International Society for Ceramics in Medicine (ISCM)

BIOCERAMICS

Volume 16

International Society for Ceramics in Medicine (ISCM)

BIOCERAMICS

Volume 16

Proceedings of the

16th International Symposium on Ceramics in Medicine

The Annual Meeting of the

International Society for Ceramics in Medicine

Bioceramics – 16

Centro de Congressos da Alfândega do Porto

Porto, Portugal

6 – 9 November, 2003

ttp TRANS TECH PUBLICATIONS LTD
Switzerland • Germany • UK • USA

Bioceramics
Volume 16

Edited by

Mário A. Barbosa
INEB-Instituto de Engenharia Biomédica
University of Porto
Portugal

Fernando J. Monteiro
INEB-Instituto de Engenharia Biomédica
University of Porto
Portugal

Rui Correia
University of Aveiro
Portugal

Betty Leon
University of Vigo
Spain

ttp TRANS TECH PUBLICATIONS

Trans Tech Publications Ltd
Brandrain 6
CH-8707 Uetikon-Zuerich
Switzerland
http://www.ttp.net

ISBN 0-87849-932-6

Volumes 254-256 of
Key Engineering Materials
ISSN 1013-9826

Covered by Science Citation Index
Full text available online at *http://www.scientific.net*

***Distributed** worldwide by*

Trans Tech Publications
Brandrain 6
CH-8707 Uetikon-Zuerich
Switzerland

Fax: +41 (1) 922 10 33
e-mail: sales@ttp.net

and in the Americas by

Trans Tech Publications
PO Box 699, May Street
Enfield, NH 03748
USA

Phone: +1 (603) 632-737
Fax: +1 (603) 632-5611
e-mail: sales-usa@ttp.net

Printed in the United Kingdom
by Hobbs the Printers Ltd

The photo on the front cover:
Rabelo boat and Porto view from Douro river
by Vitor Ferreira

Bioceramics – 16

ISCM Executive Committee 2003
International Society for Ceramics in Medicine (ISCM)

President :   M.A. Barbosa (Portugal)
Secretary General: W.   Bonfield (UK)
Treasurer: T.   Kokubo (Japan)

Members: B.   Ben-Nissan (Australia)   P.   Li (USA)
                  I.C. Clarke (USA)   A.   Moroni (Italy)
                  P.   Ducheyne (USA)   T.   Nakamura (Japan)
                  G.W. Hastings (UK)   H.   Ohgushi (Japan)
                  L.L. Hench (UK)   H.   Oonishi (Japan)
                  R.   LeGeros (USA)   L.   Sedel (France)
                                                     A.   Yli-Urpo (Finland)

International Scientific Committee

P. Abad (Colombia)   S. Aza (Spain)
R. Bizios (USA)   J.-M. Bouler (France)
J. Cavalheiro (Portugal)   G. Daculsi (France)
K. De Groot (The Netherlands)   U. Gross (Germany)
G. Heimke (Germany)   S. F. Hulbert (USA)
J. P. Lopéz (Argentina)   R. Pilliar (Canada)
J. Planell (Spain)   A. Ravaglioli (Italy)
C. Rey (France)   J. San Roman (Spain)
J.D. Santos (Portugal)   M. Vallet-Regi (Spain)
C. Zavaglia (Brazil)

Organizing Committee

Mário A. Barbosa (Portugal)   Fernando J. Monteiro (Portugal)
Chairman   Chairman

Rui Correia (Portugal)   Betty Leon (Spain)

Symposium Manager
Ana P. Filipe

BIOCERAMICS SYMPOSIA

Year	President	Site
1988	Hironobou Oonishi, M.D.	Kyoto, Japan
1989	Gunther Heimke, Ph.D.	Heidelberg, Germany
1990	Samuel F. Hulbert, Ph.D.	Terre Haute, USA
1991	William Bonfield, Ph.D.	London, UK
1992	Takao Yamamuro, M.D.	Kyoto, Japan
1993	Paul Ducheyne, Ph.D.	Philadelphia, USA
1994	Antti Yli-Urpo, Ph.D.	Turku, Finland
1995	Larry L. Hench, Ph.D.	Ponte Vedra, USA
1996	Tadashi Kokubo, Ph.D.	Otsu, Japan
1997	Laurent Sedel, M.D.	Paris, France
1998	Racquel Z. LeGeros, Ph.D.	New York, USA
1999	Hajimi Ohgushi, M.D.	Nara, Japan
2000	Antonio Moroni, M.D.	Bologna, Italy
2001	Ian C. Clarke, Ph.D.	Palm Springs, USA
2002	Besim Ben-Nissan, Ph.D.	Sydney, Australia
2003	Mário A. Barbosa, Ph.D.	Porto, Portugal

Acknowledgements

Bioceramics 16 gratefully acknowledges the support from Fundação para a Ciência e a Tecnologia, Cipan, Reitoria da Universidade do Porto, Faculdade de Engenharia da Universidade do Porto, INEB-Instituto de Engenharia Biomédica, and of the societies that endorsed the Symposium: American Academy of Orthopaedic Surgeons, Australian Society for Biomaterials Inc., Biomaterials Network – Biomat.Net, European Ceramic Society, European Federation of National Associations of Orthopaedics and Traumatology, European Society for Biomaterials, Japanese Orthopaedic Association, International Society for Technology in Arthroplasty, Italian Society for Biomaterials, Korean Society for Biomaterials, Royal Australasian College of Surgeons, Sociedade Ibérica de Biomecanica y Biomateriales, Sociedade Portuguesa de Engenharia Biomédica, Sociedade Portuguesa de Materiais, Society for Biomaterials, and The Ceramic Society of Japan.

The financial support of Fundação Calouste Gulbenkian for publication of this book is gratefully acknowledged.

PREFACE

Since their inception Bioceramics meetings have provided a treasure house of information for both scientists and clinicians in the medical field. Publication of conference books has been a major distinctive feature of Bioceramics Symposia.

This volume of the Bioceramics series includes the papers selected for the 16th International Symposium on Ceramics in Medicine, held in Porto, Portugal, from 6 to 9 November, 2003. The contents feature the following sections: Apatites, Phosphates and Glass Ceramics; Bone Grafts and Cements; Coatings and Surface Modifications; Composites and Hybrids; Dental and Orthopaedic Applications; Interactions with Cells and Tissues; Nanotechnologies Applied to Bioceramics; Porous Bioceramics; and Tissue Engineering Scaffolds. Separation of the papers in the above sections was difficult in view of the interdisciplinary nature of many papers. Others might have preferred a different separation and even other topics for the sections. However, we are convinced that the most valuable aspect of this book is the contents of the papers and not the way they have been classified.

The works reported in this book were selected after a blind peer-reviewing process of abstracts that involved over 100 scientists. The authors of selected abstracts were invited to submit the full papers, which were also reviewed, resulting in the papers selected for this book. Without the collaboration of all the reviewers this book would not have existed.

We are grateful to all the members of the Executive Committee of the International Society for Ceramics in Medicine and of the International Scientific Committee, who have contributed with their invaluable experience and suggestions.

We would like to thank all of the sponsoring institutions, organizations and companies who have supported **Bioceramics 16** both financially and otherwise.

Special thanks go to Ana Paula Filipe, who had to perform all her regular duties at INEB, while taking up the hard task of Symposium manager.

Mário A. Barbosa
Fernando J. Monteiro
Rui Correia
Betty Leon

Table of Contents

Committees   vii
Symposia   viii
Preface   ix

I.   APATITES, PHOSPHATES AND GLASS CERAMICS

Stimulation of Bone Repair by Gene Activating Glasses
L.L. Hench 3

In Vitro **Simulation of Calcium Phosphate Crystallization from Dynamic Revised**
Simulated Body Fluid
C. Deng, Y. Tan, C. Bao, Q. Zhang, H. Fan, J. Chen and X. Zhang 7

Control of Calcium Phosphate Crystal Nucleation, Growth and Morphology
by Polyelectrolytes
H. Füredi-Milhofer, P. Bar-Yosef Ofir, M. Sikiric and N. Garti 11

Fabrication and Characterization of Apatite Sintered Bodies with
c-Axis Orientation
K. Ohta, M. Kikuchi and J. Tanaka 15

Hydrothermal Preparation of Granular Hydroxyapatite with Controlled Surface
K. Ioku, M. Toda, H. Fujimori, S. Goto and M. Yoshimura 19

Influence of the Network Modifier Content on the Bioactivity of Silicate Glasses
J.P. Borrajo, S. Liste, J. Serra, P. González, S. Chiussi, B. León, M. Pérez-Amor,
H.O. Ylänen and M. Hupa 23

Textural Evolution of a Sol-Gel Glass Surface in SBF
D. Arcos, J. Peña and M. Vallet-Regí 27

Bioactive Behaviour in Biphasic Mixtures of Hydroxyapatite-sol Gel Glasses in the
System SiO$_2$-CaO-P$_2$O$_5$
J. Román, S. Padilla and M. Vallet-Regí 31

Concentrated Suspensions of Hydroxyapatite for Gelcasting Shaping
S. Padilla and M. Vallet-Regí 35

Sinterability, Mechanical Properties and Solubility of Sintered
Fluorapatite-Hydroxyapatites
K.A. Gross and K.A. Bhadang 39

Synthesis and Characterisation of Hydroxyapatite
M. Utech, D. Vuono, M. Bruno, P. De Luca and A. Nastro 43

Effect of Ag-Doped Hydroxyapatite as a Bone Filler for Inflamed Bone Defects
J.W. Choi, H.M. Cho, E.K. Kwak, T.G. Kwon, H.M. Ryoo, Y.K. Jeong, K.S. Oh
and H.I. Shin 47

Mechanical Testing of Chemically Bonded Bioactive Ceramic Materials
J. Loof, H. Engqvist, L. Hermansson and N.O. Ahnfelt 51

Assessment of Cancellous Bone Architecture after Implantation of an Injectable Bone Substitute
C.A. Davy, O. Gauthier, M.-F. Lucas, P. Pilet, B. Lamy, P. Weiss, G. Daculsi and J.-M. Bouler 55

Fabrication of Bioactive Glass-Ceramics Containing Zinc Oxide
C. Ohtsuki, H. Inada, M. Kamitakahara, M. Tanihara and T. Miyazaki 59

Apatite Hydrogel and Its Caking Behavior
Y. Yokogawa, Y. Shiotsu, F. Nagata and M. Watanabe 63

Measuring the Devitrification of Bioactive Glasses
H. Arstila, E. Vedel, L. Hupa, H. Ylänen and M. Hupa 67

Investigation of the Influence of Zirconium Content on the Formation of Apatite on Bioactive Glass-Ceramics
U. Ploska, G. Berger and M. Sahre 71

Bioactivities of a SiO_2-CaO-ZrO_2 Glass in Simulated Body Fluid and Human Parotid Saliva
P.N. De Aza and S. De Aza 75

Homogeneous Octacalcium Phosphate Precipitation: Effect of Temperature and pH
Y. Liu, R.M. Shelton and J.E. Barralet 79

Investigation of the Specific Energy Deposition from Radionuclide-Hydroxyapatite Macroaggregate in Brain Interstitial Implants
B.M. Mendes and T.P.R. Campos 83

Radiological Response of Macroaggregates Implants in an "*In Vitro*" Experimental Model
B.M. Mendes, C.H.T. Silva and T.P.R. Campos 87

Sintering Effects on the Mechanical and Chemical Properties of Commercially Available Hydroxyapatite
F.P. Cox, M.J. Pomeroy, M.E. Murphy and G.M. Insley 91

[31]P Nuclear Magnetic Resonance and X-Ray Diffraction Studies of Na-Sr-Phosphate Glass-Ceramics
R.A. Pires, I. Abrahams, T.G. Nunes and G.E. Hawkes 95

A MAS NMR Study of the Crystallisation Process of Apatite-Mullite Glass-Ceramics
A. Stamboulis, R.G. Hill, R.V. Law and S. Matsuya 99

Dry Mechanosynthesis of Strontium-Containing Hydroxyapatite from DCPD +CaO +SrO
H. El Briak-BenAbdeslam, B. Pauvert, A. Terol and P. Boudeville 103

Obtention of Silicate-Substituted Calcium Deficient Hydroxyapatite by Dry Mechanosynthesis
C. Mochales, H. El Briak-BenAbdeslam, M.P. Ginebra, P. Boudeville and J.A. Planell 107

Fabrication of Hydroxyapatite Ceramics via a Modified Slip Casting Route
E.S. Thian, J. Huang, S.M. Best and W. Bonfield 111

Newly Improved Simulated Body Fluid
H. Takadama, M. Hashimoto, M. Mizuno, K. Ishikawa and T. Kokubo 115

Preparation and Properties of Zinc Containing Biphasic Calcium Phosphate Bioceramics
A.M. Costa, G.A. Soares, R. Calixto and A.M. Rossi 119

Effects of Lead on Calcium Phosphate Ceramics Stability
E. Mavropoulos, N.C.C. Rocha, G.A. Soares, J.C. Moreira and A.M. Rossi 123

Mg-Substituted Tricalcium Phosphates: Formation and Properties
R.Z. LeGeros, A.M. Gatti, R. Kijkowska, D.Q. Mijares and J.P. LeGeros 127

Apatite Formation on HA/TCP Ceramics in Dynamic Simulated Body Fluid
Y.R. Duan, Z.R. Zhang, C.Y. Wang, J.Y. Chen and X.D. Zhang 131

Biocompatibility of Si-Substituted Hydroxyapatite
J.-H. Lee, K.-S. Lee, J.-S. Chang, W.S. Cho, Y.-H. Kim, S.-R. Kim and Y.-T. Kim 135

Process of Bonelike Apatite Formation on Sintered Hydroxyapatite in Serum-Containing SBF
T. Kokubo, T. Himeno, H.-M. Kim, M. Kawashita and T. Nakamura 139

Ceramics *In Vitro* Mineralisation Protocols: a Supersaturation Problem
P.A.A.P. Marques, M.C.F. Magalhães, R.N. Correia, A.I. Martin, A.J. Salinas and M. Vallet-Regí. 143

Characterization of Bioactive Glass-Ceramics Prepared by Sintering Mixed Glass Powders of Cerabone® A-W Type Glass/CaO-SiO₂-B₂O₃ Glass
J.H. Seo, H.S. Ryu, K.S. Park, K.S. Hong, H. Kim, J.H. Lee, D.H. Lee, B.S. Chang and C.K. Lee 147

Effect of B₂O₃ on the Sintering Behavior and Phase Transition of Wollastonite Ceramics
K.S. Park, H.S. Ryu, J.H. Seo, K.S. Hong, H. Kim, J.H. Lee, D.H. Lee, B.S. Chang and C.K. Lee 151

SiO₂- MgO-3CaO.P₂O₅- K₂O Glasses and Glass-Ceramics: Effect of Crystallisation on the Adhesion of SBF Apatite Layers
C.M. Queiroz, J.R. Frade and M.H.V. Fernandes 155

II. BONE GRAFTS AND CEMENTS

Potassium Containing Apatitic Calcium Phosphate Cements
F.C.M. Driessens and M.G. Boltong 161

HRTEM and EDX Investigation of Microstructure of Bonding Zone between Bone and Hydroxyapatite *In Vivo*
Q.Z. Chen, W.W. Lu, C.T. Wong, K.M.C. Cheung, J.C.Y. Leong and K.D.K. Luk 165

Acetabular Construction Using a Glass-Ceramic Artificial Bone Graft
S. Yoshii, M. Oka, M. Shima and T. Yamamuro 169

***In Vivo* Aging Test for Bioactive Bone Cements Composed of Glass Bead and PMMA**
S. Shinzato, T. Nakamura, K. Goto and T. Kokubo 173

In Vitro and *In Vivo* **Behaviour of Bioactive Glass Composites Bearing a NSAID**
J.A. Méndez, A. González-Corchón, M. Salvado, F. Collía, J.A. de Pedro,
B. Levenfeld, M. Fernández, B. Vázquez and J. San Román 177

Osteogenic Activity of Human Marrow Cells on Alumina Ceramics
Y. Tanaka, H. Ohgushi, S. Kitamura, A. Taniguchi, K. Hayashi, S. Isomoto, Y. Tohma
and Y. Takakura 181

In Vitro and *In Vivo* **Evaluation of Non-Crystalline Calcium Phosphate Glass as a Bone Substitute**
Y.-K. Lee, J. Song, H.J. Moon, S.B. Lee, K.M. Kim, K.N. Kim, S.H. Choi
and R.Z. LeGeros 185

Evaluation of Macroporous Biphasic HA-TCP Ceramic as a Bone Substitute
K.-S. Oh, J.L. Kim, K.O. Oh, J.W. Choi, E.K. Kwak, H.M. Ryoo, K.H. Kim
and H.I. Shin 189

Maxillary Sinus Bone Grafting with an Injectable Bone Substitute: a Sheep Study
A. Saffarzadeh, O. Gauthier, T. Humbert, P. Weiss, J.-M. Bouler and G. Daculsi 193

The Influence of Condensing Technique and Accelerator Concentration on Some Mechanical Properties of an Experimental Bioceramic Dental Restorative Material
J. Lööf, H. Engqvist, G. Gómez-Ortega, N.-O. Ahnfelt and L. Hermansson 197

A New Porous Osteointegrative Bone Cement Material
G. Georgescu, J.L. Lacout and M. Frèche 201

Low Porosity CaHPO$_4$-2H$_2$O Cement
L.M. Grover, U. Gbureck, A. Hutton, D.F. Farrar, C. Ansell and J.E. Barralet 205

Effects of Polysaccharides Addition in Calcium Phosphate Cement
T. Sawamura, M. Hattori, M. Okuyama and K. Kondo 209

Characterization of Magnetite and Maghemite for Hyperthermia in Cancer Therapy
K.M. Spiers, J.D. Cashion and K.A. Gross 213

Effects of TCP Particle Size Distribution on TCP / TTCP / DCPD Bone Cement System
S.-H. Cho, I.-S. Hwang, J.-K. Lee, K.-S. Oh, S.-R. Kim and Y.-C. Chung 217

Evaluation of Hydroxylapatite Cement Composites as Bone Graft Substitutes in a Rabbit Defect Model
M.J. Voor, J.J.C. Arts, S. Klein, L.H.B. Walschot, N. Verdonschot and P. Buma 221

Animal Study of Alpha-Tricalcium Phosphate-Based Bone Filler
J.-H. Lee, K.-S. Lee, J.-S. Chang, W.S. Cho, J.-K. Lee, S.-R. Kim and Y.-T. Kim 225

Influence of Gelatin on the Setting Properties of α-Tricalcium Phosphate Cement
A. Bigi, B. Bracci and S. Panzavolta 229

Synthesis and Comparison Characterisation of Hydroxyapatite and Carbonated Hydroxyapatite
R. Martinetti, M.F. Harmand and L. Dolcini 233

Synthesis and Properties of Bone Cement Containing Dense ß-TCP Granules
K.-S. Oh, S.-R. Kim and P. Boch 237

The Expression of Bone Matrix Protein mRNAs around Calcium Phosphate Cement Particles Implanted into Bone
M. Mukaida, K. Ohsawa, M. Neo and T. Nakamura 241

Tissue Response of Calcium Polyphosphate in Beagle Dog. Part II: 12 Month Result
S.M. Yang, S.Y. Kim, S.J. Lee, Y.K. Lee, Y.M.. Lee, Y. Ku, C.P. Chung, S.B. Han
and I.C. Rhyu 245

Thermodynamic Study of Formation of Amorphous ß-Tricalcium Phosphate for Calcium Phosphate Cements
U. Gbureck, J.E. Barralet and R. Thull 249

Effect of Albumen as Protein-Based Foaming Agent in a Calcium Phosphate Bone Cement
A. Almirall, J.A. Delgado, M.P. Ginebra and J.A. Planell 253

Construction of an Interconnected Pore Network Using Hydroxyapatite Beads
K. Teraoka, Y. Yokogawa and T. Kameyama 257

Correlating *In Vitro* Dissolution Measurement Methods Used with Bone Void Filler Bioceramics
T. Jones and M. Long 261

An Injectable Bone Void Filler Cement Based on Ca-Aluminate
N. Axén, T. Persson, K. Björklund, H. Engqvist and L. Hermansson 265

Study of Particle Size Dependant Reactivity in an α-TCP Orthophosphate Cement
C.L. Camiré, S.J. Saint-Jean, I. McCarthy, S. Hansen and L. Lidgren 269

Microporosity Affects Bioactivity of Macroporous Hydroxyapatite Bone Graft Substitutes
K.A. Hing, S. Saeed, B. Annaz, T. Buckland and P.A. Revell 273

Mechanically Induced Phase Transformation of α- and ß-Tricalcium Phosphate
J.E. Barralet, U. Gbureck, L.M. Grover and R. Thull 277

Cements from Biphasic β-Tricalcium Phosphate
K.J. Lilley, U. Gbureck, D.F. Farrar, C. Ansell and J.E. Barralet 281

Development of Novel PMMA-Based Bone Cement Reinforced by Bioactive CaO-SiO₂ Gel Powder
S.B. Cho, S.B. Kim, K.J. Cho, C. Ohtsuki and T. Miyazaki 285

Influence of Molarity and Liquid/Powder Ratio on the Synthesis of a Calcium Phosphate Cement
M.E. Murphy, R.M. Streicher and G.M. Insley 289

PLA/HAp Microsphere-Based Porous Materials for Artificial Bone Grafts
F. Nagata, T. Miyajima and Y. Yokogawa 293

Rietveld Analysis in Sintering Studies of Ca-Deficient Hydroxyapatite
L.E. Jackson, J.E. Barralet and A.J. Wright 297

Sol-Gel Derived Nano-Coated Coralline Hydroxyapatite for Load Bearing Applications
B. Ben-Nissan, A. Milev, R. Vago, M. Conway and A. Diwan 301

Hardening and Hydroxyapatite Formation of Bioactive Glass and Glass-Ceramic Cement
C.Y. Kim and H.B. Lim 305

III. COATINGS AND SURFACE MODIFICATIONS

Influence of Acetylation on *In Vitro* Chitosan Membrane Biomineralization
M.M. Beppu and C.G. Aimoli 311

Dissolution Behaviors of Thermal Sprayed Calcium Phosphate Splats in Simulated Body Fluid
K.A. Khor, H. Li and P. Cheang 315

Confocal Microscopy Characterization of *In Vitro* Tests of Controlled Atmosphere Plasma Spraying Hydroxyapatite (CAPS-HA) Coatings
K.A. Khor, M.E. Pons, G. Bertran, N. Llorca, M. Jeandin and V. Guipont 319

Effects of Some Variables on R.F. Magnetron Sputtered Hydroxyapatite Coatings on Titanium
R.A. Silva, M. Sousa, F.J. Monteiro, R. Barral, J.D. Santos and M.A. Lopes 323

The Influence of Glucose and Bovine Serum Albumin on the Crystallization of a Bone-Like Apatite from Revised Simulated Body Fluid
S.V. Dorozhkin, E.I. Dorozhkina, S. Agathopoulos and J.M.F. Ferreira 327

Synthesis and Characterization of Fluoridated Hydroxyapatite with a Novel Fluorine-Containing Reagent
K. Cheng, W. Weng, P. Du, G. Shen, G. Han and J.M.F. Ferreira 331

Novel Fluorapatite-Mullite Coatings for Biomedical Applications
J. Bibby, P.M. Mummery, N. Bubb and D.J. Wood 335

Identifying Calcium Phosphates Formed in Simulated Body Fluid by Electron Diffraction
Y. Leng, X. Lu and J. Chen 339

Novel Calcium Phosphate Fibres from a Biomimetic Process: Manufacture and Cell Attachment
E.C. Kolos, A.J. Ruys, R. Rohanizadeh, M.M. Muir and G. Roger 343

$CaO-P_2O_5$ Glass-Hydroxyapatite Thin Films Obtained by Laser Ablation: Characterisation and *In Vitro* Bioactivity Evaluation
M.P. Ferraz, F.J. Monteiro, D. Gião, B. Leon, P. Gonzalez, S. Liste, J. Serra, J. Arias and M. Perez-Amor 347

A Study of Bone-Like Apatite Formation on Calcium Phosphate Ceramics in Different Simulated Body Fluids (SBF)
Y.R. Duan, C.Y. Wang, J.Y. Chen and X.D. Zhang 351

***In Vitro* Bioactivity Study of PLD-Coatings and Bulk Bioactive Glasses**
S. Liste, P. González, J. Serra, C. Serra, J.P. Borrajo, S. Chiussi, B. León and M. Pérez-Amor 355

Calcium Phosphate Porous Coatings onto Alumina Substrates by Liquid Mix Method
J. Peña, I. Izquierdo-Barba and M. Vallet-Regí 359

Apatite Layers by a Sol-Gel Route
I. Izquierdo-Barba, N. Hijón, M.V. Cabañas and M. Vallet-Regí 363

Tribological Behaviour of Prosthetic Ceramic Materials Sliding Against Smooth Diamond-Coated Titanium Alloy
C. Met, L. Vandenbulcke, M.C. Sainte Catherine and L. Quiniou 367

The Role of Processing Parameters on Calcium Phosphate Coatings Obtained by Laser Cladding
F. Lusquiños, J. Pou, J.L. Arias, M. Boutinguiza, B. León, M. Pérez Amor, S. Best, W. Bonfield and F.C.M. Driessens 371

Effect of Surface Morphology and Crystal Structure on Bioactivity of Titania Films Formed on Titanium Metal via Anodic Oxidation in Sulfuric Acid Solution
T.-Y. Xiong, X.-Y. Cui, H.-M. Kim, M. Kawashita, T. Kokubo, J. Wu, H.-Z. Jin and T. Nakamura 375

Wollastonite Coatings on Zirconia Ceramics
J.A. Delgado, F.J. Gil, S. Martínez, M.P. Ginebra, L. Morejón, J.M. Manero and J.A. Planell 379

Tribological Study of Plasma Hydroxyapatite Coatings
J. Fernández, M. Gaona and J.M. Guilemany 383

Structure Modification of Surface Layers of Ti6Al4V ELI Implants
J. Marciniak, W. Chrzanowski, G. Nawrat, J. Zak and B. Rajchel 387

Preparation and Properties of Bioactive Calcium Phosphate Fibers
C. Klein, F.A. Müller and P. Greil 391

Consolidation of Multi-Walled Carbon Nanotube and Hydroxyapatite Coating by the Spark Plasma System (SPS)
M. Omori, A. Okubo, M. Otsubo, T. Hashida and K. Tohji 395

Influence of Carboxyl Groups Present in the Mineralising Medium in the Biomimetic Precipitation of Apatite on Collagen
E.K. Girija, Y. Yokogawa and F. Nagata 399

Apatite Deposition on Silk Sericin in a Solution Mimicking Extracellular Fluid: Effects of Fabrication Process of Sericin Film
A. Takeuchi, C. Ohtsuki, M. Kamitakahara, S.-i. Ogata, M. Tanihara, T. Miyazaki, M. Yamazaki, Y. Furutani and H. Kinoshita 403

In Vitro Surface Reactions of Bioceramic Materials
R. Gildenhaar, A. Bernstein, G. Berger and W. Hein 407

A New Procedure of a Calcium-Containing Coating on Implants of Titanium Alloy
U. Ploska, G. Berger and M. Willfahrt 411

Mechanical Evaluation of the Use of a Buffer Layer in Hydroxylapatite Coatings Produced by Pulsed Laser Deposition at High Temperature
E. Jiménez, J.L. Arias, B. León and M. Pérez Amor 415

Influence of Ca/P Ratio on Electrochemical Assisted Deposition of Hydroxyapatite on Titanium
A. Sewing, M. Lakatos, D. Scharnweber, S. Roessler, R. Born, M. Dard and H. Worch 419

Sol-Gel Apatite Films on Titanium Implant for Hard Tissue Regeneration
H.-W. Kim, H.-E. Kim and J.C. Knowles 423

Bioactive Glass Coating for Hard and Soft Tissue Bonding on Ti6Al4V and Silicone Rubber Using Electron Beam Ablation
J. Schrooten, S.V.N. Jaecques, R. Eloy, C. Delubac, C. Schultheiss, P. Brenner, L.H.O. Buth, J. Van Humbeeck and J.V. Sloten 427

Reliability Weibull Analysis for Structural Evaluation of Bioactive Films Obtained by Sol-Gel Process
A. Peláez, C. Garcia, J.C. Correa and P. Abad 431

Diamond-Like Carbon Coatings on Ti-13Nb-13Zr Alloy Produced by Plasma Immersion for Orthopaedic Applications
E.T. Uzumaki, C.S. Lambert and C.A.C. Zavaglia 435

Nano-Sized Apatite Coatings on Niobium Substrates
M.H. Prado da Silva, A.M.R. Monteiro, J.A.C. Neto, S.M.O. Morais
and F.F.P. dos Santos 439

Control of Morphology of Titania Film with High Apatite-Forming Ability Derived from Chemical Treatments of Titanium with Hydrogen Peroxide
S. Kawasaki, K. Tsuru, S. Hayakawa and A. Osaka 443

The Effect of Magnesium Ions on Bone Bonding to Hydroxyapatite Coating on Titanium Alloy Implants
P.A. Revell, E. Damien, X.S. Zhang, P. Evans and C.R. Howlett 447

Preparation of Novel Bioactive Titanium Coatings on Titanium Substrate by Reactive Plasma Spraying
M. Inagaki, Y. Yokogawa and T. Kameyama 451

Adhesion of Sol-Gel Derived Zirconia Nano-Coatings on Surface Treated Titanium
R. Roest, A.W. Eberhardt, B.A. Latella, R. Wuhrer and B. Ben-Nissan 455

Bonelike Apatite Formation on Anodically Oxidized Titanium Metal in Simulated Body Fluid
M. Kawashita, X.-Y. Cui, H.-M. Kim, T. Kokubo and T. Nakamura 459

Photocatalytic Bactericidal Effect of TiO_2 Thin Films Produced by Cathodic Arc Deposition Method
B. Kepenek, U.Ö.Ş. Seker, A.F. Cakir, M. Ürgen and C. Tamerler 463

Bonelike Apatite Formation on Synthetic Organic Polymers Coated with TiO_2
F. Balas, M. Kawashita, H.M. Kim, C. Ohtsuki, T. Kokubo and T. Nakamura 467

IV. COMPOSITES AND HYBRIDS

Organic-Inorganic Hybrids of Collagen or Biodegradable Polymers with Hydroxyapatite
M.F. Hsieh, R.J. Chung, T.J. Hsu, L.H. Perng and T.S. Chin 473

Effect of the Solvents on the Solution Mixture Derived Polylactide/Hydroxyapatite Composites
Y. Li, W. Weng, K. Cheng, P. Du, G. Shen, G. Han, M.A. Lopes and J.D. Santos 477

Bioactive Organic-Inorganic Hybrids Based on CaO - SiO_2 Sol-Gel Glasses
A.J. Salinas, J.M. Merino, N. Hijón, A.I. Martín and M. Vallet-Regí. 481

Injectable Composite Hydrogels for Orthopaedic Applications. Mechanical and Morphological Analysis
V. Sanginario, L. Ambrosio, M.P. Ginebra and J.A. Planell 485

Drug Release from Poly(D, L-Lactide) / SiO₂ Composites
M. Vaahtio, M. Jokinen, A. Rosling, P. Kortesuo, J. Kiesvaara and A. Yli-Urpo 489

Synthesis and Characterization of Hydroxyapatite on Collagen Gel
L.A. Sena, P. Serricella, R. Borojevic, A.M. Rossi and G.A. Soares 493

Preparation of Bonelike Apatite Composite Sponge
H. Maeda, T. Kasuga, M. Nogami, H. Kagami, K. Hata and M. Ueda 497

Bone-Like Apatite Forming Ability on Surface Modified Chitosan Membrane in Simulated Body Fluid
S.-H. Rhee 501

***In Vitro* of Self-Reinforced Composites of Highly Bioactive Glass Loaded Bioabsorbable Polymer**
H. Niiranen, T. Niemelä, M. Kellomäki and P. Törmälä 505

***In Vitro* Degradation of Osteoconductive Poly-L/DL-Lactide / β-TCP Composites**
T. Niemelä, M. Kellomäki and P. Törmälä 509

Water Uptake of Poly(ethylmethacrylate)/Tetrahydrofurfuryl Methacrylate Polymer Systems Modified with Tricalcium Phosphate and Hydroxyapatite
F.F. Rahman, W. Bonfield, R.E. Cameron, M.P. Patel, M. Braden, G. Pearson and S.M. Tavakoli 513

Composite Mesh Consisting of Titanium, Apatite and Biodegradable Copolymer
S. Ban, A. Yuda, Y. Izumi, T. Kanie, H. Arikawa and K. Fujii 517

Apatite-Forming Ability and Mechanical Properties of Poly(tetramethylene Oxide) (PTMO)-Ta₂O₅ Hybrids
M. Kamitakahara, M. Kawashita, N. Miyata, H.-M. Kim, T. Kokubo, C. Ohtsuki and T. Nakamura 521

Effect of Sulfonic Group and Calcium Content on Apatite-Forming Ability of Polyamide Films in a Solution Mimicking Body Fluid
T. Kawai, C. Ohtsuki, M. Kamitakahara, T. Miyazaki, M. Tanihara, Y. Sakaguchi and S. Konagaya 525

Drug Delivery Behaviour of Hydroxyapatite and Carbonated Apatite
A.J. Melville, L.M. Rodríguez-Lorenzo, J.S. Forsythe and K.A. Gross 529

Preparation of Calcium Carbonate / Poly(lactic acid) Composite (CCPC) Hollow Spheres
H. Maeda, T. Kasuga and M. Nogami 533

The Role of Phosvitin for Nucleation of Calcium Phosphates on Collagen
N. Kobayashi, K. Onuma, A. Oyane and A. Yamazaki 537

Biomimetic Coating of Laminin -Apatite Composite Layer onto Ethylene-Vinyl Alcohol Copolymer
A. Oyane, M. Uchida and A. Ito 541

Apatite Deposition on Organic-inorganic Hybrids Prepared from Chitin by Modification with Alkoxysilane and Calcium Salt
T. Miyazaki, C. Ohtsuki, M. Tanihara and M. Ashizuka 545

Bioabsorbable and Bioactive Composite Structures from SiO₂ Glassfibres and Polylactides
A.-M. Haltia, H. Heino and M. Kellomäki 549

The Degradation and Bioactivity of Composites of Silica Xerogel and Novel Biopolymer of Hydroxyproline
M. Väkiparta, M. Jokinen, M. Vaahtio, P.K. Vallittu and A. Yli-Urpo 553

Bioactive Glass (S53P4) and Mesoporous MCM-41-Type SiO$_2$ Adjusting *In Vitro* Bioactivity of Porous PDLLA
J. Korventausta, A. Rosling, J. Andersson, A. Lind, M. Linden, M. Jokinen and A. Yli-Urpo 557

Porous Body Preparation of Hydroxyapatite / Collagen Nanocomposites for Bone Tissue Regeneration
M. Kikuchi, T. Ikoma, D. Syoji, H.N. Matsumoto, Y. Koyama, S. Itoh, K. Takakuda, K. Shinomiya and J. Tanaka 561

***In Vivo* Behaviour of Bonelike®/PLGA Hybrid: Histological Analysis and Peripheral Quantitative Computed Tomography (pQ-CT) Evaluation**
J.M. Oliveira, T. Kawai, M.A. Lopes, C. Ohtsuki, J.D. Santos and A. Afonso 565

Effect of Melt Flow Rate of Polyethylene on Bioactivity and Mechanical Properties of Polyethylene /Titania Composites
H. Takadama, M. Hashimoto, Y. Takigawa, M. Mizuno and T. Kokubo 569

Preparation and Characterization of Injectable Chitosan-Hydroxyapatite Microspheres
P.L. Granja, A.I.N. Silva, J.P. Borges, C.C. Barrias and I.F. Amaral 573

***In Vitro* Mineralisation of Chitosan Membranes Carrying Phosphate Functionalities**
I.F. Amaral, P.L. Granja and M.A. Barbosa 577

***In Vitro* Bioactivity in Glass-Ceramic / PMMA-co-EHA Composites**
B.J.M. Leite Ferreira, M.G.G.M. Duarte, M.H. Gil, R.N. Correia, J. Román and M. Vallet-Regí. 581

***In Vitro* Dissolution Characteristics of Calcium Phosphate/Calcium Sulphate Based Hybrid Biomaterials**
C.P. Cleere, G.M. Insley, M.E. Murphy, P.N. Maher and A.M. Murphy 585

Hybridized Hydroxyapatite Bioactive Bone Substitutes
K. de Arruda Almeida and A.A.A. de Queiroz 589

Phase Mapping: A Novel Design Approach for the Production of Calcium Phosphate-Collagen Biocomposites
A.K. Lynn, R.E. Cameron, S.M. Best, R.A. Brooks, N. Rushton and W. Bonfield 593

V. DENTAL AND ORTHOPAEDIC APPLICATIONS

pH Changes Induced by Bioactive Glass Ionomer Cements
H. Yli-Urpo, E. Söderling, P.K. Vallittu and T. Närhi 599

Chiral Biomineralization: Epitaxial and Helical Growth of Calcium Carbonate Through Selective Binding of Phosphoserine Containing Polypeptides, and the Dental Application onto Apatite
H. Yamamoto, M. Yamaguchi, T. Sugawara, Y. Suwa, K. Ohkawa, H. Shinji and S. Kurata 603

A New Knee Prosthesis with Bisurface Femoral Component Made of Zirconia-
Ceramic (Report 2)
  T. Nakamura, E. Oonishi, T. Yasuda and Y. Nakagawa 607

Effects of Polymer Molecular Weight and Ceramic Particle Size on Flexural
Properties of Hydroxyapatite Reinforced Polyethylene
  M. Wang, L.Y. Leung, P.K. Lai and W. Bonfield 611

Diffusion of Ions from a Calcium Phosphate Cement for Dental Root Canal
Treatment and Filling
  S. Munier, H. El Briak, D. Durand and P. Boudeville 615

Biomimetic Coatings vs. Collagen Sponges as a Carrier for BMP-2: A Comparison
of the Osteogenic Responses Triggered In Vivo Using an Ectopic Rat Model
  Y. Liu, E.B. Hunziker, C. Van de Vaal and K. de Groot 619

A Randomised ROA Study of Peri-Apatite HA Coating of a Knee Prosthesis
  U. Hansson 623

Manufacturing of Paste-Made Porcelains using a Wax Solvent and Evaluation
of their Physical Properties
  K.-M. K 627

Successful Reconstruction by Myxoma-Induced Calcium Metaphosphate Ceramic-
Osteoblast Cell Granules Implantation
  S.H. Oh, J.J. Yoon, J. Park, G. Khang, H.S. Park, J.M. Rhee, J.Y. Woo, K.H. Kim
  and H.B. Lee 631

Do Calcium Phosphate Substitute Ceramics Initiate Mineralization?
  U. Gross, et al. 635

Effect of Clearance-Reduction Finishing on the Contact Mechanics of Ceramic-on-
Ceramic Hip Joint Replacements
  M.M. 639

Excellent Bone Ingrowth into HA Granules Filled in an Acetabular Massive Bone
Defect under Weight Bearing Condition
  H. Oonishi, S.C. Kim, H. Dohkawa, Y. Doiguchi, Y. Takao and K. Oomamiuda 643

Optimum Fixation at Bone / Bone Cement Interface by Interposing HA
Granules (IBBC)
  H. Oonishi, S.C. Kim, H. Dohkawa, Y. Doiguchi, Y. Takao and K. Oomamiuda 647

Biomechanical Analysis of Hydroxyapatite-Coated External Fixation Pins Removed
from Osteoporotic Trochanteric Fracture Patients
  F. Pegreffi, M. Romagnoli, A. Moroni and S. Giannini 651

A Radiological Follow-Up Study of Plasma Sprayed Bovine Hydroxyapatite
(BHA) Coatings
  S. Ozsoy, K. Altunatmaz, S. Ozyegin, F.N. Oktar, T. Yazici, O. Bayrak,
  E. Demirkesen and D. Toykan 655

Study of a Silicate Cement for Dental Applications
  S. Pereira, J. Cavalheiro, R. Branco, A. Afonso and M. Vasconcelos 659

Surface Analysis of Explanted Alumina-Alumina Bearings
  G.M. Insley and R.M. Streicher 663

Calcium
Metaphosphate
Page 631

Tricalcium-Phosphate/Hydroxyapatite Bone Graft Extender for Use in Impaction Grafting Revision Surgery – an *In Vitro* Study in Human Femora
E.H. van Haaren, T.H. Smit, K. Phipps, P.I.J.M. Wuisman, G. Blunn, G.M. Insley and I.C. Heyligers 667

Bone-Bonding Ability of Zirconia Coated with Titanium and Hydroxyapatite under Load-Bearing Conditions
T. Suzuki, S. Fujibayashi, Y. Nakagawa, I. Noda and T. Nakamura 671

Next Generation Ceramics Based on Zirconia Toughened Alumina for Hip Joint Prostheses
G.M. Insley and R.M. Streicher 675

Silica/Calcium Phosphate Sol-Gel Derived Bone Grafting Material — From Animal Tests to First Clinical Experience
T. Traykova, R. Bötcher, H.-G. Neumann, K.-O. Henkel, V. Bienengraeber and T. Gerber 679

The Effect of Dental Grinding and Sandblasting on the Biaxial Flexural Strength and Weibull Modulus of Tetragonal Zirconia
T. Kosmač 683

Rapid Prototyping Applications in the Treatment of Craniomaxillofacial Deformities - Utilization of Bioceramics
J.V.L. Silva, M.F. Gouvêia, A. Santa Barbara, E. Meurer and C.A.C. Zavaglia 687

Clinical Investigation on Bioactive Glass Particles for Dental Bone Defects
A.M. Gatti, E. Monari and L. Simonetti 691

Apatite Remineralization: *In Vivo* Long Term Study in Dental Tissue
R.L. Mourão, H.S. Mansur, F.R. Tay and L. Dlanza 695

Zirconia/Alumina Composite Dental Implant Abutments
D.-J. Kim, J.-S. Han, S.-H. Lee, J.-H. Yang and D.Y. Lee 699

The Wear Resistance Testing of Biomaterials Used for Implants
R. Sedlacek and J. Rosenkrancova 703

Finite Element Analysis of Ceramic Dental Implants Incorporated into the Human Mandible
A.H. Choi, R.C. Conway and B. Ben-Nissan 707

VI. INTERACTIONS WITH CELLS AND TISSUES

Quantitative Analysis of Osteoprotegerin and RANKL Expression in Osteoblast Grown on Different Calcium Phosphate Ceramics
C. Wang, Y. Duan, B. Markovic, J. Barbara, C. Rolfe Howlett, X. Zhang and H. Zreiqat 713

Properties of Two Biological Glasses Used as Metallic Prosthesis Coatings and after an Implantation in Body
Y. Barbotteau, J.L. Irigaray, E. Chassot, G. Guibert and E. Jallot 717

Comparative Study of the Biomineral Behavior of Plasma-Sprayed Coatings in
Dynamic Porous Environment
    Q.Y. Zhang, J.Y. Chen, C.L. Deng, Y. Cao, J.M. Feng and X.D. Zhang 721

Tissue Responses to Titanium with Different Surface Characteristics after
Subcutaneously Implanted in Rabbits
    Y. Wu, B.C. Yang, C.Y. Bao, J.Y. Chen and X.D. Zhang 725

Effect of Electrophoretic Apatite Coating on Osseointegration of Titanium
Dental Implants
    C.C. Almeida, L.A. Sena, A.M. Rossi, M. Pinto, C.A. Muller and G.A. Soares 729

The Dualism of Nacre
    E. Lopez, C. Milet, M. Lamghari, L. Pereira Mouries, S. Borzeix and S. Berland 733

Osseointegration of Grit-Blasted and Bioactive Titanium Implants:
Histomorphometry in Minipigs
    C. Aparicio, F.J. Gil, U. Thams, F. Muñoz, A. Padrós and J.A. Planell 737

Mechanism of Apatite Formation on Anodically Oxidized Titanium Metal in
Simulated Body Fluid
    H.-M. Kim, H. Kaneko, M. Kawashita, T. Kokubo and T. Nakamura 741

Osteogenetic Effect on Cortical Bone of Cultured Bone/Ceramics Implants
    N. Satoh, T. Yoshikawa, A. Muneyasu, J. Iida, A. Nonomura and Y. Takakura 745

Effect of Enamel Matrix Protein-Coated Bioactive Glasses on the Proliferation and
Differentiation of the Osteogenic MC3T3.E1 Cell Line
    S. Hattar, A. Asselin, D. Greenspan, M. Oboeuf, A. Berdal and J.M. Sautier 749

Apatite-Forming Ability of Calcium Phosphate Glass-Ceramics Improved
by Autoclaving
    T. Kasuga, T. Fujimoto, C. Wang and M. Nogami 753

Radiological and Histological Examination of Gap Healing on Plasma Sprayed HA
Coating Surface
    Y. Cao, W. Lu, B. Zhang, L.P. Wang, C.Y. Bao, B.C. Yang, C.Y. Wang, J.Y. Chen,
    J.M. Feng, S.P. Chow and X.D. Zhang 757

α-Tricalcium Phosphate Cements and the Granules to Dental Pulp and
Periapical Tissue
    M. Yoshikawa and T. Toda 761

Surface Modification of 58S Bioactive Gel-Glass with an Aminosilane
    R.S. Pryce and L.L. Hench 765

Study of Osteoblasts Mineralisation In Vitro by Raman Micro-Spectroscopy
    I. Notingher, J.E. Gough and L.L. Hench 769

Characteristic of Osteoblast Vacuole Formation in the Presence of Ionic Products
from BG60S Dissolution
    P. Valério, A.M. Goes, M.M. Pereira and M.F. Leite 773

Evaluation of Osteoblasts Viability, Alkaline Phosphatase Production and Collagen
Secretion in the Presence of TiHA
    P. Valério, F.N. Oktar, G. Goller, A.M. Goes and M.F. Leite 777

A549 Lung Carcinoma Cells: Binary vs. Ternary Bioactive Gel-Glasses
    P. Saravanapavan, S. Verrier and L.L. Hench 781

Indirect Cytotoxicity Evaluation of Soluble Silica, Calcium, Phosphate and Silver Ions
P. Saravanapavan, J. Selvakumaran and L.L. Hench 785

Bioactive Glass Coatings on Ti6Al4V Promote the Tight Apposition of Newly-Formed Bone *In Vivo*
A. Merolli, C. Gabbi, M. Santin, B. Locardi and P. Tranquilli Leali 789

Orientation of Human Osteoblast Cells on Biphasic Calcium Phosphates Tablets with Undulated Topography
E.A. dos Santos, A.B.R. Linhares, A.M. Rossi, M. Farina and G.A. Soares 793

Changes in Distant Organs in Response to Local Osteogenic Growth Factors Delivered by Intraosseous Implants: a Histological Evaluation
E. Damien and P. A. Revell 797

Experimental Study on Construction of Vascularized Bone Graft with Osteoinductive Calcium Phosphate Ceramics *In Vivo*
C. Bao, H. Fan, C. Deng, Y. Cao, Y. Tan, Q. Zhang and X. Zhang 801

Neural Cells on Iridium Oxide
I.-S. Lee, J.-C. Park, G.H. Lee, W.S. Seo, Y.-H. Lee, K.-Y. Lee, J.K. Kim and F.-Z. Cui 805

The Effect of Surface Treatment and Corrosive Etching on Flexural Strength of a Dental Porcelain
E.M. Reis, W.C. Jansen, M.M. Pereira, R. Giovani and P.R. Cetlin 809

Osteoblast Responses to Sintered and Tapecast Bioactive Glass
J.E. Gough, D.C. Clupper and L.L. Hench 813

Chondrocyte Maturation on Biphasic Calcium Phosphate Scaffold: A Preliminary Study
C.C. Teixeira, R.Z. LeGeros, C. Karkia and Y. Nemelivsky 817

***In Vitro* Mineralisation of Human Bone Marrow Cells Cultured on Bonelike®**
M.A. Costa, M. Gutierres, L. Almeida, M.A. Lopes, J.D. Santos and M.H. Fernandes 821

Biological Activity of Two Glass Ceramics in the Meta- and Pyrophosphate Region: a Comparative Study
A.G. Dias, M.A. Costa, M.A. Lopes, J.D. Santos and M.H. Fernandes 825

***In Vivo* Evaluation of Hydroxyapatite and Carbonated Hydroxyapatite Fillers**
C. Mangano, A. Scarano, R. Martinetti and L. Dolcini 829

Processing of Ca-P Ceramics, Surface Characteristics and Biological Performance
S. Cazalbou, C. Bastié, G. Chatainier, N. Theilgaard, N. Svendsen, R. Martinetti, L. Dolcini, J. Hamblin, G. Stewart, L. Di Silvio, N. Gurav, R. Quarto, S. Overgaard, B. Zippor, A. Lamure, C. Combes and C. Rey 833

Biocompatibility Evaluation of Dentine, Enamel and Bone Derived Hydroxyapatite
P. Valerio, F.N. Oktar, L.S. Ozyegin, G. Goller, A.M. Goes and M.F. Leite 837

Effects of Bioactive Glass 60S and Biphasic Calcium Phosphate on Human Peripheral Blood Mononuclear Cells
C.C.P. Silva, A. Bozzi, M.M. Pereira, A.M. Goes and M.F. Leite 841

Biological and Physical-Chemical Characterization of Phase Pure HA and SI-Substituted Hydroxyapatite by Different Microscopy Techniques
C.M. Botelho, R.A. Brooks, S.M. Best, M.A. Lopes, J.D. Santos, N. Rushton and W. Bonfield 845

Osseous Cell Response to Electrostatic Stimulations of Poled Hydroxyapatite Ceramics in Canine Diaphyses
S. Nakamura, M. Nakamura, T. Kobayashi, Y. Sekijima, S. Kasugai and K. Yamashita 849

Platelet Adhesion on Metal Oxide Layers
S. Takemoto, T. Yamamoto, K. Tsuru, S. Hayakawa and A. Osaka 853

In Vitro Cytocompatibility of Osteoblastic Cells Cultured on Chitosan-Organosiloxane Hybrid Membrane
Y. Shirosaki, K. Tsuru, S. Hayakawa, A. Osaka, M.A. Lopes, J.D. Santos and M.H. Fernandes 857

Hydroxyapatite Ceramic Particles as Material for Transfection
P. Frayssinet, E. Jean and N. Rouquet 861

Platelet Adhesion on Titania Film Prepared from Interaction of Ni-Ti Alloy with Hydrogen Peroxide Solution
S. Hayakawa, K. Tsuru and A. Osaka 865

Orthosilicic Acid Increases Collagen Type I mRNA Expression in Human Bone-Derived Osteoblasts In Vitro
M.Q. Arumugam, D.C. Ireland, R.A. Brooks, N. Rushton and W. Bonfield 869

Mechanical Properties and Biocompatibility of Surface-Nitrided Titanium for Abrasion Resistant Implant
F. Watari, Y. Tamura, A. Yokoyama, M. Uo and T. Kawasaki 873

Adhesion and Proliferation of Human Osteoblastic Cells Seeded on Injectable Hydroxyapatite Microspheres
C.C. Barrias, C.C. Ribeiro and M.A. Barbosa 877

Preliminary Radiological In Vivo Study of Calcium Metaphosphate Coated Ti-Alloy Implants
S.W. Kim, S. Oh, C.K. You, M.W. Ahn, K.H. Kim, I.K. Kang, J.H. Lee and S. Kim 881

VII. NANOTECHNOLOGIES APPLIED TO BIOCERAMICS

Preparation of Nano-HAp as Vectors for Targeting Delivery System
Y.R. Duan, Z.R. Zhang, Y. Huang and C.Y. Wang 887

Nanostructure of Hydroxyapatite Coatings Sprayed in Argon Plasma
E.I. Suvorova, Y.D. Khamchukov and P.-A. Buffat 891

Modulating the Nanotopography of Apatites
M. Bohner 895

In Vitro Evaluation of Zirconia Nanopowders
S. Braccini, C. Leonelli, G. Lusvardi, G. Malavasi and L. Menabue 899

Preparation and Characterization of Calcium Phosphate Nanoparticles
C.M. Manuel, M. Foster, F.J. Monteiro, M.P. Ferraz, R.H. Doremus and R. Bizios 903

Modulation of Defense Cell Functions by Nano-Particles *In Vitro*
  M. Lucarelli, E. Monari, A.M. Gatti and D. Boraschi 907

**Micropatterning of Apatite by Using CaO-SiO₂ Based Glass Powder
Dispersed Solution**
  T. Matsumoto, N. Ozawa and T. Yao 911

Microstructural Changes of Single-Crystal Apatite Fibres during Heat Treatment
  M. Aizawa, A.E. Porter, S.M. Best and W. Bonfield 915

Effects of Micro/Nano Particle Size on Cell Function and Morphology
  K. Tamura, N. Takashi, T. Akasaka, I.D. Roska, M. Uo, Y. Totsuka and F. Watari 919

Comparative Study of Sonochemical Synthesized ß-TCP- and BCP-Nanoparticles
  M. de Campos, F. Müller, A.H.A. Bressiani, J.C. Bressiani and P. Greil 923

**Specific Characteristics of Wet Nanocrystalline Apatites. Consequences on
Biomaterials and Bone Tissue**
  D. Eichert, H. Sfihi, C. Combes and C. Rey 927

**Influence of Water Addition on the Kinetics of Mechanochemical Synthesis of
Hydroxyapatites from DCPD+CaO**
  H. El Briak-BenAbdeslam, C. Mochales, J.A. Planell, M.P. Ginebra and P. Boudeville 931

VIII. POROUS BIOCERAMICS

**Manufacturing of Biocompatible TiO₂-Surface-Structures with a Water Based
Tape Casting**
  J. Will, S. Zuegner, H. Haugen, U. Hopfner, J. Aigner and E. Wintermantel 937

Processing and Characterisation of a Potential TiO₂ Scaffold
  H. Haugen, J. Will, A. Köhler, J. Aigner and E. Wintermantel 941

**Development of a New Calcium Phosphate Glass Ceramic Porous Scaffold for
Guided Bone Regeneration**
  M. Navarro, S. Del Valle, M.P. Ginebra, S. Martínez and J.A. Planell 945

**Fabrication and Characterization of Porous Hydroxyapatite and Biphasic Calcium
Phosphate Ceramic as Bone Substitutes**
  N. Koç, M. Timuçin and F. Korkusuz 949

Osteoinduction of Bioactive Titanium Metal
  S. Fujibayashi, M. Neo, H.-M. Kim, T. Kokubo and T. Nakamura 953

Calcium Phosphate Porous Scaffolds from Natural Materials
  L.M. Rodríguez-Lorenzo and K.A. Gross 957

Thermally Sprayed Scaffolds for Tissue Engineering Applications
  K.A. Gross and L.M. Rodríguez-Lorenzo 961

**Fabrication of α-Tricalcium Phosphate Porous Body Having a Uniform Pore
Size Distribution**
  M. Kitamura, C. Ohtsuki, S. Ogata, M. Kamitakahara and M. Tanihara 965

Bioactive Behaviors of Porous Si-Substituted Hydroxyapatite Derived from Coral
  S.R. Kim, J.H. Lee, Y.T. Kim, S.J. Jung, Y.J. Lee, H. Song and Y.H. Kim 969

Porous Bioactive Glasses with Controlled Mechanical Strength
L. Fröberg, L. Hupa and M. Hupa 973

Development of Hydroxyapatite Ceramics with Tailored Pore Structure
U. Deisinger, F. Stenzel and G. Ziegler 977

Optimising the Strength of Macroporous Bioactive Glass Scaffolds
J.R. Jones, L.M. Ehrenfried and L.L. Hench 981

Osteoblast Nodule Formation and Mineralisation on Foamed 58S Bioactive Glass
J.E. Gough, J.R. Jones and L.L. Hench 985

Macroporous Bioactive Glasses
V.J. Shirtliff and L.L. Hench 989

In Vitro and *In Vivo* **Validation of a New Bonesubstitute for Loaded Orthopaedic Applications**
G.M. Insley and R.M. Streicher 993

Porous Hydroxyapatite and Glass Reinforced Hydroxyapatite for Controlled Release of Sodium Ampicillin
A.C. Queiroz, S. Teixeira, J.D. Santos and F.J. Monteiro 997

Fabrication of Low Temperature Hydroxyapatite Foams
A. Almirall, G. Larrecq, J.A. Delgado, S. Martínez, M.P. Ginebra and J.A. Planell 1001

Osteoinductive Properties of Micro Macroporous Biphasic Calcium Phosphate Bioceramics
G. Daculsi and P. Layrolle 1005

Multi-Scale Structure and Growth of Nacre: A New Model for Bioceramics
M. Rousseau, E. Lopez, A. Couté, G. Mascarel, D.C. Smith, R. Naslain and X. Bourrat
.... 1009

In Vitro **Calcium Phosphate Formation on Cellulose – Based Materials**
L. Jonášová, F.A. Müller, H. Sieber and P. Greil 1013

Performances of Hydroxyapatite Porosity in Contact with Cells and Tissues
A. Ravaglioli, A. Krajewski, M. Mazzocchi, R. Martinetti and L. Dolcini 1017

Porous Triphasic Calcium Phosphate Bioceramics
J.S.V. Albuquerque, R.E.F.Q. Nogueira, T.D. Pinheiro da SIlva, D.O. Lima and M.H. Prado da Silva 1021

Macro Porous Hydroxyapatite with Designed Pores
E. Adolfsson 1025

Extensive Studies on Biomorphic SiC Ceramics Properties for Medical Applications
P. González, J.P. Borrajo, J. Serra, S. Liste, S. Chiussi, B. León, K. Semmelmann, A. de Carlos, F.M. Varela-Feria, J. Martínez-Fernández and A.R. De Arellano-López 1029

New Method for the Incorporation of Soluble Bioactive Glasses to Reinforce Porous HA Structures
A.F. Lemos, J.D. Santos and J.M.F. Ferreira 1033

Designing of Bioceramics with Bonelike Structures Tailored for Different Orthopaedic Applications
A.F. Lemos and J.M.F. Ferreira 1037

Combining Foaming and Starch Consolidation Methods to Develop Macroporous Hydroxyapatite Implants
A.F. Lemos and J.M.F. Ferreira 1041

The Valences of Egg White for Designing Smart Porous Bioceramics: as Foaming and Consolidation Agent
A.F. Lemos and J.M.F. Ferreira 1045

IX. TISSUE ENGINEERING SCAFFOLDS

Osteogenic Potential of Cryopreserved/ Thawed Human Bone Marrow-Derived Mesenchymal Stem Cells
M. Hirose, N. Kotobuki, H. Machida, S. Kitamura, Y. Takakura and H. Ohgushi 1051

In Vitro **Osteogenic Activity of Rat Bone Marrow Derived Mesenchymal Stem Cells Cultured on Transparent Hydroxyapatite Ceramics**
N. Kotobuki, D. Kawagoe, H. Fujimori, S. Goto, K. Ioku and H. Ohgushi 1055

The Functional Expression of Human Bone-Derived Cells Grown on Rapidly Resorbable Calcium Phosphate Ceramics
H. Zreiqat, G. Berger, R. Gildenhaar and C. Knabe 1059

Osteogenic Potential of Multi-Layer-Cultured Bone Using Marrow Mesenchymal Cells - for Development of Bio-Artificial Bone
T. Yoshikawa, J. Iida and Y. Takakura 1063

In Vitro **Study on the Osteogenetic Capacity of Expanded Human Marrow Mesenchymal Cells - for Development of Advanced Bio-Artificial Bone**
K. Miyazaki, T. Yoshikawa, K. Hattori, N. Okumura, J. Iida and Y. Takakura 1067

Osteogenic Effect of Genistein on *In Vitro* **Bone Formation by Rat Bone Marrow Cell Culture - for Development of Advanced Bio-Artificial Bone**
N. Okumura, T. Yoshikawa, J. Iida, A. Nonomura and Y. Takakura 1071

Experience of Osteogenetic Therapy with Advanceded Bio-Artificial Bone - a Study in 25 Cases
T. Yoshikawa, Y. Ueda, T. Ohmura, Y. Sen, J. Iida, M. Koizumi, K. Kawate,
Y. Takakura and A. Nonomura 1075

Fabrication of Macroporous Scaffold Using Calcium Phosphate Glass for Bone Regeneration
Y.-K. Lee, Y.S. Park, M.C. Kim, K.M. Kim, K.N. Kim, S.H. Choi, C.K. Kim,
H.S. Jung, C.K. You and R.Z. LeGeros 1079

Oriented Collagen-Based/Hydroxyapatite Matrices for Articular Cartilage Replacement
R. Zehbe, U. Gross and H. Schubert 1083

Antimicrobial Macroporous Gel-Glasses: Dissolution and Cytotoxicity
P. Saravanapavan, J.E. Gough, J.R. Jones and L.L. Hench 1087

Mineralised Membranes for Bone Regeneration
P. Mesquita, R. Branco, A. Afonso, M. Vasconcelos and J. Cavalheiro 1091

Inspired Porosity for Cells and Tissues
R. Martinetti, L. Dolcini, A. Belpassi, R. Quarto, M. Mastrogiacomo, R. Cancedda
and M. Labanti 1095

Repair of Full-Thickness Defects in Rabbit Articular Cartilage Using bFGF and Hyaluronan Sponge
T. Yamazaki, J. Tamura, T. Nakamura, Y. Tabata and Y. Matsusue 1099

Bioactive Porous Bone Scaffold Coated with Biphasic Calcium Phosphates
H.-W. Kim, H.-E. Kim and J.C. Knowles 1103

A Self Setting Hydrogel as an Extracellular Synthetic Matrix for Tissue Engineering
P. Weiss, C. Vinatier, J. Guicheux, G. Grimandi and G. Daculsi 1107

Biomimetic Apatite Formation on Chemically Modified Cellulose Templates
F.A. Müller, L. Jonášová, P. Cromme, C. Zollfrank and P. Greil 1111

Preparation of a Composite Membrane Containing Biologically Active Materials
T. Itoh, S. Ban, T. Watanabe, S. Tsuruta, T. Kawai and H. Nakamura 1115

Effect of Inorganic Polyphosphate on Periodontal Regeneration
T. Shiba, Y. Takahashi, T. Uematsu, Y. Kawazoe, K. Ooi, K. Nasu, H. Itoh,
H. Tanaka, M. Yamaoka, M. Shindoh and T. Kohgo 1119

PLGA-CMP Composite Scaffold for Articular Joint Resurfacing
J.H. Jeong, S.K. Park, D.J. Lee, Y.M. Moon, D.C. Lee, H.I. Shin and S. Kim 1123

Author Index 1127
Keyword Index 1137

I. APATITES, PHOSPHATES AND GLASS CERAMICS

Key Engineering Materials Vols. 254-256(2004) pp. 3-6
online at http://www.scientific.net
© *2004 Trans Tech Publications, Switzerland*

Stimulation of Bone Repair by Gene Activating Glasses

Larry L. Hench[1]

[1]Imperial College London, Department of Materials, Exhibition Road, London SW7 2AZ, UK.
l.hench@imperial.ac.uk

Keywords: Bioactive, Bone, Cell Cycle, Genes, Growth Factors, Osteoblasts

Abstract. Thirty-four years ago certain compositions of SiO_2-Na_2O-CaO-P_2O_5 glasses were discovered to bond to bone. These bioactive glasses are now widely used in dental and medical applications to enhance repair of bone defects. Recent molecular biology studies show that the ionic dissolution products of the glasses control the cell cycle of osteoprogenitor cells. Human osteoblasts that are not capable of differentiating into a mature osteoblast phenotype are eliminated by programmed cell death (apoptosis).

Gene micro-array analyses show that seven families of genes are up-regulated in the presence of critical concentrations of soluble Ca and Si ions released from the bioactive glasses. The activated genes include cell-signalling molecules, various growth factors unique to the osteoblast lineage, transcription factors, DNA synthesis and repair and extracellular matrix proteins. Genetic control of the osteoblast phenotype exists for human foetal stem cells as well as adult stem cells when the temporal sequence of the cell cycles are matched with the kinetics of dissolution of both melt-derived 45S5 Bioglass® or sol-gel derived 58S bioactive gel-glasses.

Introduction. In Bioceramics 7 (1994) the author proposed that *in vivo* rates of bone formation in the presence of bioactive materials depend upon the kinetics of release of ionic dissolution products [1]. Rapid dissolution of Si, Ca, P and Na ions give rise to both *intracellular* and *extracellular* responses that lead to *osteoproduction* and Class A bioactivity. Slow surface reactions with low ionic release rates result in only *extracellular* responses, *osteoconduction* and Class B bioactivity.

In Bioceramics 13 (2001) the author discussed recent results from the Imperial College Tissue Engineering and Regenerative Medicine Centre team, composed of ID Xynos, AI Edgar, L Buttery, LL Hench and JM Polak, that explained enhanced osteogenesis of Class A bioactive materials [2]. The paper shows that the time required for reaction stage 8 (attachment of osteoprogenitor cells), stage 9 (synchronised proliferation and differentiation of osteoblasts) and stage 10 (generation of extracellular matrix) was accelerated in the presence of critical concentrations of ionic dissolution products released from 45S5 Bioglass®. Bioactive control of the osteoblast cell cycle begins within the first 24-48 hours exposure to either Class A bioactive substrates, [3] or the ionic dissolution products released from the bioactive materials [4]. By 6 days the proportion of osteoblasts in synthesis phase and G2-M (mitosis) phase is greatly enhanced for the cells exposed to the bioactive stimulus [3-5]. Cells that are not capable of differentiating into a mature osteoblast phenotype are eliminated by apoptosis (programmed cell death) when exposed to critical concentrations (e.g. 15 to 20ppm of Si) of ionic dissolution products [3-5]. The bioactive stimulated osteoblasts quickly started producing osteocalcin and type I collagen and produced mineralised bone nodules [3-5].

Genetic Control of the Osteoblast Cell Cycle. Recent findings by Xynos *et al.* have shown that the bioactive shift of osteoblast cell cycle described above is under genetic control [6, 7]. Within a few hours exposure of human primary osteoblasts to the soluble chemical extracts of 45S5 Bioglass® several families of genes are activated including: genes encoding nuclear transcription factors and potent growth factors, especially IGF-II, along with IGF binding proteins, apoptosis regulators and proteins that control DNA synthesis and repair, and extracellular matrix proteins. These findings indicate that Class A bioactive glasses enhance new bone formation (osteogenesis) through a direct control over genes that regulate cell cycle induction and progression.

Transcription Factors. Entry of osteoblasts into the cell cycle (Go/G1 transition) and subsequent commencement of cell division is regulated by a family of transcription factors. These molecules provide the specific stimuli needed to develop cells of the osteoblastic phenotype. These specific proteins must be present for a bone stem cell to become a bone-growing cell. Exposing primary human osteoblast cultures to the ionic products of bioactive glass dissolution for 48 hours activates expression of a large number of transcription factors and cell cycle regulators as shown in Refs. 6 and 7 (Table 1). The transcription factors that were activated include c-jun, fra-1 and c-myc, three well characterised osteoblast transcription factors. Details of the importance of these transcription factors are described in Refs. 6-11.

For example, osteoblast proliferation and phenotypic commitment is triggered by transcription factors c-Myc and AP-1 (Table 1) but depends on successful progression through the cell. Certain cyclins are required for the progression from the G1 phase of the cell cycle to the synthesis (S) phase. These critical cyclins include cyclin D1 (G1/S specific cyclin) which phosphorylates the product of the retinoblastoma gene [9-11] resulting in the release of transcription factors important for the initiation of DNA replication. Cyclin D1 is upregulated by 400% when osteoblasts are exposed to the ionic products of bioactive glass dissolution for 48 hours (Table 1). This large increase in gene activation of cyclins demonstrates that the bioactive glass does not merely trigger the entry of osteoblasts into the cell cycle but also provides the vital stimulus needed for progression through the G1/S checkpoint, a crucial step for the successful completion of the cycle.

Two other important cell cycle regulators CDKN1A and cyclin K [12] were also activated by the ionic dissolution products of bioactive glass by 200% or more (Table 1). Both are involved in the regulation of the early stages of the mitotic cycle of the cells.

Table 1 Transcription factors and cell cycle regulators

RCL growth-related c-myc-responsive gene	500%
G1/S-specific cyclin D1 [CCND1]	400%
26S protease regulatory subunit 6A	400%
cyclin-dependent kinase inhibitor 1 [CDKN1A]	350%

DNA Synthesis and Repair. The cell cycle represents a highly synthetic phase in the life of an osteoblast. During this phase the cell produces proteins and nucleic acids that eventually result in the formation of two daughter cells. Mistakes in the synthesis of proteins and nucleic acids are quite likely, especially in the mitosis of progenitor cells of older people. In order to avoid such mistakes being passed on during cell division the cell possesses an arsenal of mechanisms that can determine whether damage is present, evaluate its extent and correct it, if feasible.

The up-regulation of DNA repair proteins by the ionic products of bioactive glass dissolution, listed in Table 2, indicates that these mechanisms are activated in human osteoblasts. At least four important genes involved in DNA synthesis, repair and recombination are differentially expressed at levels of >200% over control osteoblasts [6,7].

Table 2 DNA synthesis, repair and recombination

DNA exclusion repair protein ERCC1	300%
mutL protein homolog	300%
high mobility group protein [HMG-1]	230%
Replication factor C 38-kDa subunit [RFC38]	200%

Apoptosis Regulators. When the damage is beyond repair the cell voluntarily exits the mitotic cell cycle through death by apoptosis, programmed cell death. Apoptosis thereby prevents the creation of abnormal cells and represents a means to regulate the selection and proliferation of functional osteoblasts; i.e. osteoblasts capable of synthesising the complex array of extracellular proteins and mucopolysaccharides required to form a mineralised matrix that is characteristic of mature osteocytes. The treatment of the osteoblast cultures with the bioactive glass stimuli induced the expression of several important genes involved in apoptosis, as summarised in Table 3 [6,7]. The up-regulated genes include calpain [13] and defender against cell death (DAD1) [14]. Activation of the calpain system, a proteolytic mechanism is thought to mediate apoptotic cell death. On the other hand, DAD1, a regulator of N-linked glycosylation, is essential for cell survival since DAD1 mutation was shown by Hong *et al*. [14] to induce embryonic apoptosis in mice.

Table 3 Apoptosis regulators

defender against cell death 1 [DAD-1]	450%
calcium-dependent protease small (regulatory) subunit; calpain	410%
deoxyribonuclease II [Dnase II]	160%

Growth Factors, Cytokines, Receptors and ECM Proteins. The cell cycle does not merely provide the framework for cell proliferation but also determines to some extent cell commitment and differentiation. Bone cells cover a broad spectrum of phenotypes that include predominately the osteoblast, a cell capable of proliferating and synthesising bone cell specific products such as Type I collagen. However, a vital cellular population in bone consists of osteocytes. Osteocytes are terminally differentiated osteoblasts and are usually postmitotic and not capable of cell division. They are capable of synthesising and maintaining the mineralised bone matrix wherein they reside. Thus, osteocytes represent the cell population responsible for extracellular matrix production and mineralisation, the final step in bone development (reaction stages 10 and 11) [1,2] and probably the most crucial one, given the importance of collagen-hydroxyl carbonate apatite (HCA) bonding in determining the mechanical function of bone. Therefore, it is important to observe that the end result of the cell cycle activated by the ionic products of bioactive glass dissolution was the up-regulation of numerous genes that express growth factors and cytokines (Table 4) and extracellular matrix components (Table 5). Also, there was 700% increase in the expression of CD44 (Table 6) a specific phenotypic marker of osteocytic differentiation as shown by Hughes *et al*., 1994 [15].

Table 4 Growth factors and cytokines

insulin-like growth factor II [IGF2]; somatomedin A	320%
bone-derived growth factor 1 [BPGF1]	300%
macrophage-specific colony-stimulating factor [CSF-1; MCSF]	260%
vascular endothelial growth factor precursor [VEGF]	200%

The cDNA microarray analysis [6,7] showed that expression of a potent osteoblast mitogenic growth factor, insulin-like growth factor II (IGF-II), was increased to 320% by exposure of the osteoblasts to the bioactive glass stimuli (Table 4). This is an important finding because IGF-II is an anabolic peptide of the insulin family and constitutes the most abundant growth factor in bone and is also a known inducer of osteoblast proliferation *in vitro*. It is produced locally by bone cells and is considered to exert mostly autocrine or paracrine effects [4]. IGF-II expression is relatively high in developing bone periosteum and growth plate, healing fracture callus tissue and developing ectopic bone tissue, as reviewed by Xynos *et al*. [4, 6].

Table 5 Extracellular matrix components

matrix metalloproteinase 14 precursor [MMP14]	370%
matrix metalloproteinase 2 [MMP2]	270%
metalloproteinase inhibitor 1 precursor [TIMP1]	220%
TIMP-2 [MI]	220%
bone proteoglycan II precursor; decorin	220%

Table 6 Cell surface antigens and receptors

CD44 antigen hemoatopoietic form precursor	700%
fibronectin receptor beta subunit; integrin beta 1	600%
N-sam; fibroblast growth factor receptor1 precursor [FGFR1]	300%
vascular cell adhesion protein 1 precursor [V-CAM 1]	200%

Conclusions

The molecular biological mechanisms involved in the behaviour of bioactive glasses are finally beginning to be understood, thirty-four years after their discovery [16]. The bioactive response appears to be under genetic control. Class A bioactive glasses that are osteoproductive enhance osteogenesis through a direct control over genes that regulate cell cycle induction and progression. Cells that are not capable of forming new bone are eliminated from the cell population, a characteristic that is missing when osteoblasts are exposed to bioinert or Class B bioactive materials. The biological consequence of genetic control of the cell cycle of osteoblast progenitor cells is the rapid proliferation and differentiation of osteoblasts. The result is rapid regeneration of bone. The clinical consequence is rapid fill of bone defects with regenerated bone that is structurally and mechanically equivalent to normal, healthy bone.

Acknowledgements

The author gratefully acknowledge the support of the UK Engineering and Physics Research Council, the UK Medical Research Council and US Biomaterials Corporation.

References

[1]   L.L. Hench: Bioceramics 7, O.H. Anderson and Yli-Urpo (Eds) (Butterworth-Heinemann Ltd, Oxford, England 1994).

[2]   L.L. Hench: Key Engineering Materials Vols. 192-195 (2001), pp. 575-580.

[3]   I.D.Xynos, M.V.J. Hukkanen, J.J. Batten, I.D. Buttery, L.L. Hench and J.M. Polak: Calcif. Tiss. Int. Vol. 67 (2000), 321-329.

[4]   I.D. Xynos, A.J. Edgar, L.D. Buttery, L.L. Hench and J.M. Polak: Biochem. and Biophys. Res. Comm. Vol. 276 (2000), pp.461-465.

[5]   L.L. Hench, I.D. Xynos, L.D. Buttery and J.M. Polak: J. Mater. Res. Innovations Vol. 3 (2000), pp. 313-323.

[6]   I.D. Xynos, A.J. Edgar, L.D.K. Buttery, L.L. Hench and J.M. Polak: J.Biomed. Mater. Res. Vol. 55 (2001), 151-157.

[7]   L.L. Hench, I.D. Xynos, A. J. Edgar, L.D.K. Buttery and J.M. Polak: Proc. Int. Congr. Glass Vol. 1. (1-6 July 2001), 226-233.

[8]   A.E. Grigoriadis, Z.Q. Whang and E.F. Wagner: Cellular and molecular biology of bone, M. Noda (Ed), (Academic Press Inc., San Diego, Ca. 1996), pp. 15-24.

[9]   M. Zorning and G.I. Evan: Curr. Biol. Vol. 1 (1996), pp. 1553-1556.

[10]   B.C. Lewis, H.Shim, Q. Li, C.S. Lee, A. Maity and C.V. Dang: Mol. Cell. Biol. Vol. 17 (1997), pp. 4967-78.

[11]   M.A. Caligo, G. Cipollini, L. Fiore, S. Calvo, F. Basolo, P. Collechi, S. Pepe, M. Petrini and G. Bevilacqua: Leuk. Res. Vol. 20 (1996), pp. 161-167.

[12]   T.J. Fu, J. Peng, G. Lee, D.H. Price and O. Flores: J. Biol. Chem. Vol. 274 (1999), pp. 34527-34530.

[13]   M.K. Squier, A.C. Miller, A.M. Malkinson and J.J. Cohen: J. Cell Physiol. Vol. 159 (1994), pp. 229-237.

[14]   N.A. Hong, M. Flannery, S.N. Hsieh, D. Cado, R. Pedersen and A. Winoto: Dev. Biol. Vol. 220 (2000), pp. 76-84.

[15]   D. Hughes, D.M. Salter and R. Simpson: J. Bone Miner. Res. Vol. 9 (1994), 39-44.

[16]   L. L. Hench, R. J. Splinter, W. C. Allen, and T. K. Greenlee, Jr.: J. Biomed. Mater. Res. Vol. 2[1] (1971), pp. 117-141.

Key Engineering Materials Vols. 254-256(2004) pp. 7-10
online at http://www.scientific.net
© *2004 Trans Tech Publications, Switzerland*

In Vitro Simulation of Calcium Phosphate Crystallization

from Dynamic Revised Simulated Body Fluid

C Deng[1, 2], Y Tan[2], C Bao[2], Q Zhang[2], H Fan[2], J Chen[2] and X Zhang[2]

[1]College of Material and Bioengineering, Chengdu University of Technology, Chengdu610059, China,
dengchunlin0138@sina.com.cn
[2]Engineering Research Center in Biomaterials, Sichuan University, Chengdu610064, China

Keywords: Bone-like apatite, biphasic calcium phosphate ceramics, dynamic revised simulated body fluid (Dynamic RSBF), carbon dioxide, bone marrow stromal cells (BMSCs)

Abstract. A dynamic device was designed to study the formation of bone-like apatite on the inner surface of biphasic porous calcium phosphate in RSBF *in vitro*. The ceramics were examined with scanning electron microscopy (SEM), energy dispersive spectroscopy (EDS) and Infrared spectroscopy (IR). Spherical amorphous crystal on the surface pore walls and sheet-shaped crystals on the inner pore walls of the ceramics were found. According to co-cultured with bone marrow stromal cell (BMSCs), the cell response to the samples was investigated with SEM and MTT assay.

Introduction

Formation of bone-like apatite is one of the important subjects in osteoinductivity of bioceramics. The *in vivo* process of apatite formation is usually simulated *in vitro* by precipitation of calcium phosphates from supersaturated calcium and phosphate-containing solutions. Dynamic method, introduced by Duan et al. [1, 2] recently, is closer to physiological condition than static immersion assay. However, the pH value of simulated body fluid (SBF) changes with time after immersion, in which calcium phosphates precipitate to form non-stoichiometric hydyoxyapatite. In addition, the simulation *in vitro* is different from physiological condition *in vivo* in that CO_2 is produced in the process of metabolism in vivo [3]. To simulate the behavior of calcium phosphate ceramics veritably, a modified CO_2 supplying device was designed to adjust the pH value and to enhance buffer ability of solution and porous biphase calcium phosphate ceramics were immersed in a revised body fluid (R-SBF) introduced by Kim et al.[4]. Subsequently, cell biocompatibility was investigated by MTT test.

Materials and Methods

Biphasic porous calcium-phosphate ceramics (HA/α-TCP=60/40) were prepared by foaming with H_2O_2. The cylindrical-shaped samples ($\Phi4\times8mm$) were cleaned by ultrasonication.
4–4,5–dimethylthiazol-2-y1-2,-diphenyltetrazolium bromide (MTT) was purchased from Sigma Chemical Co. (USA), and DMEM/F12 (1:1) medium was purchased from Hyclone Co (USA).
Experiments on the formation of bone-like apatite in revised simulated body fluid (R-SBF) were performed at 37°C and a stable pH (7.4), which was controlled by filling CO_2 gas into R-SBF storage tank at a suitable rate. R-SBF flowed at a rate of 2 ml/min by a peristaltic pump (Fig. 1). HA/α-TCP was immersed for 2 to 7 days in R-SBF. After that, samples were removed, washed with distilled water for several times, then, dried at 50°C and split in halves.
The microstructures of the samples were analyzed with scanning electron microscopy (SEM). The

chemical groups of the precipitates on the surfaces were detected with Infrared spectroscopy (IR). Bone marrow stromal cells (BMSC) were isolated from the marrow of dog's femur and cultured until confluence was reached. Cells were then harvested and seeded on the immersed ceramics in 24-well plate at a concentration of 0.5×10^6 cells/cm^2. Cells seeded in wells in 24-well plate were used as Control.

Cell proliferation was determined by MTT colorimetric method after culture for 1, 2, 4 and 8 day; 3 parallel samples were used at each time point. Briefly, after removing the supernatant, ceramics(with cells) were washed with PBS, transferred into new 24-well plates containing 1ml of MTT solution(5mg/ml in PBS) and incubated for 3h at 37°C. After incubation, the supernatant was discarded, 1ml of extraction (DMSO) was added and vortexed for 5min to allow total color released from the materials, then centrifuged at 12000g for 5min. The supernatant was read at 570nm.

For SEM, at 12h, 24h, 48h, 96h, cultures were rinsed with PBS and fixed for 1h with 2.5% glutaraldehyde in 0.1mol/L phosphate buffer (pH=7.2). The cultures were subsequently rinsed and gradually dehydrated with alcohol. They were processed for critical point drying in CO_2 and then coated with gold. Samples were examined using a scanning electron microscope.

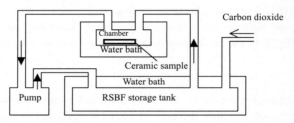

Fig. 1.Schematic drawing of dynamic immersion system

Results and Discussion

The SEM results showed that the surface pore walls of porous ceramics were completely overlaid with a dense layer of ball-like particles consisting of very tiny agglomerated crystals (Fig. 2A), and its inner pore walls were overlaid with sheet-like crystals (Fig. 2B) after immersion of 7 days in RSBF solution under dynamic condition.

The difference of shape of the formed crystal indicated that crystal environments in inner pores were different from that on surface pore. Because ions exchanged rapidly along the sample surface, local ion concentration was lower than that inside the samples. Therefore, tiny agglomerated crystals built after crystal nucleus formation. In the inner pores, solution flowed much slower than that on the surface, local ion concentration was higher and thus, sheet-like crystals grew up.

The results of EDS revealed that the precipitates consisted of calcium, phosphorus, oxygen and carbon. The existences of carbon suggested that the precipitates contained carbonate (Fig.3).

Fig 2. Typical SEM image of the precipitates formed (A) in surface pore and (B) in interior pore after immersed for 7days in RSBF with CO_2 gas.

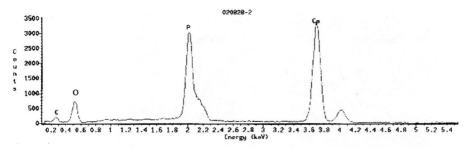

Fig. 3. EDS spectra of crystals formed on the inner pore walls of biphase calcium phosphate from dynamic RSBF with CO_2 (7 days).   Strong peaks of calcium and phosphorus and weak peaks of carbon and oxygen can be distinguished.

The IR spectroscopy suggested that characteristic peaks of CO_3^{2-}, PO_4^{3-} and OH^- appeared on the surface of the ceramics after immersion for 7 days in dynamic RSBF (Fig. 4). This was accordant with the results of EDS.

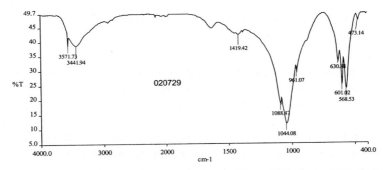

Fig. 4. IR spectra of the formed crystals in porous ceramics from dynamic RSBF (7day). The peak at 1419.42 cm^{-1} is characteristic peak of CO_2.

These results of IR, EDS and SEM indicated that the precipitates were carbonate hydroxyapatite (CHA).
When co-cultured with BMSCs, MTT assay demonstrated that the immersed ceramics and the control showed a similar trend (Fig. 5). Therefore, the precipitates can support proliferation of the BMSCs.

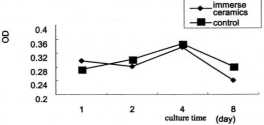

Fig. 5 The variation of OD value with time in MTT assay while osteoblasts co-cultured with ceramics.

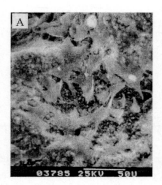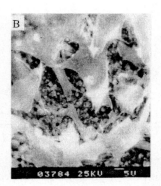

Fig.6. BMSCs were incubated with ceramics for 48h. Typical cell morphology on the surface and the pores are shown. A lot of cells proliferated on the surface could also be seen (original magnification: A =×500, B ×1000).

Fig.6 is a typical SEM image of BMSCs cultured on the surface of immersed ceramics for 48h under different multiple. The SEM results indicated that cells adhere to the surface of immersed ceramics uniformly the immersed ceramics showed excellent biocompatibility.

Conclusions

The dynamic conditions with CO_2 supplying used in Revised Simulated Body Fluid in this study are closer to physiological condition *in vivo*. Experimental results of IR, EDS and SEM indicated that the obtained precipitates were carbonate hydroxyapatite (CHA), but the shape of precipitates on surface pore walls and on inner pore walls of ceramics were different. *In vitro* data demonstrate that precipitates are suitable for cell adhesion and proliferation. The association of osteoprogenitor cells with R-SBF-immersed ceramics can provide an appropriate vehicle for repairing small bone fractures such as non-unions and cavitational defects.

Acknowledgments

This work was financially supported by the Key Basic Research Project of China (Contract No.G1999064760). The authors thank Prof. Zhao Huichuan for the IR spectroscopy test.

References

[1] Y. R. Duan, C. Y. Wang, J. Y. Chen and X. D. Zhang:The International Conference on Materials for Advanced Technologies (ICMAT 2001), July 1-6, 2001, Singapore; P2007.

[2] P Siriphannon, S Hayashi, A Yasumori, K Okada and H.Shigeo:Comparative study of the formation of hydroxyapatite in simulated body fluid under static and flowing systems. J Biomed Mater Res., Vol. 60 (2002), p. 175-185.

[3] L.Garby, J.Meldon:*The respiratory functions of blood*. Plenum Publishing Corporation, 1977.

[4] H. -M. Kim, T. Miyazaki, T. Kokubo and T. Nakamura: In: *Bioceramics 13* (Trans Tech Publication, Switzerland 2001), P47.

Key Engineering Materials Vols. 254-256(2004) pp. 11-14
online at http://www.scientific.net
© *2004 Trans Tech Publications, Switzerland*

Control of Calcium Phosphate Crystal Nucleation, Growth and Morphology by Polyelectrolytes

H. Füredi-Milhofer, P. Bar-Yosef Ofir, M. Sikiric and N. Garti

The Hebrew University of Jerusalem, Jerusalem 91904, Israel
e-mail: Helga@vms.huji.ac.il ; Pazitb@pob.huji.ac.il

Keywords: precipitation, phase transformation, crystal growth morphology, calcium hydrogenphosphate dihydrate, amorphous calcium phosphate, octacalcium phosphate, polyaspartic acid, calcium phytate, poly-l-lysine, poly-l-glutamic acid, polystyrene sulfonate.

Abstract. Crystal growth of calcium hydrogenphosphate dihydrate (DCPD), in the presence of polyaspartic acid or calcium phytate (*system A*), as well as nucleation and growth of octacalcium phosphate (OCP), in the presence of poly-l-lysine, poly-l-glutamic acid or polystyrene sulfonate (*system B*) have been investigated. In *system A* crystallization of DCPD was inhibited and the crystal growth morphology was specifically modified by preferential interaction of the respective additive with the dominant (010) crystal face. In *system B* crystals were formed via precursor phase(s) and polyelectrolytes exhibited dual action, at low concentrations inducing and at high concentrations inhibiting nucleation of the crystalline phase. Crystal/additive interactions controlling growth were nonspecific and resulted in smaller crystals with rounded edges, but with the same basic orientation as in the controls.

Introduction

Biomimetic deposition of calcium phosphate coatings onto bioinert surfaces, with the purpose to induce bioactivity and facilitate osteointegration, is rapidly gaining in importance [1-4]. In another important development, attempts are made to incorporate biologically active molecules (osteogenic agents, growth factors or drugs) into the coatings by adsorption onto preformed calcium phosphate matrices or by coprecipitation [2-4]. Of the existing methods coprecipitation seems to be preferable, but the underlying mechanisms are so far poorly understood [4]. It has, moreover, been observed [2, 4], that the properties of calcium phosphate coatings, such as thickness, crystal structure and morphology were significantly modified if protein was present in the solution during precipitation. In this context, systematic studies of the effect of polyelectrolytes (PE) on calcium phosphate crystallization seem to be of considerable importance. In a previous paper [5] we have demonstrated the dual role of PE's in amorphous calcium phosphate (ACP) – octacalcium phosphate (OCP) phase transformation, i.e. all investigated PE's at low concentrations induced and at high concentrations retarded nucleation of the crystalline precipitate. In this paper we discuss the influence of several PE's, differing in charge and molecular structure, on the rate of growth and the crystal growth morphology of calcium hydrogenphosphate dihydrate (DCPD) and OCP/apatite.

Materials and methods

Analytical grade chemicals and ultrapure water were used for all experiments. The PE's, polystyrene sulfonate, (PSS, 70000 D), poly-l-glutamic acid (PGA, 50000 D), poly-l-lysine (PLL, 50000 D) polyaspartic acid (PASP, 5000-15000 D) and calcium phytate (PHY), were purchased from Sigma and used as received. The respective molecular structures are given in Table 1. Samples were prepared by rapid mixing of equal volumes of equimolar solutions of calcium chloride and sodium phosphate at initial pH 5.5 (*system A*) and 7.4 (*system B*) respectively. In *system A* the solution of PHY or PASP was added to the phosphate solution before pH adjustment and mixing, whereas in *system B*, if not otherwise stated, solutions of PGA, PLL or PSS were added into the systems 15 min after completion of secondary precipitation, to ensure that all of the added PE was present during crystal growth. To facilitate comparison, all PE concentrations are expressed as mol monomer per liter. The reactions were followed by monitoring changes in the pH at constant temperature, 37°C (*system A*) and 25°C (*system B*), respectively. Crystals were isolated 24 h after sample preparation and characterized by optical and electron (TEM and SEM) microscopy, electron diffraction, and X-ray powder diffraction. Crystal dimensions were measured from electron micrographs taking average values from 20 – 40 measurements.

Table 1. Molecular structures of the investigated polyelectrolytes.

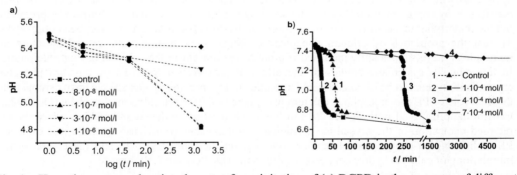

| PLL | PGA | PSS | PASP | PHY |

Results

DCPD crystals were formed without a precursor phase. In the presence of PHY (Fig. 1a) and PASP (not shown) crystallization was retarded, as indicated by the respective pH vs. time curves, the effect increasing with increasing additive concentration. In contrast, OCP and/or apatite crystals formed through precursor phase(s) in at least three distinct steps, (i) precipitation of amorphous calcium phosphate (ACP), (ii) nucleation and growth of a crystalline phase on account of the residual supersaturation and catalyzed by ACP, (secondary precipitation) and (iii) solution mediated recrystallization and growth (Fig. 1b and refs. 5, 6). PGA had a dual effect: at low concentrations it induced (curve 2 in Fig. 1), whereas at high concentrations it retarded (curve 3) or totally inhibited (curve 4) secondary precipitation (see also ref. 5). Dual action was also observed when OCP was crystallized in the presence of PLL and/or PSS (data not shown).

Fig. 1. pH vs. time curves showing the rate of precipitation of (a) DCPD in the presence of different concentrations of PHY (*system A*) and (b) OCP in the presence of different concentrations of PGA (*system B*). In both examples the additive was added before mixing of the reactants.

All investigated PE's significantly influenced crystal morphology. *System A:* In the controls DCPD crystals grew into (010) platelets approx. 100 x 200 μm (Fig. 2 a). In the presence of PHY (effective concentration range $8 \times 10^{-8} - 1 \times 10^{-5}$ mol/l) crystals appeared larger (up to 700 μm in length, Fig. 2 b), but with the same basic orientation as in the control, indicating preferential adsorption in the direction perpendicular to (010). When grown in the presence of PASP, $(5 \times 10^{-6} -$

1×10^{-5} mol/l) large (300 – 500 μm in length), fragmented DCPD crystals with a leaf-like appearance precipitated (Fig. 2 c).

System B: OCP crystals grown in the presence of PGA, PLL and PSS were generally smaller and thinner than in the controls, with rounded edges and the length/width ratio only slightly affected (Figs. 2 e, f and Table 2). Thus, growth was inhibited in all directions, due to adsorption of the PE at all active sites. Effective concentrations were at least 10x higher than for PHY and PASP (Table 2).

Fig. 2 SEM (a,b,c) and TEM (d,e,f) micrographs, showing morphological changes in *systems A* and *B.* 2 a) DCPD control; b) $1 \cdot 10^{-7}$ mol/l PHY (after ref. 7); c) $1 \cdot 10^{-5}$ mol/l PASP; d) OCP control; e) $1 \cdot 10^{-4}$ mol/l PSS; f) $7 \cdot 10^{-3}$ mol/l PSS. Aging time 24 h.

Table 2. Average particle sizes and length/width ratios of OCP crystals obtained in the controls and in the presence of different PE's.

Additive	Mw / D	c / mol/l	Length / nm	Length/width
none			557.6 ± 209.7	1.68 ± 0.55
PGA	50 000	$5 \cdot 10^{-6}$	431.9 ± 172.0	1.96 ± 0.72
		$1 \cdot 10^{-4}$	126.8 ± 28.2	2.13 ± 0.82
		$5 \cdot 10^{-4}$	129.8 ± 40.9	2.09 ± 1.41
PLL	50 000	$5 \cdot 10^{-6}$	393.9 ± 253.9	1.89 ± 0.57
		$1 \cdot 10^{-4}$	189.8 ± 110.7	2.03 ± 0.88
		$7 \cdot 10^{-3}$	165.5 ± 88.2	1.77 ± 0.65
PSS	70 000	$5 \cdot 10^{-6}$	546.4 ± 211.2	2.11 ± 0.90
		$1 \cdot 10^{-4}$	143.9 ± 43.6	1.84 ± 0.59
		$7 \cdot 10^{-3}$	148.2 ± 54.4	1.87 ± 0.52

Discussion and Conclusions

We have demonstrated two types of interactions of PE's with calcium phosphate crystals, specific and nonspecific. Specific interactions, preferentially affecting certain crystal faces, retard crystal growth in preferred directions and thus induce specific changes in the crystal growth morphology (Figs 2 b, c and refs. 7, 8). A basic requirement is that the interacting molecule is at least partially in an ordered conformation in the crystallizing solution [9]. In the course of nonspecific interactions all active sites in a crystal (faces, edges, corners) are affected and as a consequence crystals appear rounded and their sizes are reduced (Figs. 2 e, f and Table 2).

Among the additives discussed in this work (Table 1) PHY and PASP (partial beta sheet) appear in ordered conformations. PHY specifically recognizes the (010) face of DCPD (Fig. 2b), the reason being, that the distance between two phosphate groups in PHY (6.19 Å), fits fairly well the distance of two calcium atoms, exposed on the interacting crystal face (6.24 Å, [7]). Similarly, it has been shown [8], that PASP recognizes the dominant (100) face of OCP, which has been explained with the correspondence between the distances between two adjacent carboxylate groups in the beta-sheet of PASP (6.7 – 6.9 Å) and the distances between Ca atoms exposed at the (100) face of OCP (6.87 Å) However, in the case of DCPD and PASP, a structural and stereochemical fit exists between the carboxylate groups of the beta-sheet strand and the distances of neighboring calcium atoms from two adjacent layers within one Ca-HPO$_4$ bilayer (6.95 Å, [7]). Therefore, during crystallization in the presence of PASP, the polyelectrolyte is incorporated into DCPD crystals, resulting in their leaf-like appearance (Figs 2 b, c).

PGA, and PLL appear in solution as flexible, open chains with multiple charges (Table 1), while the highly charged sulfonate group of PSS is situated on a bulky benzene ring, and may therefore be less flexible. Thus, the former two molecules are likely candidates for nonspecific, electrostatic interactions, while PSS may interact both specifically and non-specifically. In this work nonspecific interactions of all three PE's with growing OCP crystals have been demonstrated (Figs. 2d-f and Table 2), while indications of preferential interactions, when crystals grow in the presence of low concentrations of PSS, still have to be confirmed.

Considerations, as outlined above, may be helpful in explaining some of the phenomena encountered in the coprecipitation of macromolecules with calcium phosphate crystals, when preparing biomimetic coatings [2, 4].

Acknowledgement. This work was supported in part by the European Commission through the SIMI project (GRDI-2000-26823). It is a pleasure to acknowledge the help with SEM and TEM by Mr. V. Gutkin and Mrs. R. Govrin-Lippman, from the Hebrew University of Jerusalem.

References

[1] Y. Abe, T. Kokubo, and T. Yamamuro, J. Mater. Sci., Mater. Med. 1 (1990), p. 233

[2] I. B. Leonor, H. S. Azevedo, C. M. Alves and R. L. Reis, Key Engineering Materials Vols. 240-242 (2003), p. 97

[3] S. Radin, J.T.Campbell and P. Ducheyne, Biomaterials 18 (1997), p. 777

[4] Y. Liu, P. Layrolle, J. de Bruijn, C. van Blitterswijk and K. de Groot, J. Biomed. Mater. Res.A Vol. 57 (2001), p. 327

[5] H. Füredi-Milhofer, P. Bar Yosef, R. Govrin-Lippman and N. Garti, Key Engineering Mat. Vols. 240-242 (2003), p. 453

[6] Lj. Brecevic and H. Füredi-Milhofer, Calc. Tiss. Res. 10 (1972), p. 82

[7] M. Sikirić, V. Babić-Ivančić, O. Milat, S. Sarig and H. Füredi-Milhofer, Langmuir 16 (2000), p. 9261; Key Engineering Mat. Vols. 192-195 (2001), p. 11

[8] H. Füredi-Milhofer, J. Moradian-Oldak, S. Weiner, A. Veis, K. P. Mintz and L. Addadi, Conn. Tissue Res. 30 (1994), p. 251

[9]L. Addadi and S. Weiner, Angew. Chemie Int. Engl. Ed. 31 (1992), p. 153

Key Engineering Materials Vols. 254-256(2004) pp. 15-18
online at http://www.scientific.net

Fabrication and Characterization of
Apatite Sintered Bodies with c-Axis Orientation

Kazushi Ohta, Masanori Kikuchi and Junzo Tanaka

Biomaterials Center, National Institute for Materials Science

1-1 Namiki, Tsukuba, Ibaraki 305-0044, JAPAN

E-mail: OHTA.Kazushi@nims.go.jp, KIKUCHI.Masanori@nims.go.jp, TANAKA.Junzo@nims.go.jp

Keywords: hydroxyapatite, sintered body, crystal orientation

Abstract. Hydroxyapatite (HAp) crystals were synthesized by the hydrolysis of $CaHPO_4 \cdot 2H_2O$ in a NaOH solution. Their aggregates formed by hydrolysis at 100°C were composed of needle-like crystals which aligned in c-axis orientation perpendicular to the wide surface of the plate-like HAp aggregate. On the other hand, the aggregates formed by hydrolysis at 25°C were composed of small crystals with random orientation. HAp sintered bodies were fabricated using these aggregates. In the HAp sintered body, the plate-like HAp aggregates formed at 100°C accumulated horizontally to the flat surface of the cylindrical sintered body and the c-axis orientation of the HAp aggregates was maintained even after sintering. On the contrary, the HAp sintered body fabricated from the aggregates formed at 25°C had no obvious crystal orientation.

Introduction

Hydroxyapatite ($Ca_{10}(PO_4)_6(OH)_2$, HAp) is a main inorganic component of hard tissues in vertebrates and is used for artificial bones [1], liquid chromatographic packings [2] and scaffolds for tissue engineering [3]. The HAp crystal has two surfaces with different electrical charges, i.e., the positively charged a-surface adsorbing negatively charged molecules and the negatively charged c-surface adsorbing positively charged molecules [2]. Therefore, regulating the morphology and agglomerate is very important to control their functions in medical and industrial use. Although many fabrication methods of HAp sintered bodies have been reported such as conventional sintering [4], hot pressing [5], spark plasma sintering [6] and microwave sintering [7], only a few reports on the control of crystal orientation in sintering are available. Fabrication of a-axis oriented HAp sintered bodies was reported recently by using a high magnetic field [8]. However, there are no reports on the fabrication of c-axis oriented HAp sintered bodies. In the present study, the c-axis oriented HAp crystals were synthesized by the hydrolysis of $CaHPO_4 \cdot 2H_2O$ in a NaOH solution and HAp sintered bodies with c-axis orientation were fabricated using the plate-like HAp crystal aggregates synthesized.

Materials and Methods

An HAp suspended solution was prepared by the hydrolysis of $CaHPO_4 \cdot 2H_2O$ (20 g; reagent grade, Kanto Chemicals) in a NaOH (reagent grade, Kanto Chemicals) solution (500 ml; 5mass%). The solution was slowly stirred at a fixed temperature either 25 or 100°C. The resultant crystal precipitates were collected by filtration, washed with distilled water (50ml, five times) and dried up at 100°C.

HAp sintered bodies were then fabricated using these aggregates. The HAp aggregates were decanted to eliminate fine particles and calcined at 750°C for 3 h. The calcined aggregates were suspended in acetone, and the suspensions were poured into metal molds of 12 mm in diameter. After natural sedimentation of the aggregates, the excess acetone was excluded from the molds by gradual pressing of the aggregates by a pushrod. The green bodies obtained were sintered at 1100 °C for 3 h; the heating rate was 3°C /min. The HAp sintered bodies fabricated from the aggregates formed at 25 and 100°C are abbreviated as HAp$_{25}$ and HAp$_{100}$, respectively.

The HAp aggregates were observed by scanning electron microscopy (SEM, JEOL JMS-5600) and field emission scanning electron microscopy (FE-SEM, HITACHI S-5000), and analyzed by X-ray diffractometry (XRD, Philips PW1700) to identify the crystal phases and to estimate the crystallite sizes by the Scherer's equation. The Ca/P molar ratios of the HAp aggregates were measured using an inductively coupled plasma spectrometer (ICP, SPS1700HVR). The mean particle sizes of the aggregates were measured with a particle size distribution analyzer (Shimadzu, SALD-2100), and the specific surface areas by the BET method (Shimadzu 2200).

Furthermore, the surfaces and the cross sections perpendicular to the flat surfaces, i.e., parallel to the pressing direction, of the cylindrical HAp sintered bodies were also observed by FE-SEM and analyzed by XRD to identify the crystal phases and to estimate the crystal orientations. The orientation index of the plane $Nh_ik_il_i$, which is defined as Eq. 1, was calculated from the XRD pattern of either the surface or the cross section.

$$N_{h_ik_il_i} \cdot \frac{R_{h_ik_il_i}}{R^0_{h_ik_il_i}} \qquad (1)$$

where $Rh_ik_il_i$ is the intensity ratio of the N plane and defined by Eq. 2, and $R^0h_ik_il_i$ is that of a random orientation obtained from JCPDS 9-432. The formula used is

$$R_{h_ik_il_i} \cdot \frac{I_{h_ik_il_i}}{I_{h_1k_1l_3} + I_{h_2k_2l_2} + I_{h_3k_3l_3} + \cdots + I_{h_nk_nl_n}} \qquad (2)$$

where $I h_ik_il_i$ is the intensity of the $h_ik_il_i$ diffraction.

Results and Discussion

The HAp aggregates obtained had plate-like form resembling the starting material CaHPO$_4$·2H$_2$O. The mean sizes of the aggregates formed by hydrolysis at 25 and 100°C were estimated as 39 and 26 μm, respectively. The XRD measurement showed that the aggregates possessed broad diffraction peaks of HAp. The crystallite sizes calculated from the 002 diffractions of the HAp aggregates formed at 25 and 100°C were estimated as 28 and 57 nm, respectively. The relative diffraction indexes of 001 and 002 of HAp aggregates increased with temperature. Fig. 1 shows the FE-SEM photographs of the HAp aggregates formed at 25 and 100°C. The aggregates

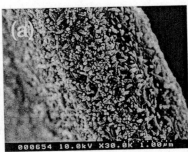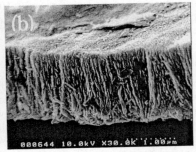

Fig. FE-SEM photographs of HAp aggregates formed by hydrolysis at (a) 25 and (b) 100°C in a 5% NaOH solution for 1 h.

formed by hydrolysis at 25°C were composed of small crystals with random orientation. On the contrary, the aggregates formed at 100°C were composed of needle-like crystals which aligned perpendicular to the developed surface of the plate-like aggregates. The results of XRD and SEM measurements indicated that the needle-like crystals in the aggregates formed at 100°C aligned in c-axis orientation perependicular to the developed surface of the plate-like aggregates.

HAp sintered bodies were fabricated using these aggregates after decantation. The mean particle sizes after sintering of the aggregates formed at 25 and 100 °C were estimated as 60 and 48 μm, respectively. The SEM photographs of the surface and cross section of HAp_{100} are shown in Fig. 2. The plate-like HAp aggregates accumulated horizontally to the flat surface of the cylindrical HAp sintered body. Fig. 3 shows the XRD patterns of the surfaces and cross sections of the HAp_{25} and HAp_{100}. The HAp_{100} was identified as a HAp single phase by XRD; however, the HAp_{25} was composed of HAp and $\beta\text{-}Ca_3(PO_4)_2$ produced by the decomposition of the HAp aggregates. The relative peak intensities of the 002 and 004 diffractions on the HAp_{100} surface were much higher than those of HAp_{25}. Fig. 4 indicates the relations between orientation and diffraction indexes of HAp_{25} and HAp_{100}. Regarding HAp_{100}, the 002 and 004 diffractions on the surface, corresponding to the c-plane, were much higher than those on the cross section whereas the 200 and 300 diffractions on the

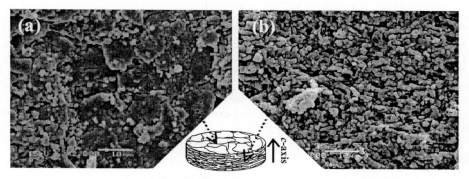

Fig. 2 SEM photographs of (a) surface and (b) cross section of HAp_{100}.

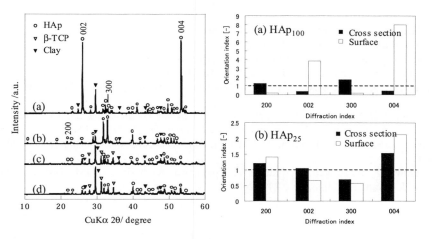

Fig. 3. XRD patterns of (a) surface and (b) cross section of HAp_{100} and (c) surface and (d) cross section of HAp_{25}. The peaks (▼) were ascribed to clay for sample fixation.

Fig. 4. Relationships between orientation and diffraction indexes of (a) HAp_{100} and (b) HAp_{25}.

surface, corresponding to the *a*-plane, were much lower. These results indicate that the *c*-axis orientation of HAp_{100} is maintained even after sintering. On the contrary, the difference of the diffraction indexes between the surface and cross section of the HAp_{25} demonstrated no tendency, i.e., the HAp sintered body had no obvious crystal orientation.

Fig. 5 shows the high magnification image of the cross section of HAp_{100}. The aggregates in the sintered body had a few connections to each other. The average grain size was about 1.5 μm. The relative density of HAp_{100} was estimated to be 81%. This low relative density in comparison to ordinary HAp sintered bodies was attributed to the restricted crystal growth in the space between the plate-like HAp aggregates in sintering as similar to that usually observed in sintering of fibrous HAp crystals [9].

Fig. 5. High magnification image of the cross section of HAp_{100}

Conclusion

HAp plate-like aggregates, whose *c*-axis of needle-like crystals was oriented perpendicular to the plane surface, were formed by the hydrolysis of $CaHPO_4 \cdot 2H_2O$ in a NaOH solution at 100 °C. HAp sintered bodies were successfully fabricated using the *c*-axis oriented HAp aggregates. The plate-like aggregates in the HAp accumulated horizontally to the flat surface of the cylindrical sintered body, and the *c*-axis orientation of the HAp aggregates was maintained even after the sintering. The HAp sintered bodies with *c*-axis orientation are expected to be useful to control biological reactions on their surfaces.

Acknowledgement

This work was supported by The Kao Foundation for Arts and Sciences.

References

[1] L. L. Hench, J. Am. Ceram. Soc., 81, 1705 (1998).

[2] T. Kawasaki, J. Chromatogr., 544, 147 (1991).

[3] H. Ohgushi and A. I. Caplan, J. Biomed. Mater. Res., 48, 913 (1999).

[4] M. Akao, H. Aoki, and K. Kato, J. Mater. Sci., 16, 809 (1981).

[5] R. Halouani, D. Bernache-Assolant, E. Champion and A. Ababou, *J*. Mater. Sci: Mater. Med., 5, 563 (1994).

[6] Y. W. Gu, N. H. Loh, K. A. Khor, S. B. Tor and P. Cheang, Biomaterials, 23, 37 (2002).

[7] Y. Fang, D. K. Agrawal, D. M. Roy and R. Roy, *J. Mater. Sci.*, 9, 180 (1994).

[8] K. Inoue, K. Sassa, Y. Yokogawa, Y. Sakka, M. Okido, and S. Asai, Key Eng. Mater., 240-242, 513 (2003).

[9] M. Aizawa, F. S. Howell, K. Itatani, Y. Yokogawa, K. Nishizawa, M. Toriyama, and T. Kameyama, J. Ceram. Soc. Jpn., 108, 249 (2000).

Key Engineering Materials Vols. 254-256(2004) pp. 19-22
online at http://www.scientific.net

Hydrothermal Preparation of Granular Hydroxyapatite with Controlled Surface

Koji Ioku[1], Manami Toda[1], Hirotaka Fujimori[1], Seishi Goto[1] and Masahiro Yoshimura[2]

[1] Division of Applied Medical Engineering Science, Graduate School of Medicine, Yamaguchi University, 2-16-1 Tokiwadai, Ube, Yamaguchi 755-8611, Japan
ioku@po.cc.yamaguchi-u.ac.jp

[2] Materials and Structural Laboratory, Tokyo Institute of Technology, 4259 Nagatsuta, Midori, Yokohama, Kanagawa 226-8503, Japan
yoshimu1@rlem.titech.ac.jp

Keywords: Hydroxyapatite, Granule, Porous Ceramics, Hydrothermal

Abstract. Microstructure designed hydroxyapatite granules from about 50 μm to 1 mm in size were prepared by hydrothermal vapor exposure method at temperatures from 105 °C to 250 °C under the saturated vapor pressure of pure water. As starting materials, powder of α-tricalcium phosphate, gelatin and a vegetable oil were used. The size of granules, the shape of particles in the granules, and the microporosity of about 0.1 μm in size of the granules were controlled.

The granular hydroxyapatite prepared at 200 °C under the saturated vapor pressure of pure water for 20 h was composed of rod-shaped crystals of about 40 μm in length with mean aspect ratio of 50. Rod-shaped hydroxyapatite crystals were locked together to make micropores of about 0.1 μm in size. It was non-stoichiometric hydroxyapatite with calcium deficient composition. This granular hydroxyapatite should have the advantage of adsorptive activity, because it has large specific crystal surface and micropores.

Introduction

Hydroxyapatite ($Ca_{10}(PO)_4(OH)_2$; HA) is known as the major component of human bones. Sintered HA with random crystal surface has already been used as a bone-repairing material which can directly bond to natural bones in bony defect. It has been known to be biocompatible and osteoconductive [1]. If materials of HA could have the tailored specific crystal surface, the HA materials should have the advantage of adsorptive activity.

The authors reported various kinds of HA materials prepared by the unique hydrothermal methods [2-11]. In the present study, HA granules with tailored crystal surface were prepared by the hydrothermal vapor exposure method [7,8,10]. The granules must be suitable as scaffold for cultured bone, for bone graft material and for drug delivery system (DDS).

Experimental

Commercial powders of α-tricalcium phosphate ($α-Ca_3(PO_4)_2$: α-TCP, Taihei Chemical Industrial Co., Ltd., Japan) were used as the starting material. After the addition of 10 mass% gelatin (Wako Chemical Co., Japan) aqueous solution, the obtained slurry was dispersed in the vegetable oil at 70 °C, and then stirred at the rate of 0-600 r.p.m. In order to control granules size. The samples were filtered off to recover, washed with ethanol, and dried at 105 °C in air. After heating at 1200 °C for 1 h in air, the samples were set in a 105 cm³ autoclave with 30 cm³ of pure water, and then they were exposed to vapor of the pure water at the temperatures from 30 °C to 250 °C under saturated vapor pressure for 20 h (Fig. 1).

The produced phases were identified by powder X-ray diffractometry with graphite-monochromatized CuKα radiation, operating at 40 kV and 20 mA (XRD; Rigaku, Geiger flex 2027, Japan). The prepared samples were dissolved in 0.1 mol.dm^{-3} nitric acid, and the chemical composition was analyzed by inductively coupled plasma spectrometer (ICP-MS; Seiko Instruments, SPQ 9000S, Japan). The microstructure of specimens was observed by scanning electron microscope (SEM; JEOL, JSM-T300, Japan).

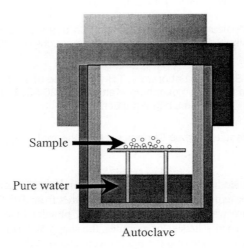

Fig. 1 Schematic illustration of the reaction apparatus.

Results and Discussion

A spherical mixture of α-TCP and gelatin was obtained (Fig. 2). Gelatin was observed among the particles of α-TCP. The size of spherical particles could be controlled by varying the stirring rate. The size of particles decreased with increasing stirring rate. Patterns of XRD for the samples are shown in Fig. 3. After heating at 1200 °C for 1 h, gelatin was burned out and no other phases than α-TCP were revealed (Fig. 3(b)). The XRD pattern of the sample after hydrothermal treatment shows high crystallinity of HA and the non-existence of phases other than HA (Fig. 3(c)).

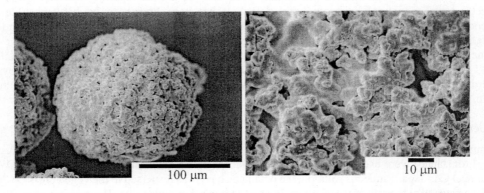

Fig. 2 Scanning electron micrographs of the spherical mixture of α-TCP and gelatin.
The stirring rate was 300 r.p.m.

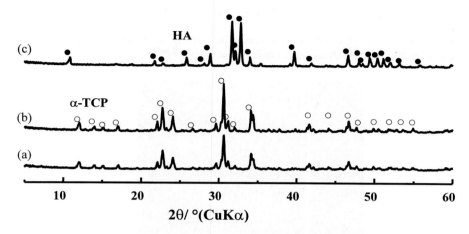

Fig. 3 Patterns of XRD of (a) spherical mixture of α-TCP and gelatin, (b) sample after heating at 1200 °C for 1 h in air and (c) sample after hydrothermal treatment of sample (b) at 200 °C under the saturated vapor pressure of water for 20 h.

Microstructure designed HA granules from about 50 μm to 1mm in size could be prepared by hydrothermal vapor exposure method at the temperatures from 105 °C to 250 °C under saturated vapor pressure of pure water. A spherical HA granule prepared hydrothermally at 200 °C is shown in Figures 4 and 5. In the case of Fig. 4, the size of HA granule was relatively large because of the stirring rate was 0 r.p.m. The granular HA was composed of rod-shaped crystals of about 40 μm in length with the mean aspect ratio of 50. Rod-shaped HA crystals were not obtained below 70 °C. These crystals were elongated along the c-axis according to TEM analysis [5]. Rod-shaped HA crystals were locked together to make micropores of about 0.1 μm in size. It was non-stoichiometric HA with calcium deficient composition of Ca/P<1.60 according to ICP-MS analysis. In the case of Fig. 5, the size of HA granule was relatively small because the stirring rate was 600 r.p.m. These granular HA must have the advantage of adsorptive activity, in comparison with other reported materials [12], because the obtained HA has large specific crystal surface and micropores.

Fig. 4 Scanning electron micrographs of the HA granule prepared hydrothermally

at 200 °C under the saturated vapor pressure of pure water for 20 h. The stirring rate was 0 r.p.m.

Microstructure designed HA granules with controlled surface must be suitable for the drug delivery system (DDS), as the scaffold of cultured bone, and as bone graft material.

10 μm

Fig. 5 Scanning electron micrograph of the HA granule prepared hydrothermally at 200 °C under the saturated vapor pressure of pure water for 20 h. The stirring rate was 600 r.p.m.

Acknowledgements

The present work has been partly supported by cooperative program in the Material and Structures Laboratory of Tokyo Institute of Technology, Japan. The authors also wish to thank Mr. S. Sasaki and Mr. K. Nakahara of Graduate School of Medicine, Yamaguchi University, Japan for their observation of the samples.

References

[1] M. Jarcho: *Clin. Orthop.*, 157, 259-278 (1981).
[2] K. Ioku, M. Yoshimura and S. Somiya: Nippon Kagaku Kaishi (J. Chem. Soc. Japan), [9], 1565-1570 (1988).
[3] K. Ioku, M. Yoshimura and S. Somiya: Bioceramics Vol.1 (Proc. 1st Int. Bioceramic Sympo.) (1989), pp.62-67.
[4] K. Ioku, M. Yoshimura and S. Somiya: Biomaterials, 11 [1], 57-61 (1990).
[5] M. Yoshimura, H. Suda, K. Okamoto and K. Ioku: J. Mater. Sci., 29 [13], 3399-3402 (1994).
[6] K. Ioku, K. Yamamoto, K. Yanagisawa and N. Yamasaki: Phosphorus Res. Bull., 4, 65-70 (1994).
[7] K. Ioku, S. Nishimura, Y. Eguchi and S. Goto: Rev. High Pressure Sci. Technol., 7, 1398-1400 (1998).
[8] K. Ioku, M. Fukuhara, H. Fujimori and S. Goto: Korean J. Ceram., 5, 115-118 (1999).
[9] K. Ioku, A. Oshita, H. Fujimori, S. Goto and M. Yoshimura: Trans. Mater. Res. Soc. Japan, 26 [4], 1243-46 (2001).
[10] K. Ioku, H. Misumi, H. Fujimori, S. Goto and M. Yoshimura, Proc. 5th Int. Conf. Solvo-Thermal Reactions (2002), pp.233-236.
[11] K. Ioku, S. Yamauchi, H. Fujimori, S. Goto and M. Yoshimura, Solid State Ionics, 151, [1-4] 147-150 (2002).
[12] V. S. Komlev and S. M. Barinov, J. Mate. Sci.: Mater. Med., 13, 295-299 (2002).

Key Engineering Materials Vols. 254-256(2004) pp. 23-26
online at http://www.scientific.net
© *2004 Trans Tech Publications, Switzerland*

Influence of the Network Modifier Content on the Bioactivity of Silicate Glasses

J. P. Borrajo[1], S. Liste[1], J. Serra[1], P. González[1], S. Chiussi[1], B. León[1], M. Pérez-Amor[1], H. O. Ylänen[2] and M. Hupa[2]

[1] Universidad de Vigo, Dpto. Física Aplicada, ETSI Industriales y Minas, Lagoas-Marcosende 9, 36200 Vigo, Spain, jpb@uvigo.es

[2] Process Chemistry Group, Combustion and Materials Research, Åbo Akademi, Piispankatu 8, 20500, Turku, Finland, heimo.ylanen@abo.fi

Keywords: bioactive glasses, network modifiers, FTIR, Raman, bioactivity.

Abstract. The influence of the substitution of calcium oxide for sodium oxide in the composition of silica-based glasses on the *in vitro* bioactivity is presented. Valuable information on the active Si-O groups present in the glasses is obtained by Fourier Transform Raman and Infrared spectroscopies. *In vitro* test analysis by Scanning Electron Microscopy and Energy Dispersive X-ray Analysis show a correlation between the network disruption induced by the modifier type and the bioactive process. It is demonstrated that glasses with high SiO_2 content can be bioactive depending on the alkali/alkali-earth modifiers ratio included into the vitreous silica network.

Introduction

Bioactive glasses are interesting materials for medical purposes due to their ability to bond chemically to living bone and soft tissues when soaked in physiological fluids [1, 2]. The clinical applications of bioactive glasses are numerous, especially in maxillofacial reconstruction, otorhinolaryngology, oral surgery and periodontal repair [1].

The bioactive silica-based glass network is basically the same of vitreous silica, where the structural units consist of slightly distorted SiO_4 tetrahedra. This structure enables the accommodation of alkali and alkali-earth cations which create non-bridging oxygen sites (Si-O-NBO) throughout the glass network [3]. When the bioactive glasses are soaked in human plasma or an analogous solution, it is known that a partial dissolution of the glass surface occurs leading to the formation of a silica-rich gel layer and, subsequently, the precipitation of a calcium phosphate film on the bioactive material takes place. The formation rate of this layer is a critical parameter, which is directly related to the type and content of the network modifiers [1].

Several spectroscopic techniques, such as Infrared and Raman spectroscopies, are sensitive to changes in the composition and the bonding configuration of the glasses, and they provide valuable information on the local structure of the silicate glasses.

Thus, the aim of this work is to evaluate, through spectroscopic techniques and *in vitro* tests, the role of content and type of network modifiers on the bioactivity of silica-based glasses.

Materials and methods

Glasses of different compositions in the quaternary system $Na_2O-CaO-P_2O_5-SiO_2$, with a systematic substitution of CaO and Na_2O concentrations, have been investigated (Table 1). Glasses were obtained by melting the appropriate quantities of analytical grade $CaCO_3$, Na_2CO_3, $CaHPO_4 \cdot 2H_2O$ and commercial Belgian quartz sand in a Pt-crucible at 1360 °C for 3 h. The glasses were cast, annealed, crushed, and remelted to improve homogeneity. In the final casting, a graphite mould of 20 mm diameter and 100 mm long was used. The test pieces were obtained by sawing discs of 2 mm thick. The discs were washed and stored in ethanol.

The structure of the glasses was studied by X-ray Diffraction (XRD). Information on the active Si-O groups present in the glasses was obtained by Fourier Transform Infrared (FTIR) and Raman spectroscopies.

Table 1. Composition of the glasses [wt %].

N series					C series				
Glass	SiO₂	P₂O₅	Na₂O	CaO	Glass	SiO₂	P₂O₅	Na₂O	CaO
N42	42	4	30	24	C42	42	4	24	30
N45	45	4	27	24	C45	45	4	24	27
N50	50	4	22	24	C50	50	4	24	22
N53	53	4	19	24	C53	53	4	24	19
N55	55	4	17	24	C55	55	4	24	17
N58	58	4	14	24	C58	58	4	24	14

The *in vitro* tests were carried out by immersion of the glasses in Simulated Body Fluid [2] (SBF) for 72 h at 37 °C ± 0.5 °C. The glass surface area to SBF volume ratio (SA/V) was 0.5 cm^{-1}. The soaked samples were analysed by Back-Scattered Electron Imaging in a Scanning Electron Microscope equipped with Energy Dispersive X-ray Analysis (BEI-SEM/EDXA).

Results and Discussion

The XRD studies reveal that the glasses are amorphous without crystallised zones detectable by this technique.

FTIR spectra of the glasses show the presence of the typical absorption bands for bioactive glasses (inset in Fig. 1). According to the literature [4], the IR bands were identified as follows: Si-O-Si(s) stretching vibration mode in the range 1000-1200 cm^{-1}, Si-O-Si(b) bending mode located around 800 cm^{-1}, Si-O-NBO (Q^3 groups) between 890-950 cm^{-1} and Si-O-2NBO (Q^2 groups) around 840 cm^{-1}. From the IR spectra, the absorbance intensity ratio of the Si-O-NBO groups and Si-O-Si(s) band have been evaluated (Fig. 1). As can be seen, with the gradual diminution of the SiO₂ content, the position of the maximum absorbance of the Si-O-Si(s) band shifts towards lower

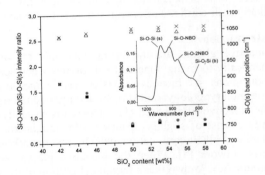

Fig.1 FTIR frequencies of the Si-O stretching vibration, for N series (Δ) and C series (X); and Si-O-NBO/Si-O-Si(s) absorbance intensity ratio, for N series (■) and C series (●), as a function of the silica content. The inset corresponds to the IR spectrum of the glass N42.

wavenumbers, until 1028 cm^{-1}, and the Si-O-NBO/Si-O-Si(s) intensity ratio increases. These effects are originated by the incorporation of the alkali and alkali-earth ions into the vitreous structure, which provokes a disruption of the network and promotes the formation of non-bridging oxygen groups [5]. Both glass series (Table 1) present the same tendency, within the experimental error, although slight differences can be observed for glasses with high silica content.

In Fig. 2 the Raman spectra of both glass series are shown. The main optical modes of the Si-O-Si groups have been identified as follows [6]: (i) the wide absorption band at 1050-1150 cm^{-1} was assigned to the asymmetric stretching mode, (ii) the very weak peak around 750 cm^{-1} was attributed to the bending vibration, and (iii) the rocking vibration was identified at 600 cm^{-1}. The Si-O-NBO group presents a Raman band at 900 cm^{-1}, and even a small peak associated to the Si-O-2NBO vibration emerges at 860 cm^{-1}.

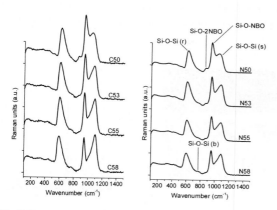

Fig 2. Raman spectra of both glass series shown in Table 1.

As can be observed in Fig. 2, the relative intensity of the NBO Raman lines increases with the incorporation of alkali and alkali-earth elements in the SiO_2 matrix. Moreover, for glasses with high silica content, the Raman spectra reveal differences between both series in the Si-O-NBO/Si-O-Si(s) intensity ratio. In particular, for samples with 62 wt% of network-forming oxide species (SiO_2 and P_2O_5) it was observed that the glass N58 presents a higher concentration of NBO groups than glass C58 and, consequently, a more disrupted silica network.

In order to study the role of the bonding configuration and composition on the bioactivity of these glasses, *in vitro* tests by soaking the glasses in SBF were carried out.

The chemical composition and the thickness of SBF developed layers were evaluated by BEI-SEM and EDX for all soaked glasses. In cross section analyses four different zones were identified: the original glass, a silica-rich layer, an intermediate film composed of silica/calcium phosphate in different proportions and, finally on the top, a calcium phosphate (CaP) rich layer. Fig. 3 shows that the thickness of the silica rich layer decreases gradually when the SiO_2 content of the glass increases. A uniform CaP layer was formed on the glasses ranging from 42 to 50 wt % of SiO_2

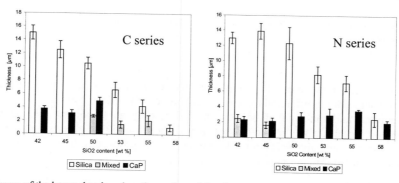

Fig. 3. Thickness of the layers developed on the surface of the glasses (Table 1) after 72 hours immersed in SBF.

content, which corresponds with high concentration of modifier oxides. Nevertheless, for glasses with low modifier content, two different behaviours depending on the glass type were found. For C series, the gradual diminution of the SiO_2-gel layer is accompanied by a vanishing SiO_2-CaP mixed layer. The CaP-rich layer was not detected in this range. Surprisingly, for N series, all glasses exhibit a bioactive behaviour and the silica-rich and CaP layers were clearly identified.

Fig. 4 shows in more detail this different behaviour between both glass series. The BEI-SEM/EDX analysis of glass C58 (micrograph a) shows some dissolution of the surface and the formation of a thin silica layer. The CaP layer is not present. Nevertheless, glass N58 exhibits the formation of both SiO_2 and CaP-rich layers (Fig. 4-b) demonstrating that a completely bioactive process is developed on the glass surface.

These different bioactive processes are in agreement with the structural properties evidenced by the spectroscopic analyses. The network disruption degree induced by the different types of network modifiers can be successfully studied by Raman and Infrared spectroscopies. Both techniques are sensitive to the presence of Si-O-NBO groups (Figs. 1 and 2), which play an important role on the bioactive process (Figs. 3 and 4). Then, the glass spectroscopic analysis could predict the final bioactivity grade of the silicate glasses.

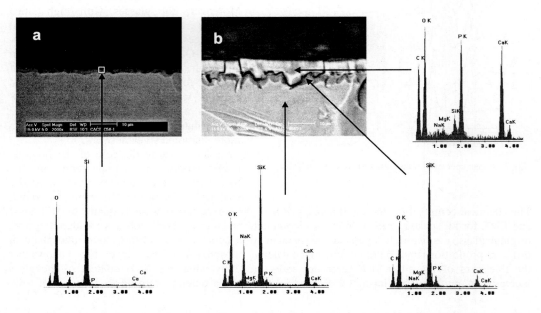

Fig. 4. BEI-SEM micrographs and EDX analysis of glasses C58 (a) and N58 (b) after 72 hours of immersion in simulated body fluid.

Conclusions

Raman and FTIR spectroscopies are complementary techniques for a better knowledge of the influence of the network modifiers on the bioactivity process of silicate glasses.

The content and the type of network modifiers play an important role on the bonding configuration and the bioactivity of the glasses. It is demonstrated that glasses with very high SiO_2 content can present a bioactive behaviour depending on the proportion between alkali and alkali-earth modifiers included into the vitreous silica network.

Acknowledgments

This work was supported by Xunta de Galicia (PGIDT99PXI32101B), Universidad de Vigo (64502I908 and 6452I106) and Ministerio de Ciencia y Tecnología (MAT2001-3434). One of the authors wishes to thank Marie Curie Fellowships Program (HPMT-CT-2001-00297).

References

[1] L.L. Hench and J. Wilson: *An Introduction to Bioceramics* (World Scientific, Singapore, 1993)

[2] H-M. Kim, T. Miyazaki, T. Kokubo, T. Nakamura: Key Eng. Mat. Vol. 192-195 (2001), p. 47.

[3] H. Scholze: *Glass: Nature, Structure and Properties* (Springer-Verlag, New York, 1990)

[4] P. Lange: J. Appl. Phys. Vol. 66 (1989), p. 201.

[5] J. Serra, P. González, S. Liste, S. Chiussi, B. León, M. Pérez-Amor, H.O. Ylänen, M. Hupa: J. Mat. Sci. Mater. Med. Vol. 13 (2002), p. 1221.

[6] P. González, J. Serra, S. Liste, S. Chiussi, B. León, M. Pérez-Amor: J. Non-Cryst. Solids. Vol. 320 (2003), p. 92.

Key Engineering Materials Vols. 254-256(2004) pp. 27-30
online at http://www.scientific.net
© *2004 Trans Tech Publications, Switzerland*

Textural Evolution of a Sol-Gel Glass Surface in SBF

D. Arcos, J. Peña and M. Vallet-Regí*

Dpto. Química Inorgánica y Bioinorgánica. Fac. Farmacia. UCM, 28040-Madrid, Spain

e-mail:vallet@farm.ucm.es

Keywords: Bioactive ceramics

Abstract. The surface area evolution of a SiO_2-CaO-P_2O_5 sol-gel glass has been followed after soaking the glass in SBF. Textural properties were studied by N_2 adsorption. The surface of the pellets prepared from this material was characterised by FTIR and SEM. Contrarily to melt derived glasses, the sol-gel glasses loss more than 50% if their surface during the first hour, mainly due to the dissolution that leads to porosity changes and for the subsequent amorphous calcium phosphate formation. The crystallisation of the amorphous CaP into hydroxycarbonate apatite leads to the recovering of high surface area values.

Introduction

It is well known that SiO_2-CaO-P_2O_5 sol-gel glasses show bioactive behaviour for a wide compositional range, due to their high surface area and porosity [1]. Since the bioactivity is a surface process, a good understanding of the surface area variations will help to tailor glasses for different applications and better "in vivo" performance.

The first stages of the bioactive process are critical for the subsequent behaviour. Actually, a faster initial ionic release will lead to a faster textural changes and OHAp formation. This process is very well known for melt-derived bioactive glasses. These materials develop a high surface area (around 100 m^2/g) when get in contact with fluids at physiological pH, as was demonstrated by Greenspan et al. [2]. The textural properties of SiO_2-CaO-P_2O_5 sol gel glasses are widely characterized [1-4]. However there is no much information about the textural changes that these glasses undergo from the very first bioactivity stages until the apatite formation on the surface [5].

In this work, we deal with the textural evolution of 58S sol-gel glass in simulated body fluid [6], as an effort to relate the surface area changes to the stages described by Hench, for the bioactivity in SiO_2 based glasses [7].

Materials and Methods

A glass of nominal composition SiO_2 58–CaO 36-P_2O_5 6 (% mol) has been synthesized by hydrolysis and polycondensation of tetraethylorthosilane (TEOS), triethylphosphate (TEP) and $Ca(NO_3)_2 \cdot 4H_2O$. In order to catalyse the reagent hydrolysis, a 2N solution of HNO_3 was added. The reagents ratio for the hydrolysis process was [mol H_2O/(mol TEOS + mol TEP)]=8. The TEP and calcium nitrate were successively added to the TEOS–H_2O–HNO_3 mixture under stirring, with 1-hour intervals between consecutive additions. After mixing all the reagents, the solutions were placed in closed Teflon® jars, where the gelation process took place at room temperature. These gels were aged at 70°C for three days and dried at 150°C for 52 hours after replacing the lids with new ones with a 1-mm hole. The dried gels were stabilized at 700°C for 3 hours and then milled and sieved, choosing the powders with sizes ranging from 32 to 68 μm. The glass powder were disk shaped by uniaxial and isostatic pressure under 5 Tons. Disks were soaked in SBF tempered at 37°C for 1, 3, 5, 7, 10, 15, 20, 30, 40, 60, 120, 240, 480, 620 and 1240 minutes, immediately soaked in acetone to stop the reaction and kept under dry atmosphere.

The calcium solubility test and pH evolution were carried out using an Ilyte Na^+, K^+, Ca^{2+}, pH system. The surface morphology was studied by scanning electron microscopy (SEM) using a JEOL 6400-LINK IN AN 1000 microscope. Fourier transform infrared (FTIR) spectra were obtained in a Nicolet Nexus spectrometer equipped with an ATR accessory. The N_2 adsorption isotherms were obtained using a Micromeritics ASAP 2010C instrument. To perform the N_2 adsorption measurements, the samples were first outgassed for 24 h at 25°C. The surface area (S_{BET}) was determined using the Brunauer-Emmett-Teller (BET) method [8]. The pore size distribution between 1.7 and 130 nm was determined from the desorption branch of the isotherm by means of the Barret-Joyner-Halenda (BJH) method [9]. Finally, the t-plot method was used to determine the presence of micropores [10].

Results and Discussion

Figure 1 shows the S_{BET} evolution of the glass as a function of the soaking time in SBF. Four stages can be clearly differenced. During the first minute, 1st stage, the glass undergoes a drastic surface decrease, that is from 138 m^2/g (original value) to 82 m^2/g, what means a 40% of surface reduction in a very short time. This is a very different behaviour compared to melt derived glasses, which have a very low surface area but develop surfaces of about 100m^2/g after being soaked in physiological simulated solutions. Afterwards, a partial surface recovering occurs between 1 minute and 10 minutes, 2nd stage, reaching a surface value of 100 m^2/g. From this point the glass begins to lose surface gradually, 3rd stage, and after one hour it has lost about the 55% of the initial surface, showing values of 62 m^2/g. Finally, the 4th stage involves the progressive surface area recovering from 1 hour until the end of the experiment, reaching values of 127m^2/g after 24 hours in SBF.

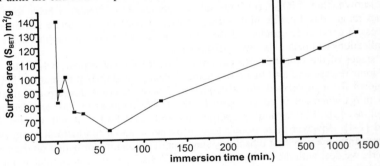

Fig. 1. Evolution of S_{BET} as a function of soaking time for 58S sol-gel glass.

In order to understand the different mechanisms during this surface evolution, we have simultaneously study the pore size distribution, pore volume and the FTIR spectra of the surface for different soaking times. Figure 2 collects the results of the most important reaction times. Before soaking (figure 2, left, 0 m.), the glass show a bimodal pore size distribution centered at 8 and 13 nm in the mesopore region. The FTIR spectra (fig. 2, right) show bands corresponding to the Si-O and P-O vibrational modes. The doublet at 580-610 cm^{-1} corresponds to crystalline phosphates, probably formed as a consequence of the high surface reactivity with the atmospheric water, or perhaps as consequence of small crystallites formation at the surface during the stabilisation process.

After one minute in SBF, the maxima of the pore size distributions are slightly shifted and the distribution shapes broadened toward higher pore size values. Moreover, the pore volume decreases, as can be seen in table 1. The Ca^{2+} solubility test (data not shown) point out that during the first minute, a burst effect for ionic dissolution (Ca^{2+} mainly) occurs. This process leads to the lost of pore volume and to increase the size of the mesopores, decreasing in this way the surface area as is observed in figure 1. At this stage, no important changes could be observed by FTIR spectroscopy.

Between 1 and 10 minutes (2nd stage of surface evolution), the ionic exchange and the incorporation of carbonates occurs (see figure 2, 3min). The formation of silanol groups from the Ca^{2+} by H^+ exchange reported by Hench, as well as the incorporation of other chemical species from the SBF, could explain the slight surface recovering during this period (2nd stage). The pore size distribution does not seem to undergo important changes at this stage.

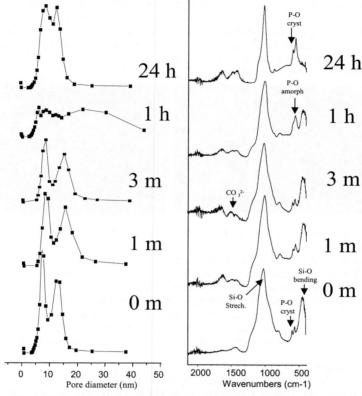

Fig.2. Pore size distribution and FTIR spectra at the different stages of S_{BET} evolution

Table I : Textural parameters of 58S sol-gel glass at different times of immersion

	S_{BET} (m^2/g)	Vp (cm^3/g)	Pore diameter (nm)
initial	138	0,38	7, 13
1m	82	0,30	8, 15
3m	90	0,30	8, 14
10m	100	0,34	8, 14
30m	74	0,30	8, 15
1h	62	0,29	10, 14, 25
2h	82	0,30	11, 14, 25
4h	108	0,34	7, 10, 13, 25
12h	116	0,36	5, 11
24h	127	0,38	5, 11

The 3rd stage of the surface evolution, that is the second surface lost happened between 10 minutes and 1 hour, takes place because of important textural and chemical changes at the surface.

At this stage the pore size show a wide maximum centered at 25 nm, whereas the distribution corresponding to the glass surface almost disappears. Moreover, the doublet at 580-610 cm^{-1} in the FTIR spectra, characteristic of crystalline phosphates, is overlapped by a new singlet characteristic of amorphous CaP. This new phase, also reported in the Hench's mechanism, is a less porous phase that cover the glass and reduces the initial surface area more than 50%.

The subsequent progressive augmentation in the S_{BET} values, reaching values of 127 m^2/g after 24 hours, is due to the CaP crystallization. The FTIR spectrum shows again the doublet assigned to crystalline CaP. The pore size distribution shows now a new bimodal pore size distribution centred at 5 and 11 nm, attributable to the porosity of the new crystallized hydroxycarbonate apatite on the glass surface. Figure 3 show the glass surface before and after soaking for 24 hours, demonstrating the presence of the new crystallized HA.

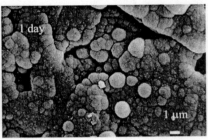

Fig. 3. Scanning electron micrographs of the glass surface before and after soaked for 1 day.

Conclusion

The bioactivity of a sol-gel glass has been examined considering the variation of the textural properties. The reactions that take place on the surface produce a drastic variation on the surface area within the first minutes of immersion. This evolution can be described as a four stages process, that is easily explained in terms of the theory for bioactive glasses: a) Lost of surface area due to the fast Ca^{2+} release, b) partial surface area restoring due to the Si-OH formation and CO$_3^{2-}$ incorporation, c) second surface area loss as a consequence of the amorphous CaP formation and d) Surface area restoring during de CaP crystallization into hydroxycarbonate apatite.

Acknowledgements

Financial support of CICYT, Spain, through research project MAT02-00025 is acknowledged. We also thank A. Rodriguez (Electron Microscopy Center, Complutense University) and F. Conde (C.A.I. X-ray Diffraction Centre, Complutense University), for valuable technical and professional assistance.

References

[1] M. Vallet-Regí, C.V. Ragel, A.J. Salinas, Eur. J. Inorg. Chem., 2003, p 1029.
[2] D.C. Greenspan, J.P. Zhong, G.P. LaTorre: Bioceramics, vol 8, (1995), p. 477.
[3] T. Peltola, M. Jokinen, H. Rahiala, E.Levänen, J.B. Rosenholm, I. Kangasniemi, A. Yli-Urpo, J. Biomed. Mater. Res., 44, (1999), p. 12.
[4] J. Zhong, D.C. Greenspan, J. Biomed. Mater. Res. (Appl. Biomater.) 53, (2000), p. 694.
[5] M.Vallet-Regi, D. Arcos and J. Perez-Pariente: J. Biomed. Mater. Res., 51, (2000), p. 23.
[6] T. Kokubo, H. Kushitani, C. Ohtsuki, S. Sakka, T. Yamamuro, J. Biomed. Mater. Res 24 (1990) p. 721.
[7] L.L. Hench, O. Andersson *Bioactive glasses* in: L.L. Hench and J. Wilson eds. *An introduction to bioceramics*, World Scientific Publish, (1993), p. 41.
[8] S. Brunauer, P.H. Emmett, E. Teller, J. Am. Chem. Soc., 60 (1938), p 309.
[9] E.P.Barret, L.J. Joyner, P.P.Halenda, J. Am. Chem. Soc., (1951), 73, 373.
[10]B.C. Lippens, B.G. Linsen, J.H. de Boer, J. Catalysis 4, (1964), p.32.

Key Engineering Materials Vols. 254-256(2004) pp. 31-34
online at http://www.scientific.net

Bioactive Behaviour in Biphasic Mixtures of Hydroxyapatite-sol Gel Glasses in the System SiO$_2$-CaO-P$_2$O$_5$

J. Román, S. Padilla and M. Vallet-Regí

Dept. Química Inorgánica y Bioinorgánica, Facultad de Farmacia, U.C.M.,
Madrid 28040, Spain. E-mail: jeromzar@farm.ucm.es

Keywords : Biphasic mixtures, hidroxyapatite, sol-gel glasses, *in vitro* assay.

Abstract. Biphasic mixtures of hidroxyapatite (HA) and sol gel glasses with a ratio of 95/5 (wt %) were prepared. The composition of the studied glasses was **55S**: 55-SiO$_2$; 41-CaO; 4-P$_2$O$_5$ (mol %) and 70S: 70-SiO$_2$; 30-CaO (mol %). The mixtures were conformed as disks by uniaxial and isostatic pressure and then heated at 1300°C for 24 h. After thermal treatment of HA, small diffraction maxima corresponding to Ca$_4$P$_2$O$_9$ (TeCP) were observed, whereas in the biphasic materials TeCP was not observed and α and β- Ca$_3$(PO$_4$)$_2$ (TCP) phases appeared, being the ratio between these two phases and HA slightly higher in the glass 70S. After soaked in SBF for 7 days the surface of biphasic mixtures was covered by an apatite like phase. In the sample HA/70S the surface was completely covered whereas in the HA/55S sample some zones of the surface were not covered.

Introduction

Hydroxyapatite (HA) has been used for many years as implant material because of its high biocompatibility and its similarity with the mineral component of the bone tissue. Nevertheless, its bioactivity is lower than other biomaterials such as bioactive glasses. In previous works the bioactivity of samples composed of HA and glasses, treated at 700 °C was studied [1 – 3]. It was stated that glasses improve the bioactivity of the HA even when a 5 wt% was added [2].

The aim of this work was to study if the biphasic mixtures maintain a bioactive behaviour when they are heated at similar temperatures to that of HA sintering. At this temperature changes in crystallinity, particle size and crystalline phases will take place both in HA and glasses, and therefore these parameters would play an important role on the bioactivity of the material.

Methods

HA was prepared by the precipitation method by reaction of Ca(OH)$_2$ (Riedel-deHaën) and H$_3$PO$_4$ (Merck), reagent grade, in aqueous solution. The H$_3$PO$_4$ solution was slowly added to the Ca(OH)$_2$ suspension, the reaction was carried out at 90°C. The HA slurry was aged for 24 h at room temperature and decanted before filtering. The filtering cake was dried at 105°C during 12 h. The solid, previously ground in a vibratory mill, was heated at 1200°C for 1 h. Finally, the HA was milled for 20 h. The HA treated at 1200°C contain 1.3 wt % of CaO.

The composition of selected glasses was **55S**: 55-SiO$_2$; 41-CaO; 4-P$_2$O$_5$ (mol %) and **70S**: 70-SiO$_2$; 30-CaO (mol %). Glasses were prepared by hydrolysis and polycondensation of stoichiometric quantities of tetraethyl orthosilicate, triethyl phosphate (for 55S glass) and Ca(NO$_3$)$_2$.4H$_2$O, as previously described [4, 5]. The dried gels were ground and sieved and the fraction of particles from 32 to 63 μm were selected.

Portions of 7.6 g of the HA and 0.4 g of glass (ratio 95:5 wt %) were mixed during 24 h in a powder mixer. Fractions of 0.5 g were used to prepare disks (13 mm in diameter x 2 mm height) by uniaxial pressure of 55 MPa and then 150 MPa of isostatic pressure. Finally, the disks were heated at 700°C/3 h and then at 1300°C for 24 h.

The materials were characterized by XRD in a Philips X'Pert MDP diffractometer (Cu Ka radiation), by FTIR in a Nicolet Nexus and by SEM-EDS using a JEOL 6400 microscope-Oxford Pentafet super ATW system.

In vitro assays were performed by soaking the disks, vertically supported in a platinum scaffold, in SBF [6] (pH = 7,30) at 37°C. The formation of an apatite-like layer on the disks surface was studied by SEM-EDS, XRD, and FTIR analysis.

Results and Discussion

The XRD pattern of HA and the biphasic mixtures, after treatment at 1300°C during 24 h, are shown in Figures 1 and 2. It was observed that the phase composition after sintering was different for HA and biphasic materials. XRD of sintered HA (Fig. 1) showed small diffraction maxima corresponding to $Ca_4P_2O_9$ (TeCP) as a consequence of the reaction between HA and CaO (present in HA at 1200°C). However, in the biphasic materials TeCP was not observed (Fig. 2) and α and β- $Ca_3(PO_4)_2$ (TCP) phases appeared. The ratio between these phases and HA was slightly higher in the glass 70S (without phosphorous). These results suggest that a reaction between glasses and HA take place producing the transformation of HA in TCP. The presence of phosphorous in the glass seems to decrease the magnitude of the HA transformation.

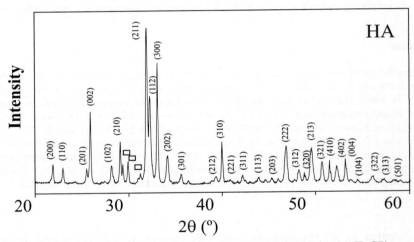

Figure 1. XRD pattern of HA sintered at 1300°C/24 h   (□-TeCP)

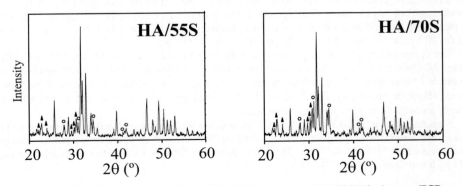

Figure 2. XRD patterns of HA/55S and HA/70S sintered at 1300°C/24 h. ($\Delta = \alpha$–TCP; O = β-TCP; no indexed maxima corresponds to HA)

In Fig. 3 SEM micrographs of the biphasic mixtures before soaking in SBF are shown. The micrographs are very similar showing a characteristic surface of compacted and sintered powder.

Figure 3. SEM micrographs of HA/55S and HA/70S sintered at 1300°C/24 h.

In vitro assays in SBF showed a very different bioactive behaviour of HA and biphasic mixtures. The HA surface shows no change after 7 days soaking in SBF, while the surface of biphasic materials was covered by a new layer formed of spherical particles. In the HA/70S sample (Fig. 4) the surface was completely covered by a new layer whereas in the HA/55S sample some zones of the surface were not covered. The SEM micrographs showed that the spherical particles were constituted by nanometric crystalline aggregates of higher size in the HA/70S sample.

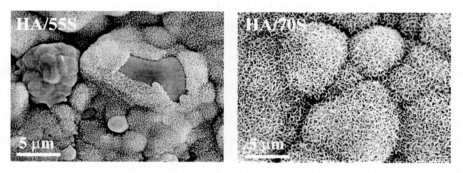

Figure 4. SEM micrographs of HA/55S and HA/70S sintered at 1300°C/24 h after 7 days Soaking in SBF.

The EDS of materials surface after 7 days in SBF showed that the layer contains calcium and phosphorous. The result of the DRX and FTIR studies (Fig. 5) showed that the layer formed after 7 days in SBF corresponds to an apatite-like phase containing carbonates ions, less crystalline and with lower crystal size than the HA present in the biphasic materials.

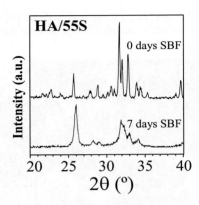
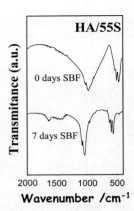

Figure 5. XRD patterns and FTIR spectra of HA/55S sintered at 1300°C/24h before (0 days) and after 7 days soaking in SBF.

Conclusions

The addition of a 5 wt % of bioactive glasses to HA provoke a considerable increment of the bioactivity, even when the mixture has been sintered at high temperature (1300°C for 24 h). The glasses provoke the transformation of HA in α and β-TCP. The transformation of HA in these phases seems to be related with the content of phosphorous in the glass.

Acknowledgment

Financial support of CICYT, Spain, through research project MAT2002-00025 is acknowledged. Authors also thank A. Rodríguez (C.A.I. electron microscopy, UCM) and F. Conde (C.A.I. X-ray diffraction, UCM) for valuable technical and professional assistance.

References

[1] A. Rámila, S. Padilla, B. Muñoz and M. Vallet-Regí M., Chem. Mater. Vol. 14 (2002), p. 2439.
[2] M. Vallet-Regí, A. Rámila, S. Padilla and B. Muñoz , J.Biomed. Mater.Res. (2003), (In press).
[3] C.V. Ragel, M. Vallet-Regí and L.M. Rodríguez Lorenzo, Biomaterials, Vol 23 (2002), p.1865.
[4] M. Vallet-Regí and A. Rámila, Chem. Mater., Vol. 12 (2000), p. 961.
[5] A. Martínez, I. Izquierdo-Barba and M. Vallet-Regí, Chem. Mat., Vol. 12 (2000), p. 3080.
[6] T. Kokubo, H. Kushitani, S. Sakka, T. Kitsugi, T. Yamamuro, J. Biomed. Mater. Res., Vol. 24 (1990), p. 721.

Key Engineering Materials Vols. 254-256(2004) pp. 35-38
online at http://www.scientific.net

Concentrated Suspensions of Hydroxyapatite for Gelcasting Shaping

Sussette Padilla and María Vallet-Regí

Dpto. Química Inorgánica y Bioinorgánica, Facultad de Farmacia, UCM, 28040, Madrid, Spain.
email: spadilla@farm.ucm.es

Keywords: Gelcasting, hydroxyapatite, suspension, colloidal processing, dispersant.

Abstract. The aim of this work was to prepare suspensions of Hydroxyapatite (HA) with a high solid content, necessary for gelcasting shaping. The HA was obtained by the precipitation method by reaction between $Ca(OH)_2$ and H_3PO_4 and it was calcined at temperatures between 900 and 1200°C for 1h. A monomeric solution containing methacrylamide and N,N'-methylenebisacrylamide was used as dispersing vehicle and Darvan 811 was used as dispersant. The monomeric solution, dispersant and HA powder were mixed in a planetary ball mill for 30 min. The influence of powder calcination temperature on phase composition, surface area, porosity, optimum concentration of dispersant and rheological behaviour were studied. Concentrated slurries (up to 60 vol.%) could be prepared using HA powder previously calcined at 1200°C.

Introduction

Janney and co-workers [1,2] developed the gelcasting method in 1990. This method allows obtaining pieces with complex shapes and high mechanical strength. In the gelcasting method, slurries with high solid content are prepared using an aqueous solution of monomers as dispersing vehicle. After polymerisation and drying, green bodies with high density and strength are obtained. Gelcasting requires slurries with good flow properties and high solid content (at least 50 vol.%). However, only solid contents lower than 50 vol.% have been achieved in most of the previous works dealing with preparation of HA slurries [3-6]. HA is a very important implantation material, because of its excellent biocompatibility and chemical composition, similar to the mineral component of the bone. However, its clinical applications have been limited due to its low mechanical properties and to the difficulty to obtain pieces with complex shapes similar to the bone defects. In this sense, the gelcasting method could be a good alternative for HA shaping. The aim of this work was to prepare suspensions of HA with a high solid content, necessary for gelcasting shaping. For this purpose the influence of calcination temperature of HA on phase composition, surface area, porosity, optimum concentration of dispersant and rheological behaviour was studied.

Materials and Methods

HA was prepared by the precipitation method by reaction of $Ca(OH)_2$ and H_3PO_4 solution. The H_3PO_4 solution was slowly added to the $Ca(OH)_2$ suspension until pH neutralization, the reaction was carried out at 90°C. The obtained HA was calcined at 900, 1000, 1100 and 1200°C for 1h. HA was characterized by FTIR, XRD, N_2 adsorption porosimetry and Hg intrusion porosimetry. The XRD patterns were register in a Philips X'Pert MDP diffractometer (Cu K_α radiation), the specific surface area was calculated using the BET method from N_2 adsorption isotherm obtained in a Micromeritics ASAP2010 analyser. The porosity was determined by Mercury intrusion porosimetry (Micromeritics AutoPore III). The molar ratio Ca/P was determined quantitatively by XR fluorescence (S4 Explorer-Brucker AXS).

The dispersing vehicle was an aqueous solution containing 15 wt% of methacrylamide (Aldrich) and N,N'-methylenebisacrylamide (Aldrich) monomers in a 6/1 ratio. Darvan 811 (D811, R.T. Vanderbilt Company, Inc.) was used as dispersant. The HA suspensions were prepared mixing the suspension vehicle, the dispersant and the HA powders in a planetary ball mill for 30 min.

The slurry viscosity was measured using a Haake ReoStress RS75 rheometer with a cone-plaque system. Measurements were performed at 20°C, in the shear rate range of 1 - 700 s^{-1}.

Results and Discussion

The raw HA is a carbonateapatite type B with a Ca/P molar ratio = 1.68. The XRD patterns of powder calcined at temperature between 900 and 1100°C correspond to a HA phase (ICDD 9-432) whereas the powder calcined at 1200°C is a mixture of HA and CaO (ICDD 43-1001) (1.3 wt-%). The CaO was formed due to the thermal decomposition of the initial carbonateapetite. The specific surface area decreased from 12 m^2/g (900°C) to 1 m^2/g (1200°C) (Table 1). The porosity of powders decreased with the calcination temperature (Table 1). The powders calcined between 900 and 1100°C are porous powders, whereas the shape of the intruded Hg volume curve of HA calcined at 1200°C is characteristic of a non-porous powder.

Table 1. Phase composition, specific surface area and pore volume of HA powder calcined at different temperatures.

Thermal treatment [°C]	Phase composition	Surface Area [m^2/g]	Vol. pore [cm^3/g]
900	HA	12	0.75
1000	HA	11	-
1100	HA	3	0.25
1200	HA + CaO	1	0.04

Slurries containing 30 vol.% of powder calcined at 900°C were prepared with different concentration of D811 (Fig. 1). Slurries with the lowest viscosity were obtained for intermediates concentrations of dispersant (≈0.9 to 3.8 mg/m^2). After that, slurries with higher concentration of solids were prepared but slurries containing 50 vol.% have a high viscosity not suitable for casting. For this reason, suspensions containing HA powders calcined at higher temperatures were prepared.

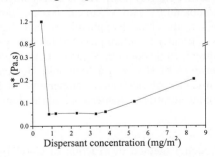

Fig. 1. Influence of Darvan 811 concentration on viscosity of slurries with 30 vol.% of OHAp calcined at 900°C.

Slurries containing 40 vol.% of HA calcined between 900 and 1200°C were prepared using a constant concentration of D811equal to 2.3 mg/m^2 (Fig. 2). Slurries with HA calcined at 900 and 1000°C have a shear thinning behaviour and show tixotropy hysteresis and their yield stress was ≈ 50 Pa. The slurries containing HA calcined at 1100 and 1200° have a very low viscosity. The shear thinning behaviour, tixotropy and viscosity decreased with the calcination temperature implied that the degree of powder agglomerate decreased when HA powders calcined at higher temperatures were used.

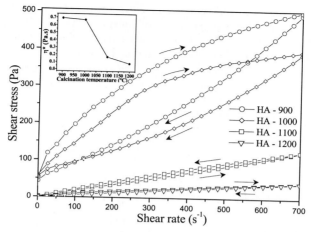

Fig. 2 Influence of calcination temperature on the rheological behaviour of HA slurries with 40 vol.% of solids.

Powder calcined at 1200°C have different phase composition to those calcined at lower temperatures, therefore different amounts of dispersant would be necessary to reach the lowest viscosity in the slurry. For this reason slurries with different concentration of dispersant were prepared using powder calcined at 1100 and 1200°C and they were compared with the slurries containing powder calcined at 900°C. In Fig. 3a the viscosity of slurries with 30 vol.% of powder calcined at 900 and 1200°C are compared and Fig. 3b shows the viscosity of slurries containing 40 vol.% of powder calcined at 1100 and 1200°C. It can be observed that the slurry with the lowest viscosity was obtained with a higher concentration of D811 when powder calcined at 1200°C was used. The D811 concentration rendered the lowest viscosity was similar for powders calcined at 900 and 1100°C, which have a similar composition. This fact could be related with the CaO presence in the powder calcined at 1200°C. In an aqueous medium CaO is transformed into $Ca(OH)_2$ which provide Ca^{2+} ions to the medium and it has been documented [7-9] that the Ca^{2+} ions enhance the adsorption of poly(acrylic acid) in slurries of alumina, titanium oxide and calcium carbonate and this effect has been related to interactions of Ca^{2+} ions with the polyelectrolyte.

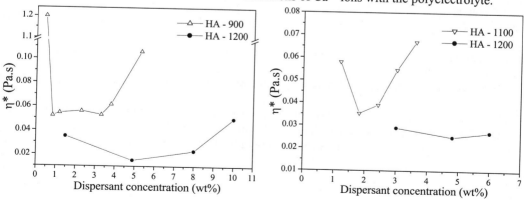

Fig. 3 Influence of Darvan 811 concentration on the viscosity of slurries with a) 30 vol.% of HA calcined at 900 and 1200°C and b) 40 vol.% of HA calcined at 1100 and 1200°C (shear rate 200 s^{-1}).

Slurries containing higher content of solids were prepared with HA calcined at 1100°C and 1200°C dry milled for 20 h, using the concentration of Darvan 811 previously determined for each powder of HA. The maximum content of solid reached in slurries with HA-1200 was 60 vol.% whereas for HA-1100 was 47 vol.%. The rheological curves of slurries containing 40, 45, 50 and 55 vol.% of HA 1200 show a Newtonian behaviour (Fig. 4) whereas the slurries with 60 vol.% and slurries containing HA calcined at 1100°C have a shear thinning behaviour.

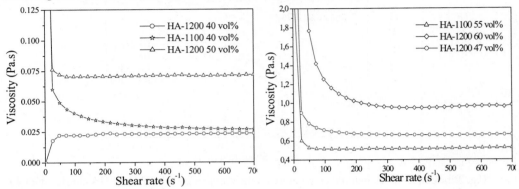

Fig. 4 Rheological curves of HA slurries containing different solid content of powder calcined at 1100 and 1200°C.

Conclusions

The calcination temperature affects the surface area, porosity and composition of HA powders, and this factors play an important role in the rheological behaviour of the suspensions. By controlling these aspects it is possible to prepare HA slurries with high content of solids. It was necessary to use powders calcined at high temperature (1100-1200°C) with low surface area and porosity in order to prepare suspensions with high content of HA and low viscosity.

References

[1] M.A. Janney: U S Patent 4894194 (1990).

[2] O.O. Omatete, M.A. Janney and R.A. Strehlow: Am. Ceram. Soc. Bull Vol. 70 (1991), p. 1641.

[3] P. Sepulveda, J.G.P. Binner, S.O. Rogero, O.Z. Higa and J.C. Brsessiani: J. Biomed. Mater. Res. Vol.50 (2000), p.27.

[4] F. Lelievre, D. Bernache-Assollant and T. Chartier. J. Mater. Sci: Mater. Med. Vol. 7 (1996), p.489.

[5] M. Toriyama, A. Ravaglioli, A. Krajewski, C. Galassi, E. Roncari, and A. Piancastelli: J. Mater. Sci. Vol.30, (1995) p.3216.

[6] H.Y. Yasuda, S. Mahora, Y. Umakoshi, S. Imazato and S. Ebisu: Biomaterials Vol.21 (2000) p.2045.

[7] L. Dupont, A. Foissy, R. Mercier and B. Mottet: J. Colloid Interface Sci. Vol.161 (1993) p.455.

[8] R.R. Vedula and H.G. Spences: Colloids Surf. Vol.58 (1991) p.99.

[9] A. Foissy, A. El Attar and J. Lamarche: J. Colloid Interface Sci. Vol.96 (1983) p.275.

Key Engineering Materials Vols. 254-256(2004) pp. 39-42
online at http://www.scientific.net

Sinterability, Mechanical Properties and Solubility of Sintered Fluorapatite-Hydroxyapatites

Kārlis A. Gross[1] and Kinnari A. Bhadang[2]

School of Physics and Materials Engineering, Building 69, Monash Uni., VIC 3800, Australia.

[1]Email: karlis.gross@spme.monash.edu.au

[2]E-mail: kinnari.bhadang@spme.monash.edu.au

Keywords: hydroxyfluorapatite, sintering, hardness, elastic modulus, fracture toughness, mechanical blends.

Abstract. Mechanical assemblies within biological systems occur widely combining different material systems and occasionally similar crystal structures. Mechanical assemblies were produced by sintering mixtures of hydroxyapatite and fluorapatite. It was found that these led to lower densities compared to solid solutions of the same composition and accompanying lower mechanical properties. Investigation of lattice parameters revealed that sintering submicron mixtures produced homogeneous mixtures. Further work on larger crystal sizes will reveal how such interfaces provide integrity for mechanically assembled apatites in biological systems and the ability to produce in-situ roughening after insertion of smooth surfaces.

Introduction

A mechanical assembly of chemically enriched apatites is a common occurrence within biological systems. Examples of these include the outer fluorapatite layer on hydroxyapatite in teeth or the chemical enrichment of inorganic crystals in bone subjected to therapeutically administered solutions or resorbable biomaterials. Mechanical blends of different apatites have not been used as biomaterials. These can be employed in biomedical devices to improve mechanical properties or administer controlled quantities of specific chemicals from a chemically mapped biomaterial surface.

The use of fluoride in apatite and bioapatites provides a means of modifying the physicochemical and mechanical properties. The natural occurrence of fluoride within apatites signifies an important function within the body. As an ion in solution, it can be used to stimulate bone growth, however, excessive quantities can lead to damaging effects. Fluoride concentrations increase in the bone with age, with apatite providing an effective storage medium. The use of fluoride for biomedical applications needs to be well designed to offer a controlled amount of this ion. The slow release of fluoride is possible by inclusion of low concentrations in bioactive materials. Slow resorption can lead to release of fluoride ions that become integrated into the surrounding tissues.

This work will investigate the sinterability, mechanical properties and solubility of sintered mechanical mixtures of fluorapatite and hydroxyapatite with reference to fluorhydroxyapatite solid solutions. Comment will be provided on the clinical usefulness of apatite ceramics produced by closely located, but chemically different apatite regions.

Materials and Methods

Hydroxyapatite and fluorapatite were synthesized using the semi-bath approach by controlled addition of an ammonium nitrate to a calcium nitrate solution. The precipitate was aged, dried in ethanol, calcined at 900 °C. Mechanical powder blends were produced by combining fluorapatite at concentrations of 0, 20, 40, 60, 80 and 100% to hydroxyapatite in an agate ball mill with ethanol, followed by drying in an oven at 120 °C. Powder blends were uniaxially compressed to form cylindrical shaped pellets, then vacuum sealed and cold isostatic pressed at 193 MPa to provide a more uniform densification of the cylindrical pellets. The pellets were sintered at 1200 °C for two hours within a tube furnace supplied with an atmosphere of air saturated with water vapour at 90 °C.

Pellets were ground on SiC paper using 500, 800, 1200 and 2400 grades followed by polishing on a short napped cloth with 1 micron diamond suspension for a period of three minutes under a load of 2.5 kg. Pellets were then assessed for density, hardness, elastic modulus and fracture toughness using micro-indentation. The degree of diffusion was assessed by examination of the lattice parameters for mixtures containing more than 40% fluorapatite.

Solubility was assessed on samples by immersion in a 0.1M potassium acetate buffer at pH 5, 37 °C for 24 hours before removal for microstructure assessment.

Results and discussion

Mechanical mixtures of 20, 40, 60 and 80% fluorapatite in hydroxyapatite powder revealed a decrease in density at a sintering time of 2 hours with a minimum at 80% fluorapatite, Fig 1. The different chemistry in crystallites has produced a lower density for mechanical blends of fluorapatite and hydroxyapatite. The tendency for the minimum to be offset towards the fluorapatite composition indicates greater difficulty in sintering of a small amount of hydroxyapatite particles with fluorapatite. This can be explained in the different packing in the unit cells. Fluorapatite is a denser structure and cannot readily accomodate the larger OH⁻ group, thus leading to a lower mobility of this ion. Conversely, movement of the smaller F⁻ ion in a less densely packed hydroxyapatite unit cell is significantly easier attributed to the small decrease in density from hydroxyapatite. The small decrease does suggest that the presence of fluoride can retard the densification of hydroxyapatite powders, and a report of fluoride concentration in hydroxyapatite formulations is a recommended addition to the requirement to report heavy metals, such as As, Cd, Hg, Pb.

Solid solutions show a different trend indicating that a minimum is only observed when fluoride and hydroxyl ions are present at similar concentrations. The lower density in the mechanical mixtures can thus be interpreted as a combination of the solid solution effect and the presence of apatites with different packing density.

The lattice parameters of the mechanical blends indicated a decrease in the a-lattice parameter, an indication that fluoride is substituted into the lattice. The lack of foreign peaks in an X-ray diffraction pattern (not shown here) indicates that diffusion between the two apatites occurs completely at sintering at 1200 °C for 2 hours.

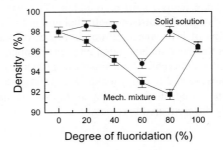

Fig 1. Density of tablets sintered at 1200 °C for 2 hours

The hardness follows the same trend in the density, where the trend is mainly attributed to the porosity within the sample. Comparison of the 20% and 100% fluoridated apatite, that exhibit comparable densities, indicated that the fluoride content gives rise to an increase in hardness, Fig 2.

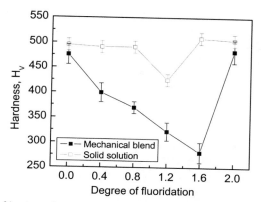

Fig 2. Hardness of hydroxyfluorapatite solid solutions and fluorapatite/hydroxyapatite blends

The elastic modulus decreased from 110MPa for mechanical blends, but a comparison of 20% and 100% fluoridated hydroxyapatite illustrates that the fluoride directly increases the elastic modulus, in agreement with data obtained from solid solutions, Fig 3. High fluoride containing apatites are thus stiffer than hydroxyapatite. A change of less than 20% occurs when all the hydroxyl groups within the lattice are replaced with fluoride groups.

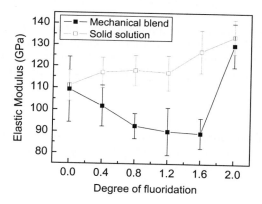

Fig 3. Elastic modulus of hydroxyfluorapatite solid solutions and fluorapatite/hydroxyapatite blends

Fracture toughness increases with fluorapatite content reaching a maximum for sintered mechanical blends at 80% fluorapatite content, Fig 4. A small addition of fluoride leads to an increase in fracture toughness, whereas complete fluoridation leads to an obvious decrease. Fluoride exhibits a larger influence on fracture toughness in comparison to hardness and elastic modulus. A total change of 50% is attained with inclusion of fluoride in the crystal structure. Although the maximum in fracture toughness occurs at a higher fluoride containing composition, a similar value suggests that fluoride is an influential element.

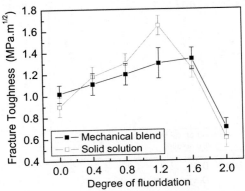

Fig 4. Fracture toughness of hydroxyfluorapatite solid solutions and fluorapatite/hydroxyapatite blends

Sintering of 20 micron sized hydroxyapatite particles together with 50 wt.% submicron fluorapatite retained rich hydroxyapatite and fluorapatite regions, with a solid solution established at the interface. The polished pellets revealed these distinctly different regions that after immersion in pH 5 buffer produced regions of different height. The interdiffused area retained more material establishing a surface covered with ridges. Fluoride is thus released from pockets of fluoride formed from the sintering process.

In-situ implant surface roughening can be produced in sintered mechanical mixtures of hydroxyapatite and fluorapatite, by selection of appropriate apatite particle sizes for improved cell contact. The interdiffusion is an attractive means to increase the fracture toughness of the sintered pellet while retaining the hardness and giving a small increase in elastic modulus.

Conclusions

Sintered pellets of fluorapatite and hydroxyapatite powder blends provides a material with a higher hardness, elastic modulus and fracture toughness. The dissolution of these materials has revealed in-situ roughening of an implanted apatite material while providing controlled fluoride release.

Key Engineering Materials Vols. 254-256(2004) pp. 43-46
online at http://www.scientific.net

Synthesis and Characterisation of Hydroxyapatite

M. Utech, D. Vuono , M. Bruno, P. De Luca and A. Nastro

Department of Pianificazione Territoriale, University of Calabria, Arcavacata di Rende, 87030

(CS) Italy, mariannautech@yahoo.it

Keywords: XRD, crystallinity, hydroxyapatite, morphology, biomaterials.

Abstract. The aim of this study is to synthesise hydroxyapatite by hydrothermal reactions and to compare the resultant pure phases obtained in static and dynamic conditions.

Introduction

The study of biomaterials begins in the XXth century, but in the last 25-30 years it shows a significant development. Polymer materials are replacing organs, but they do not stimulate the growth of bonetissue. This function is done, indeed, by ceramic materials. Those materials have biocompatibility, bioactivity and good mechanical properties. The main applications of these materials are: prosthesis, as hip prosthesis, knee prosthesis and others [1]. The use of materials like bone tissue offers good potentiality. Calcium phosphate based materials are considered an approach of bio-mimetic type. The mostly used in biomaterials are hydroxyapatite (HA), octacalcium phosphate (OCP) and ß-tricalcium phosphate (ß-TCP). In the years '90, HA was used to coat metallic prosthesis, increasing the biocompatibility with bone tissue. In 1992 Menabue et al. studied the production of bioceramics with controlled porosity [2].In 1995 Fabbri et al. synthesised hydroxyapatite-based porous aggregates [3]. In 1999 Martinetti and Nataloni studied biomimetism in maxillo facial surgery [4]. In 1996 Ito et al. synthesised HA from hydrothermal synthesis using hydroxyapatite disk as seeds [5].

Experimental

HA is synthesised using $CaHPO_4$ (Aldrich), commercial HA ($HA_{com.}$, Aldrich) and distilled water. The synthesis procedure is the following: 2.5 g of $CaHPO_4$ and 0.15 g of HA in powder are mixed in 300 ml of distilled water. The mixture is homogenised for 15 minutes and inserted into PTFE-lined Morey type inox steel autoclaves of 50 cm^3. The autoclave is put in a thermoventilated oven at 100°C. The synthesis is carried out in two ways: (1) in static conditions; (2) in dynamic conditions. The powder extracted from the autoclave at different times of synthesis is recovered by filtration, washed with distilled water and dried at about 100°C for 24 h. The samples were identified by powder XRD on a Philips PW1830 diffractometer using CuKα radiation. The scanning speed was $0.02°s^{-1}$ in the 5-45° 2θ range. The morphology of the HA is determined by scanning electron microscope MICROSPEC WDX-2A using a 25 KV accelerating potential.
Chemical microanalysis was carried out by EDS ZAF-ZAF-4/FLS analysis.

Results and discussion

The study was carried out to optimise the amounts of reagents needed to produce HA in powder. The initial system is the following (in moles): $xCaHPO_4 - yH_2O - 1.5*10^{-4} HA_{com}$ where: $1.0*10^{-2} \leq x \leq 2.6*10^{-2}$ and $11.7 \leq y \leq 21.7$. The amount of HA_{com} is considered as an optimal parameter to start the reaction and was not varied in this study.
The research is carried out using two synthesis conditions: synthesis of HA with stirring (dynamic conditions) and without stirring (static conditions). Figure 1 shows the crystallisation field of HA produced in static conditions as a function of the amounts of $CaHPO_4$ and H_2O.

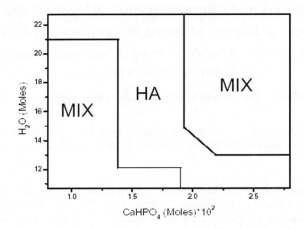

Fig. 1 – Crystallisation field of HA synthesised in static conditions.
MIX=CaHPO$_4$+ HA$_{com}$; HA= hydroxypatite

It is possible to observe that the area of crystallisation of HA is more restricted than the area in which
no reaction occured, noted as MIX in Figure 1. In this area the synthesis mixture remained unchanged
after 15 days of hydrothermal treatment in the autoclave. Figure 2 shows the X-ray diffractograms of a
dried synthesis mixture. The same spectra are obtained analysing the field where the synthesis mixture
did not react (shown in Figure 1 as mix (mixture)). The spectrum reported in Figure 2a shows two
crystalline phases: HA and not reacted CaHPO$_4$.

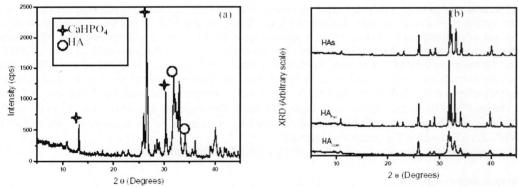

Fig. 2 – XRD of dried synthesis mixture (a) and XRD of different HA compared with synthesis HA
obtained in static conditions (b)

The field of crystallisation presents the production of HA at $1.8*10^{-2}$ moles of CaHPO$_4$ and in the range
16.7 – 21.7 moles of H$_2$O, while for lower amounts of water pure HA was not obtained. At about
$2.2*10^{-2}$ moles of CaHPO$_4$ pure HA is obtained, but only for low values of H$_2$O. This is true for
$2.6*10^{-2}$ moles of CaHPO$_4$, too. Figure 2b, reports the spectra of commercial hydroxyapatite (HA$_{com}$),
hydroxyapatite produced by Fin-ceramica Faenza (Italy - HA$_{Fin}$) and that produced in this study (HA$_s$).

The crystallinities of two HA synthetic phases (HA_{Fin} and HA_s) are comparable, while that of commercial (HA_{com}), used to start the reaction of crystallisation, has low crystallinity. Figure 3 - a shows the crystallisation field of HA obtained in dynamic conditions.

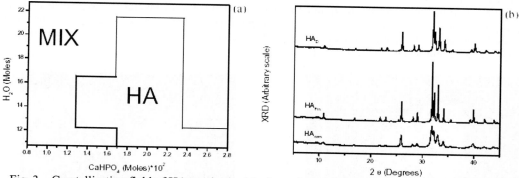

Fig. 3 – Crystallisation field of HA synthesised in dynamic conditions (a) and XRD of different HA compared with synthetic HA obtained in dynamic conditions (b).

The different synthesis conditions modify the crystallisation field. In fact, when the reaction is carried out in dynamic conditions the area of crystallisation of HA is bigger than that obtained in static conditions. HA is obtained in the of range $1.4*10^{-2} – 1.8*10^{-2}$ moles of $CaHPO_4$ and in the range of $16.7 – 21.7$ moles of H_2O. Pure HA is obtained at $2.2*10^{-2}$ moles of $CaHPO_4$ and 21.7 moles of H_2O. At $2.6*10^{-2}$ moles of $CaHPO_4$ and 11.7 moles of H_2O, HA is obtained, as it was noted in the synthesis at static conditions. Figure 3b, shows X-ray diffractograms of HA samples obtained in dynamic conditions (HAD), compared with those of Fin-ceramica and of commercial HA_{com}.

A kinetic study was carried on representative samples. This study led to the analysis of crystalline phases obtained in static and dynamic conditions. In Figure 4a the curve of crystallisation is shown as a function of time for the sample synthesised using $1.8*10^{-2}$ moles of $CaHPO_4$ and 16.7 moles of H_2O. As it can be noted, the synthesis start at about 1 day and it stop at about 70% of crystallinity, after 3 days.

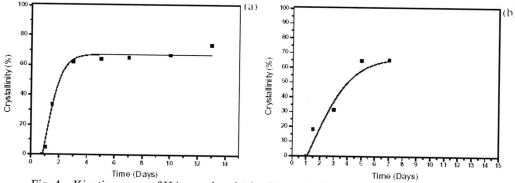

Fig. 4 – Kinetic curves of HA samples obtained in static (a) and dynamic (b) conditions.

Figure 4b shows the kinetic curve of synthesis carried out using $1.8*10^{-2}$ moles of $CaHPO_4$ and 16.7 moles of H_2O, in dynamic conditions. The synthesis starts, like in the case of static conditions, at about 1 day and stops at about 5 days at 65% of crystallinity.

The HA powders produced in static and dynamic conditions have been characterised by SEM.

Figure 5a ($1.8*10^{-2}$ moles of $CaHPO_4$ and 16.7 moles of H_2O) shows several heterogeneous crystals aggregated, obtained in static conditions, having bigger dimensions with respect to crystals shown in Figure 5b obtained in dynamic conditions. In the dynamic case, the sample presents a needle-shaped morphology, as reported by Seckler et al. [6].

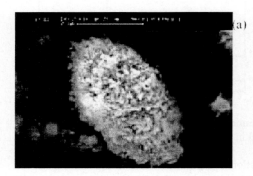 

Fig. 5 – Micrograph of synthetic HA obtained in static (a) and dynamic (b) conditions.

Conclusions

The different synthesis conditions modify the crystallinity of HA, in fact the HA powder synthesised in static conditions is more crystalline than the powder obtained in dynamic conditions.

The crystals dimension depends on the synthesis conditions: in static conditions the crystals are bigger than those needle-shaped ones obtained in dynamic conditions.

References

[1] LL. Hench : Biomaterials Vol 19 (1998), p.1269-1273

[2] L. Menabue, L. Forti et al.: Biomateriali Vol 3 / 4 (1992)

[3] Fabbri et al.:. Biomaterials, Vol 16 (1995), p.225-228

[4] A. Nataloni , R. Martinetti et al. : Atti del IX Congresso Nazionale della Società Italiana di
    Chirurgia Maxillo.facciale, Milano Marittima 1999, p. 555-559

[5] Atsuo Ito, et al: J. Crystal Growth, Vol 163 (1996), p. 311-317

[6] M.M Seckler et al.: Mat. Res. Vol 2 (1999), p.123-128

Key Engineering Materials Vols. 254-256(2004) pp. 47-50
online at http://www.scientific.net
© *2004 Trans Tech Publications, Switzerland*

Effect of Ag-Doped Hydroxyapatite as a Bone Filler for Inflamed Bone Defects

J.W. Choi[1], H.M. Cho[1], E.K. Kwak[1], T.G, Kwon[1], H.M. Ryoo[1],
Y.K. Jeong[2], K.S. Oh[2], and H.I. Shin[1]

[1]School of Dentistry, Kyungpook National University, Daegu, Korea, hishin@knu.ac.kr
[2]Nano Ceramic Center KICET, Seoul, Korea

Keywords: bone filler, antimicrobial effect, silver doped hydroxyapatite

Abstract. Silver (Ag)-doped hydroxyapatite (HAp) agglomerates containing 0.15%, 1.5% and 4.3% mole % of silver among total cations, respectively, were evaluated *in vitro* and *in vivo* to explore their potential application as a bone filler with antibacterial properties. The 0.15% Ag-doped HAp was mildly cytotoxic, whereas the 1.5% and 4.3% Ag-doped HAps were moderately cytotoxic in the standard agar overlay cytotoxicity assay. The *in vivo* test was carried out by implanting Ag-HAps in artificial bone defects at the periapical area of both mandibular 1st molar of rats and no remarkable cytotoxicity was found unlike what was observed in the *in vitro* data. All of the implanted Ag-doped HAp particles, regardless of their Ag contents, allowed appropriated cellular proliferation and favorable bone repair without remarkable inflammatory reaction through 3 week healing periods, in spite of the mild delay in organization of fibrin and inflammatory reaction with the 4.3% Ag-doped HAp at the early healing phase. They supported well new bone formation with osteointegrative and osteoconductive properties. The results suggest that HAps doped with Ag up to 4.3 % of total cations can be applied for repair of infection-associated bony defects.

Introduction

Bone defects occur in a variety of clinical situations, and their reconstruction to provide mechanical integrity to the skeleton is a necessary step in a patient's rehabilitation. Although autogenous bone graft has been considered as the gold standard for bone repair against which other skeletal substitutes are measured, the autogenous bone grafts do not solve all instances of bone deficiency due to the limited availability of autogenous bone graft materials and other drawbacks such as donor site morbidity, increased operative time, wound complications, local sensory loss, and chronic pain [1, 2]. Allograft bone has been used as an alternative but it has low osteogenicity, increased immunogenicity, and rapid resorbability. In addition, significant incidence of postoperative infection and potential risk of disease transmission arises concerns [3]. Thus, a variety of alternative bone grafts and various synthetic biomaterials have been developed to overcome these complications. Bones can regenerate themselves to repair defects up to a certain size, but if the defect is too large and severe, suitable implant materials have to be applied to facilitate the bone repair [4]. Particularly, when bone defects are associated with infection as in osteomyelitis and periapical lesions, the bone healing becomes more complicated and pathogens can not be easily removed by systemic antibiotics; thus, the application of biomaterials containing antimicrobial activity can be a promising bone healing method [5]. In this study we evaluated silver-doped hydroxyapatite fabricated in agglomerated particles with various silver ion contents as a promising bone substitute that shows appropriate antibacterial property.

Materials and Methods

HAp was fabricated by reacting solutions of $Ca(NO_3)_24H_2O$ and $(NH_4)_2HPO_4$. Subsequently, Ag ion was introduced through immersion of the prepared hydroxyapatite powder in the $AgNO_3$ solution with various concentrations. Since Ag is known to replace Ca within the lattice, the amounts of silver introduced were estimated as mole fractions among overall cations (Ca+Ag). According to the ICP(inductively coupled plasma) analysis, the mole percents of silver in HAp expressed in this standard were 0.15%, 1.50% and 4.30%, respectively. Samples were then shaped into agglomerated particles measuring 150-300μm in diameter. The biocompatibility analysis of silver-doped hydroxyapatites was performed by standard agar overlay cytotoxic assay. MC3T3-E1 cells were plated at 1.25×10^6 viable cells/6-cm tissue culture dish with 5ml α-MEM media and incubated until confluent. The culture medium was then replaced with 5ml mixed solution of 0.6% agar and 2X α-MEM supplemented with 20% FBS. After solidification, 2ml of 0.01% neutral red was added and stained for 15 mins. The test samples were place on top of the agar layer and incubated for 24 hr at 37°C in 5% CO_2. They were then implanted into 0.3 cm diameter artificial bone defects formed by a round bur in the periapical area of both mandibular 1st molar of twenty-seven rats without aseptic consideration. At 1, 2, and 3 weeks after implantation, they were recovered from 3 rats of each group and radiographs were taken and then prepared routinely for light microscopy observation.

Results and Discussion

The fabricated Ag-doped agglomerated particles were very easy to handle and can be sterilized easily with absolute ethanol and UV light.

In the agar-overlay cytotoxicity assay, as shown in Figure 1, the 0.15% Ag doped HAp revealed mild cytotoxicity, whereas the 1.5% and 4.3% Ag doped HAps were moderately cytotoxic (Fig. 1). These findings were similar to our previous data obtained from disc-shaped Ag-doped HAps with various Ag contents. They exhibited mild to moderate cytotoxicity with adequate antimicrobial effects.

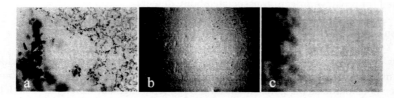

Fig. 1. Agar overlay cytotoxic assay. (a) 0.15%, (b) 1.5% and (c) 4.3% Ag doped HAps. 1.5% and 4.3% Ag-doped HAps revealed remarkable lysis of cultured neutral red stained MC3T3-E1 cells.

Table 1. Cytotoxicity determined by response index of agar overlay assay

Sample	Response Index	Cytotoxicity
Copper (Positive control)	3	Severe cytotoxic
OHP film (Negative control)	0	None cytotoxic
0.15 % Ag-doped HAp	1	Mild cytotoxic
1.5 % Ag-doped HAp	2	Moderate cytotoxic
4.3 % Ag-doped HAp	2	Moderate cytotoxic

During 3 weeks observation periods, there was no swelling or pus discharge at surgical wounds of all experimental mice and the skin wounds were completely healed at 1 week after Ag-doped HAps implantation. This indicates excellent biocompatibility and expected microbial property of Ag-doped HAps.

Radiographically the implanted Ag-doped HAps within periapical defects of mandibular 1st molar area of both sides induced gradual reduction of defect size with time, regardless of Ag contents. The defects gained similar radio density as the surrounding normal bone tissue. These findings suggest that the implanted HAps doped with up to 4.3% of total Ag cations induce active osteogenic repair without disturbance of healing cascades (Fig. 2). The Ag-doped HAps themselves also revealed similar opacity with bone tissue.

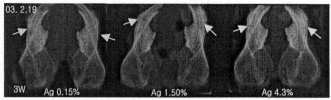

Fig. 2. Radiographs at 3 weeks after implantation of Ag-doped HAps show markedly reduced defect size with harmonized density with surrounding normal bone tissue. Arrows indicate Ag-doped HAps implanted in 1st molar periapical bone defects of rats.

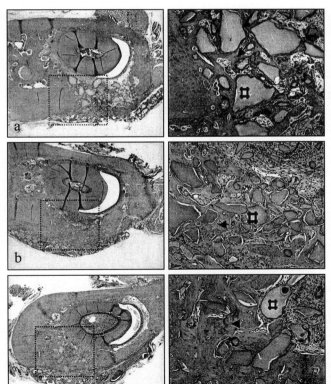

Histologically the implanted 0.15% and 1.5% Ag-doped HAps showed various sizes and shapes. They were scattered within blood vessel rich fibrous connective tissue at 1 week. Some particles located near the bone defect margins showed significant osteoconductive activity.

Fig. 3. Photomicrographs at 3 weeks after implantation of HAps doped with 0.15% (a), 1.5% (b), and 4.3% (c) Ag, respectively, into 1st molar periapical artificial bone defects of rats without aseptic consideration.
Arrow: new bone, ⌗: Ag-doped HAps. (H&E, original magnification, x40, inset, x 200).

With time the bone formation around the particles became more prominent. The Ag-doped HAp particles were well integrated with newly formed bone without fibrous tissue insertion. For 3 weeks, there was no remarkable inflammatory reaction and disturbance of periodontal space, which is very important to prevent ankylosis between bone and tooth.

The 4.3% Ag-doped HAp, which revealed moderate cytotoxicity in the agar-overlay assay, showed mild delay in organization of fibrinoid materials formed in the center of defects with a mild inflammatory reaction at the early healing phase compared to the 0.15% and 1.5% Ag-doped HAps. This may be the result of the greater cytotoxic property of HAp particles. However, they supported well new bone formation with osteointegrative and osteoconductive properties. The defects were gradually replaced by integrated HAp and new bone without any inflammatory and foreign body reactions.

Summary

In the case of bone defects associated with infections such as osteomyelitis and periapical lesions of the jaw, the removal of pathogens by systemic antibiotics is not easy, so the application of biomaterials containing antimicrobial agents can be another promising bone healing method. The Ag-doped HAps, which revealed mild to moderate cytotoxicity in an agar-overlay cytotoxicity assay, were biocompatible *in vivo* and effectively supported new bone formation without inflammatory reaction. Our data suggest that HAps doped with Ag up to 4.3% can be an effective bone filler with antimicrobial properties for infection-associated bony defects.

Acknowledgements

This work was supported by a Grant-in-Aid for Next-Generation New Technology Development Program from the Korea Ministry of Commerce, Industry and Energy (No.N11-A08-1402-07-1-3).

References

[1] W.R. Moore, S.E. Graves, and G. I. Bain: ANZ j. Surg. Vol. 71(2001), p. 354

[2] T.W. Bauer and S.T. Smith: Clin. Orthop. Res. Vol. 395(2002), p. 11

[3] J.F. Keating and M.M. McQueen: J. Bone Joint Surg[Br]. Vol 82-B(2001), p. 3

[4] I.H. Kalfas: Neurosurg Focus Vol. 10(2001), p. 1

[5] Y. Yamashita, A.Uchida, T.Yamkawa, et.al.: Int. Orthop. Vol. 22(1998), p. 247

Key Engineering Materials Vols. 254-256(2004) pp. 51-54
online at http://www.scientific.net

Mechanical Testing of Chemically Bonded Bioactive Ceramic Materials

J. Loof[1], H. Engqvist[2], L. Hermansson[3] and N. O. Ahnfelt[3]

[1]Doxa AB, Axel Johanssonsgata 4-6, 754 51 Uppsala, Sweden, jesper.loof@doxa.se

[2]Doxa AB, Axel Johanssonsgata 4-6, 754 51 Uppsala, Sweden, hakan.engqvist@doxa.se

[3]Doxa AB, Axel Johanssonsgata 4-6, 754 51 Uppsala, Sweden

Keywords: Chemically bonded ceramics, bioceramics, calcium aluminates, mechanical properties, test methods

Abstract. Most chemically bonded ceramics that are used as biomaterial are resorbable. Therefore their mechanical properties often change over time. Calcium aluminate has proven to be bioactive but non-resorbable. This opens up new application areas compared to the conventional chemically bonded ceramics, especially for load bearing applications where the long-term mechanical properties are important, e.g. in dentistry. So far no consensus has been reached upon how to test the material to obtain reliable results. This paper describes how to test hardness, expansion pressure, flexural strength and compressive strength

The delicate chemistry of the material makes mechanical testing difficult. The final microstructure develops through water uptake both from the initially added water and from the storage medium. Thus some of the materials properties will be dependent on how the material has been handled, e.g. the expansion pressure. It is therefore of outermost importance to store the material in a medium that resembles the real application as close as possible. It is also important to be very careful when preparing and polishing the samples so that no flaws or defects are introduced, especially when measuring strength and hardness.

Introduction

Chemically bonded ceramics (CBC) are widely used as biomaterials. Most CBC-biomaterials are based on CaO. The following CBC-systems have been proposed or are already used as biomaterials: Ca-phosphates, Ca-silicates, Ca-aluminates, Ca-sulphates and Ca-carbonates, see Table 1. By using chemical reactions the CBC-systems are produced at low temperatures (body temperature), which is attractive from several perspectives: cost, avoidance of temperature gradients (thermal stress), dimensional stability and minimal negative effect on the system with which the material interacts. Notable is that the hard tissue of bone (apatite) also is formed via a similar chemical reaction. The proposed CBC-systems have very good biocompatible properties.

Table 1. Chemically bonded ceramic systems

Group/name	Basic system
Calcium silicates	$CaO- SiO_2-H_2O$
Calcium aluminates	$CaO-Al_2O_3-H_2O$
Phosphates	$CaO-P_2O_5-H_2O$
Carbonates	$CaO-CO_2$

Chemically bonded ceramics constitute ceramics that are being formed due to chemical reactions [1]. Often the precursor material is a ceramic powder (e.g. Ca-silicate), which is "activated" in a liquid. A chemical reaction takes place where the initial powder is partly or completely dissolved and new phases precipitate. The precipitated phases are composed of species from both the liquid and the precursor powder. Since the precipitates have a very small grain size and a low density the porosities within the precursor powder are filled and the material hardens. The dissolution speed and the solubility products of the formed hydrate phases determine the setting time for the material. The setting time can be controlled by proper control of the precursor grain size and/or by addition of accelerating or retarding substances [2]. Since the material can be formed from a precursor powder mixed with a liquid, the material can be made mouldable simply by controlling the amount of liquid (in relation to the powder) and by the possible addition of small amounts of polymers in the liquid [3].

The Ca-phosphates, -sulphates and –carbonates are known to be resorbable when inserted in the body. Therefore their mechanical properties are not constant and change (diminish) over time. The Ca-aluminates, however, do not seem to be resorbable and the material has been proven to be bioactive and form a bond to tissue in-vivo [4]. As such its use as biomaterial differs from the other CBC-materials and thus the Ca-aluminates can be used in load bearing areas (e.g. in dentistry). The Ca-aluminates are new as biomaterials and no consensus has been reached on how the material should be stored, handled and measured in order to obtain reliable and reproducible results. The objective of this paper is to describe how to test some important mechanical properties and give indications of discrepancies between good and bad test procedures and the rationale behind the selected test procedures. The investigated properties are (including general handling and storage): hardness, flexural strength, compressive strength and expansion pressure.

Sample preparation and storage
An experimental version of a dental restorative material based on calcium aluminate was used in the tests [4]. The material was mixed with water and condensed into the desired shape for testing hardness [5], flexural strength [6], compressive strength [5] and expansion pressure [7].

Since the material is intended to be used in a 37 °C humid environment it is very important that the material is stored under similar conditions. It is also important that the storage medium has a composition resembling body fluid since the material consumes water from the surrounding. If not stored under proper conditions other precipitated phases than those found in-vivo can develop [9]. This can lead to very different test results as compared to results from samples stored properly. To elucidate the difference between dry storage, water storage and phosphate buffer storage, samples were stored under the different conditions and examined regarding hardness and expansion pressure. Hardness (Vickers 100 g load) was measured in specific PMMA-moulds resembling tooth cavities. The hardness samples were stored in 37 °C physiological phosphate buffer solution. Just prior to measurement the samples were polished to a surface roughness of 200 nm. The expansion pressure was measured using a photo elastic method; see Fig 1 [7]. In the method, Araldite moulds are filled with the material and stored in 37 °C physiological phosphate buffer solution (same as saliva). The moulds can be taken out of the storage medium and measured continuously. The samples were tested first daily and then weekly for periods up to three months to detect the maximum expansion pressure. Five hardness samples and five pressure samples were made for each test condition.

Regarding flexural strength, the influence of test method on the result for ceramic materials has been extensively investigated in the literature. Regarding CBC-systems it has been shown that the ASTM F 394 method is the most suitable [6], but depending on the purpose of the measurement different sample preparation techniques can be used. Since the results are depending on the maximum defect size, methods to fabricate samples without large defects can be beneficial when developing new formulations. In the development process an actual change in mechanical properties of the material is the aim of the measurement and not to evaluate the maximum defect size in samples produced clinically correctly.

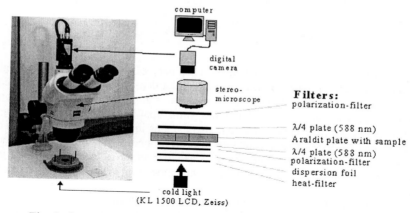

Fig. 1. Overview of the test setup for measuring expansion pressure

Applying an external pressure on the sample during setting when the material is still plastic, produces samples with no or small amounts of detrimental defects. The external pressure can be applied by using a uni-axial press with a similar or larger diameter than that of the sample. It is important that the samples are demoulded without introducing detrimental cracks; this can be achieved by using a piston with the same diameter as the sample and by gently pressing the sample out of the mould. The material should be allowed to set properly before removal from the mould. It is also very important that the samples are plane parallel and very smooth (a surface roughness below 200 nm). The flexural strength sample should be polished just prior to measurement due to possible reactions with the storage medium, which might change the surface roughness. In this study two different sample preparation techniques were used, hand condensing and external pressure. In both methods the material is first condensed into the mould and then removed from the mould or an extra pressure (approximately 100 MPa) is applied with a mechanical press as described above. The samples were stored in 37 °C physiological phosphate buffer solution for one week before measurement according to ASTM F 394 (Flexural strength testing of ceramic materials). The diameter of the samples was 5 mm and the thickness around 0.7 mm. Fifteen samples were tested for each preparation technique.

The compressive strength was measured on samples with 4.7 mm diameter and more than 7 mm in height. The difficulty in producing samples originates from producing plane parallel surface and from demoulding of the samples without introducing detrimental defects. Making the samples in plastic syringes and polish them carefully before removing the syringe can solve this. In this study fifteen samples were stored under equal conditions as in the case of the flexural strength samples and measured after one week.

Results

If stored wet the hardness increased to over 110 HV after one week, whereas if stored dry the hardness did not reach 50 HV. No difference in hardness between phosphate buffer and water storage could be found in the tests. However, unpolished samples stored in water showed low hardness due to extensive surface reactions (calcite formation [8]). The maximum expansion pressure found for the five samples was 2.4 MPa for water storage after one week and 2.0 for storage in phosphate buffer. Notably, most samples did not show any detectable pressure at all (i.e. above the detection limit of 1.75 MPa). None of the dry samples showed any pressure at all, probably due to drying shrinkage [1]. The flexural strength for hand-condensed samples reached a mean value of 65 MPa (standard deviation 11 MPa) after one week, whereas the samples pressed with an external press reached a mean of 100 MPa (s.d. 7 MPa). The compressive strength samples had a mean of 180 MPa (s.d. 12 MPa).

Discussion

It is of great importance to have insight into the chemistry of the system to understand how the material should be handled in-vitro. Since the material hardens through an acid base reaction in a dissolution precipitation reaction the surrounding medium is important for the resulting properties.

It has been proven that the Ca-aluminates form apatite on the surface when stored in phosphate buffer solutions [8]. This proves that also the ions in the storage medium can change the chemistry of the final hardened material and thus also the properties. Therefore all storage should be done in conditions that simulate the real situations as close as possible. It is also well known that CBC-systems can form different phases depending on temperature [9]. This can also have influence on the measured values.

The mechanical properties of the material are very dependent on the water to calcium aluminate ratio of the initial mix [1]. Low amount of water results in improved mechanical properties. Thus it is important to control the amount of water in the mixture and assure constant water content for all samples in a test series. Since the expansion pressure is not a common property to study, little comparable data is available for other material systems. But unpublished data show that the data in this paper is in the same range as the pressure from glass ionomers and amalgam. The dispersion in the flexural strength and compressive strength data show that the sample preparation techniques can be further improved. It is probably the removal of the samples from the moulds that introduces defects, which then affects the measurements.

Conclusions

When working with non-resorbable chemically bonded ceramics that harden in water the general handling and storage condition of the material plays a significant role. The CBC materials should be stored in a medium that resembles the real application as close as possible. The suggested methods to measure hardness, flexural and compressive strength and expansion pressure give reproducible and reliable results, but as the Ca-aluminate material studied in this work is new, further development of test procedures are necessary as new applications are found.

References

[1] F.M. Lea: LEA's Chemistry of Cement and Concrete, third ed. Edward Arnold Ltd, 1970.
[2] J. Lööf, H. Engqvist, G. Gómez-Ortega, N-O. Ahnfelt, L. Hermansson, Proceedings Bioceramics 16, Porto 2003.
[3] Lewis J. et al., J. Am.Ceram. Soc., 83 (8) 1905-13 (2000).
[4] H Engqvist, J-E Schultz-Walz, J Lööf, G A Botton, D. Mayer, M W Pfaneuf, N-O Ahnfelt, L Hermansson: accepted for publication in Biomaterials.
[5] J. Lööf, H. Engqvist, K. Lindqvist, N-O. Ahnfelt and L. Hermansson, Accepted for publication in J. Mater. Sci :Materials in Medicine.
[6] H. Engqvist, L. Kraft, K. Lindqvist, N-O. Ahnfelt and L. Hermansson, Accepted for publication in Journal of Advanced Materials.
[7] Ernst et al. Am J Dent 2000;13:69-72.
[8] L Kraft, Calcium Aluminate based Cements as Dental Restorative Materials, Ph D Thesis Dec 2002, Uppsala University, Sweden.
[9] K.L. Scrivener, J-L. Cabiron, R. Letourneux, "High-performance concretes from calcium aluminate cements", Cement and Concrete Research, 29, (1999),pp 1215-1223.

Key Engineering Materials Vols. 254-256(2004) pp. 55-58
online at http://www.scientific.net
© 2004 Trans Tech Publications, Switzerland

Assessment of Cancellous Bone Architecture after Implantation of an Injectable Bone Substitute

Catherine A. Davy[1], O. Gauthier[1, 2], M.-F. Lucas[3], P. Pilet[1], B. Lamy[4], P. Weiss[1], G. Daculsi[1], J.-M. Bouler[1]

[1] EMI 99-03, 1 Place A. Ricordeau, BP 84215, 44042 Nantes Cedex 1, France, email: catherine.davy@sante.univ-nantes.fr, jmbouler@sante.univ-nantes.fr

[2] Ecole Vétérinaire de Nantes, BP 40706, 44307 Nantes Cedex 3, France

[3] IRCCyN, UMR CNRS 6597, 1 rue de la Noë, BP 92101, 44321 Nantes Cedex 3, France

[4] LMM, 1 rue de la Noë, BP 92101, 44321 Nantes Cedex 3, France

Keywords: biomaterials, calcium phosphate ceramics, injectable bone substitute, injectable materials for bone regeneration, histomorphometry, characterization.

Abstract. This paper is aimed at describing the evolution of newly formed bone density and micro-architecture after implantation of an Injectable Bone Substitute (IBS). Cancellous bone growth (rabbit) is described as dependent of IBS formulation parameters. Possible improvements to the IBS formulation are finally discussed.

Introduction

Synthetic bone graft materials have proven useful as alternatives to autogeneous bone for repair, substitution or augmentation. Ceramic grafts are constituted of two main families: bioactive glasses and calcium phosphates. Among the latter, biphasic calcium phosphate (BCP) ceramics are made of a carefully chosen mix of calcium hydroxyapatite (HA) and beta-tricalcium phosphate (β-TCP) [1]. They are promising biomaterials, as they are resorbable and osteoconductive: BCP gradually dissolves into the body, while bone forms at the expense of the ceramic in the implanted area; moreover, those ceramics are bioactive and integrate particularly well with bone in the living body [2]: growing bone spontaneously bonds strongly to BCP as well as to the host bone. BCP ceramics are currently commercially available for orthopedic and dental applications, as blocks, particles, and also as injectable materials made of granules dispersed in a viscous polymer carrier (IBS), see for example MBCPgel™ developed by Biomatlante, Vigneux de Bretagne, France [1]. Due to their limited resistance to mechanical stresses, BCP ceramics main applications consist in filling bone defects when primary mechanical stability and contact with host bone are present [3-4]. With the purpose of improving BCP ceramics formulation, the extent of bone ingrowth has been evaluated previously for macroporous BCP structures [2], but, to our knowledge, this has not yet been thoroughly done for BCP-based IBS. The purpose of this study is therefore to characterize bone ingrowth after implantation of a BCP-based IBS and to propose possible improvements to its formulation parameters, such as the granules proportion and size. An *in vivo* experimental study on rabbit has been carried out. The originality of this paper is to propose a range of 40 histomorphometric parameters in order to fully characterize the bone mineral (the BCP granules being either included or not in this), in terms of density as well as micro-architecture. All parameters are evaluated from 2D SEM images and analyzed using data analysis techniques as proposed in [5]. Correlations are found which provide insights into the phenomenon of bone growth around the ceramic granules, so as to allow us to propose improvements to the biomaterial formulation.

Materials and Methods

In vivo **experimental study.** Bone defects were performed on 66 rabbit femurs filled for 2, 3, 6, 8 or 12 weeks with a BCP-based IBS. Three different granulometry ranges were used (grain proportion 50%): 40-80μm (small range), 80-200μm (medium range) or 200-500μm (larger range). Specimens were extracted and prepared as detailed in [6], corresponding to implant areas ranging from 20 to 70 mm². An image of the implant metallised surface was digitised (2048x2048 pixel raster in 256 grey levels) from a JEOL™ SEM (magnification x35), see Fig. 1.

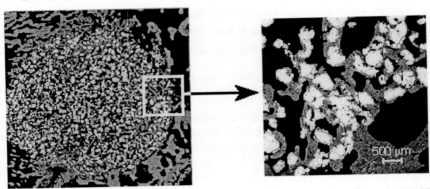

Fig. 1: B&W images of an implanted femur bone after 3 weeks, granulometry 80-200μm: calcium phosphate granules are in white, regrown and host bones are in grey.

Histomorphometry. Calculation of histomorphometric parameters was performed using a Leica Quantimet™ Q500 image analysis processor. After interactive thresholding, binary images of either the newly formed bone or the biomaterial granules were obtained and up to 40 static histomorphometric parameters were measured [7]. These include bone density (BV/TV), Parfitt's parameters (mainly the trabecular thickness TbTh, all other parameters being derived from TbTh) [8], calcified tissue skeleton characteristics (node density, free ends density, node-to-node struts mean length, etc.) [9], parameters specific to the IBS granules (density, mean grain size, mean inter-grain distance, etc.) or to their interaction with bone (mean bone/grain contact area) and also a novel parameter: the mean trabecular thickness (MTT) evaluated using the distance to the edges [10]. MTT is a direct measurement of the mean trabecular thickness whereas TbTh is derived from bone perimeter and area. Data analysis techniques [5] were then performed on the data to extract significant evolutions.

Results and discussion

All histomorphometric parameters exhibit a median very close to their mean, which is characteristic of a symmetrical distribution of data such as the Gaussian-type [5]. Several of them exhibit noticeable trends, as follows.

Mean trabecular thickness. At given granulometry and implantation time, bone density BV/TV and mean trabecular thickness MTT exhibit a very small coefficient of variation (defined as the standard deviation expressed as a percentage of the mean value), ranging from 7 to 15% only: those parameters are very robust to the biological variations from one rabbit to another. Fig. 2 displays the evolution of mean trabecular thickness evaluated (a) as suggested by Parfitt [8] (TbTh), or (b) as the MTT (average value +/- one standard deviation) vs. implantation time and for each granulometry range. According to Fig. 2(a), the mean trabecular thickness given by TbTh would be decreasing from 2 to 3 weeks and then increasing onwards. On the opposite, MTT keeps rising with the increase of implantation time. Our conclusion is that TbTh being defined as an indirect evaluation of the mean trabecular thickness does not describe correctly bone ingrowth: as direct measurement MTT shows, mean trabecular thickness rises continuously.

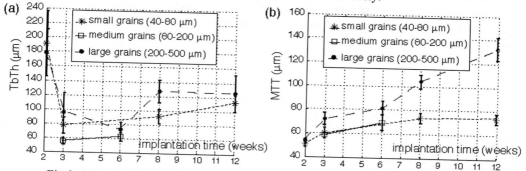

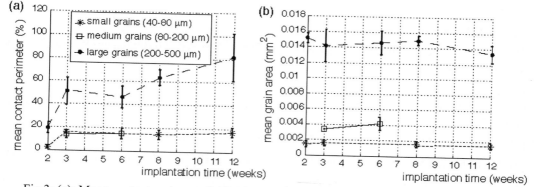

Fig.2: (a): Mean trabecular thickness measured as described by Parfitt [8] (TbTh) vs. implantation time; (b): Mean trabecular thickness measured with the distance function (MTT) vs. implantation time.

Fig.3: (a): Mean contact perimeter (MCP) between newly formed bone and ceramic granules vs. implantation time; (b): Mean grain size (MGS) vs. implantation time.

Furthermore, whereas MTT reaches an asymptote of almost 80µm (equivalent to the maximum grain size) from 3 to 8 weeks for small granulometry, MTT keeps rising for the larger granulometry range (data are insufficient to conclude for medium granulometry). Joint to the observation of the implants SEM images, this is interpreted as follows: when all grains are surrounded by newly formed trabeculae, the mean trabecular thickness reaches a maximum. Similarly, for larger granulometry, the maximum grain diameter is not overcome even at the longest implantation time available. Hence MTT increases regularly up to 12 weeks onwards in order to completely surround the biomaterial grains. This interpretation is confirmed by the evolution of the mean contact perimeter (MCP) between newly-formed bone and ceramic grains, see Fig. 3(a): whereas MCP reaches an asymptote from 3 weeks onwards for the smaller granulometry range, it keeps rising for the larger one. This bone ingrowth progression is also confirmed by Ono et al.[11] on the animal model of rat (tibiae implantation) with granules either made of apatite-wollastonite containing glass-ceramic, hydroxy-apatite or alumina (granulometry 200-355µm, no proportion granules/fluids is given). Ono et al. correlate the proportion of regrown bone to the implantation time and show that an asymptote is reached between 4 and 8 weeks for all types of ceramics. Regrown bone proportion remains constant after 8 weeks and the asymptote is reached earliest with bioactive ceramics (apatite-wollastonite and glass-ceramic or hydroxyapatite). Hence Ono et al. study exhibits very similar trends to ours: rat bone regrowth reaches an asymptote after 8 weeks implantation for a larger granulometry range (Ono et al.), whereas rabbit bone regrowth reaches an asymptote after the same implantation time for a smaller granulometry range (our study). Nevertheless, this comparison does not take into account the likely effects of the granules proportion or type.

Mean grain size and BCP resorption. The mean grain size (MGS) does not decrease significantly over the 12 weeks implantation time analyzed, see Fig.3(b). This suggests that BCP granules do not dissolve noticeably over that period of time, and that it takes them much more time to do so, as shown in [12] in the case of BCP blocks implanted on dog periodontal osseous defects.

Conclusion

This analysis has shown that the mean trabecular thickness of newly-formed bone increases after implantation of a BCP-based IBS until surrounding completely the original grains, and the smaller the granulometry range, the faster this process is achieved. This evidences the bio conductivity of IBS, which could be accelerated by increasing the available surface of the BCP grains, e.g. by optimising the grains size range and density. Besides, we have also shown no noticeable resorption of the ceramic granules up to 12 weeks implantation time. Further investigations are necessary to confirm whether the granules finally resorb and after what amount of time, or if they remain integrated into the newly formed bone architecture (contributing in that case to the overall bone mechanical strength).

Acknowledgements.

The authors are grateful to Département STIC du CNRS, INSERM and Région des Pays de la Loire for funding this project.

References

[1] G. Daculsi, O. Laboux, O. Malard and P. Weiss, J. Mater. Sci.: Mat. in Med., Vol 14 (2003), pp.195-200.

[2] T. Kokubo, H.M. Kim and M. Kawashita, Biomaterials, Vol.24 (2003), pp.2161-2175.

[3] O. Gauthier, J.M. Bouler, E. Aguado, P. Pilet and G. Daculsi, Biomaterials, Vol. 19 (1998), pp.133-139.

[4] M. Bagot d'Arc and G. Daculsi, J. Mater. Sci.: Mat. in Med., Vol 14 (2003), pp.229-233.

[5] G. Celeux, E. Diday, G. Govaert, Y. Lechevallier and H. Ralambondrainy, (Dunod Informatique 1989).

[6] O. Gauthier, J.-M. Bouler, P. Weiss, J. Bosco, E. Aguado, and G. Daculsi, Bone, Vol. 25 (1999), pp.71S-74S.

[7] D. Chappard, E. Legrand, C. Pascaretti, M. F. Baslé, and M. Audran, Microsc. Res. and Tech., Vol. 45 (1999), pp.303-312.

[8] M. Parfitt, C.H.E. Matthews, A.R. Villanueva and M. Kleerekoper, J. Clin. Invest., Vol. 72 (1983), pp. 1396-1409.

[9] J. Compston, R.W.E. Mellish and N.J. Garrahan, Bone, Vol. 8 (1987), pp.289-292.

[10] M. Coster and J.L. Chermant, Presses du CNRS (1989), pp.194-221.

[11] K. Ono, T. Yamamuro and T. Nakamura, Biomaterials, Vol. 11 (1990), pp.265-271.

[12] G. Daculsi, R.Z. LeGeros, E. Nery, K. Lynch, B. Kerebel, J. Biomed. Mater. Res., Vol.23 (1989), pp.883-894.

Key Engineering Materials Vols. 254-256(2004) pp. 59-62
online at http://www.scientific.net
© *2004 Trans Tech Publications, Switzerland*

Fabrication of Bioactive Glass-Ceramics Containing Zinc Oxide

C. Ohtsuki[1], H. Inada[1], M. Kamitakahara[1], M. Tanihara[1] and T. Miyazaki[2]

[1] Graduate School of Materials Science, Nara Institute of Science and Technology
8916-5, Takayama, Ikoma, Nara 630-0192, Japan, ohtsuki@ms.aist-nara.ac.jp

[2] Graduate School of Life Science and Systems Engineering, Kyushu Institute of Technology,
2-4 Hibikino, Wakamatsu-ku, Kitakyushu-shi, Fukuoka 808-0196, Japan

Keywords: Glass-ceramics, bioactivity, zinc, release control

Abstract. Apatite formation on glass-ceramics is an essential condition to bring about direct bonding to living bone when implanted into bony defects. Controlled surface reaction of the glass-ceramic is an important factor to govern bioactivity and biodegradation of the implanted materials. In this study, the effect of zinc oxide on bioactivity was investigated for glass-ceramics in the system $ZnO-CaO-SiO_2-P_2O_5-CaF_2$. Glass-ceramics containing apatite and wollastonite were prepared by the heat treatment of glass with compositions $xZnO \cdot (57.0-x)CaO \cdot 35.4SiO_2 \cdot 7.2P_2O_5 \cdot 0.4CaF_2$ (x=0.0, 0.7, 3.6, 7.1 or 14.2) (mol%). Apatite formation was observed on the glass-ceramics containing 0.7 mol% or less after soaking in a simulated body fluid (SBF) within 7 days, whereas it was not on the specimens containing 3.6 mol% or more. The addition of zinc oxide suppresses the dissolution of the glass phase, and makes it difficult to form the surface preferable to apatite formation. The reduced degradation can be achieved on the glass-ceramics when zinc oxide is added to bioactive compositions.

Introduction

Bioactive glasses and glass-ceramics are beneficial as bone-substitutes, because they can make a direct bond to living bone without any fibrous encapsulation when implanted into bony defects. This property is achieved by formation of a biologically active apatite layer after reaction of the glasses and glass-ceramics with surrounding body fluid. Among the bioactive glasses and glass-ceramics, glass-ceramic A-W is known as a material that shows high bone-bonding ability as well as high mechanical strength [1]. On the other hand, calcium phosphate ceramics containing zinc were recently developed because of the advantageous effect of released zinc on bone formation [2]. Zinc acts as an essential trace element that has stimulatory effects on bone formation *in vitro* and *in vivo* [3]. We expect that addition of zinc to bioactive glass-ceramic may control the reaction between the glass-ceramic and surrounding body fluid, and also that the released zinc ion from the glass-ceramic may enhance bone regeneration. In this paper, we report on the fabrication of glass-ceramics containing zinc oxide in the system $ZnO-CaO-SiO_2-P_2O_5-CaF_2$, that may produce bioactive glass-ceramics containing apatite and wollastonite.

Materials and Methods

The composition of the glass was selected from an idea on modification of glass-ceramic A-W. Glass was prepared at a composition of $xZnO \cdot (57.0-x)CaO \cdot 35.4SiO_2 \cdot 7.2P_2O_5 \cdot 0.4CaF_2$ (x=0.0, 0.7, 3.6, 7.1 or 14.2) (mol%), as shown in Table 1, by a conventional melt-quenching technique. The obtained glass was pulverized under 45 μm. Compacts of the glass-powders approximately 16 mm in diameter and 2 mm in thickness were sintered and crystallized at 930 °C for 4 hours, followed by natural cooling in the furnace. The obtained glass-ceramics were then soaked in 40 mL of a simulated body fluid (SBF) that has almost equal ion concentrations to those of human blood plasma, after the procedure reported by Kokubo *et al* [4]. The surface structures of the glass-ceramics were characterized before and after soaking in SBF, by thin-film X-ray diffraction (TF-XRD) and scanning electron microscopy (SEM). The changes in element concentrations of SBF due to soaking the glass-ceramics were examined by inductively coupled plasma (ICP) atomic emission spectroscopy.

Table 1   Compositions of the prepared glasses.

Notation	Composition / mol%				
	ZnO	CaO	SiO_2	P_2O_5	CaF_2
Zn0	0.0	57.0	35.4	7.2	0.4
Zn0.7	0.7	56.3	35.4	7.2	0.4
Zn3.6	3.6	53.4	35.4	7.2	0.4
Zn7.1	7.1	49.9	35.4	7.2	0.4
Zn14.2	14.2	42.8	35.4	7.2	0.4

Results and Discussion

Figure 1 shows TF-XRD patterns of the surfaces of the glass-ceramics before and after soaking in SBF for 7 days. Glass-ceramics containing apatite and wollastonite were obtained after the heat-treatment of the glasses at 930 °C. The relative content of wollastonite was highest at the composition of Zn7.1 among the examined specimens. Newly formed apatite characterized with broad diffraction peaks are distinctly observed for the glass-ceramics Zn0 and Zn0.7 after soaking in SBF for 7 days. Figure 2 shows SEM photographs of the surfaces of the glass-ceramics before and after soaking in SBF for 7 days. Assembles of fine particles were observed on the glass-ceramics, Zn0 and Zn0.7, while it was not observed on Zn3.6, Zn7.1 and Zn14.2. The deposition of the particles is attributed to formation of apatite crystals.These results indicate that glass-ceramics can form an apatite layer after exposure to SBF within 7 days when the glass-ceramic contains 0.7 mol% or less of zinc oxide.

Figure 3 shows the changes in element concentrations of SBF due to soaking of the glass-ceramics. The increases in silicon and calcium after the immersion of the glass-ceramics are attributed to the release of silicate and calcium ions dissolved from the glass-ceramics. The dissolution of silicate ions is distinctly reduced as the glass-ceramics contain 3.6 mol% or more of zinc oxide. Release of zinc was detected for the glass-ceramic containing 3.6 mol% or more of zinc oxide. Amounts of the released zinc are much smaller than those of silicate ions from the examined glasses. An increase in the content of zinc oxide in the glass-ceramics leads to a decrease in rate of the release of silicate, as well as an increase in that of the zinc. Decreases in calcium and phosphorus are attributed to comsumption by formation of the apatite on the glass-ceramics. A distinct decrease in phosphorus was detected due to soaking of Zn0 and Zn0.7. These results indicate that the addition of zinc oxide suppresses the dissolution of the glass-ceramics. Namely, an improvement in the chemical durability of the glass-ceramics is provided by the addition of zinc oxide, and hence decreases the rate of apatite deposition on the glass-ceramics. On the other hand, the release of zinc ions was observed when the content of zinc oxide in the glass-ceramics increased. This possibly means that the glass-ceramics enhance bone formation by released zinc ions when they have an appropriate amount of zinc oxide. Optimization of the composition is required to obtain a glass-ceramic with controlled bioactivity, apatite forming ability, as well as potential release of zinc oxide.

The results described above indicate that zinc oxide has a property that reduces the reaction between the glass-ceramics and surrounding body fluid. Zinc oxide may act as a component to reduce biodegradation of the glass-ceramics. Zinc oxide gives the glass-ceramics not only a potential on enhanced bone fomation but also reduces reaction against body fluid.

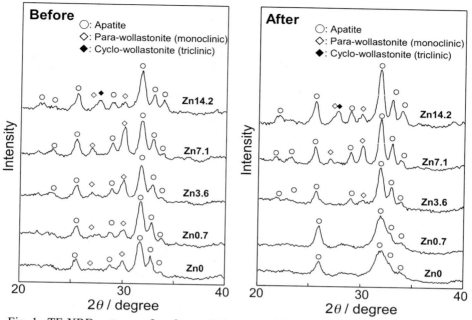

Fig. 1   TF-XRD patterns of surfaces of glass-ceramics before and after soaking in SBF
for 7 days.

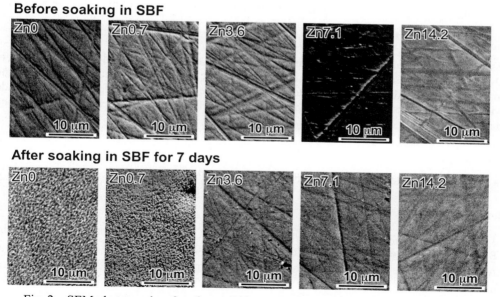

Fig. 2   SEM photographs of surfaces of glass-ceramics before and after soaking in SBF
for 7 days.

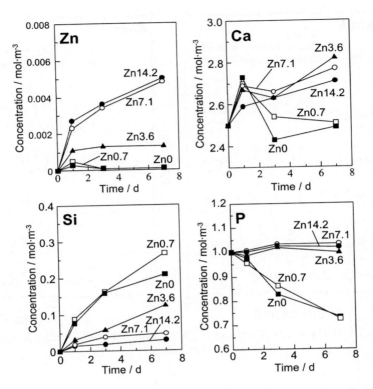

Fig. 3   Changes in element concentrations of SBF due to soaking of the glass-ceramics.

Conclusion

Glass-ceramics containing apatite and wollastonite were produced in the system $ZnO-CaO-SiO_2-P_2O_5-CaF_2$. An increase in content of zinc oxide gives an improvement in the chemical durability of the glass-ceramics to result in decrease in rate of apatite formation. Zinc oxide provides glass-ceramics with reduced biodegradation, as well as the potential enhancement of bone formation.

Acknowledgments

This study was supported by the Inamori Foundation Grant Program.

References

[1] T. Kokubo, M. Shigematsu, Y. Nagashima, M. Tashiro, T. Nakamura, T. Yamamuro, S. Higashi: Bull. Inst. Chem. Res., Kyoto Univ. Vol. 60 (1982), p.260-268.
[2] A. Ito, K. Ojima, H. Naito, N. Ichinose, T. Tateishi: J. Biomed. Mater. Res. Vol. 50 (2000), p.178-183.
[3] M. Yamaguchi, H. Oishi, Y. Suketa: Biochem. Pharmacol. Vol. 36 (1987), p.4007-4012.
[4] T. Kokubo, H. Kushitani, S. Sakka, T. Kitsugi and T. Yamamuro: J. Biomed. Mater. Res. Vol. 24 (1990), p.721-734.

Key Engineering Materials Vols. 254-256(2004) pp. 63-66
online at http://www.scientific.net

Apatite Hydrogel and Its Caking Behavior

Yoshiyuki Yokogawa[1], Yoshikazu Shiotsu[2], Fukue Nagata[1],

and Makoto Watanabe[2]

[1] Bio-Functional Ceramics Group, Ceramics Research Institute, National Institute of Advanced Industrial Science and Technology , 2266-98 Aza Anaga-hora, Ooaza Shimo-shidami, Moriyama, Nagoya, 463-8560 Aichi, JAPAN, y-yokogawa@aist.go.jp

[2] Graduate school of Engineering , Chubu University, 1200Matsumoto-cho, Kasugai, Aichi 487-8501, Japan, watanabe@isc.chubu.ac.jp

Keywords: hydroxyapatite, hydro-gel, caking, carbonate ion

Abstract. Apatite hydrogel preparation and its caking behavior was studied. Di-sodium hydrogen phosphate dodeca-hydrate and calcium chloride di-hydrate was dissolved in distilled water, and pH was adjusted to 7.40. The mixed solution was kept at room temperature for 1-21 days in air or N_2 atmosphere. With maturation period, the particle size of apatite hydrogel decreased in air, but increased in N_2. The FT-IR absorbance ratio (CO_3-band/PO_4–band) of the apatite hydrogel prepared in air increased, but those in N_2 decreased. TEM observation revealed that the crystal size of the apatite hydrogel seems to be almost similar, independent of the maturation period and atmospheric condition. So carbonate ion in apatite hydrogel is believed to strongly affect the aggregate of particles. The caking behavior of apatite hydrogel depended on the relative humidity, and the particle size. At 40-60% of relative humidity, the contraction of apatite hydrogel in volume was completed for 2 days, and the specimen set firm, stable in aqueous solution.

Introduction

Biological apatite is poorly crystalline apatite with high surface reactivity due to their nanometric crystal size and the hydrated layer rich in active environments of mineral ions on the crystal surface. Poorly crystalline apatite can be obtained by precipitation from saturated solution of calcium and phosphate [1]. Artificial hydroxyapatite (HAp) ceramics, widely employed as a biomedical implant, are generally synthesized at high temperature. A high temperature brings grain growth of ceramic, low specific surface area, and poor reactivity. The authors reported the low temperature synthesis to produce HAp porous material using the precipitates and simultaneous forming micro-size polymer previously [2]. The caking behavior depends on the particle size of precipitates, which changes with maturation period [3]. The purpose of this study is to investigate the phenomena of precipitates in solution and caking behavior during drying procedure to develop a new technique of producing an active HAp materials.

Experimental:

Di-sodium hydrogen phosphate dodeca-hydrate ($Na_2HPO_4 .12H_2O$) was dissolved in pure water, and calcium chloride di-hydrate ($CaCl_2·2H_2O$) in pure water. The latter solution was put into the former solution, and pH was adjusted to 7.40 using sodium hydroxide solution. The mixed solution was stirred at 500 rpm at room temperature in air or under N_2 atmosphere. The stirring period (maturation period) was 1-60 days. The precipitates were filtered off and washed with distilled water. The obtained gel-like materials were dried in air. The morphologies of the apatite hydrogels were observed by transmission electron microscopy (Jeol, JEM2010). All micro-FTIR spectra were recorded using samples encased in a transparent KBr. The X-ray powder diffraction patterns of the materials were recorded on a MAC Science MXP[3] diffractometer using CuKα radiation at 40 kV and

20 mA, a monochrometer. Crystallite size of specimen was calculated using apatite 002 and 310 peak width. Particle size of apatite hydrogel was measured using Particle Size Analyzer (Shimadzu, SA-CP3). The surface potential of apatite hydrogel was measured using Otsuka Electronics Laser zeta potential analyzer Leza-600. The measurements of pore size, distribution of apatite cake was carried out by Mercury Porosimetry Technique (Shimadzu, Poresizer 9310). The soaking of specimen in pure water or SBF solution [4] were also observed.

Results and Discussion

The pH value of the mixed solution was kept at around 7, but slowly changing, as the excess of phosphate and carbonate ions may act as a pH buffer. The particle size of apatite hydrogel is shown in Fig.1. With an increase of maturation period, the particle size of the gel decreased in air from 2-3 μm to 0.4 μm for 15 days of maturation period, but increased in N_2 from 1 μm to 4-5 μm for 15 days. The FTIR spectroscopy indicated that the precipitate was carbonate apatite. The ratio of the infrared absorption bands of carbonate and phosphate is shown in Fig.2. The excess carbonate ions are coming from air into solution, and carbonate ions in solution may be replaced in phosphate site of apatite. The ICP derived Ca/P ratios of apatite hydrogel formed in air increased as a function of maturation period. But the absorbance ratio of carbonate / phosphate of apatite hydrogel prepared under N_2 was slightly up with an increase of maturation period, as carbonate ions in solution was limited.

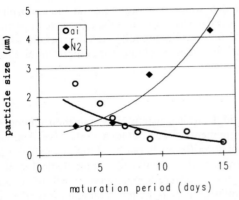

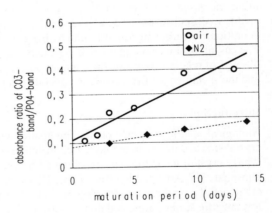

Fig. 1: Particle size of apatite hydrogel obtained in air and N_2 atmosphere as a function of maturation period

Fig. 2: FT-IR derived peak intensity of carbonate/phosphate ratio of apatite hydrogel as a function of maturation period

TEM images of the crystals in apatite hydrogel are shown in Fig.3. The crystals of apatite hydrogel formed for 1 day are angular and irregular in shape. Those formed by maturation both in air and in N_2 for several days showed rod-like shape, and seemed to grow somewhat with maturation period, but almost be the same in size, independent of maturation period and atmospheric condition. The X-ray diffraction data of the precipitates showed an apatitic pattern, and broadness of peaks were observed, which is analogous with bone minerals. The broadening of 002 and 310 lines allowed a determination of crystal dimensions. The crystallite size of the obtained gels were almost constant, independent of maturation period. So the particle size of apatite hydrogel should depend on the dispersibility of apatite crystallites, and the aggregate of apatite crystallites may be strongly affected by the carbonate ions in apatite hydrogel.

Fig.3 : TEM images of apatite hydrogel, obtained by maturation in air for 1 day (in the upper left), in air for 7 days (in the upper center), in air for 13 days (in the upper right), and in N_2 for 6 days (in the left). The specimens prepared by maturation for 1 day are with angles and irregular in shape, and those prepared by maturation for several days show rod-like shape.

The filtered apatite hydrogel was put in the plastic container of 33 mm diameter and 5mm in depth After 1 day, the specimen was taken outside plastic container, and placed on the paper. The volume changes of the specimens are shown in Fig.4. First the volume of specimen decreased exponentially, and after that the shrinkage of specimen seems to stabilize. At 80% of relative humidity, the volume of the specimen dropped from 2 to 8 days of drying period and the shrinkage seems to stabilize after 10 days of drying period. The shrinkage seems to be relatively slow for the apatite hydrogel obtained for short period maturation, so the apatite hydrogel with large particle size has slow shrinkage rate. The shrinkage behavior of specimen may involve 2 step, as re-aggregation process should be included in drying process. So at high relative humidity, the specimens with different particle size show different shrinkage behavior. At 60% of relative humidity, the volume of the specimen dropped within 2 days, and it seemed to stabilize in 3 days of drying period. The specimen set firm, and was difficult to be broken by hand. The shrinkage is so fast that the behavior seems to be similar, independent of the maturation period. At 40 % of relative humidity, the volume change of the specimens show the same manner as at 60 % of relative humidity. It is clear that the shrinkage behavior strongly depends on the relative humidity, and the particle size of apatite hydrogel affect the shrinkage behavior at high relative humidity.

The soaking of specimens in pure water or SBF solution at room temperature was conducted in order to evaluate the stability and released ions of the dried specimen in aqueous solution. Specimens of about 15 mm in diameter and 2 mm in thickness were used. With an increase in soaking time, the weight of specimen soaked in pure water was gradually decreased, which may be due to dissolution into pure water. The shape of specimens, although seems to be unchanged after soaking in pure water for 2 weeks. The weight of the specimen soaked in SBF solution increased with increasing soaking period. It may be due to the formation of new apatite layer and on/in the apatite material. The dried specimen slowly dissolves in aqueous solution, but are stable for a certain period. So the adsorptive materials included in this apatite cake may be released into solution for a certain period.

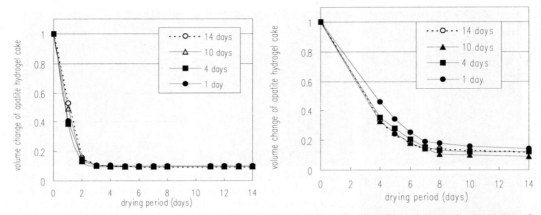

Fig.4 : The volume change of apatite hydrogel dried at 60 % (in the left) and 80 % (in the right) of relative humudity as a function of drying period.

Conclusion

Apatite hydrogel was prepared from a mixture of di-sodium hydrogen phosphate dodeca-hydrate ($Na_2HPO_4 \cdot 12H_2O$) and calcium chloride di-hydrate ($CaCl_2 \cdot 2H_2O$), and its caking behavior was studied. With maturation period, the particle size of apatite hydrogel decreased in air, but increased in N_2. The FT-IR absorbance ratio (CO_3-band/PO_4-band) of the apatite hydrogel prepared in air increased, but those in N_2 decreased. TEM observation revealed that the crystal size of the apatite hydrogel seems to be almost similar, independent of the maturation period and atmospheric condition. So carbonate ion in apatite hydrogel is believed to strongly affect the aggregation of particles. The caking behavior of apatite hydrogel depended on the relative humidity, and the particle size. At 40-60% of relative humidity, the contraction of apatite hydrogel in volume was completed in 2 days, and the specimen set firm, stable in aqueous solution. The dried specimen slowly dissolves in aqueous solution, but may be stable for a certain period. So the apatite adsorptive materials included in the apatite cake may be released into solution for a certain period.

References

[1] C.Rey, A.Hina, A.Tofighi, and M.J.Glimcher: Cells and Materials, Vol.5 (1995), p.345.
[2] Y.Yokogawa, F.Nagata, A.Hozumi, K.Teraoka, M.Inagaki, T.Kameyama, C.Rey, Bioceramics, Vol.14 (2002), p. 218 .
[3] Y.Yokogawa, Y.Shiotsu, F.Nagata, M.Watanabe, Proc.2nd Asian BioCeramics Symposium, (2002), p. 169.
[4] P.Li, C. Otsuki, T. Kokubo, K.Nakanishi, N.Soga, K.Nakamura and T.Yamamuro: J. Mater. Sci, Materials in Medicine, Vol.4 (1993) , p. 127.

Key Engineering Materials Vols. 254-256(2004) pp. 67-70
online at http://www.scientific.net

Measuring the Devitrification of Bioactive Glasses

Hanna Arstila[1], Erik Vedel[1], Leena Hupa[1], Heimo Ylänen[1] and Mikko Hupa[1]

[1] Process Chemistry Center, Åbo Akademi University, Piispankatu 8, FIN 20500, Turku, Finland

hanna.arstila@abo.fi

Keywords: bioactive glass, devitrification, crystallization, DTA, HTM

Abstract. The devitrification tendency during heat-treatment was measured for four bioactive glasses in the system Na_2O-K_2O-MgO-CaO-B_2O_3-P_2O_5-SiO_2. The measurements where performed by high-temperature optical microscopy, X-ray diffraction, scanning electron microscopy and differential thermal analysis. The combination of these methods proved to be a reliable way to define the devitrification temperatures and the composition of crystal phases formed. This information is vital in choosing glasses for implant production, especially in cases where thermal treatment is required.

Introduction

Typically, bioactive glasses have a low content of the glass forming oxides. Such glasses are prone to easily crystallize, devitrify, during the manufacturing process or any heat-treatment needed for achieving a desired form and structure of the product, e.g. an implant. The manufacturing of porous implants from melt-derived glasses through sintering is one typical example where any devitrification of the glass prevents glass working, and might decrease the bioactivity of the glass [1, 2]. Thermal studies have shown that the bioactive glass 45S5 devitrifies around 600°C. The devitrification makes it unsuitable for implant structures, which are manufactured by thermal treatments of the glass [3]. The bioactive glass S53P4 has been reported to be antimicrobial, but also this glass devitrifies easily when heat-treated [4, 5]. However, glasses referred as glass 13-93 and glass 1-98 can be sintered to porous implants by heat-treating them above glass transformation temperatures. Bioactivity of these glasses is not decreased by the sintering process, and the bioactivity can be further enhanced with surface microroughening [6]. In this work several experimental methods were applied to study the devitrification tendency of these four established bioactive glasses.

Materials

The experiments were performed with the bioactive glasses reported in literature with the codes 45S5, S53P4, 13-93 and 1-98. The experimental glasses were manufactured for laboratory use. The batches were mixtures of analytical grade raw materials and Belgian quartz sand as silica raw material. The glasses were melted in a platinum crucible for 3 hours at 1360°C, cast, annealed, crushed and re-melted for homogeneity. Small plates (14 x 19 x 0,5mm) were cut from the glasses. The rest of the glass was crushed and sieved to a powder with particle size < 45µm.

Methods

High-temperature optical microscopy (HTM). The devitrification tendency of the powdered glasses was studied with high-temperature optical microscopy (Misura 3.0 from Expert System) using the temperature range of 50°-1200°C. A constant heating rate of 20°C/min was applied during the measurement. The sample was imaged at every 5°C, and the height of the sample was measured from these images. The devitrification temperature was determined from the shrinkage graphs.

Differential thermal analysis (DTA). The devitrification temperatures were verified with a thermal analyzer TGA/SDTA851^e from Mettler Toledo. The heating rate was 20°C/min up to 1360°C. The sample size was fifteen milligrams of powdered glass. Powdered alumina was used as a reference material.

Heat-treatment and X-ray diffraction analysis (XRD). The glass plates as well as the powdered glass samples were heat-treated in an electric furnace within the devitrification range measured by HTM and DTA experiments. The samples were heated to several top temperatures with a heating rate of 20°C/min. After reaching the top temperature the samples were removed from the furnace, and allowed to cool to room temperature in static air. The samples were analyzed with XRD (X´pert by Philips) for their phase composition. The step rate for 2θ was set as 0,8°/min. The voltage of the X-ray tube was 40kV at a current of 30mA.

Scanning electron microscopy – energy-dispersive X-ray analysis (SEM-EDXA). SEM-EDXA (LEO 1530 with a Vantage EDXA analyzer from Thermo Noram) was used to study the microstructure and phase composition of both the heat-treated and untreated reference glass plates. Before the analysis the plates were coated with carbon.

Results

HTM. Two different shapes of linear shrinkage graphs were obtained from the high-temperature optical microscopy analysis. Glasses 45S5 and S53P4 gave linear shrinkage graphs where the shrinkage stabilizes before reaching 700°C. The shrinkage was only about 10%. The linear shrinkage graph obtained for glasses 13-93 and 1-98 show that the line does not stabilize until the shrinkage is over 35%. Both glasses 13-93 and 1-98 also show an extra peak between 750 and 850°C indicating intensive neck formation between the particles. The shrinkage graph of 1-98 together with the DTA graph is given in Figure 1. The horizontal plateau that indicates devitrification lies approximately at 850-1050°C. The devitrification temperatures for all the glasses were plotted from the point where the plateau begins (Table 1).

DTA. The DTA-measurements verify the HTM-results, but the devitrification temperatures obtained by the DTA are somewhat higher. Also for DTA two different forms of graphs were recorded. Glasses 45S5 and S53P4 gave similar graphs with an exothermic peak which has the onset around 650°C. This exothermic peak is a sign of devitrification. DTA graphs of glasses 13-93 and 1-98 have two exothermic peaks (Fig. 1). In these graphs the devitrification temperatures are the onset temperatures of the second exothermic peaks (Table 1). During the DTA measurements, the first peak was interpreted to indicate commencement of the sintering. Later this was confirmed by HTM and XRD.

XRD. The glasses could be divided into two categories also in respect to the XRD analysis. In XRD analysis two different crystalline phases were identified. In glasses S53P4 and 45S5 the crystalline phase formed during heating (near 650°C) was identified as sodium-calcium-silicate. Glasses 13-93 and 1-98 devitrified around 900°C and the crystal phase formed was wollastonite (Table 1).

SEM-EDXA. The crystal phases formed in the glasses were studied also by scanning electron microscopy. Glasses 45S5 and S53P4 have small angular crystals (Fig.2). According to EDXA the composition of these crystals corresponds to Na-Ca-silicate. Bigger needle-shaped crystals were found on glasses 13-93 and 1-98 (Fig.2). Analysis of the crystals proved that they contain an excess of calcium and silicon oxides, and correspond roughly to wollastonite. The composition of the glassy phase between the crystals does not differ much from the original composition of the glass.

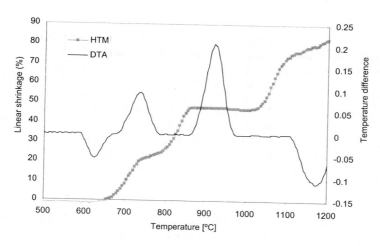

Fig. 1. High temperature optical microscopy and differential thermal analysis graphs of the bioactive glass 1-98. The devitrification temperature can be found approximately at 850°C as: I) a plateau beginning in the HTM-graph, and II) an onset of the exothermic peak in DTA-graph.

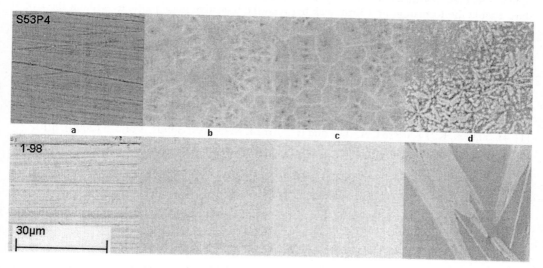

Fig. 2. Scanning electron micrographs of the heat-treated plates of glasses S53P4 and 1-98. a) Untreated plates showing scratches after polishing. Surface of the plates after heat-treatment (20°C/min) until: (b) 680°C, (c) 740°C, (d) 920°C. The magnification is the same for all the samples.

Table 1. Devitrification temperatures and crystal phases formed of the heat-treated experimental glasses as determined by HTM, DTA, XRD and SEM-EDXA.

Property measured + method	Glass 45S5	Glass S53P4	Glass 13-93	Glass 1-98
Devitrification temp, HTM [°C]	605	630	855	860
Devitrification temp, DTA [°C]	647	693	853	880
Crystal phase after heat-treatment + XRD and SEM-EDXA	Na-Ca-silicate	Na-Ca-silicate	Wollastonite	Wollastonite

Discussion

The temperature range at which the glasses are likely to devitrify during heat-treatment can be estimated from HTM and DTA-measurements. Both methods give roughly the same devitrification temperatures for one and the same glass, and the difference recorded partly depends on the interpretation of the obtained results. Different crystalline phases show typical temperature ranges for the onset of crystal formation. The crystal formation temperature for a crystalline phase varies somewhat with glass composition. The variation can be explained by the viscosity of the melts, which affects nuclei formation. The devitrification temperature as given by HTM and DTA provides a good starting point for estimating the suitability of a glass for e.g. manufacture of porous implant from crushed bioactive glass. The glasses typically suitable for sintering have a large temperature range between the glass transformation and the crystallization temperature [7].

Conclusions

The devitrification tendency of bioactive glasses can easily be verified with high-temperature optical microscopy or thermal analysis. Clear differences between established bioactive glasses were measured. Glasses 45S5 and S53P4 devitrify at low temperatures, and are thus likely to be unsuitable for heat-treatments above transformation temperatures. Glasses 13-93 and 1-98 devitrify at temperatures above 850°C. These glasses are of interest for sintering of melt-derived porous implants as their devitrification temperature is high enough to allow formation of viscous flow and, thus, enable the formation of a firm neck between the bioactive glass particles without the risk of devitrification.

Acknowledgements

Tekes, the National Technology Agency of Finland, is acknowledged for financing this study.

References

[1] H. Ylänen, K.H Karlsson, A. Itälä and H.T Aro: J. Non-Cryst Solids. Vol 275 (2000), p.107.

[2] O. Peitl Filho, G.P LaTorre and L.L Hench: J Biomed Mater Res. Vol. 30 (1996), p.509.

[3] A. El-Ghannam, E. Hamazawy and A. Yehia: J Biomed Mater Res. Vol. 55 (2000), p.387.

[4] P. Stoor, E. Söderling and J.I Salonen: Acta Odontol Scand. Vol. 56 (1998), p.161.

[5] M. Brink: Bioactive glasses with a large working range, Doctoral Thesis, Åbo Akademi University, Turku, Finland (1997).

[6] A. Itälä, J. Koort, H.O Ylänen, M. Hupa and H.T Aro: J Biomed Mater Res. *In Press*.

[7] L. Fröberg, L. Hupa and M. Hupa: Key Engineering Materials, *accepted*.

Key Engineering Materials Vols. 254-256(2004) pp. 71-74
online at http://www.scientific.net

Investigation of the Influence of Zirconium Content on the Formation of Apatite on Bioactive Glass-Ceramics

U. Ploska, G. Berger, M. Sahre

Federal Institute for Materials Research and Testing, Unter den Eichen 87, 12200 Berlin, Germany, e-mail: georg.berger@bam.de

Keywords: glass, glass-ceramics, zirconium, bioactivity, SBF

Abstract. This paper reports some investigations related to the influence of zirconium content of a bioactive glass and a glass-ceramics, respectively, on the ability to be covered with a hydroxyapatite layer when soaked into a simulated body fluid. The Zr content of the materials varied from 0 up to 10 wt%. The solubility of the materials in TRIS-HCl buffer solution is determined. A formation of hydroxyapatite on the surface of the glassy and crystallised specimens could be detected by TF-XRD. The influence of the Zr content on the shape of the spectra is discussed and also the difference between the glassy and crystallised materials.

Introduction

The addition of zirconia to hydroxyapatite (HA) markedly increases the mechanical stability of HA ceramics [1] and could enlarge the application possibilities of such ceramics. Recently presented results of animal test with zirconium containing bioceramics showed that the mineralization of the newly formed bone tissue is inhibited at the interface of the ceramic implant [2]. Compared with calcium zirconium phosphate, the implanted ceramics is characterized by a higher zirconium release [3]. Tests of the in-vitro bioactivity by soaking in simulated body fluid (SBF after Kokubo) of the bioactive glass-ceramics Ap40 (apatite and wollastonite) [4, 5] with various amounts of zirconia were carried out to proof the applicability of that method not least because of the activities for drawing up an ISO draft.

Materials and Methods

For the preparation of the glasses and glass-ceramics tricalcium phosphate, $CaCO_3$, Na_2CO_3, K_2CO_3, SiO_2, MgO, CaF_2, ZrO_2, and in two cases Al_2O_3 (reagent grade) were mixed using a wobble mixer and melted for 3 hrs at a temperature of 1550°C. The melt was poured out in a cylindrical steel mould. After quickly removing of the mould the rods were slowly cooled down to room temperature. Half of the produced glass rods were crystallized (24 hrs at 720°C and 24 h at 920°C) to obtain the glass-ceramics. Discs (about 15 mm in diameter, about 1 mm in thickness) were cut and drilled. The discs were cleaned with water and alcohol (ultrasound), dried for 1 hr at 60°C, and weighed after cooling down to room temperature. Two discs of each glassy and ceramic material were soaked in SBF solution for 1 and 4 weeks, respectively, at a temperature of 37°C. After taking out of the SBF solution the discs were rinsed with alcohol, dried, and weighed after cooling down. The surfaces of the discs were examined using the TF-XRD method. Furthermore, crushed samples (grain size range 315-400 µm) were treated with 0.2M TRIS-HCl buffer solution ($pH_{37°C} = 7.4$, 37°C, circular movement 75 r.p.m.) to determine their solubility. The concentrations of the ions leached out by this treatment were determined by means of an ICP-OES apparatus.

Results

Table: Composition of the tested materials

Sample Code	Composition in % by weight								
	SiO₂	CaO	P₂O₅	Na₂O	K₂O	MgO	CaF₂	Al₂O₃	ZrO₂
Ap40	44.30	31.90	11.20	4.60	0.20	2,80	5.00	-	-
Ap40ₖₜ₉ᵢ	43.86	31.58	11.09	4.55	0.20	2,77	4.95	0.59	0.40
Ap40ₛₜₑ	43.43	31.27	10.98	4.51	0.20	2,75	4.90	0.98	0.98
Ap40+2%ZrO₂	43.43	31.27	10.98	4.51	0.20	2,75	4.90	-	1.96
Ap40+3%ZrO₂	43.01	30.97	10.87	4.47	0.19	2.72	4.85	-	2.91
Ap40+4%ZrO₂	42.60	30.67	10.77	4.42	0.19	2.69	4.81	-	3.85
Ap40+5%ZrO₂	42.19	30.38	10.67	4.38	0.19	2.67	4.76	-	4.76
Ap40+10%ZrO₂	40.27	29.00	10.18	4.18	0.18	2.55	4.55	-	9.09

The table shows the composition of the synthesised materials (ZrO_2 and Al_2O_3 were added to Ap40). The crystallised materials with a Zr content higher than 4 wt% were not homogeneous (larger areas of two phases could be observed). The crystallised sample with 10 wt% of Zr could not be cut into discs because the produced rods had a lot of cracks.

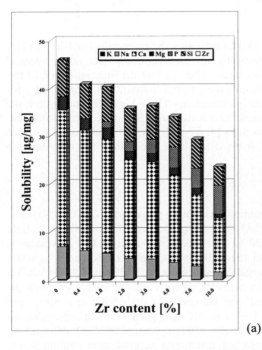 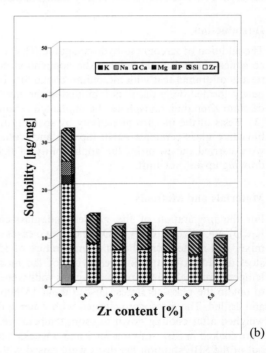

(a)   (b)

Fig. 1: Changes in solubility of Ap40 materials with various Zr content (0.2M TRIS-HCl buffer solution ($pH_{37°C}$=7.4), T = 37°C, mass-volume-ratio = 10); (a): glassy materials, (b): crystallised samples

As seen in Fig. 1a and 1b that present the results of the solubility tests with 0.2M TRIS-HCl buffer solution followed by ICP-OES analyses, the addition of ZrO_2 to Ap40 decreased the solubility of the synthesised materials. Contrary to the suggestion, the Zr release from the materials treated with

TRIS buffer solution under physiological conditions was very low. It was surprising that the Zr release from the glasses was slightly lower than from the glass-ceramics. The treatment with SBF solution under physiological conditions led to the formation of a layer on the surfaces of all materials. Already after the 7-days-treatment of the glassy samples it could be seen the change in surface appearance. After drying the soaked discs, only the surface layer of the sample Ap40 glass did not adhere well, large areas flew off. For all samples soaked in SBF solution there was observed a loss in weight. It was a little bit larger for the glassy materials than for the crystallised. The investigation of the surface layers using TF-XRD showed the formation of hydroxyapatite on almost all materials.

Discussion and Conclusion

Glassy and crystallised Ap40 are proved as bioactive materials. The addition of ZrO_2 to the composition markedly decreases their solubility. The difference is not the same for the glassy and crystalline materials, respectively. The solubility of the crystalline samples decreased most after addition of the smallest amount and changed only slightly with increasing Zr content. The addition of Zr especially influenced the release of Ca, Na, and Mg ions that became lower for more than 50 %. The formation of the surface layer on the glassy samples was slower than on the crystallised ones. On the surface of the sample Ap40 glass (7 days) no layer formation could be detected by TF-XRD but after 28 days some hydroxyapatite could be observed. The crystallised samples contained apatite and wollastonite as main phases .

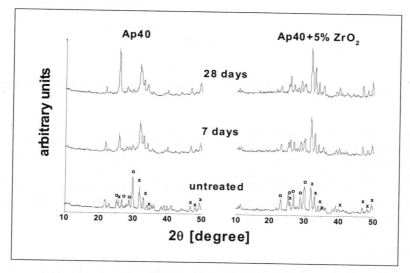

Fig. 2: TF-XRD spectra of crystallised Ap40 and Ap40+5%ZrO$_2$ samples (o - apatite, x-wollastonite)

If HA covered the surface of the samples with a thickness less than the penetration depth of the X-rays the intensity of the wollastonite peaks would be decreased as seen in Fig. 2. That difference was observed most clearly comparing the untreated sample and the 7-day sample. Between 7 and 28 days of soaking there was only a significant change in the HA peak intensities suggesting that the layers did not grow further but the crystallinity of HA was improved. These results agree with that of animal tests with another Zr containing glass-ceramics [2] about the inhibition of bone mineralization and with former results about the influence of increasing Zr content in bioactive glass-ceramics [4, 5]. Differences in the TF-XRD spectra were caused by the surface morphology of the investigated specimens. Regarding the crystalline samples, for Ap40 the grown up of HA was mainly characterized by the increase of the peak at $2\Theta \approx 24°$ whereas for the Zr containing samples the peak at $2\Theta \approx 31°$ arised. As for the glassy samples there was no difference in the appearence of the spectra concerning the detectable peaks and the spectra were similar related to a Zr amount till 5 wt%. The material with the highest Zr content showed the most structured spectra (see Fig.3). Obviously, the network of the glasses was changed with increasing Zr amount and the glass with 10 wt% might contain more crystallization nuclei responsible for the better growing up of HA.

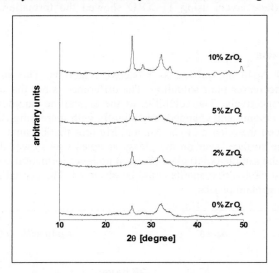

Fig. 3: Comparison of the TF-XRD spectra of Ap40 samples with various Zr content

As seen from TF-XRD spectra, the formation of HA on the surface of the specimens was influenced by the Zr content of the materials. The investigations to the theme about the test of bioactivity *in vitro* show that there are a lot of factors influencing layer formation like e.g. composition of the material, preparation of the test specimen, surface composition, solubility and so on. Before introducing an ISO draft about a test of *in vitro* bioactivity using the treatment with SBF solution the factors that influence the layer formation and their intensity have to be cleared up and also the application field for such an ISO.

Acknowledgements

A. Spitzer for the preparation of the glasses and glass-ceramics, U. Willfahrt for the tests with SBF, and S. Reetz for the TF-XRD measurement are gratefully acknowledged.

References

[1] J.A. Delgado, S. Martínez, L. Morejón, M.P. Ginebra, E. Fernández, M.T. Clavaguera-Mora, J. Rodríguez-Viejo, F.J. Gil, J.A. Planell: KEM Vol. 218-220 (2002), p. 161-164
[2] U. Gross, Ch. Müller-Mai, G. Berger, U. Ploska: KEM Vol. 240-242 (2003), p. 629-632
[3] U. Ploska: (2000) Diss. (PhD), RWTH Aachen, Germany
[4] G. Berger, R. Sauer; G. Steinborn, F.G. Wihsmann, V. Thieme, St. Köhler, H. Dressel: XV. Int. Congress on Glass, Leningrad, USSR, 3.-7.7.1989, Vol. 3a, (1989), p. 120 - 126
[5] G. Berger, V. Atzrodt: physica status solidi (a) Vol. 85 (1984), p. 9

Key Engineering Materials Vols. 254-256(2004) pp. 75-78
online at http://www.scientific.net

Bioactivities of a SiO2-CaO-ZrO2 Glass in Simulated Body Fluid and Human Parotid Saliva

P. N. De Aza[1], S. De Aza[2]

[1] Instituto de Bioingenieria. Universidad Miguel Hernandez, Edificio Torrevaillo. Avda. Ferrocarril s/n. 03202- Elche, Alicante, Spain; e-mail: piedad@umh.es

[2] Instituto de Cerámica y Vidrio, CSIC. Campus de Cantoblanco, Madrid, Spain; e-mail:aza@icv.csic.es

Keywords: Bioactivity, human parotid saliva, simulated body fluid, in vitro, glass

Abstract. A glass in the system wollastonite-zirconia, with a nominal composition of: 43.5 wt% SiO_2 -43.5 wt% CaO -13wt% ZrO_2. was obtained. This was soaked in two different solutions: simulated body fluid (SBF) and parotid human saliva (PHS), in order to compare its behaviour in a natural high protein content (PHS) medium and in an acellular protein-free solution (SBF). The SBF solution contained the same inorganic salts as in the human blood plasma, while the PHS contained both inorganic salts and various proteins. Both immersion systems were maintained for one month at 37°C.

The results have shown that the glass does not show any in vitro bioactivity when is soaked in SBF. However, a hydroxyapatite-like (HA-like) layer is formed on its surface when immersed in PHS. The interfacial reactions product was examined by Scanning Electron Microscopy (SEM), fitted with Energy-Dispersive X-say Spectroscopy (EDS). Additionally, changes in ionic concentrations in both systems and pH right at the interface glass/surrounding fluid were also determined.

The results confirm that to establish the potential bioactivity *in vitro* of a material, the physiological medium must be properly selected.

Introduction

Bioactive glasses and glass-ceramics have been widely used during the past 30 years after Hench et al [1] reported that some silicate glasses could chemically bond with living bone.

The capability of these materials for bone-bonding has been observed in vivo and is attributed to a particular modification of their surfaces, consisting of the formation of a calcium and phosphorous rich layer [2]. Direct bone apposition to implant for hard tissue replacements is essential to assure the stability of the implants clinically.

Owing the inherent difficulties of *in vivo* experiments, and since the understanding of the bioactivity demands a knowledge of the mechanisms of the (Ca-P)-rich layer formation, a significant amount of work has been carried out on the behaviour of the materials in either acellular or cellular media.[3,4]. Simulated body fluid (SBF) has been used as an *in vitro* model to study Ca-P formation on the surface of different types of biomaterials. In this test, those materials that do not present a (Ca-P)-rich layer formation on their surfaces are usually considered as non bioactives. The study examines the influence of the selected physiological medium on the *in vitro* behaviour of the material.

Materials and Methods

Keeping in mind the system $CaSiO_3-ZrO_2$, which presents a peritectic invariant point at 1467 ± 2 °C [5], the composition selected to obtain the glass was: 87 wt% $CaSiO_3$ and 13 wt% ZrO_2, which has the lowest melting temperature of the pseudobynary system.

The starting materials were high-purity ZrO_2 (99.99wt%, with low Hf content) and pseudowollastonite previously synthesized. Details of the preparation and characterization pseudowollastonite and glass can be found in previous publications [6-8].

The glass were cut by a low-speed diamond disc to prismatic pieces of 5x5x2 mm dimensions. These specimens were immersed in 50 ml of two different solutions, simulated body fluid (SBF) and human parotid saliva (PHS), at human body temperature (36.5°C) using polyethylene bottles. The immersion period of the glass was up to 4 weeks. This time span was chosen based on previous "in vitro" experiments performed in both media.[9,10].

Parotid saliva was collected from a single female subject on five occasions each lasting approximately 3 hours using a Lashley cup, after stimulating secretion with 1-2 ml of dilute lemon juice [11]. A few crystals of thymol were added as an antibacterial agent. Thymol would have no effect on the biological properties of the saliva, and no effect on salivary pH.

After the exposure period, the pellets were removed, and both media were analyzed for total Ca^{2+}, Si^{4+} and HPO_4^{2-} contents using Inductively Coupled Plasma Atomic Emission Spectroscopy (ICP-AES). The pH exactly at the pellet / human parotid saliva interface was also measured using an ion sensitive field effect transistor (ISFET-Meter) of Si_3N_4 type [12,13].

The surface structural changes of the glass and the morphology of the polished cross sections of the exposed specimens were first studied by SEM, using secondary and back scattered electron imaging and also EDS. The samples cross section were prepared by polishing up to 0.1μm diamond paste finish and cleaned gently in an ultrasonic bath, before being carbon coated.

Results and discussion

pH measurements, at the interface glass/SBF, did not show any significant changes during the period of immersion, varying from 7.25, pH of the SBF, up to 7.62. On the contrary, pH at the glass/PHS interface, varied from 8.0, pH of PHS, to a maximum of 10.0 after 10 minutes of immersion, staying constant up to one month.

Changes in the elemental ionic concentrations of PHS and SBF after one month of immersion are shown in Table I. The compositions of the original solutions are also enclosed for comparison. The chemical analyses of the PHS and SBF used in the study agree with those reported in the literature [14,15]. It was found that the silicon ionic concentration in PHS slightly increased, over the exposure time, indicating partial dissolution of the glass material. On the other hand, phosphorous and calcium ionic concentrations decreased more significantly because of the precipitation of HA-like phase on the surfaces of the pellets. However, the SBF ionic concentration did not change at all.

Table I. Concentration of Ca^{++}, Si^{4+}, $HPO_4^{=}$ in PHS and SBF before and after immersion of the glass pellets for one month

[mg/l]	Ca^{++}	Si^{4+}	$HPO_4^{=}$
Saliva control solution	64.18	35.09	441.62
Saliva after immersion	42.88	37.23	378.92
SBF control solution	100.20	-----	95.99
SBF after immersion	100. 25	-----	95.95

The surface of the glass material after one month exposure to PHS contained HA-like phase of characteristic globular morphology. The globules formed a compact and continuous layer. This is clearly presented in Fig. 1.

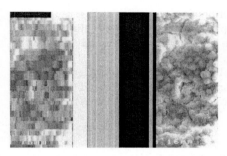

Fig. 1.- Hydroxyapatite-like layer formed on the surface of glass material after one month immersion in human parotid saliva.

Fig. 2 shows the overall microstructure of the cross-section of glass pellet after one month in PHS. The elemental X-ray maps of silicon, calcium and phosphorus are also included in the figure. The outside layer indicated a well-textured calcium phosphate phase of 10 µm average thickness on a silica rich intermediate zone.

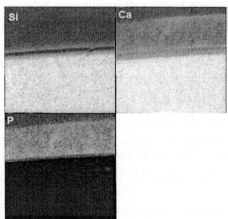

Fig.2.- X-ray maps of : Si, Ca and P, and SEM image of a cross-section of the glass pellet after soaking in human parotid saliva

Contrarily, the glass specimen soaked in SBF, did not show any morphological changes on its surface after the one month exposure. Figure 3 confirms this point. It shows the microstructure of a polished cross-section observed by SEM after 1 month immersion in SBF.

Fig.3- SEM image of a cross-section of the glass pellet after soaking in simulated body fluid

One of the reasons for the different behaviour of the glass in both media could be the distinct pH right at the interface between the glass and the respective media, being 10.0 in PHS and 7.62 in SBF. It is also possible that proteins in saliva formed a film on the glass surface, therefore influencing the transfer of ions in the same way as it happens on the enamel surface on the teeth.

Overall results suggest, that the mechanism of HA-like phase formation on the glass material in the PHS is similar to that of HA-like phase formation on other bioactive materials in simulated body [2,3]. It is understood that the HA-like layer precipitates on the surface of the glass material from the human parotid saliva due to the high pH (10.05) conditions at the glass/PHS interface, resulting from the ionic exchange of Ca^{++} from the glass network for $2H_3O^+$ from the saliva.

These results confirm those previously reported by De Aza et al [13], showing that *in vitro* bioactivity is directly related with the interfacial pH material/soaking medium. Additionally, the presence of proteins in the PHS can also play an important role in the bioactive behaviour.

Conclusions

The *in vitro* studies, carried out on the glass soaked in SBF, did not display any reactivity. There were not morphological changes observed on the surface of the glass after one month exposure to SBF. However, when the glass was exposed to PHS, an apatite-like layer was formed on its surface.

The results confirm that to establish the potential bioactivity *in vitro* of a material, the physiological medium must be properly selected.

References

[1] L. L. Hench, R.J. Splinter, W.C. Allen, T.K. Greenle: J. Biomed. Mater. Res. Symp. No. 2 (Intersciencie, New York, 1972)

[2] L. L. Hench: J. Am. Ceram. Soc. Vol. 74 (1991),p.1487

[3] C. Ohtsuki, T. Kokubo, T. Yamamuro: J Non-Crystal Solids Vol. 143(1992),p.84

[4] D. Dufrane, C. Delloye, I. Mc.Kay, P.N. De Aza, S. De Aza, Y.J. Shneider, M. Anseau: J. Mater Sci Mater Med. Vol. 14 (2003),p.33.

[5] P.N. De Aza, C. M.Lopez, F. Guitian and S. De Aza: J. Am. Ceram. Soc. Vol. 76 (1993),p. 1052

[6] P.N. De Aza, Z.B.Luklinska, M. Anseau, F. Guitian and S. De Aza: J. Microsc-Oxford Vol. 182 (1996), p. 24

[7] P.N. De Aza, F. Guitian, S. De Aza and F.J. Valle: Analyst Vol.123 (1998),p. 681

[8] P.N. De Aza, Z.B. Luklinska:. Bol. Soc. Esp. Ceram Vidrio Vol.42 (2003),p. 89

[9] P.N. De Aza, F Guitian, S. De Aza: Scripta Metall Mater Vol. 31 (1994),p.1001

[10] P.N. De Aza, Z.B. Luklinska, M.R. Anseau, F. Guitian and S. De Aza: J Dent. Vol. 27 (1999),p.107

[11] K.S. Lashley: Q. J. of Exp. Physiol. Vol.1 (1916),p. 461

[12] Merlos, I. Gracia, C. Cané, J. Esteve, J. Bartroli and C. Jimenez: *Proc. 5th Conf. on Sensors and their Applications.* (Edinburgh, UK,1991)

[13] P.N. De Aza, Guitian, F. Merlos, M. E. Lora-Tamayo and S. De Aza:. J. Mater. Sci: Mat. In Med. Vol. 7 (1996),p.399

[14] C.L.B. Larelle: *Applied oral physiology 2nd Edition.*(Wright, London 1988)

[15] J. Gamble: *Chemical Anatomy Physiology and Pathology of Extracellurar Fluid.* 6th Ed. (Harvard University Press, Cambridge 1967.)

Key Engineering Materials Vols. 254-256(2004) pp. 79-82
online at http://www.scientific.net

Homogeneous Octacalcium Phosphate Precipitation:
Effect of Temperature and pH

Y. Liu [1], R. M. Shelton [1] and J. E. Barralet [1]

[1] Biomaterials Unit, School of Dentistry, University of Birmingham, St. Chads Queensway, Birmingham, B4 6NN, UK, YXL053@bham.ac.uk

Keywords: octacalcium phosphate, crystal, precipitation, morphology

Abstract. Octacalcium phosphate (OCP) was prepared by homogenous precipitation at 40 to 60°C in the pH range 6.0 to 7.5 and depending on conditions either no precipitate was formed or a variety of morphologies of OCP were produced. At a given temperature, in low pH conditions no precipitate was formed, at higher pHs large OCP particles were precipitated which were spherical in shape with long plate-like (*LP*) crystals radiating from a central nucleus. At higher pH, within a very narrow pH range, the spherical particles were covered by short-plate (*SP*) like crystals. As the pH was raised further, aggregated spheres (*AS*) of poorly crystalline OCP were precipitated. The spherical crystals had a mean particle size range from 20 to 1000μm. FTIR spectra of *AS* were more similar to that of hydroxyapatite than OCP. As the size of the precipitated particles increased, the spectral similarity with OCP also increased.

Introduction

It has been suggested that octacalcium phosphate (OCP) $Ca_8(HPO_4)_2(PO_4)_4.5H_2O$ (Ca/P 1.33) participates in mineralized tissue formation as an initial precursor phase followed by subsequent precipitation and stepwise hydrolysis of OCP [1, 2]. Under physiological conditions, OCP is thermodynamically a metastable phase with respect to hydroxyapatite (HA). Because of the *in situ* nature of the transition from OCP to HA, oriented growth of apatite crystals may result from oriented and lengthwise growth of OCP. This may account for the characteristic packing of apatite crystals in enamel. Previous research concerning OCP precipitation has focussed on the effect of fluoride and amelogenins in order to study possible mechanisms of enamel formation and biomineralisation and there have been a number of investigations into its growth in gels in order to investigate matrix and mineral interactions [3]. OCP may be formed by controlled hydrolysis of $CaHPO_4.2H_2O$ or $\alpha\text{-}Ca_3(PO_4)_2$ [4, 5] and the OCP particles obtained using the latter method were composed of tight aggregates of strip-like microcrystals. OCP can also be precipitated as large spherical clusters of acicular crystals ~500μm in length in aqueous conditions [6]. Homogeneous precipitation is achieved by a method that relies on the retrograde solubility of OCP with temperature [7]. However, there have been no previous systemic studies on the effect of pH and temperature on the homogeneous precipitation of OCP. Here we report the morphology and size of OCP synthesised at 40 to 60°C in the pH range 6.0 to 7.5.

Materials and Methods

Sodium acetate $CH_3COONa.3H_2O$, calcium nitrate tetrahydrate $Ca(NO_3)_2.4H_2O$ and disodium hydrogen orthophosphate Na_2HPO_4 were supplied by Sigma (UK). OCP was prepared by means of homogeneous crystallization, raising the temperature of under saturated calcium phosphate solutions. To determinate the effect of pH value and temperature on the morphology of precipitation, 150ml 40mM calcium nitrate $Ca(NO_3)_2$ and 150ml 30mM disodium hydrogen orthophosphate Na_2HPO_4 were mixed in 1.5L 20mM $NaCH_3CO_2.3H_2O$ buffer using strong agitation at room temperature. After mixing, the pH of the mixture was adjusted to the desired value by addition of 20% acetic acid solution. The mixtures were warmed in a sealed container and maintained at 40, 45, 50, 55 or 60°C for 24 hours. If no precipitate was formed, the solution was maintained at the desired temperature for a further 24 hours. The precipitate obtained was filtered and dried at 37°C overnight. The same experiment was repeated at least three times for each precipitation condition. The morphology of the precipitates was determined using scanning electron microscopy (SEM) (Jeol, JSM-5300LV). Samples for SEM were dried and mounted on aluminium stubs using adhesive tabs and sputter-coated in argon atmosphere with a layer of gold-palladium. The crystal dimensions were determined according to the scale bar on electron micrographs, using Semafore software. Measurement of samples was performed on more than 100 particles and results were expressed as frequency within a size range. The density of the precipitates was determined from the average value of three samples using helium pycnometry (Accupyc 1330 Micromeritics, USA). FTIR (520 FT-IR Spectrometer, Nicolet, USA) spectra and X-ray diffraction patterns (Siemens D-5000, $2\theta=4°-80°$, step 0.02°, count time 2s) of precipitates were recorded and compared with standard JCPDS patterns to determine composition.

Results

A range of precipitates could be produced with a variety of morphologies by adjusting the pH value and temperature. Comparative micrographs illustrating the morphology of OCP crystals are presented in Figure 1. Clearly large differences in size and morphology occurred over a relatively narrow pH range and three distinct morphologies could be obtained: long plate-like (LP), (Fig. 1, A,); short plate-like (SP), (Fig. 1, B) and aggregated spheres (AS) (Fig. 1, C). At a given precipitation temperature, OCP particles were generally spherical in shape with long plate-like crystals radiating from a central nucleus at the lower pHs investigated. At higher pHs, within a very narrow pH range, the spherical particles were covered by short-plate like crystals. As the pH was raised further, aggregated spheres of poorly crystalline material were precipitated.

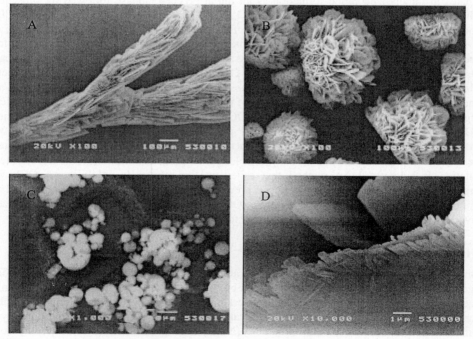

Fig. 1 The morphology of OCP A: *LP*; B: *SP*; C: *AS*; D: Submicron crystallites (Scale bar: A, B 100µm, C 10µm; D 1µm).

pH and temperature played an important role in determining the size and morphology of precipitates. The higher the solution temperature, the lower the pH required for precipitation such that it occurred at and above pH 6.6 at 40°C and 6.1 at 60°C. Above the precipitation pH, LP crystals were longer, more branched and easy to detach from the central nucleus compared with SP. As pH or temperature rose, the yield of OCP increased and SP crystals could be observed growing out from spherical nuclei. The modal size of OCP altered significantly from 20~1000 μm at pH 6.1 and 7.2 respectively. The shortening of the crystal plates resulted in a decrease in the size of spherical crystalline particles. Individual crystals appeared to consist of layers of submicron crystallites, which were arranged longitudinally (see Figure 1 D).

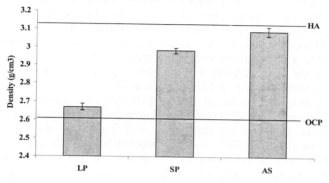

Fig. 2. The Density of OCP particles with different morphology

The density of OCP varied with size and shape. The reported theoretical densities of HA and OCP are 3.156 and 2.61g/cm^3 respectively [8]. All the precipitates' densities were intermediate between the theoretical densies of HA and OCP (see *Fig* 2). X-ray diffraction patterns of the OCP products showed a gradual reduction in intensity of peaks for OCP together with a gradual decrease in intensity of the peaks as the size of the precipitates decreased. The patterns of LP corresponded well with OCP standard (JCPDS ref: 79-0423). It was noticed that the intensities of the diffraction peaks at 2θ=4.73°(100), decreased and broadened as precipitation pH increased, indicating a less crystalline or possibly structurally imperfect material. Due to the lower crystallinity of SP and AS obtained at higher pH, no recongnizable peak could be resolved identifying the composition as containing HA.

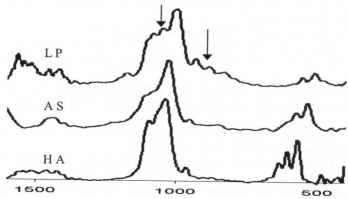

Fig. 3.FTIR spectra of LP, AS & HA

FTIR spectra of precipitates obtained at different pHs revealed that as the size of the particles increased, the structural similarity with OCP also increased and the spectra of *AS* were more similar to that of HA than OCP. Two pairs of bands, one at 1,196 and 865 cm^{-1} and another 1,297 and 910cm^{-1}, were assigned to HPO$_4$ in OCP crystal structure. The characteristic bands of OCP at 865 and 910cm^{-1} [8] became weaker, especially 910cm^{-1}, which changed dramatically as the pH increased. (see the arrows in Fig. 4). The positions and intensities of peaks in AS are very similar to HA. This indicated the possibility that apatite-like structural domains may exist in these products.

Discussion

It has been demonstrated that preparation conditions are critical in determining the size and morphology of homogeneously precipitated OCP crystals. The onset of the precipitation was related to pH and temperature. This was thought to have been a result of inverse temperature solubility for OCP [9]. The rate of precipitation was related to the supersaturation with respect to OCP in the solution. At lower pHs or temperatures, the degree of supersaturation in the initial reaction solutions was low, which led to a slow and continuous precipitation with further crystal growth. At higher temperatures or pH, the superaturation was relatively higher due to the retrograde solubility of OCP, the precipition started rapidly and the solution quickly became undersaturated. These factors modulated the nucleation and growth of OCP and thereby caused some specific growth modes, which were different from those generally observed in heterogeneous precipitation systems. Furthermore we have demonstrated that preparation conditions appear to affect the phase composition and critically, that OCP size and morphology affects phase stability in aqueous conditions. This may have occurred because the precipitate was hydrolysed to HA during the precipitation process. However, composition of the spherical particles is still ambiguous and the structural relationship between the nuclei and the plate-like crystals is intriguing and may shed light on biomineralisation mechanisms thought to involve an OCP precursor. Rod-like enamel apatite crystallite formation occurs at approximately 37°C within a relatively narrow pH range from 7.35 to 7.45 over a period of months in the human body. As biomineralisation is a complex process involving an organic template, the present results are not directly comparable with the biological model of enamel formation *in vivo*, yet provide the basis for a model with which organic/inorganic interactions may be studied. At near physiological conditions very fine spherical particles, with FTIR spectra similar to HA were obtained using this model. This study indicated that a nucleus of such a material is required for elongated macroscale OCP crystal formation to occur.

References

[1] Brown W.E., Smith J.P., Lehr J.R. & Frazier A.W., Octacalcium phosphate and hydroxyapatite, *Nature*, Vol. 196, (1962), p. 1048-1055

[2] Brown W.E., Crystal growth of bone mineral, Clin. Orthop., Vol. 44, (1966), p. 205-220

[3] Iijima M. & Moriwaki Y., Effects of ionic inflow and organic matrix on crystal growth of octacalcium phosphate: relevant to tooth enamel formation, J.Cryst. Growth, Vol. 199, (1999), p. 670-676

[4] Morden G.W. & Racz G.J., Effect of ph, calcium-concentration and temperature on the hydrolysis of dicalcium phosphate dihydrate, Canadian Journal of Soil Science Vol. 69(3), (1989), p. 689-693

[5] Monma H., Preparation of octacalcium phosphate by the hydrolysis of α-tricalcium phosphate, Journal of Materials Science, Vol. 15 (1980), p. 2428-2434

[6] LeGeros R.Z., Preparation of octacalcium phosphate (ocp): a direct fast method, Calcif. Tiss. Int. Vol. 37, (1985), p. 194-197

[7] Newesley H., Darstellung von 'oktacalcium phosphat' (tetracalciumhydrogentriphosphat) durch homogene kristallisation, Mh. Chem., Vol. 91, (1960), p. 1020-1023

[8] Elliot J.C., *Structure and Chemistry of the apaties and other calcium orthophosphates*, Elsevier Science B.V., Amsterdam, (1994), 8, 15 &20

[9] LeGeros R.Z., Preparation of octacalcium phosphate (ocp): a direct fast method, Calcif. Tiss. Int. Vol. 37 (1985), p. 194-197

Key Engineering Materials Vols. 254-256(2004) pp. 83-86
online at http://www.scientific.net

Investigation of the Specific Energy Deposition from Radionuclide-Hydroxyapatite Macroaggregate in Brain Interstitial Implants

B.M. Mendes and T.P.R. Campos

Curso de Pós Graduação em Ciências e Técnicas Nucleares - CCTN/UFMG
Av. Antônio Carlos, 6627, Pampulha Prédio PCA1, Sala 2285 CEP: 31270901
Belo Horizonte, MG, Brasil, bruno@nuclear.ufmg.br, campos@nuclear.ufmg.br

__Keywords__: hydroxyapatite, brain implants, macroaggregate, Monte Carlo, dosimetry.

Abstract. Bioceramic implants in brain can be justified by the viability of control small malignant tumors with a set of radioactive ceramics macroaggregate-type. Hydroxyapatite (HAP), a non toxic and non imunogenic bioceramic, has similar composition to the cortical bone and can incorporate radionuclides. Currently, M-hydroxyapatite macroaggregate (in which M represents ^{89}Sr, ^{90}Y, ^{165}Dy, ^{166}Ho or ^{188}Re radionuclide has been proposed. Those macroaggregates can be injected in the tumoral volume, mimicking conventional seed implants. The present paper addresses the dosimetry produced by M-HAP macroaggregates in brain interstitial implants, comparing it with ^{125}I metal seed implants. Radial Dose Profile (RDP) per unit of implant's segment has been evaluated. Bidimensional isodose curves for two distinct configurations of ^{188}Re-HAP and ^{125}I seed in multiple implants were performed based on in-house made software. Three-dimensional dose distribution is presented, from a selected ^{188}Re-HAP and ^{125}I configuration, in an anthropomorphic and anthropometrical head phantom. The present dosimetric propriety of the ^{188}Re-HAP is its major feature to approve its application at clinical oncology, establishing contrasts to ^{125}I implants.

Introduction

The radioactive seeds implants for malignant gliomas have been studied in two last decades.[1] The brachytherapeutic implants has also been used as photon megavoltage teletherapeutic booster.[2] The radioactive bioceramic implants into a brain can be proposed to control small malignant tumors. Hidroxyapatite (HAP) is the one of most interesting bioceramics for this purpose, due to its similarity in composition and chemical structure with cortex bone and to the fact that it can incorporate radionuclides. Moreover, HAP is non toxic and non immunogenic. Others bioceramic seed types for brachytherapy had already been studied in our research group.[3] Currently, new M-hydroxyapatite (M represents a metallic radionuclide) macroaggregates are being developed.

Herein, a comparative radiodosimetric analysis is presented considering ^{125}I seeds, often clinically applied to CNS brachytherapy, and hydroxyapatite macroaggregates, synthesized with incorporated high energy β-emitters, namely M-HAP, at a brain tumor implant.[4] The proposal is to verify the clinical viability of applying those bioceramic materials. The methodology is based on the investigation of the specific energy deposition from radionuclide-hydroxyapatite macroaggregate in deep brain interstitial implants through a stochastic computer code (MCNP4) and deterministic spatial dose distribution code, made in-house. The dosimetric characteristics of the present material are the major features to approve or disregard it in clinical oncology.

Methods

Radial Dose Profile (RDP) evaluation per unit of implant's segment. The RDP evaluation for M-HAP segments implanted through radio-opaque viscous gel-type medium, and for individual ^{125}I seeds were appraised through computational modeling. M-HAPs constituted by ^{89}Sr, ^{90}Y, ^{165}Dy, ^{166}Ho or ^{188}Re were studied. Figure 1 illustrates the model adopted for the radial dose profile evaluation per source segment. A 0.6mm diameter and 30mm height cylinder positioned in a sphere center represents a unitary implant's segment. Concentric cylinders 5mm height, and with 0.25mm radius increments were defined around the seed. The particles emitted (photons and electrons) absorbed dose per transition (Gy/tr) was evaluated in each radial ring utilizing the nuclear code MCNP4. The sphere and the dose cylinders are constituted by cerebral tissue.[5] The implant cylinder presents the M-HAP gel-type chemical composition. This composition varies according to the employed radionuclide. The particles emission was defined inside the central cylinder (M HAP gel-type). The β, γ and x-ray energy and emission probabilities of each radionuclide decay was considered.[6] A mathematical model was applied to discretely reconstruct the continuous betas emission spectra of the five radionuclides for the input of MCNP code.

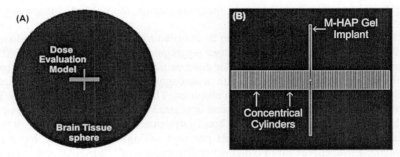

Fig. 1 – (A) MCNP4 computational model used for RDP evaluation. (B) Central region zoom.

2D isodose curves for configurations of multiple implants. Based on the RDP reproduced for each radionuclide, an in-house software generates the dose in a transversal plan, considering the M-HAP segments and ^{125}I seeds distribution. Configurations with 4, 5 and 7 implanted segments were analyzed and 2D isodose curves were generated for each configuration, looking for determination of the optimal segment arrange, preserving greater tumor dose and avoiding hot spots in implant region.

3D dose distribution, from a M-HAP implant's configuration. A brain computational phantom was developed to be executed by MCNP4 nuclear code, based on voxel model (Fig.2A). A 3.0x2.5x2.5cm tumor in left brain hemisphere was simulated. ^{188}Re-HAP cylindrical implants were placed inside de lesion site (Fig.2A). The distances between implanted segments and their number were previously defined. IMC6702 ^{125}I seeds were setup following equivalent brain implant standards (Fig.2B).

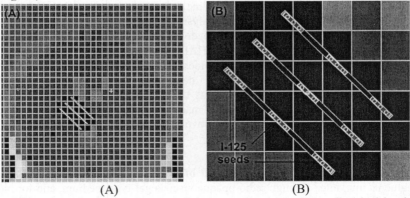

Fig. 2 – (A) Voxel model coronal slice in tumor plane. Three M-HAP cylindrical implants can be observed. (B) ^{125}I seeds placed at the model tumor.

Results and Discussion

Radial Dose Profile (RDP). The RPD per transition are presented in Figure 3. The tissue's absorbed dose for β-emitters presents profiles that diminish smoothly with the increment of the radial distance, characterizing a plateau. An abrupt fall occurs after this plateau (at a log x log graphic). Such behavior is very interesting at brachytherapy, since it permits large doses uniformly distributed in target volume, preserving adjacent healthy tissue. For ^{125}I the dose falls exponentially, obeying the physical proprieties of the gamma-rays interaction with matter. This behavior generates extremely high doses close to the seeds walls dropping by a exp(-μr) factor with radial distance. In this case, the tumoral dose varies from 100% to 1% of maximum tumor dose (MTD). The ^{90}Y and the ^{188}Re, β-emitters, with maximum energy over 2MeV, presented RDP per transition with lager plateaus, reaching up to 7-9 mm tissue-depth, which allow larger distance between segments. Beyond that, ^{188}Re emits γ and x-rays that helps monitoring leakage or absorption after implantation. The ^{188}Re half-life is 17 hours, so 90% of dose is delivered up to 22 hours. After 2 days, its activity will be decreased to insignificant levels, permitting bio-absorption and infiltration phenomena. The ^{188}Re was selected for the subsequent studies due to it's implants monitoring capability through cintilographic image.

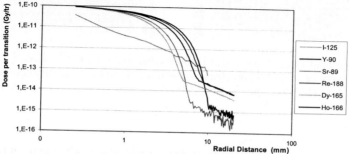

Fig. 3 – Absorbed dose per transition as function of the radial distance, in a log x log graphic, for Y-90, Sr-89, Re-188, Dy-165, Ho-166 –HAP gel-type and I-125 seed. The statistical errors were less than 5% except for ^{89}Sr and ^{90}Y in radius greater than 8mm.

Bidimensional isodoses curves from multiple segment implants. The figure 4A presents the percentage isodose curves for two distinct spatial configurations of ^{125}I seeds and ^{188}Re-HAP type gel segments. ^{188}Re-HAP segments present isodoses curves more uniformly distributed, establishing that the tumor doses varies from 100% to 10% of maximum tumor dose (MTD) (10x) and, at the same time, the healthy adjacent tissue dose fall from 10% to 1% of MTD in 2 mm, approximately. For the ^{125}I the tumor doses vary from 100% to 1% of MTD (100x).

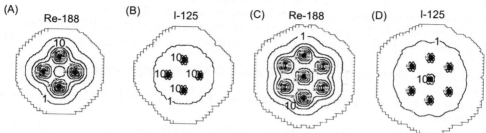

Fig. 4- Absorbed Dose bi-dimensional isodose curves (percentage of MTD) obtained for two distinct ^{188}Re-HAP gel-type segment's implants and ^{125}I seeds configurations.

Three-dimensional dose distribution in a head phantom. The Re-HAP get-type and ^{125}I seeds arrangements depicted in Fig. 4C and D were simulated in a computational head phantom, with anthropomorphic and anthropometrics equivalences to a human skull. Observing the isodose configurations (Fig.5) it is noted that ^{125}I and [188]Re presented adequate tumor dose distribution. However, the phantom discretization (5x5x5mm) was not enough, mainly in the implant proximities, for reveling the dosimetric details. Mainly for ^{125}I case, hot spots were not identified. However, healthy tissues received smaller doses when [188]Re-HAP gel was applied instead of iodine seeds.

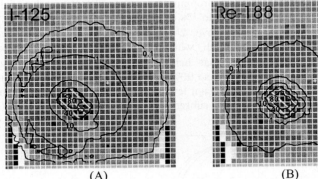

(A)   (B)

Fig. 5 – Computational three-dimensional isodoses distribution for ^{125}I seeds (A) and [188]Re-HAP gel-type implants (B), at an anthropometrics and anthropomorphic head phantom, in function of the maximum tumor dose (MTD): The ^{125}I maximum dose (100%) was $1.72\ 10^{-13}$ Gy/tr and $6.42\ 10^{-11}$ Gy/tr for [188]Re-HAP.

Conclusions

[188]Re-HAP gel-type implants present radiodosimetric advantages when compared with ^{125}I seeds, such as: i) limited dose fluctuations into the implant region which reduces the risk of sub-dosage or hot spots at tumor; ii) absolute value of MTD per transition a hundred times greater than ^{125}I one, that allows the utilization of seeds with activity 100 times lower; iii) depletion of the absorbed dose forward to normal tissue falls to 1% after 7 mm from tumor region, reducing healthy tissue damage. Future work will addresses detailed evaluations of the others M-type HAP compounds and *"in vitro"* and *"in vivo"* studies.

References

[1] K. Ulin, L.E. Bornstein, M.N. Ling, S. Saris, J.K. Wu, B.H. Curran and D.E.Wazer. (1997). International Journal of Radiation Oncology Biology Physics. Vol. 39, p. 757-767.
[2] N.J. Laperriere, P.M.K. Leung, S. Mckenzie, M. Milosevic, S. Wong, J. Glen, M. Pintilie and M. Bernstein. (1998). International Journal of Radiation Oncology Biology Physics. Vol. 41, p. 1005-1011.
[3] W.S. Roberto, M.M. Pereira and T.P.R. Campos. (2003). Key Engineering Materials. Vol. 240-242, p. 579-582.
[4] Z. Chen and R. Nath. (2003). International Journal of Radiation Oncology Biology Physics. Vol. 55, p. 825-834.
[5] ICRU-46, (1992), Report 46, International Commission on Radiation Units and Measurements, Bethesda.
[6] MIRD- http://www.nndc.bnl.gov/nndc/formmird.html. Date:11/05/2003.

Key Engineering Materials Vols. 254-256(2004) pp. 87-90
online at http://www.scientific.net
© *2004 Trans Tech Publications, Switzerland*

Radiological Response of Macroaggregates Implants in an "In-Vitro" Experimental Model

Mendes, B.M., Silva, C.H.T., Campos, T.P.R.

Curso de Pós Graduação em Ciências e Técnicas Nucleares - CCTN/UFMG
Av. Antônio Carlos, 6627, Pampulha Prédio PCA1, Sala 2285 CEP: 31270901
Belo Horizonte, MG, Brasil
bruno@nuclear.ufmg.br, campos@nuclear.ufmg.br

Keywords: radiological response, hydroxyapatite, implants.

Abstract. Hydroxyapatite macroaggregates with incorporated radionuclides, namely M-HAP, present a great potential for brachytherapy implants. HAP is biocompatible, presenting neither local nor systemic toxicity, and in some cases the organism can absorb it. At the present work, experimental models were setup to study the spatial distribution, radiological and ultrasound response of M-HAP ("M"= metallic radionuclides) macroaggregates carried by high viscosity CMC gel through interstitial implants into kidney, lung, liver, brain and muscle samples. The studied composition demonstrated high ecogenicity and easy ultrasound identification in all evaluated tissues experimental models. However, the identification of M-HAP (M=Ca) implants in X-ray anatomic image was not considered satisfactory. The Barium Sulphate addition to the gel increased considerably the radiological contrast. As conclusion, M-HAP gel presents adequate radiobiological response after brachytherapy implants. The incorporation of high atomic number (Z) radionuclides to the hydroxyapatite structure may increase the composition contrast, possibly excluding the barium sulphate after KV and mA X-ray adjustments. The M-HAP composition application through needles can be performed based on fluoroscopy or ultrasound techniques, providing adequate conditions to perform dosimetric calculations through CT, X-ray, or ultra-sound images.

Introduction

Hydroxyapatite (HAP) is a non toxic and non immunogenic bioceramic.[1] Studies report that in few milligrams it is easily absorbed by the organism.[2 HAP composition presents the chemical formula $Ca_{10}(PO_4)_6(OH)_2$. However, due to its peculiar structure, HAP can be considered as a solid solution and the Calcium ions can be substituted by others cations, for example metallic radionuclides cations.[3] Hydroxyapatite synthesis with metallic cations substituting Calcium in its structure is also possible, eliminating the need for further replacements.[4] Such characteristics evidence a great potential of the radionuclide incorporated hydroxyapatite macroaggregate M-HAP (where M is a suitable metallic radionuclide) in brachytherapy implants.

Metallic encapsulated ^{125}I and [103]Pd seeds implants occur in practical clinic for cerebral and prostatic tumor treatment.[4,5] Fluoroscopy and CT techniques are generally used in this seeds implantation and dosimetry. Those seeds presents suitable radiological identification due a metal inert core.[6] The presence of calcium atoms and β-emitters radionuclides as ^{89}Sr, ^{90}Y, [165]Dy, [166]Ho, [153]Sm and [188]Re in the HAP structure confer radiopacity to its macroaggregates, however in small amounts, location could be unpractical through conventional X-rays. The possibility of incorporating contrast inert agent still exists. In that way, the current metallic seed implantation and dosimetry techniques also could be applied for the proposed (M-HAP) macroaggregates implant.

To allow the macroaggregate implantation through thin needle, M-HAP will be mixed with a high viscosity CMC (carboxi-metil-cellulose) aqueous gel. Then, the mixture can be injected with an appropriate device. The paper will show that the CMC gel provides ecogenicity to the composition (M-HAP Gel) allowing its ultrasound visualization.

The present work evaluates the Ca-HAP [HAP-91/JHS Chemical Laboratory] macroaggregate implants (containing or not barium sulphate as X-ray contrast medium) radiological and ecogenic response in a in vitro experimental models (EM).

Methods

Experimental Model. Acrylic boxes was set up with 15x2*e* area and 15cm deep in which 2*e* is the square size taken to the radiological study. The boxes had been filled with Agar base support medium, at the density of $1.01g.cm^3$, due to its close radiological equivalence to human tissues. Before the medium solidification at laboratory temperature, animal model (pig) organs or organs fractions were added to each filled box (Fig.1), so that the tissue homogeneously occupied more than 90% of the volume. Each box had only one type of tissue. The liquid at high temperature agar-type support medium gave support to the organ and removed the empty spaces, allowing the ultrasound transmission. Image would be severely harmed in the case of air bubbles appearing between the transducer and the organ. The following organs had been evaluated: kidney, lung, liver, brain and muscle, setup the model EM-kidney, EM-lung, EM-liver, EM-brain, EM-muscle.

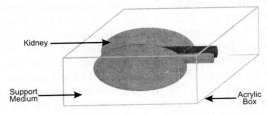

Fig. 1- Schematic drawing of the animal sample placement into the acrylic box filed with radiological tissue equivalent support medium.

Ca-HAP Macroaggregate Gel. A high viscosity gel was previously prepared adding 0.1g of carboxy-methyl-cellulose (CMC) in 10 ml of water, medium A. Medium B was obtained mixing 2.0 g of calcium hydroxyapatite [HAP-91/JHS Chemical Laboratory], in the size range of 90μm to 150μm, with 3ml of medium A. The Ca-HAP macroaggregate gel contrast medium, namely medium C, was obtained mixing 3 ml of CMC gel, 1.0 g of HAP-91 and 1.0 g of barium sulphate.

Implants and Images. 2.83μl per centimeter (needle length) of medium B was applied in the experimental models through a specific mechanical device, developed for such purpose, connected to a 0.6mm diameter needle. Medium B, without contrast, was applied in the EM-kidney, EM-brain, EM-muscle, EM-lung, EM-kidney, EM-liver experimental models. The Ca-HAP gel with the barium contrast, medium C, had been implanted only in EM-lung and EM-kidney. A copper wire was fixed at the start position of the needle application site to provide special identification for the implants without contrast, medium B, at the radiological image. A transducer of 120 elements and 7.5 MHz frequency was used in the ecographic evaluation. The X-ray equipment was set at 40 to 50 KV and 25 to 50mA current in accordance with the thickness of the box used in the experimental model. Various radiological and ultrasound shots were obtained from the experimental models, with medium B and C implants.

Results and Discussion

X-ray response. The radiological response of the implants into the EM-models filled with medium B and C may be observed in Figure 2, "a" and "b". Only two X-ray images are presented. Despite the calcium presence, the medium B cannot be easy identified. In the radiological X-ray film of the EM-kidney and EM-muscle models, the identifications of soft lines in the medium B implant sites were possible (not shown), however for the others EM-models it was unidentified. In images obtained after the X-ray film digitalization, only metallic wires can be depicted (Fig2-a). For medium C implants, even without the placement of the metallic wires at the start needle position, the location where the medium C was applied were easily spotted at the tested EM-models (fig.2-b).

The thin diameter of the implantations (0.6mm), small amounts, and the low atomic number of the macroaggregate atoms, plus 12cm tissue thickness are the main causes for the radiological contrast reduction. The implants of the medium C, using the Barium contrast, present suitable radiological identification in the EM-model tested. The high atomic number radionuclides as ^{89}Sr, ^{90}Y, ^{165}Dy, ^{166}Ho and ^{188}Re present in the M-HAP structure will improve radiological X-ray contrast.

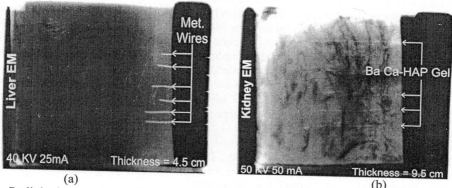

(a)   (b)

Fig. 2 – Radiological response of Experimental Models (EM) implants. (a) digitalized image of the implants of the Ca-HAP Gel without contrast (medium B) in EM-liver. (b) digitalized image of the implants of Ca-HAP Gel with Barium contrast (medium C) in EM-kidney.

Ultra-sound response. The Ca-HAP Gel ecographic response in two experimental models is shown in figure 3. Medium B and C presented high ecogenicity and notable visualization in all the soft tissues studied despite the thin diameter filled at the implants. The EM-muscle, EM-liver and EM-brain models presented suitable ecographic identification. However, EM-brain presents large amount of noise, diminishing the response. Two factors had contributed for the noise: i) the presence of air bubbles in support Agar medium amount the organ fragments; and ii) the larger reverberation phenomena produced by the plane faces of the EM acrylic box. Reverberation at the base and at the top of the image may be seen (Fig3-c). The medium B and C were fully identified in the five EM-models through ultrasound images.

(a)   (b)   (c)

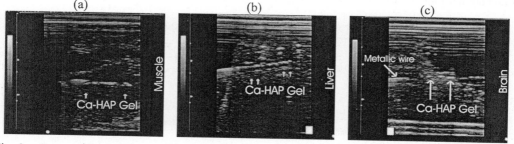

Fig. 3 – Ecographic response of the Ca-HAP Gel without contrast in EM-liver (a), EM-muscle (b) and EM-brain (c) experimental models, based on a 120 elements and 7,5 MHz transducer.

Conclusion

In the *in vitro* experimental models, Ca-HAP Gel macroaggregate presented ecographics properties that allow spatial resolution of the implants in all the studied soft tissue organs. Barium contrast supplement to the compound (medium C) allowed a good radiological X-ray identification. Hydroxyapatites containing high atomic number radionuclides as ^{89}Sr, ^{90}Y, ^{165}Dy, ^{166}Ho, ^{188}Re and ^{153}Sm substituting calcium atoms in their structure probably will improve contrast in CT and X-ray images, conditioned to KV voltage and mA current adjustments. Future work will address reduction of barium. However, some β-emitters also emits γ rays and soft X-rays allowing for spatial biodistribution monitoring through gamma camera and SPECT.

The techniques of fluoroscopy and CT clinically employed in ^{125}I and ^{103}Pd seeds implantation and post-implant dosimetry also may be used for M-HAP gel macroaggregates implants. Moreover, implant real-time procedures guided by ultrasound can be carried out for M-HAP gel in clinical practice. The main clinical features are the simplicity of the technique, easy handling, low cost, precise spatial positioning, and non-visible collateral effects.

Future work addresses *"In vivo"* implants for verifying spatial biodistribution in function at time. Dosimetric computational studies of M-HAP gel macroaggregates with β-emitters radioniclides incorporated are currently in progress.

References

[1] M. Manso, S. Ogueta, P. Herrero-Fernández, L. Vazquez, M. Langlet and J. P. García-Ruiz. Biological evaluation of aerosol-gel-derived hydroxyapatite coatings with human mesenchymal steem cells: Biomaterials. Vol. 23 (2002), p. 3985-3990.

[2] A.L. Andrade, A.P.B. Borges and S.M.C.M. Bicalho. *HAP-91 Síntese, Caracterização, Testes e Aplicações*. (2001) JHS Laboratório Químico Ltda.

[3] T. Nadari, B. Hamdi, J.M. Savariault, H.E. Feki and A.B. Salah. Substitution mechanism of alkali metals for strontium in strontium hydroxyapatite. Materials Research Bulletin. Vol. 38 (2003), p. 221-230.

[4] N.J. Laperriere, P.M.K. Leung, S. Mckenzie, M. Milosevic, S. Wong, J. Glen, M. Pintilie and M. Bernstein. Randomized study of brachytherapy in the initial management of patients with malignant astrocitoma. International Journal of Radiation Oncology Biology Physics. Vol. 41 (1998), p. 1005-1011.

[5] I. D. Pedley. Transperineal interstitial permanent prostate brachytherapy for carcinoma of the prostate. Surgical Oncology. Vol. 11 (2002), p.25-34.

[6] K. Ulin, L.E. Bornstein, M.N. Ling, S. Saris, J.K. Wu, B.H. Curran and D.E.Wazer. (1997) A technique for accurate planning of stereotactic brain implants prior to head ring fixation. International Journal of Radiation Oncology Biology Physics. Vol. 39 (1997). p. 757-767.

Key Engineering Materials Vols. 254-256(2004) pp. 91-94
online at http://www.scientific.net
© 2004 Trans Tech Publications, Switzerland

Sintering Effects on the Mechanical and Chemical Properties of Commercially Available Hydroxyapatite

F. P. Cox[1], M. J. Pomeroy[1, 2], M. E. Murphy[3] and G. M. Insley[3]

[1] Department of Material Science and Technology, University of Limerick, Limerick, Ireland
E-mail: frank.cox@ul.ie

[2] Materials and Surface Science Institute, University of Limerick, Limerick, Ireland
E-mail: michael.pomeroy@ul.ie

[3] Stryker Howmedica Osteonics, Raheen Business Park, Limerick, Ireland.
E-mail: matthew.murphy@emea.strykercorp.com
E-mail: gerard.insley@emea.strykercorp.com

Keywords: Hydroxyapatite, sintering, grain growth, microhardness

Abstract. A commercial hydroxyapatite (HA) powder has been sintered at 1150°C for time periods up to 24 hours. Rapid densification took place during the heating and cooling due to particle rearrangement. Isothermal holds of up to 6 hours resulted in densification by matter transport, with longer holds leading to the elimination of closed porosity. Microhardness values increased with sintering times as a result of porosity decreases with time although grain growth did occur. The grain growth mechanism appears to relate to lattice or surface diffusion.

Introduction

Hydroxyapatite [$Ca_{10}(PO_4)_6(OH)_2$], HA, is a primary constituent of human hard tissues like bones and teeth [1]. Synthetic HA is known to be biocompatible and forms a direct bond with bone [2]. The main problem with HA is that it lacks the mechanical properties needed for the material to be used in load bearing situations. The processing of the HA powder prior to sintering has notable effects on the sintering characteristics and the mechanical properties [3]. A large volume of work has also been devoted to the sintering of synthetic HA since the initial study by Jarcho *et al.* [2-6].

The sintering parameters (temperature, time, atmosphere, etc.) are very influential on the mechanical and chemical properties of HA [4,7]. The production of HA with improved mechanical properties has led to a number of different studies with regard to HA composites with other ceramic [8], glass [9] or polymer additions [10]. It has also induced research with respect to the specific effect(s) of each variable associated with production of synthetic HA from synthesis [11] to sintering [5]. The work reported here concentrates on the effect of one sintering variable i.e. time, on the properties of a commercially available HA material.

Materials and Methods

A commercially available HA powder (Albright and Wilson, Grade 130) was used in this study. Particle size analysis was performed on the source powder using a Malvern Zetasizer. Samples of 3 g were uniaxially pressed in a 32 mm steel die using a load of 120 MPa for a period of 20 seconds. These samples were then sintered on an alumina tile at 1150°C for sintering times of 0, 1.5, 3, 6, 12 and 24 hours using a heating rate of 5°C/min. Radial shrinkages were determined and green as well as sintered densities were calculated from geometric dimensions and mass of samples. X-Ray Diffraction (XRD) was performed using a Philips X'Pert Diffractometer with Cu-K_α radiation (λ=0.15406 nm). Data was collected over the 2-theta range of 5° to 80°, using a step size of 0.02°. All samples were polished to a 1 μm finish prior to microhardness testing and subsequently etched in a 5% citric acid solution for 2 minutes prior to microscopic analysis. Microhardness was performed using a Leco M-400-G1 Hardness Tester using a load of 500 g and an indentation time of 15 seconds. Images obtained by scanning electron microscopy (JEOL JSM-840) were used to measure the grain size using the Mendelson line intercept method [12].

Results

Fig. 1 shows that after heating samples to 1150°C and cooling without holding at 1150°C, a radial shrinkage of some 12.5% occurs. Shrinkages of 16% and 17.5% then occur after 1.5 and 3-hour isothermal holds. For longer sintering times, greater than 6 hours, shrinkage rates are extremely slow and would seem to represent pore closure. XRD analyses (Fig. 2) showed that no change in the phases present occurred with increasing sintering hold times. HA and β-TCP were the phases present and their relative intensities remained constant with increase in sintering time. This indicates that there was no change in phase chemistry during the isothermal sintering experiments.

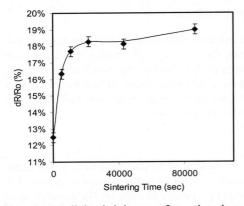

Fig. 1: Radial shrinkage after sintering at 1150°C with respect to sintering time

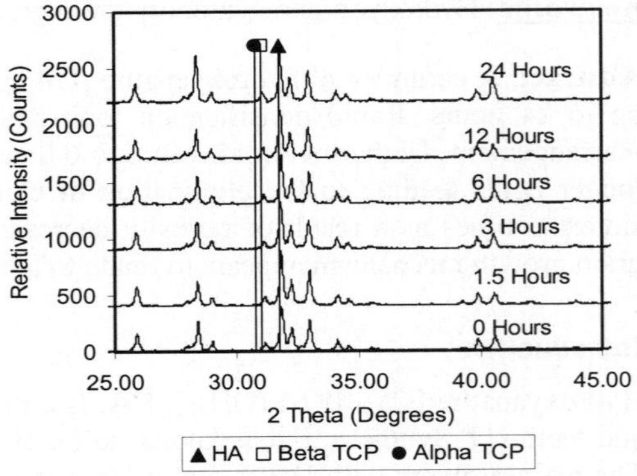

Fig. 2: XRD traces after sintering at 1150° with respect to sintering time

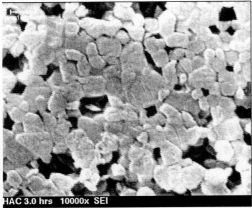

Fig. 3: SEM Image after sintering at 1150°C for 3 hours

Fig. 4: SEM Image after sintering at 1150°C for 24 hours

From the SEM micrographs shown in Figs. 3 and 4, it is seen that there is a large increase in the average grain size from approximately 1.1μm to 1.5μm after isothermal sintering for 3 hours and 24 hours respectively. This contrasts with what was observed after 0 hours where primary particle necking was seen as was the development of fine equi-axed grains. A decrease in grain growth rate can clearly be seen between 3 and 6 hours (Fig. 5) and this corresponds with the decrease in densification rate. Whilst there was a range of grain sizes observed in the microstructures after the various sintering times, no evidence of the exaggerated grain noted by Thangamani et al. [2] and Yeong et al. [13] was observed.

As might be expected, microhardness values increase with increase in radial shrinkage as comparison of Fig.1 and Fig. 5 clearly indicates. The values for samples sintered for 6 hours or more are virtually similar indicating that a decrease in porosity (pore closure) is mainly responsible for the increase in microhardness during the longer sintering times. The microhardness reaches a maximum of 3.28GPa after 24 hours of sintering at 1150°C.

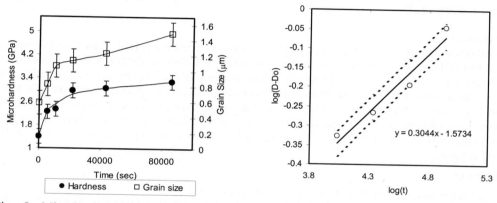

Fig. 5: Microhardness and Grain Size after sintering at 1150°C with respect to sintering time

Fig. 6: Log of the difference in grain size (D-D$_0$) versus the log of sintering time (t)

Discussion

It is seen from above that whilst changes in radial shrinkage, grain size and microhardness occurred with increasing sintering times at 1150°C, the commercial HA material formed a stable HA phase in equilibrium with approximately 6-13% β-TCP (based on X-ray intensities). Thus it can be concluded that the changes referred to occur in the absence of any phase changes. The densification of the HA material is obviously quite significant on heating to and cooling immediately and this would be expected as particle rearrangement would contribute significantly. Typical green densities for the pressed pellets were 1.64g cm^{-3}, i.e. about 52% of the theoretical density (3.16g cm^{-3}). As solid state sintering processes involving material transport ideally occur at about 70% of theoretical density then some 28% volume shrinkage (~ 11% radial shrinkage) should occur during particle rearrangement. Between 0 hours and 6 hours hold, the densities increase from 73 to 89% theoretical which is thought to occur via standard solid state sintering mechanisms. The significant fall in densification rate after the longer times as shown in Fig. 1 appears to be related to the elimination of closed porosity. The final density after 24 hours sintering is 19% radial shrinkage which corresponds to 93% of theoretical density.

Comparison of the variations in microhardness values with sintering time and the radial shrinkages (compare Figs. 5 and 1) shows that as radial shrinkage and therefore density increases, microhardness values increase also. This is in agreement with the findings of Hoepfner and Case [14] where the volume of porosity directly relates to the microhardness of HA (the lower the porosity the higher the microhardness). These authors also indicated that increases in grain size for the HA investigated had no significant effect on the microhardness. In effect, this study also indicates this, as the increased grain sizes would be expected to cause decreases in microhardness values. Accordingly, it can be concluded that whilst grain growth does occur during closed pore elimination, the parameter having most effect on microhardness is residual porosity.

For the sake of interest, the grain growth exponent was extracted using the relationship $(D-D_0) = kt^n$, where D is the measured grain size after time t, D_0 the average grain size after 0 minutes and k the rate constant. The necessary data is plotted in Figure 6 and shows that a straight line relationship holds which gives an n value of 0.3. This value which is 50% greater than that observed by Yeong et al. [13] for HA sintered at 1200°C, is typical of a lattice diffusion or vapour transport mechanism [15]. Given the slow grain growth rate observed, it appears that at 1150°C, HA can be sintered to reasonable densities after long periods. However, grain growth does occur which once densification is complete may have an adverse effect on microhardness as well as other mechanical properties.

Conclusions

A commercial HA powder was sintered at 1150°C for sintering time periods from 0 to 24 hours. It was noted that over these sintering times, the commercial HA material formed a stable HA phase in equilibrium with approximately 6-13% β-TCP. The sintering time also exhibited a significant effect on the densification with increases from 73 to 89% of theoretical density after isothermal hold times of 0 and 6 hours respectively. Given the increase in densification and associated decrease in porosity, microhardness values increased to a maximum of 3.28GPa after 24 hours sintering. A slow grain growth rate was observed but did not seem to have any significant effect on the microhardness.

References

[1] A. Posner and F. Betts: Acta Chem. Res. Vol. 8 (1975) p.273
[2] Jarcho, C.H Bolen, M.B Thomas, J. Bobick, J.F Kay and R.H Doremus: J. Mat. Sci. Vol. 11(1976), p.2027-2035
[3] N. Thangamani, K. Chinnakali and F.D. Gnanam: Ceram. Int. Vol. 28 (2002) p.355-362M.
[4] G. Muralithran and S.Ramesh: Ceram. Int. Vol. 26 (2000) p.221-230
[5] S. Raynaud, E. Champion and D. Bernache-Assollant: Biomaterials Vol. 23 (2002) p.1073-1080
[6] I.R. Gibson, S. Ke, S.M Best and W. Bonfield: J. Mat. Sci.: Mat. Med. Vol. 12 (2001), p.163-171
[7] D. Bernache-Assollant, A. Ababou, E. Champion and M. Heughebaert: J. Eur. Ceram. Soc. Vol. 23 (2003) p.229-241
[8] S. Gautier, E. Champion and D. Bernache-Assollant: J. Eur. Ceram. Soc. Vol. 17 (1997) p.1361-1369
[9] D. C Tancred, A. J Carr and B. O. McCormack: J. Mat. Sci.: Mat. Med. Vol. 12 (2001), p.81-93
[10] N.H Ladizesky, E.M Pirhonen, D.B. Appleyard, I.M Ward and W. Bonfield: Comp. Sci. Tech. Vol. 58 (1998) p.419-434
[11] Y.X Pang and X. Bao: J. Eur. Ceram. Soc. Vol. 17 (1997) p.1361-1369
[12] M.I. Mendelson: J. Am. Ceram. Soc. Vol. 52 (1969) p. 443
[13] K.C.B Yeong, J. Wang and S.C Ng: Mater. Letters Vol. 38 (1999) p.208-213
[14] T.P. Hoepfner and E.D Case: Ceram. Int. Vol.29 (2003) p.699-706
[15] R. J. Brook in F. F. Y. Wang (ed.) "Treatise on Materials Science and Technology Volume 9 – Ceramic Fabrication Processes, (Academic Press, 1976) p. 311-364

Key Engineering Materials Vols. 254-256(2004) pp. 95-98
online at http://www.scientific.net
© *2004 Trans Tech Publications, Switzerland*

[31]P Nuclear Magnetic Resonance and X-Ray Diffraction Studies of Na-Sr-Phosphate Glass-Ceramics

R.A. Pires[1], I. Abrahams[2], T.G. Nunes[1], G.E. Hawkes[2]

[1]Departamento de Engenharia de Materiais, Instituto Superior Técnico, Universidade Técnica de Lisboa, Av. Rovisco Pais, 1, 1049-001 Lisboa, Portugal, pc1539@popsrv.ist.utl.pt
[2]Structural Chemistry Group, Department of Chemistry, Queen Mary College, University of London, London E1 4NS, UK.

Keywords: glass-ceramics, phosphates, NMR, X-ray powder diffraction

Abstract. A set of Na-Sr-phosphate glass and glass-ceramic samples, with general formula xSrO:(0.55-x)Na$_2$O:0.45P$_2$O$_5$, were prepared and analysed by solid state ^{31}P nuclear magnetic resonance spectroscopy and X-ray powder diffraction. The results show the presence of Q^1 and Q^2 phosphate species in all samples. At low concentrations of Sr^{2+} ($x \leq 0.20$) the strontium is preferentially incorporated in Sr^{2+}-Q^1 crystalline phases, and only at higher Sr^{2+} concentrations are crystalline phases present which Sr^{2+} is associated with Q^2 phosphate units.

Introduction

Glasses and glass-ceramics are widely used as biomaterials, mainly as bone replacement materials [1] and as base materials for dental restorations [2, 3]. One of the main advantages of phosphate based materials, and particularly the Na-Ca-phosphate system, is their chemical relationship with hydroxylapatite, one of the main constituents of bone and teeth. However, X-ray examination of the restoration is often necessary and this demands the use of materials with a higher radiopacity. The substitution of CaO for SrO improves the material's radiopacity.

The structures of phosphate glasses and glass-ceramics are based on networks of corner sharing phosphate tetrahedra. Four types of phosphate moiety can be present, which are commonly described using Q^n nomenclature, where n represents the number of bridging oxygens (BO) of the unit. Accordingly, the (PO$_4$)$^{3-}$ (Q^0) unit has no BO, the (PO$_{3.5}$)$^{2-}$ (Q^1) unit has 1 BO, the (PO$_3$)$^-$ (Q^2) unit has 2 BO, and the (PO$_{2.5}$) (Q^3) unit has 3 BO. These phosphorous environments can be distinguished by ^{31}P solid state magic angle spinning (MAS) nuclear magnetic resonance (NMR) spectroscopy, due to their characteristic isotropic chemical shift (δ_{iso}) and chemical shift anisotropy (CSA) ranges in glass and ceramic structures [4].

In this study, a set of sodium strontium phosphate glasses and glass-ceramics, with general formula xSrO:(0.55-x)Na$_2$O:0.45P$_2$O$_5$, were synthesised. The crystalline phases were analysed by X-ray diffraction (XRD) and the phosphorous environment was probed by solid state ^{31}P MAS NMR spectroscopy.

Materials and methods

Anhydrous sodium carbonate (Analar, 99.9 %), strontium carbonate (Hopkin & Williams Ltd, 98 %) and diammonium hydrogen phosphate (BDH, 97 %) were used to prepare a set of Na-Sr-phosphate glass samples of general composition xSrO:(0.55-x)Na$_2$O:0.45P$_2$O$_5$, with x = 0.00, 0.09, 0.19, 0.29, 0.39, 0.49 and 0.55, by melt quenching. The starting materials were ground as a slurry in ethanol using an agate mortar and pestle. After drying at 70° C for 3 h the mixtures were placed in platinum crucibles and heated at 300° C for 30 min to allow the release of H$_2$O and NH$_3$, at 650° C for 30 min to allow the release of CO_2 and at temperatures between 1000-1300° C to produce the melts. After 1 h at that temperature, the melts were splat quenched in air onto a stainless steel plate.

The glass transition temperature (T_g) and crystallization temperature (T_c) were measured by differential scanning calorimetry (DSC) using aluminium pans on a Perkin-Elmer DSC 7 under nitrogen atmosphere and at a heating rate of 10° C min^{-1}.

Glass-ceramics were prepared by heating the glass samples overnight at temperatures between T_g and T_c.

The crystalline and amorphous materials were analysed by X-ray powder diffraction (XRD) on an automated Phillips PW1050/30 diffractometer, in flat plate θ/2θ geometry using Ni filtered Cu Kα radiation. Data were collected from 5 to 70° 2θ, with a step width of 0.05° and a count time of 2 s per step. Diffraction patterns were modelled by Rietveld refinement using GSAS [5].

^{31}P MAS NMR spectra of the glasses and glass-ceramics were acquired at a frequency of 242.9 MHz on a Bruker AVANCE 600; a simple single pulse, acquire sequence was used, with a recycle delay of 60 s and a pulse width of 1 μs which corresponds to a flip angle ca. 10°. 24 FIDs, each of 16384 data points were accumulated, from which, 8192 points were Fourier transformed. MAS rates were set to 12 kHz. The spectra were fitted with dmfit [6], to obtain δ_{iso}, CSA and the relative concentration of each species.

Results and discussion

^{31}P MAS NMR spectroscopy indicated the presence of two phosphorous species in the glass structure: Q^1 and Q^2. The assignment was based on the δ_{iso} and CSA of each signal. The Q^1 species, in the glass samples have δ_{iso} from, approximately, 1.2 to -6.3 ppm, while values of CSA, ranged from 130.2 to 152.7 ppm [7]. Q^2 species have δ_{iso} from, approximately, -18.5 to -22.5 ppm while values of CSA, ranged from -202.7 to 213.8 ppm [7]. The relative concentration of the two species was found to be 22 and 78 % for the Q^1 and Q^2 species respectively, which agrees with the values calculated from the equations developed by Van Wazer [4, 8].

The ^{31}P MAS NMR spectra of the glass-ceramic samples are shown in Fig. 1. Several different species are observed within the Q^1 and Q^2 δ_{iso} ranges.

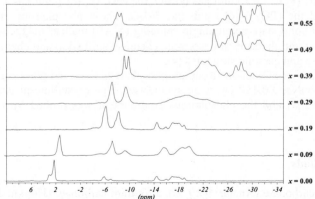

Fig. 1. ^{31}P MAS NMR spectra for glass-ceramics in the system xSrO:(0.55-x)Na$_2$O:0.45P$_2$O$_5$.

The changes in the spectra with change in the Sr^{2+} concentration reflect the stabilisation of particular crystalline phases at certain compositions. At lower Sr^{2+} concentrations (x = 0.00 to 0.19) the Q^2 band is mostly unchanged in isotropic chemical shift, whereas the Q^1 region shows species with lower frequency shifts dominating as [Sr^{2+}] increases. The sample with $x = 0.29$ shows the broadest resonances and this certainly reflects the incomplete crystallisation of this sample. The spectra with the highest concentrations of Sr^{2+} ($x = 0.39$ to 0.55) reveal a low frequency shift with [Sr^{2+}], and small changes in the Q^1 region. Turner *et al.* [9] have shown that the cation can influence the ^{31}P isotropic shift in structurally related phosphate groups, according to the factor $-Z/\sqrt{r}$, where Z is the ionic charge and r is the ionic radius. In substituting Na^+ by Sr^{2+} in the material the dominant effect is the doubling of the ionic charge, and consequently the factor is a larger number for Sr^{2+} and the negative sign results in the prediction of a shift to low frequency. Thus the changes in the spectra in Fig. 1 are consistent with the introduction of Sr^{2+} into the sodium phosphate material first producing new Sr^{2+}/Q^1 crystalline phases, and at higher concentrations producing new Sr^{2+} phases.

Fig. 2 shows the XRD powder patterns of the glass-ceramics. Several stoichiometric crystalline phases have been identified and are correlated with the composition of the parent glass. Below the $x = 0.39$ composition, sodium phosphate phases dominate the X-ray patterns, while at compositions with higher Sr^{2+} concentrations, crystalline strontium phosphates are evident.

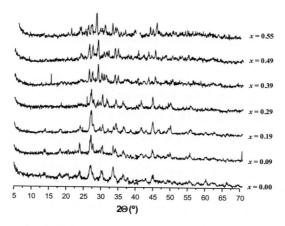

Fig.2. XRD powder patterns for glass-ceramics in the system xSrO:(0.55-x)Na$_2$O:0.45P$_2$O$_5$.

At the $x = 0$ composition $Na_3P_3O_9$ [10] and $Na_5P_3O_{10}$ [11] have been identified as the principal crystalline phases, with some evidence of a small amount of $Na_4P_2O_7$ [12]. As x increases, Sr^{2+} rich phases become evident and in the fully substituted system ($x = 0.55$) α-Sr$_2$P$_2$O$_7$ [13] and Sr(PO$_3$)$_2$ [14] have been identified. These results are consistent with the observation and variation of Q^1 and Q^2 species in the ^{31}P NMR spectra.

Theoretically, the limiting composition at which all the Na^+ ions associated with the charge balancing of the Q^1 sites is with $x \approx 0.20$. This composition is in between samples with $x = 0.19$ and $x = 0.29$, where, from the analysis of the ^{31}P MAS NMR spectra the changeover from association of Sr^{2+} with Q^1 species to Q^2 occurs.

Conclusion

The results indicate the preferential substitution of the Na^+ cations closely associated with the Q^1 phosphorous species, and only at a second stage the substitution of Q^2 associated cations occur. This conclusion is based on the ^{31}P MAS NMR spectra of the glass-ceramics, where, at lower Sr^{2+}

concentration the Q^1 species changes its surroundings, while the Q^2 species remains in a similar structural organization. This is consistent with a preferential substitution model, where the Na^+ cations are closely associated with the Q^2 phosphorous, while the Sr^{2+} cations are closely associated with the Q^1 phosphorous tetrahedron. The non-random distribution of cations in the structure of Na-Sr-phosphate glasses reported on a previous study [7] is also observed in the glass-ceramics with similar compositions.

Acknowledgments

This work was supported by the Portuguese Foundation for Science and Technology, project POCTI/33193/99. RP acknowledges a PhD research grant, PRAXIS/BD/21572/99. We gratefully acknowledge the Chemical Database Service for the ICSD database.

References

[1] T. Kokubo, H-M, Kim and M. Kawashita: *Biomaterials*, 24(13) (2003), p. 2161;

[2] A. Wilson and J. MacLean: *Glass-Ionomer Cements* (Quintessence Publishing, Chicago 1988);

[3] A. Wilson and J.W. Nicholson: *Acid-base cements – Their biomedical and industrial applications* (Cambridge University Press, Cambridge 1993);

[4] R.K. Brow: J. Non-Cryst. Solids, 263&264 (2000), p. 1;

[5] A. C. Larson and R. B. Von Dreele: Los Alamos National Laboratory Report, No. LAUR-86-748, (1987);

[6] D. Massiot, F. Fayon, M. Capron, I. King, S. Le Calvé, B. Alonso, J-O. Durand, B. Bujoli, Z. Gan and G. Hoatson: Magnetic Resonance in Chemistry, 40 (2002), p.70;

[7] R. Pires, I. Abrahams, T.G. Nunes and G.E. Hawkes: submitted to J. Non-Cryst. Solids;

[8] J.R. Van Wazer: Phosphorous and its Compounds, vol. 1, (Interscience, New York 1958);

[9] G.L. Turner, K.A. Smith, R.J. Kirkpatrick and E. Oldfield: J. Magn. Reson., 70 (1986), p. 408;

[10] H.M. Ondik: Acta Crystallogr., 18 (1965), p. 226;

[11] D.E.C. Corbridge: Acta Crystallogr., 13 (1960), p. 263;

[12] K.Y. Leung and C. Calvo: Can. J. Chem., 50 (1972), p. 2519;

[13] L.O. Hagman, I. Jansson and C. Magneli: Acta Chemica Scandinavica, 22 (1968), p.1419;

[14] M. Jansen and N. Kindler: Zeitschrift fur Kristallographie, 212 (1997), p. 141.

Key Engineering Materials Vols. 254-256(2004) pp. 99-102
online at http://www.scientific.net
© 2004 Trans Tech Publications, Switzerland

A MAS NMR Study of the Crystallisation Process of Apatite-Mullite Glass-Ceramics

A. Stamboulis[1], R.G. Hill[1], R.V. Law[2] and S. Matsuya[3]

[1] Imperial College London, Department of Materials, Prince Consort Road, London SW7 2BP, UK

[2] Imperial College London, Department of Chemistry, Exhibition Road, London SW7 2AY, UK

[3] Kyushu University, Faculty of Dental Science, Fukuoka 812-8582, Japan

e-mails: a.stamboulis@imperial.ac.uk, r.hill@imperial.ac.uk, r.law@imperial.ac.uk, smatsuya@dent.kyushu-u.ac.jp

Keywords: MAS-NMR, apatite, glass-ceramics, bioceramics, crystallisation

Abstract. The crystallisation behaviour of $4.5SiO_2$-$3Al_2O_3$-$1.5P_2O_5$-$(5-z)CaO$-$zCaF_2$ glasses with z between 0 and 3 was investigated using ^{27}Al, ^{29}Si, ^{31}P and ^{19}F Magic Angle Spinning Nuclear Magnetic Resonance (MAS-NMR) spectroscopy. The glasses showed two exothermic peaks at Tp1 and Tp2 due to crystallisation, which decreased in temperature with fluorine content. In the non-fluorine containing glass (z=0), β-tricalcium phosphate $(Ca_3(PO_4)_2$ and anorthite $(2SiO_2Al_2O_3CaO)$ were the final crystallisation products. In the fluorine containing glasses (z=2 and 3), fluorapatite (FAP) $(Ca_5(PO_4)_3)$, crystallised at Tp1 and mullite $(2SiO_23Al_2O_3)$ at Tp2. MAS-NMR results showed that a glass phase remained after crystallisation. The coordination state of the phosphate species in the glass drastically changed during crystallisation. Five coordinate Al, Al(V), in the glass containing fluoride disappeared after crystallisation of mullite. The ^{27}Al MAS-NMR spectra did not change significantly during fluorapatite crystallisation, but changed markedly on mullite formation. ^{19}F MAS-NMR demonstrated the presence of FAP in the heat treated glasses. In the glasses with a higher fluorine content than that required to convert all the Ca and P to FAP a Al-F-Ca(n) type species remained in the residual glass phase.

Introduction

There are many studies of bioactive glass-ceramics intended for application as biomaterials [1]. These include apatite glass ceramics, which are suitable for restoring and replacing hard tissues in orthopaedic and dental fields because of the good biocompatibility conferred by the apatite crystal. Apatite-mullite glass-ceramics have been developed based on SiO_2-Al_2O_3-P_2O_5-CaO-CaF_2 compositions by Hill et al. [2]. The glasses crystallise to form fluorapatite (FAP) and mullite on appropriate heat treatment. Both the apatite and mullite crystals have a needle like habit and interlock with each other giving rise to high fracture toughness values. A good example of such a glass is the LG112 glass. An implant of LG112 glass inserted in a tibia of a rat is shown in Fig.1a. There is no indication of osteointegration, in contrary there is formation of a fibrous capsular layer around the implant. The heat treated LG112 glass, on the other hand, shows excellent osteointegration and formation of mineralized bone in direct contact with the implant (Fig. 1b). A previous study [2] dealt with the effect of fluorine content of the glass on the nucleation and crystallisation behaviour by means of differential scanning calorimetry (DSC) and X-ray diffraction analysis. Two exothermic peaks were observed in the DSC curves for the glass containing CaF_2. The crystallisation temperature, Tp1 (first peak) and Tp2 (second peak) decreased with CaF_2 content. With the glass containing no CaF_2, β-$Ca_3(PO_4)_2$ (β-calcium phosphate) and $CaAl_2Si_2O_8$ (anorthite) were the final crystallisation products. On substitution of CaF_2 for CaO in the glass, $Ca_5(PO_4)_3F$ (fluorapatite) crystallised at the first exothermic peak and $3Al_2O_32SiO_2$ (mullite) at the second exothermic peak instead of β-$Ca_3(PO_4)_2$ and $CaAl_2Si_2O_8$. However, the crystallisation process was not fully understood. Magic angle spinning nuclear magnetic resonance spectroscopy (MAS-NMR) has been successfully used for investigating the structures of various P-containing silicate or aluminosilicate glasses [3]. There are also some MAS-NMR studies of the crystallisation process of these glasses [4]. In the present study ^{27}Al, ^{29}Si, ^{31}P and ^{19}F MAS-NMR analyses have been carried out to characterise the crystallisation process of the glass-ceramics.

Experimental
Preparation of the glasses. The compositions of the glasses were based on $4.5SiO_2 \ 3Al_2O_3 \ 1.5P_2O_5$ $(5-z)CaO \ zCaF_2$. The glasses were prepared by a melt quench route and have been described in detail elsewhere [2]. The glass frit was ground in a vibratory mill to produce powder (<45μm) for subsequent analysis.

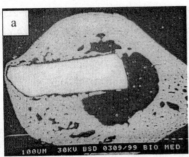
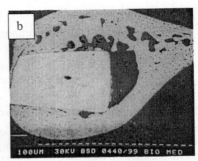

Fig. 1: a) LG112 glass implanted into a tibia of a rat. b) LG112 glass-ceramic implanted into a tibia of a rat (Courtesy Prof. Paul Hatton, University of Sheffield, UK).

MAS-NMR. MAS-NMR analyses were conducted on ^{31}P nucleus at resonance frequency of 161.98 MHz using an FT-NMR spectrometer (AM-400, Bruker, Germany). The [27]Al measurements were conducted in a 600MHz Bruker FT-NMR spectrometer at a frequency of 156.3 MHz. The ^{19}F measurements were conducted at a resonance frequency of 188.29 MHz, respectively, using an FT-NMR spectrometer (DSX-200, Bruker, Germany). Spinning rates of the sample at a magic angle were 5 kHz for the ^{31}P, 15 kHz for the ^{27}Al and ^{19}F MAS-NMR. Recycle time was 30 s for ^{31}P, 120 s for ^{19}F and 1 s for [27]Al. Reference materials for the chemical shift (in ppm) were YAG for [27]Al and 85%H_3PO_4 for [31]P. The spectra for ^{19}F were referenced to CaF_2 taken as −108 ppm.

Results and Discussion

LG116 glass (z=0)
The [27]Al spectrum of the original glass shows a broad peak around 50 ppm, which is assigned to a tetrahedrally coordinated Al, Al(IV) [5]. A large broad peak at around 55 ppm in the ^{27}Al MAS-NMR spectrum after heat treatment to Tp2 is assigned to anorthite [6], though there remains a significant amount of an Al-containing glass phase. A strong sharp peak around -10 ppm suggests the presence of octahedrally coordinated Al atom, Al(VI). The ^{31}P MAS-NMR spectrum of the glass heated up to Tp1 shows three peaks at 0, -2 and -20 ppm. The spectral pattern at 0 and -2 ppm is similar to that of β−tricalcium phosphate reported by Vogel et al. [7]. The peak at -20 ppm is probably due to a phosphate ion coordinated by one or more Al ions, which has a similar structure to aluminium phosphate described later.

LG120 glass (z=0.5)
Fig 2 shows [27]Al (Fig. 2a) and [19]F (Fig.2b) MAS NMR spectra of glass LG120 heated up to Tp1, (Tp1+Tp2)/2 and Tp2. A [27]Al spectrum of the original glass shows a large broad peak around 50 ppm as seen in the LG 116 glass. The peaks do not change after crystallisation of FAP on heating to Tp1 and (Tp1+Tp2)/2, that is the chemical environment of Al in the glass seemed to be unaffected by the crystallisation of FAP. The spectrum of the glass heated up to Tp2 shows a small shoulder peak at 39 ppm and a sharp large peak at 12 ppm. The former peak at 39 ppm was assigned to crystalline $AlPO_4$ as reported by Dollase et al. [8], though the amount and size of the crystallites was too small to be detected by X-ray diffraction analysis. The latter peak at 12 ppm corresponds to Al(V). During crystallisation and FAP formation at Tp2 there is lack of cations to charge balance Al(IV) and therefore Al in higher coordination state is promoted. The ^{31}P MAS-NMR spectrum of the original glass shows only one large peak around -6 ppm. After heat-treatment to Tp1, a strong sharp peak appeared with a chemical shift at 3 ppm assigned to the orthophosphate ion, PO_4^{3-}, in FAP. A small peak at around −29 ppm is assigned to PO_4 in crystalline $AlPO_4$ [8]. A weak broad peak is also observed, around -8 ppm, in the spectrum of glass heated to Tp1. The peak position

shifted in a more negative direction for the heat treatment at a higher temperature of Tp2. This fact suggests that the number of Al atoms coordinating around a PO_4^{3-} tetrahedron is

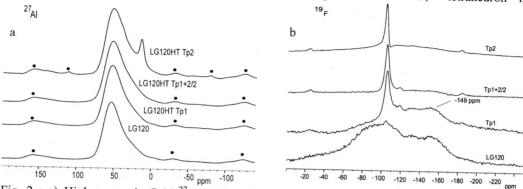

Fig. 2 : a) High magnetic field ^{27}Al MAS-NMR of LG120 and heat treated LG120 up to Tp2 (z=0.5). b) ^{19}F MAS-NMR spectrum of LG120 and heat treated LG120 up to Tp2.

increasing with heat treatment temperature. The ^{19}F NMR spectra of LG120 (Fig.1b) heat treated to Tp1, (Tp1+Tp2)/2 and Tp2 show a sharp peak at approximately -103 ppm corresponding to crystalline FAP. The initial glass exhibited two broad peaks at about -90 ppm and -150 ppm, which Zeng and Stebbins [9] assigned to F-Ca(n) and Al-F-Ca(n) respectively. The F-Ca(n) peak disappears at Tp1 and the fluorine in this environment is used preferentially in forming the FAP in contrast the Al-F-Ca(n) peak is still significant at Tp1 and only disappears at higher temperatures. This may reflect the fact that F-Ca(n) type species preferentially charge balance the non-bridging oxygens attached to the P. Little or no fluorine remains in the glass phase following crystallisation of FAP at Tp2. This is to be expected for this glass composition, since the Ca:P:F ratio is 5:3:1 corresponding to the stoichiometry of FAP.

LG26 and LG99 glasses (z=2.0 and 3.0)

The ^{27}Al spectrum of the original glass showed a large broad peak at around 50 ppm with a shoulder peak around 20 ppm and a small peak at around -5 ppm. The latter two peaks around 20 and -5 ppm were assigned to Al(V) and Al(VI) existing predominantly as fluoride/oxygen [AlF$_x$O$_y$] complex species in the original glass. Fluorine additions to the glass result in the formation of Al(V) or Al(IV) species. The peak intensity at -5 ppm decreased with increasing in heat treatment temperature up to (Tp1+Tp2)/2. Heat treatment of the glass caused crystallisation of FAP and effectively removes fluorine from the glass. It is tempting to associate the reduction in the intensity of the signal corresponding to Al(V) or Al(IV) to a reduction in the amounts fluorine complexed by Al. However, it appears to be the fluorine complexed to calcium that is used for the formation of FAP, rather than that complexed to Al. Furthermore, in the LG26 and LG99 glasses only a small fraction of the fluorine is used to form FAP. The spectrum of the glass heated up to Tp2 is quite different from those heated to lower temperatures. A shoulder peak at around 20 ppm corresponding to Al(V) also disappeared as the mullite crystallisation occurred. A broad peak around 50 ppm corresponding to a Al(IV), a shoulder at 39 ppm assigned to crystalline AlPO$_4$ and a large peak corresponding to a Al(VI) around -5 ppm and assigned to mullite, were observed in Fig. 3(a). In the ^{31}P MAS-NMR spectrum of the original glass, a broad central peak is observed around -7 ppm. The peak at 3 ppm is assigned to the orthophosphate ion, PO_4^{3-}, in FAP crystallised in the heat-treated glass. The peak around -29 ppm is also assigned to PO$_4$ in the crystalline AlPO$_4$. Besides the peak, a weak broad peak is observed around -19 ppm in the spectrum of glass heated up to Tp1. The peak position shifted in a negative direction with the heat treatment at a higher temperature. This suggests that not all the P is used to form FAP and some remains in the residual glass phase and the number of Al surrounding a PO_4^{3-} ion increases with heat treatment temperature. The ^{19}F MAS NMR spectra of LG26 and LG99 (Fig. 3b) base glasses and heat treated glasses show that F is present in at least two different environments. A small peak at -103 ppm corresponds to FAP. The amount of F in a FAP environment is very small (<1%), since the peak at −103 ppm represents a very small fraction of the total MAS-NMR ^{19}F signal. No FAP was detected by XRD as a result of the

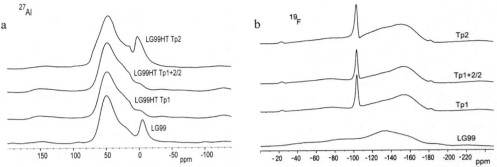

Fig. 3: a) ^{27}Al MAS-NMR spectra of LG99 base glass and heat treated up to Tp2. b) ^{19}F MAS-NMR spectra of LG99 base glass and heat treated up to Tp2.

amount and probably the size of the FAP crystals being too small. Again on heat treatment at Tp1 a sharp peak appears at -103 ppm corresponding to FAP. The peak at –150 ppm corresponding to Al-F-Ca(n) species, is much stronger than in the LG120 and LG115 glasses heat treated to Tp1 and indicates much higher fluorine content in the residual glass phase. A shoulder at –130 ppm at heat-treating up to Tp2 may reflect F changing its structural role in the residual glass network following the crystallisation of mullite. Crystallisation of mullite and a reduction of Al in the residual glass phase may force F to take up a new role. One possibility is the formation of Si-F bonds (9).

Conclusions

Substitution of fluorine for oxygen strongly influences the crystallisation process of the investigated glass compositions. A glass containing no fluorine crystallised to β-tricalcium phosphate and anorthite. Substitution of fluorine for oxygen resulted in formation of FAP instead of β-tricalcium phosphate, as a first crystallisation phase and inhibited anorthite formation. At higher fluorine contents (z>2), mullite crystallised following the crystallisation of FAP. Anorthite formation is probably hindered on crystallisation of FAP. The inhibition of anorthite favours mullite formation. Aluminium is mainly Al(IV) in the glass containing no fluorine, while fluorine additions cause the formation of Al(V) and Al(VI). These species seem to disappear on heat treatment to Tp1 and Tp(1+2)/2 but new Al(V) and Al(VI) species often appear on heat-treating to Tp2. The fluorine is present in the glasses initially as F-Ca(n) and Al-F-Ca(n) species. The F-Ca(n) species appear to be used preferentially in forming FAP.

References

[1] R.D. Rawlings, Clinical Mater. 141 (1993)55-179.
[2] A.Rafferty, A.Clifford, R.G. Hill, D. Wood, B. Samuneva, M. Dimitrova-Lukacs, J.Amer. Ceram. Soc. 83(11) (2000) 2833-2838.
[3] R. Dupree, D. Holland, M.G. Mortuza, J.A. Collins, M.W.G. Lockyer, J.Non-Cryst. Solids 112 (1989)111-119.
[4] C.M. Moisescu, T. Hoche, G. Carl, R. Keding, C. Russel, W.D. Heerdegen, J. Non-Cryst. Solids, 289 (2001) 123-134.
[5] G. Engelhardt, M. Noftz, K.Forkel, F.G. Wishmann, M. Magi, A. Samson, E. Lippma, Phys Chem Glasses 26 (1985), 157 –165,.
[6] R.J. Kirkpatrick, R.A. Kinsey, K.A. Smith, D.M. Henderson, E. Oldfield, Am. Mineral. 70 (1985) 106-123.
[7] J. Vogel, C. Russel, G. Gunther, P. Hartmann, F. Vizethum, N. Bergner, J. Mater. Sci.:Mater. Med. 7 (1996) 495-499.
[8] W.A. Dollase, L.H. Merwin, A. Sebald, J. Solid State Chem 83 (1989) 140-149.
[9] Q. Zeng, J.F. Stebbins, Am. Mineral. 85 (2000) 863-867.

Key Engineering Materials Vols. 254-256(2004) pp. 103-106
online at http://www.scientific.net

Dry Mechanosynthesis of Strontium-Containing Hydroxyapatite from DCPD + CaO + SrO

H. El Briak-BenAbdeslam, B. Pauvert, A. Terol and P. Boudeville

Faculté de Pharmacie, 15 Avenue Charles Flahault, BP 14491, 34093 Montpellier cedex 5, France,
e-mail : boudevil@univ-montp1.fr

Keywords : calcium-strontium phosphates, hydroxyapatite, strontium oxide, calcium oxide, dicalcium phosphate dihydrate, mechanosynthesis.

Abstract. By grinding mixtures of DCPD + CaO + SrO in a planetary ball mill, mixed calcium-strontium phosphates or hydroxyapatites were mechanochemically synthesized. Calcium and strontium-to-phosphate molar ratios [(Ca+Sr)/P] 1.50, 1.60 and 1.67 were tested and the relative amount of Sr and Ca was also varied. The kinetic study of the mechanochemical reaction was carried out for all the mixtures by XRD and DSC. The reaction rate was slightly lower with SrO than with CaO. XRD patterns of the powders after grinding showed an apatitic profile and after heating at 950°C for 2 h, powders were mixed Ca-Sr-hydroxyapatites or tri(Ca/Sr) phosphate or a mixture of them.

Introduction

Trace elements are an interesting alternative to growth factors in order to favor bone formation. Strontium is an element which plays a considerable part in osseous mineralization and its use is envisaged in osteoporosis treatment [1]. In this goal, some attempts were already done for introducing strontium in bioceramics [2] to enhance bone ingrowth. We showed that dry mechanosynthesis was a possible route to obtain calcium deficient hydroxyapatites (CDHAs) that can be used to prepare biphasic calcium phosphate ceramics [3] with the expected calcium-to-phosphate ratio Ca/P ± 0.01. CDAHs were prepared by grinding dicalcium phosphate dihydrate (DCPD) and calcium oxide mixtures. Given the similar chemical properties of CaO and SrO, we synthesized strontium-containing HAs or DHAs by replacing partly or completely CaO by SrO according to Eq. 1 with $0 \leq x \leq 4$, $0 \leq y \leq 4$ and $3 \leq x + y \leq 4$.

$$6\,CaHPO_4 \cdot 2H_2O + x\,CaO + y\,SrO \;\forall\; Ca_{(6+x)}Sr_y(HPO_4)_{(4-x-y)}(PO_4)_{6-(4-x-y)} + (10 + x + y)\,H_2O \quad (1)$$

Material and methods

DCPD (from Fluka) had a median particle size d_{50} of 8 µm (d_1–d_{90} = 1.6–27 µm; calculated specific surface area, 3.5 m^2 g^{-1}, Mastersizer, Malvern Instruments) and was used as received. CaO (from Aldrich) was heated at 900°C for 2 h to remove H$_2$O and CO$_2$ and stored in a vacuum desiccator. After heating, median particle size was around 7 µm (2–40 µm; calculated specific surface area = 4.3 m^2 g^{-1}). SrO was prepared by heating Sr(OH)$_2$·8H$_2$O (from Riedel-de-Haën) at 900°C for 2 h. Its XRD pattern complied with the JCPDS file 6-520 with a slight amorphous-like background.

    Mixtures (15 g) of DCPD, CaO and SrO in variable proportions depending on the desired Ca replacement by Sr and (Sr+Ca)/P ratios were ground in a planetary ball mill (Retsch Instruments: vial eccentricity on the rotating sun disc 3.65 cm) at a rotation velocity of 350 rpm with 5 balls, 2.5 cm in diameter (total mass 133 g and surface area 110 cm²). At different intervals, powder (50 mg) was taken for analysis. The DCPD content at the different intervals was determined either by XRD (Philips PW3830X, CGR horizontal goniometer, Cu K$_{\alpha 1}$ = 1.5405 Å, Ni filter), surface area of the DCPD peak at 5.80°θ (plane (0 2 0)) after baseline subtraction or by DSC (DSC6 PerkinElmer), value of the enthalpy change of the endotherm between 170-210°C corresponding to the DCPD dehydration into DCPA [4,5]. The rate constant (k) of DCPD disappearance was given by the slope of the ln(DCPD) = f(t) plot (Fig. 1) and the final reaction time (t_f) was determined with a phenolphthalein test (φφ test) [5]. After complete reaction (negative φφ test) the as-prepared powders and powders after heating at 950°C for 2 h were analyzed by XRD from 2 to 30°θ, by 0.02°θ, 5 acquisitions, acquisition delay 500 ms.

Results and Discussion

The rate constant k of DCPD disappearance and the final reaction time t_f for the different (Sr+Ca)/P ratios and CaO (x) replacement by SrO (y) are given in Table 1. The mechanochemical reaction of DCPD with SrO was slightly slower than with CaO as shown by the decrease in k and the increase in t_f when the Sr fraction y was increased. But times of grinding remained acceptable. Note that the DCPD disappearance rate seems independent on the (Sr+Ca)/P ratio (Fig. 1 and Table 1) as we observed for pure calcium apatite [5].

Table 1: Rate constants k of DCPD disappearance, final reaction times t_f and main diffraction peak angles (in °θ) for the different (Sr+Ca)/P ratios by replacing CaO (x) by SrO (y).

(Sr + Ca) / P	1.50			1.60			1.67	
y - x	0 - 3	1.5 - 1.5	3 - 0	0 - 3.6	1.8 - 1.8	3.6 - 0	0 - 4	4 - 0
k (h^{-1})	2.08	1.86	1.43	2.11	2.11	1.17	1.91	1.50
t_f (h)	2	3	3.2	3	4	5	10	18
β-TCP peak (0 2 10)	15.50	15.36	15.22	15.50	15.36	15.12		
HA peak (2 1 1)				15.84	15.78	15.62	15.84	15.60

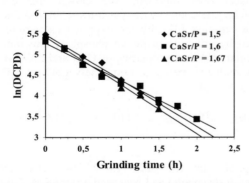

Fig. 1. Variations in ln (DCPD) with the grinding time for different (Ca+Sr)/P ratios when CaO was totally replaced by SrO.

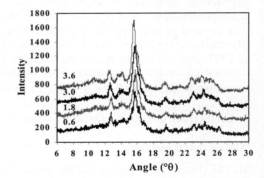

Fig. 2. XRD patterns of the ground powder when increasing the Sr fraction y indicated on each plot. (Ca+Sr)/P = 1.6.

After complete reaction (negative φφ test), whatever the (Sr+Ca)/P ratio and the amount of CaO replaced by SrO were, XRD patterns of ground powders showed the profile of an apatite with a poor crystallinity (Fig. 2). In Fig. 2, the main peak of HA (plane (2 1 1)) drifts down from 15.88°θ for y = 0 to 15.60°θ for y = 3.6. In the same way, the HA peak at 12.93°θ (plane (0 0 2)) goes down regularly to 16.60°θ for y = 3.6 due to the replacement of Ca^{2+} (ionic radii = 0.99 Å [6]) by Sr^{2+} (ionic radii = 1.12 Å).

After heating at 950°C for 2 h, the XRD pattern profiles varied with the (Sr+Ca)/P ratio. For (Sr+Ca)/P = 1.50 (Fig. 3), XRD patterns corresponded to β-TCP but all the diffraction angles were lowered when SrO was increased in the starting materials (Fig. 3, pattern 1), indicating an increase in the lattice parameters due to the partial replacement of Ca by Sr (diffraction angles reported in the two last lines of Table 1 are expressed in °θ). For example, the main diffraction peak of β-TCP decreased from 15.50°θ for β-$Ca_3(PO_4)_2$ (JCPDS file 9-169) to 15.22°θ for a theoretical formula $Ca_2Sr(PO_4)_2$.

For (Sr+Ca)/P = 1.67, XRD patterns corresponded to HA (Fig. 4) and, in the same way as for β-TCP, all the diffraction angles were lowered by replacing CaO by SrO. For $y = 4$ and thus $x = 0$, the XRD pattern of the heated product well complied with the JCPDS file 34-480 corresponding to $Ca_6Sr_4(PO_4)_6(OH)_2$ indicating the total integration of strontium in the apatitic structure. We can note the presence of a low amount of β-TCP as well in the pure Ca-HA as in the Sr-substituted HA (marked with asterisks in Fig. 4). This presence is due to the non stoichiometric amount of water in the DCPD that is nearer 1.8 than 2 moles of water [5,7]. For this reason pure HA is only mechanochemically obtained with a theoretical Ca/P ratio of 1.70 in the starting materials [3].

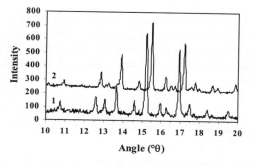

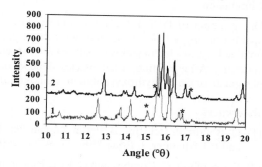

Fig. 3. XRD patterns after heating ground powders at 950 °C for 2 h. **1**: 6 DCPD + 3 SrO and **2**: 6 DCPD + 3 CaO.

Fig. 4. XRD patterns after heating ground powders at 950 °C for 2 h. **1**: 6 DCPD + 4 SrO and **2**: 6 DCPD + 4 CaO. *: β-TCP peaks.

For (Sr+Ca)/P = 1.60, β-TCP and HA mixtures were obtained (Fig. 5) with strontium integrated in the two products as shown by the shifting of their main diffraction peaks towards lower angles (Fig. 5) and reported in table 1.

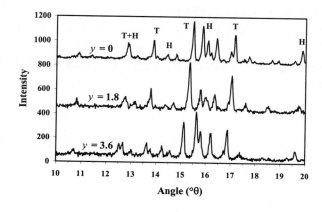

Fig. 5. Evolution in the XRD patterns of heated powders with the replacement of Ca by Sr (y indicated on each pattern). (Ca+Sr)/P = 1.60, T = tri(Ca-Sr) phosphate, H = Ca-Sr-hydroxyapatite

Yokogawa *et al.* [8] reported the obtaining of calcium-strontium apatite solid solutions over the entire compositional range for a Ca/(Ca+Sr) molar ratio of 0-1.0, by grinding for 24 h aqueous slurries of DCPD + $CaCO_3$ + $Sr(OH)_2 \cdot 8H_2O$ in variable proportions and with a molar ratio of (Sr+Ca)/P = 1.5. XRD patterns of ground powders showed effectively an apatitic profile after heating at 500°C. But we can regret that powders were not analyzed by XRD after heating over 800°C, temperature from which Ca or Sr deficient hydroxyapatites transform into tri(Ca-Sr) phosphate and Ca-Sr-HA. They also wrote "if water was not included in the ball milling process, the starting materials did not react and thus remained unchanged". This work clearly demonstrates that the mechanochemical reaction occurs even under dry conditions. A possible explanation of their remark was certainly in the low rotation velocity they used, 50 rpm, because the energy exchanged during shocks, that allows the occurring of the reaction, is proportional to the square of the rotation velocity [9].

Conclusion

Dry grinding of DCPD + CaO + SrO mixtures is a new possible route to obtain mixed Ca-Sr-apatites having a low crystallinity with the expected Sr content and (Sr+Ca)/P ratio. The powders obtained by mechanosynthesis can serve to prepare polyphasic Ca-Sr-phosphate bioceramics by sintering. After heating, final phases depend on the initial (Sr+Ca)/P ratio: either β-tri(Ca-Sr) phosphate or Ca-Sr-hydroxyapatite or a mixture of them.

References

[1] G. Boivin, P. Deloffre, B. Perrat, G. Panczer, M. Boudeulle, Y. Mauras, P. Alain Y. Tsouderos and P.J. Meunier. J. Bone Miner. Res. 11 (1996), 1302.

[2] S. Cazalbou, C. Combes and C. Rey. Bioceramics 13 (Trans Tech Publications, Switzerland 2000) 147-150.

[3] P. Boudeville, B. Pauvert, M.P. Ginebra, E. Fernandez and J.A. Planell. Bioceramics 13 (Trans Tech Publications Switzerland 2000) 115-8.

[4] S Serraj, P. Boudeville, B. Pauvert and A. Terol. J. Biomed. Mater. Res. 55 (2001) 566-75.

[5] H. El Briak-BenAbdeslam, C. Mochales, M.P. Ginebra, J. Nurit, J.A. Planell and P. Boudeville. J. Biomed Mater Res (2003) to appear.

[6] B.O. Fowler. Inorg. Chem. 13 (1974) 207-14.

[7] J.C. Elliot. Structure and Chemistry of the Apatites and other Calcium Orthophosphates. Studies in Inorganic Chemistry 18, Elsevier, Amsterdam 1994, p 29.

[8] Y. Yokogawa, M. Toriyama, Y. Kawamoto, T. Suzuki, K. Nishizawa, F. Nagata and M.R. Mucalo. Chem. Lett. (1996) 91-2.

[9] C. Mochales, H. El Briak-BenAbdeslam, M.P. Ginebra, A. Terol, J.A. Planell and P. Boudeville. Biomaterials (2003) to appear.

Key Engineering Materials Vols. 254-256(2004) pp. 107-110
online at http://www.scientific.net

Obtention of Silicate-Substituted Calcium Deficient Hydroxyapatite by Dry Mechanosynthesis

C. Mochales[1], H. El Briak-BenAbdeslam[2], M.P. Ginebra[1], P. Boudeville[2], JA Planell[1]

[1]CREB, Technical University of Catalonia (UPC), Av. Diagonal 647, 08028-Barcelona; Spain;
carolina.mochales-palau@upc.es; maria.pau.ginebra@upc.es
[2]Faculté de Pharmacie 15 Av. Charles Flahault BP14491, 34093 Montpellier; France.
boudevil@pharma.univ-montp1.fr

Keywords: silicate, substituted hydroxyapatite, mechanosynthesis

Abstract. This paper describes a new method to obtain silicate-substituted calcium deficient hydroxyapatite (Si-CDHA): mechanosynthesis. In previous studies mechanosynthesis has been showed to be an efficient method to obtain nanosized calcium deficient hydroxyapatite (CDHA). In this study, the silicate ions stemmed from calcium silicate hydrate (CSH). Both, CSH and Si-CDHA were obtained by mechanical activation in a planetary Retsch mill. Titration with 0.01M HCl of the samples of Si-CDHA obtained at different milling times demonstrated that a 9.8% w/w of silicate had been incorporated into the apatite. Moreover, XRD showed that the silicate incorporation in the apatite increased the crystallinity and the stability of the apatitic phase. These results indicated that mechanosynthesis is an effective method for incorporating silicate into CDHA. As a result a hydroxyapatite nanosized powder was obtained, with promising characteristics such as an improved bioactivity due to its calcium deficiency and silicate substitution.

Introduction

It is well-known that hydroxyapatite can incorporate different types of ionic substitutions, with the subsequent changes in its structural parameters and physico-chemical properties. Some recent studies have shown that bioactivity of hydroxyapatite is significantly enhanced by the incorporation of silicate ions into its lattice [1,2]. In previous studies mechanosynthesis has been presented as a new method to obtain nanosized hydroxyapatite with an improved control of the stoichiometry [3,4]. Even more, this technique can allow the incorporation of different ions in the apatitic structure. In this study, the introduction of silicate in a calcium deficient hydroxyapatite (CDHA) by mechanosynthesis is presented.

Materials and Methods

Two mechanochemical reactions were performed in a planetary Retsch mill, milling at 350 rpm. The first one to obtain a calcium silicate hydrate (CSH, $CaSiO_3$ with 25 wt% H_2O) and the second one to obtain the silicate-substituted hydroxyapatite.

For the first one, the reactants used were calcium hydroxide ($Ca(OH)_2$; Fluka 21181) and silicic acid hydrate(SiO_2 with 24 wt% H_2O; Fluka 60780), with a total mass of 21.5 grs.

In the second one, the reactants were adjusted to obtain a substitution of one phosphate group for one silicate group in the CDHA, having a Ca/P molar ratio of 1.6 and a Ca/(P+Si) of 1.5. For this purpose, a mixture of dicalcium phosphate dihydrate (DCPD, $CaHPO_4.2H_2O$; Fluka 21184), calcium oxide (CaO; Aldrich 24,856-8) and calcium silicate hydrate (CSH), previously obtained from the first mechanochemical reaction, were milled.

The total milled mass was 10 grams and the amount of silicate in the mixture represented a 9.8 % w/w. Powder samples were obtained after different grinding times (2, 4, 8, 12, 20, 26 and 30 h). These samples were titrated with 0.01M HCl and analyzed by X-ray diffraction before and after heating at 950°C. Titration analysis were performed with a PH-meter Tacussel PHM210 combined glass electrode Bioblock and standardized with NBS buffer PH 4, 7 and 10 (Fluka Biochemica color-coded). XRD patterns were obtained by an automatic diffractometer Philips PW 3830X with horizontal goniometer CGR, using an anticathode Cu ($K_{\alpha 1}$= 1,5405Å) and Ni filter.

Results and Discussion

CSH obtention by mechanosynthesis. CSH was obtained after 33 h of grinding calcium hydroxide and silicon oxide. The XRD patterns of the samples milled at different times (0, 1, 4, 6, 12, 16, 20, 23 and 33 h) showed the disappearance of the $Ca(OH)_2$ peaks and the appearance of an amorphous profile (Fig.1). After heating at 950 °C for 2 h a crystalline $CaSiO_3$ was formed (JCPDS 43-1460).

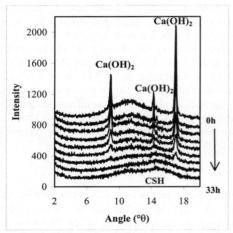
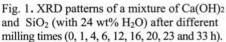
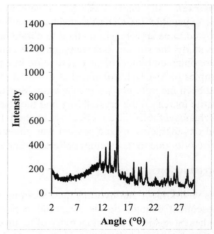

Fig. 1. XRD patterns of a mixture of $Ca(OH)_2$ and SiO_2 (with 24 wt% H_2O) after different milling times (0, 1, 4, 6, 12, 16, 20, 23 and 33 h).

Fig. 2. XRD pattern of CSH obtained after 33h of milling heated for 2h at 950°C.

Silicate incorporation into CDHA by mechanosynthesis. Before grinding only DCPD and CaO were detected by XRD. CSH was not detected since it had a very low cristallinity. After 2 h of grinding the XRD pattern showed an apatitic profile. The only change detected by XRD with increasing milling times was a progressive increase of the crystallinity of the samples (Fig.3).

The amount of silicate incorporated into the apatitic structure was calculated from the equivalent points of the titration curves of the samples at different milling times. Comparing with the titration curves of a CDHA with Ca/P molar ratio of 1.6 (which had been obtained by mechanosynthesis in a previous work [3]) and the mixture of this CDHA and CSH with a total Ca/(P+Si) molar ratio of 1.5, it could be observed that the titration curves of the samples evolved from the second one to the first one with the milling time. The results showed that the introduction of silicate in the CDHA followed an exponential behaviour. After 4 and 12 hours of grinding, the silicate introduced in the CDHA was approximately a 75 % and 96 % w/w of the initial amount of silicate in the grinded mixture, respectively. These percentages represent an incorporation of a 7.3 % and a 9.4 % w/w of silicate in the apatite powder. This amount of silicate is much higher than the values reported by other preparation techniques, such as precipitation from aqueous media or hydrothermal methods.

After 20 h of milling the titration curve of the samples remained constant, so it was assumed that at 20 h the complete amount of initial CSH (9.8% w/w) has been incorporated into the apatite. From the comparison between the titration curves of CDHA (Ca/P=1.6) and the final Si-CDHA it was observed that Si-CDHA was initially more basic than CDHA.

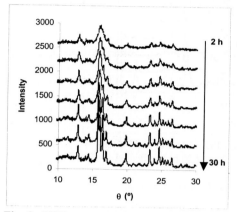

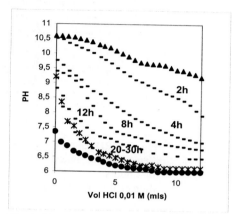

Fig. 3. XRD patterns of a mixture of DCPD, CaO and CSH after different milling times (2, 4, 8, 12, 20, 26 and 30 h).

Fig. 4. Titration curves of a mixture of DCPD, CaO and CSH after different milling times: (▪)2, 4, 8, 12, 20 and (✳) 30 h, (▲) mixture of CDHA (Ca/P=1.6) and CSH, (●) CDHA(Ca/P=1.6).

The heating of the samples at 950°C was performed for two reasons. First of all, it was necessary to be able to detect the presence of calcium silicate after different grinding times, since the CSH had an amorphous XRD profile. Secondly, to assess the effect of silicate incorporation on the stability of CDHA at high temperatures. When the samples were heated at 950°C, the phases obtained varied in function of the grinding time. It was observed that, after low grinding times, a mixture of β-TCP, CDHA and CaSiO$_3$ was obtained. However, at higher grinding times CaSiO$_3$ was not detected, the amount of β-TCP decreased and the apatitic phase increased. After 30 h of grinding, β-TCP was hardly detected, and the main phase was hydroxyapatite (Fig. 5). These results confirmed the data obtained by titration, i.e, that silicate was progressively incorporated in the apatitic phase, and in addition they suggested that this introduction of silicate in the CDHA structure stabilized the apatitic phase at high temperatures.

Stoichiometry. Taking into account the reaction mechanism proposed by Gibson *et al* [5], in which some of the phosphate ions are replaced by silicate with the simultaneous creation of hydroxyl vacancies, the stoichiometry proposed for the reaction performed in this work was:

$$5DCPD + 3CaO + 1CaSiO_3 \rightarrow Ca_9(HPO_4)(PO_4)_4(SiO_4) + 12H_2O. \qquad (1)$$

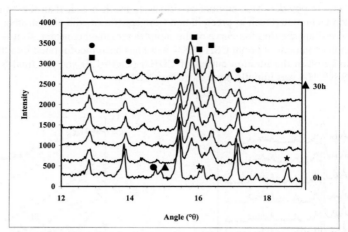

Fig. 5: XRD patterns of a mixture of DCPD, CaO and CSH after
different milling   times (0, 2, 4, 8, 12, 20, 26 and 30 h) and heated at 950 ºC
for 2 hours.
Main peaks: (●) β–TCP, (■) CDHA, (▲) CaSiO3, (★) CaO

Conclusions

Mechanosynthesis is an effective method for incorporating silicate into a calcium deficient
hydroxyapatite. As a result a hydroxyapatite nanosized powder is obtained, with promising
characteristics such as an improved bioactivity due to its calcium deficiency and silicate
substitution.

Acknowledgements

The authors thank the Science and Technology Spanish Ministry for funding this work through the
Integrated Action HF00-97.

References

[1] I.R. Gibson, K.A. Hing, P.A. Revell, J.D. Santos, S.M. Best, W. Bonfield: Key Engineering
Materials Vols. 218-220 (2002) pp. 203-206

[2] I.R. Gibson, J. Huang, S.M. Best, W. Bonfield: Bioceramics Vol. 12 (1999) pp. 191-194

[3] H. El Briak-BenAbdeslam, C. Mochales, M.P. Ginebra, J. Nurit, J.A. Planell, P. Boudeville:
Proceedings of the17th European Conference on Biomaterials , Barcelona 2002, p. P147

[4] H. El Briak-BenAbdeslam, C. Mochales, M.P. Ginebra, J. Nurit, J.A. Planell, P. Boudeville:
Proceedings of the17th European Conference on Biomaterials, Barcelona 2002, p. T22

[5] I.R. Gibson, S. M. Best, W. Bonfield : J. Biomed. Mat. Res. Vol. 44 (1999) pp. 422-428

Key Engineering Materials Vols. 254-256(2004) pp. 111-114
online at http://www.scientific.net

Fabrication of Hydroxyapatite Ceramics via a Modified Slip Casting Route

E.S. Thian, J. Huang, S.M. Best and W. Bonfield

Department of Materials Science and Metallurgy,
University of Cambridge, Pembroke Street,
Cambridge, CB2 3QZ, UK

Keywords: Hydroxyapatite, slip casting, attritor milling

Abstract. The fabrication of dense phase-pure hydroxyapatite (HA) ceramics was investigated using a modified slip casting process. With the introduction of attritor milling into the process, the mean size of the HA powder was significantly reduced to 2 μm, resulting in a high relative 'green' density of 56 %, thereby giving a high sintered density of 98 % with mean grain size of around 0.6 μm at a sintering temperature of 1200 °C. Cell culture studies revealed that cells were growing and spreading well on these HA ceramics, indicating that no contamination was introduced during the fabrication process.

Introduction

Hydroxyapatite (HA) is a biocompatible and bioactive material with a chemical composition similar to the major mineral phase of bone. HA ceramics can be formed and densified by various manufacturing processes such as conventional powder compaction [1], injection moulding [2], spark plasma sintering (SPS) [3] and pulse electric current sintering (PECS) [4]. Akao *et. al.* [1] obtained HA ceramics with a relative density of 97 % and grain size about 3 μm following sintering at 1300 °C by the compaction process. Injection-moulded HA ceramics with relative density of 98 % and grain size between 1 and 16 μm were achieved following sintering at 1300 °C by Cihlar and Trunec [2]. Nearly-full dense (99.9 %) HA ceramics were obtained using SPS at 950 °C, but with the presence of small amounts of beta-tricalcium phosphate (β-TCP) [3]. Nakahira *et. al.* [4] successfully produced HA ceramics after sintering at 700 °C with a relative density of 99.2 % and grain size of 0.3 μm using PECS. However, there is still room for improvement to form dense phase-pure HA ceramics using lower cost processes, so as to reduce the cost of biomedical devices. Slip casting represents a relatively cheap manufacturing process and also enables complex shapes to be fabricated. It is a common casting technique developed to process clay and nonclay -based ceramics into useful products. The technique is divided into three main steps: (1) grinding of raw materials into submicrometre particles with possible mixing of additives to lower the viscosity; (2) shaping of the cast; and (3) drying and sintering of the cast. The present study reports a modified slip casting process to fabricate nearly-full dense phase-pure HA ceramics.

Materials and Methods

HA powder was synthesized in-house by a precipitation reaction between commercially available calcium hydroxide ($Ca(OH)_2$) and orthophosphoric acid (H_3PO_4) solutions in a molar ratio of 10:6 [5]. The HA filter cake was ground into fine powder before being calcined at 900 °C for 2 h. 22 wt.% of HA powder and 1 wt.% of polyvinyl pyrrollindone (PVP) were mixed with deionized water to form a ceramic slurry by means of attritor milling for 3 h. The prepared slip was cast in a plaster mould with circular geometry of diameter 32 mm. The 'green' cast was then sintered at temperatures ranging from 900 to 1300 °C, at intervals of 100 °C for 2 h in air. Phase composition, morphology, Ca/P ratio and particle size of the powder or sintered HA were determined using X-ray Diffraction (XRD), Scanning Electron Microscopy coupled with Energy Dispersive Spectrometry (SEM-EDS) and Particle Size Analysis (PSA). The 'green' density was calculated by the weight and dimensional measurements, while the sintered density was determined using Archimedes' Principle.

The biocompatibility of the optimized, sintered HA was assessed using primary human osteoblast (HOB) cells, obtained from Promocell, UK. The specimens were sterilized by dry heating at 160 °C for 4 h before directly seeded with 2×10^4 cells/ml in McCoy's 5A medium containing 10 % foetal calf serum, 1 % glutamine and Vitamin C (30 µg/ml). The cells were incubated at 37 °C in a humidified atmosphere of 95 % air and 5 % carbon dioxide. On day 3, 6 and 9, the sintered HA was washed with a phosphate buffered solution after decanting the media, and was fixed with 4 % formaldehyde. The metabolic activity of the HOB cells over 3, 6 and 9 days was measured using a 10 % alamarBlue^{TM} assay. To study the morphology and attachment of HOB cells on the sintered HA, the cultures were fixed after 3 days of incubation, stained with 1 % osmium tetroxide, dehydrated in a graduated series of alcohol and then finally critical point dried. The specimen was then coated with a thin layer of carbon before it was examined using SEM-EDS.

Results and Discussion

The as-prepared HA powder consisted of crystallites ~15 nm in width and ~150 nm in length (Fig. 1a). These were agglomerates of the larger powder particles. In contrast, the calcined HA was in the form of smooth spheres of diameter ~100 nm (Fig. 1b). The energy generated during calcination has basically transformed the morphology of the powder, reducing its aspect ratio. XRD patterns (Fig. 2) showed that the uncalcined powder consisted of phase-pure HA microcrystallites, characterized by a low relative intensity and an associated broad peak width. However, the particles became more highly crystalline and increased in size after calcination at 900 °C for 2 h. EDS revealed that the measured Ca/P ratio was 1.67 ± 0.01 (mean ± standard deviation). This result was in agreement with that of the stoichiometric value of 1.67. After attritor milling for 3 h, the mean particle size of the HA powder was reduced significantly from 79 µm to 2 µm, with a well-defined narrow distribution of the particles (Table 1). The relative 'green' density of the slip-cast HA achieved in this study was 56 ± 0.2 %, which was higher than those previously reported for a slip casting process (Table 2). It was also observed that, as expected, densification increased with sintering temperature (Table 3). XRD (Fig. 2) of the sintered HA showed that HA was stable up to a temperature of 1200 °C with a Ca/P ratio of 1.67. At 1300 °C, although a slight increase in the relative density was observed, small amounts of tetracalcium phosphate (TTCP) (0.7 %) and α-TCP (4.1 %) were detected. Hence, the highest relative sintered density achieved without degrading the HA was 98 % at 1200 °C. This value was higher than those previously reported (Table 2). This result is due to the size and morphology of the HA particles achieved after attritor milling, allowing particles to pack together more efficiently, thus producing a higher 'green' density. In addition, small particles have a large surface area which tend to increase the rate of sintering. For the optimum sintering condition of 1200 °C, the grain size was in the range of 0.2 to 1.0 µm (Fig. 3a), which is likely to be beneficial in terms of mechanical properties.

Cells having visible filapodia, were seen attaching on the HA grains, with signs of extracellular matrix (ECM) synthesis (Fig. 3b). Generally, all the cells maintained their osteoblastic morphology. In addition, HOB cells were observed to proliferate with time (Fig. 4), in a similar manner to that noted for HA prepared by the conventional powder compaction. All these results indicate that nearly-full density phase-pure HA ceramics are achieved by the modified slip casting process, without contaminating the HA since the bioactive characteristics are maintained.

Table 1 Particle size distribution of HA powder (a) before- (b) after- attritor milling

	d_{10} (µm)	d_{50} (µm)	d_{90} (µm)	Span
(a)	7	79	343	0.24
(b)	1	2	4	0.67

Table 2 Comparison of various slip casting processes from the literature

Reference	[6]	[7]	[8]	[9]
HA source	commercial	in-house	in-house	in-house
Ca/P	-	-	-	1.66
Particle size (μm)	0.5 - 6	5 - 30	0.4	-
Composition (powder)	HA	HA	-	HA
Green density (%)	52.9	51.0	55.6	51.0
Sintering temperature (°C)	1300	1350	1300	1200
Sintered density (%)	94.3	94.0	96.8	96.0
Grain size (μm)	-	3 - 8	-	-
Composition (ceramics)	-	-	TCP, TTCP, HA	-

Table 3 Relative sintered density and phase composition of HA ceramics

T (°C)	900	1000	1100	1200	1300
ρ (%)	91.7	93.5	96.9	98.0	98.4
Phases	HA	HA	HA	HA	HA, TTCP, α-TCP

(a)

(b)

Fig. 1 SEM of HA powder (a) as-prepared (b) calcined

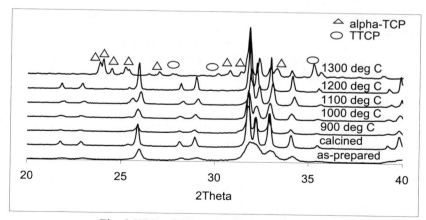

Fig. 2 XRD of HA powder and sintered HA

(a)   (b)

Fig. 3 SEM of sintered HA at 1200 °C (a) grain size (b) HOB cells

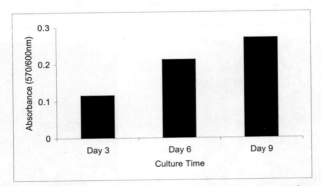

Fig. 4 Cell proliferation on sintered HA at 1200 °C versus culture time

Conclusion

Phase-pure HA ceramics with relative density of 98 % and grain size of about 0.6 μm were achieved at a sintering temperature of 1200 °C by a modified slip casting process. The bioactive characteristic of these HA ceramics was still maintained after the fabrication process, as HOB cells were observed growing and spreading readily, with the formation of bone matrix on these HA surfaces. These results demonstrate that the modified slip casting is a possible route for the fabrication of dense complex phase-pure HA ceramics for various biomedical applications.

Acknowledgements

Cambridge Commonwealth Trust (UK) and Lee Foundation (Singapore) in funding a Ph.D student (E.S. Thian). Armourers & Brasiers' Company (UK), British Alcan Aluminium (UK) and Wolfson College (Cambridge, UK) for awarding a conference travel grant.

References

[1] M. Akao, H. Aoki and K. Kato: J. Mater. Sci. Mater. in Med. Vol. 16 (1981), p. 809
[2] J. Cihlar and M. Trunec: Bioceram. Vol. 10 (1998), p. 183
[3] K.A. Khor et. al.: Bioceram. Vol. 15 (2003), p. 497
[4] A. Nakahira et. al.: Bioceram. Vol. 15 (2003), p. 551
[5] N. Patel et. al.: J. Mater. Sci. Mater. in Med. Vol. 12 (2000), p. 181
[6] C. Galassi et. al.: Proc. First Euro. Ceram. Soc. Conf. Vol. 13 (1989), p. 3.43
[7] R.R. Rao and T.S. Kannam: J. Am. Ceram. Soc. Vol. 84 (2001), p. 1710
[8] E.G. Nordstrom and K.H. Karlsson: Am. Ceram. Soc. Bull. Vol. 69 (1990), p. 824
[9] F. Lelievre et. al.: J. Mater. Sci. Mater. in Med. Vol. 7 (1996), p. 489

Key Engineering Materials Vols. 254-256(2004) pp. 115-118
online at http://www.scientific.net

Newly Improved Simulated Body Fluid

Hiroaki Takadama[1], Masami Hashimoto[1], Mineo Mizuno[1], Kunio Ishikawa[2], and Tadashi Kokubo[3]

[1] Materials Research and Development Laboratory, Japan Fine Ceramics Center (JFCC), Nagoya, Japan. takadama@jfcc.or.jp

[2] Faculty of Dental Science, Kyushu University, Fukuoka, Japan

[3] Research Institute for Science and Technology, Chubu University, Kasugai, Japan

Keywords: apatite, bioactivity, blood plasma, simulated body fluid, ion concentration, preparation process

Abstract. A newly improved simulated body fluid (n-SBF) was designed to have concentrations of ions equal to those of human blood plasma excepting that of HCO_3^-, the concentration of which is lower than the level of saturation with respect to calcite. The preparation process was modified and two different preparing processes such as powder dissolving and liquid mixing processes were attempted. The n-SBFs with nominal ion concentration can be prepared and its concentrations were unchanged over 4 weeks. The bioactive glasses showed almost same bioactivity in both n-SBFs and c-SBF. The n-SBFs is believed to produce apatite under quite similar condition to the body environment, thereby being useful for *in vitro* assessment of bioactivity of artificial material.

Introduction

A simulated body fluid (SBF) with ion concentrations approximately equal to those of human blood plasma has been used widely for *in vitro* assessment of the bioactivity of artificial materials by examining their apatite-forming ability [1, 2]. It was, however, known that the ion concentrations of conventional SBF (c-SBF) are not exactly equal to those of human blood plasma [2, 3]. It was reported that the apatite produced in c-SBF differs from bone apatite in its composition and structure, and that an apatite with a composition and structure close to those of bone apatite would be produced if the SBF could be conditioned to have ion concentrations close to those of human blood plasma [4].

Recently, several revised SBFs such as revised SBF (r-SBF) and modified SBF (m-SBF) with ion concentrations equal to and close to those of human blood plasma, respectively, have been proposed [5]. However, r-SBF and m-SBF with high HCO_3^- concentration tend to deposit calcite ($CaCO_3$) in addition to bonelike apatite especially on Ca-based bioactive materials due to the release Ca^{2+}. Therefore, these SBFs are not useful for bioactivity assessment.

Furthermore, it is known that SBF sometimes does not always show good reproducibility. One of the reasons is due to the conventional process to prepare SBF by dissolving powder. It is considered that the radical increase in the local supersaturation with respect to apatite raised by dissolving powder reagents can lead to the undesirable precipitation of apatite. In order to avoid the undesirable precipitation, some care, experience, skill or know-how is necessary to prepare SBF. It is suspected that this know-how has great effect on the reproducibility. Therefore, it is desired that stable SBF with good reproducibility and its easy preparation method with less skill should be developed.

In the present study, n-SBF was designed to have concentrations of ions equal to those of human blood plasma excepting that of HCO_3^-, the concentration of which is lower than the level of saturation with respect to calcite. Moreover, the preparation process was modified and two different preparing processes such as powder dissolving and liquid mixing processes were proposed. The n-SBFs were prepared by two different processes, respectively. For the liquid mixing process, the Ca and P solutions were prepared separately in advance and then mixed to prepare n-SBF. Ion concentration and pH of the c-SBF and n-SBFs prepared by different processes were measured and compared with those of human blood plasma.

Experimental section
Preparation of SBFs Two SBFs, i.e., c-SBF and n-SBF with different ion concentrations were prepared as shown in Table 1. Moreover, two different n-SBFs, i.e., np-SBF and nl-SBF were prepared by powder dissolving and liquid mixing processes, respectively. Table 2 lists the reagents used, their purity and the quantities used in the preparation of 1000 mL of SBFs. All reagents for the preparation of SBFs were supplied by Nacalai Tesque, Inc., Japan.

The c-SBF, abbreviated from conventional SBF, was prepared by powder dissolving process. It has ion concentrations equal to those of human blood plasma excepting that of HCO_3^- and Cl^-, and its HCO_3^- concentrations is lower and Cl^- is higher than those of human blood plasma. The c-SBF was buffered by tris-hydroxymethyl aminomethane (TRIS) and 1.0-M HCl at pH 7.40 at 36.5 °C.

The np-SBF, abbreviated from n-SBF prepared by powder dissolving process, was prepared by dissolving the reagents into the solution one by one in the similar manner to c-SBF. The np-SBF was designed to have concentrations of ions equal to those of human blood plasma excepting that of HCO_3^-, the concentration of which is set at the same as that of c-SBF. The np-SBF were buffered by 2-(4-(2-Hydroxyethyl)-1-piperazinyl)ethane sulfonic acid (HEPES) and 1.0-M NaOH at pH 7.40 at 36.5 °C. The dissolving order of was modified. The HEPES and NaOH are dissolved first of all to keep the pH constant and $NaHCO_3$ was dissolved just before the pH adjustment as shown in Table 2 to prevent the volatilization of carbon dioxide (CO_2).

The nl-SBF, abbreviated from n-SBF prepared by liquid mixing process, was prepared by mixing the Ca and P solutions prepared separately in advance. It had same ion concentrations as np-SBF. As shown in Table 1, the Ca solution was designed to have twice higher Ca^{2+} concentration than that of human blood plasma and not to contain HPO_4^{2-} and HCO_3^- to prevent the precipitation of calcite, while the concentrations of the other ions and pH were designed to be equal to those of human blood plasma. The P solution was designed to have twice higher HPO_4^{2-} and HCO_3^- concentrations than those of human blood plasma and not to contain Ca^{2+}, while the concentrations of the other ions and pH were designed to be equal to those of human blood plasma. This liquid mixing process to prepare nl-SBF can prevent the radical increase in the local supersaturation due to the powder dissolution.

Analysis of SBFs Ion concentrations of the SBFs were measured by inductively coupled plasma (ICP) emission spectroscopy (IRIS Advantage, Nippon Jarrell-Ash. Co. Japan) and ion chromatograph (DX-100, Dionex Corporation, USA). The pH of the SBFs were measured by a pH meter (Mettler-Toledo MA-235, Germany).

Soaking in SBFs The bioactive $CaO-SiO_2-Na_2O$ glasses were immersed in 20 mL of c-SBF, np-SBF and nl-SBF at 36.5°C for 12h to 5 days, respectively. After removal from solution, specimens were washed with distilled water and dried in air.

Table 1. Nominal ion concentrations of SBFs in comparison with those of human blood plasma and Ca and P solutions for preparing nl-SBF

| Ion | Concentration/mM | | | | |
| | Blood Plasma | c-SBF | np-SBF | nl-SBF | |
				Ca solution	P solution
Na^+	142.0	142.0	142.0	142.0	142.0
K^+	5.0	5.0	5.0	5.0	5.0
Mg^{2+}	1.5	1.5	1.5	1.5	1.5
Ca^{2+}	2.5	2.5	2.5	5.0	0
Cl^-	103.0	147.8	103.0	103.0	103.0
HCO_3^-	27.0	4.2	4.2	0	8.4
HPO_4^{2-}	1.0	1.0	1.0	0	2.0
SO_4^{2-}	0.5	0.5	0.5	0.5	0.5

Table 2. Reagents, their purities and amounts for preparing 1000 mL of the SBFs

Reagent*	Purity/%	c-SBF	n-SBF		
			Total	Ca solution	P solution
1.0M-NaOH	---	---	40mL	20mL	20mL
HEPES**	>99.0	---	21.183	10.591	10.591
NaCl	>99.5	8.035	5.404	2.702	2.702
NaHCO₃	>99.5	0.355	---	---	---
KCl	>99.5	0.225	0.225	0.112	0.112
K₂HPO₄·3H₂O	>99.0	0.231	0.231	---	0.231
MgCl₂·3H₂O	>98.0	0.311	0.311	0.156	0.156
1.0M-HCl	---	39 ml	---	---	---
CaCl₂	>95.0	0.292	0.292	0.292	---
Na₂SO₄	>99.0	0.072	0.072	0.036	0.036
NaHCO₃	>99.5	---	0.355	---	0.355
TRIS***	>99.0	6.118	---	---	---
1.0M-HCl	---	~0.2mL	---	---	---
1.0M-NaOH	---		~4.8mL	~2mL	~3mL

*Listed in sequence of dissolution. **HEPES: 2-(4-(2-Hydroxyethyl)-1- piperazinyl) ethane sulfonic acid. ***TRIS: Tris-hydroxymethylaminomethane.

Results and discussion

Figure 1 shows the changes in pH of the solutions in the preparation process of SBFs. In the preparing process of c-SBF, the pH became quite low after adding aqueous 1.0M-HCl. This can prevent the precipitation of apatite, but lead to the volatilization of carbon dioxide. The large pH change was involved for the pH adjustment to 7.40. On the other hand, the pH was relatively constant in the preparation process of np-SBF, because the buffer was added from the beginning. The small pH change was involved for the pH adjustment. The Ca and P solutions for the preparation of nl-SBF also showed the similar pH changes to np-SBF.

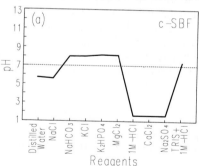
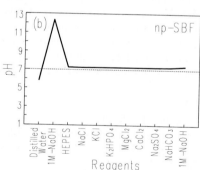

Fig. 1 The changes in pH of the solutions in the preparation processes of (a)c-SBF and (b)np-SBF.

All the as-prepared SBFs showed clear solutions without any precipitation. They kept clear even after the storage for 8 weeks at 36.5°C.

Figure 2 shows ion concentration and pH of SBFs as-prepared and after stored for 4 weeks, in comparison with those of human blood plasma. The as-prepared c-SBF showed ion concentrations almost equal to nominal values listed in Table 1, excepting those of Cl⁻ and HCO₃⁻. The measured Cl⁻ concentration was higher and HCO₃⁻ was lower than the nominal values. This is because the Cl⁻ concentration derived from 40 mL of aqueous 1.0M-HCl to lower the pH before dissolving CaCl₂ was not included in the nominal value but in the measured concentration, and that a portion of HCO₃⁻ volatilized as CO₂ gas when the pH became quite low after adding 40 mL of aqueous 1.0M-HCl in the preparing process of c-SBF. On the other hand, the as-prepared np-SBF and nl-SBF showed ion concentrations almost equal to nominal values, including those of Cl⁻ and HCO₃⁻. This indicates that the modified preparing process can prevent the volatilization of CO_2 during the preparing process. There was little difference in ion concentrations between np-SBF and nl-SBF. The c-SBF and both n-SBFs kept their ion concentrations and pH unchanged over 4 weeks at 36.5°C as shown in Figure 2.

The several bioactive CaO-SiO₂-Na₂O glasses with different compositions were soaked in the SBFs, respectively, in order to compare their apatite-forming ability. The bioactive glasses formed apatite on their surfaces within the same period in both n-SBFs and c-SBF. This indicates that the n-SBFs can be used for in bioactivity assessment of artificial materials.

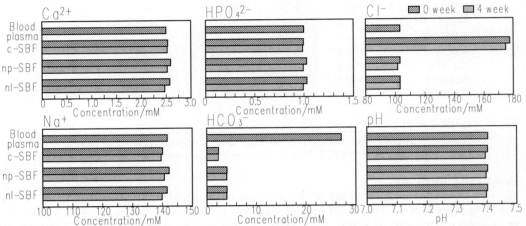

Fig. 2 The ion concentration and pH of the c-SBF, np-SBF and nl-SBF as-prepared and after stored for 4 weeks, in comparison with those of human blood plasma.

Conclusions

The np-SBF and nl-SBF with ion concentrations equal to those of human blood plasma, excepting those of HCO₃⁻, the concentration of which is equal to nominal values, can be successfully prepared. The n-SBFs kept their ion concentrations and pH unchanged over 4 weeks. The bioactive glasses showed almost same bioactivity in both n-SBFs and c-SBF. Both np-SBF and nl-SBF are believed to produce apatite under quite similar condition to the body environment, thereby being useful for *in vitro* assessment of bioactivity of artificial material and for biomimetic synthesis of bonelike apatite.

References

[1] T. Kokubo, H. Kushitani, S. Sakka, T. Kitsugi, S. Kotani, K. Oura and T. Yamamuro, *Bioceramics 1* (Ishiyaku Euro America, Tokyo 1989) p. 157-162.

[2] T. Kokubo, H. Kushitani, S. Sakka, T. Kitsugi and T. Yamamuro: J. Biomed. Mater. Res. 24 (1990) 721-734.

[3] H.-M. Kim, K. Kishimoto, F. Miyaji, T. Kokubo, T. Yao, Y. Suetsugu, J. Tanaka, T. Nakamura: J. Mater. Sci.:Mater. Med. 11 (2000) 421-426.

[4] J.L. Gamble: *Chemical anatomy, Physiology and Pathology of Extracellular Fluid, 6th Ed.*, (Harvard University Press, Cambridge 1967).

[5] A. Oyane, H.-M. Kim, T. Furuya, T. Kokubo, T. Miyazaki, T. Nakamura: J. Biomed. Mater. Res. 65A (2003) 188-195.

Key Engineering Materials Vols. 254-256(2004) pp. 119-122
online at http://www.scientific.net
© *2004 Trans Tech Publications, Switzerland*

Preparation and Properties of Zinc Containing Biphasic Calcium Phosphate Bioceramics

Andrea M. Costa[1], Gloria A. Soares[1], Reinaldo Calixto[2] and Alexandre M. Rossi[3]

[1] Dep. of Metal. and Materials Eng., UFRJ, P.O.Box 68505, Rio de Janeiro, 21941-972, RJ, Brazil, andrea@metalmat.ufrj.br

[2] Inst. Química, PUC/RJ, Rio de Janeiro, 21941-590, RJ, Brazil

[3] CBPF, Rua Dr. Xavier Sigaud, 150, Rio de Janeiro, 22290-180, RJ, Brazil

Keywords: hydroxyapatite, zinc, nanocomposite.

Abstract. Calcium deficient hydroxyapatites, CaDef-HA, have been synthesized with content of zinc varying from 1-10% mol. Zinc inhibits the apatite precipitation and reduces its crystal dimensions. Structural characterization showed that the Ca substitution by Zn reduces the hydroxyapatite thermal stability. When sintered at 1000 °C the Zn containing hydroxyapatite decomposes into a $Ca_{19}Zn_2((PO_4)_{14}$, which is more soluble than CaDef-HA. The resulting bioceramics composite may be an appropriate system for zinc liberation *in vivo* applications.

Introduction

Recent works demonstrated that osteoconductivity of calcium phosphate ceramics could be improved if these materials were doped with zinc. This metal stimulates osteogenesis by increasing bone proteins and alkaline phosphatase activity. Composites of ZnTCP and TCP (or apatite) cement with variable amounts of Zn have been processed and used as Zn carriers *in vitro* and *in vivo* studies [1-3]. These studies revealed that the release of Zn by these composites might improve human osteoblastic cells proliferation and stimulate new bone formation when implanted in femora of rabbits.

In this work we propose an alternative method to process biphasic calcium phosphate ceramics containing Zn. It consists in the preparation of a calcium deficient hydroxyapatite with a controlled amount of zinc and its sintering at temperatures higher than 1000° C. The resulting biphasic ceramics is composed by ZnTCP, $Ca_{19}Zn_2((PO_4)_{14}$, and a Zn doped hydroxyapatite, $Ca_{10-x}Zn_x(PO_4)_6(OH)_2$, CaZnHA. The structural characteristics of the composite and its dissolution behavior are being discussed in this work.

Materials and Methods

Calcium-deficient hydroxyapatite, CaDef-HA, was synthesized from drop wise addition of a $(NH_4)_2HPO_4$ aqueous solution to a $Ca(NO_3)_2$ solution at 90°C, pH = 9,0. After the addition, the solution was stirred for 4 hours at the same temperature. The precipitate was separated by filtration, repeatedly washed with deionized boiling water and dried at 100°C for 24 h. The synthesis of the CaDef-HA doped with 1, 5 and 10 % mol of Zn followed the same procedure as described before but solutions of $Zn(NO_3)_2$ and $Ca(NO_3)_2$ were used. Calcium, phosphorous and zinc contents were estimated by ICP-OES. XRD and FTIR spectroscopy were used to characterize sample mineral composition, crystallinity, lattice parameters, crystal dimension and carbonate content. Crystallite morphology were studied by transmition electron microscopy (TEM). Sample in powder and in tablets were sintered at different temperatures between 700 °C and 1000 °C, respectively.

Dissolution experiments in Milli-Q water using non heated and 1000 °C heated CaDef-HA and CaZnHA samples, in powder and tablets, were carried out in triplicate. Samples were mechanically shaken in 40 ml tubes during 7 days, and then collected, filtered using a 0.22 μm Durapore membrane Millipore and diluted in HNO_3 0.25%. The Ca and P content were then determined by ICP-OES.

Results and Discussion

Data taken from chemical analyses, Table 1, showed that non-doped sample was a calcium deficient hydroxyapatite with a Ca/P ratio of 1,60. The incorporation of Zn into the CaDef-HA structure increased the (Ca + Zn)/P molar ratio to values of 1.63, 1.64 and 1.69 for samples with 1, 5 and 10% mol of Zn, respectively. This improves in the CaZnHA stoichiometry may be attributed to the elimination of H_2O, carbonates groups and other structural defects. FTIR analyses, Fig.1a, reinforced this hypothesis because the intensities of the OH (630 and 3570 cm⁻1) and CO_3 bands (1450 and 870 cm⁻¹) decreased with the increase of the Zn content.

Table 1: Chemical composition of CaDef-HA and CaZnHA samples.

Sample	(Ca+Zn)/P	Ca/P	%Zn
0%	1.60	1.61	0
1%	1.63	1.55	1.1
5%	1.64	1.51	5.8
10%	1.69	1.35	11.7

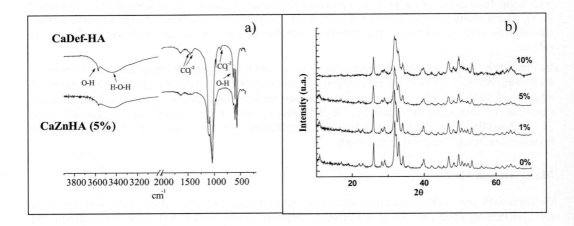

Fig 1: a) FTIR spectra of CaDef-HA and CaZnHA with 5% mol of Zn; b) DRX pattern of CaDef-HA and CaZnHA doped with 1,5 and 10 % mol of Zn.

XRD analysis showed, Fig.1b, that no other phosphate phase or zinc compound besides hydroxyapatite was produced. The unit cell parameters a=b and c varied from 0.9435 nm to 0.9422 nm and from 0.6885 nm to 0.6855 nm, respectively, for non doped to 5% zinc doped sample indicating that Zn^{2+} substitutes Ca^{2+} in the apatite structure. Zinc inhibits the apatite crystal growth because sample crystallinity and crystal mean size were strongly reduced with the incorporation of zinc into the CaDef-HA lattice. This reduction on the crystal dimensions produces a strong increase on sample surface area from 49 to 112 m^2/g and a variation on sample dissolution properties. The TEM analysis shown in Fig. 2 illustrates the diminution of the apatite crystal dimensions in a sample doped with 5 % of Zn in relation to the non-doped one.

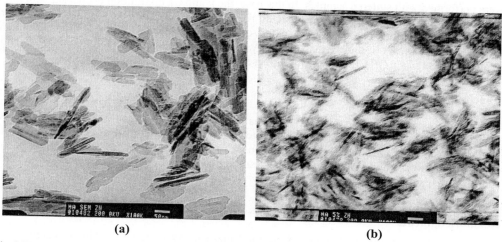

(a)   (b)

Fig.2 TEM images of a) CaDef-HA and b) CaDef-HA with 5% of Zn (magnification: 100000x)

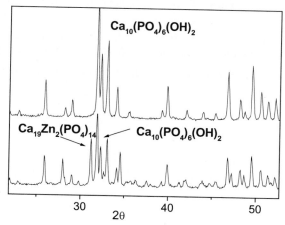

Fig. 3: DRX pattern of a) a non sintered CaDef-HA and b) a 1000 °C sintered
CaDef-HA with 5% of Zn.

Thermal treatment at 1000 °C induces the decomposition of CaDef-HA and the formation of a high crystalline β-CaTCP. The incorporation of zinc makes the apatite structure less stable than CaDef-HA. Depending on the apatite zinc content, thermal treatments at temperatures higher than 800 °C can induces the CaDef-HA decomposition and the formation of a calcium and zinc phosphate,

$Ca_{19}Zn_2(PO_4^3)_{11}$ with the same cation/anion ratio (1.3) as the β-CaTCP, Fig.3. Both phosphates are solid solutions of calcium and zinc. The decomposition of CaZnHA into CaZnTCP was confirmed by FTIR spectra because a strong band broadening in the phosphate region was observed. This effect increased with the Zn content.

Dissolution experiments in Milli-Q water using non-heated powder samples showed that zinc improves the P dissolution and Ca dissolution. This behavior changed when dissolution experiments were performed with sintered samples. In this case, the P and Ca dissolution decreased with the zinc content. The formation of a zinc hydroxide on the apatite surface after the thermal treatment could explain this behavior. The zinc release to solution was enhanced due to the dissolution of CaZnTCP, which is more soluble than CaZnHA. Therefore, its released to solution depended on the relative content of CaZnTCP and CaZnHA phases after the thermal treatment.

Conclusions

The above results suggested that thermal treatments on CaZnHA could be used to produce a biphasic ceramics with zinc uniformly distributed in both compounds: a CaZnHA and a CaZnTCP. The composition of this new biphasic composite may be controlled by choosing an adequate sintered temperature and sintered time. The resulting composite could be an alternative to ZnTCP and TCP or apatite cement used *in vivo* studies.

Acknowledgments

The work received a financial support from the Brazilian agency CAPES. This research is part of Millenium Institute for Tissue Bioengineering (IMBT), supported by CNPq.

References

[1] Ishikawa, K., Miyamoto, Y., Yuasa, T., Ito, A., Nagayama M., and Suzuki, K., Biomaterials, 23 (2002) 423-428.

[2] Ito, A., Kawamura, H., Otsuka, M., Ikeuchi, M., Ohgushi, H., Ishikawa, K., Onuma, K., Kanzaki, N., Sogo, Y. and Ichinose, N., Materials Science and Engineering C, 2002, 22: 21–25.

[3] Mayer I., Apfelbaum, F. and Featherstone J. D. B., Archs Oral Biol., 1994, 39, No1, pp 87-90.

Key Engineering Materials Vols. 254-256(2004) pp. 123-126
online at http://www.scientific.net

Effects of Lead on Calcium Phosphate Ceramics Stability

E.Mavropoulos[1], Nilce C.C.Rocha[2], Gloria A. Soares[3], Josino C.Moreira[4], and Alexandre M. Rossi [1]

[1] CBPF, Rua Dr. Xavier Sigaud, 150, Rio de Janeiro, 22290-180, RJ, Brazil, Elena@cbpf.brl

[2] Inst. Química, PUC/RJ, Rio de Janeiro, 21941-590, RJ, Brazil

[3] Dep. of Metal. and Materials Eng., UFRJ, P.O.Box 68505, Rio de Janeiro, 21941-972, RJ, Brazil,

[4] CESTEH-FIOCRUZ, R. Leopoldo Bulhões 1480, Manguinhos, Rio de Janeiro, RJ, Brazil

Keywords: Hydroxyapatite, lead, bone.

Abstract. Lead is easily incorporated in bone tissue and especially into its inorganic phase. The mechanism of lead incorporation in non-sintered and sintered synthetic deficient hydroxyapatite is studied. Aqueous solution containing Pb^{2+} promotes the hydroxyaptite dissolution by lowering its surface pH. A $Pb_{(10-x)}Ca_x(PO_4)_6(OH)_2$ and afterwards a $Pb_{10}(PO_4)_6(OH)_2$ is precipitated in sintered samples. Carbonate impurities and β-TCP can improves the formation of the $Pb_{(10-x)}Ca_x(PO_4)_6(OH)_2$.

Introduction

The interaction of metal ions with the bioceramic surface has been extensively studied because metal contamination can modify the bioceramic behavior when in contact with bone tissue [1]. Lead is one of the most toxic metals and bone is the final deposit for body lead. Recent works revealed that binding properties of bone proteins to hydroxyapatite, HA, can be affected by the presence of lead [2]. Several works studied the mechanism of Pb^{2+} removal by HA in aqueous media [3-4]. XRD analysis associated with Rietveld methodology demonstrated that lead immobilization by HA is controlled by HA dissolution and the formation of a crystalline solid solution of $Pb_{(10-x)}Ca_x(PO_4)_6(OH)_2$. Lead ions occupy preferentially the Ca(II) sites in the hydroxyapatite structure. During the reaction the new solid phase is continuously enriched by lead until its transformation into a pure lead phosphate apatite, $Pb_{10}(PO_4)_6(OH)_2$. Combination of ICP data and XRD results permitted the conclusion that lead ions are also adsorbed at HA surface. The resulting solids, CaPbHA and lead coated HA, were less soluble than the original HA.

In this work, we studied the effects of lead on the dissolution behavior of a calcium deficient hydroxyapatite and a carbonated apatite synthesized in different conditions and submitted to thermal treatments.

Materials and Methods

Stoichiometric hydroxyapatite, HA, calcium deficient hydroxyapatite, CaDef-HA, and B type carbonated apatite, CBHA, were prepared by drop wise addition of a $(NH_4)_2HPO_4$ aqueous solution to a $Ca(NO_3)_2$ or to a $Ca(NO_3)_2$ and a $Ca(CO_3)_2$ solution. Precipitation temperature was varied from 25°C to 90 °C, pH from 7 to 12, digestion time from 2 to 20 hours, addition velocity from 70 to 250 mL/h, Ca and P concentration from 0.18 to 2.0 $M.L^{-1}$ and 0.12 to 0.6 $M.L^{-1}$, respectively. The precipitates were separated by filtration and dried at 100 °C for 24 hours. The dried powders were grounded and the < 210 μm particles were separated by sieving. Aliquots of HA, CaDef-HA and CBHA samples were heated in air-atmosphere at a heating rate of 5 °C/min to temperatures up to 1000 °C. Samples were analyzed by X-ray diffractometry (XRD) and Infra Red spectroscopy

(FTIR). The Ca^{2+} and PO_4^{3-} contents and sample surface area (BET) were determined by inductively coupled plasma (ICP-OES) and N_2 adsorption-desorption isotherms, respectively. The samples (0.1 g) were added to 40 ml of $Pb(NO_3)_2.4H_2O$ aqueous solutions containing Pb^{2+} initial concentrations varying from 0.08 to 4.0×10^{-3}M. After 3, 30, 60, 180, 300, 540, 1320, 1860, 2880, 4320 and 4800 minutes, aliquots of the supernatant solution were collected, centrifuged and filtered. Three aliquots of the filtrated solution were analyzed by ICP-OES in order to determine the PO_4^{3-}, Ca^{2+} and Pb^{2+} concentrations. Dissolution experiments in Milli-Q water with hydroxyapatite and carbonated apatite samples were also performed under similar procedure. The pH was monitored during the dissolution experiments by using a pH meter Analyser-300 M.

Results and Discussion

Hydroxyapatite samples with different stoichiometries were dissolved in aqueous solutions containing a Pb^{2+} initial concentration up to 4.0×10^{-3} M. Lead was immobilized by all samples with an effectiveness that depends on apatite characteristics, solution pH, metal initial concentration and sorption time. For lead initial concentration of 4.2×10^{-4} M lead was completely removed from solution after 5 hours of reaction, Fig 1. The release of calcium to the solution was directly proportional to the lead uptake as shown in Fig. 1, whereas P was not detected in solution during the process. After dissolution in presence of lead, HA and CaDef-HA samples were studied by XRD. The DRX pattern showed the formation of the $Pb_{(10-x)}Ca_x(PO_4)_6(OH)_2$ and the $Pb_{10}(PO_4)_6(OH)_2$ for the early and later stages of dissolution, respectively, Fig.2. The solution Ca/P molar ratio of CaDef-HA increased when its dissolution occurred in presence of lead because the CaPbHA precipitation consumes phosphate ions. It was a very fast kinetical process: 4.2×10^{-4} M of lead were completely consumed in the first three minutes of reaction in order to precipitate CaPbHA. The solution pH was an important parameter that influenced Ca and P dissolution. In presence of lead the pH tends to decrease, which improves the CaDef-HA dissolution and the formation of the CaPbHA phase.

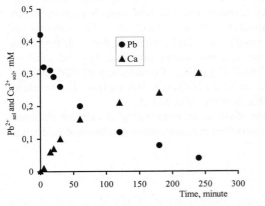
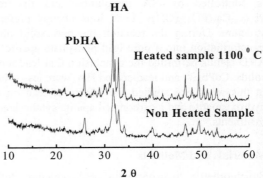

Fig.1: Ca^{2+} and Pb^{2+} in solution. Lead sorption by HA, $[Pb^{2+}]$ of 4.2×10^{-4}M.

Fig.2: XRD of CaDef-HA after Pb^{2+} sorption. $[Pb^{2+}]$ of 4.2×10^{-4}M.

Analyses by transmission electron microscopy (TEM) confirmed the XRD results and revealed that HA crystals had needle shape morphology and dimensions smaller than 150 nm before the HA dissolution. During the dissolution process, the crystal dimensions of the new CaPbHA phase increased to values up to 200 nm. The relative Pb/Ca content of CaPbHA increased with the dissolution time as confirmed by EDS analysis (results not shown). At the end of 24 hours, large particles with more than 1500 nm in length and 200 nm in width were observed, as shown in Fig. 3. These particles were mainly composed by Pb and P and their number increased with the dissolution time.

 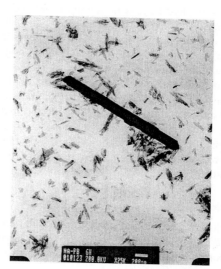

Fig.3:TEM images of a) HA sample a) before and b) after a 6 hours dissolution in an aqueous solution containing 8×10^{-4} M of lead. Magnification of a) 80000 x and b) 25000x.

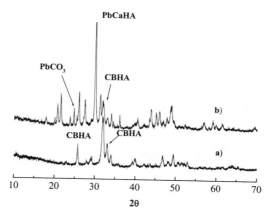

Fig.4:DRX of a CBHA a) before and b) after lead sorption. [Pb] of. 4.7×10^{-3} M

Dissolution experiments in presence of lead were also performed with eighteen hydroxyapatite samples with Ca/P molar ratio ranging from 1.53-1.75 and surface BET area from 27- 93 m^2/g. It was verified that the effectiveness of the CaPbHA precipitation was not directly controlled by Ca/P molar ratio or by apatite surface area but by the phosphorous dissolution rate of each one of the HA samples. Similar dissolution experiments as described before were performed using B type carbonated apatite samples, CBHA, containing 1-12 % weight of CO_3^{2-} ions. It was observed that the precipitation of the CaPbHA was improved when CBHA was used in the dissolution experiments containing lead but carbonate ions were not incorporated into this new structure. XRD analysis confirms that carbonate ions react with lead and precipitate the $PbCO_3$, as shown in Fig.4.

Stoichiometric hydroxyapatite samples sintered at 900 °C were dissolved in aqueous solutions without and with 8.05 $x10^{-4}$ M of Pb^{2+}. The Pb^{2+} immobilization rate by sintered samples was strongly decreased in relation to non sintered ones: sintered samples needed 80 hours to immobilized 1,87 $x10^{-3}$ M of lead while non heated samples spent only three minutes to immobilized the same amount of lead.

Calcium deficient samples decomposed into a β-TCP when sintered at temperatures higher than 900 °C. These biphasic samples were also submitted to dissolution in solutions containing lead. The chemical and XRD analysis, Fig.2, suggested that the interaction of lead with sintered CaDef-HA followed the same mechanism as non-heated samples. The PO_4^{3-} and Ca^{2+} from CaDef-HA dissolution react with Pb^{2+} and produce a CaPbHA. The effectiveness of lead removal from aqueous solution was enhanced when sintered CaDef-HA was used because β-TCP was more soluble than hydroxyapatite.

Conclusion

In conclusion, we found that the dissolution behaviors of hydroxyapatite and carbonated hydroxyapatite submitted or not to thermal treatments were strongly affected by lead. Phosphorous and carbonate from dissolution were consumed in the formation of less soluble mineral phases like CaPbHA and $PbCO_3$.

Acknowledgements

This work received financial support from the Brazilian agency CAPES. This research is part of Millenium Institute for Tissue Bioengineering (IMBT), supported by CNPq.

References

[1] Hallab, N J., Vermes,C., Messina,C., Roebuck,K.A., Glant,T.T., Jacobs,J.J., *J.* Biomed. Mater. Research,60 (3), 2002,420-433.

[2] Dowd Tl, Rosen, J.F., Mints, L., Gundberg,C.M., Biochim Biophys Acta 14, 1535 (2),2001, 153-63.

[3] Mavropoulus, H., Rossi A.M., Costa, A.M. Peres, Moreira J. C., Saldanha. M., Environmental Science and Technology, V 36, (7), (2002), 1630-1635.

Key Engineering Materials Vols. 254-256(2004) pp. 127-130
online at http://www.scientific.net
© 2004 Trans Tech Publications, Switzerland

Mg-substituted Tricalcium Phosphates: Formation and Properties

R. Z. LeGeros[1], A. M. Gatti[2], R. Kijkowska[3], D.Q. Mijares[4], J.P. LeGeros[5]

[1,4,5]New York University College of Dentistry, 345 East 24th Street, New York, New York 10010, rzl1@nyu.edu, dqm1@nyu.edu, jpl4@nyu.edu;
[2]University of Modena College of Medicine, Modena 41100 Italy, gatti@unimo.it
[3]Institute of Chemistry and Technology, Krakow, Poland, kij@chemia.pk.edu.pl

Keywords: Tricalcium phosphate, magnesium, dissolution, bone formation.

Abstract. This study aimed to investigate the formation and properties of magnesium (Mg)-substituted tricalcium phosphate, β-TCMP, its properties and potential as biomaterial for bone repair. β-TCMPs were prepared and characterized using x-ray diffraction, FT-IR and SEM. Dissolution properties were determined in acidic buffer. β-TCMP discs were implanted in surgically created holes in femoral and tibial diaphyses of rabbits.
Results demonstrated that the formation of β-TCMP and Mg incorporation in β-TCMP were dependent on reaction pH, temperature and solution Mg/Ca ratios. Sintered β-TCMP was significantly less soluble than β-TCP. Implanted unsintered β-TCMP showed osteoconductive properties associated with new bone formation. This study suggests that β-TCMP (sintered or unsintered), alone or in combination with other calcium phosphates, may be useful as biomaterials for bone repair and maybe useful in cases where slower biodegradation than that of β-TCP is desired.

Introduction

Pure beta-tricalcium phosphate (β-TCP), $Ca_3(PO_4)_2$, is obtained by sintering calcium-deficient apatite (Ca/P < 1.67) above 700°C or by solid-state reactions and cannot be obtained from solutions [1]. By itself or mixed with hydroxyapatite (HA) in biphasic calcium phosphate (BCP), β-TCP is commercially available as bioceramics for bone repair [1-3]. Magnesium (Mg)-substituted tricalcium phosphate (β-TCMP), $(Ca,Mg)_3(PO_4)_2$, occurs in pathological calcifications [1,4] and can be obtained from synthetic solutions by precipitation [1,5] or by hydrolysis of dicalcium phosphate dihydrate (DCPD), $CaHPO_4.2H_2O$, or DCPA, $CaHPO_4$ in solutions with varying Mg/Ca molar ratios [1,5,6]. β-TCMP (also referred to as whitlockite) was shown to stimulate cell proliferation and stimulate synthesis and secretion of collagenase [7]. Preliminary studies demonstrated that β-TCMP was more biodegradable than HA and F-substituted apatite [8]. The purpose of this study was to determine the factors affecting the formation of synthetic β-TCMP, its properties and ability to promote bone formation.

Materials and Methods

β-TCMPs were prepared by precipitation and hydrolysis methods. Precipitation method consisted of drop wise addition of calcium (Ca) or (Ca+Mg) solution to phosphate solution maintained at 90°C [1,5]. Hydrolysis method consisted of suspending DCPD in solutions of varying Mg/Ca molar ratio [1,6]. Initial reaction pH: 5, 7.5 and 9, reaction temperature, 90°C. The preparations were washed and dried, then characterized using x-ray diffraction, XRD (Philips X'Pert, The Netherlands), FT-IR (Nicolet 500), scanning electron microscopy (JEOL

500). β-TCP (unsubstituted) was prepared by sintering Ca-deficient apatite. Dissolution was determined in acidic buffer (0.1M KAc, pH 6, 37°C, solid/solution = 100 mg/25 ml buffer). The release of Ca ions onto the buffer with time (60 min) was monitored using a Ca-ion selective electrode.

For the *in vivo* study, β-TCMP discs were prepared, sterilized with ethylene oxide and implanted in surgically created holes (4mm dia) in femur and tibial diaphyses of two rabbits for 30 and 60 days [8]. After sacrifice the materials and surrounding tissues were explanted, fixed in 4% paraformaldehyde, dehydrated in ascending alcohol concentration and embedded in PMMA. Sections (200mm) were obtained using a diamond saw. (Accutom, Struers, Denmark) and prepared for SEM analysis (SEM XL40 Philips, The Netherlands) with energy dispersive system (EDAX) attachment for Ca and P analyses.

Results

Depending on the solution Mg/Ca molar ratio, precipitate or hydrolysis products were either biphasic calcium phosphates (BCP) consisting of mixtures of apatite and β-TCMP or of only β-TCMP. The incorporation of Mg was reflected in the shift in the x-ray diffraction peaks (Fig. 1) due to partial Mg-for-Ca substitution in tricalcium phosphate, $(Ca,Mg)_3(PO_4)_2$, causing a contraction in the unit cell dimension. Mg-incorporation in β-TCMP was dependent on solution Mg/Ca and on reaction pH. For the same solution Mg/Ca ratio, higher Mg incorporation was obtained at higher pH (7.5 and 9) compared to that obtained at low pH (pH 5). β-TCMP crystals obtained at pH 5 were larger than those obtained at pH 7.5 as shown by SEM (Fig. 2) and XRD analyses.

The extent of dissolution of sintered β-TCMP was considerably lower than that of β-TCP (Fig. 3). For this comparison, β-TCMP (at 950°C) and β-TCP obtained by sintering (at 950°C) calcium deficient apatite were used. When unsintered β-TCMPs with varying amounts of Mg concentrations were exposed to the acid buffer, the extent of dissolution was observed to increase with increasing Mg concentration. The extent of dissolution of unsintered β-TCMP was also observed to increase with decreasing crystal size.

In the *in vivo* study, the retrieved implanted unsintered β-TCMP discs have fragmented and each fragment was completely surrounded by new bone (Fig. 4). An intimate contact between the new bone and the β-TCMP fragment is observed. After 60 days, EDS of the β-TCMP surface showed similar Ca/P ratio as that of the new bone.

Discussion

In the absence of these ions, formation of pure tricalcium phosphate is not possible. For this reason, all tricalcium phosphate compounds occurring in some pathological calcifications (e.g. dental calculus, urinary stones, lungs, articular cartilage, etc) are Mg-substituted [1, 4]. This study confirmed results of previous studies that the presence of Mg in synthetic systems allows the formation of Mg-substituted tricalcium phosphate [1,5].

This study demonstrated that Mg incorporation in β-TCMP makes it more stable against acid dissolution compared to unsubstituted β-TCP (Fig. 3). However, comparing the dissolution properties β-TCMP containing different amounts of Mg showed that increasing amount of Mg increases the extent of β-TCMP dissolution. This maybe related to the effect of Mg on the crystal size of β-TCMP: the greater the incorporation, the smaller the crystal size.

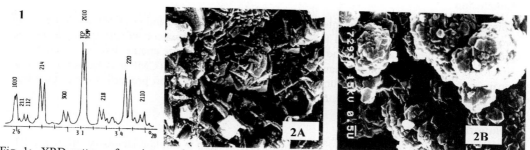

Fig. 1: XRD pattern of a mixture of unsubstituted (β-TCP) and Mg-substituted (β-TCMP) tricalcium phosphate. Partial Mg-for-Ca substitution causes shifts in the diffraction peaks reflected the decrease in the a-axis dimensions.

Fig. 2: SEM of β-TCMP crystals showing the effect of pH on crystal size: (2A) obtained from solution with pH 5, (2B) obtained from solution with pH 7.5. The Mg-incorporation in 2A is lower than that in 2B.

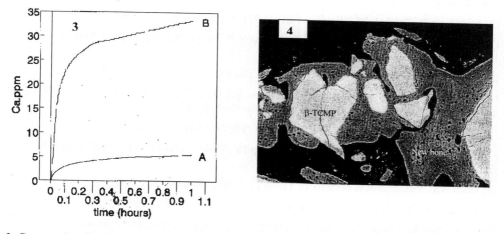

Fig. 3: Comparative dissolution β-TCP and sintered (at 950°C) β-TCMP in acidic buffer. The β-TCP was obtained by sintering (at 950oC) a calcium-deficient apatite (Ca/P = 1.5).

Fig. 4: SEM of β-TCMP surrounded with new bone after 60-day implantation in surgically created hole in rabbit femoral diaphysis.

Results of *in vivo* study demonstrated that unsintered β-TCMP was osteoconductive, i.e., serving as a scaffold or template for new bone formation. An apparent direct attachment of β-TCMP with the new bone is similar to that observed for carbonatehydroxyapatite, F-substituted apatite and calcium-deficient apatite [8]. This direct attachment is attributed to the bioactive properties of β-TCMP similar to those of other calcium phosphate biomaterials [9]. The similarity in the Ca/P ratio of the new bone compared to that of the β-TCMP after implantation suggest its transformation to bone-apatite like (i.e., carbonatehydroxyapatite) mineral similar to transformation observed with other calcium phosphates resulting from partial dissolution and reprecipitation [10].

Conclusion

This study demonstrated that unsintered BCP consisting of apatite and β-TCMP could be easily prepared. Furthermore, unsintered β-TCMP has bioactive and osteoconductive properties and promotes new bone formation and is itself transformed to bone-apatite like material. Since sintered β-TCMP is less soluble than β-TCP, it is suggested that β-TCMP may be useful in bone repair applications requiring calcium phosphate with biodegradation properties lower than that of β-TCP but higher than that of sintered HA.

Acknowledgements

This study was supported in part by L. Linkow Professorship in Implant Dentistry, NIDCR/NIH grant no. 12388 and by the Ministry of University and Scientific Research and Technology (MURST) of Italy.

References

[1] LeGeros RZ (1991). *Calcium Phosphates in Oral Biology and Medicine.* Monographs in Oral Sciences. Vol. 15. Karger, Basel.

[2] Metzger SD, Driskell TD, Paulsrud JR (1982). J Am Dent Assoc 105:1035.

[3] Schwartz C, Liss P, Jacquemaire B, Lecestre P, Frayssinet P (1999). J Mat Sci: Mat Med 10:821.

[4] Scotchford Ca, Ali SY (1995). Ann Rheum Dis 54: 339.

[5] LeGeros RZ, Daculsi G, Kijkowska R, Kerebel B (1989). In: Itokawa Y, Durlach J (eds).: *Magnesium in Health and Disease.* Libbey, New York, pp 11-19.

[6] Daculsi G, LeGeros RZ, Jean A, Kerebel B (1987). J Dent Res 66:1356.

[7] Ryan LM, Cheung HS, LeGeros RZ, Kurup IV, Toth J, Westfall PR< McCarthy GM (1999). Calcif Tiss Int 65: 374.

[8] Gatti AM, LeGeros RZ, Monari E, Tanza D (1998). Bioceramics 11:399.

[9] LeGeros RZ (2002). Clin Orthoped Rel Res 395: 81.

[10] LeGeros RZ, Daulsi G, Orly I, LeGeros JP (1991). In: *The Bone Biomaterial Interface.* University of Toronto Press, Toronto, pp 76-88.

Key Engineering Materials Vols. 254-256(2004) pp. 131-134
online at http://www.scientific.net
© 2004 Trans Tech Publications, Switzerland

Apatite Formation on HA/TCP Ceramics in Dynamic Simulated Body Fluid

Y.R. Duan[1,2,], Z.R.Zhang*[1], C.Y. Wang[2,3], J.Y. Chen[2], X.D.Zhang[2]

1.West China School of Pharmacy, Sichuan University, Chengdu 610041, China

yourongduan_2001@sina.com

2.Engineering Research Center in Biomaterials, Sichuan University, Chengdu 610064, China

jychen@scu.edu.cn

3.Yangtze River Fisheries Institute Jingzhou,Hubei Prov.434000, P.R.China

* Corresponding Author

Keywords: apatite, HA/TCP ceramics, dynamic, simulated body fluid (SBF)

Abstract. In vitro study of biomaterials in SBF is a very important approach to understand the bioresponse of implants *in vivo*. This study aimed at exploring the effect of dynamic SBF flowing at normal physiological rate of body fluid in skeletal muscle upon the formation of bone-like apatite on the surface of pores and dense HA/TCP calcium ceramics. Results demonstrated that in normal physiological rate, the surface of dense ceramic can not be found the bone-like apatite. Results showed that bone-like apatite formation could only be found in the internal pores of the materials when SBF flowing at physiological rate was coordinated with that of in vivo implantation of calcium phosphate ceramics: most of the ectopic bone formation was detected inside the pore of the porous calcium phosphate ceramics and no new bone was found on surface of dense ceramics. This result demonstrated that dynamic model used in this study was better than usually static immersion model in mimicking the physiological condition of in vivo formation of bone-like apatite. Dynamic SBF method is very useful for understanding the *in vivo* formation of bone-like apatite and the mechanism of ectopic bone formation of calcium phosphate ceramics.

Introduction

Calcium phosphate biomaterials were reported to be osteoinductive; they can induce bone formation in extraskeletal sites without additional osteogenic cells or bone morphogenetic proteins. An essential requirement for an calcium phosphate ceramics to induce bone form is the formation of a bonelike apatite layer on its surface in body environment. Bone-like apatite formation on implants is believed to be crucial to the osteoinductivity of calcium phosphate ceramics [1]. *In vitro* immersion study can simplify the complicated interaction of many factors *in vivo* to study single factor's contribution to bone-like apatite formation on calcium phosphate ceramics so as to explore the mechanism of osteoinduction. Dynamic SBF is better in mimicking the living body fluid than static SBF to show the relation between apatite layer formation and osteoinduction in biomaterials than that from static SBF immersion experiment *in vitro*. In this work, we investigated the apatite formation on the surfaces of porous and dense hydroxyapatite (HA)/tricalcium phosphate (TCP) (HA/TCP, 70/30) in SBF flowing at physiological rate.

Materials and methods

Fabrication of porous materials: Porous are cylinders of 4 mm diameter×8 mm length. They were prepared in our laboratory by foaming with H_2O_2 and sintering at 1200 ℃. The porosity was 50 ~ 60 % and pores were interconnected. Dense calcium phosphate ceramics disks with diameter of 3mm was fabricated by pressing HA/TCP powder in 15 MPa for 3 min, the green body was then sintered at 1200℃. Disks were sandblasted with Al_2O_3 particles, the surface roughness was 3.5-4.0 μm.

Preparation of SBF: One liter of SBF was prepared by dissolving NaCl 7.995g, NaHCO$_3$ 0.353g, KCl 0.224g, K$_2$HPO$_4$ .3H$_2$O 0.228g, MgCl$_2$·6H$_2$O 0.305g, CaCl$_2$ 0.227g, and Na$_2$SO$_4$ 0.0710g into distilled water and adjusting pH with Tris and HCl to 7.4 [2]. The temperature of SBF was maintained at 36.5±1 °C.

The equipment for *in vitro* circulation: The dynamic SBF equipment with circulation device is shown in Fig.1. The dynamic condition was realized by controlling the SBF flowing in/out of the sample chamber of 100ml. The flow rate of 2ml/100ml.min (*i.e.* SBF in the sample chamber of 100ml was refreshed at a rate of 2ml/min by flowing in/out) is closed to that in human muscle environment [3]. The SBF in storage tank was replaced with fresh one in every two days in order to keep the solution composition stable.

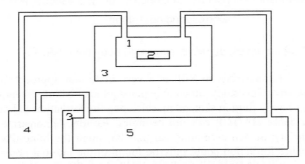

Fig.1 Schematic diagram of flow chamber system
1.Sample chamber; 2. Samples; 3. Water bath; 4. Pump; 5. SBF storage tank.

Immersion: Calcium phosphate ceramics materials were immersed in SBF in the dynamic equipment shown in Fig.1. The experimental conditions and methods of immersion as follow: immersion for 14 days in SBF flows at rate 2 ml/100ml.min and immersion for 14 days in static SBF.

Surface analysis: The morphology of the sample surfaces was examined by a scanning electron microscope (SEM). The chemical composition of the surface after immersion was determined by reflective infrared spectroscopy (RIR).

Results and Discussion

Dense HA/TCP ceramics: No significant changes were detected after the dense HA/TCP ceramics immersion in static or dynamic SBF for 14 days. A layer of newly formed materials can be found to cover the surface of the sandblasted dense HA/TCP after immersion in static SBF for 14 days (Fig.2b). The apatite can be also seen on the surface of dense HA/TCP ceramics immersion in dynamic 1.5SBF (the concentration of Ca^{2+} and HPO$_4^{2-}$ in 1.5 SBF is 150% of that in SBF) for 7 days (Fig.2c). Dense materials with a smooth surface have a relatively low solubility and thus the local concentration of Ca and P near the surface is relatively low. Owing to roughness of the surface, materials have a larger surface area and high solubility. Therefore, the local concentration of Ca and P is relatively higher when compared with smooth surfaces. As a result, the crystal nucleus formation on rough surface is faster than that on smooth surface. The local concentration of Ca and P is higher and the speed of nucleus formation is faster. Therefore, on the rough surface bone-like apatite formation is faster than on smooth surface. On the contrary, in dynamic SBF, ions on the surface of materials diffused rapidly and the concentration of Ca and P can not easily reach the threshold for nucleus formation or in another word, it will take a relatively long time to reach the threshold. In our experimental condition, there was no crystal nucleus formation on both rough and smooth material surface in dynamic SBF. When increase the concentration of Ca and P of SBF, nucleus can formed on rough surface. The results also demonstrated that local ion concentration in solution near the nucleation site on the samples played a key role in nucleation.

a b c

Fig 2 SEM picture of the dense surface of HA/TCP: (a) sandblasted surface before immersion; (b) sandblasted surface after immersion in static SBF for 14days(×10k); (c)dense surface after immersion in 1.5SBF flows at rate 2ml/min for 7 days (X1000).

Porous HA/TCP ceramics: Fig. 3 showed the examination results of porous HA/TCP ceramics by SEM. The apatite layer cannot be formed on the cylinder surfaces when SBF flows at rate 2 ml/100ml.min, but it appears on the pore wall (Fig.3c). The experimental results that apatite formation in both dynamic SBF and osteoinduction in animal experiment appeared only in the pores of samples [4] confirmed the direct relation between them. The coincidence between dynamic SBF and animal experiment results indicated that the dynamic SBF simulated biological environment better than static SBF. The result that the apatite formed on the surface (Fig.3b), when the flow rate was 0 ml/min, was in agreement with other results obtained from immersion studies in static SBF [5-6].

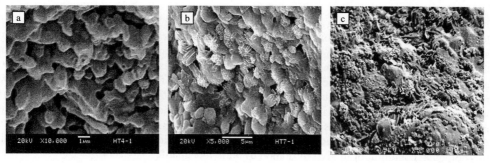

Fig.3 SEM micrographs of calcium phosphate ceramics before and after immersion for 14 days in SBF: (a) the outer surface of HA/TCP before immersion; (b) apatite formed on the surface in static SBF; (c) Apatite formed in the internal surface of pores in SBF flow at rate 2ml/100ml.min(X1000).

When ceramic materials contacted SBF, many events took place. The first step of adhesion was that Ca^{2+}, HPO_4^{2-}, PO_4^{3-} ions were released from the ceramic surface and entered the SBF solution. The ions were attracted by static electricity and adhered to the surface. The dissolved ions could supply sufficient Ca^{2+}, HPO_4^{2-}, PO_4^{3-} in microenvironment and resulted in crystal nucleation on the surface.Once crystal nucleus were formed on the surface of biomaterials, even distributed bone-like apatite could form. In our study, when SBF flowed at physiological rate of tissue fluid in muscle, crystal nucleus could not form on the surface of materials, but crystal nucleus could form inside the porous ceramics. Because SBF solution flowed slower inside the porous ceramics than outside, higher local concentration of Ca and P could be maintained inside the ceramics. This was why the apatite formed inside the porous calcium phosphate.

Microenvironment with a supersaturation concentration of calcium and phosphate ions was crucial for apatite to nucleate and grow in SBF. This result that bone-like apatite formed only on the walls of the porous calcium phosphate ceramics but not on the surface of the ceramics was consistent with the results observed in *in vivo* implantation experiment in animal body. Dynamic SBF is a good method to understand the mechanism of bone induction of calcium phosphate ceramics.

Fig.4 shows XPS C1s spectrum for sandblast dense HA/TCP after immersion in SBF flows at rate 2ml/100ml.min. XPS analysis results showed that the flake-like structure grown on the surface was composed of 4 elements, i.e Ca, P, C, O. C1s which was divided into 3peaks located in 384.6Ev, 286.5eV, 287.998eV and respectively corresponding to contamination carbonate, C-C, C-O and O-C-O bond.

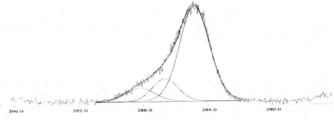

Binding energy/eV
Fig.4 XPS C1s spectra for HA/TCP after immersion in SBF

The surface structure was scraped and collected carefully after confirmed by SEM to have significant changes on the surface and then subjected to IR analysis. Result showed that there were specific peaks for CO_3^{2-} in 1458cm^{-1}, 1420cm^{-1} and 873cm^{-1}. The SEM photos, XPS and IR results of the flake-like structure are similar to bone apatite. so it is called bone-like apatite.

Conclusion

Results demonstrated that in normal physiological rate, on the surface of dense ceramics bone-like apatite can not be found. The results also demonstrated that local ion concentration in solution near the nucleation site on the samples played a key role in nucleation. Results showed that bone-like apatite formation could only be found in the internal pores of the materials when SBF flowing at physiological rate was coordinated with that of *in vivo* implantation of calcium phosphate ceramics: most of the ectopic bone formation was detected inside the pore of the porous calcium phosphate ceramics and no new bone was found on surface of dense ceramics. This result demonstrated that dynamic model used in this study was better than usually static immersion model in mimicking the physiological condition of in vivo formation of bone-like apatite.

Acknowledgment

This work was financially supported by the key basic research project of China, contract No.G1999064760

References

[1] de Bruijn JD, Yuan HP, van Blitterswijk CA et al. *Bone Engineering* edited by Davies JE.(Toronto, Canada), p.421-431.

[2] Kokubo T, Ito S, Yamamura T, et al. J. Biomed. Mater. Res. 24(1990), p.331.

[3] Wu X. *Normal Physiological Data of Human body*, (Scientific Dissemination Publishing Company, Beijing1987), p. 68.

[4] Zhang XD, Yuan HP, de Groot K. "Notebook: Workshop1#, Biomaterials With Intrinsic Osteoinductivity," The 6th World Biomaterials Congress. 2000.

[5] Kokubo T, Materials Science Forum. Vol. 293(1999), p. 65.

[6] Bouler JM, Daculsi G. Key Engineering Materials. Vol. 192-195 (2001), p. 119

Key Engineering Materials Vols. 254-256(2004) pp. 135-138
online at http://www.scientific.net
© 2004 Trans Tech Publications, Switzerland

Biocompatibility of Si-Substituted Hydroxyapatite

Jong-Heon Lee[1*], Kang-Sik Lee[1], Jae-Suk Chang[1], Woo Shin Cho[1]
Younghee Kim[2], Soo-Ryong Kim[2], and Yung-Tae Kim[1]

[1] Department of Orthopedic Surgery, ASAN Medical Center, Seoul 138-736, Korea
[2] Ceramic Building Materials Department, Korea Institute of Ceramic Engineering and Technology, Seoul 153-023, Korea
*Corresponding Author, e-mail: jongheon@amc.seoul.kr

Keywords: Si-substituted hydroxyapatite, bone substitutes, biocompatibility, cell proliferation, rabbit, implantation study

Abstract. The 2wt% Si-substituted hydroxyapatite (Si-HA) powder was prepared by a precipitation method and sintered to form a disk-shaped Si-HA (98% dense, 10 mm diameter and 1.5 mm thick) specimen at 1200°C for 3 hrs. After polishing the surfaces of specimen with the 2000-grit SiC metallographic paper, the cell proliferation experiment was performed as *in vitro* test at various times from 3 to 24 hrs. In addition, for *in vivo* study, four Si-HA specimens (1 mm x 1 mm x 10 mm) were implanted into the paravertebral muscle of a rabbit. As the results, symptoms of hemorrhage, necrosis and discolorations were not macroscopically observed on the surrounding areas of the implants, and typical foreign body reaction, without neutrophil, was revealed from microscopical observation. Therefore, the biocompatibility of Si-HA seems to be good and it is a feasible material as bone substitute.

Introduction

Hydroxyapatite (HA) has been of considerable interest as a biomaterial for implants and bone augmentations since its chemical composition is close to that of bone [1-3]. Previous studies have demonstrated that its bioactivity is strongly dependent on a number of both physical and chemical factors and is significantly enhanced by substitution of silicon (Si) ions into phosphate (P) sites in HA lattice [4]. The presence of silicon in HA has shown to play an important role on the formation of bone [5]. To study it, Si-substituted hydroxyapatite (Si-HA) has been synthesized by several methods [6,7] but its biocompatibility was not fully reported yet. The purpose of this study was to evaluate the biocompatibility of Si-HA through *in vitro* test of cell proliferation and *in vivo* study of implantation into the paravertebral muscle of a rabbit.

Materials and Methods

Preparation of 2wt% Si-substituted Hydroxyapatite

The 2wt% Si-substituted hydroxyapatite (Si-HA) was prepared using a precipitation method. Tetraethyl orthosilicate $Si(OC_2H_5)_4$, the source of silicon, was first mixed with H_3PO_4. The mixture was added to $Ca(OH)_2$ slurry while stirring at room temperature in a reactor. The reacted mixture was aged without stirring for a day. The aged precipitate was then dried at 90°C and heated at 1200°C for 3 hrs. The Si-HA precipitate was uniaxially pressed to form a disk-shaped compact (10 mm diameter and 1.5 mm thick) and the compact was then pressed at 200MPa by cold isostatic pressing (CIP). The chemical compositions of Si-HA was analyzed by inductively coupled plasma (ICP) emission spectroscopy and atomic absorption spectroscopy (AAS). The density of the disk-shaped compact was determined by the Archimedes method. Cell proliferation test was conducted on the surface of compact mechanically polished with the 2000-grit SiC metallographic paper and observed by an optical microscope.

Cell Proliferation

MG63 osteoblast-like cells, a continuous line derived by human osteosarcoma (Korean Cell Line Bank, Seoul, Korea), were used. They were cultured in Dulbecco's modified Eagle's medium (DMEM) containing 10% fetal bovine serum (FBS)(HyClone, Logan, Utah) and 0.5% antibiotics at 37°C in a humidified atmosphere of 95% air and 5% CO_2. For cell proliferations, 2×10^4 cells/ml were placed both in the polystyrene plate containing the Si-HA substrate and in polystyrene control plate. The numbers of cells in the trypsinized plates were counted by a hemocytometer after 3, 6 and 24 hrs of incubation time.

Si-HA Implantation

A New Zealand white rabbit weighing average of 2.5kg (3-4 months old) was preanesthetized with an atropine (0.5 mg/kg) and then anesthetized with ketamin (35 mg/kg) and xylazine (5 mg/kg) intramuscularly. After general anesthesia, four strips of the Si-HA (1x1x10 mm) were implanted into the paravertebral muscle on one side of the spine by means of a hypodermic needle (15-gauge), 2.5 to 5 cm from the midline and parallel to the spinal column, and about 2.5 cm apart from each other. Two strips of USP Negative Control Plastic RS were implanted into the opposite muscle of the rabbit in a similar way of the Si-HA samples. The rabbit was sacrificed after 120 hrs. The sites of Si-HA samples and controls were macroscopically observed for hemorrhage, necrosis and discolorations. Microscopical observation was conducted on the areas surrounding both sample and control to observe biocompatibility of Si-HA samples.

Results and Discussion

The bulk density and the chemical compositions of Si-HA were 95% and $Ca_{10}(PO_4)_{5.31}(SiO_4)_{0.69}(OH)_{1.31}$, respectively. Figure 1 shows the morphology of the attached cells on the samples with a passage of time. The average size of cells was 30µm with homogeneous distribution after 3 hrs. Their sizes were increased to 70µm after 6 hrs, but they were difficult to be distinguished after 24 hrs by their connections. The increases in the cells size, for those attached on to the specimens were proportional to the increase of time, indicating good biocompatibility of Si-HA.

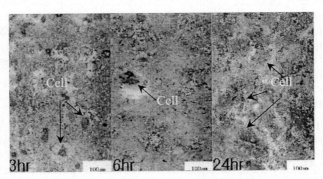

Fig. 1. Cell number and size were increased with a passage of time.

The numbers of cells proliferated in the Si-HA sample plate and in the control plate were counted as shown in Figure 2. With increaseing time, the number of cells in both plates was gradually increased up to 12×10^4 cells/ml for the control plate, and 33×10^4 cells/ml for the sample plate. The degree of cell proliferation in the sample plate was higher than that in the control one. These results indicate that the Si-HA sample is better than the control in terms of biocompatibility. From the implantation experiment, symptoms of hemorrhage, necrosis and discolorations were not observed on the sites surrounding the implanted Si-HA samples from macroscopical observation as shown in Fig. 3.

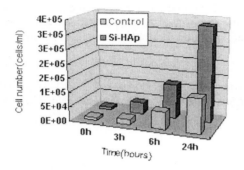

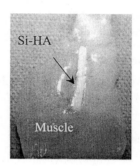

Fig.2. Cell proliferation in time-course.   Fig. 3. Si-HA implant.

Figure 4 shows the microscopical observation on implanted sites by an optical microscope. Fibroblasts are surrounding the implanted areas of both USP negative RS control (Figure 4 (a, b)) and Si-HA sample (Fig. 4 (a, b)), and a few giant cells appear in the fibroblast around the control, which may disapper in a few days. The typical foreign body reaction, without neutrophil and macrophage,

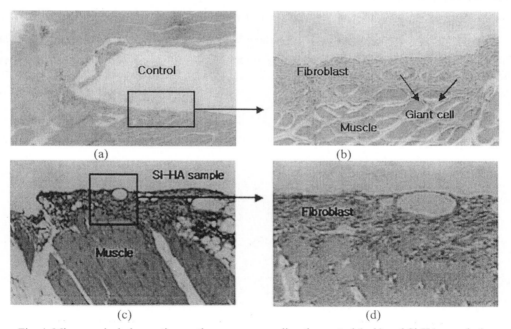

(a)   (b)

(c)   (d)

Fig. 4. Microscopical observation on the areas surrounding the control (a, b) and Si-HA sample (c, d), showed similar findings; fibroblast proliferation without any inflammatory cells, except a few giant cells.

was revealed on the control and sample. It indicates that the control and sample may be biocompatible.

Conclusion

The degree of cell proliferation and the cell size attached on the 2wt% Si-HA gradually increased with time. Symptoms of hemorrhage, necrosis and discolorations were not macroscopically observed in the surrounding area of the implanted Si-HA. Neutrophil and macrophage did not appear on the implanted sites of Si-HA sample. Based on these results, the 2wt% Si-substituted hydroxyapatite has good biocompatibility and seems to be an useful material for implants and bone substitutes.

Acknowledgments

This work was supported by Next Generation Project (N11-A08-1402-06-1-3) of Ministry of Commerce, Industry and Energy, and by the ASAN Institute for Life Sciences (No. 2003-088), Seoul, Korea.

References

[1] J.D. de Bruijn, C.A. van Blitterswijk, J.E. Davies: J. Biomed. Mater. Res. Vol. 29, No. 1 (1995), pp. 89-99.

[2] M. Jarcho: Dent. Clin. North Am. Vol. 36, No. 1 (1992), pp. 19-26.

[3] M. Okumura, H. Ohgushi, Y. Dohi, T. Katuda, S. Tamai, H.K. Koerten, S. Tabata: J. Biomed. Mater. Res. Vol. 37 (1997), pp. 122-129.

[4] I.R. Gibson, K.A. Hing, P.A. Revell, J.D. Santos, S.M. Best, W. Bonfield: Key Eng. Mate. Vols. 218-220 (2002), pp. 203-206.

[5] E.M. Carlisle: Science, Vol. 167 (1970), pp. 179-180.

[6] I.R. Gibson, S.M. Best, W. Bonfield: J. Biomed. Mater. Res. Vol. 44, No. 4 (1999), pp. 422-428.

[7] P.A.A.P. Marques, M.C.F. Magalhaes, R.N. Correia, M. Vallet-Regi: *Bioceramics 13th* (2000), pp. 247-250.

Key Engineering Materials Vols. 254-256(2004) pp. 139-142
online at http://www.scientific.net
© 2004 Trans Tech Publications, Switzerland

Process of Bonelike Apatite Formation on Sintered Hydroxyapatite in Serum-Containing SBF

Tadashi Kokubo[1], Teruyuki Himeno[2], Hyun-Min Kim[3], Masakazu Kawashita[2] and Takashi Nakamura[4]

[1]Research Institute for Science and Technology, Chubu University, 1200 Matsumoto-cho, Kasugai-shi, Aichi 487-8501, Japan

kokubo@isc.chubu.ac.jp

[2]Department of Material Chemistry, Graduate School of Engineering, Kyoto University, Sakyo-ku, Kyoto 606-8501, Japan

[3] Department of Ceramic Engineering, School of Advanced Materials Engineering, Yonsei University, 134, Shinchon-dong, Seodaemun-gu, Seoul 120-749, Korea

[4]Department of Orthopaedic Surgery, Graduate School of Medicine, Kyoto University, Sakyo-ku, Kyoto 606-8507, Japan

Keywords: Sintered hydroxyapatite, bonelike apatite, serum, SBF, zeta-potential, TEM, EDX

Abstract. It has been revealed that even sintered hydroxyapatite forms a bonelike apatite layer on its surface in the living body and bonds to surround bone through the apatite layer. In the previous study, process on bonelike apatite formation on sintered hydroxyapatite was investigated in a protein-free acellular simulated body fluid (SBF) with ion concentrations nearly equal to those of the human blood plasma. In the present study, the effect of addition of serum to SBF on the process of bonelike apatite formation was investigated by using transmission electron microscopy attached with energy dispersive X-ray spectroscopy (TEM-EDX) and laser electrophoresis spectroscopy. In this solution similar to in the original SBF, the sintered hydroxyapatite formed the bonelike apatite on its surface via formation of Ca-rich hydroxyapatite and amorphous calcium phosphate with low Ca/P ratio. The rate of these structural changes in the serum-containing SBF was, however, lower than that in the serum-free SBF. This was attributed to lower negative charging of the surface of the sintered hydroxyapatite at the initial stage due to adsorption of positively charged globulin and lower Ca^{2+} ion concentrations in SBF due to its combination with negatively charged albumin, in serum-containing SBF.

Introduction

It has been revealed that even sintered hydroxyapatite forms a bonelike apatite on its surface in the living body and bonds to the surrounding bone through the apatite layer, similar to Bioglass® and glass-ceramic A-W [1].

How is the bonelike apatite layer formed on the sintered hydroxyapatite? In the previous study, this process was investigated in a protein-free acellular simulated body fluid (SBF) with ion concentrations nearly equal to those of the human blood plasma [2]. As a result, it was found that the sintered hydroxyapatite is initially negatively charged on its surface and hence combines with positively charged Ca^{2+} ion in SBF to form a calcium-rich hydroxyapatite. As the Ca^{2+} ions are accumulated, its surface is positively charged and hence combines with negatively charged phosphate ions to form an amorphous calcium phosphate with low Ca/P ratio. This calcium phosphate is metastable and hence eventually transforms into a stable bonelike apatite.

Actual body fluid contains not only inorganic components but also various kind of organic components. What kind of effect do these organic components have on the formation of the bonelike apatite? In the present study, the process of bonelike apatite formation on a sintered hydroxyapatite was investigated in a serum-containing SBF by means of transmission electron microscopy (TEM) attached with energy dispersive X-ray analysis (EDX) and measurement of laser electrophoresis spectroscopy.

Experimental

Hydroxyapatite (HA: Mitsubishi Materials Co., Japan) sintered at 800°C was ground and sieved into particles less than 5 μm in size. The HA particles 50 mg in mass were immersed in 120 mL of simulated body fluid (SBF) containing 25wt% bovine calf serum (Hyclone Inc., Utah, USA) with pH 7.44 at 36.5°C, where the serum contains proteins in 6.8 g/dl, and albumin and globulin in α-1, α-2, β and γ type amount 47.3, 5.4, 14.8, 20.4 and 12.1% of it. After some given periods, the surface structure of the HA particles was observed by TEM (JEM-2000FXIII, Jeol Co., Tokyo, Japan) and EDX (Voyager III, Noran Instruments, Inc., Middleton, USA). For the TEM-EDX, the HA particles were gathered from the SBFs by dipping a 200-mesh nylon TEM grid covered with polyvinaylformal film. The serum-containing SBFs (SBF-s) dispersed with HA particles for different periods were filled in a high-purity silica glass cell, measured with laser electrophoresis spectroscopy (ELS9000K, Otsuka Electronics Co., Osaka, Japan) to determine the zeta potential of the HA surface.

Results and discussion

TEM-EDX showed the presence of distinct particles of HA with Ca/P ratio of 1.67 before soaking in SBF-s.

After soaking in SBF-s for 6 h, aggregates of small particles, which give an electron diffraction pattern ascribed to crystalline apatite but high Ca/P ratio of 1.69, were observed, as shown in Fig. 1.

After 24 h, a homogeneous amorphous surface layer, which gives no distinct electron diffraction pattern and low Ca/P ratio of 1.45, was observed, as shown in Fig. 2.

After 72 h, aggregates of tiny particles, which give a weak diffraction pattern ascribed to apatite and somewhat higher Ca/P ratio of 1.53, were observed as shown in Fig. 3.

After 336 h, needlelike nano-sized particles, which give a distinct diffraction pattern ascribed to crystalline apatite and Ca/P ratio of 1.62, almost equal to that of the bone mineral, were observed, as shown in Fig. 4.

Figure 5 (a) shows the variation of zeta potential of HA particles as a function of soaking time in SBF-s. In the same figure, variation Ca/P ration of HA particles was also plotted. For reference, variation of zeta potential and Ca/P ratio of HA particles in SBF, which are reported in the previous paper [2], was reproduced in Fig 5 (b).

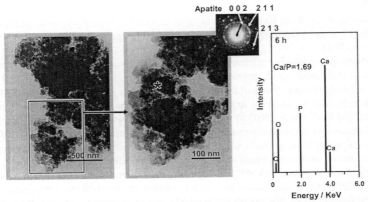

Fig. 1 TEM-EDX pictures of HA particles after soaking in SBF-s for 6 h.

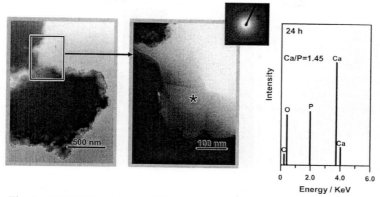

Fig. 2 TEM-EDX pictures of HA particles after soaking in SBF-s for 24 h.

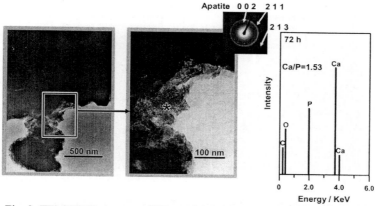

Fig. 3 TEM-EDX pictures of HA particles after soaking in SBF-s for 72 h.

It is clear from Fig 5 that HA particles show an initial increase, subsequent decrease and final convergence to the level of bone mineral in its Ca/P ratio as well as in the zeta potential in SBF-s, as in SBF.   They also form the bonelike apatite on their surfaces via formation of a Ca-rich hydroxyapatite and subsequent formation of an amorphous calcium phosphate in SBF-s, as in SBF. The magnitudes of the variations in the Ca/P ratio and zeta potential and the rate of the structural changes in SBF-s are, however, not so high as in SBF. In X-ray photoelectron spectroscopy (XPS: Model MT 5500, ULVAC-PHI Co. Ltd, Chigasaki, Japan) of the surface of a plate of the same HA, a fairly strong N1s peaks was detected from its surface after soaking in SBF-s for 6 h. This indicates that some proteins such as positively charged globulins are adsorbed on its surface in SBF-s to reduce its surface negative charge.

On the other hand, Ca^{2+} ion concentration in SBF-s, which was measured by Ca^{2+} ion electrode (CA-135B, Toa Electronics, Ltd, Tokyo, Japan) was 1.7 mM. This value is much lower than that in SBF, 2.1 mM. This indicates that some calcium ions are combined with negatively charged proteins such as albumin, to reduce concentrations of the free Ca^{2+} ion in SBF-s. Both of these factors might give lower magnitude of the variation in Ca/P ratio and zeta potential, and lower rate of structural changes in SBF-s, in comparison with those in SBF.

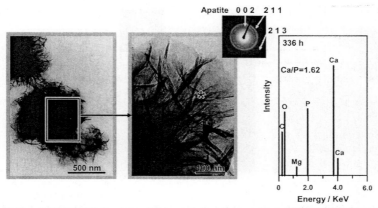

Fig. 4 TEM-EDX pictures of HA particles after soaking in SBF-s for 336 h.

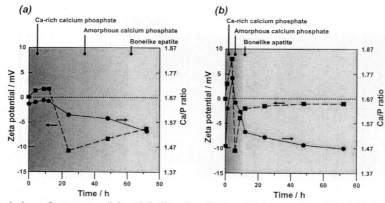

Fig. 5 Variation of zeta potential and Ca/P ratio of HA particles as a function of soaking time in SBF-s (a) and SBF (b).

Conclusions

Sintered hydroxyapatite forms a bonelike apatite on its surface via formation of Ca-rich hydroxyapatite and subsequent formation of an amorphous calcium phosphate with low Ca/P ratio in serum-containing simulated body fluid, as in serum-free sintered body fluid. The magnitude of variation in Ca/P ratio and zeta potential, and rate of structural changes in the process described above are, however, lower in serum-containing simulated body fluid than in serum-free simulated body fluid. This is attributed to the adsorption of positively charged proteins to the surface of HA and combining of Ca^{2+} ions in SBF-s with negatively charged proteins.

References

[1] M. Neo, T. Nakamura, C. Ohtsuki, T. Kokubo and T. Yamamuro (1993) J. Biomed. Mater. Res., 27, 999-1006.
[2] T. Himeno, H.-M. Kim, H. Kaneko, M. Kawashita, T. Kokubo and T. Nakamura (2003) *Bioceramics*, 15, pp 457-460, Trans. Tech. Publication.

Key Engineering Materials Vols. 254-256(2004) pp. 143-146
online at http://www.scientific.net
© *2004 Trans Tech Publications, Switzerland*

Ceramics *In Vitro* Mineralisation Protocols: a Supersaturation Problem

P.A.A.P. Marques[1], M.C.F. Magalhães[1], R.N. Correia[1], A. I. Martin[2], A.J. Salinas[2], M. Vallet-Regí[2]

[1] CICECO, University of Aveiro, 3810-193 Aveiro, Portugal; paulam@dq.ua.pt
[2] Dpt. Inorganic Chemistry, Faculty of Pharmacy, Complutense University, E-28040 Madrid, Spain

Keywords: Mineralisation, hydroxyapatite, sol-gel glass, *in vitro* protocols, supersaturation.

Abstract. Two ceramics: a $25CaO-2.5P_2O_5-72.5SiO_2$ mol% sol-gel glass (S72.5P2.5) and a commercial hydroxyapatite (HAP), both bioactive in Kokubo's Simulated Body Fluid (SBF), were studied according to two recently proposed *in vitro* protocols. The first one (*SBF-dynamic*) avoids the variations of the ionic concentration in the assay solution with a continuous renewal of SBF. The second protocol, uses a carbonated simulated inorganic plasma (*CSIP*), prepared by substituting the tris(hydroxymethyl aminomethane)/HCl buffer (TRIS) of SBF for a CO_2/HCO_3^- buffer, whereby a physiological concentration of HCO_3^- in solution is reached (24 to 27 mM) whereas the pH is maintained between 7.3 and 7.4. After the *in vitro* immersions, S72.5P2.5 exhibited a hydroxycarbonate apatite (HCA) surface layer in both cases, but some differences were found depending on the *in vitro* protocol. In *CSIP*, a higher HCA crystallization rate was initially observed (1 day *vs* 3 days in *SBF-dynamic*). Nevertheless, after 7 days in *SBF-dynamic*, the new layer was thicker (6 μm *vs* 1 μm) and presented a Ca/P molar ratio lower than in *CSIP* (1.7 *vs* 2.1). In the case of HAP ceramic, a HCA layer was observed only in *CSIP*. In *SBF-dynamic* the local supersaturation degree of the solution and/or the system dynamics were not appropriated for HCA layer formation.

Introduction

In vitro studies in synthetic solutions are currently practised as a first stage in the assessment of the *in vivo* surface behaviour of implantable materials [1,2]. Probably, the most popular solution to study the *in vitro* bioactivity of materials is the Simulated Body Fluid K#9 (SBF) proposed by Kokubo *et al* in the early nineties [3]. This aqueous solution contains a concentration of inorganic ions almost equal to human plasma and has been widely reported because it constitutes a fast, easy and convenient method for the evaluation of new candidates to bioactive implant materials. In comparison with *in vivo* conditions, SBF is lower in HCO_3^-: 4.2 mM (vs. 27 mM in plasma).

Moreover, SBF is buffered at plasmatic pH with the non-biological TRIS, whereas CO_2/HCO_3^- is the main buffer in plasma. All these factors, together with the possible complexation of Ca^{2+} ions by TRIS [4], affect HCA layer formation since they modify the solution composition and, thus, the solution supersaturation. Another point worth noting is that mineralisation under continuous flow - as occurs in the body - may differ from that in confined media.

Recently, some authors [5-8] identified limitations in conventional SBF to test the *in vitro* bioactivity, and proposed alternatives to maintain the initial composition of solution during the whole assay, and to obtain an inorganic composition exactly equal to plasma, namely concerning $[HCO_3^-]$. In the present work two ceramics: a $25\%CaO-2.5\%P_2O_5-72.5\%SiO_2$ sol-gel glass and a commercial hydroxyapatite (HAP) (Captal®), both bioactive in static SBF but with different HCA layer formation kinetics, were studied following two recently proposed *in vitro* protocols [5,6]: (1) *SBF-dynamic*, where materials are kept under a continuous flow of conventional SBF. (2) *CSIP*, a static solution buffered at pH 7.3-7.4 with CO_2/HCO_3^- gas mixture, the concentration of $[HCO_3^-]$

being maintained between 24 and 27 mM. The comparative results will allow evaluating the capability of these *in vitro* systems to test the bioactivity of ceramics with very different dissolution properties. In addition, new insights in the apatite formation process under *in vitro* conditions are expected by comparing the characteristics of the HCA layer formed in each case.

Materials and Methods

HAP ceramic discs (9 mm diameter, 1 mm thick) and sol-gel glass discs (12 mm diameter, 2 mm thick) were obtained. The sol-gel glass pieces (72.5% SiO_2–25% CaO–2.5% P_2O_5, mol%) were synthesized as described elsewhere [6] by heating discs prepared by uniaxial and cold isostatic pressing of dry gel powders. Commercial HAP (Captal®) discs were sintered and polished. Mineralisation tests were conducted for periods ranging from 3 hours to 7 days at 37°C. In *SBF-dynamic* the conventional SBF was pumped at 1 mL/min; in *CSIP* the pH of 7.3-7.4 was controlled by an appropriated mixture of CO_2 /N_2 gases continuously bubbled in flasks containing 40 mL of solution. Simultaneously, a physiological concentration of HCO_3^- (24 to 27 mM) is reached in *CSIP*. After the incubation, the immersed discs were rinsed with distilled water and acetone and air-dried. Sample surfaces were analysed by XRD (low angle incidence), FTIR (reflectance mode) and SEM-EDS. Time variations of the Ca(II), P(V), Si(IV) and H_3O^+ concentrations in the solutions were monitored by selective electrode measurements and UV-Vis and ICP spectroscopies.

Results

Table 1 summarizes results obtained by XRD, FTIR and SEM-EDS for discs of commercial HAP and S72.5P2.5 sol-gel glass after 7 days in both conditions.

Table 1. Characteristics of the newly formed HCA layer after 7 days

	Commercial HAP	S72.5P2.5 sol-gel glass
SBF-dynamic	No layer	Layer ≈ 6 μm thick Ca/P = 1.70 ± 0.11
CSIP	Microcrystalline layer ≈ 1 μm thick with plate like morphology Ca/P molar ratio = 1.70 ± 0.11 % Mg = 1.27 ± 0.25; % Na = 1.94 ± 0.43	Layer ≈ 1 μm thick Ca/P molar ratio = 2.10 ± 0.12 $CaCO_3$ present

The ionic compositional variations in the assay solutions were:
- In *SBF-dynamic,* HAP immersion caused no variation; however, the soaking of S72.5P2.5 glass slightly changed the solution composition during the first 3 hours of assay, rising pH (7.37 → 7.47), [Ca(II)] (2.23 → 3.02 mM) and [Si(IV)] (0 → 0.11 mM), and decreasing [P(V)] (0.96 → 0.89 mM). Nevertheless, for higher soaking times, these parameters returned to the initial values in SBF.
- In *CSIP,* for S72.5P2.5 glass, a careful regulation of CO_2 partial pressure was required to maintain the pH constant, due to the dissolution properties of this material. After 7 days, Ca(II) and Si(IV) concentrations in solution increased (2.5 → 3.1mM and 0 → 2.1mM, respectively, and P(V) decreased (1.0 → 0.06mM). For HAP after 7 days of immersion in *CISP* the chemical analysis of the solution showed a continuous decrease in [Ca(II)] and [P(V)] (2.5 → 1.3 mM and 1.0 → 0.25 mM, respectively) due to the HCA precipitation.

Therefore, HAP was more reactive in *CSIP* than in static SBF [9] but showed no reactivity in *SBF-dynamic* (Table 1). However, S72.5P2.5 glass was highly reactive in both systems. In the case of HAP immersed in *CSIP*, FTIR showed only carbonated apatite groups (Fig. 1(a)), and XRD showed very broad apatite peaks. A plate-like morphology was evidenced by SEM (Fig. 1(b)).

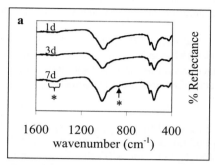
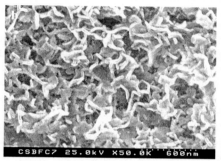

Fig. 1. (a) FTIR spectra after 1, 3 and 7 days of immersion of HAP in *CSIP* (* indicates CO_3^{2-} groups), (b) SEM image of HAP after 7 days of immersion in *CSIP*.

For S72.5P2.5 gel glass immersed in *CSIP* a higher initial rate of carbonate apatite crystallization was observed by XRD (1 day vs. 3 days in *SBF-dynamic)*. (See Figure 2a). However, after 7 days in dynamic mode, the layer was thicker (6 μm vs. 1 μm). For higher soaking times, a layer of flake-shaped particles covered the surfaces in both cases and FTIR showed carbonate apatite, possibly mixed with octacalcium phosphate (OCP).

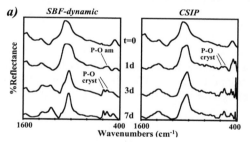
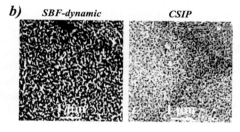

Fig. 2. (a) FTIR spectra of S72.5P2.5 sol-gel glass before and after soaking for 1, 3, and 7 days in both protocols (b) SEM micrographs of S72.5P2.5 after soaking for 7 days in both protocols.

Discussion

Simulated inorganic plasma solutions used are supersaturated with respect to OCP and HAP. These solutions may be metastable for practically infinite time periods and may return to equilibrium only when a disturbance occurs, like the introduction of seed crystals, especially those of the corresponding phosphates which increase the calcium and phosphate content of the aqueous medium and/or the pH.

The two ceramics tested in the present study presented a remarkable difference in their solubility behaviour, being the sol-gel glass much more soluble than the HAP. In this sense the glass causes greater variations in the solution supersaturation values, namely by increasing Ca(II), P(V) concentrations and pH. The *in vitro* results obtained allow us to establish that:

- In *SBF-dynamic* a thicker layer with a lower Ca/P molar ratio is obtained. In fact, *CSIP* is almost a static method because H_3O^+ and HCO_3^- are the only ions renewed. Therefore, when the calcium and phosphate ions coming from the solution or from the partial dissolution of glass are exhausted, the layer ceases to grow. On the contrary, in *SBF-dynamic* new ions that can be incorporated to the glass are continuously added.

- In *CSIP* an initial higher rate of HCA crystallization is observed. This is due to the higher increase of the local supersaturation as a consequence of the static character of *CSIP* compared with *SBF-dynamic*.
- The higher value of the Ca/P molar ratio after 7 days in *CSIP* can be explained by the co-precipitation of calcium carbonate, as the XRD and FTIR studies point out. At the same time this could justify the relatively low increase of calcium concentration in *CISP* solution, compared with the results obtained where this glass was studied in static SBF in a previous work [10].

Conclusions

The HCA layer on commercial HAP will probably be formed in *CSIP* by epitaxial growth; but in *SBF-dynamic*, the solution flow retards the kinetics of mineralisation. Contrary to HAP, the gel glass is highly soluble in both media, rising pH, and Ca(II), P(V) and Si(IV) ionic concentrations; in *SBF-dynamic* solution flow removes the lixiviated ions from near the surface and active growth sites will form, resulting in a HCA layer. For glass in static *CSIP* the solution becomes supersaturated and the buffer capacity available is not sufficient to prevent pH rise and $CaCO_3$ precipitation occurring together with that of HCA.

Regarding the advantages and limitations of both protocols, *CSIP* is a static system and, therefore, with higher tendency to include impurities in the layer (Mg, Na) and, eventually to co-precipitate calcite if material dissolution causes a pH rise that overrides the capacity of the buffer. In *SBF-dynamic* the continuous ion supply allows to maintain constant local conditions for layer growth, although in materials with lower reactivity, like HAP, local supersaturation may be not be enough for HCA layer formation.

Acknowledgements

P. Marques thanks the Portuguese Foundation for Science and Technology (FCT) for the doctoral grant PRAXIS XXI/BD/16157/98. Financial support of CICYT through MAT2002-0025, MAT2001-1445-C02-01, and the Spanish-Portuguese Integrated Action HP2000-051 is acknowledged.

References

[1] M. Vallet-Regí: Mater. J. Chem. Soc. Dalton Trans. Vol. 2 (2001), p. 97.
[2] L.L. Hench: J. Am. Ceram. Soc. Vol. 81 (1998), p. 1705.
[3] T. Kokubo, H. Kushitani, S. Sakka, T. Kitsugi, T. Yamamuro: J. Biomed. Mater. Res. Vol. 24 (1990), p. 721.
[4] J. Hlaváč, D. Rohanová, A. Helebrant: Ceram Silik Vol. 38 (1994), p.119.
[5] P.A.A.P. Marques, M.C.F. Magalhães, R.N. Correia: Biomaterials. Vol. 24 (2003), p. 1541.
[6] I. Izquierdo-Barba, A.J. Salinas, M. Vallet-Regí: J. Biomed. Mater. Res. Vol. 51 (2000), p. 2301.
[7] H.M. Kim, T. Miyazaki, T. Kokubo, T. Nakamura: Key Eng. Mater. Vol. 192 (2000), p. 47.
[8] A.Oyane, K Onuma, A. Ito, H.M.Kim, T. Kokubo, T.Nakamura: J. Biomed. Mater. Res. Vol. 64 (2003), p. 339.
[9] P.A.A.P. Marques, A.P. Serro, B.J. Saramago, A.C. Fernandes, M.C.F. Magalhães, R.N. Correia: J. Mater. Chem. Vol. 13 (2003), p.1484.
[10] A.J. Salinas, A.I. Martin, M. Vallet-Regi: J. Biomed. Mater. Res. Vol. 61 (2002), p. 524.

Key Engineering Materials Vols. 254-256(2004) pp. 147-150
online at http://www.scientific.net
© 2004 Trans Tech Publications, Switzerland

Characterization of Bioactive Glass-Ceramics Prepared by Sintering Mixed Glass Powders of Cerabone® A-W Type Glass/CaO-SiO₂-B₂O₃ Glass

J.H. Seo[1], H.S. Ryu[1],K.S. Park[1], K.S. Hong[1], H. Kim[1],J.H. Lee[2], D.H. Lee[2], B.S. Chang[2], C.K. Lee[2]

[1] School of Material Sci. & Eng., Seoul Nat'l Univ., Seoul 151-742, Korea, seoscy1@snu.ac.kr

[2] Department of Orthopedic surgery, Seoul Nat'l Univ. Seoul 110-744, Korea

Keywords: Bioactive, biodegradable, bioceramics, SBF, mechanical property

Abstract. From our previous work, CaO-SiO₂-B₂O₃(CSB, CaO 45.8, SiO₂ 45.8, B₂O₃ 8.4mol%) glass-ceramic was proven to show fast dissolution rate *in-vitro* and *in-vivo* and Cerabone® A-W (Cera) hardly degraded in SBF. To control the biodegradation rate, we prepared glass-ceramics by mixing glass powders of Cera and CSB. The glass-ceramics sintered 800°C for 2h showed dense microstructure and were composed of crystalline β-wollastonite, apatite and a residual glass matrix. So mixing CSB with Cera enabled low temperature crystallization of β-wollastonite. In addition, as the amount of CSB glass powder increased, rate of dissolution and mechanical properties increased, but bioactivity slightly decreased.

Introduction

Bioactive glass-ceramics have been extensively studied for better bioactivity than hydroxyapatite or other calcium phosphate compounds[1-3]. Especially Kokubo et al. reported apatite–wollastonite containing glass-ceramics(Cerabone® A-W) with high mechanical strength and it has been widely used as bone replacement[4].

Recently, biodegradation of bioactive materials became important for bone substitutes. For complete new bone replacement, rate of bone ingrowth must be similar with that of dissolution of implant. So, biodegradation rate of ideal bone graft must be controlled. The present authors found that glass-ceramics in the CaO-SiO₂-B₂O₃ system were not only non-toxic, bioactive and biodegradable but also osteoconductive. However, the glass-ceramics presented so fast dissolution rate *in-vitro* and *in-vivo* that significant bone growth could not be anticipated[5].

In this study, novel biodegradation controllable glass-ceramics prepared by sintering mixed glass powders of Cerabone® A-W type glass and CaO-SiO₂-B₂O₃ glass were investigated.

Materials and Methods

Table.1 The chemical compositions of Cerabone-AW® and CSB glass

Sample Name	CaO(wt%)	SiO₂(wt%)	P₂O₅(wt%)	B₂O₃(wt%)	MgO(wt%)	CaF₂(wt%)
Cera	44.9	34.2	16.3	-	4.6	0.5
CSB	43.4	46.6	-	10		

Table. 2 Mixing ratios of prepared glass-ceramics

Sample Name	Cera(wt%)	CSB(wt%)
C3	75	25
CB	50	50
B3	25	75

The compositions of Cerabone® A-W type glass(Cera) and $CaO-SiO_2-B_2O_3$ glass(CSB) are presented in Table 1. Cera and CSB were prepared by the same method described in the literature [4,5]. Then the glass powders of Cera and CSB were mixed in weight ratios of 3:1(C3), 1:1(CB), 1:3(B3) by ball-milling for 24h(Table.2). The shrinkage behaviors of glass powder compacts were observed by dilatometry. The bulk density and open porosity of the samples sintered at 750~900°C for 2h were determined using Archimedes' method and the phases of sintered sample were identified by XRD.

Bioactivity and biodegradability were investigated by *in-vitro* test. Mirror-polished specimens were soaked in simulated body fluid(SBF) for 5 to 60 days. To evaluate biodegradation behavior, their weight losses as a function of soaking time were also investigated. Calcium and phosphorous ion concentrations in SBF were measured by ICP(Inductively Coupled Plasma).

Results and Discussion

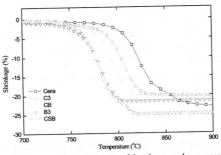

Fig. 1. Thermal shrinkage curves of the glass powder compacts   Fig. 2. XRD patterns of specimens sintered at 800°C for 2h.

Fig. 1 shows the shrinkage behavior of pellets made of five types of glass powders. As the amount of CSB increased, the onset temperature of shrinkage decreased. The bulk density and open porosity of the sintered samples(Figure not shown) confirmed the effect of mixing of Cera and CSB. The samples of Cera sintered at 900°C for 2h exhibited the maximum bulk density while samples of C3, CB, and B3 sintered at 800°C for 2h exhibited the maximum bulk density.

Fig. 2 presents XRD patterns of the samples sintered at 800°C for 2h. β-wollastonite crystals starts to be precipitated (shifted to high angle) while those of C3, CB and B3 were fully precipitated. Especially XRD pattern of C3 sintered at 800°C for 2h was almost the same as that of Cera sintered at 1000°C for 2h. Thus, mixing Cera with CSB glass powder enabled low temperature cystallization of β-wollastonite. It is known that β-wollastonite crystals can effectively increase the mechanical properties of A-W glass-ceramics[4], but β-wollastonite crystals of Cera were fully precipitated when sintered at 1000°C for 2h. However, those of C3, CB and B3 were fully precipitated when sintered at 800°C for 2h. Therefore C3, CB and B3 could be sintered at lower temperature(800°C) without decrease in mechanical properties.

Table 3. Mechanical properties of examined glass-ceramics

	Compressive Strength(MPa)	Bending Strength(MPa)	Fracture Toughness(MPa·m$^{1/2}$)
Cera	1050±32	162±6	1.54±0.07
C3	1124	163±10	1.82±0.01
CB	1295	189±24	2.21±0.02
B3	1388	200±13	2.55±0.05
CSB	2826± 186	212±11	3.16±0.16

Table 3 shows the mechanical properties of the examined glass-ceramics. Cera sintered at 1000°C for 2h and the others sintered at 800°C for 2h. All the mechanical properties were increased as the amount of CSB increased. Therefore this confirms the effect of mixing Cera and CSB.

Fig. 3 SEM photographs of the surface of glass-ceramics soaked in SBF for 1 day.

In Fig. 3, after soaking in SBF for 1 day, Hydroxy Carbonate Apatite(HCA) layer covered completely the surface of C3 and the half of surface of CB, however in B3, HCA started to be nucleated. This result is confirmed in thin-film XRD patterns of samples soaked for 1 day(Fig.4). C3 glass-ceramics showed the strongest apatite peaks among the three glass-ceramics after 1day, but intensity of apatite slightly decreasd in CB and more decreased in B3. However all samples were covered completely with HCA layer after 3 days soaking in SBF(Figures are not shown). In our previous works, thin HCA layer covered on surface of CSB after 3 days soaking in SBF, and thick HCA layer was present after 5days[5]. And thick HCA layer covered completely the surface of Cera after 1days soaking in SBF. So this also confirms the effect of mixing Cera and CSB.

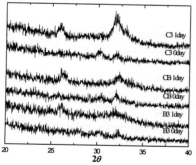

Fig. 4 Thin film XRD patterns of C3, CB and B3 soaked in SBF for 1 day.   Fig. 5 Weight losses of samples as a function of soaking time in SBF

Fig. 5 shows weight loss as a function of soaking time. Cera hardly degraded in SBF and CSB degraded the most between the five kinds of glass-ceramics. The rate of dissolution is increased proportionally to the content of CSB. In fig. 6, variation of calcium and phosphorous ion concentrations of the SBF after soaking C3, CB and B3 for various periods are presented. This shows the same result as the weight loss in fig.5. The rate of calcium ion released increased with the amount of CSB. The phosphrous ions were not detected in all three samples after 60 days soaking in SBF and their consumption rate also changed with the amount of CSB. Therefore we can control the rate of dissolution by controlling mixing ratios of Cera and CSB.

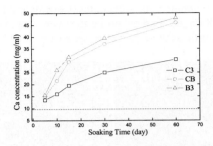 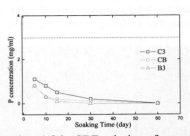

Fig. 6 Variation of Calcium and Phosphorous concentrations of the SBF solution after soaking the C3, CB and B3 for various periods

Conclusions

 Biodegradation-controllable glass-ceramics were fabricated by sintering the glass powder compact of mixed Cerabone® A-W and CSB. As the amount of CSB glass increased, the shrinkage and crystallization started at lower temperature. The main crystalline phases of the glass-ceramics were wollastonite and apatite as in Cerabone® A-W. However, as opposed to Cerabone® A-W, C3, CB and B3 showed crystallization of β-wollastonite at lower temperature(800°C) and thus the authors could sintered C3, CB and B3 at lower temperature without decreasing the mechanical properties. From *in-vitro* test, HCA was formed after 1day on the suface of C3, CB and B3, however, coverage and thickness of HCA were dependent on the composition of the glass-ceramics. The rate of dissolution increased with the amount of CSB. In short, all properties in C3, CB and B3 showed combined effects of Cera and CSB. These glass-ceramics make it possible to overcome drawbacks of both Cerabone® A-W and CaO-SiO_2-B_2O_3 glass-ceramics.

Acknowledgement

This work was supported by a Grant-in-Aid for Next-Generation New Technology Development Programs from the Korea Ministry of Commerce, Industry and Energy (No. N11-A08-1402-08-1-3).

References

[1] Hench LL: J. Am. Ceram. Soc. Vol. 81 (1991), p.1497

[2] Hench LL, Wilson JW: *Introduction to bioceramics* (World Scientific,1984), p41.

[3] Hench LL, Paschall HA: J. Biomed. Mater. Res. Vol. 5 (1974), p49.

[4] T. Kokubo, S. Ito, M. Shigematsu, S. Sakka, and T. Yamamuro: J. Mater. Sci. Vol 20 (1985), p. 2001.

[5] H. S. Ryu, J.H. Seo, H. Kim, K.S. Hong, D.J. Kim, J.H. Lee, D.H.Lee, B.S. Chang, C.K. Lee and S.S.Chung: J. Biomed. Mater. Res., (Accepted)

Key Engineering Materials Vols. 254-256(2004) pp. 151-154
online at http://www.scientific.net
© 2004 Trans Tech Publications, Switzerland

Effect of B₂O₃ on the Sintering Behavior and Phase Transition of Wollastonite Ceramics

K.S Park[1], H.S Ryu[1], J.H Seo[1], K.S Hong[1], H Kim[1], J.H Lee[2], D.H Lee[2], B.S Chang[2], C.K Lee[2]

[1] School of Material Sci. & Eng., Seoul Nat'l Univ., Seoul 151-742, Korea, nugo1@snu.ac.kr

Department of Orthopedic surgery, Seoul National University, Seoul 110-744, Korea

Keywords: wollastonite, phase transition, sintering

Abstract. By adding B_2O_3 powder into alpha wollastonite ceramics synthesized by solid-state reaction, the sintering temperature of wollastonite ceramics lowered by 300°C and temperature that exhibit maximum density decreased by about 400°C. In addition, we found that liquid-phase sintering observed in wollastonite-B_2O_3(WB) ceramics is due to melting of calcium borate which is confirmed by DTA measurement. Phase transition between alpha wollastonite and beta wollastonite is observed in WB ceramics, while no phase transition is observed in alpha wollastonite. In WB ceramics, beta wollatonite started to appear at 900°C, and above 950°C, phase transition from alpha phase to beta phase took place and only alpha wollastonite remained above 1100°C. Especially, the sample sintered at 950°C was composed of almost beta wollastonite in the case of W5B. In addition, we observed abnormal grain growth due to liquid phase in the sintered sample.

Introduction

Since the discovery of Bioglass® by Hench in 1972[1], various materials including glasses[2,3], sintered hydroxyapatite[4], glass ceramics[5,6], composite materials[7] have been studied for biomedical applications. Among them, wollastonite ceramics have been reported to have good bioactivity and biocompatibility [8]. Furthermore, because it has good mechanical property, there are possible applications of wollastonite ceramics for the manufacture of artificial bone and dental root. However in spite of its availability of as a biomaterial, its sintering temperature is very high. And study about two polymorphisms of wollastonite ceramics, alpha phase and beta phase, is not enough. In this study, to lower sintering temperature of wollastonite, B_2O_3 was added to pure alpha wollastonite ceramics, and we studied the phase transition and the microstructure.

Method

To obtain pure alpha wollatonite ceramics, a mixture of $CaCO_3$ (99.99%, high purity chem., Japan) and SiO_2(99.9% high purity chem., Japan) with a CaO/SiO_2 molar ratio equal to one was prepared by ball-milling for 24h in ethanol media and then the mixture was calcined at 1300 °C for 12h in a Pt crucible. Calcined wollastonite powder was confirmed to be pure alpha phase by XRD, and pulverized by additional ball milling for 24h. By adding 5 and 10wt% B_2O_3 powder (99.9% high purity chem., Japan) to pure wollastonite powder and mixing them, W5B and W10B were obtained. The compositions are presented in table 1. These three powders including pure wollastonite were granulated and compacted into pellets. The shrinkage behaviors of the pellets were observed using a dilatometry (DIL 402C, Netsch, German) and the crystalline behavior was observed by differential thermal analysis (DTA, SDT 2960, TA inst., USA). The pellets of WB ceramics were sintered at 800~1100 °C for 2h, and those of wollastonite were sintered at 1000~1400°C for 2h. The bulk density and open porosity of sintered samples were determined using Archimedes' method. The phases of sintered samples were confirmed by XRD (M18XHF-SRA, Mac Sci., Japan) and their mirror-polished surfaces were observed by SEM (JSM-5600, Jeol, Japan).

Table. 1 The compositions of WB ceramics

wt%	wollastonite	B_2O_3
W5B	95	5
W10B	90	10

Results and Discussion

Fig. 1 shows the shrinkage behaviors of the samples and DTA curve of W10B sample. While wollastonite ceramics shrank at about 1200°C, WB ceramics shrank at about 900°C. That is, by addition of small amount of B_2O_3, the onset temperature is lowered by about 300°C. The ledge (↑) observed at about 700°C in WB ceramics is due to crystallization of calcium borate as confirmed in XRD patterns of sample quenched at 550 and 800°C which are presented in Fig. 2. While only alpha wollastonite exists at 550°C, calcium borate peak is observed at 800°C. Trace of beta wollastonite peak is observed at 1000°C.

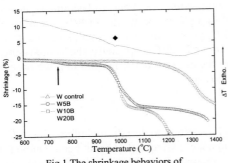

Fig.1 The shrinkage behaviors of
Wollastonite and WB ceramics

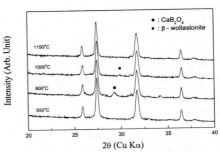

Fig. 2 XRD patterns of sample quenched
at various temperature of W10B

The endothermic peak at 985°C (indicated by an arrow) by melting of calcium borate in DTA curve of W10B sample. This agrees with the starting point of shrinkage of W10B sample, which shows that the liquid-phase sintering observed at WB ceramics is due to melting of calcium borate.

The bulk density and open porosity of the sintered samples in Fig. 3 confirmed the effect of addition of B_2O_3. While the samples of pure wollastonite ceramics sintered at 1400°C for 2h exhibited maximum bulk density, the samples of W5B sintered at 950°C for 2h exhibited maximum bulk density.

XRD patterns of W5B samples sintered at various temperature for 2h are shown in figure 4. Pure beta wollastonite is difficult to obtain because calcined wollastonite is alpha phase and can not be transformed into beta phase for another heat treatment as shown at Fig. 5. Even if alpha wollastonite powder calcined at 1300°C was sintered at 900°C for 20h, not significant difference was observed. In addition, second phases such as α-quartz and β-Ca_2SiO_4 were observed when calcinations is carried out at 1000~1200°C.

The sample sintered at 800°C for 2h consisted of alpha wollastonite and calcium borate. Beta wollastonite started to appear at 900°C and the sample sintered at 950°C was almost all composed of beta wollatonite. Above 950°C, phase transition from alpha phase to beta phase took place and only alpha wollastonite remained above 1100°C. So it is remarkable that the sample of W5B sintered at 950°C for 2h was composed of almost only beta wollatonite. In the case of W10B, the pattern of XRD was similar to that of W5B.

In Fig. 6, SEM photographs of thermally etched wollastonite and WB ceramics are presented. Pure wollastonite (A) was sintered at 1300°C and W5B, W10B were sintered at 950°C, 1000°C, respectively. In wollatonite ceramics, liquid phase exists at grain boundary and grain size is below 10. In WB ceramics, liquid phase is more abundant than wollastonite ceramics, consequently, abnormally grown grains are observed.

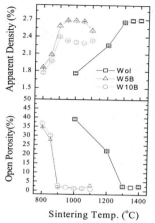

Fig. 3 Bulk density and open porosity of wollastonite and WB ceramics

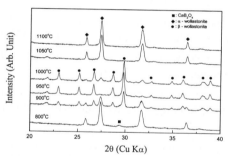

Fig. 4 XRD patterns of W5B sintered for 2h

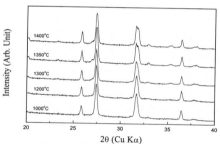

Fig. 5 XRD patterns of wollastonite sintered at various temperature for 2h

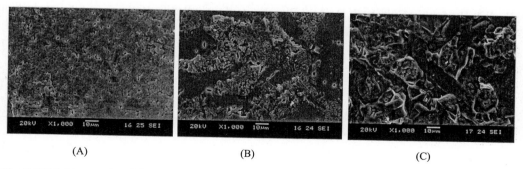

(A)   (B)   (C)

Fig. 6 SEM photographs of thermally etched sintered samples of (A) pure alpha wollastonite (B) W5B (C) W10B

Conclusion

In this study, we added B_2O_3 into alpha wollatonite ceramics. By addition of B_2O_3, the sintering temperature of wollastonite ceramics are lowered. Because melting of calcium borate helps sintering of WB ceramics. We could observe the phase transition between alpha phase and beta phase of the mixture, consequently, almost pure beta wollastonite could be obtained. In addition, we observed abnormal grain growth due to liquid phase in the sintered sample.

Acknowledgement

This study was supported by a Grant-in-aid of the Next-Generation New Technology Development Programs, Ministry of Commerce, Industry & Energy, Republic of Korea (No. N11-A08-1402-08-1-3).

References

[1] L.L. Hench, R.J. Splinter, W.C. Allen and T.K. Greenle: Bonding mechanisms at the interface of ceramic prosthetic materials, J. Biomed. Mater. Res., Vol. 2(1972), p. 117-141

[2] Y. Ebisawa, T. KoKubo, K. Ohura, T. Yamamuro: Bioactivity of $CaO-SiO_2$-based glasses: In vitro evaluation, J. Mater. Sci: Mater. Med., Vol. 1(1990), p. 239-244

[3] K. Ohura, T. Nakamura, T. Yamamuro, T. Kokubo, Y. Ebisawa, Y. Kotoura, M Oka: Bone –bonding ability of P2O5-free CaO•SiO2 glasses, J. Biomed. Mater. Res., Vol. 25(1991), p. 357-365

[4] M. Jarcho, J.F. Kay, K.I. Gumaer, R.H. Doremus, H.P. Drobeck: *Tissue, cellular and subcellular events at a bone-ceramic hydroxyapatite interface*, J. Bioeng., Vol. 1(1977), p. 79-92

[5] T.Kokubo, H.Kushitani, C. Ohtsuki, S.Sakka and T.Yamamuro: Chemical Reaction of Bioactive Glass and Glass-ceramics with a Stimulated Body Fluid, Materials in Medicine, Vol. 3(1992), p. 79-83

[6] T. Yamamuro: *Reconstruction of the Iliac Crest with Bioactive Glass-ceramic Protheses in Handbook of Bioactive Ceramics*, Vol 1, Edited by T. Yamamuro, L.L.Hench, J.Wilson, CRC Press, Boca Raton, (1990) 335-42

[7] P. Ducheyne, W.V. Raemdonck, J.C. Heughebaert, M. Heughebaert: Structural analysis of hydroxyapatite coating on titanium, Biomaterials, Vol. 7(1986), p. 97-103

[8] P.N De Aza, F. Guitian, S. De Aza: Bioactivity of wollastonite ceramics: in vitro evaluation, Scripta Materialia, Vol. 8 (1994), p. 1001-1005

Key Engineering Materials Vols. 254-256(2004) pp. 155-158
online at http://www.scientific.net

SiO$_2$- MgO-3CaO·P$_2$O$_5$- K$_2$O Glasses and Glass-Ceramics: Effect of Crystallisation on the Adhesion of SBF Apatite Layers

C.M. Queiroz, J.R. Frade, M.H.V. Fernandes

Ceramics and Glass Engineering Department (CICECO)

University of Aveiro, 3810-193 Aveiro, Portugal

Keywords: glass-ceramics, crystallisation, adhesion, apatite layer, scratch test, bioactivity.

Abstract: In this work glasses with nominal molar compositions 0.45SiO$_2$–(0.45-x)MgO–xK$_2$O–0.1(3CaO.P$_2$O$_5$) (x=0 and 0.09) were heat-treated according to adequate time-temperature schedules to obtain glass-ceramics (GCs) with phosphate phases. The GCs were immersed for different periods of time in SBF at 37°C. The morphology, crystallinity, and adhesion of the apatite layers thus formed are compared and discussed for the glasses and the GCs. Adhesion measurements were performed for the GCs using the scratch test technique. Distinct adhesion behaviours of the newly formed apatite layers to the substrate were observed when the GCs composition is K-free or when it contains K. Adhesion parameters for the apatite coating on the K-containing GCs showed to be affected by the immersion time in SBF but were not time-dependent in the case of the K-free GCs.

Introduction

One of the main reasons for the development of GCs for biomedical applications [1] was the possibility of improving the mechanical properties of their parent glasses. Crystallization often decreases the *in vitro* bioactivity of the base material in terms of the apatite layer thickness [2,3]. However it was found that the apatite layers precipitated on the materials, during the *in vitro* acellular tests in Kokubo´s simulated body fluid (SBF) [4], were much more adherent to the GCs than to the glasses. So far the observed differences are not yet completely understood and few explanations are found in the literature.

When the built up apatite coatings are poorly adherent or become quickly absorbed by the patient's body the substrate bare can be left behind, the loosening of the implant can occur and eventually, further revision surgery will be needed.

In a previous work Queiroz et al [5] have developed potentially bioactive glasses of the Si-Mg-Ca-P-K system with poorly attached bone-like apatite layers. In the present study it was intended to improve the apatite coating adhesion by properly heat treating the parent glasses. The relatively high MgO content is believed to contribute for the reinforcement of the layer, since Mg is a conspicuous constituent of the hard bone tissue [6] and it has been successfully used in the osteoporosis treatment. K is expected to promote the biomineralization since it was found to be involved in the process of dentine mineralization [7]. Apatite layers precipitated on the GC samples after immersion in SBF were studied by scratch tests and the adhesion parameters were used to discuss the observed differences.

Materials and Methods

Bulk annealed glass samples of the intended compositions (0.45SiO$_2$–(0.45-x)MgO–xK$_2$O–0.1(3CaO.P$_2$O$_5$), x=0 and 0.09) were prepared as reported elsewhere [8]. Well-crystallized white GCs were then produced from the crystallization of the bulk glass samples, heated at 5 °C.min^{-1} and held for 2 hours at the peak temperatures suggested by the DTA runs (880 °C and 820 °C, respectively, for the K-free and for K-containing composition). X-ray diffraction (XRD), Fourier

held for 2 hours at the peak temperatures suggested by the DTA runs (880 °C and 820 °C, respectively, for the K-free and for K-containing composition). X-ray diffraction (XRD), Fourier transform infrared spectroscopy (FTIR) and scanning electron microscopy (SEM) were used to characterize the GCs, before and after exposure to the SBF solution.

The in vitro bioactivity of the crystallized glasses was assessed by soaking the heat-treated samples in SBF. Prismatic specimens ($10 \times 10 \times 1 mm^3$) with optical polished surfaces were immersed into 10ml SBF until 30 days at 37°C. Analysis of the liquids comprised pH measurements and determination of the concentration of Ca^{2+}, P^{5+}, Si^{4+}, Mg^{2+} and K^+ by inductively coupled plasma – optical emission spectroscopy (ICP).

The adhesion of the apatite layer to the GC substrates was evaluated using scratch tests performed on a Csem-Revetest scratcher with a Rockwell C diamond stylus (hemispherical tip; 200µm diameter), under linearly increasing load (F_n from 10 to 60N) and a specimen translation speed of 5mm min^{-1}, thus yielding 5mm scratch channels. The adhesion was assessed from both *in situ* transverse (wear) force measurements and by light optical microscopy imaging of the coating losses on the scratch tracks, searching for evidence of debris flaking from the apatite layer or massive losses were the coating fails catastrophically [9,10].

Results and Discussion

Heat treated samples exhibited the presence of witlockite-type orthophosphate phases, $Ca_7Mg_2(PO_4)_6$ for GCs with x=0 and $Ca_9MgK(PO_4)_7$ for GCs with 0.09, as major crystalline phases.
Immersion in SBF gave rise to a rapid initial leaching of Si^{4+}, Mg^{2+} and K^+ from the glass and the GCs samples, particularly for the higher K_2O contents, resulting in a favourable alkaline reaction, responsible for the precipitation of needle-like crystals that tend to become thicker as time increases, covering the whole surface (Fig. 1). The increase in pH was found higher for the glasses than for the GCs samples. From the ICP analysis it become apparent that a calcium-phosphate precipitate was probably forming on the GCs surfaces after immersion in SBF. FTIR and low angle XRD spectra confirmed the formation of an apatite layer over the parent glasses and the respective GCs.

SEM observations showed that thick and regular SBF apatite layers precipitated on the parent glass samples, although weakly attached or even detached from them. By the contrary, apatite deposits on the GCs were thinner and not uniformly distributed over the whole surface, except for the higher immersion times. Improved adhesion was however detected for the GCs, as discussed next.

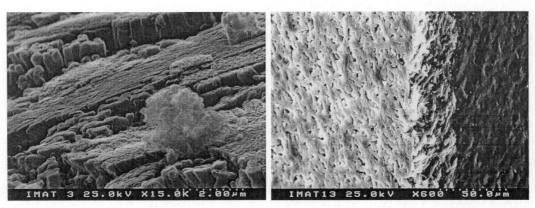

Fig. 1. SEM-micrograph of the whitlockite-apatite layer and the corresponding onset of apatite deposition, for a plate sample of composition $0.45SiO_2–(0.45-x)MgO–xK_2O–0.1(3CaO.P_2O_5)$ with x= 0.09, soaked in SBF for 7 days (left) and 15 days (right).

The scratch plots (Fig 2) showed that the apatite precipitates were deeply anchored to the GCs substrates and the films generally fail under a 25-40N load range, just prior to the GCs substrates themselves (typically above 40N load). Under substrate failure both the noise level of the transverse force (F_t) and the acoustic emission show higher contributions, while the F_t versus F_n slope also tends to increase. For the K-enriched GCs the slope increases with the soaking period while for the K-free GCs this is not apparent and lower slopes are observed. These results suggest that the composition of the glassy degradable phase is of major importance for controlling the adhesion apatite/GCs substrates.

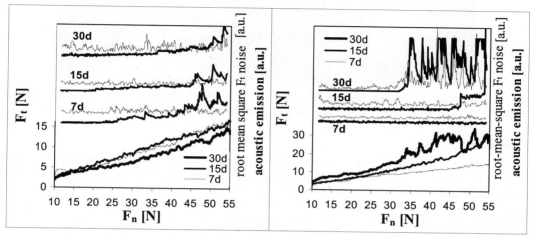

Fig. 2. Scratch tests for GCs with x=0 (left) and 0.09 (right), soaked in SBF for 7, 15 and 30 days. The transverse (F_t) versus perpendicular force (F_n) plots are displayed on the bottom part of the graphics. Plots on the top represent both the acoustic emission and the high-frequence component of F_t (noise) as obtained through fast Fourier analysis on the frequence domain.

The formation of the apatite precipitates on the GC substrate seems to be the result of two mechanisms acting simultaneously: (i) the intersticial amorphous phase is easily leached out from the GCs microstructure thus leaving out the interlocked whitlockite structure, and (ii) the apatite crystals, formed with the contribution of SBF ions, nucleate and grow over the less soluble whitlockite structure, possibly starting at nucleation sites (silanol groups Si-OH) at the glass phase dissolving level [11].

The improved adherence of the apatite layer to the K-containing GC, compared to the K-free GC, may be attributed to the presence of K, a quite effective modifier in the glassy network, which is known to promote degradation in aqueous solutions and allow the control of biodegradability [12]. Under physiological conditions the dissolution of the glass interstitial phase could be expected to reduce the gradient in elastic module between the GC implant and the bone, thus reducing the difference on their interfacial response to applied stress [13], although GC implants have been so far restricted to statically loaded implants due to the mismatch of elastic module [14].

The above results were paralleled by optical microscope observation of the samples. Fig. 3 shows the difference between the tracks produced in the scratch tests on the apatite layers precipitated over the K-free and over the K-containing GCs, after 15 days immersion in SBF. In accordance with the plots shown in Fig.2, failure of the layer on the K-free GC is occurring earlier and in a more catastrophic way.

In the above discussion the differences in the glassy phase compositions, and the effect of surface topography and roughness on the adherence of apatite layers to GC substrates have been considered. Further studies are needed on the relationship between chemical and crystallographic characteristics of the phases precipitated *in situ* within the heat treated glasses and the apatite-like phases growing on the samples after SBF immersion.

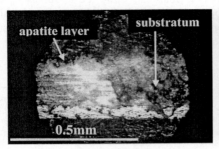 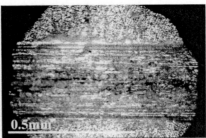

Fig. 3.   Optical microscope photographs of wear tracks and debris formed on apatite layers deposited on K-free (left) and on K-containing (right) GCs after 15 days of immersion in SBF.

Conclusions

Apatite layers precipitated on GCs samples, after immersion in SBF, exhibited improved adhesion to the base material, compared with the apatite layers deposited on the respective parent glasses. Adhesion of SBF apatite layers to the base GCs showed to be lower for the K-free GCs than for the K-containing GCs. For the latter the immersion time in SBF was seen to affect the adhesion behaviour, in opposition to the K-free GCs, for which adhesion did not change with immersion time. Parameters such as composition of the glassy and crystalline phases, topography and local roughness of the virgin material and of the precipitated apatite, helped to explain those findings.

Acknowledgements

The authors wish to thank Dr. A.J.Fernandes (Physics Department, University of Aveiro, Portugal) for technical assistance in the scratch tests. C.M.Queiroz acknowledges the Portuguese Foundation for Science and technology (FCT, Portugal) for financial support (grant SFRH/ BD/ 1243/ 2000).

References

[1] L. L. Hench: J. Am. Ceram. Soc. Vol. 81(7) (1998) p.1705.
[2] O. Peitl, E. D. Zanotto and L. L. Hench: J. of Non-Cryst. Solids Vol. 292 (2001) p. 115.
[3] Oliveira, J. M., Correia, R. N., Fernandes, M. H.: Biomaterials Vol. 16(11) (1995) p. 849.
[4] T. Kokubo, H. Kushitani, S. Sakka, T. Kitsugi and T. Yamamuro: J. Biomed. Mater. Res. Vol. 24 (1990) p.721.
[5] C.M.Queiroz, S. Agathopoulos, J.R.Frade and M.H.Fernandes, Mater. Sci. Forum (2003) *in press.*
[6] T. Aoba, E. C. Moreno and S. Shimoda: Calcif. Tissue Int. Vol. 51(2) (1992), p. 143.
[7] H. P. Wiesmann, U. Plate, K. Zierod and H. J. Hohling: J. Dent. Res. Vol. 77(8) (1998), p.1654.
[8] C.M.Queiroz, M.H.Fernandes and J.R.Frade: Phys. Chem. of Glasses Vol. 43C (2002) p. 281.
[9] J. Sekler, P. A. Steinmann and H. E. Hintermann: Surf. and Coat. Tech., Vol. 36 (1988) p. 519.
[10] B. Bhushan, X. Li: J. Mater. Res. Vol. 12 (1997), p. 54.
[11] T. Kokubo, H.-M. Kim, M. Kawashita and T. Nakamura: J. Aust. Ceram. Soc. Vol. 36(1) (2000) p. 37.
[12] G. Berger, R. Sauer and G. Steinborn: *Proceedings of the 4th International Otto-Schott-Coloquium* (Jena, 1990), p. 50.
[13] L. L. Hench: J. Am. Ceram. Soc. Vol. 74(7) (1991) p. 1487.
[14] G. Steinborn, G. Berger, G. Neumann and W. Knöfler: *Proceedings of the 4th International Otto-Schott-Coloquium* (Jena, 1990), p. 53.

II. BONE GRAFTS AND CEMENTS

Key Engineering Materials Vols. 254-256(2004) pp. 161-164
online at http://www.scientific.net

Potassium Containing Apatitic Calcium Phosphate Cements

F.C.M.Driessens[1] and M.G.Boltong[2]

[1]University of Nijmegen, Pr. Beatrixlaan 3, NL-6109 AH Ohe en Laak, the Netherlands

[2]Calcio B.V.,Ohe en Laak, The Netherlands, trudydriessens@hotmail.com

Keywords: Potassium containing apatite, calcium phosphate cement, calcium deficient hydroxyapatite, calcium potassium phosphate, calcium potassium sodium phosphate

Abstract: Previously it has been shown that potassium containing nanoapatites can precipitate in the range $0.8 < Ca/P \leq 1.5$, when mixtures of MCPM, CPP, CPSP and α-TCP are mixed with water. In this study it is found that similar cement-like products are formed by using powders of CPP, CPSP and/or α-TCP and mixing them with phosphoric acid or acetic acid or citric acid. Also, α-TCP can be used with a solution of KH_2PO_4 and K_2HPO_4. Another possibility is a combination of CPP and/or CPSP and α-TCP and KH_2PO_4. Setting times varied from 2 to 15 min. Compressive strengths after 3 days of immersion in 0.9% saline solution at 37°C could go up to 35 MPa. In all these cases the cements contained potassium containing apatite in combination with calcium deficient hydoxyapatite.

Introduction

Calcium phosphate cements CPC's consist of a powder and a liquid. The powder consists of a mixture of solid compounds part of which can dissolve in the liquid upon mixing. The liquid then becomes supersaturated and a precipitate is formed which contains at least one calcium phosphate. By entanglement of the crystals of the precipitate the mass of cement gains strength. In Table 1 the calcium phosphates involved in this study are mentioned.

Table 1. Names, abbreviations and formulas of some calcium phosphates

Abbreviation	Name	Formula
MCPM	monocalcium phosphate monohydrate	$Ca(H_2PO_4)_2.H_2O$
CPP	calcium potassium phosphate	$CaKPO_4$
CPSP	calcium potassium sodium phosphate	$Ca_2KNa(PO_4)_2$
KCA	potassium containing apatite	$Ca_5K_{4-x}Na_x(HPO_4)_4(PO_4)_2(H_2O)$
CDHA	calcium deficient hydroxyapatite	$Ca_9(HPO_4)(PO_4)_5(OH)$
HA	hydoxyapatite	$Ca_{10}(PO_4)_6(OH)_2$
α-TCP	alfa-tertiary calcium phosphate	α-$Ca_3(PO_4)_2$
DCP	dicalcium phosphate	$CaHPO_4$
TTCP	tetracalcium phosphate	$Ca_4(PO_4)_2O$

Brown and Chow [1] found that CPC's of HA are formed as follows

$$2\, DCP + 2\, TTCP + H_2O \rightarrow HA \qquad (1)$$

We found [2] that CPC's of CDHA are formed by the reaction

$$3\, \alpha\text{-}TCP + H_2O \rightarrow CDHA \qquad (2)$$

Also, CPC's consisting of mixtures of HA and CDHA can be formed from reactions of mixtures of DCP, TTCP and α-TCP. Moreover these apatites can incorporate sodium and carbonate ions so that they cover the range $1.50 \leq Ca/P < 1.85$ [3].

Later on, we synthesized KCA either from a combination of MCPM with CPP or from a combination of MCPM with CPSP. Such cements set upon mixing with water.[4]. Initial setting times varied from 8 to 14 min and final setting times from 18 to 26 min.. The compressive strength after soaking for 18 h in 0.9% saline solution at 37°C did not exceed 13 MPa.

By combination of CPP, CPSP and MCPM in the cement powders it was possible to obtain mixed cements of the general formula mentioned in Table 1 for KCA. Upon soaking for several weeks in saline solutions or simulated body fluids, which were refreshed every other day, such cements were transformed into nearly pure CDHA and thus had lost K^+, Na^+ and phosphate.

Whereas the setting of HA cements as well as that of CDHA cements is practically completely inhibited by Mg^{2+} ion concentrations, as occurring in body fluids, this is not so with the setting of KCA cements [5]. Another difference is the pH range in which these cements precipitate. HA cements precipitate in the range $10 < pH < 11.5$, CDHA cements in the range $8 < pH < 9.5$ and KCA cements in the range $6 < pH < 7.5$. The pH of a slurry of CPP or of CPSP is close to 8, so that from these slurries no spontaneous setting of KCA is expected to occur. But the use of weak acids might be sufficient to bring the pH into the desired range, so that setting might then occur. It was the first purpose of the present study to investigate this aspect.

In a previous study [5] we found that mixed type KCA and CDHA cements could be obtained from combinations of CPP, CPSP, MCPM and α-TCP. Initial setting times t_I varied from 1.5 to 26 min and final setting times t_F from 5 to 53 min. The compressive strength after soaking in 0.9% saline solution at 37°C for 18 h could go up to 47 MPa. The solubility behaviour of these mixed type KCA and CDHA cements was completely determined by their KCA component. Therefore, the precipitate could not be a solid solution, but should be characterized as a conglomerate of molecular domains of KCA and CDHA within the nanocrystals [5].

Theoretically, as far as their composition is concerned, it should be possible to obtain such mixed type KCA and CDHA cements also from   combinations either of CPP, CPSP, α-TCP and KH_2PO_4 or of α-TCP, K2HPO4 and KH_2PO_4. This might occur by the reactions

$$2\ CPP + \alpha\text{-}TCP + 2\ KH_2PO_4 + H_2O \rightarrow KCA \qquad (3)$$

(or the equivalent reaction with CPSP) and

$$5\ \alpha\text{-}TCP\ +\ 4\ K_2HPO_4 + 4\ KH_2PO_4 + 3\ H_2O \rightarrow 3\ KCA \qquad (4)$$

The second purpose of the present investigation was to study the feasibility of these reactions (3) and (4) and the properties of the obtained cements.

Materials and Methods

CPP was prepared by heating a 1:2 M mixture of K_2CO_3 and DCP at 1000°C for 2 h and quenching, and subsequent milling in a ball mill. CPSP was prepared by heating a 1:1:4 M mixture of K_2CO_3, Na_2CO_3 and DCP at 1000°C for 2h, followed by quenching and milling. α-TCP was prepared by heating a 1:2 M mixture of $CaCO_3$ and DCP at 1350°C for 2h, followed by quenching and milling. PHA (precipitated HA) was used as obtained.

The following weak acids were selected for testing, whether CPP and/or CPSP in combinations with slightly acidic solutions could result in setting: acetic acid, citric acid, lactic acid and phosphoric acid. They were diluted down to the desired concentration.

The compounds K_2HPO_4 and KH_2PO_4 were used as obtained.

Results

The data of Table 2 show that some acids are suitable to make cements out of combinations of CPP and α-TCP or of CPSP and α-TCP. Initial and final setting times are mostly in the desired range and the compressive strength after soaking in a 0.9% saline solution for 18 h or longer at 37°C was fairly high.

Table 2. Initial setting time t_I (min), final setting time t_F (min) and compressive strength CS (MPa) of cements obtained from different powder-liquid combinations

Powder	Liquid	L/P	t_I	t_F	CS
20g CPP+60g α-TCP+1.6gPHA	2N acetic acid	0.5	3.5	9	24
	1N H_3PO_4	0.4	7	12	24
	1.5N citric acid	0.35	7.5	24	23
	2N lactic acid	0.5	>20	-	-
20g CPSP+60g α-TCP+1.6gPHA	2N acetic acid	0.45	7	11	26
	1N H_3PO_4	0.4	7	16	30
	1.5N citric acid	0.35	7	18	34
	2N lactic acid	0.5	>20	-	-

Much more work has still to be done to optimize these cement formulations.

In order to test the occurrence of reaction (3) we prepared a powder containing 15g CPP, 2g KH_2PO_4, 60g α-TCP and 1.6g PHA. Taking water as the cement liquid and a water/powder ratio of 0.4 ml/g, we obtained a t_I of 4.5 min and an 18h compressive strength of 14 ± 2 MPa. With CPSP in stead of CPP t_I was 6.5 min and the 18h compressive strength 15 ± 3 MPa.

The occurrence of reaction (4) was tested by making a mixture of 60g α-TCP, 22g KH_2PO_4 and 8g K_2HPO_4. Taking water as cement liquid, the resulting t_I was 1.5 min and t_f 7.5 min at a W/P ratio of 0.35. Similarly, with 65g α-TCP and a cement liquid of 2M K_2HPO_4 + 0.75M KH_2PO_4 at a L/P ratio of 0.45 t_I was 9.5 min and the 24h strength 15 MPa.

Conclusions

1.Mixtures of CPP, and/or CPSP and α-TCP and PHA react with aqueous solutions of some weak acids to form calcium phosphate cements

2.Mixtures of CPP and/or CPSP and α-TCP react with KH_2PO_4 and water to form calcium phosphate cements

3.Mixtures of α-TCP with K_2HPO_4 and KH_2PO_4 and water react to form calcium phosphate cements

4.In view of the previous studies it is concluded that these cements present potassium containing apatite in combination with calcium deficient hydroxyapatite

References

[1] W.E.Brown and L.C.Chow, J.Dent.Res., Vol. 62(1983), p. 672

[2] F.C.M.Driessens, Fourth Euro Ceram, Vol. 8(1995), p.77-83

[3] Y.Miyamoto, T.Toh, K.Ishikawa, T.Yuasa, M.Nagayama and K.Suzuki, J. Biomed. Mater. Res., Vol. 54(2001), p. 311-319

[4]F.C.M.Driessens, E.A.P. de Maeyer, E.Fernandez, M.G.Boltong, G.Berger, R.M.H.Verbeeck, M.P.Ginebra and J.A.Planell, Bioceramics , Vol 9(1996), p. 231-234

[5]F.C.M.Driessens, M.G.Boltong, E.A.P. de Maeyer, R.Wenz, B.Nies and J.A.Planell, Biomaterials Vol. 23(2002), p. 4011-4017

Key Engineering Materials Vols. 254-256(2004) pp. 165-168
online at http://www.scientific.net
© 2004 Trans Tech Publications, Switzerland

HRTEM and EDX Investigation of Microstructure of Bonding Zone between Bone and Hydroxyapatite *In Vivo*

Q.Z. Chen[1], W.W. Lu[2], C.T. Wong[2], K.M.C. Cheung[2], J.C.Y. Leong[2] and K.D.K.Luk[2]

[1] Department of Mechanical Engineering, The University of Hong Kong, Hong Kong, P.R. China, email: qzchena@hkucc.hku.hk

[2] Department of Orthopaedic Surgery, The University of Hong Kong, Hong Kong, P.R. China, email: wwlu@hkusua.hku.hk

Keywords: Bone, Hydroxyapatite, Microstructure, Interface, Transmission electron microscope

Abstract. An investigation of the microstructure of the bonding region between hydroxyapatite (HA) and host bone *in vivo* has been undertaken using high-resolution transmission electron microscope (HRTEM). The TEM observation showed that the bonding region is bone-like, but not identical to bone; and that it is nanostructured with crystallites smaller than 10 nm. A continuous structure had been established crossing the interface between HA and bone-like region at 3 months after implantation. EDX analysis showed that the bone-like zone had higher levels of calcium and phosphate than bone. In addition, there was an amorphous layer on the surface of HA particles at this stage. These findings indicate that the transformation from crystalline to amorphous HA had occurred prior to HA biodegradation, and that the mechanism of the formation of bonding zone might have involved the dissolution of amorphous HA in surrounding solution and precipitation of nanocrystalline HA in the over-saturated solution. The good mechanical properties of the bonding region can be attributable to its nanostructure, high levels of calcium and phosphate, and chemical bonding crossing the interface between the bonding-zone and HA particles.

Introduction

Calcium phosphates have been extensively investigated as important raw materials used for the repair of bone defects [1]. The attractiveness of calcium phosphates is their ability to become chemically and thus tightly bonded to bone, particularly hydroxyapatite (HA) [2, 3, 4, 5]. The mechanism of bone-ceramics bonding, however, is not yet completely clear. There have been a number of SEM and TEM investigations on the interfaces between bone and HA [2-5, 6, 7, 8, 9], but they delivered conflicting findings. An amorphous bonding zone was reported first by Jarcho [2], and later by others [3-5]. Tracy and Doremus [6] found that the bone at the HA ceramic surface was the same as normal bone. Daculsi [8] and other subsequent authors [5, 9] observed, between bone and HA, a layer of biological apatite different from bone apatite. Based on these inconsistent findings, different mechanisms have been proposed for bone-ceramic bonding [4-5, 8-9]. It is therefore necessary to carry out a careful high resolution TEM investigation on the interface of bone and HA. This study investigated the structure of the bonding zone between HA and host bone, with the aim of elucidating the bonding mechanism.

Materials and experimental procedures

The study was conducted *in vivo* on white rabbits. The bioactive bone cement contained a filler blend of mainly strontium-containing hydroxyapatite (Sr-HA) and a resin blend of bisphenol A diglycidylether dimethacrylate (Bis-GMA) and triethylene glycol dimethacrylate (TEGDMA). Twenty-one New Zealand white rabbits were used. The rabbits were 4 to 6 months old and weighed from 2.5 to 3.9 kg. The animals were operated on under general anaesthesia. A cavity of 3 mm in diameter and 12 mm in depth was drilled in ilium. The bioactive bone cement was injected into the ilium and observed for 1, 3 and 6 months, respectively. The animals were killed after the observation periods. The undecalcified bulk samples were dried and embedded. TEM foils were prepared by ultra-thin sectioning using a diamond knife and were examined using a Tecnai 20 microscope at 200 kV and energy-dispersive X-ray (EDX) analysis was carried out in a Philip CM 30 microscope at 200 kV.

Results and discussion

Collagen fibres were formed on and parallel to the surface of bone at 1 month after surgery (Fig. 1a). A number of multi- and mono-nuclear cells were located at the bone surface. Cement debris frequently appeared around some cells (Fig. 1b). They might have been phagocytosed and dissolved by these cells [10].

At 3 months after surgery, direct contact between bone and cement was established (Fig. 2a). However, there was a bone-like zone between the typical matured bone and cement, which was not identical to bone. HRTEM observation revealed that the bone-like region was nanostructured (Fig. 2b). Most nanocrystallites were 1-2 nanometres in diameter. There was an amorphous layer on the surface of the crystalline HA particle (Fig. 2b). This suggests that the transformation of crystalline HA into amorphous HA had occurred prior to its biodegradation. It is thus hypothesized that the bonding region is formed through following processes: (i) firstly the crystalline HA transforms into amorphous HA, (ii) then the amorphous HA dissolves into the surrounding solution, resulting in over-saturation; and (iii) finally the nanocrystallites are precipitated from the over-saturated solution. It is well known that the crystalline state is more stable than the amorphous state in terms of thermodynamic energy. Hence, the amorphous HA is probably an intermediate stage in the transformation from crystalline HA to nanostructured HA. The energy barrier can be surmounted with the assistance of the biological system. It is likely that the transition from crystalline to amorphous HA is the controlling factor in the process of surface biodegradation of crystalline HA, as it is energy-increasing, and thus the slowest step. This finding can explain several previously reported phenomenon, e.g. the amorphous HA dissolves faster than crystalline HA [11]. Carbonate substitution has been shown to cause a reduction in the crystallinity of HA [12], and carbonated HA biodegrades faster than pure HA [13].

Table 1 Chemistry of bone, bonding zone and HA (at %, by EDX in TEM)

	Bone	Bonding zone	Hydroxyapatite
O	37.27	29.01	23.81
P	21.94	23.35	28.43
Ca	40.81	47.41	46.23
Sr		0.23	1.54

It was also found that at 3 months a continuous structure at the atomic level had been built up crossing the interface between HA and bonding zone (Fig. 3), indicating that a chemical bonding had been established. EDX analysis revealed that the bonding zone had higher levels of calcium and phosphate than bone (Table 1). It can be concluded that the nanostructure, higher levels of calcium and phosphate in the bonding zone and the chemical bonding crossing the interface were responsible for the good combination of strength and toughness in the bonding region [10].

At 6 months after implantation, the matured bone contacted the cement directly (Fig. 4). Eventually HA particles were completely surrounded by bone matrix. Mechanical locking, therefore, should be the main mechanism of load transfer between bone and cement at this stage.

Conclusions

The transformation of crystalline HA into amorphous HA occurred prior to its biodegradation, which was the controlling factor of the biodegradation rate of crystalline HA. The formation of the bonding region involved the transformation of crystalline HA into amorphous HA, which dissolved into the surrounding solution, resulting in over-saturation, and precipitation of nano-crystallites from the over-saturated solution.

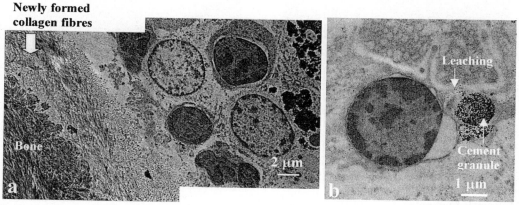

Fig. 1. (a) New collagen fibres were formed at 1 month after implantation. (b) Cement granules were present near the nucleus of a cell.

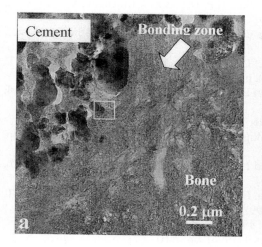
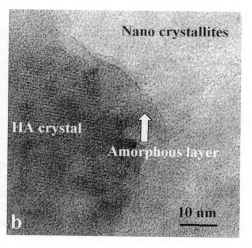

Fig. 2 HA particles contacted bone directly at 3 months after surgery. (a) The bonding zone between cement and bone was bone-like, but not identical to bone. (b) HRTEM images of the framed position in (a).

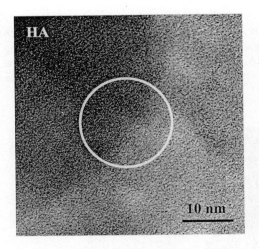

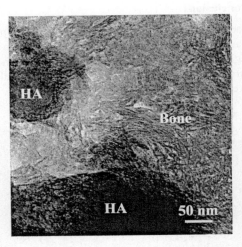

Fig. 3. Continuous structure was observed
crossing the interface of HA and bonding
zone at 3 months after implantation.

Fig. 4. HA particles were completely surrounded
by matured bone at 6 months after
implantation.

Acknowledgements

The authors thank Dr. Alfonso W.H. Ngan for his helpful comments on the paper.

References

[1] M. Jarcho: Clin. Orthop. Rel. Res., Vol. 157 (1981), pp. 259-278

[2] M. Jarcho, J.F. Kay, K.I. Gumar, R.H. Doremus, and H.P. Drobeck: J. Bioengin., Vol. 1 (1977), pp. 79-92

[3] H.W. Denissen, K. de Groot, P. Kakkes, A. van den Hooff, and P.J. Klopper: J. Biomed. Mater. Res., Vol. 14 (1980), pp. 713-721

[4] M. Neo, S. Kotani, Y. Fujita, T. Nakamura, T. Yamamura, Y. Bando, C. Ohtsuki, and T. Kokubo: J. Biomed. Mater. Res., Vol. 26 (1992), pp. 255-267

[5] J.D. de Bruijn, J.S. Flach, K. de Groot, C.A. van Blitterswijk and J.E. Davies: Cells Mater., Vol. 3 (1993), pp. 115-127

[6] B.M. Tracy and R.H. Doremus, J. Biomed. Mater. Res., Vol. 18 (1984), pp. 719-726

[7] G.L. de Lange, C. de Putter, and F.L.J.A. de Wijs: J. Biomed. Mater. Res., Vol. 24 (1990), pp. 829-845

[8] G. Daculsi, R.Z. LeGeros, M. Heughebaert, and I. Barbieux: Calcif. Tissue Int., Vol. 46 (1990), pp. 20-27

[9] Y. Okada, M. Kobayashi, H. Fujita, Y. Katsura, H. Matsuoka, H. Takadama, T. Kokubo, and T. Nakamura: J. Biomed. Mater. Res., Vol. 45 (1999), pp. 277-284

[10]U. Gross and V. Strunz: J Biomed. Mater. Res. 19 (1985), pp. 251-271

[11] R. Z. LeGeros, and J. P. LeGeros: Bioceramics 9 (1996), pp. 7-10.

[12] A. Milev, G.S.K. Kannangara and B. Ben-Nissan: Bioceramics 15 (2002), pp. 481-484

[13] T. Sakae, A. Ookubo, R.Z. LeGeros, R. Shimogoryou, Y. Sato, S. Lin, H. Yamamoto, and Y. Kozawa: Bioceramics 15 (2002), pp. 395-398

Key Engineering Materials Vols. 254-256(2004) pp. 169-172
online at http://www.scientific.net
© *2004 Trans Tech Publications, Switzerland*

Acetabular Construction Using A Glass-Ceramic Artificial Bone Graft

Satoru Yoshii[1], Masanori Oka[2], Mitsuhiro Shima[1] and Takao Yamamuro[3]

[1] Institute of Biomedical Engineering, Kansai Denryoku Hospital, Fukushima 2-1-7, Fukushima-ku, Osaka 553-0003 Japan, K-20433@kepco.co.jp

[2] Department of Tissue Regeneration, Field of Clinical Application, Institute for Frontier Medical Sciences, Kyoto University, 53 Syogoin-Kawahara-cho,Kyoto 606-8507 Japan

[3] Department of Orthopaedics, Faculty of Medicine, Kyoto University, 54 Syogoin-Kawahara-cho, Kyoto, 606-8507 Japan

Keywords: artificial bone, glass, ceramic, acetabular dysplasia, bone construction.

Abstract. We have developed a block of glass-ceramic to augment the dysplastic acetabulum. Four cases of acetabular dysplasia underwent implantation of a block of glass-ceramic on the lateral surface of the ilium, a pelvic bone, just above the hip joint. In these cases, the mean Harris hip score was 46±8 points pre-operatively and 85±13 points 4 to 7 years post-operatively. Our procedure is unique in that we use a bioactive glass-ceramic block as an artificial bone to construct the acetabular bone. The results appear to be satisfactory.

Introduction

Failure of normal acetabular development is inevitable in congenital dislocation of the hip if it remains unrecognized until late infancy or childhood. The diminished weight-bearing area and increased shear stress cause attrition of the hyaline cartilage, and ultimately degenerative arthrosis and pain [1]. Inadequately supported subluxation of the hip often progresses into a painful and disabling condition in the young adult. Operations on the pelvis to increase the coverage of the femoral head are useful for patients with congenital dysplasia of the hip [2,3]. By increasing the coverage of the femoral head an adequate support to the subluxated femoral head is obtained and this may provide lasting symptomatic relief [4,5]. However, after these operations weight bearing must be avoided for a long time until the graft has become stable. Moreover a thick augmentation requires an abundant bone graft. We have developed a block of bioactive glass-ceramic to solve these problems.

Materials and Methods

Characteristics of the implant. The chemical composition of the glass-ceramic was MgO 4.6, CaO 44.9, SiO_2 34.2, P_2O_5 16.3, CaF_2 0.5 in terms of weight ratio. We made the glass-ceramic by compacting glass powder of the above composition under a pressure of 40 MPa and heating it to 1050°C for sintering and crystallization. The resulting glass-ceramic had high bending strength and a compressive strength of 196 MPa and 1076 MPa, respectively. The Young's modulus was 117600 MPa, and the porosity was 0.7%. The previous studies proved that the glass-ceramic had good bone-bonding capability (3.3 MPa in dog at one year post-operatively) and suggested that it could be used in block to reconstruct an acetabular defect [6,7]. We made a block of the glass-ceramic that measured 30x27.5x20mm (Fig. 1).

Patients. Between June 1996 and November 1998, four cases of acetabular dysplasia had acetabular augmentation using the glass-ceramic block. Pre-operatively, the patients were aggressively treated conservatively with rest, cane and medication. However, all patients had severe pain at times and pain at rest. They had marked limitation of activities. The mean Harris hip score [8] was 46±8 points pre-operatively.

Operation. The lateral iliac surface was exposed subperiosteally over an area of 3x3cm just above the hip joint. The lateral surface of the ilium was planed using an ilium plane of 30mm in diameter. The glass-ceramic block was placed on the hip joint capsule above the center of the femoral head and fixed to the lateral surface of the ilium with two titanium screws. Four weeks after the operation, the cane was discarded and the patients returned to their daily living.

Results

Radiographic Study. Pelvic radiographs were made pre-operatively and at 1-year interval postoperatively under weight-bearing condition. The 4 to 7 years' follow-up radiographs have revealed that the glass-ceramic implants have been securely fixed on the iliac surface. No translucent zone between the implant and the iliac surface, or migration of the implant has been found (Fig. 2). The mean coverage of the femoral head has increased by 30%, compared with the pre-operative measurements (Table 1). The mean center-edge angle of Wiberg has increased by 35 degrees.

Hip Scores. The patients had no or mild pain of operated hip joints in their activity of daily living at 4 to 7 years post-operatively, except case 1. The Harris hip score was determined for all patients at 1-year interval post-operatively. The mean Harris hip score was 85±13 at 4 to 7 years post-operatively (Table 1). The mean Harris hip score has improved 39 points 4 to 7 years postoperatively. The pain score has improved 24 points and the gait score has improved 16 points.

Discussion

The results of our procedure appeared to be satisfactory. The Harris hip score improved by 39 points 4 to 7 years after the operation. Especially, the pain score and gait score improved markedly. The advantages of this surgical procedure are 1) correct and solid fixation of the implant, 2) a shorter operation time, 3) less blood loss since iliac crest is not used, 4) customized sizing of glass-ceramic block to achieve coverage, 5) improved stability of the joint, 6) quantum improvement in Harris hip scores, 7) rapid clinical recovery without requirement for cast or a brace. The bonding that occurred between bone and the glass-ceramic implant was thought to be chemical and biological [6]. According to a previous study, the bonding between bone and the glass-ceramic implant is considered to progress between 2 weeks and 2 months after the implantation [7].
Construction of the acetabulum has been achieved effectively using the artificial bone. The medium-term functional results of the hip joints are excellent. Using the artificial bone one can build a solid and functional bony structure of any shape at any place, as it may be wished.

Table 1 Radiological and clinical results.

Patient	Age	Sex	Coverage of The femoral head	Harris hip score	Pain score	Gait score
1	41	F preop.	60%	51	10	20
		7-yr p.o.	100%	68	20	25
2	54	F preop.	57%	54	10	17
		6-yr p.o.	86%	93	40	30
3	21	F preop.	77%	37	10	10
		6-yr p.o.	100%	96	44	33
4	43	F preop.	51%	41	10	10
		4-yr p.o.	76%	83	30	30
mean±S.D.	40±14	preop.	61±11%	46±8	10±0	14±5
		postop.	91±12%	85±13	34±11	30±3

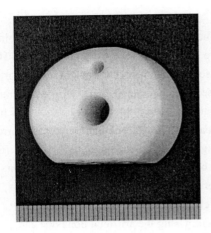

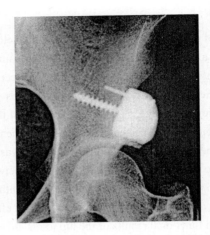

Fig. 1 A block of the glass-ceramic 30x27.5x20mm. It has 2 holes, one 6.5mm and one 2.8mm in diameter.

Fig. 2 A radiograph 6 years after the operation. The coverage of the femoral head has increased by 23%. New bone formation is found under the implant.

References

[1] B.R.T. Love, P.M. Stevens and P.F. Williams: J. Bone Joint Surg. Vol. 62 (Br) (1980),   p.321

[2] D.M. Bosworth, J.W. Fielding, T. Ishizuka and R. Ege: J. Bone Joint Surg. Vol. 43 (Am) (1961) p.93

[3] D. Wainwright: J. Bone Joint Surg. Vol. 58 (Br) (1976) p.159

[4] F. König: Verh. Dtsch. Ges. Chir. Vol. 20(1891) p.75

[5] R.E. White Jr. and F.C. Sherman: J. Bone Joint Surg.   Vol. 62 (Am) (1980) p.928

[6] S. Yoshii, Y. Kakutani, T. Yamamuro, T. Nakamura, T. Kitsugi, M. Oka, T. Kokubo and M. Takagi: J. Biomed. Mater. Res. Vol. 22 (1988) p.327

[7] S. Yoshii, T. Yamamuro, T. Nakamura, M. Oka, H. Takagi and S. Kotani: J. Appl. Biomat. Vol. 3 (1992) p.245

[8] W.H. Harris: J. Bone Joint Surg. Vol. 51(Am) (1969) p.737

Key Engineering Materials Vols. 254-256(2004) pp. 173-176
online at http://www.scientific.net
© 2004 Trans Tech Publications, Switzerland

In Vivo Aging Test for Bioactive Bone Cements Composed of Glass Bead and PMMA

S. Shinzato[1], T. Nakamura[2], K Goto[3] and T. Kokubo[4]

[1] Department of Orthopaedic Surgery, Moriyama Municipal Hospital, Moriyama 4-14-1, Moriyama City, Shiga, 524-0022, Japan, shinzato@mbox.kyoto-inet.or.jp

[2, 3] Department of Orthopaedic Surgery, Graduate School of Medicine, Kyoto University, Kawahara-cho 54, Shogoin, Sakyo-ku, Kyoto 606-8507, Japan

[4] Department of Material Chemistry, Graduate School of Engineering, Kyoto University, Yoshida-honmachi, Sakyo-ku, Kyoto 606-8501, Japan

Keywords: aging test; bending strength; bioactive bone cement; glass; polymethyl methacrylate

Abstract. The degradation of a bioactive bone cement (GBC), consisting of bioactive MgO-CaO-SiO_2-P_2O_5-CaF_2 glass beads and high-molecular-weight polymethyl methacrylate (hPMMA) was evaluated in an *in vivo* aging test. Hardened rectangular specimens were prepared from two GBC formulations (containing 50%w/w [GBC50] or 60%w/w [GBC60] bioactive beads) and a conventional PMMA bone cement control (CMW-1). Initial bending strengths were measured using the three-point bending method. Specimens of all three cements were then implanted into the dorsal subcutaneous tissue of rats, removed after 3, 6 or 12 months, and tested for bending strength. The bending strengths (MPa) of GBC50 at baseline (0 months), 3, 6 and 12 months were 136±1, 119±3, 106±5 and 104±5, respectively. Corresponding values were 138±3, 120±3, 110±2 and 109±5 for GBC60, and 106±5, 97±5, 92±4 and 88±4 for CMW-1. Although the bending strengths of all three cements decreased significantly ($p<0.05$) from 0 to 6 months, those of GBC50 and GBC60 did not change significantly thereafter, whereas that of CMW-1 declined significantly between 6 and 12 months. Thus, degradation of GBC50 and GBC60 does not appear to continue after 6 months, whereas CMW-1 degrades progressively over 12 months. Moreover, the bending strengths of GBC50 and GBC60 (especially GBC60) were significantly higher than that of CMW-1 throughout. We believe that GBC60 is strong enough for use under weight-bearing conditions and that its mechanical strength is retained *in vivo*.

Introduction

Polymethyl methacrylate (PMMA) bone cement is widely used for the fixation of prostheses or as a bone substitute in orthopedics. However, conventional bone cement is unable to bond to bone and has poor mechanical properties. Recently, bioactive bone cements have been developed to overcome these disadvantages. We have developed a new bioactive bone cement (GBC) that showed excellent osteoconductivity and strong mechanical strength [1-8]. The purpose of the present study was to evaluate the degradation of GBC in an *in vivo* aging test [9,10].

Materials and methods

Preparation of the three types of cement

Preparation of bioactive glass beads: Powder with a nominal composition (%w/w) of MgO, 4.6; CaO, 44.7; SiO_2, 34.0; P_2O_5, 16.2 and CaF_2, 0.5 was melted in a SiC furnace at 1500°C for 4 h. The melt was quenched between water-cooled steel rollers, formed into ribbons and pulverized in an alumina-ball mill to produce a glass powder. This was subsequently softened using a burner and quenched to produce spherical glass beads [2,3]. The density of the bioactive glass beads was 2.9 g/cm³, the average particle size was 3.2 μm and the specific surface area was 0.7 m²/g. The bioactive glass beads were treated with γ-methacryloxy propyl trimethoxy silane (Nippon Unicar

Co. Ltd., Tokyo) at a ratio of 0.2%w/w, then dried at 110°C for 2 h. A polymerization initiator, benzoyl peroxide (Wako Pure Chemical Industry, Osaka, Japan), was added to the treated glass bead filler at a ratio of 1.0%w/w. GBC formulations containing 50 or 60%w/w of the bioactive glass bead filler were prepared; the resulting mixtures were designated GBC50 and GBC60, respectively.

Preparation of hPMMA powder: hPMMA powder was synthesized by suspension polymerization [4]. The particles were spherical, with an average diameter of 5 μm (S.D., 2 μm) and an average molecular weight of 270,000. The proportions of hPMMA powder used in GBC50 and GBC60 were 25 and 19%w/w, respectively.

A liquid was prepared from methyl methacrylate (MMA) monomer (Wako). A polymerization accelerator, N,N-dimethyl-p-toluidine (Wako), was dissolved in the MMA monomer at a concentration of 2.0%w/w. The proportions of MMA monomer liquid used in GBC50 and GBC60 were 25 and 21%w/w, respectively.

GBC50 and GBC60 were prepared by mixing the appropriate amount of liquid with the premixed treated glass bead filler and hPMMA powder for 1 min; polymerization occurred within 5–6 min.

Conventional PMMA cement (CMW-1, DePuy International Ltd., Blackpool, UK) was prepared by mixing the liquid and powder components for 1 min; polymerization occurred within 5–6 min.

Preparation of cement specimens

Blocks of polymerized cement (50x80x4 mm) were cut into rectangular specimens (20x4x3 mm) using a slicing machine (model TSK-40205M; Tokyo Seiki Co., Tokyo, Japan) and their surfaces were abraded with No. 2000 alumina paper. They were cleaned in a distilled-water-filled ultrasonic cleaner for 10 min and sterilized conventionally with ethylene oxide gas.

Initial mechanical testing of the cement specimens

The bending strength of GBC50, GBC60 and CMW-1 were measured to provide baseline values. After soaking in phosphate-buffered saline solution (PBS) for one day, their bending-strength was tested on a materials testing machine (model AG-10TB; Shimadzu Co., Kyoto, Japan), run at a cross-head speed of 0.5 mm/min, using the three-point bending method [3-6].

***In vivo* aging test for the cement specimens**

Eight-week-old male Wistar rats weighing approximately 200g were used. Surgery was conducted under general anesthesia (Nembutal: 40 mg/kg body weight). The dorsum of each rat was shaved and sterilized with 70% ethanol. Four symmetrical 1 cm incisions were made through the dorsal skin of each rat and a rectangular cement specimen was implanted into the subcutaneous tissue through each incision [9,10]. The skin incisions were closed with nylon sutures. A total of 27 rats were used, of which 9 received GBC50, 9 received GBC60 and 9 received CMW-1. Thus, 36 specimens of each cement were implanted.

At each assessment time-point (3, 6 and 12 months after implantation), 12 specimens were removed. After macroscopic examination and soaking in physiological saline, bending-strength tests were performed using the three-point bending method as described above for the initial mechanical test. After mechanical testing, the specimens were embedded in conventional resin (Epofix; Struers Co., Copenhagen, Denmark). New surfaces were created 2 mm below each break point by abrasion and polishing with No. 4000 alumina paper. The specimens were then subjected to SEM observation.

Results

By macroscopic examination, No surface changes or surface breakages were observed on the any cement specimens taken from the rats.

Figure 1 shows the results of bending strength tests. The values (MPa, n = 12) for GBC50 at the initial condition (0 months), 3, 6 and 12 months were 136±1, 119±3, 106±5 and 104±5, respectively. Corresponding values were 138±3, 120±3, 110±2 and 109±5 for GBC60, and 106±5,

97±5, 92±4 and 88±4 for CMW-1. Although the bending strengths of all three cements decreased significantly ($p<0.05$) from 0 to 6 months, those of GBC50 and GBC60 did not change significantly thereafter, whereas that of CMW-1 declined significantly between 6 and 12 months. Thus, degradation of GBC50 and GBC60 does not appear to continue after 6 months, whereas CMW-1 degrades progressively over 12 months. Moreover, the bending strengths of GBC50 and GBC60 (especially GBC60) were significantly higher than that of CMW-1 throughout.

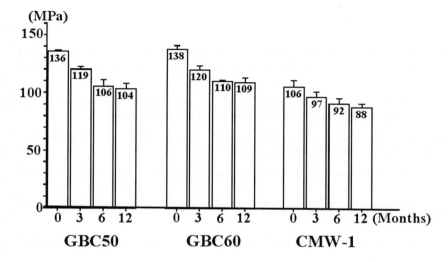

Fig. 1. Bending Strengths (MPa) for GBC50, GBC60, and CMW-1 at 0, 3, 6, and 12 months after implantation. n = 12.

By SEM observation, disappearance of some of the glass beads from the cement surfaces was observed with GBC50 and GBC60 at 6 and 12 months after implantation.

Discussion

GBC have shown greater bending strength than CMW-1 from 0 to 12 months because it contains spherical glass beads with a smaller particle diameter. The spherical shape and smaller particle size improves the filling effects of GBC [3,6]. Moreover, the use of an appropriate amount of silane coupling agent and an appropriate amount of bioactive glass beads were believed to have increased and retained bending strength [4,5,7].

Based on these test results, we believe that both GBC50 and GBC60 are strong enough for use under weight-bearing conditions and that their mechanical strength is retained *in vivo*. GBC60 may be more resistant to cement degradation than GBC50 because its bending strengths were significantly higher than those of GBC50 at 6 and 12 months, even though there was no significant difference at 0 and 3 months.

In previous studies, we packed GBC into the intramedullary canals of rat tibiae and found that it had excellent osteoconductivity [2-8]. The reasons for this were that the glass beads had high bioactivity and the hPMMA powder had a high molecular weight (270,000) [2-4,11]. The glass beads are derived from a mother glass of AW-GC [12,13], a well known bioactive material, while the hPMMA has low solubility in the MMA monomer liquid used to make up the cement, resulting in extensive exposure of the bioactive glass beads at the cement surface [2-4]. These exposed glass beads are then able to bind to bone.

This study showed that the degradation of GBC may not proceed rapidly. However, its dynamic fatigue behavior will need to be studied before widespread application in the clinical setting.

Conclusions

The present study revealed that the degradation of GBC does not appear to continue after 6 months *in vivo*, whereas CMW-1 continues to degrade progressively over 12 months. Moreover, the bending strengths of both GBC50 and GBC60 (especially GBC60) were significantly higher than that of CMW-1 throughout the study. We therefore believe that GBC60 is strong enough for use under weight-bearing conditions and that its mechanical strength is retained *in vivo*.

References

[1] Nakamura T, Kato H, Okada Y, Shinzato S, Kawanabe K, Tamura J, Kokubo T. In: Giannini S, Moroni A, editors. *Bioceramics Vol 13*. Bologna, Italy: Trans Tech Publications Ltd; 2000. p 661-664.

[2] Shinzato S, Nakamura T, Kokubo T, Kitamura Y. In: Giannini S, Moroni A, editors. *Bioceramics Vol 13*. Bologna, Italy: Trans Tech Publications Ltd; 2000. p 665-668.

[3] Shinzato S, Kobayashi M, Mousa WF, Kamimura M, Neo M, Kitamura Y, Kokubo T, Nakamura T.: J Biomed Mater Res, Vol. 51 (2000), p. 258-272.

[4] Shinzato S, Nakamura T, Kokubo T, Kitamura Y. J Biomed Mater Res, Vol. 54(2001), p. 491-500.

[5] Shinzato S, Nakamura T, Kokubo T, Kitamura Y. J Biomed Mater Res, Vol. 55 (2001), p. 277-284.

[6] Shinzato S, Nakamura T, Kokubo T, Kitamura Y. J Biomed Mater Res, Vol. 56 (2001), p. 452-458.

[7] Shinzato S, Nakamura T, Kokubo T, Kitamura Y. J Biomed Mater Res, Vol. 59(2002), p. 225-232.

[8] Shinzato S, Nakamura T, Kawanabe K, Kokubo T. J Biomed Mater Res Part B 2003;65B:262-27.

[9] Kitsugi T, Yamamuro T, Nakamura T Kakutani Y, Hayashi T, Ito S, Kokubo T, Takagi M, Shibuya T. J Biomed Mater Res, Vol. 21(1987), p.467-484.

[10] Shinzato S, Nakamura T, Kawanabe K, Kokubo T. J Biomed Mater Res Part B 2003;*in press*.

[11] Mousa WF, Kobayashi M, Shinzato S, Kamimura M, Neo M, Yoshihara S, Nakamura T. Biomaterials, Vol. 21(2000), p. 2137-2146.

[12] Kokubo T, Ito S, Shigematsu M, Sakka S, Yamamuro T. J Mater Sci, Vol. 20(1985), p. 2001-2004.

[13] Nakamura T, Yamamuro Y, Higashi S, Kokubo T, Ito S. J Biomed Mater Res, Vol. 19(1985), p.685-698.

Key Engineering Materials Vols. 254-256(2004) pp. 177-180
online at http://www.scientific.net
© 2004 Trans Tech Publications, Switzerland

In Vitro and *In Vivo* Behaviour of Bioactive Glass Composites Bearing a NSAID

J. A. Méndez[1], A. González-Corchón[1], M. Salvado[2], F. Collía[2], J. A. de Pedro[2], B. Levenfeld[3], M. Fernández[1], B. Vázquez[1] and J. San Román[1]

[1]Instituto de Ciencia y Tecnología de Polímeros, CSIC. C/ Juan de la Cierva, 3 28006-Madrid. Spain. icth341@ictp.csic.es.

[2]Facultad de Medicina, Universidad de Salamanca, C/ Alfonso X el Sabio, s/n. Campus Miguel de Unamuno, 37007-Salamanca, Spain

[3]Dpto Ciencia de Materiales e Ingeniería Metalúrgica, Universidad Carlos III. C/ Butarque, 15. 28911-Leganés. Madrid. Spain.

Keywords: PMMA, bioactive bone cement, bioactive glass, fosfosal, hydroxyapatite.

Abstract. Bioactive acrylic formulations based on PMMA, bioactive glasses in the system SiO_2-CaO-Na_2O, and the drug fosfosal, have been effective in providing the formation of an apatite layer on their surface *in vitro*. The *in vivo* behaviour studied in the femur of rabbits showed a clear integration between this material and the distal end of the femoral bone shaft after 12 weeks, without a defined structural interface bone-cement. Osseous neoformation was detected by the existence of areas of osteoid around the pellet of cement.

Introduction

Since it was demonstrated that certain compositions of silicate glasses had the ability of bonding to bone these materials have been used to repair and reconstruct bone tissue[1]. Bioactive glasses [2,3,4] and other ceramics such as hydroxyapatite [5] or glass ceramic [6] have been used to formulate acrylic bone cements in order to overcome the lack of adhesiveness to bone. The acrylic component of the cement can provide initial support and the bioactive component may allow the release of ions conducive to osteogenesis. Formulations based on bisphenol-a-glycidyl dimethacrylate (Bis-GMA) as acrylic resin have shown good results [7,8]. Also, bioactive polymethyl methacrylate (PMMA) bone cements have been prepared in presence of bioactive glasses and the results showed a direct contact between bone and both PMMA and the glass beads [9]. Other bioactive bone cements have been developed to be used as controlled delivery systems of antibiotics [10].

This paper reports on the preparation and characterisation of bioactive PMMA cements formulated with bioactive glass in the system SiO_2-Na_2O-CaO, in the presence or absence of a non-steroidic anti-inflammatory drug (NSAID) bearing phosphate groups, in order to study the influence of the phosphate groups in the formation of an apatite-like layer. The drug used was fosfosal, the sodium salt of 2-phosphonoxibenzoic acid [11]. Setting of cements was evaluated by curing parameters measurements. *In vitro* behaviour was studied in simulated body fluid (SBF) and the evolution of the surface was analysed by microscopic and spectroscopic techniques. The bone bonding ability of the cement *in vivo* was studied by intraosseous implantation of the cements in rabbits.

Materials

PMMA beads (Industrias Quirúrgicas de Levante) [12], methyl methacrylate, MMA, (Acros Organics), fosfosal (Laboratorios Uriach & Cia), $NaNO_3$ (Fluka), $Ca(NO_3)_2 \cdot 4H_2O$ (Fluka), $Si(OC_2H_5)_4$ (TEOS) (Fluka) and HNO_3 (Panreac) were used as received. Benzoyl peroxide, BPO, (Fluka) was recrystallized from methanol. 4-N,N-dimethylaminobenzyl alcohol, DMOH, was prepared previously [13].

Methods

The bioactive glasses (BV) were prepared by the sol-gel process using a 2N HNO_3 solution as the catalyst. The glass obtained had a nominal composition (wt-%) of 45.0 SiO_2, 27.5 CaO, 27.5 Na_2O. The particle size distribution was determined by laser ray scattering using a Microtrac SRA-150[©], Leeds & Northrup, North Wales, PA analyser. A Gaussian distribution was obtained with an average particle size of 40 μm. The bioactive bone cements were formulated using a solid: liquid ratio of 1.7:1. The liquid component consisted of MMA and DMOH (1 wt-%) as activator of reduced toxicity, in all cases. The solid component consisted of PMMA beads, the bioactive glass, BV, (40 wt-%), the drug fosfosal, FS, (20 or 30 wt-%) and BPO (1.5 wt-%) as the initiator. The exothermic polymerisation temperature profile was registered automatically at 25°C using a mould described in a previous paper [8]. Curing parameters were determined according to the international standard specification (ISO 5833) [14]. Discs of 15 mm diameter and 1 mm thickness were immersed in simulated body fluid (SBF) at 37°C. The morphology of the surface of samples after immersion in SBF was examined by ESEM, using a Philips XL 30 microscope. The composition of the surface was analysed by ATR-FTIR spectroscopy (Perking Elmer Spectrum One) and by X-Ray diffraction (Philips-MRD). Intraosseous implantation was studied in New Zealand rabbits. A bone defect in the femoral condyle was created by using a slow-speed drill and the bioactive cement BV/FS-40/30 was injected and allowed to cure inside. Animals were sacrificed at 12 weeks. The femoral condyle was cut longitudinally and then was included in methacrylate [15]. 5μm sections were obtained with a microtome MICROM-HS. The sections were stained with Von Kossa and Goldner techniques and observed with a light microscope Nikon Microphot FXA.

Results and Discussion

The drug-loaded bioactive cements provided an increase of the dough and setting times with respect to the PMMA control. Also, a decrease of the maximum temperature in approximately 10°C was measured for the formulation containing bioactive glass, and in 20°C for those incorporating both bioactive glass and fosfosal components. Values of curing parameters are summarised in table1

Table 1 Values of curing parameters of the bioactive acrylic formulations prepared in this work.

CEMENT	t_{dough} [min] (s.d.)	$t_{setting}$ [min] (s.d.)	$t_{working}$ [min] (s.d.)	T_{max} [°C] (s.d.)
PMMA	4.3 (0.1)	10.7 (0.0)	6.4 (0.1)	83.1 (3.4)
BV-40	6.3 (0.4)	15.5 (0.7)	9.3 (0.3)	71.8 (0.3)
BV/FS-40/20	10.8 (0.4)	24.0 (0.9)	13.3 (0.5)	61.2 (2.8)
BV/FS-40/30	11.6 (0.1)	26.9 (0.1)	15.4 (0.0)	56.7 (0.8)

The ATR-FTIR spectra of the surface of any cement before immersion showed the typical bands assigned to the polymer PMMA. After 4 days in SBF, the spectra of BV/FS-40/20 and BV/FS-40/30 cements showed a new broad band at 1025 cm^{-1} which was assigned to the vibration of the phosphate groups of hydroxyapatite. The XRD pattern of the deposited layer exhibited the characteristic peaks at $2\theta=26°$ and $32°$ attributable to hydroxyapatite. ESEM photographs of the surfaces of these cements showed spherical particles containing tiny crystals which correspond to the apatite identified by XRD. Apatite started to precipitate after 4 days of immersion in SBF and the spherulites increased in both number and size with immersion time. However, for the cement BV-40 no deposition of hydroxyapatite was detected during the time of the experiment (13-15 days). This fact indicates that the phosphate groups derived from fosfosal participate in the apatite nucleation accelerating the process, although its detailed mechanism is not yet known.

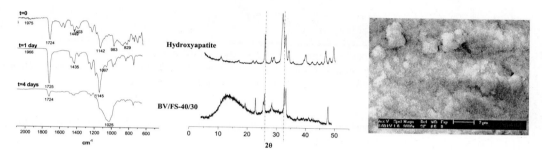

Fig. 1. ATR-FTIR spectra (right) and XRD pattern (middle) of the cement BV/FS-40-30 after 4 days of immersion in SBF, and ESEM photograph (left) of the cement after 15 days of immersion .

In vivo experiments were carried out by intraosseous implantation of the cement BV/FS-40/30 in the femur of rabbits. Figure 2 shows the histological results after 12 weeks of implantation. In the right figure, a panoramic view showing the integration between the bioactive cement and the femoral bone shaft, along with normal haematopoietic bone marrow without alterations can be seen. In the left figure the cortical femoral bone in black (upper) and the relationship between the cement and neoformed bone (down) are shown. In general terms, the cement was not totally reabsorbed, but there existed a clear integration between bone and this material, without a clear structural bone-cement interface. The haematopoietic bone marrow picture showed no distribution alterations nor cytological changes. Osseous neoformation was detected by the existence of areas of osteoid around the pellet of cement. The inflammatory reaction was very weak, almost limited to the presence of scanty multinucleated giant cells as well as tangles of reticular connective cells.

Conclusions

The presence of phosphate groups provided by fosfosal accelerates the precipitation of an apatite like layer on the surface of a bioactive bone cement formulated with polymethyl methacrylate and phosphorous free silicate glasses when it was soaked in SBF. The *in vivo* study showed a clear integration between bone and this material after 12 weeks, without a defined structural interface bone-cement.

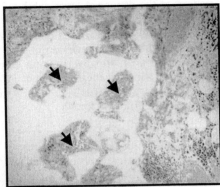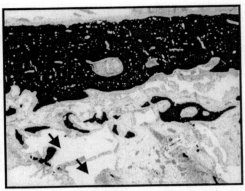

Fig. 2. Histological analysis results. Right: the arrows show proliferated connective tissue island between clear areas of bioactive cement. Goldner technique (10x). Left: the arrows point at the clear cement surrounded by bone forming tissue. Von Kossa technique (10x)

Acknowledgements
Financial support from the *Comision Interministerial de Ciencia y Tecnologia*, CICYT (MAT2002-04147-C02-02) is grateful acknowledged.

References

[1] L. L. Hench, R. J. Spinter, W. C. Allen, T.K. Greenlee. J. Biomed. Mater. Res. Vol. 2, (1971) p. 117.
[2] J. T. Heikkila, A. J. Aho, I. Kangasniemi and A. Yli-Urp. Biomaterials Vol.17 (1996), p.1755.
[3] S. Shinzato, T.Nakamura, T. Kokubo and Y. Kitamura: J. Biomed. Mater. Res. Vol. 59 (2002), p. 225.
[4] J. Raveh, H. Stich, P. Schawalder, C. Ruchti and H. Cottier: Acta Otolaryngol. Vol. 94 (1982), p. 371.
[5] A. Castaldini and A. Cavallini: Biomaterials Vol. 6 (1985), p. 55.
[6] W. Hennig, B. A. Blencke, H. Broemer, K. K. Deutscher, A. Gross and W. Ege: J. Biomed. Mater. Res. Vol. 13 (1979), p. 89.
[7] M. Saito, A. Maruoka, T. Mori, N. Sugano and K. Hino: Biomaterials Vol. 15 (1993), p. 156.
[8] J. Tamura, K. Kawanabe, T. Yamamuro, T. Nakamura, T. Kokubo, S. Yoshihara and T. Shibuya: J. Biomed. Mater. Res. Vol, 29 (1995), p. 551.
[9] S. Shinzato, M. Kobayashi, W. F. Mousa, M. Kamimura, M. Neo, Y. Kitamura, T. Kokubo and T. Nakamura: J. Biomed. Mater. Res. Vol. 51 (2000), p. 258.
[10] R. P. Del Real, S. Padilla, M. Vallet-Regí. J. Biomed. Mater. Res. Vol. 52 (2000), p. 1.
[11] M. Vallet-Regí, M. Gordo, C.V. Ragel, M.V. Cabañas and J. San Román. Solid State Ionics Vol. 101-103, (1997), p. 887.
[12] B. Pascual, B. Vázquez, M. Gurruchaga, I. Goñi, M. P. Ginebra, F. J. Gil, J. A. Planell, B. Levenfeld and J San Román. Biomaterials Vol. 17 (1996), p. 509.
[13] C. Elvira, B. Levenfeld, B. Vázquez and J San Román. J. Polym. Sci. Polym. Chem. Vol. 34, (1996), p. 2783.
[14] International Standard ISO 5833. Implants for Surgery-Acrylic Resins Cements. (1992).
[15] R. Baron, A. Vignery, L. Neff, A. Silverglate, A. Santa Maria. Processing of undecalcified bone specimens for bone histomorphometry. In: R. Recker, ed. Bone Histomorphometry: Techniques and Interpretation. Boca Raton, Florida, CRC Press, Inc., (1983) p. 37.

Key Engineering Materials Vols. 254-256(2004) pp. 181-184
online at http://www.scientific.net
© 2004 Trans Tech Publications, Switzerland

Osteogenic Activity of Human Marrow Cells on Alumina Ceramics

Yasuhito Tanaka[1], Hajime Ohgushi[2], Shigeyuki Kitamura[2], Akira Taniguchi[1], Koji Hayashi[1], Shinji Isomoto[1], Yasuaki Tohma[1], and Yoshinori Takakura[1]

[1] Department of Orthopaedic Surgery, Nara Medical University, Kashihara, Nara, 634-8522, JAPAN, e-mail: yatanaka@naramed-u.ac.jp

[2] Tissue Engineering Research Center (TERC), National Institute of Advanced Industrial Science and Technology (AIST), Amagasaki site, 3-11-46 Nakouji, Amagasaki City, Hyogo 661-0974, JAPAN, e-mail: hajime-ohgushi@aist.go.jp

Keywords: Alumina ceramic, human, marrow cell, stem cell, artificial joint

Abstract. To apply osteogenic potential of marrow mesenchymal cells to tissue engineered artificial joint replacements, we performed experimental studies using human bone marrow cells. We cultured fresh human marrow cells to expand the number of mesenchymal cells (primary culture), then cultured the mesenchymal cells on culture dish as well as alumina ceramics surface (subculture). The subculture was done in the presence of vitamin C, glycerophosphate and dexamethasone. After 8 to 10 days in the subculture, mineralized areas began to appear and the areas become obvious in 15 to 19 days on culture dish and alumina ceramics. After 14 days, high alkaline phosphatase activities were detected on both culture substrata. On the basis of experimental results, alumina ceramic total ankle prostheses loaded with the human mesenchymal cells were used for the replacement of osteoarthritic ankle joints. The present paper describes the in vitro osteogenic potential of human mesenchymal cells on alumina ceramics and focuses on preliminary results of the tissue engineered ankle joint replacements.

Introduction

The integrity of the interface between host bone and the joint prosthesis is very important and the failure of the interface (loosening) continues to be the leading cause of total joint replacement failure. It is well known that stromal or mesenchymal cells reside in bone marrow and the cells have the capability to differentiate into osteoblasts, which fabricate bone matrix in culture conditions. The in vitro osteoblasts/matrix constructs are called regenerative cultured bone. The in vitro bone formed on the surface of ceramic artificial joints can be expected to prevent the aseptic loosening [1]. In the present paper, we address the issue using the tissue engineered approach.

Materials and Methods

Cell culture

Three ml of fresh marrow cells was aspirated from anterior iliac crest of the osteoarthritic patients and cultured in a humidified atmosphere of 95% air with 5% CO_2 at 37°C. After six to 11days of the primary culture, adherent cells were released from the substratum and applied on the ceramic surface (1×10^5cells/ cm^2) as well as culture dish and further cultured in the medium supplemented with 10 mM Na beta-glycerophosphate, 82 μg/ml vitamin C phosphate (L-Ascorbic Acid Phosphate Magnesium salt n-Hydrate) and with or without 100nM dexamethasone (Dex).

Ceramics

Polycrystalline alumina ceramics disks and alumina ankle prosthesis were obtained from Kyocera Corp., Kyoto, Japan. The base of the disc was made by alumina of 93% purity. The surface of the ankle prosthesis was coated with alumina beads. The diameters of the beads were 710-850 micron. To enable the use of an optical microscope to observe the cultured cells on the ceramic, we also used transparent single-crystal alumina disks (30-mm diameter, 1.5-mm thick) that were manufactured using the Czochralski crystal growth technique.

ALP staining of the human marrow cell culture

For the Alkaline phosphatase (ALP) staining, the subcultured cell layers were washed twice with PBS (-) (phosphate-buffered saline without Ca^{2+} and Mg^{2+}) and fixed with 4% paraformaldehyde in PBS (4°C, 10 min). Then, 2 mL of the substrate solution, which consisted of 1 mg of Naphthol AS-MX phosphate sodium salt and Fast Red Violet B salt (Nacalai Tesque, Inc., Kyoto, Japan) dissolved in 2 mL of 0.056-M 2-amino-2- methyl propanol (AMP) buffer (PH 9.9), was added to the culture well.

Visualization of mineralized matrix by fluorescence emission during subculture

During the subculture periods, to enable the detection of the mineralized extracellular matrix of the cultured cells, 1 μg/mL of calcein (Dojindo Laboratories, Kumamoto, Japan) was added to the culture wells whenever the culture medium was renewed. Prior to the assay, the subcultured cell layers were washed twice with PBS (-), after which 2 mL of maintenance medium was added. The fluorescence of the calcein in the mineralized extracellular matrix of the cultured cells was visualized and observed by using a fluorescence microscope (IX 70, OLYMPUS Co., Ltd., Japan) [2].

Measurement of DNA content and ALP activity of the 14-day subcultured cells

The subcultured cell layers were washed twice with PBS (-) and collected by scraping with 1 mL of 10-mM Tris-buffer (pH 7.4, 1-mM EDTA, 100-mM NaCl). The cell suspension was then sonicated, after which 20 μL of the sonicated cell suspension was used to quantify the DNA content. The DNA content was quantified using Hoechst 33258 (Molecular Probes) [3].

To measure the alkaline phosphatase (ALP) activity, the same sonicated cell suspension was centrifuged at 13,000 g for 1 min at 4 centigrade degrees. An aliquot (20 μL) of the supernate was assayed for ALP activity using a p-nitrophenyl phosphate substrate (Zymed Laboratories Inc., CA, USA). The activity was represented by p-nitrophenol, which was released after incubation for 30 min at 37 centigrade degrees.

Total ankle replacement

An alumina ceramic total ankle prosthesis loaded with cultured cells were used for the replacement of the osteoarthritic ankle in a 58-year-old male patient. The ethics committee in our university approved this procedure for the clinical application. We informed very well and received her and her families' consent in writing. The operation was performed with a standard manner [4]. Clinical and radiographic results were investigated 3 months after surgery.

Results

Cultured cells grew well on the alumina ceramics as well as culture dish surface and showed abundant mineralization evidenced by calcein uptake and alizarin red staining in the culture with Dex (Fig. 1,2). ALP activities of the cells cultured on the alumina ceramics and culture dish were less than 0.06 μmole / 30min / μg DNA in the absence of Dex but significantly higher of these cells in the presence of Dex and these activities were in the range of 0.15 to 0.4 μmole / 30min / μg DNA.

Clinical AOFAS (American Orthopaedic Foot Ankle Surgery) scores were used to evaluate the clinical performance. Both pain and function scores were significantly improved after the surgery. Radiographic findings showed early bone fixation around the cell seeded areas even 2 months after the operation.

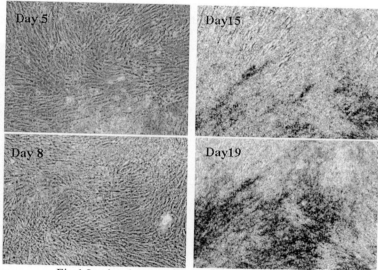

Fig.1 In vitro bone formation on alumina ceramic

Human marrow cell culture was done in the presence of Dex on transparent alumina ceramic disks. After 8 to 10 days in the subculture, mineralized areas began to appear and the areas become obvious as seen in black areas in Day 15 and 19.

Phase contrast microscopy   Fluolescent microscopy

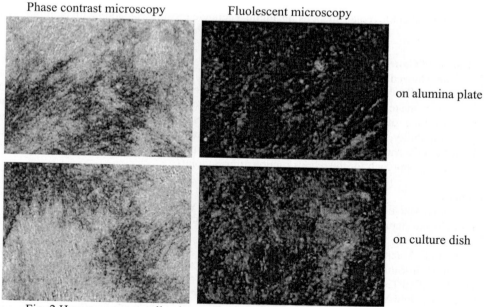

on alumina plate

on culture dish

Fig. 2 Human marrow cell culture on alumina and culture dish

To demonstrate mineralization of the cultured hBMMCs over time during the subsequent culture period, we added calcein to the culture medium whenever the medium was renewed. Calcein has been found to be specifically incorporated and deposited into extracellular bone matrix as evidenced by co-staining with Alizarin Red S [2]. After 14 days of the culture with Dex, the mineralized areas were seen under phase contrast microscopy as black areas (left) as well as under fluorescent microscopy as white areas (right). The mineralization of the culture was well seen on both alumina ceramics and culture dish.

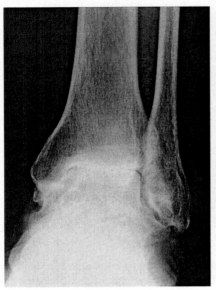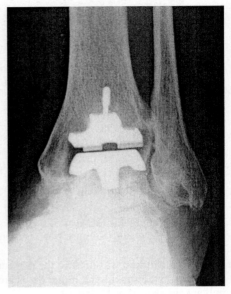

Fig. 3 Total ankle replacement using alumina ceramic prosthesis loaded with cultured cells
Pre-op   Post-op 4 weeks

Discussion and Conclusion
Present data showed that alumina ceramic is qualified material for culturing human marrow mesenchymal cells, which finally differentiate into osteoblasts. Importantly, the differentiated osteoblasts can make bone matrix consisting of hydroxyaptatite crystal on the alumina ceramic surface. Coverage of osteoblast/bone matrix on the ceramic surface vests the ceramics with osteogenic capability, which demonstrate the early bone fixation around the ceramic total ankle joint. All of these results show our method using patient marrow mesenchymal cells and alumina ceramics is the promising approach for the reconstruction arthritic joints.

References
[1] Ohgushi H, Machida H, Ikeuchi M, Tateishi T, Tohma Y, Tanaka Y, Takakura Y. Marrow
    mesenchymal stem cells cultured on alumina ceramics. From basic science to clinical applications.
    Bioceramic 2002; 5: 651-654.
[2] Uchimura E, Machida H, Kotobuki N, Kihara T, Kitamura S, Ikeuchi M, Hirose M, Miyake J,
    Ohgushi H. In-Situ Visualization and Quantification of Mineralization of Cultured Osteogenetic
    Cells. Calcified Tissue International 2003; in press.
[3] Ohgushi H, Dohi Y, Yoshikawa T, Tamai S, Tabata S, Okunaga K, Shibuya T. Osteogenic
    differentiation of cultured marrow stromal stem cells on the surface of bioactive glass ceramics. J
    Biomed Mat Res 1996; 32: 341-348.
[4] Takakura Y, Tanaka Y, Sugimoto K, Tamai S, Masuhara K. Ankle Arthroplasty. A Comparative
    Study of Cemented Metal and Uncemented Ceramic Prosthesis. Clin Orthp 1990; 252:209-216.

Key Engineering Materials Vols. 254-256(2004) pp. 185-188
online at http://www.scientific.net
© 2004 Trans Tech Publications, Switzerland

In Vitro and In Vivo Evaluation of Non-Crystalline Calcium Phosphate Glass as a Bone Substitute

Y.-K. Lee[1,3], J. Song[1], H.J. Moon[1], S.B. Lee[1], K.M. Kim[1], K.N. Kim[1,3], S.H. Choi[2,3] and R.Z. LeGeros[4]

[1] Department and Research Institute of Dental Biomaterials and Bioengineering, Yonsei University College of Dentistry, 134 Shinchon-dong, Seodaemun-ku, Seoul 120-752, Korea, leeyk@yumc.yonsei.ac.kr

[2] Department of Periodontics, Yonsei University College of Dentistry, Seoul 120-752, Korea

[3] Brain Korea 21 Project for Medical Science, Yonsei University, Seoul 120-752, Korea

[4] New York University College of Dentistry, New York, NY 10010, USA, rzl1@nyu.edu

Keywords: Bone, calcium phosphate, non-crystalline, osteoblast cell, rat

Abstract. The purpose of this study was to evaluate the bone formability of calcium phosphate glass in vivo as well as in vitro. We prepared calcium phosphate glass in the system $CaO-CaF_2-P_2O_5-MgO-ZnO$ through the conventional melting process. Pre-osteoblastic MC3T3-E1 cells were cultured onto the calcium phosphate glass in α-MEM with β-glycerophosphatase and ascorbic acid. Calcium phosphate glass particles were transplanted onto the critical-sized calvarial defects of Sprague-Dawley rats. The alkaline phosphatase activity in the experimental group was enhanced by the calcium phosphate glass significantly at 10-18 days after incubation than that of the control group ($p<0.05$). The promotion of bone-like tissue formation by the calcium phosphate glass was observed after 7 days and thereafter. In vivo test, new bone was formed in the upper side of the defects as well as the defect margin and dura mater. Experimental group always exhibited significantly higher values in the length, area and density of the newly formed bone than that of the control group ($p<0.05$). The results of the present study indicate that the prepared calcium phosphate glass affected osteogenesis by increasing collagen synthesis and calcification of the extracellular matrix in vitro and promoted new bone formation in the calvarial defects in the Sprague-Dawley rats.

Introduction

The treatment of the bone defects resulting from trauma, neoplasm, surgery, or infection is one of the major concerns in dentistry. The major goal is the functional, esthetical regeneration of supporting structures already destructed by disease. Transplantation technique have been used to provide a scaffold for bone regeneration, to augment bony defects resulting from trauma or surgery, to restore bone loss caused by dental disease, to prevent the collapse the alveolar ridge in recent extraction sites, to replace bone loss by periodontal disease, to augment the alveolar ridge in implant surgery. There are autogeneous, allogenic, xenogenic and alloplastic bone-grafts in transplantation. Among the alloplastic bone-graft materials, hydroxyapatite and tricalcium phosphate has been received much attention since the mineral phase of the hard tissue has been identified by XRD as having an apatite structure, idealized as calcium hydroxyapatite. They are bioactive, in general, bond to surrounding osseous tissue and enhance bone tissue formation. There have been some difficulties to synthesize the biological apatite, since minor elements have been associated with the biological apatites besides the major calcium and phosphate ions. Moreover, they are not adequate for the filling of large bony defects due to their low degradation rate. Calcium phosphate glass obtained from the system $CaO-P_2O_5-Al_2O_3$ was firstly developed as a dental crown material since it is easily fused and has a low viscosity in the molten state. This material can be expected to extend application field to biomaterials for hard tissue repair because of excellent bioactivity and great biodegradation rate.

Methods

Calcium phosphate glass with Ca/P ratio 0.6 was prepared from the system $CaO-CaF_2-P_2O_5$. A molar ratio of CaO/CaF_2 was fixed as 9. MgO and ZnO were added 1% in weight percent, respectively. Mixed batch was melted in a platinum crucible at 1250°C and quenched to the room temperature. As-quenched glass was cut into 10×10×1 mm for *in vitro* test and milled to 400 µm of mean diameter for *in vivo* test.

Newborn C57B/6 rat calvaria-derived MC3T3-E1 cells were employed for *in vitro* test. The cells were maintained in α-MEM supplemented with 10% fetal calf serum, 100 U/*ml* penicillin and 100 µg/*ml* streptomycin in a humidified 5% CO_2 balanced-air incubator at 37°C, the media being changed every other day. In order to induce the spontaneous differentiation into osteoblasts and mineral deposition of MC3T3-E1 cells, 10 mM β-glycerophosphate as well as 50 µg/*ml* ascorbic acid was added to the culture medium. MC3T3-E1 cells were seeded into 6-well tissue culture plates at a density of $1.0×10^5$ cells/well in the differentiation medium with calcium phosphate glass or polystyrene and incubated for 2, 4 and 6 days. After each incubation period, the number of cells in each tissue culture plate was directly counted with the aid of a hemocytometer under light microscope. In order to evaluate the initial differentiation of MC3T3-E1 cells quantitatively, alkaline phosphatase activity was determined every other day during 20 days. The optical density at 405 nm was measured at both 0.5 and 2.5 min after mixing and compared with the value of a series of p-nitrophenol standards. On day 7, 12, 17 and 21, the cultures in the plates were stained for 1 hr with 0.1% Alizarin red S to detect the calcium precipitates and followed by subsequent staining with 0.1% light green SF yellowish for 30 min. Dual-stained samples were observed under light microscope.

In vivo bone formability was carried out using 60 Sprague-Dawley rats. Routine dental infiltration anesthesia (2% lidocaine hydrochloride with 1/100,000 epinephrine) was used at the surgical sites. The calvarial skin was incised along the midline. The critical-sized defects in 8 mm diameter were produced surgically using trephine bar with extreme care to avoid injury to the dura mater. The produced defects were filled with the prepared calcium phosphate glass powder, whilst sutured without grafting anything in the control group. They were sacrificed at 2, 4 and 8 weeks after transplantation. The histologic, histomorphometric and radiodensitometric analyses were performed. The statistical significant differences of the results of the *in vitro* and *in vivo* tests were analyzed using an ANOVA with Kruskal-Wallis test and Mann-Whitney test at a level of 0.05.

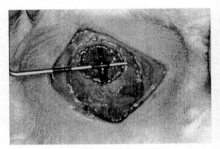
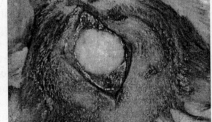

Fig. 1. Calvarial defect formation.   Fig. 2. Sample transplantation.

Results

Change of cell number was exhibited in Fig. 3. Cell number was increased with increasing incubation time. They were increased dramatically till the 4th day and the growth rates were somewhat diminished after the 4th day. The level of alkaline phosphatase activity was shown in Fig. 4. At the same incubation time, the experimental group showed significantly higher activity than that of in the control group during 10th-18th days (p<0.05).

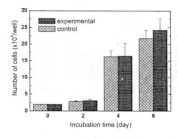
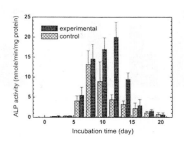

Fig. 3. Number of MC3T3-E1 cells.   Fig. 4. ALP activity of MC3T3-E1 cells.

Based on the appearance of MC3T3-E1 cells under optical light microscopy, various stages were identified (Fig. 5). These were composed of growth stage in Fig. 5(a), confluent monolayer stage in Fig. 5(b), and multilayer stage with nodule formation in Fig. 5(c).

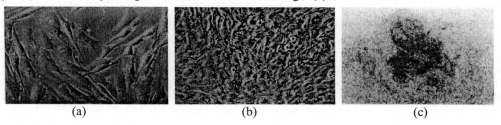

(a)   (b)   (c)

Fig. 5. Bone-like tissue formation of MC3T3-E1 cells in the experimental group
with incubation time at (a) 2($\times$100), (b) 4($\times$100), and (c) 12 days($\times$40).

Histoligical results of *in vivo* test were represented in Fig. 6. New bone was formed triangular shape around the margin of defect and in the deep layer of the dura mater in the control group. The center of defects was filled with loose connective tissue. The margin of new bone was filled with osteoid and lots of osteoblasts was arranged its forward after 2 weeks. After 8 weeks, however, osteoblasts were seen on newly formed bone while no osteoid was seen. In the experimental group, the connective tissue was observed denser and regularly than that of the control group. Many blood vessels were scattered and the grafted particles were remained after 8 weeks. Giant cells were observed. New bone was formed in the upper side of the defects as well as the defect margin and dura mater.

Histomorphometric results of *in vivo* test were shown in Fig. 7. All the length, area and density of the new-formed bone were increased with increasing transplantation time. Experimental group always exhibited the significantly higher values than that of the control group in every healing time ($p<0.05$).

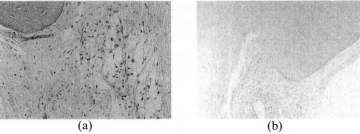

(a)   (b)

Fig. 6. Histologic mircophotographs after 8 weeks of transplantation with magnification
of $\times$200 for (a) control and (b) experimental groups, respectively.

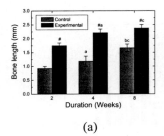 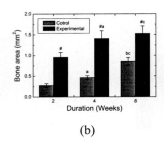 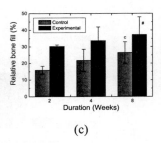

       (a)   (b)   (c)

Fig. 7. Histomorphometric amalysis of newly formed bone (a) length, (b) area, and (c) density.
      (# : statistically significant difference from the control group, $p < 0.05$
      a : statistically significant difference between 2 and 4 weeks, $p < 0.05$
      b : statistically significant difference between 4 and 8 weeks, $p < 0.05$
      c : statistically significant difference between 2 and 8 weeks, $p < 0.05$)

Discussion and Conclusion

In order to investigate the potential for the prepared calcium phosphate glass in hard tissue repair, we studied its influence on the differentiation and calcification *in vitro* and on bone formation *in vivo*. MC3T3-E1 cell and Sprague-Dawley rats were employed in this study. Owen *et al.* classified the progress of cell differentiation into 3 stages. The initial stage is characterized as cell proliferation and a high level of type I collagen gene expression, biosynthesis, and secretion, but the cells remain undifferentiated. Down regulation of the replication and expression of the differentiated osteoblast function is characterized as the second stage. The final stage of MC3T3-E1 maturation begins at about the 20th day and is defined by matrix calcification associated with progressive increases in extracellular matrix accumulation and alkaline phosphatase activity. Gerstenfeld *et al.* and Siffert reported that alkaline phosphatase activity is associated with the maturation of the matrix. Although the results in this study could not be compared directly with those of previous reports, it should be observed that a greater amount of bone nodule was formed in the experimental group than in the control group.

Critical-sized defect may be defined as the smallest size that will not heal spontaneously during the lifetime of animal. Tagaki and Urist determined that 8 mm diameter defects created in the calvaria of Sprague-Dawley rats and reported no further healing after 12 weeks. Schmitz *et al.* reported that the biologic inertness of the skull as compared to other bones could be attributed to a poor blood supply and a relative deficiency of bone marrow. In contrast to many long bones that contain a primary nutrient artery, there is no primary nutrient artery in the human calvaria. The middle meningeal artery provides the main cranial blood supply.

The results of the present study indicate that the prepared calcium phosphate glass affected differentiation and calcification of the pre-osteolastic MC3T3-E1 cells *in vitro* and promoted new bone formation in the calvarial defects in the Sprague-Dawley rats.

Acknowledgements

This study was supported by a grant of the Korea Health 21 R&D Project, Ministry of Health & Welfare, Republic of Korea (01-PJ5-PG3-20507-0105).

References

[1] T.A. Owen et al., J. Cell Physiol. Vol. 143 (1990), p. 420.
[2] L.C. Gerstenfeld *et al.*, Dev. Biol. Vol. 122 (1987), p. 49.
[3] R.S. Siffert, J. Exp.Med. Vol. 93 (1951), p. 415.
[4] K. Tagaki and M.R. Urist, Ann. Surg. Vol. 196 (1980), p. 100.
[5] J.P. Schmitz *et al.*, Acta Anat. Vol. 138 (1990), p. 185.

Key Engineering Materials Vols. 254-256(2004) pp.189-192
online at http://www.scientific.net
© *2004 Trans Tech Publications, Switzerland*

Evaluation of Macroporous Biphasic HA-TCP Ceramic as a Bone Substitute

K.S. Oh[1], J.L. Kim[2], K.O. Oh[2], J.W. Choi[3], E.K. Kwak[3],
H.M. Ryoo[3], K.H. Kim[3], and H.I. Shin[3]

[1]Nano Ceramic Center KICET, Seoul, Korea
[2]Oscotec, Chunan, Korea
[3]School of Dentistry, Kyungpook National University, Daegu, Korea, hishin@knu.ac.kr

Keywords: Biphasic HA-TCP ceramic, bone substitute, macroporous structure

Abstract. To improve the efficiency of osteogenic repair, we developed macroporous biphasic HA-TCP ceramic as a bone substitute and evaluated its efficiency by evaluation of cellular toxicity, cellular attachment and proliferation rate, and osteogenic supportive effect. The biphasic hydroxyapatite (HA)- tricalcium phosphate (TCP) ceramic with macroporous structure has excellent biocompatibility and allows for favorable cellular attachment with acceleration of cellular proliferation and osteogenic differentiation support as well. Our data suggest that the macroporous biphasic HA-TCP ceramic can be a promising scaffold for scaffold–guided tissue regeneration technology, which has the potential to maximize the repair of large bone defects.

Introduction

Bone defects occur in a variety of clinical situations, and their reconstruction to provide mechanical integrity to the skeleton is a necessary step in the patient's rehabilitation. Bones can regenerate themselves to repair defects up to a certain size but if the defect is too large and severe the suitable implant materials have to be applied to facilitate the bone repair [1]. Although the autograft is a current gold standard for bone reconstruction, there are limitations in the quantity of bone available. To overcome the limited availability of natural bone graft materials for hard tissue repair and replacement, various synthetic biomaterials have been developed. Ideally, synthetic bone graft substitutes should be biocompatible, show minimal fibrotic reaction, undergo remodeling and support new bone formation. Also, they should have the strength similar to that of the cortical/cancellous bone being replaced [2]. A number of ceramic analogs of bone such as hydroxyapatite and tricalcium phosphate have shown promise because of their exceptional tissue compatibility and direct osteointegration with native bone [3]. They, however, have no osteoinductive/ostoegenesis properties that facilitate bone healing in numerous situations whereby healing normally would not occur [4]. As part a line of efforts to improve the efficiency of osteogenic repair, we developed macroporous biphasic HA-TCP ceramic as a bone substitute and evaluated its efficiency by determining its cellular toxicity, cellular attachment and proliferation rate, and osteogenic supportive effect.

Materials and Methods

The biphasic HA-TCP ceramic with macroporous structure was prepared according to N. Korvak's description with few modifications. The biocompatibility analysis of biphasic HA-TCP ceramic was performed by the standard agar overlay cytotoxic assay [5]. For agar overlay assay MC3T3-El cells were plated at 1.25×10^6 viable cells/6-cm tissue culture dish with 5ml α-MEM media and incubated until confluent. Culture medium is then replaced with 5ml mixed solution of 0.6% agar and 2X α-MEM supplemented with 20% FBS. After solidification, 2ml of 0.01% neutral red was added and stained for 15 mins. The test samples were place on top of the agar layer and incubated for 24 hr at 37°C in 5% CO_2. The cells were then scored for cytotoxic effect. The cellular attachment and proliferation rates were analyzed by a MTT assay [5] and by SEM observation. The cellular attachment on test discs was allowed by incubation of 1.2×10^5 cells/well for 4 hours at 37ºC in 5% CO_2 using a 48 well plate. The attached cells were reacted with 3.75mM ρ-nitrophenyl-N-acetyl β-D-glucosaminide (Sigma N9376) and 50nM citrate buffer (pH10.4) containing 0.25% Triton-X 100 for 1 hour and then read the absorbance at 405nm using a plate reader. The effect of macroporous biphasic HA-TCP ceramic on osteoblastic differentiation was evaluated by bone nodule formation assay [6] and by an analysis of an osteoblast related gene expression pattern using the RT-PCR method [7].

Results and Discussion
The agar-overlay assay revealed that macroporous HA-TCP ceramic caused neither distaining nor lysis of cultured neutral red stained of MC3T3-E1 cells as the negative control OHP film, indicating noncytotoxic (Fig. 1).

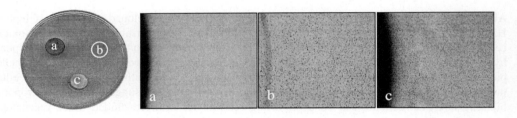

Fig. 1. Agar overlay assay of (a) copper; positive control, (b) OHP film; negative control, and (c) macroporous HA-TCP ceramic disc.

As shown in Fig. 2, HA-TCP ceramic discs showed a relatively favorable cellular attachment rate in the early period, culturing with 24 hours. The comparative percentage of the attached cell at the bottom of the well (indicated as control) versus within the HA-TCP ceramic (indicated as control) was around 50%. The attached cell number on HA-TCP ceramic discs, however, increased with time. These findings suggest that the porous frame of macroporous biphasic HA-TCP ceramic works well as a scaffold for the attachment of of primary calvarial osteoblastic cells and allows their active proliferation (Fig.2).

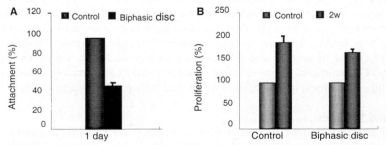

Fig. 2. Rate of cellular attachment and proliferation of cultured of primary calvarial osteoblastic cells on HA-TCP ceramics analyzed by MTT assay.

The fabricated biphasic HA-TCP ceramic revealed a well-organized macroporous structure with a porosity of 89%. The pore size was around 200um in diameter. These porous structures were large enough to allow the ingrowth of cells and blood vessels. The porous frame of macroporous biphasic HA-TCP ceramic revealed well attached of primary calvarial osteoblastic cells with filed appearance, indicating osteoblastic characteristics (Fig.2).

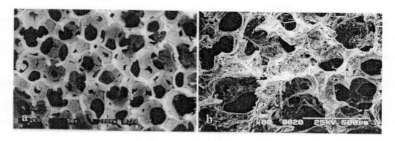

Fig. 3. Scanning Electron Micrographs of macroporous biphasic HA-TCP ceramic before (a) and after (b) cellular attachment and proliferation. The cultured primary calvarial osteoblastic cells were well attached and proliferated along the inner surfaces of evenly distributed pore structures measuring around 200um in diameter.

The cultured primary calvarial osteoblastic cells with macroporous biphasic HA-TCP ceramics in osteogenic media for 3 weeks formed numerous mineralized nodules, indicating functional osteoblastic differentiation of primary calvarial osteoblastic cells (Fig. 3).

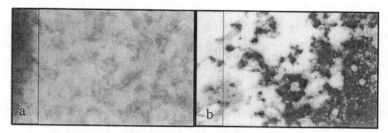

Fig. 4. Bone nodule forming assay. The cultured MC3T3-E1 cells with macroporous biphasic HA-TCP ceramics in α-MEM supplemented with 10% FBS, 50μg/ml ascorbic acid, 10^{-7}M dexamethasone and 10^{-2}M β-glycerophosphate for two weeks formed numerous nodules which were stained as red by Alizarin red. The dotted box indicates area of HA-TCP ceramic disc. (Original magnification, x100)

Furthermore, and and the two week-cultured MC3T3-E1 cells with macroporous biphasic HA-TCP ceramics revealed a favorable expression of osteoblast related genes such as alkaline phophatase (ALP), type I collagen(Col I), osteopontin(OPN), and osteocalcin(OC) mRNAs in RT-PCR analysis. At 1 wewk, the expression of ALP and OC mRNAs was downregulated compared to control but that was more upregulated(Fig. 4). This finding indicates that the culured primary calvarial osteoblastic cells proliferating within macroporous biphasic HA-TCP ceramics do not obtain fully differentiated osteoblastic characterstrics at 1 week but rapidly differentiate into functional osteoblastic cells within 2 weeks.

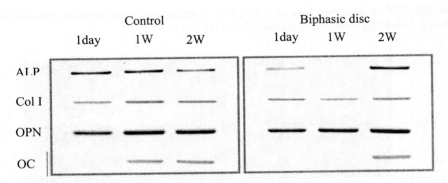

Fig. 5. Expression pattern of osteoblast related mRNAs in cultured MC3T3-E1 cells
with macroporous biphasic HA-TCP ceramics.

Summary

Our findings suggest that the biphasic HA-TCP ceramic with macroporous structure has excellent
biocompatibility and that it allows for favorable cellular attachment with an acceleration of cellular
proliferation and osteogenic differentiation support as well. Thus, with the controlled
biodegradability, the biphasic HA-TCP ceramic can be a promising scaffold for scaffold–guided
tissue regeneration technology, which has the potential to maximize the repair of large bone defects.

Acknowledgement

This work was supported by a Grant-in-Aid for Mission Oriented Basic Research Programs from the
Korea Science and Engineering Foundation (R01-2001-000-00429-0).

References

[1] H. Ohgushi, T. Yoshikawa, H. Nakajima, et al.: J. Biomed Mater Res. Vol. 44 (1998), p. 381

[2] W.R. Moore, S.E. Graves, and G.I. Bain: AnZ J. Surg. Vol. 71 (2001), p. 354

[3] J.E. block and M.R. Thorn: Calcif. Tissue Int. Vol. 66 (2000), p. 234

[4] T.W. Bauer and S.T. Smith: Clin. Orthop. Res. Vol. 395 (2002), p. 11

[5] G. Sjogren , G. Sletten, J.E. Dahl: J. Prosthet. Dent. Vol. 84 (2000), p. 229

[6] H. Chang, T.Y. Jin, W.F. Jin, S.Z. Gu, Y.F. Zhou: Biomed Environ Sci. Vol. 16 (2003), p.83

[7] L.F. Eng, Y.L. Lee, G.M. Murphy, A.C. Yu: Prog. Brain Res. Vol. 105(1995), p. 219

Key Engineering Materials Vols. 254-256(2004) pp. 193-196
online at http://www.scientific.net
© 2004 Trans Tech Publications, Switzerland

Maxillary Sinus Bone Grafting with an Injectable Bone Substitute: a Sheep Study

Afchine. Saffarzadeh[1], O. Gauthier[1, 2], T. Humbert[2], P. Weiss[1], JM. Bouler[1], G. Daculsi[1]

1 Centre de recherche sur les matériaux d'intérêt biologique, EMInserm 99-03, faculté de chirurgie dentaire, 1 place Alexis Ricordeau, 44042 Nantes, France

e-mail : afchine.saffarzadeh@sante.univ-nantes.fr

2 Ecole nationale vétérinaire de Nantes, BP 40706, 44307 Nantes, France

Keywords: sinus bone grafting, injectable bone substitute, biphasic calcium phosphate ceramic

Abstract. The aim of this animal study was to verify the osteoconductivity and the biofunctionnality of an Injectable Bone Substitute (IBS) used to perform maxillary sinus bone augmentation. Autologous bone graft was used as control. Histological and quantitative analysis using Scanning Electron Microscopy (SEM) performed after three months, showed new bone formation in tested sinuses. The amount of newly formed bone was significantly higher for sinuses grafted with IBS.

Introduction

Maxillary sinus bone grafting has frequently been used for reconstruction of the severely atrophied posterior maxilla prior to the placement of dental implant [1]. Autogenous bone grafts harvested from the iliac crest or the calvarian bone are considered as the gold standard material to perform sinus grafting.

However, the use of autogenous bone requires an additional surgical site and leads to morbidity.

Calcium phosphate ceramics have been widely reported as bone substitutes for orthopaedic, periodontal and dental applications . Their biological behaviour is directly related to their physico-chemical characteristics. Recently, the combination between hydroxyapatite (HA) and beta tricalcium phosphate (βTCP) has provided biphasic calcium phosphates (BCP) with controlled bioactivity. The association of BCP granules with a hydrophilic polymer provided an injectable "ready to use" bone substitute (IBS) [2].

The aim of this animal study was to evaluate the biological properties and the biofunctionality of this injectable bone substitute (IBS). This IBS was injected into maxillary sinus of sheep and compared with an autogenous bone graft.

Materials and methods

Injectable bone substitute. The biomaterial used for this study was obtained by the association of a calcium phosphate mineral phase and a hydrophilic polymer.

The ceramic was a biphasic calcium phosphate (BCP) with a 60/40 HA/βTCP ratio. This biphasic ceramic was granulated and sifted to obtain 200-500 μm diameter granules (MBCP^{TM}, Biomatlante, France).

A cellulose derivated polymer (Methyl-Hydroxy-Propyl-Cellulose HPMC) was used regarding to its biocompatibility and rheologic properties that confer injectability to the final composite.

A 3% solution of HPMC was prepared by dissolving HPMC powder in bidistilled water under stirring for 48 hours.

The Injectable Bone Substitute (IBS) (MBCP Gel[TM], Biomatlante, France) was obtained by mixing a 3% HPMC solution with the 200-500 µm BCP granules in a 50/50 weight ratio [3]. The resulting composite was then packed in "ready to use" 10 ml glass flasks and steam sterilized at 121°C for 20 minutes.

Animal experiments. Six adult female sheeps, with a mean weight of 64 kg, were used for maxillary sinus grafting. All animals were bred in an approved farm, kept at the National Veterinary school of Nantes according to the european community guidelines for the care and use of laboratory animals (E.D 86/609/EEC). Animals were quarantined for 10 days prior to the surgical procedure to check their general statement and ensure the absence of general or infectious disease.

Surgical procedures. All surgical procedures were performed under general anesthesia induced with intravenous diazepam (1mg/kg) and ketamine (8 mg/kg) and followed by volatile anesthesia with halothane. The facial bony antral wall was exposed over a 4 cm long incision above the facial tubercle. The masseter muscle was detached and a bony window sized 3 x 1.5 cm was created with a round dental bur and then removed using an osteotome. The sinus membrane was detached and gently displaced dorsocranially in order to create an intra sinusal bony defect.

An autologous corticocancellous bone graft was simultaneously harvested from the homolateral distal femoral epiphysis site. The bone graft was crushed with a rongeur and then packed into the sinus to achieve sinus lift procedure in the control side. The controlateral sinus (tested side) was grafted using IBS. An Ependorf syringe was filled with homogenized IBS and injected into the sinus in a retrograde manner. The composite was then sligthly packed in order to perfectly fit with the anatomic borders of the sinus cavity.

Three months after surgery, the animals were sacrified by an intravenous injection of overdosed sodium pentobarbital.

Sample preparation. Maxillary sinus segments were removed and then fixed in 10 % formalin solution, dehydrated in ascending graded ethanol and pure acetone, impregnated and embedded in a glycomethylmetacrylate resin.

Histological evaluation. Both tested and control sinuses were studied through, scanning electron microscopy (SEM). Each resin block was cut in different sections along the coronoapical axis of molars and perpendicularly to the dorsocranial axis of the filled area.

Three sections of 500 µ m thick were obtained using a diamond saw. The six surfaces obtained from each sinus were polished and gold-palladium coated prior to their observation with a 20 kV scanning electron microscope using backscattered electrons.

Quantitative evaluation. In order to compare bone ingrowth in both sinuses grafted with IBS and control ones, a semi automatic image analyser (Quantimet Q500, Leica) was used to mesure different parameters from the SEM images of the implanted area.

The limits of the grafted area were determined by the operator. Then, for each sample, the surfaces occupied by BCP granules, soft tissues, newly formed bone and total mineralised tissues (BCP + bone), were delimited. Their respective surfaces were automatically calculated in mm^2 before their expression as a percentage of the total analysed surface.

For the tested sinuses, SEM images were analysed to determine the graft mean height and the bone ingrowth mean height. Each image was first calibrated to calculate the exact milimeter value of a pixel. Then, on each calibrated image, the operator determined two areas corresponding to the whole grafted area and the bone ingrowth area. The respective mean heights of these surfaces were determined in pixels before their expression in milimeters according to the calibration values.

Statistical analysis. Differences in bone ingrowth between tested sinuses and control ones were studied for statistical purposes with an Anova test. P values < 0.05 were considered statistically significant

Results and discussion

Clinical results. No major surgical complication occurred during the implantation procedures. Wound healing occurred normally 10 days after surgery in all animals without any suture dehiscence or any signs of sinus infection.

Histological results. Scanning electron microscopy evaluation showed new bone formation in both tested and control sinuses. Bone formation seemed to be more important in peripheral areas of tested sinuses. In some cases, bone trabeculaes network created a spongious bony bridge between the lateral and the medial sinus walls including BCP granules in close contact with the newly formed bone. These results demonstrated the osteoconductive properties of IBS and confirmed results obtained from previous animal (rabbit) study [4]. In the central area of tested sinuses, bone formation seemed to be less important but it was also organized around the BCP granules without any gap at the interface between bone and ceramic (Fig. 1).

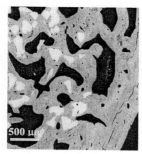
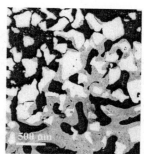

Fig 1: SEM images from the peripheral (left) and central (right) areas of tested sinuses. BCP appears in white, bone in grey and non mineralised tissue in black

All tested and control sinuses were considered for the bone ingrowth evaluation. Quantitative measurements of newly formed bone were provided from 12 sinuses (6 tested and 6 control sinuses). After 3 months, the mean rate of newly formed bone was significantly higher in tested sinuses than in control ones ($P = 0.003$). Thus, newly formed bone represented 18.9% ± 5.4 of the surface of tested sinuses whereas in control ones, 12.9% ± 7.9 of the surface was occupied by newly formed bone (Table 1).

Table 1: bone ingrowth percentage after three months

	Tested sinuses (IBS) (n=6)	Control sinuses (autogenous bone) (n=6)
No mineralised tissue (%)	53,6 ± 5,8	84,9 ± 15,1
Newly formed bone (%)	18,9 ± 5,4	12,9 ± 7,9

For the tested sinuses, calibrated SEM images allowed to measure the mean height of the grafted area and the mean height of bone colonisation into this grafted area. Three months after implantation, the mean height of the grafted area was 5.2 ± 1.3 mm and the mean height of bone colonisation was 2.9 ± 1.1 mm representing 55.3% of the mean height of the grafted area.

The osteoconductive properties of this biomaterial (IBS) have been previously demonstrated in a rabbit study [4]. However, this was the first time that this biomaterial confirmed its biofunctionnality in a specially developed large animal model to reproduce an human surgical indication.Currently, the IBS (MBCP Gel™) is used for an orthopaedic clinical trial.

Conclusion

The present animal study confirmed that an IBS composed of biphasic calcium phosphate and a cellulose derivative polymer was suitable to support bone ingrowth when used for maxillary sinus grafting.

Many experiments have been conducted to study the bone ingrowth after sinus bone grafting with different biomaterials. Thus, some authors report primate or canine models [5, 6,7]. Sheep studies have also been conducted to evaluate new bone formation after maxillary sinus grafting [8, 9]. In this study, the sheep model was choosen because of the anatomic similarity with human sinus.

Further studies will have to investigate the dental implant osseointegration into previously augmented sinuses with IBS.

References

[1] P. Boyne, R. James: J Oral Surg Vol. 38 (1980), p. 613-616

[2] G. Daculsi: Biomaterials Vol. 19 (1998), p. 1473-1478,

[3] G. Daculsi, P. Weiss, J. Delecrin, G. Grimandi, and N. Passuti: CNRS patent WO 95/21634 (1995)

[4] O. Gauthier, J.M. Bouler, P. Weiss, J. Bosco, G. Daculsi, and E. Aguado: J Biomed Mater Res Vol. 47 (1999), p 28-35

[5] M.D. Margolin, A.G. Cogan, M. Taylor, D. Buck, T.N. Mc Allister, C. Toth, and B.S. Mc Allister: J Periodontol Vol. 69 (1998), p 911-919

[6] O. Gauthier, D. Boix, G. Grimandi, E. Aguado, J.M. Bouler, P. Weiss, and G. Daculsi: J Periodontol Vol. 70 (1999) p.375-383

[7] A.C. Wetzel, H. Stich, and R.G. Cafesse: Clin Oral Impla Res Vol. 6 (1995) p. 155-163

[8] P. Bravetti, H. Membre, L. Marchal, R. jankowski : J Oral Maxillofac Surg Vol. 56 (1998) P ; 1170-1176

[9] R. Haas, K. Donath, M. Fodinger, and G. Watzek: Clin Oral Impla Res Vol. 9 (1998) p. 107-116

Key Engineering Materials Vols. 254-256(2004) pp. 197-200
online at http://www.scientific.net
© 2004 Trans Tech Publications, Switzerland

The Influence of Condensing Technique and Accelerator Concentration on Some Mechanical Properties of an Experimental Bioceramic Dental Restorative Material

Jesper Lööf[1], Håkan Engqvist[2], Gunilla Gómez-Ortega[3], Nils-Otto Ahnfelt[3], Leif Hermansson[3]

[1] Doxa AB, Axel Johanssonsgata 4-6, 754 51 Uppsala, Sweden, jesper.loof@doxa.se
[2] Doxa AB, Axel Johanssonsgata 4-6, 754 51 Uppsala, Sweden, hakan.engqvist@doxa.se
[3] Doxa AB, Axel Johanssonsgata 4-6, 754 51 Uppsala, Sweden

Keywords: Calcium aluminate, dental restorative, mechanical properties, condensing techniques

Abstract. In this paper the influence of condensing technique and accelerator concentration on hardness and dimensional stability of an experimental version of the bioceramic filling material Doxa™ T is tested. In addition a simple test of the setting time depending on accelerator concentration has also been done together with a test showing the temperature increase experienced in the bottom of a cavity when condensing with ultrasonic tips and varying the amplitude. In order to achieve early age hardness (after 4 hours) an addition of lithium to the hydration liquid is necessary if condensing is made manually. The ultrasonic vibrations bring the filler grains closer together yielding a denser end product where the maximum pore size is decreased as compared with a manually condensed sample. When using ultrasonic condensing no lithium is needed. The ultrasonic tip also transfers extra energy, mainly in the form of heat, to the material, which may speed up the hydration giving improved early age properties. Condensing with the ultrasonic device also gives higher long-term (64 days) hardness than the manual condensing. The samples condensed with the ultrasonic device and with 0 and 18mM LiCl both showed a hardness of about 150 HV(100g) after 64 days. Both for manual and ultrasonic condensing the highest hardness is achieved by using no or low additions of lithium. Regarding the dimensional stability no differences could be seen either for the different accelerator concentrations or for the different condensing techniques. The setting time is roughly the same for the three different accelerator concentrations and undetectable with this method when not using any lithium. The temperature rise in the filling when condensing with the ultrasonic device is dependent on the amplitude of the vibrations and the contact time between tip and filling.

Introduction

As a more biocompatible and environmentally friendly alternative to the filling materials present on the market, a bioceramic material based on calcium aluminate has been developed. The material's main constituents are calcium aluminate and filler and the material is chemically bonded through a dissolution precipitation process of the calcium aluminate in water [1]. It has been proven that the material is bioactive and bonds to the tooth [2] and that it has sufficient mechanical properties in order to be used as a dental restorative [3] The material needs to be condensed in the cavity using either manual or ultrasonic aided condensing and in order to achieve sufficient early age mechanical properties of the material an accelerator should be added. In this paper the influence of condensing technique and accelerator concentration on hardness and dimensional stability is tested. In addition a simple test of the setting time depending on accelerator concentration has also been done together with a test showing the temperature increase experienced in the bottom of a cavity when condensing with ultrasonic tips and varying the amplitude.

Materials and methods

The material used was an experimental version of the bioceramic material Doxa™ T. Three accelerator solutions with different concentrations of LiCl; 0, 18 and 35 mM were tested. For manual condensing (MC) regular amalgam pluggers were used and for the Ultrasonic condensing (US) an Amdent US 30 equipped with amalgam condensing tips were used. The hardness was tested according to the Vickers method with a load of 100g performed on a Buehler micromet 2100 series hardness tester. The samples were made from 3mm tablets that were hydrated and then condensed into holes of 3mm in diameter in acrylic blocks. The blocks were stored in water at 37°C and taken out for measuring after 1hour, 4 hours and after 1, 7, 14 and 64 days. Before measuring, the samples were polished with as fine as 4000 grit grinding paper. The dimensional stability was measured as restrained linear expansion using the method described in [4]. The samples were made from 3mm tablets with the two different condensing techniques. The expanders were stored in water at 37°C and taken out for measuring at 7, 14 and 64 days. The method has an uncertainty of ~0.05%. The setting time was measured by placing a thermocouple (type K) in the middle of a hydrating sample and recording the temperature changes. This was done for accelerator concentrations of 0, 18. 35 and 75 mM LiCl without condensing. The temperature rise in the filling using ultrasonic condensing at different amplitudes was measured by inserting a thermocouple (type K) through a narrow hole into the bottom of a cavity made in a plastic tooth, recording the temperature changes during condensation. An initial period of condensing was followed by a cool off period and then a second condensing period. Polyethylene foil was used to keep the material in place and the tip was in constant contact with the foil. The tablet and the liquid were both at room temperature (approx. 25° C) before the test started.

Results and discussion

The initial hardness development (up to 24 hours) showed large differences both depending on condensing technique and Li-content (Fig. 1a.) Using only water and manual condensing did not give any hardness at all during the first hours. Thus, with the MC condensing technique it is clear that the Li concentration must be at least 18 mM in order to get a sufficient hardness after 1 hour. Studying the US samples they showed a higher hardness compared to the MC samples independently on Li content (after 4 hours). After 64 days the US samples still gives a significantly higher hardness than the MC, see Fig. 3. The hardest MC sample has roughly the same hardness as that of the lowest US condensed. The influence of Li-concentration on hardness declines after the first days. Thereafter the samples respective position seems independent of the Li concentration and more influenced by the condensing technique.

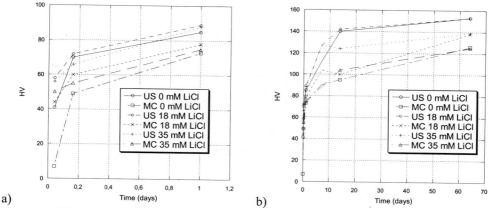

Fig 1: a) : Vickers hardness up to 1 day comparing the two condensing techniques, Manually condensed (MC) and Ultrasonically condensed (US), and different accelerator concentrations. b) Vickers hardness up to 64 days comparing the two condensing techniques

The condensing technique clearly has impact on the resulting hardness of the material. When using the ultra sonic device the hardness after 64 days is significantly higher than that with the manual method. The most obvious difference can be seen if the two samples with 0 Li concentration are compared after 1 hour. With the US condensing the hardness had reached 40 HV and with MC condensing the hardness was less then 10 HV. This implies that the ultra sonic device transfer energy to the material which speeds up the hydration process. This energy is mainly in the form of heat as shown by the heat transfer study. The setting time for high alumina cements is in general decreased with increasing temperature [1]. The higher long-term hardness of the material condensed by US implies that the US technique gives a more dense body than with manual condensing. The US vibrations bring all the grains closer together resulting in a denser and more compact body as compared with MC condensed samples. When using US the material is vibrated and "floats" more or less out into the cavity space enabling also the very small spaces and undercuts to be filled with material ensuring that the cavity is totally filled with material.

Regarding the dimensional stability differences could be seen neither for the different accelerator concentrations nor for the different condensing techniques with this method.

The setting time is approximately the same for the three different accelerator concentrations and undetectable when not using any lithium see Fig 2.

Lithium salts have been used as accelerators for high alumina cements (HAC) for a long time. More recently Matusinović et al have made a rather extensive study on how lithium salts affect the hydration of HAC's [5].The lithium immediately reacts and precipitates as a lithium hydrometaaluminate hydrate ($LiH(AlO_2)_2$. These hydrates then function as heterogeneous precipitation sites for the calcium aluminate hydrates and helps to supress the kinetic nucleation barrier connected with the precipitation. This lithium salt action suppresses the induction period of the cement hydration and accelerates the setting thus resulting in a high early strength. If the addition of lithium salts is too high, this results in lowered long-term strength [5].

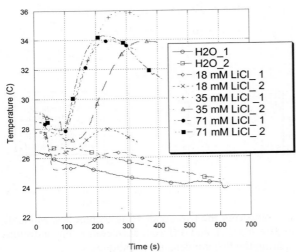

Fig 2: Setting time for different concentrations of LiCl in the hydration liquid, 1 and 2 denotes two different samples with the same concentration

Conclusions

Condensing with aid of ultrasonic vibrations clearly enhances the hardness. The addition of accelerator is necessary if the material is condensed manually. The lithium addition should be kept low to give the best long-term properties and the setting time measurements show that setting occurs after approximately the same time independently of the accelerator concentrations tested. The best hardness for calcium aluminate bonded restorative materials can be achieved with no lithium salt added, and if condensing with the ultrasonic device is used.

References

[1] F.M. Lea: LEA's Chemistry of Cement and Concrete, third ed. Edward Arnold Ltd, 1970

[2] H Engqvist, J-E Schultz-Walz, J Lööf, G A Botton, D. Mayer, M W Pfaneuf, N-O Ahnfelt, L Hermansson, accepted for publication in J. Biomaterials.

[3] J. Lööf, H. Engqvist, K. Lindqvist, N-O. Ahnfelt and L. Hermansson, accepted for publication in J. Mater. Sci :Materials in Medicine

[4] L. Kraft, L. Hermansson, G. Gomez-Ortega: *Shrinkage of Concrete- Shrinkage 2000* PRO 17 RILEM, 16-17 October 2000, ed by V. Baroghel-Bouny and P.-C. Aitcin, pp 401-413

[5] T. Matusinović, N. Vrbos, D. Čurlin: Ind. Eng. Chem. Res. 1994, 33, 2795-2800

Key Engineering Materials Vols. 254-256(2004) pp. 201-204
online at http://www.scientific.net

A New Porous Osteointegrative Bone Cement Material

G. Georgescu, J.L. Lacout and M. Frèche

CIRIMAT,UMR 5085,ENSIACET-INPT,Physico-Chimie des Phosphates,118 route de Narbonne, 31077 Toulouse, France. E-mail : Jeanlouis.Lacout@ensiacet.fr

Keywords: calcium phosphate cements, porosity, porogenic agents, resorbable bone substitute.

Abstract. A new way to create macropores in calcium phosphate cements has been developed. The method consists in adding a mixture of porogenic agents to the initial solid phase of the cement. The reaction of the effervescent mixture which forms CO_2 bubbles occurs in the first moments following mixing of the cement. Apparent porosity values showed a drastic increase both in macropores with an average size of 0.5-3.5 mm and in total porosity (- even higher than 75). Only, due to the increase of porosity, the compressive strength of the porous cement decreases significantly.

Introduction

Apatitic cements are being increasingly used for orthopaedic purposes. Calcium phosphate cements show several advantages with respect to other materials that are used for bone repair. They are easy to shape and to place in the surgical site. Therefore, they are very effective in filling bone defects with an irregular shape. On the other hand, the resorption of calcium phosphate cements in vivo is very slow. In view of this, two types of resorption can be distinguished: passive and active [1]. Passive resorption is due to the dissolution of the material into the body fluids and it depends on the final components of the set cement. The rate of this type of resorption is determined by the porosity of the samples, ionic substitutions, crystallinity and pH of the cement tissue interface [2]. Active resorption is due to cellular activity and is, therefore, related to passive resorption. The osteoclastic cells produce a pH close to 5.5. Usually this type of resorption only occurs on the cement surface because the pores - present in the cement do not allow the penetration of cells or blood vessels through the material [3-5]. Several attempts have already been made to improve the resorption behaviour of apatitic cements by increasing the porosity of the material. So far, the most extensive experiments have been done by mixing the calcium phosphate cements (CPC) with crystals of the right dimensions of highly soluble and non - toxic compounds, such as sucrose [6] or mannitol [7] . Addition of sucrose or mannitol requires dissolution of these components after application and setting of the cement in the bone defect in order to get macroporosity. Increasing the porosity of a phosphocalcium cement consists in creating pores either during the preparation of the cement by means of porogenic substances (- initial porosity -) or after implantation during maturation in the surgical site by elimination of mineral or organic particles that are more soluble than the cement. Therefore, the aim of this work has been the development of another method to increase the macroporosity of CPCs , which is based on the use of a porogenic agent during the step of cement preparation.

Materials and Methods

Starting calcium phosphate cement. The cement used in this study is manufactured under the name of CEMENTEK® (Teknimed – France) [8] . It is composed of a solid phase made up of two calcium phosphates- : tetracalcium phosphate (- TTCP – $Ca_4(PO_4)_2O$ – 49%) and α- tricalcium phosphate (α-TCP- $Ca_3(PO4)_2$- 38% -) with additionally, sodium glycerophosphate (NaGP- $Na_2C_3H_5(OH)_2PO_4$- 13% -). The solid phase is designed to be mixed with a liquid phase. This acidic phase was prepared from lime (- $Ca(OH)_2$ – 32%) and phosphoric acid (H_3PO_4 – 68%). The liquid to solid weight ratio (L/S) was equal to 0.43. After mixing these two phases, a succession of acid-base and dissolution - precipitation reactions takes place in order to arrive at the final cement structure. This structure corresponds to hydroxyapatite with an atomic Ca/P ratio of 1.635.

Preparation of porous calcium phosphate cement. To create porous cement, a mixture of porogenic agents (citric acid – 46.65% w/w, sodium bicarbonate $NaHCO_3$ –39.45% w/w, and anhydrous sodium carbonate Na2CO3- 13.90% w/w) was added to the above mentioned solid components of the starting powder. In the presence of water this mixture provides carbonic acid (- which decomposes into H_2O and CO_2 -). Five samples of every composition were removed from the moulds after 7 days of maturation at 37°C in wet medium and then dried at the room temperature for 2 days. In order to evaluate the mechanical properties of the cements a Hounsfield Series S apparatus was used to measure the compressive strength (C). The X-ray diffraction patterns of the cements were recorded by CPS 120 INEL, Co-radiation, $\lambda = 1.78892\text{Å}$, 30mA, 40kV , duration of measurement: 2θ- 10-70°).Different samples of the cement were prepared by varying the amount of porogenic mixture in the range 0-20% and the L/S ratio in the range 0.27-0.43.

Experiments

Setting time. First the initial setting time of the apatitic cements, which corresponds to the evolution of a paste mixture towards a hard product, was determined. The determination was carried out by choosing the moment when the cement ceases to be malleable. The various formulations showed an unquestionable effect of the liquid/solid ratio, but did not show any significant effect of the amount of porogenic agent above 5%. The values of the setting time are reported in table 1.

Table 1: Initial setting time

	Standard	5% porogenic agent			10% porogenic agent			20% porogenic agent		
L/S	0.43	0.33	0.41	0.43	0.27	0.39	0.43	0.33	0.39	0.43
Setting time τ [min]	25	12	15	>50	10	13	>50	12	16	>50

Porosity. Previous studies of the liquid/solid ratio for the effervescent cement were made and we observed that a sufficiently low L/S ratio provided a homogeneous distribution of the pores in the total volume of the final material. The evaluation of the pore dimensions and of the pore distribution in the samples, was done using an optical microscope. The apparent porosity (closed and connected porosity) was also determined by measuring the density of the samples. The values of the apparent porosity are reported in table 2 and some photographs of samples with different % of porogenic agent are presented in figure 1.

Table 2: Variation of the porosity of the samples with the amount of porogenic agent and the liquid/solid ratio

Amount of porogenic agent	Liquid/Solid ratio	Porosity [%]
0 %	0.43	50
5 %	0.33	65.3
	0.41	68.7
	0.43	71.6
10 %	0.27	71.8
	0.39	75.5
20 %	0.35	74.1
	0.39	76.6

The porosity of the samples increases with the amount of porogenic agent and also with the liquid/solid ratio.

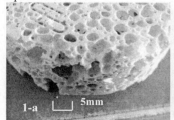

Fig. 1 : Photographs of different samples studied in our work: 1-a: 20% effervescent and L/S = 0.39; 1-b: 10% effervescent and L/S = 0.27; and 1-c: 5% effervescent and L/S = 0.37.

The photographs show that for 5 % effervescent agent, the pores are small and not interconnected, but for 10 or 20% effervescent mixture the pores are large and numerous. Their size range was 0.5 – 3.5 mm and the total porosity from 65 to75% .

X-ray diffraction shows that independently of L/S ratio and of the amount of porogenic agent, all the materials progressively evolved towards a non-stoichiometric hydroxyapatite. After one week the evolution was complete. Therefore, no extra phases appeared due to the process of creating pores inside the material. Figure 2 presents the X-ray diffraction diagrams for three formulations studied of the phosphocalcium cements.

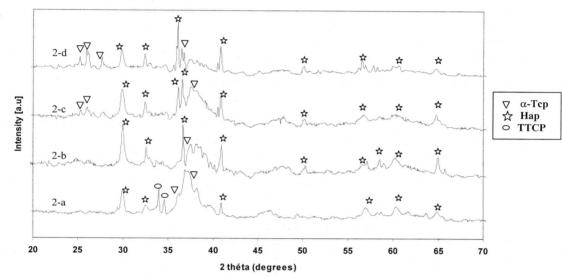

Fig 2. X-ray diffraction patterns for the samples after 7 days in a wet medium : 2-a: standard cement; 2-b: 5 % effervescent agent; 2-c: 10% effervescent agent; 2-d: 20% effervescent agent.

Mechanical properties. The compressive strengths values are listed in table 3.

Table 3: Variation in compressive strength with the amount of porogenic agent and the liquid/solid ratio.

Amount of porogenic agent	Liquid/ Solid ratio	Compressive strength [Mpa]	Porosity [%]
0 %	0.43	20	50
5 %	0.33	6.3	65.3
	0.41	2.2	68.7
	0.43	2.4	71.6
10 %	0.27	4.1	71.8
	0.39	2.4	75.5
20 %	0.35	1.4	74.1
	0.39	1.2	76.6

The mechanical properties decreased drastically with the increase of the porosity. They also decreased with the number and the dimensions of the pores which explains the differences observed between the 10% and 20% samples. A significant decrease with respect to the standard samples (20 MPa for Cementek) can be noticed in all the cases (4 MPa for 10% porogenic agent and 0.27 L/S ratio).

Discussion

The addition of a porogenic agent to the phosphocalcium cement can have several effects: an effect on the chemical composition and the crystallisation state of the solid phase, an effect on the mechanical-physical properties and an effect on the biological behaviour. It is obvious that these various effects are strongly interdependent.

The chemical composition is slightly modified by the addition of sodium ion brought by the porogenic agent. The location of this element inside the apatitic structure or not was not determined, but does not influence the formation of apatite . On the other hand, the presence of CO_2 leads to the formation of a slightly carbonated B-type apatite, highlighted by infra-red spectrometry. The state of crystallisation of apatite, after 7 days of maturation in a wet medium, increased with the amount of porogenic agent (fig.2) - this can be due to the creation of inter-connected pores which allows easy diffusion of solution through the mass of the solid and thus a good and quick chemical maturation.

The contribution of the porogenic agent and the formation of bubbles, lead to a significant modification of the rheology of the cement during mixing: there is in particular a strong reduction in viscosity. The elimination of CO_2 bubbles is then fast before the hardening of the material and the formation of large-sized pores is limited. It was thus necessary to decrease the liquid/solid ratio, in order to increase the viscosity of the mixture, to limit the displacement of the bubbles, to enable the paste to harden and pores to be formed. The strong reduction of setting time with the addition of porogenic agent must be mainly related to the decrease imposed to the L/S ratio.

The mechanical properties depend at the same time on the nature of the solid phase which forms the walls of the pores and on the number, the dimension and the distribution of these pores. The chemical and crystallographic modifications on the solid phase are too weak to cause a significant effect on the mechanical properties. On the other hand, porosity has a strong influence. A drastic reduction occurred in the compressive strength, related primarily to the increase in total porosity when 5 % of porogenic agent was added. The increase of the amount of porogen (10 and 20 %) did not modify the total porosity (- which remained equal to 75 %) but led to coalescence of the bubbles to form large pores resulting in an additional reduction of the compressive strength of the final material.

The biological properties are not discussed here, but it can be pointed out that the presence of inter-connected pores, and the formation a B-type carbonated apatite seems very compatible with the preparation of a biointeractive material.

Conclusion

This work leads to a convenient formulation for porous phosphocalcium cement. The addition of 10 % of CO_2 porogenic agent with a L/S ratio of 0.27 generated a total porosity of 75 % with a pore size in the range 0.5-2 mm. The formation of the bubbles outside the biological site, before the hardening step, the setting time and the mechanical properties seem to be convenient for application as a filling biomaterial. The appearance of the phosphocalcium cement was similar to trabecular bone which suggests excellent osteointegration properties.

References

[1]   R.P.del Real, J.G.C.Wolke, M.Vallet-Regi, J.A.Jensen : Biomaterials, 2002 ;vol.23, 3673-80.
[2]   R.Z. LeGeros, J.R.Parsons, G.Daculsi, F.Driessens, D.Lee: Biodegrad-Bioresorp,1988;268-71.
[3]   R.Holmes: Plast Reconstruction Surgery 1979;63:623-33.
[4]   K. Shimasaki, V.Mooney: Journal Orthop Res 1985;3:301-10.
[5]   P.S. Eggli, W.Muller,R.K Schenk: Clin Orthop 1988;232;127-38.
[6]   S.Takagi, L.C.Chow: Journal Mater Sci: Mater Med 2001;12:135-9.
[7]   M.Markovic, S.Takagi, L.C.Chow: Bioceramics 13; 2001, p. 773-6.
[8]   J.L.Lacout ,Z. Hatim, M. Frèche, French Patent N° 9803459, 1998.

Key Engineering Materials Vols. 254-256(2004) pp. 205-208
online at http://www.scientific.net
© *2004 Trans Tech Publications, Switzerland*

Low Porosity CaHPO$_4$.2H$_2$O Cement

L.M. Grover[1], U. Gbureck[2], A. Hutton[1], D.F. Farrar[3], C. Ansell[3] and J.E. Barralet[1]

[1] Biomaterials Unit, School of Dentistry, University of Birmingham, St Chad's Queensway, Birmingham, England, B4 6NN;LXG865@bham.ac.uk

[2] Department of Functional Materials in Medicine and Dentistry, University of Wurzburg, Pleicherwall 2, D-97070, Wurzburg, Germany.

[3] Smith and Nephew Group Research Centre, Heslington Science Park, York, YO10 5DF, England.

Keywords: Artificial bone grafts, mechanical properties, bone cements, injectable materials for bone regeneration

Abstract. Compaction can be used to reduce the porosity of cements thereby improving their mechanical properties. In this study a brushite calcium phosphate cement that set following the mixture of monocalcium phosphate monohydrate (MCPM), β-tricalcium phosphate (β-TCP) and water was uniaxially compacted at pressures of up to 100 MPa prior to setting. The resultant cements exhibited wet compressive strengths in serum of up to 43% higher than uncompacted cement. Compaction reduced the extent of the setting reaction from 95% with no compaction to 78 % after 100 MPa of compaction. Although, compaction inhibited the setting reaction it was an effective means of improving the compressive strength of brushite cement and may provide a method to control the degradation rate in vivo.

Introduction

Calcium phosphate cements can be moulded to match irregular contours using finger pressure and set at physiological temperature to form a hardened matrix consisting predominantly of either hydroxyapatite (HA) or brushite. HA cements are typically stronger than brushite cements and have attracted more research activity. Studies have shown that although HA cements are initially resorbed, following implantation, they can persist for upwards of a year [1]. In contrast, brushite cements are largely resorbed [2].

A means by which the compressive strength of brushite cement can be improved is by reducing relative porosity. The relative porosity of a ceramic material, such as cement, is related to its strength by an inverse exponential relationship [3]. Porosity reduction of cements can be accomplished by reducing the water demand of the system or by compaction. The water demand can be reduced by altering the particle size distribution of the reactants so that they pack with a high efficiency or by using dispersants thereby preventing agglomeration. Compaction has previously been shown to effectively reduce the porosity of apatitic cement and consequently improve its compressive strength [4,5]. In this study a brushite cement slurry was compacted at pressures of 10, 25, 50 and 100 MPa. The effects of compaction at these pressures on compressive strength, relative porosity, and phase composition were investigated.

Methods and Materials

β-TCP was synthesised by heating a mixture of CaCO$_3$ (Merck) and CaHPO$_4$ (Mallinckrodt Baker) with an overall calcium to phosphate ratio (Ca:P) of 1.5 to 1050 °C for 24 h, prior to quenching at room temperature in a dessicator. The product was verified as being phase-pure highly crystalline β-TCP by X-ray diffraction (XRD; D5005, Siemens). After grinding using a pestle and mortar, the powder was passed through a 355 μm sieve, subsequently 75 g of powder was milled using a planetary ball mill (PM400 Retsch) at 200 rpm in 500ml agate jars each containing 4 agate balls (30 mm) for 1 h.

Ground β-TCP was then combined with MCPM (Mallinckrodt Baker) and tri-sodium citrate (Mallinckrodt Baker) at a molar ratio of 9:9:2; the resultant mixture was homogenised using a vibrating platform. The powder component was then mixed with water at a powder to liquid mixing ratio (P:L) of 3.3 g/ml to form a slurry, which was poured into a stainless steel split mould of 8 mm in diameter. The plunger was placed into the mould and the slurry was compacted at a cross-head speed of 1 mm/min using a Universal testing machine (Instron 1158) at pressures of 10, 25, 50 and 100 MPa. The amount of cement compacted was altered so that each cylindrical sample had an aspect ratio of 2:1. Following compaction, each sample was left in the mould at 37 °C for 1 h, after which the sample was carefully removed and placed into bovine serum (Sigma-Aldrich) containing 0.1wt% sodium azide (Sigma-Aldrich) and stored at 37 °C for 24 h. Prior to testing in compression dimensional measurements were made. The samples were tested in compression at a cross-head speed of 1 mm/min using a Universal testing machine (Instron 5544). The fragments were subsequently retrieved and dried at 37 °C until no further weight loss was observed. The strut densities of the dried cement fragments were then measured using helium pycnometry (Accupyc 1330). The relative porosity of the cement was calculated from apparent and strut densities.

In order to determine phase composition, tested samples were ground to a fine powder, placed in a specimen holder and then analysed using a diffractometer. The resultant diffraction patterns were compared with reference patterns for brushite (PDF Ref. 09-0077) and β-TCP (PDF Ref. 09-0169). Phase compositions of the cements were determined using Rietveld refinement phase analysis.

Results

Compaction of the cement mixes did result in an increase in the wet strength of the set material. For example, after compaction at 100 MPa mean compressive strengths of ~25MPa were measured, compared to ~13MPa for the uncompacted samples (Fig. 1).

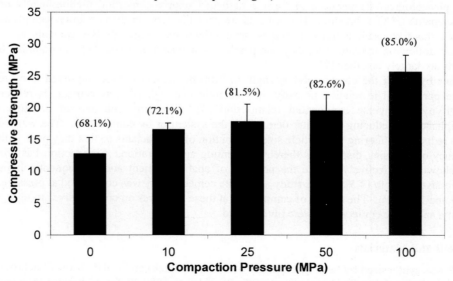

Fig. 1. The effect of compaction on compressive strength, relative density shown in brackets.

As compaction pressure increased the measured porosity in the cements decreased such that 85% dense cement was formed following compaction at 100MPa. Even uncompacted cement contained little porosity (~32%) and this is thought to stem from the consumption of water during this hydraulic reaction (Fig. 1).

Table 1 shows the densities of the cement reaction products. Two main features are to be noted, firstly the agreement between the density measured by pycnometry and that calculated from phase composition is generally very good, (< 3%) providing independent verification of the phase analysis data and secondly density appeared to increase with compaction pressure, suggesting that the amount of unreacted ßTCP, (ρ = 3.0 g cm^{-3}) was higher in the compacted material.

Indeed, although an increase in compressive strength was noted when the cement was compacted the percentage conversion of β-TCP to brushite was reduced (Fig. 2). When made at ambient pressure 92wt% (95mol%) brushite was formed yet with 100 MPa of compaction a yield of only 65wt% (78mol%) was obtained (Fig. 2). At compaction pressures of between 10-25 MPa, although the reduction in the porosity was ~10% this did not result in a marked increase in compressive strength possibly due to the lower degree of reaction (Fig.2).

Table 1. The calculated and measured strut densities of compacted cements.

Compaction Pressure [MPa]	Strut Density [g/cm^3]	Calculated Density [g/cm^3]
0	2.38	2.39
10	2.43	2.42
25	2.40	2.48
50	2.44	2.47
100	2.56	2.58

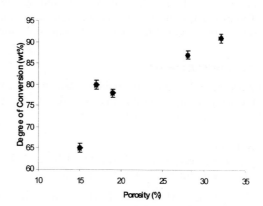

Fig. 2. Degree of conversion – porosity relationship.

Discussion

Brushite cement differs notably from apatite cements since the product of this cement system occupies a markedly smaller volume than the reactants. This occurs because a considerable volume of water (ρ = 1.0 g cm^{-3}) is incorporated into the brushite phase ((ρ = 2.3 g cm^{-3}). Although gross dimensional change may not occur, the shrinkage of the matrix would cause the formation of ~11% porosity in the set cement, if the cement were stoichiometric. However, the cement formulation that was used was not stoichiometric as it contained excess water. The probable setting reaction is described in Eqn. 1.

$$Ca_3(PO_4)_2 + Ca(H_2PO_4)_2.H_2O + 9.4\ H_2O \rightarrow 4\ CaHPO_4.2H_2O + 2.4\ H_2O \qquad [1]$$

This excess water would not participate in solid formation and itself would constitute a further ~11% porosity. The soluble tri-sodium citrate occupied ~9% of the set cement volume and is thought to have dissolved in the ageing medium. Hence the total porosity from contraction, excess water and retardants was calculated to be ~31% for the uncompacted cement which correlated well with measurements (Fig. 1).

On compaction water was ejected from the mould and consequently the porosity of the set cement was reduced.   Although compaction did lead to an increase in the compressive strength of the cement, it was not as marked as may have been expected. Chow et al. [4] demonstrated that following the compaction of an apatitic cement to only 700 kPa, the increase in the compressive strength when compared with non-compacted cement was ~29% and in another study a compaction pressure of 106 MPa resulted in an increase of ~850% [5]. Such dramatic gains were not noted following the compaction of this cement system; this is probably due to the reduction in the extent of reaction with an increase in compaction pressure. Compaction inhibited the extent of reaction by reducing the amount of water present for the reaction. However, cements were immersed in an aqueous ageing medium following setting. Clearly the diffusion of water through the low porosity cement was limited. Long term in vitro studies are required to determine whether the reaction can proceed any further.

Conclusion

Although the compaction of brushite cement appeared to reduce the extent of the reaction, it was shown to be an effective means by which the compressive strength of the cement could be increased. The reason why reaction did not persist in aqueous conditions is not yet understood. The production of cements at given densities offers the possibility of tailoring cement degradation rate to that required for specific clinical applications.

Acknowledgements

We acknowledge the financial support of the EPSRC (studentship) and Smith and Nephew Group Research Centre, York, UK (CASE award) (Mr. L.M. Grover) and provision of a Nuffield undergraduate bursary (A. Hutton).

References

[1] D. Knaack, M.E.P. Goad, M. Ailova, C. Rey, A. Tofighi, P. Chakravarthy, D. Lee: J Biomed Mater Res Vol 43 (1998), p. 399-409.

[2] K. Ohura, M. Bohner, P. Hardouin, J. Lemaitre, G. Pasquier, B. Flautre: J Biomed Mater Res Vol 30 (1996), p. 193-200.

[3] E. Ryshkewitch: J Am Ceram Soc Vol 36 (1953), p. 65-68.

[4] L. C. Chow, S. Hirayama, S. Takagi, E. Parry: J. Biomed. Mater. Res. (Appl. Biomater.) Vol 53 (2000), p. 511-517.

[5] J.E. Barralet, T. Gaunt, A.J. Wright, I.R. Gibson, J.C. Knowles: J Biomed Mater Res (Appl Biomaterials) Vol 63 (2002), p. 1-9.

Key Engineering Materials Vols. 254-256(2004) pp. 209-212
online at http://www.scientific.net

Effects of Polysaccharides Addition in Calcium Phosphate Cement

T.Sawamura[1], M.Hattori[2], M.Okuyama[3] and K.Kondo[4]

[1]R & D Center, NGK Spark Plug Co., Ltd., 2808 Iwasaki, Komaki, Aichi 485-8510,Japan,
t-sawamura@mg.ngkntk.co.jp

[2]The same address as first author's, m-hattori@mg.ngkntk.co.jp

[3]The same address as first author's, m-okuyama@mg.ngkntk.co.jp

[4]The same address as first author's, ka-kondo@mg.ngkntk.co.jp

Keywords: Calcium Phosphate Cement, Polysaccharides, decay, hydration reaction,
Hydroxyapatite.

Abstract In preparation of Calcium Phosphate Cement (CPC) made from tetracalcium phosphate (TeCP) and dicalcium phosphate anhydrous (DCPA), the effects of the addition of polysaccharides on the decay in a simulated body fluid (SBF) and the hydration reaction of CPC mixing body were studied. The aqueous solutions of polysaccharides including Dextran (DX), Dextran Sodium Sulfate (DS), Sodium Hyaluronate (HS), and Chondroitin Sodium Sulfate (CS) were evaluated as mixing liquids for CPC powder. When each CPC paste, prepared by mixing CPC powder with the mixing liquid, was soaked into SBF, CPC paste with DX decayed after soaking, similarly to CPC paste with the distilled water. On the other hand, the addition of HS, CS or DS was found to decrease the decay behaviour in SBF. The addition of polysaccharides also showed the effects on the hydration of CPC paste. For compressive strength, CPC paste with DS or the distilled water showed approximately 50 MPa after 8h. However CPC paste with CS or HS took 24h to reach 50MPa. Moreover, by XRD and SEM, it was found that CPC paste with DS had high hydration reaction compared with CS. The differences in CPC characteristics were most likely due to the molecular weight of each polysaccharide and the viscous and adhesive characteristics of these aqueous solutions.

Introduction

Calcium Phosphate Cement (CPC) attracts considerable interest as a new type of bone substitute, because of the formability, injectability, setting property of CPC paste, and high bioactivity of the resultant CPC consisting of low crystalline bone-like apatite [1-3].
However, when CPC paste obtained by mixing CPC powder and distilled water was soaked in a simulated body fluid (SBF), an issue may exist that is, the decay of CPC paste in SBF before setting. It is pointed out that non-reacted CPC powder may cause inflammation in the living body and therefore the decay of CPC paste may be one important issue in clinical use [4,5].
In this study, to overcome this issue, the effects of the addition of polysaccharides into CPC on the decay in SBF and the reactivity of CPC were studied.

Materials and Methods

CPC powder was prepared from Tetracalcium phosphate (TeCP; $Ca_4(PO_4)_2O$) and dicalcium phosphate anhydrous (DCPA;$CaHPO_4$). Commercially available DCPA powder was milled with the distilled water for 10h by using an alumina ball mill, and dried at 120°C for 20h. The obtained DCPA showed the mean particle size of 0.6μm and the specific surface area (BET) of 4.0m^2/g. TeCP was prepared from dicalcium phosphate dihydrous (DCPD;$CaHPO_4.2H_2O$) and calcium carbonate($CaCO_3$) .   A mixture of DCPD and $CaCO_3$ at a molar ratio of 1/1 was fired at 1550°C for 10h in a electrical furnace, quenched at room temperature, and was followed by crushing down to the mean particle size of 200μm.   CPC powder was prepared by mechanically mixing the TeCP and DCPA at a molar ratio of 1/1 with automatic mortar.   The mean particle size and the specific surface area of the CPC powder thus obtained were 6.3μm and $2.5m^2/g$, respectively.   Polysaccharides used as additives in this study were Dextran (DX), Dextran Sodium Sulfate (DS), Sodium Hyaluronate (HS), and Chondroitin Sodium Sulfate (CS), and the aqueous solutions of each polysaccharide were used as mixing liquid for CPC powder.

CPC powder were mixed with each mixing liquid at a liquid-to-powder ratio (L/P) of 0.26, and CPC pastes were obtained. The CPC pastes just after mixing were molded (12mm in diameter and 8mm in height) and soaked immediately in SBF [6] at 37°C, and the status after 10min soaking was observed by naked eyes to evaluate the decay behavior.   For the compressive strength of the CPC pastes, the molded CPC pastes (6mm in diameter and 12mm in height) were kept in incubator for 1h at 37°C and 95% relative humidity, soaked in SBF for 23h at 37°C, and were subjected to the measurement according to JIS (Japanese Industrial Standard) No.6602.   Crystalline phases of the CPC pastes soaked in SBF at 37 °C for 8-72h were identified by an X-ray diffraction pattern (XRD) and the microstructures of these CPC pastes were observed by a scanning electron microscope (SEM).   CPC paste mixed with distilled water was evaluated simultaneously and the results were used as comparison.

Results

Figure 1 shows the decay behavior in SBF of CPC pastes with various polysaccharides.   CPC paste with DX decayed after soaking in SBF, similarly to CPC with the distilled water. However, the addition of HS or CS appeared to decrease the decay behavior, and moreover CPC paste with DS didn't decay at all and maintained the shape as molded.

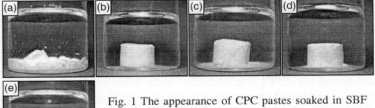

Fig. 1 The appearance of CPC pastes soaked in SBF for 10min, with (a) DX, (b) DS, (c) HS and (d) CS and (e) without additive.

Figure 2 shows the compressive strength of CPC pastes with various polysaccharides in SBF. The compressive strength increased with soaking time in SBF, and the strength for all evaluated CPC paste reached approximately 50 MPa.   However

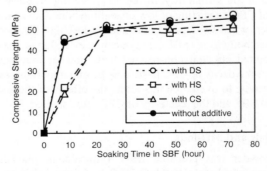

Fig. 2 The compressive strength for CPC pastes with or without polysaccharides, as a function of soaking time in SBF.

major differences were observed in the time to reach 50MPa between the CPC pastes. For CPC paste with DS, the strength reached 50MPa after 8h, as early as CPC paste with the distilled water. For CPC paste with CS or HS, however, the strength after 8h was 20MPa and it took about 24h to reach 50MPa, the addition of CS or HS appeared to retard the hardening behavior.

Figure 3 shows the change in XRD patterns for CPC pastes with DS, CS or distilled water for several soaking times in SBF. CPC paste just after mixing consisted of TeCP and DCPA. For CPC pastes with DS or distilled water, HAp phase appeared after 8 h and gradually increased with time, accompanied with the decrease of TeCP and DCPA phases. In addition, DCPA phase disappeared

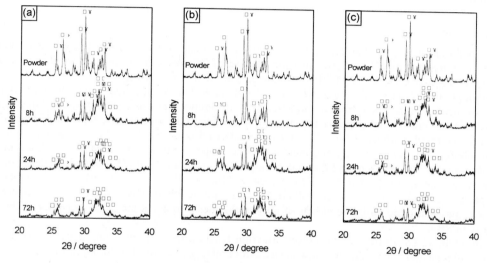

Fig. 3 X-ray diffraction patterns of CPC pastes with (a) DS and (b) CS and (c) without additive.
(□)hydroxyapatite,(○)TeCP,(▼)DCPA

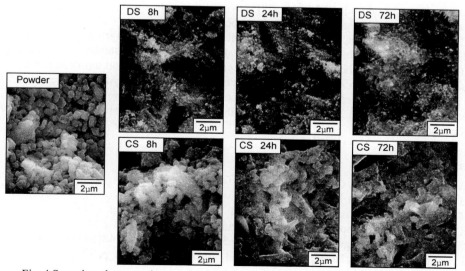

Fig. 4 Scanning electron micrograph of fracture surface for CPC pastes with DS and CS for different soaking time.

completely and low crystalline HAp and TeCP existed after 72h, as shown in Fig. 3(a) and (b).   On the other hand, for CPC pastes with CS, no HAp was observed after 8h and major DCPA was detected even after 24 h.   After 72h soaking, HAp increased slightly and DCPA remained in CPC paste, as shown in Fig. 3(c).   It appeared that the addition of CS delayed the hydration reaction.

Figure 4 shows the scanning electron micrograph of fracture surface of the CPC paste with DS and CS.   For CPC paste with DS, many precipitated HAp crystals were observed and their amount increased with soaking time. However, for the CPC paste with CS, DCPA particles of starting powder remained after 8h and small amount of crystalline HAp were observed with voids after 72h of soaking.

Discussion

The addition of polysaccharides into CPC showed the major effects on the decay in SBF and the hydration reaction of CPC.   Especially, the addition of DS was found to prevent the decay behavior without negative effects on hydration reaction and strength of CPC paste.

Although one of the possibilities to be considered was pH of mixing liquid, the measured pH was almost neutral in the range from 6.3 to 6.9, and no difference was observed between mixing liquids. The effect to prevent decay behavior was most likely due to the viscous and adhesive characteristics of mixing liquids with DS, HS or CS. It was considered that these polysaccharides associated to each particle of CPC during setting and maintained the shape in SBF.   In the case of sodium alginate addition into CPC, it is reported that sodium alginate formed a gel with Ca ion and the decay of CPC could be restrained [5].   DS, CS and HA behaved like a sodium salt, similarly to sodium alginate, especially DS, which is the sodium sulfate of DX. Therefore DS may have similar effects to form gel. These may cause the desirable effects of polysaccharides to prevent the decay behavior of the CPC paste in SBF.

In addition, DS did not disturb transformation reaction to HAp, probably because DS had low molecular weight and could be easily released from mixing body to SBF during setting.

Conclusion

It was found that the addition of DS into CPC gave the desirable effects on properties of CPC paste, such as non-decay in SBF and good hardening reaction.

For CPC prepared from TeCP and DCPA, the effects of the addition of various polysaccharides (DX, DS, HS and CS) on decay in a SBF and reactivity of mixing body were studied.   CPC paste with DS didn't decayed in SBF and held the sufficient compressive strength in short term compared with other polysaccharides-added CPC.   Moreover, this paste showed sufficient hydration reaction similarly to CPC paste with the distilled water.   The differences in properties were most likely caused by the molecular weight of each polysaccharide and the viscous and adhesive characteristics of these aqueous solutions.

References

[1] W. E. Brown and L. C. Chow: US Patent No. 4,612,053(1986)
[2] H. Monma : FC Report.12(1988), p. 475
[3] B. R. Constantz, I. C. Ison, M. T. Fulmer et al: Science. 267(1995), p. 1796
[4] K. Ishikawa, Y. Miyamoto, M. Kon, et al: Biomaterials. 16(1995), p. 527
[5] Y. Miyamoto, K. Ishikawa, M. Takechi, et al: J Biomed Mater Res. 48(1999), p. 36
[6] T. Kokubo, S. Ito, Z.T. Huang, et al: J Biomed Mater Res. 24(1990), p. 721

Key Engineering Materials Vols. 254-256(2004) pp. 213-216
online at http://www.scientific.net
© *2004 Trans Tech Publications, Switzerland*

Characterization of Magnetite and Maghemite for Hyperthermia in Cancer Therapy

K.M. Spiers[1,2], J.D. Cashion[2] and K.A. Gross[2]

[1]School of Electrical and Computer Systems Engineering, Monash University, 3800, Australia.
[2]School of Physics and Materials Engineering, Monash University, 3800, Australia.
Email: kathryn.spiers@spme.monash.edu.au, john.cashion@spme.monash.edu.au,
karlis.gross@spme.monash.edu.au.

Keywords: Magnetite, maghemite, hyperthermia, X-ray diffraction, Mössbauer spectroscopy, hysteresis heating

Abstract. Submicrometre particles of magnetite and maghemite were produced for potential use for hyperthermic treatment of cancer. X-ray diffraction and Mössbauer effect measurements showed that the samples were of good crystallinity and SQUID magnetization measurements showed that magnetite should produce the larger heating effect.

Introduction

Hyperthermia for treatment of tumours may be induced in tissue by applying an alternating magnetic field to a region implanted with magnetic materials. Small magnetic particles or spheres deposited in tissue will cause heating through hysteresis loss. This form of therapy requires a magnetic material with the optimum heat evolution.

Little attention has been placed on the synthesis and characterization of iron oxide-based magnetic particles for cancer therapy. A detailed knowledge of the magnetic particles is a key aspect in optimizing the magnetic properties for heat production. This work will detail the synthesis of magnetite and maghemite powders, discuss their differentiation using X-ray diffraction and Mössbauer spectroscopy and their heat generation characteristics using SQUID magnetometry.

Methods

To synthesize magnetite, 560 ml of 0.5 M $FeSO_4 \cdot 7H_2O$ was dissolved in deionized water and heated to 90 °C. All deionized water used for the synthesis of magnetite was previously bubbled with N_2 to minimize any dissolved gases that may disrupt the synthesis reaction. A mixture of 3.33 M KOH with 0.27 M KNO_3, made to a volume of 240 ml, was placed in a stoppered separating funnel above the reaction vessel and added to the air-tight glass vessel containing the ferrous sulphate continuously purged with N_2. The resultant precipitate was washed in deionized water, in ethanol, and then dried at 80 °C in flowing N_2 [1] to produce submicrometre particles. Maghemite and haematite were produced by heating magnetite powder in flowing air in a tube furnace for 6 h at different temperatures. X-ray diffraction (XRD) was performed on the powder samples using a CuK_α radiation source within a 2θ range of 15 to 65 degrees. Mössbauer spectroscopy was performed on magnetite powders heat treated at 220 °C, 250 °C and 280 °C. SQUID magnetometry was conducted at 40 °C, as appropriate for in-vitro applications, on two samples pressed into cylindrical pellets at a pressure of approximately 700kPa. Samples included synthesized magnetite and maghemite produced from heat-treatment of magnetite in air at 280 °C.

Results and Discussion

Characterization of the heat-treated magnetite powders by X-ray diffraction showed the oxidation of magnetite to maghemite above 200 °C, producing a pure maghemite at 280 °C, Fig. 1. The maghemite retains the spinel structure and cubic unit cell of magnetite, and the slight shift in the X-ray peaks of the maghemite to larger 2θ angles is due to the decrease in its unit cell size resulting from oxidation of Fe^{2+} ions and the introduction of cation vacancies. Heating in air at 450 °C produces a mixture of maghemite and haematite, with pure haematite resulting at 900 °C, Fig. 1. The X-ray diffraction patterns of haematite are clearly distinct from those of magnetite and maghemite, reflecting the change in structure from the cubic magnetite and maghemite to the rhombohedral haematite. The transformation of maghemite to haematite is isochemical, and there is a direct orientation relationship between the two, with the (311), (422) and (511) planes of maghemite corresponding to the (110), (116) and (122) planes (indexed on the hexagonal unit cell) of haematite respectively.

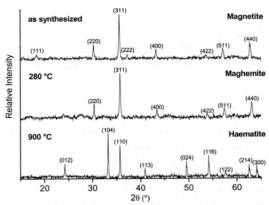

Fig. 1. XRD patterns of synthesized magnetite and after heat treatment in air at 280 °C and 900 °C

The ^{57}Fe Mössbauer spectrum of the as-synthesized magnetite sample at room temperature is shown in Fig. 2a. It has been fitted to two sextets, the outer one representing the Fe^{3+} ions and the inner one for the electron-hopping Fe^{2+}-Fe^{3+} ions. In stoichiometric magnetite, the former would all be in tetrahedral sites and the latter in octahedral sites, with the relative area ratios being 1:2. The relative areas in our spectrum are 1:1.35, with the linewidths being considerably broadened to give a full-width half maximum for the outer lines of the outer sextet of 0.37 mm/s and similarly for the inner sextet of 0.55 mm/s. These changes are caused by oxidation of some of the Fe^{2+} which causes their contribution to the spectrum to be shifted from the inner to the outer sextet. A significant vacancy contribution is then incorporated on the octahedral sites, so that both the octahedral and tetrahedral ions can then have a variety of cation neighbour configurations, which leads to the observed line broadening. The measured hyperfine fields were 49.2 T and 46.2 T, which are very close to the literature values of 49.0 T and 46.0 T respectively, but slightly enhanced due to the higher fraction of the iron in the ferric state.

The room temperature spectrum of the 280 °C heat-treated maghemite sample (Fig. 2b) is notable in being more symmetrical than other spectra in the literature. Fits were attempted with two sextets, combined as an inner and outer sextet or a left and right hand sextet. Most authors have preferred the inner-outer combination (e.g. [2]) because of obvious asymmetry in their spectra, but Pollard [3] preferred the left-right combination. In the absence of constraints on the line intensities, the left-right arrangement produced the most physically acceptable fit to our spectrum and it is this which is the fitting model used, and we obtained values of 49.4 T (280 °C sample), 49.7 T (250 °C) and 49.6 T (220 °C) which are all well within the main literature range of (48.9-50.0) T, showing that the samples are of very good quality. Values below this range would be indicative of poor crystallinity.

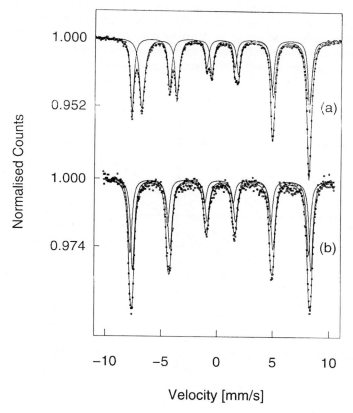

Fig. 2. Room temperature Mössbauer spectra of (a) magnetite and (b) maghemite

SQUID magnetometry was used to ascertain the hysteresis loop area for each material. The samples were magnetized to 24 kA/m (300 Oe) before measurements were taken along the hysteresis loop defined by applied magnetic fields of ±24 kA/m (±300 Oe).The resultant hysteresis loops are shown in Fig. 3.

The internal magnetic field acting on the samples will be smaller than the applied magnetic field, due to demagnetizing effects. The effective heating power, P, due to hysteresis may be determined according to Eq. 1,

$$P = f \oint H_{int} \, dM_m \, ,$$

(1)

where f is the frequency of the applied magnetic field, H_{int} is the internal magnetic field, M_m is the magnetization per unit mass of the material, and $\oint H_{int} \, dM_m$ is the area of the hysteresis loop per unit mass of the material. The internal magnetic field may be determined according to Eq. 2,

$$H_{int} = H_{app} - NM_v, \qquad\qquad (2)$$

where H_{app} is the applied magnetic field, N is the demagnetizing factor, and M_v is the magnetization per unit volume of material. The demagnetizing factor depends on the dimensions of the samples. The measured samples were cylindrical and of identical diameter; however, as the magnetite sample was shorter than the maghemite sample, its demagnetizing factor is larger. The hysteresis loops of Fig. 3. are very similar in area, but after the demagnetization factor corrections have been made, the corrected loop shows that the magnetite will produce greater hysteresis heating for a given applied field.

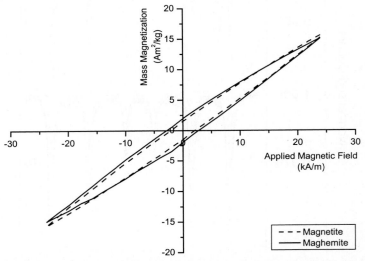

Fig. 3. Hysteresis loops of magnetite and maghemite for applied magnetic fields of ±24 kA/m (±300 Oe)

Conclusions

Submicrometre particles of magnetite and maghemite were synthesized from solution. X-ray diffraction and Mössbauer spectroscopy showed that the samples were of good quality but there was some oxidation apparent in the magnetite. SQUID magnetization measurements showed that the magnetite will give a larger hysteresis heating for a given applied magnetic field.

Acknowledgements

The Australian Research Council is acknowledged for funding towards a QEII Fellowship for Dr. Kārlis A. Gross.

References

[1] U. Schwertmann and R.M. Cornell, *Iron Oxides in the Laboratory* (VCH, Weinheim 1991), p. 111.
[2] G.M. da Costa, E. De Grave, L.H. Bowen, R.E. Vandenberghe, and P.M.A. de Bakker, Clays and Clay Min. 42 (1994), p. 628.
[3] R.J. Pollard, Hyperfine Interactions 41 (1988), p. 509.

Key Engineering Materials Vols. 254-256(2004) pp. 217-220
online at http://www.scientific.net

Effects of TCP Particle Size Distribution on TCP / TTCP / DCPD Bone Cement System

Sang-Hwan Cho[1,2], In-Soo Hwang[1], Jong-Kyu Lee[1], Kyung-Sik Oh[1],

Soo-Ryong Kim[1] and Yong-Chae Chung[2]

[1]Korea Institute of Ceramic Engineering and Technology, Seoul 153-801, Korea,
cementlab@kicet.re.kr

[2]Department of Ceramic Engineering, Han-Yang University, Seoul 133-791, Korea,
yongchae@hanyang.ac.kr

Keywords: calcium phosphate, TCP granule, bone cement, compressive strength

Abstract. In this work, properties of tricalcium phosphate (TCP)-tetracalcium phosphate (TTCP)-dicalcium phosphate dehydrate (DCPD) bone cement containing dense TCP granules were studied. 5 groups of cement were prepared with respect to the blending ratio between granular and powdery TCP. We intended to improve the workability maintaining the compressive strength through the addition of TCP granules. Even if the best compressive strength of specimen was observed in the cement without granule at a power to liquid (P/L) ratio of 2.5, comparable compressive strength (58.36 MPa) could be also obtained in the cement by replacing half of powdery TCP with granules in weight. However, the granular cement exhibited proper workability unlike a powder one. In granular cement, P/L ratio could be further increased from P/L ratio (3.5) with optimum strength, while such an increase of P/L ratio from optimum (2.5) was not possible in the cement without granules.

Introduction

Self-setting calcium phosphate cements are useful for various applications such as orthopaedic surgery and dentistry. Calcium phosphate cements have excellent biocompatibility and less heat evolution compared with conventional PMMA bone cement. However, the mechanical properties need to be improved. It is well known that the particle size and the activity of the starting materials are critical parameter influencing the mechanical properties (setting time and compressive strength) of hydraulic cements [1]. The calcium phosphate cement is typically composed of α-tricalcium phosphate (α-TCP), TTCP (tetracalcium phosphate) and DCPD (dicalcium phosphate dehydrate). Among them, α-TCP is the least soluble and thus controls the overall hardening reaction. Therefore, in this work reactivity was controlled through the introduction of granular α-TCP. Addition of granules in calcium phosphate cement affects on powder to liquid ratio (P/L). With the introduction of granules, the volume of liquid necessary for reaction can be reduced and results in the decrease of porosity. Compressive strength, setting time, phase evolution, and morphology were examined as a function of the amount of granules.

Materials and Methods

α-TCP and TTCP were prepared by calcination of the mixture of calcium hydrogen-phosphate (DCPA) and calcium carbonate at 1500°C for 6 h in furnace, followed by quenching at room temperature. Subsequently obtained powders were blended with DCPD. 0.5 mole of $NH_4H_2PO_4$ and 0.3 mole of citric acid were dissolved in a liter of deionized water to prepare a setting agent [2,3]. The TCP granules were synthesized through sintering of green bodies at 1350°C and subsequent grinding and sieving. The composition of TCP/TTCP/DCPD system bone cement was determined at Ca/P=1.50 according to result in previous work [1]. The average particle size of TTCP and DCPD was 9 μm and 16 μm, respectively. The hydration reaction was performed for 7 days in simulated body fluid at 37°C and RH 100%. 5 groups of cement were prepared with respect to the blending ratio between granular and powdery TCP. The weight percent of granules were 0, 20, 50, 80 or 100 and designated as R, G20, G50, G80 and G100, respectively, as summarized in *Table 1*. The powder to liquid (P/L) ratio was changed from 2.0 to 5.0 depending on the state of mixture. The compressive strength of hardened cement was measured for cube specimens with edge of 1 cm using universal testing machine (INSTRON 4204, maruto) at a compression speed of 0.5 mm/min. The setting time was also determined using Vicat needle. The samples were characterized with X-ray diffractometer and scanning electron microscopy.

Table 1. Designation of sample with the mixing ratio (wt %) between granular and powdery TCP

	TCP type and size range	
	granule 45~90 µm	Powder 9~11 µm
G100	100	0
G80	80	20
G50	50	50
G20	20	80
R	0	100

Results

Table 2 shows the compressive strength with the percentage of granules. The compressive strength of specimen R (P/L ratio=2.5) after curing 7 d in SBF solution was 61.0±3 MPa, which was the best in this work. But the workability of specimen R (P/L ratio=2.5) was not proper for handling and further increase of P/L ratio was impossible. Similar value of compressive strength (58.36 MPa) was also observed in G50 with the P/L ratio of 3.5. However the workability of specimen G50 (P/L ratio=3.5) was much improved as can be convinced from the fact that the P/L ratio in G50 could be further increased to 4.0. The setting time was tested for the specimen with best compressive strength at each composition. It is well known that a proper setting time is about 8~12 min including mixing time. As shown in Table 2, the initial setting time of all compositions was about 10 min, except for G100.

Table 2. Compressive strength and setting time of bone cement prepared

	Compressive strength(Mpa)							Setting Time (min)
	2.0	2.5	3.0	3.5	4.0	4.5	5.0	
G100	-	-	-	-	27.20	30.28	28.91	15
G80	-	-	-	39.20	47.53	34.25	-	11
G50	-	34.40	41.31	58.36	39.41	-	-	10
G20	-	39.79	51.89	✗	-	-	-	9
R	45.00	61.00	✗	-	-	-	-	11

Fig. 1 shows the X-ray diffraction pattern of specimens with the change of TCP particle size after curing 7d in SBF. The specimens R and G20 were composed of HAp phase after hardening. However, specimen G50 consisted of TCP and HAp, while G80 and G100 were composed of only TCP. The result in Fig 1 suggests that the transformation to HAp during setting reaction is not complete for the

higher granule contents. To increase strength of cements, two aspects need to be considered. First, the setting reaction should be carried out homogenously in the overall specimen leaving the unreacted part as a minimum. Second, the specific strength of final product needs to be as high as possible. Table 2 shows that compressive strength requires an optimum condition in P/L ratio except the groups with smaller granular contents (R, G20). With the increase of P/L ratio, the initial increase of strength can be attributed to the decrease of porosity, while final decrease of strength is ascribed to the insufficient homogeneity in reaction throughout the specimen.

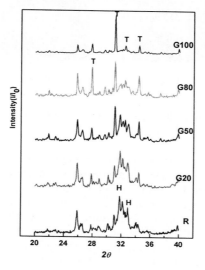

Fig. 1 XRD patterns of Bone Cement (7days)
(T: α-Tricalcim phosphate and H: Hydroxyapatite)

Fig. 2 shows the SEM photographs of specimens with the change of TCP particle size after curing 7 d in SBF solution. In specimens G20 and G50, petal-like HAp could be observed on the surface of the hardened body even though some TCP granules are still remained. For specimens G80 and G100, the HAp was hardly visible and granular morphology of TCP was dominant.

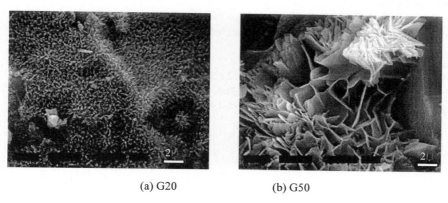

(a) G20   (b) G50

(c) G80   (d) G100

Fig. 2. SEM photographs (5000) of Bone Cement in SBF solution (7days)

Discussion and conclusions
The setting and hardening properties of calcium phosphate cements are due to the progressive dissolution of the TCP phase and the formation of an entangled network of calcium-deficient hydroxyapatite (CDHAp) crystals. With the introduction of granular TCP, P/L ratio could be effectively increased and thus the advantage in workability could be realized while maintaining the similar level of compressive strength. However, further increase of TCP granules rather decreased the compressive strength in spite of the decrease of porosity. It was attributed to the incomplete reaction in forming HAp that has the most superior specific strength in the system. The relative chemical inertness of granules might have prevented the reaction. The study has shown the two important factors in synthesizing self-setting cements with superior strength when introducing TCP granules. First, the P/L ratio should be determined in consideration of not only minimization of porosity but also completeness of reaction. Second, even though the introduction of granules is positive in enhancing the workability, they need to be controlled so that sufficient conversion to HAp can be ensured.

Acknowledgements
This work was supported by a Grant-in-Aid for Next Generation New Technology Development Programs from the Korea Ministry of Commerce, Industry and Energy (No. N11-A08-1402-06-1-3).

References
1. J. K. Lee et al. Key Eng. Mater. Vols. 240-242 (2003), p. 377-380
2. K. Kurashina et al. J. Mater. Sci.: Mater. In Med. 6 (1995), p. 340-47
3. Honglian Dai et al. Key Eng. Mater. Vols. 192-195 (2001), p. 821-24

Key Engineering Materials Vols. 254-256(2004) pp. 221-224
online at http://www.scientific.net
© 2004 Trans Tech Publications, Switzerland

Evaluation of Hydroxylapatite Cement Composites as Bone Graft Substitutes in a Rabbit Defect Model

MJ Voor[1], JJC Arts[2], S Klein[1], LHB Walschot[2], N Verdonschot[2], P Buma[2]

[1]Orthopaedic Bioengineering Laboratory, Department of Orthopaedics Surgery, University of Louisville, Louisville, KY, USA, 40292 mike.voor@louisville.edu
[2]Orthopaedic Research Laboratory,Department of Orthopaedics,UMC St Radboud Nijmegen, P.O. Box 9101,6500 HB Nijmegen,the Netherlands

Keywords: bone graft, hydroxylapatite cement, cancellous, incorporation, bone graft sustitute

Abstract. To evaluate *in vivo* performance of hydroxylapatite cement (HAC) as a porous bone graft substitute, HAC was mixed (1:1 volume ratio) with either porous calcium phosphate GRANULES (80% TCP, 20% HA) or defatted morsellized cancellous bone allograft (MCB) and implanted bilaterally in rabbit distal femurs. Groups with EMPTY defects and impacted MCB were used for reference. After eight weeks, one femur from each pair was examined histologically. Contralateral specimens and time-zero specimens were used for mechanical indentation tests.

Histology showed that the EMPTY defects were filled with newly formed osteopenic bone after eight weeks. The MCB defects showed complete remodelling of graft into vital bone. Incorporation of the HAC/MCB composites was extensive but not completed, while minimal new bone ingrowth into the HAC/GRANULE composites was found. Indentation tests showed that both cement composites were significantly stronger than EMPTY defects, incorporated MCB and intact bone.

Composites of HAC and porous biomaterials can maintain their relatively high strength over eight weeks *in vivo*, but their incorporation into a new bony structure is slowed by the synthetic materials. The HAC/MCB composite showed more favourable strength and incorporation.

Introduction

The ideal bone graft material should be implantable through a minimal surgical exposure, conform to fill irregular defects, be as rigid and strong as intact bone for immediate load bearing, promote new bone formation and incorporation by the host, and be fully synthetic. Injectable hydroxylapatite cement (HAC) is currently being evaluated for a number of limited clinical applications [6], but it is not quickly replaced by living host bone nor is it sufficiently strong in any loading mode except compression. These characteristics limit its use to confined locations, non-load bearing situations, or fracture applications only with supplemental fixation [6]. One approach to improving HAC would be to accelerate the process of new bone formation within the cement. Several studies have looked at inducing porosity within the cement to achieve this goal [3,8-10]. Unfortunately, when porosity is generated by creating bubbles in an otherwise single-phase cement, the initial strength, even in compression, is significantly reduced [3,8]. Other attempts have used a second phase material that is rapidly absorbed or dissolved to create pores while maintaining short-term strength [1,10,11]. The objective in this study was to determine if adding porous particles to HAC could improve its incorporation without compromising strength.

Materials and Methods

A total of twenty-four rabbits weighing an average of 3040 g (2300-4200 g) were used for both mechanical and histological evaluation. The university ethical committee approved all procedures. Bilateral cancellous bone defects in the form of cylindrical holes 10 mm deep and 5.5 mm in diameter were drilled from the lateral direction in the distal femora [2-4]. A polyethylene ring and plug were used to seal the defect at the lateral femoral cortex (Fig. 1).

Fig. 1. Graft specimens (8 mm x 5.5 mm dia.) were placed in rabbit lateral femoral condyle through a threaded alignment ring (5 x 8 mm).

Three materials were used in various combinations to fill the defects: BoneSource Classic HAC (Stryker), BoneSave GRANULES (Stryker), and defatted morselized cancellous bone allograft (MCB) of approximately 2 mm particle size obtained from four donor rabbits. HAC is an equimolar combination of tetra calcium phosphate $[Ca_4O(PO_4)2]$ (TTCP) and dicalcium phosphate anhydrous $[CaHPO_4]$ (DCPA). The particle size of the TTCP is approximately 15 μm while the particle size of the DCPA is approximately 1.5 μm. The size of the particles of the two components is important, in both the absolute sense and relative to each other, in producing the desired setting reaction. These components react in an aqueous environment to form hydroxylapatite $[Ca_{10}(PO_4)6(OH)2]$. A setting time of approximately five minutes can be achieved by mixing the cement powder with a 0.25 mol/L solution of Na_2HPO_4 [1]. BoneSave GRANULES are made up of 80% TCP $[Ca_3(PO_4)2]$ and 20% HA $[Ca_{10}(PO_4)6(OH)2]$. The granules used in this study have a nominal surface porosity of 50% and were sorted into sizes ranging from 1.5 to 2.5 mm diameter.

The four rabbit groups were: empty defect control (EMPTY), impacted MCB allograft (MCB), HAC mixed with MCB allograft in an approximate 1:1 ratio (HAC/MCB), HAC mixed with porous BoneSave granules in an approximate 1:1 ratio (HAC/GRANULES). Five femurs in each group were randomly (left or right) designated for histological examination. The five contralateral femurs from each group were used for mechanical indentation testing of the graft sample *in situ*.

For mechanical testing, the polyethylene ring was used for positioning, alignment and fixation of the distal femur on a custom loading pillar using plaster (Fig. 2). Immediately prior to testing, each embedded specimen was placed in a lathe and the surface plaster removed as well as the medial aspect of the femur to reveal the distal femoral cancellous bone of the specimen and the included graft material. The cylindrical indenter (1.5 mm diameter) was fixed on the hydraulic actuator (MTS). The specimen was supported by a 2kN capacity load cell operating in the 1 kN range (Eaton). The indenter was pressed into the exposed surface of the graft material at a rate of 0.1 mm/s to a distance of 2 mm and the maximum load was determined.

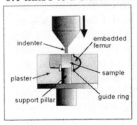

Fig. 2. After removing the surface of the plaster and part of the femur to reveal the exposed graft material, an indentation test was performed using a 1.5 mm diameter indenter tip at a displacement rate of 0.1 mm/s.

The specimens were cut along the long axis of the graft sample and half of each specimen was prepared for undecalcified histology. The other half was decalcified in EDTA. Undecalcified specimens were embedded in poly-methyl-methacrylate and sectioned using a diamond blade saw (Leica SP1600) to a thickness of 20 μm. Contiguous sections were stained with hematoxylin and eosin (HE) or left unstained for fluorescence microscopy. Decalcified sections (7 μm) were stained with HE and TRAP (Tartrate Resistant Acid Phosphate) to visualize osteoclasts. Specimens were examined qualitatively and quantitatively for the extent of bone ingrowth and the presence of ceramic or bone graft.

Statistical analyses were performed as ANOVA with post-hoc t-tests (Tukey) to determine significant differences between groups for mechanical strength and quantitative histology measures.

Results

The indentation test results (Fig. 3) showed that both HAC/GRANULES and HAC/MCB were stronger than intact cancellous bone initially and after eight weeks *in vivo* (p<0.05). There was no difference between HAC groups. The HAC groups were stronger than the EMPTY defect and MCB groups at eight weeks (p<0.05). The MCB was much stronger initially than after being *in vivo* for eight weeks (p<0.05) but after eight weeks *in vivo* was not significantly different from intact bone.

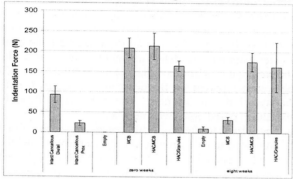

Fig. 3. The indentation strengths of the various materials tested showed that the initial strength of all three of the implanted materials exceeded that of intact cancellous bone. After eight weeks *in vivo* the strength of the impacted allograft MCB reduces to that of intact bone. Solid bars indicate mean with error bars representing SEM.

Histology showed that EMPTY defects were healed after eight weeks, but only sparse thin trabecular structures were observed within some of the defect sites (Fig. 4a). The MCB group showed complete incorporation and remodelling to a new trabecular structure with organization resembling intact bone (Fig. 4b). The HAC/MCB group showed extensive incorporation and remodelling of the regions originally occupied by the bone particles (Fig. 4c). The cement was also partially resorbed with osteoclasts active adjacent to the HAC where there was actively incorporating MCB. The surface of the HAC was osteoconductive and had new bone attached to it along the original defect margin. The HAC/GRANULE group showed little resorption or remodelling activity (Fig. 4d). There was a small amount of bone ingrowth as observed on fluorescence microscopy, but in general the HAC/GRANULE cylinders remained intact.

A. EMPTY   B. MCB   C. HAC/MCB   D. HAC/GRANULES

Fig. 4. Histology shows little or no new bone in the EMPTY group, complete remodelling in the MCB group, extensive remodelling in the HAC/MCB group, and little activity in the HAC/GRANULE group.

There was excellent bone attachment to the surface of the HAC cylinder along the periphery of the original defect. Also, when there were porous granules in contact with this original outer surface or when cracking of the HAC occurred, some bone ingrowth and granule resorption did occur. The results of the quantitative histology showed no significant differences between groups because of the amount of variation within each group. Two of the EMPTY defects healed well, while two remained empty. Similarly, two HAC/GRANULE specimens showed substantial bone ingrowth and remodelling, while the other three did not.

Discussion and Conclusions

The impacted MCB group showed full incorporation after eight weeks. This was in contrast to the EMPTY group which showed filling of the defect with osteopenic bone having very low indentation strength at eight weeks. The addition of porous particles as a second phase to the bulk HAC improved its rate of incorporation without compromising its compressive strength. This was especially true in the HAC/MCB group. By eight weeks, nearly all of the MCB was replaced by living bone, yet the indentation strength and stiffness did not decrease. The HAC and the MCB formed continuous bonds with the original margins of the drilled defect and it is assumed that over time the material will become stronger in other loading modes such as tension, shear, and bending. There was evidence of direct osteoclast resorption of the HAC followed by deposition of new bone. The HAC/GRANULE group was also osteoconductive and showed signs of new bone formation in some of the pores and cracks, but overall it was not very active. In a study of different formulations of calcium phosphate cements in the cancellous bone of goats, Ooms et al. [5] also found that bone covered the surface of pores in contact with outer margin of the defect at eight weeks. Also, as the HAC cracks and allows osteoconduction to occur along its surfaces, bone begins to fill in the voids in the material. This result is similar to the results reported by Boyde et al. [7] in which bulk hydroxylapatite ceramic was used in a diaphyseal defect in sheep. The HAC is sufficiently osteoconductive to allow rapid filling of cracks and exposed pores with new bone.

An interesting observation from this study is related to the basic differences between the two HAC groups. The MCB particles are organic compared to the synthetic BoneSave granules and the MCB pores are more extensive and interconnected compared to the BoneSave granules. Excellent osteoconduction was observed in all three materials; HAC, MCB, and BoneSave. MCB, however, was much more rapidly resorbed and replaced by living bone. A combination of the organic nature of the MCB and its interconnected porosity was most likely responsible for this effect. In conclusion, mixing MCB allograft with HAC in a 1:1 ratio is a simple and effective way to accelerate the incorporation of injectable hydroxylapatite cement without compromising its compressive load bearing capacity.

References

[1] Xu, H.H.K., Quinn, J.B., Takagi, S., Chow, L.C., Eichmiller, F.C. "Strong and Macroporous Calcium Phosphate Cement: Effects of Porosity and Fiber Reinforcement on Mechanical Properties." J. Biomed Mater Res 57 (2001) 457-66,

[2] Friedman RJ, An YH, Ming J, Draughn RA, Bauer TW. Influence of biomaterial surface texture on bone ingrowth in the rabbit femur. J Orthop Res. May;14(3) (1996) 455-64..

[3] Bauer TW, Togawa D, An YH, Woolf SK, Hawkins M, Edwards B. Evaluation of resorption rates of five injectable bone cements using a rabbit femoral defect model. Orthopaedic Research Society. Poster 738, February 2002.

[4] An YH, Friedman RJ (ed.). Animal Models in Orthopaedic Research. CRC. Boca Raton, 1999.

[5] Ooms EM, Wolke JG, van der Waerden JP, Jansen JA, Trabecular bone response to injectable calcium phosphate (Ca-P) cement. J Biomed Mater Res 61(1) (2002) 9-18.

[6] Larsson S and Bauer TW, Use of injectable calcium phosphate cements for fracture fixation: a review. Clin Orthop 395 (2002) 23-32.

[7] Boyde A., Corsi A, Quarto R, Cancedda R, Bianco P, Osteoconduction in large macroporous hydroxyapatite ceramic implants: evidence for a complementary integration and disintegration mechanism. Bone 6 (1999) 579-89.

[8] del Real RP, Wolke JG et al., A new method to produce macropores in calcium phosphate cements. Biomaterials. 23(17) (2002) 3673-80.

[9] Barralet JE, Grover L, Gaunt T, Wright AJ, Gibson IR, Preparation of macroporoous calcium phosphate cement tissue engineering scaffold. Biomaterials 23(15):3063-72.

[10] Nilsson M, Fernandez E, Sarda S, Lidgren L, Planell JA, Characterization of a novel calcium phosphate/sulphate bone cement. J. Biomed Mater Res 61(4) (2002) 600-7.

[11] Xu HH, Quinn JB. Calcium phosphate cement containing resorbable fibers for short-term reinforcement and macroporosity. Biomaterials 23(1 (2002) 193-202.

Key Engineering Materials Vols. 254-256(2004) pp. 225-228
online at http://www.scientific.net

Animal Study of Alpha-Tricalcium Phosphate-Based Bone Filler

Jong-Heon Lee[1*], Kang-Sik Lee[1], Jae-Suk Chang[1], Woo Shin Cho[1], Jong-Kyu Lee[2], Soo-Ryong Kim[2], and Yung-Tae Kim[1]

[1] Department of Orthopedic Surgery, ASAN Medical Center, Seoul 138-736, Korea

[2] Ceramic Building Materials Department, Korea Institute of Ceramic Engineering and Technology, Seoul 153-023, Korea

*Corresponding Author: Jong-Heon Lee (e-mail: jongheon@amc.seoul.kr)

Keywords: Alpha tricalcium phosphate, bone filler, animal study, rabbit

Abstract. The viscous α-tricalcium phosphate (TCP)-based bone fillers were inserted into the drill holes (5mm in diameter) made through the lateral femoral condyles of 18 New Zealand white rabbits. The rabbits were sacrificed at 1, 3 and 6 weeks respectively and the interface layers were then examined by scanning electron microscope. Histopathological analysis was performed on the areas surrounding the inserted sites. The new bony formation did not begin after one week. At 3 weeks, bony ongrowth to the inserted bone filler could be found, and the new bone surrounded the bone filler at 6 weeks. In conclusion, the rate of new bone ongrowth increased with a passage of time from 1 week through 6 weeks. The difference in the amount of new bone ongrowth however was not clearly shown by varying the percentage of α-TCP between 60 and 75% in the bone filler, unlike the effect of time passage to the rate of new bone ongrowth.

Introduction

Autologous bone transplants are highly biocompatible, very well tolerated by the human body and therefore do not show signs of rejection after implantation. Allografts are transplants that stem from other human beings and their limited availability is reduced as compared to the autografts. However, they have potential problems: for autografts, donor site complications, prolonged operation time and higher costs; for allografts, the risk of disease and infection transmitted from the donor, such as Hepatitis or HIV. Their problems therefore lead to the use of synthetically produced ceramics [1]. Calcium phosphate bone cements (CPBC) have been developed by Brown and Chow [2] and used as bone fillers. They have attracted as substitutes to fill the sites of bone defects in orthopedic surgery due to their good biocompatibility and capacity to be replaced by new bone [3-6]. However, their clinical applications are limited due to slow resorption *in-vivo,* poor mechanical strength, unsuitable setting time and the biological properties. *In-vivo* studies have been recently conducted to evaluate the biocompatibility [7,8] and the mechanical strength of CPBC [9] but the properties of CPBC are still insufficient for safe application. The objective of the present study is to investigate the new bony growth at the interfaces between the bone fillers and the cancellous bones of the lateral femoral condyles of New Zealand white rabbits.

Materials and Methods

Materials

Calcium hydrogen phosphate ($CaHPO_4$) and calcium carbonate ($CaCO_3$) were used as the starting materials to produce α-TCP and tetracalcium phosphate (TTCP) by solid-state reaction. The α-TCP-based bone filler powders with the average particle size of 7.5μm were then prepared after mixing the three components, α-TCP, TTCP and dicalcium phosphate dihydrate (DCPD, 99.9% pure, research grade, Aldrich Co.), based on the compositions listed in Table 1. Distilled water containing 0.5M of ammonium dihydrogen phosphate was used as the liquid phase. Hydroxyapatite was the desired end product after hydrolysis of the α-TCP-based bone filler powders mixed with the liquid phase at the powder-to-liquid (P/L) ratio of 1.7 g/ml at 36.5°C.

Table 1. Classification of the alpha-TCP-based bone fillers

Sample ID	Compositions of bone fillers, %		
	α-TCP	TTCP	DCPD
A	60	20	20
B	65	15	20
C	75	20	5

Animal Study

Eighteen New Zealand white rabbits weighing average of 2.5 kg (3-4 months old) were used. They were preanesthetized with an atropine (0.5 mg/kg) and then anesthetized with ketamin (35 mg/kg) and xylazine (5 mg/kg) intramuscularly. Drill holes (5 mm in diameter and 10 mm in length) were made through both the right and left lateral femoral condyles of rabbits and hemostasis was conducted for 3 minutes by filling and compressing the holes with gauze. Immediately after removing the gauze, the enough amount of bone filler powder mixed at the P/L ratio of 1.7 g/ml was inserted into the holes. For each sample, two rabbits were used and euthanized at each time period of 1, 3 and 6 weeks after operation, respectively. The bony ongrowth at the interface layers between the α-TCP-based bone fillers and the cancellous bones of the lateral femoral condyles were examined by the back scattered mode of scanning electron microscope (SEM).

Histopathological Analysis

After sacrifice of the rabbits, the femoral condyles were retrieved. The specimen containing the sample site and the area surrounding it was cut and immediately placed in a 70% ethyl alcohol fixative solution for 7 days. Then, dehydration of tissues commenced in 70% ethyl alcohol and continued in 80%, 90% and 100% ethyl alcohol. The specimen was soaked in each solution for more than 2 hours, which was followed by soaking twice in 100% ethanol for more than 4 hours. It was immersed in propylene oxide for 2 hours and then progressively embedded in epoxy type resin. Post-hardening of the polymerization of the resin was done at 35°C for 24 hours, then at 60°C for 4 days. A rotating diamond saw, using glycerol as cooling fluid, cut approximately 0.2mm thick disks from the specimen, perpendicular to the longitudinal direction of insertion. Each disk was fixed with epoxide resin on glass slides, polished to a thickness of approximately 30-60µm and then stained by the modified Goldner's method, using Weigert's hematoxylin solution, Ponceau - Fuchsin - Azophloxin solution, light green in 0.3% acetic acid and Phosphomolybdic acid-Orange G solution. The stained section was examined by a Nikon UFX-DX transmitted light microscope.

Results and Discussion

At one week, all samples A, B and C did not show signs of hemorrage, necrosis and discolorations on the femoral condyles surrounding the inserted bone fillers through macroscopic observation. As shown in Figure 1, SEM observations showed that for all samples, the new bone ongrowth was not found and the boundaries between the bone filler and the cancellous bone were clearly revealed. SEM observations after 3 and 6 weeks of bone filler insertion are shown in Figure 2 and 3, respectively. New bone ongrowth interfaces were visible at 3 weeks (Figure 2). After 6 weeks, new bone seems to grow among the particles of bone filler so that they were partially surrounded by new bone on the boundary (Figure 2). In addition, the amount of bone ongrowth seems to be larger at 6 weeks (Figure 3) than that at 3 weeks. However, Samples A, B and C which were respectively mixed with 60, 65 and 75% of α-TCP at the present study, did not show the clear evidence of difference in bony ongrowth activity as a function of α-TCP amount.

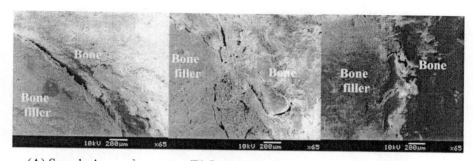

(A) Sample A   (B) Sample B   (C) Sample C

Fig. 1. SEM observations on the bone filler insertion sites after one week (back scattered images).

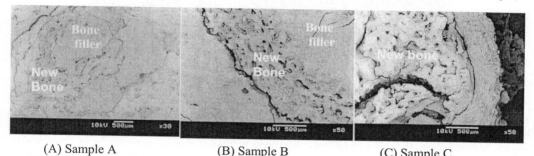

(A) Sample A   (B) Sample B   (C) Sample C

Fig. 2. SEM observations showing new bony ongrowth at 3 weeks (back scattered images).

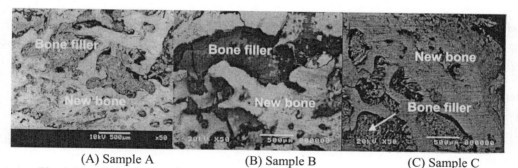

(A) Sample A   (B) Sample B   (C) Sample C

Fig. 3. New bony formation observed by SEM after 6 weeks (back scattered images).

From the histopathological observation of Sample C after 1 and 6 weeks follow-up studies, as shown in Figure 4, almost no new bone formation was observed on the boundary between the bone filler and the cancellous bone at a week (Figure 4(A)). As time goes on, bony ongrowth increased and the large amount of mineralized new bone was surrounding the inserted bone filler at 6 week (Figure 4(B)). Nevertheless, Samples A, B and C did not show the remarkable differences in bone formation process at the same time of follow-up study. This indicates that the differences of the amount (between 60 and 75%) of α-TCP in the bone fillers (Table 1) seem not to affect on the bone formation activity if the follow-up time is same.

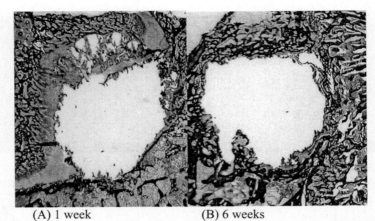

(A) 1 week   (B) 6 weeks
Fig. 4. Histopathological observation of Sample C at 1 and 6 weeks follow-up studies
(Goldner's Staining: black contrast is minalized bone; gray contrast is osteoid).

Conclusion

At one week of insertion, the new bony ongrowth was not observed and the boundaries between the inserted bone filler and the cancellous bone are clearly revealed. As time goes on, bony ongrowth began to form at 3 weeks and the bone filler particles were surrounded by the new bone ongrowth at 6 weeks. In conclusion, the rate of new bone ongrowth increases as time increases from 1 to 6 weeks. However, the effect of the amount of α-TCP in bone filler was not clearly observed on the process of new bone formation at the same period of follow-up study.

Acknowledgments

This work was supported by Next Generation Project of Ministry of Commerce, Industry and Energy, and by the ASAN Institute for Life Sciences (No. 2003-088), Seoul, Korea.

References

[1] A. Gisep, Injury: Int. J. Care Injured, Vol. 33 (2002), p. S-B88.
[2] W.E. Brown and L.C. Chow: U.S. Patent No. 4518430, 1985.
[3] E. Fernandez, F.J. Gil, M.P. Ginebra, F.C.M. Driessens, J.A. Planell: J. Mater. Sci. Mater. Med. Vol. 10 (1999), p. 169.
[4] E. Fernandez, F.J. Gil, M.P. Ginebra, F.C.M. Driessens, J.A. Planell: J. Mater. Sci. Mater. Med. Vol. 10 (1999), p. 177.
[5] K. Ishikawa and K. Asaoka: Biomaterials, Vol. 16 (1995), p. 527.
[6] F.C.M. Driessens, M.G. Boltong, M.I. Zapatero, R.M.H. Verbeeck, W. Bonfield, O. Bernudez, E. Fernandez, M.P. Ginebra and J.A. Planell: J. Mater. Sci. Mater. Med. Vol. 6 (1995), p. 272.
[7] J.G.C. Wolke, E.M. Ooms and J.A. Jasen: Key Eng. Mate. Vols. 192-195 (2001), p. 793.
[8] C. Faldini, A. Moroni, M. Rocca, S. Stea, E. Donati, M. Mosca and S. Giannini: Key Eng. Mate. Vols. 192-195 (2001), p. 805.
[9] H. Yamamoto, S. Niwa, M. Hori, T. Hattori, K. Sawai, S. Aoki, M. Hirano and H. Takeuchi: Biomaterials. Vol. 19 (1998), p. 1587.

Key Engineering Materials Vols. 254-256(2004) pp. 229-232
online at http://www.scientific.net

Influence of Gelatin on the Setting Properties of α-Tricalcium Phosphate Cement

Adriana Bigi, Barbara Bracci, Silvia Panzavolta

Department of Chemistry "Ciamician", University of Bologna, via Selmi 2, 40126 Bologna, Italy,
bigi@ciam.unibo.it

Keywords: cement, α-tricalcium phosphate, gelatin, compressive strength, X-ray diffraction

Abstract. This study investigates the effect of gelatin on the setting time, compressive strength, phase evolution and microstructure of calcium phosphate cement. Different cement formulations were prepared from suspensions of α-$Ca_3(PO_4)_2$ in gelatin aqueous solutions. The cements, enriched with $CaHPO_42H_2O$, and prepared with a liquid/powder ratio of 0.3 mL/g, were soaked in simulated body fluid (SBF) for different periods of time up to 21 days. The results indicate that the presence of gelatin accelerates the setting reaction, and improves the mechanical properties of the cement.

Introduction

Calcium phosphates are widely employed for the preparation of bone cements, thanks to their good biocompatibility and bioactivity, which make these compounds particularly suitable for hard tissue replacement [1]. The product of the setting reaction is quite often calcium deficient hydroxyapatite (CDHA), which exhibits structural and chemical properties similar to biological apatite. A variety of different compositions have been proposed for calcium phosphate cements (CPCs) leading to CDHA as the final setting product [2,3]. In particular, a number of organic and polymeric additives have been used with the aim to improve the mechanical properties of CPCs [4,5]. Thanks to its relatively high solubility, α-tricalcium phosphate, α-$Ca_3(PO_4)_2$ (α-TCP), is one of the main components of several bioactive calcium phosphate cements. We have investigated the effect of gelatin on the setting properties of cement constituted of α-TCP enriched with $CaHPO_42H_2O$ (DCPD).

Materials and methods

α-TCP was obtained by solid state reaction of a mixture of $CaCO_3$ and $CaHPO_4 \cdot 2H_2O$ in the molar ratio of 1:2 at 1300°C for 5 hours [6]. Gelatin-α-TCP powders with different gelatin content (5, 10, 15 and 20% wt) were prepared by grinding and sieving the solid compounds obtained by casting gelatin aqueous solutions containing α-TCP. 5% wt of $CaHPO_42H_2O$ were added to the cement powders before mixing with the liquid phase, with a liquid/powder ratio of 0.3 mL/g. Teflon moulds were used to prepare cement cylinders 6 mm in diameter and 12 mm high. Soaking was performed in simulated body fluid (SBF) at 37°C [7] for different periods of time up to 21 days.

Initial and final setting times were determined by the Gillmore method. Total porosity was evaluated by measurement of the dimensions, and of the wet and dry weights of the cements after 7 days of storage in SBF solution. The specimens were removed from SBF after different times of soaking, from 4 h to 21 days, de-moulded and immediately submitted to compression tests by using a 4465 Instron testing machine, equipped with a 1KN load cell. At least six specimens for each incubation time were tested at a crosshead speed of 1 mm/min. X-ray diffraction analysis was carried out by means of a Philips PW 1050/81-powder diffractometer equipped with a graphite monochromator in the diffracted beam. CuKα radiation was used (40 mA, 40 kV). Morphological investigation of the dried films before and after treatment with the calcifying solutions was performed using a Philips XL-20 Scanning Electron Microscope. The samples were sputter-coated with gold prior to examination.

Results and discussion

Control cement prepared without gelatin exhibits initial and final setting times of 5 and 14 minutes respectively. As gelatin content increases from 5 to 20% wt, both initial and final setting times slightly decrease, so that the values recorded for the sample containing 20% wt of gelatin are 4 and 10 minutes. The presence of gelatin does not affect the X-ray diffraction pattern of the cement powders that exhibit the diffraction reflections characteristic of α-TCP together with the diffraction reflections at 11.7 and 20.9° of 2θ, due to DCPD. The diffraction peaks characteristic of DCPD are no longer appreciable in the patterns recorded after just 4 h of aging, while the relative intensities of the diffraction peaks characteristic of apatite increase with respect to those due to α-TCP as a function of the soaking time.

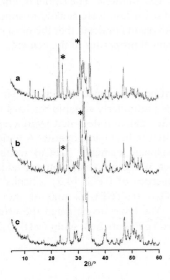

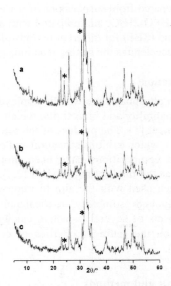

Fig. 1. Powder X-ray diffraction patterns of the control cement after different times of soaking in SBF. (a) 1 day, (b) 2 days, (c) 7 days. The main diffraction peaks of α-TCP are indicated with (∗).

Fig. 2. Powder X-ray diffraction patterns of the samples after two days of soaking in SBF. (a) control cement, (b) cement containing 5% wt of gelatin, (c) cement containing 20% wt of gelatin. The main diffraction peaks of α-TCP are indicated with (∗).

Figure 1 reports the powder X-ray diffraction patterns of the control cement after different times of soaking in SBF solution. The incubation time necessary for the complete conversion of α-TCP into apatite decreases on increasing gelatin content, so that the setting reaction of the cement containing 20% wt of gelatin is almost complete after two days of soaking in SBF. The comparison of the X-ray patterns reported in fig. 2 clearly indicates that gelatin accelerates the conversion of α-TCP into CDHA.

The lattice constants of the apatitic phase obtained after setting do not show appreciable variations as a function of gelatin content of the cements, and assume mean values of $a = 9.42(1)$ Å and $c = 6.88(1)$ Å. The control cement exhibits poor mechanical properties even after setting, as shown by the low value of the compressive strength of the samples aged in SBF at 37°C for 7 days (Table 1).

Table1. Mean values of compressive strength (σ_c) and total porosity of the cements aged in SBF for 7 days.

Gelatin content (% wt)	σ_c (MPa)	Porosity (%)
0	2.0 ± 0.3	59
5	2.3 ± 0.3	50
10	6.5 ± 1.2	43
15	11.0 ± 1.2	38
20	14.0 ± 0.9	37

The compressive strength of the cements at different gelatin contents increases as the setting reaction proceeds. Table 1 reports the mean values of the compressive strength of the samples soaked in SBF for 7 days as a function of their gelatin content. It is evident that the increase of gelatin content provokes a significant increase of the compressive strength. The values of total porosity of the aged cements reported in the same Table 1, suggest a close relationship between the improvement of the mechanical properties and the reduction of the total porosity recorded on increasing gelatin content. A similar relationship has been previously reported for apatitic calcium phosphate cements of different total porosity obtained by compaction of the cements [8]. The data support the hypothesis that porosity is a key factor for the control of the mechanical properties of cements [8,9].

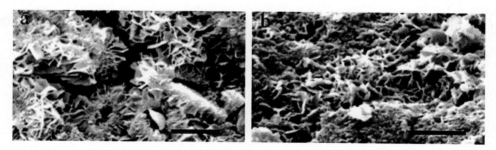

Fig. 3. SEM micrographs of the fracture surfaces of two samples soaked for 7 days in SBF. (a): control cement, bar = 5 μm; (b) cement containing 20% wt of gelatin, bar = 10 μm.

The SEM micrographs of the cements after 7 days of soaking in SBF show that the fracture surfaces of the cements are covered with entangled plate-like crystals (Figure 3). However, the fracture surface of the gelatin-α-TCP cements appear much more compact and uniform than that of the control cement, in agreement with the reduction of the total porosity of the cements induced by the presence of gelatin.

Conclusions

The results of this work indicate that the presence of gelatin improves the setting properties of the cement. With respect to the control sample, gelatin cements exhibit a more compact microstructure, a faster rate of conversion into CDHA, reduced total porosity and significantly greater compressive strength. The addition of gelatin to α-TCP based cements can be utilized to modulate the setting properties of the cements, and to improve their mechanical properties.

Acknowledgements

This research was carried out with the financial support of MURST, and of the University of Bologna (Funds for Selected Research Topics).

References

[1] E. Fernández, F.J. Gil, M.P. Ginebra, F.C.M. Driessens, J.A. Planell, S.M. Best: J. Mater. Sci. Mater. Med. Vol. 10 (1999), p. 223.
[2] W. Suchanec, M. Yoshimura: J. Mater. Res. Vol. 13 (1998), p. 94.
[3] F.C.M. Driessens, M.G. Boltong, E.A.P. de Maeyer, R. Wenz, B. Nies, J.A. Planell: Vol. 23 (2002), p. 4011.
[4] I. Khairoun, F.C.M. Driessens, M.G. Boltong, J.A. Planell: Biomaterials Vol.20 (1999), p. 393.
[5] L.A. Dos Santos, L.C. De Oliveira, E.C.S. Rigo, R.G. Carrodeguas, A.O. Boschi, A.C.F. DeArruda: Bone Vol. 25 (1999), p. 99.
[6] A. Bigi, E. Boanini, R. Botter, S. Panzavolta, K. Rubini: Biomaterials Vol. 23 (2002), p. 1849.
[7] A. Bigi, E. Boanini, S. Panzavolta, N. Roveri: Biomacromolecules Vol. 1 (2000), p. 752.
[8] J.E. Barralet, T. Gaunt, A.J. Wright, I.R. Gibson, J.C. Knowles: J. Biomed. Mater. Res. Vol.63 (2002), p. 1.
[9] R.P. del Real, J.G.C. Wolke, M. Vallet-Regi, J.A. Jansen: Biomaterials Vol. 23 (2002), p. 3673.

Key Engineering Materials Vols. 254-256(2004) pp. 233-236
online at http://www.scientific.net
© 2004 Trans Tech Publications, Switzerland

Synthesis and Comparison Characterisation of Hydroxyapatite and Carbonated Hydroxyapatite

R. Martinetti[1], M.F. Harmand[2], L. Dolcini[1]

[1] FIN-CERAMICA FAENZA s.r.l., Faenza, Italy, roberta.martinetti@fin-ceramicafaenza.com

[2] LEMI, Technopole Montesquieu, Martillac, France, lemi@atlantel.fr

Keywords: Bioactive ceramics, artificial bone graft

Abstract. The aim of the present study was to compare two calcium phosphate biomaterials regarding bone biomimetic characteristics in term of chemical and biocompatibility, in particular cell metabolic activity and mitochondria activity were investigated. Hydroxyapatite (HA) and Carbonated Hydroxyapatite (CHA) powders were both synthetized by wet method; chemical, morphological and biocompatibility evaluations were performed. Using the Balb 3T3 test system CHA was shown non-cytotoxic while HA was slightly cytotoxic. Using human osteogenic bone marrow cells (HOBMC), both powders are non-cytotoxic. Moreover mitochondria activity (MTT test) was not modified. Thus CHA and HA powders could be both suggested for bone filling material development ; in particular CHA due to his bone biomimetic chemical composition could have higher development than HA for bone repair or hard tissue engineering.

Introduction

The requirement of bone-like substitutes is growing up mainly in the orthopaedic, maxillofacial and neurosurgery field ; this need to replace a portion of lost bone has induced the development of synthetic calcium phosphates more similar to biological apatite such as carbonated hydroxyapatite (CHA) with respect to the more used stoichiometric hydroxyapatite (HA).

Literature data have shown that CHA has higher *in vitro* solubility that HA while comparative data in term of compatibility degree are not reported. This study reports the comparative data in term of physico-chemical characterisation and cytocompatibility behaviours; this data may contribute for the selection of calcium phosphates powders to develop biomimetic medical devices for bone replacements.

Materials and Methods

HA and CHA were synthetised by wet methods using Ca, P and, in particular for CHA, CO_3^{2-} sources. The powders were investigated by X-Ray diffraction analysis (CuKα radiation, Rigaku Miniflex), IR spectroscopy (Perkin Elmer FT-IR mod.1600 spectrometer, spectral range 4000-400cm^{-1}, Kbr pellets as supports), elemental analysis (Ca, P, Na, Mg, Pb, Hg, As, Cd and C) (ICP), morphology (SEM, Leica, Cambridge), specific surface area (B.E.T. method) and particle-size distribution.

Powders cytotoxicity was assessed, in direct contact (25 mg/culture well) and using an extract (0.2 g/mL – 120 h – 37° C according to ISO 10993-12 [1]), according to ISO 10993-5 [2] following 24 and 72 h incubation periods, using two methods: cell count (dead and living cells are counted using a hemocytometer) and MTT test [3].

Two cell lines were used, Balb/c 3T3 and human osteogenic bone marrow cells (3rd passage) (HOBMC).

Results

The close processing routes produced CHA and HA powders with similar mean particle size and surface area while morphology (Figure 1 and 2) was quite different even if the powder shows homogeneous aggregates constituted with nanoparticles. X-ray diffraction analysis of powders revealed no foreign phases besides CHA and HA for respectively powders (Figure 3 and 4); traces showed also that powders are characterised by low crystallinity degree: respectively 30 vol % for CHA and 60 vol % for HA powders. The carbonate content of the synthetic powder was determined from the c:a ratio (lattice parameters): the c:a tends to 0,75. IR spectra show that CHA were B-type and that the carbonate bands were produced between 1410 and 1460 cm^{-1} and at 870cm^{-1} in agreement with literature data [4, 5]

Table 1 report also comparative data in term of elemental analysis.

Fig. 1. HA powder morphology

Fig. 2. CHA powder morphology

Table 1.

Samples	Ca / P	CO$_3$ [wt %]	Na [%]	Mg [%]	Pb [ppm]	Hg [ppm]	As [ppm]	Cd [ppm]
HA	1,66	-	0,000233	0,00185	< 1	< 1	< 1	< 1
CHA	1,90	14	2,5	-	< 1	< 1	< 1	< 1

Two methods were used in this study to assess cytotoxicity at two levels: cell viability and growth using cell count and cell metabolic activity through the conversion of MTT (bromure of 3-{4,5-dimethylthiazol-2-yl}- 2,5 - diphenyl tetrazolium) to formozan crystals by mitochondrial succinate dehydrogenase. Two tests systems were used: first a transformed cell line (Balb/c 3T3), thereafter HOBMC, which are cells responsible for bone formation and repair, in order to confirm the results obtained using a transformed cell line with a cell line arising from the future implantation site of bone filling materials (cytocompatibility study).

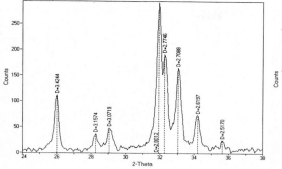

Fig. 3. X-ray HA powder

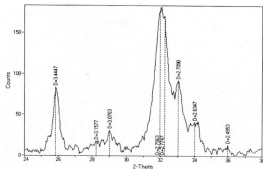

Fig. 4. X-ray CHA powder

In direct contact, CHA powder did not modify Balb/c 3T3 viability and growth (Figure 5) nor MTT test data. At the contrary, HA inhibited slightly cell growth (- 39 %, $P < 0.01$) at 24 h, whereas cell growth was restaured at 72 h, together with MTT test data. This effect is a cytostatic effect rather than a cytotoxic effect taking into account cell morphology (Figure 6), and could be the consequence of early pH changes in micro-areas of contact between powder grains and cells.

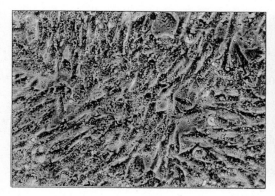

Fig. 5. Balb 3T3 morphology in contact with HA powder – 72 h

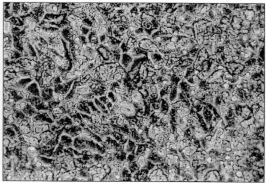

Fig. 6. Balb 3T3 morphology in contact with CHA powder – 72 h

Using liquid extracts no cytotoxic or cytostatic effect was found whatever the test material and incubation time period.

HOBMC were shown less sensitive to cytotoxic or cytostatic effect, since no inhibition of cell growth, or MTT conversion effectiveness was observed whatever the test material, the evaluation procedure, and the incubation period (Figures 7 and 8).

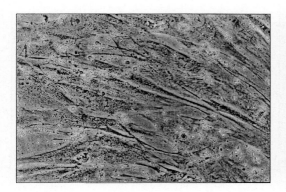 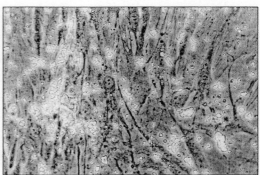

Fig. 7. HOBMC morphology in contact with HA   Fig. 8. CHA HOBMC morphology in contact
                powder – 72 h   with CHA powder – 72 h

Discussion and Conclusions

With regards to cytotoxicity results, especially when using human osteogenic cells from bone marrow, which are cells responsible for bone formation and repair, no cytotoxic or cytostatic effect was observed, whatever the incubation time and the cytotoxicity assessment procedure, through viability and growth or metabolic activity. This first step in biocompatibility testing validates the use of CHA and HA powders for bone filling material development.

Acknowledgments

The authors thank Dr. G.C. Celotti (ISTEC-CNR, Faenza) for X-ray diffraction analysis.
This work was carried out within the frame of an European Community Project: BRIMAS (Contract N°: BRPR 98 0707 Project N°: BE97-5086).

References

[1] ISO 10993-12. Biological evaluation of medical devices – Part 1.2 Samples preparation and reference materials (1996)
[2] ISO 10993-5. Biological evaluation of medical devices – Part 5. Tests for cytotoxicity evaluation (1999)
[3] T. Mosmann, Rapid colorimetric assay for cellular growth and survival: Application to proliferation and cytotoxic assays. J. Immunol. Methods, Vol. 65, (1983), p. 55-63
[4] R.Z. LeGeros, M. Tung, Carves Res., Vol. 17. (1983), p. 19-29
[5] Y.Doi, T.Shibutan, Y.Mortiwaki, Y.Iwayama, J. Biomed.Mater.Res. Vol. 39 (1998), p.03-610

Key Engineering Materials Vols. 254-256(2004) pp. 237-240
online at http://www.scientific.net
© 2004 Trans Tech Publications, Switzerland

Synthesis and Properties of Bone Cement Containing Dense β-TCP Granules

Kyung-Sik Oh[1], Soo-Ryong Kim[1] and Philippe Boch[2]

[1]Korea Institute of Ceramic Engineering and Technology, Seoul 153-801, Korea
[2]ESPCI 10 Rue Vauquelin,75005 Paris, France
ksoh@kicet.re.kr

Keywords: calcium phosphate, bone cement, β-TCP, granules

Abstract. Effort was made to enhance the compressive strength and to suppress temperature rise during setting of β-Ca$_3$(PO$_4$)$_2$ (TCP) based cement through the control of powder to liquid ratio. To increase the powder to liquid ratio, highly dense granules of β-TCP with the size around 300 μm were prepared and blended with monocalcium monophosphate (MCPM) and calcium sulfate hemihydrate (CSH). Due to the decrease of the liquid necessary for setting, compressive strength was remarkably increased from 3.9 to 20.6 MPa. The use of granular β-TCP was also affirmative in heat evolution, since temperature rise was virtually negligible during setting.

Introduction

Calcium phosphate bone cements have merits of excellent biocompatibility and mild heat evolution compared with the conventional polymer based cement like PMMA. Among the various compositions of calcium phosphate cements, β-TCP based bone cement is unique in absorbability due to the metastable end product. For the number of calcium phosphate cements based particularly on α-TCP and tetracalcium phosphate (TeCP), hydroxyapatite (HAp) is the end product with the variation in the degree of crystallinity and carbonation [1,2]. Since HAp is the least soluble in most physiological conditions among calcium phosphates, it is hardly resorbed and replaced with natural bone. On the contrary, hydroxyapatite does not participate during setting of β-TCP based cement and the chronic side effects can be prevented. A shortcoming of β-TCP based cement is relatively poor compressive strength around 5 MPa [3], while that of α-TCP or TTCP based cement is ranged near 50 MPa [1]. The increase of strength can be effectively achieved through the decrease in porosity, since theoretical prediction indicates that strength is proportional to the inverse of porosity [4]. Less liquid as possible should be used to decrease the porosity. In this work, highly dense and coarse (~300 μm) granules of β-TCP were prepared and introduced to decrease the volume of liquid necessary for setting. Difficulty in obtaining dense β-TCP is the limitation of sintering temperature. Densification should be completed below the transformation temperature to α-TCP, since intervention of α-TCP is not desirable due to its chemical reactivity and fragility. Therefore dense green body is desirable to be sintered under this limited sintering temperature. Colloidal sedimentation was adopted to prepare such a dense green body.

Materials and Methods

β−TCP (Ca$_3$(PO$_4$)$_2$) powder was synthesized based on the N. Kirvak's recipe [5]. 0.4 M of Ca(NO$_3$)$_2$ and 0.14 M of (NH$_4$)$_2$HPO$_4$ solutions were reacted at 40°C. The volumes of solutions were prepared so that molar ratio of Ca/P is 1.4. The precipitate was separated from supernatant and dried in the oven. Through the calcinations at 900°C for 1 h, the precipitate transformed to β−TCP. The slurry was prepared through dispersion of β−TCP powder in distilled water with 0.2 ml of Darvan C per gram of powder. Cloudy state of the slurry was verified to ensure dense settling of sediments. Decomposable net was introduced during sedimentation for the ease of breakdown after sintering. After drying, the sediment was fired at 1000°C for preliminary consolidation. The fired sediments were crushed and sieved to obtain granules of sizes between 300~450 μm. The collected granules were fired again at 1180°C for further densification. 41.7 g of dense β-TCP granules were mixed with 13.0 g of MCPM and 10.4 g of CSH according to K.Ohura's recipe [3]. Distilled water kept at 36.5°C was used as a setting agent. Effect of the volume of setting agent on hardening and workability was observed. Between cement mixture with and without dense granules, compressive strength, setting time and temperature change were compared. The compressive strength was measured using an INSTRON 4204 at a crosshead speed of 0.5 mm/min and the setting time was measured with Vicat needle. The temperature rise during setting was measured with thermocouple embedded in the mixture.

Results

The β-TCP slurry used in the experiment was sufficiently stable and took more than 24 h for complete sedimentation. The density of the green body was measured to be 1.69 g/cm^3 corresponding to a relative density of 55%. As a result of sintering at 1180°C for 2 h, density as high as 3.04 g/cm^3 (relative density: 99%) was attained. The sintered sediment was broken down and ground to round granules as shown in Fig.1.

Fig. 1 β–TCP granule used in the preparation of bone cement

The least amount of setting agent to obtain suitable workability was 0.6 ml/g for cement with powdery β-TCP with the average particle size of 1.5 μm. In this state, the mixture of cement and setting agent could be injected or poured as long as several minutes after mixing. However, when powdery β-TCP was replaced with granular one, the amount of setting agent necessary for obtaining similar workability drastically decreased to 0.2 ml/g. With the decrease of the amount of liquid necessary, the density of the hardened body increased from 1.19 to 1.93 g/cm^3 and porosity decreased from 43.9% to 23.0% as summarized in Table 1. The decrease of porosity with the use of granular β-TCP led to the drastic increase of the compressive strength from 3.9 to 20.6 MPa.

Table 1. Properties of bone cement with and without granular β-TCP

	Cement with powdery β-TCP	Cement with granular β-TCP
Least volume of liquid to achieve injectability (ml/g of power)	0.6	0.2
Density (g/cm^3)	1.19	1.93
Porosity (%)	43.7	22.9
Compressive Strength (Mpa)	3.9±0.6	20.6±1.8
Maximum Temperature rise (°C)	3.4	0.1

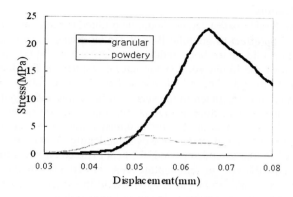

Fig.2 Displacement and development of stress for hardened β-TCP cement

Figure 2 shows an example of typical fracture behavior of each sample. In case of a sample with granular β-TCP, the stress increased steeply till the distinct point of fracture. Thus the fracture of granular sample shows the typical brittle fracture pattern of hard materials. On the contrary, the point of fracture is not clearly visible with the powdery sample. It can be realized that the powdery β-TCP cement is still relatively soft even after hardening. The phase analysis after hardening for one week showed that both of the samples were composed of β-TCP and DCPD as also observed by Ohura et al. [3]. Even if a considerable increase of compressive strength of β-TCP based cement has been realized, it is much inferior to α-TCP and TeCP based cements. One of the reasons for the difference is from the specific strength of end products, since HAp is much stronger than $CaHPO_4 2H_2O$ (DCPD) of β-TCP based cement.

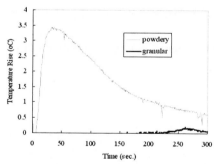

Fig.3 Temperature rise during the setting of cement containing powdery or granular β-TCP

Figure 3 shows the temperature change with the hardening of each sample. The powdery sample exhibited a steep rise in temperature to 3.4°C during initial 30 seconds. On the contrary, the temperature change of the granular sample was practically negligible. Maximum increase of 0.1°C was recorded after 4 min and 20 sec. The change of time for peak temperature was in correspondence with the change of setting time measured by Vicat needle. The setting time determined by the Vicat needle was 5 min. for powdery cement, while the granular one had extended setting time of 8 min.

Discussion and Conclusions

The major drawbacks of β-TCP based cement could be bridged in many aspects taking advantage of dense β-TCP granules. The effect of granular β-TCP can be discussed in mechanical and chemical aspects. First, granulation into dense body reduced the amount of liquid required for wetting the particles. The reduction of the liquid volume for setting resulted in the decrease of porosity and increase of compressive strength. Second, it also led to the moderation of the reactivity. With granular β-TCP, the reaction is restricted to the surface of granules and might gradually progress to the core over an extended period. Such a gradual reaction can contribute in preventing sudden hardening of cement, which is not desirable for practical usage. Slow hardening can provide the practician with more margin to avoid irreversible over hardening. At the same time, since hardening reaction is completed within the extended period, the heat evolved during setting is dispersed over a longer period suppressing the rise in temperature very effectively. In the practical aspects, the cement developed from the work can be particularly useful in surgeries related in nervous systems, since nervous tissue is critically vulnerable to the rise of temperature.

Acknowledgements

This work was supported by a Grant-in-Aid for Next-Generation New Technology Development Program from the Korea Ministry of Commerce, Industry and Energy (No.N11-A08-1402-07-1-3)

References

[1] L.C.Chow, S.Takagi, P.D.Constantino, C.D.Friedman, Mat.Res.Soc.Symp.Proc. Vol. 179 (1991), p.3

[2] Z.Hatim, M.Freche, A.Kheribech, J.L.Lacout, Ann.Chim.Sci.Mat. Vol. 23 (1998), p.65

[3] K.Ohura,M.Bohner,P.Hardouin,J.Lemaitre,G.Pasquier and B.Flautre: J.Biomed.Mat.Res Vol. 30 (1996), p.193

[4] J.C.Le Huec, T.Schaeverbeke, D.Clement, J.Faber and A.Le Rebeller: Biomaterials Vol. 16 (1995), p.113

[5] N.Kirvak and A.C.Tas: J.Am.Ceram.Soc Vol. 81 (1998), p.2245

Key Engineering Materials Vols. 254-256(2004) pp. 241-244
online at http://www.scientific.net
© 2004 Trans Tech Publications, Switzerland

The Expression of Bone Matrix Protein mRNAs Around Calcium Phosphate Cement Particles Implanted into Bone

Masashi Mukaida[1] , Kunitaka Ohsawa[2] , Masashi Neo[3] ,Takashi Nakamura[3]

[1] Department of Orthopedic Surgery Graduate School of Medicine, Kyoto University,54 Kawahara-cho,Shogoin,Sakyo-ku,Kyoto 606-8507,Japan
mmasashi@kuhp.kyoto-u.ac.jp
[2]Shiga Medical Center for Adults,5-4-30 Moriyama,Moriyama city,Shiga 524-8524 ,Japan
Kunitaka5@aol.com
[3] Department of Orthopedic Surgery Graduate School of Medicine, Kyoto University,54 Kawahara-cho,Shogoin,Sakyo-ku,Kyoto 606-8507,Japan

Keywords: in situ hybridization, calcium phosphate cement , bone formation, biocompatibility

Abstract. Tissue response around calcium phosphate cement(CPC) particles (150-300μm in diameter) implanted into rat tibiae was analyzed by *in situ* hybridization with digoxigenin-labeled procollagen α1(I) (COL), osteopontin (OPN), and osteocalcin (OC) mRNA probes. Specimens were collected at 3,5,7, and 10 days after the operation. New bone was formed centripetallyand all three kinds of mRNA were expressed in activated osteoblasts. A COL signal was expressed most strongly and widely, and was detected at the peripheral region of the hole at day 3. The other two mRNAs were also expressed in bone forming osteoblasts by day 7. In the earlier cell reaction stage, OPN mRNA was seen exclusively in the cells on the particles, and an OPN signal was detected not only in COL-positive cells, but also in COL-negative cells.

In the previous studies, we have shown tissue response around β-tricalciumphosphate (β-TCP) particles using *in situ* hybridization with COL, OPN, ON, and OC mRNA probes. In the present study, the temporal and spatial patterns of cellular response on calcium phosphate cement surface were similar to that on β-TCP. The COL- and OPN-positive cells were speculated to be osteoblasts and to reflect active bone formation on the surface of the biomaterial, whereas the COL-negative OPN-positive cells were thought be macrophages and to reflect foreign body reaction. Expression of these OPN mRNAs induced by implantation of these bioactive materials may play a role in bone formation on the materials and in determining their biocompatibility.

Introduction

Tissue response to implanted bioactive materials is one of the most important themes in the field of biomaterials. Differences in these bioactive materials are reflected in differences in the rate of bone formation on their surfaces. However the reasons for these differences are not yet understood.

Ohgushi et al. found that all bioactive materials had an ability to differentiate stromal bone marrow cells into osteoblasts on their surface, whereas nonbioactive materials did not [1]. But most of these previous findings have been based on histological and histomorphometric analysis or *in vitro* analysis, and little is known about the series of events occurring at the cellular level *in vivo*.

In previous studies, we have shown the temporal and spatial patterns of cellular response at the bone-bioactive material interface using *in situ* hybridization [2-6]. In the present study, we investigated the effects of calcium phosphate cement (CPC) particles on bone formation using *in situ* hybridization with digoxigenin-labeled procollagen α 1(I) (COL), osteopontin (OPN), and osteocalcin (OC) mRNA probes.

Materials and methods
Materials
Calcium phosphate cement particles were made from BIOPEX (Mitsubishi Pharma Ltd., Osaka,

Japan). The cement powder consisting of α-TCP(75%), DCPD(5%), TeCP(18%), and HA(2%) was mixed with water containing sodium succinate (12%) and sodium chondroitin sulphate (5%) as described in the manufacturer's instructions. After the cement had hardened, it was pulverized and the particles (150-300 μm in diameter) were collected.

Before this study, the particles were examined by using X-ray diffraction (XRD). The XRD demonstrated the low crystallinity of the apatite, which was recognized as type B carbonate apatite (low crystalline hydroxyapatite).

Animal surgery and tissue preparation

Sixteen mature male Wistar rats, with an average weight of about 320g, were used. Animal surgery was performed according to the procedures described in previous experiments [5-6].   Briefly a hole 3 mm in diameter was drilled bilaterally in the medial cortex of tibiae just distal to the proximal epiphyseal plate, and calcium phosphate cement particles were implanted in the hole. Four rats from each group were killed at 3,5,7 and 10 days after surgery. At killing, the rats were fixed by perfusion with 4% paraformaldehyde solution in phosphate-buffered saline (PFA/PBS), pH 7.4, from the left ventricle of heart. Horizontal segments of the tibiae, about 3-mm thick, including the holes, were excised immediately and further fixed in the same solution for 24 h at 4°C. The specimens were dehydrated in graded ethanol series and defatted with chloroform. They were then rehydrated and decalcified in 20% ethylenediaminetetraacetic acid solution, pH 7.4, for 14 days at 4°C. After this, the specimens were dehydrated in serial ethanol and chloroform, then embedded in paraffin. The implanted CPC was almost completely lost during the decalcification. Horizontal sections, 4 μm thick, were cut on a microtome and mounted on aminopropyl-trithoxysilane-coated slides (DAKO Japan Co., Ltd., Kyoto, Japan). These were stored at 4°C until hybridization. Serial sections were used for histological examination and *in situ* hybridization.

In situ **hybridization**

The following complementary DNA (cDNA) clones were used as hybridization probes. Rat procollagen α1(I) cDNA containing a 1.3-kb fragment, mouse ON containing a 0.9-kb fragment; mouse OPN cDNA containing a 0.9-kb fragment, and rat OC cDNA containing a 0.47-kb fragment were used. There is a nucleotide sequence homology between mouse and rat ON and OPN, so these mouse probes were used to detect rat mRNA. The specificity of these probes was confirmed. Digoxigenin (DIG)-labeled single-strand RNA probes were prepared for hybridization using a DIG RNA labeling kit (Boehringer Mannheim biochemical, Mannheim, Germany) according to the manufacturer's instructions.

In situ hybridization was performed according to the method described in previous experiments. Briefly, sections were deparaffinized, rehydrated, and then fixed with 4% PFA/PBS before and after treatment with pronase.   After immersion in 0.2N HCl, they were acetylated in 0.25% acetic anhydride diluted in 0.1M triethanolamine. Fifty microliters of hybridization buffer containing approximately 5 ng /μL DIG-labeled probe was applied to each section, which then was covered with Parafilm. Hybridization was performed at 60°C for 18 h in a moisture chamber saturated with 50% formamide. After hybridization, the sections were washed according to the method described previously. The signals were detected using a nucleic acid detection kit (Boehringer Mannheim Biochemica). Positive signals were detected as a blue color. After fixation in 4% PFA/PBS, the slides were counterstained with safranin 0.

Results

Positive blue signals were observed only in the sections hybridized with the antisense probes.

New bone was formed centripetally, and all kinds of mRNA were expressed in activated osteoblasts.

At day 3, spindle-shaped cells in the space between the particles were COL positive in the peripheral region. Very few OPN and OC signals were observed around the hole.

At days 5 and 7, in the central region, which represented an earlier stage of osteoblast differentiation, some of the spindle-shaped cells on and among the particles were COL-positive and OPN-positive

[Fig 1a]. And some OPN-positive round-shaped cells were observed on the particles where no COL signal was detected [Fig 1b]. In the peripheral region, which represented later stage of osteoblast differentiation, new bone was formed directly from the surface of particles. COL-positive cells were detected at the surface of particles and in and on woven bone among the particles [Fig 2a]. OC [Fig 2c] mRNAs were detected in a pattern similar to that of the COL signal. OPN [Fig 2b] signals were also detected in the COL-positive cells, but were limited to the bone-forming front. In this region, there were fewer COL-negative OPN-positive cells than at the earlier stage [Fig 2b].

At day 10, woven bone reached the central region and new trabecular bone formation progressed. Direct bone formation on CPC particles was observed more widely than before. There were fewer COL-negative OPN-positive, round-shaped cells than that of at days 5 and 7.

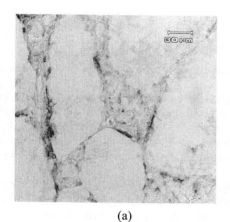 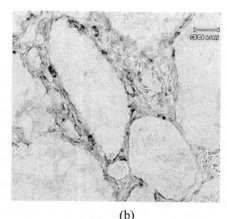

(a)   (b)

Fig 1. Bone formation around CPC at day 5. (a)Localization of COL mRNA. (b)Localization of OPN mRNA.

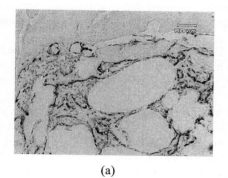

(a)

Fig 2. Bone formation around CPC at day 7
(a) Localization of COL mRNA
(b) Localization of OPN mRNA
(c) Localization of OC mRNA

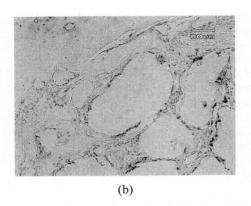

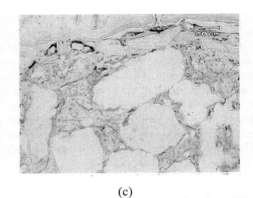

(b)   (c)

Discussion

To investigate the cellular reaction around bioactive materials at the molecular level, we used *in situ* hybridization with mRNAs for bone matrix proteins.

In the previous studies, Neo et al. studied *in vivo* osteoblast reaction on HA, A-W GC, and β-TCP using *in situ* hybridization with procollagen $\alpha 1$(I) probe. They concluded that no qualitative differences were demonstrated between HA, A-W GC, and β-TCP. But they used bulk materials, so more information may be obtained by furtther investibation using particulate materials, because the differece in bioactivity is amplified when bioactive materials are compared in the form of particulates.

Ohsawa et al. have shown tissue response around β-tricalciumphosphate (β-TCP) particles using *in situ* hybridization with COL, OPN, ON, and OC mRNA probes. The COL- and OPN-positive, spindle-shaped cells were speculated to be osteoblasts and to reflect active bone formation on the surface of the biomaterial, whereas the COL-negative OPN-positive, round-shaped cells were thought be macrophages and to reflect foreign body reaction.

In the present study, the temporal and spatial patterns of cellular response on calcium phosphate cement surface were similar to that on β-TCP.

Sugimoto et al. found that after mixing and implantation CPC changed into hydroxyapatite which is similar to the natural bone mineral [7]. In the present study the XRD showed that the CPC particles exhibit a low crystalline hydroxyapatite pattern.

So the similar patterns of cellular response between β-TCP and CPC demonstrate no difference in biological reactions between β-TCP and low crystalline hydroxyapatite.

References

[1] Ohgushi H, Okumura M, Caplan AI, et al. J Biomed Mater Res 24 (1990) 1563-1570.
[2] Neo M, Voigt CF, Gross UM, et al. J Biomed Mater Res 30 (1996) 485-492.
[3] Neo M, Voigt CF, Gross UM, et al. J Biomed Mater Res 39(1998)1-8.
[4] Neo M, Voigt CF, Gross UM, et al. J Biomed Mater Res 39(1998) 71-76.
[5] Ohsawa K, Neo M, Nakamura T, et al. J Biomed Mater Res 52(2000) 460-466.
[6] Ohsawa K, Neo M, Nakamura T, et al. J Biomed Mater Res 54(2001) 501-508.
[7] Sugimoto T, Niwa S, Aizawa Y, et al. Bioceramics 12 (1999) 533-536.

Key Engineering Materials Vols. 254-256(2004) pp. 245-248
online at http://www.scientific.net

Tissue Response of Calcium Polyphosphate in Beagle Dog
Part II: 12 Month Result

S.M. Yang[1,4], S.Y. Kim[2], S.J. Lee[3], Y.K. Lee[4] Y.M. Lee[1], Y. Ku[1], C.P. Chung[1], S.B. Han[1], I.C. Rhyu[1]

[1]. Dept.of Periodontology, Seoul National University, Seoul, South Korea,
icrhyu@snu.ac.kr
[2]. Dept. of Material Engineering Young-Nam Univ. Taegu, South Korea,
sykim@yn.ac.kr
[3]. Dept. of Pharmacy, Ewha Women's University, Seoul, South Korea,
sjlee@ewha.ac.kr
[4]. Dept. of Periodontology, Samsung Medical Center, School of Medicine, Sungkyunkwan University, Seoul, South Korea,
pkoyang@smc.samsung.co.kr

Keywords: Calcium polyphosphate, chitosan, Na_2O, osteoconduction, biodegradation

Abstract. CPP is chemically similar to natural bone and has higher stiffness. As one of our previous studies has shown, CPP had osteoconductivity and biocompatibility. The purpose of this study was to compare the new bone formation in CPP granules between 3 and 12 months and to evaluate CPP as bone substituting materials.

Introduction

Various types of bone grafts have been used in the treatment of periodontal intraosseous defects. Autogenous bone grafts appear to possess a good osteogenic potential. [1]. Allografts get much wider availability and little patient morbidity associated with graft procurement. But they have problems in clinical use. One possible approach toward addressing the respective problems inherent in autograft and allograft material is employing tissue engineering which is actively researched today.

Calcium phosphates are generally considered materials of choice as bone substitutes. While calcium phosphate ceramics meet some of the needs for bone replacement they are limited by their inherent stiffness, brittleness and low fatigue properties relative to bone [2] and are generally not resorbed during bone remodeling. The condensed phosphates are very numerous and exist both as crystalline salts and as amorphous glasses. Among the many candidates for bioabsorbable or transient implants, the use of bioabsorbable Calcium Polyphosphate (CPP) has twofold advantages over that of the other bioabsorbable polymeric materials, that is, CPP is chemistry similar to natural bone and has higher stiffness. As our previous study showed, CPP might be good scaffold for tissue engineering of bone tissue and regardless of its additive components, had osteoconductivity and biocompatibility [3]. Chitosan is a biodegradable cationic polysaccharide composed of N-acetylglucosamine residues which is known to accelerate wound healing and bone formation [4]. The purpose of this study was to compare the new bone formation in CPP granules between 3month and 12 month and to evaluate CPP as bone substituting materials.

Methods

The protocol details of this study have previously been reported [5]. Therefore only a brief description is included. Interconnected porous calcium ployphosphate (CPP) blocks were prepared by condensation of anhydrous $Ca(H_2PO_4)_2$ (Duksan Chemical Co.,Inc.) to form non-crystalline $Ca(PO_3)_2$. From the latter, a homogenous melt was created by thermal treatment, quenched in distilled water, and the block was then milled to produce CPP powder with addition of Na_2O. The CPP granule with chitosan was prepared by mixing into each powder of CPP, $CaSO_4$, and chitosan in 5% chitosan solution as binder, according to weight ratio 5:1:1. After getting CPP granules with chitosan which size was 300-500um, animal model prepared. The 10 teeth were extracted and as premolar teeth are biradicular, 20 alveolar extraction sites were available for bone filling. After meticulous check of extraction sites, Extraction sites were grafted with biomaterial or left unfilled; i.e, the mesial socket of a tooth was left unfilled and the distal socket filled with the composite biomaterial. The CPP granule with chitosan and CPP granule with Na_2O were injected into the extraction socket in a retrograde manner from the bottom of the socket to the top of the alveolar crest. The animals were sacrificed 12 month after implantation. Both treated and control mandibular and maxillary sites were histologically evaluated with light microscopy. The independent samples t-test and repeated measure ANOVA were utilized to evaluate differences between groups.

Results

Like 3 month results, all control and experimental sites healed uneventfully with no clinical evidence of inflammatory response to the CPP implants and DFDB in 12month. Histologically, the amounts of newly formed bone appear to be increased 12 month specimens with CPP granules, with Na_2O and control sites as compared 3 month. However, the tissue appears to demonstrate adipose tissues between newly formed bone and CPP granules and some of the CPP granules were simply surrounded with fibrous connective tissue. At 12 month time period, irregular surface of CPP granules and adipose tissues unlike 3month showed that resorption or dissolution process of the CPP granules took place laying down of mineralized bone matrix simultaneously and mesenchymal cells differentiated to not osteoblasts but fat cells. (Fig1).

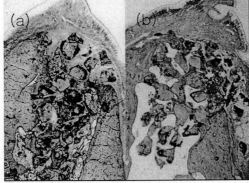

Fig.1. Bone formation around CPP granules with Na_2O. (a) 3month (b) 12month.

No adverse gingival tissue reaction was noted in areas of CPP granules contact, that is, relatively smooth appearing surface which contacted directly newly formed bone around CPP granules. The bone showed large marrow spaces composed of adipose tissues with no evidence of active bone formation. Unlike 3 month results, there was little osteoclastic activity evidenced by few multinucleated giant cells, but osteoblast-like cells were lining newly formed bone as like 3 month (Fig2) At 12 month, the one-way ANOVA showed that all the treatments produced statistically significant higher gain in new bone formation than did the control groups like 3 month results($p<0.05$) (table 1). For both filled and control extraction sites, the rate of newly formed bone was significantly higher in mandibular than maxillary sites. For implanted sites comparisons, the analysis showed that there is significant difference between experimental sites with implant material in contrast to 3 month results (Fig3). In experimental sites, CPP granules with Na_2O are showed increase of new bone formation to 3 month results and significant difference in DFDB, CPP granules with chitosan groups. But between the CPP granules with chitosan and DFDB, there is no significant difference. There is no significant difference between 3month and 12 month in experimental sites with implant material.

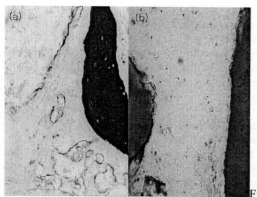

Fig.2. Tissue response of CPP in extraction site. A few inflammatory cell infiltrated in 12 month. (a) 3 month (b) 12 month (H&E stain x400)

Table1. % area of New Bone Formation in experimental sites

Group	Mean % of regenerated area	
	3month	12 month
1	44.04±4.7*	58.87±2.0*
2	62.64±6.1#	69.21±3.0#
3	66.48±1.0#	70.16±2.8#
4	80.46±5.7#	82.07±7.5#

* : Mean ± S.D.

: There were significant differences % area of new bone formation between group 1 and the other groups($p<0.05$). Group1 : control, Group2 : demineralized freeze-dried bone, Group3 : CPP chitosan granule, Group 4: CPP granule with Na_2O

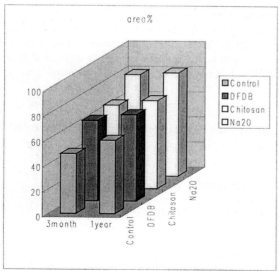

Fig. 3. Area % of new bone formation

Discussion and Conclusions

Bone-substituting materials should fulfill several requirements such as biocompatibility, osteogenecity, malleability, biodegradability[5]. However, these materials revealed some drawbacks including bone resorption, immune response, disease transmission, low biodegradability, poor adaptation. In this point, regardless of additional component, CPP granules might exhibit biocompatible, highly osteoconductive characteristics. But they exhibited very slow degradation rate. That may pose a problem for the replacement of these devices with new bone and may alter the mechanical properties of the newly formed bone.

Since chitosan can be well adapted to inorganic materials such as calcium phosphate, tricalcium phosphate and hydroxyapatite containing chitosan paste has been introduced as a self-setting material for bony lesion [6]. Also in present study CPP granules were employed for enhancement of material properties and osteocondutivity of chitosan matrix. In histologic and histomorphometric finding, newly formed osseous tissues infiltrated into and were mingled with the matrices being degraded simultaneously, and no significant differences between CPP granules with Na_2O and CPP granules with chitosan were found in matrix degradation but the amount of induced new bone by the two is significant. These findings indicate that the addition of chitosan to CPP granules probably does not evoke any adverse effects to degradation of the matrix or healing of defect, and in accordance with our former results[5] does not serve any additional osteoconductive effect, that is, chitosan degrades over time and is possibly replaced by bone slowly. Therefore there is little increase of newly formed bone and large marrow space in 12 month compared in 3 month.

Like other study[7], the CPP granules with Na_2O were incorporated into the surrounding tissue by both bone ingrowth into the implant volume and by apposition of bone against the implant surface. With increasing time of implantation, % newly formed bone increased. Newly formed bone increase slightly, suggesting that the CPP granules with Na_2O degrade over time and is replaced by bone. In histologic findings, surface contact (apposition) between bone and CPP granules with Na_2O was greater than ingrowth of bone into CPP granules with Na_2O. This suggests that CPP granules with Na_2O may be degrading (albeit slowly), and there is a lag time before new bone growth enters the recently degraded area. However, the data presented here do not provide sufficient evidence for this hypothesis. Another possibility is that bone preferentially adheres to the CPP granules with Na_2O, leaving a slightly higher porosity in the implant than in surrounding tissue.

In this study, regardless of its additive components, CPP granules provided necessary structural properties, showed increase of newly formed bone in 12 month, and were biocompatible, but they exhibited very slow degradation rate and replacement by bone. CPP is available in various forms and its degradation rate is controllable, so in future study, it would be useful to determine if the various forms of CPP and additive components allow full bone replacement and eventual return of normal metabolic remodeling.

Acknowledgment

This study was supported by a grant of the Korea Health21 R&D project, Ministry of Health & Welfare, Republic of Korea (HMP-99-E-10-0003)

References

[1] S.J. Froum, R. Thaler, I.W. Scopp, S.S. Stahl: J Periodontol, Vol 46, (1975), p. 515-521
[2] I.L. Hench: Biomaterial Science, Vol 76, (1996)
[3] S.M. Yang, S.Y. Kim, S.J. Lee, Y. Ku, Han SB, C.P. Chung, I.C. Rhyu: Key Engineering Materials, Vol 218-220 (2002), p. 657-660
[4] R.A. Muzzarelli: Carbohydr Polym, Vol 20, (1993) p. 7-16
[5] R.A. Yukna: J Periodontol, Vol 65, (1994), p.   342-349
[6] M. Maruyama, M. Ito: J Biomed Mater Res, Vol 32 (1996), p. 527-532
[7] C.R. Nunes, S.J. Simske, R. Sachdeva, L.M. Wolford: J Biomed Mater Res, Vol 36, (1997) p. 560-563

Key Engineering Materials Vols. 254-256(2004) pp. 249-252
online at http://www.scientific.net

Thermodynamic Study of Formation of Amorphous ß-Tricalcium Phosphate for Calcium Phosphate Cements

U. Gbureck[1], J. E. Barralet[2], R. Thull[1]

[1] Department for Functional Materials in Medicine and Dentistry, University of Würzburg, Pleicherwall 2, D-97070 Würzburg, Germany, e-mail: uwe.gbureck@fmz.uni-wuerzburg.de

[2] Biomaterials Unit, School of Dentistry, University of Birmingham, St. Chad's Queensway, Birmingham, B4 6NN, UK

Keywords: tricalcium phosphate, phase transformation, thermodynamic aspects

Abstract. The reactivity and setting properties of calcium phosphates in water are considerably determined by the particle size / specific surface area and the crystallinity of the compounds. Wet or dry grinding is a common method for adjusting the particle size of calcium phosphate compounds, but this process can also lead to a partial amorphous transformation of the materials. In this work we could show that this mechanically induced phase transformation of ß-tricalcium phosphate is strongly increasing the formation enthalpy ΔH of the compound up to 420 KJ/mol, which is a result of the formation of intrinsic defects within the crystal lattice. The increase of ΔH is linearly correlated to the loss of crystallinity as well as the weight loss during heating. X-ray diffraction analysis showed, that the amorphous fraction is responsible for a setting reaction of mechanically activated ßTCP to calcium deficient hydroxyapatite.

Introduction

The reactivity of calcium phosphates in self setting hydroxyapatite (HA) cements is normally adjusted via the particle size and hence the specific surface area. Reactants for HA cements are crystalline calcium phosphates like tetracalcium phosphate (TTCP) [1-3], α/ß-tricalcium phosphate (TCP) [4-6] as single component cements or combined with slightly acidic compounds like dicalcium phosphate anhydride (DCPA) or dicalcium phosphate dihydrate (DCPD). After mixing with an aqueous liquid phase, all cement types form a supersaturated solution with regard to HA or calcium deficient HA, which is precipitated and leads to the hardening of the cement structure. Rate limiting parameters of the setting reaction are mainly the rates of dissolution (kinetic solubility) of each cement component as well as the degree of supersaturation of the cement liquid (thermodynamic solubility). In the case of highly crystalline compounds, the thermodynamic solubility is equivalent to the solubility product and the rate of dissolution can be ascribed by means of mathematical models as a function of the specific surface area, e.g. the hydrolysis of αTCP hydration to CDHA [7]. The most common method for adjusting the particle size of calcium phosphates is grinding in solid or liquid phase. Besides a reduction in particle size, a crystalline to amorphous mechanically induced phase transformation may occur during prolonged ball milling as previously shown [8]. The mechanically induced phase transformation is related to the formation of defects within the crystal lattice and it is linked to an increase of the solubility and the formation enthalpy of the compounds. In this work we studied in detail the thermodynamic aspects of the mechanical activation of ß-tricalcium phosphate by means of differential scanning calorimetry (DSC) and thermogravimetric (TG) measurements.

Materials and Methods

β-Tricalcium phosphate (βTCP) was synthesized by heating a mixture of monetite (DCPA; Mallinckrodt Baker, Griesham, Germany) and calcium carbonate (CC; Merck, Darmstadt, Germany) to 1050°C for 24 h followed by quenching to room temperature in a desiccator. The sintered cake was crushed with a pestle and mortar and passed through a 355μm sieve. Milling was performed in a planetary ball mill (PM400 Retsch, Germany, diameter) at 250 rpm in 99.9% ethanol. The ground powders were dried in a vacuum oven at 60°C.

X-ray diffraction patterns of the starting ßTCP, ground materials and set cements were recorded on a diffractometer D5005 (Siemens, Karlsruhe, Germany). Data were collected from $2\theta = 20\text{-}40°$ with a step size of 0.02° and a normalized count time of 1s/step. The relative crystallinity of the mechanically activated ßTCP samples were calculated on X-ray diffraction patterns according to the areas of the strongest ßTCP reflection peak at 31.1° for ground samples compared to the crystalline raw material. Crystal sizes and quantitative phase compositions of the materials were calculated by means of total Rietveld refinement analysis with the TOPAS software (Bruker AXS, Karlsruhe, Germany). Differential scanning calorimetry (DSC) and thermogravimetric measurements (TG) were performed (Model STA 409, Netsch, Germany) at a heating rate of 20°C/min up to 1500°C.

Results

It was found that a loss of crystallinity of the ßTCP occurred as grinding time increased. Thermal analysis showed that this loss of crystallinity was accompanied by a strong increase in the formation enthalpy ΔH. Typical DSC heating curves are displayed in Figure 1 for mechanically activated ß-TCP. During heating of the mechanically activated ßTCP to 1500°C, firstly an endothermic reaction between 440-620°C with an enthalpy of about 30-60 KJ/mol, followed by a strong exothermic peak between 620 – 1100°C with an formation enthalpy of up to 420 KJ/mol were observed after 24h grinding (Figure 1).

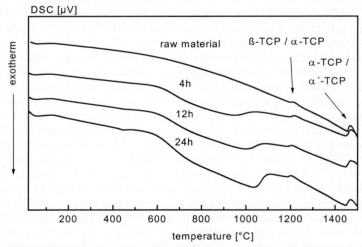

Fig. 1: Typical DSC heating curves of ßTCP after mechanical activation in ethanol for various time periods

The first process can probably be related to the formation of a decomposition product indicating that substantial phase transformation had occurred since these peaks were absent from the unmilled ßTCP. It was found that the first peak was mostly independent of the activation (grinding) time and relative crystallinity of the materials. The second is determined by recrystallisation and crystal growth of the ßTCP and the healing of intrinsic defects within the crystal lattice. In this case a linear relationship between the crystallinity and the enthalpy was found (Figure 2). Furthermore an increase in weight loss of the mechanically activated ßTCP occurred with prolonged grinding. The total weight loss of the raw material was about 0.6% while the 24h activated material lost about 2.5% of its weight when heated up to 1500°C (Figure 3), probably due to the formation of calcium pyrophosphate as it is known for crystalline ßTCP in literature [9]. Weight loss occurred mainly in the temperature range between 400 – 1200°C, such that an evaporation of adsorbed solvent can be excluded as explanation.

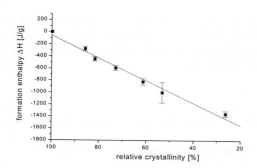
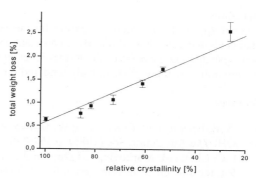

Fig. 2: Correlation between relative crystallinity and formation enthalpy

Fig. 3: Correlation between relative crystallinity and weight loss

While the crystalline ßTCP raw material showed no setting reaction with water or sodium phosphate solution, a hardening of the mechanically activated ßTCP occurred within 5-16 min when 2.5% Na_2HPO_4 solution is used as cement liquid. X-ray diffraction analyses of the set cements indicated that the amorphous fraction within the materials is responsible for the primary setting reaction and hardening of the cements, while the crystalline fraction remained unreacted (Table 1).

Table 1: HA content of mechanically activated ßTCP after 24h setting with 2.5% Na_2HPO_4

Grinding time [h]	1	4	12	24
HA content [%] after 24h setting	19.2	41.4	85.8	92.7

Discussion

Up to now it was believed that only particle size / specific surface area of calcium phosphate particles were responsible for reactivity. Recently we could show that prolonged grinding of ßTCP additionally results in a mechanically induced formation of amorphous CaP. This transformation drastically increases the reactivity and solubility of the compound, which can not be solely correlated to a reduction of the particle size, but to the presence of the amorphous TCP fraction. The resulting materials showed such a high reactivity that a setting reaction to CDHA occurred while highly crystalline ßTCP is nearly unreactive when combined with a neutral aqueous phase:

$$3 \text{ MA } Ca_3(PO_4)_2 \quad + \quad H_2O \longrightarrow Ca_9(PO_4)_5(HPO_4)OH$$

In this work we could show that the formation of amorphous ßTCP by means of high energy ball milling led to a strong increase in the formation enthalpy ΔH of the material, which is responsible for the setting reaction in the presence of an aqueous phase. The energy input into the material during grinding is linearly related to the loss of crystallinity; maximum ΔH values of up to -420 KJ/mol (24h grinding) were obtained which are an order of magnitude higher than other phase transitions in the TCP system, e.g. the ßTCP / αTCP transition at about 1125°C (3.7 –8.0 kJ/mol) or the formation of the high temperature modification α'TCP at about 1460°C (10.1 - 24.5 kJ/mol). For comparison the total formation energy of crystalline ßTCP is 4109.9 kJ/mol [10]. Due to the absence of compositional changes during grinding, the resulting vacancies and defects within the crystal lattice leave the material stoichiometric.

Conclusions
The novel approach to cement formation by mechanical activation of calcium phosphates and a fundamental understanding of the parameters that control it offer a way of making a variety of new cement systems. The increase of the formation enthalpy of the compounds enables setting reactions, which are normally thermodynamically hindered for crystalline materials.

References

[1]   W.E. Brown, L.C. Chow, in: P.W. Brown, editor. Cements Research Progress. Westerville, OH: The Am Ceram Soc (1986), p.352-379.
[2]   L.C. Chow, M. Markovic, S. Takagi in: L.J. Struble, editor. Cements Research Progress. Westerville, OH: The Am Ceram Soc, (1997), p. 215-238.
[3]   L.C. Chow: J Ceramic Soc Japan Vol. 99 (1991), p. 954-964.
[4]   M.P. Ginebra, E. Fernandez, M.G. Boltong, F.C.M. Driesens, J.A. Planell, in: L. Sedel, C. Rey, editors. Bioceramics 10. New York: Elsevier Science, (1997). p. 481-484.
[5]   M.P. Ginebra, E. Fernandez, F.C.M. Driessens, M.G. Boltong, J. Muntasell, J. Font, J.A. Planell, J. Mater. Sci.: Mater. Med., Vol 6 (1995), p. 857-860.
[6]   M. Ginebra, E. Fernandez, M.G. Boltong, F.C.M. Driessens, J. Ginebra, E.A.P. De Maeyer, R.M.H. Verbeeck, J.A. Planell, J. Dent. Res. Vol. 76 (1997), p. 905-912.
[7]   M.P. Ginebra, E. Fernandez, F.C.M. Driessens, J.A. Planell, J. Am. Ceram .Soc., Vol. 82(10) (1999), p. 2808-12.
[8]   U. Gbureck, O. Grolms, J.E. Barralet, L.M. Grover, R. Thull, Biomaterials Vol. 34(24) (2003), p. 4123-4131.
[9]   R.Z. LeGeros, G. Daculsi, R. Kijkowska, B. Kerebel, in: Y. Itokawa, J. Durlach, editors. Magnesium in Health disease, John Lilbey (1989), p.11-19.
[10]   J.C. Elliott: Structure of the apatites and other calcium orthophosphates, Elsevier, Amsterdam (1994).

Key Engineering Materials Vols. 254-256(2004) pp. 253-256
online at http://www.scientific.net

Effect of Albumen as Protein-based Foaming Agent in a Calcium Phosphate Bone Cement

A. Almirall[1], J.A. Delgado[1], M.P. Ginebra[1] and J.A. Planell[1]

[1] Research Centre in Biomedical Engineering, Biomaterials Division. Dept. Materials Science and Metallurgy. Technical University of Catalonia (UPC) , Av. Diagonal 647, 08028, Barcelona, Spain, e-mail: Amisel.Almirall@upc.es; Maria.pau.ginebra@upc.es

Keywords: calcium phosphate cement, macroporosity, injectable bone graft, foams, scaffolds

Abstract. In this work, the preparation and characterization of a macroporous α-tricalcium phosphate (α-TCP) cement with albumen as foaming agent is discussed. X-ray diffraction (XRD) and infrared (IR) analysis of the samples reveal that the conversion of α-TCP to calcium deficient hydroxyapatite (CDHA) was not affected by the addition of albumen in the cement paste. SEM observations showed the formation of spherical macropores with diameters between 100 and 500 μm. The use of phosphate solutions, which have an accelerating effect of the setting reaction, influenced the foam stability, reducing the macroporosity of the set cements.

Introduction

Calcium Phosphate Cements (CPC) have the ability to be slowly replaced by bone [1,2]. However, one of the weak points in the biological/clinical performance of apatitic CPC is their slow rate of resorption. The introduction of macroporosity in CPC would be a way to facilitate bone ingrowth and to accelerate the transformation of the synthetic biomaterial in newly formed bone tissue. In this context, one of the challenges in the investigation on CPC's lies in the introduction of macroporosity without loosing their setting properties and injectability. Up to now, some porogenic agents have been suggested, such as mannitol [3], sucrose [4], frozen sodium phosphate solution particles [5], or some surfactant molecules [6].

In this study a new approach is proposed, taking advantage of the good emulsifying and foaming properties of some proteins. Specifically, the great foaming capacity of albumen, the protein mixture derived from the egg white, which contains ovalbumin, conalbumin, ovomucoid, lysozome, globulins and ovomucin, is used to foam a calcium phosphate cement paste.

Materials and Methods

The α-tricalcium phosphate (α-TCP) was obtained in the laboratory as described elsewhere [7]. The powder was grounded in an agate ball mill to a median size of 6.21 μm, as measured by laser diffraction. A 2 wt% of precipitated hydroxyapatite (PHA, Merck) was added as a seed material. This powder was mixed with a liquid phase, which was either distilled water or a 2,5 wt% disodium hydrogen phosphate solution, in a liquid to powder ratio of 0.35 ml/g. The phosphate solution was used since it is reported that it accelerates the setting reaction of the CPC [8]. The resulting paste was further mixed with different proportions of an albumen foam (5 and 10 wt%), which was prepared from dehydrated egg white (Igreca, France). The albumen powder was dissolved in water in the ratios 1:7 to give a 12 wt% solution, which is the natural concentration of fresh egg white. The foam was generated by mechanical stirring. The compositions of the different formulations studied are reported in Table 1.

Table 1. Different formulations studied in this work.

Code	Weight % of Na$_2$HPO$_4$ in the liquid phase of the cement	Weight % of albumen foam
AF-5W	0	5
AF-10W	0	10
AF-5A	2,5	5
AF-10A	2,5	10

The foamed cement samples were immersed in Ringer's solution at 37°C for 7 days. The phase composition was determined by X-ray diffraction (XRD) and infrarred spectroscopy (IR), the microstructure was analysed by Scanning Electron Microscopy (SEM). The macroporosity was evaluated by an indirect method, proposed by Takagi and Chow [3], and based in the measurement of the apparent density of samples. This parameter was measured by immersion in Hg, applying the Archimedes principle.

The pore size distribution was evaluated by Image Analysis (Omnimet 1.5) on micrographs obtained by light microscopy. The compressive strength was evaluated in a Universal Testing Machine using cylindrical specimens (diameter = 6 mm and heigh = 12 mm) at a crosshead speed of 1 mm.min^{-1}. Five specimens were tested for each formulation.

Results and Discussion

The structural integrity of the cement was strongly influenced by the amount of the proteinaceous foam incorporated in the cement paste. The results showed that there is a limit in the amount of foam added to the CPC, which ensures the spatial continuity of entangled CDHA crystals, responsible for the mechanical consolidation of the cement. Thus, when the amount of albumen foam was 10wt% (AF-10W and AF-10A), many defects and flaws were formed and the samples fractured in small fragments. In contrast, the samples prepared with 5 wt% albumen foam (AF-5) maintained their foamed structure and were able to set after immersion in Ringer's solution, showing a good structural integrity.

XRD and IR patterns revealed that the hydrolysis of α-TCP to calcium deficient hydroxyapatite (CDHA) was not hindered by the presence of albumen in the cement. As it can be observed in Fig. 1, the only phase detected after 7 days of immersion in Ringer's solution at 37°C was CDHA.

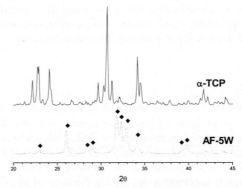

Fig 1. XRD pattern of α-TCP powder and AF-5W sample after 7 days in Ringer's solution
(♦ CDHA)

The IR spectra suggested that the albumen was eliminated during ageing in Ringer's solution, or it was present at very low amounts after 7 days of soaking.

SEM microstructures for the specimens containing 5 wt% of albumen are shown in Fig 2. When water was used as the liquid phase spherical macropores were formed, with diameters between 100 and 600 µm (Fig. 2 b). However, when the phosphate solution was used, the amount of macropores obtained was drastically reduced, as shown in Figure 2 a). The characteristic needle-like small crystals of CDHA obtained through cementitious low-temperature reactions were observed in all formulations studied, being these low crystallinity and high specific surface, important advantages of the apatites formed in CPC.

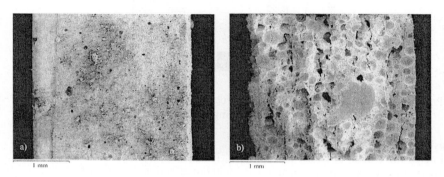

Fig 2. SEM micrographs (fracture surfaces) of AF-5 after 7 days in Ringer's solution, a) with the disodium hydrogen phosphate solution and b) with water as liquid phase

The apparent density, the macro, micro and total porosity and the compressive strength of the specimens containing 5wt% of albumen are summarized in Table 2.

Table 2. Apparent density, porosity and mechanical properties of the cements prepared with 5 wt% of the albumen foam. Standard deviation between brackets.

Series	Apparent density (g/cm^{-3})	Total porosity (%)	Macroprosity (%)	Microporosity (%)	Compressive strength (MPa)
AF-5W	0.91 (0.03)	71.0 (0.4)	53.6 (0.9)	17.4 (0.2)	2.5 (0.4)
AF-5A	1.46 (0.06)	53.4 (1.8)	23.1 (3.0)	30.3 (2.1)	29.6 (5.9)

The macroporosity was strongly reduced when the phosphate solution was used. As expected, the introduction of macroporosity into the CPC produced a decrease of the compressive strength.

The pore size distribution obtained by Image Analysis from light microscopy images for formulations with 5% of albumen is shown in Fig 3. The materials with disodium hydrogen phosphate as the liquid phase (AF-5A) presented almost all pores in the range of 100 to 300 µm. However, the samples with water (AF-5W) had a large quantity of pores between 300 and 400 µm, being this size within the range considered as adequate for the colonisation of surrounding bone tissue.

All the results described indicate that there is an interaction between the disodium hydrogen phosphate solution and the albumen foam. Indeed, both the total macroporosity and the pore size were reduced when a phosphate solution was used. This result suggests that the phosphate solution reduces the foam stability of albumen. Actually, it is well known that the foam stability of albumen is very sensitive to pH [9], being reduced at high pH values. Thus, the incorporation of disodium hydrogen phosphate, which increases the pH of the cement during setting [10], results in a decrease of the foam stability.

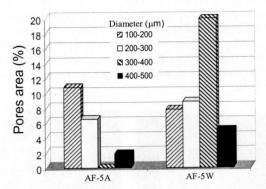

Fig 3. Pore size distribution represented as area occupied by pores (%) for AF-5A and AF-5W after 7 days in Ringer's solution

Conclusions

According to the results obtained in this work, it can be concluded that albumen can be used in appropriate amounts as foaming agent in CPC without affecting their *in situ* setting ability. The use of phosphate solutions, which have an accelerating effect of the setting reaction, influences the foam stability, reducing the macroporosity of the set cements. The size of the macropores obtained in the cement containing a 5wt% of the albumen foam and water as the liquid phase was adequate to allow the colonisation by the surrounding bone tissue. Further studies are being performed in order to assess the biological behaviour of these materials.

Acknowledgements

The authors thank the Science and Technology Spanish Ministry for funding this work through project CICYT MAT2002-04297. Further, the authors are grateful to Ms. Marsal for their assistance in SEM observations.

References

[1] B.R. Constantz, B.M. Barr, I.C. Ison, M.T. Fulmer, J. Baker, L. McKinney, S.B. Goodman, S. Gunasekaren, D.C. Delaney, J. Ross and R.D. Poser: J. Biomed. Mater. Res. (Appl. Biomater.) Vol. 43 (1998), p. 428

[2] EM Ooms, JGC Wolke, MT Van de Heuvel, B Jeschke, JA Jansen: Biomaterials 24(2003) p. 989

[3] M. Markovic, S. Takagi and L.C. Chow: *Bioceramics 13* (Trans Tech Publications Ltd., 2001), p. 773

[4] S. Takagi and L.C. Chow: J. Mater. Sci.: Mater. Med Vol. 12 (2001), p. 135

[5] J.E. Barralet, L. Grover, T. Gaunt, A.J. Wright and I.R. Gibson: Biomaterials Vol. 23 (2002), p. 3063

[6] S. Sarda et al., J. Biomed. Mater. Res. Vol. 65A (2003) p. 216

[7] M.P. Ginebra, E. Fernandez, E.A.P. De Maeyer, R.M.H. Verbeeck, M.G. Boltong, J. Ginebra, F.C.M. Driessens, J.A. Planell: J. Dent. Res. Vol. 76 (1997), p.905.

[8] M.P. Ginebra, E. Fernández, F.C.M. Driessens, J.A. Planell: *Bioceramics 11* (World Scientific Publishing Co, 1998), p. 243.

[9] J.F. Zayas: *Functionality of Proteins in Food* (Springer-Verlag, 1997)

[10] M.P. Ginebra: PhD Thesis, Universitat Politècnica de Catalunya, Barcelona (1997)

Key Engineering Materials Vols. 254-256(2004) pp. 257-260
online at http://www.scientific.net

Construction of an Interconnected Pore Network
Using Hydroxyapatite Beads

Kay Teraoka[1], Yoshiyuki Yokogawa[1] and Tetsuya Kameyama[1]

[1] 2266-98, Anagahora, Shimoshidami, Moriyama, Nagoya, Aich, Japan, ok-teraoka@aist.go.jp,
y-yokogawa@aist.go.jp, t-kameyama@aist.go.jp

Keywords: bone regeneration, hydroxyapatite, porous ceramics

Abstract. A new method for fabrication of macro porous grafts for bone regeneration was presented in this study. Spherical hydroxyapatite ceramics 1 mm in diameter with a cylindrical through-hole 300 µm in diameter (HA beads) were fabricated as components of a bone graft. By integrating the HA beads, the through-holes and inter-bead spaces were connected to each other, forming a single interconnected network. The integrated HA beads showed macro porosity of $47.7 \pm 1.9\%$. The interconnected macro spaces network performed remarkably concerning bone regeneration. The through-holes conducted bone formation during the 7-day-long animal test using rabbits.

Introduction

Porous biomaterials are attracting much attention as bone regenerative implants [1-3]. Pores are considered to allow cell migration and humoral transmission to take place in implants. In particular, macro pores (> 100 µm) are reported to be effective for osteoconduction [1, 4, 5]. As well as the size of individual pores, efficiency of porous biomaterials highly depends on the interconnectivity of the pores. Perfectly interconnected macro pores would allow cellular activities to take place throughout implants, which results in faster bone regeneration. However, no conventional porous bone grafts have interconnected pore network.

We are trying to fabricate a perfect macro porous bone graft by integrating small components made with biomaterials. In this study, spherical hydroxyapatite (HA) ceramics with a through-hole (HA beads) were prepared as the component. By integrating the HA beads, the through-holes and inter-bead spaces were connected to each other, forming a single interconnected network. The interconnected macro space network was evaluated by micro X-ray CT analysis. Bone regeneration in the interconnected macro space network was evaluated by animal tests.

Materials and Methods

Spherical HA ceramics beads 1 mm in diameter with a cylindrical through-hole 300 µm in diameter were fabricated as a model of the HA beads. Fabrication of the HA beads is described elsewhere [6]. Briefly, HA beads were prepared by sintering ϕ 1.8 mm gelatinous HA spheres that consisted of HA powder and sodium alginate. Before sintering, each gelatinous HA sphere was provided with a ϕ 500µm through-hole obtained by using a trimming tool. The HA beads were ultrasonic-washed for 15 min and rinsed 10 times with ultra-pure water. The through-hole is designed as an osteoconductive macro space. Porosity and specific surface area of the HA beads are measured by mercury intrusion employing an AutoPore IV 9500 (Shimadzu, Japan). The HA beads are integrated in a ϕ 5 x 5 mm cylindrical space. Morphology and an interconnectivity of the through-hole of the integrated HA beads was evaluated by image analysis based on a µ-CT data set of the integrated HA beads. Bone regeneration in the integrated HA beads was evaluated by an animal test using two male New Zealand white rabbits at the age of 12 weeks. Bone defects (ϕ 5 × 5 mm) were created on the proximal epiphysis of the tibiae. One hundred HA beads were implanted in the defect. After 7 days, the implanted HA beads and surrounding tissue were cut into tissue sections, and examined histologically.

Results and Discussion

The HA beads fabricated in this study were nearly spherical, but had a slightly oblate shape (Fig. 1). The average lengths of the major and minor axes were 1.03 mm and 0.87 mm, respectively. The HA beads were porous having a porosity and a specific surface area of the HA beads of 25.2 % and 0.27 m^2/g, respectively. The HA beads absorbs azo dyes due to the capillary action, implicating good interconnectivity of the micro pores in the HA beads.

The μ-CT analysis of the integrated HA beads proved that the through-holes and the gaps between the HA beads are composing a single interconnected network. The arrangement of the HA beads approached hexagonal closest packing in certain regions. The porosity of the integrated HA beads was 47.7±1.9%.

Fig. 1. A photomicrograph of the HA beads fabricated in this study.

The HA beads were appropriately implanted in the ϕ 5×5 mm bone defect without inflammatory responses, or degeneration and necrosis of the surrounding tissue. All implanted HA beads remained at the implantation site during the implantation period. Fibrous tissue formation was observed all over the implantation site. The implanted HA beads were infiltrated by cells and fibrous tissue. Remarkable new bone formation was found in the through-hole of the HA beads that locates on the edge of the implantation site (Fig. 2). The new bone was accompanied with cell layers along the outlines. The cells directly on the outlines were considered osteoblasts. Spindle cells along the osteoblast layer were considered primitive mesenchimal cells. The above findings are considered favorable bone repair in the through-hole. In contrast, new bone formation in the gaps between the HA beads was minimal and localized on the bead surfaces. Mostly the HA beads were surrounded by an osteoid layer, presumed to be a precursor of new bone formation. Therefore, the interconnected macro space network of the integrated HA beads could be filled with new bone at a relatively early stage.

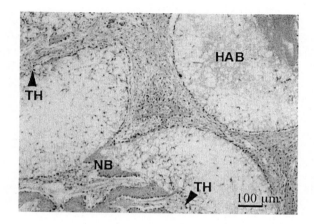

Fig. 2. A photomicrograph of the implanted HA beads located along the boundary of the implantation site. Remarkable new bone formation occurred in the through-hole of the HA beads.
HAB: HA bead, TH: Through-hole, NB: New bone

Conclusion

This study demonstrated a new method for constructing porous grafts for bone regeneration by integrating HA beads that served as an osteoconductive space. By employing the present method, a perfectly interconnected macro space network could be generated. The macro space network performed remarkably concerning bone regeneration. The through-holes conducted new bone formation during the 7-day-long animal test using rabbits. It is therefore concluded that the integrated HA beads can function as an effective bone graft.

References

[1] L.L. Hench: J. Am. Ceram. Soc. Vol. 81 (1998), p. 1705
[2] R.A. Ayers, S.J. Simske, C.R. Nunes and L.M. Wolford: J. Oral. Maxillofac. Surg. Vol. 56 (1998), p. 1297
[3] C.R. Quinones, M.B. Hurzeler, P. Schupbach, A. Kirsch, P. Blum, R.G. Caffesse and J.R. Strub: Clin. Oral. Implants. Res. Vol. 8 (1997), p. 487
[4] J.E. Aubin and F. Liu.: *Principles of bone biology* (Academic Press, San Diego 1996).
[5] J.T. Triffitt, R.O.C. Oreffo: *Molecular and cellular biology of bone* (JAI, London 1998).
[6] K. Teraoka, Y. Yokogawa and T. Kameyama: Biomaterials (submitted).

[illegible] HA beads could be filled with new bone at a relatively early stage.

Fig. 2 A light micrograph of the implanted HA beads located along the boundary [illegible] the implantation site [illegible] new bone formation occurred in the through-holes of the HA beads.
HAB: HA bead, [illegible], NB: New bone

Conclusion

This study demonstrated a new method for constructing [illegible] bone [illegible] [illegible]

References

[1] [illegible]
[2] [illegible]
[3] [illegible]
[4] [illegible]
[5] [illegible]
[6] [illegible]

Key Engineering Materials Vols. 254-256(2004) pp. 261-264
online at http://www.scientific.net
© 2004 Trans Tech Publications, Switzerland

Correlating In-vitro Dissolution Measurement Methods Used With Bone Void Filler Bioceramics

Tracey Jones and Marc Long

Smith & Nephew, Inc., 1450 Brooks Road, Memphis, TN 38116, USA,
tracey.jones@smithnephew.com

Keywords: dissolution, tricalcium phosphate, calcium, bone

Abstract. In-vitro dissolution of bone void filler bioceramics (tricalcium phosphate) was conducted using a novel protocol that measured the gravimetric weight loss of ceramic granules as a function of time until full dissolution in an acidic buffered solution (pH=4.0) was achieved. This study concluded that gravimetric measurements using this novel protocol were comparable to weight loss measurements using spectroscopy (ICP and colorimetric) methods to characterize the in-vitro dissolution behavior of bioceramics.

Introduction

Bioceramic materials are increasingly used as synthetic bone void fillers (BVF) to support bone healing in orthopaedic trauma applications. BVF bioceramics may be used alone, as scaffolds, or as carriers for bioactive agents [1-3]. The dissolution or resorption rate of these materials after implantation is a key performance parameter to support new bone growth [4]. In-vitro dissolution testing of BVF, including dissolution in various solutions and resorption in cell culture, can be used as an indicator of in-vivo performance. During the bone resorption process multinucleated osteoclast cells attach to the bone mineral surface and create a sealed acidic environment with low pH. The acidic pH is a critical factor in locally creating resorption lacunae and removing the bone mineral matrix [5]. Previous in-vitro dissolution studies aiming to simulate the in-vivo cellular resorption mechanism for different calcium phosphate bioceramics concluded that the low pH in-vitro degradation pattern was similar to that observed in-vivo. These protocols have typically studied the level of calcium or phosphate ion release using atomic spectroscopy techniques, but only through incomplete dissolution of the materials [6,7]. A novel in-vitro dissolution protocol has been developed that measures the weight loss of bone void filler ceramics as a function of time to full dissolution [8]. Dissolution is conducted in an acidic buffered solution at pH=4.0 to mimic the acidic environment associated with bone resorption. This study compared measurements using this gravimetric weight loss method to those using ICP (Inductively Coupled Plasma) and colorimetric spectroscopy methods to demonstrate the validity of the dissolution protocol.

Materials and Methods

A preliminary study was conducted to establish the correlation between the two measurement techniques. A larger and final study was then conducted to confirm the correlation between gravimetric and spectroscopy weight loss measurements. The in-vitro dissolution behavior of tricalcium phosphate (TCP) ceramic materials was evaluated using both methods. Sample materials consisted of shaped granules that were produced from TCP powder blends using a dry powder compaction process. The granules used in the preliminary study were composed of 89.0wt.% TCP powder (Berkeley Advanced Biomaterials, San Leadro, CA) with 11.0wt.% processing aids. These TCP granules were then sintered at 1150°C for 2 hours to produce full β-phase. The sintered granules were approximately 5mm from tip to tip. The granules used in the final study were composed of 97.5wt.% amorphous calcium-deficient precursor TCP powder (Plasma Biotal, Tideswell, UK) and 2.5wt.% processing aids. These TCP granules were then sintered at 900, 1200, and 1350°C for 2 hours. The various sintering temperatures were selected to produce TCP granules with various densities and phase compositions and assess their effect on in-vitro dissolution. The sintered granules were approximately 4mm from tip to tip. TCP granule density and phase composition were measured using dry powder pycnometry and x-ray diffraction quantitative analysis (Rietveld method), respectively.

The in-vitro dissolution protocol was designed to produce continuous and full dissolution of the samples by allowing the ceramic materials to dissolve without reaching their solubility limits [8]. Each sample in the preliminary study consisted of two TCP granules (~110mg) placed in Nalgene® bottles to which 44mL of KHP (Potassium Hydrogen Phthalate) acid buffer solution (pH=4.01; Orion brand) was added. Each sample in the final study consisted of five TCP granules (~98mg) placed in Nalgene® bottles to which 40mL of KHP acid buffer solution (pH=4.01) was added. For both studies the samples were then placed in an orbital shaker set to 37°C and 50 RPM. The test was run until full dissolution of the granules was observed with samples collected daily. Both the 400:1 ratio (volume of buffer in mL: weight of granules in g) and the daily replacement of the acid buffer solution had been defined to provide an environment that would allow the ceramic materials to dissolve without reaching their solubility limit. Triplicate samples were collected daily for analyses. These were rinsed with a controlled amount of distilled water (preliminary study: 8±0.2mL; final study: 20±0.2mL), filtered onto filter papers (1.2 μm pore size) and dried overnight at 40°C to steady-state weight. The filtered buffer solution from the same samples was collected for spectroscopy measurements. The buffer solution was replaced in all remaining samples each day. The test ran for 15 days for the preliminary study and 8 days for the final study. The gravimetric weight of the sample was measured using a precision scale (0.1mg resolution and accuracy). The equivalent weights of calcium (Ca) and phosphorus (P) dissolved were computed from the weight loss values adjusted for the stoichiometric composition of TCP, $Ca_3(PO_4)_2$, as follows:

$$Mass\ Loss_{[Ca]} = Mass\ Loss_{[TCP]}.(3/1).(Atomic\ Weight_{[Ca]} / Atomic\ Weight_{[TCP]}) \qquad \text{Equation 1}$$

$$Mass\ Loss_{[P]} = Mass\ Loss_{[TCP]}.(2/1).(Atomic\ Weight_{[P]} / Atomic\ Weight_{[TCP]}) \qquad \text{Equation 2}$$

The Ca and P concentrations in the collected solution were measured (mg/L) using ICP atomic emission spectroscopy and colorimetric methods respectively (EPA method 6010B and 365.2 respectively). The Ca concentration was measured using standard ICP techniques upon an initial acid digestion of the filtrate solution using nitric and hydrochloric acid. The P concentration was measured using colorimetric means. The total Ca and P contents (in mg) were computed from the ICP and colorimetric measurements by multiplying the concentration levels by the total volume of filtered solution. The filtered solution was comprised of the buffer solution and the controlled amount of rinsing distilled water.

Weight loss for both methods was plotted as a function of time (days) for each condition. The number of days to full dissolution was determined when 95% weight loss was reached. Above this weight loss, the samples were mostly fine powders and the accuracy and precision of the measurements decreased. Weight loss values for Ca and P computed from spectroscopy measurements were correlated to their equivalent weight loss values from gravimetric measurements using regression analyses (Statgraphics Plus Version 4.0) for weight loss data up to full dissolution only.

Results and Discussion

In the preliminary study, the TCP granules fully dissolved within 12 days (Fig. 1). Cumulative spectroscopy weight loss values of Ca and P contents dissolved in the buffer solution showed a similar behavior as a function of time (Fig 1). Dissolution measurements derived from gravimetric and spectroscopy weight loss techniques correlated very well (R^2=0.995) with a one to one relationship (0.98 slope, p<0.01) between the two methods (Fig 2). This correlation was found to be independent of the intrinsic slight variability in the amount of buffer solution and rinsing water.

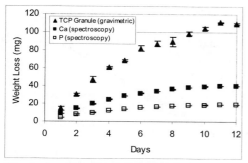

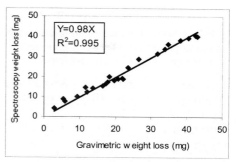

Fig. 1. In-vitro dissolution weight loss measurements (gravimetry and spectroscopy) as a function of time (n=3).

Fig. 2. Correlation between gravimetry and spectroscopy weight loss measurements.

The TCP granules used in the final study exhibited greater density values with increasing sintering temperature (Table 1). Sintering at 900°C produced a mixture of 7% HA-like phase (untransformed original precursor phase) and 93% β-TCP. Sintering at 1200°C produced 100% β-TCP phase. Sintering at 1350°C resulted in the formation and retention of 21% α-TCP phase with the remaining 79% β-TCP phase (Table 1).

Table 1. Density and phase composition of TCP granules.

Sintering Condition	Density [g/cm^3]	Phase Composition
900°C / 2hours	1.53 ± 0.09	93% β-TCP + 7% HA-like
1200°C / 2hours	2.67 ± 0.09	100% β-TCP
1350°C / 2hours	2.67 ± 0.17	79% β-TCP + 20% α-TCP

The TCP granules sintered at 900 °C, 1200 °C, and 1350 °C dissolved within 6, 8, and 7 days, respectively (Fig. 3, Standard Deviations < 5% of Mean). Cumulative spectroscopy weight loss values of Ca and P contents dissolved in the buffer solution showed a similar behavior as a function of time (Fig. 4, Standard Deviations < 7% of Mean). The linear correlation between Ca and P contents derived from gravimetric weight loss measurements and Ca and P contents measured by spectroscopy was observed at each sintering condition (R^2>0.997) with correlations close to a one to one relationship (slope = 1.06-1.12) between the two methods (Table 2). When data for all three conditions were combined, the correlation remained consistent (R^2=0.998) with a one to one relationship (1.08 slope, p<0.01) between the two methods (Fig. 5).

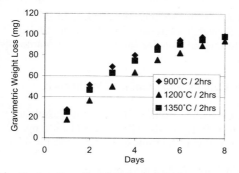

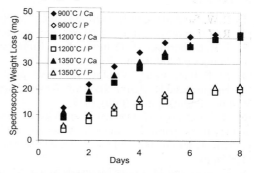

Fig. 3. In-vitro dissolution weight loss gravimetric measurements of TCP granules sintered at various temperatures (n=3).

Fig. 4. In-vitro dissolution weight loss from cumulative spectroscopy measurements for TCP granules sintered at various temperatures (n=3).

Table 2. Correlation factor and relationship factor (slope, p<0.01) between gravimetric and spectroscopy weight loss measurements.

Sintering Condition	Correlation Factor (R^2)	Relationship (slope)
900°C/2hrs	0.997	1.07
1200°C/2hrs	0.998	1.12
1350°C/2hrs	0.999	1.06

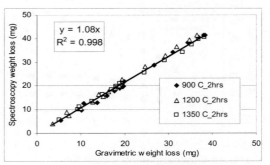

Fig. 5. Correlation between gravimetric and spectroscopy weight loss measurements for Ca and P (all three sintering temperatures combined).

The excellent correlation between the two measurement techniques demonstrates that the in-vitro dissolution protocol used in this study is comparable to protocols using spectroscopy techniques in previous studies. The novel protocol provided additional benefits as full dissolution of the test samples was achieved in a continuous manner throughout the bulk of the bioceramic granules. This novel protocol might then be applicable to screen bone void filler ceramic materials as well as granules or products of various forms prior to in-vivo implantation. In addition to comparing two methods of dissolution measurements, this study also showed that the TCP shaped granules fully dissolved and may therefore be an acceptable bioresorbable bone void filler scaffold.

Conclusions

Based on the results of this study, the in-vitro dissolution protocol based on gravimetric weight loss measurements correlated well with spectroscopy techniques to screen bone void filler ceramic materials and products. In-vitro dissolution may be used to design bone substitute materials with in-vivo resorption rates tailored to maximize the quality of the new bone formed.

Acknowledgements

The authors thank Lela Granberry, Carie Alley, Hilary Hornbuckle, Jeff Holbrook, Mike Cooper, Ed Margerrison, and Trevor Allen for their assistance in this study.

References

[1] R.W. Bucholz: Clinical Orthopaedics and Related Research, Number 395 (2002), pp. 44.

[2] R. Z. LeGeros: Clinical Orthopaedics and Related Research, Number 395 (2002), pp. 81.

[3] J. R. Lieberman, A. Daluiski, T. A. Einhorn: The Journal of Bone & Joint Surgery, Volume 84-A, Number 6 (2002), pp. 1032.

[4] M. Bohner: Injury, Volume 31 (2000), pp. S-D37.

[5] N. Sims and R. Baron: *Skeletal Growth Factors*, E. Canalis, ed., Lippincott Williams & Wilkins, Philadelphia (2000), pp.1.

[6] R.Z. LeGeros, G. Daculsi: *Handbook of Bioactive Ceramics*, Volume II, Calcium Phosphate and Hydroxyapatite Ceramics, T. Yamamuro, L. Hench, J. Wilson, eds., CRC Press (1990), pp. 17

[7] H.K. Koerten and J. Van der Meulen: Journal of Biomedical Materials Research, Volume 44 (1999), pp. 78

[8] T. Jones and M. Long: Materials Science Forum, Vols. 426-432 (2003), pp. 3037

Key Engineering Materials Vols. 254-256(2004) pp. 265-268
online at http://www.scientific.net
© *2004 Trans Tech Publications, Switzerland*

An Injectable Bone Void Filler Cement Based on Ca-Aluminate

Niklas Axén, Tobias Persson, Kajsa Björklund, Håkan Engqvist and Leif Hermansson

Doxa AB, Axel Johanssons gata 4-6, SE-751 26 Uppsala,Sweden, Niklas.Axen@doxa.se

Keywords: Bone cement, bone void filler, Ca-aluminates, biocompatibility, osteoporosis

Abstract. This article describes a novel orthopaedic cement based on calcium aluminate. Calcium aluminate is a chemically curing ceramic, which upon mixing with water hardens through chemical reactions between the ceramic powder and water. The curing process leads to the formation of hydrate crystallites, which build up the matrix in the hence formed solid. Due to the higher amounts of water involved in the hydration of calcium aluminate, dense and strong bodies of low residual porosity can be formed. Aspects of rheology, mechanical properties and biocompatibility of the calcium aluminate cement are addressed in this paper.

Introduction

As a result of the high frequency of osteoporosis, particularly among the ageing populations of well-fare countries, the number of complicated and difficult to treat spinal and metaphyseal extremity fractures in patients with osteoporotic bone is increasing steadily. There is a need for better treatment options to complement the conventional surgery procedures based on metal implant fixation elements.

Treatments of osteoporotic fractures need to provide a stable mechanical fixation of the fracture that enables rapid rehabilitation without pain and with a limited risk for mechanical failure prior to healing of the fractured bone. In addition, the surgery should be simple and safe and ideally require no or a minimum of open surgery.

A major pathway in the search for new treatments of osteoporosis related fractures, is to explore injectable and in-situ curing bone graft materials (orthopaedic cements) that may both provide the necessary early fixation of the fracture and contribute to the strength of the surrounding porous bone. Several new types of orthopaedic cements for these purposes are commercially available, e.g. for spinal vertebral-body fractures, near-joint extremity fractures or cranio-facial restoration. Most of these products are based on calcium phosphates or calcium sulphates, but also traditional acrylic bone cement (polymethyl-methacrylate or PMMA) is used. So far there is however no product that fulfils all necessary criteria for these types of treatments.

Orthopaedic cements for treatment of osteoporotic fractures should be injectable with established systems of syringes and needles, cure within a number of minutes without excessive heat generation, possess the biocompatibility or bioactivity to favour bone regeneration, and be strong and stable enough to allow for early and active loading.

This work describes a novel orthopedic cement based on calcium aluminate for orthopaedic applications. Calcium aluminate is used as a powder, which upon mixing with water hardens through a reaction between the powder and the water. Due to the high amounts of water involved in the hydration, a high degree of mouldability is achievable, and a dense body of low residual porosity and high strength is possible to form. Aspects of rheology, biocompatibility and various mechanical and physical properties of the new material are addressed in this work.

Materials and Methods

Materials preparation. In this work an orthopaedic cement composition based on calcium aluminate as the only hydrating component is described. The hydraulic system is composed of a mixture of $CaO \cdot Al_2O_3$ and $CaO \cdot 2Al_2O_3$. The molar ratio of calcium to aluminium oxide is approximately 70/30. In this work, calcium titanate ($CaO \cdot TiO_2$) was used as an inert-filler for purposes of microstructure optimisation. The calcium titanate was found to disperse well into the formed hydrates and to influence the mechanical properties favourably. In addition, the oxide power mix contained traceable amounts of calcium and silicon oxides.

During hydration, calcium and aluminate $Al(OH)_4^-$ ions are released and the pH of the solution increases due to formation of hydroxide ions, OH^-. A number of different hydrates can be formed. However, at 37 °C $3CaO \cdot Al_2O_3 6H_2O$ (katoite) and γ-$Al(OH)_3$ (gibbsite) are the stable end products of the hydration [1].

The materials preparation includes mixing of the powder ingredients by milling using silicon nitride balls in a polyethylene container with iso-propanol as milling liquid. After mixing, the alcohol is evaporated and polymeric residues are removed in a furnace at 400 °C.

To prepare samples, the powder is mixed with water in two steps. First, 9.5 ml of deionised water is added to a batch of 30 g of the powder mix, and a paste is prepared by mixing with a propeller. Second, 0.8 ml of a 8.5 gr/l LiCl solution is added and quickly mixed. This procedure allows for sufficient time to mix powder and water into a smooth paste without having to take curing into account initially. After the accelerator solution is added, the paste is mouldable for approximately 15 minutes at room temperature, thereafter it solidifies. The water to hydraulic cement ratio was kept at 0.38, as counted by weight. The material samples were cured at 37 °C in a phosphate buffer solution (Dulbecco, Sigma Aldrich). The material was evaluated regarding aspects of importance to an orthopaedic cement.

Temperature generation. Approximately 5 cm^3 of cement were placed in a heat conducting aluminium container directly after the addition of the accelerator. The container was placed in a water bath holding 37 °C with a thermocouple element placed in the centre of the volume. The temperature rise was recorded as a function of time. Comparison is made to PMMA bone cement.

Mechanical strength. The sample test geometry for compression and compressive fatigue was selected according to the ISO-5833.2 standard, i.e. cylinders with a diameter of 6 mm and 12 mm in height. The samples were manufactured by injection of the paste with a 1 ml syringe into moulds prepared in silicone rubber. The material is cured in the mould at 37 °C; the mould is thereafter cut open to remove the sample. Compression strength, elastic modulus and compressive fatigue were tested with an Instron® machine.

To evaluate the flexural strength, a biaxial testing procedure described in the ASTM standard F394 was used. The test explores circular disks of diameter 12.0 mm and thickness 1.4 mm, which are loaded at one point at the centre disk from the one side, and supported at three points located along a circle of diameter 8 mm from the other side.

Expansion. Dimensional stability was evaluated using the acrylic split pin method as described in [2]. This method uses 200x30x4 mm acrylic plates which are split with a central slit running from one side along the plate to a hole at the opposite side. The paste is placed in the hole and expansion is measured as the opening of the slit with an optical microscope. The expanders with sample material are stored as described above at 37 °C in saturated humidity.

Biological evaluations. The following in vitro biocompatibility tests, selected following to the ISO 10993 standard, were performed.

In vitro cytotoxicity assay. The material was tested in cultured mammalian cells (L929 mouse fibroblasts) according to the methods described in the ISO 10993-5 guideline. The paste was mixed and an extract of the paste was prepared during the curing process by placing disks of the paste in cell culture medium at 37 °C for 24 hours. The ratio of paste to medium was 0.2g/ml. The extract was adjusted from pH 9.4 to pH 7.4, filtered to ensure sterility and used to treat the L929 cells. The cured samples were also tested in direct contact with separate cell cultures.

Ames test. The material was tested in the Ames test using Salmonella typhimurium strains TA 102, TA 100, TA 98, TA 1537 and TA 1535. Modifications included the preparation of extracts of the test article in sterile saline and dimethyl sulphoxide (70 °C for 24 hours using a ratio of 0.2 g of the test material/ml). The extracts were tested in a plate incorporation test and a pre-incubation test. Treatments were performed both with and without metabolic activation (S-9 mix). Triplicate plates were prepared at each test point.

Delayed contact hypersensitivity. The dermal sensitising potential of a 0.9% NaCl extract and a sesame oil extract of the material was investigated according to ISO 10993 part 10. The Guinea-Pig Maximisation test was used. The prepared paste was placed directly in the 0.9% NaCl and the sesame oil. The ratio of paste to extracting medium was 0.2g/ml and the extraction period was 72 hours at 37 °C. The study comprised induction, (intradermal injections and a topical application one week apart), and challenge, (Hill Top chamber topical application 4 weeks after the intradermal injection), phases.

Results and discussion

The described composition forms a mouldable paste, which can be injected through 12-14 G needles into e.g. the cancellous bone of spinal vertebrae, see Fig. 1. Test results for the materials evaluation are summarised in Table 1, and commented further below.

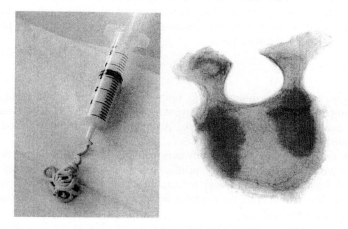

Fig. 1. Illustration of the paste (left), and the paste injected into a spinal vertebra (right).

Compared with commercial PMMA bone cement, the cured calcium aluminate based material has a slightly higher compressive strength, but lower flexural strength. Compared with the calcium phosphate cement (the table provides typical values for the commercial Biobone or Norian products) the mechanical strength is significantly higher. In compressive fatigue the calcium aluminate material is comparable to a PMMA material, and exceeds the calcium phosphate materials by about one order of magnitude.

The calcium aluminate based material expands slightly over time. The majority of the approximately 0.5% expansion occurs within the first 4-5 weeks.

Table 1. Summary of results for the calcium aluminate based cement and comparison with selected literature data for a calcium phosphate cements and PMMA bone cements.

Property	Calcium aluminate	PMMA	Ca-phosphates
Biocompatibility	Good	Poor	Good
Setting temperature [°C]	~40	~70	~37
Dimensional stability [%]	0.5 expansion	1-2 shrinkage	~0
Long term chemical stability	High	High	Resorbable in bone
Compression strength [MPa]	150-180	80-120	20-50
Flexural strength [MPa]	30-50	50-60	~5
Comp. fatigue, 10^6 cycles [MPa]	60	60	5
Elastic modulus [GPa]	10 –12	1-3	~5

The evaluation of the biological compatibility supports the use of the material for implantation. The increased pH-value however during curing leads to cytotoxicity in an extract in-vitro test unless pH adjustment is performed. Animal studies show no cell necrosis, which indicates that a living body buffers the pH increase. The diluted extract showed no to slight toxicity, (cytotoxicity grade 0 to 1), the pieces of cured material showed no toxicity, (cytotoxicity grade 0).

Extracts of the material are not mutagenic in the Ames test. The extracts were also not toxic to test bacteria at any dose level in this test. No biologically significant increases in the number of colonies, compared to the negative control values were observed on plates treated with the extracts.

There was no evidence of delayed contact hypersensitivity after treatment with undiluted 0.9% NaCl and sesame oil extracts of the material.

Conclusion

The scientific literature on calcium aluminate as biomaterials is scarce, but generally describes favourably the biocompatibility of the calcium aluminate systems [3,4,5]. Calcium aluminate based materials have been evaluated for oral applications. A dental product based on Ca-aluminate was launched on the Swedish market in 2000.

This article shows that calcium aluminate powders can be used as base material to develop orthopaedic cements with promising property profiles, suitable for spinal vertebra compression fracture, metaphyseal fractures treatment and for cranio-facial applications. Properties such as biocompatibility, injectability, moderate temperature increase curing and high mechanical strength are highly suitable for these applications.

The described calcium aluminate based material is highly mouldable, applicable to orthopaedic cavities with standard syringes and needles. The paste cures within about 5 minutes at 37 °C and develops strength values comparable to PMMA bone cement. The cured material is long-term stable and shows promisingly good biocompatibility.

References

[1] A.M. Neville, "*High-alumina cement*" in Properties of Concrete, Fourth Edition, p. 91-103 Prentice Hall.
[2] L. Kraft, L. Hermansson, G. Gomez-Ortega, *Proceedings of Shrinkage of Concrete* – Shrinkage 2000, Pro 17, RILEM, 16-17 Oct. 2000, Ed. by V. Baroghel-Bouny and P-C. Aitcïn, p. 401-413.
[3] L Kraft, *Calcium Aluminate based Cements as Dental Restorative Materials*, Ph D Thesis Dec 2002, Uppsala University, Sweden.
[4] M.L. Roemhildt, T.D. McGee and PS.D. Wagner, J. Mater. Sci.: Materials in Medicine 14 (2003) 137-141
[5] H. Engqvist, J-E Schulz-Walz, J Lööf, G A Botton, D. Mayer, M W Pfaneuf, N-O Ahnfelt, L Hermansson: accepted for publication in Biomaterials.

Key Engineering Materials Vols. 254-256(2004) pp. 269-272
online at http://www.scientific.net

Study of Particle Size Dependant Reactivity in an α-TCP Orthophosphate Cement

C.L.Camiré[1], S. Jegou Saint-Jean[2], I. McCarthy[1], S.Hansen[2], L.Lidgren[1]

[1]Department of Orthopaedics, Lund University Hospital, 221-85 Lund, Sweden.
Christopher.Camire@ort.lu.se

[2]Department of Materials Chemistry, LTH, Box 124, 221-00, Lund, Sweden.
Simon.Jegou_Saint-Jean@materialkemi.lth.se

Keywords: Tricalcium phosphate, bone void filler, calcium deficient hydroxyapatite

Abstract: Calcium phosphates have been of great interest in the field of medicine for many decades due to their biocompatibility, hardening properties and diverse areas of application. In this study two different alpha phase tricalcium phosphate powders were produced of distinctly different particle size distributions. These two powders were then tested for reactivity and phase evolution. X-ray diffraction was used to characterise the beginning and final products and the exotherm were measured with an isothermal calorimeter. The powders were then viewed relative to their starting properties and differences were drawn. Fine material exhibited a much larger exotherm, which was not directly proportional to the difference in surface area of the particulate. The difference has been attributed to the variant concentrations of nucleation sites assisting the ionic dissolution and precipitation reaction. For proper and ultimately efficient use of this material in the clinic it remains important to fully understand and have control over reactivity. Higher success rates and more reliable fracture fixation will result.

Introduction

Calcium phosphates, which have the ability to precipitate in solution and form apatites, have become particularly promising in the field of medicine. One quite widely accepted material is alpha phase tricalcium phosphate (α-TCP) which is an interesting osteoconductive biomaterial that has been shown to be applicable for bone void filling and fracture fixation. Upon hydration with fluids such as water or organic salts in solution it undergoes a mildly exothermic 2-step reaction dissolving from α-TCP and precipitating to form calcium deficient hydroxyapatite (CDHA) [1].

As the skeletal metabolic system produces a porous matrix with a mineral phase of apatite, synthetic apatites have good biocompatibility, as they can be removed by osteoclast action and replaced by normal bone. Precipitates such as calcium deficient hydroxyapatite (CDHA) show closer stoichiometric relationships to human bone than does hydroxyapatite (HA) which makes α-TCP one of the more promising compounds [2]. Mechanical characteristics of the cement remain central, while other factors such as porosity, composite and pharmaceutical additions grow in relevancy. A better understanding of hydration will yield a cement that can be more optimally designed for specific application in the clinic. In order for this material to be used more effectively, factors controlling the rate of hydration need to be understood in detail. The aim of the study was to observe the difference in reactivity of two distinct particle size distributions of α-TCP using isothermal calorimetry, which was verified by X-ray diffraction analysis.

Materials and Methods

Alpha-TCP was produced on site by mixing Calcium Carbonate (Sigma-Aldrich, catalog number C-4830) and Calcium Phosphate (Sigma-Aldrich, catalog number C-7263) at a 2:1 molar ratio. The materials were mixed vigorously in a mortar and pestle for 15 minutes and then placed in a platinum crucible for heat treatment. The material then underwent a heating cycle ending at 1300°C for 2 hours.   After heat treatment and quenching at room temperature the material was then milled utilising a Retsch S100 centrifugal agate ball mill. This material was then divided and half of the batch was then further milled for one hour more at high revolutions per minute, resulting in   α-TCP powders of two distinctly different particle size distributions from the same source of α-TCP. Both materials after grinding were assessed utilising a laser diffraction particle size analyser from Beckman Coulter (Miami, FL) model LS13320 system equipped with a Universal Liquid Module. Sufficient sample size meeting the requirements for the laser diffraction unit was met then samples were disbursed in ethanol, sonicated and mixed with a magnetic stirring device during measurement. Particle characterisation was achieved yielding a size profile along with median and mean statistics. The particle reactive surface area was then measured by BET nitrogen gas surface analysis (Micrometrics model ASAP 2400), the samples had been degassed overnight.   Isothermal calorimetry tests were performed at room temperature by hydrating 3-gram samples with a liquid powder ratio of 1:3 using a 2.5% Na_2HPO_4 solution. Calorimetry measurements were taken in millivolts per second utilizing PICO software (www.pico.com) every 15 seconds until the reaction reached zero.   The curves were then analysed with MatLab (Math Works Version 6.5.0.18913a Release 13).   These samples were hydrated for four weeks time and tested until they no longer registered thermal activity for a period of 1 week.   These samples were then removed and set in acetone for further analysis. X-ray diffraction was performed for characterisation of the beginning materials as well as on the final products obtained after testing. Testing was performed on a Siemens D-500 X-ray diffraction unit.

Results

Particle Size and Area. Particle size analysis yielded a mean diameter of 9.866 µm with a median of 5.559 µm for the fine particulate, while the coarse powder yielded a mean of 29.920 µm while the median was 28.700 µm. BET gas surface analysis resulted in values of surface area of 0.2601 +/- 0.0013 m^2/g for the coarse powder and 2.5306 +/- 0.0043 m^2/g for the powder with smaller particle distribution sizes.   Through observing that the Langmuir equation resulted in a non-linear distribution, it was concluded that the surface contained little if no micro pores between 1 and 30 Å.

Calorimetry. The following reaction curves (Fig.1) were acquired during measurement and the integration of these curves yielded 131.10 kJ/mol for the finely ground powder and 54.99 kJ/mol for the coarse powder. Integrating the curves segment-wise with the first minimum representing the end of the primary reaction, the following amounts were determined: for the fine powder the primary reaction yielded 2.84 kJ/mol and the secondary hydration yielded 128.26 kJ/mol. This value compares quite well with studies performed by TenHuisen and Brown [3]. For the coarse powder the primary reaction yielded 2.32 kJ/mol and the secondary reaction yielded 52.66 kJ/mol.

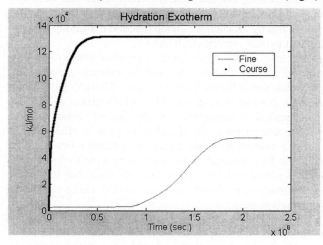

Fig.1: Heat evolution curves for the hydration

X-ray Diffraction Analysis (XRD). As the material undergoes the transition from α-TCP to CDHA the corresponding XRD patterns (Fig.2) change accordingly, α-TCP peaks are reducing while CDHA peaks are growing. Peaks however are changing at different linear rates and are initiated according to the material lattice structure change, which occur at variant times. In order to quantify the resulting material composition and register the remaining α-TCP left in the material the standard external method was utilised [4]. Significant peaks with no minimum, overlap or correspondence with

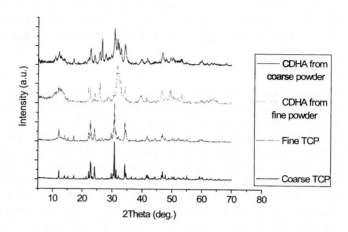

Fig.2: Evolution of XRD pattern from α-TCP to CDHA

other peaks present in α-TCP or CDHA were chosen for analysis (Table 1). Peaks were referenced with JCPDS cards 23-359 and 9-432 respectively. The specific peaks were removed from the pattern and then integrated. The following equation (Eq.1) was then employed, leading to the comparison of the weighted ratio (Eq.2), thus the resulting values seen below (Table 1).

$$R(\%) = A_f / A_o \qquad\qquad (1)$$

$$WR(\%) = [\ \Sigma A_f / \Sigma A_o\ ] \qquad\qquad (2)$$

A_f is the area of the peak in final form and A_o is the area of the peak before reaction. Because the peak intensities are changing at different linear rates (refer to Af/Ao Table 1) it was not possible to determine the exact percentage of α-TCP left in both samples. However, comparing the weighted ratio obtained for fine and coarse α-TCP revealed the amount of reacted material of the fine α-TCP to be about 2.61 times higher than that of the coarse α-TCP.

Table 1: Table of characteristic α-TCP peaks chosen for analysis of material transformation from α-TCP to CDHA and calculations exhibiting the decrease in area under the significant α-TCP peaks.

	2Theta	Coarse			Fine			Coarse/Fine
		Ao	Af	Af/Ao	Ao	Af	Af/Ao	
peak1	22.20	126.275	22.850	0.181	93.450	3.425	0.037	
peak2	24.10	179.475	40.500	0.226	119.225	2.825	0.024	
peak3	29.64	98.275	2.250	0.023	92.925	4.000	0.043	
peak4	30.74	720.950	27.550	0.038	18.000	0.000	0.000	
sum		1124.975	93.150		323.600	10.250		
WR(%)				0.083			0.032	2.614

Discussion and Conclusion

It may be assumed that surface area is a reliable method of indirectly determining degree or intensity of reactivity. If so the hydration exotherm would reflect the magnitude 10 difference seen in the surface area data. During primary reaction the fine material yields nearly the same amount of energy as compared to the coarse material. Post primary hydration of the fine material yields more than double the amount of exothermic energy. The reactivity of α-TCP could be more accurately understood by application of an equation modelling the velocity of the reaction inward into the particles, which will yield information on depth of reaction, and volume of reacted material to determine the energy output of such a reaction. In deriving an equation to cope with the situation we can refer to work performed in the cement industry [5] and observe that the degree of hydration (α) can consequently be calculated according to the below equation (Eq.3)

$$\alpha = 1 - [1 - (kt / r_o)]^3 \qquad\qquad (3)$$

where k is the linear velocity of the reaction front, radius r_o and time t. The cubic nature of the equation is due to the volumetric circumstance. The k term is a chemical constant, which given the same conditions should be the same for all α-TCPs. Utilising t and r_o from the hydration curves and laser diffraction analysis, and setting the completeness of reaction equal to the exotherm showing the fine particulate reacting in full we see the k term resulting in 0.0217 μm/hr.

Clearly particulate surface area has a large influence on reactivity however hydration exotherm could not be directly accurately determined by surface area analysis. It remains critical to observe how much remaining α-TCP is present in the material in order to quantify the reaction as a control. According to XRD analysis there is over 2.6 times the amount of α-TCP remaining in the coarse material than in the fine material. This could have a close relation to the full exotherm being less than half as large.

Secondary hydration shows a larger difference in resulting exotherm while primary a smaller. The primary reaction releases heat when fluid contacts the particulate surface initiating ion release into the surrounding fluid. Continuation of this reaction may be more influenced by the number of nucleation points releasing ions. Fracture sites, edges and corners, which are present at higher concentrations due to smaller particle size, may explain the differences in primary and secondary hydration exotherm. As well as the capabilities of the environment surrounding the particulate, accessibility to fluid for transfer and precipitation of ions remains ultimately important for the reaction of such a material.

Acknowledgements
We would like to thank the Swedish Research Council and the Medical Faculty of Lund University, as well as the UPC in Barcelona Spain for providing testing facilities.

References

[1] Ginebra M.P., Fernandez E., Driessens F.C.M., and Planell J.A.: Modelling of the hydrolysis of alpha-tricalcium phosphate. Journal of the American Ceramic Society , Vol 82, (1999), p. 2808-2812.

[2] Driessens F.C.M., Boltong M.G., Zapatero M.I., Verbeeck R.M.H., Bonfield W., Bermudez O., Fernandez E., Ginebra M.P., and Planell J.A. In-vivo behavior of 3 calcium –phosphate cements and a magnesium phosphate cement. Journal of Materials Science-Materials in Medicine: Vol 6, (1995), p. 272-278.

[3] TenHuisen K.S. and Brown P.W.: Formation of calcium-deficient hydroxyapatite from alpha-tricalcium phosphate. Biomaterials, Vol 19, (1998), p. 2209-2217.

[4] Fernandez E., Ginebra M.P., Boltong M.G., Driessens F.C.M., Ginebra J., DeMaeyer E.A.P., Verbeeck R.M.H., and Planell J.A.: Kinetic study of the setting reaction of a calcium phosphate bone cement. Journal of Biomedical Materials Research , Vol 32, (1996), p. 367-374.

[5] Evju C., Hansen E., and Hansen S.: Simulating the hydration of cementitious phases with an oscillating rate of reaction. Cement and Concrete Research, Vol 29, (1999), p. 1513-1517.

Key Engineering Materials Vols. 254-256(2004) pp. 273-276
online at http://www.scientific.net
© *2004 Trans Tech Publications, Switzerland*

Microporosity Affects Bioactivity of Macroporous Hydroxyapatite Bone Graft Substitutes

K. A. Hing[1], S. Saeed[2], B. Annaz[1], T. Buckland[3], P. A. Revell[2]

[1]IRC in Biomedical Materials, Queen Mary University of London,

Mile End Road, London E1 4NS, UK, k.a.hing@qmul.ac.uk.

[2]Osteoarticular Research Group, Dept. of Histopathology, University College London,

Royal Free Campus, NW3 2PF, UK.

[3]Apatech Ltd., IRC in Biomedical Materials, Queen Mary University of London,

Mile End Road, London E1 4NS, UK.

Keywords: Hydroxyapatite, bone graft substitute, macroporosity, microporosity, accelerated osteoconduction

Abstract. This paper describes an investigation into the influence of microporosity on early osseointegration within porous hydroxyapatite scaffolds. Two batches of phase pure porous hydroxyapatite were produced with total porosities of approximately 80%, but with varied levels of microporosity such that the strut porosity of the two batches were 10 and 20%. Cylindrical specimens 4.5mm in diameter were implanted in the distal femur of 6 month New Zealand White rabbits and retrieved for histological and histomorphometric analysis at 1 and 3 weeks. Optical microscopy demonstrated variation in the degree of capillary penetration and bone morphology within scaffolds at 1 and 3 weeks, respectively. Moreover, histomorphometry demonstrated that there was significantly more bone ingrowth within HA80-2 scaffolds and that the rate of bone formation within these scaffolds was significantly faster. These results indicate that the bioactivity of porous hydroxyapatite scaffolds may be improved by increasing the level of microporosity within the ceramic struts.

Introduction

The combined affects of an ageing population and greater expectations in quality of life have resulted in an increasing global demand for orthopaedic implants for the replacement or augmentation of damaged bones and joints. In bone grafting current 'gold standards' include the use of autograft (living bone from the patient) or allograft (dead, sterilised bone from bone banks) but these methods are increasingly recognised as non-ideal due to limitations in supply and consistency [1]. Porous ceramics have been considered for use as bone graft substitutes in the treatment of bone defects for over 30 years [2]. In particular, calcium phosphates such as hydroxyapatite (HA) have been promoted as a result of their osteoconductive properties. However, while it is well recognised that both the rate of integration and the final volume of regenerated bone may be primarily dependent on various features of the macro-porosity, such as volume fraction, pore size and pore connectivity [3], recent in vitro studies have demonstrated bone cell sensitivity to the level of microporosity within the ceramic struts [4]. The aim of this study was to investigate the influence of microporosity on early osseointegration within porous HA (PHA) scaffolds.

Materials and Methods

Two batches of phase pure PHA were produced using a novel slip foaming technique [5], both had total porosities of approximately 80%, but varied in the volume fraction of porosity distributed

(a)   (b)   (c)

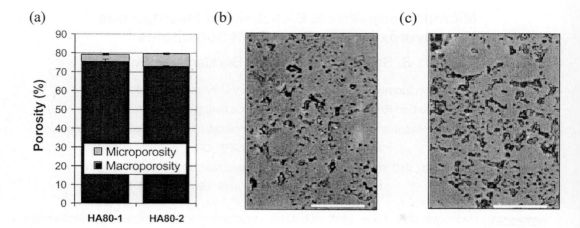

Fig 1 – (a) Porosity distribution within batches HA80-1 and HA80-2. Micropore morphology within the struts of batches (b) HA80-1 and (c) HA80-2.   Bar = 50 μm

between their macropore (>50μm) and micropore (<20μm) populations (Fig. 1). This variation was such that the strut porosity of batches HA80-1 and HA80-2 were 10 and 20%, respectively.

Morphological characterisation of both the macro- and microporosity was performed through a combination of immersion densitometry and image analysis of serial sections using a Zeiss Axioskop optical microscope linked to a KS300 image analyser. Macropore size, interconnection size and connectivity index, a measure of inter-pore connectivity, were all measured. Both the open and closed % of microporosity within the struts was quantified.

Cylindrical specimens 4.5mm in diameter were implanted in the distal femur of 6 month New Zealand White rabbits and retrieved for histological and histomorphometric analysis at 1 and 3 weeks. The volume of new bone ingrowth was calculated using a Weibel grid and the mineral apposition rate (MAR) was determined through the administration of fluorochrome labels at 1 and 2 weeks and measurement of the inter-label distance using a Zeiss Axioskop optical microscope with a UV light source, linked to KS300 image analyser.

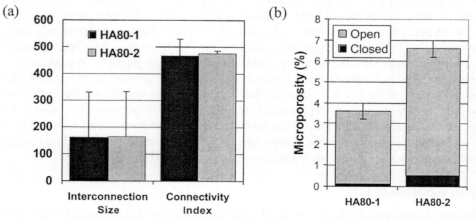

Fig 2 – (a) Macropore interconnection data and (b) porosity distribution of open and closed microporosity within batches HA80-1 and HA80-2.

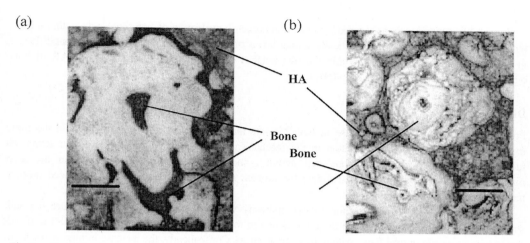

Fig 3 – Variation in ingrowth morphology within batches (a) HA80-1 (Goldner's trichrome) and (b) HA80-2 (Toluidine blue).   Bar = 100 μm

Results

There was no significant difference in the mean pore interconnection size or connectivity index of the two batches (Fig 2a). Modal values of macropore size were 150 and 175μm for batch HA80-1 and HA80-2, respectively, while 93 and 88% of the total microporosity within the struts of the two batches was confirmed as being interconnected with the macroporosity (Fig 2b) and significantly different in volume between the two batches (P<0.05).

At one week, osseointegration of the PHA scaffolds was well progressed, bone apposition was evident within peripheral porosity and neovascularisation was evident throughout HA80-2, but only within the deep porosity of HA80-1. However, at 3 weeks there was a distinct variation in the morphology of bone ingrowth within the two scaffolds, with both 'free standing' trabecluae and regions of direct bone apposition on the scaffold struts appearing thicker within HA80-2 (Fig. 3).

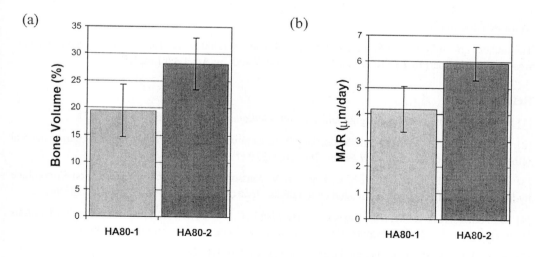

Fig 4 – Variation in the (a) volume and (b) rate of bone ingrowth within the two batches of PHA scaffolds.

These observations were confirmed by histomorphometric evaluation of bone ingrowth within the two batches (Fig. 4). Despite having statistically similar levels of total porosity there was a significant difference in the amount of bone ingrowth within the two batches of PHA ($P<0.05$), and in the MAR of bone deposited between weeks 1 and 2 ($P<0.01$).

Discussion

Possible explanations for the increase in both the rate and volume of bone ingrowth within the more microporous scaffolds include; variation in the mechanical environment within the scaffolds acting via mechano–transduction on local bone cell activity, and/or increased permeability within the more microporous scaffolds acting to up-regulate the process of osseointegration through enhanced nutrient transfer and induction of angiogenesis.

An increase in the volume fraction of microporosity within HA80-2 scaffolds will reduce the bulk strength and modulus values despite the equivalence in total porosity as compared to HA80-1 scaffolds due to the increase in critical flaws within the ceramic struts. Consequently, the increase in ingrowth volume and change in morphology from HA80-1 to HA80-2 could result from mechanical mediation of ingrowth where more numerous, thicker trabeculae develop to augment the mechanical properties of the weaker structure.

Revascularisation is known to be a prerequisite to bone formation in fracture repair. The disparity in capillary penetration within the scaffolds at 1 week suggest that the relative rate of development of the vascular network within the PHA scaffolds may have contributed to the observed variation in bone apposition at these early time points.

Hence it would appear that the increase in microporosity could act on the extrinsic bioactivity of the scaffold by its influence on both the bulk mechanical properties of the scaffold and its struts, in addition to the rate of vascularisation.

Conclusions

The results of this study indicate that the bioactivity of PHA scaffolds can be improved by increasing the level of microporosity within the ceramic struts.

Acknowledgements

The authors gratefully acknowlegde the expertise of Tom MacInnes and Taneisha McFalern. This work was supported in part by EPSRC grant No. AF/99/0077 and by ApaTech Ltd, UK.

References

[1] R.R. Betz. Limitations of autograft and allograft. *Orthopedics* 25(5) S561-S570 (2002)

[2] J.J Klawitter, S.F. Hulbert. Application of porous ceramics for the attachment of load bearing internal orthopedic applications. *J Biomed Mater Res* 5:161-229 (1971)

[3] J.H. Kühne, R. Bartl, B. Frish, C. Hanmer, V. Jansson, M. Zimmer. Bone formation in coralline hydroxyapatite. Effects of pore size studied in rabbits. *Acta Orthop Scand.* 65(3):246-252 (1994)

[4] B. Annaz, K.A. Hing, M. Kayser, T. Buckland, L. Di Silvio. In-Vitro Assessment of Porous Hydroxyapatite as a Bone Substitute. *Proc 17th Meeting European Soc Biomater.* (2002)

[5] K.A. Hing, W. Bonfield, Foamed Ceramics, Pat No. GB99/03283.

Key Engineering Materials Vols. 254-256(2004) pp. 277-280
online at http://www.scientific.net

Mechanically Induced Phase Transformation
of α- and β-Tricalcium Phosphate

J. E. Barralet [a], U. Gbureck [b], L. M. Grover [a], R. Thull [b]

[a] Biomaterials Unit, School of Dentistry, University of Birmingham, St. Chad's Queensway,
Birmingham, B4 6NN, UK; e-mail: J.E.Barralet@bham.ac.uk

[b] Department for Functional Materials in Medicine and Dentistry, University of Würzburg,
Pleicherwall 2, D-97070 Würzburg, Germany

Keywords: calcium phosphate cement, X-ray diffraction, mechanical properties, tricalcium phosphate, phase transformation

Abstract. The reactivity of calcium phosphates for self-setting bone cements is normally adjusted by altering the particle size and hence the specific surface area of the compounds. In this report we show that prolonged high energy ball milling of tricalcium phosphates (ß-TCP, α-TCP), led to mechanically induced phase transformation from the crystalline to the amorphous state with first order reaction kinetics. The process increased the thermodynamic solubility compared to the unmilled materials and accelerated the normally slow reaction with water to form single component cement systems. By using a 2.5% Na_2HPO_4 solution setting times were reduced to 5-16 minutes rather than hours. X-ray diffraction analyses indicated that the amorphous fraction within the materials was responsible for the primary setting reaction and hardening of the cements, while the crystalline fraction remained unreacted (ß-TCP) or converted only slowly to hydroxyapatite (α-TCP). Mechanically activated α- and β-TCP cements were produced with compressive strengths of up to 50 – 70 MPa and initial setting times between 5-16 min. In the powder:liquid (P:L) ratio range investigated, mechanical properties were found to be linearly related to activation time for ß-TCP, because of a higher degree of conversion to hydroxyapatite after setting. In contrast mechanical strengths of α-TCP cements decreased with prolonged ball milling, probably because of a much faster setting reaction of the mainly amorphous materials with the formation of smaller HA-crystals and a hence less homogeneous cement microstructure.

Introduction

Apatitic calcium phosphate cements (CPC) are usually based on acid base reactions between several calcium orthophosphate combinations and set *in situ* in the presence of an aqueous phase. Cement formation is based on the pH dependent solubilities of the calcium phosphates and the reaction product hydroxyapatite [1]. Reactants for HA cements are crystalline calcium phosphates like tetracalcium phosphate (TTCP) [2, 3, 4] or α/ß-tricalcium phosphate (TCP) [5, 6, 7] combined with slightly acidic compounds like dicalcium phosphate anhydride (DCPA) or dicalcium phosphate dihydrate (DCPD). The preparation of CPC involves several manufacturing steps, like sintering, grinding or mixing the dry powder components. Grinding in solid or liquid phase is a common method for adjusting the reactivity of these compounds by altering their particle size and hence their specific surface area [8]. Besides a decrease of the particle size however, we have found that this process leads to a mechanically induced phase transformation from the crystalline to the amorphous state with a higher solubility and reactivity of the amorphous fraction [9]. In this work we compared the influence of the grinding process on the phase composition, setting and mechanical properties of α and β TCP cements.

Materials and Methods
α or ß-TCP was synthesised by heating a mixture of monetite (DCPA; Mallinckrodt Baker) and calcium carbonate (CC; Merck) to 1050°C (β) or 1400°C (α) followed by quenching to room temperature in a desiccator. The sintered cake was crushed and sieved. Amorphous transformation was performed in a planetary ball mill (PM400 Retsch) at 250 rpm with 500ml agate jars, 200 agate balls (10mm) and a load of 125g TCP and 125ml ethanol (99.9%), per jar for times up to 24 hours. Phase analyses were performed by means of X-ray diffraction (Siemens D5005). The relative crystallinity of the ground TCP samples were calculated on x-ray diffraction patterns according to the areas of the TCP reflection peaks at 31.1° (β-TCP) and 30.7° (α-TCP) after milling compared to the peak areas of the highly crystalline raw materials. Crystal sizes and quantitative phase compositions of the materials were calculated by means of total Rietveld refinement analysis with the TOPAS software (Bruker AXS, Karlsruhe, Germany). As references the system internal database structures of ß-TCP and HA were used together with a *Chebychev* forth order background model and a Cu K_a emission profile [10]. Cements were formed using a 2.5% Na_2HPO_4 solution as liquid phase at a P/L ratio of 2.5. Wet compressive strengths were determined using cylindrical samples with an aspect ratio of 2 after 24h setting (Zwick 1440).

Results
It was found that a continuous loss of crystallinity of both the TCPs occurred as grinding time increased, while the particle sizes reached a limit of about 3μm (β-TCP) and 6μm (α-TCP) after 2-4h. The relative crystallinity calculated on from X-ray diffraction data, are shown in Fig. 1.

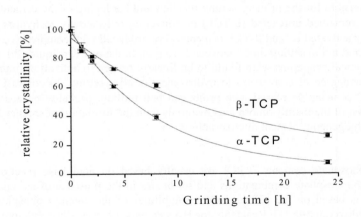

Fig. 1: Relative crystallinities of TCPs after grinding for up to 24h

Using water only as liquid phase, no setting reaction and hardening of the highly crystalline *β*-TCP and only a slow setting of the *α*-TCP raw material occurred within 8-12 h. This hardening time subsequently decreased with prolonged grinding to about 2h for the MA *α*-TCP ground in ethanol for 24 hours and about 5h for the MA *β*-TCP. The use of a 2.5% Na_2HPO_4-solution as liquid results in cement pastes with initial setting times from 10-16min (1h milling) to 5min (24h milling) for both the *α*- and *β*-modification.
Cements formed with 2.5% Na_2HPO_4 solution results in wet compressive strengths as shown in Fig. 2. Strength of ß-TCP cement increased with grinding time from 7 MPa after 1h to over 50 MPa after 24h. In contrast strengths of cements made from amorphous αTCP showed the opposite effect.

The relatively crystalline material ground for only 1h formed a cement with 70MPa strength which fell to 40 MPa after 24h milling.

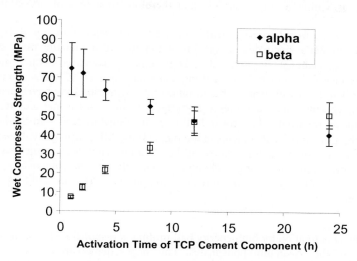

Fig. 2: Wet strengths of amorphous TCP cements after 24h at 37°C

The phase compositions of mechanically activated TCPs after 24h setting with 2.5% Na_2HPO_4 solution at 37°C are given in Table 1. While the MAα-TCP showed for all grinding levels a nearly complete hydrolysis reaction to HA, the degree of reaction of the MAβ-TCP samples continuously increased with prolonged grinding time from only 19% after 1h to nearly 93% after 24h milling.

Table 1: Phase compositions of MATCPs after setting for 24h according to Rietveld refinement analysis of X-ray diffraction data

Grinding time [h]	1	2	4	8	12	24
HA content of MA α-TCP cement [%]	91.7	-	97.4	-	-	98.7
HA content of MA β-TCP cement [%]	19.2	25.0	41.4	59.7	85.8	92.7

Discussion

Up to now it was believed that grinding of calcium phosphates only results in a decrease of particle size and therefore a higher reactivity. In this work we could show that in addition a mechanically induced phase transformation of both α and ß TCPs occurred by means of high energy ball milling. Following mechanical activation, the setting reaction of TCP to calcium deficient hydroxyapatite can be accelerated to the point where clinically useful setting cement systems can be formed:

$$3 \text{ ß-Ca}_3(PO_4)_2 + \quad H_2O \longrightarrow Ca_9(PO_4)_5(HPO_4)OH \quad (1)$$

The effect of mechanical activation could not be attributed only to a decrease of the particle size but to a mechanically induced phase transformation from the crystalline to the amorphous state. This phenomenon is already known for a number of compounds in the field of metallurgy and ceramics

[11]: at a critical particle size, the milling energy can no longer reduce particle size, but leads to the formation of defects within the crystal lattice. This process results in the formation of nanocrystalline domains within a single particle and therefore to a loss of crystallinity of the material as a whole.

Mechanical properties were found to be linearly related to activation time for ß-TCP up to a limiting strength of ~50 MPa since degree of conversion to hydroxyapatite after setting also increased. In contrast, mechanical strengths of α-TCP cements decreased with prolonged ball milling, probably because of a much faster setting reaction since α TCP has a much higher solubility than ß TCP hence a lower degree of amorphous transformation may have been necessary for cement formation. The maximum compressive strengths of the cement matrices reported here are higher than previously reported values for α TCP cements. Ginebra et al. and Sarda et al. [12] found compressive strengths of α-TCP cements in the range 38-60 MPa; strength values for a single component ß-TCP cement system are not reported in the literature. However, comparisons between values are difficult since cement mixing technique, mould design, precompaction, flaw size, powder: liquid ratio and porosity can all influence strength [13, 14], in addition to reactant crystallinity as reported here.

Conclusion

The effect of mechanical activation by high energy ball milling is responsible for an increase of the hydrolysis reactivity of the tricalcium phosphates. This paper demonstrates that mechanically induced changes in crystallinity represent a valuable processing route by which to modify the behaviour of bioceramics.

References

[1] L. C. Chow, S. Takagi, K. Ishikawa : W.E. Brown, E. Constanz, editors. *Hydroxyapatite and Related Materials* (CRC-Press, USA 1994) p. 127.

[2] W.E. Brown, L.C. Chow: P.W. Brown, editor. *Cements Research Progress* (The Am Ceram Soc USA 1986) p.352.

[3] L.C. Chow, M. Markovic, S. Takagi: L.J. Struble, editor. *Cements Research Progress* (The. Am. Ceram. Soc. USA 1997) p. 215

[4] L.C. Chow, J. Ceramic Soc. Vol. 99 (1991), p. 954

[5] F.C.M. Driessens, M.G. Boltong, J.A. Planell, O. Bermudez, M.P. Ginebra, E. Fernandez: P. Ducheyne, D. Christiansen, editors. *Bioceramics 6* (Butterworth-Heinemann UK 1993). p. 469

[6] M.P. Ginebra, E. Fernandez, M.G. Boltong, F.C.M. Driesens, J.A. Planell: L. Sedel, C. Rey, editors. *Bioceramics 10.* (Elsevier Science USA 1997) p. 481

[7] M. Ginebra, E. Fernandez, M.G. Boltong, F.C.M. Driessens, J. Ginebra, E.A.P. De Maeyer, R.M.H. Verbeeck, J.A. Planell, J. Dent. Res. Vol 76 (1997) p. 905

[8] M. Otsuka, Y. Matsuda, Y. Suwa, J.L. Fox, W.I. Higuchi, J Biomed Mater Res Vol 29 (1995) p.25

[9] U. Gbureck, O. Grolms, J.E: Barralet, L.M. Grover, R. Thull, Biomaterials Vol 24 (2003) p. 4123

[10] *TOPAS Tutorial Quantitative Analysis, Users Manual,* Bruker AXS, Karlsruhe, 2001.

[11] A.W. Weeber, H. Bakker, Physica B Vol 153 (1988) p.93

[12] S. Sarda, E. Fernandez, M. Nilson, M. Balcell, J.A. Planell., J Biomed Mater Res Vol. 61 (2002) p. 653

[13] J.E. Barralet, T. Gaunt, A.J. Wright, I.R. Gibson, J.C. Knowles, J Biomed Mater Res Vol 63 (2002) p.1

[14] K. Ishikawa, K. Asaoka, J. Biomed. Mater. Res. Vol 29 (1995) p. 1537.

Key Engineering Materials Vols. 254-256(2004) pp. 281-284
online at http://www.scientific.net
© 2004 Trans Tech Publications, Switzerland

Cements from Biphasic β-Tricalcium Phosphate

K.J. Lilley[1], U. Gbureck[2], D.F. Farrar[3,] C. Ansell[3] and J.E. Barralet[1]

[1] Biomaterials Department, School of Dentistry, University of Birmingham, B4 6NN, UK; e-mail
j.e.barralet@bham.ac.uk

[2] Department of Functional Materials in Medicine and Dentistry, University of Würzburg,
Pleicherwall 2, D-97070

[3] Smith and Nephew Group Research Center, Heslington Science Park, York, YO10 5DF, UK.

Keywords: Calcium phosphate cement, β-tricalcium phosphate, brushite, mechanical properties

Abstract. Brushite cements are three component systems whereby phosphate ions and water react
with a calcium phosphate to form cement. When ß-tricalcium phosphate (ß-TCP) is the calcium
phosphate component, brushite is the main product. However, ß-TCP may be fabricated with non-
stoichiometric calcium phosphate (Ca:P) ratios to form biphasic mixtures of ß-TCP and ß-dicalcium
pyrophosphate or hydroxyapatite. These biphasic mixtures were considered as cement reactants in
this study. Altering Ca:P ratio extended setting time and altered the composition, structure and
behavior of cements. The use of a biphasic reactant enables the means of controlling setting time
and increasing strength of biodegradable brushite cements.

Introduction

Brown and Chow formed apatitic cements by mixing at least two calcium phosphates with a liquid
component [1]. Apatitic cements are used as bone substitute materials because they set and harden
at physiological temperature and are osteoconductive [2]. However, apatitic cements are poorly
resorbed *in vivo* [3]. In 1987 Lemaitre *et al.* [4] formed a cement where the most abundant calcium
phosphate product was brushite. These brushite cements are more resorbable than apatitic cements
in vivo [3] and hence offer a route to a resorbable ceramic cement bone graft.

Brushite cements are three component systems whereby phosphate ions and water react with a
calcium phosphate to form $CaHPO_4.2H_2O$. ßTricalcium phosphate (ßTCP) has commonly been
used as a calcium phosphate component to form brushite cement [4].

Interestingly, ßTCP (calcium phosphate ratio (Ca:P) = 1.5) may be fabricated with non-
stoichiometric Ca:P ratios to form a biphasic mixture of ßTCP and another calcium phosphate.
When calcium deficient (Ca:P < 1.5) a mixture of ßTCP and ß-dicalcium pyrophosphate (ßDCPP) is
present. When phosphate deficient (Ca:P > 1.5) mixtures of ßTCP and hydroxyapatite (HA) are
formed. Altering the stoichiometry of the calcium phosphate was hypothesized to alter the quantity
of phosphate ions and hence phosphoric acid concentration needed to form brushite.

Biphasic mixtures of calcium phosphate have not previously been assessed as brushite cement
reactants. In this study we have examined these biphasic ceramics as cement components. We
investigated the effect of Ca:P ratio and acid concentration on cement setting times, composition
and compressive strength.

Methods and Materials

Cements were formed with non-stoichiometric β-TCP and either 2 M or 3 M phosphoric acid
(Sigma, Dorset, UK) /50 mM sodium citrate (Sigma, Dorset, UK) solutions. Cements were formed
at room temperature with a powder to liquid ratio of 1.5 g/ml.
ßTCP preparation. Monetite (DCPA; Mallinckrodt Baker, Griesham, Germany) and calcium
carbonate (CC; Merck, Darmstadt, Germany) were mixed to give theoretical Ca:P ratios of between

1.25 and 1.6. These mixtures were heated to 1050 °C for 24 h and then quenched at room temperature in a desiccator. The resulting sintered cake was ground to a particle size of <355 μm and milled. Milling was performed in a planetary ball mill (PM400 Retsch, Germany, diameter: 400 mm bidirectionally at 200 rpm) with 500ml agate jars, 4 agate balls (30mm) and a load of 75g ßTCP per jar for 1h.

XRD Analysis. X-ray diffraction patterns of β-TCP and set cements were recorded on a diffractometer D5005 (Siemens, Karlsruhe, Germany). Data was collected from $2\theta = 20\text{-}40°$ with a step size of 0.02° and a count time of 5-10s/step. The phase composition was checked by means of JCPDS reference patterns for ßTCP (PDF Ref. 09-0169), HA (PDF Ref. 09-0432), brushite (PDF Tef. 09-0077) and ßDCPP (PDF Ref. 09-0346). Crystal sizes and quantitative phase compositions of materials were calculated by means of total Rietveld refinement analysis with TOPAS software (Bruker AXS, Karlsruhe, Germany). As references, the system internal database structures of ßTCP, HA, brushite and ßDCPP were used together with a Chebychev forth order background model and a Cu Ka emission profile.

Density. Relative density was calculated from the strut density measured using He pycnometry (Accupyc 1330, Micromeritics, UK) and apparent density.

Setting time. Initial and final setting times were measured using the Gilmore needle technique [5]. The time passed between cement mixing and the needle not making an impression on the cement surface was measured. This elapsed time represented setting time. Setting time was measured at room temperature.

Compressive strength. Cements were stored at room temperature for 24 h before testing. Cement samples with a diameter of 6 mm and a height of 12 mm were loaded under compression at a crosshead speed of 1 mm/min until failure using a universal testing machine (Instron, 5544, Bucks, UK).

Results

Phase analysis of x-ray diffraction patterns of cement reactants prepared in this study showed heating mixtures of DCPA and CC with varied theoretical Ca:P ratio produced biphasic mixtures of β-TCP and β-DCPP or HA. When the theoretical Ca:P ratio was between 1.25 and 1.4 the biphasic mixture was made up of ßTCP and ßDCPP, however, when the Ca:P ratio was 1.5 only ßTCP was present, and when the ratio was 1.6 ; ßTCP and HA were present. When biphasic cement reactants were added to 2 M or 3 M H_3PO_4 solutions a slurry was formed that set to a cement. These cements were triphasic and consisted of brushite, ßTCP, ßDCPP or HA, Table 1.

Table 1, Phase composition and relatve density of cements formed with β-TCP of various Ca:P ratios and 2 M or 3 M phosphoric acid

Ca:P	H_3PO_4 [M]	Cement composition [%]				Relative Density [%]	H_3PO_4 [M]	Cement compostion [%]				Relative Density [%]
		Brushite	ßDCPP	ßTCP	HA			Brushite	ßDCPP	ßTCP	HA	
1.25	2	51.21	34.87	13.92	0	51.3	3	65.60	30.48	3.92	0	55.0
1.30	2	52.39	26.83	20.78	0	52.0	3	65.59	25.43	8.99	0	59.5
1.35	2	50.96	20.08	28.95	0	67.3	3	66.04	22.58	11.39	0	63.5
1.40	2	48.79	11.01	40.21	0	73.4	3	50.39	11.63	37.98	0	64.9
1.50	2	46.26	2.23	51.51	0	50.5	3	53.33	3.33	43.34	0	55.5
1.60	2	54.89	0	18.27	26.84	64.2	3	55.69	29.40	14.91	29.40	51.4

The proportion of ßTCP in cement increased as the theoretical Ca:P ratio of biphasic cement reactants was increased whilst the proportion of ßDCPP decreased, Table 1. Cement formed from biphasic calcium phosphate of theoretical Ca:P ratio 1.6 contained HA and no ßDCPP, Table 1.

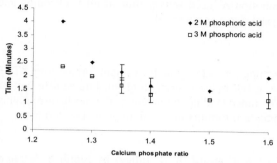

Fig. 1. Effect of Ca:P ratio of biphasic cement reactans on initial setting time of cements.

Setting time was retarded from 2 mins to 4 mins as the Ca:P ratio was decreased, Figure 1. Increasing the concentration of H_3PO_4 in the liquid component of the cement from 2 M to 3 M reduced setting time, Figure 1.

Compressive strength of cement increased as the theoretical Ca:P ratio of biphasic cement reactants was decreased, Figure 2. Increasing the concentration of H_3PO_4 in the liquid component of the cement from 2 M to 3 M increased compressive strength at all Ca:P ratios of ßTCP, Figure 2. There appeared to be a correlation between brushite crystal size in cement and strength but the reason why reactant Ca:P ratio altered crystal size was not determined, Figure 3.

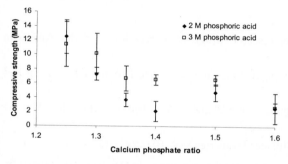

Fig. 2. Effect of Ca:P ratio of biphasic cement reactans on compressive strength of cements.

Discussion

The strength of cement has an inverse exponential relationship to porosity [6] only if all other factors are constant. Crystal size, degree of reaction and flaw size and /or distribution all affect the strength of cement. As compressive strength, was not related to density, (Table 1, Figure 2), variation in degree of reaction and crystal size, may have been the cause of variation in strength (Figure 3).

It is probable that the secondary phase in non-stoichiometric ßTCP acted as an inert filler in the cement since ßDCPP and HA are relatively insoluble. Although pyrophosphate ions have previously been shown to retard setting [7], in this system they are present as the insoluble calcium salt, β-DCPP, thus it seems improbable that pyrophosphate ions participated significantly in the setting reaction. The decrease in setting time observed as the Ca:P ratio of cement reactants was reduced, (Figure 1), may have been as a result of a

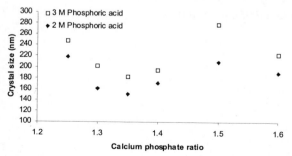

Figure 3. Effect of Ca:P ratio of biphasic cement reactans on brushite crystal size in set cements

higher extent of reaction as evidenced by lower amounts of residual ßTCP or ßTCP dissolution may have been physically retarded by the insoluble ßDCPP matrix.

Our discovery that cements formed from reactants of Ca:P ratio 1.25 were more than twice as strong (12.4 ± 2.4 MPa) as cements formed from stoichiometric ßTCP is highly significant. It appeared that ßTCP did not appear to contribute to strength in cements formed from reactants of lower Ca:P ratios. These cements contained less unreacted ßTCP, whilst the proportion of brushite in cements remained reasonably constant (46-52%) as reactant Ca:P ration varied, since the quantity of phosphate ions in the liquid phase was insufficient for complete reaction. As shown in previous studies [8], increasing phosphoric acid concentration reduced setting time at all Ca:P ratios and increased compressive strength and brushite crystal size was observed.

Conclusion

Biphasic TCP offers a novel means to control setting time of degradable cement formed from ßTCP and ßDCPP in a brushite matrix. The presence of HA reduced strength of brushite cement but the cause of this effect has yet to be determined. By understanding the nature of the mechanism giving rise to mechanical strength we may engineer optimised formulations with improved performance.

Acknowledgements
We gratefully acknowledge the provision of a CASE award (K. J. Lilley) by Smith & Nephew group research center, York, UK.

References

[1] Brown W.E. and Chow L.C.: Cements research progress, Proceedings of the American Ceramics Society (1986), p. 352-379

[2] Hwang J.J.H., Siew C., Robinson P., Gruninger S.E., Chow L.C. and Brown W.E.: J Dent Res Vol. 65 (1986), p. 195

[3] Nilsson M., Fernández E., Sarda S., Lidgren L. and Planell J.A.: J Biomed Mater Res Vol. 61 (2002), p. 600-607

[4] Lemaitre J., Mirtchi A. and Mortier A.: Silicates Industries Vol. 10 (1987), p. 141-146

[5] American national standards institute/American dental association. JADA Vol. 101 (1980), p. 669-671

[6] Ryshkewitch E.: J Am Ceram Soc Vol. 36 (1953), p. 65–68

[7] Mirtchi A.A., Lemaitre J., Munting E.: Biomaterials Vol. 10 (1989), p. 634-638.

[8] Bohner M., Lemaitre J.: Third Euro Ceramics (1995), p. 95-100

Key Engineering Materials Vols. 254-256(2004) pp. 285-288
online at http://www.scientific.net

Development of Novel PMMA-Based Bone Cement Reinforced by Bioactive CaO-SiO$_2$ Gel Powder

[1]S.B. Cho , [1]S.B. Kim, [1]K.J. Cho, [2]C. Ohtsuki, [3]T. Miyazaki

[1]Korea Institute of Geoscience and Mineral Resources (KIGAM),
Kajung-dong 30, Yusong-ku, Taejon 305-350, Korea, sbcho@kigam.re.kr
[2]Graduate School of Material Science, Nara Institute of Science and Technology,
8916-5, Takayama-cho, Ikoma-shi, Nara 630-0101, Japan
[3]Graduate School of Life Science and Systems Engineering, Kyushu Institute of Technology,
1-1, Sensui, Tobata, Kitakyusyu, Fukuoka 804-8550, Japan

Keywords: bioactive bone cement, PMMA, apatite formation, SBF

Abstract. PMMA cement was reinforced by 20CaO-80SiO$_2$ gel in order to induce bioactivity, and its apatite forming ability was investigated using simulated body fluid. The gel containing 20CaO-80SiO$_2$ was easily prepared by sol-gel method. When the gel powders replaced 50wt% of the PMMA, the cement was hardened after mixing for 7 min. Conventional PMMA cement did not form apatite on its surface, but the novel PMMA-based bone cement reinforced by sol-gel derived 20CaO-80SiO$_2$ gel formed bonelike apatite on its surface after soaking in SBF7.25 for 2 weeks. From these results, it is suggested that novel PMMA-based bone cement can bond to living bone and then it can be used for bioactive bone cement.

Introduction

Bone cement consisting of poly(methyl methacrylate) (PMMA) powder and methyl methacrylate(MMA) liquid, in which they are mixed and polymerized, is clinically used for the fixation of implants such as artificial hip joint. Significant problems on the PMMA bone cement are caused by loosening at the interface between bone and cement, since the cement does not show bone-bonding, i.e. bioactivity. Many attempts have been done to induce bioactivity in PMMA bone cement. It has been proposed that the CaO-SiO$_2$ system can be a good basic composition of bioactive glasses and glass-ceramics, since the dissolved Ca ion from the matrix increases the degree of supersaturation and the hydrated silica developed on them induce apatite formation. In the present study, novel PMMA-based bone cement that was reinforced with bioactive CaO-SiO$_2$ powder was developed and its apatite forming ability was investigated.

Methods

As a bioactive powder, 20CaO-80SiO$_2$ gel was prepared by the hydrolysis and polycondensation of tetraethoxysilane (TEOS, Si(C$_2$H$_5$O)$_4$). Polyethylene glycol (M.W. 10,000) and calcium nitrate (Ca(NO$_3$)$_2$·4H$_2$O) were dissolved into ion-exchanged distilled water, and concentrated nitric acid was added. Then TEOS was added to the above solution under vigorous stirring. After 5 min., the solution was transferred into a plastic Petri dish with its top tightly sealed, and was kept at 40°C in an air-circulating oven for gelation. After it was aged for another 18h, the obtained wet gel was immersed in aqueous nitric acid solution for 6 hours. The nitric acid solution was renewed every 2h. Then, the obtained wet gel was dried at 40°C for 6days, it was heated at 600°C at a heating rate of 100°C/h, and held at the temperature for 2h, then allowed to cool to room temperature. After heat treatment, the gel was powdered by planetary ball mill.

The PMMA powder was mixed with 20CaO-80SiO$_2$ gel powder at 50wt% of the powder. Benzoyl oxide was then added to the powder as a polymerization initiator. MMA liquid was mixed with N,N-dimethyl-p-toluidine as a polymerization accelerator. The powder was mixed with the liquid at a powder to liquid ratio of 1g/ 0,5 g at room temperature. The paste was shaped to cylindrical specimens and then the setting time of the cement was measured. Apatite forming ability of the cement was investigated using simulated body fluid (SBF) with ion concentrations nearly equal to those of the human blood plasma (Na^+ 142.0, K^+ 5.0, Mg^{2+} 1.5, Ca^{2+} 2.5, Cl^- 147.8, HCO_3^- 4.2, HPO_4^{2-} 1.0, SO_4^{2-} 0.5 mmol/dm^3)[1]. Before and after soaking in SBF, the surfaces of the specimen were investigated by SEM observation, X-Ray diffraction and FT-IR diffusive reflection spectroscopy.

Results

The structural characteristics and their apatite forming ability of the sol-gel derived $CaO-SiO_2$ system were well described in a previous study [2]. According to the results of the X-ray diffraction for the sol-gel derived $CaO-SiO_2$ system, the gels were amorphous after heat treatment at 600°C. However, SEM observation results showed that the structural morphology changed with their composition. Higher than 15wt% CaO, the gels having interconnected micrometer-ranged pore structure were obtained, whereas CaO content exceed 20wt%, the interconnected pore structure was destroyed and then changed to agglomeration of spherical powder. Regarding the bioactivity of $CaO-SiO_2$ gels it was shown that the induction period of apatite formation for the $CaO-SiO_2$ gels increased with the CaO content. The $20CaO-80SiO_2$ gels heat-treated at 600°C formed bonelike-apatite on their surfaces within 1 week soaking in SBF7.25. Furthermore, it is thought that the spherical powder positively affects the rheological properties of the cement. Based upon these results, the present authors used $20CaO-80SiO_2$ gel for the induction of bioactivity on PMMA-based cement in this study.

When the gel powders replaced all the PMMA powder, the cement was not hardened. However, we could obtain hardened cement by replacement of 50wt% of the PMMA by the gel powder after mixing for 7 min. Figure 1 shows that fractured surface of novel PMMA-based cement in which 50wt% of PMMA was replaced. The result of SEM observation for the cement showed that the gel powder was well dispersed and remained in PMMA cement matrix.

As a essential requirement to bond to living bone, artificial materials have to form the bonelike apatite layer on their surfaces *in vivo* or *in vitro*[3,4]. It is well known that conventional PMMA cement did not form apatite on its surface *in vivo* as well as *in vitro*. Figure 2 shows that Thin Film-XRD patterns of the novel PMMA-based cement reinforced by $20CaO-80SiO_2$ gel before and after soaking in SBF for 2 weeks. Before soaking in SBF, only a broad reflection peak due to amorphous $20CaO-80SiO_2$ gel is detected, but new diffraction peaks due to apatite appears on the surface of PMMA-based cement after 2 weeks in SBF, as shown in Fig. 2. According to SEM observation as shown in Fig. 3, platelike particles were deposited on the novel cement surface. The morphology of the apatite crystals induced on the surface of the cement is very similar to that of crystals on the surfaces of various bioactive glass, glass-ceramics. From these results, it is suggested that the novel PMMA-based cement reinforced by sol-gel derived $20CaO-80SiO_2$ gels bond to living bone when they are implanted in the body environment and then can be used for bioactive bone cement.

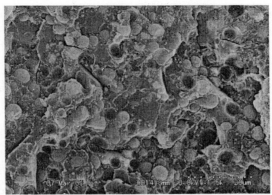

Fig. 1. Fractured surface of the novel PMMA-based bone cement reinforced by sol-gel derived CaO-SiO$_2$ powder.

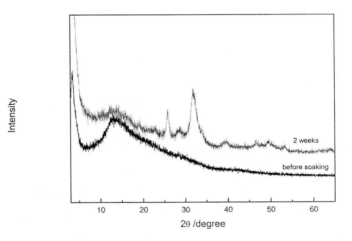

Fig. 2. TF-XRD patterns of the PMMA-based bone cement before and after soaking in SBF.

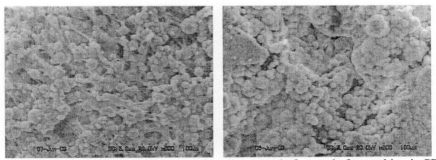

Fig. 3. SEM photographs of the PMMA-based bone cement before and after soaking in SBF.

Acknowledgements
This work was supported by grant No.R01-2002-00256 from the Korea Science & Engineering Foundation.

References

[1] S.B. Cho, K. Nakanishi, T. Kokubo, N. Soga, C. Ohtsuki, T. Nakamura, T. Kitsugi, and T. Yamamuro: J. Am. Ceram. Soc., Vol. 78, (1995), p. 1769-1774.

[2] S.B Cho, S.B. Kim, K.J. Cho, Y.J.Kim, T.H. Lee, Y. Hwang: Proc. 2nd Asian BioCeramics Symposium, (2002), p. 77-80.

[3] T. Kokubo, Biomaterials, Vol. 12, (1991), p. 155-163.

[4] T. Kokubo, H. Kushitani, S. Sakka, T. Kitsugi and T. Yamamuro, J. Biomed. Mater. Res., Vol. 24, (1990), p. 721-734.

Key Engineering Materials Vols. 254-256(2004) pp. 289-292
online at http://www.scientific.net
© 2004 Trans Tech Publications, Switzerland

Influence of Molarity and Liquid/Powder Ratio on the Synthesis of a Calcium Phosphate Cement

M. E. Murphy[1], R. M. Streicher[2] and G. M. Insley[3]

[1] Stryker Howmedica Osteonics, Raheen Business Park, Limerick, Ireland.
matthew.murphy@emea.strykercorp.com
[2] Stryker SA, Florastrasse 13, CH-8800 Thalwil, Switzerland.
robert.streicher@emea.strykercorp.com
[3] Stryker Howmedica Osteonics, Raheen Business Park, Limerick, Ireland.
gerard.insley@emea.strykercorp.com

Keywords: Calcium phosphate, cement, mechanical properties.

Abstract. Current calcium phosphate based cements (CPCs) for use as bone substitutes provide the surgeon with a material that is readily formable and with mechanical properties similar to that of human bone. However, a better understanding of the influence of the component materials used is required in order to further develop these cements. This study investigates the influence of varying the molarity and liquid-to-powder (L:P) ratio of the sodium phosphate (Na_2HPO_4) accelerator on the mechanical, chemical and microstructural properties of a CPC.

Introduction

Early studies with calcium phosphates reported on a self-hardening CPC that contained an equimolar mixture of finely ground tetracalcium phosphate (TTCP) and dicalcium phosphate anhydrous (DCPA) as the starting phases. When mixed with water, the cement forms hydroxyapatite (HA) as the end product. The use of a concentrated phosphate solution as the cement liquid, e.g. Na_2HPO_4, has been reported to influence the setting time of CPCs [1, 2]. This paper reports the influence of Na_2HPO_4 on the mechanical and chemical properties of a CPC.

Materials and Methods

Reagent grade TTCP and DCPA were blended together (using an equimolar ratio of 2.69) to obtain an homogenous powder mixture. This powder was mixed using (a) L:P ratios of 0.6 to 0.8 with a 3 M solution of Na_2HPO_4 and (b) varying molarity of Na_2HPO_4 (0.25 to 3 M) and a fixed L:P of 0.65.

The setting time of the cement samples was measured using the Gilmore needle method (ASTM C266), using a needle with a tip diameter of 1 mm. The mix time was recorded as the time taken to form a smooth homogeneous cement, whilst the work time was the time with which the cement could be worked without any crumbling of the cement (internal standards). For the compressive strength measurement, the specimens were compressed in a universal testing machine (Instron, USA) at a constant crosshead speed of 1 mm/min. Scanning electron microscopy (SEM, Hitachi S-3500N VPSEM) was used to examine the cement's fracture surfaces after 24 hrs of setting.

Results and Discussion.

Set-times. Figure 1(a) displays the influence of varying the L:P ratio (using 0.25 M Na_2HPO_4) on the mix, work and setting time of a CPC.

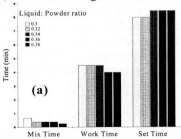
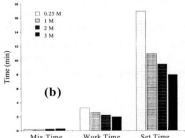

Fig. 1 Plot displaying the (a) influence of the liquid powder ratio and (b) variation in molarity of the Na_2HPO_4 on the mix, work and setting time of a CPC

From Figure 1(a), a comparable relationship between the mixing time and working time can be observed. That is, as the L:P ratio was increased, the mixing time tended to decrease as more liquid was available to wet the powder to produce a cement with a putty-like consistency. As the L:P ratio increased, the working time tended to decrease, this can be explained by the presence of excess Na_2HPO_4 which acts as a accelerator in the dissolution/precipitation reaction of the cement's initial powder components [1]. An inverse relationship can be observed with respect to the setting time for the higher L:P ratios, i.e. as the L:P ratio increased, the setting time also tended to increase due to the presence of excess liquid which also has a corresponding adverse effect on the mechanical strength of the cement, a topic that will be discussed later [2].

Figure 1(b) displays the susceptibility of the mix, work and setting times of CPCs to the molarity of the liquid used to mix the dry components, i.e. Na_2HPO_4 with molarities varying from 0.25 to 3 M, using a L:P ratio of 0.65. The mix time was found to increase slightly with increasing molarity solutions due to a more rapid dissolution of the solids when higher molar solutions are used [3]. However, due to this occurrence, any available high molarity solution will be tied into the dissolution of the dry components, hence giving the appearance of a drier cement with a longer mixing time compared to one mixed with a lower molarity solution.

Nurit *et al.* [2] reported that the cement's working and setting time depends on the rate at which supersaturation is reached, and that this depended upon a number of factors, including the molarity of the mixing liquid. From figure 1(b), it can be observed that as the molarity of the Na_2HPO_4 was increased for a set L:P ratio, the working and setting time decreased correspondingly. It is the entanglement of the precipitated HA crystals that provide the cement cohesion and its mechanical strength and that the cement setting time depends on the rate at which supersaturation is reached [3]. The increased setting time for the 0.25 molar solution correlates to a slower precipitation that is due to the rate limiting nature of dissolution of the reactants which prevents the supersaturation of PO_4 ions. Since the function of the Na_2HPO_4 accelerator is to impart PO_4 ions to the medium to create sufficient supersaturation thus speeding up the precipitation reaction and thereby bringing down the setting time [4]. It therefore stands to reason that if the cement reaches supersaturation more rapidly with a higher molarity solution, that the working and setting times will correspondingly decrease with further increases in the molarity used. These findings correlate to a similar study on CPCs that were mixed using citric acid as the mixing liquid with various molarities [1].

Compressive Strength. It is the formation of HA that is the key factor in the hardening and the mechanical strength of CPCs [3]. Once the dry powder components are combined with the mixing liquid (Na_2HPO_4), the TTCP and DCPA will dissolve into solution, form a supersaturated solution and the result of which is the precipitation and growth of HA crystals. As the HA crystals grow, they will become interlocked, as a result, the microstructure will become denser with respect to time and the mechanical strength of the cement will correspondingly increase [5].

Figure 1(a) has shown the benefit of a using a higher L:P ratio resulting in a wetter cement so as to improve the workability of a CPC. The mechanical strength, i.e. compressive strength, was therefore similarly examined for various L:P ratios. so as to determine the influence, if any, that varying the L:P ratio had on the strength of the resultant cement. From figure 2(a), an initial value of ~14.5 MPa can be observed for a CPC using a L:P ratio of 0.6. This value increases to ~18.5 MPa when the L:P ratio is increased to 0.65. However, further increases in the L:P ratio resulted in a continual drop in the compressive strength. From figure 2(a), the variation in the L:P ratio with respect to the compressive strength can be seen to set a limitation on the compressive strength value [6].

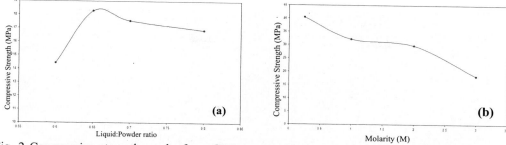

Fig. 2 Compressive strength results for a CPC varying (a) the L:P ratio and (b) the molarity on the compressive strength

With respect to the influence of the molarity of the Na₂HPO₄ on the compressive strength of CPCs, results from earlier studies in this area have indicated that the concentration of the liquid component of the cement strongly influences both the mechanical properties and biocompatibility of the cement [7]. From figure 2(b), it can be observed that as the molarity of the Na₂HPO₄ was increased from 0.25 M to 3 M, the compressive strength tended to decrease at what can be considered an almost linear rate from ~40 MPa to ~18 MPa. Such a decrease is understandable when the corresponding SEM of the fracture surfaces of these samples are examined (see SEM section). From SEM studies, it was found that higher the molarity Na₂HPO₄ resulted in a coarser microstructure that had less entanglement of the precipitated crystals that provide cohesion and its mechanical strength to the cement. Therefore, if the more concentrated molarities result in a coarser microstructure with less interlocking, or 'entanglement', per unit area compared to that of a lower concentration and finer microstructure cement, it is understandable that the mechanical properties decreased as a result [2].

SEM. Any change in the mechanical strength can be related to the kinetics of recyrstallization of the starting powders to form HA, where the mechanical strength increases according to the progress of recyrstallization [8]. The fracture surface of CPCs with increasing L:P ratio can be seen in figures 3(a), (b) and (c). From these micrographs, both plate and rod-like crystals can be observed. However, from these micrographs, it would appear that the rod-like crystals tend to increase in size with increasing L:P ratio. This is most probably linked to the fact that any excess liquid in the setting cement acts as a void in the final set cement, therefore, allowing for crystal growth through these 'voids'. Hence the coarser rod-like microstructure for the higher L:P ratio's. As was stated earlier, the mechanical strengths of CPCs are derived primarily from the interlocking of the HA crystals. Thus, for the higher L:P ratio samples, a coarser microstructure would explain the lower compressive strength values observed in figure 2(a) [9].

Comparison of the original powder with the surface of the set cements revealed an interlocked structure forming the resultant set cement, as can be seen in the SEM micrographs, see figures 3 to 4, inclusive. Needle-like crystals, which are typical of HA, were observed in all of the set specimens. In addition to needle-like crystals, plate-like crystals were also observed in the set cement. The degree of needle and platelet formation appears to be dependent on the starting liquid molarity, i.e. it appears that the higher the molarity of the Na₂HPO₄ used, the 'coarser' the resulting microstructure. Another point to note is the presence of what appears to be particles in figure 4 (d) (ringed areas). The phase of these particles are unknown, however, the presence of these particles in the 3 molar solution cement and not in any of the other samples indicates an inherently poorer mechanical bonding/strength compared to the other samples, this statement being supported by the mechanical properties obtained for these samples (see figure 2 (a) and (b)) [5].

Fig. 3 SEM micrographs of the fracture surfaces of a CPC, 24 hours after mixing using 3 molar
Na$_2$HPO$_4$ solution and (a) 0.6, (b) 0.7 and (c) 0.8 L:P ratio

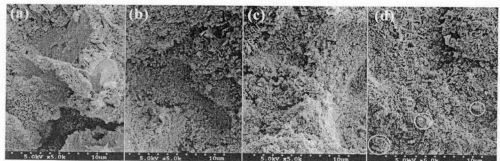

Fig. 4 SEM micrographs of the fracture surface of a CPC, 24 hours after mixing, using (a) 0.25 (b)
1 (c) 2 and (d) 3 molar solutions of Na$_2$HPO$_4$

Conclusions

Varying the molarity and L:P ratio can have a large effect on the compressive strength and HA
conversion of the CPC studied in this work. This study has also shown how the mechanical,
chemical and microstructural properties of CPC cements can be modified by varying the molarity
and/or L:P ratio of the mixing liquid.

Acknowledgements

The authors would like to acknowledge the contributions of Cillian Cleere and Peter Maher of the
Biomaterials Department, Stryker Howmedica Osteonics, Limerick, to this work.

References

[1] Y. Doi, Y. Shimizu, Y. Moriwaki, M. Aga, H. Iwanaga, T. Shibutani, K. Yamamoto, Y.
     Iwayama: Biomat. Vol. 22 (2001), p.847-854.
[2] J. Nurit, J. Margerit, A. Terol and P. Boudeville: J. Mat. Sci.: Mat. Med. Vol. 13 (2002),
     p.1007-10014.
[3] L. C. Chow, S. Takagi, P. D. Costantino and C. D. Friedman: Mat. Res. Soc. Symp. Proc. Vol.
     179 (1991), p.3-24.
[4] M. Komath, H. K. Varma and R. Sivakumar: Bull. Mat. Sci. Vol. 23 (2000), p.135–140.
[5] Y. Miyamoto, K. Ishikawa, H. Fukao, M. Sawada, M. Nagayama, M. Kon and K. Asaoka:
     Biomat. Vol. 16 (1995), p.855-860.
[6] K. Ishikawa: Key Engineering Materials Vol. 240-242 (2003), p.369-372.
[7] A. Yokoyamaa, S. Yamamotoa, T. Kawasakia, T. Kohgob and M. Nakasuc: Biomat. Vol. 23
     (2002), p.1091-1101.
[8] H. Yamamoto, S. Niwa, M. Hori, T. Hattori, K. Sawai, S. Aoki, M. Hirano and H. Takeuchi:
     Biomat. Vol. 19 (1998), p.1587–1591.
[9] H. H. K. Xu and J. B. Quinn: Biomat. Vol. 23 (2002), p.193-202.

Key Engineering Materials Vols. 254-256(2004) pp. 293-296
online at http://www.scientific.net
© 2004 Trans Tech Publications, Switzerland

PLA/HAp Microsphere-based Porous Materials
for Artificial Bone Grafts

Fukue Nagata, Tatsuya Miyajima, Yoshiyuki Yokogawa

Ceramics Research Institute, National Institute of Advanced Industrial Science and Technology,
2266-98 Anagahora, Shimoshidami, Moriyama, Nagoya 463-8560 JAPAN, f.nagata@aist.go.jp

Keywords: microsphere, porous material, hydroxyapatite, poly(D,L-lactide)

Abstract. Porous materials have been fabricated from surfactant-free PLA/HAp microspheres by a filtration method under reduced pressure. In order to form porous disks from microspheres, dissolved PLA was used as a glue to bond neighboring microspheres at ambient temperature. The concentration of PLA solution was an important factor of mechanical properties and morphologies of the resulting discs. High concentration of PLA solution induced dissolution of the microspheres, consequently collapsed porous structures. Three-dimensional porous materials were fabricated with PLA solutions adjusted low concentrations enough to maintain the morphology of the microspheres, containing two distributions of pores on the orders of 10 and 500 μm. They would be promising biodegradable bone grafts for bone repair.

Introduction

Poly(α-hydroxyl acid) such as poly (D,L-lactide), PLA, has received significant attention for biomedical applications because it has biodegradable and biocompatible characters. Formation of porous materials using PLA microspheres would have the possibility to obtain easily three-dimensional designed materials for bone grafts [1]. However, most of PLA microspheres are prepared by emulsion method with the aid of surfactants, which remain in the final products [2]. The residual surfactants are also suspected of a disadvantageous effect to human body such as an allergy-like reaction because of their non-biodegradability. In order to solve this problem, several researchers have reported surfactant-free preparations of PLA or PLGA microspheres [3-5]. We have shown a novel method to form surfactant-free PLA microspheres covered with hydroxyapatite (HAp) using interfacial interaction between inorganic-organic materials [6]. Since PLA/HAp composite has been also found to counter a pH drop by PLA degradation, which is another disadvantage of pure PLA, porous materials using the surfactant-free PLA/HAp microspheres would be useful for artificial bone grafts. Previously, we have reported an injection method to fabricate porous materials from the PLA/HAp microspheres, the microspheres, however, were deformed after the method because of an injecting pressure. In this paper, we report a filtration method under reduced pressure in order to maintain morphology of microspheres which determine pore sizes in porous materials.

Methods

Surfactant-free PLA/HAp microspheres were fabricated by a solvent evaporation method. The organic phase composed of PLA (Mw = 20000; WAKO) and dichloromethane was poured in various concentrations of $(CH_3COOH)_2Ca$ aqueous solution. The mixture was stirred to yield emulsion. $(NH_4)_2HPO_4$ aqueous solution was added into the emulsion. The suspension was stirred for 24 - 72 h at room temperature until dichloromethane was removed thoroughly. The suspension consisting of microspheres and ionic solution was washed to eliminate excess ions. In order to fabricate porous material using microspheres, 2.5 mL of the resulting microspheres were mixed with various amount of water and 5.0 mL of PLA solution saturated in ethanol. PLA concentrations were adjusted by the volume ratio of water and PLA dissolved in ethanol. After various reaction times, the mixtures were filtrated under reduced pressure to form discs.

The morphology of the products were observed using a scanning electron microscope (SEM, HITACHI, S3000) and an optical microscope (Nikon SZX-1500). Energy-dispersive X-ray micro analyzer (EDX, HORIBA, EMAX-7000) was used for elemental analysis. The phases of the microspheres were analyzed by powder X-ray diffraction (XRD, MAC science, MXP3). Binding sites between PLA and calcium phosphate were analyzed by Fourier transform infrared (FT-IR, JASCO, FT/IR-8000S).

Results and Discussion

Fig. 1 shows an SEM image of microspheres as prepared. The microspheres were approximately 100 μm in size and precipitates were observed on the surface of microspheres. The precipitates were identified as calcium phosphate by EDX analysis. An optical microscope image shows that the microspheres were well dispersed in water (Fig. 2). Fig. 3 shows the XRD pattern of the composite microspheres. A broad peak around 15° was ascribed to the amorphous structure of the PLA. The peaks of the XRD pattern were in good agreement with the JCPDS value of HAp but they were not sharp, which indicated the precipitate on the microspheres to be incompletely crystallized HAp. The results from FT-IR revealed that HAp would be precipitated on the carboxyl end groups of PLA, which have the ability to induce the precipitation of calcium phosphate. Owing to the precipitated HAp acting as a stabilizer in oil-in-water emulsion, PLA and HAp composite microspheres were obtained without surfactant.

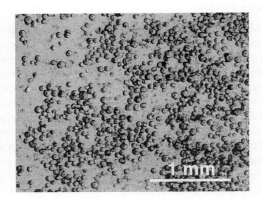

Fig. 1   An SEM image of microspheres as prepared.

Fig. 2   An optical microscope image of microspheres as prepared.

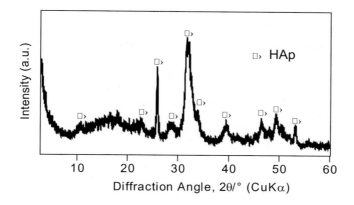

Fig. 3. XRD pattern of the microspheres prepared without surfactants

PLA solutions were used for fabrication of porous disks from the microspheres, being expected to play a role as a glue which binds neighboring microspheres. Mechanical properties and morphologies of the resulting discs were strongly affected by the volume ratio of PLA solution to water in the mixture before the filtration. Large volume ratios of PLA solution to water caused some discs to exhibit a remarkable elasto-plastic behavior. In particular, when the volume ratio of PLA solution to water was more than 5 to 2, the resulting discs were easily bent by finger without any cracking. The discs had many closed pores ranging in diameter from 0.2 - 1.0 mm caused by evaporation of ethanol, though no microsphere was observed in the discs. These results indicated that high concentration of PLA solution would lead to dissolution of the microspheres, and thereby the PLA characteristic would be predominant over the composites characteristic. On the other hand, the microspheres maintained their spherical morphology in the discs under small PLA volume ratio conditions. For example, when the volume ratio of PLA solution to water was 5 to 15, the resulting discs consisted of aggregates of the microspheres, having bi-modal distribution of pores (Fig. 4a, 4b).

Pores in the order of 10 μm were made by the spaces between adjacent microspheres bonded to each

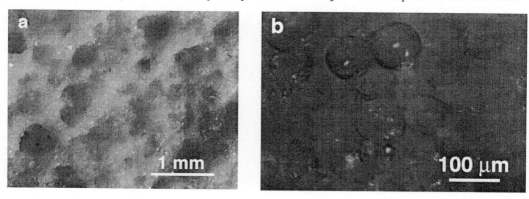

Fig. 4   Optical microscope images of porous disks fabricated from the microspheres, showing bi-modal distribution of pores.

other at their contact points (Fig. 4b). Pores of approximately 500±300 μm in diameter (Fig 4a) were formed by the evolution of bubbles arising from ethanol during filtration.

Conclusions

In this paper, we have shown that it is possible to fabricate porous materials from PLA/HAp microspheres by the filtration method under reduced pressure. The volume ratio of PLA solution and water in the mixture was an important factor to control mechanical properties and morphologies of porous materials

References

[1] M. Bordenm, M. Attaiwa, Y Khan and C.T. Laurencin: Biomaterials Vol. 23 (2002), p. 551

[2] M.F. Zambaux, F. Bonneaux, R. Gref, P. Maincent, E. Dellacherie, M.J. Alonso, P Labrude and C. Vigneron: J. Controlled Release Vol. 50 (1998), p. 31

[3] A. Carrio, G. Schwach. J. Coudane, M. Vert: J. Controlled Release Vol. 37 (1995), p. 113

[4] R. Gref, V. Babak, P. Bouillot, I. Lukina, M. Bodorev and E. Dellacherie: Colloids Surf. Vol. 143 (1998), p. 413

[5] Y. I. Jeong, C. S. Cho, S. H. Kim, K. S. Ko, S. I. Kim, Y. H. Shim and J. W. Nah: J. Appl. Polym. Sci. Vol. 80 (2001), p. 2228

[6] F. Nagata, T, Miyajima, Y. Yokogawa: Chem. Lett. Vol. 32 No.9 (2003), p. 2

Key Engineering Materials Vols. 254-256(2004) pp. 297-300
online at http://www.scientific.net
© 2004 Trans Tech Publications, Switzerland

Rietveld Analysis in Sintering Studies of Ca-Deficient Hydroxyapatite

L.E. Jackson[1], J.E. Barralet[2] and A.J. Wright[1]

[1] School of Chemistry, University of Birmingham, Edgbaston. Birmingham B15 2TT, UK.

[2] Biomaterials Unit, School of Dentistry, University of Birmingham, B4 6NN, UK

(emails; LEJ820@bham.ac.uk, j.e.barralet@bham.ac.uk, a.j.wright@bham.ac.uk)

Keywords: Rietveld analysis, hydroxyapatite, beta-tricalcium phosphate, alpha-tricalcium phosphate, biphasic calcium phosphates.

Abstract. Rietveld analysis of X-ray powder diffraction data is a powerful tool that can quantitatively determine sample phase composition and stoichiometry of the phases. It is known that calcium deficient hydroxyapatite (HA) will decompose into β-TCP and HA on heating yet precise quantification has not been attempted. In this study a precipitated calcium deficient HA was sintered in air at temperatures of between 600 and 1500°C. Quantification of phase composition surprisingly indicated the continued decomposition of HA with increasing sintering temperature. The presence of both polymorphs of tricalcium phosphates (α-TCP and β-TCP) were detected and interestingly, α-TCP was observed to be the first decomposition product and not β-TCP as has been suggested in other studies.

Introduction

Calcium phosphate based materials are of great importance in a diverse range of fields involving chemistry, materials, biology and medicine. In particular, insoluble HA, ($Ca_{10}(PO_4)_6(OH)_2$) which is the main inorganic component of bone and teeth, and more soluble tricalcium phosphate (TCP, $Ca_3(PO_4)_2$) have been widely studied due to their osteoconductivity and their bioresorbability respectively.

It is known that the mineral component of bones and teeth, when sintered, produce a biphasic mixture of HA and TCP, suggesting that despite having carbonate for phosphate substitutions, this mineral is a calcium deficient HA. Extensive research has been devoted to producing ceramics with various quantities of both HA and TCP (known as biphasic calcium phosphate, BCP) [1, 2] whose resorption rate can be controlled by manipulation of the HA/TCP ratio. [3-6]

This study used Rietveld analysis [7] to probe the stability of calcium deficient HA when it is subjected to a range of sintering conditions. The phases produced at each sintering temperature were characterised using powder X-ray diffraction (XRD) and the exact phase and sample composition was determined by Rietveld refinement using the GSAS software. [8]

Materials and Methods

A precipitation reaction based on one reported by Nelson and Featherstone [9] was used for the preparation of HA. A 300 mM solution of analytical grade tri-ammonium orthophosphate was dripped into a 500 mM continuously stirred solution of analytical grade calcium nitrate 4- hydrate. The temperature was maintained at approximately 70° C using a thermostatically controlled hot plate. The pH of the calcium nitrate solution was maintained to above 11 using ammonia solution. The resultant precipitate was then aged for 24 hours and was then filtered and washed 5 times with distilled water to remove any unwanted impurities. The solid was then dried at 95° C for 24 hours in an oven. The samples were then intimately ground and sintered for 12 hours in air at temperatures from 600° C up to 1500° C (heating rate of 10°C/min). The resulting powders were cooled in a controlled manner at $\leq 10°$C/min and studied using XRD. Diffraction data were collected using a Siemens D5000 diffractometer, with a Ge primary beam monochromator providing Cu Kα1 radiation, from 5-100° 2θ. All XRD data obtained were analysed using the Rietveld method to obtain accurate phase and sample composition. The three phases observed were HA, α-TCP and β-TCP and the starting structural model for each phase was obtained from literature reports. [10, 11, 12] In all of the Rietveld analyses, backgrounds were modeled by a linear interpolation function and peak shapes by a pseudovoigt function.

Table 1 Summary of the results of the Rietveld analyses

Sinter Temp. [° C]	Phase composition (Weight % / Mole %)	Lattice Parameters [Å]	Refined Calcium Content	Refinement fit parameters			
				x^2	Rwp [%]	Rp [%]	$R_{F}2$ [%]
600	HA (100 / 100)	a= 9.4197(1), b= 6.8870(1)	9.92(3)	3.037	4.04	3.05	2.89
700	HA (98.988 / 96.80)	a= 9.4197(1), b= 6.8799(1)	9.94(3)	3.039	3.88	2.87	4.74
	α-TCP (1.12(1)/ 3.20)	a= 12.89(1), b= 27.277(9) c= 15.223(5), γ= 126.19(5)					
800	HA (83. 659 / 61.25)	a= 9.4077(1) b=6.8731(1)	9.92(3)	2.253	2.93	2.21	5.08
	α-TCP (15.0(2) / 35.69)	a= 12.867(1), b= 27.239(1) c= 15.1968(8), γ= 126.176(7)					
	β-TCP (1.29(4) / 3.07)	a= 10.4247(4), c= 37.322(2)					
900	HA (83.71 / 61.39)	a= 9.41022(9) b=6.87368(7)	9.92(3)	2.082	2.57	1.91	7.71
	α-TCP (8.2(1) / 19.35)	a= 12.868(2), b= 27.245(2) c= 15.200(1), γ= 126.18(1)					
	β-TCP (8.1(1) /19.26)	a= 10.4246(3), c= 37.324(1)					
1000	HA (75.73 /49.08)	a= 9.4093(2) b=6.8732(1)	9.91(3)	2.123	2.89	2.10	4.93
	β-TCP (24.3(2) / 50.92)	a= 10.4157(2), b= 37.3004(8)					
1100	HA (76.20 / 49.81)	a= 9.4359(1), b= 6.89341(8)	9.89(3)	1.977	4.24	3.15	6.28
	α-TCP (1.46(7) / 3.09)	a= 12.855(1), b= 27.364(2) c= 15.245(1), γ= 126.271(6)					
	β-TCP (22.3(2) / 47.10)	a= 10.4459(2), b= 37.488(1)					
1200	HA (73.71 / 46.44)	a= 9.4038(1) b=6.8710(1)	9.93(3)	2.119	3.63	2.55	4.95
	α-TCP (26.3(2)/ 59.14)	a= 12.837(1), b= 27.291(9) c= 15.1954(8), γ= 126.308(6)					
1300	HA (69.15 / 40.86)	a= 9.41432(9) b=6.87971(8)	10	2.001	3.27	2.31	6.30
	α-TCP (30.9(2) / 59.14)	a= 12.859(1), b= 27.304(1) c= 15.2101(6), γ= 126.259(6)					
1400	HA (63.97 /35.43)	a= 9.4145(1) b=6.8800(1)	10	1.883	3.89	2.87	4.89
	α-TCP (36.0(2) / 64.57)	a= 12.8637(9), b= 27.303(1) c= 15.2122(6), γ= 126.19(5)					
1500	HA (51.40 / 24.67)	a= 9.4168(1) b=6.8820(1)	10	1.704	3.32	2.45	5.30
	α-TCP (48.6(2) / 75.33)	a= 12.8676(6), b= 27.3101(7) c= 15.2186(4), γ= 126.243(3)					

Results and Discussion

The results of the Rietveld analyses are summarized in Table 1, with a section of a typical Rietveld profile shown in Fig. 1, indicating the excellent agreement obtained between observed and calculated data. The relative phase compositions of sintered samples are plotted in Fig. 2 and demonstrate a decrease in HA content at 800°C as α-TCP was first formed, then again at 1000°C as the total TCP content increased and then finally between 1200 and 1500°C. We observed increasing amounts of α-TCP at the higher sintering temperatures, but unusually, we observed the

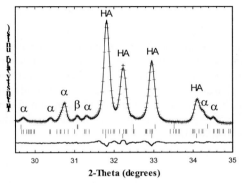

Fig. 1   Multiphase Rietveld analysis of X-ray diffraction data for sample sintered at 800°C (lines for calculated and difference; crosses for observed data). Reflection positions of α-TCP, β-TCP and HA indicated by labels and tick marks.

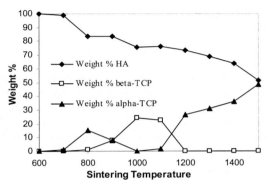

Fig. 2 Phase composition (weight %) of sintered samples.

presence of significant amounts of α-TCP at 800-900°C, before any significant β-TCP formation (see Fig. 1). The α-TCP phase is usually only observed above 1125°C, when it forms principally through the conversion of β-TCP [13]. The observation of α-TCP at lower temperatures is unusual but this may be a consequence of the sensitive nature of Rietveld analysis that is able to deconvolute the diffraction pattern to accurately account for all intensity, something not possible by conventional "fingerprint" analysis.

The lattice parameters of the HA showed an anomalously high volume at the sintering temperature

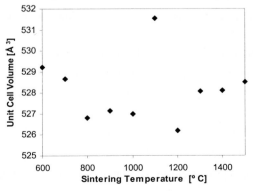

Fig. 3 Variation in unit cell volume of HA component as a function of sintering temperature (note; error bars too small to view)

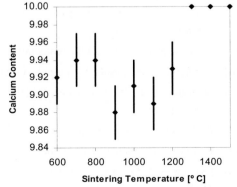

Fig. 4 Variation in calcium content of HA component as a function of sintering temperature (contents within error of 10 were fixed at this value)

of 1100°C (Fig. 3). Repeated data collection and analysis confirmed the validity of this volume increase which appears to coincide with the decomposition of α-TCP and increasing content of β-TCP. It should also be noted that the HA is non-stoichiometric up to 1300°C (see Fig. 4). At all of the sintering temperatures, the refinement was sufficiently stable to allow for the full variation of the HA atomic parameters. These atomic coordinates were found not to vary significantly from those of the original model [10] but Ca occupancies were found to vary significantly and are plotted in Fig. 4. No evidence was found for the reduction in occupancy of the hydroxyl sites in HA, as has been reported to occur in the formation of oxyapatite [14]

Conclusions

The application of Rietveld analyses on diffraction data has now become a very powerful tool in the study of materials. Although a number of previous studies have reported estimated sample compositions from X-ray diffraction studies, the results of Rietveld analyses of XRD data presented here have been able to quantitatively determine the phases present within the samples and also the actual composition of these phases. This is particularly important in accessing the calcium content of the HA phases and determining any potential bioactivity. Heat treatment of calcium deficient HA may be a route to producing novel bi- and triphasic ceramics.

References

[1]   R. F. Ellinger, E. B. Nery and K. L. Lynch: J. Periodont. Restor. Dent. Vol 3 (1986), p 223.

[2]   E. B. Nery, K. K. Lee and S. Czajkowski: J Periodontol. Vol 61 (1990), p 737.

[3]   R. Z. Legeros, E. Nery, G. Daculsi, K. Lynch and B. Kerebel: *"Third World Biomaterials Congress"* (1998), abstract 35.

[4]   R. Z. Legeros and G. Daculsi, in *"CRC Handbook of Bioactive Ceramics"* (CRC Press, Boca Raton, 1990), p 17.

[5]   G. Daculsi, R. Z Legeros, E. Nery, K. Lynch and B. Kerebel: J. Biomed. Mater. Res Vol.23 (1989), p 883.

[6]   E. B. Nery, R. Z. Legeros, K. L. Lynch and J. Kalbfleisch: J Periodontol Vol. 63 (1992), p 729.

[7]   H. M. Rietveld: Acta crystallogr., Vol. 22 (1967), p 151.

[8]   A. C. Larson and R. B. Von Dreele, General Structure Analysis System, Los Alamos National Laboratory, (1994).

[9]   D. G. A. Nelson and J.D.B. Featherstone, Calcified Tissue Int.Vol. 34 (1982), p S69.

[10]   K. Sundarsanan and R. A. Young: Acta Crystallogr B Vol. 24 (1982), p 1938.

[11]   M. Mathew, L. W. Schroeder, B. Dickens, and W. E. Brown: Acta Crystallogr. B Vol. 32 (1992), p 120.

[12]   B. Dickens, L. W Schroeder and W. E Brown: J. Solid State Chem. Vol 10 (1974), p 232.

[13]   W. Fix, H. Heyman and R. Heinke: J Am Ceram Soc. Vol 52 (1969), p 346.

[14]   J.M. Zhou, X.D. Zhang, J.Y Chen, S.X. Zeng and K. Degroot: J. Mater. Sci. Vol 4 (1993), p 83.

Key Engineering Materials Vols. 254-256(2004) pp. 301-304
online at http://www.scientific.net
© *2004 Trans Tech Publications, Switzerland*

Sol-Gel Derived Nano-Coated Coralline Hydroxyapatite for Load Bearing Applications

B. Ben-Nissan[1], A. Milev[1], R. Vago[2], M. Conway[3] and A.Diwan[4]

[1]Department of Chemistry, Materials and Forensic Science University of Technology, Sydney PO Box 123, Broadway, NSW, Australia 2007
[2]Institute for Applied Bio Sciences, Ben-Gurion University of the Negev, Beer Sheva, 84105 Israel

[3]Sydney Eye Hospital, University of Sydney, Clinical Ophthalmology, Sydney, NSW 2000, Australia
[4] St.George Hospital, Orthopaedic Research Centre, UNSW, Sydney, NSW 2052, Australia
b.ben-nissan@uts.edu.au

Keywords: hydrothermal conversion, coral, Hydroxyapatite, sol-gel, mechanical properties, nano-coating

Abstract. Current bone graft materials are mainly produced from coralline hydroxyapatite (HAp). Due to the nature of conversion process, commercial coralline HAp has retained coral or $CaCO_3$ and the structure possesses nanopores within the inter pore trabeculae resulting in high dissolution rates. Under certain conditions these features reduce durability and strength respectively and are not utilised where high structural strength is required. To overcome these limitations, a new-patented coral double-conversion technique has been developed.

The current technique involves two-stage application route where in the first stage complete conversion of coral to pure HAp is achieved. In the second stage a sol-gel derived HAp nano-coating is directly applied to cover the micro and nano-pores within the intra pore material, whilst maintaining the large pores. Biaxial strength was improved due to this unique double treatment. This application is expected to result in enhanced durability and longevity due to monophasic hydroxyapatite structure and strength in the physiological environment.

It is anticipated that this new material can be applied to load bearing bone graft applications where high strength requirements are pertinent.

Introduction

Synthetically derived materials have been used as bone grafts in an attempt to overcome the limitations of autograft and allografts. Coral mineral (aragonite or calcite forms of calcium carbonate) has had considerable success considering its porous structure (which ranges from 150-500μm) is similar to cancellous bone and is one of a limited number of materials that will form chemical bond with bone and soft tissues *in vivo* [1]. Studies indicate that a favourable pore size and microstructural composition are important factors facilitating in-growth of fibrovascular tissue or bone from the host [2]. However unconverted coral (calcium carbonate) is unsuitable for most long term implant purposes due to its very high dissolution and poor longevity and stability. Current commercially available technology using hydrothermal exchange achieves conversion of coral to coralline hydroxyapatite (HAp), termed partially converted HAp. However, fully converted coralline hydroxyapatite has the advantage of retaining a favourable pore size and bioactivity, with improved durability compared to native or partially converted coral, allowing it to remain stable under the harsh conditions encountered in the human body [3]. Unfortunately, the poor mechanical properties (brittle nature, low biaxial stress) of the synthetic bone graft produced limit their use in load bearing applications (long bones, spine and mandible and many more). This significantly influences strategies in patient care including surgical technique, recovery, rehabilitation and cost of such procedures.

Researchers first evaluated coral as a potential bone graft substitute for humans in 1974 [4].   More recent studies have demonstrated that coral can be converted into monophasic hydroxyapatite by a low temperature hydrothermal process, providing a bioactive material similar to the mineral content of teeth and bone [5]. The hydroxyapatite produced from coral as the starting material, has been characterized using different coral sources and the pore sizes, for the synthesized products, were measured using optical and electron microscopy and porosimeter techniques [5-7]. The pore size and distribution -specifically their interpore micro and nanopores- and 3-D structure is an important parameter that influences the strength and the rate of new bone growth [2,7].

The present investigation was carried out to determine whether the porous structure of Australian coral after double conversion treatment was in fact strong enough to replace bone in an implant, and if so, will it be able to replace a load bearing bone whilst adequately interacting with its surrounding environment.

Materials and Methods

Australia has rich variety of corals, however, their potential for use in implants has not been adequately investigated. In this study, a new species of Australian coral was used for hydroxyapatite conversion. This particular species, similar to the genus *Goniopora* was chosen because of its favourable strength and structural properties [3].

Hydrothermal treatment: The coral was obtained from the Australian Great Barrier Reef and were supplied by the Australian Marine Institute, Townsville, Qld. The coral was shaped in the form of a block and was treated with boiling water and 5% NaClO solution.

Hydrothermal conversion was carried out in a Parr reactor (Parr Instrument Company, USA) with a Teflon liner at 250°C and 3.8MPa pressure with excess $(NH_4)_2HPO_4$. The product was then washed and dried at 70°C.

Sol-Gel coating: In our previous work on the sol-gel processes using alkoxide or metal salt precursors it has been demonstrated that a period up to 48h at 70°C is required for the production of monophasic hydroxyapatite [8]. The precursor solution was formed by mixing calcium diethoxide $[Ca(OEt)_2]$ or calcium acetate mono hydrate $[Ca(OAc)_2.H_2O]$ with ethylene glycol and acetic acid 1:1 mol ratio glycol/acetic acid. A stoichiometric amount (Ca/P ration 1.67) of diethyl hydrogen-phosphonate $[(C_2H_5O)_2P(O)H]$ was added after the complete dissolution of calcium precursor. The resulting clear sol was aged at 70°C up to 48h in a closed vial. Coatings were formed using these solutions followed by subsequent heat treatments. If the solution was used earlier than 24h at that temperature the presence of calcium oxide in addition to hydroxyapatite phase was detected. Detailed study of the reactions that occur at that temperature revealed a number of compounds that have been formed. The application of mainly solution-NMR techniques revealed that these compounds were glycol mono- and diacetate, ethyl acetate, ethanol. Most importantly three major phosphorous containing compounds have been identified. All of them were phosphonates having direct P-H bonds [9].

Characterisation: Mechanical testing involved a standard 4-point bend test according to ASTM C1161 to measure the flexural strength and flexural modulus of the coral. Comparative biaxial strength tests were also carried out. Fracture surfaces were then viewed using stereo optical microscopy as well as scanning electron microscopy. SEM was performed on a LEO-Supra55VP. Samples to be analysed by X-Ray diffraction (XRD) with the Siemens D-5000 (Karlsruhe, Germany) were prepared by crushing a small amount of each coral (natural, converted and converted and coated) using a mortar and pestle. The XRD scan was carried out from 20.0 to 60.0 in 0.020 steps at a step time of 2.0s. Thermogravimetric and differential analyses (TGA/DTA) were performed on a TA Instruments SDT 2960 simultaneous DTA-TGA at a heating rate of 10°C/min.

Results

Characterisation studies of the natural and converted corals using XRD, SEM, DTA/TGA, NMR and Raman spectroscopy were reported in previous publications [3,5,8].

Mechanical Testing: Mechanical testing was performed in directions longitudinal and tangential to the net direction of pore travel. Due to the nature of the coral skeleton, pores do not follow a straight path, but curve and flow depending on the corals position in the reef with respect to other polyps and natural structures, so a specimen with uniform pore direction was sometimes difficult to achieve. A summary of the mechanical results is given in Table 1 and a comparison with natural bone is made.

Table 1. Flexure strength and E_{flex} results of the 4- point bend test in both tangential and longitudinal directions for the coral investigated.

		Flexure Modulus [GPa]	Flexure Strength [MPa]
	Tangential	20 ± 5	6.7 ± 0.7
Coral	Longitudinal	11 ± 3	4.1 ± 0.5
Bone	Cancellous	0.3 ± 0.2	3.1 ± 1.7
	Cortical	$7 - 18$	$40 - 160$

The specimens utilised displayed relatively consistent pore direction and showed distinct directional properties similar to that of wood or bone. In comparison to bone, which has a natural collagen matrix providing toughness, coral is a brittle material and has less energy absorption capacity than bone. Coral displays a high E_{flex} of ~14GPa due to the extremely low flexibility. It is a brittle material and the rough surface and micro interpores allows easy propagation of cracks.

The SEM images of the new species used shows a much larger pore size and greater pore interconnectivity giving it a lower overall porosity and hence a greater strength than other corals investigated. The pore size of the coral investigated in the new species is between 250-500μm with interconnections of about 150μm. In detailed analysis of the interpore structure shows very porous (nano and micro) interconected structure (Fig.1).

The effectiveness of coralline hydroxyapatite as a bone graft substitute depends on its strength. Hydrothermal conversion retains the structure of the coral while altering the chemical composition from aragonite to hydroxyapatite. It is desirable to use a coral species with the highest possible strength and a large pore size to enable bone ingrowth to occur quickly. It was reported by Elsinger *et al.* [10] that coralline hydroxyapatite bone graft substitutes surpassed cancellous bone strength after 6 months due to infiltration of bone making the graft an effective hybrid material. This is a desirable property because it gives excellent early structural support.

This currently patented technique [11], involves two-stage production route. In the first stage a complete conversion of coral to monophasic hydroxyapatite is achieved. In the second stage a sol-gel derived HAp nano-coating is directly applied to cover the micro and nano-pores within the intra pore trabeculae material (Fig.2), whilst maintaining the useful large pores. In this current study it is shown that the biaxial strength could be improved due to this unique double treatment (Table 2).

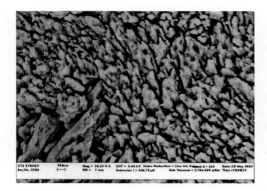

Fig.1: Detailed surface of the natural coral.   Fig. 2: Surface of the converted and HAp coated coral.

Table 2. Biaxial strength of coral, hydrothermally converted, and converted and HAp coated coral.

	Biaxial Strength (MPa)
Coral	6.5 ±2.9
Converted Coral	7.6 ± 1.4
Converted and Hap Coated Coral	13.3 ± 6.5

This new application is expected to generate, enhanced durability and strength in the physiological environment through the elimination of intra-trabecular pores and high stability due to complete conversion to monophasic hydroxyapatite. It is anticipated that this new material can be applied to bone graft applications where high strength requirements and longevity are pertinent.

Acknowledgments
The authors would like to thank Dr. R. Wuhrer from the UTS-MAU for his contribution during the SEM studies. Financial support provided to Prof. Ben-Nissan by the Japan Society for the Promotion of Science (JSPS) under the JSPS/Australian Academy of Science Fellowship Program for Research in Japan scheme is also gratefully acknowledged.

References
[1] B. Ben-Nissan and G. Pezzotti: J. Ceram. Soc. of Japan, 110 [7] (2002), p. 601.

[2] J.H. Kühne, R. Bartl, B. Frisch, C. Hammer, V. Jansson, M. Zimmer, Acta Orthopaedia Scandinavia; 65 (3) (1994), p. 246.

[3] J. Hu, R. Fraser, J. Russell, R. Vago and B. Ben-Nissan, J. Mater. Sci. Technol., 16, (6) (2000), p. 591.

[4] D.M. Roy, S.K. Linnehan, Nature; 247 (1974), p. 220.

[5] J. Hu, J.J. Russell, B. Ben-Nissan and R. Vago, J. Mater. Sci. Letters, 20 (1) (2001), p. 85.

[6] J. Pena, R. LeGeros, R. Rohanizadeh, J. LeGeros. Key Eng. Mater. Vol., 192-195 (2001), p. 267.

[7] R.A. White, E.W. White, R.J. Nelson, Biomat, Med. Dev, Art. Org, 7(1) (1979), p. 127.

[8] B. Ben-Nissan, D. D Green, G.S.K.Kannangara, C.S. Chai, A. Milev. J. Sol-Gel Scien. Techn., 21 (2001), p. 27.

[9] A. Milev, G.S.K. Kananngara, and B. Ben-Nissan, in *Bioceramics 15*, edited by B. Ben-Nissan, D. Sher, W. Walsh (Trans. Tech. Publications, Switzerland, 2003) p. 481.

[10] E. C. Elsinger and L. Lear, Journal of Foot and Ankle Surgery, 30 (1996), p. 396.

[11] B. Ben-Nissan *et al.* PCT -WO 02/40398 A1 (May 23, 2002)

Key Engineering Materials Vols. 254-256(2004) pp. 305-308
online at http://www.scientific.net
© *2004 Trans Tech Publications, Switzerland*

Hardening and Hydroxyapatite Formation of Bioactive Glass and Glass-Ceramic Cement

Cheol Y. Kim and Hyong Bong Lim

Department of Ceramic Engineering, Inha University 253 Yonghyun-dong, Nam-ku, Incheon, 402-751, Korea, E-mail: cheolkim@inha.ac.kr

Keywords: Bone cement, glass, glass-ceramics, DCPD, hydroxyapatite,

Abstract. Several glasses and glass-ceramics have been prepared to be used as bioactive cements. The self-setting of the cement when mixed with phosphoric acid is due to the development of dicalcium phosphate dihydrate($CaHPO_4 \cdot 2H_2O$: DCPD). The setting time of glass cements was much longer than that of the glass-ceramic cements. The slower setting in the glass cement is due to the slower leaching of Ca^{2+} ions from the glass. The DCPD transformed into a hydoxyapatite when the cement was soaked in simulated body fluid (SBF). No hydroxyapatite was observed when the glass-ceramics containing apatite crystal (33P9C) was soaked in SBF even for 4 weeks. The compressive strength of the cement is strongly dependent on the conversion behavior of DCPD to hydroxyapatite

Introduction

Ceramic bone cement can be classified as crystalline calcium phosphate cements (CPCs) and glass cements in the system of CaO-SiO_2-P_2O_5 glass [1]. Since Brown and Chow reported the first CPCs in 1987 [2], many different calcium and phosphate-containing compounds have been investigated as potential CPC materials. CPCs are excellent bone substitutes: they are biocompatible, resorbable, and osteoconductive. However their low strength and brittle nature limit their potential use to non-load-bearing applications [3].

Bioactive bone cement based on CaO-SiO_2-P_2O_5 glass was introduced by Kokubo [4]. He mixed the glass powder with ammonium phosphate solution to make a cement. When implanted, however, the glass cement releases lots of ammonia gas, and it might be harmful to living cells [1].

In the present study, two kinds of glasses with different compositions and the corresponding glass-ceramics are prepared. The primary objective of this study is to examine the behavior of hydroxyapatite formation on the hardened cement when soaked in a simulated body fluid (SBF). Hardening time and compressive strength depending on glass composition and hydroxyapatite formation are also studied.

Materials and Methods

The compositions of the glasses for this study are shown in Table 1. Appropriate amounts of raw materials, from the reagent grades of SiO_2, $CaCO_3$, and H_3PO_4, were weighed and mixed in gyroblender for 2h. Then, each glass batch was melted in Pt-Rh crucible at 1500°C for 4 h. The glass melt was quenched on stainless steel and milled into a size of less than 40μm in a planetary agate mill. Crystallized cement was obtained by heat-treatment of the glass powder at 950°C for 4h. Different crystalline phases are observed depending on glass composition as shown in Table 1.

0.5 gram of powder was mixed with 0.5 cm³ of phosphoric acid for 30sec, and the paste was molded in a syringe under vacuum. The hardened cement was soaked in SBF at 37°C for various times. The initial and final setting times for the cement paste were determined by Gilmour needles method. Needles with 113.9 gram and 453.4 gram of weight for initial and final setting times, respectively, were applied on the mixed cement paste.

The phase changes of the bioactive cement during setting and reacting with SBF were examined by thin film XRD and SEM. The compressive strength of the cement measured by a universal tester.

Results and Discussion
Crystallization of the Glass and Hardening of the Cements
When the glass powder was fired at 950°C for 4 hours, each glass has different crystalline phases. 45P2G glass crystallized into α-wollastonite, and 33P9G glass into oxyapatite.

Initial and final setting times were measured by Gilmour needles method. The results are shown in Table 1. Setting time of the glass cement was longer than that of glass-ceramic cement. The setting time of the bioactive cements is closely related to the forming behavior of a dicalcium phosphate dihydrate(CaHPO$_4$·2H$_2$O : DCPD). The DCPD is formed by reacting phosphoric acid with Ca^{2+} ion, which is provided from the cement. It is known that α-wollastonite reacts with acid and releases Ca^{2+}ions easily, and this shortens the setting time in the glass-ceramic cements. The glass-ceramics 33P9C showed longer setting time than 45P2C because of the oxyapatite crystal, which delayed the DCPD formation.

Table 1. Compositions and setting time of glasses and glass-ceramics.

Smple	Composition [mol.%]			Setting Time [min.]	
	CaO	SiO$_2$	P$_2$O$_5$	Initial Time	Final Time
45P2G	53.2	44.7	2.1	4	15
33P9G	58.2	32.6	9.2	1.5	7
45P2C	cystalline phase: α-wollastonite			1	3
33P9C	cystalline phase: apatite			2.5	5.5

Hydroxyapatite Formation on the Hardened Cements.
Hardened 45P2G and 45P2C were immersed in SBF for various times, and the results are shown in Fig.1 (a) and (b). DCPD crystal developed on both samples just after setting, and the DCPD converted into hydroxyapatite after reaction in SBF. For the conversion, it took 9h for the glass sample (45P2G), but 24h for the glass-ceramic sample (45P2C), which has α-wollastonite phase. Fig.1(b) shows that the α-wollastonite crystal was unchanged even after 1 week of reaction [5,6].

Next, 33P9G and 33P9C were reacted in SBF for various times, and the results are shown in Fig.1 (c) and (d). All DCPD disappeared in 9h of reaction for the glass sample (33P9G) and hydroxyapatite replaced the DCPD. The DCPD and apatite crystals, however, stays unchanged even after 4 weeks of reaction, and no hydroxyapatite was observed for the glass-ceramic sample (33P9C) [5]..

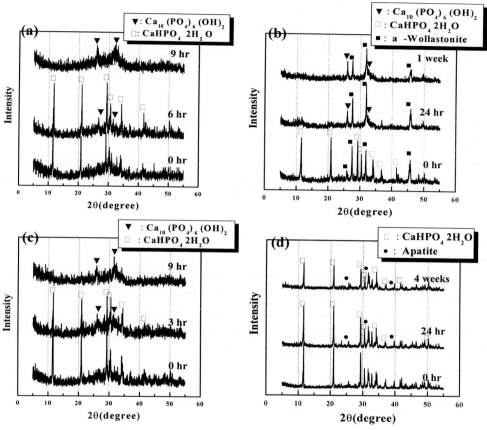

Fig. 1. XRD patterns for hardened glasses and glasses-ceramics in SBF for various times. (a) 45P2G, (b) 45P2C, (c) 33P9G, (d) 33P9C.

Compressive Strength

Compressive strength for all samples was measured after reaction in SBF for various times, and the results are shown in Fig 2. The compressive strength increased with reaction time, and became almost constant after 3 days of reaction. 45P2C shows higher compressive strength than 45P2G. The compressive strength of the cement is closely related to the transformation behavior of DCPD to hydroxyapatite, and this transformation was easier for 45P2G than for 45P2C

The hydroxyapatite formation rate in 33P9G was very fast and it shows high compressive strength. The corresponding glass-ceramic cement (33P9C), which contains oxyapatite crystals, showed very low compressive strength. No conversion of DCPD into hydroxyapatite in this sample was observed even after one month of reaction. This indicates that the formation of hydroxyapatite plays an important role to enhance the strength.

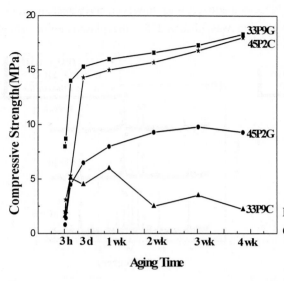

Fig. 2. Compressive strength of the Cement as a function of soaking in SBF.

Conclusion

Setting time is faster in glass-ceramic cement than in glass cement, which indicates that the formation of DCPD is faster in glass-ceramic cement. When reacted in SBF, the DCPD transforms into hydroxyapatite in the cements. No hydroxyapatite development is observed for the sample containing apatite (33P9C) even for longer exposure times. Higher compressive strength is observed in the cements which developed hydroxyapatite.

References

[1] M.Bohner: Injury, Int. J. Care Injured 31(2000) S-D37-47

[2] C.Chow, Shozo Takagi: J. Res. Natl. Stand. Technol., Vol. 106 (2001), 1029-1033

[3] E.Charriere, S.Terrazzoni, C.Pittet, Ph. Mordasini: Biomaterials, Vol. 22 (2001) 2937-2945

[4] Tadashi Kokubo, Sataru Yoshihara: J. Am. Ceram. Soc., Vol 74(7) (1991) 1739-1741

[5] Mukesh Kumar, Jing Xie, Krishnan Chittur: Biomaterials, Vol.20 (1999) 1389-1399

[6] P.N. De Aza, Z.B. Luklinska, M.R. Anseau: J. Dentistry, Vol.27(1999) 107-113

III. COATINGS AND SURFACE MODIFICATIONS

Key Engineering Materials Vols. 254-256(2004) pp. 311-314
online at http://www.scientific.net
© 2004 Trans Tech Publications, Switzerland

Influence of Acetylation on *In Vitro* Chitosan Membrane Biomineralization

Marisa Masumi Beppu[1] and Cassiano Gomes Aimoli[2]

[1] Faculdade de Engenharia Química, Universidade Estadual de Campinas, CP 6066, Campinas SP 13083-970, Brazil, beppu@feq.unicamp.br

[2] Faculdade de Engenharia Química, Universidade Estadual de Campinas, CP 6066, Campinas SP 13083-970, Brazil, c008313@dac.unicamp.br

Keywords: Chitin, chitosan, biomineralization, calcification, acetylation.

Abstract. In a previous work of this research group, we studied the *in vitro* calcification of dense and porous chitosan membranes. Chemical modifications had been promoted, but further investigations were needed to better understand the role of some chemical groups in the process of calcification. In the present study, we proposed the acetylation of the already-molded chitosan membranes, producing a "pseudo-chitin", to be used in calcification. The acetylated chitosan was submitted to mineralization by soaking the membranes in simulated body fluids (SBF) with 1x and 1.5x the concentration of ions found in human serum, for 7 days at 36.5°C. Morphological characterization was performed using SEM and compositional analyses were done using SEM-EDX and FTIR-ATR techniques. The results showed that acetyl groups induce calcification, forming deposits that present a Ca:P ratio different of those formed on pristine chitosan.

Introduction

Chitosan is recently being appointed in many studies as a promising biomaterial. A sub-area of application that deserves attention is the use of chitosan as organic matrix to produce biocomposites through biomimetic processes.

The biomimetic process is an alternative of ceramic synthesis that avoids usual difficulties found in sol-gel processes such as particle aggregation. In these processes, the inorganic phase is obtained under an organic matrix control, which regulates the shape, size and orientation of deposits and hence determines the structural and mechanical characteristics of the final material [1]. For application of these processes in the biomaterial field, there is still a need to better understand how to control mineralization as it may be wanted in some uses, such as in ostheogenic implants, or completely undesired in others, as it is the major reason for some medical device failure, such as in cardiac devices.

This laboratory has been studying chitosan as a potential biomaterial and for bioseparation processes since 1997. Previous studies have appointed that some chemical treatments, such as crosslinking with glutaraldehyde, PAA adsorption, etc, on chitosan changed the way it induces *in vitro* calcification [2-4].

In this work we expanded the investigation, proposing the acetylation of the already-molded chitosan membranes, producing a "pseudo-chitin", to be used in calcification. This modification was inspired in the fact that chitin, along with calcium carbonate, produces nacre, one of the most resistant biocomposites found in nature. This treatment would allow increasing the n-acetyl groups of organic matrix without loosing the moldability of chitosan. The membranes underwent a calcification process and the deposits were analyzed chemical and morphologically.

The results would allow us to compare the differences between a highly acetylated chitosan and a deacetylated chitosan or, in other words, the influence between amino and acetyl groups in biomineratization.

Methods
Dense and porous membranes were obtained by coagulation of an acidic chitosan solution, using a
basic solution. These membranes were chemically modified by immersion into a 0.6% methanol-
acetic anhydride (v/v) solution for 24h [5]. The membranes were then washed and kept in Milli-Q
water at 4°C. After this process, the membranes presented a reduction in area and solubility in
acidic solutions. Potentiometric titration and thermal analyses (DSC and TGA) were conducted to
check the blockage of amino groups.

In vitro calcification experiments were done as follows: the substrates were soaked in
simulated body fluids (SBF) with 1x and 1.5x the concentration of ions found in human serum,
during 7 days at 36.5°C. Effects of pH, calcifying medium composition and physical (porosity,
permeability) and chemical characteristics of substrates on calcification quality and degree were
observed.

Morphological characterization was performed using SEM and compositional analyses were
done using SEM-EDX on deposits and FTIR-ATR techniques on chitosan films.

Results and Discussion
It was observed from titration results that acetylation was promoted extensively on chitosan
membranes. Potentiometric titration showed that pristine chitosan presented a deacetylation degree
of 81% and that 24h-acetylated chitosan (pseudo-chitin) presented a near-zero value for
deacetylation.

FTIR-ATR spectra in Fig.1 indicate that the chains on the membrane surface are acetylated
within minutes of reaction. The reaction causes a retraction of membranes probably due to the
change in solubility.

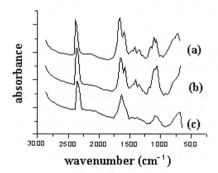

Fig. 1- FTI-ATR spectra of (a) 24-hour-acetylated chitosan (b) 2-min-acetylated chitosan
(using 0,03 ml acetic anhydride in methanol) and (c) natural chitosan.

24h-acetylated chitosan (pseudo-chitin) underwent calcification. After a 7-day-period,
extensive calcification could be observed and the morphology of deposits was different from those
observed on precipitates obtained by sol-gel process, without the presence of an organic matrix.

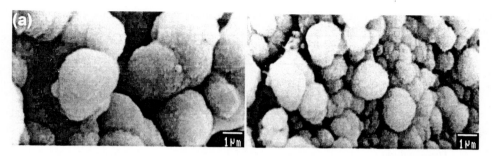

Fig. 2: SEM micrographs of porous acetylated chitosan membranes after calcification at (a) pH 7.4 and (b) pH 7.8.

This fact confirms the influence of matrix on the mineralization process. The pH also influenced the morphology: at lower pH, the sphere-like precipitate units showed to be bigger, which suggests that at pH 7.4, growth would be more favoured than nucleation when compared to pH 7.8 (Fig.2).

The size of calcification clusters showed to be bigger on acetylated than on pristine chitosan. This fact could be clearly observed on dense membranes (Fig. 3).

Fig. 3: SEM micrographs of dense chitosan membranes after calcification at pH 7.4 (a) natural and (b) acetylated.

EDX analyses of calcification deposits were conducted and the results are depicted on table 1. The Ca:P ratio showed to be influenced by acetylation only on dense membranes. This fact is associated with the importance of micromechanical characteristic of their surfaces: flat membranes have different influence on calcification when compared to porous membranes with higher surface area.

However, for definition on the mechanisms involved in both kinds of acetylated membranes, further studies are being conducted in this laboratory.

Chemical and morphological analyses showed that the nature and intensity of calcification depended on: *1)* porosity and surface irregularities of membranes; *2)* acetylation that introduces carboxyl groups that adsorb calcium ion [6] and *3)* Composition and pH of fluid used for calcification experiments.

These results show the influence from both chemical and micromechanical characteristics of surface on *in vitro* calcification.

Table 1: Ca:P molar ratios determined by EDX on calcifications formed on chemically modified chitosan membrane samples.

Sample	% calcium atoms	% phosphorus atoms	Description of analysed deposit	Ca:P molar ratio
Dense acetylated membrane pH 7.4	51.72	42.70	cluster	1.21
Dense acetylated membrane pH 7.8	51.62	41.74	cluster	1.24
Dense natural membrane pH 7.4	51.29	24.37	cluster	2.10
Dense natural membrane pH 7.8	62.99	30.64	cluster	2.06
Porous acetylated membrane pH 7.4	29.14	13.10	cluster	2.22
Porous acetylated membrane pH 7.8	48.21	16.96	cluster	2.84
Porous natural membrane pH 7.4	36.66	16.94	cluster	2.16
Porous natural membrane pH 7.8	45.79	19.00	cluster	2.41

Conclusion

Acetylation of chitosan promoted a different kind of *in vitro* calcification when compared to pristine chitosan, mainly on dense membranes. The presence of carboxyl groups may increase the ionotropic effect of groups on substrate surface. However, further investigations shall be done in order to check whether the calcification follows a heterogeneous nucleation process or a homogeneous nucleation followed by surface anchorage. At this point, there is a noticeable influence of membrane micromechanics and solution pH on calcification morphology.

References

[1] P.Calvert and P. Rieke: Chem. Mater. Vol 8 (1996) p.1715.

[2] M.M. Beppu and C.C. Santana: Key Eng. Mater. Vol 125-129 (2000). p.34.

[3] M.M. Beppu and C.C. Santana: Mater. Sci. Vol 5 (2002), p.47.

[4] M.M. Beppu: *Estudo da calcificação in vitro da quitosana* (PhD thesis – FEQ – State University of Campinas. Brazil 1999)

[5] S. Hirano et al.: *Industrial Polysaccharides* (Elsevir Science Publishers. Amsterdam, 1987)

[6] S. Zhang and K.E. Gonsalves: J. Appl. Polym. Sci. Vol 56 (1995) p.687.

Key Engineering Materials Vols. 254-256(2004) pp. 315-318
online at http://www.scientific.net

Dissolution Behaviors of Thermal Sprayed Calcium Phosphate Splats in Simulated Body Fluid

K.A. Khor[1], H. Li[1] and P. Cheang[2]

[1] School of Mechanical and Production Engineering, Nanyang Technological University, Nanyang Avenue, Singapore 639798, mkakhor@ntu.edu.sg

[2] School of Materials Engineering, Nanyang Technological University, Nanyang Avenue, Singapore 639798

Keywords: Dissolution; Calcium phosphate; Splat; *In vitro*; Thermal spray

Abstract. This study aims at revealing the dissolution behavior of individual calcium phosphate (CP) splats after incubation in simulated body fluid (SBF) over various periods. The CP splats were prepared using both plasma spraying and high velocity oxy-fuel (HVOF) technique. The *in vitro* dissolution rates (defined here as the ratio of dissolved area to overall area of a splat) together with the morphological changes of the splats indicated that the extent of hydroxyapatite (HA) transformation to other CP during plasma spraying was more extensive than during HVOF spraying. It was found that 2 hours of *in vitro* incubation resulted in complete dissolution of the plasma sprayed CP splats; whilst the dissolution rate of the HVOF sprayed CP splats significantly depended on the melt states. For fully melted HVOF splats, complete dissolution occurred after 4 hours' incubation. The present results further confirmed that HA decomposition predominantly occurred within the melted part of the sprayed particles. In other words, there could be a relationship between melt states of HA particles during the spraying and phase composition of the resultant splats. The temperatures of the sprayed HA particles were also measured before their impingement on the titanium alloy substrates prior to forming splats.

Introduction

Calcium phosphate (CP) coatings deposited on titanium alloy implants have shown promising effects on rapid bone remodeling and suitable functional life in orthopedic and dental applications. It has also been recognized that precipitation of a bone-like apatite layer on the CP matrix *in vitro* was directly related to dissolution of the phases within the coating. In order to understand the *in vitro* behavior of the coatings, the clarification of their dissolution behavior is essentially required. To date, it was claimed that different phases within the CP family exhibited distinctive different structures, and thus, could lead to varying biological responses in simulated body fluid (SBF) [1]. The dissolution rate of monophasic CP ceramics increases in the following order: HA<CDHA<OHA<β-TCP<α-TCP<TTCP [1]. Furthermore, the influence of crystallinity of HA coatings has been clarified through *in vitro* test by immersing HA coating with miscellaneous amorphous calcium phosphate (ACP) content in SBF [1,2]. However, due to the high temperatures attained by the HA powders during thermal spraying, phase composition of the coatings is heterogeneous, which makes precise characterization of the dissolution elusive. Conversely, thermal sprayed coating composed of a layered structure, which is effectively an accumulation of individual solidified splats. Researchers have extensively conducted the study on thermal sprayed splats [3-5]. Generally, splat formation is an isolated event, which means minor influence would be exerted by the subsequent splat on the phases of the prior deposited splat. Therefore, the overall *in vitro* behavior of a bulk HA coating should be directly related to that of individual HA splats. A good understanding of the *in vitro* behavior of a single HA splat would significantly contribute to the knowledge on dissolution/precipitation mechanism of HA coatings. In the present study, the dissolution behavior of individual CP splats *in vitro* was characterized. The splats were deposited using both plasma spraying and HVOF onto polished Ti-6Al-4V substrates.

Experimental Details

In-house synthesized HA powders (via wet chemical synthesis and spray drying) were utilized as feedstock for both HVOF and plasma spray processes. A fully computerized HV2000 HVOF system (*PRAXAIR, USA*) with a nozzle diameter of 19 mm and d.c. plasma spray system (Model 4500 *PRAXAIR, USA*) were utilized for the splats deposition. The spray parameters of both the HVOF spraying and plasma spraying are listed in Table 1. The HA splats were collected on polished Ti6Al4V plate substrates (through polishing with 1 µm diamond paste). The powders with a wide range of sizes were sprayed to prepare splats with different morphology on the polished substrates. A filter plate was placed between the substrate and flame/arc to facilitate the collection of single HA splats, and, several holes of 1 mm in diameter were drilled on the plate.

Table 1 Spray parameters for the HA splats preparation.

Plasma spray gun	HVOF spray gun
Model: SG-100, Praxair, USA	Model: HV2000, Praxair, USA
Net energy: 12 kW	O_2: 283 l/min
Ar: 30.6 l/min	H_2: 566 l/min
He: 22.6 l/min	Carrier gas (Ar): 19 l/min
Carrier gas (Ar): 10 l/min	Powder feed rate: 8 g/min
Powder feed rate: 8 g/min	Spray distance: 250 mm
Spray distance: 120 mm	

During *in vitro* test, the splats were incubated in the SBF solution for various periods. The substrate on which the HA splats were deposited, and, used for *in vitro* tests was of the dimension $12 \times 12 \times 2$ mm^3 in width, length and width, respectively. The Kokubo SBF (pH = 7.40) [6] was used for the *in vitro* incubation. The solution is composed of 142.0 mM Na$^+$, 5.0 mM K$^+$, 1.5 mM Mg^{2+}, 2.5 mM Ca^{2+}, 147.8 mM Cl$^-$, 4.2 mM HCO$_3^-$, 1.0 mM HPO$_4^{2-}$, and 0.5 mM SO$_4^{2-}$. The *in vitro* test was conducted in a continuously stirred bath containing distilled water with a stable temperature of 37°C. Each sample was incubated in 70 ml SBF contained in a polyethylene bottle. Once the sample was taken out from the solution, it was washed in distilled water and subsequently dried at ambient temperature. Topographical features of the splats were observed by scanning electron microscope (SEM, JEOL JSM-5600LV). The ImagePro image analysis software was used for the quantitative determination of the dissolution rates after incubation of the splats in the SBF. In the present study, in order to further explain the differences in melt state of the powders during HVOF and plasma spraying, the temperatures of the sprayed HA particles were measured using SprayWatch-2i (Oseir, Finland).

Results and Discussion

Figure 1 shows the typical HA splats deposited by plasma spraying and HVOF, which indicates a fully melted state of the plasma sprayed HA particle (Fig. 1a) and partially melted state (Fig. 1b) and fully melted state (Fig. 1c) of HVOF HA particles depending on their particle size.

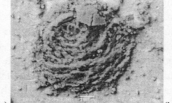

Fig. 1 Typical topographical morphology of plasma sprayed HA splat (a) and HVOF HA splats (b,c), showing the different melt state of the particles depending on particle size and spray methods.

Due predominantly to the significant differences in temperatures of the heating source, it is not surprising that the melt state of the HA powders are different during HVOF and plasma spray, respectively. The surface temperatures of the HA particles during plasma spraying (Fig. 2) were measured using a CCD camera system (SprayWatch-2i, Oseir, Finland). It was found that the temperatures of the HVOF sprayed HA powders in-flight are far lower than 2000°C, which is the lowest value to be detectable using the current CCD camera. Figure 2 also shows the influence of plasma spray parameters on the in-flight

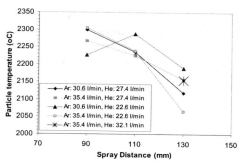

Fig. 2 Temperatures of in-flight HA particles during plasma spraying, showing influence of spray parameters on the particle temperatures (power: 12 kW)

temperatures of the HA powders. The highest temperature attained by the HA powders during plasma spraying is ~ 2300°C. It has been extensively reported that decomposition of HA started at the temperatures above 1000°C; therefore, significant phase transformation from HA to other CP phases would take place during plasma spraying as a result of the high temperatures attained by the powders.

Microstructure changes of the splats induced by the *in vitro* ageing are shown in Fig. 3. It showed that, for the typical plasma sprayed HA splats, the relative dissolved areas increased with the soaking time.

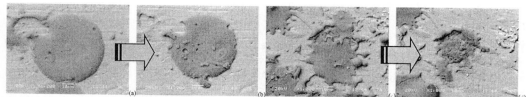

Fig. 3 Typical microstructure changes of plasma sprayed HA splats (a-as-sprayed splat, b-after 1 hr ageing; c-as-sprayed splat, d-after 1.5 hr ageing)

It was found that 2 hours *in vitro* soaking has induced complete dissolution of the plasma sprayed HA splats. However, for the HVOF sprayed splats, the dissolution rate (defined here as the ratio of dissolved area to overall area of a splat) is lower than those deposited by plasma spraying. And, the rate shows a dependence on the melt state of the particles. Figure 4 shows the dissolution rates of both the plasma sprayed and HVOF sprayed HA splats over 72 h *in vitro*. It is noted that partially melted powders show a remarkably low dissolution rate. The *in vitro* dissolution rates (Fig. 4) together with the microstructure changes of the splats (Fig. 3) reflect the various phases within different locations of the splats. It has been determined that the dissolution rate of different CP phases in the SBF was different [1]. The current dissolution results of the splats demonstrate that the peripheral parts of the splats dissolved preferentially into the SBF, which indicate that various phases exist within different parts of individual HA splats. It is noted that even fully melted HA splats deposited through plasma spray (45~75 μm) and HVOF (20~45 μm), respectively, apparently exhibited different dissolution behavior (Fig. 4). The present results suggested that splats with seemingly similar melt state offer differing dissolution rates *in vitro*. This is perceived to be largely contingent on the phases present within the splats. For plasma sprayed HA splats, 2 h of *in vitro* immersion resulted in total dissolution of the phases while a similar outcome took 4 hrs for HVOF sprayed HA splats (fully melted). For the partially melted HVOF splats, precipitation appeared after

24 hrs *in vitro*, and the dissolution reached a relative stable state after 14 hrs, which indicates decidedly the low dissolvability of crystalline HA. The present results further confirmed that HA decomposition occurred mainly within the melted part of the sprayed particles. This is consistent with previous studies on phase transformations of HA at elevated temperatures. [7,8] In other words, there exists a relationship between melt states of HA particles during spraying, and phase composition of the resultant splat. A previous study [9] also show that, compared to crystalline HA, TCP, TTCP, and ACP dissolved preferentially in the SBF.

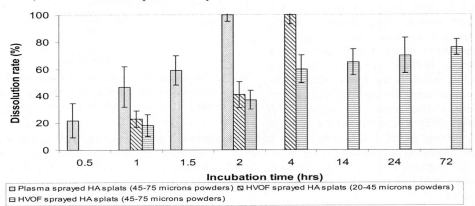

Fig. 4 In vitro dissolution rates of the HA splats prepared by plasma and HVOF spraying different HA powders, showing remarkable influence of spray processes and melt state of the powders on the dissolution rate of resultant splats.

Conclusion

The present study investigated the dissolution behavior of thermal sprayed (plasma and HVOF) individual HA splats, and their dissolution rates were revealed. The results showed that the fully melted HA splats deposited by d.c. plasma spraying dissolved completely after 2 hrs immersion in SBF. Conversely, a fully melted HA splats produced by HVOF spray showed a slower dissolution rate, and their dissolution behavior was influenced significantly by the melt state of the corresponding particles. The present results also suggested that HA decomposition occurred predominantly within the melted portion of the sprayed particles. It is found that the dissolution test can be effectively utilized with other techniques to further investigate the process/ microstructure/phase composition relationships in thermal sprayed HA.

References

[1] P. Ducheyne, S. Radin, L. King: J. Biomed. Mater. Res., 27 (1993), 25-34.
[2] Y. Harada, J.T. Wang, V.A. Doppalapudi, A.A. Willis, M. Jasty, W.H. Harris, M. Nagase, S.R. Goldring: J. Biomed. Mater. Res., 31 (1996), 19-26.
[3] L. Bianchi, A.C. Leger, M. Vardelle, A. Vardelle, P. Fauchais: Thin Solid Films, 305 (1997), 35-47.
[4] G. Montavon, S. Sampath, C.C. Berndt, H. Herman, C. Coddet: J. Therm. Spray Technol., 4 (1995), 67-74.
[5] P. Gougeon, C. Moreau: J. Therm. Spray Technol., 10 (2001), 76-82.
[6] T. Kokubo, H. Kushitani, S. Sakka, T. Kitsugi, T. Yamamuro: J. Biomed. Mater. Res., 24(1990), 721-724.
[7] C. Liao, F. Lin, K. Chen, J. Sun: Biomaterials, 20 (1999), 1807-1813.
[8] J. Zhou, X. Zhang, J. Chen, S. Zeng, K. de Groot: J. Mater. Sci. Mater. Med., 4 (1993), 83-85.
[9] K.A. Khor, H. Li, P. Cheang, S.Y. Boey: Biomaterials, 24 (2003), 723-735.

Key Engineering Materials Vols. 254-256(2004) pp. 319-322
online at http://www.scientific.net
© *2004 Trans Tech Publications, Switzerland*

Confocal Microscopy Characterization of *In Vitro* tests of Controlled Atmosphere Plasma Spraying Hydroxyapatite (CAPS-HA) Coatings

K.A. Khor[1], M. Espanol Pons[1], G. Bertran[2], N. Llorca[2], M. Jeandin[3]
and V. Guipont[3]

[1] School of Mechanical & Production Engineering, Nanyang Technological University, 50 Nanyang Avenue, 639798 SINGAPORE, mkakhor@ntu.edu.sg, p43690802w@ntu.edu.sg

[2] Universitat de Barcelona. Metallurgical and Chemical Engineering Department, Barcelona E-08028 SPAIN, llorca@material.qui.ub.es, bertran@material.qui.ub.es

[3] Center for Plasma Processing (C2P), École Nationale Supérieure des Mines de Paris, Evry, FRANCE, vincent.guipont@mat.ensmp.fr, michel.jeandin@ensmp.fr

Keywords: *In vitro* test, CAPS coatings, Confocal Laser Scanning Microscope, interconnected defects, fluorescence mode, prosthesis, bone growth.

Abstract. This paper aims to introduce the confocal microscope in its fluorescent mode of operation, as a powerful tool able to investigate how the interconnected defects networks existing in as-sprayed plasma coatings evolve upon immersion in simulated body fluid (SBF). The Controlled Atmosphere Plasma Spraying (CAPS) system has been used to produce a set of coatings with tailored composition and density. This has been accomplished varying the chamber pressure (100, 150, 200 and 250 KPa) and the atmosphere (air and argon) surrounding the gun during the spraying process. For the characterization of the in vitro CAPS-HA coatings the conventional Scanning Electron Microscope (SEM) technique has been used in addition to the Confocal Laser Scanning Microscope (CLSM).

Introduction

The success in the clinical field of biomedical prosthesis with plasma-sprayed HA coatings on metal implants depends on the quality of the coatings. Detailed study of the coating's properties and its microstructure can be a good indicator of subsequent behavior in the body. Many papers dealt with the characterization of plasma-sprayed HA coatings, and despite all the effort being put in this field there still remain some important aspects of the coatings which have not yet been addressed [1].

The interconnected defects network is for instance one of them. Plasma sprayed coatings are well known for their anisotropic structure arising from the building up process of solidified droplets. Pores formed depend on the type of powder, and the degree of melting of each particle upon impinging the substrate. Microcracks arise too, as a mechanism for relieving residual stresses. From the mechanical point of view, when high strength coatings are needed, the presence of these defects may be detrimental. However, as HA has the ability of bonding directly to bone (bioactivity) [2] as well as to encourage bone already being formed to adhere to its surface (osteoconductivity) [3], the presence of an interconnected network of defects (when pores and microcracks are connected together) in the coating will promote greater fixation of the new bone to the prosthesis. Nevertheless one of the problems that can be sought is that if defects interconnection reaches the substrate there would arise problems of metal-ion release from the metallic implant. Hence the importance of studying the evolution of the coating's defects networks when immersed in simulated body fluid at different duration.

Experimental Materials and Procedures

The Controlled Atmosphere Plasma Spraying (CAPS) system has a 18 m³ chamber which operates in a controlled atmosphere of either air or argon, or nitrogen from 0.1 KPa to 350 KPa. A conventional 55 kW DC plasma gun with F4-MB nozzle and 6 mm anode diameter was used to spray the powder. The experiments were carried out at chamber pressures of 100, 150, 200 and 250 KPa in an argon atmosphere [4]. Detailed information on the spray conditions and the raw powder that have been used can be found in a previous paper [4]. For the *in vitro* test, SBF has been prepared following Kokubo's protocol [5] and the samples have been separately immersed in the solution keeping constant the ratio of surface coating to the SBF volume. A water bath has been used to keep the temperature constant at 36.5°C and different samples have been removed at 2, 4, 7, 14, 21 and 28 days intervals for analysis. The calcium ion concentration from the solutions has been measured with the Inductive Coupled Plasma-Atomic Emission Spectrometer (ICP-AES) Perkin Elmer 400. Rietveld analysis has been applied using Rietquan Quantitative Analysis software package to determine the amount of the different phases and the amorphous content on the coatings. The coatings microstructure has been evaluated with the JEOL JSM-5600 LV Scanning Electron Microscope (SEM) and the defects interconnectivity has been studied with the Leica TCS4D Confocal Laser Scanning Microscope (CSLM) in its fluorescent mode of operation. This has been accomplished impregnating the samples with a low viscosity resin (Spurr's kit from Fluka, Switzerland and Sigma, St. Louis, USA), which contained a fluorescent dye under vacuum (10^{-2} Pa) to ensure maximum resin penetration [6].

Results and Discussion

To understand the *in vitro* behavior of CAPS-HA coatings it should be kept in mind that coatings sprayed at high pressures (150, 200 and 250 KPa versus 100 KPa) renders coatings with increased percentages in the amount of amorphous phase and other crystalline calcium phosphate phases and also, to a marked decrease in the percentage of internal defects as it has been evaluated in a previous paper. This occurs from concentration of the plasma energy within a smaller volume at high pressure thus, improving the heat transfer ability from the plasma to the particles [8]. When post spray heat-treatment (e.g. 800°C, 1 h) is carried out in the CAPS-HA coatings the coating becomes fully crystalline with only HA phase and new microcracks, mainly oriented perpendicular to the coating's surface, arises in the coatings as the mechanism for relieving the residual internal stresses (being more important in coatings sprayed at 200 and 250 kPa). Cracks randomly oriented are also noticeable and these are attributed to volume changes due to phase transformation during the HT process. The typical SEM microstructure of CAPS as sprayed and HT coating is shown in Fig.1.

Fig.1, SEM micrographs for the as-sprayed and post heat-treated CAPS coatings sprayed at 150kPa.

From these images it is impossible to know whether the cracks and porosity that can be distinguished are interconnected or not, and, consequently, it is not possible to assess either the presence or the complexity degree of an interconnected network of defects within the coating. This is of great importance for two main reasons: First, as mentioned already in the introduction, due to the presence of interconnected defects in the coating there is the possibility of substrate corrosion by the physiological fluid. Second, as reported by other authors [8], the generation of many other calcium phosphate phases in as-sprayed coatings, which are more soluble than HA in physiological fluid, can be responsible of delamination of the coating because they are located preferentially at the coating-substrate interface. However, if precipitation of new bone takes place fast enough, depending on the coating composition, this can be impeded or at least minimized, especially in the first instance. Therefore, the importance of following the mechanisms aforementioned implies the necessity of looking for a technique, which will be able to assess the presence and changes of the interconnected defects in this kind of samples, especially during *in vitro* testing. The Confocal Scanning Laser Microscopy (CSLM) in its fluorescent mode was used in this application. As reported previously, this technique enables the possibility of studying the interconnected defects of thermal sprayed coatings by impregnating the coatings with a low viscosity resin mixed with a fluorescent dye under vacuum [6].

Immersion in SBF of all these coatings, as-sprayed and post-spray HT CAPS-HA samples sprayed at the different conditions resulted in diverse behavior. Whereas all as-sprayed coatings, irrespective of the spraying condition, yielded precipitation of a poorly crystalline apatite layer that is easily detected in the SEM micrographs (and assessed as well through X-ray diffraction), none of the microstructures for the heat-treated samples apparently changed after immersion in SBF. However, analysis of the calcium ion concentration in the solutions indicated that an uptake of calcium ions from the solution took place throughout the in vitro test. The calcium ion concentration was always below 100 ppm, which was the original calcium concentration in the SBF solution. The answer to this discrepancy has been found through the confocal studies: Fig.2 shows the confocal images of two sets of coatings, as-sprayed and post spray heat-treated, sprayed at 150 kPa before and after being immersed in SBF at different immersion periods.

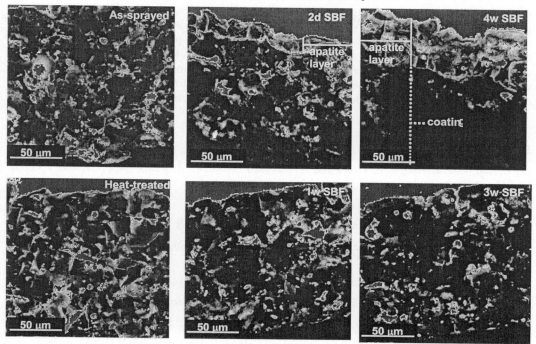

Fig.2, Confocal images of the as-sprayed coating cross-section (on the top) and post heat-treated coatings before and after being immersed in SBF.

Confocal images of heat-treated samples shows that as the immersion time increases the fluorescent intensity decreases, meaning that the coating's defects are being filled up, and there is no space left for the resin to penetrate. Thus, despite the SEM images not showing any changes after the *in vitro* test, the confocal study does reflect traces of the calcium ions. All the microcracks and pores in the coating are the most preferred sites where precipitation of apatite can easily occur as they are recessed regions where the requisite supersaturation level for precipitation can be easily attained. The same trend occurs in the as-sprayed coatings but the presence of soluble phases (e.g. amorphous phase and other calcium phosphate phases) greatly increases the supersaturation level of the solution thus, resulting in an accelerated precipitation of a thick layer instead of discrete precipitation in the coating's internal defects. In this case, the images show that most of the fluorescent intensity is concentrated at the coating's surface where the apatite layer precipitates. This is the perceived outcome because this layer of gel-like material upon drying (the required step prior to the impregnation of the samples) would leave behind gaps or fissures when the water evaporated. These gaps or fissures were subsequently filled with the resin. Moreover, among all the as-sprayed coatings studied none of them failed by delamination or disintegration during the *in vitro* test. This indicates that the phases are homogeneously distributed and also implies that the precipitated layer acts as a barrier impeding the fluid from passing through it. In addition, the time required for the open defects network to close up is due to differences between heat-treated and as-sprayed samples' microstructure. The defects network for the as-sprayed coating sprayed at 150 KPa becomes closed in less than a month, varying the time with their composition and spraying pressure, whereas for the analogous heat-treated coating the time taken is longer, always beyond one month, hence, increasing the chance of ions leaching from the substrate.

Conclusions

Confocal microscopy in its fluorescent mode has confirmed the presence of interconnected defects network in the CAPS-HA coatings, and its evolution throughout the *in vitro* test. As-sprayed and post-spray heat-treated CAPS-HA coatings showed diverse behavior during the *in vitro* test.

Whereas the post-spray heat-treated samples gave discrete precipitation exclusively at the recessed regions in the coatings, pores, and microcracks, and thus tended to seal the coating as the immersion period progressed, the as-sprayed samples yielded a homogenous precipitated layer on top of the coatings, which was found to act as a barrier impeding further precipitation on the defects of the coatings. In less than a month the defects network connectivity in the as-sprayed coatings was completely concealed.

References

[1] L. Sun, C.C. Berndt, K.A. Gross and A. Kucuk: J. Biomed. Mat. Res. (Appl. Biomat.) Vol.58 (2001), p.570

[2] L.L. Hench: J. Am. Ceram. Soc. Vol.74 (1991), p.94

[3] J. Black: Biological performance of materials (New York, 1999)

[4] V. Guipont, M. Espanol, F. Borit, N. Llorca-Isern, M. Jeandin, K.A. Khor: Mat. Sci. Eng A. Vol. A325(1-2) (2002), p.9

[5] T. Kokubo: Protocol for preparing SBF, Dep.of Mat. Chem., Kyoto Univ., Japan.

[6] N. Llorca-Isern, M. Puig, M. Espanol: J. of Therm. Spray Technol. Vol. 8(1) (1999), p. 73

[7] Sodeoka, M. Suzuki, K. Ueno: J. of Therm. Spray Technol. Vol. 5(3) (1996), p. 277

[8] L. Nimb, K. Gotfrendsen, J. Steen Jensen: Acta Orthop. Belgica, Vol. 59(4) (1993), p. 333

Key Engineering Materials Vols. 254-256(2004) pp. 323-326
online at http://www.scientific.net
© 2004 Trans Tech Publications, Switzerland

Effects of some variables on R.F. Magnetron Sputtered Hydroxyapatite Coatings on Titanium

R.A.Silva[1,2], M.Sousa[1], F.J.Monteiro[2,3], R. Barral[1], J.D Santos[2,3] and M.A.Lopes[2, 3]

1-ISEP- Instituto Superior de Engenharia do Porto, Rua Dr.António Bernardino de Almeida,431,4200- 072 Porto, Portugal, ras@isep.ipp.pt
2-INEB- Instituto de Engenharia Biomédica- Laboratório de Biomateriais, Rua do Campo Alegre,823, 4150-180 Porto, Portugal
3- Faculdade de Engenhria da Universidade do Porto, Dep. Engenharia Metalúrgica e Materiais, Rua Dr. Roberto Frias, 4200-465 Porto, Portugal, fjmont@fe.up.pt

<u>Keywords</u> : Hydroxyapatite, thin films; RF sputtering

Abstract. RF sputtering may be used to produce thin, well adherent bioactive coatings with long term stability to ensure contact with bone tissue. RF Sputtered Hydroxyapatite (HA) coatings were produced onto Ti at 200 and 150 W. S.E.M. analysis indicated that films obtained with 200 W were thicker than those with 150 W. Significant variations were observed, depending on the location of the substrates with respect to the sample holder in the sputtering chamber, with thickness ranging from 1.8 to 3.2 μm. EDS analysis of 200 W films revealed the presence of Ca and P, in ratios typical of HA. XPS of thinner films obtained at 150 W, showed Ca and P, but also some carbonate probably partially substituting OH^-.
Anodic polarisation curves of samples coated at 200 W and 150 W and for different sputtering times, compared to Ti substrate, indicate that HA films were protective and that the degree of protection increased with film thickness, i.e., with power intensity and time of deposition.

Introduction

Hydroxyapatite $(Ca_{10}(PO_4)_6(OH)_2)$ - (HA) is a major constituent of bone and dental tissues[1-2]. Previous studies have established HA as a material of choice for many applications involving bone repair and bone-implant contact.
Plasma spraying is still the most popular method for coating Ti implants, due to its concept simplicity, and the ability to form thick layers (50 – 200 μm) of HA [3]. However, the films are neither fully crystalline nor dense, and their composition is heterogeneous [4,5]. In-vitro [6] and in-vivo [7] studies have highlighted degradation at the coating-substrate interface, partial film dissolution and release of debris.
Sputtering is well known for its ability to yield dense, uniform, smooth and adherent films, with thickness control, depositing films with nearly bulk-like properties, which are predictable and stable [8-10].

Materials and methods

Targets were prepared from HA (Plasma Biotal Ltd), compressed in a mould at $1.4x10^6$ Pa, and sintered in air at 1200 °C, with heating rate of 4°C/ min, and natural cooling inside the furnace. Plates of Ti c.p.15 x 15 x 1 mm were wet polished, and sonicated in de-ionised water and acetone for five minutes.
RF Sputtering power varied between 150 W and 200 W, argon pressure varied between 0.3-0.5mbar for 150 W and was kept at 0.03 mbar for 200 W. Deposition time varied between 2 to 6 hours. Distance between target and sample was 42 mm.
SEM/EDS was used for surface characterisation and chemical analysis. XPS was applied to determine finer surface chemical composition of the coatings.

Anodic polarisation was applied as a means to evaluate the degree of corrosion protection that was provided to Ti by the sputtered films. Electrochemical polarisation curves were obtained after 1 hour immersion in 0.15 M NaCl, varying the potential between –0.5 and 1.5 V, at 0.25 mV/s sweep rate. Data from the curves as i_{corr}, Ba, Bc and E(I=0) were calculated. Saturated calomel electrode (SCE) was the reference electrode and a platinum plate was the auxiliary electrode.

Results

SEM studies allowed determination of the films chemical composition and surface morphology, evaluating the homogeneity of the films, depending on the location of the substrates within the samples' holder.

A spectrum of a film formed after the substrate was located at the centre of the holder for two hours under 200 W is shown in Fig.1a:

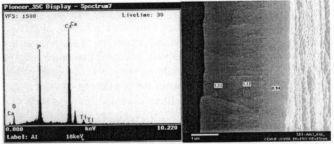

Fig 1a- EDS spectrum of sample located at the centre of the holder, 2 hours, 200W.
Fig. 1b- Cross section of coating of Fig 1a.

The film is composed of three layers with different thickness, as marked in figure1b. The overall thickness is around 3.2μm. Fig. 2a presents the spectrum of a film that was located at approximately 30 mm from the holder centre, where a much thinner film was obtained. Ca to Ti ratio shows a strong signal from the substrate, as a result of reduced film thickness.

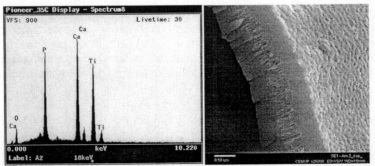

Fig 2a- EDS spectrum of sample located distant from the centre of the holder, 2 hours, 200 W.
Fig. 2b- Cross section of the coating of Fig 2a.

The SEM image, Fig.2b, confirmed the thinner three layers present in the cross section with an overall thickness of 1.8μm, as a sandwich-like structure with two dense and compact layers, with undefined structures at this magnification, separated by an intermediate columnar grained layer.

XPS was used to characterise the films surface when RF sputtering energy was varied between 150 W and 200 W. Both films presented similar Ca/P ratios, and the C 1s peak de-convolution revealed the presence of carbonate.

Films obtained with 150 W were thinner and discontinuous, revealing the presence of Ti from the substrate.

Fig 3, shows a XPS spectrum of a sample obtained with 150 W sputtered for 4 hours.

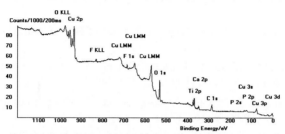

Fig. 3- XPS Survey spectrum of a sample treated for 4 hours at 150W.

Electrochemical characterisation was performed to evaluate the protective behaviour of the sputtered HA films. Ti presented the most regular passivation region, but the HA coatings curves, although comparatively depressed, do not show sudden current density increases, meaning that no corrosion phenomena occurred. The results of polarization curves are given in table 1, where the smallest values of i_{corr} correspond to the thicker HA films. This means that the HA coatings are not detrimental, in terms of corrosion, for Ti and on the contrary they are protective.

Table 1. Anodic polarization results

Sample	$E_{corr} = f(t)$ (mV)	E(I=0) (V)	i_{corr} (x10^{-3}) ($\mu A/cm^2$)	Ba (V/decade)	Bc (V/decade)
Ti c.p.	-300.00	- 0.2904	111.10	0.2141	0.1776
4h,150W	56.00	-0.7995	18.42	0.1094	0.1398
6h,150W	9.00	0.1125	36.85	0.4363	0.4274
2h,200W	- 57.00	-0.1271	3.62	0.2767	0.1617

Conclusions

The thickness of the HA films depends on the power and location on the holder. Three distinct layers with different thicknesses compose the films. Thicker films are more homogeneous and dense in terms of morphology. Concerning corrosion resistance, the HA films are not detrimental to Ti; on the contrary, they are protective. Thicker films show lower corrosion rates, i_{corr}, thus being more protective.

Acknowledgements

The authors acknowledge project POCTI/CTM/35478/2000 for funding this work.

References

[1] K.A. Hing, S.M. Best, K.E Tanner, W. Bonfield, and P.A Revell: Journal of Materials Science: Materials in Medicine, Vol. 8 (1997) p. 731-736,

[2] S. Raynaud, E Champion, D. Bernache-Assolant: and D. Tetard, Journal of Materials Science: Materials in Medicine, Vol. 9, (1998), p. 221-227

[3] J Weng, X. Liu, X. Zhang, and K. de Groot: Journal of Biomedical Materials Research, Vol. 30, (1996) p. 5-11,.

[4] K.A. Gross, V. Gross, and C.C Berndt,.Journal of the American Ceramic Society, Vol. 81, 1, (1998), p. 106-112,

[5] J.T Edwards, J.B. Brunski, H.W. and Higuchi, Journal of Biomedical Materials Research, Vol. 36, (1997), p. 454-468,

[6] C.Y. Yang, R.M.Lin, B.C. Wang, T.M. Lee, E. Chang, Y.S. Hang, and P.Q. Chen: Journal of Biomedical Materials Research, Vol. 37, (1997), p. .335-345

[7] M. Ogiso, M. Yamamura, P.T. Kuo, D. Borgese and T. Matsumoto: Journal of Biomedical Materials Research, Vol. 39 (1998), p. 364-372

[8] K. Van Dijk, H.G. Schaeken, J.G.C Wolke, C.H.M. Maree, F.H.P.M. Habraken and J.A. Jansen: In P. Vincenzini, (ed.) *Materials in Clinical Applications*, , (1995), p. 249-256

[9] M. Yoshinari, T. Hayakawa, J.G.C. Wolke, K. Nemoto and J.A. Jansen: Journal of Biomedical Materials Research, Vol. 37, (1997), p. 60-67,.

[10] J.G.C. Wolke, K. de Groot and J.A. Jansen: Journal of Biomedical Materials Research, Vol. 39, (1998), p. 524-530

Key Engineering Materials Vols. 254-256(2004) pp. 327-330
online at http://www.scientific.net
© 2004 Trans Tech Publications, Switzerland

The Influence of Glucose and Bovine Serum Albumin on the Crystallization of a Bone-Like Apatite from Revised Simulated Body Fluid

S.V. Dorozhkin[1], E.I. Dorozhkina[1], S. Agathopoulos[2] and J.M.F. Ferreira[2]

[1] Research Institute of Fertilizers, Kudrinskaja sq. 1-155, 123242 Moscow D-242, Russia
sedorozhkin@yandex.ru

[2] Department of Ceramics and Glass Engineering, University of Aveiro, 3810-193 Aveiro, Portugal

Keywords: revised simulated body fluid (rSBF), bovine serum albumin (BSA), glucose, crystallization, calcium phosphates, bone-like apatite, biomineralization.

Abstract. Revised simulated body fluid (rSBF) was modified by addition of bovine serum albumin (BSA) and glucose in physiological amounts. Precipitation experiments were carried out in sealed plastic containers. The influence of both components on precipitation and crystallization of calcium phosphates from supersaturated solution equal to 4 times the ionic concentration of rSBF was studied under physiological conditions (solution pH = 7.40 ± 0.05, temperature 37.0 °C ± 0.2 °). The experimental results showed that BSA was co-precipitated with calcium phosphates, but it evidently hindered the crystallization of the precipitates. Glucose showed negligible influence on crystallization of calcium phosphates.

Introduction

In vitro investigations of biomineralization are very useful for the simulation of the actual *in vivo* conditions. The liquid medium for mineralization studies introduced by the research group of Prof. Kokubo in the early 90's, named as simulated body fluid (SBF) [1], has become the most popular simulating solution. The drawbacks of SBF have been recently overcome by introducing a revised SBF (rSBF) whose ionic composition is precisely equal to the human blood plasma [2]. However, neither SBF nor rSBF contain biologically associated organic compounds, such as carbohydrates, proteins, etc., which are always present in biological liquids (e.g. blood, serum, saliva, etc.). Therefore, significant chemical and structural differences have been reported between biological apatites in calcified tissues and bone-like apatites precipitated from SBF and rSBF. From literature it is known that carbohydrates have a minor influence [3, 4], while proteins have a major effect [5 – 7] on dissolution and precipitation of calcium phosphates. These investigations were done in various buffered solutions; however, no experiments with rSBF containing glucose and albumin have been reported yet. This work aims at closing the gap in the knowledge with regards to the influence of BSA and glucose on the precipitation and crystallization of calcium phosphates from rSBF.

Materials and experimental procedure

The precipitation experiments were carried out in plastic containers. The solution of rSBF was prepared by dissolving the inorganic salts in double-distilled water, as described earlier [2]. To accelerate precipitation and increase the amount of precipitates, the experiments were performed with solutions containing 4 times the ionic concentration of rSBF (4rSBF). Hence, the amount of the inorganic salts was taken in quadruplicate. To avoid immediate precipitation, calcium and magnesium cations were separated from hydrogenphosphate and hydrogencarbonate anions by preparing two different solutions, named as 4rSBF-Ca and 4rSBF-PO$_4$, respectively. Bovine serum albumin (BSA) and glucose were dissolved in both 4rSBF-Ca and 4rSBF-PO$_4$ in the following amounts: 4 g/l for glucose and 2.5 to 80 g/l for BSA, respectively. 4 g/l corresponds to 4 times of the physiological concentration of glucose, while 2.5–80 g/l corresponds to 3–100% of the physiological concentration of BSA [8]. Preheated (37.0 °C ± 0.2°) solutions of 4rSBF-Ca and

4rSBF-PO$_4$ (250 ml of each) containing either BSA or glucose were quickly mixed in sterilized plastic containers. The containers were immediately sealed and kept under gently shaking at 37.0 °C ($\pm$ 0.2°) for 3 days. Then, the suspensions were filtered; the precipitates were washed with water to remove the excess of rSBF, and dried overnight at 37.0 °C ($\pm$ 0.2 °). The precipitates were analyzed with SEM, chemical analysis (EDS), XRD, and IR-spectroscopy. Results from similar experiments with 4rSBF containing neither BSA nor glucose were used as a control.

Results and discussion

Characteristic SEM images of precipitates formed from 4rSBF in the presence of glucose and different amounts of BSA after 3 days are shown in Fig. 1. Their spectra from XRD and IR analyses are plotted in Fig. 2. The spectra of the precipitates formed in the presence of glucose are omitted because there was no difference between them and those corresponding to the precipitates formed without glucose. The chemical analyses (by EDS) of the precipitates formed from 4rSBF containing either glucose or different amounts of BSA are shown in Fig. 3.

In general, the presence of glucose has apparently negligible effect in the chemical composition, crystalline state and microstructure of the precipitates, whereas BSA has evidently a stronger influence. According to the SEM images (Fig. 1), the presence of glucose influences neither the size nor shape of the precipitates (Figs. 1a and 1b). On the other hand, BSA strongly influenced both the size and the shape of precipitates (Figs. 1c, 1d and 1e). In particular, the higher concentration of the dissolved BSA, the smaller dimensions of the precipitates.

According to Fig. 3, the precipitates contained carbon, oxygen, phosphorus and calcium as the main chemical elements together with traces of sodium and magnesium. It is well-known that the chemical composition of precipitates from SBF and rSBF can be described as non-stoichiometric sodium- and magnesium-containing carbonate apatite [1, 2]. The present measurements agree fairly well with those earlier findings.

The influence of glucose and BSA on the chemical composition of precipitates is clearly illustrated with the results of EDS (Fig. 3). There is generally a negligible effect of glucose on the chemical composition of the precipitates, since the vast amount of the ions of calcium, magnesium, phosphate and carbonate was independent on presence of glucose (Fig. 3a). In the presence of BSA, the intensity of all peaks (other than that of carbon, whose intensity increases) gradually decreases over BSA increasing (Fig. 3b). This finding can be attributed to both co-precipitation of calcium phosphates with BSA (carbon peak increases), and an inhibiting effect of BSA on crystallization of calcium phosphates (for instance, due to the chelating properties of BSA).

The experimental results from XRD and IR-spectroscopy revealed that the presence of glucose in 4rSBF influenced neither crystal structure (XRD data) nor the IR-spectra of the precipitates (data are not shown). In the case of presence of BSA, even in small concentrations (e.g., 6% of the physiological value) in rSBF, there was a strong influence on both the crystalline structure and the chemical composition of the precipitated calcium phosphates (Fig. 2). The crystallinity of the precipitates progressively decreased with increasing concentration of dissolved BSA, which can be seen as intensity decreasing of the diffraction patterns (Fig. 2a). In general, the XRD data clearly point out a strong negative influence of the amount of dissolved BSA on the crystallinity of the precipitates formed from rSBF.

The results of IR-spectroscopy (Fig. 2a) disclose that even in the presence of 5 g/l of dissolved BSA, a calcium phosphate – BSA composite (co-precipitate) was formed. This conclusion can be supported by the fact that the typical absorption bands of BSA (those around 2900, 1650, and within 1400 – 1570 cm^{-1}) are seemingly superimposed with that of Ca-P. Meanwhile, at the spectrum without BSA (i.e. 0 BSA), the absorption bands at 1650 and within 1400 – 1570 cm^{-1} might be due to the incorporation of carbonates, those present in rSBF. Hence, the IR data do not allow distinguishing between the two possible options.

Fig. 1. Microstructure of calcium phosphate precipitates formed from 4rSBF containing different amounts of BSA and glucose, after 3 days at 37 °C: (a) 4rSBF, (b) 4rSBF + 4 g/l of glucose, (c) 4rSBF + 5 g/l BSA, (d) 4rSBF + 20 g/l BSA, (e) 4rSBF + 80 g/l BSA.

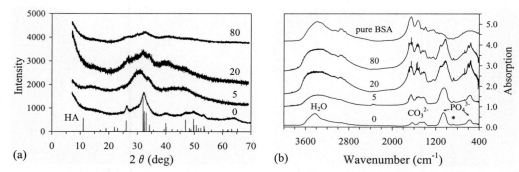

(a)   2 θ (deg)

(b)   Wavenumber (cm^{-1})

Fig. 2. XRD (a) and IR (b) spectra of the precipitates formed from 4rSBF containing different amount of the dissolved BSA (in g/l), after 3 days in 37 °C. The X-ray diffraction patterns of well crystallized hydroxyapatite (HA) and the IR-spectrum of pure BSA are also plotted for comparison purposes.

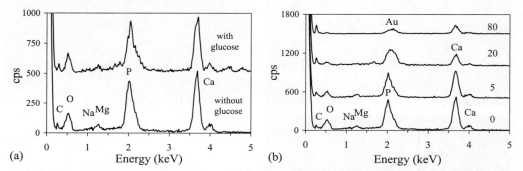

(a)   Energy (keV)

(b)   Energy (keV)

Fig. 3. Chemical analysis (by EDS) of the precipitates formed from 4rSBF in the presence of (a) glucose and (b) different amounts of BSA (in g/l), after 3 days at 37 °C.

Conclusions

There is negligible effect of glucose but very strong influence of BSA on the precipitation and crystallization of calcium phosphates from rSBF. BSA co-precipitates with calcium phosphate but increasing quantity of dissolved BSA results in deceasing crystallinity of the Ca-P precipitates.

References

[1] T. Kokubo, H. Kushitani, S. Sakka, T. Kitsugi, T. Yamamuro: J. Biomed. Mater. Res. Vol. 24 (1990), p.721.

[2] H.M. Kim, T. Miyazaki, T. Kokubo, T. Nakamura: In: S. Giannini, A. Moroni, editors. Bioceramics 13. Zurich: Trans Tech Publications Vol. 192-195 (2001), p.47.

[3] K.K. Makinen, E. Soderling: Calcif. Tissue Int. Vol. 36 (1984), p. 64.

[4] T. Matsumoto, M. Okazaki, M. Taira, J. Takahashi: Caries Res. Vol. 34 (2000), p. 26

[5] S. Shimabayashi, Y. Tanizawa, K. Ishida: Chem. Pharm. Bull. (Tokyo), Vol. 39 (1991), p. 2183.

[6] D.T. Wassell, R.C. Hall, G. Embery: Biomaterials Vol. 16 (1995) 697.

[7] H.B. Wen, J.R. de Wijn, C.A. van Blitterswijk, K. de Groot: J. Biomed. Mater. Res. Vol. 46 (1999) 245.

[8] I.R. Kupke, B. Kather, S. Zeugner, Clin. Chim. Acta Vol. 112 (1981) 177.

Key Engineering Materials Vols. 254-256(2004) pp. 331-334
online at http://www.scientific.net
© 2004 Trans Tech Publications, Switzerland

Synthesis and Characterization of Fluoridated Hydroxyapatite with a Novel Fluorine-Containing Reagent

Kui Cheng[1], Wenjian Weng[1*], Piyi Du[1], Ge Shen[1], Gaorong Han[1] and J.M.F.Ferreira[2]

[1] Department for Materials Science and Engineering, Zhejiang University, Hangzhou 310027 P.R.China, Wengwj@zju.edu.cn

[2] Department of Ceramic and Glass Engineering, University of Aveiro, Aveiro 3810, Portugal, jmf@cv.ua.pt

Keywords: Sol-gel, ammonium hexafluorophosphate, fluoridated hydroxyapatite;

Abstract. A sol-gel method is utilized to synthesize a fluoridated hydroxyapatite (FHA) phase. Calcium nitrate tetrahydrate, phosphoric pentoxide and ammonium hexafluorophosphate were used as precursors. The Ca, P and F precursors were mixed under designed proportions to form solutions with a Ca/P ratio of 1.67. In order to obtain FHA phase with various fluorine contents, different amounts of ammonium hexafluorophosphate were added in the Ca-P mixed solutions. The Ca-P-F mixed solutions were quickly evaporated on a hot plate to become powders, then the powders were further calcined at a temperature up to 900°C, and the FHA phase could be obtained. The fluorine content in the FHA phase could be tailored by the varying the amount of ammonium hexafluorophosphate added.

Introduction

Hydroxyapatite (HA) has been widely investigated because of its resemblance to natural bone minerals. However, being an inorganic material, the high elastic modulus and low fracture toughness limited its application as bulk materials especially under load bearing conditions. A practical way to utilize HA is to function as coatings or films on medical metals which have excellent mechanical properties. Thus, the combination of the metal and HA can meet both demands of bioactivity and good mechanical properties [1]. Further investigation revealed that the HA coatings show relative large dissolution under physiological or simulated physiological conditions, which may be caused by both intrinsic solubility and impurity phases [2]. Therefore, the long-term effectiveness of the coatings was deteriorated. Fluoridated hydroxyapatite (FHA) is regarded as a good alternative for pure hydroxyapatite because of its lowered solubility, which may improve the long-term effectiveness [2].

Since the most prevalent way to prepare coatings on medical metals is plasma spray which needs synthetic powders, many efforts are made to synthesize FHA phase powders, like solid reaction [3], pyrolytic method [4] and co-precipitation [5]. Recently, sol-gel method is also reported [6,7].

In our previous work [7], FHA could be obtained from the systems containing fluorine-containing reagent and Ca-P precursors. However, either corrosive or low efficient reagent was used. In this work, a novel fluorine-containing reagent with higher efficient and less corrosive was adopted to synthesize FHA.

Experimental

Phosphoric pentoxide (P_2O_5, Riedel-deHaen) was dissolved in absolute ethanol (Merck) to a concentration of 1mol/L and refluxed for 24 hours. Calcium nitrate tetrahydrate ($Ca(NO_3)_2 \cdot 4H_2O$, Reidel-deHaen) was also dissolved in absolute ethanol to form 2mol/L solutions. A designed amount

---

* Corresponding author. Tel.: +86-571-87952324; Fax: +86-571- 87952341. Email address: wengwj@zju.edu.cn

of NH_4PF_6 was added into the Ca-P mixed solution to form a Ca-P-F mixed solution with a Ca/P ratio of 1.67. The adding amount of NH_4PF_6 in each sample was listed in Table 1. As-prepared mixed solution was quickly evaporated on a stainless steel plate held at $150°C$ and then calcined at $900°C$. The calcined powders were characterized by XRD (Rigaku D/Max RA) and FTIR (Nicolet Avatar 360) for phase purity. The fluorine content in the powders was determined by fluorine sensitive electrode method. ^{19}F liquid NMR analysis (BRUKER AMX300. ^{19}F NMR, 282.4MHz, $C_6F_6(-163ppm)$ as reference compound) was also performed on the Ca-P-F mixed solution to understand the FHA formation mechanism.

Table 1. The sample composition table

Sample annotation	HNF-1	HNF-2	HNF-3	HNF-4	HNF-5	HNF-6
$[PF_6^-]$/Ca ratio in the solution	1/90	2/90	3/90	4/90	5/90	6/90

Results

The as- prepared mixed solutions with NH_4PF_6 were clear. After hours (> 6h), depending on the adding amount of NH_4PF_6, they became cloudy or even precipitated. The powders were prepared when the solutions remained clear.

As shown in the XRD results, the peaks attributed to apatite phase could be observed with powders using this novel fluorine-containing reagent (Fig. 1). After the adding amount of NH_4PF_6 increased up to F/Ca=0.2, pretty pure apatite phase could be obtained (Fig. 1.c). The pure phase remained with even higher adding amount of NH_4PF_6 (Fig. 1.d, e, f). The chemical analysis result of the fluorine content in the powders was given in Fig. 2. It was found the fluorine content increased with the adding amount of NH_4PF_6 at first, while the fluorine content almost kept unchanged after the F/Ca ratio in the Ca-P-F mixed solution reached 0.2. Some detailed information on the structure of the apatite phases was given by the FTIR results in Fig. 3. The results showed typical apatite spectra except that the librational band of OH in HA phase located at $630cm^{-1}$ was totally missing, which indicated that the amount of OH in apatite structure was reduced. Further considering the decreasing tendency of lattice parameter a with increasing fluorine content (as shown in Fig.4), the powders obtained were actually hydroxyapatite/fluorapatite solid solutions.

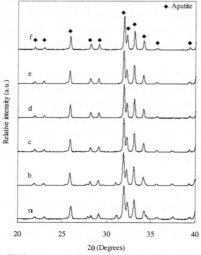

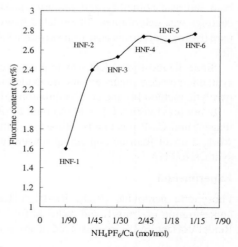

Fig 1 XRD patterns of samples with different adding amount of NH_4PF_6
a: HNF-1, b: HNF-2, c: HNF-3, d: HNF-4, e: HNF-5, f: HNF-6

Fig 2 Fluorine contents of samples with varied NH_4PF_6 amount in the mixture

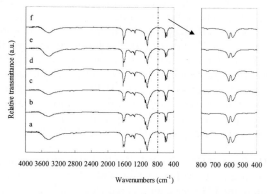
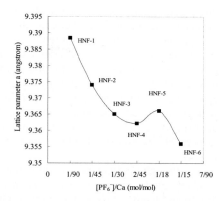

Fig 3 FTIR spectra of samples with different amounts of NH₄PF₆
a: HNF-1, b: HNF-2, c: HNF-3, d: HNF-4, e: HNF-5, f: HNF-6

Fig 4 Lattice parameter a of samples with different fluorine content

Discussion

As shown in the experimental results, NH₄PF₆ is effective for F incorporation in synthesis of FHA phase. However, when the NH₄PF₆ content in the Ca-P-F mixed solution is high (HNF-3, HNF-4, HNF-5, HNF-6), the fluorine content in the powders (Fig. 2) and their lattice parameter α (Fig. 4) remained unchanged with the adding amount. The phenomenon should be related to the F incorporation process of NH₄PF₆ into the apatite.

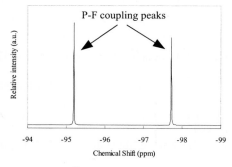
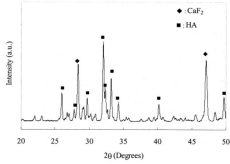

Fig. 5 ^{19}F liquid NMR results of the Ca-P precursors with NH₄PF₆

Fig. 6 Phase composition of the precipitates

The ^{19}F liquid NMR spectrum (Fig. 5) of the clear Ca-P-F mixed solution (HNF-6) shows two peaks with almost the same intensity. The two peaks could be attributed to P-F coupling peaks which are characteristic peaks of PF₆⁻ ion [8]. Hence, PF₆⁻ radicals seem to be quite stable in the clear solution. After long time aging (>6h), precipitation occurs in this clear mixed solution. It is found that the precipitates consist of calcium fluoride and apatite as shown in Fig. 6. Therefore, the F incorporation process of NH₄PF₆ into apatite could be suggested as follows. The PF₆⁻ ions are stable in the clear mixed solution at first. Then, since there exists water in the mixed solution (brought by the Ca(NO₃)₂·4H₂O), the PF₆⁻ begins to hydrolyze slowly and release free F⁻ ions, which react with Ca ions and cause the solution to be cloudy or precipitate. When the clear solution is quickly evaporated on a hot plate, the quick evaporation may prevent calcium fluoride from growing well, so that only tiny calcium fluoride particles with high reactivity are formed. In the following calcining process, these particles react with other calcium and phophate and form the FHA phase.

It can be concluded from the process above that the efficiency for NH_4PF_6 to incoporate F into apatite depends on hydrolysis degree of NH_4PF_6. Since the H_2O in the mixed solution comes from the calcium nitrate reagent, the amount is limited. In the Ca-P-F mixtured solutions some phosphate precursors also undertake hydrolysis (into H_3PO_4 finally), which can be proved by the existence of apatite phase in the precipitates. Therefore, the proportion of hydrolyzed NH_4PF_6 species decreases with an increase in the adding amount of NH_4PF_6 because H_2O from the calcium nitrate reagent is unchanged in the Ca-P-F mixted solution. Consequently, imcomplete hydrolysis of NH_4PF_6 is responsible for no expected high fluorine content in powders HNF-4, HNF-5 and HNF-6 (Fig. 2 and Fig. 4).

Although the fluorine content in powders HNF-1 and HNF-2 is relative low, the OH librational band (should be located at $630cm^{-1}$) in their FTIR spectra are still almost invisible. The disappearance of $630cm^{-1}$ band indicates the length of OH chain in the FHA structural is reduced, implying that the FHA phase synthesized in this work has pretty good fluorine distribution uniformity as a solid solution because the existence of librational band of OH is related to the length of OH chain in the apatite structure [9].

Conclusion

In general, FHA phase could be synthesized with this novel fluorine-containing reagent. The fluorine incorporation mechanism of NH_4PF_6 is described as a 3-step process: the hydrolysis of PF_6^- and the releasing of F^-; the reaction between Ca and F; the development of FHA phase. The FHA phase obtained has pretty uniform fluorine distribution as a solid solution. Therefore, NH_4PF_6 could be a good fluorine-containing precursor for the synthesis of fluoridated hydroxyapatite by sol-gel method.

References

[1] P. Ducheyne, W. Van. Raemdonck, J. C. Heughebaer and M. Heughebaert, Biomaterials, Vol. l7 (1986) p.97

[2] S. Overgaard, M. Lind, K. Josephsen, A. B. Maunsbach, C. Bnger and K. Soballe, J. Biomed. Mater. Res, Vol. 39 (1998) p.141

[3] J. C. Elliott: *Studies in Inorganic Chemistry 18 : Structure and Chemistry of the Apatites and Other Cacium Orthophosphates* (Elsevier, Amosterdam, 1994).

[4] U. Partenfelder, A. Engela and C. Russel, J. Mater. Sci.: Materials in Medicine, Vol. 4 (1993) p. 292

[5] K. A. Gross, J. Hart and L. M. Roderiguez-Lorenzo, Key Engineering Materials, Vol. 218-220 (2002) p. 165

[6] M. Cavalli, G. Gnappi, A. Montener, D. Bersani, P. P. Lottici, S. Kaciulis, G. Mattogno and M. Fini, Journal of Materials Science, Vol. 36 (2001) p.3253

[7] Kui Cheng, Ge Shen, Wenjian Weng, Gaorong Han, J.M.F. Ferreira and Juan Yang, Materials Letters, Vol. 51 (2001) p.37

[8] D.G. Gorenstein: *Phosphorus-31 NMR: Principle and Application* (Academic Press, New York, 1984).

[9] F. Freund and R. M. Knobel, J. C. S. Dalton, (1977) p.1136

Key Engineering Materials Vols. 254-256(2004) pp. 335-338
online at http://www.scientific.net
© 2004 Trans Tech Publications, Switzerland

Novel Fluorapatite-Mullite Coatings for Biomedical Applications

J. Bibby[1], P.M. Mummery[1], N. Bubb[2], and D.J. Wood[2]

[1]Manchester Materials Science Centre, University of Manchester and UMIST, Manchester, England.

[2]Leeds Dental Institute, University of Leeds, Leeds, England

Jennifer_bibby@hotmail.com

Keywords: Fluorapatite, mullite, glass ceramic ,bioactivity, SBF, osteoblasts

Abstract. The bioactivity of cast fluorapatite glass-ceramic LG112 was investigated using SBF and a culture of human osteoblasts. The glass-ceramic underwent different heat treatments at 800 and 1000˚C in both air and in a vacuum. The bioactivity was determined using SBF and osteoblast cell culture, examined using SEM, using Ti6Al4V and unheated glass as controls. The glass was successfully deposited as a coating by electrophoretic deposition and by magnetron sputter coating.

Introduction

Apatite-mullite glasses and glass-ceramics have been developed from dental ionomer glasses, by careful heat treatment. These have an advantage over other bioactive glasses as the interlocking needle shaped crystals have high fracture toughness [1]. The mix of the bioactive fluorapatite and the inert mullite, makes a bioactive glass ceramic with a low dissolution rate. The fluorine, in the fluorapatite, forms denser bone at the interfaces, and, in dental applications, can prevent caries. LG112, has an advantage over other apatite-mullite glass-ceramic systems as it has the same calcium to phosphorus ratio, 1:1.67, as hydroxyapatite, the mineral phase in bone. It is designed to have a similar thermal expansion coefficient as titanium, which makes it a suitable material for coating titanium and titanium alloy implants.

Glass-ceramics have brittle properties and therefore are more successfully used as a coating. Many coating systems have been tried using bioactive materials, including dipping, blasting, plasma spray, sol gel, enameling, sputter coating and electrophoretic deposition with varying degrees of success [2,3,4,5,6,7]. There has been some discussion over the bioactivity of similar glass-ceramics [8,9]. This paper will be examining the bioactive properties of cast LG112 and investigating the possibilities of using this glass as a coating using electrophoretic deposition (EPD) and magnetron sputter coating.

Methods

Glass Production. Apatite-mullite glass ceramics are based on the Al_2O_3-SiO_2- P_2O_5-$CaCO_3$-CaF system. Both the glass and the glass-ceramic were used for testing the bioactivity and for coating by EPD. The glasses were produced by weighing the constituents and placed in a rotary mill for one hour. The resulting mixed powder was subsequently placed in a $250cm^3$ mullite crucible and heated in at 1550°C for two hours. The molten glass was then quenched in room temperature water. The cooled glass frit was then placed in a grinding dish (ring and puck) set at low amplitude and ground for 17 minutes. Grinding homogenises the glass. This powder was remelted at 1550˚C for two hours. A cylindrical graphite mould was preheated at 580˚C for 20 minutes. The molten glass was poured into this mould and annealed at 580˚C for 2 hours. 580˚C is 50˚ below the glass transition temperature of this glass and it has been found, by trial and error, to be the best annealing temperature. The final glass rod was cut into 3mm discs using an Accutom 5 cutting machine.

Heat Treatment. The samples to be tested in SBF were heated in air at 1000°C for 1 hour and furnace cooled. Samples for cell culture were also heated at 800°C and 1000°C for 1 hour in air and furnace cooled.

SBF. A solution of Kokubo's formulation simulated body fluid (SBF) with an ion concentration (moles/litre) of $Na+$ 142, $K+$5.0, Mg 2+ 1.5, Ca 2+ 2.5, $Cl-$ 147.8, HCO_3- 4.2, HPO_42- 1.0, SO_42- 0.5. The cast samples were soaked in this solution for 1, 6, 24, 48 and 72 hours in a carbon dioxide atmosphere at 38°C.

Cell Culture. Confluent osteoblast cells were trypsinised and resuspended in Dulbeccos Eagle Media. 2cl of this cell solution was put in cell culture wells containing the glass/glass-ceramic/titanium alloy samples as well as two empty wells to check for confluency and contamination. The cells were incubated until there was confluency in the empty wells. This took 2 days. For examination in the SEM, the cells were dried out using graded ethanol and fixed, followed by a gold coating.

SEM. A JOEL SEM and a Phillips 525 SEM with an EDAX attachment were used.

XRD. A Phillips X'pert thin film XRD was used to measure the crystallinity.

EPD. The powder from the ring and puck mill was crystallised in air at 1000°C. As this caused the powder to sinter, this was further ground in an attritor mill for 6 hours. The final particle size was 0.5µm. Successful EPD coatings were deposited on to a Ti6Al4V electrode by suspending 3wt% of the glass ceramic in distilled water. The conditions used were 10V, 2A, for 5 minutes and a distance of 1.5cm. The coating was then dried for 1 hour at 70°C, followed by sintering for 1 hour at 1000°C in a vacuum, to prevent oxidation of the substrate.

Sputter Coating. Polished Ti6Al4V was sputter coated using a magnetron sputter coater for 2 hours in an argon atmosphere. A 10cm square target was made by pressing and sintering the powdered glass.

Results and Discussion

Heat treatment: Figure 1 shows the formation of the expected needle like structure. These needles have formed star shaped clusters approximately 5µm across. The glass-ceramic had crystallised to fluorapatite and mullite, which was confirmed by XRD. It also showed evidence of other phases such as anorthite and calcium alumina silica. The extra phases are due to the loss of silicon tetrafluoride on melting. This gives an unstoichiometric composition, therefore not just fluorapatite and mullite are formed.

SBF: After 24 hours in SBF, there were some areas that showed signs of nucleation of a hydroxyapatite (HA) layer. Other areas of the same sample were completely smooth.

Fig. 1 Glass heated at 1000°C for 1 hour no days in SBF.

Fig. 2 Glass heated at 1000°C, 24 hours in SBF, some signs of HA layer formation

Fig. 3 Smooth area on same sample

Samples taken from the SBF at different times, also showed a great variation in surface texture. EDAX taken from both the rough areas and the smooth areas showed very similar results. Two days is obviously not long enough to give a complete HA layer.

Cell Culture: After 2 days, the osteoblasts were confluent on the polystyrene cell wells. After drying, there were a number of healthy cells adhering to both the plain glass sample and the sample heated at 800° C in a vacuum for 1 hour. As shown in the figures below:

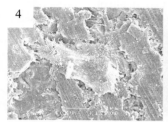  

Fig. 4 A micrograph showing osteoblast cell on the glass
Fig. 5 A micrograph showing osteoblast cells on the surface of LG112 heated at 800°C for 1 hour.
Fig. 6 A micrograph showing osteoblast cells on the surface of a Ti6Al4V sample

Even on the plain, as cut, uncrystallised glass, the cells have spread out, with plenty of fibrils, as shown in Figure 4. This indicates that the cells are healthy. The growth of cells appeared to be localised, some areas with many cells, others with none. This may just be because 2 days isn't sufficient time for cells to be evenly spread across the surface.

In figure 5, a disc of the glass ceramic that had been heated at 800°C for one hour, also shows good attachment of healthy cells. After 2 days, the control, the titanium alloy, showed only a few round, unhealthy cells, as shown in figure 6. This shows that, even after only 2 days, this glass-ceramic shows good bioactivity.

EPD: The EPD conditions chosen for coating titanium proved to be successful. This can be seen in figure 7, below. It can be seen that the coating covers the substrate completely and the underlying substrate cannot be seen.

Fig. 7 A micrograph of the unsintered EPD coating

The titanium alloy electrodes were weighed before and after coating and, as a known area was coated, an approximate thickness could be calculated. Over 5 samples, the average thickness was approximately 100μm. SEM of the coating (not shown here) showed good sintering of the powder, with necking between the particles.

Sputter Coating: The titanium substrate was successfully coated with the glass. A slight variation of thickness could be seen as a variation in colour across the sample. This indicates that the thickness of the coating is approximately the wavelength of light. It can be assumed that this coating is amorphous, as the coating is too thin for any crystallinity to be detected by XRD. SEM of the sputter coating showed that the coating was extremely flat. The few scratch marks remaining on the surface of the polished substrate appeared to have been filled.

Conclusions

It can be seen that LG112 can easily be cast and is bioactive in both SBF and osteoblast cell culture. Upon heating it crystallises to fluorapatite and mullite as well as other phases, such as anorthite and calcium alumina silica, due to the change in stoichiometry due to the loss of silicon tetrafluoride. As the coefficient of thermal expansion of LG112 is similar to that of Ti6Al4V, it can be used to successfully coat titanium alloys by both EPD and magnetron sputter coating.

References

[1] A. Clifford, R. Hill J. Non. Crystal. Sol. Vol. 196 (1996) p.346-351

[2] C. Jana, P. Wange, G. Grimm, W. Gotz: Glass Science and Technology – Glastechntsche Berichte 68,4 (1995) p.117-122

[4] J.M Gomez-Vega, E. Saiz, A.P. Tomsia: Journal of Biomedical Materials Research 46:4 (1999) p.549-559

[5] D. C. Greenspan Medical and Diagnostic Industry Magazine March 1999 Special Section

[6] M Gomez-Vega, E.Saiz et al Advanced Materials 12:12 (2000) p.894-898

[7] L. Sun, C. C. Berndt, K. A. Gross, A. Kucuk Journal of Biomedical Materials Research 58:5 (2001) p.570 – 592

[8] A. Boccaccini, I. Zhitomirsky Current Opinion in Solid State and Materials Science 6 (2002) p.251-260

[9] X. Liu, C. Ding, Z. Wang Biomaterials 22 (2001) p.2007-2012

[10] S. Agathopoulos, D.V. Tulyaganov, P.A.A.P. Marques, M.C. Ferro, M.H.V. Fernandes, R.N. Correia: Biomaterials, Vol. 24 (2003) P.1317-1331

Key Engineering Materials Vols. 254-256(2004) pp. 339-342
online at http://www.scientific.net

Identifying Calcium Phosphates Formed in Simulated Body Fluid by Electron Diffraction

Yang Leng[1], Xiong Lu[1] and Jiyong Chen[2]

[1]Department of Mechanical Engineering, Hong Kong University of Science & Technology
Hong Kong, China, meleng@ust.hk
[2]Engineering Research Center in Biomaterials, Sichuan University, China, jychen@scu.edu.cn

Keywords: Calcium Phosphates, TEM, electron diffraction, simulated body fluid.

Abstract. Calcium phosphate precipitation on hydroxyapatite/tricalcium phosphate ceramics and alkaline treated titanium surfaces were examined using the single crystal diffraction in transmission electron microscope. We found that the precipitation turns out to be of octacalcium phosphate, instead of "bone-like apatite". The crystal structure similarity between hydroxyapatite and octacalcium phosphate could cause the confusion and misidentification of precipitation. We also found that thin film X-ray diffraction spectroscopy could also lead to misidentification octacalcium phosphate as hydroxyapatite because of the preferential orientation of crystal precipitates.

Introduction

Bioactivity of bioceramics and other orthopedic materials is characterized by the capability of forming bioactive calcium phosphates (BCP) in vitro and in vivo. It is widely believed that the BCP formed on the bioceramics and other bioactive metal surfaces in simulated body fluid (SBF) is hydroxyapatite (HA) with carbonate content and certain degree of calcium phosphate deficiency. Note that the definition of apatite is twofold: 1) the chemical formula of $Ca_{10}(PO_4)_6 X_2$ where X is OH for hydroxyapatite; and 2) the hexagonal crystal structure [1]. Although it is the thermodynamic stable phase in physiological environment, hydroxyapatite might not be the calcium phosphate phase to form in supersaturated solution as in SBF. Octacalcium phosphate (OCP) and dicalcium phosphate dehydrate (DCPD) have been reported as intermediate phases or precursor of forming apatite in precipitation because of kinetic reasons [2-4]. Note that OCP has crystal structure, which can be considered as the one with alternative stacking of apatite layers and hydrated layers along [100] direction [3]. Great caution should be taken in differentiating HA and OCP, particularly using the X-ray diffraction spectroscopy (XRD) because their high intensity peaks overlap in the 2θ rang of 20 ~ 40°. We found that the thin film XRD technique may also misidentify phases due to the preferential orientation of precipitation on titanium surfaces. The most reliable method of identifying calcium phosphate precipitates is the single crystal diffraction. We managed to extract the single crystalline precipitates from biphasic HA/TCP and alkaline treated titanium surfaces. From the electron diffraction pattern of single crystalline precipitates, we found that the commonly believed "bone-like apatite" turns out to be OCP.

Experimental

Two different bioactive surfaces for the calcium phosphate precipitation were examined: biphasic HA/TCP ceramics and alkaline treated titanium. The HA/TCP ceramics were prepared from HA and TCP powder with weight ratio of 70 to 30. The uniformly mixed powder was foamed with H_2O_2 and sintered at 1200 °C. The porosity of bulk BCP was 50-60 % and the pores in the samples were interconnected. The titanium was commercial pure titanium (CP Ti) (Baoji Special Iron and

Steel Co. Ltd., Shangxi, China). The alkali treatment was performed by immersing titanium plates in 100mL of 10M NaOH aqueous solution at 60°C for 24 h. After the alkali treatment, the titanium plates were gently washed with distilled water and dried at 40°C for 24 hr in an air atmosphere. The alkali-treated plates were then heated to 600 °C at a rate of 5 °C /min and kept at 600 °C for 1 hour and cooled to room temperature in the furnace.

Each liter of simulated body fluid (SBF) was prepared by dissolving 7.995g of NaCl, 0.353g of $NaHCO_3$, 0.224g of KCl, 0.228g of $K_2HPO_4.3H_2O$, 0.305g of $MgCl_2.6H_2O$, 0.227g of $CaCl_2$, 0.0710g of Na_2SO_4 into distilled water. The pH value of 7.4 was maintained by adjusting amounts of Tris (tris-hydroxymethylaminomethane) and HCl.   After 14 days immersion in SBF, the HA/TCP and titanium specimens were rinsed with distilled water and dried at 50°C.

The precipitates on specimen surfaces were carefully separated from the substrates by ultrasonic vibration. Then, the precipitates were collected using copper meshes with carbon film for TEM examinations. The precipitates, at least in one dimension were sufficiently thin, so that no further thinning processing was necessary for TEM sample preparation.

Results

HA/TCP. 10 precipitates on HA/TCP surfaces were examined and their typical TEM image is shown in Fig. 1a. The diffraction patterns indicate that the flake-like precipitates are single crystals (Fig. 1b).   The single diffraction pattern in Fig. 1b was identified as OCP with **B** = [110], which implies the OCP flake surface is parallel to the (110) plane. The pattern could be misidentified as HA with **B** = [010] because both of them show a plane spacing of 0.68 nm and the OCP pattern looks like orthogonal as that of HA (actually, the angle in OCP pattern is 90.3°).   The main difference in the patterns however is that the OCP pattern reveal the ($\bar{1}$10) plane with a plane spacing of 0.938nm, which is unique among all the calcium phosphates. No hydroxyapatite (HA) precipitate was found in all the examined samples.

Alkaline Treated Titanium. The crystalline precipitates extracted from alkaline treated titanium surfaces exhibit different morphology and size from those on HA/TCP surfaces. The precipitate shape is neither flake nor plane-like. Figure 2a shows the bright field image of precipitates in much smaller size than those on HA/TCP surfaces. Figure 2b shows a single diffraction pattern of one precipitate in the field of Fig. 2a. The diffraction pattern in Fig. 2b was identified as **B** = [$0\bar{1}2$] of OCP, showing (300) and (210) planes with an interplanar angle of 73.08°. Totally, we identified 11 OCP diffraction patterns, while no HA pattern was found. The OCP precipitates were rod-like crystals and their diffraction patterns exhibit various orientations, such as **B** = [$\bar{1}\bar{1}1$], [255], etc.

Discussion

We noted the possibility of misidentifying OCP as HA in thin film XRD. For example, Fig. 3 is a thin film XRD spectrum of calcium phosphate precipitates on the alkaline-treated titanium, which has been identified as OCP by electron diffraction. The XRD spectrum is more likely to be identified as that of HA, due to similar peak positions of OCP and HA in the 2θ range of 20 to 40°. The main differences between the OCP and HA exist in the angle of 4 to 20°, mainly the OCP (100) peak at 2θ = 4.738°, and OCP ($\bar{1}$10) peak at 2θ = 9.417°. The thin film XRD spectrum however cannot reveal those peaks when the OCP precipitates orientate their crystallographic c-axis perpendicular to the titanium surface.   Without electron diffraction, it would not be possible for us to know that the XRD pattern in Fig. 3 is that of OCP.

We found that the ring pattern of electron diffraction, as shown in Fig. 4, might also lead to confusion because of two reasons: 1) the rings of OCP (100) and other low index (hk0) being not visible, and 2) the most visible two OCP rings of (002) and ($\overline{4}02$) being almost identical to those of HA (002) and (211). Actually, Fig. 4 is the ring pattern of OCP, because we identified individual OCP precipitates in the same field by single diffraction patterns. In summary, the commonly believed apatite formation on bioceramics and on bioactive titanium should be further examined with caution. In this study, the hard evidence provided by single crystal diffraction indicates that the BCP precipitation in SBF, under our experimental conditions, is OCP. Thus, the osseointegration mechanisms of bioactive orthopedic materials should be investigated in the case of absence of bone-like apatite formation.

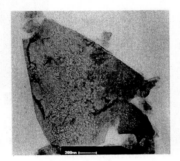 

Fig. 1 a) TEM bright field image of OCP precipitates on HA/TCP; b) its diffraction pattern with **B** = [110].

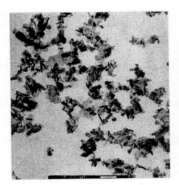 

Fig. 2 a) TEM bright field image of OCP precipitates on alkali-treated titanium surface; b) one of single crystals diffraction pattern with **B** = [$0\overline{1}2$].

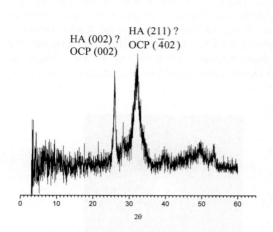

Fig.3 Thin film XRD spectrum of precipitates on alkaline treated titanium from

Fig. 4 OCP precipitates on HA/TCP surfaces and their ring patterns.

Acknowledgement

The work was financially supported by the Research Grants Council of the Hong Kong SAR government and by an internal fund of Hong Kong University of Science and Technology.

References

[1] J. C. Elliot, *Structure and Chemistry of the Apatite and Other Calcium Orthophosphates*, Elsevier, 1994.

[2] Newman WF, Neuman MW. *The Chemical Dynamics of Bone Mineral*. University of Chicago Press, Chicago 1958.

[3] Brown WE, Eidelman N, Tomazic B. Octacalcium phosphate as precursor in biomineral formation. Adv Dent Res 1987;1:306-313.

[4] LeGeros RZ, Daculsi G, Orly I, Abergas T, Torres W. Solution-mediated transformation of octacalcium phosphate (OCP) to apatite. Scan Microscopy 1989;3:129-138.

Key Engineering Materials Vols. 254-256(2004) pp. 343-346
online at http://www.scientific.net

Novel Calcium Phosphate Fibres from a Biomimetic Process: Manufacture and Cell Attachment

E.C. Kolos[1], A.J. Ruys[1], R. Rohanizadeh[2], M.M. Muir[2] and G. Roger[3]

[1] Centre of Advanced Materials Technology, University of Sydney, Sydney, Australia,
elizabeth.kolos@aeromech.usyd.edu.au; a.ruys@aeromech.usyd.edu.au

[2] Skin and Bone Laboratory, Department of Physiology. University of Sydney, Sydney, Australia

[3] ASDM, Australian Surgical Design and Manufacture, St Leonards, Sydney, Australia

Keywords: Biomimetic, simulated body fluid (SBF), calcium phosphate, fibres, cell attachment.

Abstract. Biomimetic method has been used extensively to coat many materials in the attempt to make them bioactive. Using a biomimetic method, combustible fibres were coated with a calcium phosphate apatite. Phosphorylation was employed as a pretreatment and simulated body fluid (SBF) as the growth solution. Burnout of the combustible fibres at varying temperatures produced biomimetically coated calcium phosphate hollow fibres. There was generally increased sintering and crystallisation with increasing temperature. Cell studies of the sintered fibres showed successful cell adhesion of osteoblast cells. Results were presented with SEM, XRD and EDS. This research could lead to applications of calcium phosphate fibres in tissue scaffolds and internal bandages, and possibly as composites.

Introduction

Biomimetic method has been used extensively to coat many materials in the attempt to make them bioactive. This method comprises two-steps of pre-treatment and deposition. Kokubo *et al.* in 1982 [1] developed apatite wollastonite (AW) glass-ceramic that exhibited good bone bonding ability, with no foreign body infection, and was seen as a load bearing bioactive bone substitute. Kokubo *et al.* [2] completed tests in an acellular simulated body fluid. In 1991, it was then found that any material and any shape could achieve a bioactive coating including alumina, cotton and carbon [3]. M.R. Mucalo *et al.* [4] continued this work coating cellulose and cotton in 1993, bamboo was coated by S.H. Li *et al.* [5] and chitosan was coated by Y. Yokogawa *et al.* [6] in 1997. Thus a bioactive fibrous material using a biomimetic approach was achieved. Using electrophoretic deposition to coat hydroxyapatite on carbon, I. Zhitomirsky [7] burnt out the carbon to achieve a hollow hydroxyapatite fibre. With the background of both biomimetic coating and burnout, the following paper will look at the experimental approach to coating cotton with a calcium phosphate apatite, and the burning out of the combustible substrate cotton to achieve calcium phosphate fibres.

Experimental Procedure

Preparation of Calcium Phosphate Fibre Manufacture
Phosphorylation Treated Cotton

Phosphorylation of cotton samples were carried out following the preparation reported by Inagaki *et al.* [8]. Cotton pieces were placed in a round bottom flask equipped with a thermometer, mechanical stirrer, condenser and N_2 gas inlet tube. Urea dissolved in dimethyl formamide (DMF) was added to the flask and heated to 130°C, upon which phosphorous acid (H_3PO_3) was added and heated to 145°C. The reaction was allowed to reflux for thirty minutes. Cotton fibres were then washed repeatedly in distilled water and dried in an oven at 50°C.

Ca(OH)₂ Treatment

The phosphorylated cotton was soaked (without stirring) in a saturated solution of $Ca(OH)_2$ (pH ~ 11-12) in closed screw-top glass bottle for periods of up to 8 days. The $Ca(OH)_2$ solution was renewed every 4 days. Upon completion of the soaking period the samples were subsequently filtered, rinsed thoroughly with distilled water and dried in an oven at 50°C.

Growth of Calcium Phosphate

1.5 times the concentration of Simulated Body Fluid (1.5SBF) was prepared. SBF simulates the ionic concentration of blood plasma. The pH was measured and adjusted to pH 7.4 with tris(hydroxymethyl)aminomethane (($CH_2OH)_3CNH_2$) and dilute hydrochloric acid (HCl). Samples of pre-treated cotton were placed in 1.5SBF in closed screw-top glass jars and re-buffered to pH of 7.4. The glass jars were immersed in a shaking water bath at 36.5°C was for two weeks. The 1.5SBF solution was renewed every two days to maintain pH = 7.4 and ion concentration. Upon completion of soaking period, samples were washed with distilled water and dried in air before further examination. A 5.0SBF solution was prepared in a similar way.

Heat treatment – Cotton Burnout

To determine both the effect of sintering behaviour at various temperatures on the calcium phosphate apatite, and to burn out the cotton substrate, coated cotton fibres were fired in air to 950, 1150 and 1250°C at a rate of 100°C per hour.

Experimental Procedure of Cell Cultures of Sintered Fibres

The sintered calcium phosphate using 1.5SBF and the 5.0SBF were weighed to 3µg and autoclaved for 45 minutes at 125°C. Both the 1.5SBF and the 5.0SBF groups were seeded with human derived MG63 cells at a density of 50000 cells per ml cell media. The media of the cells was replaced after 4 days. After one week, the media was pipetted out of each well and the specimens were rinsed twice with Phosphate Buffer Solution (PBS). The specimens were then fixed with 2% glutaraldehyde, rinsed with PBS, dehydrated in a series through a graded series of ethanol and critical point dried. Finally they were mounted on aluminium plates and coated with a thin layer of gold.

Results and Discussion

Fig. 1 of coated cotton with 1.5SBF and 5.0SBF shows fairly uniform coverage of calcium phosphate on cotton fibres with no heat treatment. **Fig. 2** shows the heat treatment study on 1.5SBF at 950, 1150 and 1250°C. Heat treatment at 950°C shows the cotton substrate has been burnt out, however very little sintering of the calcium phosphate phase has occurred. With heat treatment at 1150°C there seems to be necking within the coat, the earlier stage of sintering. While with heat treatment at 1250°C there is a considerable amount of sintering, resulting in sintered porosity. There is also a reduction in surface area at 1250°C compared with heat treatment at 950°C. It is also evident from these SEM pictures that the fibres maintain a tubular morphology. At 950°C, the cotton has burnt out leaving a fairly thick walled (approximately 1µm) hollow fibre. Increasing the sintering temperature to 1150°C, the tubular morphology is maintained but at 1250°C sintering occurs such that the hollow fibres start to open to take the form of tapes.

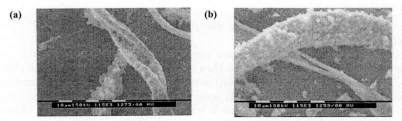

Fig. 1 – Calcium Phosphate prepared (a) in 1.5 times concentration of SBF; (b) in 5.0 times concentration of SBF

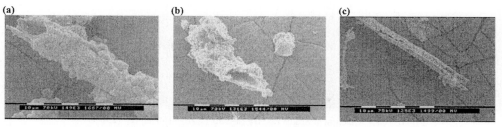

Fig. 2 – Heat Treatment Study for 1.5SBF (a) 950°C; (b) 1150°C; (c) 1250°C

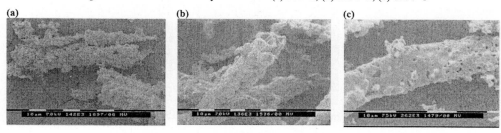

Fig. 3 – Heat Treatment Study for 5.0SBF (a) 950°C; (b) 1150°C; (c) 1250°C

XRD analysis confirmed the calcium phosphate phase was low crystalline apatite. XRD plots show the crystallinity of the apatite phases generally increasing with increasing heat treatment temperature for 5.0SBF, but 950°C treatment of 1.5SBF samples has higher crystallinity than the 1150°C. This may be an experimental error and requires further investigation. The main calcium phosphate phase present at the increasing heat treatment temperature is whitlockite, a decomposing phase of hydroxyapatite, which, in accordance with the literature, is expected [10].

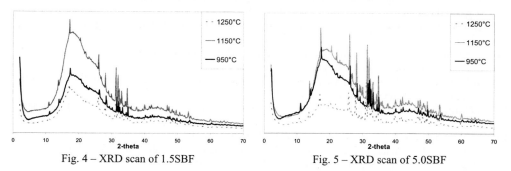

Fig. 4 – XRD scan of 1.5SBF   Fig. 5 – XRD scan of 5.0SBF

Samples tested in cell cultures were sintered at 950, 1150, and 1250°C. It was observed that all samples were favourable to cell attachment and the sintering temperature did not directly influence the proliferation and attachment of the cells. The purpose for cell cultures was to test the general biocompatibility of the fibres, but also to see if any residue of cotton from the burnout stage could deleteriously affect the biocompatibility. The osteoblasts grew along length of the fibres and did not attempt to enter the hollow area. They were well attached and spread over the surface with contact to other cells. Thus the results show the biomimetically produced hollow fibres are biocompatible with no cytotoxic effects. Further studies will be required to indicate the role of surface topography, calcium phosphate crystallinity, phase composition and concentration of SBF on the attachment and growth of anchorage-dependent cells.

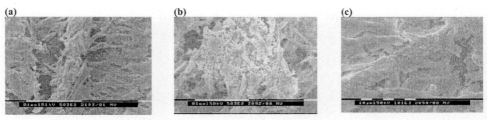

Fig. 6 – Cell Attachment Study 1.5SBF (a) 950°C; (b) 1150°C; (c) 1250°C

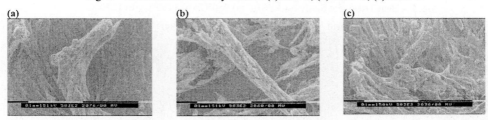

Fig. 7 – Cell Attachment Study 5.0SBF (a) 950°C; (b) 1150°C; (c) 1250°C

Acknowledgements

The authors would like to thank Tony Romeo from the Electron Microscope Unit, University of Sydney, for assistance in biological preparation of SEM samples.

References

[1] T. Kokubo, M. Shigematsu, Y. Nagashima, M. Tashiro, T. Nakamura, T. Yamamuro, S. Higashi, Bull. Inst. Chem. Res., Kyoto Univ., 60 [3-4], 260-268, (1982).

[2] T. Kokubo, T. Hayashi, S. Sakka, T. Kitsugi, T. Yamamuro, J. Ceram. Soc. Japan, 95 785-791, (1987).

[3] T. Kokubo, K. Hata, T. Nakamura, T. Yamamuro, Bioceramics Proceedings of the 4th International Symposium on Ceramics in Medicine, 4 113-120, (1991).

[4] M.R. Mucalo, Y.Yokogawa, M. Toriyama, T. Suzuki, Y. Kawamoto, F. Nagata, K. Nishizawa, J. Mater. Sci. Mater. Med., 6 597-605, (1995).

[5] S.H. Li, Q. Liu, J. de Wijn, B.L. Zhou, K. de Groot, J. Mater. Sci. Mater. Med., 8[9], 543-549, (1997).

[6] Y. Yokogawa, J.P. Reyes, M.R. Mucalo, M. Toriyama, Y. Kawamoto, T. Suzuki, K. Nishizawa, F. Nagata, T. Kamayama, J. Mater. Sci. Mater. Med., 8[7] 407-412, 1997.

[7] Zhitomirsky I., J. Europ. Ceram. Soc., 18, 849-856, (1998).

[8] N. Inagaki, S. Nakamura, H. Asai, K. Katsuura, J. App. Poly. Sci., 20 2829-2836 (1976).

[9] A.J. Ruys, C.C. Sorrell, A. Brandwood, B.K. Milthorpe, J. Mater. Sci. Lett, 14 744-747 (1995).

[10] A.J. Ruys, G.N. Ehsani, B.K. Milthorpe, and C.C. Sorrell, International Ceramic Monographs, 191 106-110 (1994).

Key Engineering Materials Vols. 254-256(2004) pp. 347-350
online at http://www.scientific.net

CaO-P₂O₅ Glass-Hydroxyapatite Thin Films Obtained by Laser Ablation: Characterisation and *In Vitro* Bioactivity Evaluation

M. P. Ferraz[1,2], F. J. Monteiro[1,3], D. Gião[1,3,4], B. Leon[4], P. Gonzalez[4], S. Liste[4], J. Serra[4], J. Arias[4] and M. Perez-Amor[4]

[1]INEB, Instituto de Engenharia Biomédica, Laboratório de Biomateriais, Rua Campo Alegre 823, 4150-180 Porto, Portugal, mpferraz@ufp.pt

[2]Universidade Fernando Pessoa, Faculdade de Ciências da Saúde, Rua Carlos da Maia 296, 4200-150 Porto, Portugal

[3]Universidade do Porto, Faculdade de Engenharia, Departamento de Engenharia Metalúrgica e Materiais, Rua Roberto Frias, 4200-465, Porto,Portugal

[4]Universidad de Vigo, Departamento de Física Aplicada, Lagoas, Marcosende, 9, 36280 Vigo, Spain

Keywords: CaO-P₂O₅ glass, hydroxylapatite thin films, laser ablation, *in vitro* bioactivity.

Abstract. Hydroxyapatite (HA) coatings have been applied to improve adhesion of non-cemented implants to host bone. Plasma spraying is the most common technique leading to thick calcium phosphate films (>120μm). Pulse laser deposition (PLD), is a possible alternative method to obtain thin (<10 μm), well adherent hydroxyapatite (HA) films. Similarly to synthetic HA, biological apatites contain Ca^{2+}, PO_4^{3-} and OH^-, but also several trace ions, like Na^+, Mg^{2+}, K^+ and F^-, which may be introduced by CaO-P₂O₅ glasses. In this study, calcium phosphate coatings based on HA and glass modified HA were applied by PLD onto Ti-6Al-4V, using deposition times of 3 hours.

SBF immersion up to 1 week was used to test the films bioactivity. PLD thin films before and after SBF immersion were observed by SEM/EDS and analysed by XPS and XRD. PLD thin films presented columnar cross section structures, independently of the coatings' chemical composition. After SBF immersion, apatite films formed on PLD coatings, both of HA and HA+1.5% glass, did not present the usual morphology of immersion films, but appeared to replicate the previous films. The main difference between HA and modified HA coatings could be seen in the XPS analyses at short immersion periods. Natural apatite was calcium deficient with a Ca/P of ±1.2-1.3. The results seem to indicate that modified HA coatings with lower Ca/P ratio induced earlier formation of natural apatite.

Introduction

Calcium phosphate ceramics have been often applied as coatings for implants, in most cases using thermal plasma spraying processes, leading to thick calcium phosphate films (>120μm). Recently various attempts have been done to produce thin (<5 μm) hydroxyapatite (HA) coatings. Pulse laser deposition (PLD), is based on the ablation of a target by a pulsed laser beam, producing a plasma plume and ablating products which are transferred to the substrate, forming a film, being a possible method to obtain thin film coatings. Previous works indicated that HA films obtained by PLD were thin, continuous, adherent to the substrate and presented adequate biocompatibility and bioactivity [1,2].

Similarly to synthetic HA, biological apatites contain Ca^{2+}, PO_4^{3-} and OH^-, but the inorganic part of bone also contain several trace ions, particularly Na^+, Mg^{2+}, K^+ and F^-, which may be introduced through CaO-P₂O₅ glasses. In an attempt to increase bioactivity and simultaneously maintain the bonding strength levels of HA, modified HA coatings were prepared based on PLD technique, applied to targets obtained by HA/CaO-P₂O₅ glass homogeneous mixtures. Due to the presence of the glass in the target the modified HA layer was expected to induce a rapid initial response, as previously observed in coatings obtained by plasma spraying [3].

In this study, calcium phosphate coatings based on HA and glass modified HA (HA+1.5G) have been applied by PLD on Ti-6Al-4V substrates, aiming at evaluating their possible use as biomaterials and compare their bioactivity.

Materials and Methods

A $CaO-P_2O_5$ glass containing 35, 35, 20 and 10 mol% of P_2O_5, CaO, Na_2O and K_2O respectively was prepared as previously described. The modified HA preparation method has been described elsewhere [4]. Glass addition of 1.5 wt% to HA (Pure, fully crystalline Ha, Ca/P= 1.67, by Plasma Biotal Ltd, UK) was used.

Ti-6Al-4V was used as substrate for these experiments. PLD coating technique applied to calcium phosphates was described elsewhere [2]. Coatings were produced at 460°C in a reactive atmosphere of water vapour, at constant pressure of 0.45 mbar, with a focussed ArF laser beam (193nm), operating at 10Hz and 200mJ/pulse. Targets were made of mixtures of HA and 1.5 wt% $CaO-P_2O_5$ glass, and coatings were produced using several deposition times (0.5, 1, 2 and 3 hours).

SBF immersion for several periods up to 1 week was used to test bioactivity of films obtained with 3 hour of deposition, as they presented the desired thickness. PLD thin films before and after immersion in SBF were observed by SEM/EDS and analysed by XPS and XRD. For the S.E.M analysis of cross sections, samples were bent up to 10° max, in a specific jig, to disrupt the coated layers and expose, under similar conditions, the cross-sections.

Results and Discussion

Under the above conditions, PLD treatment formed thin films on Ti-6Al-4V with columnar cross section structures, independently of the coating chemical composition, as found in previous works on films obtained with HA targets [2].

Coating thickness depends on the deposition time and not on the chemical composition of the target, being thicker with longer deposition times (Fig 1).

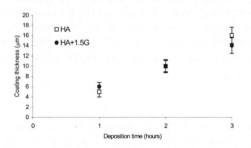

Fig. 1. Coating thickness *versus* deposition time

The surface morphologies are also similar for both types of targets, and none of them shows the droplet morphology, very often found in laser ablation studies of other materials. In spite of these common features and the very slight variation in composition of the targets (1.5%), the PLD coatings produced from the HA and the composite target have respectively Ca/P ratios of 1.65 ± 0.04 and 1.52 ± 0.03. These differences resulting from the ablation process are also clear from the XRD patterns, depicted in Fig. 2.

In both cases the only crystalline phase observed is HA, no amorphous phase could be detected, but the coating deposited from the composite target shows preferential growth in the (002) and (112) directions. From these results, one can conclude that within the processes taking place during pulsed laser deposition, namely ablation, transport through a plasma plume and surface reactions on the substrate, the 1.5% of calcium phosphate glass added to the HA plays a modifying role in the deposited material.

Coatings are calcium deficient hydroxylapatite, in which the Na and K ions supplied by the glass are substituting Ca in the HA structure, and are responsible for the observed preferential orientation

After SBF immersion, chemical changes clearly indicate the transformations taking place at the

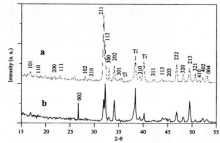

Fig. 2. XRD patterns of the coatings produced from HA (a) and HA+1.5BG (b) targets.

surface (Fig 3). However the apatite film formed on PLD coatings (both HA and HA/1.5% glass (HA/1.5BG)) does not present the usual morphology of immersion films, but instead it replicates the previous films (Fig 4a/4c and Fig 5a/5c). A possible explanation for this fact might be a substitution process taking place instead of the usual precipitation normally occurring on HA surfaces when immersed in SBF. The main difference between HA and HA/1.5BG coatings could be seen for short immersion periods both on the XPS analyses (Fig 3), SEM (Fig4b and 5b) and XRD.

As it can be seen on Fig 3, HA at 3 days immersion has nearly the same Ca/P ratio as the HA+1.5BG at 1 day of SBF immersion. The corresponding SEM images (I and II respectively) show a very similar morphology, indicating that HA+1.5BG induces a faster process than HA.

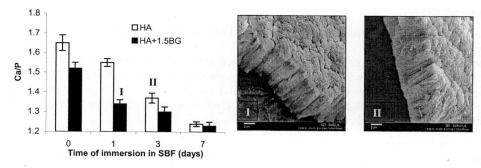

Fig. 3- Ca/P relation versus immersion time in SBF obtained by XPS. (I) SEM insert of HA+1.5BG 1 day immersed in SBF, (II) SEM insert of HA 3 day immersed in SBF.

Natural apatite is calcium deficient with a Ca/P of ±1.2-1.3. The results seem to indicate that HA+1.5BG coating with lower Ca/P ratio induced earlier formation of natural apatite. This is confirmed by SEM, once that at 1 day of SBF immersion, the HA+1.5BG samples present a morphology similar to that observed after 7 days immersion (Fig5a/5b). On the contrary, for the same immersion period, HA is still undergoing the surface modification process deriving from the apatite film being formed and apparently substituting the previous PLD film, although also replicating the previous coating layer with the formation of the new apatite film (Fig 4a/4b). Different tilting angles were used in these observations, eventually inducing misleading thickness evaluations.

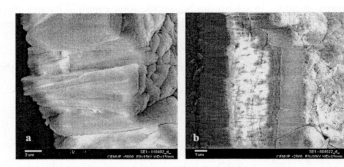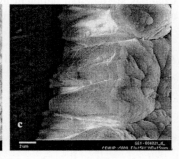

Fig 4 – SEM of HA coating (a), after 1day (b) and 7 days (c) of SBF immersion. Tilting angle was not kept constant.

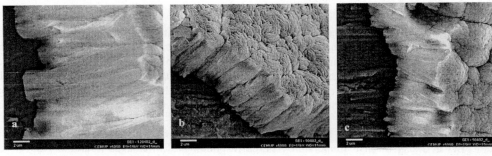

Fig. 5 – SEM of HA+1.5BG coating (a), after 1day (b) and 7 days (c) of SBF immersion. Tilting angle was not kept constant.

Conclusions

PLD thin films presented columnar cross section structures, independently of the coatings' chemical compositions.

The apatite film formed on both HA and modified HA PLD coatings did not present the typical morphology of immersion films, but instead it replicated the previously existing PLD films.

The main difference between HA and HA/1.5BG coatings could be seen for short immersion periods, where HA+1.5BG coating with lower Ca/P ratio induced earlier formation of natural apatite, and better defined crystallinity.

Acknowledgements

The authors wish to thank F.C.T. project POCTI/CTM/35478/2000, and CRUP "Acções Integradas Portugal/ Espanha E-17/20" for their financial support, and Socrates/ Erasmus programme, for Ms Gião's grant provided during her training period at University of Vigo.

References

[1] J.L.Arias et al :J Mater. Sci. Mater.Med.Vol. 8 (1997), p. 873-876.
[2] C. Peraire *et al* : 13th European Conference on Biomaterials, Göteborg, Sweden, 4-7 Sept. 1997, p. 75.
[3] M. P. Ferraz, F. J. Monteiro, J. D. Santos: J Biomed Mater Res, Vol. 45 (1999), p.373.
[4] M. P. Ferraz, M. H. Fernandes, A. Trigo-Cabral, J. D. Santos, F. J. Monteiro: J. Mat Sci: Mat. Med, Vol.10(9) (1999), p.567.

Key Engineering Materials Vols. 254-256(2004) pp. 351-354
online at http://www.scientific.net
© *2004 Trans Tech Publications, Switzerland*

A Study of Bone-Like Apatite Formation on Calcium Phosphate Ceramics in Different Simulated Body Fluids (SBF)

Y.R. Duan[1,2], C.Y. Wang[1,3], J.Y. Chen*[1], X.D.Zhang[1]

1. Engineering Research Center in Biomaterials, Sichuan University, Chengdu 610064, China
jychen@scu.edu.cn
2. West China School of Pharmacy, Sichuan University, Chengdu 610041, China
yourongduan_2001@sina.com
3. Yangtze River Fisheries Institute Jingzhou,Hubei Prov.434000, P.R.China

Keywords: Bone-like apatite, calcium phosphate ceramics, simulated body fluid

Abstract. Five kinds of simulated body fluids were prepared according to the Ca^{2+} ions concentration in body fluids of human, dog, pig, rabbit and monkey so that the different biological environments in different animals were simulated. Results showed that Ca^{2+} can induce bone-like apatite formation in a dose-dependent manner; there is a threshold of Ca^{2+} local concentration for the formation of bone-like apatite. The threshold of Ca^{2+} for bone-like apatite formation in static stimulated body fluid (SBF) is different from dynamic SBF. The threshold of static SBF is lower than that of dynamic SBF. The threshold of Ca^{2+} local concentration is 0.2459g/L in static SBF and 0.3392g/L in dynamic SBF.

Introduction

Ten years ago, several research groups [1-3] reported ectopic formation of bone in porous calcium phosphate ceramics (CaP) implanted in muscle or subcutis of animals. de Bruijn [4] considered that formation of bone-like apatite on the surface of calcium phosphate ceramics was a prerequisite for osteoinduction. The factors affecting bone-like apatite formation on surface of calcium phosphate ceramics are not clear yet. They seem to be both material-dependent and animal-dependent. X. Zhang [5] has reported that the osteoinductivity of Ca-P biomaterials varied among different kinds of animals. The body fluids of different animals have different Ca^{2+} ion concentration. Ca^{2+} can induce bone-like apatite formation in a dose-dependent manner; there is a threshold of Ca^{2+} local concentration for the formation of bone-like apatite. In this study, a series of SBF were prepared in order to reveal the threshold of Ca^{2+} using a biomimetic method. The threshold of Ca^{2+} for bone-like apatite formation in static stimulated body fluid (SBF) is different from dynamic SBF.

Materials and Methods

Materials: Each sample cylinder shape with 4 mm in diameter and 8 mm in length were prepared by sintering a green block of biphasic hydroxyapatite/calcium phosphate(HA/TCP=70/30) at 1200^0C.

Preparation of SBF: The ion composition of SBF was nearly the same as that of human plasma [6]. 1 liter SBF was prepared by dissolving NaCl 7.995g, NaHCO$_3$ 0.353g, KCl 0.224g; K$_2$HPO$_4$.3H$_2$O 0.228g, MgCl$_2$.6H2O 0.305g, CaCl$_2$ 0.227g, Na$_2$SO$_4$ 0.0710g into distilled water, and buffering the solution to pH7.4 with tris-hydroxymethyl-aminomethane and hydrochloric acid.

Preparation of SBFa: A series of SBF (SBFa1~SBFa5) with various Ca^{2+} ion concentration and same ion composition of SBF were prepared. The Ca^{2+} ion concentration agreed with that of body fluids of human, dog, pig, rabbit and monkey [7]. The Ca^{2+} concentration is shown in Table 1. The preparation method of SBFa was similar to SBF.

Table 1 Ca^{2+} concentration of simulation body fluid (mg/dl)

	Human	Monkey	Pig	Dog	Rabbit
SBFa	SBFa1	SBFa2	SBFa3	SBFa4	SBFa5
Ca^{2+}	11.0	11.6	10.6	15.0	14.0

Preparation of SBFs: The seven kinds of simulated body fluid (SBFS1~SBFS7) were prepared by changing the Ca^{2+} ion concentration and keeping constant the other ion concentrations and composition of SBF. The Ca^{2+} concentration is shown in Table 2. A similar method of preparation SBF was used to prepare the SBFs.

Table 2.Ca^{2+} concentration of SBFs (g/L)

SBFs	SBFS1	SBFS2	SBFS3	SBFS4	SBFS5	SBFS6	SBFS7
Ca^{2+}	0.1837	0.2148	0.2459	0.2770	0.3081	0.3392	0.3703

Immersion of samples in static SBFa: Each sample vessel contained 100ml of SBF and 3 samples were immersed in it for 14 days. The sample vessel was placed in a bath shaker in order to keep the solution homogeneous. The SBFa was changed every other day. Samples were rinsed several times with distilled water after immersion and dried at 50°C in dry oven.

Immersion in dynamic SBFs: In vitro experiment with dynamic SBF to mimic the flowing of body fluid *in vivo* is of great significance. The cycling equipment of SBF was shown in figure 1. Each time 3 samples were immersed in SBF solution of the sample chamber. The volume of the sample chamber was 100ml. The flow rate of SBF solution flowing through the sample chamber was expressed as refresh ratio of the SBF solution in the chamber by flowing. The flow rate of 2ml/100ml.min meant that 2ml among 100 ml solution in sample chamber flowed out and 2ml of solution from the store tank was pumped into the chamber at the same time. The flow rate 2ml/100ml.min was nearly the same as that of body fluid in muscle.

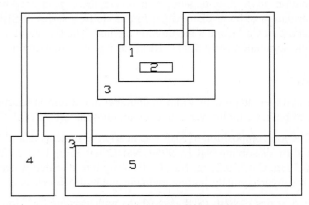

Fig.1 Schematic diagram of flow chamber system
1.Sample chamber; 2. Sample; 3. Water bath; 4. Pump; 5. SBF storage tank.

Characterization of the surface of samples: The morphology of the sample surfaces and pore wall of cross section was examined by a scanning electron microscope (SEM). The chemical composition of the surface after immersion was determined by reflective infrared spectroscopy (RIR).

Results and Discussion

After 14-day immersion in SBFa, bone-like apatite formed on the pore wall with micro pores (Fig.2). The bone-like apatite grew faster in SBF of dog and rabbit than in other SBF. This sequence was consistent with the sequence of the Ca^{2+} concentration in these animals. In animal experiments the same growth speed sequence of the apatite were also observed: bone-like apatite formed faster in specimens implanted in dog and rabbit than that in other animals [8]. But the sequence of bone-like apatite amount formed in the specimens is different from the sequence of osteoinductivity of biphasic calcium phosphate ceramics implanted in these animals. The results showed that the difference of ion concentrations in body fluids of different kinds of animals had significant effect on the formation of bone-like apatite in porous calcium phosphate ceramics. The bone-like apatite layer formed on the materials is a prerequisite for their osteoinduction but osteoinduction of biomaterials may also be affected by other material factors and physiological factors.

Fig.2. The picture of SEM for HA/TCP before and after immersion in static SBF for 14days:(a)before immersion; (b) after immersion in SBFS2; (c) after immersion in SBFS3.

  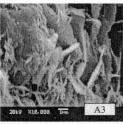

Fig.3. The picture of SEM for HA/TCP after immersion in dynamic SBFS6 for 14days:(A1) surface of samples; (A2) concave of surface; (A3) wall of porous

Table 3 Morphology of the porous ceramics after immersion in static SBFs and dynamic SBFs for 14days

SBFs	SBFs1	SBFs2	SBFs3	SBFs4	SBFs5	SBFs6	SBFs7
static	No changes	crystal nuclei	Flake	Flake	Flake	Flake	Flake
dynamic	No changes	No changes	No changes	No changes	No changes	Flake	Flake

The surface morphology of samples was observed with SEM(Fig.3). After 14-day immersion in static SBFs3, SBFs4, and SBFs5, bone-like apatite formed on the pore wall while no apatite could be observed on the pore wall immersion in dynamic SBFs5. The bone-like apatite formed on the pore wall immersion in dynamic SBFs6 for 14days could be observed. Results are shown in table3.
The chemical compositions of the sample surface after immersion were analyzed with infrared spectroscopy (IR). Result showed that before immersion, a peak specific for HPO_4^{2-} has appeared at 873 cm^{-1}; after immersion, two peaks, which represent CO_3^{2-}, appeared at 1458 cm^{-1} and 1420 cm^{-1}, while the peak at 873 cm^{-1} , specific for HPO_4^{2-}, also appeared. This result indicated that the CO_3^{2-}

ions in the solution substitute OH⁻ and PO₄³⁻. This apatite formed in SBFs is similar to the inorganic phase of bone, so it was called bone-like apatite.

When SBF is static, Ca^{2+} and HPO_4^{2-} released from the material surface cannot easily disperse and the resultant relatively high concentration of Ca^{2+} and HPO_4^{2-} near the surface of the samples may reach the threshold of nucleation. After nucleation, growth of crystal consumes great amount of ions from the SBF solution, which results in the decrease of ion concentration near the sample surface to a level lower than standard SBF. When SBF is dynamic, Ca^{2+} and HPO_4^{2-} ions dissolved from sample surface can easily leave the sample surface to enter the SBF solution under the function of ions concentration-gradient-driven dispersion and stress-gradient-driven transportation. Ca and P concentration near the sample surface can only be slightly higher than that in solution and thus, ions can not easily accumulate on the surface. The threshold for nucleation is not easily reachable.

This nucleation theory indicates that increasing the solution supersaturation and reducing the net interfacial energy are advantageous to the heterogeneous nucleation[9]. Therefore, the bone-like apatite grew faster in static SBFs than in dynamic SBFs.

Conclusion

After 14-day immersion in SBFa, bone-like apatite formed on the pore wall with micro pores could be observed. The bone-like apatite grew faster in SBF of dog and rabbit than in other SBF. This sequence was consistent with the sequence of the Ca^{2+} concentration in these animals.

Ca^{2+} can induce bone-like apatite formation in a dose-dependent manner; there is a threshold of Ca^{2+} local concentration for the formation of bone-like apatite. The threshold of Ca^{2+} for bone-like apatite formation in static stimulated body fluid (SBF) is different from dynamic SBF. The threshold of static SBF is lower than that of dynamic SBF. The threshold of Ca^{2+} local concentration is 0.2459g/L in static SBF and 0.3392g/L in dynamic SBF. This result demonstrated that local ion concentration and microporous structure of pore wall are key factors affecting the formation of bone-like apatite on calcium phosphate ceramic.

Acknowledgment

This work was financially supported by the key basic research project of China, contract No.G1999064760.

References

[1] X.D. Zhang, in Trans.of 19th annual meeting of the society for biomaterials, USA, April 1993, p299.

[2] M. Heughebeart, M. Ginest, G. Bonel, et al, J Biomed. Mater. Res, Appl Biomater 22(1988): 257-268.

[3] C. Klein and K. de Groot and W. Chen and Y. Li and X. Zhang, Biomaterials 35(1994): 31-34T.

[4] J. de Bruijn, H.Yuan, R Dekker, P. Layrolle, K. de Groot, in Bone Engineering, Ed. by J.E.Davies, Em Squared Inc., 287 Garden Avenue, Toronto, Canada, (2000): 421-431.

[5] Z. Yang, H. Yuan, X. Zhang, Biomaterials 17(1996): 2131-2137.

[6] Kokubo T, Ito S,Yamamuro T. Ca, P-rich layer formed on high-strength bioactive glass-ceramic A-W. J Biomed. Mater Res 24(1990): 331-343.

[7] H. Wei☐in Medicine experiment zoology (in Chinese) ☐Science Technology publisher, Beijing, China (1998) :240.

[8] Y.R.Duan ,J.Y.Chen, X.D.Zhang, Journal of Materials Science Letters 21(2002): 775–778.

[9] Hyun-Man Kim, Y. Kim, J. S. Ko, Biomaterials 21(2000): 1129-1134.

Key Engineering Materials Vols. 254-256(2004) pp. 355-358
online at http://www.scientific.net
© *2004 Trans Tech Publications, Switzerland*

In Vitro Bioactivity Study of PLD-Coatings and Bulk Bioactive Glasses

S. Liste, P. González, J. Serra, C. Serra, J. P. Borrajo, S. Chiussi, B. León
and M. Pérez-Amor

Universidad de Vigo, Dpto. Física Aplicada, ETSI Industriales y Minas, Lagoas-Marcosende 9,
36200 Vigo, Spain, sliste@uvigo.es

Keywords: Bioactive glass, laser ablation, coatings, *in vitro* test, FTIR, XPS, SEM, EDX.

Abstract. Bioactive glass coatings have been obtained by Pulsed Laser Deposition (PLD) from bulk glasses of different compositions in the system SiO_2-Na_2O-K_2O-CaO-MgO-P_2O_5-B_2O_3. A comparative study of the *in vitro* bioactive behaviour of the PLD-coatings and bulk glasses was carried out. Fourier Transform Infrared and X-ray Induced Photoelectron Spectroscopies show that the bonding configuration of the bulk glasses is not congruently transferred to the coatings during the ablation process. These results are in agreement with the *in vitro* tests that show a similar bioactive process but a different bioactivity grade between the bulk glasses and the corresponding PLD coatings. The composition and bonding configuration of the bioactive coatings grown by PLD should be carefully tuned in order to obtain an adequate biological response.

Introduction

Bioactive glasses are known to be materials that have the ability to form an intimate bond with living tissues. *In vitro* and *in vivo* studies show that the formation of new tissue occurs through the growth of a biologically active apatite layer on the glass surface [1,2].

The bioactivity of these glasses is based in the existence of alkali and alkali earth cations, e.g., Na^+, Ca^{2+} in the silica network, which results in a disruption of its continuity and promotes the formation of non-bridging silicon-oxygen groups (Si-O-NBO). This is a key step of the complex chemical process, which can be abridged in various phases [1]: Rapid exchange of alkali or alkali earth elements with H^+ or H_3O^+ from the solution, loss of soluble silica in form of $Si(OH)_4$ and formation of Si-OH (silanols), condensation and repolymerization of a SiO_2-rich layer, migration of Ca^{2+} and PO_4^{3-} groups to the surface and growth of an amorphous CaO-P_2O_5 rich film.

A promising application of the bioactive glasses is to coat metallic prosthesis, which could be a route to combine both mechanical resistance and bioactivity in one material. A bioactive coating with a good adhesion to the metal forms a chemical bond with the surrounding tissues, it prevents the formation of a dense fibrous tissue capsule on the metal and it promotes an intimate contact between bone and prostheses.

Pulsed Laser Deposition (PLD) is an alternative coating technique [4], which involves four main stages: i) the laser interaction with the target (bulk bioactive glass), ii) expansion of the ablated products (plume), iii) interaction of the plasma plume with the substrate, and iv) nucleation and growth of the coating on the substrate surface. Compared with conventional film deposition techniques, e.g. sputtering, plasma spraying, enamelling and sol-gel, PLD method presents several advantages, such as, materials with high melting-point can be deposited, no contamination is present, coatings can be prepared in a reactive environment and ability to transfer the stoichiometry of very complex materials to the coating. In short, the congruent ablation of the target is a crucial characteristic of this method due to the important role of the film composition in the bioactive behaviour of these glasses.

In this work *in vitro* tests by immersion in Simulated Body Fluid (SBF) were performed to study the bioactive behaviour of several compositions of bulk bioactive glasses and the corresponding coatings grown by PLD.

Materials and Methods

Bioactive silica based glasses of different compositions (Table 1) in the system SiO_2- Na_2O-K_2O-CaO-MgO-P_2O_5-B_2O_3 have been studied.

Table 1. Composition of bioactive glasses (wt%).

Glass	SiO₂	Na₂O	K₂O	CaO	MgO	P₂O₅	B₂O₃
BG42	42	20	10	20	5	3	-
BG50	50	15	15	15	2	-	3
BG55	55	21	9	8	2	4	1
BG59	59	10	5	15	5	3	3

The coatings were grown on Ti and Si substrates in a high vacuum chamber using an ArF excimer laser (λ=193 nm). The laser was operated at a repetition rate of 10 Hz providing an energy density of 4.17 J/cm^2. The substrates were kept at a constant temperature of 200°C during the film growth [6].

The bioactivity study was carried out by immersion of the bulk and coatings in SBF with ion concentrations and pH nearly equal to those of human plasma [2]. The sample surface area to solution volume (SA/V) ratio was 0.5 cm^{-1}. The flasks (sterilised polystyrene) containing the solution and the specimens were maintained at 36.5 °C for 72 hours in an incubator.

The glass bonding configuration was observed by Fourier Transform Infrared (FTIR) and X-ray induced Photoelectron Spectroscopies (XPS). After *in vitro* test, the layers and the bulk glasses were analysed by Scanning Electron Microscopy (SEM) and Energy Dispersive X-ray spectrometry (EDS).

Results and discussion

Previous work [5] indicates that in the PLD coating technique the stoichiometry of the bulk glass used as ablation target is congruently transferred to the coating. Nevertheless, FTIR and XPS analyses show important differences between the bulk and the film bonding configuration. Fig. 1 shows the FTIR spectra of a typical bioactive glass target and the corresponding coating. The main peaks of bioactive glass are observed [6,7]: a) a band at 1000-1200 cm^{-1} assigned to the Si-O-Si asymmetric stretching vibration, b) a weak band around 750 cm^{-1} associated to the Si-O-Si bending vibration, and c) an additional band at 900-950 cm^{-1} corresponding to the non-bridging silicon-oxygen groups (Si-O-NBO). Important variations in the IR bands can be observed: a) the peak associated to the Si-O-Si stretching vibration shifts to lower wavenumbers and its intensity increases, and b) the band intensity assigned to the Si-O-NBO groups decreases. These results demonstrate that the glass bonding configuration is not correctly transferred to the PLD coatings.

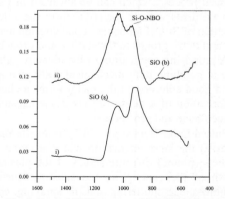

Fig. 1: Typical FTIR spectra of i) BG50 bulk glass used as ablation target and ii) the corresponding PLD coating (C50).

In order to clarify the chemical bonding of the coatings X-ray induced Photoelectron Spectroscopy (XPS) analysis [8] were carried out. The O_{1s} and Si_{2p} photoelectron spectra of different bioactive glass coatings are shown in Fig. 2. When the silica content decreases, both photopeaks shift towards lower binding energy values. The O_{1s}

photopeak can be used to provide useful information on the bonding states of the oxide ions in the silicate glasses. The curve fitting of a typical O_{1s} photoelectron spectrum for bioactive glass coating corresponding to BG42 is shown in Fig. 3. Two well-resolved peaks around 532 and 530 eV associated with the bridging silicon-oxygen groups (BO) and non-bridging silicon-oxygen groups (NBO), respectively can be observed. Important variations in the intensity of the BO and NBO bands with the silica content of the bioactive glasses are observed. When the content of the network modifiers (Na^+, K^+, Ca^{2+} and Mg^{2+}) in the glass increases the peak intensity corresponding to BO groups decreases due to the formation of NBO groups and vice versa.

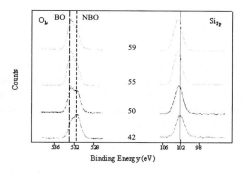

Fig. 2. XPS binding energies of the Si_{2p} and O_{1s} photoelectrons for different bioactive glass coatings.

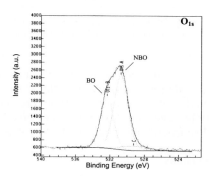

Fig. 3. Deconvolution of the O_{1s} photoelectron band for bioactive glass coating C42.

In Table 2 are summarised the chemical binding energies of the different coatings. An increase of the binding energies associated to Si_{2p}, O_{1s} (NBO) and O_{1s} (BO) photopeaks with the increase of the glass silica content is observed. These results are associated with a change in the chemical environment of the silica structural unit. The vitreous silica structure is formed by SiO_4 tetrahedra connected by BO groups. If network modifiers (Na^+, K^+, Ca^{2+}, Mg^{2+}) are added to the silica network, the ions induce structural changes leading to the breaking of Si-O-Si groups between adjacent tetrahedra. The charge is compensated by the formation of NBO groups, which are associated with a nearby alkali or alkali earth ion.

Table 2. Chemical shift of the peaks corresponding to Si_{2p} and O_{1s} (NBO and BO).

	Si2p	O1s (NBO)	O1s (BO)
Coating 42	102.08	530.73	532.33
Coating 50	102.44	530.82	532.53
Coating 55	102.65	531.08	532.67
Coating 59	102.72	531.06	532.64

SEM/EDS analyses of the *in vitro* tested bulk glasses and of the corresponding PLD coatings after immersion in SBF are shown in Figs. 4 and 5. The bulk glasses (Fig. 4) present the three expected layers (CaP-rich, CaP-SiO_2 mixture and SiO_2-rich), which corresponds to different phases of the chemical routes of the bioactivity process explained previously. When the silica content of the glass decreases, the thickness of the CaP-rich layer increases and, therefore, the glass bioactivity increases due to the presence of the network modifiers.

The coatings grown by PLD (Fig. 5) show a similar behaviour than the bulk glasses (Fig. 4) but two important differences can be observed: a decrease of the layer thickness and a shifting of the *in vitro* bioactivity behaviour with respect to the bulk glass. It should be noted that a similar bioactivity grade was found for the bulk BG50 and the coating BG42.

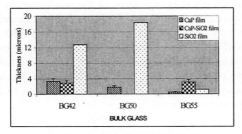 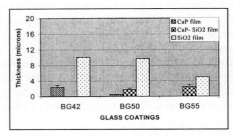

Fig. 4. Thickness of CaP, CaP-SiO$_2$ and silica-
rich layers of the bulk bioactive glasses.

Fig. 5. Thickness of CaP, CaP-SiO$_2$ and silica-rich
layers of the bioactive glass coatings.

This phenomenon corroborates that the bonding configuration of the bulk glass is not correctly transferred to the coating during the ablation process, and this effect plays an important role in the bioactivity of the materials. Therefore, the growth of thin film bioactive materials should be carefully controlled because its composition and bonding configuration are key factors for the development of biomedical products with an adequate biological response.

Conclusions

FTIR and XPS analyses demonstrate that in the laser ablation process the glass bonding configuration is not congruently transferred to the coatings. This result is in agreement with the *in vitro* bioactivity studies. Composition and bonding configuration of the coatings should be controlled in order to obtain an adequate biological response.

Acknowledgements

The work was supported by Xunta de Galicia (PGIDT02PXIC30302PN), Universidade de Vigo (64502I908 and 6452I106) and Ministerio de Ciencia y Tecnología (MAT2001-3434). The authors thank Dr. H. Ylänen for providing some glasses and Ángeles Fernández for her collaboration during the experimental work.

References

[1] L.L. Hench and J. Wilson: *An Introduction to Bioceramics* (World Scientific, Singapore, 1993).
[2] T. Kokubo, M. Tanahashi, T. Yao, M. Minoda, T. Miyamoto, T. Nakamura and T. Yamamuro: Bioceramics Vol. 6 (1993), p. 327.
[3] L. D'Alessio, R. Teghil, M. Zaccagnino, I. Zaccardo, D. Ferro and V. Marotta: Appl. Surf. Sci. Vol. 138-139 (1999), p. 527.
[4] J. Serra, P. González, S. Chiussi, B. León and M. Pérez-Amor: Key Engineering Materials Vol. 192-195 (2001), p. 635.
[5] S. Liste, P. González, J. Serra, J.P. Borrajo, S. Chiussi, B. León, M. Pérez-Amor, J. García López, F.J. Ferrer, Y. Morilla and M.A. Respaldiza: Thin Solid Films (2003), in press.
[6] P. González, J. Serra, S. Liste, S. Chiussi, B. León and M. Pérez-Amor: Vacuum Vol. 67 (2002), p. 647.
[7] J. Serra, P. González, S. Liste, S. Chiussi, B. León, M. Pérez-Amor, H.O. Ylänen and M. Hupa: J. Mat. Sci: Mat. Med. Vol. 13 (2002), p. 1221.
[8] J. Serra, P. González, S. Liste, C. Serra, S. Chiussi, B. León, M. Pérez-Amor, H.O. Ylänen and M. Hupa: J. Non-Cryst. Solids (2003), in press.

Key Engineering Materials Vols. 254-256(2004) pp. 359-362
online at http://www.scientific.net
© 2004 Trans Tech Publications, Switzerland

Calcium Phosphate Porous Coatings onto Alumina Substrates by Liquid Mix Method

J. Peña, I. Izquierdo-Barba, M. Vallet-Regí *

Dpto. Química Inorgánica y Bioinorgánica. Fac. Farmacia. UCM, 28040-Madrid, Spain

e-mail:vallet@farm.ucm.es

Keywords: Coating technologies-chemical and physical

Abstract. The objective of this work was to cover alumina substrates with a calcium phosphate layer by the Liquid Mix method. This method was employed to prepare solutions with Ca/P ratios of 1.5 and 1.67 in which Al_2O_3 were dip coated 1, 5 or 10 times.. Depending on the precursor solution employed β-TCP (1.5) or HA/β-TCP (1.67) layers were obtained after annealing at 800 or 1000°C,respectively. All the coatings yield a porous morphology that increases with their thickness.

Introduction

Ceramic materials such as alumina (Al_2O_3) or yttria-stabilized zirconia (YSZ) have high wear resistance and high strength, but forms no chemical or biological bond between tissues and their surfaces. The concept of coating these materials with calcium phosphates lies on the synergic combination of the mechanical benefits of the former with the biocompatibility of hydroxyapatite (HA) or β-tricalcium phosphate (β-TCP). These two calcium phosphate phases are implanted both as single phases or forming biphasic mixtures that combine the higher resorbability of β-TCP with the superior biocompatibility of HA. The coatings were prepared by the dip-coating method starting from solutions obtained from the Liquid Mix Method. This synthetic route, based on the Pechini´s technique [1], commonly used to obtain different oxides [2,3] has been successfully applied to the preparation of calcium phosphates [4] and is adapted in this work to the preparation of coatings of hydroxyapatite and β-tricalcium phosphate.

Materials and Methods

Briefly (fig. 1), the metallic salts ($CaNO_3.4H_2O$ and $H_2NH_4PO_4$), with molar Ca/P ratios of 1.5 and 1.67, were dissolved in 0.2M citric acid (CA) solutions in water previously heated on a stirring plate. After thorough mixing of the solution, the volume was reduced by slow heating on a hot plate with continuous stirring, at this moment ethylene glycol (EG) (at a 1:1 molar ratio with CA) is added to one of the solutions. The so obtained solutions were diluted in ethanol (1:2) and immediately employed to cover polycrystalline alumina substrates (10 x 10 x 0,65 mm) by the dip-coating method. Dipping was carried out by means of an outfit designed and implemented at the Condensed Matter Physics Department of Universidad de Cádiz. The substrates were extracted at a rate of 1000 μm/s, dried in air and annealed at a rate of 1°C/min at 400°C for 12 hours. This procedure was repeated 1, 5 or 10 times to yield the 1c, 5c and 10c coatings, respectively.

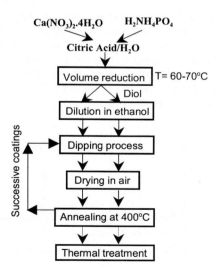

Fig. 1: Synthetic route employed

The layers obtained from the 1.5 and 1.67 solutions were annealed at 800 or 1000°C, respectively. The surface of the coatings was studied by XRD, FTIR and AFM. Rheological measurements were carried out to characterize the precursor solutions.

Fourier transform infrared (FTIR) spectra were obtained in a Nicolet Nexus spectrometer equipped with a Smart Golden Gate ATR accessory. The crystallinity of the materials was analyzed by X-ray diffraction (XRD) in a Philips X-Pert MPD diffractometer equipped with a thin film attachment. The fixed incidence angle was 1° for all samples. Surface morphology was analyzed by scanning in a JEOL 6400 electron microscopy. The Ca and P content was determined by energy dispersive spectroscopy (EDS) using a LINK 10000 analyzer on samples covered with graphite. The topography of the samples was obtained by using an Autoprobe-CP SFM from Park Scientific Instr working in contact mode. Cantilevers with a 0.4 N/m force constant and conical tip were used.

Results and Discussion

The solutions prepared were freshly employed as they degrade: i.e. precipitate, few hours after its preparation. Viscosity measurements were carried out on both types of solutions before and after ethanol addition. These measurements show a Newtonian-like behavior in all cases with similar viscosity values for LM (24 mPa) and CIT (18 mPa) solutions. Ethanol addition leads to a decrease on these values down to a unique value of 10 mPa.

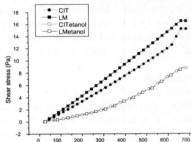

Fig. 2: Viscosity properties of the solutions.

The two precursor solutions (with and without ethylene glycol) yield coatings with no apparent differences, this fact can be explained considering the similarity in their rheological properties between these two solutions after its dilution with ethanol and the complete elimination of the precursors with the thermal treatment.

Single phase β-TCP coatings are obtained from the Ca/P=1.5 solutions after a thermal treatment at 800°C (fig 3a). The annealing temperature was chosen considering the results previously obtained for the obtention of β-TCP in the form of powder [2].

In a similar way, the coatings obtained from a solution with a Ca/P=1.67 were annealed at 1000°C. However, it is not possible to obtain single phase hydroxyapatite annealing at 1000°C (fig. 3b); longer annealing times do not produce any evolution in the biphasic -HA/β-TCP (40/60%wt.)-coatings obtained. In fact, thermal treatments at higher temperatures (up to 1300°C) does not produce any variation in the HA/β-TCP proportion; this stabilization effect can be attributed to the alumina substrate.

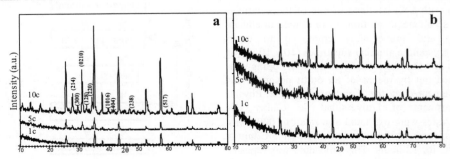

Fig. 3: X-ray diffraction of layers prepared from 1.5 (a) and 1.67 (b) Ca/P solutions

The SEM and AFM characterization is summarized in figure 4, where the results obtained for the Al_2O_3 substrate and for the β-TCP layers (1c and 5c) prepared without the addition of ethylene glycol are collected. All types of coatings are characterized by a porous morphology that increases with the number of layers. This porosity seems to be caused by the elimination of the organic precursors. The number of coatings does not only affect the porosity of the annealed layers, it also determines their roughness and particle size. The AFM characterization allows to quantify the diminution in the roughness (rms) and in the particle size with the number of layers deposited.

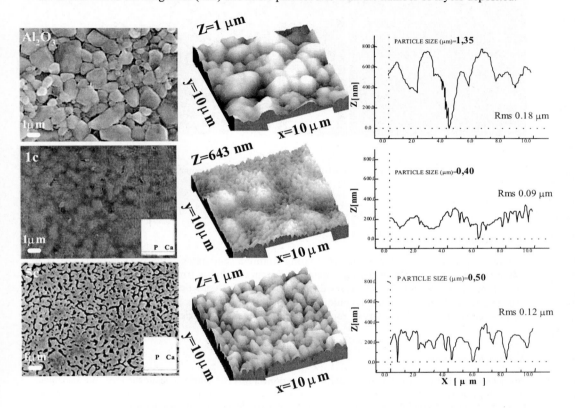

Fig. 4 SEM and AFM micrographs and profiles for the Al_2O_3 substrate and β-TCP layers (1c, 5c) obtained without the addition of ethylene glycol.

In all cases, single coatings (1c) morphology are strongly influenced by the substrate as can be deduced from SEM and AFM results that allow to observe the underlying irregular surface of the Al_2O_3 substrate. In fact, the roughness profile is composed by sharp peaks and valleys attributable to the particles of the calcium phosphate layer and by less defined curves caused by the coarse grains of Al_2O_3. Higher thickness, five coating cycles, causes not only an increase on the porosity of the samples –attributing this term to the cracks generated by the elimination of the precursors– but also an augmentation on the particle size (0.12 μm). In this case the topography of the substrate has been completely covered.

Ten coating cycles yield a more porous films, as can be explained by the higher quantity of precursors to be degassed, but, unfortunately, present detached zones (fig. 5) due to the excessive number of coatings which do not adhere strong enough to constitute an unique layer.

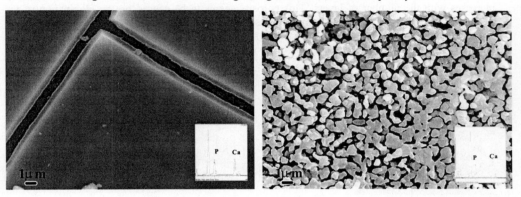

Fig. 5. SEM micrograph of a 10c layer before (a) and after annealing at 1000°C (b).

Conclusions

In this work we have succeeded in coating alumina substrates with β-TCP or HA/β-TCP (40/60) porous layers. These coatings present different roughness (rms), particle size and thickness as a function of the number of layers deposited. Both the synthetic route employed, Liquid Mix Technique, and the deposition technique, dip-coating, are costless and allow to cover substrates with a certain degree of porosity or roughness and complex geometries.

Acknowledgements

Financial support of CICYT, Spain, through research project MAT02-00025 is acknowledged. We also thank A. Rodriguez (Electron Microscopy Center, Complutense University) and F. Conde (C.A.I. X-ray Diffraction Centre, Complutense University), for valuable technical and professional assistance.

References

[1] M.P. Pechini, U.S. Patent No. 3330697, 1967.
[2] M. Vallet-Regí *Perspectives in Solid State Chemistry*, Ed. K.J. Rao, Narosa Pub. House, 1995, pp-37
[3] L-W. Tai, P.A. Lessing, J. Mater. Res. 7, 2 (1992) 502-510.
[4] J.Peña, M. Vallet-Regi, J. Eur. Ceram. Soc. 23 (2003) 1687.

Key Engineering Materials Vols. 254-256(2004) pp. 363-366
online at http://www.scientific.net

Apatite Layers by a Sol-Gel Route

I. Izquierdo-Barba, N. Hijón, M.V. Cabañas and M. Vallet-Regí*

Departamento de Química Inorgánica y Bioinorgánica. Facultad de Farmacia.

Universidad Complutense. 28040-Madrid. Spain. Email: ibarba@farm.ucm.es

Keywords: calcium phosphate, sol-gel, thin films

Abstract. Carbonate-hydroxyapatite coatings were deposited by a sol-gel method using sols with different characteristics, i.e. with different ethanol content. In calcium phosphate film preparation, an important factor in determining the composition of the layers is the aging time of the sol, which in turn depends on the precursor-sol characteristics. The shortest time to synthesize hydroxyapatite was observed when an aqueous system was used, but the sols containing ethanol are stable for a longer time, and more homogeneous coatings are obtained.

Introduction

Ceramic coatings over metallic substrates combine the good mechanical properties of the metals with the excellent biocompatibility and bioactivity of the ceramics. Different methods have been developed to coat metallic implants with calcium phosphate layers [1]. The sol-gel route is a wet chemical method which does not require high vacuum nor high temperatures, and is considered to be one of the most flexible techniques for the preparation of high quality thin films [2,3]. However, one of the main problems of this method when applied to calcium phosphates is the aging time of the precursor-sol, which influences the phase composition of the coatings, as well as its textural properties, homogeneity, etc.

In this work, the influence of the sol characteristics in the preparation and features of hydroxyapatite (HA) coatings onto Ti6Al4V substrates using the dip-coating technique is studied.

Materials and Methods

For the preparation of sols, triethyl phosphite, $P(OCH_2CH_3)_3$, was dissolved in ethanol (EtOH) or hydrolyzed in water (in a 1:4 ratio). A calcium nitrate solution, $Ca(NO_3)_2.4H_2O$, aqueous or ethanolic, was added to the phosphite sol, in a Ca/P ratio equal to 1.67 (Figure 1). The mixed sol solution was agitated during 15 minutes and aged at 60°C in an oven during different times before being used to make coatings. The pH and viscosity of the sols were continuously monitored throughout the aging period.

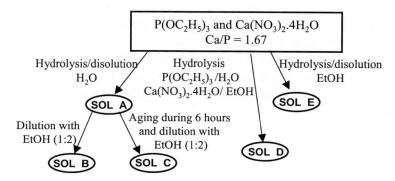

Fig. 1 Schematic diagram showing the different precursor-sols used for coating preparation.

For the preparation of films, the Ti6Al4V substrates were dip coated with the precursor-sol solution, with a withdrawal speed varying from 200 to 2500 µm/s. The coatings were dried and annealed in air at 550°C for 10 minutes. Figure 2 shows a scheme of the different steps used in film preparation by the dip-coating method. The coatings were characterized by X-ray diffraction (XRD) and Fourier Transform Infrared (FTIR) spectroscopy. The surface of the films was characterized by Scanning Electron Microscopy (SEM) and Scanning Force Microscopy (SFM).

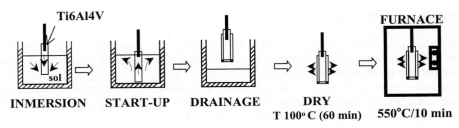

Fig. 2 Scheme followed for the preparation of coatings by the dip-coating technique

Results and Discussion

A continuous pH measurement of the precursor-sol solutions shows a decrease of pH values as the aging time increases, with a larger decrease for the totally aqueous system.

XRD patterns show different crystalline phases, for the same aging time, as a function of characteristics of the precursor-sols. The minimum aging time to obtain pure HA layers, 6 hours, is observed when the sol A is used. At this time, sol A and C lead to single apatite phase coatings, whereas $CaCO_3$ (calcite) is observed when sol B is used (Figure 3a). For this aging time, it was not possible to obtain coatings using the sols with higher ethanol content (sol D and E) because no adherence was observed between the sol-gel and the substrate at this aging period.

Also, the XRD study shows that in presence of ethanol, the aging time to deposit monophasic HA films is higher. The HA coatings have similar XRD patterns, irrespective of the diluting medium employed.

It is interesting to remark, from a practical point of view, that sol A, aqueous system, is the less stable; after 12 hours, it is not possible to obtain thin films, because powder is formed onto the substrate, whereas more than 8 days are needed to observe powder when the sol E was used.

Figure 3b shows FTIR spectra of pure HA films deposited by using sol A. Similar spectra were observed in coatings deposited by using the other sols. The FTIR spectra show characteristic apatite absorption bands [4]. The most intense bands, in the ranges 550-610 and 950-1100 cm^{-1}, can be assigned to the major absorption modes of the phosphate groups. The bands at 872, 1446 and 1455 cm^{-1} correspond to CO_3^{2-} groups. These bands are assigned to a B-type carbonate substituting for phosphate groups in the apatite structure [5]. Then, the obtained coatings can be considered as a carbonate-hydroxyapatite.

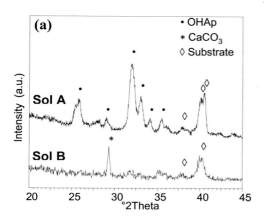
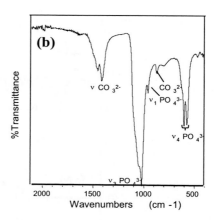

Fig. 3 (a) XRD patterns corresponding to coatings deposited by using sol A and B at 6 hours of aging time; (b) FTIR spectra corresponding to a HA coating deposited by using sol A.

Film characterization by SEM and SFM reveals slight differences in textural properties, depending on the characteristics of the sol. The study by SEM shows that all the obtained films are dense and homogeneous. The SEM micrographs show that the coatings deposited using the aqueous system, sol A, are constituted by a smooth surface, similar to that observed for the uncoated metal substrate (Figure 4), but the EDS analysis allows to distinguish the presence of calcium and phosphorous on the substrate coated with sol A. Films with higher roughness seem to be deposited when the ethanol content into the precursor-sol increases (Figure 4).

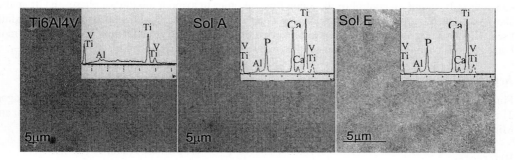

Fig. 4 SEM micrographs corresponding to uncoated substrate and HA coatings deposited by using an aqueous system, sol A, and ethanolic system, sol E.

The advantages in the use of SFM lay on the real 3D information offered, which allows to study quantitatively the evolution of the surface roughness versus the characteristics of precursor-sol. Figure 5 shows the profiles obtained by SFM, corresponding to uncoated Ti6Al4V and coatings deposited from sol A and E. In agreement with the SEM results, the SFM study shows that, in the case of sol A, the obtained coatings are very dense with a roughness value close to 4nm; the films deposited by using the ethanolic sol are more rough and porous, with a pore size in the range between 0.35 and 0.15 nm.

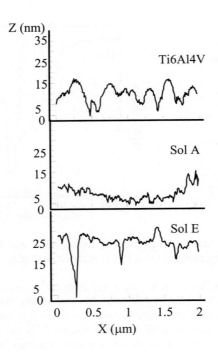

Fig. 5 Profiles corresponding to
uncoated Ti6Al4V and coatings
deposited from sol A and sol E.

Conclusions

The preparation of monophasic HA coatings onto metallic substrates, working with aqueous or ethanolic sols, implies a control of the aging time, in order to avoid a mixture of phases. FTIR studies show that the films deposited correspond to a carbonate-HA type B, independently of the medium used. Higher aging times are needed to obtain pure HA using EtOH containing sols, although more homogeneous coatings are then obtained; this, in turn, allows coating larger surfaces more easily.

Acknowledgements

Financial support of CICYT (Spain) through Research Project MAT2002-0025 is acknowledged. The XRD, SEM and SFM measurements were performed at C.A.I Difracción de Rayos X and Microscopia Electrónica, Universidad Complutense, respectively.

References

[1]. W. Suchanek and M. Yoshimur, J. Mater. Res. vol.13 (1998), p. 94-117.
[2]. W. Weng, J. Am. Ceram. Soc. vol. *82*, (1999), p. 27-32.
[3]. I. Izquierdo-Barba, M. Vallet-Regí, J.M. Rojo, E. Blanco and L. Esquivias. J. Sol-gel Science and Technology, vol. 26 (2003), p. 1179-1182.
[4]. B.O. Fowler, Inorganic Chemistry, vol. 13 (1974), p. 194-206.
[5]. J.C. Elliot, *"Structure and Chemistry of the Apatites and other Calcium Orthosphosphates"*, *Studies in Inorganic Chemistry* 18, (Ed. Elsevier, Amsterdam 1994).

Key Engineering Materials Vols. 254-256(2004) pp. 367-370
online at http://www.scientific.net
© 2004 Trans Tech Publications, Switzerland

Tribological Behaviour of Prosthetic Ceramic Materials Sliding Against Smooth Diamond-Coated Titanium Alloy

C. Met, L. Vandenbulcke, M.C. Sainte Catherine[#] and L. Quiniou[#]

LCSR – CNRS et Université d'Orléans, 45071 Orléans cedex 2, France,
met@cnrs-orleans.fr, vanden@cnrs-orleans.fr

[#] DGA / CTA, 16 bis av. Prieur de la Côte d'Or, 94114 Arcueil cedex, France,
Marie-Christine.Sainte-Catherine@dga.defense.gouv.fr

Keywords: Nano-smooth surfaces, diamond coatings, Ti-6Al-4V alloy, alumina, zirconia, pin-on-disc friction tests, coefficients of friction, wear rates, Ringer's solution, synthetic serum

Abstract. Duplex coatings with an external nano-smooth fine-grained diamond layer, a thin titanium carbide interlayer and a carbon diffusion layer have been deposited by PACVD on titanium alloy at 600°C. Rotating pin-on-disc friction tests have been carried out at room temperature in ambient air, in Ringer's solution and synthetic serum to approach the in-vivo wear conditions, with hemispherical pins fabricated from alumina and zirconia balls and diamond-coated Ti-6Al-4V samples as the discs. The friction coefficients and wear rates of the different materials allow one to compare their performance and to demonstrate the potential of the nano-smooth diamond coatings for bio-mechanical applications.

Introduction

While ceramic parts are usual components of orthopaedic implants, the use of ceramic coatings on metallic alloys is restricted to diamond-like-carbon on stainless steel or titanium alloy and oxidized zirconium. This seems due to the risk of delamination of the coatings. A very strong adhesion on a usual substrate is therefore the first condition that a ceramic coating must satisfy. A very good tribological behaviour of this coating is obviously also required against biomaterials.

Nano-smooth diamond coatings on titanium alloys are presented here. These coatings have potential applications in the field of prostheses because of their high resistance to corrosion and wear. Their nano-smooth surface might be interesting because it gives low friction coefficient and low wear of various counterface materials in ambient air. Their friction behaviour against various alloys and against UHMWPE has already been reported in ambient air and in solutions [1,2]. Their tribological performance when sliding against ceramic counterfaces is reported here.

Materials and Methods

Very adherent nano-smooth fine-grained diamond (SFGD) coatings have been deposited by a two-step microwave plasma-assisted chemical vapor deposition (PACVD) process on titanium alloys or titanium-coated substrates at 600°C. The procedure allows to avoid the incorporation of important amount of hydrogen and oxygen at the interface while an oxygen-containing mixture can be used for the deposition of the diamond coating at moderate temperature [3]. These coatings have been characterized by visible and UV Raman spectroscopy, X-ray diffraction, SEM, TEM / EELS. Their surface morphology and roughness have been studied by AFM and the interfaces and titanium carbide interphase with the substrate have been characterized by XPS and SIMS [4]. Their intrinsic mechanical properties have been determined by nanoindentation and Brillouin light scattering [5,6].

The friction tests were carried out using a rotating pin-on-disk tribometer with a fixed pin. Diamond-coated Ti-6Al-4V samples were used as the discs. The counterface materials were hemispherical pins, 6 mm in diameter, fabricated from commercial alumina (Inceral®) and zirconia (Prozyr®) balls. The radius of curvature of the pin tip was 14 mm, corresponding to ceramic heads for hip prostheses of 28 mm in diameter. The sliding velocity was 0.1 m.s^{-1} and the normal applied load was 13 N, leading to a contact pressure of about 450-500 MPa, the influence of the coating being negligible. Besides the measurement of the pin height decrease during the test, the final specific wear rates of the pins were determined from the diameter of the wear scars. The tests were conducted at room temperature in various media, in ambient air at relative humidity in the 40-60% range, in the Ringer's solution and a synthetic serum (Plasmion® from Rhône Poulenc which contains 25 g/l proteins) to approach the in-vivo wear conditions. After the friction tests, the pin tips and the disc tracks were investigated by AFM and SEM.

Results

Coating Characteristics. Contrary to the classical polycrystalline diamond coatings, the SFGD coatings exhibit a smooth surface, in the 15-35 nm (rms) range as a function of the deposition conditions, the lowest surface roughness being obtained when some more sp^2-hybridized carbons are incorporated in the coatings. However the hardness and the Young's modulus are always high, in the 70-90 GPa and 600-900 GPa ranges respectively [5,6].

Another important characteristic of these coatings is due to the relatively high diffusion coefficients of carbon in titanium and in titanium carbide at moderate temperature that permits to obtain a duplex coating with an external diamond layer, a thin titanium carbide interlayer and a carbon diffusion layer in the α (and β) solid solution(s) [1]. The formation of this thin titanium carbide phase and of the diffusion layer in the substrate allows a very strong bonding of the diamond layer to be obtained as will be shown in the next section.

Adherence of the Coatings. Figure 1 shows the surface of a SFGD coating after a Rockwell C indentation test under heavy load (1470 N). Only radial and orthoradial cracks are observed without delamination. The same test carried out on a poly-crystalline coating (Fig. 2) allowed to evaluate the stresses by Raman spectro-scopy as a function of the reduced distance from the center of the indent. The results are reported on fig. 3.

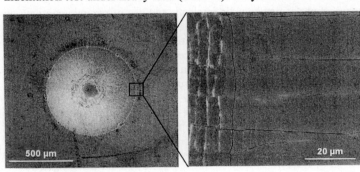

Fig. 1: SEM images of the indent on the SFGD coating.

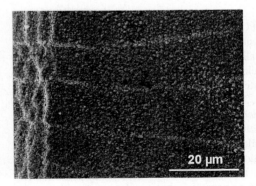

Fig. 2: SEM image of the indent edge on a polycrystalline coating.

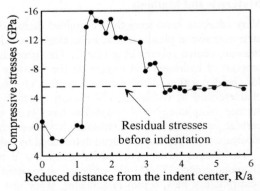

Fig. 3: Variations of the diamond coating stresses versus the distance from the indent center (a=indent radius).

While a relaxation of the residual stresses is observed inside and near the edge of the indent, high compressive stresses, up to 16 GPa, are observed which then decrease to the actual value in the coating. According to the VDI guidelines 3198 (1991), this test permits to conclude that the adherence of the coating is high. Therefore such duplex coating with an internal diffusion layer, a strongly bonded carbide interphase and an external diamond-based layer appears very attractive for applications requiring no risk of delamination.

Tribological Tests.   Figure 4 shows the variations of the coefficient of friction recorded in ambient air, in the Ringer's solution and the synthetic serum when zirconia and alumina pins respectively are sliding against polished (final roughness of about 10 nm as measured by AFM) SFGD coatings deposited on Ti-6Al-4V discs. In ambient air, the friction coefficients are in the 0.15-0.2 range for both pin materials. They are much lower in the solutions, especially for alumina, lower than 0.03 in the Ringer's solution, but a little bite more in the synthetic serum, about 0.04.

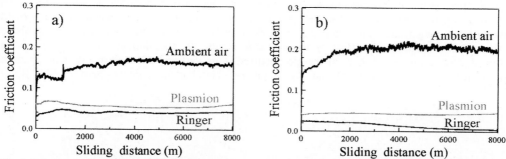

Fig. 4: Friction coefficients versus sliding distance of: a) zirconia and b) alumina on SFGD coatings (sliding speed = 0.1 ms^{-1}, normal load = 13 N) in different environments.

The wear rates being very low, they could not be measured during the test. The final mean wear rates deduced from the diameter of the wear scars or from the pin profiles are reported in figure 5.

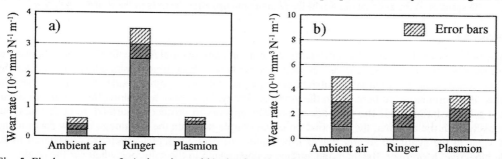

Fig. 5: Final wear rates of: a) zirconia and b) alumina pins after sliding on SFGD coatings for 8 km (sliding speed = 0.1 ms^{-1}, normal load = 13 N). Note that the wear scales of a) and b) differ by a ten factor.

In ambient air, the wear rates are of the same order of magnitude for both zirconia and alumina pins (few 10^{-10} mm^3.N^{-1}.m^{-1}). A coarse micro-polishing of the pin occurred during the friction tests that eliminated the initial scratches of the polished balls which can be observed on figure 6a.

In the Ringer's solution, wear rates of the zirconia counterface are much higher, about 3×10^{-9} mm^3.N^{-1}.m^{-1}, as can be seen on figures 5a and 7a. This could be attributed to a tribo-corrosion effect of this saline solution, the friction tests producing some pulling out of the grains, as observed on the AFM images of the wear scar (Fig. 6b). For the alumina pins, such phenomenon was limited. As a result, the final wear rates of the alumina pins are low, about 2×10^{-10} mm^3.N^{-1}.m^{-1} (Figs 5b and 7b).

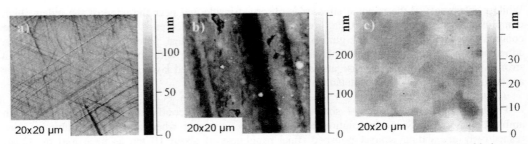

Fig. 6: AFM images of: a) an alumina pin before the friction test; b) a zirconia pin after the friction test carried out in the Ringer's solution (note the very different z range {300 nm} in that case); c) an alumina pin after the friction test carried out into the synthetic serum.

In the plasmion solution, the counterface surface damages are encountered for the zirconia pins only, however they are less sizeable than in the Ringer's solution. The zirconia wear rate is of the same order of magnitude than in ambient air, about 5×10^{-10} mm^3.N^{-1}.m^{-1}. This could be attributed to the protective third body effect of the synthetic serum. On the alumina pins, no damage was observed. Only a micro-polishing of the pin occurred (Figs. 6c) and a low wear rate was obtained, as in the Ringer's solution (Figs. 5b and 7b), in spite of a high hertzian contact pressure (480 MPa).

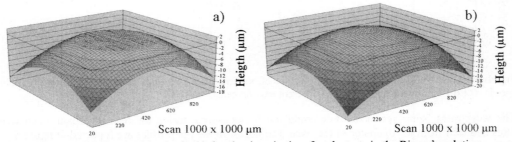

Fig. 7: 3D scans of a) the zirconia pin tip, b) the alumina pin tip, after the tests in the Ringer's solution.

Conclusion

While unidirectional sliding of counterparts do not simulate accurately the final results obtained in multi-directional motion encountered in various prostheses, these first results allow to consider the SFGD coatings as a potentially outstanding material for use against oxide ceramics, especially against the alumina tested here. With the zirconia (Prozyr®), a tribo-corrosion effect seems to occur in the saline solution while the presence of proteins appears to protect its surface. In both cases, but especially with the alumina, the diamond on oxide couple presents the advantage of low friction and low wear in the synthetic serum, that is low amount of ceramic debris which are considered to be little bioactive. Moreover, the diamond surface is undamaged in the solutions whatever the oxide.

References

[1] L. Vandenbulcke, M.I. De Barros, C. Met, G. Farges, Key Engineer. Mat., 218-220 (2002) 595.
[2] C. Met, L. Vandenbulcke, M.C. Sainte Catherine, Wear xxx (2003) xxx (in press).
[3] M.I. De Barros, L. Vandenbulcke, Diamond Relat. Mater., 9 (2000) 1862.
[4] M.I. De Barros, V. Serin, L. Vandenbulcke, G. Botton, P. Andreazza, M.W. Phaneuf, Diamond Relat. Mater., 11 (2002) 1544.
[5] P. Djemia, C. Dugautier, T. Chauveau, E. Dogheche, M.I. De Barros, L. Vandenbulcke, J. Appl. Phys., 90 (2001) 3771.
[6] L. Vandenbulcke, M.I. De Barros, Surf. Coat. Technol., 146-147 (2001) 417.

Key Engineering Materials Vols. 254-256(2004) pp. 371-374
online at http://www.scientific.net
© 2004 Trans Tech Publications, Switzerland

The Role of Processing Parameters on Calcium Phosphate Coatings Obtained by Laser Cladding

F. Lusquiños[1], J. Pou[1], J.L. Arias[1], M. Boutinguiza[1], B. León[1], M. Pérez Amor[1], S. Best[2], W. Bonfield[2], F.C.M.Driessens[3]

[1]Dpto. Física Aplicada, Universidade de Vigo, Lagoas-Marcosende 9, E-36200 Vigo, Spain, jpou@uvigo.es

[2] Dpt. Of Materials Science and Metallurgy, Univ. of Cambridge, Great Britain

[3] University of Nijmegen, Calcio B.V., The Netherlands

Keywords: calcium phosphate, laser cladding, coatings

Abstract. Laser surface cladding has become an extensively technique used in metallurgical applications in order to improve surface properties of materials. We have proposed this technique in the field of biomaterials to coat the surface of titanium alloy substrates used in orthopaedical implants with a calcium phosphate (CaP) bioceramic to promote the growth of the bone when the implant is inserted in the body.

An exhaustive study about the influence of the processing parameters on the CaP coating properties was carried out. To tackle this subject, an appropriate selection of the relevant processing parameters was done.

Introduction

Several techniques have been proposed to apply bioceramic coatings on metallic substrates, such as electrophoretic deposition [1], magnetron sputtering deposition [2], ion-beam deposition [3] and pulsed laser deposition [4] and the widely used method commercially available plasma-spray [5]. In the last years, a novel technique in the biomaterials field has been introduced: laser surface cladding [6]. This surfacing technique is well-known in the metallurgical field as a way to improve the mechanical and physico-chemical properties of the surfaces of the materials.

Laser cladding technique is a material processing technique in which a laser is used as a heating source to melt the precursor material (Hydroxyapatite) to be cladded onto the substrate (Ti6Al4V). A gas jet containing fine HA powder is directed via an off-laser beam axis nozzle through the path of the laser beam, which is focused slightly above the workpiece. The powder heats up and this laser/particles stream is traversed across the workpiece. A small melt-pool is formed on the surface of the workpiece where the molten particles land to form a strip upon cooling, about several hundreds of microns thick, after the laser beam has moved on. The characteristics of a built up strip mainly depends on the processing parameters. Therefore, these parameters governing the cladding process have to be studied carefully. They play an important role in determining the clad profile, the dilution of the metallic substrate, the cracking due to thermal stresses, composition and microstructural evolution from the top of the coating to the substrate. Dilution is an important factor and a desired range should be set to assure the strength of the chemical bonding between the coating and the substrate avoiding the presence of the metallic substrate in the top layers of the coating. It is not possible to predict the influence of a single parameter on the cladding process. In general, several parameters have to be varied simultaneously to obtain the desired characteristics of the deposited coating [7].

Screening experiments and investigation works of other authors [8,9] directed us to reduce the processing parameters to three of them: powder feed rate, laser beam power density and CNC table feed rate. The influence of these parameters in the properties of the clad coatings are discussed.

Materials and Methods

Materials

Ti alloy (Ti6Al4V) plates with a thickness of 6 mm and dimensions of 50x50 mm^2 were used as substrates with as-received surface finishing. The hydrosyapatite powder (LAH-97) was supplied by Fin-Ceramica Faenza (Italy) having a particle size distribution ranging from 50 to 500 µm. The Ca:P molar ratio value of 1.68 confirm the quasi stechiometric composition of the HA, which, at the same time, presents a high crystallinity ($\geq 95\%$) as it was demonstrated by X-ray diffraction (XRD).

Methods

The powder blowing technique was used to produce the coatings by laser cladding. A pulsed Nd:YAG laser (Rofin Sinar RSY 500P) guided by optical fibre was used. The precursor powder (hydroxyapatite) was supplied by a pneumatic conveyor specifically designed to keep an uniform mass flow. The stream of the powder was injected in the molten pool by an off-axis nozzle. The relevant processing parameters was reduced to the laser power (300-480 W), the powder mass flow (7,0-20 mg/s) and the transversal speed of the substrate (0,5-1,5 mm/s). The lower value of transverse speed interval was theoretically obtained to allow melting of the titanium alloy surface and the upper value guaranteed the melting of the titanium alloy surface for the maximum laser power density. The influence of these parameters on the geometrical (height, width and aspect ratio), finishing and compositional properties of CaP strips obtained were studied by analytical techniques (scanning electron microscopy, energy dispersive X-ray spectroscopy and X-ray diffraction).

Results

Geometrical properties of ceramic coatings (strip dimensions: width, height, dilution)

The width of the ceramic strips is directly and inversely proportional to laser power density and the transverse speed respectively, being negligible the role of powder mass flow due to the slight attenuation of the laser beam power. These results can be clearly seen in fig. 1 (a).

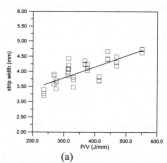
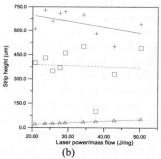

(a)   (b)

Fig 1 (a). Lineal relation between width of strip and lineal power density. Fig1 (b).Lineal relation between height of strip and energy per mass of powder for different CNC transverse speed(+: 0,8 mm/s, □: 1mm/s, Δ: 1,2 mm/s)

The height of the strips depends on the powder mass flow and laser power density if the laser power density is high enough to melt the substrate and powder mass flow. This behaviour can be clearly seen in fig. 1(b).

These results must be taken into account to define the overlapping of strips to coat a surface. Therefore, the aspect ratio (width/height) has to reach the value of 5 in order to avoid inter-strips pores.

The geometrical dilution grows with the laser power density and decreases with the powder mass flow and transverse speed as reported by many authors[7]. Its behaviour is shown in figure 2.

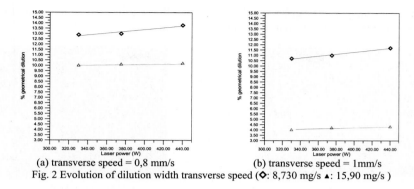

(a) transverse speed = 0,8 mm/s (b) transverse speed = 1mm/s

Fig. 2 Evolution of dilution width transverse speed (◆: 8,730 mg/s ▲: 15,90 mg/s)

Finished surface

The roughness of the surface and the number of cracks per lenth unit were studied, verifying their growing with the transverse speed due to the thermal stresses caused by the rapid thermal solidification and the dissimilar expansion coefficient of substrate and ceramic coating. Also, the powder mass flow plays a significant role in the roughness if it reachs a value that provokes insufficient melting of the substrate and powder by the laser beam.

Composition

In order to understand the composition of the coating it is essential to use high temperature phase diagrams of ceramists where the phases obtained by the heating of HA are compiled [10]. From XRD spectra, it can be concluded the presence of different phases at the surface of the coating: α-TCP, $Ca_2P_2O_7$, HA (fig. 3); and HA glass was revealed from the EDX analysis.

The presence of $CaTiO_3$ increases with depth from the top of the coating. An inverse behavior can be seen for the evolution of calcium phosphate phases. The interface of the Ti alloy and the coating is enriched in calcium titanates and compounds from Titanium and Phosphorous (fig. 4) which reveals the chemical bonding between coating and substrate. This fusion bonding is responsible of the strong adhesion of the coating.

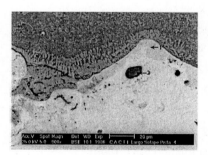
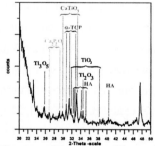

(a) SEM image of interface (1:Ti_3P_2, 2:$Ca_3Ti_2O_7$, 3:$CaTiO_3$) (b) XRD pattern of the surface of the coating with 1° grazing angle

Fig. 3 (a) Composition of the interface and (b) Phases on the top of the coating by XRD analysis by grazing angle

The presence of the different calcium phosphates can be attributed to the non uniform temperature field in the molten pool, and the presence of HA is assured if the laser beam power is not high enough to melt the HA particles during flight and the transverse speed is high enough to melt partially the substrate and the impinging powder cloud. Therefore, the higher contents of HA are obtained with increasing CNC transverse speed.

Finally, the different in-depth microstructures were found. A refined structure was the result of the higher CNC transverse speed and lower laser beam power. In this way a rapid solidification takes place. Moreover refined grains can be found on the top of the coating, and an increase of grains size is observable at the bottom of the coating (figure 4).

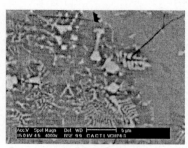

(a) surface   (b) 200 μm in depth

white zones: CaTiO₃ and gray zones: calcium phosphates

Fig. 4 Evolution of in-depth microstructure for a strip (transverse speed = 0,6 mm/s, powder mass flow = 8,750 mg/s, laser beam power density = 3500W/cm²)

Conclusions

The laser cladding technique produced an homogeneous calcium phosphate layer onto the surface of a titanium alloy without previous treatment of the surface.

Different phases of calcium phosphate were found including tricalcium phosphate (α-TCP), HA and a calcium phosphate glass with the same Ca/P ratio than HA. Both components (HA and HA glass) are localized on the surface of the coating. Strong fusion bonding between the CaP coating and the titanium alloy has been achieved through the formation of calcium titanates and titanium phosphates at the coating/substrate interface.

Literature References

[1] J.G.C. Wolke, et al.: J. Biomed. Mat. Res. Vol. 28 (1994), p.1477

[2] T.S.Chen et al.: J. Mater. Res. Vol. 9 (1994), p. 1284

[3] M. Shirkhazadeh et al.: Mater. Lett. Vol.18 (1994), p. 211

[4] C.M. Cottell et al.: J. Appl. Biomat. Vol.3 (1992), p. 87

[5] M.J.Filiaggi et al.: J. Biomed. Mater.Res. Vol.27 (1993), p. 191

[6] F. Lusquiños et al.: J. Appl. Phy. Vol. 9 (2001), p. 4231

[7] M.F. Schneider: "Laser cladding with powder" PhD.Thesis (1998), Univ. of Twente

[8] R. Mallikharjua et al.: http://web.umr.edu/~landersr/

[9] P.A. Vetter et al: Proc SPIE Vol. 2207 (1994), p. 452

[10] F.C.M. Driessens, "Biominerals" (CRC Press, Boca Raton (USA), 1990)

Key Engineering Materials Vols. 254-256(2004) pp. 375-378
online at http://www.scientific.net
© *2004 Trans Tech Publications, Switzerland*

Effect of Surface Morphology and Crystal Structure on Bioactivity of Titania Films Formed on Titanium Metal via Anodic Oxidation in Sulfuric Acid Solution

Tian-Ying Xiong[1], Xin-Yu Cui[1,2], Hyun-Min Kim[3], Masakazu Kawashita[2], Tadashi Kokubo[4], Jie Wu[1], Hua-Zi Jin[1] and Takashi Nakamura[5]

[1]Institute of Metal Research, Chinese Academy of Sciences,
72 Wenhua Road, Shenyang, 110016, P.R. China

tyxiong@imr.ac.cn

[2]Department of Material Chemistry, Graduate School of Engineering, Kyoto University,
Sakyo-ku, Kyoto 606-8501, Japan

[3]Department of Ceramic Engineering, School of Advanced Materials Engineering, Yonsei University
134 Shinchon-dong, Seodaemun-gu, Seoul 120-749, Korea

[4]Research Institute for Science and Technology, Chubu University,
1200 Matsumoto-cho, Kasugai-shi, Aichi 487-8501, Japan

[5]Department of Orthopaedic Surgery, Graduate School of Medicine, Kyoto University,
Sakyo-ku, Kyoto 606-8507, Japan

Keywords: titanium metal, anodic oxidation, sulfuric acid solution, apatite, simulated body fluid

Abstract. Porous titania layers composed of rutile and anatase were formed on the surface of titanium metals by anodic oxidation at voltages of 100, 150 and 180 V in H_2SO_4 solutions, with concentrations of 0.5, 1.0 and 2.0 *M*. Pore size increased with increasing applied voltage from 100 to 180 V under a constant H_2SO_4 concentration of 0.5, 1.0 or 2.0 *M*. The relative amount of rutile to anatase and the apatite-forming ability of titanium metals increased with increasing applied voltage from 100 to 180 V under a constant H_2SO_4 concentration of 0.5, 1.0 or 2.0 *M* and increasing H_2SO_4 concentration from 0.5 to 2.0 *M* under a constant applied voltage of 100, 150 or 180 V. Titanium metal anodically oxidized at 180 V in 2.0*M*-H_2SO_4 solutions showed extremely high apatite-forming ability in simulated body fluid (SBF).

Introduction

Some ceramics such as Bioglass®, sintered hydroxyapatite, glass-ceramic A-W form bonelike apatite layer on their surfaces, and bond to living bone through the bonelike apatite layer. They are called bioactive ceramics, and have been used as important bone repairing materials. They have, however, lower fracture toughnesses and higher in elastic moduli than the human cortical bone, and hence they can not be used under load-bearing conditions. Metals such as titanium metal and its alloys have high fracture toughness, but they do not bond to living bone. Recently, titanium and its alloys were able to bond to living bone, when they were previously subjected to chemical and subsequent heat treatments [1,2].

An essential requirement for an artificial material to bond to living bone is the formation of a bonelike apatite layer on its surface in a body environment. Recently, it has been shown that titania gel with anatase structure showed a high apatite-forming ability in a body environment [3]. Anodic oxidation of a metal is a useful technique for obtaining a metal oxide layer on the metal surface. Therefore, the titania layer with anatase structure can be formed on the surface of tough titanium by anodic oxidation, and hence the anodically oxidized titanium is expected to show a high apatite-forming ability in body environment.

The present authors investigated the apatite-forming ability of titanium anodically oxidized, using various kinds of electrolytes, and revealed that an electrolyte of H_2SO_4 solution is most useful for inducing apatite-forming ability on titanium metal [4]. In the present study, titanium metals were subjected to anodic oxidation at various applied voltages in H_2SO_4 solutions with various concentrations and their apatite-forming ability was investigated in simulated body fluid (SBF).

Experimental

Rectangular specimens of titanium metal (purity: 99.9%, Kobe Steel Co. Ltd., Kobe, Japan) with 10x10x1 mm in size were abraded with #400 diamond plate, and then washed with pure acetone, ethanol and ultra-pure water in an ultrasonic cleaner.

The specimens were subjected to anodic oxidation at room temperature for 1 min under the applied voltages of 100, 150 and 180 V. Sulfuric acid (H_2SO_4) solutions with concentrations of 0.5, 1.0 and 2.0 M were used as electrolyte. After the anodic oxidation, the specimens were gently washed with ultra pure water, and dried at 40°C for 24 h.

The specimens were immersed in 30 mL of an acellular SBF [5] with ion concentrations (Na+ 142.0, K^+ 5.0, Ca^{2+} 2.5, Mg^{2+} 1.5, Cl^- 147.8, HCO_3^- 4.2, HPO_4^{2-} 1.0, SO_4^{2-} 0.5 mM) and pH nearly equal to those of human bone at 36.5°C for 3 and 7 days. After removal from solution, specimens were washed with distilled water and dried in air at 40°C for 24 h. The specimens were analyzed by thin-film X-ray diffraction (TF-XRD, Rigaku RINT-2500) and field-emission scanning electron microscopy (FE-SEM, Hitachi S-4700).

Results and discussion

Figure 1 shows TF-XRD patterns of the surfaces of the specimens anodically oxidized at different applied voltages in H_2SO_4 solutions with different concentrations. It can be seen from this figure that titania layer composed of a large amount of rutile and a small amount of anatase was formed on the surface of the specimens by anodic oxidation. The relative amount of rutile to anatase increased with increasing concentration of H_2SO_4 and applied voltage.

Figure 2 shows FE-SEM photographs of the surfaces of specimens anodically oxidized at different applied voltages in H_2SO_4 solutions with different concentrations. The specimens showed little change in their surface morphology, when they were anodically oxidized at an applied voltage

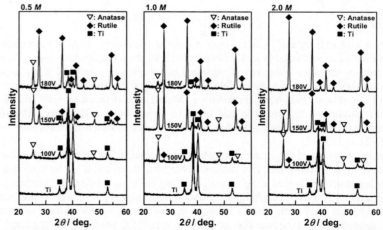

Fig. 1 TF-XRD patterns of the surfaces of the specimens anodically oxidized at different applied voltages in H_2SO_4 solutions with different concentrations.

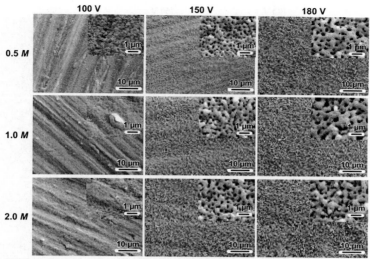

Fig. 2   FE-SEM photographs of the surfaces of the specimens anodically oxidized at different applied voltages in H_2SO_4 solutions with different concentrations.

of 100 V, whereas they originated porous structure on their surfaces, when they were anodically oxidized at higher applied voltages, above 150 V. This indicates that porous titania layers were formed on the surface of the specimens after the anodic oxidation at above 150 V. The pore size slightly increased with increasing applied voltage up to 180 V.

Figure 3 shows TF-XRD patterns of the surfaces of the specimens anodically oxidized at different applied voltages in H_2SO_4 solutions with different concentrations, and subsequently soaked in SBF for 7 days. It can be seen from this figure that the specimens anodically oxidized at 150 and 180 V and those anodically oxidized at 100 V in $2.0M$-H_2SO_4 solution formed apatite on their surfaces, whereas those at 100 V in 0.5 and $1.0M$-H_2SO_4 solutions did not form it on their surfaces in SBF after 7days.

Figure 4 shows FE-SEM photographs of the surfaces of the specimens anodically oxidized at

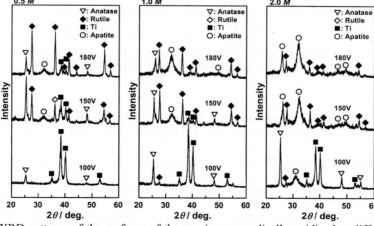

Fig. 3   TF-XRD patterns of the surfaces of the specimens anodically oxidized at different applied voltages in H_2SO_4 solutions with different concentrations and subsequently soaked in SBF for 7 days.

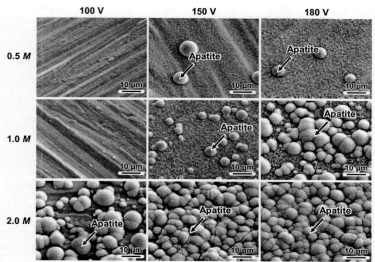

Fig. 4   FE-SEM photographs of the surfaces of the specimens anodically oxidized at different applied voltages in H$_2$SO$_4$ solutions with different concentrations and subsequently soaked in SBF for 3 days

different applied voltages in H$_2$SO$_4$ solutions with different concentrations, and subsequently soaked in SBF for 3 days. Apatite was formed on the surfaces of the specimens anodically oxidized at 100 V in 2.0M-H$_2$SO$_4$ solution and at 150 and 180 V in 0.5 to 2.0M-H$_2$SO$_4$ solutions within 3 days in SBF. The apatite-forming ability of the specimens increased with increasing concentration of H$_2$SO$_4$ and applied voltage.

In conclusion, titanium with high apatite-forming ability can be obtained by the anodic oxidation at 180 V in 2.0M-H$_2$SO$_4$ solution.

Conclusions

Porous titania layers composed of rutile and anatase were formed on the surface of titanium metals by anodic oxidation at voltages of 100, 150 and 180 V in H$_2$SO$_4$ solutions with concentrations of 0.5, 1.0 and 2.0 M.   Titanium anodically oxidized at 180 V in 2.0M-H$_2$SO$_4$ solutions showed extremely high apatite-forming ability in SBF.   Thus obtained titanium with high apatite-forming ability is useful as bioactive metals for bone-repairing.

References

[1] H.-M. Kim, F. Miyaji, T. Kokubo and T. Nakamura (1996) J. Biomed. Mater. Res., 32, 409-417.
[2] W.-Q. Yan, T. Nakamura, M. Kobayashi, H.-M. Kim, F. Miyaji and T. Kokubo (1997) J. Biomed. Mater. Res., 37, 267-275.
[3] M. Kawashita, X.-Y. Cui, H.-M. Kim, T. Kokubo and T. Nakamura (2003) Bioceramics, 16, in press.
[4] M. Uchida, H.-M. Kim, T. Kokubo, S. Fujibayashi and T. Nakamura (2003) J. Biomed. Mater. Res., 64A, 164-170
[5] T. Kokubo, H. Kushitani, S. Sakka, T. Kitsugi and T. Yamamuro (1990) J. Biomed. Mater. Res., 24, 721-734.

Key Engineering Materials Vols. 254-256(2004) pp. 379-382
online at http://www.scientific.net
© *2004 Trans Tech Publications, Switzerland*

Wollastonite Coatings on Zirconia Ceramics

J.A. Delgado[1], F.J. Gil[2], S. Martínez[3], M.P. Ginebra[2], L. Morejón[1], J.M. Manero[1] and J.A. Planell[2]

[1] Centro de Biomateriales, Universidad de La Habana, C. Habana, 10400, Cuba, e-mail: jose.angel.delgado@upc.es

[2] CREB, Dept. Materials Science and Metallurgy, Technical University of Catalonia (UPC) , Av. Diagonal 647, 08028, Barcelona, Spain, e-mail: francesc.xavier.gil@upc.es

[3] Dept. Cristallografia, Mineralogia I Dip. Minerales, Fac Geología, Universitat de Barcelona, 08028-Barcelona, Spain

Keywords: wollastonite, zirconia, bioactive materials, simulated body fluid

Abstract. A wollastonite coating on two different zirconia substrates was studied. The results obtained in the preparation and micro-structural characterization are discussed. The observations by scanning electron microscope (SEM) showed a good deposition of the coating on both zirconia substrates. The presence of peaks attributable to wollastonite and cubic, tetragonal and monoclinic phases of the zirconia at the X-ray diffraction (XRD) patterns were identified. After two weeks in SBF, the observations by environment scanning electron microscopy (ESEM) and SEM revealed the formation of apatite layer which covered fully the samples. The chemical composition of the layer was confirmed by energy-dispersive-spectrometry (EDS).

Introduction

The use of bioactive coatings on zirconia ceramics is an attractive alternative to prepare implant materials which combines the good mechanical properties of the zirconia with the ability to bond with the surrounding tissue owing to the coatings [1]. Wollastonite ($CaSiO_3$) ceramics was studied as material to substitute bone because of its good bioactivity and biocompatibility [2-3]. Some authors reported that the rate of apatite layer formation on the surface of wollastonite ceramics is faster than those of the other biocompatible materials in simulated body fluid solution (SBF) [4]. On the other hand, the zirconia ceramics are already tested in animal models and used in clinical applications and no local or systemic adverse reactions related to the materials were detected [5]. In this work, wollastonite coatings on zirconia were prepared, the mineralogical and chemical compositions, the microstruture and the changes occurring after aging in SBF were studied.

Materials and Methods

The wollastonite powder was obtained from a natural source. Two commercial zirconia ceramics were used as substrates (magnesia-partially-stabilized zirconia (Mg-PSZ) and ytria-partially-stabilized zirconia (Y-PSZ)). The ceramics were cut into small bars and covered with a alcoholic suspension of powdered wollastonite, the suspension also contained some amount of polyvinylalcohol as binder. The thermal treatments were carried out at 1350°C for 2 hours in air. The cross-sections of the covered samples were polished and analysed at the interface by EDS and SEM. SBF solution was prepared by dissolving reagent-grade chemical and buffered to pH 7.4 at

37°C [6]. The samples were immersed in the SBF solution and maintained at 37°C for two weeks. The changes at the surfaces were investigated by ESEM and EDS.

Results and Discussion

The XRD patterns of coated samples showed peaks corresponding to wollastonite and cubic and tetragonal zirconia for Y-PSZ substrate, also small signals attributable to monoclinic phase were detected. These results demonstrated that the thermal treatment did not affect appreciably the mineralogical composition of the Y-PSZ substrate. However, the XRD results for Mg-PSZ revealed the amount considerably of monoclinic zirconia (Fig. 1).

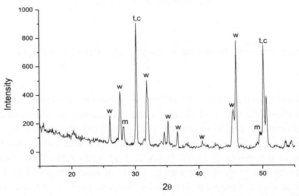

Fig 1. XRD pattern of wollastonite coating on Y-PSZ substrate: (w) wollastonite, (t) tretagonal, (c) cubic and (m) monoclinic zirconia.

A good deposition of the coating on both zirconia substrates was observed by SEM, no cracks at the interface between coating and substrate were detected, also the SEM observation revealed the presence of several pores in the wollastonite coating. This porosity could be induced by an insufficient sintering temperature for the wollastonite. The thickness of the coating was about 200 μm (Fig. 2). The EDS spectra showed some diffusion of silicon from the coating to the zirconia.

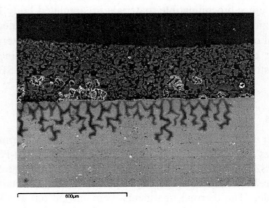

Fig 2. SEM photographs of the wollastonite coating on Y-PSZ substrate.

The SEM view of the surface of covered sample immersed in SBF solution for two weeks is shown in Fig. 3. The surface of the wollastonite coating after immersion was completely covered by apatite layer. The formation of apatite on the surface of the coating was confirmed by EDS.

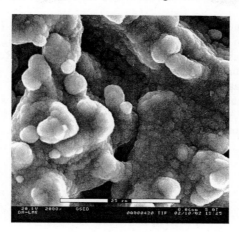

Fig 3. SEM photograph of the surface of wollastonite coating on Y-PSZ substrate soaked in SBF for two weeks.

The rate of formation of apatite layer after aging in SBF for two weeks was very similar for both zirconia substrates. The calcium/phosphate (Ca/P) ratio obtained from EDS results was higher than (1.67) which is the value for stoichiometric apatite. This result should be related with the formation of the carbonated apatite. The formation of the carbonated apatite layer is associated with the precipitation of calcium, phosphate, carbonate and hydroxyl ions from the SBF solution on the silica rich surface provided by the wollastonite coating [7]. The presence of carbonate in the layer indicated that apatite similar to the mineral phase of bone was formed on the surface of coated materials.

Conclusions

Considering the results obtained in this work, we concluded that this kind of wollastonite seems to be an adequate coating to provide bioactivity to zirconia ceramics. The presence of tetragonal phase found on coated samples of Y-PSZ is important to maintain the good mechanical properties of the zirconia. The formation of carbonated apatite layer with chemical composition similar to bone apatite on the coating surfaces, after aging in SBF predicts that these materials may chemically bind to living bone. Finally, more studies have to be done to determine the adherence of the coatings.

Acknowledgments

This work was supported by the Scientific Cooperation with Latin America Program of the AECI and Ministry of Education, Culture and Sports, Spain.

References

[1] A. Krajewski, A. Ravaglioli, M. Mazzochi and M. Fini: J. Mater. Sci. Mat in Med. Vol. 9 (1998), p. 309

[2] P.N. De Aza, F. Guitian and S. De Aza: Scr. Metall. Mater. Vol. 31 (1994), p. 1001

[3] P.N. De Aza, F. Guitian and S. De Aza: J. Microsc. Vol. 182 (1996), p. 24

[4] P. Siriphannon, S. Hayashi, A. Yasumori and K. Okada: J. Mater. Res. Vol. 14 (1999), p. 529

[5] X. Liu and C. Ding: Biomaterials Vol. 23 (2002), p. 4065

[5] C. Piconi and G. Maccauro: Biomaterials Vol. 20 (1999), p. 1

[6] T. Kokubo, H. Kushitani, H. Sakka, T. Kitsugi and T. Yamamuro: J. Biomed. Mater. Res. Vol. 24 (1990), p. 721

[7] X. Liu and C. Ding: Biomaterials Vol. 23 (2002), p. 4065

Key Engineering Materials Vols. 254-256(2004) pp. 383-386
online at http://www.scientific.net

Tribological Study of Plasma Hydroxyapatite Coatings

J. Fernández, M. Gaona and J. M. Guilemany

CPT Thermal Spray Centre. Materials Engineering. Dept. Enginyeria Química i
Metal.lúrgia. Universitat de Barcelona. C/Martí i Franquès, 1. E-08028 Barcelona (Spain).
e-mail: cpt@material.qui.ub.es

Keywords: Hydroxyapatite, plasma spraying, adhesive strength, fretting wear.

Abstract. Stainless steel 316L AISI were coated with hydroxyapatite by Air Plasma Spraying
(APS). Different coatings were obtained by changing the stand-off distance spraying. The coatings
were subjected to bond test and ball-on-disc test. It was found that lower spraying stand-off
distances show slightly better adhesion at the interface substrate/HAp and the main wear
mechanism is the brittle cracking of the coating.

Introduction

Hydroxyapatite (HAp) coatings on prosthetic devices have been used for several years. The
combination of excellent mechanical properties of metals with the osteoconductive properties of
HAp make this kind of implants widely used in orthopaedic surgery. The ideal HAp coating in
biomedical applications would be the one with a strong cohesive strength, good adhesion to the
substrate, optimal degree of porosity to enhance bone integration, a high degree of crystallinity and
high chemical and phase stability [1]. The most used coating method is plasma spraying due to the
rapid application and economic process. However, the main drawback of this technique is the low
stability of the coating in a load-bearing situation after long-term implantation. The stability seems
to be potentially weak due to the failure of the HAp/metallic interface rather than the HAp/bone
interface one [2].
The aim of the present study was to evaluate the tribological properties of different plasma-sprayed
HAp coatings under different stand off distance spraying conditions.

Materials and Methods

HAp powder used was obtained from Plasma Biotal (Captal®30). The particle size was determined
by a Beckman Coulter LS Particle Size Analyser. The mean size of HAp particles was
32.59 ± 3.083 μm. and a normal size distribution was observed. The Scanning Electron Microscopy
micrograph, shown in Fig.1., indicates that the particles were crushed and sintered. The XRD
pattern of the powder is shown in Fig 2., where a high degree of cristallinity of the powder is
corroborated.

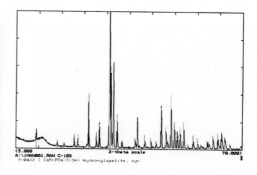

Fig1: SEM of HAp powder particles

Fig 2: XRD pattern HAP powder

Austenitic stainless steel 316L AISI was used as a substrate. The substrates were cleaned and grit blasted with Al_2O_3 under 6 kg cm^{-2} air pressure before spraying.

The air plasma spraying (APS) conditions were the same for every coating, except the stand-off distance (SOD), which was varied from 80 mm to 120 mm. The spraying parameters are summarized in table 1:

Table 1: APS spraying parameters employed (Plasma-Technik A-3000s).

Primary gas (Ar), flow rate [l min^{-1}]	50
Secondary gas (H), flow rate [l min^{-1}]	5
Powder carrier gas [Ar], flow rate [l min^{-1}]	3.65
Arc Current (A)	500
Arc Voltage (V)	54-56
Stand-off distance [mm]	80, 100, 120
Torch speed [mm s^{-1}]	1250

Coating characterization was carried out using Bright Field Optical Microscopy and Scanning Electron Microscopy. Image analysis was used in order to quantify the porosity and thickness of every coating. XRD was performed to determine crystalline and amorphous phases content.

Friction and wear test were carried out using a Ball on Disc instrument following the ASTM G99-90 under lubricated conditions (0.9 % NaCl) and loads of 5 N and 10 N. Tests were run using a Al_2O_3 balls at a sliding velocity of 131 r.p.m. for 19894 cycles. The friction coefficient was recorded continuously throughout the test. Scanning White Light interferometry and Scanning Electron Microscopy were used to determine the wear loss of the coatings estimating the traces of the surface profiles across the wear tracks and the kind of wear that took place.

Bond Strength was measured using tensile test (ASTM F1147-99). A thin layer of bonding glue was applied onto a coated and an uncoated grit-blasted surface of two rod 25.4 mm diameter. Both rods were then joined using compressive stress while curing at 180° in an oven. The pressure of the rods is then released and the resin-bonded rods were pulled out using a tester applying a load at a rate of 0.018 mm/min. Fracture surfaces were observed using a binocular lens with Image Analysis to determine what type of failure mechanism took place (cohesive, adhesive or fracture through the glue).

Results
Coating characterization

Fig. 3a and 3b show the free surface for different SOD HAp coatings. Pores and cracks are clearly visible. The pores were formed as a result of the poor bonding between adjacent splats, whereas microcracking arises from shrinkage of the splat during quenching and subsequent differential thermal contraction between substrate and coating.

   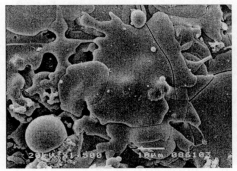

Fig 3 (a) | Fig 3 (b)

Fig 3. SEM micrographs of HAp coatings of the free surface (a) SOD 80mm, (b) SOD 120 mm.

Lower amount of small Hap particles are found on the surface when sprayed at greater SOD. Therefore they do not posses enough kinetic energy to overcome longer distance. Also particles, which are not enough fixed on the surface, are removed. Particles of different size, which are not deformed or only slightly deformed after their impingement on the substrate, could be fixed on the surface due to the roughness and also due to superimposing by the other particles. Coatings were growing by a gradual, non-uniform superposition of particles of different type. A greater deformation of a particle after its impingent on the substrate, reflex a greater probability of its fixing on the substrate. No big differences were observed in the porosity and roughness of the different coatings (table 2) for the working stand-off distances.

Table 2: Coating properties for different SOD spraying.

	Thickness (μm)	Porosity	Roughness (μm)
SOD 80 mm	96.65 ±9.90	18-20 %	5.80 ± 0.51
SOD 100 mm	128.38 ± 6.42	17-20 %	5.48 ± 0.49
SOD 120 mm	103.25 ± 4.85	15-19 %	5.43 ± 0.52

Bond test

The bond strength data are shown in Fig 4, where each value is the average of 3 data. With an increase of the SOD, the bond strength decreases. The highest value was obtained for the coating, which was sprayed at lower SOD (80mm). This coating presents the thinner thickness and the highest roughness value. No splats were observed on the surface. It could be supposed that the adhesion is mostly governed by physical phenomena of the less molten particles that reach the substrate surface during the spray process.

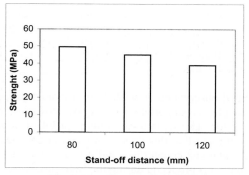

Fig 4: Bond strength vs SOD

The area percentage of adhesive failure in each coating is reported in Table 3. The coating at SOD 80 mm shows a high adhesion to the steel substrate.

Table 3: Percentage of failure for every coating.

SOD 80	10 % Adhesive	90 % Glue
SOD 100	13 % Adhesive	87 % Glue
SOD 120	27 % Adhesive	73 % Glue

Ball-on-Disc under lubricated conditions test

The wear track studied by Interferometry and SEM observation are shown in Fig 5 and Fig 6. Table 4 shows the changes of the friction coefficient and the wear loss of the substrates for the tests carried out under 5 N and 10 N loads. Lower coefficients of friction were obtained under lubricated conditions for 5N load experiments. NaCl 0.9 % can be retained in the porous structure of the coating in order to maintain low coefficients of frictions [3].

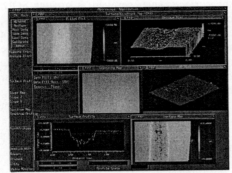

Fig 5: Wear track of a HAp polished coating of   Fig 6: Wear track of the HAp coating. Note the
SOD 80 mm tested against alumina in   material deposited on the coating surface
lubricated conditions

Table 4: Friction parameters for HAp coating under 5 N and 10 N loads

	5N		10N	
	Coefficient of friction	Coating Volume loss (mm^3)	Coefficient of friction	Coating Volume loss (mm^3)
SOD 80 mm	0.360	0.126	0.372	0.304
SOD 100 mm	0.403	0.290	0.424	0.520
SOD 120 mm	0.390	0.203	0.501	0408

SEM observations revealed not important scratches indicating that the abrasive wear process does
not play an important role (Fig 6). However, there is an important cracking pattern on the coating
surface. The wear track contains a lot of deposited material for the coatings. The EDS analysis
confirmed that this deposited material contains only Ca, P and O, therefore it does not come from
the alumina counterface and it can be concluded that no material transference from the ball to the
coating is produced during the Ball-on-Disc test
Splats morphology studies must be carried on to determine the impact mechanism affecting the
adhesion and efficiency of the process.

Conclusions
Reasonable high coating strengths were obtained. The highest value was obtained for the coating
which was sprayed at lower SOD (80mm). This coating shows the high roughness and thinness. The
SEM observation confirms the present of no splats on its surface. It is supposed that the adhesion is
mostly governed by physical phenomena of the less molten particles that reach the substrate surface
during the spray process. Although no big different values were observed for the different spraying
conditions.
From the SEM observation of the surface after the Ball-on-Disc test, it can be concluded that the
main wear mechanism is the brittle cracking of the coating promoted by its high porosity and
brittleness of HAp coatings.

Acknowledgements
The authors want to thank to the Generalitat de Catalunya for funding through the project
2001SGR00145 and the grant 2003FI 00384.

References
[1] Y.C. Tsui, C. Doyle, T.W. Clyne. Biomaterials 19 (1998) 2015-2029.
[2] B.Y. Chou, E. Chang. Surface and Coating Technology 153 (2002) 84-92
[3] Y. Fu, A.W. Batchelor, K.A. Khor. Wear 230(1999) 98-102.

Key Engineering Materials Vols. 254-256(2004) pp. 387-390
online at http://www.scientific.net
© *2004 Trans Tech Publications, Switzerland*

Structure Modification of Surface Layers of Ti6Al4V ELI Implants

J. Marciniak[1], W. Chrzanowski[1], G. Nawrat[1], J. Zak[1] and B. Rajchel[2]

[1] Biomedical Engineering Center, Silesian University of Technology,

ul. Akademicka 2a, 44-100 Gliwice, Poland, veq@polsl.gliwice.pl

[2] Institute of Nuclear Physics, ul. Radzikowskiego 152, Kraków, Poland

Boguslaw.Rajchel@ifj.edu.pl

Keywords: biomaterials, corrosion, surface modification, carbon coating.

Abstract. In the work, the influence of electrochemical modifications of Ti6Al4V ELI on the in vitro corrosion resistance was studied. The specimens with carbon layer produced with the use of IBAD method on the passivated specimens were also examined. The characteristics of the grinded, the electropolished, the electropolished and electrochemically passivated and with the passive-carbon layer specimens were evaluated by the potentiodynamic tests in Tyrode's solution. A flexibility of the passive layer was examined in potentiodynamic tests of the bended specimens. Results show that electropolishing and anodic oxidation in the examined solutions increased the in vitro corrosion resistance of the alloy. The corrosion potential of the polished and passivated specimens is much higher then the grinded samples. The increase of passive current density was not observed for the polished and passivated specimens. The passive layer was uniform that was proved by the AFM investigation. The tests proved high corrosion resistance of the passivated specimens that were bended (up to 90°) to evaluate the flexibility of passive layer. The passive-carbon layer ensures corrosion resistance, which is not as high as for passivated samples.

Introduction

Titanium and its alloys are corrosion resistant biomaterials that are characterized by a wide passive range [1, 2, 3]. The breakdown potentials are higher then membrane potentials of tissue in a living body (0,2-0,45V)[4]. It should be assumed that the loss of passivity in the electrochemical fluids and tissue system is rather impossible. However, a passive layer can be mechanically or chemically damaged. Metallic surface is then uncovered and corrosion processes are initiated. Currently the scientific research is focused on surface layer modification techniques in order to increase the corrosion resistance of implants. It should be said that implants undergo plastic deformation. For this reason prepared layers should be characterized by the plastic deformation ability. The aim of the work was to work out conditions of producing passive-carbon layers on Ti6Al4V ELI alloy surface and the evaluation of the corrosion resistance.

Materials and methods

Ti6Al4V ELI was used in the research. Chemical composition and mechanical properties met the ASTM standard [5]. Surface preparation involved: grinding, electrochemical polishing, anodic oxidation and carbon layer creation with the use of double beam method - Ion Beam Assisted Deposition (IBAD) [6]. Electrochemical polishing was carried out in the bath composed of: sulfuric acid (20÷70% vol.) + hydrofluoric acid (20÷50% vol.) + ethylene glycol (10÷20% mass) + acetanilide (20% mass). Anodic oxidation was carried out in the solution of $CrO_3 - 100÷200g/litre$. The specimens were passivated at different potentials. The corrosion resistance of layers was evaluated by potentiodynamic method in the Tyrode's solution (36,6±1°C and pH=6,9÷7,5). Non-deformed and deformed in transverse bend test passive layers were evaluated. Passive-carbon layers were tested comparatively in non-deformed state. Topography of the surface was evaluated with the use of an AFM method. Chemical constitution and thickness of the layers was estimated in Rutherford Backscattering Spectroscopy (RBS) and Particle Induced X-ray Emission (PIXE) method.

Results

Results of the pitting corrosion tests are compiled in the table 1. Ti6Al4V ELI alloy with grinded surface had the corrosion potential in the range of E_{cor}=+50÷+59 mV, the breakdown potential was in the range of E_B =+1540÷+1980 mV. Electrochemical polishing caused the increase of the corrosion potential to E_{cor}= +112÷+125 mV and breakdown potential to E_B=+2240÷+2410 mV – table 1. For the polished and passivated electrochemically specimens the corrosion potential increased to E_{cor}=+342÷+402 mV. The increase of anodic current density in the investigation range up to +5V was not observed for the passivated specimens.

Passivated samples were then deformed in transverse bend test (0÷90°). The decrease of the corrosion potential to E_{cor}=+211÷+273 mV was observed for the specimens deformed about 10°. The breakdown potential was in the range of E_B=+4710÷+4920 mV. The increase of the deformation angle to 45° caused further decrease of the corrosion and breakdown potentials to the value of E_{cor}= +70÷+78 mV and E_B= +4300÷+4918 mV. For the deformation angle 90° assigned potentials were: E_{cor}=+25÷+35, E_B=+3750÷+4280 mV – table 1, fig. 1.

Investigation of topography with the use of the AFM method did not reveal any damages of the passive film. Roughness of the layer did not exceeded R_a=27 nm – fig. 2.

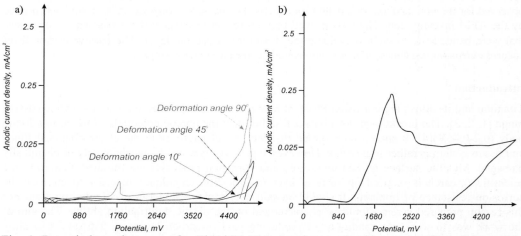

Fig. 1. Potentiodynamic curves for Ti6Al4V ELI specimens a) with passive film, deformed in different deformation angle, b) with passive-carbon layer

The carbon layer was deposited on the passive surface with the use of the IBAD method. Corrosion potential of the passive-carbon layer was in the range of E_{cor}=+163÷+193 mV. Slight increase of current density in the range of about +1360÷+1448 mV was observed but when the potential reached about +2000 mV the decrease of current density was observed. The change of the polarization direction caused a rapid decrease of the current value. Analyzes carried out by proton and ion He$^+$ beam of energy 2000 keV proved the presence of 1 µm thick carbon layer.

Conclusion

In the work electrochemical polishing and passivation technology for Ti6Al4V ELI alloy was elaborated. The test revealed that the passive film is uniform (fig. 2) and ensured the pitting corrosion resistance of the analyzed alloy. For the polished and passivated specimens the increase of the pitting corrosion resistance was observed in refer to grinded specimens. The corrosion potential increased by about 300 mV – Table 1. Any changes of the current density on the potentiodynamic curves were not also observed for the polished and passivated specimens. The passive-carbon layers also ensure high corrosion resistance.

Table1. Results of potentiodynamic investigation

Specimens	Angle, °	Corrosion potential E_{corr}, mV	Breakdown potential, E_B, mV
Girding	0°	50÷59	1540÷1980
Polished electrochemically	0°	112÷125	2240÷2410
Polished electrochemically and passivated	0°	342÷402	-
Polished electrochemically and passivated	10°	211÷273	4710÷4920
	45°	70÷78	4300÷4910
	90°	25÷35	3750÷4280
With the passive-carbon layer	0°	+163÷+193	+1360÷+1448

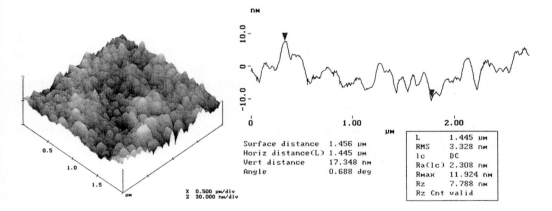

Fig. 2. The passive film topography of the Ti6Al4V ELI alloy

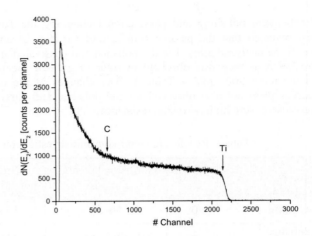

Fig. 3. RBS spectrum recorded for Ti6Al4V ELI covered by thin C_{IBAD} coating layer. For determination of the Ti and C depth distribution the beam of He^+ ions at energy of 2000 keV was used. The C and Ti surface position are marked.

The alloy is often used as biomaterial, that is deformed during the implantation and the working life so, flexibility of the surface layer is essential. Additionally the corrosion resistance tests carried out on deformed specimens proved flexible properties of the layer. The specimens deformed even up to 90° had the higher breakdown potential E_B=+3750÷+4280 mV – Table 1, fig. 1) then polished ones (E_B=+2240÷+2410 mV – Table 1).

References

[1] J. P. Simpson: *Electrochemical behavior of titanium and titanium alloys with respect to their use as surgical implant materials.* In: Christel P., Meunier A., Lee A. J. C.: *Biomedical and Biomechanical Performance of Biomaterials.* Elsevier, Amsterdam 1986, p. 63-68.

[2] U. Zwicker :*Titan and Titanlegierungen.* Springer, Berlin Heidelberg New York 1974, ISBN 3-540-05233.

[3] J. Marciniak: Biomaterials. Wydawnictwo Politechniki Slaskiej, Poland, Gliwice 2002.

[4] J.Wirz, F. Schmidli, S. Steinmann, R. Wall: Aufbrennlegierungen im Spaltkorrosionstest. Swieitz. Monatsschr. Zahnmedizin 97, 1987.

[5] ASTM-F136-84 (1984, USA).

[6] L. Hanley , S. B. Sinnott: The growth and modification of materials via ion–surface processing Surface Science Vol. 500 (2002), p. 500–522.

Key Engineering Materials Vols. 254-256(2004) pp. 391-394
online at http://www.scientific.net
© *2004 Trans Tech Publications, Switzerland*

Preparation and Properties of Bioactive Calcium Phosphate Fibers

Caroline Klein, Frank A. Müller and Peter Greil

University of Erlangen-Nuremberg, Department of Materials Science – Glass and Ceramics,
Martensstr. 5, 91058 Erlangen, Germany, fmueller@ww.uni-erlangen.de

Keywords: hollow fibers, tricalcium phosphates, hydroxyapatite, brushite, electrodeposition

Abstract. Ca-deficient carbonated hydroxyapatite (HCA) and brushite ($CaHPO_4 \cdot H_2O$, DCPD) were precipitated under galvanostatic conditions from an electrolyte solution containing calcium nitrate and ammonium dihydrogen phosphate with a molar ratio adjusted to Ca:P = 1.67. Carbon multifilament fibers were used as deposition electrodes (cathode). Tailorable surface morphologies could be achieved by varying the deposition parameters. After calcination in air and burnout of the carbon core above 800 °C hollow fibers were obtained. Single phase β- tricalcium phosphate ($Ca_3(PO_4)_2$, TCP) was formed after calcination at 1100 °C and α-TCP at 1200 °C, respectively. Thermal induced phase transformations of the deposited phases into other calcium phosphates were examined.

Introduction

Calcium phosphate-based powders, granules and coatings are currently used for a number of dental and skeletal prosthetic applications due to their excellent biocompatibility and osteointegration properties [1]. TCPs and mixtures with hydroxyapatite ($Ca_{10}(PO_4)_6(OH)_2$, HAP), so called biphasic calcium phosphate (BCP), have found application as bone cements and as bone implant material, respectively. TCP exists in three polymorphs, β-TCP below 1180 °C, α-TCP between 1180 and 1400 °C and α`-TCP above 1470 °C [2]. The order of relative solubility has been suggested as α-TCP > β-TCP >> HAP [3]. Bioactive as well as bioresorbable ceramic fibers are of particular interest for reinforcement of osteosynthetic biopolymer matrix composites [4]. HAP hollow fibers have already been prepared by electrophoretic deposition on carbon fibers [5]. In the present study we produced hollow calcium phosphate fibers with tailorable surface microstructure and controllable phase composition by electrolytic deposition onto carbon template fibers.

Materials and Methods

Polyacrylnitril derived carbon fibers (Tenax Fibers, Germany) with a diameter of 7 μm and a length of 2 cm were used as deposition electrodes (cathode) under galvanostatic conditions. An aqueous electrolyte solution with a Ca:P-ratio of 1.67 corresponding to HAP was prepared by stirring a solution of 0.042 mol/l $Ca(NO_3)_2 \cdot 4H_2O$ and 0.025 mol/l $NH_4H_2PO_4$ at room temperature for 24 h [6]. Ethanol was added to reduce H_2 evolution at the deposition electrode [7]. A solution temperature of 40 °C, deposition times up to 120 minutes and a constant current between 20 and 100 mA were used. After calcium phosphate deposition the coated fibers were dried at room temperature for 24 hours. Subsequently the carbon core was burnt out in a temperature range between 550 and 800 °C. The resulting hollow fibers were calcinated at elevated temperatures up to 1200 °C for 2 h. Coating morphology and microstructures were analyzed by scanning electron microscopy (Philips XL 30, Stereoscan MK II). Average crystal sizes and fiber diameters were determined from SEM micrographs. The specific surface area was measured by a gas adsorption method in krypton atmosphere (Sorptomatic 1990). Crystalline phases were examined by x-ray diffraction analysis (Siemens D-500) of the removed and grinded coatings. The diffractometer was equipped with a heating system to allow in situ measurements at elevated temperatures up to 1450 °C. Spectra were taken every 50 °C after temperature has equilibrated for 30 min. Fourier transform infrared (FT-IR) spectra were obtained in a Nicolet Impact 420 spectrometer over the 400 – 2000 cm^{-1} region using KBr : calcium phosphate powder mixtures with a weight ratio of 300:1.

Results and Discussion
Homogenous, crack free calcium phosphate coatings with an average layer thickness of 35 µm and a specific surface area of 67 m^2/g were achieved by electrolytic deposition of a calcium and phosphate containing solution under galvanostatic conditions, when a deposition current of 100 mA was applied for 60 minutes. XRD of the deposited phases correspond to DCPD and HAP with low crystallinity. XRD data suggest that in the initial stage mainly DCPD was deposited serving as nucleation site for the subsequent growth of HAP crystals. Increasing current and thus the higher quantity of available calcium and phosphate ions lead to a transition from insular crystal clusters to a tabular habit of the HAP phase (fig. 1), due to changes in the nucleation of calcium phosphates. Differences in surface morphology can be explained due to a faster rate of nuclei growth to a critical level and a faster crystal growth at increasing current.

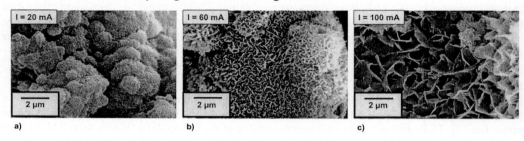

Fig. 1: SEM micrographs of calcium phosphate layers deposited under different currents.

The mass of the deposited layer increased linearly with deposition time and current. In a previous paper it was demonstrated that precipitation of calcium phosphates on the cathode is induced by the supersaturation of calcium phosphate salts due to increased pH value [8]. Thus, the electrochemical deposition process includes the electrochemical half reaction, the acid-base reaction and the calcium phosphate precipitation reaction. First water and nitrate are reduced at the cathode surface producing hydrogen gas, nitrite and hydroxide ions, as shown in the following:

$$2\,H_2O + \quad 2e^- \qquad\qquad \Leftrightarrow \quad H_2 \uparrow + \quad 2\,OH^- \tag{1}$$

$$NO_3^- + \quad H_2O + \quad 2e^- \qquad \Leftrightarrow \quad NO_2^- + \quad 2\,OH^- \tag{2}$$

Due to Faraday`s laws:

$$Q = I \cdot t = \frac{m}{M} \cdot z \cdot N_a \cdot e = \frac{m}{M} \cdot z \cdot F, \quad \rightarrow \quad m = \frac{I \cdot t \cdot M}{z \cdot F} \tag{3}$$

where Q is the charge, I the current, t the deposition time, m the mass, M the molecular mass, z the charge number, N_a the Avogadro constant, e the electrical charge of an electron and F the Faraday constant (F = 9.648*10^4 C/mol), the amount of OH$^-$ ions on the surface of the cathode can be controlled by controlling current and deposition time. The hydroxide ions generated at the surface may react with dihydrogen phosphate and subsequently with hydrogen phosphate ions according to

$$OH^- + \quad H_2PO_4^- \qquad \Leftrightarrow \quad H_2O + \quad HPO_4^{2-} \tag{4}$$

$$OH^- + \quad HPO_4^{2-} \qquad \Leftrightarrow \quad H_2O + \quad PO_4^{3-} \tag{5}$$

Taking into account the equilibrium constant K

$$K = \frac{c(H_2O) \cdot c(HPO_4^{2-})}{c(OH^-) \cdot c(H_2PO_4^-)} \qquad (6)$$

an increase of OH⁻-concentration is supposed to increase the concentration of the end products.

$$Ca^{2+} \quad + \quad HPO_4^{2-}+ \quad 2\,H_2O \qquad \Leftrightarrow \qquad CaHPO_4 * 2\,H_2O \qquad (7)$$

$$(10\text{-}x\text{-}y)\,Ca^{2+}+x\,NH_4^{+}+(6\text{-}y)\,PO_4^{3-}+y\,CO_3^{2-}+2\,OH^- \Leftrightarrow Ca_{10\text{-}x\text{-}y}(NH_4)_x(PO_4)_{6\text{-}y}(CO_3)_y(OH)_2 \quad (8)$$

The reactions 7 and 8 depend on the solubility product L_p.

$$xa \quad + yb \qquad \Leftrightarrow \qquad ab \qquad (9)$$

$$L_p(ab) = \left[c(a)\right]^x \cdot \left[c(b)\right]^y \qquad (10)$$

If the right side of equation (10) exceeds L_p, the precipitation will be spontanous. Current and deposition time control the pH at the surface, and thus the equilibrium in equation (4) and (5). The concentrations of HPO_4^{2-} and PO_4^{3-} influence the equilibrium in equation (7) and (8) and thus the deposition rate of nonstoichiometric HAP and brushite is increasing linearly with current and time.

The FT-IR spectra of the unsintered sample is shown in fig. 2. The band at 1645 cm⁻¹ indicates DCPD present in the deposited layer. Bands at 1087 and 1040 cm⁻¹ are assigned to the triply degenerated asymmetric stretching modes, v_{s3a} and v_{s3b}, and at 962 cm⁻¹ to the nondegenerated symmetric stretching mode v_1 of the P-O binding, bands at 602 and 574 cm⁻¹ to the triply degenerated bending mode v_{4a} and v_{4b} and the band at 462 cm⁻¹ to the stretching mode v_{2b} of the O-P-O bonds of the phosphate group of hydroxyapatite. Furthermore there is evidence of CO_3^{2-} ions present in the crystals. It is detected from the peaks at 1550, 1460 and 1420 and 872 cm⁻¹ corresponding to carbonate ions in type B sites of the HAP crystal, which replace parts of the PO_4^{3-} ions [6]. The peak at 1380 cm⁻¹ can be assigned to the N-H deformation vibration of the ammonium groups. According to [2] precipitated HAP easily adsorbs soluted ions and HAP is known to form defect structures with carbonate, ammonium and other ions.

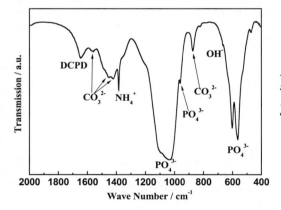

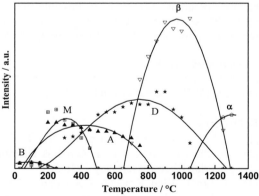

Fig. 2: FT-IR spectra of the electrolytically deposited layer.

Fig. 3: XRD intensities of characteristic peaks of calcium phosphate phases in correlation to temperature. B: brushite (040), A: apatite (002), M: monetite (041), D: dicalcium diphosphate (008), β: β-TCP (300), α: α-TCP (170).

Thermal treatment of the coated carbon fibers resulted in a series of phase transformations (Fig. 3). At 200°C brushite converted into monetite under loss of crystal water. Monetite transformed into dicalcium diphosphate at 400 °C and subsequently into β-TCP at 1100 °C. HAP transformed into β-TCP at 750 °C, which indicates it to be Ca-deficient, because stoichiometric HAP is known to transform into tetracalcium phosphate and α-TCP at temperatures above 1200 °C [2]. β-TCP remained stable until 1200 °C where it transformed into the high-temperature α-TCP phase.

The average layer thickness of fibers deposited for 60 minutes and sintered at 1100 °C for 2 h reached 13 μm with an average outer diameter of 33 μm and a mean particle size of 1 - 2 μm. The specific surface area amounts 1.08 m^2/g. Fig. 4 shows typical SEM micrographs of a β-TCP hollow fiber.

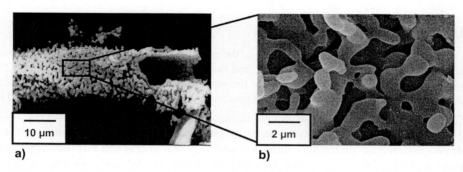

a)   b)

Fig. 4: SEM micrographs of electrolytically deposited β-TCP hollow fibers, sintered at 1100 °C for 2h.

Conclusion

Resorbable TCP hollow fibers have been prepared by electrolytic deposition of Ca-deficient HAP and brushite on C-fiber templates. The microstructure of electrodeposited TCP could be controlled by adjusting deposition parameters and thus nucleation and grain growth conditions. With increasing current the surface roughness increased. Subsequent fiber burnout and thermal treatment resulted in hollow TCP fibers. Possible applications of electrodeposited hollow fibers include resorbable scaffolds for tissue engineering, carriers in drug delivery systems and reinforcement of resorbable polymers in load bearing orthopedic applications.

Acknowledgement
Financial support from VW-stiftung under contract no. I / 73 774 and Verbund der Chemischen Industrie (VCI) is greatfully acknowledged.

References
[1]   L.L. Hench: J. Am. Ceram. Soc. Vol. 81/7 (1998), p. 1705
[2]   E. Pietsch and J. Meyer, Gmelins Handbuch der Anorganischen Chemie, Calcium Teil B (Verlag Chemie Germany 1956/57)
[3]   F. Lin, C. Liao, K. Chen, J. Sun and C. Lin: Biomat. Vol. 22 (2001), p. 2981
[4]   T. Kasuga, Y. Ota, M. Nogami and Y. Abe: Biomat. Vol. 22/1 (2001), p. 19
[5]   I. Zhitomirsky, L. Gal-Or, A. Kohn and M.D. Spang: J. Mater. Sci. Vol. 32 (1997), p. 803
[6]   M. Shirkhanzadeh: J. Mater. Sci.: Mater Med. Vol. 6 (1995), p. 90
[7]   C. Jim-Shone, J. Horng-Yih and H. Min-Hsiung: J. Mater. Sci.: Mater. Med. Vol. 9 (1998), p. 297
[8]   S. Ban and S. Maruno: J. Biomed. Mater. Res. (1998), p. 387

Key Engineering Materials Vols. 254-256(2004) pp. 395-398
online at http://www.scientific.net

Consolidation of Multi-Walled Carbon Nanotube and Hydroxyapatite Coating by the Spark Plasma System (SPS)

M. Omori[1], A. Okubo[1], M. Otsubo[2], T. Hashida[2] and K. Tohji[2]

[1] Institute for Materials Research, Tohoku University, 2-1-1 Katahira, Aoba-ku,Sendai 980-8577, Japan, email: mamori@imr.tohoku.ac.jp
[2] Graduate School of Engineering, Tohoku University, Aoba, Aramaki, Aoba-ku, Sendai, 980-8577, Japan

Keywords: carbon nanotube, multi-walled carbon nanotube, consolidation of carbon nanotube, spark plasma system (SPS), spark plasma sintering (SPS), hydroxyapatite coating

Abstract. A multi-walled carbon nanotube (MWNT) was mixed with phenol resin and consolidated by the spark plasma system (SPS). Properties of the MWNT consolidated at 1200°C at 120 MPa were as follows: bulk density was 1.74 g/cm^3; apparent porosity was 16.3%; Young's modulus was 11.1 GPa. Hydroxyapatite was coated on the consolidated MWNT at 1000°C and 120 MPa by SPS, using the suspension prepared from 6 moles of $CaHPO_4 \cdot 2H_2O$ and 4 moles of $Ca(OH)_2$.

Introduction

Carbon nanotubes (CNT) consist of single-walled carbon nantubes (SWNT) and multi-walled carbon nanotubes (MWNT) and have been attracting a lot of attention since their discovery [1]. It is said that molecular circuit devices will be fabricated from SWNT [2]. MWNTs are composed of various kinds of tube diameters and a number of carbon network layers. CNT are still expensive, but the cost of their fabrication will surely decreased in the near future. Low-cost CNTs will be used for fillers of composites and starting materials to produce structural and/or functional compacts. Graphite is a hard-to-sinter material, and its powder can only be sintered at very high temperatures under pressing [3]. The sintering ability of CNT is the same as that of graphite, and advanced techniques are needed to consolidate it at lower temperatures, before the transformation into graphite. The spark plasma system (SPS) has been developed for sintering of metal and ceramics in the plasma and electric field [4, 5], and it is used for consolidation of various kinds of materials such as metals, ceramics and polymers [6]. The bioactivity of graphite is not excellent. The best way to increase the bioactivity of the consolidated MWNT is deposition of hydroxyapatite (HA) films on it. Plasma spraying is widely used for manufacturing HA coating on Ti or Co-Cr-based implants. However, a multitude of phase changes occurs at high temperatures at the plasma spraying process [7]. Two compounds of 6 moles of $CaHPO_4 \cdot 2H_2O$ and 4 moles of $Ca(OH)_2$ reacted at 150°C to produce HA and H_2O by the hydrothermal hot-pressing method [8]. This reaction is able to apply to HA coating on biomaterials because the reaction product is only HA except for H_2O.

In this paper, the MWNT was mixed with phenol resin in ethanol. After removing the ethanol and decomposing the phenol resin by heating, the mixture of the MWNT and the amorphous carbon was consolidated by SPS. The consolidated MWNT was dipped in the suspension of 6 moles of $CaHPO_4 \cdot 2H_2O$ and 4 moles of $Ca(OH)_2$. The two compounds reacted and bonded to the consolidated MWNT at 1000°C at 120 MPa by SPS.

Experimental procedures
Consolidation of MWNT

CNT used for the consolidation was MWNT (NanoLab Inc., USA, 80% purity). The MWNT was purified to remove a metal catalyst using a solution of 50% HNO_3. Phenol resin was dissolved in ethanol. The MWNTs were put in the ethanol solution. After evaporating ethanol, the phenol resin film on the MWNT was decomposed at about 200°C in air. The coated MWNTs were put in a graphite die and set in the spark plasma system (SPS) (Sumitomo Coal Mining, Japan, SPS-1050). The consolidation was carried out between 1000°C and 1600°C at 120 MPa in a vacuum. In case of the consolidation at 1000°C, the consolidation temperature was raised as follows: heating rate from 0°C to 900°C at 100°C/min, from 900°C to 980°C at 20°C/min, from 980°C to 1000°C at 5°C/min and holding time at 1000°C for 5 min.

The microstructure of the consolidated MWNT was analyzed by a transmission electron microscope (JEOL, Japan, JT-007). The polished surface of the consolidated MWNT was observed with an optical microscope (Nikon, Japan, N-01). X-ray diffraction (XRD) was carried out on the MWNT and the consolidated one using Cu Kα line by an X-ray diffractometer (Rigaku, Japan, Rotaflex, RU-200B). Density of the consolidated disk was determined based on Archimedes' principle using water. Elastic modulus of disk samples (3 mm in thickness and 20 mm in diameter) was measured by a pulse-echo overlap ultrasonic technique, using an ultrasonic detector (Hitachi Kenki Co. Ltd., Japan, ATS-100) and a storage oscilloscope (Iwasaki Tsushinki Co. Ltd., Japan, DS6411).

Coating of HA on the consolidated MWNT

$CaHPO_4 \cdot 2H_2O$ (6 moles) and $Ca(OH)_2$ (4 moles) powders (Wako Pure Chemical Ind., Japan, reagent grade) were used to form HA films. These powders were suspended in distilled water using glycolic acid (Wako Pure Chemical Ind., Japan, reagent grade). The consolidated MWNT (1 x 1 x 5 mm^3) was dipped in the suspension and dried. The coated MWNT was put in the graphite die with carbon powders and set in SPS. The coating of HA was carried out at 1000°C at 120 MPa in a vacuum. The heating rate was controlled as follows: from 20°C to 900°C at 100°C/min, from 900°C to 980°C at 20°C/min and from 980°C to 1000°C at 5°C/min. The holding time at 1000°C was 5 min.

Results and discussion
Consolidation of MWNT

An X-ray diffraction pattern of the MWNT was similar to that of graphite (small squares in Fig. 1). As shown in the transmission electron micrograph (TEM) of Fig. 2, the MWNT consisted of varied tube diameters. The dispersion of the diameter was mainly from 10 nm to 50 nm, and a thick MWNT of 200 nm in diameter was found among them.

The MWNT had about 20% amorphous carbon and was not consolidated by SPS. This amorphous carbon did not enhance the consolidation of the MWNT by SPS. A phenol resin was added to achieve the consolidation, and its carbon residue was of about 20%. The optical micrograph revealed that the MWNT consolidated with the 30% phenol resin contained coarse pores. There were no coarse pores in the MWNT consolidated with 50% phenol resin. It was considered that the addition of the 50% phenol resin was adequate to obtain the dense compact. The X-ray diffraction pattern of the MWNT consolidated at 1000°C was similar to those of the MWNT and graphite. When the consolidation was carried out at

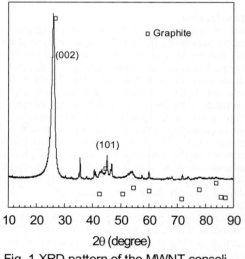

Fig. 1 XRD pattern of the MWNT consolidated at 1600 °C.

1200°C and at 1400°C, the (101) peak was strong compared with other ones. The pattern of the MWNT consolidated at 1400°C was slightly different from that of the MWNT and graphite. The difference was emphasized on the pattern of the MWNT consolidated at 1600°C, but the intensity of the (101) peak decreased, as shown in Fig. 1. The new diffraction peaks did not correspond to those of graphite. Fig. 3 shows the TEM image of the MWNT consolidated at 1600°C with 30% phenol resin. The MWNT was partly decomposed and converted into different

Table 1 Density and mechanical properties of the consolidated MWNT

Consolidation temperature (°C)	Bulk density (g/cm³)	Apparent porosity (%)	Closed porosity (%)	Young's modulus (GPa)	Poisson's ratio
1000	1.67	16.7	9.6	3.05	- 0.62
1200	1.74	16.8	6.4	11.1	0.074
1400	1.73	15.6	8.1	10.1	0.034

compounds from graphite. MWNT is not decomposed until 2400°C by heating [9]. The SPS process consists of some effects such as the spark plasma, electric field and others, and carbon fibers are decomposed into powders by the SPS treatment [6]. The carbon network of the MWNT should be partly decomposed by the spark plasma and resulted in new structures different from graphite.

The density and mechanical properties are shown in Table 1. The low bulk density of the MWNT consolidated at 1000°C indicated that the consolidation was not accomplished. The bulk density of 1.74 g/cm^3 was not high and depended on the tube structure because there were no coarse pores. The apparent porosity was almost the same for all consolidated MWNT. The closed porosity was calculated from the apparent porosity and theoretical density of graphite (2.266 g/cm^3). Since the theoretical density of the MWNT must be lower than that of graphite, the closed porosity should decrease less than the values indicated in Table 1. Young's modulus and Poisson's ratio were measured on the surface where the MWNT was aligned parallel to the pressing direction of SPS. Young's modulus of the MWNT consolidated at 1000°C was lower than that of the ones consolidated

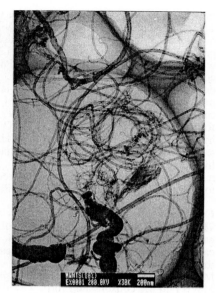

Fig. 2 TEM image of MWNT.

1200°C and 1400°C. The Young's modulus of 11.1 GPa of the consolidated MWNT was not low, considering the density of the consolidated MWNT and that of 16 GPa of a commercial graphite product with high density of 2.0 g/cm^3. Given that, the Young's modulus of human bone is 7 - 30 GPa [9], the consolidation of the MWNT could produce the material with the low modulus. Poisson's ration was negative for the MWNT consolidated at 1000°C, and was very little for the ones consolidated at 1200°C and 1400°C. The negative Poisson's ratio indicated that the bond between the MWNTs was not completed at 1000°C, which is consistent with the Young's modulus of 3.05 GPa.

The bending strength of the consolidated MWNT was measured by a three-point bending test method, but it was not obtained. The sample for the bending test was curved by stress at first, and then it was fractured. Bending strength of ductile materials like metals cannot be measured. The consolidated MWNT behaved in the same way as metal for bending test. Bending strength of human bone is 50 – 150 MPa [9], and that of the consolidated MWNT seemed to be lower than it. It is obvious the strength of the consolidated MWNT is not enough for application to human bone at the moment.

Fig. 3 TEM image of the MWNT consolidated at 1600 °C.

Coating of HA on the consolidated MWNT

CaHPO$_4$·2H$_2$O and Ca(OH)$_2$ were easily suspended in water compared with HA powder because they contained an OH group or H$_2$O. The mirror surface of the consolidated MWNT was inadequate for HA coating. A coarse surface was prepared by polishing with SiC powders of 64 □m. Two suspended compounds of 6 moles of CaHPO$_4$·2H$_2$O and 4 moles of Ca(OH)$_2$ were coated on the consolidated MWNT and reacted at 1000°C and 120 MPa by SPS. The reacted film was identified as HA by X-ray diffractometry. Fig. 4 shows an optical micrograph of the consolidated MWNT coated with HA. The HA film did not contain cracks. The coated HA seemed bonded with the rough surface of the consolidated MWNT, although the bonding strength was not determined. The bonding was partly based on the anchor effect of the coarse surface. It was not clear whether the chemical bond was associated with the bonding. The compounds of 6 moles of $CaHPO_4 \cdot 2H_2O$ and 4 moles of $Ca(OH)_2$ were allowed to dehydrate and produce HA at high pressure, such as 120 MPa. HA was not formed at 1000°C and 10 MPa from those compounds by SPS. The reaction temperature decreased from 1000°C to 600°C with increasing the pressure from 120 MPa to 300 MPa [13], but the film coated at the low temperature peeled. HA can be formed from $CaHPO_4$ or $CaHPO4 \cdot 2H_2O$ at 200°C by hydrothermal technique [10-12]. The reaction temperature by SPS was higher than those of hydrothermal and hydrothermal hot-pressing methods. The hydrothermal reaction requires water to complete the formation of HA, but SPS does not.

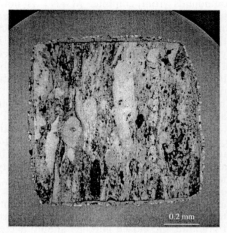

Conclusions
The MWNT containing about 20% of amorphous carbon was not consolidated by SPS. Amorphous carbon transformed from the phenol resin enhanced the consolidation. The structure of the consolidated MWNT was almost the same as that of graphite. The MWNT in the consolidated form was aligned in the

Fig. 4 Optical micrograph of the consolidated MWNT coated with HA.

direction parallel to the (101) plane of graphite. The consolidated MWNT was decomposed at temperatures higher than 1400°C. Properties of the MWNT consolidated at 1200°C and 120 MPa were as follows: density was 1.74 g/cm3; apparent porosity was 16.8%; Young's modulus was 11.1 GPa; Poisson's ratio was 0.074.
HA was coated on the consolidated MWNT at 1000°C at 120 MPa, using the suspension prepared from 6 moles of $CaHPO_4 \cdot 2H_2O$ and 4 moles of $Ca(OH)_2$. HA was not cracked and tightly covered the surface of the consolidated MWNT.

Acknowledgments
This study is supported by Research on Advanced Medical Technology in Health and Labour Sciences Research Grants from Ministry of Health, Labour and Welfare of Japan. The authors are thankful to Yoshihiro Murakami, Shun Ito, Yuichiro Hayasaka and Yoshiyuki Sato for the measurements of X-ray diffraction patterns and observations of transmission electron microscope and optical microscope.

References
[1] S. Iijima: Nature, Vol. 354 (1991), p. 56.
[2] L. Chico, L. X. Menedic, S. G. Louie and M. L. Cohen: Phys. Rev., B54 (1996), p. 2600.
[3] H. Boeder and E. Fitzer: Carbon, Vol. 8 (1970), p. 453.
[4] K. Inoue: US Patent, No. 3,241,956 (1966).
[5] K. Inoue: US Patent, No. 3,250,892.
[6] M. Omori: Mater. Sci. Eng., A287 (2000), p. 183.
[7] S. R. Radin and P. Ducheyne: J. Mater. Sci.: Mater. Med. Vol. 3 (1992), p.   33.
[8] K. Hosoi, T. Hashida, H. Takahashi, N. Yamazaki and T. Korenaga: J. Am. Ceram. Soc., Vol. 79 (1996), p. 2771.
[9] A. Bougrine, N. Dupont-Pavlovsky, A Naji, J. Ghanbaja, J. F. Mareche and D. Billaud: Carbon, Vol. 39 (2001), p.685.
[10] L. H. Hench: J. Am. Ceram. Soc., Vol. 81 (1998), p. 1705.
[11] A. P. Peroff and A. S. Posner: Science, Vol. 124 (1958),p. 583.
[12] E. Hayek, W. Boehler, J. Lechleitner and H. Petter: Z. Anorg. Allge. Chem., Vol. 295 (1958), p. 241.
[13] M. Omori, T. Onoki and T. Hashida: Unpublished data

Key Engineering Materials Vols. 254-256(2004) pp. 399-402
online at http://www.scientific.net
© *2004 Trans Tech Publications, Switzerland*

Influence of Carboxyl Groups Present in the Mineralising Medium in the Biomimetic Precipitation of Apatite on Collagen

E.K.Girija, Y.Yokogawa and F.Nagata

Biofunctional Ceramics Research Group, Ceramics Research Institute,
National Institute of Advanced Industrial Science and Technology,
2266-98, aza Anaga-hora, Ooaza Shimo-Shidami, Moriyama-ku,
Nagoya 463-8560, Aichi Japan.
girijae@yahoo.com

Keywords: Collagen, hydroxyapatite, simulated body fluid, biomimetics, bovine serum albumin, polyacrylic acid

Abstract. Calcium phosphate was precipitated by biomimetic method on collagen fibrils from conventional and revised simulated body fluid (C and R SBF) solutions in presence of bovine serum albumin (BSA) and polyacrylic acid (PAAc). The precipitates formed on the collagen fibrils were analyzed using SEM, TF-XRD and FT-IR. Partial modification in the morphology of the precipitates was observed. An increase in concentration of BSA and PAAc in the SBF solution reduced the crystallinity of the precipitates. The precipitates formed were B-type carbonate apatite. Presence of chelation complex of Ca and carboxyl groups was also identified.

Introduction

Mineralized tissues such as bone and teeth are biologically produced composites of collagen and hydroxyapatite (HAP). The carboxylate group of collagen is one of the key factors for the nucleation of HAP (1). The non-collagenous proteins present in the body fluids was found to play potential roles in the mineralization process *in vivo* (2). Also compounds containing COOH groups are known to be effective in the formation of apatite in the body environment. In this study *in vitro* investigation of the effect of BSA and PAAc [having carboxyl moieties] dissolved in the mineralizing solution (SBF) in the formation of apatite on collagen fibrils was carried out.

Experimental

Description of the experimental procedure for preparing cross-linked collagen gel (from Type I collagen solution) and C and R-SBF solutions were given in our earlier report (3). Cross-linked collagen gels were incubated in BSA (1, 5 and 10 mg/ml) solublised C-SBF for 48 hours with the renewal of the medium after 24 hours (hereafter referred as initial and renewed solution). PAAc (molecular weight 25,000) was directly mixed to R-SBF solution (0.1 and 1.0 mg/ml) and the gels were incubated for 1, 3, 5 and 7 days. Control experiments were done without BSA in C-SBF and PAAc in R-SBF for comparison. The pH values of the solutions were measured for both cases. Incubation of the gels were done at 36.5°C. Then the gels were washed with water and dried by critical point drying and a thin film of collagen gel was obtained. pH measurements were done using Mettler Toledo MP220 pH meter. SEM was performed using a Hitachi S-3000N scanning electron microscope. Thin film X-ray diffraction (TF-XRD) pattern of the gels were recorded on a MAC Science MXP3VAII diffractometer using monochromated CuKα radiation at 40 kV and 20 mA. Micro FT-IR spectra was recorded on a Jasco Micro-FT-IR Jansen MFT-2000 Fourier transform infrared spectrometer by encasing the sample in a transparent KBr matrix.

Results and Discussion

pH measurements had shown that pH of the renewed C-SBF solution was higher than the pH of the initial C-SBF solution irrespective of the presence or absence of BSA. Hence it is clear that the maximum precipitation occurs in first day of soaking. Incubation of collagen gel in R-SBF resulted in a decrease in pH with an increase in the incubation time in the absence of PAAc. This is an indirect proof of apatite precipitation, that involves consumption of OH ions, and hence pH decreases. Dissolution of PAAc into R-SBF solution resulted in slight increase of the pH initially and then became constant. Formation of a chelation complex from the Ca and carboxyl groups and hence the inhibition effect exerted on apatite formation might be the reason for this increase in pH.

SEM analysis

Samples mineralised in the absence of BSA and PAAc had shown the formation of spherical precipitates on the reconstituted collagen fibrils. The flakes forming the spherulites could be observed in the precipitates from C-SBF (Fig.1a). Plain coating covering the fibrils was observed randomly either alone or mixed with tiny flakes and spherical precipitates when BSA was present in the mineralising medium (Fig. 1 b and c). The amount of plain coating formation increased with an increase in BSA concentration. Such plain coating covering the fibril was also observed when PAAc was dissolved in R-SBF (Fig.1d). When BSA or PAAc was solublised in the medium calcium complexation and apatite formation occurs in the medium and gets deposited along the fibrils. Amino groups along the collagen fibrils may also try to form spherical precipitates.

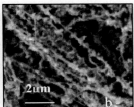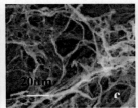

Fig. 1 SEM micrographs, [a] without BSA in C-SBF, [b] and [c] with BSA in C-SBF and [d] with PAAc in R-SBF.

TF-XRD analysis

TF-XRD patterns of samples are given in Fig.2 A and B. and the patterns could be assigned to HAP (4). Intensity of the most intense broad peak at 31.74° and the broad peak at 26° decreased with increase in concentration of BSA. For 10mg/ml of BSA peak at 26° disappeared and intensity of that at 31.7° was reduced by 7 times than those of first two cases. Increase in concentration of PAAc from 0.1 to 1.0 mg/mL has also shown a decrease in the crystallinity of the precipitates formed. The non-apatite extra peak at 27.38° might represent traces of calcium carbonate formed (5).

FT-IR results

Collagen samples soaked in C and R-SBF (Fig.3.A(a) and B(a)) has shown the γ_4 and γ_3 vibrations of PO_4 at 566 and 610 cm^{-1} and in the region 920-1190 cm^{-1} (6). The absorption bands in the spectral region 1500 to 1300 cm^{-1}are associated with the presence of carbonate. The carbonate bands at 875, 1420 and 1465 cm^{-1} indicated that B-type carbonate substitution has occurred which is

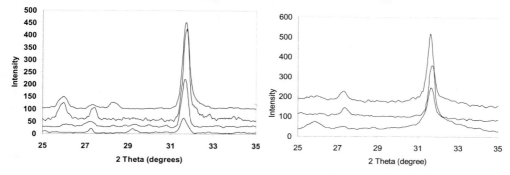

Fig.2A.TF-XRD patterns of samples from BSA absent and present at 1, 5 and 10 mg/ml concentrations in C-SBF (from top to bottom respectively)

Fig.2.B.TF-XRD patterns of the samples from PAAc absent and present at 0.1 and 1 mg/ml concentrations in R-SBF (from top to bottom respectively)

characteristic of calcified bone apatite (7). The band at 1660 cm^{-1} arises from the C=O of the peptide bond of the collagen. There is an additional band at 1560 cm^{-1} in the spectra of the samples mineralised from BSA and PAAc solublised solutions (Fig.3.A(b-d) and B(b-c)). This additional band and the band at 1420 cm^{-1} can be assigned to the chelation complex formed from the carboxlyate group of the additive and the Ca ion from the SBF solution and the band at 1660 cm^{-1}arise from the amide I band of BSA (8,9,10). The samples formed in presence of BSA had shown an increase in the intensity of the bands in the spectral region 1200 – 1375 cm^{-1}that might suggest the incorporation of carboxyl groups into the calcium phosphate phase. Also a decrease in the sharpness and intensity of the phosphate bands are observed with the addition of BSA and PAAc .

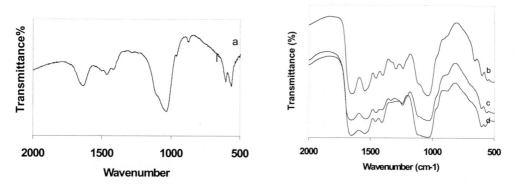

Fig.3(A) FT-IR spectra for BSA absent (a) and present at 1 (b), 5 (c) and 10 (d) mg/ml concentrations in C-SBF

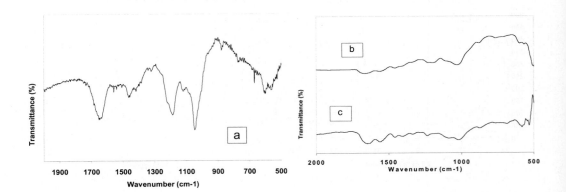

Fig.3.(B) FT-IR spectra of samples from PAAc absent (a) and present at 1.0 (b) and 0.1 mg/ml
(c) concentrations in R-SBF

Conclusions

Presence of BSA and PAAc containing carboxyl groups in the mineralising medium (C and R -SBF)
affected the crystallinity and partially the morphology of the precipitates formed on the collagen fibrils.
A chelation complex formed from the carboxlyate group of the additive and the Ca ion from the SBF
solution is found to be present in the precipitates. The precipitates formed were B-type carbonate
apatite

Acknowledgement

EKG is grateful to AIST for giving Post Doctoral Fellowship to carry out the work

References

[1] S.H.Rhee, J.D.Lee and J.Tanaka, J. Am. Ceram. Soc. 83 (2000) 2890.
[2] J.D.Termino, A.B.Belcourt, K.M.Conn and H.D.Kleinman, J. Biol. Chem. 256 (1981) 10408.
[3] E. K. Girija, Y.Yokogawa and F. Nagata, Chemistry Lett. (2002) 702
[4] JCPDS # 9-432.
[5] JCPDS # 41-1475
[6] A.M.Jacqueline, V.D.Houwen, G.Cressey, B.A.Cressey and E.V.Jones, J. Cryst.
     Growth 249 (2003) 572.
[7] M.Vignoles, G.Bonel, D.W.Holcomband R.A.Young, Calcif. Tissue Int. 43,
     (1988) 33.
[8] Y. E. Greish and P. W. Brown, J. Mater. Res. 14 (1999) 4637.
[9] Y. Yokogawa, F. Nagata, M. Toriyama, K. Nishizawa and T. Kameyama, J.Mater.Si. Lett. 18
     (1999) 367.
[10] H.Zeng, K.K.Chittur, W.R.Lacefield. Biomaterials 20 (1999) 377.

Key Engineering Materials Vols. 254-256(2004) pp. 403-406
online at http://www.scientific.net
© 2004 Trans Tech Publications, Switzerland

Apatite Deposition on Silk Sericin in a Solution Mimicking Extracellular Fluid: Effects of Fabrication Process of Sericin Film

Akari Takeuchi[1], Chikara Ohtsuki[1], Masanobu Kamitakahara[1],
Shin-ichi Ogata[1], Masao Tanihara[1], Toshiki Miyazaki[2], Masao Yamazaki[3],
Yoshiaki Furutani[4] and Hisao Kinoshita[4]

[1] Graduate School of Materials Science, Nara Institute of Science and Technology,

8916-5 Takayama, Ikoma, Nara 630-0192, Japan

ta-akari@ms.aist-nara.ac.jp

[2] Graduate School of Life Science and Systems Engineering, Kyushu Institute of Technology,

2-4 Hibikino, Wakamatsu, Kitakyushu, Fukuoka 808-0196, Japan

[3] Kyoto Prefectural Institute for Northern Industry, Mineyama, Naka, Kyoto 627-0011, Japan

[4] CENTMED Inc., 2-185-2 Issha, Meitou, Nagoya, Aichi 465-0093, Japan

Keywords: hydroxyapatite, simulated body fluid, biomimetic process, sericin

Abstract. An apatite-organic polymer hybrid is expected as a novel material for medical application, because it shows high biological affinity and high flexibility. In order to fabricate this type of hybrid, a biomimetic process was proposed, in which a bone-like apatite layer can be coated onto organic substrates by using a simulated body fluid (SBF) at ambient conditions. Potential of induction of heterogeneous nucleation of apatite on substrate materials is an important parameter to achieve a successful coating of apatite. Recently, it was reported that sericin, a protein existing on the surface of raw silk fiber, showed apatite-forming ability in 1.5SBF which has 1.5 times the ion concentrations of SBF. In the present study, the structural effect of sericin on its apatite-forming ability was investigated in 1.5SBF. Apatite was deposited on sericin with high molecular weight and β-sheet structure in 1.5SBF after 7 days.

Introduction

Hydroxyapatite is a kind of calcium phosphate that has high biological affinity to bone. Sintered hydroxyapatite is now widely used as bone substitute. However, the application of hydroxyapatite is restricted because it does not have enough mechanical properties to bear a large load. Therefore, a hydroxyapatite-organic polymer hybrid is expected as a novel material for medical application, because it shows high biological affinity and high flexibility. Hydroxyapatite coating on organic polymers is an attractive way to develop such hybrid materials. Kokubo and his colleagues previously proposed a biomimetic process to deposit a hydroxyapatite layer on various substrates, such as organic polymers. In this process, a bone-like apatite layer can be coated onto organic polymer substrates by either using a simulated body fluid (SBF) with ion concentrations nearly equal to those of human blood plasma, or using a more concentrated fluid under mild conditions [1]. Potential of induction of heterogeneous nucleation of apatite on the substrate materials is an important parameter to achieve a successful coating of apatite layer. Apatite crystals can spontaneously grow in the simulated body fluid (SBF), since SBF is supersaturated with respect to hydroxyapatite. We recently reported that sericin, a protein existing on the surface of raw silk fiber, is able to induce apatite deposition in 1.5SBF, which has 1.5 times the ion concentrations of SBF [2]. This means that sericin has a suitable composition and/or structure for inducing apatite nucleation in body environment. In this study, we have investigated the effective structure of sericin needed to deposit apatite crystals in 1.5SBF. Some sericin films were prepared from a solution containing sericin extracted under various conditions. The capacity of the sericin films for apatite deposition was examined by soaking them in 1.5SBF.

Materials and Method

Four types of sericin were obtained by extraction from raw silk fiber under various conditions. Raw silk fiber was autoclaved in ultra pure water at 105 or 120°C for 1 hour. The resultant solutions were denoted as 105-0d or 120-0d respectively. These solutions were subsequently stored at 4°C for 2 weeks, and then they were denoted as 105-2w or 120-2w. The molecular weight and secondary structure of sericin were analyzed by gel permeation chromatography (GPC), circular dichroism (CD) spectroscopy and Fourier-transform infrared (FT-IR) spectroscopy. The solution containing sericin was cast in a Petri dish to form a film of sericin at the bottom of the dish, followed by drying at room temperature. The cast films were exposed to 1.5SBF (ion concentrations: $Na^+ = 213.0$, $K^+ = 7.5$, $Mg^{2+} = 2.3$, $Ca^{2+} = 3.8$, $Cl^- = 221.7$, $HCO_3^- = 6.3$, $HPO_4^{2-} = 1.5$, and $SO_4^{2-} = 0.8$ mol/m^3). The solution was buffered at pH = 7.25 using 75 mol/m^3 of tris(hydroxymethyl) aminomethane and appropriate amount of hydrochloric acid using the method reported by Kokubo et al [3]. The temperature of the solution was maintained at 36.5°C. A volume of 15 mL of 1.5SBF was put into a Petri dish coated with the sericin film, and kept for 7 days at 36.5°C. The surfaces of the films before and after soaking in 1.5SBF were observed under a scanning electron microscope (SEM), and characterized using thin-film X-ray diffraction (TF-XRD).

Results and Discussion

Figure 1 shows GPC profiles of the sericin solutions prepared under various conditions. The most frequent distribution in molecular weight appeared approximately at 134, 106, 42 and 43 kDa for 105-0d, 105-2w, 120-0d and 120-2w respectively. This indicates that higher molecular weight of sericin was obtained by extraction at 105°C than at 120°C. Both the extracted solutions at 105 and 120°C did not show significant changes in the molecular weight after storage at 4°C for 2 weeks. Figure 2 shows CD spectra of the sericin solutions prepared under various conditions. The CD spectrum of 105-0d was assigned to a random coil structure from the typical negative peak near 198 nm. After the storage for 2 weeks at 4 °C, the intensity of such negative peak decreased and the intensity of a negative peak at 217 nm increased. This means that β-sheet structure became dominant during the storage of the solution [4-5]. This behaviour of the sericin extracted at 105 °C was similar to that of the solution extracted at 120 °C (Fig.2 (b)). In addition, FT-IR spectra of sericin films prepared from each solution showed that sericin structure in film was the same as in aqueous solution. Figure 3 shows SEM photographs of the surfaces of sericin films after soaking in 1.5SBF for 7 days. Deposition of spherical particles was observed on the surface of 105-2w film whereas it was not noticed on the surfaces of the three other films. From XRD analysis, particles on the surface of 105-2w film were characterized as low-crystalline apatite. These results indicate that sericin with high molecular weight and β-sheet structure induces apatite deposition in 1.5SBF.

Previous studies showed that carboxyl groups may induce apatite nucleation in a solution mimicking body fluid. Sericin contains approximately 20 mol% of acidic amino acid that gives carboxyl groups in its structure. The high content of carboxyl groups plays an important role in apatite nucleation. The present results indicate that heterogeneous nucleation of apatite on silk sericin is governed not only by the content of carboxyl group included in it but also by its secondary structure. Sericin with high molecular weight and specific structure dominant in β-sheet provides effective surface for apatite nucleation in 1.5SBF. Arranged carboxyl groups can provide effective site for heterogeneous nucleation of apatite. Consequently, sericin with appropriate structure may satisfy the requirement to induce apatite deposition and can be useful as a polymer material in the fabrication of apatite-polymer hybrid through biomimetic processes.

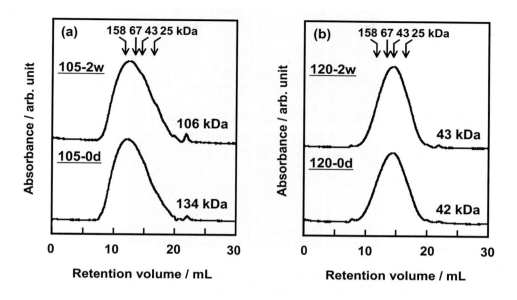

Fig. 1. GPC profiles of sericin extracted at: (a) 105 °C and (b) 120 °C.

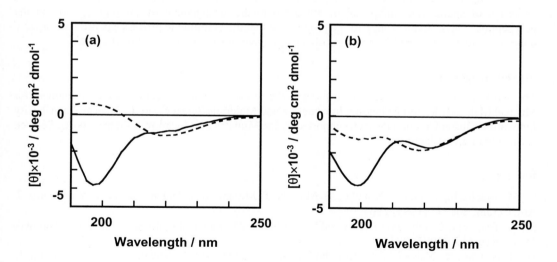

Fig. 2. CD spectra of sericin solution obtained under various conditions..
(a) —— : 105-0d, ----- : 105-2w, (b) —— : 120-0d, ----- : 120-2w

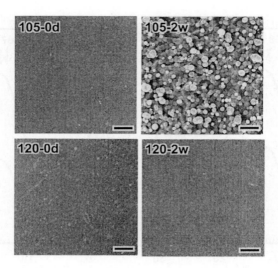

Fig. 3. SEM photographs of the surfaces of sericin films after soaking
in 1.5SBF at 36.5°C for 7 days. Bar indicates 20 μm.

Conclusion

Sericin can induce apatite deposition on its surface in a biomimicking physiological solution, such as
1.5SBF, when it is dominated by the β-sheet specific structure and high molecular weight. The
induction of apatite nucleation can be attributed to the arrangement of carboxyl groups on the
polypeptide. These findings support the proposition that the biomimetic process can produce organic
polymer coated with bone-like apatite utilizing a polypeptide containing carboxyl groups.

Acknowledgement

This work was conducted under the auspices of the research project, "Technology Development for
Medical Materials Merging Genome Information and Materials Science", in the Kansai Science City
Innovative Cluster Creation Project, supported by the Ministry of Education, Culture, Sports, Science
and Technology of Japan.

References

[1] M. Tanahashi, T. Yao, T. Kokubo, M. Minoda, T. Miyamoto, T. Nakamura and T. Yamamuro: J.
Am. Ceram. Soc. 77 (1994) 2805.
[2] A. Takeuchi, C. Ohtsuki, T. Miyazaki, H. Tanaka, M. Yamazaki and M. Tanihara: J. Biomed.
Mater. Res. 65A (2003) 283.
[3] S. B. Cho, K. Nakanishi, T. Kokubo, N. Soga, C. Ohtsuki, T. Nakamura, T. Kitsugi and T.
Yamamuro: J. Am. Ceram. Soc. 78 (1995) 1769.
[4] N. Greenfield and G. D. Fasman: Biochemistry 8 (1969) 4108.
[5] V. Madison and J. Schellman: Biopolymers 11 (1972) 1041.

Key Engineering Materials Vols. 254-256(2004) pp. 407-410
online at http://www.scientific.net
© *2004 Trans Tech Publications, Switzerland*

In-vitro Surface Reactions of Bioceramic Materials

R. Gildenhaar², A. Bernstein¹, G. Berger², W. Hein¹

¹University of Halle, Department of Orthopaedics, Magdeburger Str. 22, 06097 Halle/Saale,
Germany, renate.gildenhaar@bam.de
²Federal Institute for Materials Research and Testing, Unter den Eichen 87, 12205 Berlin,
Germany

Keywords: ceramics, coating, biomaterial, bone, biocompatibility, surface reaction

Abstract. Bioceramics used as coatings show different biocompatibility and bioactive behaviour in relation to their chemical and morphological composition. Hydroxyapatite of low crystallinity can be detected by XRD on the surface of different plasma sprayed bioceramic materials after treatment in SBF solution. Solubility and cell adhesion tests as well as test of toxicity on discs, however, did not show similar effects. Especially the crystallographic composition and also surface morphology of the material determined the release of ions into the solutions which could be responsible for cytotoxic reactions as well as for the ability of the hydroxyapatite formation. However, the reaction mechanisms which lead to a formation of apatite layers seems to be of different nature. Moreover, there is also the question if such an apatite formation in vitro can indicate the material's ability to favour osteogeneses.

Introduction

Whereas the primary fixation mode of cementless prostheses is mechanical and depends on a physical interlock between the implant and the bone, the secondary fixation is biological and achieves osseointegration at the implant-bone interface by means of bone growth onto or into the substrate. Bioactive ceramics such as hydroxyapatite (HA) promote and enhance biological fixation. Therefore, these materials are widespreadly used as interfacing osseoconductive coatings of metallic surgical implants [1, 2].
Solubility investigations of bone substitution materials lead to a differentiation into resorbable and long-term stable materials. There is still a discussion on the desired longevity of the coatings. One theory is that coatings are only required till the initial phase of osseointegration. Hence, fully resorbable coatings are to be preferred. The ingrown new bone will easily compensate the loss of the coating, and the titanium substrate will eventually establish a strong, direct bond to bone by contact osteogenesis. On the other hand, stable coatings will provide an essential role as a strong semipermanent connection between the implant and tissue. Such a less resorbable coating requires an optimum between stable resorption rate, flexural strength and adhesive strength of the coating. Their biorelevant function is bonding osteogenesis.
The in vivo performance of such coatings depends on a large array of factors. The most notable ones are coating thickness, chemical composition, crystallinity, phase purity, cohesive and adhesive strengths, and resorption behavior. In this work, the biocompatible and bioactive behaviour of four plasma sprayed bioceramic materials are examined. The cytotoxicity in vitro was determined on discs by agar diffusion and filter tests and the microculture tetrazolium (MTT) assay. After surface characterization by scanning electron microscopy (SEM) cell proliferation parameters were measured. Solubility investigations were carried out on plasma sprayed coatings as well as on discs.

Materials

Plasma sprayed coatings and discs were produced using four calcium phosphate ceramic powders (see Table I). The samples S1 and S2 are characterized by a high alkali content in relation to HA (S3) and the sample S4 contains no alkali but remarkable quantities of zirconium. The materials were synthesized by a melting process and crushed to a grain size of 50-125μm for the plasma spraying process.

Table I: Composition of powders

Sample Code	Chemical composition (% by weight)					Crystal phases
	CaO	P_2O_5	K_2O	Na_2O	MgO	
S1	30,7	43,1	14,3	9,4	2,5	$Ca_2KNa(PO_4)_2$
S2	38,0	46,2	8,6	5,7	1,5	$Ca_3(PO_4)_2$, $CaK_2P_2O_7$ $Ca_5Na_2(PO_4)_4$
S3	55,8	42,4				$Ca_5(PO_4)_3OH$
S4	31,1	42,9	4,5 CaF_2	21,4 ZrO_2		$Ca_5(PO_4)_3F$, $CaZr_4(PO_4)_6$

TiAl6V4 alloy of the company DePuy Motech, Inc. was used as substrate material for biomedical applications. The alloy was roughened with corundum granules of the grain size 0,5-1mm with a jet pressure by 3 bar. The plasma coating process took place on a PTM 1000 plant, equipped with a Sulzer Metco F4MB electronic torch. For a layer thickness of 120-180 μm four plasma-over runnings were necessary.

Discs were prepared from powders by pressing and temperature treatment for both solubility tests and investigations of proliferation of human osteoblasts.

For solubility investigation coated targets as well as pure ceramic discs were stored in simulated body fluid (SBF) after KOKUBO and in 0,2M TRIS-HCl buffer solution (TRIS) ($pH_{37°C}=7.4$), respectively. The treatment was carried out over a period of 24 weeks. The solutions were renewed weekly. At the same time, the targets and discs were dried and weighed. After 24 weeks the surfaces of the layers were examined by X-ray diffraction and by Scanning Electron Microscopy (SEM).

Human osteoblasts were cultured on discs. To determine cell proliferation, the cells were detached with trypsin/EDTA after 3, 7, 15, and 19 days [3, 4]. Osteoblasts were obtained by outgrowth from small explants of human bone tissue (tibia) from healthy patient undergoing treatment for prosthesis. Bone chips were grown in Dulbecco´s modified Eagle medium (DMEM) without calcium, supplemented with 10% fetal calf serum (FCS), 1% penicillin , and 1 % streptomycin. The chips were incubated at 37°C in a humidified 5% CO_2 atmosphere. Cells began to migrate from the chips within 10-14 days. When the cells reached confluence, they were transferred to medium. The osteoblastic phenotype of outgrowing cells was characterized by the presence of alkaline phosphatase and the elaboration of a type I collagen matrix.

For the agar diffusion test, the filter test, and for the MTT assay material granules (315-400μm) were used. The ceramic granules were sterilized at 200°C for 4h. Extracts were prepared by storing the ceramics in water (0,4g/l). The extracts were tested by the MTT assay. MG63, an osteosarcoma cell line (ATCC), was employed as cell line for the agar diffusion test, the filter test, and for the MTT assay. MG63 cells were maintained in Dulbecco´s modified Eagle medium (DMEM) supplemented with 10% fetal calf serum, penicillin (100U/ml) and streptomycin (100μg/ml). The agar diffusion test and the filter test were performed as recommended by ISO 10993-5 and described in the literature. [5]

Concerning the MTT assay the ability of living cells is employed to reduce a water-soluble yellow dye, 3-(4, 5-dimethyl-thiazol-2-yl)-2,5-diphenylterazolium bromide (MTT) to a water-insoluble purple formazan product by mitochondrial succinate dehydrogenases [5].

Results and discussion

The four materials employed in this study showed a good biocompatible behaviour.
On the plasma sprayed samples a significant change in surface composition could be observed during storing in SBF and in TRIS, respectively. Fig. 1 shows the change of surface structure of sample S1.
After treatment with TRIS the surface of sample S1 did not change but in SBF hydroxyapatite of low crystallinity can be examined by XRD. Similar performance was observed regarding the materials S3 and S4. In contrast to that, the sample S2 showed a very different behaviour. After treatment with SBF no surface change is noticeable but after storing in TRIS the formation of $Ca_2P_2O_7$ could be observed.

a)   b)   c)

Figure 1: SEM photographs of plasma sprayed coating S1; a) untreated, b) storing in TRIS for 28 weeks, c) after treatment with SBF for 28 weeks

If hydroxyapatite was formed during the sample treatment in SBF a significant mass accumulation could be found whereas during the treatment with TRIS a mass loss was observed as it is shown in Fig. 2. The plasma sprayed samples showed surface changes caused by the medium influence. However, using the same treatment for discs these changes could not be found. Variations in the amorphous part and in porosity between the disc materials and the plasma sprayed coatings could be the reason for the different behaviour. The amorphous part as well as the porosity may increase the solubility and also may influence the chemical reactions.

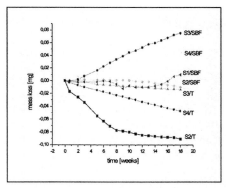

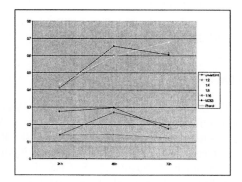

Figure 2: Mass loss of plasma sprayed S1 bioceramics in SBF and in TRIS

Figure 3: MTT test of ceramic (absorbance measured at 550nm)

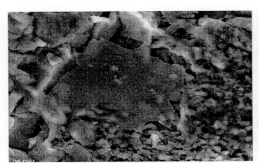 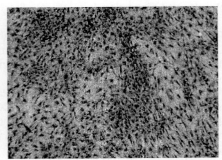

Fig.4: ESEM picture of disc-surface of   Fig. 5: Light micrograph of osteoblast –
material S4   growth on S4 (toluidine blue colouring)

Further investigations are required to determine the role of ion levels in the culture medium and of solution-induced surface transformations on cellular adhesion, metabolism and proliferation. The cells treated with extracts of ceramics in MTT-test showed an activation of mitrochondrial activity strongly dependent on concentration. For all tested materials the growth curves showed a similar kinetics as it is shown in Fig. 3 for material S1. For both MTT-test and filter test, cytotoxic reactions were not detected at every point in time.

A different growth pattern of the osteoblasts was observed on the ceramics after 10 days of incubation. Primary human osteoblast-like cells were used which provide a better representation of the situation in vivo than commercially available cell lines.

The four materials employed in this study had surface characteristics that were similar to those of corresponding clinical implants (Fig.4). The surface roughness of all materials was very similar. It was tried to minimize the influence of the surface roughness on the cellular reaction.

It can be concluded that the general toxicity in vitro of the calcium phosphate ceramics depends on the solubility of the materials. Solubility himself is a function of material composition and morphology as well as dependent on solution composition and solution volume. Further investigations are necessary to determine the role of ion levels in the culture medium and of solution-induced surface transformations on cellular adhesion, metabolism and proliferation.

Acknowledgements

Financial support by the German Research Foundation (DFG grant no. BE1339/10-1 and BE1964/2-1) is gratefully acknowledged.

References

[1] J.F. Osborn: *Neuere Ergebnisse in der Osteologie* (Springer, Germany, 1989, p. 358-364)
[2] G. Willmann: Mat.-wiss. u. Werkstofftech., Vol. 30 (1999), p. 317-325
[3] C. Knabe, W. Ostapowicz, R.J. Radlanski, R. Gildenhaar, G. Berger, R. Fitzner,
    U. Gross: J. Mater. Science - Materials in Medicine Vol. 9 (1998), p. 337-345
[4] C. Schmidt, A. Ignatius, L. E. Claes: Biomed. Mater. Res. Vol. 54 (2001), p. 209-215
[5] A. Ignatius, C. Schmidt, D. Kaspar, L.E: Claes: J. Biomed. Mater. Res. Vol. 55 (2001),
    p. 285-294

Key Engineering Materials Vols. 254-256(2004) pp. 411-414
online at http://www.scientific.net
© 2004 Trans Tech Publications, Switzerland

A New Procedure of a Calcium-Containing Coating on Implants of Titanium Alloy

U. Ploska[1], G. Berger[1], M. Willfahrt[2]

[1] Federal Institute for Materials Research and Testing, Unter den Eichen 87, 12200 Berlin, Germany, georg.berger@bam.de
[2] Merete Medical GmbH, Alt-Lankwitz 102, 12247 Berlin, Germany, mwillfahrt@merete.de

Keywords: TiAl6V4, calcium titanate, surface layer, salt melt

Abstract. The paper describes a new method for converting the existing oxide layer of implants of titanium and titanium alloys in calcium containing surface layers. The layers were generated by dipping the implants in a salt melt consisting of calcium nitrate at temperatures of 520°C until 560°C for 2 until 4 hrs. The melt reacted with the surface layer of the implants by forming calcium titanate surface layers. Besides, the thickness of the TiO_2 layer of the implants became larger. Several methods for the determination of the layer thickness were investigated related to their suitability. The micro hardness of the layer and the tribological behaviour were investigated. The presented procedure can be used for generating calcium containing surface layers on implants of titanium alloys even with complicated forms.

Introduction

Implants of titanium or titanium alloys are proved as bioinert. For improving the biological properties the implants were coated to cause a contact between implant and bone without connective tissue. Several coating techniques are well known, e.g. the procedure of plasma spraying [1], CVD [2], electrolytic depositing [3], ion implantation [4] or biomimetic methods [5]. Most of the generated coatings are porous and their adhesive strengths are not always sufficient. Plasma spraying is not suitable if the samples had irregular forms. For the deposition of hydroxyapatite coatings on titanium implants by biomimetic methods, it is necessary to modify the sample surface before soaking in simulated body fluid to improve the deposition of hydroxyapatite. This paper describes a new method for covering titanium alloys with calcium containing surface layers by treating the samples with a salt melt.

Materials and Methods

The treatment of the TiAl6V4 samples with the salt melt was carried out in a gas-tight furnace in a nitrogen atmosphere. For the salt melt calcium nitrate tetra hydrate (p.a., Merck GmbH, Germany) was used. The procedure was carried out in a TiAl6V4 crucible. After dehydration of the salt it was melted in the furnace at a temperature of 580°C. Samples like screws and cylinders were blasted with glass balls or Ap40 granulate. The samples were then scoured using the weakly alkaline universal cleaner Tickopur R33 (Dr. H. Stamm GmbH, Germany) in an ultrasound bath, cleaned with water by ultrasound, dried on air, and then dipped into the salt melt at temperatures of 520-560°C. When the treatment time was over the sample were taken out of the melt and put in hot water in an ultrasound bath to remove adhering parts of the salt melt on the sample surfaces followed by cleaning with diluted HCl (1+9) in an ultrasound bath. Finally the samples were washed with distilled water and dried on air. For the determination of the layer thickness disks were cut from TiAl6V4 rods, drilled and polished before they were scoured. The time of treatment with the salt melt was varied between 0.5 and 16 hrs. The structure of the surface layers was identified by thin film X-ray diffraction (TF-XRD) measurements. Depth profiles were obtained by Auger electron spectrometry (AES) and the layer thickness was first determined by scanning electron microscopy (SEM) investigation. Several other methods like cross cut, sphere cut, ellipsometry and eddy current were used for checking their suitability as process controlling. The micro hardness was determined and the tribological behaviour of the covered samples compared with that of untreated samples.

Results

The procedure is suitable to generate a surface layer on TiAl6V4 implants by the treatment of the samples with a salt melt consisting of calcium nitrate. The layer had a deeper grey colour than the untreated samples (Fig. 1a). Fig. 1b shows coated screws as used in surgery for fixing of bone.

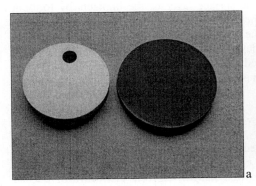 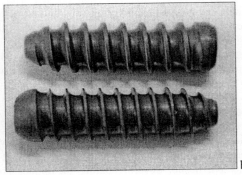

Fig. 1 a: Comparison of untreated titanium alloy (left) and coated titanium alloy (right)
Fig. 1 b: Coated screws as used in surgery as example of complicated forms

By means of TF-XRD measurements, the crystalline phase of the surface layer was shown to be the CaO-rich calcium titanate ($Ca_4Ti_3O_{10}$). The layer was generated after 30 min but their depths varied in a large range. Treatment times of 2 hrs until 8 hrs lead to layers with increased depth and decreased differences in thickness.

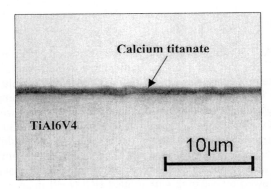

Fig. 2: Light microscopic picture of a cross cut view (treatment time 16 hrs)

Fig. 3 und 4 show the AES measurement of 2 samples soaked for 4 hrs and 16 hrs, respectively, into the salt melt. By means of the AES measurement, the thickness of a calcium titanate layer (treatment time 16 hrs) was estimated to be 1.2 µm. Using higher treatment times like 16 hrs the thickness of calcium titanate layer was increased.

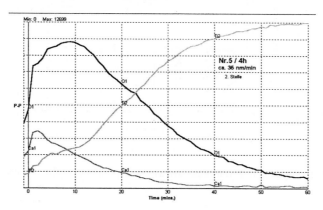

Fig. 3: AES spectrum of a sample soaked for 4 hrs in the calcium nitrate melt

The thickness of that calcium titanate layer was about 1 µm thick and grew when the time of treatment was longer than 8 hrs. The thickness of the intermediate TiO$_2$ layer is broadened compared to that of untreated TiAl6V4.

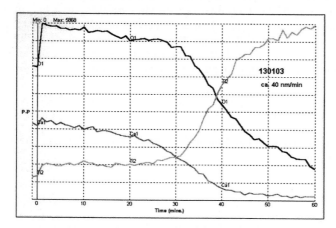

Fig. 4: AES spectrum of a sample soaked for 16 hrs in the calcium nitrate melt

The surface layers were elastic (average hardness 4.3 GPa, average e-modulus 128 GPa). The determination of the layer thickness by means of eddy current is suitable for process controlling. In Fig. 5 the variation of the layer thickness with treatment time is represented inclusively the errors of measurement. It is obvious that there is no significant difference in the layer thickness if compared the treatment times of 2 until 8 hrs.

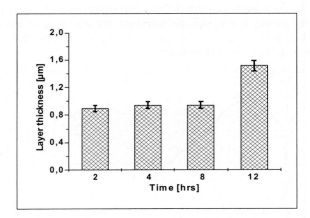

Fig. 5: Comparison of the layer thickness dependent on the treatment time

Discussion and Conclusion

As shown by the determination of the micro hardness, the surface layers have an elastic behaviour. No cracks are obvious at the area of the intender's smash even not at the borderline of the smash. A reaction time of 2 or 8 hrs is sufficient to generate the calcium titanate surface layer. There is no significant difference in layer thickness when treated from 2 up to 8 hrs. Longer reaction times (>8 hrs) led to a slightly larger thickness of the calcium titanate layer but also to higher thickness of the TiO_2 intermediate layer resulting in a loss of adhesive strength. For process controlling, methods based on cross cutting or eddy current seemed to be suitable. Irregular forms can be coated well because the whole surface area of the samples is in contact with the melt. The presented procedure can be used for generating calcium containing surface layers on implants of titanium alloys even with complicated forms.

Acknowledgements

Financial support by the German Ministry of Economics and Labour (VI A 2 – 15/02) is gratefully acknowledged.

References

[1] Y.W. Gu, K.A. Khor, P. Cheang: Biomaterials, Vol. 24 9 (2003), p. 1603
[2] S. Rupprecht, A. Bloch, S. Rosiwal et al.: International Journal of oral max impl Vol. 17 6 (2002), p. 778
[3] L.A. de Sena, M.C. de Andrade, A.M. Rossi et al.: Journal of Biomedical Materials Research, Vol. 60 1 (2002), p. 1
[4] M.A. de Maeztu, J.I. Alava, C. Gay-Escoda: Clinical oral implants research, Vol. 14 1 (2003), p. 57
[5] T. Kokubo, H.M. Kim, M. Kawashita: Biomaterials, Vol. 24 13 (2003), p. 2161.

Key Engineering Materials Vols. 254-256(2004) pp. 415-418
online at http://www.scientific.net
© *2004 Trans Tech Publications, Switzerland*

Mechanical Evaluation of the Use of a Buffer Layer in Hydroxylapatite Coatings Produced by Pulsed Laser Deposition at High Temperature

E. Jiménez, J.L. Arias, B. León, M. Pérez-Amor

Universidade de Vigo, Departamento de Física Aplicada, Lagoas-Marcosende, 9, 36200 Vigo, Spain,
jlarias@uvigo.es

Keywords: Hydroxylapatite coatings pulsed laser deposition, mechanical properties.

Abstract. Hydroxylapatite is used as coating on metallic implants because its composition and structure is similar to those of bone. As an alternative to plasma spraying, pulsed laser deposition has been successfully applied. Nevertheless, to obtain a high crystalline coating at temperatures higher than 550 °C are needed, leading to a deterioration of the coating-to-substrate adhesion. Some coatings were produced by pulsed laser deposition at 650 °C with a previous buffer layer deposited al lower temperature (460 °C). The combination of this buffer layer with a high water vapour flow (60 Pa l/s) clearly improved the mechanical behaviour of the system coating-metal.

Introduction

Hydroxylapatite (HA) is a calcium phosphate used as coating on metallic orthopaedic and dental implants, because its composition and structure are similar to those of bone [1]. As an alternative to plasma spraying, the commercial coating technique, pulsed laser deposition (PLD) allows a perfect control of composition and crystallinity of the coating [2, 3, 4] and shows an optimal osteoconduction in animal models [5]. Nevertheless, when a certain application requires a very high crystallinity, the high temperatures needed (>550 °C) provoke the deterioration of the coating-substrate interface, due to the formation of TiO_2 [2,4]. In the present study a HA buffer layer is produced by PLD, at lower temperatures (460 °C), in order to avoid the TiO_2 formation and to obtain a very crystalline coating with an appropriate mechanical response.

Materials and Methods

The beam of an ArF laser (λ = 193 nm) was focused on a HA sintered target to achieve an energy density of 1.2 Jcm^{-2}. The ablated material was deposited onto a grade 3 pure titanium substrate situated parallel to the target at 4.8 cm from it, and kept at 650° C. This laser ablation process took place in a water vapour pressure of 60 Pa with a flow of 30 Pa l/s.

Two different type of coatings (1 μm thick) were obtained, with and without a previous buffer layer. This buffer layer (25 nm thick) was deposited with the same energy density as the coatings, but at a lower substrate temperature of 460 °C and in a total water pressure of 45 Pa with a flow of 25 Pa l/s. These are the conditions to obtain a carbonated HA (C-HA) coating with a high coating-to-substrate adhesion.

The crystallinity and phase composition of these coatings were determined by X-ray diffraction (XRD) and infrared spectroscopy (FT-IR). X-ray photoelectron spectroscopy (XPS) was carried out to identify the presence of some phases in the coatings and at the coating-substrate interface.

The coating-to-substrate adhesion was evaluated by micro-scratch testing, while the hardness and elastic modulus of the coatings were determined by progressive nanoindentation along their whole depth. The fatigue resistance of coatings was studied by impact tests.

Results and Discussion

The XRD patterns of both types of coatings (Fig. 1) produced without and with buffer layer show that they are constituted by a HA phase. Their FT-IR spectra (Fig. 2) not only confirm this point with the presence of the phosphate and hydroxyl vibrational bands in the positions characteristic of HA, but also allow us to say that they correspond to carbonated hydroxylapatite (C-HA) since the vibrational bands of carbonate corresponds to a carbonate substitution for phosphate in the HA. The XRD pattern of the coatings produced without buffer layer (Fig. 1.a) presents clearly a TiO_2 phase (rutile), formed at the substrate surface. The use of the buffer layer avoids the formation of TiO_2 (Fig. 1.b), but it provokes the formation of a new phase with a diffraction peak at 37.2° 2θ, and the duplication of the diffraction peaks of titanium (Ti^x).

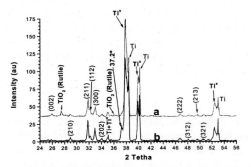

Fig. 1. XRD patterns of the HA coatings produced by PLD (a) without buffer layer and (b) with buffer layer.

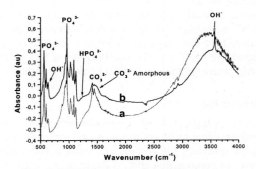

Fig. 2. FT-IR spectra of the HA coatings produced by PLD (a) without buffer layer and (b) with buffer layer.

This diffraction peak at 37.2° can be assigned to either CaO or a titanium hydride ($TiH_{0.71}$). Although the absence of the main peak for CaO at 53.9° in the XRD pattern (Fig. 1.b) seems to exclude this possibility, an study by XPS of the binding energy environment of Ca in the whole coating thickness has been performed to clarify this point. The XPS spectra (Fig. 3.a) show that the binding energy of Ca only corresponds to that of HA. Therefore, the diffraction peak at 37.2° should be assigned to a titanium hydride ($TiH_{0.71}$).

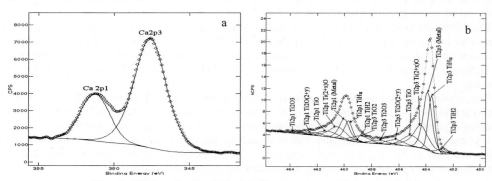

Fig. 3. XPS analysis of the binding energy enviroment of: (a) Ca inside the HA coating, and (b) Ti in the coating-sustrate

The analysis of the coating-substrate interface by XPS (Fig. 3.b) shows the presence of traces of oxygen and a shoulder at the metallic Ti 2p energy peak, probably due to this titanium hydride $TiH_{0.71}$ (with an energy shift lower than that of TiH_2). Therefore, the additional Ti XRD peaks (Ti^x in Fig. 1.b) are probably due to a stressed Ti phase produced by the diffusion into the substrate of both oxygen and hydrogen.

Although the TiO_2 formation is avoided with the use of a buffer layer, the coating-to-substrate adhesion is not clearly improved. The scratch testing gave a critical load at which the coating detached from the substrate of 3-6 N for the coatings produced with buffer layer as compared to the values of 2-4 N obtained for those without. This low adherence can be mainly due to the brittle nature of the titanium hydride.

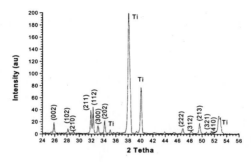

Fig. 4. XRD patterns of the HA coatings produced by PLD
with buffer layer and a water vapour flow of 60 Pa l/s.

In order to reduce the formation of these phases at the interface, the coatings with a previous buffer layer were produced in the same conditions but with a higher water vapour flow of 60 Pa l/s. The aim of these procedure was to drag out the hydrogen, produced by the water dissociation at the surface during the coating deposition. As it can be observed in the XRD of these coatings (Fig. 4) both the titanium hydride ($TiH_{0.71}$) and the stressed titanium phase (Ti^x) are not present. The disappearance of these phases when the water vapour flow is increased led to values of the critical load higher than 30 N (the evaluation limit of the scratch tester).

The progressive nanoindentation allowed to determine the Young modulus and the hardness of the coatings as functions of the penetration depth (Fig. 5). The coatings produced with a previous buffer layer with different water vapour flows gave similar values for the hardness (1.8 GPa) and Young modulus (65 GPa) inside the coating. However, at the interface, the values of the hardness (2.4 for 30 Pa l/s and 3.4 for 60 Pa l/s) and Young modulus (2.4 GPa for 30 Pa l/s and 3.4 GPa for 60 Pa l/s) are very different. Therefore, the presence of the titanium hydride and the stressed titanium phase clearly modify the mechanical properties of the coating-substrate interface.

The impact tests gave a much higher fatigue resistance for the coatings produced with buffer layer and a water vapour flow of 60 Pa l/s. These coatings either failed after 75-100 cycles of impacts with 16 mN load with a spherical diamond (25 mm radius) or did not fail after the complete test (150 cycles). While, the coatings produced with buffer layer and a flow of 30 Pa l/s already failed after 25 cycles.

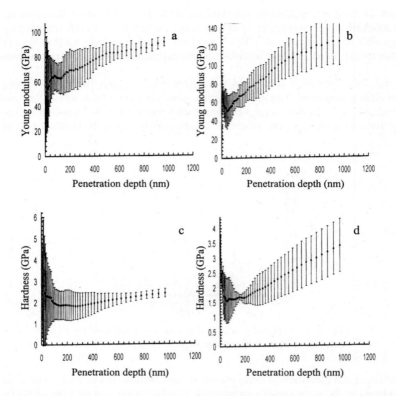

Fig. 5. Nanoindentation tests of the coatings produced by PLD with a buffer layer. Young modulus and hardness are given as functions of the penetration depth: (a) Young modulus of the coating produced at a water vapour flow of 30 Pa l/s, (b) Young modulus of the coating produced at a water vapour flow of 60 Pa l/s, (c) hardness of the coating produced at a water vapour flow of 30 Pa l/s, and (d) hardness of the coating produced at a water vapour flow of 60 Pa l/s.

References

[1] L.L. Hench, Ö. Anderson: in L.L. Hench, J. Wilson editors: *An Introduction to Biocer*amics (World Scientific, Singapore 1993), p. 239.

[2] C.M. Cotell, D.B. Chrisey, K.S. Grabowski, J.A. Sprague: J. Appl. Biomater. Vol. 3 (1992), p. 87.

[3] J.M. Fernández-Pradas *et al*:_Appl. Surf. Sci. Vol. 195 (2002), p. 567.

[4] J.L.Arias *et al* :J Mater. Sci. Mater.Med.Vol. 8 (1997), p. 873.

[5] C. Peraire *et al*: Proc. 17th European Conf. Biomater. (2002), p. T83.

Key Engineering Materials Vols. 254-256(2004) pp. 419-422
online at http://www.scientific.net
© 2004 Trans Tech Publications, Switzerland

Influence of Ca/P Ratio on Electrochemical Assisted Deposition of Hydroxyapatite on Titanium

A. Sewing[1], M. Lakatos[2], D. Scharnweber[2], S. Roessler[2], R. Born[2], M. Dard[1] and H. Worch[2]

[1] Biomet Merck BioMaterials GmbH, Frankfurter Str. 129, D-64271 Darmstadt, Germany,
Andreas.Sewing@biomet-merck.com

[2] Max-Bergmann-Zentrum für Biomaterialien, Technische Universität Dresden,
Institut für Werkstoffwissenschaft, Budapester Str. 27, D-01062 Dresden, Germany,
Dieter.Scharnweber@mailbox.tu-dresden.de

Keywords: Calcium phosphate, hydroxyapatite, electrochemistry, coating.

Abstract. An electrochemical assisted process under cathodic polarization was used to produce hydroxyapatite (HAP) coatings on titanium surfaces. Within the parameter field of current density, polarization time and for Ca/P ratios of the electrolyte of 1, 1.33 and 1.67 the deposited calcium phosphate phases were identified by FTIR and SEM. It was found that the transformation of amorphous calcium phosphate (ACP) to HAP with needle like morphology takes place at higher current densities and after longer times when the Ca/P ratio of the electrolyte is moved away from the stoichiometric 1.67 ratio. The Ca/P ratio of the coatings were analyzed by ICP-OES showing that additional calcium is stored in the layer. Electrolyte speciation and migration processes near the Ti surface are discussed including layer morphology to propose amorphous calcium hydroxide to be responsible for the Ca/P values of around 2.

Introduction

Different methods for coating of metal implants with calcium phosphate phases (CPP) beside the most widely used plasma spray technique have been developed over the last decade. Aim is to overcome the weaknesses of plasma spray coatings caused by high temperature process conditions resulting in coating properties that deviate from the mineral phase of bone with respect to crystal structure and solubility. Biomimetic approaches using a precipitation of the CCP from solution result in coating properties that are closer to mimic the surrounding tissue at the implant site. Electrochemical assisted processes have the advantage of possible process control and higher efficiency over pure precipitation processes. Our process leads to a direct formation of HAP via an amorphous phase at process conditions near physiological pH and temperature [1]. In this paper we report on the influence of Ca/P ratio in the electrolyte on CCP composition of the layer as a function of current density and time.

Material and Methods

Ti6Al4V disks (ASTM F136) with diameter of 10 mm and 2 mm thickness and polished surfaces were used as substrates. CPP coatings have been prepared using an electrochemical assisted process under cathodic polarization of the sample in a $Ca^{2+}/H_xPO_4^{(3-x)-}$ solution near physiological conditions (pH 6.4, 37 °C), which is described in detail in ref. [1]. Ca/P ratio of the electrolyte was set to 1.67, 1.33, and 1 by choosing the right molar ratio of $CaCl_2$ and 0.02 M $NH_4H_2PO_4$ and diluting equal volumes at a ratio of 1:20 with distilled water. Samples have been prepared in galvanostatic mode for current densities of -5, -10 and -15 mA/cm² and for polarization times of 10, 20, 30, 45, 60, 90 and 120 minutes. For each parameter setting two sets of samples were prepared, leading to 12 samples available for characterization. SEM (DSM 982 Gemini, Carl Zeiss) was used to characterize the morphology of the CCP. Chemical composition was analyzed by FTIR (Perkin Elmer FTS 2000). Five spectra were taken from each sample to prove homogeneity. Characteristic spectra were analyzed for peak position and intensity and compared to reference spectra. The Ca/P ratio of the resulting layer was measured by ICP-OES (Varian Vista-PRO). Therefore the coating

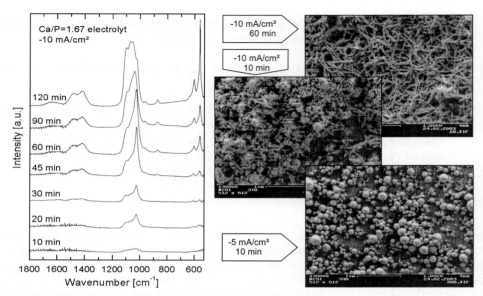

Fig. 1: (Left) FTIR spectra for different polarization times of coatings produced at a current density of 10 mA/cm² in a Ca/P=1.67 electrolyte. (Right) Selected SEM micrographs of ACP, intermediate and HAP phase (all at same magnification of 30000).

the resulting layer was measured by ICP-OES (Varian Vista-PRO). Therefore the coating was dissolved from the substrate in 0.1 M nitric acid. Six samples were measured per deposition parameter setting and values were corrected for background signals of blank samples and chemicals.

Results and Discussion

Electrochemical assisted CPP deposition is based on the diffusion and migration processes in the electrolyte and the pH dependent solubility of CPP. During cathodic polarization of a metal in an aqueous electrolyte alkalization close to the surface takes place resulting in the precipitation of an initial amorphous CPP [1]. Fig. 1 shows selected SEM micrographs of the phase formation taken from samples prepared in an electrolyte with Ca/P ratio of 1.67. The morphology of the amorphous phase is characterized by spheres with diameters up to 300 nm which cover the surface already after short periods of time (-5 mA/cm², 10 min). With increasing current density or time a second needle like phase appears. The growth of this phase is accompanied by the dissolution of the ACP spheres ending up in a complete transformation into the needle like nanocrystalline phase that was identified as HAP by FTIR. The HAP crystals form a porous structure with crystallites up to 500 nm in length and diameters below 60 nm, as shown in Fig. 1 for -10 mA/cm² and 60 min polarization time.

The transformation into the crystalline phase was also monitored by FTIR. On the left hand side of Fig. 1 the development of the spectra with time for a current density of -10 mA/cm² in a Ca/P = 1.67 electrolyte is shown. The initial amorphous phase is characterized by a broad band around 3400 cm^{-1}, indicating high water content, and a second broad band around 1050 cm^{-1} in the region of the ν_1 and ν_3 stretching vibrations of the P-O bond of the phosphate groups. With increasing polarization time the peaks of the triply degenerated ν_3 asymmetric stretching mode (1000-1100 cm^{-1}) and ν_4 bending mode (550-620 cm^{-1}) of the phosphate groups appear as well as the weaker symmetric stretching mode ν_1 at 962 cm^{-1}. The relative peak intensities of the ν_3 mode are changing with increasing polarization time indicating further crystal growth and higher degree of order. After 120 min of polarization relative peak intensities are reached that are comparable to spectra known from technical HAP.

The formation of HAP is accompanied by the increasing intensity of the bands at 873, 1420 and 1470 cm^{-1} assigned to carbonate. The incorporation of carbonate ions (CO_3^{2-}) may take part during the formation of the CPP layer or be caused by reaction of hydroxyl ions with carbon dioxide from air during storage of the samples. The OH libration band at 630 cm^{-1}, characteristic for HAP, was not detected in our samples. The OH stretching mode at 3570 cm^{-1} was preferentially detected in samples with thicker layers corresponding to longer polarization times or higher current densities.

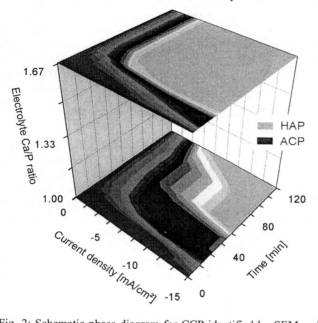

Fig. 2: Schematic phase diagram for CCP identified by SEM and FTIR as a function of current density and time for Ca/P ratios of the electrolyte of 1 and 1.67.

In Fig. 2 the phases identified by SEM and FTIR are summarized in a schematic phase diagram as a function of current density and polarization time for the Ca/P ratios of the electrolytes of 1 and 1.67. Pure amorphous phases are found for a Ca/P ratio of 1.67 only for polarization times below 20 min and current densities below -10 mA/cm². In a small parameter field an intermediate state is present, where amorphous spheres are found as well as HAP needles. Above about 30 min polarization time and current densities of at least -5 mA/cm² only the crystalline HAP phase is found. For lower Ca/P ratios longer times are needed to get a complete transformation. At an electrolyte Ca/P ratio of 1 the pure HAP phase is found above polarization times of 60 min for -15 mA/cm², while for current densities of -10 around 100 min and for -5 mA/cm² more than 120 min are needed for complete conversion. The stability of the ACP phase is only slightly affected by the lower Ca/P ratio, whereas the transformation of the intermediate state is extended over a much wider parameter field.

The influence of the Ca/P ratio of the electrolyte on the Ca/P ratio of the coating was measured by ICP-OES. The Ca/P ratio of the coating was expected to reflect the phases identified above with values of 1.67 for HAP and around 1.5 for an ACP phase supposed to consist mainly out of nm-size $(Ca_3(PO_4)_2)_n$ clusters [1, 2]. Nevertheless only weak dependencies on the process parameters were found with typical values around 2 for the coating Ca/P ratio. In Fig. 3 the Ca/P ratio of the coatings as a function of polarization time is shown for the Ca/P = 1 electrolyte, because the curves for the different current densities show most pronounced differences for this electrolyte. Not taking into account the values for low polarization times, which are influenced by low layer thickness, the Ca/P values show a slight increase with increasing polarization time (> 45 min). Also for higher current densities higher Ca/P values are reached. The comparison of the different electrolytes for a given current density shows that for polarization times above 60 min the Ca/P value of the coating increases with decreasing Ca/P value of the electrolyte. This is indicating that additional Ca must be stored in the coating beside the HAP phase and that this process is more pronounced for higher current densities, longer polarization times and lower Ca/P ratio of the electrolyte. As HAP is the phase

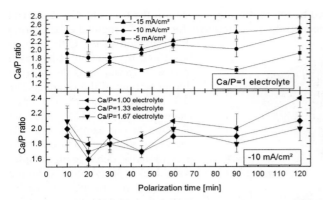

Fig. 3: Ca/P ratios of the coating as a function of time. (Above) for different current densities in Ca/P=1 electrolyte. (Below) for constant current density and different electrolytes.

with highest Ca/P ratio and lowest solubility in aqueous solutions at ambient temperatures the increase in Ca/P ratio can only be caused by a non-phosphate phase. Most probably due to ion content of the electrolyte is $Ca(OH)_2$. With minor probability $CaCO_3$ is possible, not excluding a transformation from $Ca(OH)_2$ to $CaCO_3$ after storage of the samples in air.

Contrary to Ca^{2+} ions that independent on pH are present in the electrolyte, PO_4^{3-} ions ($pK_{s3} = 12.3$) which can react directly to form ACP/HAP, have to be formed in significant concentration by dissociation from $H_xPO_4^{(3-x)-}$ ions. Consequently except from very close to the sample surface where a sufficient high pH is reached during cathodic polarization in other parts of the pH gradient a high surplus of Ca^{2+} ions over the necessary PO_4^{3-} concentration for ACP/HAP precipitation exists. This situation is enhanced due to migration of the ions under the electrochemical conditions that further increase the surface concentration of the positively charged Ca ions at the cathode and decreases the amount of negatively charged phosphate ions.

The increase of Ca content in the layer with increasing polarization time and therefore with increasing layer thickness shows that the process is also influenced by the morphological development of the coating. With increasing surface coverage the current density is more and more concentrated on porous pathways within the structure of the growing layer, leading to a local enhancement of the current density within pores. This will enhance the migration process of Ca^{2+} ions accompanied by a further increase of pH and cause the deposition of amorphous $Ca(OH)_2$ when the corresponding solubility product of $3.0 \cdot 10^{-5}$ mol^3 l^{-3} [3] is reached. The deposition as amorphous well distributed $Ca(OH)_2$ phase was not detectable by FTIR, but a transformation into $CaCO_3$ after storage in air may contribute to the carbonate bands, as shown above.

Conclusion

An electrochemical assisted deposition of HAP can be achieved for Ca/P ratios of the electrolyte away from the stoichiometric value of 1.67, but longer polarization times and higher current densities are needed. This wider parameter field offers the possibility to produce coatings with a higher solubility than HAP under more stable conditions. The coatings contain an additional amount of Ca which is stored in the layer as $Ca(OH)_2$. This process becomes more pronounced the more the CPP precipitation is hindered by none stoichiometric Ca/P ratio of the electrolyte.

References

[1] S. Roessler, A. Sewing, M. Stölzel, R. Born, D. Scharnweber, M. Dard and H. Worch: J. Biomed. Mater. Res. Vol 64A (2003) p. 655.

[2] R.Z. LeGeros: Calcium phosphate in oral biology and medicine (Karger AG, Switzerland 1991); S. Koutsopoulos: J. Biomed. Mater. Res. Vol 62 (2002) p. 600.

[3] H. Remy: Lehrbuch der Anorganischen Chemie (Akademische Verlagsanstalt Geest & Portig, Germany 1950).

Key Engineering Materials Vols. 254-256(2004) pp. 423-426
online at http://www.scientific.net
© *2004 Trans Tech Publications, Switzerland*

Sol-Gel Apatite Films on Titanium Implant for Hard Tissue Regeneration

Hae-Won Kim[1,2,#], Hyoun-Ee Kim[1] and Jonathan C Knowles[2]

[1] School of Materials Science and Engineering, Seoul National University, 151-742, Seoul, Korea,
[2] Biomaterials and Tissue Engineering, Eastman Dental Institute, University College London,
WC1X 8LD, London, UK
[#] Corresponding author, email: hkim@eastman.ucl.ac.uk

Keywords: Sol-gel coating, hydroxyapatite, fluor-hydroxyapatite, titanium implant

Abstract. Titanium (Ti) dental implant was coated with bioactive hydroxyapatite (HA) and fluorine substituted hydroxyapatite (FHA) layers via sol-gel method. Sol and coating parameters were carefully controlled to obtain a pure and dense coating layer through the complex shaped surface. Under processing conditions of low sol concentration, high spinning speed, and slow heating/cooling rate, a homogeneous and dense apatite layer was formed. The obtained HA and FHA coating layers showed different dissolution rate. HOS cells grew and spread similarly on both coating samples. The ALP activities of the cells on both coatings expressed to higher degrees compared to those on pure Ti, suggesting an improved cell differentiation with apatite coatings.

Introduction

Titanium (Ti) gained much interest as hard tissue implant due to its biocompatibility [1]. The osseointegration of bone and Ti implant can be improved by a surface treatment, such as sand blasting, acid etching, oxidation, and hydroxyapatite (HA) coatings [2,3]. Among those, the HA coatings are widely used to enhance bone ingrowth due to their osteoconductivity and bioactivity [4-6]. Currently, most HA coatings have been performed by a plasma-spraying method [6]. However, the plasma-sprayed coating easily resulted in phase and structural inhomogeneity and reduction in coating-substrate interfacial strength.

Sol-gel coating method has the benefits of phase and structural homogeneity due to low temperature processing. Moreover, it is simple, cost efficient, and beneficial for complex shaped material [7,8]. The purpose of this study was to obtain apatite coatings onto a Ti implant by a sol-gel method. Processing conditions were carefully tailored to obtain a dense and uniform coating layer. Two types of apatites (hydroxyapatite and fluor-hydroxyapatite) were coated. The phase and microstructure as well as the *in-vitro* dissolution behaviors and cellular responses of the coatings were examined.

Experimental Procedures

Precursors for HA and FHA coatings were calcium nitrate (CN), triethyl phosphite (TEP), and ammonium fluoride (AF). CN and TEP were dissolved in ethanol and distilled water separately at a ratio of *Ca/P*~1.67. After stirring for 24 h, each solution was mixed and stirred, and then aged for 3 days to make HA sol. For FHA sol, a controlled amount ([P]/[F] = 6) of AF was dissolved into the TEP solution prior to mixing CN and TEP solutions. The sol concentration was varied from 0.2 to 2 M to determine an optimal condition for a uniform coating. As a coating substrate, Ti dental implant of screw and root type was used. After dipping into the apatite sols, the implant was spin coated at different spinning rate (500 to 4000 rpm). The coated sample was oven dried, and then heat-treated at 500 °C for 1 h. Heating and cooling rates were varied from 1 to 20 °C/min.

Phase change of the coating layer was characterized with X-ray diffraction (XRD) patterns. Surface and fractured cross-section morphologies were observed with scanning electron microscope (SEM). Dissolution behavior of the films was investigated with incubation in a physiological saline solution. At each time period, the sample was removed and the Ca^{2+} ion concentration dissolved from the film was measured using Inductively Coupled Plasma-Atomic Emission Spectroscopy (ICP-AES).

Human osteosarcoma HOS cells were plated on the coated samples at a density of 2×10^4 cells/ml. Pure Ti was also tested for comparison. The cells were cultured in an incubator for 3 and 7 days. At each culture period, the cell layers were detached using trypsin-EDTA solution. The cells were gathered, centrifuged, and counted using a hemocytometer. Cell morphology was observed using SEM after fixation, dehydration, and critical point drying. Cell differentiation was assessed by alkaline phosphatase (ALP) activity. After culturing for 10 days, the HOS cells were trypsinized, resuspended with Triton X-100, and disrupted by cycling freezing/thawing process. After centrifugation, the cell lysates were reacted with p-nitrophenyl phosphate at pH 10.3 for 60 min. The result-out color product, p-nitrophenol, was measured at 410 nm using a spectrophotometer.

Results and Discussion

The phase of the sol-gel coatings on Ti was observed with XRD, as shown in Fig. 1. Typical apatite peaks were clearly observed. There were no other peaks than apatite, showing that little reaction between coating and substrate occurred.

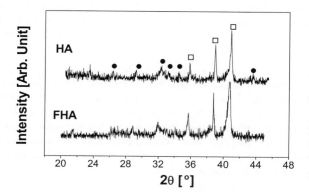

Fig. 1. XRD patterns of HA and FHA coatings on Ti (●: Apatite, □: Ti)

The typical microstructures of the HA coated Ti implant obtained under controlled coating conditions were represented in Fig. 2. The sol concentration was low (0.5 M), spinning speed was high (3000 rpm), and heating/cooling rate was slow (1 °C/min). The coated surface apparently maintained the initial shape of the dental implant, i. e., the screw type (Fig. 2A). The coating layer was highly dense and uniform through the surface. The thin coating layer apparently reflected the roughness and morphology of the implant surface, which was sandblasted and acid-etched (Fig. 2B). From a fractured cross section view, the layer thickness was observed to be approximately 1-2 μm (Fig. 2C).

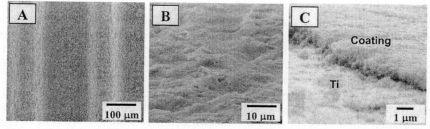

Fig. 2. HA coating morphologies on Ti dental implant obtained at low sol concentration of 0.5 M

At high sol concentration (2 M), uniform coating layer could not be achievable, as shown in Fig. 3. Coating layer was uneven depending on the implant position (Fig.3A). Cross-section view shows the formation of thick coating layer within the valleys (as indicated by arrows) even though relatively thin coating was formed on the hills (Fig. 3B). The layer thickness in the valleys was as high as ~50 μm. The coating morphology obtained at low spinning speed (< 1000 rpm) also showed a similar morphology to that obtained at higher sol concentration (data not shown here).

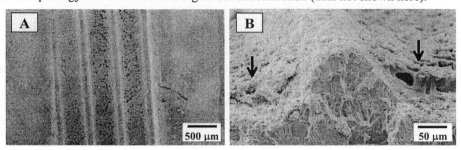

Fig. 3. HA coating morphologies on Ti dental implant obtained at high sol concentration of 2 M

As observed in Figs 2 and 3, highly dense and homogeneous apatite layers were obtained under controlled processing conditions, such as sol concentration, spinning conditions, and heat treatment history. Since the Ti implant has a complex geometry, i. e., a screw and root type with hill and valley structures, a special care should be taken to form a uniform layer. Under coating at high sol concentration or at low spinning speed, the uniform coating was impossible due to the excess apatite sol existing in the valleys, which could not be removed due to high viscosity. The low heating and cooling rates were needed for a dense structure without cracking and void by alleviating thermal mismatch and fluctuation.

The obtained HA and FHA coating layers had different dissolution rate as shown in Table 1. The concentration of Ca ions released from the HA coating was higher than those from the FHA coating. Such a dissolution difference facilitates functional apatite coatings in term of solubility control. Importantly, the F^- ions can be beneficially useful as dental restorations in bacteria-existing situations [8].

Table 1. Ca^{2+} ion concentration released from apatite coatings on Ti

Incubation periods (day)		1	3	7	14
Ca^{2+} concentration	HA	1.89	3.45	4.35	5.81
(ppm)	FHA	1.01	2.22	2.63	3.67

Osteoblast-like cells proliferated well on both coating samples, as presented in Fig. 4. The cells grew actively and spread in intimate contact with the coatings. The cell proliferation numbers and ALP activity on the coatings were quantified as presented in Table 2. The cells numbers were much higher at 7 days when compared to 3 days, indicating a favorable cell viability. There was little significance difference among the coatings and pure Ti. However, ALP activity of the cells expressed much higher (by a factor of ~2) on both the coatings compared to the pure Ti, suggesting that the cells differentiated more actively on the coating samples [9]. There was little difference between HA and FHA coatings. The bioactive coatings are reportedly beneficial to up-regulate the osteoblast cell differentiation [7]. At this point, the exact role of each ion in the apatite coatings could not be elucidated, however, there should favorably be ion-related cell responses with CaP coatings [10,11]. Since the ALP assay do not give the whole knowledge on the *in vitro* behavior of the osteoblast cells, the evaluations by other bone-associated markers (osteocalcin, bone sialoprotein, and type I collagen etc.) are needed.

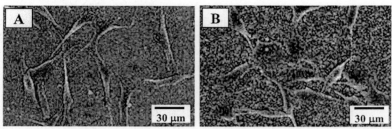

Fig. 4. HOS cell growth morphologies on (A) HA and (B) FHA coated Ti implant after culturing for 5 days

Table 2. Cell numbers and ALP activities on apatite coated- and pure-Ti

		Coating		Pure-Ti
		HA	FHA	
Cell numbers (x10^5 / cm^2)	3 days	0.65 (±0.06)	0.69 (±0.04)	0.58 (±0.07)
	7 days	3.99 (±0.42)	3.90 (±0.41)	4.08 (±0.48)
ALP activity (µmol/hr/mg)		0.34 (±0.13)	0.35 (±0.05)	0.18 (±0.04)

Summary and Conclusions

Hydroxyapatite (HA) and fluorine substitute hydroxyapatite (FHA) layers were coated on a Ti dental implant by a sol-gel spinning method. Coating parameters, such as sol concentration, spinning speed, and thermal history were carefully controlled. A phase-pure, thin (1-2 µm), and dense layer was obtained on the complex shaped implant. The apatite coatings had favorable cell proliferation and superior differentiation behaviors to pure Ti.

Acknowledgement

This work was supported by the Post-doctoral Fellowship Program of Korea Science & Engineering Foundation (KOSEF)

References

[1] P. Tengvall, I. Lundström: Clin. Mater. Vol. 9 (1992), p. 115-34.

[2] B. Kasemo, J. Lausmaa: CRC Crit. Rev. Biocomp. Vol. 2 (1986), p. 335-80.

[3] J. F. Kay JF: Dent. Clin. North. Am. Vol. 36 (1992), p. 1-18.

[4] K. de Groot K, R. G. T. Geesink, C. P. A. T. Klein, P. Serekian: J. Biomed. Mater. Res. Vol. 21 (1987), p. 1375-87.

[5] S. D. Cook, K. A. Thomas, J. F. Kay, M. Jarcho: Clin. Orthop. Vol. 230 (1988), p. 303-12.

[6] D. M. Liu, Q. Yang, T. Troczynski, W. J. Tseng: Biomaterials Vol. 23 (2002), p. 1679-87.

[7] H.-W. Kim, Y.-M. Kong, C.-J. Bae, Y.-J. Noh, H.-E. Kim: Biomaterials (In print).

[8] E. C. Moreno, M. Kresak, R. T. Zahradnik: Nature Vol. 247 (1974), p. 64-5.

[9] S. Ozawa, S. Kasugai: Biomaterials Vol. 17 (1996), p. 23-9.

[10] H.-W. Kim, Y.-J. Noh, Y.-H. Koh, H.-E. Kim, H.-M. Kim, J. Ko: Biomaterials Vol. 23 (2002), p. 4113-21.

[11] P. J. Marie, M. C. de Vernejoul, A. Lomri: J. Bone. Miner. Res. Vol. 7 (1992), p. 103-13.

Key Engineering Materials Vols. 254-256(2004) pp. 427-430
online at http://www.scientific.net

Bioactive Glass Coating for Hard and Soft Tissue Bonding on Ti6Al4V and Silicone Rubber Using Electron Beam Ablation

Jan Schrooten[1], Siegfried V.N. Jaecques[2], Rosy Eloy[3], Claire Delubac[3], Christoph Schultheiss[4], Patrice Brenner[4], Lothar H.O. Buth[4], Jan Van Humbeeck[1] and Jos Vander Sloten[2]

[1] Department of Metallurgy and Materials Engineering, Katholieke Universiteit Leuven, Kasteelpark Arenberg 44, B-3001 Leuven, Belgium, Jan.Schrooten@mtm.kuleuven.ac.be

[2] Division of Biomechanics and Engineering Design, Department of Mechanical Engineering, Katholieke Universiteit Leuven, Celestijnenlaan 200A, B-3001 Leuven, Belgium, Siegfried.Jaecques@mech.kuleuven.ac.be

[3] Biomatech, Z.I. de l'Islon, Rue Pasteur, F-38670 Chasse-Sur-Rhône, France

[4] Forschungszentrum Karlsruhe, Institute for Pulsed Power and Microwave Technology, Postfach 3640, D-76021 Karlsruhe, Germany

Keywords: Bioactive glass, soft tissue bonding, electron beam ablation, *in vitro*, cell culture

Abstract. The adhesion of soft tissue to any commonly used implant material is poor. Therefore percutaneous implants are an infectious passage of bacteria into the body. To address this problem, a method that allows the growth of living soft tissue on medical implants ranging from metals to temperature sensitive polymers is developed. The implant surface is to be coated with bioactive glass (BAG) by applying electron beam ablation (ELBA). A *proof of principle* of the ELBA potential is given by coating Ti6Al4V and silicone rubber (poly-dimethylsiloxane, PDMS) substrates with thin BAG layers (~10 μm thickness). The BAG coatings are amorphous, sufficiently adherent and showed the desired bioactive *in vitro* dissolution behaviour in Hanks' solution. Finally, *in vitro* fibroblast cell adhesion experiments were performed. Compared to an uncoated material both the coated PDMS and Ti6Al4V samples showed good cell adhesion and proliferation. These results support the hypothesis that a BAG coating deposited by ELBA can enhance the tissue bonding potential of many types of implants, including those made of temperature sensitive materials.

Introduction

In contrast to the bone bonding potential of existing implants, the adhesion of soft tissue to any commonly used implant material is poor. Therefore an important problem of percutaneous implants is the infectious passage of bacteria into the body at the interface between tissue and implant [1]. Subcutaneous implants, e.g. access ports for injections used in chemotherapy, suffer under low fixation and should be anchored with the surrounding soft tissue.

The main objective of this research is to develop a method, which allows the growth of living soft tissue on medical implants ranging from metals to temperature sensitive polymers. The surface of the implant is to be coated with bioactive glass (BAG), which has proven bonding potential to hard and soft tissue [2]. A new coating procedure based on electron beam ablation (ELBA) will be used [3]. A first goal was to deliver a *proof of principle* of the ELBA technique by coating Ti6Al4V and silicone rubber (poly-dimethylsiloxane, PDMS) substrates with micrometer thin BAG layers.

Materials and Methods

The ELBA coating technique [3] uses a pulsed high power electron beam that hits a BAG target and evaporates a thin layer in an explosive manner, which is called ablation. Vapour and fine molten particles condense at the surface of the substrate (Ti6Al4V or PDMS) and build an adherent and dense layer. BAG targets (Fig. 1a) (⌀30mm, thickness d = 5-10mm), produced by conventional melting as generally applied to produce bulk BAG material [4], with a composition shown in Table 1, were used. The BAG composition is within the soft and hard tissue bonding compositional area of bioactive glass [1, 5]. The following substrates were used: (i) Ti6Al4V, cylindrical disks (∅10mm, d = 1mm) (Fig. 1b), with a glass-blasted surface finish prior to coating and (ii) PDMS cylindrical disks (∅10mm, d = 1mm), in as-received condition.

Fig. 1: Bioactive glass target (a), Ti6Al4V substrates (b)
and coated substrates (c, right) in comparison with an uncoated one (c, left).

Several properties of the coated samples (Fig. 1c) were evaluated. The composition was measured by SEM-EDX. The amorphous character was determined by grazing angle X-ray diffraction (XRD), with angles of incidence varying from 0.5 to 5°. Coating thickness was measured on cross sections by SEM imaging and by theoretically calculating the thickness from the change in mass. The coating roughness was measured using optical 3D profiling. Adhesion strength was determined in a pull-off test. To evaluate the bioactive potential of the coating, short term (24h) *in vitro* dissolution kinetics in Hanks' balanced salt solution (HBSS) were performed. Afterwards, the reacted samples were subjected to XRD and SEM and the compositional changes in the Hanks' solution were analysed by inductively coupled plasma-atomic emission spectroscopy (ICP-AES). Finally, cell adhesion was tested with L929 mouse fibroblast cells (ISO-10993-5 standard). Tissue culture polystyrene (TCPS) was used as a control. The number of cells per ml medium in contact with the TCPS was taken as 100%. The coated samples were incubated for 24 and 48 hours. Afterwards the adherent cells were trypsynized, a supravital staining was performed and the cells were counted.

Results and Discussion

Fig. 2 shows a detail of a BAG coating on a Ti6Al4V substrate and an overview of a BAG coating on a PDMS substrate. The coatings on both substrates are similar. The composition of these BAG coatings was changed through the ELBA process as shown in Table 1. This final composition is still within the boundaries of the soft and hard tissue bonding area of bioactive glasses [1, 4].

Table 1: BAG target and coating composition

[wt%]	SiO_2	CaO	P_2O_5	Na_2O
Target	47	33	2	18
Coating	43	37	3	16

The deposited BAG layers had a thickness ranging from 4 to 14µm depending on the ELBA process parameters. The XRD results are shown on Fig. 3 and indicate that an amorphous structure was formed, showing an amorphous peak similar to that of molten glass [6]. The adhesion strength decreased with increasing the coating thickness, for example 6.8MPa for 2.4µm thickness, versus 2.5MPa for 8.3µm. However, the adhesion strengths obtained with the thickest coatings are still sufficient for the intended use [7, 8].

Fig. 2: SEM images of a BAG coating on a Ti6Al4V substrate (left, detail) and on a PDMS substrate (right, overview)

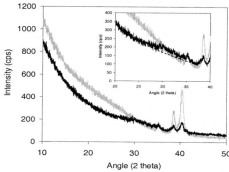

Fig. 3: XRD image of Ti6Al4V (grey) and a BAG coated Ti6Al4V substrate (black) taken at a grazing angle incidence of 0.5°. The inlay shows more in detail the amorphous BAG peak.

The bioactive potential of the deposited coatings was evaluated by *in vitro* dissolution tests in HBSS and by cell culture tests. A strong indication of the reaction behaviour of the BAG can be seen on Fig. 4, showing crystalline phases formed on the surface of a BAG target stored under ambient atmosphere for a short period. The *in vitro* dissolution experiments in HBSS showed a compositional change that is expected for BAG, related to leaching, dissolution and reprecipitation of Ca and P [1, 4]. After these *in vitro* tests the coating roughness (Ra) was significantly decreased (from 882nm to 797nm, $p<0.05$ for Ti6Al4V substrates and from 1470nm to 531nm, $p<0.01$ for PDMS substrates) and crystalline phases were formed.

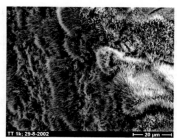

Fig. 4: SEM images showing the presence crystalline phases on top of a BAG target after a short storage in normal conditions

The fibroblast cell adhesion results are shown in Table 2. There is good cell adhesion and proliferation on the treated PDMS samples and also on the treated Ti6Al4V samples, but on a lower level. After 48 hours the coating was completely covered with fibroblasts. After 24 hours the coating on the PDMS samples showed cracks. The cell adhesion on these samples was heterogeneous, which was probably due to the degradation of the coating. This was not the case for the coated Ti6Al4V samples. The coating thickness after testing varied between 5 and 12μm.

Table 2: Cell culture results of BAG coatings after 24 and 48 hours

		Negative control	PDMS	Titanium
24h	**Results [cells/mL]**	1.2×10^4	8.95×10^4	1.45×10^4
	Percentage of the negative control [%]	100	750	120
48h	**Results [cells/mL]**	7.5×10^3	9.65×10^4	4.1×10^4
	Percentage of the negative control [%]	100	800	340

Conclusions

The electron beam ablation (ELBA) coating process can deposit thin bioactive glass (BAG) oatings on metal (Ti6Al4V) and polymer (PDMS) substrates. During the ELBA process the amorphous character and the reaction behaviour are preserved, but the BAG composition changes slightly (deviations of 1-4% in composition between BAG in ELBA target and BAG deposited as coating). In comparison to cell growth on untreated TCPS control surfaces, ELBA-coated Ti6Al4V-alloy and PDMS samples showed an increased cell population at the surface after 48 hours. The reported BAG coatings produced by ELBA have the ability to promote cell growth. These results will be used as a basis for investigating other BAG compositions more rigorously and understanding the reaction mechanism of thin BAG coatings with relation to soft and hard tissue bonding.

Acknowledgements

The presented research is part of EU FP5 Growth project G5RD-CT-2001-00533 INCOMED "Innovative Coating of Temperature Sensitive Medical Implants with Biofunctional Materials using Electron Beam ablation".

References

[1] E.A. Ross, C.D. Batich, W.L. Clapp, J.E. Sallustio and N.C. Lee, Kidney Int. 63 (2003), p. 702.
[2] J. Wilson and D. Nolletti, "Bonding of Soft Tissue to Bioglass", in *Handbook of Bioactive Ceramics*, Ed. T. Yamamuro, L.L. Hench & J. Wilson. (CRE Press, Boca Raton, FL 1990).
[3] G. Müller, M. Konijnenberg, G. Krafft and C. Schultheiss, "Deposition by means of Pulsed Electron Beam Ablation", in: *Science and Technology of Thin Films*, ed. F.C. Matacotta & G. Ottaviani (World Scientific Publishing, 1995).
[4] E. Verne, C.V. Brovarone, C. Moisescu, E. Ghisolfi and E. Marmo, Acta Mat. 48 (2000), p. 4667.
[5] Hench LL and Andersson O, Bioactive glasses, in *An Introduction to Bioceramics*, Ed. L.L. Hench , J. Wilson (World Scientific Publishing, Singapore 1993).
[6] J.A. Helsen, J. Proost, J. Schrooten, G. Timmermans, E. Brauns and J. Vanderstraeten, J. Eur. Ceram. Soc. 17 (1997), p. 147.
[7] T. Kameyama, M.Ueda , K.Onuma, A.Motoe, K.Ohsaki, H.Tanizaki and K.Iwasaki, Proc. of ITSC 1995, p.187.
[8] E. Barth and H. Herø, Biomaterials 7 (1986), p. 273.

Key Engineering Materials Vols. 254-256(2004) pp. 431-434
online at http://www.scientific.net
© *2004 Trans Tech Publications, Switzerland*

Reliability Weibull Analysis for Structural Evaluation of Bioactive Films Obtained by Sol-Gel Process

Alejandro Peláez[1,2], Claudia Garcia[2], Juan Carlos Correa[2] and Pablo Abad[2]

[1] Instituto de Ciencias de la Salud. CES. Facultad de Odontología.
[2] Universidad Nacional de Colombia sede Medellín, Facultad de Ciencias ,
A.A. 3840 Medellín, Colombia. continuo@unalmed.edu.co

Keywords: Thin bioactive film, sol-gel, Weibull analysis, flexural strength

Abstract. Gaussian distributions are the most frequently reported method used for the evaluation of the mechanical properties of materials, although these distributions only characterize the statistical behavior of the metallic materials used as substrates. However, it is the mechanical flaw in the case of glass coating the one that finally prevails as agent of strange body. Therefore, the analysis should be centered in the asymmetric statistical distribution that characterizes the answer of the bioceramic coating.

In this study the analysis of reliability of Weibull is used with the objective of verifying if a glass substratum with a bioactive layer obtained by sol - gel method acquires a smaller variation range in the resistance values to the fracture, and therefore, its structural reliability is bigger than that of the glass without coating.

Thirty samples per group was prepared and tested. The experimental group was formed by microscope colorless glass slides of 75 mm x 25 mm x 1 mm coated with a suspension of a sol of silica and bioactive glass particles, and the control group was formed by the uncoated microscope glass slides.   The three point flexure test was used to measure the fracture force of two groups. The fracture stress values were analyzed by two parameter Weibull analysis to determine the modulus values (m) and 5% probabilities of failure.

Finally, results showed that the flexural strength average of the coated samples have lower values whose difference is not statistically significant with respect to the uncoated samples. Moreover, the value of the variance for the coated samples is much smaller and the Weibull modulus is higher for the coated samples than for the uncoated samples.   The coated samples show more reliability and the Weibull analysis of the strength data characterizes more completely the fracture potential of this material. A higher value of Weibull modulus ensures fewer fatal flaws, a smaller error in strength estimation and a greater practical reliability. The application of a bioactive particle containing coating, improves the reliability of the failure strength of the glass substrate in the lower values of the curve.

Introduction

One of the most important challenges in the biomaterials area research is the development of materials with good mechanical properties and with good bioactivity and biocompatibility.[1-3] For this purpose, today there are a lot of glass and ceramic coatings which make the substrates more bioactive and more biocompatible.

The evaluation of the mechanical properties in metallic materials coated with ceramic films, presents a typical gaussian distribution doubt to the influence of the metallic substrates [4]. In fact, in these case the evaluation has been done over the metal and no over the coating. The mechanical properties of ceramics and glasses present asymmetric distributions as a result of random locations of the flaws.

Weibull risk of rupture analysis is a widely accepted model for material and structural evaluation in which the probability of failure P_f, is given as [5]:

$$P_f = 1 - \exp\left(-\int_s \left(\frac{\sigma}{\sigma_0}\right)^m dS\right)$$ (1)

Where σ = stress; σ_0 = a characteristic strength (stress level at which 63.2 % of the specimens have failed); s = the surface (or volume) under stress; and m = the Weibull modulus, is a constant which determines the slope of the distribution.   A high Weibull modulus is desirable for all materials since it indicates and increased homogeneity in the flaw population and a more predictable failure behavior.

The purpose of the present study was, using the reliability Weibull analysis, to verify that glass substrate with a thin sol-gel bioactive film have a smaller range of variation of the values of failure strength and therefore greater structural reliability than the uncoated glass.

Materials and methods

The present experiment consisted in a series of failure test performed on microscope glass slides of 7.5 cm x 2.5 cm (Knittel Glaser, Germany). Two groups of glass slides were evaluated. The control group consisted in thirty glass slides as delivered (not coated).   The experimental group was compounded of thirty specimens which were coated by a dipping method with a sol-gel suspension consisting in a hybrid sol and 10% of hydroxyapatite particles.

The hydroxyapatite particles were prepared by the precipitation route using as initial aqueous solutions   tetra-hydrated calcium nitrate $(Ca(NO_3)4H_2O)$ (Aldrich) and ammonic phosphate $(H_2(PO_4)NH_4)$ (Aldrich) in concentrations 1M and 0.48M, respectively.

The sols were prepared by the method of acid catalysis in a single stage. Tetraethylorthosilicate (TEOS, Merch) and methyltriethoxisilane (MTES, Aldrich) were selected as silica precursors for the sol, prepared in alcoholic media using HNO_3 0.1 N and acetic acid as catalysts. The molar ratio TEOS:MTES was fixed as 40:60, the H_2O:(TEOS+MTES) ratio equal to 2:1 and the water:acetic acid ratio was 7:1, to give a final concentration of SiO_2 of 200 g/l. The suspensions were prepared by adding 10 % by weight of particles of hydroxyapatite to the solution, and treating them with a high shear mixer, (Heidolph, Germany), during 6 minutes. As a dispersant was used a phosphate ester (Emphos PS21, Whitco, Chem, USA) 2% by weight with respect to the solid.

The microscope glass slides was coated by dip coating with a withdrawal rate of 4 cm/min. The coatings were dried at room temperature and heat treated at 450°C for 30 min. The integrity and characteristics of the coatings were evaluated by optical microscopy (Olympus BX41, USA).

The three point flexure test was used to measure the fracture force for the glass slide specimens. The tests were made using a Universal testing machine (Test Resources model 650R) operated with a displacement rate of 1 mm/min. The failure load was recorded and the flexural strength calculated according with the procedure reported previously by Catell et al. [6].

Weibull distribution was used to calculate fracture probabilities of each group of samples as a function of applied stress [7]. The Kolmogorov-Smirnov test was applied, in order to verify if the Weibull model be an adequate distribution for the present experimental data. This test compares the empirical distribution of the data with the theoretical distribution and considers the maximum distance between both functions. If this distance is not greater than a threshold value, then we can use the theoretical model as an appropriate model for the data. The statistical estimation of the Weibull parameters is calculated by the maximum likelihood method [8]. Maximum likelihood estimators are optimum in the sense that they are asymptotically efficient and normally distributed. This property allows us to construct tests that are adequate for the questions we want to answer. Several tests for judging if there are statistical differences among the two Weibull models were applied. One of the tests that we want to verify is if $\sigma 1 = \sigma 2$. The parameter σ in the Weibull distribution is called *the characteristic life*, since it is always the $100(1-1/e) \sim 63.2$th percentile and it has the same units as the original variable [9]. To test this hypothesis, we use the statistic test that follows a standard normal distribution.

Results

The optical microscope evaluation of the coatings showed no defects and the estimated average size particles was 5 μm. No zones of the surface richer in particles than others were found.

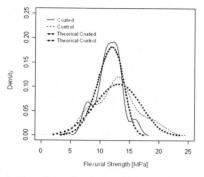

Fig. 1. Plot of experimental Weibull distribution for the coated and uncoated samples (control) and their corresponding theoretical Weibull distribution

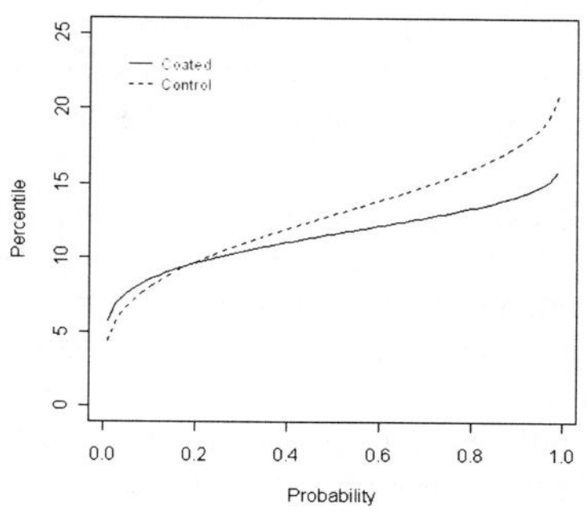

Fig. 2. Curves of percentiles for the flexural strength of the coated and control samples

The Kolgomorov-Smirnov test verify that the Weibull model is an adequate distribution for the present experimental data; for the coated samples, we obtained a p-value = 0.8893, and for the Control samples we obtained a p-value = 0.6802; therefore, we consider that the Weibull model is an appropriate choice (Fig. 1).

In Table 1 we summarize the mean flexural strength, the Weibull modulus (m) and the characteristic strength (σ) for the two groups of samples studied. The characteristic strength value of the control samples is higher than that of the coated samples, and their statistical difference is significant, with a z-value z = -2.283857 and a p-value = 2 P (Z > |z|) = 0.02237991.

In order to evaluate the difference between the 5% percentiles for the flexural strength in both distributions, we construct a 95% bootstrap confidence interval. The obtained values were (-1.949125 , 2.664825); this shows that there is a statistically significant difference. We performed a similar test for the 10% percentile and obtained the interval (-1.459000, 2.123437); this shows that there is not a statistically significant difference (Fig. 2).

Table1. Values of Flexural Strength,   Weibull Modulus and Characteristic strength of the two groups of samples

Samples	N	Flexural Strength [MPa]		m	s.e. m	σ	s.e. σ
		Mean	Variance				
Coated	30	11.46072	4.886918	6.0298884	0.8281905	12.3502548	0.3949355
Control	30	12.84014	13.44576	3.9221031	0.5444550	14.1818415	0.6979849

Discussion

As can be seen in the table 1, the flexural strength average of the coated samples is lower than that of the uncoated samples but the difference is not statistically significant. This means that the coating does not seem to play an important role in the improvement of the flexural mechanical behavior (modulus of rupture) of the samples, and this can be explained by the fact that the coating thickness is too small in comparison with the substrate thickness.

Moreover, the value of the variance for the coated samples is much smaller, and the Weibull modulus is higher for the coated samples than for the control samples. This indicates that the coated samples show more reliability. In this case, the Weibull analysis of the strength data characterizes more completely the fracture potential of this material. A higher value of Weibull modulus ensures fewer fatal flaws, a smaller error in strength estimation and a greater practical reliability. To explain these results, we can propose some hypotheses: (a) it is probable that the higher reliability and the statistically significant difference for lower data values by the coating action, can be due to an more even distribution during the initial loading and to the healing of the superficial irregularities present in the glass slides or (b) the coatings have lower thermal contraction than the glass slide and, therefore, these coatings might be working in compression, which may imply an improvement in the mechanical behavior. This is seen only in the lower extreme of the data distribution (Fig. 2) because the coating thickness is very small as compared with that of the substrate.

We do not have yet a satisfactory hypothesis for the behavior of the higher data values, that show lower failure strength values for the coated samples. More detailed studies are necessary in order to explain the failure mechanisms of different substrata coated with thin films. For example, it could be interesting to determine the influence of sol-gel coatings on dental ceramics.

Acknowledgement

We area grateful to Professor Ramiro Restrepo for his comments on the manuscript. This work was supported by Young Researcher Program of Colciencias, Universidad Nacional de Colombia Sede Medellín, and Biomaterials Laboratory of CES and EIA, Colombia.

References

[1] P. Galliano: J. Sol-Gel Sc. Tech. Vol. 13 (1998), p. 723.
[2] T. Kokubo: J. Ceram. Soc. Jap. Vol. 99 (1991), p. 937.
[3] J. Gallardo and P. Galliano: J. Sol-Gel Sc. Tech. Vol. 19 (2000) p. 107..
[4] J. Tinschert, D. Zewz, R. Marx, K.J. Anusavice: J. Dent. Vol. 28. (2002), p. 529.
[5] J.E. Ritter: Dent. Mater. Vol. 11 (1995), p. 147.
[6] M.J. Cattell, R.L. Clarke and J.R. Lynch: J. Dent.Vol. 25 (1997), p. 399.
[7] S.S. Scherrer, W.G. de Rijk, U.C. Berlser, J.M. Meyer: Dent. Mater. Vol. 10 (1994), p. 172.
[8] Azzalini, A. Statistical Inference Based on the Likelihood. (Chapman & Hall, London 1996)
[9] Nelson, W. Applied Life Data Analysis. (John Wiley & Sons, New York 1982)

Key Engineering Materials Vols. 254-256(2004) pp. 435-438
online at http://www.scientific.net
© *2004 Trans Tech Publications, Switzerland*

Diamond-like Carbon Coatings on Ti-13Nb-13Zr Alloy Produced by Plasma Immersion for Orthopaedic Applications

E. T. Uzumaki[1], C. S. Lambert[2] and C. A. C. Zavaglia[1]

[1] Department of Materials Engineering, Mechanical Engineering Faculty, State University of Campinas, Campinas/SP, Brazil, P.O. Box 6122, 13083-970, emilia@fem.unicamp.br

[2] Gleb Wataghin Physics Institute, State University of Campinas, Campinas/SP, Brazil

Keywords: diamond-like carbon, coatings, orthopaedic implants, biomaterials

Abstract. Diamond-like carbon (DLC) films have been intensively studied with a view to improving orthopaedic implants. Various techniques have been employed to manufacture DLC films, and recently, the plasma immersion process has been used to provide non-line-of-sight deposition on three-dimensional pieces with complex shapes. In this method, the whole surface of the target is coated, even without moving the sample, and without an intermediate layer. DLC films were deposited on a silicon wafer and Ti-13Nb-13Zr alloy substrates, using the plasma immersion process. The films were analysed by Raman spectroscopy, atomic force microscopy (AFM), and nanoindentation. As examples, uniformly DLC coated orthopaedic implants (knee implant and femoral head) are shown.

Introduction

Diamond-like carbon (DLC) films are often considered a suitable coating material for orthopaedic applications. It has proven characteristics, such as hardness, wear resistance, low friction coefficient and biocompatibility that improve the properties of solid and articulated implants [1,2].

DLC coatings can be deposited using various techniques, such as plasma enhanced chemical vapour deposition, magnetron sputtering, laser ablation, and others [3]. Recently, the plasma immersion process was used to deposit DLC films with superior adhesion properties [4,5]. In the plasma immersion process, a developing technique for surface modification of three-dimensional components, pulsed high negative voltage is applied to the target, producing a plasma, and the total surface of the target is coated, even without moving the sample, and without an intermediate layer.

Titanium alloys are widely used as biomaterials, because they have good corrosion resistance and desirable mechanical properties close to those of bone [6]. However, titanium alloys have relatively poor resistance against wear and a high friction coefficient. A DLC film could be used to protect titanium-based implants, improving wear behaviour and surface hardness.

Ti-6Al-4V alloy has been extensively used for many years as an implantable material mainly in the application of orthopaedic prostheses. However, the toxicity of Vanadium and neurological disorders associated with Aluminium, have also created problems for biological applications, and new types of alloys have been developed [6,7]. Ti-13Nb-13Zr alloy has been proposed as an alternative to the Ti-6Al-4V because of its superior corrosion resistance and biocompatibility.

In this study, DLC coatings were deposited on Ti-13Nb-13Zr alloy substrates using the plasma immersion process. We characterized the films obtained using nanoindentation, Raman spectroscopy, and atomic force microscopy (AFM). The study is in an initial phase and other analyses, as characterization of mechanical and tribological properties are under way.

Materials and Methods

Two different substrates (silicon wafer and Ti-13Nb-13Zr alloy) of 100x150x1 mm were used for the investigation. Before the coating deposition, the substrates were polished to a 1 μm finish with diamond paste, and ultrasonically cleaned in acetone for 20 minutes. The substrate surface was sputter cleaned with Ar plasma before deposition, to remove surface contamination. After Ar sputtering, DLC deposition was performed by immersion in a methane plasma. Films of DLC were also deposited on knee implant made of Co-Cr-Mo, and femoral head made of Ti-6Al-4V (prostheses manufactured by Microsteel Ind. Com. Ltda, Sumaré/SP, Brazil).

Our deposition equipment possesses a vacuum system, a power source and a reactor. The power source has a pulsed DC negative output and the reactor contains two electrodes. The cathode is the substrate to be coated. After pumping the system down to 10^{-6} Torr, methane is introduced until reaching the working pressure, and a potential difference of 2.000 V between the electrodes (300 mA) is used. The temperature of the piece was kept at 350 °C for 2 h. The surface morphology was analysed by atomic force microscopy (AFM), the structure of the films was evaluated by Raman spectroscopy and the hardness was obtained by nanoindentation.

Results

The DLC coating grew homogeneously on the substrate without defect formation (Fig. 1). The thickness of coatings was in the range of 0.7-1.0 μm. The Raman spectrum showed structures typical of DLC films, exhibiting two Raman features: the D peak located at about 1350 cm^{-1}, and the G peak positioned at approximately 1580 cm^{-1} (Fig. 2). Hardness, measured by a nanoindenter, ranged from 16.4-17.6 GPa.

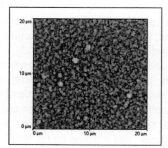

Fig. 1. Surface morphology of DLC film (not polished) by AFM.

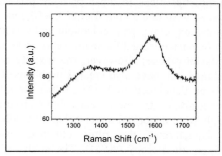

Fig. 2. Raman spectrum of DLC coating.

Good adhesion was obtained by depositing DLC films directly onto the prosthesis and substrates using plasma immersion equipment. Fig. 3 shows the knee implant with (a) bright plasma luminescence around, and (b) DLC film. A hip prosthesis made of Ti-6Al-4V with DLC coated femoral head can be seen in Fig. 4. This technique has shown potential for improving medical devices with irregular geometries.

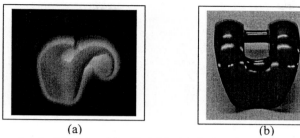

(a)   (b)

Fig. 3. Knee implant (made of Co-Cr-Mo): (a) during deposition, and (b) with DLC film.

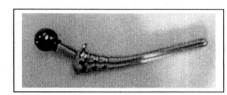

Fig. 4. Hip prosthesis (made of Ti-6Al-4V) with DLC coated femoral head.

Discussion and Conclusions

The plasma immersion process is a very attractive method for surface treatment of biomedical implants, and may be advantageous for the manufacture of DLC coatings on articulated orthopaedic implants. Films with good uniformity, without delamination, and good adherence were obtained. To complete our research an extended series of tests will be performed, such as wear and friction studies, corrosion resistance and biological response.

Acknowledgements

The authors are very grateful to CAPES for financial support, to Microsteel Ind. Com. Ltda (Sumaré, SP, Brazil) for prostheses, to Laboratório de Interfaces (IFGW/UNICAMP), and Grupo de Propriedades Ópticas (IFGW/UNICAMP).

References

[1] V.M. Tiainem: Diamond Rel Mater Vol. 10 (2001), p. 153.

[2] D. Dowling, P. V. Kola, K. Donnelly, et al: Diamond Rel Mater Vol. 6(1997), p. 390.

[3] A. Grill: Diamond Rel Mater Vol. 8 (1999), p. 428.

[4] K. C. Walter, M. Nastasi; C. Munson: Surf Coat Technol Vol. 93 (1997), p. 287.

[5] S. Miyagawa, S. Nakao, K. Saitoh, et al: Surf Coat Technol Vol.128-129 (2000), p. 260.

[6] M. Long, H. J. Rack: Biomaterials Vol. 19 (1998), p. 1621.

[7] K. Wang: Materials Science and Engineering A Vol. 213 (1996), p. 134.

Key Engineering Materials Vols. 254-256(2004) pp. 439-442
online at http://www.scientific.net
© *2004 Trans Tech Publications, Switzerland*

Nano-Sized Apatite Coatings on Niobium Substrates

M.H. Prado da Silva[1], A.M.R. Monteiro[2], J.A.C. Neto[2], S.M.O. Morais[2] F.F.P. dos Santos[2]

[1] Universidade Federal do Ceará, Centro de Tecnologia, Deptº Eng. Mecânica e de Produção, Bl.714, Campus PICI, 60455-760, Fortaleza, CE, mhprado@ufc.br

[2] Universidade Federal do Ceará, Centro de Tecnologia, Deptº Eng. Química, Bl.709, Campus PICI, 60455-760, Fortaleza, CE, ancelitamonteiro@yahoo.com.br

Keywords: Niobium, nano-sized apatite, monetite, hydroxyapatite, bioactivity.

Abstract. In the present study an apatite coating was produced on niobium substrates. This coating consists of nano-sized needles of carbonate substituted apatite. The nano-sized needles re-precipitate from the conversion of monetite in an alkali solution. Bioactivity tests showed that the partially substituted hydroxyapatite is highly bioactive.

Introduction

Titanium is the most widely used metal in the fabrication of metallic implants. Recent studies have pointed niobium as an alternative metallic biomaterial for applications where structural stability is desired [1-5]. Surface treatments used to improve mechanical bone apposition consist of creating suitable roughness for improving interlocking of newly formed bone to the implant surface. Calcium phosphate coatings are an alternative surface treatment to minimize the time for osseointegration to occur [6-8]. This practice intends to reduce the time between surgical interventions in the case of two-stage surgeries. In the present study, calcium phosphate coatings were produced on niobium surbstrates. This research is based in previous studies on electrolytically coating titanium substrates with monetite and subsequent hydrothermal conversion to hydroxyapatite [7, 8]. The present method is a hydrothermal treatment in two steps: in the first one, a monetite coating is produced and then converted to hydroxyapatite. It was found that during the conversion process, the metallic substrate is bioactivated. This finding points to an intimate bonding between the metallic niobium substrate and the final bioceramic coating, that consists of partially substituted carbonate apatite.

Materials and Methods

In this study, niobium sheets ($10 \times 10 \times 2 mm^3$) were hydrothermally coated with monetite and hydroxyapatite. The starting solution, hereby called clear solution, has the composition: 0.3M H_3PO_4, 0.5M $Ca(OH)_2$, 1M CH_3CHCO_2HOH. The monetite coating was produced by the immersion of the specimens in the solution at 80°C during 1 hour. The sheets were then removed, washed in deionised water and dried in an oven at 100°C during 1 hour. Hydroxyapatite coatings were produced by alkali conversion of the monetite coatings by immersion of the coated sheets in a solution of NaOH with pH=12.5 during 24 hours at 60°C. Specimens were then removed from the alkali solution, washed in deionised water and dried in an oven at 100°C overnight. Scanning electron microscopy (SEM) analysis was used to assess the morphology of the coatings, with a scanning electron microscope Philips model XL-30. Phases were identified by X-ray diffraction analysis (XRD) with a Philips X'Pert X-ray Diffraction System operating with CuKa (l=0,1542nm) with 40kV and 40mA. Fourier transform infra-red spectroscopy (FTIR) (IR- Prestige-21 Shimatzu

Co) analysis was used as a complementary tool to the XRD analysis. Samples were prepared with KBr and analysed in absorbance. The bioactivity of the produced coating was assessed by immersion of the coated samples in simulated body fluid (SBF) prepared according to Kokubo and co-workers [9].

Results

SEM analysis showed that the hydrothermal coatings in the clear solution at 80°C consisted of crystals with a parallelepiped-like morphology, as described in Figure 1a. XRD analysis identified this coating as being monetite. After conversion in the alkali solution, SEM analysis revealed the presence of nano crystals of hydroxyapatite grown from the precursor monetite phase, Figure 2. XRD analysis confirmed the presence of hydroxyapatite. Infra-red spectroscopy analysis identified the presence of carbonate groups in the final coating, as presented in Figure 3.

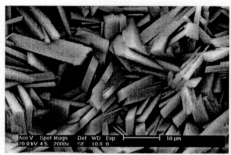

Fig. 1 – Morphology of the monetite crystals.

Fig. 2 – Nano-sized apatite coating produced by alkali conversion from monetite.

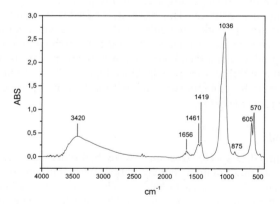

Fig. 3 – FTIR analysis showing carbonate groups at 1656, 1461, 1419 and 875 cm-1.

The final coating appeared to be highly bioactive. After 6 days incubated in SBF, the specimens were coated with bone-like apatite. Figure 5 presents a detail of the bone-like apatite precipitated from SBF. Figure 6 illustrates how the niobium substrate without precursor monetite coating is coated with bone-like apatite after being incubated during 25 days in SBF.

Fig. 4 – SEM of sample soaked in SBF during 6 days.

Fig. 5 – Sample incubated in SBF during 9 days.

Fig. 6 – Detail of bone-like apatite precipitation on niobium substrate after 25 days in SBF.

Discussion

In previous studies, the author has produced monetite coatings with the same solution in an electrolytic coating process on titanium substrates. Similar results have been obtained without the presence of voltage and current. It is well established that the stability of calcium phosphates is dependent on pH value and temperature. In this study, the use of temperature was enough for reproducing results where voltage and current were used to produce monetite coatings. The conversion of monetite to hydroxyapatite showed similar results when compared to hydrothermal conversion on titanium substrates. A new observed feature is the bioactivation of the niobium substrate during the alkali conversion of monetite to hydroxyapatite. This feature indicates a good bonding between the substrate and the coating and points to a good bone-bonding ability.

Conclusion

The obtained monetite coating consisted of parallelepiped-like crystals on niobium substrates. This result is similar to previous studies where the author produced similar coatings on titanium substrates. After hydrothermal conversion, nano-sized hydroxyapatite coatings were produced. This result is also similar to the ones obtained in previous studies. The hydrothermal method appears as a simple alternative to the electrolytic method, with an advantage of bioactivating the previously inert substrate.

Acknowledgements
The authors would like to thank CNPq through PROFIX programme 540191/-1-9, CBMM for providing metallic niobium for this study, Andre Galdino (DEMP-UFC) and Joao Batista A. S. Junior and Leonia Gonzaga (Polymer lab., dep. organic and inorganic chemistry, UFC), for their help in the XRD and FTIR analyses.

References

[1] S.V.M. de Moraes, K.M. Ribeiro, T. Ogasawara, G.A. Soares, Trans. of the 15th Symp. on Apatite, Toquio, Japao (1999), p. 41-44.

[2] M.C. de Andrade, M.H. Prado da Silva, S.V.M. de Moraes, T. Ogasawara, G.A. Soares, I Workshop de Biomateriais, Campinas, Brazil (1999), p. 5.

[3] DeGroot, K., Bioceramics, Vol. 11 (1998), p. 41.

[4] M.C. de Andrade, M.H. Prado da Silva, S.V.M. de Moraes, K.M. Ribeiro, I.R. Gibson, S.M Best, T. Ogasawara, G.A. Soares, IV Forum Nacional de Ciência e Tecnologia em Saúde e XVI Cong. Bras. de Eng. Biomedica, Curitiba, Brazil, (1998) p. 7.

[5] M. H. Prado da Silva, J.A.C. Neto, F.F.P. dos Santos, R.O.Feitosa, T.D. Pinheiro da Silva, D.O. Lima, J.I.L.Almeida Jr., "Recobrimento nanocristalino de hidroxiapatita em substratos de nióbio", Proceedings of the 1st Brazilian Meeting in Medical Physics, Porto Alegre, Brazil, 2003.

[6] I.R. Gibson, M.H. Prado da Silva, S.M. Best and W. Bonfield, "Electrolytically Deposited and Chemically Converted Hydroxyapatite Coatings". Proc. of the European Ceramic Society, Brighton, U. K., (1999).

[7] M.H. Prado da Silva, G.A. Soares, C.N. Elias, I.R. Gibson, S.M. Best, Key Eng. Mater. Vol. 192 (2000) p.59.

[8] M.H. Prado da Silva, PhD Dissertation, COPPE/UFRJ, Brazil (1999) p.166.

[9] M.H. Prado da Silva J.H.C. Lima, G.A. Soares, C.N. Elias, M.C. de Andrade, S.M. Best, I.R. Gibson, Surf. Coat. Tech., Vol. 137 (2001), p. 270.

Key Engineering Materials Vols. 254-256(2004) pp. 443-446
online at http://www.scientific.net
© *2004 Trans Tech Publications, Switzerland*

Control of Morphology of Titania Film with High Apatite-forming Ability Derived From Chemical Treatments of Titanium with Hydrogen Peroxide

S. Kawasaki, K. Tsuru, S. Hayakawa and A. Osaka

Biomaterials Laboratory, Faculty of Engineering, Okayama University,

3-1-1, Tsushima, Okayama-shi, 700-8530, Japan

(corresponding author. osaka@biotech.okayama-u.ac.jp)

Keywords: Titanium, Titania, Anatase, Morphology, Topography, Chemical and Thermal Treatment, Apatite Formation, Hydrogen Peroxide

Abstract. We prepared titanium substrates with sub-micron scale pores and without sub-micron scale pores by means of chemical treatments with hydrogen peroxide and subsequent thermal treatments. Sub-micron scale pores were observed on the surface of the titanium substrate treated with 3 mass% H_2O_2 at 90°C for 1 h(H1) by scanning electron microscopy. On the other hand, the sub-micron scale pores were not observed on the substrate treated with 3mass% H_2O_2 at 90°C for 72 h(H72). The surface roughness, Ra of H1 and H72 substrates was 27 and 14 nm, respectively. Both substrates showed high *in vitro* apatite-forming ability in simulated body fluid (SBF of Kokubo's recipe). The apatite-forming ability of H72 substrate with low Ra value was higher than that of H1 with high Ra value.

Introduction

Titanium and its alloys are used as important materials at orthopedic and dental fields because of their high corrosion resistance and biocompatibility. However, these materials cannot bond to human bone directly. In order to overcome this problem, several chemical treatment methods have been reported for providing bioactivity to titanium surface [1-2]. Osaka et al. [4-6] already reported that titania film derived from the treatment of titanium substrates with a hydrogen peroxide solution and subsequent thermal treatment deposited a bone-like apatite layer in simulated body fluid (SBF of Kokubo's recipe [3]) within 3 days, which has inorganic ion concentration similar to the human plasma. Since these surfaces have many sub-micron scale pores, the morphology is strongly related to the size, number and distribution of surface sub-micron scale pores. Recently, the surface topography is considered to be an essential factor to control the apatite-forming ability or the adhesion of an osteoblastic cell.

In this study we tried to control the surface morphology of titania films derived from chemical treatment of titanium with H_2O_2 solution and subsequent thermal treatment, and examined their *in vitro* apatite-forming ability.

Materials and Methods

Titanium substrates of 10x10x0.1mm in size were cut from a sheet of commercially available pure titanium (cp Ti) and washed with acetone for three times 5 min in an ultrasonic cleaner. The titanium substrates were then immersed in 20 ml of 3 mass% H_2O_2. After being kept at 90°C for 1, 12, 24, 36, 48, 60 and 72 h, they were washed with distilled water for 5 min in an ultrasonic cleaner. Subsequently, these substrates were heated at 400°C for 1h in an electric furnace. After the H_2O_2 and thermal treatments, the *in vitro* formation of apatite on the titanium substrates was examined by soaking them in 20ml of SBF at 36.5°C and pH 7.4 for 3 days. Their surface structures were investigated by thin film X-ray diffraction (TF-XRD: Cu Kα), scanning electron microscopy (SEM) and atomic force microscopy (AFM). The concentration of the Ti (IV) species dissolved from titanium substrates into H_2O_2 solutions was measured by means of inductively coupled plasma emission spectroscopy (ICP).

Results

Fig.1 shows the TF-XRD patterns of the substrates treated with 3 mass% H_2O_2 solutions for 1, 12, 24, 36 and 72 h (a), and subsequently heated at 400°C for 1 h (b) (denoted hereafter as H1, H12, H24, H36 and H72 substrate, respectively). From Fig.1 (a), H24 substrate showed the diffraction of anatase at ~25° in 2 θ (JCPDS 21-1272). H36 and H72 substrates showed the diffraction of anatase and rutile at ~27° (JCPDS 21-1276). The diffraction patterns of H36 and H72 substrates were almost the same. From Fig.1 (b), all substrates showed the diffractions of anatase and rutile. The intensities of anatase and rutile diffraction peaks increased due to heat treatment at 400°C for 1h except for H72 substrate.

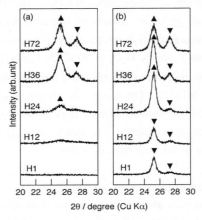

Fig.1 TF-XRD patterns for H1, H12, H24, H36 and H72 substrates treated with the 3mass% H_2O_2 solutions (a) and subsequently heated at 400°C for 1h (b). ▲ : Anatase,▼ : Rutile

Fig.2 shows the SEM photographs of H1, H12, H24, H36 and H72 substrates after the chemical and heat treatments. Sub-micron scale pores were formed on the surface of H1 substrate which has the same morphology as those observed on titania film in previous study [4-6]. The estimated pore size was about 200 nm in diameter. On the other hand, the other substrates did not show any sub-micron scale pores. No clear difference in the surface morphology was seen between H12 and H24 substrates, while the titania gel layer of H36 and H72 substrates seems to be thicker than that of H12 and H24 substrates.

Fig.3 shows the AFM topographies of H1 and H72 substrates. The roughness, Ra of H1 substrate was 27 nm, while H72 substrate was 14 nm, indicating that the surface of H1 substrate was rougher than that of H72 substrate.

Fig.4 shows the TF-XRD patterns of H1and H72 substrates after soaking in SBF for 3days. The diffractions at 26° and 32° in 2 θwere due to apatite on both substrates. The apatite peak intensity of H72 substrate was larger than that of H1 substrate. This indicates that H72 substrate was more effective than H1 substrate in terms of *in vitro* apatite formation.

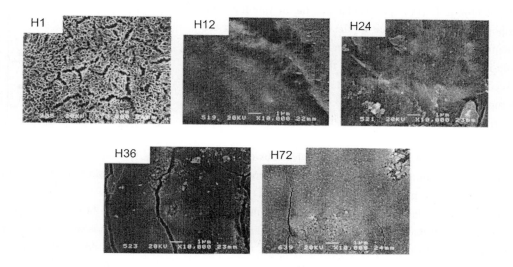

Fig.2 SEM photographs for H1, H12, H24, H36 and H72 substrates treated with the 3mass% H_2O_2 solutions and subsequently heated at 400°C for 1h.

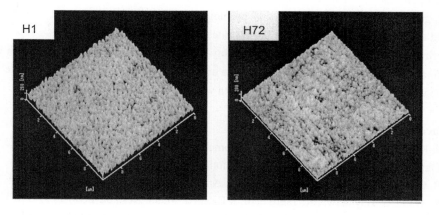

Fig.3 AFM topographical surveys for H1 and H72 substrates treated with the 3mass% H_2O_2 solutions and subsequently heated at 400°C for 1h.

Discussion

Shibata et al. [6] reported that the titania gel formation and Ti(IV) species dissolution occurred competitively during the reaction between H_2O_2 solution and titanium substrate. Thus, the change in the morphology and crystalline phase formed on titanium substrates can be outlined below. (1) Titania gel with sub-micron pores formed on titanium substrate within 1 h (Fig.2, H1). (2) The crystallization of titania gel layer occurred after 24 h. (3) The dissolution of Ti (IV) species completely deposited as titania gel on the surface of titanium substrate within 36 h, on the basis of the results of ICP analysis(not shown here). The deposition of Ti (IV) species changed the morphology and topography of the surface of the titania gel layer on the titanium substrate, resulting in the disappearance of sub-micron scale pores in the titania gel layer. Accordingly, it was found that the chemical treatment temperature and time are key factors to control the morphology and nm-scale topography of the titania gel layer.

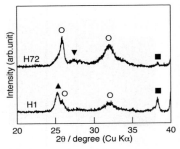

Fig.4 TF-XRD patterns for H1 and H72 substrates treated with the 3mass% H_2O_2 solutions, subsequently heated at 400°C for 1h and soaked in SBF for 3days. ▲ : Anatase, ▼ : Rutile, ○ : Apatite, ■ : Titanium

As shown in Fig.4, H1 and H72 substrates showed high *in vitro* apatite-forming ability. Li et al. [7-8] reported that the titania film prepared by sol-gel method deposited apatite on the surface after soaking in SBF within 2 weeks. They suggested that the Ti-OH group in the titania gel was the key factor to deposit apatite by providing negatively charged sites. Wang et al. [4-5] suggested that the two types of negatively charged sites such as Ti-OH and TiOOH groups were present on the titania gel derived from the chemical treatment of titanium with hydrogen peroxide solution and that the thermal treatment at 400°C eliminated the Ti-O-O-Ti and TiOOH groups from the titania gel and promoted the formation of anatase. Therefore, it is expected that the surfaces of H1 and H72 substrates must be rich in Ti-OH groups.

We confirmed high *in vitro* apatite-forming ability of the titanium substrates treated with hydrogen peroxide and subsequently heat treatment. The apatite-forming ability of H72 substrate with low Ra value was higher than that of H1 with high Ra value.

Reference

[1] HM. Kim, F. Miyaji, T. Kokubo et al., J. Biomed. Mater. Res., 32 (1996) 409-417.

[2] C. Ohtsuki, H. Iida, S. Hayakawa et al., J. Biomed. Mater. Res., 35, (1997), 39-47.

[3] S. B. Cho, K. Nakanishi, T. Kokubo et al., J. Am. Ceram. Soc., 78 (1995), 1769-1774.

[4] X-X. Wang, S. Hayakawa, K. Tsuru et al, J. Biomed. Mater. Res., 52(2000), 171-176.

[5] X. X. Wang, S. Hayakawa, K. Tsuru and A. Osaka, Biomaterials, 23 (2002), 1353-1357.

[6] K. Shibata, K. Tsuru, S. Hayakawa and A. Osaka, , Bioceramics, 15 (2002), 55-58.

[7] P. Li, C. Ohtsuki, T. Kokubo, K. Nakanishi et al., J. Biomed. Mat. Res., 28 (1994), 7-15.

[8] P. Li, I. Kangasniemi, K. de Groot, and Kokubo T., J. Am. Ceram. Soc., 77 (1994), 1307-1312.

Key Engineering Materials Vols. 254-256(2004) pp. 447-450
online at http://www.scientific.net

The Effect of Magnesium Ions on Bone Bonding to Hydroxyapatite Coating on Titanium Alloy Implants

P A Revell [1], E Damien [1], X S Zhang[1], P Evans [2], C R Howlett [3]

[1] Osteoarticular Research, Dept. of Histopathology, Royal Free Campus,

UCL, London NW3 2PF, UK   e.mail: prevell@rfc.ucl.ac.uk

[2] ANSTO, Lucas Heights, NSW 2234, Australia   email: pev@ansto.gov.au

[3] School of Pathology, University of New South Wales, PO Box 1,

NSW 2033, Australia e mail: R.Howlett@unsw.edu.au]

Keywords: Bone bonding; bioactive; hydroxyapatite; magnesium

Abstract. After 6 weeks implantation in the NZW rabbit, statistically significant enhanced bone bonding, measured as interfacial shear strength, has been demonstrated with Mg ion beam embedded HA coated TiAlV cylinders compared with an ordinary HA coating ($p < 0.05$, n=7, student 't' test). The results are in keeping with previous studies of the effects of magnesium on bone cell activity.

Introduction

Hydroxyapatite (HA) applied by plasma spraying is the most commonly used bioactive coating used in orthopaedics [1- 4]. The amount of bone apposition in relationship to retrieved human HA-coated implants varies between 32 and 78 per cent [5], while experimental studies have shown that bone will grow into HA-plasma sprayed slots in metal implants placed in canine and rabbit bone [6,7] . That there are different osseous responses due to the porosity, crystallinity and indeed, chemical composition of different hydroxyapatite layers is apparent. Ion beam implantation might also have effects on the biological properties of hydroxyapatite coatings. Ion beam implantation is a method whereby the surface atomic layers of any material substrate may be altered without changing its bulk properties. Previous work with non-bioactive surfaces has shown that Mg ion implantation in metal, ceramic (alumina) and polymer changes the adhesiveness of osteoblasts to these materials as well as their functional synthetic activity [8-10].

It was hypothesised that ion beam embedding of Mg ion into an HA coating would enhance bone growth and attachment. In order to study this, identically prepared HA coated implants which were Mg ion implanted or not were inserted in rabbit femoral bone. The biological response in terms of the bone bonding effect, measured biomechanically, was evaluated and the results are reported here.

Materials and Methods

Preparation of implants. Cylindrical TiAlV alloy implants (3.5 x 8mm;diam x length) were plasma sprayed with hydroxyapatite over the total curved surface and at one end to a thickness of $50\mu m$ (Plasma Biotal, UK). Some of these coated implants were left without further treatment (HA group) while identically prepared cylinders were additionally ion beam implanted with Mg ions on the coated surfaces (Mg-HA group), using a MEVVA ion source and ion beam energies of 23keV to give a Mg ion concentration of 1 x 10^{17} ions/cm^2 . Implants were ethylene oxide (EtO) treated with residual EtO levels of <3 ppm. (Moorfields Hospital, London).

Experimental implantation procedure. Mature female NZW rabbits (2.7 ± 0.2 kg) were anaesthetised with Hypnorm/Diazepam induction and maintenance was with 1.5 to 2% halothane in 25% oxygen/75% nitrogen. Cylinders (HA, Mg-HA) were implanted bilaterally into the lower femur using an intraarticular approach in which the femoral condyles were exposed and a slow speed saline cooled drill (AO, Switzerland) used to make a 3.5mm diam hole from the intercondylar notch into the lower femur to a depth of 8-10mm. The implants were press fitted into this hole and the joint and skin closed. Post-operative care was provided by an analgesic (Temgesic) and antibiotic (enrofloxacin) both injected subcutaneously. All animals were fully mobile within 24 hours of surgery. There were no post-operative complications and all animals remained in a normal healthy condition until the experiments were terminated.

Experimental design. Fourteen rabbits received bilateral implants with HA-coated cylinders on one side and Mg-HA coated cylinders in the contralateral lower femur. Implants with surrounding bone were retrieved after 3 and 6 weeks (7 implants in each group). They were immediately dissected down to the bone and sectioned at right angles to the long axis of the implant as determined radiologically. A plane parallel cut was made using an Exakt saw to produce a cuboid of bone 15 x 15 x 10mm which was 10mm thick and had the implant centrally placed [11-12]. The sample was kept moist while being transferred to the biomechanical testing laboratory. Push-out tests were performed using an Instron 4464 testing machine (High Wycombe, UK) to measure the interfacial shear strength (ISS) between bone and materials. The specimens were accurately orientated perpendicular to the plane parallel surfaces and the push-out tests performed at a cross-head speed of 1mm/min in a hydrated environment. Maximum ISS was calculated for each specimen by examination of the stress/strain plots, using the following formula :

$$ISS = \frac{Lp}{\pi \emptyset L} \quad (MPa) \qquad (1)$$

where Lp is the initial peak load required to dislodge the implant (in Newtons), Ø is the diameter of the implant (in millimetres) and L is the length of the specimens (in millimetres). 1 mega-Pascal (MPa) is $9.87 \ kg/cm^2$. The means and standard deviations for the ISS values were calculated for each group and the significance of the differences between the groups calculated using the student 't' test.

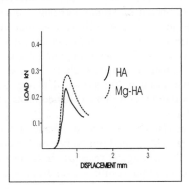

Fig.1 Displacement curves for push out test on HA and Mg-HA implants in the same animal

Table 1. Interfacial shear strengths for HA coated and Mg-HA coated metal implants in rabbit bone for 3 and 6 weeks. (Means and SDs) . There were 7 implants in each group at each time.

Duration of implant-ation	HA Implant	Mg-HA Implant	Student 't' test
3 weeks	2.46±1.05 (MPa)	3.31±1.24 (MPa)	not significant
6 weeks	1.97+1.17 (MPa)	3.54+1.41 (MPa)	p<0.05

Results

Strain data were recorded during the push out tests performed at a cross head speed of 1mm/min. Typical displacement curves for HA and Mg-HA implants from the same animal are shown in figure 1. The mean and standard deviations for each group (HA vs. Mg-HA) were calculated for 3 and 6 weeks of implantation. The results are shown in table 1. Although there was an increase in ISS after 3 weeks implantation of Mg-HA compared with HA, this did not reach statistical significance. However, at 6 weeks there was a statistically significant increase in interfacial shear strength for Mg ion embedded HA coated implants compared with HA coated controls (Mg-HA vs HA; p<0.05, students 't' test).

Discussion

Bioactive coating of non-cemented joint prostheses is now well established, and achieved mostly by the provision of a layer of HA applied to the metal surface by plasma spraying. Changes to the chemistry of the HA coating, in a previous study, by the implantation of magnesium ions at a dose of 1×10^{17} ions/cm^2 produced a significant increase in bone growth into coated slots after both 3 and 6 weeks when compared with HA controls [13]. The ion beam implantation method enables the alteration of the surface chemistry of the HA-coating without changes to the hydroxyapatite bulk properties. At 23 keV average energy, it is presumed that Mg ions will not have penetrated beyond 60nm from the surface of the HA-coating and that the Mg ions remain as isolated atomic forms with no incorporation into the molecular structure by bonding.

Mg ion implanted HA coatings showed a significantly greater bone bonding strength at 6 weeks as evaluated by calculating interfacial shear stress after push out tests. Push out testing has been employed before as a means of assessing the mechanical strength of bonding between bone and biomaterial in vivo [11,14-16]. The reproducibility of ISS testing is sometimes questioned. Dhert and colleagues showed that the configuration of the testing rig, the cortical bone thickness and the stiffness of the implant could all affect the values obtained for interfacial stress distribution [17]. However, the model used in the present experiment has been employed in previous studies, with identical implant dimensions, site of implantation, pre-test preparation of samples, conditions of testing and testing machine. The test results obtained are in line with those found previously using HA-coated cylinders [11,18].

The point of failure at the interfaces between titanium alloy, HA-coating and bone is between ceramic and metal [4]. Failure in the present study was less consistently defined occurring also in the bone and HA although splitting between ceramic and metal was the most conspicuous (data not shown). An analysis of the fracture mechanics of push out data obtained using implants of the same dimensions and placed at the same site in rabbit bone has been reported [12]. HA-coated TiAlV alloy and epoxy resin with an adhesively bonded HA layer were compared, using the known material properties of the metal, epoxy, HA-coatings and bone, and the elastic-plastic energy release rate was calculated for the different interfaces. The mathematically predicted crack propagation patterns coincided with the experimental findings, the point of failure for HA-coated TiAlV being between these two materials while that for the epoxy system was between the coating and the bone, or within the bone itself [11,12]. In the present study, obvious failure occurred between hydroxyapatite and metal but fractures were observed in adjoining osseous trabeculae with occasional splits in the hydroxyapatite and the bonded bone in the Mg ion implanted group. It is interesting that the increased bone ongrowth was sufficient to give an increase in interfacial shear strength for the Mg ion implanted HA coatings.

It has been suggested that amorphous HA coatings are better than crystalline with respect to bone growth and the incorporation of magnesium may have the effect of changing the crystallinity of the HA-coating. But recent results using Zn ion beam implantation failed to reveal any difference in ISS between the HA coated and Zn-HA coated Ti alloy (data not shown), suggesting that any change in crystallinity by the ion bombardment is unlikely to have an effect on the bone bonding effect of the HA.

Having demonstrated increased bone ingrowth and improved bonding with Mg ion implanted HA-coatings, the question arises as to the mechanisms whereby this enhancement occurs. Magnesium is present in small amounts in normal bone, amounting to 0.5 to 1% of bone ash [19]. It is the fourth most abundant cation in the human body and not a trace element [20]. There are species differences in the magnesium content of bone. Magnesium deficiency in man is related to the development of osteoporosis [21-23]. Studies in the rat indicate that magnesium deficiency has an effect on both osteoclastic and osteoblastic activity with uncoupling of bone formation and resorption leading to net bone loss [24]. Magnesium supplementation in oophorectomised rats resulted in increased bone formation, reduced bone resorption and increased bone strength [25]. In rabbits the minimum daily requirement for magnesium is above 10mg/100g dry weight feed, below which level various pathological conditions have been recorded [26]. The feed of the animals in this present study contained magnesium at approximately 4 times this level. It is unlikely, therefore, that any magnesium deficiency was present in the animals.

References
[1] R.G.T.Geesink:*Hydroxyl-apatite coated hip implants* (PhD thesis, Maastricht University 1988)
[2] K.DeGroot: *Ceramics of calcium phosphates* (CRC Press. Boca Raton, Florida 1982)
[3] K.DeGroot, R.G.T.Geesink, C.R.A.T.Klein, P.Serekian: J.Biomed.Mater.Res;21(1992):1375-81
[4]W.J.A. Dhert: Medical Progress through Technology;20 (1994):143-154
[5]T.W.Bauer, R.C.T.Geesink, R.Zimmerman, J.T.McMahon:J.Bone Jt.Surg;73A: (1991) 1439-1452
[6] P.K.Stephenson, M.A.R.Freeman, P.A.Revell, J.Germain, M.Tuke, C.J. Pirie: J. Arthroplasty; 6(1991):51-58
[7] P.A.Revell, K.E.Tanner, W.Bonfield: J. Pathol; 181(Suppl.) 12A (1997)
[8] K.S.Grabowski KS, C.R.Gosset, F.A.Young, J.C.Keller:Mat.Res.Soc.Symp.Proc;110 (1987): 697-702.
[9] J-S.Lee, M.Kaibara, M.Iwaki, H.Sasebe, Y.L.Suzuki, M.Kusakabe: Biomaterials;14(1993) :958-960
[10] C.R.Howlett, H.Zreiqat, R.O'Dell, J.Noorman, P.Evans, B.A.Dalton, J.McFarland, J.G.Steele: J. Mater Sci: Mater. Med; 5(1994):715-722
[11] X.S.Zhang, P.A.Revell, S.L.Evans, M.A.Tuke, P.J.Gregson: J.Biomed.Mater.Res;46(1999) :279-286
[12] J.I.Thompson, P.J.Gregson, P.A.Revell: J. Mater. Sci:Mater. Medicine; 10(1999):863-868
[13]X.S.Zhang, P.A.Revell, P.Evans, K.E.Tanner, C.R.Howlett: Trans. Soc. Biomat;21(1998):187

[14] B.C.Wang, T.M.Lee, E.Chang, C.Y. Yang: J. Biomed. Mater. Res;27(1993):1315-1327

[15] T.Inadome, K.Hayashi, Y.Nakashima, H.Tsumura,Y.Sugioka: J.Biomed.Mater.Res;29(1995): 19-24

[16] S.D.Cook, K.A.Thomas, J.F.Kay, M. Jarcho: Clin. Orthop. Rel. Res; 232(1988):225-243
[17] W.J.A.Dhert, C.C.P.M.Verheyen, L.H.Braak, J.R.de Wijn, C.P.A.T.Klein, K.de Groot, P.M. Rozing: J. Biomed. Mater. Res; 26(1992):119-130
[18] M.C.Blades, D.P.Moore, P.A.Revell, R.Hill: J.Mater.Sci: Mater.Med; 9(1998): 701-706
[19] S.Wallach: Magnes. Trace Elem; 9(1990):1-14
[20] G.A. Quamme: Magnesium; 5(1986):248-72
[21] R.K.Rude, M.E.Kirchen, H.E.Gruber, M.H.Meyer, J.S.Luck, D.L.Crawford: Magnes. Res; 12(1999):257-267
[22] J.Durlach, P.Bach, V.Durlach, Y.Rayssiguier, M.Bara, A.Guiet-Bara: Magnes. Res.; 11(1998): 25-42
[23] L.Cohen: Magnes. Res.;1(1988) :85-7
[24]R.K.Rude, M.E.Kirchen, H.E.Gruber, A.A.Stasky, M.H.Meyer: Miner Electrolyte Metals; 24(1998):314-20
[25]Y.Toba, Y.Kajita, R.Masuyam, Y.Takada, K.Suzuki, S.Aoe: J. Nutr.; 130(2000):216-20
[26] H.O.Kunkel, P.B.Pearson: J.Nutri;36(1948):657-666

Key Engineering Materials Vols. 254-256(2004) pp. 451-454
online at http://www.scientific.net
© *2004 Trans Tech Publications, Switzerland*

Preparation of Novel Bioactive Titanium Coatings on Titanium Substrate by Reactive Plasma Spraying

Masahiko Inagaki[1], Yoshiyuki Yokogawa[2] and Tetsuya Kameyama[3]

[1] 2266-98 Anagahira Shimoshidami Moriyama-ku, Nagoya, 463-8560, JAPAN, National Institute of Advanced Industrial Science and Technology (AIST), m-inagaki@aist.go.jp

[2] 2266-98 Anagahira Shimoshidami Moriyama-ku, Nagoya, 463-8560, JAPAN, National Institute of Advanced Industrial Science and Technology (AIST), y-yokogawa@aist.go.jp

[3] 2266-98 Anagahira Shimoshidami Moriyama-ku, Nagoya, 463-8560, JAPAN, National Institute of Advanced Industrial Science and Technology (AIST), t-kameyama@aist.go.jp

Keywords: Reactive plasma spraying, titanium, titania, titanium nitride, coatings, bioactivity

Abstract. A simple treatment method using radio-frequency reactive plasma spraying (rf-RPS) was studied to induce bioactivity of titanium (Ti) coatings. Ti coatings were deposited on Ti substrates by an rf-RPS method using a thermal plasma of Ar gas containing 1-6% N_2 and/or O_2 at a input power of 16 kW. Composition change of coating's surface during soaked in a simulated body fluid (SBF) was examined by micro Fourier transform infrared spectroscopy and thin film X-ray diffraction. Ti coatings prepared with Ar-O_2 and Ar-N_2-O_2 plasma formed apatite after 3 days of soaking in 40 ml SBF. This indicates that such coatings have the ability to form a biologically active bone-like apatite layer on the surface. In the XRD patterns for both Ti coatings, minute peaks ascribable to TiO_2 (anatase and rutile phase) were commonly observed. On the other hand, composition change of coating's surface cannot be observed for Ti coating sprayed with Ar-N_2 plasma after 7 days of soaking in SBF.

Introduction

Surface modification of metallic titanium by chemical treatments [1,2] and ion implantation [3] has been studied to induce bioactivity for implant materials. Such surface treated titanium formed apatie while soaked in a simulated body fluid (SBF). Such layer also induced bone conduction and directly bond to bone. In this paper, we propose a novel bioactive coating method using radio frequency reactive plasma splaying (rf-RPS). This method is based on the plasma-enhanced reaction between Ti particles and a plasma gas during plasma spraying of titanium powder.

Methods

Ti coatings were deposited on titanium substrates (0.8 mm thick, ASTMB348-GR2) by a RF-RPS method. A thermal plasma of Ar gas containing 1-6% N_2 or O_2, was generated at 4 MHz frequency at a RF input power of 16 kW. Ti powder (average particle size = 68±15 mm, 99.9%) were used as raw materials. A simulated body fluid (SBF) was prepared as described in the literature [4], and pH was adjusted to 7.4 at 36.5 °C. Ti coated samples and non-coated samples were soaked in 40 ml SBF for 1, 3, 5 and 7 days at 36.5 °C. After soaking in SBF, the samples were rinsed with Milli-Q water and ethanol, and then dry in atmosphere. These samples were evaluated by micro Fourier transform infrared reflection spectroscopy (micro FT-IR) (MICRO-20, JASCO co.) and thin film X-ray diffraction (TF-XRD) (MPX3, MAC Science, Japan). The morphologies of the coatings before and after soaking were also observed by scanning electron microscope (SEM) (S3000N, Hitachi) and confocal scanning laser microscope (VK-8510, Keyence).

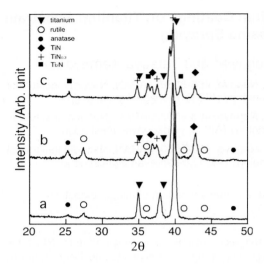

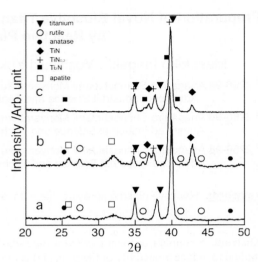

Fig. 1 TF-XRD patterns for Ti coatings prepared with Ar-O$_2$ (a), Ar-N$_2$-O$_2$ (b) and Ar-N$_2$ (c).

Fig. 2 TF-XRD patterns for Ti coatings prepared with Ar-O$_2$ (a), Ar-N$_2$-O$_2$ (b) and Ar-N$_2$ (c), after soaking in SBF for 7 days.

Results and Discussion

Figure 1 shows TF-XRD patterns for Ti coatings prepared with different plasma gas compositions. When sprayed with Ar-O$_2$ (Fig. 1a) and Ar-N$_2$-O$_2$ (Fig. 1b) plasma, minute XRD peaks ascribable to TiO$_2$ (rutile and anatase phase) were commonly observed. Minute XRD peaks ascribable to titanium nitrides (TiN, Ti$_2$N and titanium-nitrogen solid solution) were also observed in Ti coatings sprayed

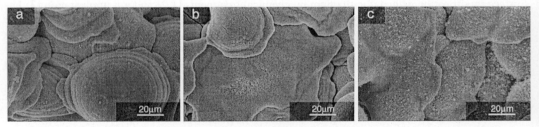

Fig. 3 typical SEM images obtained for surface of Ti coatings prepared with Ar-O$_2$ (a), Ar-N$_2$-O$_2$ (b) and Ar-N$_2$ (c).

Fig. 4 typical SEM images obtained for surface of Ti coatings prepared with Ar-O$_2$ (a), Ar-N$_2$-O$_2$ (b) and Ar-N$_2$ (c) plasma after soaking in SBF for 7 days.

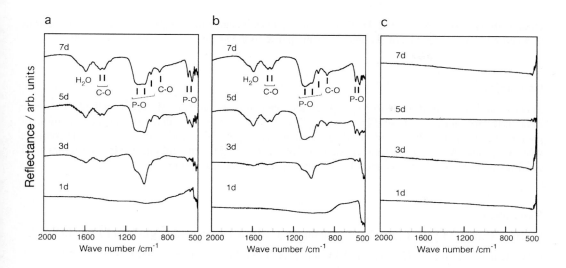

Fig. 5 FT-IR spectra for Ti coatings prepared with Ar-O$_2$ (a), Ar-N$_2$-O$_2$ (b) and Ar-N$_2$ (c) plasma after soaking in SBF for different periods.

with Ar-N$_2$-O$_2$ (Fig. 1b) plasma. TF-XRD patterns for Ti coatings prepared with Ar-N$_2$ (Fig. 1c) plasma indicates that titanium nitrides were formed in such coatings. However, XRD peaks ascribable to titanium oxides were not observed in Ti coatings prepared with Ar-N$_2$ (Fig. 1c) plasma.

Figure 2 shows TF-XRD patterns for Ti coatings prepared with different plasma gas composition, after soaking in SBF for 7 days. XRD peaks ascribable to (002) and (211) diffraction of apatite were observed in Ti coatings sprayed with Ar-O$_2$ and Ar-N$_2$-O$_2$ plasma.

Figure 3 shows typical SEM images obtained for surface of Ti coatings prepared with Ar-O$_2$, Ar-N$_2$-O$_2$ and Ar-N$_2$ plasma. On the confocal scanning laser microscope observation, the surface roughness (R$_a$) of these Ti coatings before soaking in SBF was 12.5±0.1 μm, 9.1±0.7 μm and 10.6±0.4 μm, respectively. Figure 4 shows typical SEM images obtained for surface of Ti coatings prepared with different plasma gas composition after soaking in SBF for 7 days. It can be seen the surface of coatings sprayed with Ar-O$_2$ and Ar-N$_2$-O$_2$ plasma covered with apatite particles after soaking in SBF. Therefore, it is indicated that Ti coatings sprayed with oxygen containing plasma have the ability to form a biologically active bone-like apatite layer on the surface. Kaneko et al. [6] suggested that the apatite formation on anatase was initiated by combination of Ti-OH end on anatase and calcium ions in SBF. An rf-RPS of Ti particles with oxygen containing plasma yielded the anatase phase that was inducing apatite nucleation. SEM images of Ti coating sprayed with Ar-N$_2$ plasma were unchanged from those before soaking. Thus titanium nitrides phases have no ability to form a biologically active bone-like apatite layer on the surface.

Figure 5 shows the FT-IR spectra for Ti coatings prepared with different plasma gas compositions, after soaking in SBF for different periods. In the FT-IR spectra for Ti coating sprayed with Ar-O$_2$ and Ar-N$_2$-O$_2$ plasma, absorption band of the PO$_4$ group at 570 cm^{-1}, 600 cm^{-1}, 960 cm^{-1}, 1040 cm^{-1} and 1090 cm^{-1} and absorption band of the CO$_3$ group at 870 cm^{-1}, 1420 cm^{-1} and 1460 cm^{-1} ascribable to carbonated apatite [5] were observed after 3 days of soaking in SBF. However, any absorption band cannot be observed for up to 7 days of soaking in FT-IR spectra for control (Ti substrate) and Ti coating sprayed with Ar+6% N$_2$ plasma. From the changes of FT-IR spectra of our Ti coatings by soaking in SBF, it can be seen that the apatite-forming ability is not much difference between Ti coating sprayed with Ar-O_2 and Ar-N_2-O_2 plasma. Thus it seems that titanium nitrides in Ti coatings sprayed with Ar-N_2-O_2 plasma is inert to form apatite in the same manner as metal titanium.

Conclusions

Surface modification of Ti deposits by plasma-enhanced reaction is an effective method to provide bioactivity for plasma sprayed Ti coatings. Our results clearly revealed that the composition of plasma gas during rf-TPS played a crucial role in improving the bioactivity of Ti coatings. Ti coatings with Ar-O_2 and Ar-N_2-O_2 plasma can deposit apatite after 3 days of soaking in SBF. In contrast, Ti coatings with Ar-N_2 plasma cannot deposit apatite for 7 days of soaking.

References

[1] H.M. Kim, F Miyaji, T Kokubo, T Nakamura, J. Biomed. Mater. Res. Vol 32, (1996), 409-417.

[2] C. Ohtsuki, H Iida, S Hayakawa, A Osaka, J. Biomed. Mater. Res. Vol 35, (1997), 39-47.

[3] T Hanawa, Y Kamiura, S Yamamoto, T Kohgo, A Amemiya, H Ukai, K Murakami K Asaoka, J. Biomed. Mater. Res. Vol 36, (1997), 131-136.

[4] HM Kim, F. Miyaji, T Kokubo, T Nakamura, J. Biomed. Mater. Res Vol 45, (1999), 100-107.

[5] I Rehman, W Bonfield, J. Mater. Sci. Mater. Med Vol 8, (1997) 1-4.

[6] H Kaneko, M Uchida, HM Mim, T Kokubo, T Nakamura, Key Eng. Mater. Vol 218-20, (2002) 649-652.

Key Engineering Materials Vols. 254-256(2004) pp. 455-458
online at http://www.scientific.net
© 2004 Trans Tech Publications, Switzerland

Adhesion of Sol-Gel Derived Zirconia Nano-Coatings on Surface Treated Titanium

R. Roest[1], A.W. Eberhardt[2], B.A. Latella[3], R. Wuhrer[4], B. Ben-Nissan[1]

[1] Department of Chemistry, Materials and Forensic Science, University of Technology, Sydney, PO Box 123, Broadway, N.S.W., 2007, Australia
[2] Department of Biomedical Engineering, University of Alabama at Birmingham, Birmingham, AL, 35294, USA
[3] Materials and Engineering Sciences, ANSTO, PMB 1, Menai, N.S.W., 2234, Australia
[4] Microstructural Analysis Unit, University of Technology, Sydney, PO Box 123, Broadway, N.S.W., 2007, Australia

Keywords: Adhesion, phosphate treatment, nanocoatings, zirconia, sol-gel, titanium, anodizing

Abstract. The morphology, adhesion and tribological properties of the zirconia sol-gel coatings on phosphate treated, anodized and un-treated titanium surfaces were investigated. The anodization of titanium involves the formation of a thin, compact, oxide layer, which improves the wettability for further coating. This process involves the conversion of the rutile structure of the original titanium oxide into a mostly crystalline anatase structure. The samples were anodized in sulphuric and phosphoric acid at varying concentrations. The samples were anodized at differing currents and differing time periods ranging from 10 to 30 minutes. Phosphate adsorption treatment involves soaking samples in 10% H_3PO_4 solution for 10 minutes.

These samples were spin coated with zirconia, yielding 100 nm thick films. The nanocoatings were prepared by alkoxide sol-gel chemistry, using techniques and protocols developed in an earlier work and were examined with x-ray diffraction, and scanning electron microscopy.

Interfacial and adhesion properties were measured using a micromechanical tensile test. The tribological properties were investigated using an Orthopod machine, with commercial grade ultra-high molecular weight polyethylene (UHMWPE) pins (3/8 inch diameter) that can articulate in number of different combinations against opposing coated and control specimens. The UHMWPE pins were used in a bovine serum environment. The amount of the wear was measured gravimetrically and wear features were observed using SEM.

Introduction

Surface coatings offer the possibility of modifying the surface properties of an surgical component and thereby achieving improvements in performance, reliability and biocompatibility. Zirconia films (both nano and macro) produced by different methods are of great interest for variety of biomedical and engineering applications. As a result, a wide range of different deposition techniques for zirconia have been proposed. Many of the common thermal barrier coatings and corrosion resistant coatings use plasma or thermal spraying, while several more recent techniques have also been used including physical vapour deposition, sputtering, thermal and electron beam evaporation, plasma MOCVD, electrochemical vapour deposition, and sol-gel processing [1]. The term sol-gel is currently used to describe any chemical procedure or process capable of producing ceramic oxides, non-oxides and mixed oxides from solutions.

In this current work we aimed to modify pure titanium metal surfaces by first anodizing and then phosphotazing with specific phosphate adsorption and photocatalysis treatments to improve the wettability. The specimens were then coated with a sol-gel derived alkoxide based zirconia for increased biocompatibility and adhesion. After characterization, in this preliminary work, tensile adhesion and tribological pin-on-disc wear tests were carried out for their assessment of interfacial adhesion properties.

Materials and Methods

The titanium samples, both CP Ti and Ti alloy, were machined from a titanium rod without lubricant and with a HSS tool, then polished on a Leco Auto polisher to a 1 micron finish, before the final polish in a vibratory polisher with an alpha alumina to 0.5 micron finish. Samples were then degreased in MEK and finally cathodically cleaned in an 10% sodium hydroxide solution for 1 minute. The micro-adhesion and tribology samples were punched out using a flywheel punch and die from a 1mm thin CP Ti sheet and then put through the same polishing processes as the other titanium samples.

The titanium alloy samples were anodized in a phosphoric and sulphuric solution at varying concentrations. The samples were anodized at differing currents and differing time periods ranging from 10 mins to 30 mins based on an earlier work [2]. Phosphate adsorption treatment involves soaking samples in 10% H_3PO_4 solution for 10 minutes then rinsing immediately in pure distilled water then drying at 70^0 C. Photocatalysis treatment consists of the treatment of the anatase layer with a UV wavelength of approximately 380nm for 1 hour. Exposing the catalyst to UV generates an excited state on the surface, which is able to initiate subsequent processes like redox reactions and molecular transformations. These samples were then coated with alkoxide-derived zirconia. These coatings were applied by sol-gel spin coating methods, using techniques and protocols developed in an earlier work and were examined with x-ray diffraction and scanning electron microscopy [3,4].

Tensile tests on titanium samples were conducted using a specially designed micromechanical tensile tester equipped with a 2500N capacity load cell. The flat "dog-bone-shaped" converted and coated titanium samples were tested in tension at a rate of 0.005 mm/s using a specially designed high-stiffness micromechanical testing device positioned directly under the objective lens of an optical microscope (Zeiss Axioplan) at a fixed magnification of 200x (Fig.1). This allowed direct observation of crack initiation, and evolution and debonding of the thin films on the titanium specimens. The applied load and the imposed displacement were recorded during the tests (every 2 s). Simultaneously, optical images of the coated surface were captured every 2 s using a MTI analog camera with image analysis software (Scion Image, NIH). After testing, the samples were imaged with a LEO SUPRA55VP, scanning electron microscope (SEM).

The Ortho-POD (AMTI, Watertown, MA) was programmed carefully not to touch the edges of the disks. A square wear path was chosen for tests. A cycle time of 666ms (1.5 Hz) was used. During testing, the materials were immersed in diluted bovine serum (50% serum, 50% deionized water). The tests were run for 500,000 cycles, with the serum changed every 125,000 cycles. (The control pins were soaked in bovine serum while each test was running).

Cobalt chromium was the metallic chosen for the first wear test. Of the four disks, two disks had a zirconia coating and two disks had a zirconia coating combined with a phosphate adsoption treatment. A titanium substrate was chosen for the second wear test. Of the four disks, two disks had a zirconia coating and two disks had a zirconia coating combined with phosphate a UHMWPE sorption treatment. All of the polyethylene samples were each weighed three times using an AG245 scale (Mettler Toledo, 0.00001 gm resolution), and were averaged before and after testing.

Results and Discussion

Adhesion is a complex phenomenon related to physical effects and chemical reactions at the interface. Adhesive forces are set up as the coating is applied to the substrate and during curing or drying. The magnitude of these forces will depend on the nature of the surface and the coating. The micro adhesion tester was set up with an optical microscope and digital camera so that at any point the force at that point of time on the sample can be calculated (Fig.1).

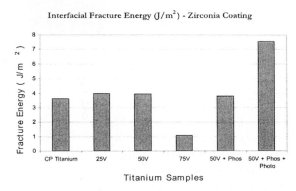

Interfacial Fracture Energy (J/m^2) - Zirconia Coating

Figure 1. Micro adhesion tester.

Figure 2. A graph of interfacial fracture energy of zirconia coatings on titanium substrates.

The results of the adhesion tests for CP Ti (S1), 25V(S2), 50V(S3), 75V(S4) anodized and 50V anodized and phosphate treated prior to sol-gel zirconia coating (S5) and 50V anodizing prior to phosphate treatment, photo catalysis reaction and phosphate treatment prior to zirconia coating (S6), have shown that the last combination of treated sample (S6) samples had higher interfacial energy (Fig.2), and fracture energy (Fig.3).

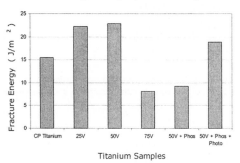

Fracture Energy of Film (J/m^2) - Zirconia Coating

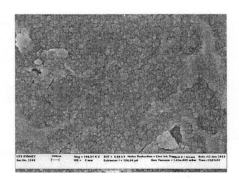

Figure 3. Measured fracture energies of the samples

Figure 4. SEM image of the Sample S6 after tensile test.

Phosphate and photocatalysis treated and zirconia coated sample shows higher fracture energy (K_{Ic}) in comparison to the untreated samples. SEM image of the sample S6 is shown on Figure 4.

Ortho-Pod results although only partially completed at this stage shows, impressively low wear rates on anodized titanium substrates with phosphate conversion, photocatalysis and zirconia coating in comparison to other combinations (Figure 5).

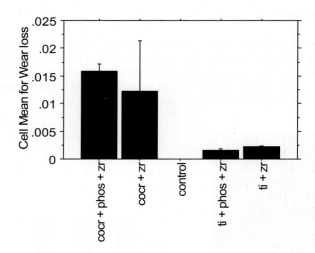

Fisher's PLSD for Wear loss
Effect: Material
Significance Level: 5 %

	Mean Diff.	Crit. Diff.	P-Value	
cocr + phos + zr, cocr + zr	.004	.011	.4159	
cocr + phos + zr, control	.016	.011	.0116	S
cocr + phos + zr, ti + phos + zr	.014	.011	.0173	S
cocr + phos + zr, ti + zr	.014	.011	.0205	S
cocr + zr, control	.012	.011	.0303	S
cocr + zr, ti + phos + zr	.011	.011	.0474	S
cocr + zr, ti + zr	.010	.011	.0574	
control, ti + phos + zr	-.002	.011	.7200	
control, ti + zr	-.002	.011	.6147	
ti + phos + zr, ti + zr	-.001	.011	.8813	

Figure 5. Wear loss measurements of various coated substrate groups. The final mass loss data was plotted on a *Statview*. S next to the table indicates that a significant difference was observed between those particular mass losses, as determined by ANOVA ($p < 0.05$).

Acknowledgements:
The authors gratefully acknowledge support from the Australian Academy of Science and the wear testing by Somaieh Kaschef, Yundan Cheng and Desmin Milner from the UAB.
References:

[1] M. J. Paterson, P.J.K. Paterson and B. Ben-Nissan, J. Mater.Res.,13, 2 , (1998) p.388.
[2] R.Roest and B.Ben-Nissan, "Surface Modification of Anodized Titanium for Calcium Phosphate Coatings", in the *Proceedings of the Engineering Materials 2001*, 23-26 September 2001, Melbourne, Australia. (Eds.) E. Pereloma and K. Raviprasad, (2001), p. 115.
[3] M. J. Paterson and B.Ben-Nissan, Surf. & Coat. Technol., 86-87 (1996), p. 156.
[4] M. Anast, J. Bell, T. Bell and B. Ben-Nissan, J. Mater. Sci. Lett., 11, (1992), p.1483.

Key Engineering Materials Vols. 254-256(2004) pp. 459-462
online at http://www.scientific.net
© *2004 Trans Tech Publications, Switzerland*

Bonelike Apatite Formation on Anodically Oxidized Titanium Metal in Simulated Body Fluid

Masakazu Kawashita[1], Xin-Yu Cui[1,2], Hyun-Min Kim[3], Tadashi Kokubo[4]
and Takashi Nakamura[5]

[1]Department of Material Chemistry, Graduate School of Engineering, Kyoto University,

Sakyo-ku, Kyoto 606-8501, Japan

kawashita@sung7.kuic.kyoto-u.ac.jp

[2]Institute of Metal Research, Chinese Academy of Sciences, Shenyang, 110016, P.R. China

[3] Department of Ceramic Engineering, School of Advanced Materials Engineering,
Yonsei University, 134, Shinchon-dong, Seodaemun-gu, Seoul 120-749, Korea

[4] Research Institute for Science and Technology, Chubu University,
1200 Matsumoto-cho, Kasugai-shi, Aichi 487-8501, Japan

[5]Department of Orthopaedic Surgery, Graduate School of Medicine, Kyoto University,
Sakyo-ku, Kyoto 606-8501, Japan

Keywords: titanium metal, anodic oxidation, apatite, simulated body fluid, rutile,

Abstract. Porous titania layers which were mainly composed of rutile and anatase were formed on the surface of titanium metals by the anodic oxidation in H_2SO_4 and Na_2SO_4 solutions. Titanium metals subjected to the anodic oxidation in CH_3COOH and H_3PO_4 solutions formed porous titania layers which were essentially amorphous and contained only a small amount of anatase. Titanium metals anodically oxidized in H_2SO_4 and Na_2SO_4 solutions formed bonelike apatite on their surfaces in simulated body fluid (SBF) within 1 and 3 days, respectively. It was found that the rutile precipitated on their surfaces had (101) plane parallel to the substrate. These results indicate that not only anatase but also (101) plane of rutile can show high apatite-forming ability in SBF. On the contrary, specimens anodically oxidized in CH_3COOH and H_3PO_4 solutions did not form apatite even after the soaking in SBF for 7 days. This might be attributed to the formation of only a small amount of anatase on their surfaces.

Introduction

Some ceramics such as Bioglass®, sintered hydroxyapatite, glass-ceramic A-W can bond to living bone through a bonelike apatite layer, which is formed on their surfaces. They are called bioactive ceramics, and have been used as important bone repairing materials. They are, however, lower in fracture toughnesses and higher in elastic moduli than the human cortical bone, and hence can not be used under load-bearing conditions. Metals such as titanium metal and its alloys have high fracture toughnesses, but they do not bond to living bone. Recently, titanium metal and its alloys can bond to living bone, when they are previously subjected to chemical and subsequent heat treatments [1,2].

An essential requirement for an artificial material to bond to living bone is the formation of a biologically active bonelike apatite layer on its surface in a body environment. Recently, it has been shown that titania gel with anatase structure showed an excellent apatite-forming ability in a body environment [3]. Anodic oxidation of a metal is well known to *in situ* produce a metal oxide layer on the surface of metal. Therefore, the titania layer with anatase structure can be formed on the surface of tough titanium metal by the anodic oxidation, and hence the anodically oxidized titanium metal is expected to show a high apatite-forming ability in a body environment. In the present study, Ti metals were subjected to anodic oxidation under various conditions and their apatite-forming ability was investigated in simulated body fluid (SBF).

Experimental

Rectangular specimens of titanium metal (purity: 99.9%, Kobe Steel Co. Ltd., Kobe, Japan) with $10 \times 10 \times 1$ mm^3 in size were abraded with #400 diamond plate, and then washed with pure acetone, ethanol and ultra-pure water in an ultrasonic cleaner.

The specimens were subjected to anodic oxidation at room temperature for 1 min under the applied voltages of 100, 150 and 180 V. Sulfuric acid (H_2SO_4), acetic acid (CH_3COOH), phosphoric acid (H_3PO_4) and sodium sulfate (Na_2SO_4) solutions with concentrations of 0.5, 1.0 and 2.0 M were used as electrolyte. After the anodic oxidation, the specimens were gently washed with ultra pure water, and dried at 40°C for 24 h.

The specimens were immersed in 30 mL of an acellular SBF [4] with ion concentrations (Na^+ 142.0, K^+ 5.0, Ca^{2+} 2.5, Mg^{2+} 1.5, Cl^- 147.8, HCO_3^- 4.2, HPO_4^{2-} 1.0, SO_4^{2-} 0.5 mM) and pH nearly equal to those of human bone at 36.5°C for 3 and 7 d. After removal from solution, specimens were washed with distilled water and dried in air at 40°C for 24 h. The surface structural changes of the specimens were analyzed by thin-film X-ray diffraction (TF-XRD, Rigaku RINT-2500) and field-emission scanning electron microscopy (FE-SEM, Hitachi S-4700).

Results and discussion

Figure 1 shows TF-XRD patterns of the surfaces of the specimens anodically oxidized at various voltages in different electrolytes. The specimens formed an essentially amorphous titania layers

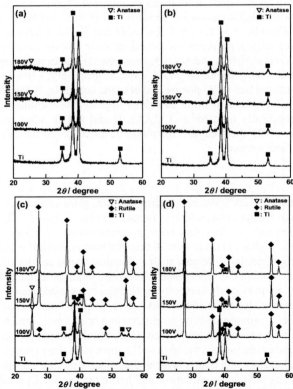

Fig. 1 TF-XRD patterns of the surfaces of the specimens anodically oxidized at various voltages in different electrolytes ((a): 2.0M-H_3PO_4, (b): 2.0M-CH_3COOH, (c): 2.0M-H_2SO_4, (d): 1.0M-Na_2SO_4).

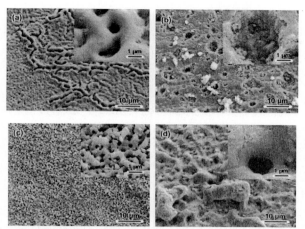

Fig. 2 FE-SEM photographs of the surfaces of the specimens anodically oxidized at 180 V in different electrolytes ((a): $2.0M$-H_3PO_4, (b): $2.0M$-CH_3COOH, (c): $2.0M$-H_2SO_4, (d): $1.0M$-Na_2SO_4).

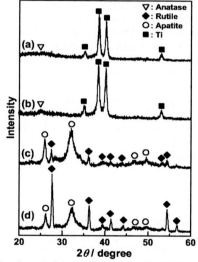

Fig. 3 TF-XRD patterns of t the surfaces of the specimens anodically oxidized at 180 V in different electrolytes and subsequently soaked in SBF for 7 days ((a): $2.0M$-H_3PO_4, (b): $2.0M$-CH_3COOH, (c): $2.0M$-H_2SO_4, (d): $1.0M$-Na_2SO_4).

containing a small amount of anatase on their surfaces in H_3PO_4 and CH_3COOH solutions (Figs. 1-(a) and (b)). In contrast, they formed titania layers with essentially rutile structure on their surfaces in H_2SO_4 and Na_2SO_4 solutions and the relative amount of rutile to anatase increased with increasing applied voltage (Figs. 1-(c) and (d)).

Figure 2 shows FE-SEM photographs of the surfaces of the specimens after the anodic oxidation at 180 V in different electrolytes. It can be seen from this figure that porous titania layers were formed irrespective of the kind of electrolytes, although morphological differences were observed.

Figure 3 shows TF-XRD patterns of the surfaces of the specimens anodically oxidized at 180 V in different electrolytes and subsequently soaked in SBF for 7 days. No change in diffraction patterns was observed for the specimens anodically oxidized in H_3PO_4 and CH_3COOH solutions (Figs. 3-(a)

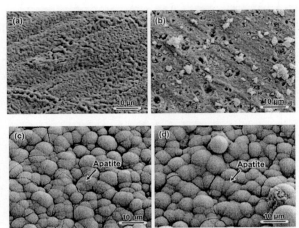

Fig. 4 FE-SEM photographs of the surfaces of the specimens anodically oxidized at 180 V in different electrolytes and subsequently soaked in SBF for 3 days ((a): $2.0M$-H_3PO_4, (b): $2.0M$-CH_3COOH, (c): $2.0M$-H_2SO_4, (d): $1.0M$-Na_2SO_4).

and (b)), whereas new peaks ascribed to apatite were observed for the specimens anodically oxidized H_2SO_4 and Na_2SO_4 solutions (Fig. 3-(c) and (d)). Figure 4 shows FE-SEM photographs of the surfaces of the specimens anodically oxidized at 180 V in different electrolytes and subsequently soaked in SBF for 3 days. After the soaking in SBF, a dense and uniform bonelike apatite layer was formed on the surfaces of the specimens after the anodic oxidations in H_2SO_4 and Na_2SO_4 solutions within 3 days, whereas no change was observed on the surfaces of the specimens after the anodic oxidations in H_3PO_4 and CH_3COOH solutions.

The present results indicate that the specimens anodically oxidized in H_3PO_4 and CH_3COOH solutions show no apatite-forming ability, whereas those in H_2SO_4 and Na_2SO_4 solutions show excellent apatite-forming ability in SBF. It was found from TF-XRD patterns of the surfaces of the specimens that the rutile precipitated on their surfaces had (101) plane parallel to the substrate (Fig. 1). This indicates that not only anatase but also (101) plane of rutile can show high apatite-forming ability in SBF.

Conclusions

Porous titania layers which were mainly composed of rutile and anatase were formed on the surface of specimens by the anodic oxidation in H_2SO_4 and Na_2SO_4 solutions. Specimens anodically oxidized in H_2SO_4 and Na_2SO_4 solutions formed bonelike apatite on their surfaces in SBF within 3 days. It is expected that the present anodic oxidation is useful for obtaining bioactive metals for bone-repairing.

References

[1] H.-M. Kim, F. Miyaji, T. Kokubo and T. Nakamura: J. Biomed. Mater. Res., 32 (1996) 409-417.
[2] W.-Q. Yan, T. Nakamura, M. Kobayashi, H.-M. Kim, F. Miyaji and T. Kokubo: J. Biomed. Mater. Res., 37(1997), 267-275.
[3] M. Uchida, H.-M. Kim, T. Kokubo, S. Fujibayashi and T. Nakamura: J. Biomed. Mater. Res., 64A(2003), 164-170
[4] T. Kokubo, H. Kushitani, S. Sakka, T. Kitsugi and T. Yamamuro: J. Biomed. Mater. Res., 24 (1990) 721-734.

Key Engineering Materials Vols. 254-256(2004) pp. 463-466
online at http://www.scientific.net
© 2004 Trans Tech Publications, Switzerland

Photocatalytic Bactericidal Effect of TiO$_2$ Thin Films Produced by Cathodic Arc Deposition Method

B. Kepenek[1], U. Ö. Ş. Seker[2], A. F. Cakir[1], M. Ürgen[1], C. Tamerler[2]

[1] İstanbul Technical University, Faculty of Chemical and Metallurgical Eng., Dept. of Metallurgical & Materials Eng., 34469, Maslak-Istanbul / TURKEY. kepenekba@itu.edu.tr

[2] İstanbul Technical University, Faculty of Science and Letters, Dept. of Molecular Biology and Genetics, 34469, Maslak-Istanbul / TURKEY.

Keywords: Titanium dioxide, anatase, bactericidal effect, photocatalytic activity, cathodic arc deposition

Abstract. Photocatalytic TiO$_2$ thin films were successfully obtained by cathodic arc deposition technique. The films were deposited on stainless steel (AISI 304) sheets at different substrate temperatures using a titanium metal target in 100 % O$_2$ atmosphere. Applied magnetic focusing provided anatase crystal structure at all substrate temperatures. The bactericidal effect of the films was monitored by determining the survival ratio of *Escherichia coli* K12 strain on the films after UV treatment.

Introduction

TiO$_2$ thin films are commonly used in various fields, especially on purification and treatment of water and air, since they exhibit strong oxidation power to breakdown organic compounds when exposed to ultraviolet light; either from natural sunlight or commercial lamps. Thus photocatalysis is expected to find applications in materials possessing antibacterial functions.

Conventional disinfection methods such as chlorination and ozonation produce undesired byproducts like trihalomethanes (THM), and other carcinogenic byproducts. UV disinfection is not completely effective alone, due to long processing times, and the long treatment times may result in mutations of the undesired organisms which makes them more resistant to UV [1].

Different techniques have been employed to produce TiO$_2$ thin films such as sol-gel [2], sputtering [3], CVD [4], atomic layer deposition [5], plasma immersion ion implantation [6], etc. However, no work has been reported on the production of TiO$_2$ thin films on stainless steel substrates via cathodic arc deposition method.

Since anatase exhibits the highest photocatalytic activity among the three distinct TiO$_2$ phases; anatase, rutile and brookite, the aim of this study was to produce TiO$_2$ thin films in anatase form on stainless steel substrates by cathodic arc deposition process and to monitor their bactericidal effect as a result of its photocatalytic properties.

Experimental

Stainless steel (AISI 304) sheets having dimensions of 25 mm x 40 mm x 1 mm were used as substrate materials. The surfaces of the specimens were prepared by standard metallographical techniques. Before coating, the samples were degreased ultrasonically in hot alkaline baths and then rinsed in deionized water and dried in propanol.

The TiO$_2$ films were deposited in 100 % O$_2$ environment using a pure Ti cathode. The O$_2$ gas flow was adjusted by independent mass flow controllers and the total pressure was maintained at 7.5 mTorr by setting the total gas-flow rate to 60 sccm (in full) during the deposition process.

Coating was conducted in an Arc-PVD unit (Model NVT-12, Novatech-SIE, Moscow). Experimental conditions were as follows: arc current, 50 A; cathode substrate distance, 250 mm; deposition time, 6 min and 10 min. Samples were produced with magnetic focusing having a magnitude of 7.92 mT.

The crystalline structures of films were evaluated by Phillips PW 3710 Grazing Angle X-Ray Diffractometer with a CuKα source under an applied voltage of 40 kV and a current of 40 mA. Scanning rate was 0.02 degree/sec. Because coating thicknesses were in the range of nanometers, grazing incidence was chosen as $\theta = 0.5$ ° and scan range $2\theta = 10\text{-}90$ °.

Cultures of *Escherichia coli* K12 were grown aerobically in 20 ml nutrient broth at 37 °C on a rotary shaker (200 rpm, Φ15 cm) for 15 hours. After incubation overnight *E.coli* cells were harvested by centrifugation at 7000 x g for 5 minutes at 4 °C, the supernatant was discarded and the cell pellet was resuspended in 25 ml PBS. Serial dilutions were done to achieve 10^4 and 10^5 colony forming units (CFU) ml^{-1}.

5 ml of the *E. coli* cell culture was pipetted on the TiO_2-coated metal surface placed in a cubical dish. The coated samples were illuminated by a UV source with the center wavelength of 365 nm positioned 5 cm above the coated metal surface. An amount of 100 μl aliquots of serially diluted suspensions was spread on the nutrient agar plates at 10 minutes intervals. The inoculated nutrient agar plates were incubated at 37°C for 18 h. The experiment was carried in duplicates. After incubation the survival cell numbers were determined counting the colonies on agar plates. The survival ratios of the cells were calculated.

Results and discussion

Structural Characterization. XRD patterns of crystal structures belonging to the films prepared with magnetic focusing on stainless steel substrates are shown in Fig. 1. Films prepared with magnetic focusing at all temperatures exhibited anatase structure. Magnetic focusing results in high concentration of plasma per unit area. Thus, deposition rate of the film increases. However, this increase in the deposition rate affects nucleation process and causes formation of defective structures. Consequently, the metastable phase, anatase is deposited.

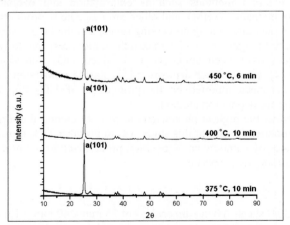

Fig. 1 XRD patterns of prepared films.

Bactericidal effect of films. Bactericidal effect of the coatings was monitored through the survival ratios of the *E. coli* K12 strains. *E .coli* K12 strains possess recA gene which is responsible for the UV resistance [7]. In order to examine the bactericidal effect, TiO_2 coated stainless steel samples produced at different substrate temperatures and deposition time – 375 °C (10 minutes), 400 °C (10 minutes), 450 °C (6 minutes) - were used. The death of the cells could be clearly observed on the Petri dishes, which were inoculated after different UV processing times on a TiO_2 coated stainless steel (Fig. 2). The response of uncoated stainless steel to UV irradiation was also monitored as a control group.

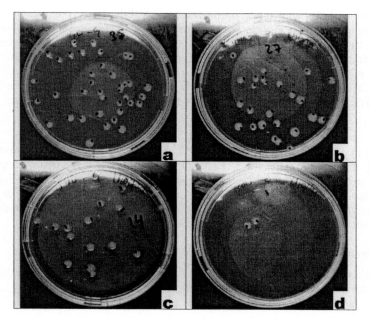

Fig. 2 The bactericidal effect of coating due to their photocatalytic activities caused a gradual decrease in the total cell number. a) 10 min UV treatment, b) 20 min UV treatment, c) 40 min UV treatment, d) 80 min UV treatment.

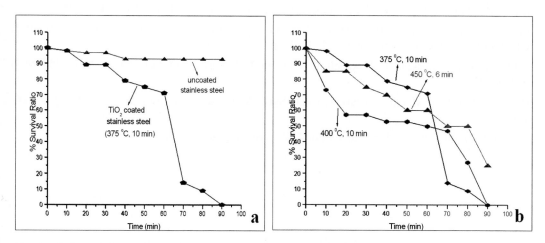

Fig. 3, a) Comparison of the bactericidal activities of the TiO_2 coated and uncoated stainless steel under UV irradiation, b) Effect of thin films on *E. coli* K12 cells survival as a function of time.

In Fig. 3a the bactericidal effects of the TiO_2 coated material and uncoated stainless steel are given. It is clearly seen that the death rate on uncoated stainless steel was low with respect to TiO_2 coated stainless steel. It is concluded that TiO_2 coating creates a bactericidal effect on the stainless steel. The bactericidal effect of each coating is represented in Fig. 3b. The survival ratios reflected that films prepared at substrate temperatures 375 °C and 400 °C showed close photocatalytic activities (deposition time; 10 minutes). Therefore, it can be stressed that the substrate temperature during the deposition has no remarkable effect on bactericidal effect of the films. The film prepared at 450 °C substrate temperature was coated only for 6 minutes hence thinner coating was obtained. Its lower activity may be due to the lower thickness of the coating hence its lower surface area because of its porous columnar structure. As a common point for all of the coatings, after 60th minutes there is a sharp decrease in the total cell number. It is clear that the coatings deposited for 10 minutes have lower time to kill all the bacteria in the suspension. The total killing time for these two coatings is 90 minutes and that for the coating deposited for 6 minutes is 20 min longer or more.

Conclusion

TiO_2 thin films in anatase crystal structure were deposited on stainless steel substrates irrespective of the substrate temperatures. Results of bactericidal effect test showed that thickness of coatings hence surface area is an important factor determining the activity of TiO_2 coatings. Due to the photocatalytic activity of the coatings a gradual decrease in the initial cell number was observed. Surface characteristics affected the bactericidal effect of the films. TiO_2 thin films with high photocatalytic activity enable safer and rapid disinfection.

Acknowledgements

The authors acknowledge the support given by T.R. Prime Ministry the State Planning Organization through the Advanced Technologies in Engineering project.

References

[1] R. Sommer, M. Lhotsky, T.Haider, A.Kabaj, J.Food Protect, Vol. 63 (2000), p. 1015.

[2] J. Yu, X. Zhao, Q. Zhao, Materials Chemistry and Physics, Vol. 69 (2001), p. 25-29.

[3] T. M. Wang, S. K. Zheng, W. C. Hao, C. Wang, Surf. and Coat. Tech., 155, 141 – 145

[4] C. J. Taylor, D.C. Gilmer, D.G. Colombo, G.D. Wilk, S.A. Campbell, J. Roberts, W.L. Gladfelter, J. Am. Chem. Soc., Vol. 121 (1999), p. 5220-5229.

[5] M. Ritala, M. Leskela, E. Nykanen, P. Soininen, L. Niinisto, Thin Solid Films, Vol. 225 (1993), p. 288-295.

[6] G.Thorwarth, S. Mandl, B. Rauschenbach, Surf. And Coat. Tech., Vol. 128-129 (2000), p. 116-120.

[7] J.G. Peak, M.J. Peak, Photochem. Photobiol., Vol. 36 (1982), p. 103

Key Engineering Materials Vols. 254-256(2004) pp. 467-470
online at http://www.scientific.net
© *2004 Trans Tech Publications, Switzerland*

Bonelike Apatite Formation on Synthetic Organic Polymers Coated with TiO$_2$

F. Balas[1,2], M. Kawashita[2], H. M. Kim[3], C. Ohtsuki[1], T. Kokubo[4], T. Nakamura[5]

[1] Graduate School of Material Science, Nara Institute of Science and Technology (NAIST), Nara, Japan, balas@ms.aist-nara.ac.jp

[2] Graduate School of Engineering, Kyoto University, Kyoto, Japan

[3] Department of Ceramic Engineering, Yonsei University, Seoul, South Korea

[4] Research Institute for Science and Technology, Chubu University, Kasugai, Japan

[5] Graduate School of Medicine, Kyoto University, Kyoto, Japan

Keywords: bioactive materials, coatings, composites, in vitro, polymers, surface modification

Abstract. Synthetic organic polymers were modified with silane-coupling agents, subsequently coated with hydrated titania abundant in Ti-OH group, to produce novel materials with bone-bonding property, i.e. bioactivity. Formation of bonelike apatite layer on the modified polymers was examined using a simulated body fluid (SBF). The apatite formation in SBF was induced on the surface of silane-coupled and TiO$_2$-coated substrates, within two days, irrespective of the types of polymers among PET, EVOH, Nylon 6 and Nylon 6/6. Consequently, these surface modifications provide organic polymers with the inductive effect of apatite nucleation in SBF, regardless of their structures or compositions.

Introduction

Bone is a natural composite in which inorganic nanometric apatite crystals are deposited on organic collagen fibers fabricating into a three dimensional structure [1]. Composite materials consisting of apatite nanocrystals and polymer fibers that mimicks bone structure attracted much attention, because they may open new avenues for novel bone-reparing materials. Such nanoapatite-polymer fiber composite could show bone-bonding ability through the apatite layer deposited on surface, as well as similar mechanical properties of those of human bones. Biomimetic processing utlizing an acellular simulated body fluid (SBF) was applied to produce a composite that has nanoapatite crystals on polymer substrates. On this process, the formation of apatite layer is triggered by heterogenous nucelation of apatite on the polymer surface in SBF. In the last years, it was revealed that several functional groups such as Si-OH, Ti-OH, Ta-OH, Nb-OH or Zr-OH [2] are able to induce the apatite nucleation in SBF. Among them, Ti-OH shows high apatite-forming ability when it has arranged structure of anatase [3]. Therefore, hydrated titania gel with anatase-like structure has a benefit on induction of bonelike apatite formation. Furthermore, the formation of apatite on the surface of hydrated titania gel is easier to control, since its solubility is quite low in SBF and body environment.

Apatite formation on ethylene-*co*-vinyl alcohol (EVOH) substrate in SBF was sucessfully achieved by modification of its surface by silane-coupling and subsequent TiO$_2$-coating [4]. This method has the chance to be spreaded over several kinds of organic polymer substrates for medical applications. In this study, the surface modification was applied to four different types of organic substrates to investigate the effect of the polymer substrates on the apatite formation in SBF.

Experimental Specimen preparation

Four kinds of organic polymers given in Table 1 were used in the present study. Samples were obtained as plates after compression moulding of pellets of the polymers, under a pressure of 9 MPa for 10 minutes at several temperatures (T$_P$). Plates were cut and abraded with #400 diamond pad to obtain polished plates of 10x10x1 mm^3. Specimens were ultrasonically washed in 2-propanol and dried at 40°C.

Table 1. Data of the polymer substrates used in this study

Polymer	Code	T_P [°C]	Manufacturer
Poly(ethylene terephatalate)	**PET**	270	Toyoboseki
Ethylene-*co*-vinyl alcohol (32% ethylene)	**EVOH**	210	Kuraray
Poly(6-caprolactam)	**Nylon 6**	230	Scientific Polymer Products
Poly(hexamethylene adipamide)	**Nylon 6/6**	230	Scientific Polymer Products

Silane coupling. PET and EVOH were silane-coupled by reaction with 3-isocyanatopropyl triethoxysilane in toluene at 50°C for 5 hours, using dibutyltin dilaurate as catalyst [5]. Nylon 6 and Nylon 6/6 were treated with 3-glycidoxypropyl trimethoxysilane in *N,N*-dimethyl acetamide at 50°C for 5 days in presence of triethylene diamine [6]. After reactions, specimens were soaked in a 0.05M-HCl solution at 40°C for 12 hours, washed with water and vacuum dried at 25°C for 24 hours.

TiO$_2$-coating. The titania (TiO$_2$) precursor solution was prepared according to a previous report [4]. A solution with molar ratio of Ti(OC$_3$H$_7$)$_4$/H$_2$O/CH$_3$CH$_2$OH/HNO$_3$ = 1.0/1.0/9.25/0.1, was obtained after mixing stoichiometric amounts of the above-mentioned reagents. Specimens were immersed in the solution at 25°C for 24 hours and withdrawn at 2 cm/min, dried in air at 80°C for 10 minutes, soaked in TiO$_2$ solution again and withdrawn at same rate. This process was performed once for every specimen and then dried in oven at 80°C for 24 hours. Finally, samples were immersed in 0.1M-HCl for 8 days at 80°C, followed by washing with distilled water.

SBF Soaking. Specimens were immersed in 30 mL of SBF [7]. SBF (Na$^+$ 142.0, K$^+$ 5.0, Ca^{2+} 2.5, Mg^{2+} 1.5, Cl$^-$ 147.8, HCO$_3^-$ 4.2, HPO$_4^{2-}$ 1.0, SO$_4^{2-}$ 0.5 mol·m^{-3}) has ion concentrations and pH nearly equal to those of human blood plasma at 36.5°C. After soaking for 2 and 7 days, specimens were removed from the solution and washed with distilled water, followed by drying in air at 40°C for 24 hours.

Specimen analysis. Specimens were analyzed by thin-film X-ray diffraction (TF-XRD, Rigaku RINT2500), field-emission scanning electron microscopy (FE-SEM, Hitachi S-4700), energy dispersive X-ray spectroscopy (EDX, Horiba EMAX) and X-ray photoelectron spectroscopy (XPS, Shimadzu ESCA-3200). Adhesion of deposited TiO$_2$ and apatite layers to organic substrates was checked after detaching test using Scotch® tape.

Results and discussion

Figure 1 shows FE-SEM images of surface of specimens before and after soaking in SBF for 2 and 7 days. Deposition of spheres consisting of fine scale-like particles are observed on all the examined specimens silane-coupled and TiO$_2$-coated specimens. The deposited particles consist of bonelike apatite, as confirmed from the results of TF-XRD shown on Figure 2. Typical peaks at 26° and 32° assigned to bonelike apatite are detected for the specimens soaked in SBF for 7 days, while these peaks are not distinct for the EVOH specimens soaked of 2 days. The analysis by EDX also confirmed that the surface of the specimens showed compositions rich in calcium and phosphorus, after soaking in SBF for 2 and 7 days. This indicates that the formed particles on the surface of the specimens are calcium phosphate compounds. All these results support that the bonelike apatite can formed on the surface of the modified polymer in SBF, regardless of the substrates compositions. Deposition of the apatite crystals was initiated on the surface within 2 days after soaking in SBF.

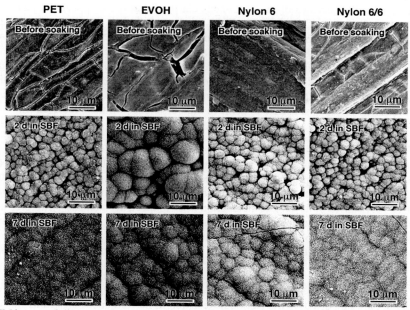

Fig. 1. FE-SEM images of silane-coupled and TiO₂-coated polymer specimens before and after soaking in SBF for 2 and 7 days

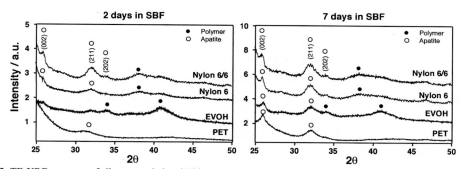

Fig. 2. TF-XRD patterns of silane-coupled and TiO₂-coated polymer specimens after soaking in SBF for 2 and 7 days

Although the modified substrate can form apatite layer on their sufaces in SBF, formation of layers seems to depend on the employed substrate, as seen on FE-SEM images of the surface of the specmens soaked in SBF for 2 days. The apatite layer is formed by aggregates of spheres with sizes ranging from 5 to 8 μm for PET, Nylon 6 and Nylon 6/6. In contrast, the apatite layer on EVOH is formed by larger spheroidal aggregates (10 to 15 μm) that are spreaded on all the surface. As the immersion time increases, the deposited apatite layers become thicker and denser on all the examined specimens, since the FT-XRD peaks became clear to show the existence of apatite phase after soaking for 7 days. It is assumed from the results that the number of nucleation sites of the apatite on the modified substrate of EVOH is smaller than those of PET, Nylon 6 and Nylon 6/6, since less nucleation sites on the substrate allow larger area for growing the apatite crystals.

FE-SEM images and EDX spectra obtained after the peeling-off tests using Scotch® tape demonstrate a good integration of apatite layers to the organic substrates regardless of their chemical compositions and structures. The coated TiO₂ may have enough interaction not only with the substrate but also to the apatite layer.

XPS analysis of the surfaces of the specimens clarified that there appear the peaks assigned to titanium core-level spectra (Ti2p) after the TiO_2-coating, while it is hard to detect the existence of titania phase by TF-XRD. XPS of Ti2p and oxygen core-levels (O1s) show binding energies associated with those found for anatase [8], for all the studied specimens. This means that all the surface of the polymer could be covered with Ti-OH groups with anatase-like structure. The concentration of the anatase-like structure is suggested to be lower on the surface of EVOH specimen than PET, Nylon 6 and Nylon 6/6. Optimization of the thickness and concentration of anatase structure is required for each substrate to achieve high efficiency on their fabrication.

Conclusions

Titanium oxide (TiO_2) coatings could be achieved on silane-coupled PET, EVOH, Nylon 6 and Nylon 6/6 after a sol-gel dipping technique and hot HCl treatment. Thus TiO_2-coated polymer substrates induced the nucleation and growth of a bonelike apatite layer on specimen surfaces after immersion in SBF up to 7 days. The nucleation was induced by anatase structure, although the concentration of anatase structure may be affected by the types of the organic substates. Surface modification with silane-coupling and subsequent TiO_2-coating leads novel methods for production of apatite nanocrystals-organic polymer, that can be applied to various kinds of organic substrates.

Acknowledgements

This work is in part supported by the National Research & Development Programs for Medical and Welfare apparatus entrusted from the New Energy and Industrial Technology Development Organization (NEDO).

References

[1] J.B. Park and R.S. Lakes: Biomaterials (Plenum Press, New York 1992)

[2] T. Kokubo, H.M. Kim and M. Kawashita: Biomaterials, 24 (2003) 2161.

[3] M. Uchida, H.-M. Kim, T. Kokubo, S. Fujibayashi and T. Nakamura: J. Biomed. Mater. Res. 64A (2003) 164.

[4] A. Oyane, M. Kawashita, T. Kokubo, M. Minoda, T. Miyamoto and T. Nakamura: J. Ceram. Soc. Japan 110 (2002) 248.

[5] J. Wen and G.L. Wilkes: Polym. Bull. 37 (1996) 51.

[6] Y.W. Park and J.E. Mark: Colloid. Polym. Sci. 278 (2000) 665.

[7] T. Kokubo, H. Kushitani, S. Sakka, T. Kitsugi and T. Yamamuro: J. Biomed. Mater. Res. 24 (1990) 721.

[8] E.C. Plappert, K.H. Dahmen, R. Hauert and K.H. Ernst: Chem. Vapor Dep. 5 (1995) 79.

IV.   COMPOSITES AND HYBRIDS

Key Engineering Materials Vols. 254-256(2004) pp. 473-476
online at http://www.scientific.net
© *2004 Trans Tech Publications, Switzerland*

Organic-Inorganic Hybrids of Collagen or Biodegradable Polymers with Hydroxyapatite

M.F. Hsieh[1], R.J. Chung[2], T.J. Hsu[3], L.H. Perng[4] and T.S. Chin[5]

[1] M.F. Hsieh, Biomedical Engineering Center, Industrial Technology Research Institute, 195, Sec.4, Chung Hsing Road, Chu Tung, Hsinchu 31041, Taiwan, Republic of China, [1] MFHsieh@ITRI.ORG.TW

[2-3,5] Department of Materials Science and Engineering, National Tsing Hua University, Hsinchu 30043, Taiwan, Republic of China, [2] d907501@oz.nthu.edu.tw, [3] g913536@oz.nthu.edu.tw, [5] tschin@mse.nthu.edu.tw

[4] L.H. Perng, Department of Chemical Engineering, Cheng Shiu University, Niauu-Sung, Kaohsiung 83305, Taiwan, Republic of China,[4] cd7543@ms41.hinet.net

[5] Affiliated also with National Lien-Ho Institute of Technology, Miaoli, 36000, Taiwan, Republic of China

Keywords: Hydroxyapatite, Biomimetic, PLGA, Collagen, Organic-inorganic hybrids.

Abstract. In this research we utilize three polymers, including collagen, diblock PEG-PLGA copolymer and triblock PEG-PLGA-PEG copolymer for biomimetic hydroxyapatite (HAp) preparation. Average crystallite sizes of precipitated HAp from polymer solutions were estimated from XRD patterns to be ca. 23 nm for those in collagen-HAp hybrids and 30 nm for those without polymer addition. Scanning electron micrographs showed that the organic macromolecules induced HAp with regular plate-like shape and nano-sized structure. Thermogravimetric analyses showed 2-3 wt% polymer content in the hybrids. Solid state NMR revealed polymer incorporation in HAp crystal. The hybrids containing these biodegradable polymers can be processed to various morphologies for tissue engineering applications.

Introduction

Biomimetic syntheses of organic-inorganic hybrid draw a lot of research interests. Among those novel attempts, self-assembly of hydroxyapatite (HAp) in calcium and phosphate-containing solutions gains a lot of attention because organic-inorganic HAp hybrids (org-HAp) can be tailor-made. These hybrids are categorized as biomimetic HAp or bio-inspired HAp, since preparation methods are adopted from natural development of mammalian bones and teeth. Organic templates, ranging from macromolecules of various shapes, linear, branched or dendrite[1-3], to simple molecules like citric acid [4], affect precipitation of HAp crystals. In previous reports, HAp crystals with a c-axis-preferred orientation were obtained using collagen and chondroitin sulfate fibrils [5]. These macromolecular fibrils provide their axial space for nucleation of HAp when local calcium and phosphate ion concentrations exceed solubility products of HAp. It may be noted that HAp has been used as gene delivery vehicle [6]. Positively charged Ca^{2+} sited on HAp surface was proposed as binding sites of phosphate linkages on DNA chains. However, other than electrostatic interaction, research on how the van der Waal (VDW) interaction affects HAp precipitation is still in its infancy. But some literature, focused on surface energy, have coined ways to explore this VDW interaction [7]. The aim of this study is to investigate org-HAp formation in the presence of collagen or biodegrabale PEG-PLGA-PEG copolymer, that is a non-ionic, hydrophilic PEG-hydrophobic PLGA alternating triblock polyester, and to discuss the mechanism of org-HAp formation based on structural analyses of org-HAp.

Experimental

Preparation of polymers:

The three polymers used in the present study were PEG-PLGA, PEG-PLGA-PEG and collagen. The synthesis of PEG-PLGA-PEG copolymer (molecular weight=3.9KDa) was described elsewhere

[8]. In brief, a ring opening polymerization of glycolide (GA) and lactide (LA) by monomethoxy poly(ethylene glycol), PEG, followed by a coupling reaction, was used to produce a tribolck poly(ethylene glycol)-poly(DL-lactic acid-co-glycolic acid)-poly(ethylene glycol). Feeding molar ratio of LA/GA was 78/22. Reagents used in this study were reagent grade and were used without further purification. Collagen was extracted from porcine skin.

Preparation of HAp:

Ca(NO$_3$)$_2$·4H$_2$O, (NH$_4$)$_2$HPO$_4$ and polymers were dissolved separately in acetic acid, and then mixed in a 500mL flask resulting a solution of Ca/P molar ratio 1.67, that corresponds to the ratio of stoichiometric HAp. 10% NH$_4$OH was added slowly through a peristaltic pump, until pH 7.5 was reached and steady. The solutions were kept still at 37°C for 48 h. The precipitates were centrifuged at 5000 rpm, washed with distilled water and then freeze dried. The resultant powders containing different polymers were abbrevatied as col-HAp, di-HAp and tri-HAp. HAp with no polymer addition was also precipitated for comparision. Thermogravimetry (TGA), field emission scanning electron microscopy (FE-SEM), energy dispersive x-rays spectroscopy (EDX), x-ray diffraction (XRD), and magic angle spinning solid state NMR (MAS SS-NMR) were used for structural analyses of these powders.

Results and Discussion

Thermal stability of precipated HAp powders:

TGA analyses showed 2-3% weight loss difference between org-HAp and HAp. Weight loss of the org-HAp from 150°C to 800°C was 6-7%, that can be mainly attributed to thermal decomposition of the polymer, while less than 4% ocurred for HAp. It should be noted that reasonable weight loss can occur even when pure HAp is heated above 600°C. The results evidence the incorporation of biopoymers into the HAp powders, which is more noticeable for triblock PEG-PLGA-PEG polymer than for the other two polymers.

FE-SEM morphology of precipated HAp powders:

Fig. 1 presents the FE-SEM micrographs of the powders. The micrographs of org-HAp display nano-grains in a dense and regular form. However, in the absence of polymer, irregular HAp nano-crystals, with surface feature of whiskers and irregular thin plates with average radius less than 100nm, were obtained. Col-HAp has extended thin sheets. Morphology of di-HAp resembles short sticks also less than 100nm. Tri-HAp has extended sheets while is composed of short sticks. Ca/P ratio measured by EDX equiped along with SEM showed Ca/P ratios to be around 1.67 for all powders.

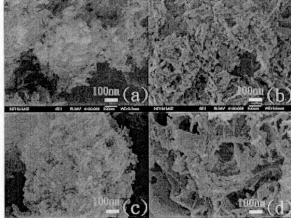

Fig. 1. FE-SEM micrographs of synthesized powders (a) HAp (b) col-HAp (c) di-HAp (d) tri-Hap

XRD analyses of HAp precipitated powders:

The mean crystallite sizes were calculated using the Scherrer equation, $d_{002}=K./B_{002}\cos..$ d_{002} is the mean crystallite size along the [002] direction in Å, K is the shape factor, B is the broadening of the diffraction line measured at half of its maximum intensity, λ is the wavelength of X-ray and θ is the Bragg's diffraction angle. This calculation ignores line broadening due to strain in the crystallites. For the shape factor K, a widely accepted value, i.e., $K=0.9$, has been chosen. Powder XRD patterns of HAp prepared were shown in Fig. 2 and Table 1 lists the calculated crystallite sizes of org-HAp. Precipitated powders are all less than 30nm, and addition of polymer apparently decrease the crystalite sizes by 10.5% to 21.9%. XRD patterns of both org-HAp and HAp exhibit broadened characteristic peaks of hydroxyapatite phase, further indicating partial crystallinity of HAp.

Table 1. Crystallite sizes of the precipitates.

	Crystallite size (nm)	a (Å)	c (Å)
HAp	30.33	9.41	6.86
col-HAp	23.33	9.38	6.84
di-HAp	27.14	9.27	6.83
tri-HAp	25.48	9.26	6.81

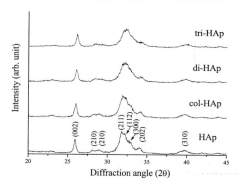

Fig. 2. XRD diffraction patterns of org-HAp and HAp powders.

^{13}C and ^{31}P MAS SS-NMR examination of precipitated HAp powders:

Fig. 3 shows the ^{13}C MAS SS-NMR spectra of the precipated powders. The evidence of the polymer incorporation in HAp powders can be confirmed in Fig. 3, because there is no significant carbon peak in Fig. 3a (HAp) but Figs. 3b-d reveal major ^{13}C peaks of PEG-PLGA, PEG-PLGA-PEG and collagen. Chemical shifts of Figs. 3b-d are assigned according to literature [9-10].^{31}P MAS SS-NMR spectra (not shown here) displayed two phosphate species for all HAp precipitates; the peak located at 2.8 ppm is assigned to HAp and a weak peak located at 0.9 ppm as a shoulder may explain the interaction of calcium phosphate and polymers. To manifest this weak interaction, application of $^{1}H\rightarrow^{31}P$ cross polarization MAS SS-NMR technique to the precipitated HAp powders resulted in enhanced intensity of the weak peak. Accordingly, the local interaction between HAp phase and polymer may happen in the present study.

PEG-PLGA-PEG is a linear polymer with alternating polarity. In aqueous solution, this polymer can self-organize in stable micelle forms of a few hundred manometers [11]. Hence, self-assembly of HAp using this template polymer can be envisaged by a decrease in surface energy of micelle once supersaturated Ca^{2+} and PO_4^{3-} ions being selectively absorbed by functional groups, such as hydrophilic PEG segment. However, we do not exclude the possibility of PLGA segments as ion absorption sites. It is also possible that carbonyl group (C=O), having partly ionic nature may attract Ca^{2+} ions and initiate HAp precipitation [12]. So the inner and outer spaces of micelle can accommodate HAp nuclei. It is interesting to note that morphology of di-HAp and tri-HAp was transformed from plate-like to oval shape when acetone or ethanol was used as dispersion media during SEM sample preparation. The selective dissolution for polymers but not for HAp ceramic implies that one can manipulate microstructure of di-HAp and tri-HAp, and apply this technique to produce tissue engineering scaffolds with controlled structures.

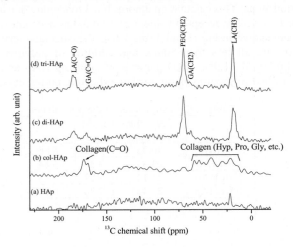

Fig. 3. ^{13}C SS-NMR spectra of the precipated powders: (a) Hap, (b) col-Hap, (c) di-Hap, (d) tri-Hap.

Conclusions

Precipitated HAp particles having regular nano-grain shape have been synthesized in the presence of biodegradable polyesters and collagen. Micelle surface as well as inner space of polyesters in an aqueous solution may serve as nucleation sites for Ca^{2+} and PO_4^{3-} ions. ^{31}P MAS SS-NMR spectra of all HAp precipitated revealed the local interaction between HAp and polymers. Therefore, this local interaction resulted in precipitated HAp crystallite size reduction by 10.5% to 21.9%.

Acknowledgements

The authors are grateful for sponsorship of this research by the National Science Council of the Republic of China NSC 91-2216-E-230-001, and also thank Miss Feng in Instrument Center of National Tsing-Hua University for taking NMR spectra.

References

[1] Q. Liu, L. Ding, F.K. Mante, S.L. Wunder and G.R. Baran: Biomaterials, 23 (2002) 3103-3111.
[2] S.I. Stupp and P.V. Braun, Science, 277 (1997) 1242-1248.
[3] A.J. Khopade, S. Khopade and N.K. Jain, Int. J. Pharm., 241 (2002) 145-154.
[4] R. Tang, M. Hass, W. Wu, S. Gulde anf G.H. Nancollas, J. Colloid. Interf. Sci., 260 (2003) 379-384.
[5] S.H. Rhee, Y. Suetsugu and J. Tanaka, Biomatieral, 22 (2001) 2843-2847.
[6] Y.W. Yang and J.C. Yang, Biomaterials, 18 (1997) 213-217.[7] N. Spanos, V. Deimede and P.G. Koutsoukos, Biomaterials, 23 (2002) 947-953.
[8] B. Jeong, Y.H. Bae and S.W. Kim, J. Biomed. Mater. Res., 50 (2000) 171-177.
[9] N. Dastbaz, D.A. Middleton, A. George and D.G. Reid, Mol. Simul., 22 (1999) 51-55.
[10] D. Huster, J. Schiller and K. Arnold, Magnet. Reson. Med. 48 (2002) 624-632.
[11] B. Jeong, Y.H. Bae and S.W. Kim, Colloids and Surfaces B, 16 (1999) 185-193.
[12] W. Zhang, Z.L. Huang, S.S. Liao and F.Z. Cui, J. Am. Cerm. Soc., 86 (2003) 1052-1054.

Key Engineering Materials Vols. 254-256(2004) pp. 477-480
online at http://www.scientific.net

Effect of the Solvents on the Solution Mixture Derived Polylactide/hydroxyapatite Composites

Yanbao Li[1], Wenjian Weng[1*], Kui Cheng[1], Piyi Du[1], Ge Shen[1], Gaorong Han[1], M. A Lopes[2,3] and J. D. Santos [2,3]

[1]Department of Materials Science and Engineering, Zhejiang University, Hangzhou, 310027, China, wengwj@zju.edu.cn

[2] FEUP - Faculdade de Engenharia da Universidade do Porto, DEMM, Rua Dr. Roberto Frias, 4200-465 Porto, Portugal.

[3] INEB - Instituto de Engenharia Biomédica, Lab. de Biomateriais, Rua do Campo Alegre, 823, 4150-180 Porto, Portugal.

Keyword: hydroxyapatite, polylactide, composite, solvent, dispersion.

Abstract. Polylactide/hydroxyapatite composites were prepared by a solution mixture method. Chloroform and tetrahydrofuran were used as solvents. The effect of the solvents on the polylactide/hydroxyapatite composites was investigated by TEM, mechanical properties, sedimentation time and UV-Vis transmission spectra. The results indicate that the polylactide/hydroxyapatite prepared using tetrahydrofuran as solvent has better dispersion of the hydroxyapatite filler in the matrix and larger bend strength than the one using chloroform as solvent. This is ascribed to the proper interaction between the solvent and hydroxyapatite filler.

Introduction

Hydroxyapatite (HA) has osteoconductivity, leading to direct bonding between HA and the bone. However, HA ceramics has been restricted to applications involving non-load or low load bearing because of its poor mechanical properties, especially their brittleness [1]. An approach to overcoming this limitation is the combination of polymers with the ceramics to form composite materials [1,2]. If the polymer is biodegradable, the composite will be more close to meet the requirements of the current clinical applications. Polylactide/HA composite is one of most potential biomaterials with a biodegradable characteristic [3,4]. The mechanical properties of the polylactide/HA composites can be controlled by both intrinsic properties of the components and microstructures of the composite [3,4]. For the latter, the dispersion and agglomeration of HA fillers in the polylactide matrix have a great influence on mechanical properties.

A solution-mixture method has advantage of involving simple equipment and well-dispersing of inorganic fillers in the matrix, and the method is widely employed to prepare composite biomaterials [2].

In this paper, a solution-mixture method was adopted to prepare the polylactide/HA composites; the effect of solvents on dispersion degree of fillers in the polylactide/HA composite and mechanical properties was investigated.

Materials and methods

HA was prepared by a precipitation method at 40°C using reagent grade $Ca(OH)_2$, $(NH_4)_2HPO_4$ as the starting materials, then the HA precipitates were heat-treated at 800 °C for 3h. The chloroform or tetrahydrofuran was used as a solvent to prepare the composites. HA powders were put into the solvent by ultrasonic agitation to form a homogeneous suspension, and an appropriate amount of

* Corresponding author. Tel.: +86-571-87952324; Fax: +86-571- 87952341. Email address: wengwj@zju.edu.cn

Poly(l-lactide) (PLA, molecular weight 200 kDa) (PLA/HA=90/10 weight ratio) was dissolved in the suspension. Finally PLA/HA/solvent mixture was formed and poured into a glass dish to evaporate the solvent at 40 °C for 48h, and composite sheets were obtained. Bulk composites were prepared by hot-pressing the sheets at 180 °C under 40MPa. The samples for mechanical properties were cut into $2 \times 3 \times 40$ mm^3 by diamond knife. The bend strength was tested using universal test machine (Reger RG2000). The dispersion of HA in the PLA was observed by transmission electron microscopy (TEM). TEM observation was conducted on a JEM200 CX transmission electron microscope, using an accelerating voltage of 160kv. The Archimedes' method was employed to determine the density of the bulk composites. HA powders were shaken and ultrasonically dispersed in chloroform or tetrahydrofuran at solids loadings of 1g/1L. The sedimentation time was defined as the time required for all particles to settle out of solution, leaving a completely clear supernatant. 10ml HA/solvent suspension obtained from the upper part of the HA/solvent suspension after 20h sedimentation was used to characterize the dispersion property of the HA powders in a Cary 100 Bio UV-Vis spectrophotometer.

Results

The HA powders prepared by wet-chemical method have an average size of 92nm and are used as fillers in this work. Fig.1 shows TEM photos of the dispersion of HA powders in chloroform or tetrahydrofuran solvent. HA powders are dispersed homogeneously in the tetrahydrofuran and produce a little agglomeration in the chloroform. Fig.2 shows the dispersion of HA fillers in the PLA matrix obtained through different solvent. The HA fillers are well dispersed in the composite prepared from the PLA/HA/tetrahydrofuran system (refer as COM1) while there is obvious agglomeration in the composite from the PLA/HA/chloroform system (refer as COM2). The relative densities of the sample COM1 and COM2 are 98.5% and 96.3%, respectively. The bend strengths of the sample COM1 and COM2 are 225.14 and 145.28MPa, respectively. Sample COM1 with better dispersion of the fillers has better mechanical properties than Sample COM2.

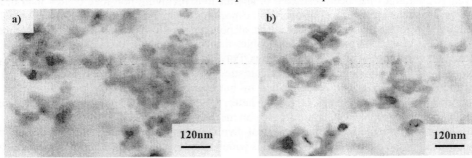

Fig.1 TEM photo of hydroxyapatite dispersed chloroform (a) and THF (b) solutions.

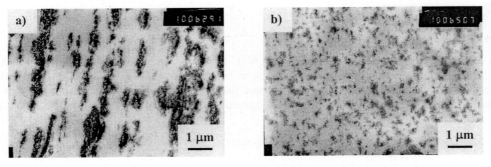

Fig.2 TEM photo of hydroxyapatite dispersed in the PLA matrix
by chloroform (a) and THF (b).

Discussion

The sedimentation time for HA powders dispersed in chloroform (30h) is shorter than that in tetrahydrofuran (70h), and UV-Vis transmittance of the upper part of 20h aged HA/chloroform suspension is higher than that of the HA tetrahydrofuran suspension (fig.3). This result indicates that HA powders are better dispersed in the tetrahydrofuran than in chloroform. The TEM results also demonstrate that the dispersion of HA powders in tetrahydrofuran is better than in chloroform (fig.1).

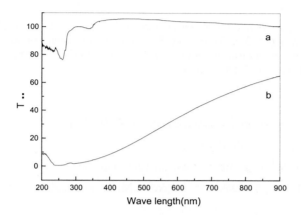

Fig.3 UV-Vis transmission spectra of the hydroxyapatite dispersed in the chloroform (a) and tetrahydrofuran (b) after sedimentation for 20h.

The difference in the dispersion of HA powders in the solvents could result from difference in the compatibility between HA powders and the solvent. The surfaces of HA powders have OH groups, which tend to interact with polar groups in solvent molecules through hydrogen bond [5]. Hence, an increase in polarity of a solvent favors the compatibility between HA powders and the solvent, and as a result, the HA powders can be well dispersed in the solvent. Since there is an oxygen atom in tetrahydrofuran ring molecular structure ($\bigcirc$), the relative dielectric constant of tetrahydrofuran (7.52) is 1.57 times as that of chloroform (4.8)[6], that is, the polarity of tetrahydrofuran is larger than that of chloroform. Therefore, HA powders can be dispersed well in tetrahydrofuran (fig.1b) than in chloroform (fig.1a).

When PLA is added in HA/solvent suspension, PLA becomes soluble molecules with high molecular weights, resulting in increase of viscosity of PLA/HA/solvent mixture. The viscosity increase is greatly helpful to maintain the dispersion status of HA fillers in the solvent, and could be considered that the dispersion of HA fillers in PLA matrix strongly depends on the HA filler dispersion in the solvent after the solvent is removed. Hence, HA fillers in tetrahydrofuran-derived composite have better dispersion than that in chloroform derived composite (Fig.2).

Since the compatibility between HA powders and chloroform is not good, HA powders in chloroform suspension tend to agglomerate. There exist easily voids in the agglomerated HA powders, that could remain in the composite, causing the chloroform derived composite (COM2) to have a lower relative density than tetrahydrofuran derived composite (COM1). The void existence deteriorates the composite microstructure, and weakens the mechanical property with bend strength of 145.28MPa, much lower than tetrahydrofuran-derived composite (COM1) with bend strength of 225.14MPa.

Conclusion

When tetrahydrofuran and chloroform are selected as solvent for preparing PLA/HA composites by a solution mixture method, the solvent have a significant influence on the HA filler dispersion in PLA. Since tetrahydrofuran has a larger polarity than chloroform, tetrahydrofuran is easily compatible with HA fillers, leading to a well dispersed HA suspension. Further, HA fillers can be dispersed well in PLA matrix, and the resulting composite has a good mechanical property.

Acknowledgments

The authors would like to thank the research fund of Science and Technology Department of Zhejiang province, China (11106225) and the FCT for the financial support through the project entitled "*Interactive calcium-phosphate based materials prepared by post-hybridisation and in situ hybridisation*", Ref. PCTI/1999/CTM/35516.

References

[1] L. L. Hench: Journal of American Ceramic Society, 81(1998), p.1705

[2] M. Wang: Biomaterials, 24(2003), p.2133

[3] J. Hao, M. Yuan and X. Deng: Biomaterials, 22 (2001), p.2867

[4] N. Ignjatovic, S. Tomic, M. Dakic,M. Miljkovic,M. Plavsic and D. Uskokovic: Biomaterials, 20(1999) , p.809

[5] D. N. Misra: Langmuir 4(1988) , p.953

[6] J. A. Dean: *Lange's Handbook of Chemistry* (15th Edition) (McGraw-HILL, INC. New York 1999)

Key Engineering Materials Vols. 254-256(2004) pp. 481-484
online at http://www.scientific.net
© 2004 Trans Tech Publications, Switzerland

Bioactive Organic-Inorganic Hybrids Based on CaO - SiO$_2$ Sol-Gel Glasses

A.J. Salinas, J.M. Merino, N. Hijón, A.I. Martín and M. Vallet-Regí

Department of Inorganic and Bioinorganic Chemistry; Faculty of Pharmacy; Complutense University; Madrid. E-28040-Madrid; Spain. salinas@farm.ucm.es

Keywords: organic-inorganic hybrids, sol-gel, CaO–SiO$_2$, PDMS, PVA, *in vitro* bioactivity

Abstract. Bioactive organic-inorganic hybrids, CaO–SiO$_2$–poly(dimethylsiloxane) (PDMS) and CaO–SiO$_2$–poly(vinyl alcohol) (PVA), were synthesized by a sol-gel method and characterized before and after soaking in Kokubo´s Simulated Body Fluid (SBF). The synthesis conditions to obtain monoliths were established in both series. CaO–SiO$_2$–PDMS hybrids presented elastic behavior when TEOS:PDMS molar ratio was under 3.5, and they showed in vitro bioactivity when Ca(NO$_3$)$_2$.4H$_2$O:TEOS molar ratio was close to 0.10. Moreover, the relative amount of H$_2$O used for TEOS hydrolysis also played a role in their mechanical behavior and in vitro bioactivity. In the CaO–SiO$_2$–PVA hybrids, the presence of PVA favored both the synthesis of crack-free monoliths as well as a marked degradation, around to 50 wt-%, for the first 6 h in SBF. All the studied compositions presented in vitro bioactivity. In addition, PVA contents played a major role in the in vitro behavior.

Introduction

The sol-gel method is a suitable way to prepare bioactive and biocompatible glasses which can be used as implant materials [1]. Furthermore, the soft thermal conditions of this method allow to achieve organic-inorganic hybrids with tailored properties for new clinical applications [2-4]. In the present work, CaO–SiO$_2$–PDMS and CaO–SiO$_2$–PVA hybrids were synthesized and characterized. CaO–SiO$_2$ was the inorganic constituent because of the well-known bioactivity of gel glasses in this system [5]. The biocompatible polymers PDMS and PVA were chosen due to the elastic properties of PDMS and the broad uses of PVA in drug delivery systems. The aim of this work was to obtain monolithic organic-inorganic hybrids, showing in vitro bioactivity in SBF [6] based on CaO–SiO$_2$ bioactive sol-gel glasses modified with PDMS and PVA.

Materials and Methods

The starting materials were reagent grade tetraethyl orthosilicate (TEOS) (Aldrich), Ca(NO$_3$)$_2$.4H$_2$O (Fluka), as SiO$_2$ and CaO precursors, and PDMS (Aldrich) and PVA (Aldrich). The studied compositions and the synthesis methods are shown in Table 1 and in Figures 1 and 2. In the CaO–SiO$_2$–PDMS hybrids, together with the compositional variations showed in Table 1, the effect of other synthesis parameters was studied including: kind of PDMS, OH- or CH$_3$- terminated, the catalyst, HNO$_3$ or HCl, and the aging and drying conditions (see Fig.1).

For the syntheses of CaO–SiO$_2$–PVA hybrids, measured amounts of PVA were dissolved in water at 80°C under reflux and pH was set at 0.5 with HNO$_3$. The gel ageing took place at 60°C for 48 h in a polystyrene multi-dwell sealed container and the drying at 60°C for 24 h after removing the sealant tape allowing gas leaking (Fig. 2).

The hybrids were characterized by CHN elemental analysis, TG/DTA, DSC, FTIR, SEM-EDS, Hg porosimetry and N$_2$ adsorption. In vitro assays were carried out in SBF. After soaking up to 7 days, the ionic variations in solution and modifications on the hybrids surfaces were analyzed.

Table 1. Compositions of the organic-inorganic hybrids.

		TEOS:PDMS molar	Ca(NO₃)₂:TEOS molar	H₂O:TEOS molar	TEOS:PVA weight
CaO–SiO₂–PDMS	Series A	6 - 12	0.03 - 0.20	3	-
	Series B	1 - 6	0.05 - 0.33	2-4	-
CaO–SiO₂–PVA		-	0.33	8	72, 144, 288, 722

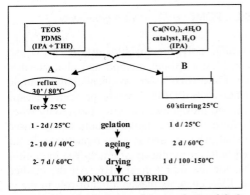

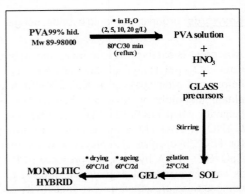

Fig. 1. Schematic depiction of the synthesis of CaO–SiO₂–PDMS hybrids. In A series, lower in PDMS, the hydrolysis conditions were stronger and drying ones softer than in B series, high in PDMS.

Fig. 2. Schematic depiction of the synthesis of CaO–SiO₂–PVA hybrids. (*) Changes with respect to conditions used in the synthesis of CaO–SiO₂ sol-gel glasses [5].

Results and Discussion

CaO–SiO₂–PDMS hybrids. In the hybrids of the Series A, low in PDMS, the hydrolysis conditions were strong, 30 minutes at 80°C under reflux, whereas ageing and drying conditions were soft, 40°C and 60°C respectively (see Fig. 1). Hybrids with around a 30% of porosity and high BET surface area that increased from 489 m²/g, when Ca(NO₃)₂:TEOS = 0.20, to 693 m²/g, when the molar ratio was 0.03, were obtained. These hybrids contained micropores (< 2 nm) and showed two maxima in the pore size distribution at 3.4 and 1.4 nm. An apatite-like layer was formed on the hybrids surface after 7 days in SBF. After the in vitro assays, both cracks and CaCl₂ (chlorine coming from catalyst) on the hybrids with higher CaO contents (Ca(NO₃)₂:TEOS > 0.15) were detected.

For the synthesis of CaO–SiO₂–PDMS hybrids of the Series B, high in PDMS, the hydrolysis conditions were gentle, just one hour stirring at room temperature. However, the samples were dried at temperatures up to 150°C (Fig. 1). For a constant molar ratio H₂O:TEOS:Ca(NO₃)₂ = 4:1:0.05, the porosity increased from 51% to 78%, when the PDMS content increase (TEOS:PDMS decreasing from 6 to 3). For values of this ratio between 3.5 and 2, the hybrids showed high porosity and elasticity to compression, but for the highest PDMS content (TEOS:PDMS = 1), the porosity was 12%. The pore size distribution was also dependent on the PDMS content. For TEOS:PDMS = 2 to 3, most pores were between 6 and 60 μm, when this molar ratio was between 4 and 5 most porosity was under 2.5 μm, and for TEOS:PDMS = 6, under 0.25 μm. On the other hand, as the proportion of water decreased (H₂O:TEOS = 2), the porosity of the hybrids decreased and their rigidity increased. Moreover, for high CaO contents, Ca(NO₃)₂:TEOS > 0.20, some calcium was detected outside the hybrid network. Some hybrids of this series with Ca(NO₃)₂:TEOS ratio between 0.05 and 0.10, showed in vitro bioactivity that was more marked for lower water contents H₂O:TEOS = 2 (see Figures 3 and 4).

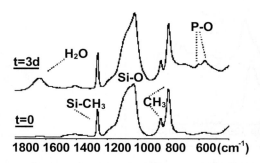

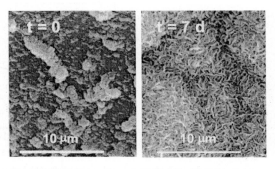

Fig. 3. FTIR spectra of a CaO-SiO$_2$-PDMS hybrid (Ca(NO$_3$)$_2$:TEOS=0.1, TEOS:PDMS=3, H$_2$O:TEOS =2) before and after soaking in SBF.

Fig. 4. SEM micrographs of a CaO-SiO$_2$-PDMS hybrid (Ca(NO$_3$)$_2$ TEOS= 0.1, TEOS:PDMS = 3, H$_2$O:TEOS = 2) before and after soaking in SBF.

CaO–SiO$_2$–PVA hybrids. The presence of PVA greatly favored the formation of transparent monoliths. In Figure 5, a picture of one CaO-SiO$_2$-PVA hybrid is shown. For all the PVA contents studied, transparent, crack-free monoliths were obtained. Thus, the presence of PVA in the hybrids let us to obtain monoliths even drying at atmospheric pressure. The in vitro studies showed the marked solubility of these materials, close to 50% in weight (Fig. 6), and the formation of an apatite-like layer, after 1 day of immersion in SBF (see Figures 7 and 8).

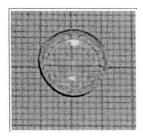

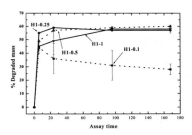

Fig. 5. Photograph of a CaO-SiO$_2$-PVA hybrid (TEOS:PVA= 72).

Fig. 6. Degradation (dissolution) behavior of the CaO-SiO$_2$-PVA hybrids with soaking time in SBF.

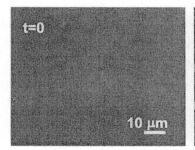
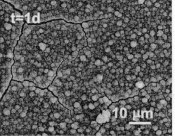

Fig. 7. SEM micrographs of a CaO-SiO$_2$-PVA hybrid (TEOS:PVA= 288) before and after soaking in SBF.

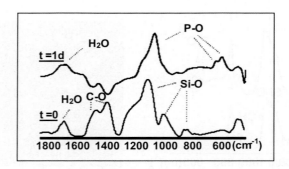

Fig. 8. FTIR spectra of a CaO-SiO_2-PVA hybrid (TEOS:PVA= 144) before and after soaking in SBF.

Conclusions

Homogeneous CaO–SiO_2–PDMS and CaO–SiO_2–PVA hybrids were obtained as monoliths. CaO–SiO_2–PDMS hybrids showed elastic behavior, when TEOS:PDMS molar ratio was close to 3, and in vitro bioactivity, for Ca$(NO_3)_2$:TEOS ratios between 0.05 and 0.10. However, the amount of water used for the TEOS hydrolysis play a major role in both mechanical behavior and in vitro bioactivity of these hybrids. The presence of PVA in the CaO–SiO_2–PVA system makes easy the obtention of crack-free monoliths. CaO–SiO_2–PVA hybrids are also bioactive as the FTIR and SEM results pointed out. Moreover, these hybrids are substantially degraded (up to 50 weigth-% of the mass) when soaked in SBF and this aspect influences the formation process of the apatite-like layer.

Acknowledgements

Financial support of CICYT though MAT2002-0025 and MAT2001-1445-C02.-01, is acknowledged.

References

[1] Vallet-Regí M, Ragel CV, Salinas AJ, Eur J Inorg Chem. 2003, 1029-43.

[2] Tsuru K, Aburatani Y, Yabuta T, Hayakawa S, Ohtsuki C, Osaka A, J. Sol-Gel Sci Tech 2001, 21, 89-96

[3] Kamitahara M, Kawasita M, Miyata N, Kokubo T, J. Mater Sci-Mater M 2002, 13, 015-1020

[4] Pereira APV, Vasconcelos WL, Orefice RL, J. Non-Cryst Solids 2000; 273 180-185.

[5] Martínez A, Izquierdo-Barba I, Vallet-Regi M, Chem Mater 2000, 12, 3080-3088.

[6] Kokubo T, Kushatani H, Sakka S, Kitsugi T, Yamamuro T. J Biomed Mater Res 1990; 24 721-734.

Key Engineering Materials Vols. 254-256(2004) pp. 485-488
online at http://www.scientific.net
© 2004 Trans Tech Publications, Switzerland

Injectable Composite Hydrogels for Orthopaedic Applications. Mechanical and Morphological Analysis

V. Sanginario(1), L. Ambrosio(1), M.P. Ginebra(2) and J.A. Planell(2)

(1)Institute of Composite Biomedical Materials - CNR and Interdisciplinary Resarch Center on Biomaterials, University of Naples "Federico II" Piazzale Tecchio 80, 80125, Napoli, Italy, Tel: +390817682513 Fax: +390817682404 sangival@yahoo.it, ambrosio@unina.it

(2)Department of Materials Engineering and Metallurgy, University Polytechnic of Catalu nya, Avda. Diagonal 647, 08028 Barcelona, Spain, Tel-Fax: +34934016706 maria.pau.ginebra@upc.es

Keywords: composite, bone cement, injectability

Abstract. The large number of orthopaedic procedures performed arthoscopically each year, has lead to a great interest in injectable biodegradable materials for bone regeneration. In this study a composite hydrogel has been prepared with Calcium Phosphate particles (H-cement) reinforcing hydrophilic matrix of Poly-VinylAlcohol (PVA). The composite material has demonstrated good mechanical properties, in the range of spongious natural bone and they can be employed as a substitute material. Mixing the ceramic phase with the organic matrix, an easy-handling injectable paste has been obtained. The composite hydrogel is also a self-setting system: the α-TCP setting and hardening reaction occurs thanks to the water contained in the PVA matrix showing higher mechanical properties than the system prepared with the traditional methodology.

Introduction

Since the 1980's a new class of ceramic biomaterial has been developed, named Calcium Phosphates (CPs). The interest with using CPs is due to their similarity to the mineral component of human bone tissue. CPs are non-toxic, biocompatible and not are recognised as foreign materials in the body, furthermore they exhibit bioactive behaviour [1]. CPs are integrated into bone tissue through the same process which regulates the remodelling of healthy bone by resorption and deposition of bone mineral. The intimate physico-chemical bonds between CPs implants and bone leads to osteointegration [2]-[8]. CPs are also known to support osteoblasts adhesion and proliferation [5], [9], [10]. The main characteristic of these ceramic materials is the possibility to develop a self-setting system which is mouldable during the handling and it hardens with time when in contact with water. This system, referred as Calcium Phosphate Bone Cements (CPBCs), is an injectable bone substitute that can be used to fit irregular defects or to reintegrate damaged tissues. Planell et al.[11] demonstrated that, in order to have an injectable cement with suitable mechanical properties, the ceramic powder (P) has to be mixed with a water solution (L) of NaH2PO4 (2.5%wt) using L/P=0.35. The obtained paste must be kept in the same water solution L to allow the setting and the hardening of the cement.

Setting and hardening mechanism are a microstructural evolution, which changes the rounded α-TCP particles into needle-like Calcium Deficient Hydroxyapatite (CDHA) crystals [11], as shown in reaction (1):

$$3Ca_3(PO4)2+H2O \rightarrow Ca(HPO4)(PO4)5OH \quad (1)$$

The use of these self-setting CPBCs is limited by their low compressive strength, relative to natural bone [1]. As reported in many works [1],[7], when an organic phase is added, mechanical properties improve, reducing cement brittle behaviour. The presence of the organic phase produces a material that is structurally more similar to natural bone tissue. In fact human bone may be considered as a composite in which the organic phase is Collagen and the ceramic phase is Hydroxyapatite (HA). In this work an hydrogel Poly(VinylAlcohol) was used as the organic phase. Hydrogels can be defined as aqueous gel networks, that are able to swell and retain a large amount of water [12]. They have some unique properties that make them highly biocompatible. Firstly, they have a low interfacial tension with the surrounding biological fluids and tissues. This minimises the driving force for protein adsorption and cell adhesion [13],[14]]. Secondly, because of its very high water content, the hydrogel surface is highly hydrophilic and it is able to simulate some properties of natural tissues, thus making it biocompatible [[15],[17]]. Thirdly, the soft nature of hydrogels minimises mechanical and frictional irritation to the surrounding tissues [18]. For our purpose, the PVA main characteristic is its water solubility. In fact, mixing water PVA solution with the ceramic powder, it is possible to have an injectable composite, without using any toxic substances, which hardens in a clinically acceptable period of time. The composite had demonstrated enhanced mechanical properties with respect to the cement alone. It is also possible to demonstrate that the water content of the polymeric solution is enough to allow the hardening of the ceramic phase, in this sense the system is referred as a self-setting system. In order to verify the microstructure during the setting and hardening process, E.S.E.M. micrography was performed. The E.S.E.M. technique was chosen because it allows the microstructure analysis of wet specimens, showing swelling of the polymeric phase. Mechanical tests have been performed on different kinds of specimens in order to compare composite behaviour with the cement plain one.

Materials and Methods

PVA (Aldrich Chemical Company, Inc. Mw=85,000-146,000) was dissolved in distilled water at T= 80°C to obtain a solution 10% by weight of PVA. The PVA/water solution was then mixed for two minutes with the cement powder (H-cement=98%wt α-TCP, 2%wt HA) in order to achieve a paste which was subsequently injected in a Teflon dish to produce cylindrical specimens (D=5 mm, L=10 mm). Two types of specimens were prepared: the first, named Hydrate (id), the composites were immersed in an aqueous solution of NaH_2PO_4 (2.5%wt). The second type, named self setting system (ss), where the composite was kept in a dry state, in order to permit the setting reaction with no external water. The temperature was fixed at 37°C. To compare mechanical properties, H - cem specimens were prepared without the polymer phase, using the appropriate quantity of the aqueous solution of NaH_2PO_4 (2.5%wt) to obtain a L/P ratio at 0.35 [11]. The composite specimens were analysed after seven days of reaction. After drying, the specimens where tested with a dinamometric testing machine (Instron 4020), a standard testing method for compressive properties was applied (ASTM D 695). After testing, the specimens were analysed with E.S.E.M. in the swollen phase.

Results and Discussion.

Table 1 lists the mechanical properties of the cement (100%wt H-cem) compared with the composites (H-cem/PVA 93/7) prepared by both the hydrate technique and self-setting. All specimens were allowed to react for 7 days at 37°C. In Table 1: σ[MPa] is the Maximum Compressive Strength, E [GPa] is the Elastic Modulus, εm[mm/mm] is the maximum strain, εu[mm/mm] is the strain at the break- point and T [MPa] is the toughness. In Figure 1 the representative stress-strain curves of the prepared materials are reported.

Table 1: Compressive properties of H-cem and PVA reinforced with 93%wt of H-cement, 37C, 7 days of reaction.

Specimens	E [GPa]	σ[MPa]	εm[mm/mm]	εu[mm/mm]	T[MPa]
100%wt Hcem(id)	0.8±0.2	23±3	0.02±0.01	0.04±0.01	0.5
93%Hcem+ 7%PVA(id)	1.1±0.2	29±3	0.03±0.01	0.07±0.01	1.3
93%Hcem+ 7%PVA(ss)	1.0±0.3	43±4	0.04±0.01	0.07±0.01	1.9

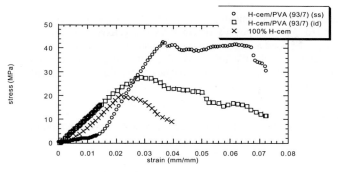

Fig. 1: Compressive properties of H-cem and PVA reinforced with 93%wt of H-cement, 37C, 7 days of reaction.

The results show that both composites have higher mechanical properties than H - cement. A thougher material was produced by the addition (7% by weight) of polymeric phase. In fact the brittle nature of the cement was reduced by the addition of the organic phase. All the composites present higher mechanical properties at compression. With regard to the sample preparation, it is clear that the self-setting specimens showed greater maximum compressive strengths and Elastic Moduli than the Hydrate ones, but toughness remained the same. This is due to an increased interaction between the two phases. In fact the water necessary for the setting and hardening reactions must be released by the polymeric solution. The presence of the negative charge on the PVA chains produces an attractive effect on the neighbouring Ca++ ions released during the setting reaction. PVA enhances the mechanical properties of the cement, acting at microstructural level. Further investigations are necessary to understand if this process influences crystallinity of the inorganic phase.

Figure 2 shows E.S.E.M. micrographs of H- cem/PVA (93/7, id) composite at different times of reaction: 2 hours (a), 1 day (b) and 7 days (c).

(a)   (b)   (c)

Fig. 2: E.S.E.M. analysis of H-cem/PVA (93/7, id) in swollen phase.

After two hours the composite is still mouldable and can be easily injected by hand. In figure 2(a), the α-TCP rounded particles are evident and close to the PVA matrix, the setting process had not begun as no needle-like CDHA crystals were observed. After 1 day of reaction, 2(b), CDHA crystals had formed among the PVA chains. The crystals had a thin needle-like morphology and they formed a diffuse network over the entire surface.

As the reaction proceeded, the crystals grew in both size and number, as shown in figure 2(c), forming agglomerates entrapped and held together by the hydrogel. From the picture it seems that the PVA chains act as preferential sites for the deposition of the ceramic phase.

Conclusion

In order to improve the mechanical properties of CPBCs (H-cem) an organic phase (PVA water solution) was added. The composites H-cem/PVA (93/7 % by weight) showed higher mechanical properties and it can be used a self-setting system, where the setting and hardening processes occurs due to the water retained in the PVA solution. The self-setting system presented higher compressive properties than the hydrate one. E.S.E.M. analysis showed the interactions of the CDHA crystals and the polymer phase.

Acknowledgement

This work has been funded by the 5th European Framework Program, Contract $N° G5RD-CT-2000-00267$, "Disc".

References

[1] R.Mickiewicz, A.M.Mayes, D.Knaack.Polymer Calcium Phosphate cement composites for bone substitute. Wiley Interscience. DOI: 10;1002/jbm 10222.

[2] K.C.Dee, R.Bizos:Mini Review: Proactive biomaterials and bone tissue engeneering.Biotech, Bioeng, 50 (1996) 438-442

[3] P.Weiss, F.Millot, G.Dacalusi,G.Grimandi, In vitro evaluation of a new injectable calcium hosphate material. J.Biomed Mater Res 39 (1998) 660-666.

[4] O.Gauthier,G.Bosco,J.M.Bouler,P.Weiss,E.Aguado,G.Dacalusi:*Kinetic study of bone ingrowth and ceramic resorption associated with the implantation of different injectable bone substitute* Wiley Pubbl. (1999) 28-35

[5] K.de Grooth: Clinical application of calcium phosphate biomaterials, a review. Ceram int 19(1993) 363-366.

[6] P.Ducheyne, Q.QiuBioactive ceramics: the effects of surface reactivity of bone formation and bone cell function. Biomaterials 19 (1999) 2287-2303.

[7] G.Dacalusi: Biphasic calcium phosphate concept applied to artificial bone, implant coating and injectable bone substitute. Biomaterials 19 (1999) 1473-1478.

[8] K.A.Hing, S.M.Best, W.Bonefield: Characterization of porous Hydroxyapatite. Jour of Mat. Sci:Mat in Med 10 (1999) 135-145.

[9] Osteoblast adhesion on biomaterials. K.Anselm . Biomaterials 21 (2000) 667-681

[10] J.E.Davies: In vitro modelling ofthe bone-implant surface. Anat. Rec 245 (1996) 426-445.

[11] E.Fernandez, F.J.Gil, M.P.Ginebra, F.C.M.Driessens, J.A.Planell, S.M.Best. Setting reaction and hardening of an apatitic Calcium Phosphate cement.. J. Dent. Res 76(4) April 1997 905-912 .

[12] M.G.Cascone, S.Maltinti, N.Barbani :Effect of Chitosan and Dextran on the properties of Poly(VinylAlcohol) hydrogels. , Jour. Of Mat Sci.: Mat in Med 10 (1999) 431-435

[13] O. Wichterele and D.Lim, Nature 185 (1960) 117.

[14] M.S.Jhon, J.D.Andreade. Jour Biomed. Mater. Res. 7 (1973) 509

[15] A.S.Hoffman : *Polymers in medicine surgery* edited by R.L.Kronenthal et al. Plenium Press, New York 1975 p.33

[16] D.S.Brook: *Properties of biomaterials in physiological environment*. CRC Press, Boca Ranton, FL 1980

[17] J.D.Andrade :*Hydrogels for medical and related applications*.American Chemical Society,Washington,1976.

[18] B.D.Rotnerin: *Biocompatibility of clinical implant materials*. Vol II, edited by D.F.Williams CRC Press, Boca Ranton,1981

Key Engineering Materials Vols. 254-256(2004) pp. 489-492
online at http://www.scientific.net
© 2004 Trans Tech Publications, Switzerland

Drug Release from Poly(D, L-lactide) / SiO2 Composites

Minna Vaahtio[1,4], Mika Jokinen[1,4], Ari Rosling[2], Pirjo Kortesuo[3], Juha Kiesvaara[3] and Antti Yli-Urpo[5]

[1]Institute of Dentistry & Biomaterials Research, University of Turku, Lemminkäisenkatu 2, FIN-20520 Turku, Finland
minna.vaahtio@utu.fi and mika.jokinen@utu.fi
[2]Laboratory of Polymer Technology, Åbo Akademi University, Piispankatu 8, FIN-20500 Turku, Finland
[3]Orion Corporation, Orion Pharma, P.O. Box 425, FIN-20101 Turku, Finland
[4]Turku Centre for Biomaterials, Itäinen Pitkäkatu 4 B, FIN-20520 Turku, Finland
[5]Department of Prosthetic Dentistry, Institute of Dentistry, University of Turku, Lemminkäisenkatu 2, FIN-20520 Turku, Finland

Keywords: Drug release, silica gel microparticles, toremifene citrate, poly(D,L-lactide)

Abstract. The aim was to develop a biodegradable carrier system for toremifene citrate based on poly(D,L-lactide) and sol-gel derived SiO_2. Two molecular weights of P(D,L-LA) (LMW 130 000 g/mol and HMW 240 000 g/mol) were used in the composites. The release rate of toremifene citrate from P(D,L-LA)/SiO_2 composites was evaluated by in vitro dissolution tests. It was shown that it is possible to prepare a controlled release system of toremifene citrate by adding SiO_2 particles and/or pores to the used polymer. Release of toremifene citrate can be adjusted from 30 days to 6 months.

Introduction

Polylactides have been used as matrix materials for drug release [1], but the clear and sudden changes in structure, e.g., steep decrease in molecular weight caused by enhanced autocatalytic degradation [2], is a problem with respect to the controlled release of drugs. Biodegradable, sol-gel derived SiO_2 is known to be biocombatible and it has been used for controlled drug delivery as such [3]. In this work toremifene citrate (TC) was used as a model drug in polymer/silica gel composites. Toremifene citrate is an antiestrogenic compound that has been used in the systemic treatment of hormone-dependent breast cancer. Local hormone therapy after breast cancer surgery could provide targeted and long-lasting disease control.

The purpose of this study was to develop a controlled release formulation of toremifene citrate using biodegradable delivery systems based on poly(D, L- lactide) and sol-gel derived SiO_2 composites. Also the effect of pore forming CO_2 treatment and addition of a fast dissolving mesoporous SiO_2 (MCM-41) on composites was investigated. Furthermore, the effect of the molecular weight of poly(D,L- lactide) on release rate of toremifene citrate was studied.

Materials and Methods

Silica sol was prepared in a mole ratio of TEOS: H_2O: HCl, 1.0: 14.2: 0.0096. Toremifene citrate (Orion Pharma Ltd, Turku, Finland) was added into the clear hydrolysed silica sol at room temperature. The pH of the sol was adjusted to 2.4 before spray drying with a mini spray dryer (B-191, Büchi Labortechnik AG, Switzerland). The theoretical drug concentration in the silica sol was 2 wt.%, corresponding to 13,4 wt.% drug in spray dried microparticles (MP). Pore structure of calcified, TEOS-derived MCM-41-type SiO_2 was modified by cetyltrimethylammonium bromide/dimethylhexadecyl amine ratio.

HMW and LMW P(D,L-LA) polymers were prepared from racemic lactide (Purac) recrystallised once from ethylacetate. The racemic D,D- and L,L-lactide was polymerised in melt using tin octoate as catalyst at 140 °C for 5 hours under N_2 atmosphere and mechanical stirring. Solvent casted PDLLA composite films were prepared by dissolving PDLLA in chloroform mixed with SiO_2 microparticles and drug. They contained 40 or 60 wt.% of toremifene citrate containing SiO_2

microparticles, corresponding to 6 and 9 wt.% drug content, or 6 wt.% of pure toremifene citrate. Two composites contained 6 wt.% TC and SiO_2 (20 wt.% microparticles or 5 wt.% MCM-41) separately in the polymer bulk. The HMW P(D,L-LA) composites were treated with CO_2 at 50 bar for 5, 20 or 36 hours. Composites without CO_2 treatment were used as controls. The fast dissolving mesoporous SiO_2 (MCM-41) was added to LMW P(D,L-LA) composites. Molecular weights were obtained by Size Exclusion Chromatography (SEC). The morphology of the porous LMW P(D,L-LA) / SiO_2 composites were detected by a scanning electron microscopy (SEM; JEOL, Model JSM-5500, Tokyo, Japan).

The release profiles (n=3) of toremifene citrate were studied using a shaking water bath at 37°C (75 shakes per minute). Simulated body fluid (SBF, pH 7.4) [4] containing 0.5 wt.% sodiumlaurylsulphate (SDS, Sigma) was used as a dissolution medium. The volume of the dissolution medium was 50 ml and the weight of the specimens films was approximately 50 mg with varying specimen volume (35 – 210) mm^3 depending on the influence of the CO_2 treatment on the composite specimen size and morphology. Alternatively, a 5 ml sample or the whole sample solution was removed from each flask and replaced immediately with an identical volume of fresh medium. The absorbance values of the dissolution samples were measured on a UV-Visible spectrophotometer (Shimadzu, UV-1601) at the maximum absorbance for toremifene citrate (A_{278}).

Results and Discussion

The molecular weights of pure polymers were 240 000 g/mol for HMW P(D,L-LA) and 130 000 g/mol for LMW P(D,L-LA). Molecular weights of HMW P(D,L-LA) specimens before and after dissolution are shown in Table 1.

Table 1. Molecular weights of HMW P(D,L-LA) specimens before and after dissolution.

Composite	Before dissolution			After dissolution			
	M_w	M_n	MWD	Time	M_w	M_n	MWD
P(D,L-LA) +	[g/mol]	[g/mol]		[days]	[g/mol]	[g/mol]	
Pure HMW P(D,L-LA)	240 000						
TC (6%), 20h, 50 bar	220 100	136 900	1,61	169	23 500	15 200	1,54
TC MP (40%), 20h, 50 bar	197 400	129 800	1,52	169	93 400	65 200	1,43
TC (6%), 5h, 50 bar	203 600	112 100	1,82	169	28 400	16 200	1,75
TC MP (40%), 5h, 50 bar	214 300	129 100	1,66	169	98 600	64 400	1,53
TC (6%), without pores	210 000	119 600	1,68	105	16 700	8 800	1,88
TC MP (40%), without pores	202 000	116 100	1,74	126	143 700	85 800	1,67
TC MP (60%), 36h, 50 bar	247 400	166 100	1,48				

Table 2. Changes in molecular weights of some LMW P(D,L-LA) specimens during dissolution.

Composite	Before dissolution			After dissolution			
	M_w	M_n	MWD	Time	M_w	M_n	MWD
P(D,L-LA) +	[g/mol]	[g/mol]		[days]	[g/mol]	[g/mol]	
Pure LMW P(D,L-LA)	130 000	82 000	1,60				
TC MP (40%)	97 298	55 218	1,76	60	52 112	28 933	1,80
TC (6%) + MCM-41 (10%)				60	68 015	40 286	1,69
TC (6%)				2	67 504	38 676	1,75
				8	59 899	34 673	1,73
				13	44 702	25 208	1,77
				27	19 182	9 992	1,92
TC MP (40%) + MCM-41 (5%)				2	124 752	71 503	1,75
				8	112 916	70 309	1,61
				27	90 763	58 569	1,55
				46	92 367	56 667	1,63
				60	78 317	49 651	1,58

The changes in molecular weights for selected LMW P(D,L-LA) specimens during dissolution are shown in Table 2. Generally, silica-containing and/or specimens treated with CO_2 degraded slower than the polymer/drug specimens. Degradation of polymer in LMW P(D,L-LA) specimens was significantly faster compared to that in HMW P(D,L-LA) specimens.

Toremifene citrate (TC) release results from P(D,L-LA)/drug polymer and the P(D,L-LA)/silica gel composite matrices are shown in Fig. 1. The toremifene citrate release was clearly slower from both polymers and composites made with HMW P(D,L-LA) than from composites with LMW P(D,L-LA), as shown in Fig. 1b and Fig. 1c. The summary of the cumulative release profiles of toremifene citrate from polymer and composite matrices are shown in Fig. 1a.

(a)

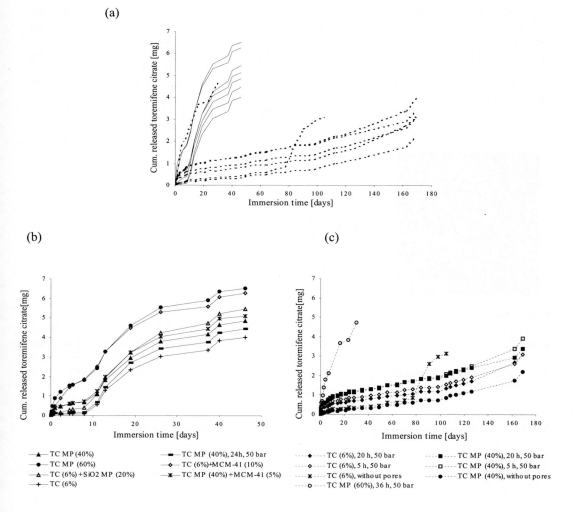

(b)   (c)

Fig.1. The summary of toremifene citrate release profiles are shown in Fig.1(a) and in more detail in Fig.1 (b) for LMW P(D,L-LA) specimens and in Fig.1 (c) for HMW P(D,L-LA) specimens.

Toremifene citrate was released at a constant rate obeying square root of time kinetics from the SiO_2 containing HMW P(D,L-LA) composites and from the CO_2 treated HMW P(D,L-LA) composites (Fig. 1c). The release was faster from the SiO_2 containing composites than from pure P(D,L-LA) polymer. In both cases the CO_2 treatment speed up the release of toremifene citrate. The time of pore forming CO_2 treatment had no significant effect on the release of toremifene citrate. The release rate was fastest from composites (both HMW & LMW P(D,L-LA)) containing 60 wt.% TC-containing SiO_2 microparticles. The release of TC showed an abrupt change from the untreated HMW P(D,L-LA) polymer containing 6 wt.% of toremifene citrate (Fig. 1c) due to a typical significant change in the molecular weight [5]. This undesired effect was not observed for SiO_2 containing composites or for CO_2-treated, porous HMW P(D,L-LA) within 6 months of immersion in SBF. This is due to the faster out-diffusion of polylactide degradation products.

All LMW P(D,L-LA) specimens showed a some kind of lag phase until they released TC faster after 8 days. Within that time the molecular weight of the polymer specimen containing 6 wt.% TC, which represents a case where the lag phase influence is stronger, was decreased about 50 % (Table 2.). Composites containing 6 wt.% TC and SiO_2 (as microparticles or as MCM-41) separately in the polymer bulk showed interesting results. The release of TC was faster than that of the others (except the composite containing 60 wt.% TC-containing SiO_2 microparticles that have the release of the same order of the magnitude), although the amount of SiO_2 was clearly lower (20 wt.% & 5 wt.%). This provides possibilities to reduce the amount of SiO_2 whether it is important to use as much as possible polymer in the composite, e.g., in order to optimise the mechanical properties or to work up the implant.

Scanning electron micrographs of the porous LMW P(D,L-LA)/silica gel composites are shown in Fig. 2. This image shows the formed porous structure of the CO_2- treated composite.

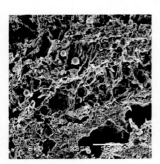

Fig. 2. SEM cross-section image of a CO_2-treated LMW P(D,L-LA)/SiO_2 composite containing 40 wt.% TC silica gel microparticles. The image showing the presence of pores.

Conclusion

In summary, it was shown that incorporation of SiO_2 particles or CO_2 treatment resulted in controlled release pattern of toremifene citrate. Release of toremifene citrate can be adjusted from 30 days to 6 months by adding silica and/or pores to the biodegradable polymeric carrier system. The steep change in structure observed for pure HMW P(D,L-LA) was avoided using SiO_2 and/or CO_2 treatment.

Acknowledgements

The National Technology Agency of Finland (TEKES) is acknowledged for financial support.

References

[1] C.G. Pitt, A.R. Jeffcoat, R.A. Zweidinger and A. Schindler: J. Biomed. Mater. Res. Vol. 13 (1979), p. 497

[2] M. Malin, M. Hiljanen-Vainio, T. Karjalainen and J. Seppälä: J. Appl. Polym. Sci. Vol. 59 (1996), p. 1289

[3] P. Kortesuo, M. Ahola, M. Kangas, I. Kangasniemi, A. Yli-Urpo and J. Kiesvaara: Int. J. Pharm. Vol. 200 (2000), p. 223

[4] T. Kokubo, H. Kushitani, S.Sakka, T.Kitsugi, T. Yamamuro: J. Biomed. Mater. Res. Vol. 24 (1990), p. 721

[5] J. Rich, P. Kortesuo, M. Ahola, A. Yli-Urpo, J. Kiesvaara and J. Seppälä: Int. J. Pharm. 212 (2001), p. 121

Key Engineering Materials Vols. 254-256(2004) pp. 493-496
online at http://www.scientific.net
© 2004 Trans Tech Publications, Switzerland

Synthesis and Characterization of Hydroxyapatite on Collagen Gel

Lídia A. Sena[1], Patrícia Serricella[2], Radovan Borojevic[2], Alexandre M. Rossi [3]
and Gloria A. Soares[1]

[1] Metallurgy and Materials Dep., UFRJ, P.O.Box 68505, Rio de Janeiro, 21941-972,
RJ, Brazil, lidia@metalmat.ufrj.br, gloria@ufrj.br

[2] ICB/UFRJ, Rio de Janeiro, 21941-590, RJ, Brazil, radovan@iq.ufrj.br

[3] CBPF, Rua Dr. Xavier Sigaud, 150, Rio de Janeiro, 22290-180, RJ, Brazil, rossi@cbpf.br

Keywords: Hydroxyapatite, collagen, nanocomposite.

Abstract. Bone is a complex natural composite with the mineral phase (mainly hydroxyapatite, HA) arranged in a specific way on the type-I collagen fibers. Tissue engineering tries to mimic this structure by carrying out the hydroxyapatite synthesis in the presence of collagen. Several techniques like electron microscopy, x-ray diffraction and infrared spectroscopy were employed to characterize the materials produced. The synthesis conditions allow the production of a calcium-deficient hydroxyapatite exhibiting nanometric crystals. The presence of collagen had little influence on the hydroxyapatite synthesis, slightly reducing the cristallinity of apatite phase present in the composite.

Introduction

Synthetic calcium phosphate (CaP) materials have been widely used in medical and dental applications as they possess the ability to promote cellular function, osteoconductivity and, in certain circumstances, may become osteoinductive [1]. Third-generation biomaterial are being developed to stimulate specific cellular responses [2]. For bone tissue engineering, researches concerning scaffolds design have been outstanding [3, 4]. One option deals with scaffolds composed by synthetic or natural hydroxyapatite ($Ca_{10}(PO_4)_6(OH)_2$, HA) and collagen (Col). The main reason of this choice is the preponderance of these components in the bone tissue, as bone structure is composed by nanocrystals of HA with the c-axis aligned to the Col fibres in a specific way [5]. The aim of this work was to synthesise nanocrystalline hydroxyapatite in the presence of collagen gel in order to produce a HA-Col composite suitable to be used in bone reconstruction.

Methods

The synthesis methodology followed the route used by Kikuchi et al. [5]. The composite was prepared by co-precipitation method using a HA-Col weight ratio of 80:20. The $Ca(OH)_2$ (99,6 mM) suspension and H_3PO_4 (59,64 mM) solution were simultaneously dropped with aid of peristaltic bombs into a reaction vessel with 25 mL of milliQ water which was vigorously stirred. Natural type collagen I was used in the gel form and was added in the H_3PO_4 solution. The synthesis temperature was 40 °C and the pH was adjusted to the range of 8-9 with ammonia. After ageing the precipitates for 18 hours, the material was filtered and lyophilised. Following the same route, another synthesis was prepared without the presence of collagen to compare to the HA-Col composite. The materials obtained were evaluated by using scanning electron microscopy (SEM, Zeiss, DMS 940A), transmission electron microscopy

(TEM, JEOL, 2000 FX), X-ray diffractometry (XRD, Rigaku, Miniflex) and diffuse refletance infrared Fourier-transformed spectroscopy (DRIFT, Spectra Tech).

Results and Discussion

The lyophilised HA-Col composite consisted in an homogenous powder as observed by SEM (Figure 1). This material seems to be suitable to produce tablets or discs by using uniaxial or isostatic pressure without calcination step. Calcination step changes material characteristics but is fundamental to increase the mechanical strength of green material [6]. The employment of collagen even in denatured form may provide moldability and mechanical resistance to the composite. TEM micrography of HA-Col composite (Figure 2) showed HA nanocrystals (not clearly resolved at 50,000x of magnification) precipitated on collagen gel and exhibiting some degree of orientation similar to that observed by other authors [5, 7].

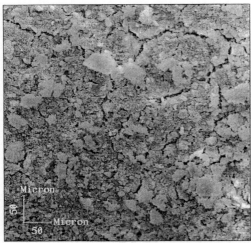

Fig.1 SEM micrograph of HA-Col composite; 200x.

Fig.2.TEM image and electron diffraction of the HA-Col composite, 50,000x.

XRD analysis (Figure 3) showed a general pattern typical of calcium-deficient hydroxyapatite (CD-HA, $Ca_{10-x}(HPO_4)_x(PO_4)_{6-x}(OH)_{2-x}$) for both materials [7]. Nanocrystals size, determined by Debye-Sherrer formula, seems not to be influenced by collagen presence as both materials exhibit length and diameter of 21 nm and 6.4-6.6 nm, respectively, similar to that observed in natural bone. By TEM analysis HA crystals with needle-like appearance were observed with length ranging from 20 to 100 nm (Figure 4). The comparison of the XRD pattern of HA synthesised with and without collagen showed that collagen in the reaction vessel had little influence on the hydroxyapatite synthesis, with the HA-Col composite presenting slight lower crystallinity.

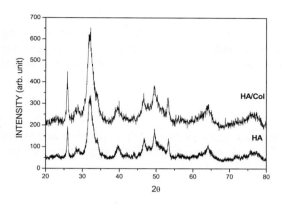

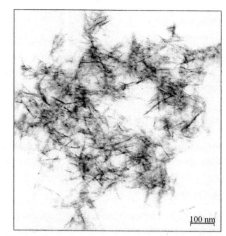

Fig. 3   XRD patterns of HA-Col composite and HA.

Fig. 4   Individual crystals of HA (original magnification: 80,000x).

The DRIFT spectrum of HA synthesised without collagen (Figure 5a) showed typical bands of structural OH⁻ (3571 cm^{-1}), PO$_4^{3-}$ (1088, 1025 and 963 cm^{-1}). The bands observed at 1650 and 3440 cm^{-1} corresponds to adsorbed water in the material. On HA-Col spectrum (Figure 5b) the characteristic bands of the collagen was additionally identified. Amide I, II and III are the major bands characteristic of collagen. The decrease in the intensity of the C=O stretching vibration of $-$ COO$^-$ groups at 1652 cm^{-1} was observed and this is considered an evidence of the chemical interaction between hydroxyapatite and collagen fibres [8,9]. The presence of CO$_3^{2-}$ bands at 1416 and 873 cm^{-1} indicates that both HA formed are compatible to a carbonated-substituted apatite [10]. The carbonate group is usually incorporated into the solution from the air during material preparation.

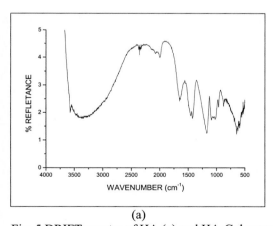

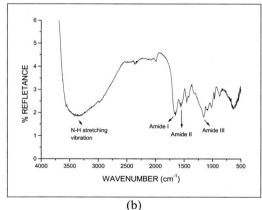

(a)   (b)

Fig. 5 DRIFT spectra of HA (a) and HA-Col composite (b).

Conclusions

The obtained results allow us to conclude that is quite easy to produce HA-Col composites by using a co-precipitation route. Moreover the collagen did not significantly affect the hydroxyapatite characteristics. Therefore, this route may be useful to produce HA-Col composite for several applications in bone reconstruction.

Acknowledgements

Financial support was received from several Brazilian agencies like CAPES, FAPERJ and FUJB. This research is part of Millenium Institute for Tissue Bioengineering (IMBT), supported by CNPq.

References

[1] R. Z. LeGeros: Clinical Orthopaedics and Related Research, Vol. 395 (2002), p. 81-98.
[2] L.L. Hench and J. M. Polak: Science 295 (2002), p. 1014-1017.
[3] K.J.L. Burg; S. Porter; J. F. Kellam: Biomaterials, Vol. 21 (2000), p. 2347-2359.
[4] D. A. Parry: Biophysical Chemistry, Vol. 29 (1988), p. 195-209.
[5] M. Kikuchi, Y. Suetsugu, J. Tanaka, S. Itoh, S. Ichinose, K. Shinomiya, Y. Hiraoka, Y. Mandai, S. Nakani: Bioceramics, Vol. 12 (1999), p. 393-396.
[6]] J.F. Oliveira, P.F. Aguiar, A.M. Rossi, G.A. Soares: Artif.l Organs, V.27 (2003), p. 386-391.
[7] H. Aoki: *Medical Application of Hydroxyapatite* (Tokyo, EuroAmericana, 1994).
[8] M. C. Chang, J. Tanaka: Biomaterials, Vol. 23 (2002), p. 4811-4818.
[9] S.-H. Rhee, J.D. Lee, J. Tanaka: J. Am. Ceram. Society, Vol. 83, 11 (2000), p. 2890-2892.
[10] J. C. Elliott: *Structure and Chemistry of the Apatites and other Calcium Orthophosphates* (Amsterdam, Elsevier Science, 1994).

Key Engineering Materials Vols. 254-256(2004) pp. 497-500
online at http://www.scientific.net
© *2004 Trans Tech Publications, Switzerland*

Preparation of Bonelike Apatite Composite Sponge

H. Maeda[1], T. Kasuga[1], M. Nogami[1], H. Kagami[2], K. Hata[3] and M. Ueda[4]

[1] Graduate School of Engineering, Nagoya Institute of Technology

Gokiso-cho, Showa-ku, Nagoya 466-8555, Japan

hirotaka@zymail.mse.nitech.ac.jp

[2] Nagoya University School of Medicine

[3] The Genetic Regenerative Medical Center, Nagoya University Hospital

[4] Nagoya University Graduate School of Medicine

[2-4] Tsurumai-cho 65, Showa-ku, Nagoya 466-8550, Japan

Keywords: Bonelike apatite, porous composite, Poly(lactic acid), calcium carbonate, simulated body fluid

Abstract. A novel sponge, coated with bonelike hydroxycarbonate apatite (HCA) on its skeleton surface, was derived via a particle-leaching technique combined with a biomimetic processing. In the present work, a compact consisting of calcium carbonate / poly(lactic acid) composite (CCPC) and sucrose formed by hot-pressing, was soaked in the simulated body fluid at 37 °C. Within initial 1 h, the sucrose was completely dissolved out, resulting in the formation of large-sized pores in the compact, and subsequently, after 3 d of soaking, bonelike HCA formed on the skeleton consisting of CCPC. The formed sponge has numerous, large pores of 450 to 580 μm in diameter, which are connected with channels having a diameter in the range of 70 to 120 μm, as well as a high porosity of 89 %.

Introduction

When osteoconducting materials, which have the ability to directly form a chemical bond with living bone, such as hydroxyapatite and ß–tricalcium phosphate ceramics, are implanted in living tissue, new bone forms around the materials via cell attachment, proliferation, differentiation and extracellular matrix production and organization. Recently, much attention has been paid to hydroxycarbonate apatite (HCA) as a novel biomaterial, since HCA is very similar to the apatite in terms of living bone in its chemical composition and structure[1] and shows effective compatibility in cell attachment, proliferation, and differentiation on the material[2], *i.e.,* osteoconductivity, as well as good bioresorbability[3]. We expect that the sponges composed of a bonelike HCA skeleton can be applied to bone-fillers or scaffolds for tissue engineering. A preparation method for bonelike HCA using simulated body fluid (SBF), a tris-buffer solution with inorganic ion concentrations almost equal to those of human plasma, is known as a biomimetic method. This method has an advantage over conventional methods in that materials can be coated with nano-sized HCA without heating.

In our earlier work a compact of calcium carbonate / poly(lactic acid) composite (CCPC) consisting of 25 wt% poly(lactic acid) (PLA) and 75 wt% calcium carbonate was reported to form bonelike HCA on its surface even after 3 h of soaking in SBF at 37 ºC[4]. The rapid formation of the HCA was suggested to originate from the integration of PLA having carboxy groups bonded with calcium ions for the HCA nucleation and a large amount of calcium carbonate (vaterite) having an ability to effectively increase the supersaturation of the HCA. We believe that various novel biomaterials can be prepared using CCPC. In the present work we prepared the novel sponge composed of a bonelike HCA composite skeleton utilizing CCPC and SBF.

Methods

The preparation procedure of calcium carbonates containing vaterite was described in the previous paper[4]. CO_2 gas was blown for 3 h at a flow rate of 300 mL/min into the suspension consisting of 7.0 g of $Ca(OH)_2$ in 180 mL of methanol at 0 °C in a Pyrex® beaker. The resultant slurry was dried at 70 °C in air, resulting in fine-sized powders. 2.0 g of PLA (with a molecular weight of 160 ± 20 kDa, determined by gel permeation chromatography) was dissolved in 10 mL of methylene chloride at room temperature. The calcium carbonate powders were added to the PLA solution and then the mixture was stirred. The weight ratio of $CaCO_3$/PLA was 1/2. We have already reported that the CCPC containing $\sim$ 30 % calcium carbonate has an excellent mechanical properties with a high HCA-forming ability in SBF[5]. For example, the composites containing 20 $\sim$ 30 % calcium carbonate show the bending strength of 40 $\sim$ 60 MPa and Young's modulus of 3.5 $\sim$ 6 GPa, and they exhibit a good ductility. Moreover, CCPC containing 30 % calcium carbonate has a high HCA-forming ability in SBF.

The sponge in the present work was prepared using a conventional particle-leaching technique. This method results in a porous structure of poorly interconnected pores. In the present work the interconnective pore structure of the sponge was developed utilizing sucrose as a sacrificial phase. Sucrose particles, which were sieved with the opening from 0.5 mm to 1.0 mm, were added to the CCPC slurry. The weight ratio of CCPC/sucrose was 1/6. The mixture slurry was stirred and cast into a stainless steel die, and then dried in air for solidification. After that, the product in the die was heated at 180 °C and uniaxially hot-pressed under a pressure of 40 MPa. When the product of CCPC and sucrose was hot-pressed at 180 °C, the sucrose particles melted (it begins to melt at 160 °C), leading to adjacent particles connected each other. As a result, both the sucrose and CCPC phases were unified into an interconnecting three-dimensional network. After the hot-pressing treatment, the specimen was cut in methanol with a diamond saw.

The hot-pressed sample was soaked in a SBF (consisting of 2.5 mM of Ca^{2+}, 142.0 mM of Na^+, 1.5 mM of Mg^{2+}, 5.0 mM of K^+, 148.8 mM of Cl^-, 4.2 mM of HCO_3^-, 1.0 mM of HPO_4^{2-}, and 0.5 mM of SO_4^{2-}) that included 50 mM of $(CH_2OH)_3CNH_2$ and 45.0 mM of HCl at pH 7.4 at 37 °C. After soaking, the sample was removed from the SBF, gently washed with distilled water, and dried at room temperature. Our strategy for the preparation of the sponge composed of CCPC skeleton with a bonelike HCA coating is to leach out the sucrose phase and simultaneously to form bonelike HCA on the composite skeleton.

The crystalline phases in the sponges are identified by X-ray diffraction analysis (XRD). Their morphology was observed by scanning electron microscopy (SEM).

Results and Discussion

Figure 1 shows XRD patterns before and after soaking the compact of CCPC/sucrose mixture in SBF. The XRD patterns show that the peaks corresponding to sucrose and vaterite disappear after 1 h. In the XRD pattern after 3 d of soaking, the peaks corresponding to apatite crystal can be seen. Almost no significant change in the peaks corresponding to aragonite is observed before and after the soaking.

Figure 2 shows the SEM photographs of the sample after 3 d of soaking in SBF. The SEM photograph shows that the sponge has numerous, large pores of 450 to 580 μm in diameter and large interconnected channels of 70 to 120 μm. The sponge has a continuous open foam with a 3D interpenetrating network of struts and pores. After 3 d of soaking, the surface of the CCPC skeleton is covered with the numerous leaf-like deposits, which are bonelike HCA, judged from the morphology and the XRD patterns. The porosity was estimated to be 89 % by comparison of the bulk density with the theoretical true density. The theoretical density of CCPCs was estimated using the value of 1.27 g/mL for PLA and 2.65 g/mL for calcium carbonate.

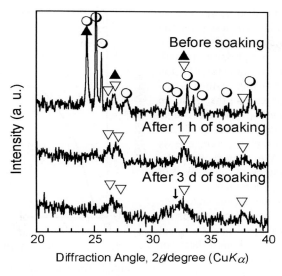

Fig. 1. XRD patterns of before and after soaking the compact of the CCPC/sucrose mixture in SBF. (O)sucrose, (▲)vaterite, (▽)aragonite, (⬇)apatite.

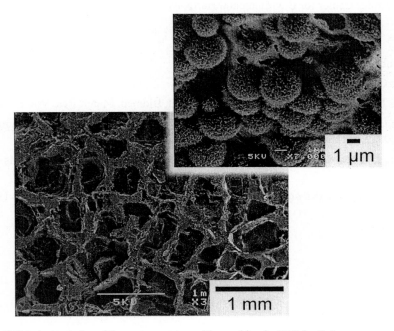

Fig. 2. SEM photographs of the sponge prepared by soaking in SBF for 3 d.

Figure 3 shows a typical stress-stain curve measured by the compressive tester for the sample prepared by soaking in SBF for 3 d. The sponge shows the compressive strength of ~1.5 MPa and the maximum strain for the fracture of the sponge is ~15 %, which is two orders of magnitude larger than that of dense hydroxyapatite ceramics[6]. The curve also shows that the fracture proceeds gradually beyond the maximum stress; the sponge leads to ductile fracture. We expect that the sponges in the present work are not broken in normal handling during operations.

The formed interconnected pore size of the sponge is effective for bone ingrowth or for cell delivery. The sponges may be great potential candidates as bone-fillers or scaffolds for tissue engineering.

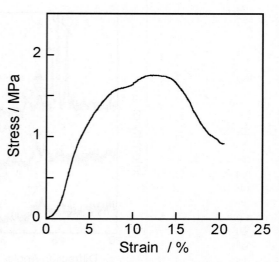

Fig. 3. Typical stress-strain curve of the sponge prepared by soaking for 3 d.

Acknowledgements

This work was supported in part by a Grant-in-Aid for Scientific Research from Japan Society for the Promotion of Science (No. 14380398) and grant from the NITECH 21st Century COE Program "World Ceramics Center for Environmental Harmony".

References

[1] T. Kokubo: Acta Mater. Vol. 46 (1998), p. 2519.

[2] Y. Doi: Cells and Materials Vol. 7 (1997), p. 111.

[3] Y. Doi, T. Shibutani, Y. Moriwaki, Y. Iwayama: J. Biomed. Mater. Res. Vol. 39 (1998), p. 603.

[4] H. Maeda, T. Kasuga, T. Nogami, Y. Hibino, K. Hata, M. Ueda, Y. Ota: Key Engng. Mater. Vol. 240-242 (2002), p. 163.

[5] T. Kasuga, H. Maeda, K. Kato, T. Nogami, K. Hata, M. Ueda, Y. Ota: Biomaterials Vol. 24 (2003), p. 3247.

[6] H. Aoki: *Medical Applications of Hydroxyapatite* (IshiyakuEuroAmerica, Japan 1994), p. 286.

Key Engineering Materials Vols. 254-256(2004) pp. 501-504
online at http://www.scientific.net
© *2004 Trans Tech Publications, Switzerland*

Bone-like Apatite Forming Ability on Surface Modified Chitosan Membrane in Simulated Body Fluid

Sang-Hoon Rhee

Department of Dental Biomaterials Science, College of Dentistry, Seoul National University,

Seoul 110-749, Korea, rhee1213@snu.ac.kr

Keywords: chitosan, apatite, surface modification, bioactivity, simulated body fluid

Abstract. A chitosan membrane modified with silane coupling agent and calcium salt was prepared and evaluated as a bioactive guided bone regeneration material. The chitosan membrane, which contained calcium nitrate tetrahydrate, was prepared and it was further subjected to a surface-modification with 3-isocyanatopropyl triethoxysilane (IPTS). Besides, a chitosan membrane which contained calcium nitrate tetrahydrate was also prepared as a control. Two membranes were exposed to a simulated body fluid (SBF) for 1 week. The SBF exposure led to the deposition of a layer of apatite crystals on the surface of the chitosan membrane, which contained calcium salt and modified with IPTS. In contrast, the chitosan membrane which contained only calcium salt did not show the apatite-forming ability. It implies that silanol groups and calcium salt acted as nucleation sites and accelerator for the formation of apatite crystals, respectively. Therefore, this chitosan membrane modified with IPTS and calcium salt is likely applied to a bioactive guided bone regeneration material because of its apatite-forming ability.

Introduction

Chitosan is a widely used material for the application as biomaterial because of its excellent biocompatibility and mechanical properties. Therefore, the chitosan or its hybrid materials with other natural or synthetic polymers are extensively studied for the applications as artificial skins, cartilages, or bones. However, the chitosan or its hybrids with other polymers have no apatite-forming ability in vitro or in vivo conditions. Thus, many investigations have already been made to develop a bioactive chitosan system for the application as bone restorative materials through the hybridization with chitosan and calcium phosphates. Unfortunately, however, they show good biocompatibility but could hardly show bone-like apatite forming ability in the SBF [1].

The silane coupling agent, 3-isocyanatopropyl triethoxysilane (IPTS), has already been used to modify the non-bioactive ethylene vinyl alcohol copolymer [2] or silicone [3] membranes combined with tetraethyl orthosilicate to give the bioactivity. However, there has been no report to use IPTS alone without tetraethyl orthosilicate for giving bioactivity to originally non-bioactive materials. The purpose of this study is to develop a chitosan membrane, which has an apatite-forming ability in the SBF through the compositional and surface modifications together. A chitosan membrane containing calcium salt was made and subjected to a surface-modification with IPTS because the calcium ion and silanol group are known to act as an accelerator and a nucleation site for the formation of apatite crystals, respectively.

Materials and Methods

The chitosan membrane containing calcium salt was prepared by dissolving chitosan powders to the 3 % acetic acid solution with the concentration of 4 wt % and calcium nitrate tetrahydrate with the concentration of 0.5 wt % to the chitosan. The solution was poured into a Petri's dish and then dried in an ambient condition for about 1 week for obtaining a membrane.

The surface modification of the chitosan membrane (1.5 cm ×1.5 cm × 0.01 cm in size) with silane coupling agent was accomplished through the reaction with the 3-isocyanatopropyl triethoxysilane

(IPTS; Aldrich Chem. Co. Inc., WI) using dibutyltin dilaurate (Aldrich Chem. Co. Inc., WI) as a catalyst in dry toluene and N_2 atmosphere for 24 hours. The concentration of IPTS in toluene was 0.5 M and the molar ratio of dibutyltin dilaurate to IPTS was 0.005. Following reaction, it was vigorously rinsed using acetone several times and then dried at room temperature. Subsequently, the chitosan membrane modified with calcium salt and silane coupling agent together was hydrated in 0.001 M HCl solution for 0.5 hours, rinsed with ion-exchanged distilled water several times, and then dried in an ambient condition. Hereafter, the chitosan membrane containing calcium salt alone and that containing calcium salt and surface modified with silane coupling agent will be referred to as specimen C and S, respectively.

The bioactivity of the specimens was assessed by evaluating their ability to form apatite on their surfaces in simulated body fluid (SBF). The SBF was prepared by dissolving reagent grade NaCl, $NaHCO_3$, KCl, $K_2HPO_4 \cdot 3H_2O$, $MgCl_2 \cdot 6H_2O$, $CaCl_2$, and Na_2SO_4 in ion exchanged distilled water [4]. Their ionic concentrations were Na^+ 142, K^+ 5.0, Mg^{2+} 1.5, Ca^{2+} 2.5, Cl^- 147.8, HCO_3^- 4.2, HPO_4^{2-} 1.0, SO_4^{2-} 0.5 (in mM). The solution was buffered at pH 7.4 with tris(hydroxymethyl) aminomethane $((CH_2OH)_3CNH_2)$ and 1 M hydrochloric acid (HCl) at 36.5 °C. Non-sterilized square shaped two specimens 1 cm × 1 cm × 0.02 cm in size were cut and then soaked in 30 mL of the SBF at 36.5 °C for various periods of time. After soaking, they were removed from the fluid and gently rinsed with ion-exchanged distilled water, and then dried at room temperature.

The surface of specimens C and S were analyzed by X-ray photoelectron spectroscopy (XPS; Model) and attenuated total reflection mode Fourier transformed infrared spectroscopy (FTIR-ATR; Thunderdome Swap-Top Module, Thermo Spectra-Tech, CT), respectively. The microstructure and phase of the crystals formed on the specimens were analyzed by scanning electron microscopy (SEM; JSM-840A, JEOL, Tokyo, Japan) and thin film X-ray diffractometry (TF-XRD; D8 Discover, Bruker, Germany) with an angle of 2° to the direction of incident X-ray beam, respectively.

Results

Figure 1 shows the XPS survey spectra of specimens C and S. Peaks ascribed to Si2s, Si2p, Ca2p at about 153, 102, 347 eV, respectively, were observed in specimen S but not in specimen C. It means that silane-coupling agent was successfully bonded to chitosan membrane. However, the silanol or siloxane group originated from silane-coupling agent was not detected in FTIR-ATR spectra and it is believed to result from its small amount of deposition.

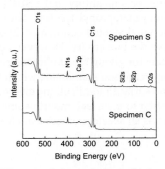

Fig. 1. XPS survey spectra of specimens C and S.

Figure 2 shows the SEM photographs of specimens (a) C and (b) S after soaking them into the SBF for 1 week. No apatite crystals were observed to occur except for some surface scratches on the surface of specimen C but apatite layer, which is described later, was observed to occur on the surface of specimen S.

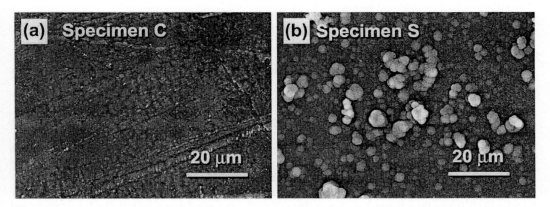

Fig. 2. SEM photographs of (a) specimen C and (b) S after soaking in SBF at 36.5 °C for 1 week.

Figure 3 shows the TF-XRD results for specimens (a) C and (b) S after soaking them in the SBF for 1 week. No peaks assigned to apatite crystal were observed in specimen C except for a peak, which is likely to have originated from chitosan phase. However, apatite peaks denoted by the '●' symbol were observed to occur in specimen S.

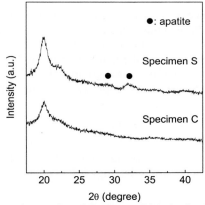

Fig. 3 TF-XRD patterns of specimens C and S after soaking in the SBF for 1 week at 36.5 °C.

Discussion

Some functional groups, such as Si-OH, COOH, NH_2, or OH, are known to induce the formation of apatite crystals in the SBF [5,6]. Chitosan has one amine group and two hydroxyl groups in its structure but it has no ability to induce the formation of apatite crystals in the SBF as shown in this work. Therefore, some surface modification, which can give nucleation sites for the apatite crystals, is required for using chitosan as a bioactive bone repairing materials.

In this work, the surface modification which gives the silanol group as a nucleation site for the formation of apatite crystals was performed using 3-isocyanatopropyl triethoxysilane. Two hydroxyl groups in a chitosan structure are likely to react with the isocyanate group in the IPTS and formed urethane linkage as in the following reaction (1).

$$\text{chitosan-OH} + \text{OCN(CH}_2)_3\text{Si(OEt)}_3 \xrightarrow{\text{dibutyltin dilaurate}} \text{chitosan-OC(O)NH(CH}_2)_3\text{Si(OEt)}_3 \quad (1)$$
$$\text{IPTS}$$

Subsequent hydrolysis of triethoxysilane end-capped chitosan membrane produced the silanol groups on its surface as in the following reaction (2).

$$\text{chitosan-OC(O)NH(CH}_2)_3\text{Si(OEt)}_3 \xrightarrow[\text{H}_2\text{O}]{\text{HCl}} \text{chitosan-OC(O)NH(CH}_2)_3\text{Si(OH)}_3 \quad (2)$$

When the chitosan membrane, which has silanol group on its surface and contains calcium salt, is soaked in the SBF, the silanol group acts as nucleation site and the dissolved calcium ion accelerates the formation of apatite crystals by increasing the ionic activity product of apatite crystals in the SBF. Indeed, apatite crystals did not form on the surface of chitosan membrane which contained only calcium salt while they formed on the surface of chitosan membrane which had silanol group and contained calcium salt, together. It means that the nucleation site and accelerator are required together for the formation of apatite crystals on the chitosan membrane in the SBF.

From the results, this chitosan membrane has a potential for the application as a guided bone regeneration membrane because of its apatite-forming ability and biocompatibility.

Conclusion

The bioactive chitosan membrane is newly developed for the application as a guided bone regeneration material. Calcium salt containing chitosan membrane was modified with silane coupling agent for providing the nucleation site for the formation of apatite crystals. Compositionally and surface modified chitosan membrane showed the apatite-forming ability in the SBF within one week. Its bioactivity is likely to originate from the silanol group and calcium salt because they have been known to act as nucleation site and accelerator, respectively, for the formation of apatite crystals in the SBF.

Acknowlegement

This work was supported by a Grant-in Aid for International Science & Technology Coopeartion Program from the Ministry of Science and Technology (MOST) (No. 05000020-02A0100-01610).

References

[1] H. K. Varma, Y. Yokogawa, F. F. Espinosa, Y. Kawamoto, K. Nishizawa and F. Nagata, T. Kameyama: Biomaterials 20 (1999), p. 879.

[2] A. Oyane, M. Minoda, T. Miyamoto, R. Takahashi, K. Nakanishi, H.-M. Kim, T. Kokubo and T. Nakamura: J. Biomater. Mater. Res. 47 (1999), p. 367.

[3] A. Oyane, K. Nakanishi, H.-M. Kim, F. Miyaji, T. Kokubo, N. Soga and T. Nakamura: Biomaterials 20 (1999), p. 79.

[4] T. Kokubo, H. Kushitani, S. Sakka, T. Kitusgi and T. Yamamuro: J. Biomed. Mater. Res. 24 (1990), p. 721.

[5] M. Tanahashi and T. Matsuda: J. Biomed. Mater. Res. 34 (1997), p. 305.

[6] S.-H. Rhee and J. Tanaka: Biomaterials 20 (1999), p. 2155.

Key Engineering Materials Vols. 254-256(2004) pp. 505-508
online at http://www.scientific.net
© *2004 Trans Tech Publications, Switzerland*

In Vitro of Self-Reinforced Composites of Highly Bioactive Glass Loaded Bioabsorbable Polymer

H. Niiranen, T. Niemelä, M. Kellomäki and P. Törmälä

Institute of Biomaterials, Tampere University of Technology, P.O.Box 589, 33101 Tampere, Finland

henna.niiranen@tut.fi

Keywords: Composite, bioactive glass, bioabsorbable polymer

Abstract. The aim of the study was to manufacture composites of bioabsorbable polymer P(L/DL)LA 70:30 and bioactive glass 13-93 with a high glass filler content. The effect of glass content to the degradation was investigated by studying mechanical properties and polymer matrix degradation. The mechanical properties were evaluated by 3-point bending and shearing. Polymer matrix degradation during manufacturing and *in vitro* was measured by gel permeation chromatography (GPC) and gas chromatography (GC). Also the water absorption (WA) and weight loss (WL) of composites were studied. The increase in glass amount was found to influence the molecular weight decrease during the extrusion but not to have an effect on the lactide monomer content of bioabsorbable polymer matrix. The mechanical properties were strongly dependent on the glass filler amount. *In vitro* degradation of composites was dependent on the glass content and the composite structure formed during self-reinforcing.

Introduction

The bioabsorbable, osteoconductive composite materials for guided tissue regeneration can be produced by combining bioactive glass with bioabsorbable polymer matrix [1]. In this study we have manufactured composites using twin screw extruder which allows higher filler content compared to single screw compounding applied previously [2]. The degradation of self-reinforced, highly filled composites is presented in this paper.

Materials and Methods

The bioabsorbable polymer used as a matrix material for composite rods of SR-PLA70+BaG was: a lactide stereocopolymer, poly(L/DL)lactide 70:30, with the trade name Resomer LR 708 (Boehringer Ingelheim, Ingelheim am Main, Germany). The polymer is medical grade, has an inherent viscosity of approximately 6.1-6.3 dl/g and has a residual monomer content of less than 0.5 % (both given by manufacturer). The bioactive filler for composites was bioactive glass 13-93 with composition of 6 wt-% Na_2O, 12 wt-% K_2O, 5 wt-% MgO, 20 wt-% CaO, 4 wt-% P_2O_5, 53 wt-% SiO_2 (Vivoxid Ltd, Turku, Finland). The glass spheres had the size distribution of 50-125 μm. The bioactive glass filler content of composites was 30, 40 or 50 wt-%.

The non-reinforced composite billets were extruded using twin screw extruder (Mini ZE 20*11.5 D). The PLA70+BaG30 and PLA70+BaG40 composites were further self-reinforced (SR) to the draw-ratio of 3.5. For PLA70+BaG50 rods the draw-ratio was 2.0. Phosphate-buffered saline (PBS) (pH 7.4) was used as *in vitro* solution. The follow-up times were 0, 1, 3, 6, 12, 24, 30, 36, 48 and 52 weeks. All the specimens were sterilized using gamma irradiation (minimum dose 25 kGy) prior to the hydrolysis. The composites were mechanically tested by three-point bending and shearing using Instron materials testing machine (Instron 4411, Instron Ltd, High Wycombe, England). The mechanical results are means of four or six parallel samples. The changes in viscosity average molecular weight (M_v) were measured by gel permeation chromatography, GPC (Waters, Milford, Massachusetts, USA). Narrow polystyrene standards were used for calibration and chloroform was used as a solvent and eluent. M_v was calculated using Mark-Houwink parameters $\alpha = 0.73$ and $K = 5.45 * 10^{-4}$. M_v results are means of three parallel samples with two injections. The gas chromatography, GC, (Trace GC, Thermo Finnigan, CE Instruments, Italy) was used to determine the lactide monomer content. The monomer contents are averages of two parallel samples. The percentages of water absorption (WA) and weight loss (WL) were determined from the wet, dry and initial weights of the specimens. The results are averages of three specimens.

Results

The increase in glass content interferes with the self-reinforcing process, thus affecting the maximum achievable draw-ratio and initial mechanical properties. [3] The composites and initial mechanical properties are presented in Table 1.

With all composites bending strength started to decrease after 12 weeks hydrolysis, as presented in a Fig 1.

Table 1. Materials and initial mechanical properties. B.S= bending strength, S.S= shear strength.

Composite	Glass content [wt-%]	Draw-ratio	Diameter [mm]	Initial B.S. [MPa] (STD)	Initial Bending Modulus [GPa] (STD)	Initial S.S [MPa] (STD)
SR-PLA70+BaG30	30	3.5	3.1	89 (6)	2.2 (0.2)	69 (3)
SR-PLA70+BaG40	40	3.5	3.1	70 (4)	2 (0.3)	55 (4)
SR-PLA70+BaG50	50	2.0	4.2	58 (11)	1.8 (0.4)	39 (7)

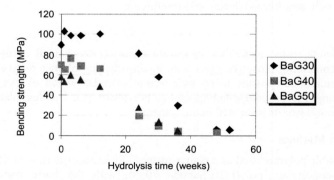

Fig.1. Bending strength retention of the composites *in vitro*.

The glass addition did not affect the lactide monomer content during processing. After extrusion and self-reinforcing the monomer content was less than 0.05 % for all studied composites. Gamma irradiation had no effect on the monomer content.

The change in molecular weight of polymer was studied after processing, after gamma irradiation and during the hydrolysis. The effect of processing and gamma irradiation on M_v is presented in a Fig 2. With all composites the polymer M_v had decreased approximately 90 % of the initial M_v of the raw material after Gamma irradiation. The decrease in M_v during *in vitro* hydrolysis is presented in a Fig 3. The M_v of BaG30 was 50 % of the initial value (processed and gamma irradiated) at approximately 23 weeks *in vitro*. With BaG40 and BaG50 the M_v was reduced to 50 % of the initial after 19 and 18 weeks respectively.

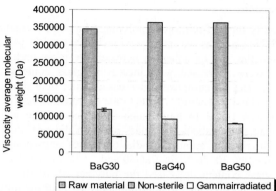

Fig. 2. Change in viscosity average molecular weight (M_v) of composites in processing and gamma irradiation.

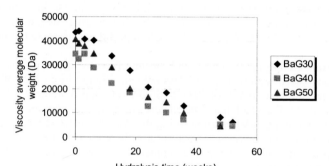

Fig. 3. Decrease in viscosity average molecular weight (M_v) during *in vitro* hydrolysis.

The WA of composites was dependent on the glass content. With BaG50 the WA was greatest during the whole follow-up period. At the 48th *in vitro* week the WA was 46 % for BaG30, 62 % for BaG40 and 76 % for BaG50. At the same time the WL was 4 % for BaG30, 11 % for BaG40 and 13 % for BaG50. The WL of composites as a function of viscosity average molecular weight is presented in a Fig 4.

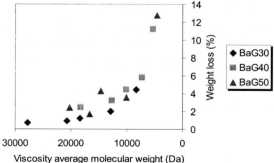

Fig. 4. Weight loss of composites as a function of viscosity average molecular weight (M_v).

Discussion and Conclusions

The glass particles modify the self-reinforced polymer structure by creating pores around the glass particles along the drawing axis as we have previously reported [1, 4]. Due to this porosity in a composite structure and the loss of adhesion between the glass particles and matrix, the initial mechanical properties were decreased with increasing glass content. *In vitro*, the mechanical properties of BaG40 and BaG50 composites decreased faster than the properties of BaG30. Thus the glass content and composite structure affects to the degradation rate of composites.

The glass addition was found to have some influence on the polymer degradation during processing. According to GC analysis, the glass addition did not increase the lactide monomer content during the extrusion or gamma irradiation. After extrusion the remaining M_v of the composites was 35 % for BaG30, 26 % for BaG40 and 22 % for BaG50 of the M_v of raw materials. Thus the increase in glass content decreased the polymer M_v of the extruded composites. After Gamma irradiation the M_v was reduced to approximately 10 % of the initial with all the composites studied. Our previous studies with single screw extrusion showed that the remaining M_v of PLA70 was 52 % after extrusion and 14 % after gamma irradiation.

Molecular weight (M_v) decrease *in vitro* was slightly faster with BaG40 and BaG50 compared to BaG30. However, comparing to our previous results, the polymer degradation is not faster with composites than with PLA70. [5] Thus the glass addition did not promote the polymer degradation *in vitro*.

Due to the porous composite structure, and the interfaces of glass and polymer matrix, water absorption (WA) increased with increasing glass content. Weight loss (WL) of BaG40 and BaG50 was slightly greater compared to WL of BaG30. Slight WL, approximately 4-5 %, existed with M_v values greater than 10000 Da which may indicate erosion type degradation.

Thus it is concluded that the degradation of highly bioactive glass loaded composites is strongly related to the self-reinforced composite structure. For the composites having sufficient bioactivity and mechanical properties the optimal filler content may be between 15 to 30 weight-percent depending on the application.

References

[1] M. Kellomäki et al.: Biomaterials 21 (2000), p. 2495-2505

[2] H. Niiranen, P. Törmälä: Key Engineering Materials Vols. 192-195 (2001), p. 721-724

[3] T. Paatola, H. Niiranen, P. Törmälä: Vincenzini, P. & Barbucci, R. (Eds.). CIMTEC 2002 - 6th International Conference "Materials in Clinical Applications", 10th International Ceramics Congress and 3rd Forum on New Materials, Florence, Italy, July 14 – 18th. (2002), p. 171 – 174

[4] H. Niiranen. P. Törmälä: Neenan, T., Marcolongo M. & Valentini, R (Eds.). Biomedical materials-drug delivery, Implants Tissue Engng. Vol 550 (1999), p. 267-272

[5] H. Niiranen et al., submitted to J Mater Sci Mater Med

Key Engineering Materials Vols. 254-256(2004) pp. 509-512
online at http://www.scientific.net
© *2004 Trans Tech Publications, Switzerland*

In vitro Degradation of Osteoconductive Poly-L/DL-lactide / β-TCP Composites

T. Niemelä, M. Kellomäki and P. Törmälä

Tampere University of Technology, Institute of Biomaterials, P.O Box 589,
FIN-33101 Tampere, Finland, tiiu.niemela@tut.fi

Keywords: β-tri calcium phosphate, composite, osteoconductive, polylactide

Abstract. Combining an osteoconductive ceramic with a bioabsorbable polymer and processed using self-reinforcing may create an interesting composite material. In this study *in vitro* degradation of osteoconductive poly-L/DL-lactide / β- tricalcium phosphate composites were reported. The composites were compared to plain poly-L/DL-lactide 70/30. The studied materials were characterized determining the mechanical properties, molecular weight, dimensions, mass and structural changes in phosphate buffered saline (PBS, pH 7.4) at 37 °C for up to 87 weeks. 20 wt-% β-TCP addition decreased the mechanical properties compared to plain poly-L/DL-lactide 70/30 but the composites retained their original shape longer than plain poly-L/DL-lactide 70/30. It was concluded that the studied composite material is a novel implant material potential for fracture fixations.

Introduction

Combining an osteoconductive ceramic with a bioabsorbable polymer has been studied to improve the osteoconductivity of degradable implant materials [1, 2]. On the other hand, self-reinforcing process improves the mechanical properties of materials [3]. Combination of these two methods gives interesting materials and structures. The aim of this study was to evaluate the mechanical properties and the *in vitro* degradation behaviour of composite rods made of poly-L/DL-lactide 70/30 with 20 wt-% of β-TCP.

Materials and methods

β-tricalcium phosphate (β-TCP) particles (CAM Implants B.V., Leiden, The Netherlands) were compounded with high molecular weight (inherent viscosity of approximately 6.3 dl g^{-1}) poly-L/DL-lactide with an LL/DL dimer ratio 70/30 (Boehringer Ingelheim, Ingelheim am Rhein, Germany). β-TCP was used as sintered powder with particle size distribution 50 – 125 μm, Ca/P ratio 1.5 and density 85 – 90 % of dense β-TCP.

The non-reinforced composite rods were manufactured by extrusion using twin screw extruder (Mini ZE 20*11.5 D). The extruded composite rods contained 20 wt-% of β-TCP and the diameter was 5.4 – 5.8 mm. To increase mechanical properties of the composites, the rods were self-reinforced (SR) using solid-state die-drawing [3, 4] to the draw ratio 3.5 – 4 and diameter 2.9 – 3.0 mm. The SR-rods were cut to desired lengths, washed with alcohol, dried in vacuum, packed and finally sterilized with gamma irradiation (25 kGy). To find out the effect of the β-TCP filler the plain poly-L/DL-lactide 70/30 rods were manufactured similarly for comparison (DR 3.5 – 4 and diameter 2.7 – 2.9 mm). The studied samples are shown in Table 1.

Table 1. Studied samples.

Composite	β-TCP content	DR	Diameter
SRPLA70	0 wt-%	3.5 – 4	2.7 – 2.9 mm
SRPLA70TCP20	20 wt-%	3.5 – 4	2.9 – 3.0 mm

Table 2. Testing parameters for mechanical testing.

	Bending test [5]	
	Crosshead speed	5 mm / min
	Bending span	48 mm
	Shear test [6]	
	Crosshead speed (tension)	10 mm / min

The composites were immersed in phosphate buffered saline (PBS, pH 7.4) at 37 °C for up to 87 weeks (plain poly-L/DL-lactide up to 36 weeks). The buffer solution was changed every two weeks. The studied composites were characterized determining the mechanical properties, molecular weight, dimensions, masses and structural changes. Dry samples (four parallel) were mechanically tested (Table 2) at room temperature by three-point bending and shearing (Instron 4411, Instron Ltd., High Wycombe, England). Weight average molecular weight (M_w) of the rods was studied using conventional gel permeation chromatography (GPC) with narrow polystyrene standards using chloroform as solvent and eluent. The equipment consisted of differential refractometer detector (Waters 410 RI) and HPLC-pump (Waters 515). The GPC columns were PLgel 5 µm Guard and two PLgel 5 µm mixed-C. The concentration of injected sample was 0.1 mass%, injection volume and eluent flow rate 150 µl and 1 ml min⁻¹, respectively. Data are means of two repeated injections. β-TCP particles were filtered off before GPC analysis. For determining changes in diameter, length and mass, all specimens were measured and weighted dry before *in vitro* testing. During *in vitro* experiments, specimens were removed from the buffer solution and dried. After drying the specimens were measured and weighed again. The weight measurements were done using a Mettler Toledo AG245 and the results are average of three specimens. Structural changes were studied using scanning electron microscope (SEM, JEOL T100, Tokyo, Japan).

Results

The compounding and self-reinforcing of the composites was successful. Some problems were noticed when manufacturing the plain poly-L/DL-lactide 70/30 rods. The self-reinforcing increased the mechanical properties of the samples, for example for the SRPLA70TCP20 composites the bending strength increased from 88 ± 5.5 MPa to 116 ± 8.5 MPa and shear strength from 36 ± 0.5 MPa to 79 ± 6.5 MPa. For SRPLA70 rods properties were 168 ± 4.5 MPa and 108 ± 1.5 MPa, respectively. The plain poly-L/DL-lactide 70/30 rods had better initial mechanical properties but the strength was lost earlier than for the composites, as seen in the Fig. 1a. The mechanical properties of the composite remained unchanged for up to 24 weeks, and after 52 weeks *in vitro* composites still had 65–75 % of the initial mechanical properties. The plain poly-L/DL-lactide 70/30 rods had lost approximately 35 % of the initials strengths already during the first 24 weeks *in vitro*. The mechanical properties were completely lost at 75th week with the composites and approximately at 48th week with the plain polylactide.

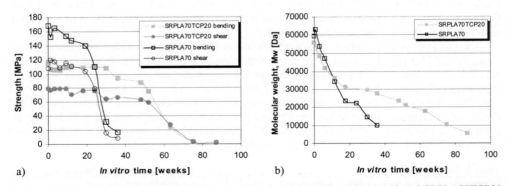

Fig 1.   a) Decrease of the bending and shear strength of gammasterilized SRPLA70 and SRPLA70TCP20 rods (diameter 2.7 – 3.0 mm) during *in vitro* (PBS). Each data point represents the mean of four independent measurements. b) Decrease of the weight average molecular weight (M_w) of gammasterilized SRPLA70 and SRPLA70TCP20 rods during *in vitro* (PBS). Data are means of two repeat injections.

Before processing of the samples, the matrix polymer had a weight average molecular weight (M_w) of about 380 kDa. The extrusion, self-reinforcing process and gamma irradiation decreased the M_w of matrix polymer to about 65 kDa in the case of SRPLA70 and to about 55 kDa in the case of SRPLA70TCP20 composites. At the beginning of *in vitro testing,* M_w decreased linearly in both samples. After 12 weeks the decreasing differed. Although the SRPLA70 had higher initial M_w, the M_w decreased more rapidly, being about 15 % of the initial value after 36 weeks *in vitro*, whereas in SRPLA70TCP20 composites it took about 80 weeks to decrease to the same level.

The composite rods maintained their size and shape practically unchanged for up to 52 weeks *in vitro* whereas the dimensions of the SR-rods of plain poly-L/DL-lactide 70/30 changed already after 12 weeks *in vitro* (Fig 2a). After 52 weeks the composite rods started to show relaxation. The diameter of the rods increased and the length decreased. At the week 75 changes in dimensions were about 60 %. After that the changes balanced. The weight of the SRPLA70 rods did not change at all during the first 36 weeks *in vitro*, while the weight of the SRPLA70TCP20 composites remained unchanged during the first 52 weeks. After that the weight started to decrease steadily. Fig 2b shows the weight average molecular weight (M_w), bending strength and weight of the SRPLA70TCP20 composites during *in vitro*. From that figure the order in which the properties start to decrease during *in vitro* can be well observed. The M_w started to decrease already at the beginning, bending strength after 36 weeks and weight only after 52 weeks.

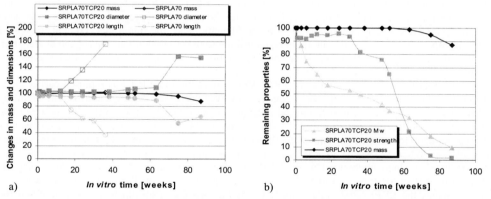

a)   *In vitro* **time [weeks]**   b)   *In vitro* **time [weeks]**

Fig 2. a) Changes in weight and dimensions of gammasterilized SRPLA70 and SRPLA70TCP20 rods (diameter 2.7 – 3.0 mm) during *in vitro* (PBS). b) Weight average molecular weight (M_w), bending strength and weight of the gammasterilized SRPLA70TCP20 composites. All properties are marked as 100 % at the beginning of hydrolysis.

When studying the structural changes using scanning electron microscope the relaxation was observed at 75 weeks *in vitro*. The internal fibrous structure of the composite after self-reinforcing and gamma irradiation is shown in the Fig 3a. It was clearly noticed from the longitudinal section when the composites had completely lost their orientation after 75 weeks *in vitro*.

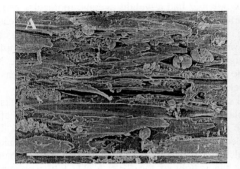

Fig 3.   a) The internal structure of the SRPLA70TCP20 composite after self-reinforcing and gamma irradiation (at the beginning of the *in vitro* studies). b) The internal structure of the SRPLA70TCP20 composite at 75 weeks *in vitro*. Scale bar 1000 μm.

Discussion and Conclusions

In the present study two different materials were studied *in vitro* in PBS. One material was plain poly-L/DL-lactide 70/30 and the other was composite of β-TCP (20 wt-%) and poly-L/DL-lactide 70/30. Both materials were self-reinforced to increase the mechanical properties [3]. The processing of the plain poly-L/DL-lactide 70/30 was problematic and the monomer content increased significantly. All impurities, like monomers, contribute to the strength retention and the degradation of the material [7]. The final diameter and draw ratio have an effect on the initial mechanical properties [7].

The initial mechanical properties of SRPLA70 were at the same level as reported earlier [1, 8, 9]. Due to the monomer content, the properties however decreased slightly rapidly. The blending with 20 wt-% β-TCP decreased the initial mechanical properties but the composites retained the strength level longer than plain poly-L/DL-lactide 70/30. The molecular weight of the composites decreased also slower than the M_w of SRPLA70. The decreasing of the M_w was the first sign of degradation. When M_w has decreased about 50 % the strength started to decrease. The weight of the composites remained unchanged up to 52 weeks. The composite rods also kept their original shape longer than SRPLA70. The SRPLA70 started to relax and swell already after 12 weeks *in vitro* whereas the SRPLA70TCP20 composites maintained the shape up to 52 weeks. When the mechanical properties of studied composites were compared to those earlier reported [1], the initial strength values were relatively similar. However, the studied composites retained the mechanical properties and shape longer than the materials reported earlier [1].

Blending 20 wt-% of β-TCP to bioabsorbable polymer and the self-reinforcing process together created an osteoconductive composite with porous structure. The porous structure is potentially beneficial for the bone tissue ingrowth. To conclude, the studied composite material is a novel implant material potential for fracture fixations.

References
[1] Ignatius A.A. et al. J Biomater Sci. Polymer Eds, 12, pp. 185-194. 2001.
[2] Kellomäki M. et al. Biomaterials, 21, pp. 2495-2505. 2000.
[3] Törmälä P., Clinical materials, 10, pp. 29-34. 1992.
[4] Niiranen H. et al. MRS Fall Meeting, Boston, USA, 1998.
[5] SFS-EN ISO 178. "Plastics. Determination of flexural properties". 1997.
[6] BS 2782: Part 3: Method 340B. "Determination of shear strength of sheet material". 1978.
[7] Törmälä P. et al. Proc Instn Mech Engrs, Vol 212, Part H. pp 101-111. 1998.
[8] Pohjonen T. et al. 13th European Conference on Biomaterials, Göteborg, Sweden. 88. 1997.
[9] Pohjonen T. et al. 13th European Conference on Biomaterials, Göteborg, Sweden. P43. 1997.

Key Engineering Materials Vols. 254-256(2004) pp. 513-516
online at http://www.scientific.net
© *2004 Trans Tech Publications, Switzerland*

Water Uptake of Poly(ethylmethacrylate)/Tetrahydrofurfuryl Methacrylate Polymer Systems Modified with Tricalcium Phosphate and Hydroxyapatite

F.F Rahman[1], W. Bonfield[1], R.E. Cameron[1], M.P.Patel[2], M.Braden[2], G. Pearson[2] and S.M.Tavakoli[3].

[1]Department of Material Science & Metallurgy, University of Cambridge, Cambridge, CB2 3QZ, UK, ffr20@cam.ac.uk
[2]IRC in Biomedical Materials, Department of Biomaterials in Relation to Dentistry, St Bartholomew's and Royal London School of Medicine and Dentistry, Queen Mary, University of London, London, E1 4NS UK, m.patel@mds.qmul.ac.uk
[3]TWI, Granta Park, Great Abington, Cambridge, CB1 6AL, UK, smtavakoli@twi.co.uk

Keywords: Diffusion, tricalcium phosphate, hydroxyapaptite, methacrylate polymers, bone growth, water uptake

Abstract. This work involves the incorporation of two bioactive fillers, tricalcium phosphate (TCP) and hydroxyapatite (HA) in a heterocyclic methacrylate polymer system (PEM/THFM) to create a bioactive material for encouraging bone regeneration and to provide a potential scaffold for bone growth.
The parent polymer system exhibits high water uptake. The effect of the addition of TCP and HA on this water uptake process has been studied. The pattern of water uptake of the TCP composite is similar to that of the parent polymer. In the case of the HA composite significantly higher water uptake is observed. The uptake is lower in phosphate buffered saline for all three systems. The water uptake process is complex and may be strongly influenced by water soluble species.

Introduction

Calcium phosphate bioceramics have created an interest as implant materials in orthopaedic surgery and dentistry because of their biocompatibility with hard living tissues. TCP has an important role as a resorbable bioceramic due to its high solubility and bioactivity. In terms of the mechanical and biological properties, HA is an appropriate reinforcement for organic polymers. HA is known to be osteoconductive and osteophillic leading to bone formation and the in-growth of bone-forming cells [1].
A heterocyclic methacrylate polymer system (PEM/THFM) had originally been developed to be used as hearing aid and temporary crown and bridge materials in dentistry. This system was compared to existing PMMA based bone cements because it exhibited a much lower shrinkage [2], a low reaction exotherm [3] and better biological properties [4]. It was also shown to interface effectively with subchondral bone, resulting in mechanical stability with favourable uptake kinetics [5]. It has been shown to support the growth of bone and cartilage [6]. Other features of this system are that it exhibits distinctive water uptake properties, permitting rapid water absorption initially followed by a slower prolonged uptake [7]. The system delivers drugs in a controlled manner [8]; biologically active recombinant human bone morphorgenetic protein –2, released from this system, resulted in new bone formation, in the periodontium of a rat model [9].
This paper discusses the incorporation of TCP and HA into a heterocyclic methacrylate polymer system which is based on poly(ethyl methacrylate) polymer powder mixed with tetrahydrofurfuryl methacrylate monomer (PEM/THFM) and polymerized at room temperature, to provide a potential scaffold for bone growth.

Materials and Methods

TCP was produced via an aqueous precipitation method [10], using a reaction between calcium hydroxide $(Ca(OH)_2)$, and phosphoric acid (H_3PO_4). The precipitation reaction produced a gelatinous precipitate which was dried at 60°C overnight, sintered at 1100°C for 4 hours and cooled to room temperature at 5°C/min.

HA was produced in a similar manner to TCP, however the H_3PO_4 was added drop wise to an agitated suspension of $Ca(OH)_2$ in water [11]. The reaction was modified by the addition of ammonium hydroxide (NH_4OH) in order to keep the pH of the reaction within the alkaline region, so that the gelatinous precipitate formed hydroxyapatite after sintering. The HA precipitate was also dried at 60°C overnight, sintered at 1200°C for 4 hours and cooled to room temperature at 5°C/min. The average particle size of the TCP and HA powder was 28 μm.

PEM/THFM (control) specimens were produced, by mixing 5g PEM (containing 0.6% residual benzoyl peroxide) with 3ml THFM containing 2.5% N,N- dimethyl-p-toluidine (v/v). The mixture was poured into rectangular polyethylene moulds and cured (~30mins) under pressure (~2.5 bar). Specimens were cut and polished using 4μm silicon carbide paper to 30x10x10 mm³.

For inclusion of active species, TCP and HA respectively, were ball milled with the PEM powder (a maximum weight fraction of HA or TCP powder equivalent to 40% on the PEM powder), prior to being mixed with the monomer, and prepared as PEM/THFM.

Water absorption measurements were carried out gravimetrically. Each sample was initially weighed to an accuracy of ± 0.0001g and specimens were immersed in distilled water (DW-100ml) and phosphate buffered saline (PBS-100ml) at 37°C (3 samples per system). At noted intervals, each specimen was taken out of the bottle, blotted on filter paper to remove excess water, weighed and returned to the bottle. On the first day, several readings were taken. The uptake of water was recorded until there was no significant change in weight (i.e. equilibrium was attained).

In order to examine the surface characteristics of the samples produced, cured square sections (10x10x10 mm³) were mounted on aluminium stubs, gold sputter-coated in an atmosphere of argon and then examined in the Jeol 250 scanning electron microscope.

Results

Scanning Electron Microscopy (SEM) of the novel HA/ TCP filled polymer composites showed an homogenous distribution of particles throughout the matrix (Fig. 1), (c.f. Fig. 1(B) control with (A) and (C)). In Fig. 1 (C), the HA particles can be seen distinctively on the surface.

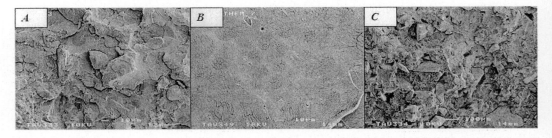

Fig. 1:   SEM micrographs of A) PEM/THFM/TCP, B) PEM/THFM and C) PEM/THFM/HA
         (x 500 magnification)

Fig. 2 shows the water uptake of the PEM/THFM /TCP system in DW which is similar to PEM/THFM in DW. It can also be seen that at any time point beyond $t^{1/2}$= 50 mins$^{1/2}$, the water uptake is lower in PBS. In contrast to these results the incorporation of HA in PEM/THFM substantially increases water uptake (Fig. 3) (from 2% (control) to 8% (PEM/THFM/HA) at $t^{1/2}$ = 100 min$^{1/2}$ in DW).

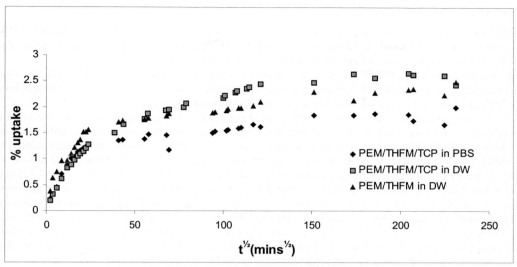

Fig. 2: Water uptake of PEM/THFM/TCP in DW and PBS

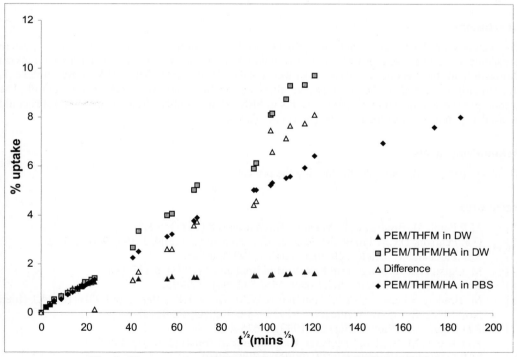

Fig. 3: Water uptake of PEM/THFM/HA system in DW and PBS.

Discussion

While the effect of the inclusion of TCP on water uptake is slight, the effect of the inclusion of HA is considerable. This result is in marked contrast to earlier work, where HA incorporated in the PEM-n-butyl methacrylate system showed only neglible effects on water uptake [12].

The water uptake of the PEM/THFM system in the presence of HA is enhanced and prolonged, with equilibrium not being attained over the period of the experiments. The uptake of this system is less in PBS, but has the same characteristics to the parent system as shown in Fig. 3.

Hence water uptake in each of the systems studied consists of a two-stage process similar to that published in previous work [7,13]. The first stage represents water saturation of the matrix and is Fickian in nature. This is followed by a secondary process of droplet growth as a consequence of osmotic pressure, which appears to be water uptake associated with the filler. This effect is demonstrated by subtracting the data collected during immersion in distilled water for PEM/THFM from that of PEM/THFM/HA. This procedure gives a $t^{1/2}$ linear plot, with an intercept on the time axis $t^{1/2} = 23.6$ mins$^{1/2}$ and a slope of 0.088% min$^{1/2}$. The uptake of the PEM/THFM/HA system in PBS is similar in nature (Fig. 3), but the uptake has decreased. The corresponding data in PBS has an intercept of $t^{1/2} = 10$min$^{1/2}$ and a slope of 0.0446% min$^{1/2}$. These results are consistent with earlier findings [14]. They also indicate the importance of the molarity of the external solution on the prospective implantation site of these systems with respect to the extracellular fluid.

The high water uptake associated with the HA polymer system is being investigated further. Also the effect of the water uptake on the mechanical properties of all the composites is being studied.

Conclusions

The kinetics of the water uptake of PEM/THFM/TCP is similar to that of the parent system consisting of a two-stage process, the first involves rapid saturation of the matrix by Fickian diffusion and the second involves a slower uptake. With PEM/THFM/HA, water uptake is substantially enhanced in the second stage although the first stage is similar to the control. The second process is reduced in all three systems in PBS. Hence, either of the novel composites could be used as potential scaffolds for bone growth at this stage.

Acknowledgements

Thanks to the ESPRC and TWI for funding the project.

References

[1]   M.P.Patel, M. Braden, K.W.M Davy: Biomaterials 8 (1987), p 53.
[2]   G.J. Pearson, D.C.A Picton, M. Braden, C. Longman : Int. End. J. 19 (1986), p 121.
[3]   M.P.Patel and M. Braden: Biomaterials 12 (1991), p 645.
[4]   M .Ogiso et al : J.Long Term Effects Med Implants 2 (1993), p 137-148.
[5]   S. Downes et al: J. Mater. Sci. Mater. Med 5 (1994), p 88-95.
[6]   M. Braden, S.Downes, M.P. Patel and K.W.M. Davy. UK Patent Appl. GB210734.1 (filed Nov 1992), European Application. 92923088.6, PCT/GB92/02128.
[7]   P.D Riggs et al: Biomaterials 20 (1999), p 435-441.
[8]   P.D. Riggs, M. Braden, M.P.Patel: Biomaterials 21 (no 4) (2000), p 345-351.
[9]   M.P.Patel et al: Biomaterials 22, (2001), p 2081-2086.
[10]   M.Akao, H. Aoki, K, Kato and A.J, Sato: J. Mater Sci 17 (1982), p343-346.
[11]   E. Hayek and H. Newesley: Inorganic Synthesis 7 (1963, p 121-128
[12]   S.Deb, M. Braden and W. Bonfield: Biomaterials 16 (1995), p 1095-1100.
[13]   M.P. Patel et al: J. Mater. Sci. Mater. Med 10 (1998), p 147- 151
[14]   R.M. Sawtell, S. Downes, R.L. Clarke, M. Braden and M.P.Patel: J. Mater. Sci. Mater. Med 8 (1997), p 667-674.

Key Engineering Materials Vols. 254-256(2004) pp. 517-520
online at http://www.scientific.net
© *2004 Trans Tech Publications, Switzerland*

Composite Mesh Consisting of Titanium, Apatite and Biodegradable Copolymer

S. Ban, A. Yuda, Y. Izumi, T. Kanie, H. Arikawa, K. Fujii

Kagoshima University Graduate School of Medical and Dental Sciences
8-35-1, Sakuragaoka, Kagoshima, 890-8544, Japan
sban@denta.hal.kagoshima-u.ac.jp

Keywords: apatite coating, composite, titanium mesh, biodegradable copolymer

Abstract. Apatites were formed on pure titanium mesh using a hydrothermal-electrochemical method. The mesh covered with apatites was dipped in the solution of dichloromethane and poly-lactic acid/poly-glycolic acid copolymer (PLGA) and dried in air. SEM observations showed that the apatites were well embedded in the PLGA film having many pores left by the evaporation of dichloromethane and the tips of the apatites were mostly exposed. These results imply that PLGA does not interfere with the bioactivity of apatite and maintain the mechanical strength of the coating during operation and initial stage of bone formation.

Introduction

Titanium mesh has been used in oral and maxillofacial surgery for the reconstruction of large and small bone defects [1-4]. Titanium mesh has excellent mechanical properties in terms of stiffness and elasticity, indicating easy to handle during surgery. An additional advantage is that the titanium mesh can be provided with calcium phosphate coatings [5-7]. It is generally assumed that calcium phosphate coatings enhance the bone formation process. We have been studying the hydrothermal-electrochemical apatite coating on titanium plate [8, 9]. Recently, we reported that apatites were successfully deposited on titanium mesh using hydrothermal-electrochemical method [10]. On the other hand, synthetic polymers such as poly-lactic acid, poly-glycolic acid, and poly-lactic acid / poly-glycolic acid copolymer (PLGA) are frequently used as biodegradable membranes in dental and medical applications. These polymers are absorbed after regeneration of the target tissue. The reasons for using biodegradable membranes are the elimination of a second stage operation, their low rate of exposure, and the high probability of healing with a complete restitution, *ad integrum* even after an exposure [11]. When biodegradable polymers are used as scaffold in guided tissue regeneration (GTR) and alternative materials, it is necessary that they are compatible with the mechanical properties of target tissue and their decomposition rates correspond to the regenerating speed of tissue. It has been reported that the composite membranes consisting of apatite particles and biodegradable polymer were prepared to improve their mechanical strength and decomposition rate [12-14]. In the present study, the PLGA was used to protect the apatites on the titanium mesh from damages during operation. The objective of this preliminary study was to characterize the surface of the composite mesh and to discuss the possible application of hydrothermal-electrochemical apatite coating for the production of bioactive composite mesh used as a scaffold for bone regeneration.

Materials and Method

#80 mesh woven with 120-μm diameter titanium fiber was cut to 10 x 10 mm using conventional scissors. Surface area of the 10-mm square titanium mesh was derived to be 2.475 cm^2, from measurement data of fiber size and the mesh pattern.

Fig.1 shows a schematic illustration of the preparation procedure used in this study. Apatites were formed on the pure titanium mesh using a hydrothermal-electrochemical method in an autoclave. The electrolyte was prepared by dissolving 137.8 mM of NaCl, 1.67 mM of K_2HPO_4, and 2.5 mM of $CaCl_2 \cdot 2H_2O$ into distilled water. The solution was buffered to pH 7.2 at room temperature with 50 mM tris(hydroxymethyl)-aminomethane $[(CH_2OH)_3CNH_2]$ and an adequate amount of HCl. The electrolyte was maintained at 100°C and a constant direct current at 12 mA/cm^2 was loaded for 1hr. After loading of the constant current, the mesh was rinsed with distilled water and dried at 37°C in air.

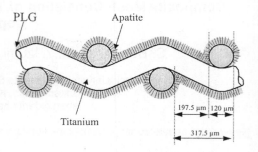

Fig.1 Schematic illustration of the composite membrane

The titanium mesh after the hydrothermal-electrochemical deposition was dipped in the copolymer solution prepared with PLGA and dichloromethane (CH_2Cl_2), and dried at 37°C for 24 hours in air. The surface of the composite mesh was characterized by X-ray diffractometry (XRD) and scanning electron microscopy (SEM).

Results and Discussion

Fig. 2 shows SEM micrographs of the surface of the titanium mesh without any treatments. The mesh was fabricated from 120-μm diameter fibers and the distance between the woven fibers was about 320 μm. The surface of the titanium fiber was in a rough state with scales. It seems to be due to the ductile work for its production.

Fig. 3 shows SEM micrographs of the surface of the titanium mesh coated with apatites using hydrothermal-electrochemical method. SEM micrograph showed that the mesh was homogeneously coated with needle-like or plate-like deposits. The edge shapes of the needles were hexagonal flat planes. The widths of the needles were 0.5–1 μm and the lengths were 7–10 μm. The needles were well oriented to the substrate. The thickness of the thin plate was about 0.1 μm and the widths were 30-80 μm.

Fig. 2 SEM micrographs of the surface of the titanium mesh.

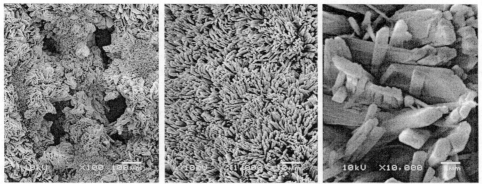

Fig. 3 SEM micrographs of the surface of the titanium mesh coated with apatite using hydrothermal-electrochemical method.

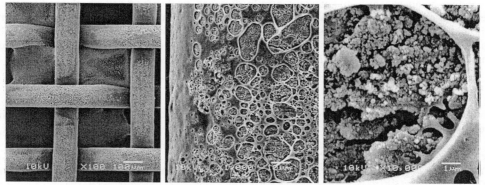

Fig. 4 SEM micrographs of the surface of the titanium mesh coated with PLGA.

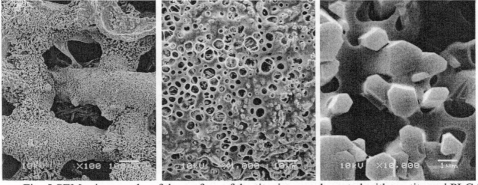

Fig. 5 SEM micrographs of the surface of the titanium mesh coated with apatite and PLGA.

XRD patterns of the deposits demonstrated that the deposits were hydroxyapatite and the diffraction peak corresponding to (002) of hydroxyapatite [$Ca_{10}(PO_4)_6(OH)_2$, referred to HA] of the deposits had a large intensity ratio compared to the standard intensity ratio of the random HA. It suggests that the deposits are identified as HA crystal rods grown along the c-axis and perpendicular to the substrate and the thin plates are also apatite, which was derived from octacalcium phosphate [15]. Our previous study [16] demonstrated that plate-like apatite was more frequently formed when the current density was lower, the electrolyte temperature was lower, and the loading time was shorter.

Fig. 4 shows SEM micrographs of the surface of the titanium mesh coated with PLGA film. The surface of the titanium mesh was covered with porous PLGA film. It seems that these pores, 1-10 μm, were formed due to the evaporation of dichloromethane used as solvent. The pore size could be regulated by the concentration of PLGA/dichloromethane solution and the dry condition.

Fig. 5 shows SEM micrographs of the surface of the titanium mesh coated with apatite and PLGA. SEM observations showed that most of the apatites were well embedded in the PLGA film having many pores by the evaporation of dichloromethane and the tips of the apatites were mostly exposed. Our previous study revealed that the exposed HA particles on the surface of the HA-glass composite act as nucleation sites for precipitation in physiological environment [17]. Furthermore, we confirmed that the coating film of this composite was not macroscopically damaged by bending and usual handling for the experimental procedure. These results imply that the PLGA will not interfere with the bioactivity of apatite and maintain the mechanical strength of the coating film during operation and initial stage of bone formation. A more elaborate study with objective quantitative and qualitative parameters is required to evaluate its biodegradation rate and bioactivity in vitro and in vivo.

Acknowledgements

This work was partially supported by a Grant-in-Aid for General Scientific Research from the Ministry of Education, Culture, Sports, Science and Technology of Japan.

References
[1] Y.C.G.J. Paquay, J.E. de Ruijter, J.P.C.M. van der Waerden and J.A. Jansen: Biomaterials Vol.18 (1997), p.161
[2] T. von Arx and B. Kurt: Clin. Oral Impl. Res. Vol.10 (1999), p.24
[3] I.P. Janecha: Arch. Otolaryngol. Head Neck Surg. Vol.126 (2000), p.396
[4] T. Schug, H. Rodemer, W. Neupert, and J. Dumbach. J. Cranio-Maxillo. Surg. Vol.28 (2000), p.235
[5] J.A. Jansen, J.P. van de Waerden, J.G. Wolke and K. de Groot: J. Biomed. Mater. Res. Vol25 (1991), p.973
[6] J.W.M. Vehof, P.H.M. Spauwen, and J.A. Jansen: Biomaterials Vol.21 (2000), p.2003
[7] T. Kim et al.: J. Biomed. Mater. Res. Part B: Appl. Biomater. Vol.64B (2003), p.19
[8] S. Ban and S. Maruno: Jpn. J. Appl. Phys. Vol.32 (1993), p.1577
[9] S. Ban and S. Maruno: J. Biomed. Mater. Res. Vol.42 (1998), p.387
[10] A. Yuda, Y. Izumi, and S. Ban: J. J. Dent. Mater. Vol.22 (2003), p.72 (in Japanese)
[11] E. Fontana, P. Trisi and A. Piattelli: J. Periodontol. Vol.65 (1994), p.658
[12] A. Piattelli, M. Franco, G. Ferronato et al..: Biomaterials Vol.18 (1997), p. 629
[13] P. Cerrai, G.D. Guerra, M. Tricoli et al.: J. Mater. Sci.: Mater. Med. Vol.10 (1999), p. 677
[14] K. Kesenci, L. Fambri, C. Migliaresi et al.: J. Biomed. Sci.: Polymer Edn. Vol.11 (2000) p.617
[15] S. Ban and S. Maruno: Biomaterials Vol.19 (1998), p1245
[16] S. Ban and J. Hasegawa: Biomaterials Vol.23 (2002) p.2965
[17] S. Ban, S. Maruno, H. Iwata, and H. Itoh: J. Biomed. Mater. Res. Vol.28 (1994), p.65

Key Engineering Materials Vols. 254-256(2004) pp. 521-524
online at http://www.scientific.net
© *2004 Trans Tech Publications, Switzerland*

Apatite-Forming Ability and Mechanical Properties of Poly(tetramethylene oxide) (PTMO)-Ta₂O₅ Hybrids

M. Kamitakahara[1,2], M. Kawashita[2], N. Miyata[2], H.-M. Kim[3], T. Kokubo[4], C. Ohtsuki[1] and T. Nakamura[5]

[1] Graduate School of Materials Science, Nara Institute of Science and Technology
8916-5, Takayama, Ikoma, Nara 630-0192, Japan, kamitaka@ms.aist-nara.ac.jp
[2] Department of Material Chemistry, Graduate School of Engineering, Kyoto University
Yoshida, Sakyo-ku, Kyoto 606-8501, Japan
[3] Department of Ceramic Engineering, School of Advanced Materials Engineering, Yonsei University,
134, Shinchon-dong, Seodaemun-gu, Seoul 120-749, Korea
[4] Research Institute for Science and Technology, Chubu University
1200 Matsumoto-cho, Kasugai, Aichi 487-8501, Japan
[5] Department of Orthopaedic Surgery, Graduate School of Medicine, Kyoto University
Shogoin, Sakyo-ku, Kyoto 606-8507, Japan

Keywords: apatite, mechanical properties, hybrid, PTMO-Ta₂O₅.

Abstract. Calcium-free poly(tetramethylene oxide) (PTMO)-Ta₂O₅ hybrids were prepared by a sol-gel method from triethoxysilane-functionalized PTMO (Si-PTMO) and tantalum ethoxide (Ta(OEt)₅) with Si-PTMO/Ta(OEt)₅ mass ratio of 30/70, 40/60 and 50/50 (hybrids PT30Ca0, PT40Ca0 and PT50Ca0, respectively). A calcium-containing PTMO-Ta₂O₅ hybrid was prepared from Si-PTMO, Ta(OEt)₅ and CaCl₂ with Si-PTMO/Ta(OEt)₅ weight ratio of 40/60 and CaCl₂/Ta(OEt)₅ mole ratio of 0.15 (hybrid PT40Ca15). Crack-free transparent monolithic hybrids were obtained for all the compositions except for PT30Ca0. Hybrids PT40Ca0 and PT50Ca0 formed apatite on their surfaces in a simulated body fluid (SBF) within 14 d. Hybrid PT40Ca0 showed higher mechanical strength and larger strain to failure than human cancellous bone. Its mechanical strength was a little increased by soaking in SBF. Hybrid PT40Ca15 formed apatite on its surface in SBF within 3 d. This hybrid, however, showed a significant decrease in mechanical strength after soaking in SBF. Bioactive flexible materials of calcium-free PTMO-Ta₂O₅ hybrids are expected to be useful for bone repair.

Introduction

Bioactive ceramics are used clinically as important bone-repairing materials. They are, however, brittle and hence limited in their applications. It is desirable to develop new types of flexible bioactive materials. The present authors previously revealed that CaO-free polydimethylsiloxane (PDMS)-TiO₂ [1] and poly(tetramethylene oxide) (PTMO)-TiO₂ [2] hybrids in which anatase is precipitated by hot-water treatments show apatite-forming ability on their surfaces in a simulated body fluid (SBF) and flexibility. These hybrids are expected to form an apatite layer on their surfaces in the living body and bond to living bone through the apatite layer. Their mechanical strengths, however, are decreased by the hot-water treatments for precipitation of anatase. A sol-gel-derived Ta₂O₅ gel shows apatite-forming ability in SBF even in the amorphous state [3]. Therefore, PTMO-Ta₂O₅ hybrids prepared by a sol-gel method can be expected to form apatite in SBF without any subsequent treatment. In the present study, PTMO-Ta₂O₅ hybrids without and with addition of

CaO were prepared by a sol-gel method, and their apatite-forming ability in SBF and mechanical properties were examined before and after soaking in SBF.

Materials and Methods

The mixture of poly(tetramethylene oxide) (PTMO, HO-(-CH$_2$CH$_2$CH$_2$CH$_2$O-)$_n$-H, molecular weight = 1000) and 3-isocyanatopropyltriethoxysilane (IPTS, (C$_2$H$_5$O)$_3$SiCH$_2$CH$_2$CH$_2$NCO), whose molar ratio was [PTMO/IPTS] = [1/2], was stirred at 70 °C for 5 d under a nitrogen atmosphere to obtain triethoxysilane functionalized poly(tetramethylene oxide) (Si-PTMO). This was mixed with tantalum ethoxide (Ta(OEt)$_5$) and CaCl$_2$ at the ratios given in Table 1. The obtained solutions were kept at 40 °C for 3 w for gelation and drying. The obtained hybrids were soaked in SBF with pH 7.40 and ion concentrations (Na$^+$ 142.0, K$^+$ 5.0, Ca^{2+} 2.5, Mg^{2+} 1.5, Cl$^-$ 148.8, HCO$_3^-$ 4.2, HPO$_4^{2-}$ 1.0, SO$_4^{2-}$ 0.5 mM) nearly equal to those of the human blood plasma [4] at 36.5 °C. The surface structural changes of the specimens due to soaking in SBF were analyzed by thin-film X-ray diffraction (TF-XRD) and scanning electron microscopy (SEM). The changes in the element concentrations of SBF due to soaking of the specimens were measured by inductively coupled plasma (ICP) atomic emission spectroscopy. The mechanical properties were measured before and after soaking in SBF with an Instron-type testing machine using a three-point bending test.

Table 1. Starting compositions of prepared hybrids

Notation	Composition				
	Si-PTMO/Ta(OEt)$_5$ (mass ratio)	CaCl$_2$/Ta(OEt)$_5$ (mole ratio)	H$_2$O/Ta(OEt)$_5$ (mole ratio)	HCl/Ta(OEt)$_5$ (mole ratio)	IPA/Ta(OEt)$_5$ (mole ratio)
PT30Ca0	30/70	0	2.5	0.05	30
PT40Ca0	40/60	0	2.5	0.05	30
PT50Ca0	50/50	0	2.5	0.05	30
PT40Ca15	40/60	0.15	2.5	0.05	30

Results and Discussion

Crack-free transparent monolithic discs approximately 40 mm in diameter and 2–4 mm in thickness were obtained for all the compositions except for PT30Ca0. Hybrid PT30Ca0 formed many cracks during preparation. All the as-prepared hybrids were amorphous.

Fig. 1 shows TF-XRD patterns of the surfaces of the hybrids before and after soaking in SBF for various periods. Figure 2 shows SEM photographs of the surfaces of the hybrids before and after soaking in SBF for various periods. Apatite was formed on the surfaces within 14 d for hybrids PT40Ca0 and PT50Ca0 in SBF. The amount of apatite formed on hybrid PT40Ca0 was larger than that of hybrid PT50Ca0. It is considered that the Ta$_2$O$_5$ component in hybrids induced apatite nucleation on their surfaces. Hybrid PT40Ca15 formed apatite within 3 d. This is because incorporated calcium ions were released into SBF and accelerated apatite nucleation by increasing the ionic activity product of apatite. The release of calcium ions from hybrid PT40Ca15 into SBF was confirmed by ICP.

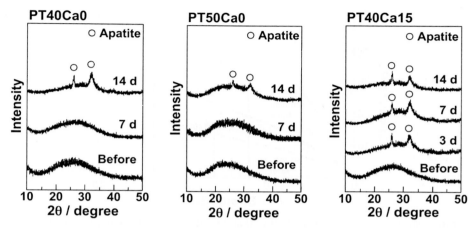

Fig. 1 TF-XRD of surfaces of hybrids before and after soaking in SBF for various periods.

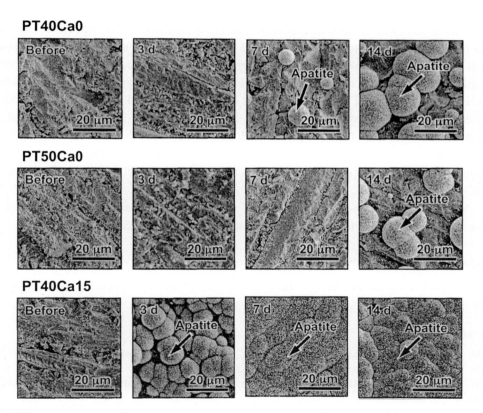

Fig. 2 SEM photographs of surfaces of hybrids before and after soaking in SBF for various periods.

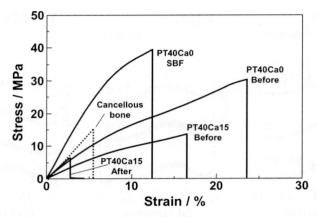

Fig.3 Stress-strain curves of hybrids PT40Ca0 and PT40Ca15 before and after soaking in SBF.

Fig. 3 shows stress-strain curves of hybrids PT40Ca0 and PT40Ca15 before and after soaking in SBF. Hybrid PT40Ca0 showed higher mechanical strength and larger strain to failure than those of human cancellous bone. Its mechanical strength was a little increased by soaking in SBF, whereas its strain to failure was decreased. Hybrid PT40Ca15 showed mechanical strength lower than those of hybrid PT40Ca0 even before soaking in SBF, and showed a significant decrease in mechanical strength and strain to failure by soaking in SBF. It is considered that the incorporated calcium ions are easily released via exchange with hydronium ions in SBF and water molecules also easily enter into the network of calcium-containing hybrid to break its network. It has been revealed that calcium-containing PDMS-CaO-SiO$_2$-based and PTMO-CaO-TiO$_2$ hybrids show a significant decrease in mechanical strength in SBF [5]. The degradation mechanism of hybrid PT40Ca15 in SBF is considered to be similar to that of the calcium-containing PDMS-CaO-SiO$_2$-based and PTMO-CaO-TiO$_2$ hybrids.

Although the apatite-forming ability of hybrid PT40Ca0 is lower than that of the calcium-containing hybrid, its mechanical properties were improved in SBF.

Conclusion

Bioactive flexible materials of calcium-free PTMO-Ta$_2$O$_5$ hybrids are expected to be useful for bone repair.

References

[1] M. Kamitakahara, M. Kawashita, N. Miyata, T. Kokubo, T. Nakamura: *Bioceramics Vol. 14* (Trans Tech Publications, Switzerland 2001), p.633-636.
[2] M. Kamitakahara, M. Kawashita, N. Miyata, T. Kokubo, T. Nakamura: Biomaterials Vol. 24 (2003), p.1357-1363.
[3] T. Miyazaki, H.-M. Kim, T. Kokubo, H. Kato, T. Nakamura: J. Sol-Gel Sci. Tech. Vol. 21 (2001), p.83-88.
[4] T. Kokubo, H. Kushitani, S. Sakka, T. Kitsugi and T. Yamamuro: J. Biomed. Mater. Res. Vol. 24 (1990), p.721-734.
[5] N. Miyata, M. Kamitakahara, M. Kawashita, T. Kokubo and T. Nakamura: *Bioceramics Vol. 15* (Trans Tech Publications, Switzerland, 2002), p.943-946.

Key Engineering Materials Vols. 254-256(2004) pp. 525-528
online at http://www.scientific.net
© *2004 Trans Tech Publications, Switzerland*

Effect of Sulfonic Group and Calcium Content on Apatite-Forming Ability of Polyamide Films in a Solution Mimicking Body Fluid

T. Kawai[1], C. Ohtsuki[1], M. Kamitakahara[1], T. Miyazaki[2], M. Tanihara[1],
Y. Sakaguchi[3], S. Konagaya[3]

[1] Graduate School of Materials Science, Nara Institute of Science and Technology (NAIST),
8916-5, Takayama, Ikoma, Nara 630-0192, Japan, kaw-taka@ms.aist-nara.ac.jp
[2] Graduate School of Life Science and Systems Engineering, Kyushu Institute of Technology,
2-4, Hibikino, Wakamatsu, Kitakyushu, Fukuoka, 808-0196, Japan
[3] Toyobo Research Center Co., Ltd., Katata, Ohtsu, Shiga 520-0292, Japan

Keywords: Polyamide film, sulfonic group, apatite, simulated body fluid, calcium ion, biomimetic process.

Abstract. Apatite-polymer hybrids have attractive features as novel bone substitutes since they may show bone-bonding ability and mechanical performances analogous to those of natural bone. We previously reported that polyamide films containing sulfonic group ($-SO_3H$) and calcium chloride can effectively deposit apatite on their surfaces in a solution (1.5SBF) mimicking body fluid. 1.5SBF has 1.5 times ion concentrations of a simulated body fluid (SBF). In the present study, the apatite-forming ability of polyamide films with different amounts of sulfonic groups and $CaCl_2$ was investigated in 1.5SBF. It was found that the polyamide films showed the ability of apatite formation, which was increased with increasing amounts of not only calcium chloride but also sulfonic group. Adhesion of apatite layer to the organic polymer may be enhanced by increasing the amounts of sulfonic group in the polymer. These results indicate that increasing sulfonic groups could accelerate the rate of heterogeneous nucleation of apatite on its surface. The interaction between sulfonic group and apatite crystals constitute a tight attachment.

Introduction

Hydroxyapatite ($Ca_{10}(PO_4)_6(OH)_2$) shows high biological affinity to living bone. Synthetic hydroxyapatite is now widely used as a substitute material for filling bony defects. However, the body of sintered hydroxyapatite does not have good enough mechanical properties for extending applications on medical fields. Development of apatite-polymer hybrids has attractive features as novel bone substitutes, because such hybrids may show bone-bonding ability and mechanical performances analogous to those of natural bone.

Kokubo *et al.* proposed a biomimetic process that utilizes a reaction between bioactive glass and a simulated body fluid (SBF), in order to deposit a hydroxyapatite layer on organic substrates [1]. This process can also be used for producing a polymer substrate covered with bone-like apatite. Bone-like apatite layer is a key substance to bring about direct bonding of artificial materials to living bone. It is therefore expected that the coating of bone-like apatite through the biomimetic method will produce organic-inorganic hybrids that can show direct bonding to living bone, i.e. bioactivity, in addition to specific mechanical properties arising from organic materials.

In the biomimetic process utilizing a simulated body fluid, apatite deposition on an organic polymer can be triggered both by the existence of a specific functional group effective for induction of heterogeneous nucleation of apatite, and by an increased concentration of calcium ions in the surrounding fluid. Tanahashi *et al.* reported that carboxyl (-COOH) groups play an effective role in heterogeneous nucleation of apatite in the body environment [2]. It is reported that polyamide films containing carboxyl groups had the ability to induce deposition of apatite crystals on their surfaces in 1.5SBF that has ion concentrations 1.5 times those of SBF, when the films were incorporated with calcium chloride ($CaCl_2$). We recently found that the same type of polyamide film can also deposit apatite on its surface when it was modified with sulfonic (-SO_3H) groups and calcium chloride [3]. This means that sulfonic group also acts as a functional group effective on apatite formation in body environment, although the detailed behavior of apatite formation on the polyamide film was not clarified. In the present study, apatite formation on polyamide films with different amounts of sulfonic groups and $CaCl_2$ was investigated in 1.5SBF to obtain detailed finding on fabrication of apatite-polyamide hybrids.

Materials and Methods

Polyamides shown in Fig. 1 were prepared according to a literature method [4]. The polyamides, S(0), S(20) and S(50), have 0, 20 and 50 % of sulfonic group in the polymer molecule, respectively. One gram of each polyamide powder was dissolved in 10 mL of *N,N*-dimetyl-acetamide, both with and without $CaCl_2$. $CaCl_2$ was added to the polyamide in various mass ratios of $CaCl_2/(Polyamide+CaCl_2)$ = 0, 0.10, 0.20 and 0.40. The mixture was stirred for 12 hours and then coated onto flat glass plates, using a bar coater. The solution coated on the glass plate was dried in a vacuum oven at 60°C under 133 Pa for 8 hours. The obtained polymer films were then removed from the glass plates and cut into 10x10 mm^2 sections. Polyamide (S(x)) films were prepared by modification with y mass% of $CaCl_2$ to a given total of S(x) and $CaCl_2$, and hereafter, were denoted as S(x)Ca(y).

The obtained films were then soaked in 1.5SBF (Na^+ 213.0, K^+ 7.5, Mg^{2+} 2.3, Ca^{2+} 3.8, Cl^- 221.7, HCO_3^- 6.3, HPO_4^{2-} 1.5, and SO_4^{2-} 0.8 $mol·m^{-3}$). The pH of the solution was buffered at 7.40 using 75 $mol·m^{-3}$ of *tris*(hydroxymethly)aminomethane and an appropriate volume of 1 $mol·dm^{-3}$ HCl solution. After soaking at 36.5°C for a given period, the films were removed from the solution, washed with ultrapure water, and then dried at room temperature. The surfaces of the films were characterized both before and after soaking in 1.5SBF, using thin-film X-ray diffraction (TF-XRD) and scanning electron microscopy (SEM). Composition of the specimens was measured by energy dispersive X-ray microanalyzer (EDX).

Results and Discussion

TF-XRD patterns of the surfaces of S(50)Ca(y) films (y=0, 10, 20 and 40) after soaking in 1.5SBF for 7 days showed peaks ascribed to apatite in the patterns of the S(50)Ca(20) and S(50)Ca(40), whereas no apatite peaks for S(50)Ca(0) and S(50)Ca(10). Figure 2 shows TF-XRD patterns of S(x)Ca(20) and S(x)Ca(40) films (x = 0, 20 and 50) after soaking in 1.5SBF for 7 days. Peaks assigned to apatite were detected at 2θ = 26 and 32° for the films irrespective of the content of sulfonic groups when the films

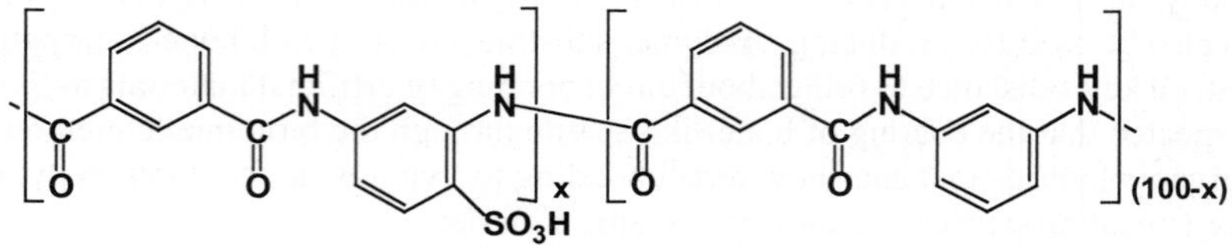

Fig. 1 Structural formula of polyamide S(x).

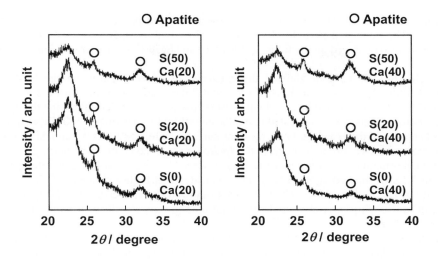

Fig. 2 TF-XRD patterns of the surfaces of S(x)Ca(20) and S(x)Ca(40) films after soaking in 1.5SBF for 7 days.

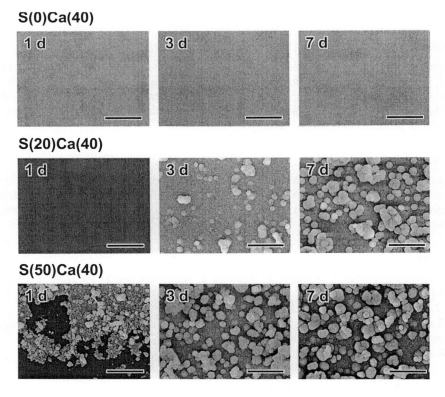

Fig. 3 SEM photographs of the surfaces of S(x)Ca(40) films after soaking in 1.5SBF for various periods. (Scale bar indicates 25μm.)

respectively. The results of EDX distinctly showed that the particles consisted of calcium and phosphorus. The deposited particles were attributed to hydroxyapatite. Formation of the apatite particles were also observed on the films S(20)Ca(20) and S(50)Ca(20) after soaking in 1.5SBF for 3 days.

Although TF-XRD patterns showed the existence of hydroxyapatite on the S(0)Ca(20) and S(0)Ca(40) films after soaking in 1.5SBF for 3 and 7 days, almost all the surfaces looked smooth under SEM because the formed layer of hydroxyapatite was easily peeled off during the operation for SEM observation. This result suggested that adhesion of hydroxyapatite to S(0)Ca(y) films was too weak to keep the hydroxyapatite layer on their surfaces, while S(20)Ca(y) and S(50)Ca(y) could show enough adhesive strength to keep the hydroxyapatite layer on their surfaces during SEM observation. Formation of hydroxyapatite was observed on S(50)Ca(40) in shorter periods after soaking in 1.5SBF than on S(50)Ca(20). This indicates that higher contents of $CaCl_2$ resulted in higher rates of apatite formation. S(50)Ca(40) also showed shorter periods for the apatite formation than S(20)Ca(40). (Fig.3). Higher contents of sulfonic groups also lead to higher rates of apatite formation in 1.5SBF.

The present results indicate that existence of sulfonic group accelerated apatite formation on the polyamide film when it contained the same amount of $CaCl_2$. Apatite deposition could be detected by TF-XRD even when the polyamide did not contain sulfonic groups but did contain large amounts of $CaCl_2$. However, the adhesive strength between apatite layer and polyamide film without sulfonic groups was too low to be held on the surface under SEM observation. This means that the interaction between the sulfonic group and hydroxyapatite crystals may provide increased strength in adhesion of the apatite layer to the substrates.

Conclusions

Polyamide containing sulfonic groups and calcium chloride formed an apatite layer on its surface in 1.5SBF. The rate of apatite formation increased with increasing amounts of sulfonic group as well as calcium chloride. Sulfonic groups in polyamide gave an increase in adhesive strength between the apatite layer and the substrate. Organic polymers modified with sulfonic groups and calcium chloride are candidate materials to produce apatite-polymer hybrids.

Acknowledgements

This work was conducted under the auspices of the research project, "Technology Development for Medical Materials Merging Genome Information and Materials Science", in the Kansai Science City Innovative Cluster Creation Project, supported by the Ministry of Education, Culture, Sports, Science and Technology of Japan.

References

[1] M. Tanahashi, T. Yao, T. Kokubo, M. Minoda, T. Nakamura and T. Yamamuro: J. Am. Ceram. Soc. 77 (1994) 2805.
[2] M. Tanahashi and T. Matsuda: J. Biomed. Mater. Res. 34 (1997) 305.
[3] T. Kawai, T. Miyazaki, C. Ohtsuki, M. Tanihara, J. Nakao, Y. Sakaguchi and S. Konagaya: Key engineering Materials. 240-242 (2003) 59.
[4] S. Konagaya and M. Tokai: J. Appl. Polym. Sci. 76 (2000) 913.

Key Engineering Materials Vols. 254-256(2004) pp. 529-532
online at http://www.scientific.net
© *2004 Trans Tech Publications, Switzerland*

Drug Delivery Behaviour of Hydroxyapatite and Carbonated Apatite

A.J. Melville[1], L.M. Rodríguez-Lorenzo[1], J.S. Forsythe[1] and K.A. Gross[1]

[1]School of Physics and Materials Engineering, Building 69, Monash Uni., VIC 3800, Australia.

e-mail: karlis.gross@spme.monash.edu.au

Keywords: Carbonated apatite, hydroxyapatite, ibuprofen, drug adsorbance, drug release rate.

Abstract. The effectiveness of biomaterials is being increased through the design of substrates capable of allowing slow release of therapeutic drugs or chemicals to increase the function of implanted devices. This investigation showed carbonated apatite to have a higher adsorbance of Ibuprofen than hydroxyapatite, and furthermore to release this drug at a faster rate into a tris buffer solution. It is suggested that carbonated apatite may also serve as a higher adsorbance substrate than hydroxyapatite when loaded with other therapeutic drugs.

Introduction

Calcium phosphate materials are currently being investigated as potential drug delivery systems, such as carriers for antibiotics [1] due to their osteoconductive, biocompatible and bioresorbable behaviour [2]. The aim of this study is to investigate the drug delivery capabilities of hydroxyapatite and carbonated apatite when loaded with Ibuprofen, a non-steroidal anti-inflammatory drug.

Materials and Methods

Hydroxyapatite (HAP) and 5wt% carbonated apatite (CAP) were synthesised by a precipitation method, whereby a 1L solution containing (0.6M) $(NH_4)HPO_4$ (Riedel-de Haën, extra pure) for hydroxyapatite and $(NH_4)HPO_4/NaHCO_3$ (Selby-Biolab, analytical reagent) for carbonated apatite was added to 1L of 1M $Ca(NO_3)_2.4H_2O$ (Riedel-de Haën, analytical reagent) solution in 10ml quantities using a 736 GP Titrino and 700 Dosino (Metrohm, Switzerland) over a period of approximately 8hrs. HAP synthesis was conducted with (high purity) nitrogen gas bubbling through the solution to prevent carbonation. Both reactions were conducted in a polypropylene vessel supported in a water bath at 37°C, and the pH was maintained at 9.4 through the addition of an ammonium hydroxide solution. After completion of the reaction, the precipitate was aged for 24hrs, filtered, washed twice with deionised water and once with ethanol, and then dried at 80°C for 24hrs.

To prepare porous apatite discs, 1g of polylactic acid (PLA) was dissolved in chloroform overnight, then half of the solution was thoroughly mixed to form a slurry with 2.5g of HAP and half with 2.5g CAP. After the chloroform had evaporated the mixtures were filtered 3 times with ethanol, then dried for 2h at 100°C. Discs were prepared in a uniaxial press, whereby 1g of HAP, HAP/PLA, CAP, and CAP/PLA were pressed at 127MPa for 30s to form 2 discs of each composition. The discs were then sintered at 900°C at a heating rate of 1°C/min. HAP sintering was conducted in a dry air atmosphere, and CAP in a dry CO_2 atmosphere.

An Ibuprofen solution was prepared by dissolving 0.8g of Ibuprofen (Aldrich) in 100ml of ethanol. One disc of each composition was immersed in 25ml of this solution, and the four vessels then placed under vacuum to impregnate the Ibuprofen solution into the pores of the apatite discs. The vacuum reached 200mbar for 5 minutes, after which it was slowly released and the vessels left still for a further 5 minutes. The discs were then dried for 30 minutes at 50°C, well below the melting temperature of Ibuprofen of 71°C [3].

Characterisation of each composition included thermal behaviour, chemical and phase analysis, density, porosity, crystalline defect examination, and drug release rate measurement. The thermal behaviour of PLA was investigated using Thermal Gravimetric Analysis (TG92 Setaram) (TGA), measured in both air and CO_2 atmospheres at a heating rate of 1°C/min. The specific surface area of each apatite was measured using Micromeritics Gemini 2360 Brunauer-Emmett-Teller (BET) gas adsorption apparatus, with nitrogen gas as the adsorbate. The samples were degassed overnight using a Micromeritics Flowprep 060 at 100°C. Chemical bonding was determined from Fourier Transform Infra Red (FTIR), using a Perkin Elmer Spectrum GX infrared spectrometer. Spectra were collected for both apatite powders between 4000 and 400cm^{-1}, with a resolution of 2cm^{-1} at an interval of 0.5cm^{-1}. X-ray Diffraction (XRD) data was analysed to determine the phase composition of the apatites, using a step of 0.02° between 10° and 90°. Calcium content of the dried apatite powders was measured using an Atomic Absorption Spectrometer (Varian), and phosphorous content using a UV-visible Spectrophotometer (Varian). Calcium was analysed at 422.7nm using an air-acetylene flame, and phosphorous at 420nm. Carbon content was measured by The Campbell Microanalytical Laboratory (University of Otago, New Zealand) using a CHN elemental analyser. Density was quantified using the Archimedes principle. Crystalline defects were examined via Transmission Electron Microscopy (TEM), using a Phillips CM20 for hydroxyapatite and JEOL 2011 for carbonated apatite, and porosity was observed from Scanning Electron Microscopy using a Hitachi S570 (SEM). The drug release rate was determined using UV-visible Spectrophotometry (Varian) measured at 37°C. Spectra were collected between 600 and 220nm, and the absorbance of Ibuprofen at λ =264nm examined. The Ibuprofen-loaded apatite discs were placed in 50ml of pH 7.3 0.1M tris(hydroxymethyl)aminomethane solution, and positioned in an Environmental Shaker (ES-20) set at 100rpm and 37°C. The amount of Ibuprofen released into the tris buffer was measured at 15 minutes after the disc was immersed in the solution, then every 30 minutes for 4hrs, and every hour for a further 3hrs.

Results

To ensure complete polymer burnout from the apatite discs, TGA was conducted in atmospheres identical to that to be used for sintering. It can be seen in Fig. 1 that PLA was completely removed by 450°C in both atmospheres when heating at a rate of 1°C/min, thus implying that a sintering temperature of 900°C would be satisfactory. The specific surface area, Ca/P ratio and carbon content of the apatite compositions examined are presented in Table 1. Density measurements revealed that the discs loaded with PLA were less dense, whereby HAP

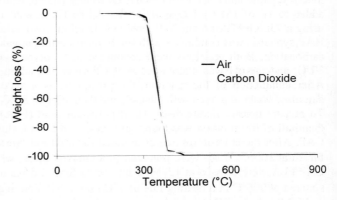

Fig. 1. TGA curves of PLA in air and CO_2

was 1.452, HAP/PLA 1.369, CAP 1.469 and CAP/PLA 1.181g/cm^3. Ca/P ratios showed that stoichiometric HAP was produced, and a ratio of 1.8 confirmed B-type carbonate substitution for CAP. CAP is seen to contain 4.597wt% carbonate within its structure, and is thus closely matched to natural bone. Characterisation of the apatites demonstrated that each was composed of a single apatitic phase, and that additional CaO and β-TCP phases were not present as seen in Fig. 2.

FTIR spectra of the dried powders displays water remaining from synthesis that was later removed during sintering. Carbonate modes present are also visible as indicated by arrows in Fig. 3.

SEM results showed that significantly more porosity was present in the composite discs as compared to the pure apatite discs. Pore sizes were measured to be on average between 0.2µm and 0.5µm for all discs, with some larger pores randomly dispersed throughout (Fig. 4).

TEM results confirmed that the samples possessed the hexagonal apatitic structure, complementing the measured lattice parameters of a = 9.427Å and c = 6.896Å for hydroxyapatite measured by Rietveld analysis from X-ray diffraction data. Fig. 5 highlights the difference in crystallite structure between the differently chemically substituted samples.

Ibuprofen release rates are illustrated in Fig. 6. It can be seen that CAP displayed the highest release rate of Ibuprofen over the time period measured, followed by HAP/PLA, and HAP. Measurements of drug loading have shown that CAP had the highest adsorbance of Ibuprofen of 4.07mg, followed by HAP with 3.70mg and HAP/PLA with 3.20mg, which would correspond to a final molarity of 3.95×10^{-4}M for CAP, 3.55×10^{-4}M for HAP and 3.10×10^{-4}M for HAP/PLA if all of the Ibuprofen were released into solution.

Table 1. Characterisation of HAP and CAP.

	HAP	CAP
Specific surface area [m²/g]	60.01	53.38
Ca/P ratio	1.7	1.8
Carbon content [wt%]	-	4.597

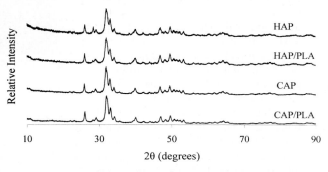

Fig. 2. XRD patterns of HAP, HAP/PLA, CAP & CAP/PLA.

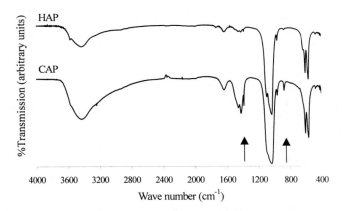

Fig. 3. FTIR spectra of dried HAP and CAP powders.

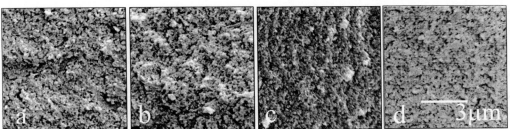

Fig. 4. SEM pictures of sintered (a)HAP, (b)HAP/PLA, (c)CAP and (d)CAP/PLA discs.

Discussion.

CAP displayed the highest uptake of Ibuprofen, and also the fastest release rate (Fig. 6). Furthermore, only a small difference was noted between the uptake of HAP and the HAP disc prepared with PLA. Porosity of the discs was calculated as 54.42% for CAP, 56.64% for HAP/PLA and 54.01% for HAP. This suggests that the uptake and release of Ibuprofen is primarily related to the surface of the discs and not the extent of porosity available to contain additional Ibuprofen. This is confirmed when noting that 88.8% for CAP, 92.7% for HAP/PLA and 85.6% for HAP of Ibuprofen present in the pellets was released over the first 7hrs. Such a fast release would imply that the drug was adsorbed more onto the surfaces rather than contained within the disc, as this would require a longer period to diffuse out into solution. TEM results (Fig. 5) show that CAP possessed more lattice defects, primarily in the form of grain boundaries, and hence a higher surface energy. This would account for the increased adsorbance of Ibuprofen on the surface of CAP as compared to HAP.

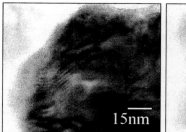
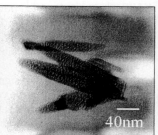

Fig. 5. TEM pictures of (a)CAP and (b)HAP dried powders.

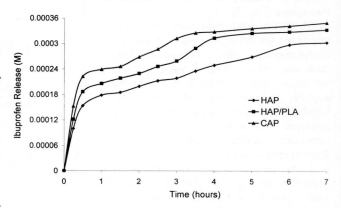

Fig. 6. Ibuprofen release from HAP, HAP/PLA and CAP discs in 0.1M tris buffer at 37°C

Conclusion

Ibuprofen was found to adsorb to a larger extent on CAP than HAP surfaces, and is also released from CAP at a faster rate, hence it is the better apatite if faster drug delivery of Ibuprofen is desired. Future work can involve increasing the porosity fraction of the discs, and loading with a higher drug concentration.

Acknowledgements

The authors gratefully acknowledge the assistance of Christian Maunders, Chamini Mendis, Silvio Mattievich and Timothy Scott. This work was supported by a Monash Research Fund grant and Australian Research Council grants F10017027 and A10017174.

References

[1] M. Otsuka, Y. Matsuda, D. Yu, J. Wong, J.L. Fox, and W.I. Higuchi: Chem. Pharm. Bull. Vol. 38 (1990), p. 3500.

[2] S. Kimakhe, S. Bohic, C. Larrose, A. Reynaud, P. Pilet, B. Giumelli, D. Heymann, and G. Daculsi:J. Biomed. Mater. Res. Vol. 47 (1999), p. 18.

[3] S. Ladrón de Guevara-Fernández, C.V. Ragel and M. Vallet-Regí: Biomaterials Vol. 24 (2003) p. 4037.

Key Engineering Materials Vols. 254-256(2004) pp. 533-536
online at http://www.scientific.net
© 2004 Trans Tech Publications, Switzerland

Preparation of Calcium Carbonate / Poly(lactic acid) Composite (CCPC) Hollow Spheres

H. Maeda, T. Kasuga, M. Nogami

Graduate School of Engineering, Nagoya Institute of Technology

Gokiso-cho, Showa-ku, Nagoya 466-8555, Japan

hirotaka@zymail.mse.nitech.ac.jp

Keywords: Hollow sphere, calcium carbonate, Poly(lactic) acid, composite, apatite

Abstract. Novel hollow spheres were prepared using calcium carbonate / poly(lactic acid) composite (CCPC). The CCPC slurry was drop-wise added to 1 % poly(ethylene glycol) (PEG) aqueous solution with stirring for 1 d. Numerous CO_2 gas generated from calcium carbonates in CCPC during the process which contributes to the formation of the hollow sphere. The size of the spheres is around 0.8 ~ 1.5 mm in diameter. The spheres have thicknesses of the shells in the range of 50 to 300 μm.

Introduction

Materials having a high bioactivity or bioresorbability play an important role in the recovery of bone defects. In our earlier work a composite consisting of calcium carbonate (vaterite) and poly(lactic acid) (PLA) (denoted by CCPC) was reported to form bonelike hydroxycarbonate apatite (b-HA) on its surface even in several hours of soaking in simulated body fluid (SBF)[1, 2]. The rapid formation of b-HA was suggested to originate from the integration of PLA having carboxy groups bonded with calcium ions for the b-HA nucleation and a large amount of vaterite having an ability to effectively increase the supersaturation of the b-HA. We believe that various biomaterials such as bone-fillers or bone plates can be prepared using PLA and calcium carbonate.

In general, PLA microspheres are prepared by an oil-in-water emulsion method with the aid of surfactants [3]. It was also reported that PLA composite microspheres can be prepared without surfactants [4]. Formation of porous materials using biodegradable microspheres would have the possibility to obtain easily three-dimensional designed materials for bone grafts[5]. In the present work hollow spheres of CCPC were prepared for developing novel biomaterials that may make it possible to deliver cells or drugs in addition to a conventional application such as bone fillers.

Methods

The preparation procedure of calcium carbonates consisting of vaterite was described in the earlier papers [1, 2]. CO_2 gas was blown for 3 h at a flow rate of 300 mL/min into the suspension consisting of 7.0 g of $Ca(OH)_2$ in 180 mL of methanol at 0 °C in a Pyrex® beaker. The resultant slurry was dried at 70 °C in air to prepare fine-sized powders. The BET surface area was measured to be ~40 m^2/g. Figure 1 shows a scanning electron micrograph (SEM photographs) and XRD pattern of the powders. Secondary spherical particles of 0.5~1 μm diameter consist of agglomeration of primary particles of ~100 nm diameter. The particles are suggested to be vaterite crystals because they have a tendency to form in fine-sized spherical shapes[6]. X-ray diffraction analysis shows that the calcium carbonate powders obtained in the present work consist predominantly of vaterite with small amounts of aragonite and calcite.

The molecular weight of the received PLA was determined to be 160 ± 20 kDa by gel permeation chromatography. 0.5 g of PLA was dissolved in 10 mL of methylene chloride, which includes dilute hydrochloric acid as a stabilizer, at room temperature. The calcium carbonate powders were put into

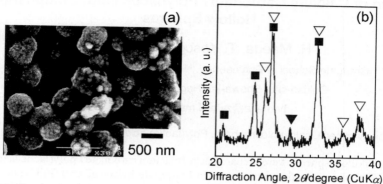

Fig. 1. (a) SEM photograph and (b) XRD pattern of the calcium carbonate prepared in the present work. (■) vaterite, (▽) aragonite, and (▼) calcite.

the above PLA solution and then the mixture was stirred. The weight ratio of CaCO₃/PLA was 1/2. The small sphere in the present work was produced by an oil-in-water emulsion method[3]. The CCPC slurry was drop-wise added to 1 % poly(ethylene glycol) (PEG) aqueous solution with stirring for 1 d. The small spheres were isolated by filtration, washed with ethanol, and dried at 40 °C. In our preliminary experiments, the PEG solution percolates gradually through the sphere. As a result, the shape of the sphere is effectively maintained wihtout fracture.

In the present work the apatite-forming ability of a small sphere was examined. The small spheres were soaked under static conditions in a 50 mL of SBF (consisting of 2.5 mM of Ca^{2+}, 142.0 mM of Na^+, 1.5 mM of Mg^{2+}, 5.0 mM of K^+, 148.8 mM of Cl^-, 4.2 mM of HCO_3^-, 1.0 mM of HPO_4^{2-}, and 0.5 mM of SO_4^{2-}) that included 50 mM of $(CH_2OH)_3CNH_2$ and 45.0 mM of HCl at pH 7.4 at 37 °C. After the soaking, the samples were removed from the SBF and were gently washed with distilled water. They were dried at room temperature.

The crystalline phases of the small spheres before and after the soaking in SBF were identified by X-ray diffractometer (XRD). Their morphology of the surface and cross-section face was observed by scanning electron microscopy (SEM).

Results

The hollow spheres of 0.8 ~ 1.5 mm diameters were obtained in the present work. Figure 2(a) shows a typical SEM photograph of the hollow sphere. Figure 2(b) shows the exterior surface of the sphere with high magnification. Needle-like crystals, a typical shape of aragonite crystals, are seen. As shown in Figure 2(c), the sphere has a shell with a thickness in the range of 50 to 300 μm. The shell includes numerous pores, which are smaller than 100 μm in diameter (Fig. 2(c)). The pores below several micrometers in diameter are also seen at the inside surface of the shell (Fig. 2(d)).

Figure 3 shows an XRD pattern of the hollow spheres prepared in the present work. It showed that the hollow spheres consist of crystalline PLA and CaCO₃ (mainly aragonite). A large amount of vaterite in calcium carbonates as the starting material disappeared after the spheres formed.

In XRD patterns for samples soaking in SBF for 1 week, the peaks corresponding to apatite crystal can be seen. Figure 4 shows the SEM photograph of the exterior surface of the hollow sphere after 1 week of soaking in SBF. In the SEM photograph, the exterior surface of the sphere is partially covered with the leaf-like nanodeposits, which are b-HA, judged from the morphology and the XRD pattern. SEM observation showed that no deposit formed newly on the inside surface.

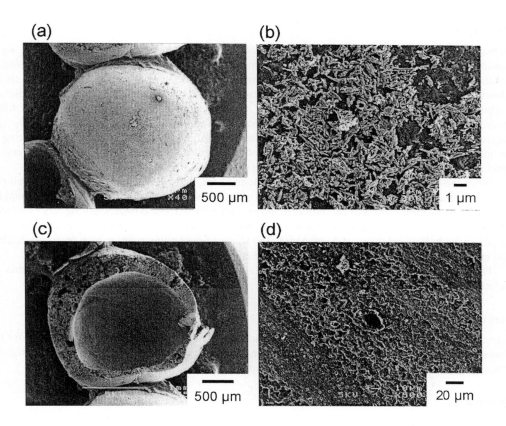

Fig. 2. SEM photographs of the small sphere. (a, b) exterior, (c) fracture face, and (d) inside surface of the shell.

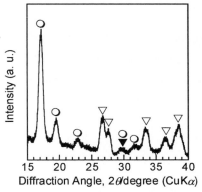

Fig. 3. XRD pattern of the small sphere.
(○) PLA, (▽) aragonite, and (▼) calcite.

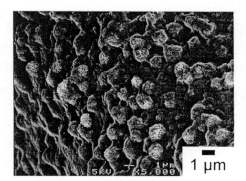

Fig. 4. SEM photograph of the exterior surface of the small sphere after 1 week of soaking in SBF at 37 °C.

Discussion

When the CCPC slurry is cast into the PEG aqueous solution, numerous droplets of CCPC form automatically. The formed spheres are 0.5 ~ 1.0 mm in diameter at this time. As methylene chloride solution gradually releases from the surface of the droplets, the interior of the droplets starts to set slowly from the surface towards the inside, resulting in formation of the thin shells. Calcium carbonate (vaterite) powders in interior phase, which has not set yet, would react with dilute hydrochloric acid included in methylene chloride solution to generate numerous CO_2 gas microbubbles. Gradually, these microbubbles gather and unify to grow into large bubbles, resulting in an increase in the diameter of the spheres. Then the generated numerous CO_2 gas in the spheres is gradually emitted through connective pores in the shells. As a result, the hollow spheres form.

When the CCPC slurry was prepared using calcite or aragonite as calcium carbonates, no hollow spheres formed. It is important for the preparation of hollow spheres that numerous CO_2 gas is rapidly generated by decomposition of calcium carbonates. Since the vaterite powders having large surface area and high solubility would react readily with dilute hydrochloric acid in methlyne chloride solution, the hollow spheres can be prepared. The shape of the hollow spheres may be controlled by the amount and particle sizes of calcium carbonates.

The hollow spheres may be great potential candidates as cell- or drug-delivery systems or injectable bone fillers.

Acknowledgements

This work was supported in part by a Grant-in-Aid for Scientific Research from Japan Society for the Promotion of Science (No. 14380398) and grant from the NITECH 21st Century COE Program "World Ceramics Center for Environmental Harmony".

References

[1] H. Maeda, T. Kasuga, T. Nogami, Y. Hibino, K. Hata, M. Ueda, Y. Ota: Key Engng. Mater. Vol. 240-242 (2002), p. 163.

[2] T. Kasuga, H. Maeda, K. Kato, T. Nogami, K. Hata, M. Ueda, Y. Ota: Biomaterials Vol. 24 (2003), p. 3247.

[3] *JPN patent application* H7-108165 (1996).

[4] F. Nagata, T. Miyajima, Y. Yokogawa: Key Engng. Mater. Vol. 240-242 (2002), p. 167.

[5] M. Bordenm, M. Attaiwa, Y. Khan, C. T. Laurencin: Biomaterials Vol. 23 (2002), p. 551.

[6] T. Yasue, Y. Koijma and Y. Arai: Gyp. Lime, Vol. 247 (1993), p. 471 (in Japanese).

Key Engineering Materials Vols. 254-256(2004) pp. 537-540
online at http://www.scientific.net
© 2004 Trans Tech Publications, Switzerland

The Role of Phosvitin for Nucleation of Calcium Phosphates on Collagen

Naoko Kobayashi[1], Kazuo Onuma[2], Ayako Oyane[2] and Atsushi Yamazaki[3]

[1] Major in Mineral Resources Engineering and Materials Science and Engineering, Mineral Resources Engineering, Mineral Resources Engineering Specialization, Science and Engineering, Waseda University, Japan, woodstock@ruri.waseda.jp

[2] National Institute of Advanced Industrial Science and Technology, Japan, k.onuma@aist.go.jp, a-oyane@aist.go.jp

[3] Department of Resources and Environmental Engineering, Science and Engineering, Waseda University, Japan, ya81349@waseda.jp

Corresponding author: Dr. Kazuo Onuma

Keywords: collagen, calcium phosphate, AFM, phosvitin

Abstract. The role of phospho-protein in bone formation was investigated in this study. A phosvitin layer was formed on a collagen sheet by soaking the sheet in a fluid containing phosvitin of 0.04wt% and various concentrations of dimethylsuberimidate. The dimethylsuberimidate concentration was changed from 0.04 to 0.3wt% to find the appropriate concentration for phosvitin-collagen layer formation. The composite layer was soaked in physiological solutions and the nucleation process of calcium phosphates was observed by using atomic force microscopy (AFM). It was found that small hydroxyapatite (HAP) crystals were formed throughout the layer with their a-axes arranged perpendicular to it. On the other hand, HAP with its c-axis perpendicular to the layer was formed when phosvitin was not present. This makes it clear that the roles of phosvitin are to enhance nucleation of HAP, and to control the orientation of HAP crystals on collagen in the same way as that observed in human bone.

Introduction

Unusual bone metabolism diseases such as osteoporosis have been found to be particularly prevalent in aged persons and in persons who have an unbalanced diet. A great many people who suffer from such diseases need to use high-quality artificial bones that can be used safely for long periods [1-3]. Human bone is a composite of nano-sized calcium phosphate crystals and collagen having both high strength and flexibility. The calcium phosphates are carbonated hydroxyapaite (HAP) crystals, and they regularly form their a-axes perpendicular to collagen fibers. For more than 30 years, many attempts have been made to investigate the formation mechanism of bone structure in nanoscale, but no final conclusions have yet been reached. [4-6] We believe that the development of artificial bone that has the same structure and chemical composition as those of human bone will help to satisfy an important medical need, and that it cannot be doubted that fundamental investigations of the bone formation process in nanoscale are necessary to achieve this. One of the important issues concerning the bone formation process that has been clarified through previous works is the role of phospho-protein for nucleation of HAP [7, 8]. A central topic that has been discussed is that phospho-proteins exist on the collagen, and that they play a "glue" role in binding HAP crystals to the collagen. Until now, however, no detailed mechanism of phospho-proteins has been clarified. To address this issue, we investigated the effect of phospho-proteins on nucleation of HAP on collagen under a pseudo-physiological condition in the present study. The morphology, chemical composition, and orientation of deposited HAP were characterized by using atomic force microscopy (AFM), X-ray photoelectron spectroscopy (XPS), and thin-film X-ray diffraction (TH-XRD).

Experimental

1. Sample preparation

A type I collagen sheet was soaked in a solution containing phosvitin and dimethylsuberimidate for a few days to form a collagen-phosvitin composite layer. 0.04 wt% phosvitin and 0.04 - 0.3 wt% dimethylsuberimidate were separately dissolved in 200 mM of tris solution at pH 8.5, and both solutions were stirred for one night at 4°. After both solutions were mixed with the same volume, the collagen sheet was soaked in the latter solution for a few days. The sheet was then removed from the solution and washed with ultra pure water for one day to remove the unbounded phosvitin on collagen completely. This was done because the isolated phosvitin in the latter solution strongly inhibits nucleation and growth of HAP [9]. The collagen-phosvitin composite layer was then soaked in a physiological solution (hereafter referred to as simulated body fluid, or SBF) for 1-5 days. The SBF composition was 140 mM of NaCl, 1 mM of $K_2HPO_4.3H_2O$, and 2.5 mM of $CaCl_2$, and was buffered by 50 mM of tris and 1 N HCl at pH 7.4 [10].

2. Observation and characterization of precipitates

Samples were taken from the SBF and washed with ultra pure water before making observations and characterizations. We used AFM to observe precipitates on the composite layer, TH-XRD to characterize precipitates, and XPS to search for an appropriate concentration of dimethylsuberimidate before soaking the samples in a composite layer in SBF.

Results and discussion

1. Estimation of bound phosvitin on collagen by XPS

The amount of bound phosvitin was analyzed by using XPS (Figs. 1a, b, and c). Measured spectra were fitted according to the Gaussian–Lorentzian equation as follows,

$$B(X_i, Q) = H \left\{ PG \cdot e^{-\ln 2 \left[\frac{2(X_i - PP)}{FWHM} \right]^{-2}} + \frac{1 - PG}{1 + \left[\frac{2(X_i - PP)}{FWHM} \right]^2} \right\}$$

where $B(X_i, Q)$ is a functional value at the energy point X_i that is the binding energy value for the data point i, PP is the binding energy at the center of a peak, H is the height of a peak at its center, FWHM is the full width at the half maximum of a peak, and PG is a proportion of the Gaussian part in the equation.

After fitting, curves were normalized by using the Si intensity, which was measured for silica glass and used as the base line of a spectrum.

In the present study, the dimetylsuberimidate concentrations used were 0.04, 0.1, 0.15, 0.2, 0.25, and 0.3 wt%. From 0.04 to 0.15 wt%, the amount of P atoms on the collagen sheet was very small, which indicates that the amount of dimethylsuberimidate was not enough to bind phosvitin to the collagen molecules. In contrast, at 0.3wt% of dimethylsuberimidate, the number of P atoms gradually increased through 3 days of soaking, but after 3 days it decreased, as shown in Fig.1a. This feature was not observed at 0.2 wt %, in which case the number of P atoms continuously increased throughout 5 days of soaking. We therefore set the dimethylsuberimidate concentration at 0.2 wt% to form a phosvitin-collagen composite layer.

The reason the critical concentration of dimethylsuberimidate for P atoms in the composite layer was observed is as follows.

Dimethylsuberimidate is a cross-linker, and has two $R-CH_3$ groups that react with $R-NH_2$ of protein. Therefore, when the excess dimethylsuberimidate exists and is freely solved in the solution, the groups easily bind to NH_2 groups of phosvitin, which are already bound to the collagen. This means that dimethylsuberimidate covers some parts of the phosvitin molecules and decreases the number of effective binding sites for calcium phosphates. The results seen in Figs.1b and c are consistent with this idea.

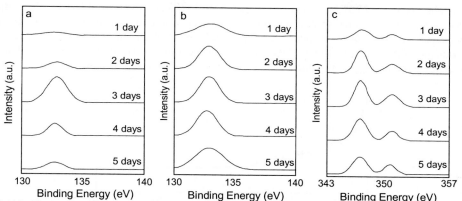

Fig. 1. (a) Change in XPS spectrum of P atoms with soaking time on phosvitin-collagen layer at the concentration of 0.3 wt% of dimethylsuberimidate. (b) Change in P atoms after soaking in SBF under the same condition as (a). (c) Change in Ca atoms after soaking in SBF under the same condition as (a).

2. AFM observations

Figure 2 shows AFM images of the phosvitin-collagen composite layer (Fig.2a), precipitates nucleated on the composite layer (Fig.2b), and precipitates nucleated on the collagen layer without phosvitin (Fig.2c). It was found that calcium phosphate particles with 20-30 nm diameters were regularly formed over the whole surface of the composite layer. The precipitates are densely packed. On the other hand, calcium phosphates nucleated irregularly on the surface when phosvitin was not bound to the collagen. The area where nucleation took place was limited, and precipitates tended to grow normally.

Fig. 2. (a) Collagen-phosvitin composite layer. (b) Calcium phosphates nucleated on (a). (c) Calcium phosphates nucleated on collagen without phosvitin. All images were taken at $1 \mu m \times 1 \mu m$ scale.

The fast nucleation rate in the presence of phosvitin is attributed to two effects. The first of these is that PO_4 groups in phosvitin are likely to bind calcium ions following HAP forms. The second is that the surface energy, γ, of the phosvitin-collagen layer probably decreases compared to that of the collagen layer. Because the activation energy of nucleation is inversely proportional to γ^3, even a slight change in γ drastically affects the nucleation rate.

3. TH-XRD analysis

A precipitate was characterized by using TH-XRD. In the absence of phosvitin, a sharp intense peak was observed at around 31.7°, which corresponded to (002) reflection of HAP . This indicates that HAP crystals arrange themselves so that their c-axes are perpendicular to the collagen layer. On the other hand, in the presence of phosvitin, three peaks were observed as shown in Fig. 2. These peaks coincide with those of (100), (200), and (300) reflections of HAP. This indicates that HAP crystals arrange themselves so that their a-axes are perpendicular to the composite layer.

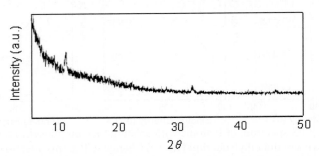

Fig. 3. TH-XRD pattern of precipitates (soaked in phosvitin solution for 5 days and then soaked in SBF for 5 days)

It is not yet clear why HAP crystals arrange themselves so that their a-axes are perpendicular to the composite layer. The reason may relate to the distance between PO_4 groups of phosvitin molecules.

Conclusion

The roles of phosvitin are summarized as follows.
  (1) Nucleation rate of calcium phosphate is greatly accelerated in the presence of phosvitin.
  (2) Orientation of precipitated calcium phosphate is controlled by phosvitin. The arrangement whereby the a-axis is perpendicular to the collagen surface is the same as that in the human bone structure.

References

[1] S. Itoh, M. Kikuchi, Y. Koyama, K. Takakuda, K. Shinomiya and J. Tanaka: Biomaterials Vol.23-19 (2002), p. 3919
[2] Myung C. Chang and J. Tanaka: Biomaterials Vol. 23-18 (2002), p. 3879
[3] M. Kikuchi, S. Itoh, S. Ichinose, K. Shinomiya and J. Tanaka: Biomaterials Vol. 22-13 (2001), p. 1705
[4] J. Christoffersen, M. R. Christoffersen, S. B. Christensen and G. H. Nancollas: Journal of Crystal Growth Vol. 62-2 (1983), p. 254
[5] R. Bareille, M. H. Lafage-Proust, C. Faucheux, N. Laroche, R. Wenz, M. Dard and J. Amedee: Biomaterials Vol.21-13 (2000), p. 1345
[6] I. Ono, M. Inoue and Y. Kuboki: Bone Vol.19-6 (1996), p. 581
[7] T. Saito, A. L. Arsenault, M. Yamauchi, Y. Kuboki and M. A. Crenshaw: Bone Vol.21-4 (1997), p. 305
[8] R. A. Gelman, K. M. Conn and J. D. Termine: Biochimica et Biophysica Acta (BBA) Vol.630-2 (1980), p. 220
[9] F. C. Church, C. W. Pratt, R. E. Treanor and H. C. Whinna: FEBS Letters Vol. 237-1-2 (1998), p. 26
[10] I. B. Lenor, A. Ito, K. Onuma, N. Kanzaki and R. L. Reis: Biomaterials Vol. 24-4 (2003), p. 579

Key Engineering Materials Vols. 254-256(2004) pp. 541-544
online at http://www.scientific.net
© *2004 Trans Tech Publications, Switzerland*

Biomimetic Coating of Laminin-Apatite Composite Layer onto Ethylene-Vinyl Alcohol Copolymer

Ayako Oyane[1], Masaki Uchida[2] and Atsuo Ito[3]

[1] Research Center of Macromolecular Technology, National Institute of Advanced Industrial Science and Technology, 2-41-6 Aomi, Koto-ku, Tokyo, Japan, a-oyane@aist.go.jp

[2] Institute for Human Science and Biomedical Engineering, National Institute of Advanced Industrial Science and Technology, 1-1-1 Higashi, Tsukuba, Ibaraki 305-0046, Japan, uchida-m@aist.go.jp

[3] Institute for Human Science and Biomedical Engineering, National Institute of Advanced Industrial Science and Technology, 1-1-1 Higashi, Tsukuba, Ibaraki 305-0046, Japan, atsuo-ito@aist.go.jp

Keywords: Apatite, laminin, biomimetic method, ethylene-vinyl alcohol copolymer (EVOH) , composite, skin terminal

Abstract. A laminin-apatite composite layer was formed on the surface of ethylene-vinyl alcohol copolymer (EVOH) by a biomimetic process, in which the EVOH substrate was alternatively soaked in calcium and phosphate solutions to make apatite nuclei on its surface, and then soaked in a laminin-containing calcium phosphate solution to grow a laminin-apatite composite layer on the nuclei. An epithelial-like cell, BSCC93 strongly adhered to the surface of thus-prepared specimen. The present biomimetic process would be effective in producing a new skin terminal with improved epithelial adhesion.

Introduction

A skin terminal is a medical device providing a percutaneous pathway for fluids, electricity or other functions between outside of the body and internal tissues, organs or implanted devices. Essential and the most important requirement for skin terminals is prevention of epidermal downgrowth and bacterial infection. Although dense hydroxyapatite skin terminal has demonstrated long-term prevention of epidermal downgrowth and bacterial infection, its mechanical properties are far dissimilar to those of skin tissue [1, 2]. For example, dense hydroxyapatite has extremely high Young's modulus and brittle nature compared with skin tissue. On the basis of this background, an unbrittle material with low Young's modulus matching to skin tissue, good biocompatibility and strong skin-adhesiveness, is desired to be developed.

A polymeric material with a laminin-apatite composite layer on its surface should be a promising candidate as an ideal skin terminal, because apatite provides the material with good biocompatibility [3], whereas laminin, which is a cell adhesion molecule for epithelial cells, provides the material with strong skin-adhesiveness [4]. As a polymeric material, ethylene-vinyl alcohol copolymer (EVOH) is intrinsically useful, since EVOH is chemically stable and its mechanical properties can be considerably varied by changing the monomer ratio, i.e., the ratio of ethylene and vinyl alcohol segments [5]. In the present study, coating of a laminin-apatite composite layer onto the EVOH was attempted by the following biomimetic process. First, the EVOH substrate was alternatively soaked in calcium and phosphate solutions to introduce apatite nuclei on the EVOH surface [6]. Second, the specimen was soaked in a laminin-containing calcium phosphate solution which is supersaturated with respect to apatite to grow a laminin-apatite composite layer on the nuclei. Adhesiveness of an epithelial-like cell, BSCC93, to the surface of thus-prepared specimen was examined in comparison with that of the untreated EVOH and the apatite-coated EVOH as controls.

Materials and Methods

Preparation of specimens. A substrate of EVOH ($1 \times 10 \times 10 \, \text{mm}^3$) with quoted ethylene content of 32 mol%, generously supplyed by Kuraray Co. Ltd. Japan, was obtained by hot-pressing at 210°C. The EVOH substrate was abraded with a #2000 sandpaper, ultrasonically washed with aceton and ethanol, and then dried at 100°C under vacuum for 24 h.

Biomimetic process. The specimen of EVOH was soaked in a 20 mL of 200 mM-$CaCl_2$ aqueous solution for 10 s, dipped in ultra pure water and then dried. The specimen was then soaked in a 20 mL of 200 mM-$K_2HPO_4 \cdot 3H_2O$ aqueous solution for 10 s, dipped again in ultra pure water and then dried. The above alternate soaking in calcium and phosphate solutions was repeated three times. The specimen was subsequently soaked in 3 mL of a calcium phosphate (CaP) solution (denoted as EVCP) or the CaP solution suplemented with 40 μg/mL of laminin (denoted as EVCPL) at 25°C for 24 h. The CaP solution was prepared by dissolving reagent-grade NaCl (142 mM), $K_2HPO_4 \cdot 3H_2O$ (1.50 mM) and $CaCl_2$ (3.75 mM) in ultra pure water, and buffering the solution at pH 7.40 at 25°C with tris(hydroxylmethyl)aminomethane and 1 M-HCl.

Surface characterization. Surface structural changes of the specimen due to the biomimetic process were examined by a scanning electron microscope (SEM: XL30 ESEM FEG, FEI Company Japan Ltd., Japan), a thin-film X-ray diffractometer (TF-XRD: RINT 2000, RIGAKU Co., Japan) with CuKα and an X-ray photoelectron spectrometer (XPS: QUANTUM 2000, ULVAC-PHI Co. Ltd., Japan) with AlKα.

Cell-adhesiveness assay. The specimens of EVOH, EVCP and EVCPL were soaked in 2 mL of Dulbecco's modified eagle medium containing 3.0×10^5 cells of BSCC93 (RIKEN CELL BANK, Japan). After incubating at 37°C under 5% CO_2 for 2 h, the non-adherent cells were removed from the specimens by aspirating the medium and washing with PBS. The number of the cells adhered to the specimen surface was counted after being detached from the specimen using a EDTA-trypsin solution. For each kind of the specimen, six substrates were used to obtain mean value and standard deviation of the number of the adhered cells. The data was assessed using an one-way ANOVA followed by the Fisher's protected least significant difference test. Differences at $p < 0.01$ were considered to be statistically significant.

Results and Discussions

Figure 1 shows the SEM photographs of surfaces of the EVOH, EVCP and EVCPL. After the biomimetic process, a layer with a flake-like structure was newly observed on the surfaces of both EVCP and EVCPL.

Figure 2 shows the TF-XRD patterns and XPS spectra for these specimens. On the XRD patterns (Fig.2, left), peaks ascribed to low-crystalline apatite were observed for the EVCP and EVCPL, but they were not observed for the EVOH. This result indicates that the surface layers on the EVCP and EVCPL, which were observed in Fig.1, are low-crystalline apatite. Peak intensity of the apatite was slightly lower for the EVCPL than that for the EVCP. This might be due to an inhibitory effect of laminin on the apatite growth in the laminin-containing CaP solution. On the XPS spectra (Fig.2, right), peaks ascribed to calcium and phosphorous were detected for the EVCP and EVCPL, but they were not detected for the EVOH. These peaks are due to the apatite layer on the surface of the specimen, according to the XRD results (see Fig.2, left). For the EVCPL, a peak due to nitrogen, which is a component of laminin, was also detected in addition to peaks of calcium and phosphorus. Therefore, the apatite layer formed on the EVCPL contains laminin.

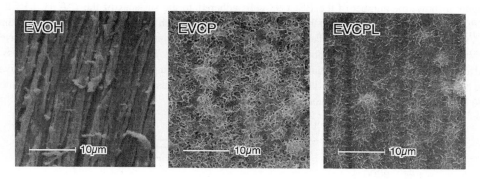

Fig 1.   SEM photographs of surfaces of the EVOH, EVCP and EVCPL.

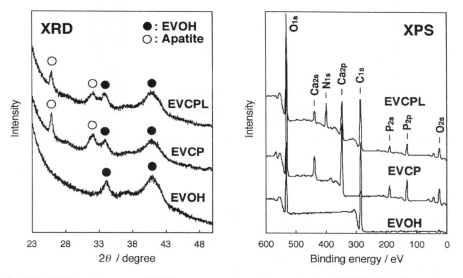

Fig 2. TF -XRD patterns and XPS spectra of surfaces of the EVOH, EVCP and EVCPL.

It was found from the above results that the apatite layer and the laminin-apatite composite layer were successfully formed on the surfaces of the EVCP and EVCPL, respectively, by the present biomimetic process. Since untreated EVOH does not form apatite on its surface in the CaP solution regardless of the laminin concentration (not shown), the alternate soaking process is prerequisite for the apatite formation on the EVCP and EVCPL. The mechanism of the apatite formation can be explained as follows. After the alternate soaking in calcium and phosphate solutions, the EVOH formed apatite nuclei or precursors of apatite on its surface. In the case of EVCP, the nuclei or the precursors spontaneously grow into the apatite layer by consuming calcium and phosphate ions in the CaP solution, since the solution is highly supersaturated with respect to apatite. In the case of EVCPL, the spontaneous growth of apatite and the attachment of laminin onto the surface of the growing apatite proceed simultaneously in the laminin-containing CaP solution to produce the laminin-apatite composite layer on the specimen surface.

Figure 3 shows the number of the BSCC93 cells adhered to the surfaces of the EVOH, EVCP and EVCPL. It is clear from the figure that the BSCC93 cells adhered much more to the EVCPL surface than to the EVOH surface and the EVCP surface. It is considered that the laminin in the surface composite layer on the EVCPL induced the strong cell adhesion to the specimen surface. The present biomimetic process was found to be useful for coating of a laminin-apatite composite layer onto a polymer surface without inactivation of the protein. A new skin terminal with strong adhesiveness to skin tissue as well as good biocompatibility would be developed by the present biomimetic process.

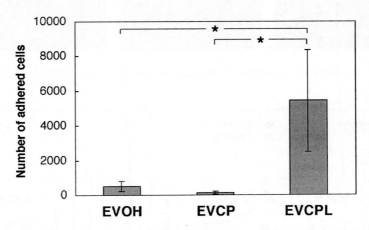

Fig 3. Number of BSCC93 cells adhered to the surfaces of the EVOH, EVCP and EVCPL (* : p<0.01, n=6)

Conclusions

The laminin-apatite composite layer was successfully formed on the EVOH substrate by the present biomimetic process. The epithelial-like cell, BSCC93, strongly adhered to the surface of thus-prepared specimen. This kind of composite material has great potential as a skin terminal with improved adhesiveness to skin tissue as well as good biocompatibility.

Acknowledgements

The authors thank to Keitaro Usui M.D., Department of Dermatology, Jichi Medical School, for his work in the origination of the epithelial-like cell, BSCC93.

References

[1] H. Aoki, M. Akao and Y. Shin: Med. Prog. Technol. Vol. 12 (1987), p.213.
[2] H. Aoki, Science and medical applications of hydroxyapatite, Takayama Press, Tokyo 1991, pp. 152
[3] M. Jarcho, J.L. Kay, R.H. Gumaer and H.P. Drobeck: L. Bioeng. Vol. 1 (1977), p. 79.
[4] G.R. Martin and R. Timpl: Annu. Rev. Cell Biol. Vol. 3 (1987), p. 57.
[5] S. Yamashita, S. Nagata and K. Takakura: Kobunshi Ronbunsyu Vol. 36 (1979), p. 257.
[6] T. Taguchi, A. Kishida and M. Akashi: Chem. Lett. Vol. 8 (1998), p. 711.

Key Engineering Materials Vols. 254-256(2004) pp. 545-548
online at http://www.scientific.net
© 2004 Trans Tech Publications, Switzerland

Apatite Deposition on Organic-inorganic Hybrids Prepared from Chitin by Modification with Alkoxysilane and Calcium Salt

T. Miyazaki[1], C. Ohtsuki[2], M. Tanihara[2] and M. Ashizuka[1]

[1]Graduate School of Life Science and Systems Engineering, Kyushu Institute of Technology, 2-4 Hibikino, Wakamatsu-ku, Kitakyushu-shi, Fukuoka 808-0196, Japan
tmiya@life.kyutech.ac.jp

[2]Graduate School of Materials Science, Nara Institute of Science and Technology, 8916-5, Takayama-cho, Ikoma-shi, Nara 630-0192, Japan

Keywords: Organic-inorganic hybrid, chitin, alkoxysilane, calcium salt, bioactivity, simulated body fluid, apatite.

Abstract. So-called bioactive ceramics have been attractive because they spontaneously bond to living bone. However, they are much more brittle and much less flexible than natural bone. Previous studies reported that the essential condition for ceramics to show bioactivity is formation of a biologically active carbonate-containing apatite on their surfaces after exposure to the body fluid. The same type of apatite formation can be observed even in a simulated body fluid (Kokubo solution) with inorganic ion concentrations similar to those of human extracellular fluid. Organic-inorganic hybrids consisting of organic polymers and the essential constituents of the bioactive ceramics, i.e. silanol (Si-OH) group and calcium ion (Ca^{2+}), are useful as novel bone substitutes, owing to bioactivity and high flexibility. In the present study, organic-inorganic hybrids were synthesized from chitin by modification with glycidoxypropyltrimethoxysilane (GPS) and calcium chloride ($CaCl_2$). Their apatite-forming ability was examined in Kokubo solution. Homogeneous bulk gel was obtained when mass ratio of chitin to the total of GPS and chitin was 0.25 or more and molar ratio of $CaCl_2$ to GPS was 0.25 or more. The prepared hybrids formed apatite on their surfaces in Kokubo solution within 7 days. This indicates that modification with alkoxysilane and calcium salt provides to chitin-based biomaterials with bone-bonding ability.

Introduction

Several kinds of ceramics such as Bioglass®, sintered hydroxyapatite and glass ceramics A-W, are known to exhibit specific biological affinity. Namely, they have the potential to bond to living bone when implanted into a bony defect [1]. They are called bioactive ceramics and have been already subjected to clinical applications as bone substitutes in orthopedic surgery, neurosurgery, dentistry and so on. However, their higher brittleness and lower flexibility than natural bone limits their clinical application to low load conditions. Materials exhibiting both high bioactivity and high flexibility are required for novel bone substitutes.

The prerequisite for artificial materials to show bioactivity is formation of a bone-like apatite layer on their surfaces in the body environment [2]. The same type of apatite formation can be observed even in a simulated body fluid (Kokubo solution) with inorganic ion concentrations similar to those of human extracellular fluid [3]. Fundamental studies on the mechanism of apatite formation on bioactive glass and glass-ceramics show that silanol (Si-OH) group and calcium ion (Ca^{2+}) are essential components to induce apatite deposition in the body environment [4]. Design of organic-inorganic hybrids by modification of these components with flexible organic polymers is expected to solve the problem of the bioactive ceramics.

In this study, we synthesized organic-inorganic hybrids from chitin by modification with alkoxysilane and calcium salt, which provides a Si-OH group and Ca^{2+}, respectively. Chitin is a natural organic polymer known as a constituent of club shell. Its chemical structure is shown in Fig. 1.

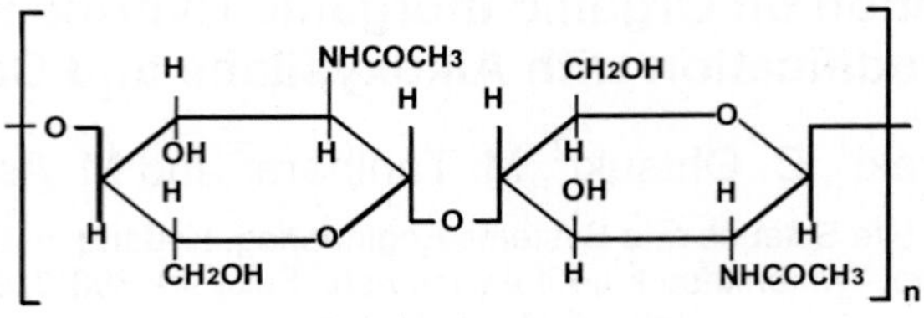

Fig. 1. Structural formula of chitin.

It has attractive features for implant materials such as low toxicity and high chemical durability [5]. Ability of apatite deposition on the hybrids was examined in Kokubo solution.

Materials and Methods

Compositions of the prepared specimens are shown in Table 1. Lithium chloride (LiCl) was dissolved in N,N-dimethylacetamide (DMAc). Chitin powder was then dissolved in the solution. After stirring for 3 h, glycidoxypropyltrimethoxysilane (GPS, $CH_2(O)CHCH_2O(CH_2)_3Si(OCH_3)_3$) and calcium chloride ($CaCl_2$) were added into the solution. The prepared sol solution was subsequently poured into a glass dish and dried at 100°C for given periods within 24 h. The mass ratio of chitin to the total of GPS and chitin ranged from 0.10 to 0.50, and the molar ratio of $CaCl_2$ to GPS also ranged from 0.10 to 0.50. Schematic representation of the synthesis of the organic-inorganic hybrids is shown in Fig. 2. The obtained homogeneous bulk gels were then soaked in Kokubo solution at pH 7.40 at 36.5°C for various periods up to 7 d. Surface structural changes on the hybrids after soaking in Kokubo solution were examined by scanning electron microscopy (SEM) and thin-film X-ray diffraction (TF-XRD).

Results

The appearance of the specimens after drying is summarized in Table 2. Homogeneous bulk gel was obtained when the mass ratio of chitin to the total of GPS and chitin was 0.25 or more and the molar ratio of $CaCl_2$ to GPS was 0.25 or more. The gels showed such high flexibility as to be easily bended. Fig. 3 shows SEM photographs of the specimens of the bulk gel soaked in Kokubo solution for 7 d. Deposition of fine particles was observed on the surfaces of all the specimens after soaking. Fig. 4 shows TF-XRD patterns of the specimens soaked in Kokubo solution for 7 d. Broad peaks assigned to apatite were observed at 26 and 32° in 2θ for all the specimens after soaking. These results indicate that the prepared organic-inorganic hybrids induce apatite deposition on their surfaces in Kokubo solution.

Table 1. Compositions of the hybrids examined in this study.

Sample	$\dfrac{\text{Chitin}}{\text{Chitin+GPS}}$ (mass ratio)	$\dfrac{\text{CaCl}_2}{\text{GPS}}$ (molar ratio)
S90C10Ca05	0.10	0.50
S75C25Ca05	0.25	0.50
S50C50Ca05	0.50	0.50
S75C25Ca025	0.25	0.25
S75C25Ca01	0.25	0.10

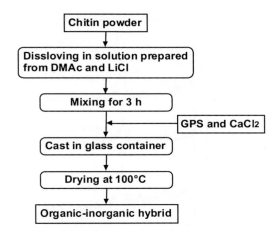

Fig. 2. Schematic representation of synthesis of organic-inorganic hybrids.

Table 2. Appearance of the prepared hybrids.

Sample	Appearance of the sample
S90C10Ca05	No bulk gel
S75C25Ca05	Homogeneous bulk gel
S50C50Ca05	Homogeneous bulk gel
S75C25Ca025	Homogeneous bulk gel
S75C25Ca01	No bulk gel

Fig. 3. SEM photographs of the surfaces of the hybrids, which were soaked in Kokubo solution for 7 days.

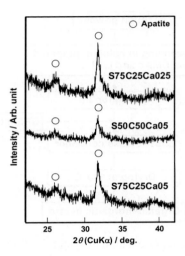

Fig. 4. TF-XRD patterns of the surfaces of the hybrids, which were soaked
in Kokubo solution for 7 days.

Discussion and Conclusions

The results obtained in this study show that organic-inorganic hybrids can be synthesized from chitin by modification with appropriate amounts of GPS and $CaCl_2$. The prepared hybrids have the potential to induce deposition of apatite in simulated body environment. Si-OH groups, which were formed on the surfaces of the hybrids by hydrolysis of GPS, trigger heterogeneous apatite nucleation. This apatite nucleation is accelerated by released Ca^{2+} ions by increasing the degree of supersaturation with respect to the apatite in the body fluid-the same mechanism of apatite formation as on bioactive glasses and glass-ceramics.

Calcium ions incorporated within the hybrids have an additional role. When the molar ratio of $CaCl_2$ to GPS was decreased to 0.10, bulk gel could not be formed after drying (see Table 2). This indicates that Ca^{2+} has an important role not only in accelerating nucleation of the apatite by the release into surrounding fluid but also in supporting formation of bulk gel. Namely, the incorporated Ca^{2+} would cross-link the polymer chain of chitin, or the network of silica gel formed by hydrolysis and polycondensation of GPS.

Consequently, chemical modification of chitin with alkoxysilane and calcium salt is quite effective for development of organic-inorganic hybrids with the ability to deposit apatite. The obtained chitin-based hybrids are expected to bond to bone in the living body.

Acknowledgements

This study was supported by Saneyoshi Scholarship Foundation, Japan. One of the authors (CO) acknowledges the support by Grant-in-Aid for Scientific Research ((B)13450272), Japan Society for the Promotion of Science.

References

[1] L.L. Hench: J. Am. Ceram. Soc. Vol. 81 (1998), p. 1705.
[2] H.-M. Kim: J. Ceram. Soc. Japan Vol. 109 (2001), p. S49.
[3] T. Kokubo, H. Kushitani, S. Sakka, T. Kitsugi and T. Yamamuro: J. Biomed. Mater. Res. Vol. 24 (1990), p. 721.
[4] C. Ohtsuki, T. Kokubo and T. Yamamuro: J. Non-cryst. Solids Vol. 143 (1992) p. 84.
[5] E. Khor and L.Y. Lim: Biomaterials Vol. 24 (2003), p. 2339.

Key Engineering Materials Vols. 254-256(2004) pp. 549-552
online at http://www.scientific.net
© *2004 Trans Tech Publications, Switzerland*

Bioabsorbable and bioactive composite structures from SiO₂-glassfibres and polylactides

Anna-Maija Haltia[1], Harri Heino[2] and Minna Kellomäki[3]

[1] Tampere University of Technology, Institute of Biomaterials, P.O. Box 589, FIN-33101 Tampere, Finland, anna-maija.haltia@tut.fi

[2] Linvatec Biomaterials Ltd, P.O. Box 3, FIN-33720 Tampere, Finland, hheino@linvatec.com

[3] Tampere University of Technology, Institute of Biomaterials, P.O. Box 589, FIN-33101 Tampere, Finland, minna.kellomaki@tut.fi

Keywords: Bioactive, bioabsorbable, composite, biaxial orientation

Abstract. In the current study three-layer membrane composites were developed and studied. Membranes were made of biodegradable polymers and bioactive SiO₂-glass fibres based on sol-gel method and prepared with dry spinning technique. Membranes were compression moulded and further processed by biaxial orientation for malleability. Sol-gel fibres, previously found loosely attached onto the membrane, were bound on the membrane with a separate layer of polymer.

Introduction

In clinical work, for example in cranial applications, lack of malleability at the room temperature is considered as a disadvantage for especially larger plates. The different preparation methods for bioabsorbable plates and membranes, and their use have been reported, for example, by Serlo *et al.* and Kellomäki *et al.* [1-3]. Ideally, the plates and membranes should be bent and reformed into required shape in operation room without heating, in such a way, that plate follows the form of bones' surfaces and guides the bone regeneration. However, during the reshaping the plate or membrane should also maintain its mechanical strength. In order to attain the adequate mechanical properties bioabsorbable materials can be reinforced with solid state deformation techniques such as oven drawing [4].

Calcium phosphates and materials having ability to form CaP precipitations *in vivo* have been studied to increase the osteoconductivity and bioactivity of the devices. One of these materials is the sol-gel derived silica gel [5-10]. The use of sol-gel-derived silica fiber as a bioactive component in composite membrane has been earlier reported by Haltia *et al.* and Peltola *et al.* [11-12]. Here one new preparation technique for composite membranes and plates is reported in which the osteoconductivity and bioactivity of the SiO₂-fibres were combined with enhanced malleability of membrane.

Materials and methods

SiO₂-fibres were prepared by compounding tetraethyl orthosilicate (TEOS), ethanol, deionised water and HNO₃ using the sol-gel method and dry spinning technique. The spinning process of the fibres has previously been illustrated in detail [5-7].

A poly(L/D,L)lactide 70/30 (PLA70, Resomer LR 708 i.v. 6.7 dl/g) membrane was the layer to give the structural strength for the membrane. It was prepared by extruding PLA70 granules to the billet at 170°C in the barrel and at 212°C in the die by twin screw extruder (Mini ZE 20 * 11.5 D) and by compression moulding at 165°C. The final thickness of PLA 70 membrane was 1.1 mm. Another polymer layer of poly(L/D)lactide 50/50 (PLA50, Resomer R 207 i.v. 1.6 dl/g) was compression moulded from flocky powder at 140°C and the thickness of moulded membrane was about 0.4 mm. The lamellar composite structure of PLA70, PLA50 and staple SiO_2-fibres was formed by compression moulding at 170°C layers in the mentioned order. In this way, the composite membranes had thicknesses of 1.2-1.4 mm. The composite membranes were orientated biaxially (Lab Stretcher Karo IV equipment, VTT Chemical Technology) at 73°C to draw ratio approximately 2.9 (both orientation directions). In the orientation the membranes were fixed between clips and the carriage took the membrane into the oven for preheating and orientation. The preheating time was 360 s and stretching speed 0.254 m/min. The maximum draw stresses ranged between 3.7 and 5.4 MPa depending the fibre content of membrane and orientation direction. PLC controlled the machine and the computer enabled the following of orientation and data logging. The total fibre content of the composite membranes was between 5.5 and 8.0 wt-%, but surface was practically completely covered. The thickness of the membranes varied between 0.3-0.4 mm. For comparison, the orientated membrane without SiO_2-fibres were also prepared. Polymers belong to the portfolio of Boehringer Ingelheim, Ingelheim am Rhein, Germany.

The membrane samples were tensile tested in both orientation directions (MD = machine direction and TD = transverse direction) using an Instron 4411 (Instron, Ltd., High Wycombe, England) materials testing instrument at crosshead speed of 10 mm min[-1] and a gauge length 17 mm. The flexibility of membranes (50 x 10 mm^2) has also studied with three point bending test using an Instron 4411 materials testing instrument. The microstructure of membranes (gold coated) were characterised using a JEOL JSM-T100 (JEOL LTD, Tokyo, Japan) scanning electron microscopy (SEM) at 15 kV acceleration voltage.

Results

At biaxial orientation no delamination of the double layer film occurred and staple SiO_2-fibres remained attached on the membrane with PLA50-layer. SEM examination shown that the SiO_2-fibres became almost completely covered by PLA50 layer (Figures 1a-b). In some places the SiO_2-fibres had only partially covered with PLA50, and then the fibres were totally in the surface of composite membrane, but the fibres seemed nevertheless attached to the composite.

a)   b)   c)

Fig. 1: The PLA50 surface of composite membranes. In a) and b) the membranes with the SiO_2-fibres, and in c) pure polymer composite.

The elongation of membranes with no biaxial treatment was less than 30 % while for the biaxial oriented SiO_2-fibre membranes the multidirectional alignment of molecular chains increased the elongation at break to 40-90 % (Fig. 2). The deviation of strain results was relatively high due to the uneven distribution of SiO_2-fibres on the structure. The other variables in these membranes might be the ratio of polymers, the thickness of membranes and the settlement of fibres to polymers. The biaxially oriented, layered membrane with no SiO_2-fibres had strain at break from 70 to 120 %. Several factors, such as distribution of SiO_2-fibres, orientation degree and alignment of polymer layers, may influence the strain properties of composite. The strength of SiO_2-composite membranes remained at orientation and the strength was regulated by PLA70-layer (Fig. 3). Three point bending tests were performed although samples were thin. The membranes were highly flexible due to the thinness and biaxially orientated structure, and were able to bend without breaking and sign of deformation.

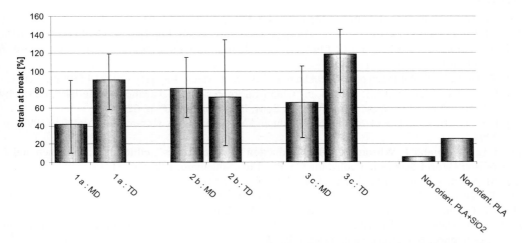

Fig. 2: Tensile strain of orientated and non-orientated membranes. *1a* and *2b* are biaxially orientated composite membranes with SiO_2-fibres, *3c* is biaxially orientated composite membrane without SiO_2-fibres. *Non orient. PLA* and *Non orient. PLA+SiO2* are non-orientated polymer membrane with and without SiO_2-fibres.(MD and TD).

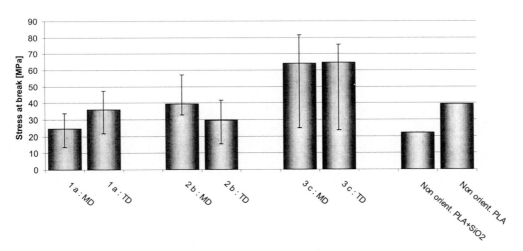

Fig. 3: Tensile stress of orientated membranes in two orientation directions (MD and TD) and stress of non-orientated membranes with and without SiO_2-fibres.

Discussion and Conclusions

The results showed that the handling properties of a composite of SiO2-fibres and two polylactide layers can be improved by biaxial orientation and by bonding fibres with thin and flexible polymer layer onto the structural layer. The faster bioabsorbing polymer layer also controls the contact of SiO_2-fibres with tissue fluid, and thus regulates the bioactivity of membrane *in vivo*. The lamellar glass/polymer composite membrane is applicable to, for example, guided tissue growth applications.

References

[1]   W. Serlo, et al.: Scand. J. Plast. Reconstr. Hand Surg. Vol. 35 (2001), p. 285.
[2]   M. Kellomäki, S. Paasimaa and P. Törmälä: Proc. Inst. Mech. Eng [H] Vol. 214 (2000), p. 615.
[3]   R. Suuronen, I. Kallela and C. Lindqvist: The Journal of Cranio-Maxillofacial Trauma Vol. 6 (2000), p.19.
[4]   P. Törmälä, T. Pohjonen and P. Rokkanen: Proc. Inst. Mech. Eng [H] Vol. 214 (1998), p. 101.
[5]   R. Li, A.E. Clark and L.L. Hench, in "Chemical Processing of Advanced Materials", edited by L.L. Hench and J.K. West (John Wiley & Sons, Inc., USA, 1992), p. 627.
[6]   P. Li, C. Ohtsuki, T. Kokubo, K. Nakanishi and N. Soga: J. Am. Ceram. Soc. Vol. 75 (1992), p. 2094.
[7]   L.L. Hench and J. Wilson, in "An Introduction to Bioceramics", edited by L.L. Hench and J. Wilson (World Scientific Publishing Co. Pte. Ltd., Singapore, 1993), p. 21.
[8]   M. Jokinen, et al.: J. Eur. Ceram. Soc. Vol. 20/11 (2000), p.1739.
[9]   T. Peltola, et al.: Biomaterials Vol. 22/6 (2001), p. 589.
[10]   T. Peltola, et al.: J. Biomed. Mat. Res. Vol. 54/4 (2001), p. 579.
[11]   A.-M. Haltia, M. Vehviläinen and M. Kellomäki: 17th European Society for Biomaterials Conference, Barcelona, Spain (2002), P203.
[12]   T. Peltola, et al.: Bioceramics 15, Key Engineering Materials Vol. 240-242 (2003), p. 156.

Key Engineering Materials Vols. 254-256(2004) pp. 553-556
online at http://www.scientific.net
© *2004 Trans Tech Publications, Switzerland*

The Degradation and Bioactivity of Composites of Silica Xerogel and Novel Biopolymer of Hydroxyproline

M. Väkiparta[1], M. Jokinen[1], M. Vaahtio[1], P.K. Vallittu[2] and A. Yli-Urpo[2]

[1]University of Turku, Institute of Dentistry, Turku Centre for Biomaterials, Itäinen Pitkäkatu 4B, FIN-20520 Turku, Finland, marju.vakiparta@utu.fi and minna.vaahtio@utu.fi

[2]University of Turku, Institute of Dentistry, Department of Prosthetic Dentistry and Biomaterials Research, Lemminkäisenkatu 2, FIN-20520 Turku, Finland

Keywords: Novel biopolymer, silica xerogel, bioactivity, degradation.

Abstract. Biodegradable composites of silica xerogel and hydroxyproline monomer or poly(hydroxyproline) were made by the sol-gel method. Composites contained up to 30 wt-% of monomer/polymer. The fast dissolution of prolines is combined with dissolving silica matrix that makes the composites potential bioactive porogens in bone implants. At the same time, water-soluble prolines with about 34-fold difference in molecular weight are studied as model molecules for controlled release of biologically active agents from silica xerogels. Composites were immersed in SBF for three weeks to investigate the degradation of the silica xerogel matrix, the release of monomer/polymer and formation of CaP layer on the surface of the material. The results for silica xerogels containing 30 wt-% of prolines are promising with respect to their potential use as bioactive porogens in bone. The release results showed both dependency and independency between the matrix structure & degradation and release of prolines showed the sensibility of the sol-gel process and pointed out the detailed design of synthesis for silica xerogels used in the controlled release of biologically active agents.

Introduction

One of the properties that promote artificial materials to bond to living tissue, especially to bone, is the presence/formation of carbonate-containing hydroxyapatite on the surface of the material. Another desired property is the presence/formation of macropores that promote tissue ingrowth. A bioactive hydroxyapatite layer can be formed on sol-gel-prepared silica gels *in vitro* [i] and *in vivo*. [ii] Recent development in bioresorbable polymers has produced new types of polymers like biodegradable poly(hydroxyproline). [iii] Hydroxyproline-based polymers have been used as porogens for biostable materials (e.g. bone cements) in order to promote tissue bonding to the material. [iv] Therefore, composites of silica xerogel and poly(hydroxyproline) may provide biodegradable material with improved bioactivity, especially in bone tissue due to the stimulative effect of dissolved silicon on bone-forming cells, osteoblasts. [v] Many biologically active agents (drugs) are water-soluble and the silica xerogel properties (e.g., structure and its influence e.g., on diffusion) are often dependent on the amount of water used in the synthesis. [vi] Therefore, hydroxyproline (M_w = 130 g/mol) and poly(hydroxyproline) (M_{vis} = 4 400 g/mol) serve as good model molecules to study the release of biologically active agents from silica xerogel. The aim of this work was to prepare composites of silica xerogel and prolines to be used as bioactive porogens and model matrices for release of water-soluble molecules of different size.

Materials and methods

The composites were made by the sol-gel method. Tetraethylortosilicate (TEOS), deionised water (H_2O: TEOS mole ratio was 14.2:1 for Si1 and 3:1 for Si2) and acid catalyst (HCl) was used. After hydrolysis of TEOS, powder of experimental linear poly(hydroxyproline) (P) synthesized by Puska et al. [iv] or powder of hydroxyproline monomer (M, H6002, Sigma) was added. Mixture of TEOS and P (SiP) or M (SiM) was polycondensated at 40 °C and 40 % relative humidity for 18 hours to disc shape (Ø 5 mm, thickness 2 mm), and dried at 40 °C and 40 % relative humidity to constant weight. The quantity of the M varied between 5 and 30 wt-% and P from 7 to 30 wt-% in the air-dried silica xerogel. Test specimens (n = 3) were immersed in simulated body fluid (SBF; pH 7.4) at +37 °C for three weeks. Control materials were pure silica xerogels (Si1 & Si2). Silica, phosphorus and hydroxyproline concentrations in SBF were detected by UV-VIS spectrophotometer (Shimadzu, UV-1601) [vii,viii,ix]. The amount of calcium was measured by atomic absorption spectrophotometer (AAS, Perkin-Elmer 460) and UV-VIS spectrophotometer from SBF. The morphology of precipitations of apatite like minerals (HA) on the surface of composites was examined with scanning electron microscope (SEM, Stereoscan 360, Cambridge).

Results and discussion

Degradation of the silica xerogel varied depending both on the silica xerogel and composite compositions (Fig. 1). The difference between pure silicas (controls) is clear; Si1 degrades about 10 times faster than Si2. The quantity of M and P in the silica xerogel affected also the degradation. As the quantity of M or P was 30 wt-% (Si1M30 and Si1P30), the silica xerogel degraded faster than controls and composites with lower quantities of M and P. However, with 30 wt-%, there was no significant difference between M- and P-containing composites. Also Si1P7 composites degraded a bit faster than the control, but the difference between Si1M5 and control was insignificant. In summary, the Si1 composites seem to degrade faster mainly due to presence of M or P in the structure. More M or P in the structure makes the structure more heterogeneous. Another possibility is the interaction between M or P and silica during the sol-gel process, but it is less likely, because no influence is observed with a smaller amount of M (5 wt-%). Si2 degrades clearly slower and the influence of the composite structure is also visible. Composite containing 2.5 wt-% M (Si2M2.5) degrades clearly faster than the control and also 2.5 wt-% P (Si2P2.5) had a clearer influence than in the case of Si1. This depends mainly on the difference in the silica xerogel matrix. The amount of water in Si2 synthesis is clearly lower than in that of Si1 that makes the Si2 structure denser. In other words, additional components, M and P, disturb more a denser Si2 structure than a more porous Si1.

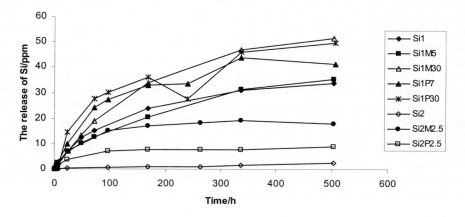

Fig. 1: The release of Si from composites containing monomer M or polymer P.

There are also some significant differences in M and P release results (Fig. 2). As the Si1 composites containing 30 wt-% M or P are compared to each other, it is shown that M is released a bit faster than P. The results show that size of the molecule is not important. Although P has a larger Mw, it is still strongly water-soluble. The rapid release is either due to fast degradation to monomers or smaller oligomers in the presence of water and rapid diffusion through the silica xerogel matrix or due to the gradient formation near the silica xerogel matrix or due to both reasons. Neither retards Si2 matrix significantly the release of M although the matrix is denser and degrades more slowly. However, Si2 matrix completely retards the release of P in time period of three weeks. In these cases it is more likely that the lower amount of water affected the distribution of M and P in the silica xerogel matrix. The used water amount in the sol was near the solubility limit of 2.5 wt-% M and P. More difficult dissolution and the amounts near the solubility limit makes that M and P separate easier and form gradient structures. M enriched near to the surface caused a rapid release despite of the slow degradation of the matrix. However, although the solubility limit is near and the gradient formation is likely, the molecular weight of P is high enough for a dense Si2 structure to be able to retard its release. The release from Si1 composites containing 5 wt-% M and 7 wt-% P are a bit slower than that of other composites. The main reason is the same as in the degradation of silica; the smaller amount of M and P does not make the silica xerogel matrix structure so heterogeneous.

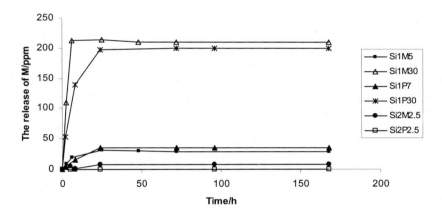

Fig. 2: The release of M (ppm) from composites containing monomer M or polymer P.

Also calcium phosphate formation was observed for the composites. Ca concentration decreased about 20 % in SBF for Si1 composites containing M, and about 5 % for Si1 composites containing P.
SEM micrographs and EDX analysis supported these results showing precipitation of calcium phosphate (CaP) on the surface of both composites (Fig. 3). In Si1M5, the precipitation is different from that found in Si1P30. In Si1M5 the layer is rougher while the shape of CaP is spherical, but smooth in Si1P30 (Fig. 3). The released amount of M, that is much greater for Si1P30, seems to affect the morphology of the formed CaP.

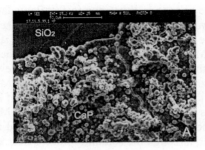 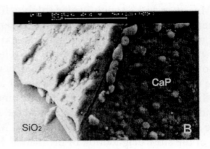

Fig. 3: SEM micrographs of A) composite Si1M5 (original magnification 500x; bar = 50 μm) and B) composite Si1P30 (original magnification 2000x; bar = 20 μm) after three weeks in SBF.

Conclusions

In conclusion, composites of the silica xerogels and hydroxyproline or poly(hydroxyproline) can be prepared containing up to 30 wt-% of prolines. The results for silica xerogels containing 30 wt-% of prolines are promising with respect to their potential use as bioactive porogens in bone. The prolines are rapidly released from the composites and the concentration levels of dissolving Si are within the levels that are potentially bioactive in bone. In addition, Si1 composites form a CaP layer on their surfaces. The proline release results showing both dependency and independency on the matrix structure & degradation, showed the sensibility of the sol-gel process. This points out that the silica xerogel synthesis steps have to be considered in detail, as the matrices for the controlled release of biologically active agents are designed.

Acknowledgement

The National Technology Agency of Finland (TEKES) is gratefully acknowledged for financial support.

References

[1] S-B. Cho, K. Nakanishi, T. Kokubo, N. Soga, C. Ohtsuki, T. Nakamura, T. Kitsugi and T. Yamamuro: J. Am. Ceram. Soc. 78(7) (1995), p.1769
[2] P. Li, X. Ye, I. Kangasniemi, J.M.A. de Blieck-Hogervorst, C.P.A.T. Klein and K. de Groot: J. Biomed. Mat. Res. 29 (1995), p. 325.
[3] Y. Lim, Y.H. Choi and J. Park: J. Am. Chem. Soc. 121 (1999), p.5633
[4] M.A. Puska, A.K. Kokkari, T.O. Närhi and P.K. Vallittu: Biomaterials 24 (2003), p. 417
[5] L.L Hench, J.M. Polak, L.D.K. Buttery and I.D. Xynos: Patent WO0204606, 2002.
[6] P. Kortesuo, M. Ahola, M. Kangas, A. Yli-Urpo, J. Kiesvaara and M. Marvola: Int. J. Pharm. 221 (2001), p. 107.
[7] O.G. Koch and G.A. Koch-Dedic: *Handbuch der Spureanalyse* (Spinger-Verlag, Berlin 1974).
[8] L.F. Leloir and C.E. Cardini: *Methods in Enzymology III* (Academic Press, New York 1972).
[9] M. Stegemann: Z. Physiol. Chem. 311(41) (1958).

Key Engineering Materials Vols. 254-256(2004) pp. 557-560
online at http://www.scientific.net
© *2004 Trans Tech Publications, Switzerland*

Bioactive Glass (S53P4) and Mesoporous MCM-41-type SiO$_2$ Adjusting *In vitro* Bioactivity of Porous PDLLA

Joni Korventausta[1], Ari Rosling[2], Jenny Andersson[3], Anna Lind[3], Mika Linden[3], Mika Jokinen[1], and Antti Yli-Urpo[4]

[1] Turku Centre for Biomaterials and Department of Prosthetic Dentistry, Institute of Dentistry, University of Turku, Lemminkäisenkatu 2, FIN-20520 Turku, Finland, email: joni.korventausta@utu.fi

[2] Laboratory of Polymer Technology, Åbo Akademi University, Piispankatu 8, FIN-20500 Turku, Finland

[3] Department of Physical Chemistry, Åbo Akademi University, Porthaninkatu 3-5, FIN-20500 Turku, Finland

[4] Department of Prosthetic Dentistry, Institute of Dentistry, University of Turku, Lemminkäisenkatu 2, FIN-20520 Turku, Finland

Keywords: Multicomposites, *in vitro*, ion release, CaP formation.

Abstract. SiO$_2$-based bioceramics, MCM-41-type SiO$_2$ and the bioactive glass S53P4, in composites with poly(D,L)lactide were studied in the simulated body fluid. The parameters controlling ion dissolutions and calcium phosphate formation were studied and the data was used to create multicomposites with locally varying properties (*e.g.*, CaP formation on the other side, uninhibited silica dissolution and possibility to drug release from the other side of composite)

Introduction

Bioactive and biocompatible bioceramics, such as bioactive glass (BAG) or sol-gel derived SiO$_2$, are widely studied materials but they have weaknesses, such as brittleness that limits their potential use in clinical applications. The utilisation of bioceramics can be improved by forming a composite with load-bearing or moldable polymers. An important property, the possibility to vary the dissolution properties of biologically important ions (dissolving SiO$_2$, Ca^{2+}, and PO$_4^{3-}$ with respect to the bone growth) depends on the specific properties of bioceramics and composite properties. The properties of BAGs are fixed to the composition, but the sol-gel derived SiO$_2$ provide more possibilities to adjust individual ion dissolutions by controlled pore structures and freely adjustable SiO$_2$ dissolution, with or without Ca^{2+} and PO$_4^{3-}$. The properties of polymers, such as porosity, provide additional parameters to adjust the dissolution properties and calcium phosphate (CaP) formation on composites.

The object of this work is to study biologically important processes (*i.e.* SiO$_2$ dissolution and CaP formation with respect to the bone formation) of BAG-PDLLA (poly(DL)lactide) and MCM-41-SiO$_2$-PDLLA composites in simulated body fluid (SBF). The composite parameters studied were content of bioceramics in composite, composite porosity, and bioceramics itself. The data is combined with earlier results on corresponding bioceramics (calcium and phosphorus releasing silica (CaPSiO$_2$) and pure silica (baSiO$_2$)) in corresponding composites to prepare and study new multicomposites with local functionalities.

Materials and methods

SiO$_2$ applied for composite studies was calcified MCM-41 with d-spacing modified by cetyltrimethylammonium bromide/dimethylhexadecyl amine ratio. The ratio was 1.0 and d-spacing 88.4 Å. BAG was S53P4. For multicomposite studies, calcium phosphate doped silica (CaPSiO$_2$) and pure silica (baSiO$_2$) were prepared as described earlier [1].

The bioceramic-PDLLA composites were made by a casting technique. The pores were generated in composite using high pressure (50 bar/24 h) CO_2, which let off rapidly. The multi-composites ([baSiO$_2$(40)-PDLLA(60, porous)]-[PCL]-[CaPSiO$_2$(40)-PDLLA(60, porous)]) were prepared by heating PCL polymer and bioceramic-PDLLA composites after which the composites were pasted on the plastified PCL.

The SBF test specimens containing 50 mg bioceramics each in 50 ml of SBF were tested in a shaking water bath at 37 °C. Three parallel samples were taken at each time point. Silica, calcium, and phosphate ion concentrations were measured as a function of immersion time. The silica and phosphate ions were measured with UV-Vis spectrophotometer, and the calcium concentrations were measured by AAS. The CaP formation properties of the multicomposites were studied by SEM-EDX (JEOL JSM-5500).

Results

The calcium phosphate (CaP) formation was detected by monitoring phosphorus concentration of SBF. Pure MCM-41-SiO$_2$ formed CaP within 5 days, but in the MCM-41-SiO$_2$-PDLLA composites the CaP formation was blocked (Fig. 1) independently on the varying the porosity of the composite or the MCM-41-SiO$_2$ content between 20 and 60 wt-% in the composite. However, these parameters had an effect on the silica dissolution.

The BAG-PDLLA composite formed CaP as expected, but the formation was much slower than on the pure BAG surface (Fig. 1). The CaP-free surface had a major role on the CaP formation. The composite containing 60 % BAG formed a CaP layer, which inhibited effectively the calcium dissolution and also further CaP formation. Non-porous composite had a low calcium dissolution and low CaP formation. The most intense CaP formation was on the porous composite containing 40 % of BAG, which had less CaP on the composite surface than other BAG-PDLLA composites studied. The amount of CaP on the surface can be established from calcium or silica (Table 1) dissolution results. The dissolution of calcium and silica increased as the composite surface contained less CaP. In MCM-41-SiO$_2$-PDLLA composites the inhibiting effect on the dissolution was not observed because no CaP was formed.

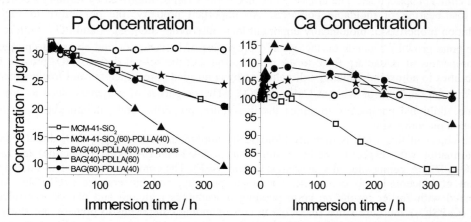

Fig. 1 Evolution of the phosphorus and calcium concentration in SBF as a function of immersion time for the MCM-41-SiO$_2$, MCM-41-SiO$_2$-PDLLA composite, and BAG-PDLLA composite.

The silica dissolution profile of MCM-41-SiO$_2$-PDLLA composites obeyed following guidelines: the more MCM-41-SiO$_2$ in the composite, the faster and more silica dissolution and the porous composite resulted in higher dissolution and dissolution rate (Table 1). The CaP formation regulated clearly the silica dissolution of BAG-PDLLA composites. The silica dissolution rate was fastest and the dissolution level highest for the composite (40 % BAG) having the least amount of CaP on the surface. The non-porous BAG-PDLLA composite dissolved 60 % slower and reached a constant level that was only 50 % of that of the corresponding porous composite with same preparation parameters.

Table 1 SiO_2 dissolution rates and concentration levels in SBF of MCM-41-SiO_2, MCM-41-SiO_2-PDLLA composite, and BAG-PDLLA composite.

Material (wt-% in composite in brackets)	SiO_2 dissolution rate [ppm/h]	Constant level of SiO_2 concentration [µg/ml]
MCM-41-SiO_2	9.77	145
MCM-41-SiO_2(20)-PDLLA(80), no pore	0.18	38
MCM-41-SiO_2(20)-PDLLA(80)	0.29	43
MCM-41-SiO_2(40)-PDLLA(60)	3.65	155
MCM-41-SiO_2(60)-PDLLA(40)	10.49	147
BAG(40)-PDLLA(60), no pore	0.68	39
BAG(40)-PDLLA(60)	1.69	78
BAG(60)-PDLLA(40)	1.06	55

The SEM-EDX analysis of the multicomposite revealed fast formation of CaP on the CaPSiO$_2$-PDLLA side in SBF (Fig. 2). CaP can be seen as white or light grey in the SEM graphs. After 4 days of immersion the CaP layer covered fully the CaPSiO$_2$-PDLLA side and the baSiO$_2$-PDLLA side of the multicomposite had some precipitated CaP. After 7 days of immersion the baSiO$_2$-PDLLA side had evidently CaP not only on shear plane but also on the surface of the composite, however less than on the CaPSiO$_2$-PDLLA side. Within 2 weeks the CaP layer was formed effectively enough to break up the CaPSiO$_2$-PDLLA and PCL bonding. CaP was spread on the PCL surface and the major part of the baSiO$_2$-PDLLA surface was covered with CaP.

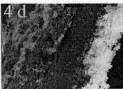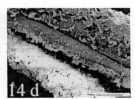

Fig. 2 SEM graphs of the [baSiO$_2$(40)-PDLLA(60, porous)]-[PCL]-[CaPSiO$_2$(40)-PDLLA(60, porous)] multicomposite shear plane as a function of immersion time in SBF. The scale bar is 500 µm.

Discussion

The MCM-41-SiO_2 is not effective enough for CaP nucleation in the composite. Though, CaP was formed on pure MCM-41-SiO_2 without polymer in the normal SBF, *i.e.* without any extra calcium. The MCM-41-SiO_2-PDLLA composite formed CaP in multiple ion concentrations of SBF's, with at least 1.2 x SBF (not shown). These results are in accordance with our previous results on other pure SiO_2:s in composites [2]. CaP is formed on the BAG-PDLLA composites, which release extra calcium. This result supports the need of extra calcium for the acceleration of the CaP formation.

The silica and calcium dissolution results revealed that the easier the flux of fluid in the composite structure the higher the dissolution. The flux can be increased by using more porous composites, higher ceramic content in the composite, and by using ceramics that form less CaP.

The results achieved on the bioceramic-polymer composites enable us to design multicomposites where the properties of the structure can be varied widely. In the studied multicomposite (Fig. 2 and schematic structure in Fig. 3) it is shown that CaP is formed also on the baSiO$_2$-PDLLA side of the multicomposite (although not so extensively as on CaPSiO$_2$-PDLLA side of the multicomposite). This is in accordance with the fluid intermediative CaP formation, a biomimetic process introduced earlier by Kokubo's group [3]. Although the phosphate concentration is decreased quite effectively due to the CaP formation on the CaPSiO$_2$-PDLLA side of the multicomposite, the CaP is formed also on the baSiO$_2$-PDLLA side of the multicomposite (in our study SBF is not changed to a fresh one during the experiment).

The properties of the multicomposites can be adjusted to fulfil various demands for implants, *e.g.*, mechanical properties combined with specific properties with respect to the different tissues that will be in contact with the implant. For example, the introduced multicomposite provides possibilities to design an implant where mechanical properties depend mostly on the bulk polymer with no tissue contact, the hard tissue contact is favoured on the other side of the implant and (uninhibited) drug and silica release on the other side. In addition, the results provide possibilities to design structures that have various bioresorption rates in different parts of the multicomposite that may be needed, *e.g.*, in tissue guiding devices.

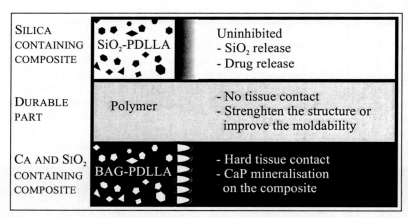

Fig. 3.   Schematic picture of the multifunctional composite composition.

Conclusion

The silica release and calcium phosphate formation can be effectively adjusted by the chosen material combinations. The bioceramics properties determine the major properties of whole composite structure with respect to bioactivity and the composite parameters (porosity, bioceramics content) provide possibilities to fine adjustments. It was shown that it is possible to prepare multicomposite that have different functions depending on location within the composite.

Acknowledgement
The National Technology Agency of Finland (TEKES) is acknowledged for financial support.

References
[1] Peltola T, Jokinen M, Rahiala H, Levänen E, Rosenholm JB, Kangasniemi I, Yli-Urpo A. J Biomed Mater Res 1999;44:12-21.
[2] Korventausta J, Jokinen M, Rosling A, Peltola T, Yli-Urpo A, Calcium phosphate formation and ion dissolution rates in silica gel-PDLLA composites, Biomaterials, accepted (3/2003).
[3] Tanahashi M, Yao T, Kokubo T, Minoda M, Miyamoto T, Nakamura T, Yamamuro T. J Am Ceram Soc 1994;77:2805.

Key Engineering Materials Vols. 254-256(2004) pp. 561--564
online at http://www.scientific.net
© *2004 Trans Tech Publications, Switzerland*

Porous Body Preparation of Hydroxyapatite/Collagen Nanocomposites for Bone Tissue Regeneration

M. Kikuchi[1,2], T. Ikoma[1,2], D. Syoji[3], H.N. Matsumoto[4] Y. Koyama[2,4], S. Itoh[2,4], K. Takakuda[2,4], K. Shinomiya[2,4] and J. Tanaka[1,2]

[1] Biomaterials Center, National Institute for Materials Science, 1-1, Namiki, Tsukuba, Ibaraki 305-0044, Japan, KIKUCHI.Masanori@nims.go.jp, IKOMA.Toshiyuki@nims.go.jp, TANAKA.Junzo@nims.go.jp

[2] CREST, Japan Science and Technology Corp., Kawaguchi Center Building, 4-1-8 Honcho, Kawaguchi, Saitama 332-0012, Japan

[3] PENTAX Corp., 2-36-9,Maeno-cho,Itabashi-ku, Tokyo 174-8639, Japan, daisuke.shoji@aoc.pentax.co.jp

[4] Tokyo Medical and Dental University, 1-5-45 Yushima, Bunkyo-ku, Tokyo 113-8510, Japan, hnm.mech@tmd.ac.jp, koyama.mech@tmd.ac.jp, itoso.gene@tmd.ac.jp, takakuda.mech@tmd.ac.jp, shinomiya.orth@tmd.ac.jp

Keywords: Porous body, hydroxyapatite/collagen nanocomposite, bone tissue regeneration, elastic property

Abstract. Elastic porous bodies were fabricated from the self-organized hydroxyapatite/collagen nanocomposite fibers by lyophilization with the use of collagen as a binder. The porous bodies obtained rapidly absorbed water droplet on them and composed of interconnected pores. The pore size was depending on freezing temperatures. The porous bodies demonstrated sponge-like elastic property when wetted and 70 kPa in tensile strength. The cell attachment and resorption with osteoclastic cells for the porous bodies examined by animal tests were the same as those for the compact bodies of hydroxyapatite/collagen nanocomposite. These mechanical and biological properties provide easier handling to operator and better cell invasiveness in both *in vitro* and *in vivo* than those of "hard" porous ceramics and "soft" collagen sponges.

Introduction

Hydroxyapatite (HAp) and β-tricalcium phosphate are widely used as bone fillers and bioactive coating on metals due to their osteoconductivity [1]. Recently, these materials are also used as a scaffold material for bone tissue engineering to regenerate critical bone defects by diseases and injuries. Although both experimental and clinical trials of bone tissue engineering using calcium phosphate ceramics indicated excellent results in bone regeneration [2], the remaining of the ceramics is still regarded as a problem for clinical use, because mechanical properties of ceramics are very different to those of bones and induced fractures around the ceramics. High-soluble calcium compounds, *i.e.*, coral and synthetic calcium carbonates [3] and calcium sulfates [4], were also applied as scaffold materials as well as bone filler, their solubility is too high to cooperate with bone formation because their dissolution generally takes place without biological reactions. Biodegradable synthetic polymers, such as poly-L-lactide based polymers [5], and composites of the polymers and bioactive ceramics have been developed [6, 7]; however, their bioactivity and biocompatibility are still not enough for tissue engineering use [7]. Biopolymers, mainly collagen, are excellent materials for tissue engineering due to their biocompatibility [8], cell affinity and biodegradability. However, their mechanical strengths do not allow easy handling by physicians.

Although, Chen *et al.* prepared porous poly-L-lactide-co-glycolide (PLGA) coated with collagen sponges and demonstrated high cell affinity on it [9], the problem of the acidity by decomposing of PLGA has not been solved yet.

In the last several years, the authors developed a novel artificial bone material composed of bone-like-oriented HAp nanocrystals and collagen molecules and reported their mechanical properties and excellent biological reactions, *i.e.*, they have similar mechanical properties to bone and incorporate bone-remodeling process to regenerate new bone [10-15]. In this paper, the authors describe preparation of porous body of this hydroxyapatite/collagen (HAp/Col) bone-like nanocomposites for bone tissue regeneration.

Materials and Method

The HAp/Col composite fibers for the porous bodies were synthesized from 100 cm3 of 400 mM Ca suspension and 200 cm^3 of 120 mM H$_3$PO$_4$ solution with addition of 2 g of atelocollagen (extracted from porcine dermis) at a titration rate of 15 cm^3/min and reaction temperature of 40°C as reported previously [15]. The composite fibers obtained were filtered and lyophilized. The composite fibers, collagen solution, H$_2$O and Dulbecco's phosphate buffered saline were mixed well at mixing ratios yielding final porosities of 90 and 95%, and stored 37°C to allow gellation of collagen solution. The gelled mixtures were frozen at -10, -20, -30, -40 and -80°C respectively and lyophilized followed by cross-linkage. The porous bodies obtained were cut into test pieces and examined by scanning electron microscopy, tensile strength test and animal tests using rats and Beagles. All animal tests were carried out under the "Guidelines of Tokyo Medical and Dental University for Animal Care" conformable to the "NIH guidelines for the care and use of laboratory animals" (NIH Publication #85-23 Rev. 1985). All operations and treatments were carried out by veterinarians, and animal boarding management was carried out by veterinarians and animal health technicians.

Results and Discussion

The structure of porous bodies observed by SEM shown in Fig. 1 indicated that mean pore size was increased with increasing freezing temperature. The pore in the porous body prepared by lyophilization is generally formed by removal of ice crystals grown between the substances, in this case, the HAp/Col fibers, *i.e.*, the pore size depends on growth behavior of the ice crystals. The rapid decreasing of temperature of the gelled mixture resulted in nucleation of large amounts of ice crystals and ice crystals between the fibers became smaller. Consequently, larger amounts of smaller pore were formed in the porous body frozen in lower temperature in comparison to that prepared at higher freezing temperature. In fact, in the porous bodies frozen at -30 and -40°C porosity existed in the thick walls consists of larger pores than that prepared at -20°C. The heterogeneous structure in the porous body prepared at -30 and -40°C could be formed by an exclusion of the fiber/small-ice conjugates by fusion of small ice crystals rapidly grown from outside to inside. The exclusion also could explain laminar structures around the walls.

The tensile strength of wet porous bodies at porosity of 95% was around 76±14 kPa instead of 32±9 kPa for the porous body without binder. The porous bodies showed sponge-like elastic behavior against compression, stretching and cutting with scalpel as shown in Fig. 2. These mechanical properties were suitable for both surgical and scaffold use, and enough to easy handling for operators.

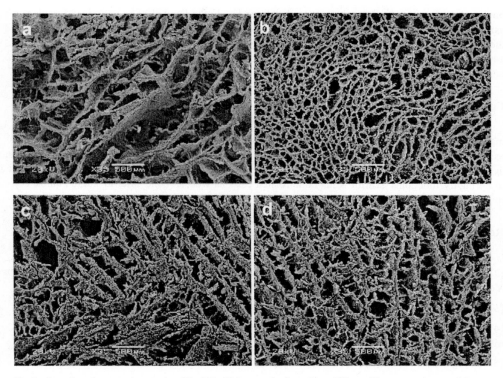

Fig. 1. Scanning electron micrographs of the HAp/Col porous bodies. Bar = 500µm. The preparation conditions are -10, b: -20, c: -30 and d: -40°C.

time course

Fig.2 Sponge-like elastic property of the wet HAp/Col sponge. The sponge deformed and water retained in the sponge was squeezed out. The sponge recovers the original shape with absorbing the water.

The cell adhesion on the surface of porous body was similar to that of collagen sponges. The porous body implanted into bone tissue demonstrated that good cell invasion. In the same way as the dense body of the HAp/Col composites, they show resorption by osteoclasts followed by osteogenesis by osteoblasts, *i.e.*, the materials retained the property to incorporate into bone remodeling process. The resorption and absorption rates of the porous body should be higher than those of the dense bodies due to larger surface area to react with cells and body fluid and must be optimized for the use as bone filler and scaffold materials to cooperate with new bone formation. Fortunately, the rates of the composites are easily controlled by cross-linkage ratio of collagen molecules as mentioned in ref. 14, the porous HAp/Col nanocomposites will expected to be good bone fillers instead of autogenous bone and good scaffold for bone tissue regeneration.

Conclusion

The materials demonstrate excellent elastic properties compared to both ceramics and collagen materials. This sponge-like property is utilized for both bone filler and scaffold materials for its easiness of handling, due to its neither brittleness in porous ceramics nor softness in collagen sponges, as well as cell/body fluid infiltration with repeating deformation and reformation. They are expected to be highly functional scaffolds for bone-related cells as well as bone filler to induce quick cell invasion and substitution with new bone in suitable period.

References

[1] Aoki H: *Medical Applications of Hydroxyapatite*, (Ishiyaku-Euro America, Tokyo, 1994).

[2] Yoshikawa T, Ohgushi H, Nakajima H, Yamada E, Ichijima K, Tamai S, Ohta T: Transplantation Vol. 69 (2000); p. 128.

[3] Fuller DA, Stevenson S, Emery SE: Spine. Vol. 21 (1996) p. 2131.

[4] Kim CK, Chai JK, Cho KS, Moon IS, Choi SH, Sottosanti JS, Wikesjo UM: J Periodontol. Vol. 69 (1998) p.1317.

[5] Hasegawa Y, Sakano S, Iwase T, Warashina H: J Biomed Mater Res. Vol 63 (2002), p. 679.

[6] Kikuchi M, Suetsugu Y, Tanaka J, Akao M: J. Mater. Sci.: Mater. In Med. Vol 8 (1997), p. 361.

[7] Sikinami Y, Okuno M: Biomaterials Vol. 20 (1999), p. 859.

[8] Mizuno S, Glowacki J: Biomaterials. Vol. 17 (1996), p. 1819.

[9] Chen G, Ushida T, Tateishi T: J. Biomed. Mater. Res. Vol. 51 (2000), P.273.

[10]Kikuchi M, Itoh S, Ichinose S, Shinomiya K, Tanaka J: Biomater. Vol. 22 (2001), p. 1705.

[11]Itoh S, Kikuchi M, Takakuda K, Koyama Y, Matsumoto HN, Ichinose S, Tanaka J, Kawauchi T, Shinomiya K: J. Biomed. Mater. Res. Vol. 54 (2001), p. 445.

[12]Itoh S, Kikuchi M, Takakuda K, Nagaoka K, Koyama Y, Tanaka J, Shinomiya K: J. Biomed. Mater. Res. Vol. 63 (2002) p. 507.

[13]Itoh S, Kikuchi M, Koyama Y, Takakuda K, Shinomiya K, Tanaka J: Biomater. Vol. 23 (2002) p. 3919.

[14]Kikuchi M, Matsumoto HN, Yamada T, Koyama Y, Takakuda K, Tanaka J: Biomater. in print.

[15]Kikuchi M, Itoh S, Matsumoto HN, Koyama Y, Takakuda K, Shinomiya K, Tanaka J: Key-Eng. Mater. Vol. 240-242 (2003) p.567.

Key Engineering Materials Vols. 254-256(2004) pp. 565-568
online at http://www.scientific.net
© *2004 Trans Tech Publications, Switzerland*

In Vivo Behaviour of Bonelike®/PLGA Hybrid: Histological Analysis and Peripheral Quantitative Computed Tomography (pQ-CT) Evaluation

J.M. Oliveira[1,2], T. Kawai[3], M.A. Lopes[1,2], C. Ohtsuki[3], J.D. Santos[1,2] and A. Afonso[4]

[1] INEB - Instituto de Engenharia Biomédica, Lab. de Biomateriais, Rua do Campo Alegre, 823, 4150-180 Porto, Portugal, jmoliv@fe.up.pt, malopes@fe.up.pt, jdsantos@fe.up.pt
[2] FEUP - Faculdade de Engenharia da Universidade do Porto, DEMM, Rua Dr. Roberto Frias, 4200-465 Porto, Portugal
[3] NAIST - Nara Institute of Science and Technology, 8916-5 Takayama, Ikoma, Nara 630-0192, Japan
[4] FMDUP - Faculdade de Medicina Dentária da Universidade do Porto, Rua Dr. Manuel Pereira da Silva, 4200-393 Porto, Portugal

Keywords: Bonelike®, PLGA, pQ-CT, histological analysis, *in vivo* studies

Abstract. The present work consisted in assessing the *in vivo* biological behaviour of novel Bonelike®/PLGA hybrid materials through subcutaneous and tibiae implantations for a 4 week period using an animal model. No sign of inflammation was observed as a result of subcutaneous implantation, which reveals that these materials were well-tolerated by the soft tissues. pQ-CT analysis demonstrated that the implanted materials were highly osteointegrated in bone tissue. SEM analysis at bone defect area revealed that Bonelike® resorption occurred after 4 weeks of implantation. Although the osteoconductive properties of Bonelike® have been reported in literature, the present work allowed to demonstrate that the silane-coupling agent and the PLGA treatments do not significantly modify the biological performance of Bonelike®. This result is strongly encouraging since Bonelike®/PLGA hybrid materials are being developed for bone tissue healing/regeneration purposes with simultaneous delivery of therapeuthic molecules.

Introduction

Synthetic polymers, such as poly(D,L-lactide-co-glycolide) (PLGA) have been used to deliver therapeutic molecules for medical applications of tissue regeneration [1, 2]. The great attention given to these materials is mainly related to the fact that it is possible to control their biodegradation by changing the lactide/glycolide ratio and molecular weight [3] and therefore designing biomaterials structures that deliver therapeutic molecules in a controlled manner [4].

Bonelike® is highly bioactive *in vivo* as it has been reported in literature [5-7] and has been designed to mimic the mineral part of bone tissues. Hybrid structures of Bonelike® and PLGA will allow to combine the advantageous of both materials in terms of bioactivity and drug deliver capability and therefore extend the medical applications of Bonelike®.

This work aims at analysing the *in vivo* osteoconductivity of Bonelike®/PLGA hybrids after 4 weeks implantation period using qualitative histology and peripheral Quantitative Computed Tomography (pQ-CT).

Materials and Methods

Implant materials

Bonelike® block samples were prepared by wet mixing, in methanol, hydroxyapatite (HA), $Ca_{10}(PO_4)_6(OH)_2$, (Plasma Biotal-P201 R) with 4wt% of $CaO-P_2O_5$-based glass. The slip was dried and sieved to less < 75μm and mixed powders were uniaxially pressed at 288MPa followed by sintering at 1300°C with a ramp rate of 4°C/min and 1h dwell time, as previously described by Santos *et al* [8].

Bonelike® granules were also prepared for tibiae implantation with size ranging from 150 to 500µm using standard milling and sieving techniques. Granules were degreased, washed with deionised water and silanized (γ-methacryloxypropyltrimethoxysilane), as previously reported [9]. The silane-treated samples were *post*-hybridised with a 5wt% PLGA/ethyl lactate solution using a solvent evaporation method [7].

Implantation technique
Prior to implantation, Bonelike® samples were sterilized by soaking in several ethanol solutions and dried in vacuum. The *in vivo* studies were performed at the Japan SLC Inc. (Japan) using White male Japanese rabbits (~2kg). Bonelike® block samples were implanted subcutaneously and Bonelike® granules implanted in the tibiae using standard aseptic procedures. Animals were sacrificed with an overdose of general anaesthetic after the 4 weeks of implantation and Bonelike® samples retrieved from the implantation sites.

Computed Tomography and Histological analysis
The implanted area was sectioned, washed with physiological saline solution, fixed in 10% formalin/phosphate buffered solution and dehydrated with a graded series of alcohol solutions (70, 80, 90 and 100%).
Peripheral Quantitative Computed Tomography (pQ-CT) of retrieved samples was carried out using XCT Research SA+ equipment (Stratec, Germany) and Scion Image Software. After assessment of bone structure using pQ-CT, the implants were embedded in methylmethacrylate resin, cut perpendicularly to the tibiae axis with a diamond blade microtome (Struers Accutom), hand-ground down to 70-80µm thick and stained with Goldner-Trichrome for histological examination under light microscopy. Retrieved samples were also analysed using SEM microscope (JEOL JSM-6301F) for bone/implant interface characterisation.

Results and Discussion
No sign of inflammatory response was observed for both Bonelike® and Bonelike®/PLGA hybrid blocks implanted in back of rabbits, as it may be seen in Fig. 1. This result clearly indicates that both implanted materials are well-tolerated by the soft tissues.
Both the cross-section and the 3-D pQ-CT images revealed callus formation for both implanted materials with excellent osteointegration after 4 weeks implantation as depicted in Fig. 2 (A) and (B).

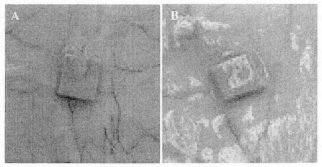

Fig. 1- Bonelike® (A) and Bonelike®/PLGA hybrid (B) subcutaneously implantated for 4 weeks period. No sign of inflammatory response was observed.

Both the cross-section and the 3-D pQ-CT images revealed callus formation for both implanted materials with excellent osteointegration after 4 weeks implantation as depicted in Fig. 2 (A) and (B).

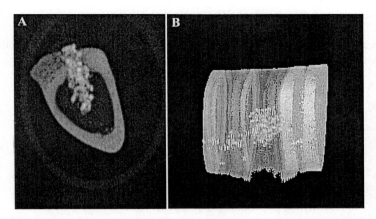

Fig. 2- pQ-CT scans of implantation sites after the 4 week implantation period for Bonelike®/PLGA hybrid: cross-section image (A) and 3-D image (B).

The histological analysis confirmed extensive new bone formation apposed on Bonelike® granules with several stages of calcification (Fig. 3). Additionally, several areas of bone remodelling were also detected throughout the samples in the cortical area defect with evidences of vascularization *de novo* bone.

These results seem to demonstrate that silane and PLGA treatments did not significantly modify the osteoconductive properties of Bonelike®, which that have been observed previously in both animal experimentation [7] and clinical practice [6].

SEM studies also revealed that new bone ingrowth occurred at resorption/degradation areas in Bonelike®/PLGA hybrid granules (Fig. 4B).

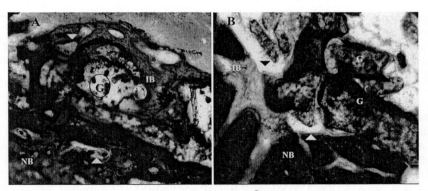

Fig. 3- Light microscopy photographs of Bonelike® granules (G) implanted in rabbit tibiae for 4 weeks at the bone defect area. Extensive new bone formation among granules may be seen for: Bonelike® (A) and Bonelike®/PLGA hybrid (B). Newly formed mineralised bone (NB) and Immature bone "osteoid" (IB) and areas of bone remodelling (arrowhead). (Undecalcified section, Goldner-Trichrome stain; original magnification x100).

The chemical composition of Bonelike® mimics the mineral phase of bone tissues, and also comprises some percentage of β-tricalcium phosphate (TCP), $Ca_3(PO_4)_2$, and α-TCP [9]. Therefore,

the *in vivo* resorption of Bonelike® may be explained by the presence of these biodegradable phases.
This study revealed that Bonelike®/PLGA hybrids can effectivelly support bone remodelling with simultaneous material resorption and extensive new bone formation.

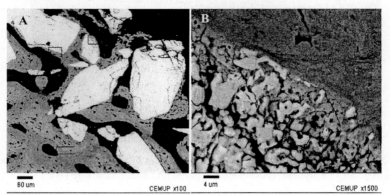

Fig. 4- Backscattered scanning electron image of Bonelike®/PLGA hybrid granules after a 4 week period: bone defect area (A) and showing areas of Bonelike® resorption with new bone ingrowth (NB).

Conclusion

The *in vivo* evaluation of Bonelike®/PLGA hybrid demonstrated that the silane-coupling agent and the PLGA surface treatment do not modify the osteoconductive properties of Bonelike®. Therefore, these hybrid materials can be used as bone allograft capable of delivering therapeutic molecules.

Acknowledgements

Authors would like to acknowledge the financial support of FCT (Fundação para a Ciência e Tecnologia) through the project "*Interactive calcium-phosphate based materials prepared by post-hybridisation and in situ hybridisation*", Ref. PCTI/1999/CTM/35516, Rotary Club de Caldas das Taipas and Mrs. Ana Mota for her technical assistance in the histological studies.

References

[1] E. Solheim, E. Pinholt, G. Bang *et al:* Journal Neurosurgery 76(1992), 275-279.
[2] C.T. Laurecin, M.A. Attawia, L.Q. Lu, M.D. Borden, H.H. Lu, W.J. Gorum, J.R. Lieberman: Biomaterials 22 (2001) 1271-1277.
[3] F.F. Nielsen, T. Karring, S. Gogolewski: Acta Orthop Scandinavica; 63, (1992) 66-69.
[4]G. Jiang, B.H. Woo, F. Kang, J. Singh, P.P. DeLuca: Journal of Controlled Release 79 (2002) 137-145.
[5] F. Duarte, M.A. Lopes, J.D. Santos: European Cells and Materials (*in press*).
[6] F. Duarte, A. Afonso, J.D. Santos: Advanced Materials Forum (*in press*).
[7] J.M. Oliveira, M.A. Lopes, T. Kawai, C. Ohtsuki, J.D. Santos, A. Afonso: Advanced Materials Forum (*in press*).
[8] J.D. Santos, J.K. Campbell, G.H. Winton: European Patent N° WO0068164 (2000).
[9] J.M. Oliveira, T. Miyazaki, M.A. Lopes, C. Ohtsuki and J.D. Santos: Biomaterials (*submitted*).

Key Engineering Materials Vols. 254-256(2004) pp. 569-572
online at http://www.scientific.net
© *2004 Trans Tech Publications, Switzerland*

Effect of Melt Flow Rate of Polyethylene on Bioactivity and Mechanical Properties of Polyethylene /Titania Composites

Hiroaki Takadama[1], Masami Hashimoto[1], Yorinobu Takigawa[1], Mineo Mizuno[1], and Tadashi Kokubo[2]

[1] Materials Research and Development Laboratory, Japan Fine Ceramics Center (JFCC), Nagoya, Japan. takadama@jfcc.or.jp

[2] Research Institute for Science and Technology, Chubu University, Kasugai, Japan

Keywords: apatite, bioactivity, composite, simulated body fluid, titania (anatase), polyethylene

Abstract. Bioactive bone-substitutes with mechanical properties analogous to those of natural bone are strongly desired to be developed. In the present study, HDPE/TiO$_2$ composites were prepared from titanium dioxide (TiO$_2$) nano-powder with anatase structure and high density polyethylene (HDPE) with different melting flow rate (MFR) through a batch-kneader mixing. The composite with a uniform dispersion of TiO$_2$ powder were prepared. The composite prepared from HDPE of MFR=8 shows the highest bending strength (about 50 MPa) and Young's modulus (about 7.5 GPa) within the range of the mechanical properties of human cortical bone. The composites formed apatite on their surfaces in a simulated body fluid within 7 days. Therefore, these PE/TiO$_2$ composites with such mechanical properties and bioactivity are considered to be useful as bone-repairing materials.

Introduction

Recently, many kinds of bioactive materials have been developed [1,2]. Most of them are ceramics and metallic materials. The mechanical properties of them were, however, quite different from those of human cortical bone. Especially, their Young's moduli are quite higher than that of human cortical bone [3]. This is a critical problem, since high young's modulus are liable to induce resorption of the surrounding bone because of stress shielding. Therefore, new bioactive bone-substitutes with mechanical properties analogous to those of natural bone and bone-bonding ability i.e. bioactivity, are strongly desired to be developed.

For this purpose, it is considered that inorganic-organic composites are suitable, as natural bone is composed of inorganic apatite and organic collagen. Many researchers have tried to design various bioactive inorganic-organic composites with mechanical properties close to those of natural bone. One of the typical inorganic-organic composites is hydroxyapatite reinforced HDPE composite (HAPEX®) [4], which has been already clinically used. However, HAPEX® was not designed to have high bending strength or Young's modulus but to have high fracture toughness. Though some of its mechanical properties are already desirable for clinical use, its bending strength and Young's modulus are lower than those of human cortical bone. Moreover, bioactive inorganic-organic composites with mechanical properties equal to those of natural bone other than HAPEX® have not been developed yet.

Recently, Ti-OH groups in a titania gel with an anatase structure have been found to be remarkably effective in inducing apatite formation [5]. Therefore, TiO$_2$ with anatase structure can induce apatite formation in the body environment. In addition, it shows higher mechanical properties than hydroxyapatite, which has been clinically used for bone substitutes. On the other hand, polyethylene shows good biocompatibility. From these findings, the composite of titanium oxide and polyethylene is expected to show mechanical properties analogous to those of natural bone and high bioactivity.

In our previous reports, bioactive UHMWPE/TiO$_2$ composites were prepared by hot pressing the powders of ultra-high molecular weight polyethylene (UHMWPE) and TiO$_2$ [6]. The mechanical properties of UHMWPE/TiO$_2$ composites were, however, lower than those of human cortical bone, as it was quite difficult to disperse a large amount of TiO$_2$ nano-powder uniformly in UHMWPE matrix due to the quite high viscosity of UHMWPE even at relatively high temperature. In the present study, to improve the mechanical properties, HDPE with higher mechanical properties and lower viscosity than UHMWPE was chosen, and HDPE/TiO$_2$ composites were prepared from HDPE of different melting flow rates (MFR) and TiO$_2$ nano-powder through a batch-kneader mixing. The effect of melt flow rate of HDPE on mechanical properties and bioactivity of HDPE/TiO$_2$ composites were investigated.

Materials and Methods

Sample preparation TiO$_2$ powder with an anatase structure and an average size of 200 nm (Ishihara Sangyo Kaisha Ltd., Osaka, Japan) was added into melted HDPE with different MFR of 0.3, 8.0, 20 and 40 (Japan Polyolefins Co., Ltd., Tokyo, Japan), respectively and then mixed under shear stress by a batch-kneader to prepare the inorganic-organic compounds. The MFR means the weight of a material that is forced by compression through a die orifice in 10 minutes under specific temperature and pressure defined for each polymer [g/10min]. The TiO$_2$ content was set at 40 volume percent. Then, the prepared compounds were hot-pressed at 230 °C for 1 h under the pressure of 2.5 MPa to form the HDPE/TiO$_2$ composites.

SBF soaking The HDPE/TiO$_2$ composites of $10 \times 10 \times 4$ mm^3 in size after polishing using sandpapers were soaked in 20 mL of a simulated body fluid (SBF) with ion concentrations (Na$^+$ 142.0, K$^+$ 5.0, Mg^{2+} 1.5, Ca^{2+} 2.5, Cl$^-$ 147.8, HCO$_3^-$ 4.2, HPO$_4^{2-}$ 1.0, SO$_4^{2-}$ 0.5 mM) nearly equal to those of human blood plasma [7] at 36.5°C for 7 days. After removal from SBF, the composites were washed with distilled water and dried in air.

Mechanical tests A universal electromechanical testing machine (Model 5582, Instron Corporation, MA, USA) was used to perform the three-point bending. The HDPE/TiO$_2$ composites were cut and then polished using sandpapers to a size of $45 \times 10 \times 4$ mm^3. A load was applied over 30 mm span at the center of the 45×10 mm^2 surface, using a cross-head speed of 1.0 mm/min at room temperature. The bending strength and Young's modulus of the HDPE/TiO$_2$ composites were calculated. As a reference, HAPEX® was also tested in three-point bending under the same condition.

Sample analysis The distribution of TiO$_2$ powder in the composites were observed by a field emission scanning electron microscope (FE-SEM: JSM-6330F, JEOL DATUM Co. Ltd., Nagoya, Japan). The surfaces of the HDPE/TiO$_2$ composites before and after soaked in SBF were analyzed by thin-film X-ray diffraction (TF-XRD: Model-265A, Rigaku Corporation, Osaka, Japan).

Results and discussion

Optimizing the mixing speed and temperature of batch-kneader and adding the TiO$_2$ nano-powder gradually into melted HDPE provided the compounds with a uniform dispersion of TiO$_2$ powder in HDPE matrix. Though it was quite difficult to disperse TiO$_2$ nano-powder of 40 volume percent in UHMWPE uniformly in the previous work, this batch-kneader mixing can disperse more than 40 volume percent of TiO$_2$ powder in HDPE uniformly. These compounds were hot-pressed to prepare the HDPE/TiO$_2$

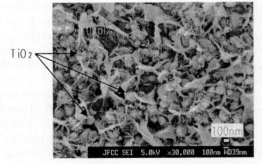

Fig. 1 FE-SEM image of the fracture surface of the HDPE/TiO$_2$ composite prepared from HDPE of MFR=8.

composites. Their fracture surfaces were observed by FE-SEM. Regardless of MFR, all the prepared HDPE/TiO$_2$ composites showed a uniform distribution of TiO$_2$ nano-powder in the composites as shown in Figure 1.

The HDPE/TiO$_2$ composites prepared from HDPE of different MFR were polished by sandpapers and then soaked in SBF. Regardless of MFR, all the composites formed the apatite layer on their surfaces within 7 days as shown in Figure 2, since TiO$_2$ with anatase structure induces apatite formation.

Figure 3 shows the representative load-displacement curves as obtained from three-point bend testing of the HDPE/TiO$_2$ composites prepared from HDPE of different MFR in comparison with HAPEX®. The composites prepared from HDPE of MFR=8 and 20 showed the highest breaking load. The displacement at the break point decreased and the composites became more brittle with the increase in MFR. The reason is that the amount of HDPE of low molecular weight became larger with the increase in MFR. Regardless of MFR, all the composites had larger effective fracture energy estimated by work of fracture method than HAPEX®. This may be due to the fact that the mechanical properties such as bending strength and Young's modulus of TiO$_2$ are higher than those of hydroxyapatite and that a nano-powder was used as a filler for

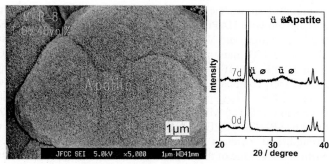

Fig. 2 FE-SEM image and TF-XRD pattern of the surface of the HDPE/TiO$_2$ composites prepared from HDPE of MFR=8 after soaked in SBF for 7 d.

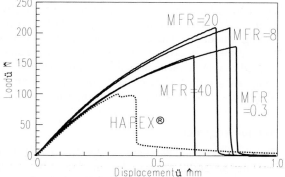

Fig. 3 The load-displacement curves obtained from three-point bend testing of HDPE/TiO$_2$ composites and HAPEX®.

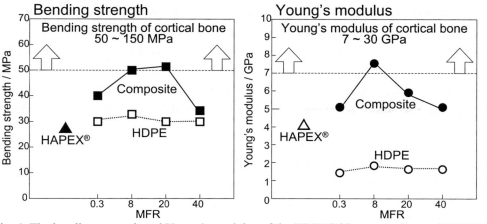

Fig. 4. The bending strength and Young's modulus of the HDPE/TiO$_2$ composites and HAPEX®.

this composites, while micron-powder was used for HAPEX®. Since materials which do not break off but deform to some extent and absorb the fracture energy even under highly loaded condition are desirable for bone substitutes, this composites are useful as bone-repairing materials.

Figure 4 shows the bending strength and Young's modulus of the HDPE/TiO$_2$ composites in comparison with HAPEX®. Regardless of MFR, all the HDPE/TiO$_2$ composites showed the higher bending strength and Young's modulus than those of HDPE itself. This indicates that the uniformly distributed TiO$_2$ nano-powder reinforced the HDPE matrix. The HDPE/TiO$_2$ composites prepared from HDPE of MFR=8 and 20 showed the highest bending strength of about 50 MPa. The HDPE/TiO$_2$ composites prepared from HDPE of MFR=8 showed the highest Young's modulus of about 7.5 GPa. These values were almost equal to those of human cortical bone, since human cortical bone shows 50-150 MPa for the bending strength and 7-30 GPa for the Young's modulus. From these results, the HDPE/TiO$_2$ composites prepared from HDPE of MFR=8 shows the highest bending strength and Young's modulus, within the range of the mechanical properties of natural human cortical bone. One of the reasons is that the HDPE/TiO$_2$ composites prepared from HDPE of MFR=8 shows the highest packing density and crystallinity of HDPE in the composites.

Conclusions

The HDPE/TiO$_2$ composites with the uniform dispersion of TiO$_2$ nano-powder in HDPE matrix were prepared by hot-pressing the compounds prepared by batch-kneader mixing under shear stress. The HDPE/TiO$_2$ composite prepared from HDPE of MFR=8 showed the highest bending strength and Young's modulus. These values were almost equal to those of human cortical bone. The composites showed the large effective fracture energy estimated by work of fracture method. The composites formed apatite on their surfaces in a simulated body fluid within 7 days. Therefore, the HDPE/TiO$_2$ composites with the mechanical properties almost equal to those of human cortical bone and bioactivity were prepared. These composites are considered to be useful as bone-repairing materials.

Acknowledgements

This work is in part supported by the National Research & Development Programs for Medical and Welfare apparatus entrusted from the New Energy and Industrial Technology Development Organization (NEDO) to the Japan Fine Ceramics Center.

References

[1] T. Kokubo, M. Shigematsu, Y. Nagashima, M. Tashiro, T. Nakamura, T. Yamamuro and S. Higashi: Bull. Inst. Chem. Res., Kyoto Univ. Vol. 60 (1982) 260-268.

[2] M. Jarcho, J.L. Kay, R.H. Gumaer and H.P. Drobeck: J. Bioeng. Vol. 1 (1977) 79-92.

[3] L.L. Hench and J. Wilson: *An Introduction to Bioceramics* (World Scientific, Singapore 1993) p. 1-24.

[4] M. Wang, D. Porter and W. Bonfield: Brit. Ceram. Trans., 94 (1994) 91-95.

[5] M. Uchida, H.-M. Kim, T. Kokubo, S. Fujibayashi and T. Nakamura: J. Biomed. Mater. Res. 64A (2003) 164-170.

[6] H. Takadama, M. Hashimoto, Y. Takigawa, M. Mizuno, Y. Yasutomi and T. Kokubo: *Bioceramics 15* (Tran Tech Publication, Switzerland 2002) p. 951-954.

[7] T. Kokubo, H. Kushitani, S. Sakka, T. Kitsugi and T. Yamamuro: J. Biomed. Mater. Res. 24 (1990) 721-734.

Key Engineering Materials Vols. 254-256(2004) pp. 573-576
online at http://www.scientific.net
© *2004 Trans Tech Publications, Switzerland*

Preparation and Characterization of Injectable Chitosan-Hydroxyapatite Microspheres

P.L. Granja,[1] A.I.N. Silva,[1,2] J.P. Borges,[2] C.C. Barrias,[1,3] and I.F. Amaral[1,3]

[1] INEB - Instituto de Engenharia Biomédica, Laboratório de Biomateriais, R. Campo Alegre 823, 4150-180 Porto, Portugal, pgranja@ineb.up.pt
[2] Dept. Ciência dos Materiais, Faculdade de Ciências e Tecnologia, Universidade Nova de Lisboa
[3] Universidade do Porto, Faculdade de Engenharia, Dept. Eng. Metalúrgica e de Materiais

Keywords: chitosan, hydroxyapatite, microsphere, injectable, composite, bone regeneration.

Abstract. The combination of chitosan and hydroxyapatite (HAp) in the form of injectable, porous and biodegradable structures seems to be an interesting route to promote localized bone regeneration, especially with the incorporation of cells or cell-targeted molecules.

In the present work, chitosan-HAp microspheres were prepared and characterized in terms of size, morphology, water sorption and structure. Chitosan-HAp porous microspheres were successfully prepared using tripolyphosphate as coagulating agent. The size increased and the water sorption decreased with increasing HAp contents. The ceramic particles were well embedded and homogeneously distributed within the polymer matrix.

Introduction

A number of orthopedic diseases could benefit from minimally invasive surgical techniques allowing less patient discomfort. Injectable microspheres provide such possibility due to their ability to fill-in defects, as well as the easy incorporation of proteins and cells, thus facilitating the regeneration of bone tissue [1].

Chitosan is a natural polymer which has been receiving increasing attention for biomedical applications due to its biodegradability, coupled with biocompatibility and regenerative properties [2]. Hydroxyapatite (HAp) is a bioactive ceramic widely used for orthopedic applications due to its similarity to the mineral constituent of hard tissues [3]. The combination of these two materials in the form of injectable, biocompatible, porous and biodegradable structures seems to be an interesting route to promote localized bone regeneration, especially with the incorporation of cells or cell-targeted molecules.

In the present work, chitosan-HAp microspheres carrying different amounts of HAp were prepared and characterized in terms of size, morphology, water sorption and structure. The interaction between polymer and ceramic was investigated as well as the influence of the chitosan acetylation degree.

Materials and Methods

Materials. Squid chitosan, kindly donated by France-Chitine (France), presenting a viscosity average molecular weight of 2.48×10^6 Da, and a molar fraction of *N*-acetylated units of $[F_A]$ of 0.30, as determined by FTIR according to Baxter *et a* [4], was used. Deacetylated chitosan was prepared by heterogeneous deacetylation in 50% (w/v) NaOH, at 70°C, under N_2 atmosphere, resulting in a $[F_A]$ of 0.14, as determined by FTIR according to Brugnerotto *et al* [5]. Commercial HAp (Plasma Biotal, UK) was used. Granulometric analysis of HAp particles revealed that 90% of them were smaller than 6.49 μm, with a volume average diameter of 3.328 μm.

Preparation of composite microspheres. 1% (w/v) chitosan solutions in 1% (v/v) CH_3COOH were prepared. Composites were obtained by adding HAp powder (1 to 30% w/v) under stirring, until homogenization. After degassing under reduced pressure, beads were formed by dropping the mixture through a syringe at a rate of 40 mL·h^{-1}, into a 5% (w/v) sodium tripolyphosphate (TPP) solution, according to Mi *et al* [6]. The size was controlled by applying a coaxial air stream. The microspheres were cured during 30 min, washed with water, frozen at -70°C and lyophilized.

Characterization of composites. Microsphere diameter was assessed by optical microscopy. Water sorption was determined by incubating a known weight of dried samples in bidistilled water and placing them under reduced pressure for 30 min, in order to assure complete hydration. When equilibrium was reached samples were finally weighed and the weight gain was calculated. The surface and cross-sectional morphology of Au sputtered samples was analysed by scanning electron microscopy (SEM). The structure was analysed by Fourier transform infrared (FTIR) spectroscopy.

Results and Discussion

Chitosan solutions dropped into TPP solutions formed gel beads instantaneously. According to Mi *et al* [6], at the pH of TPP solution, both OH$^-$ and tripolyphosphoric ions in TPP solution diffuse into chitosan droplets during the curing period, to react with NH$_3^+$ functionalities, by deprotonation or ionic crosslinking, respectively. The beads will preferentially be formed by coacervation together with slight ionic crosslinking.

The average diameter of freeze-dried microspheres ranged from ca. 220-250 μm for chitosan to 400-530 μm for chitosan-30%HAp, as assessed by optical microscopy (Fig. 1). A lower degree of acetylation promoted a decrease in microsphere diameter. Heterogeneous deacetylation is known to lead to a decrease in molecular weight, thus to less viscous solutions. The decrease in microsphere diameter of deacetylated samples is probably due to this phenomenon.

The water sorption of chitosan microspheres was of about 400%, whereas in chitosan-HAp composite microspheres it decreased with increasing HAp content (Fig. 2), becoming lower than 100% for the higher amounts of HAp tested. No significant differences were found in water sorption as a function of the degree of acetylation.

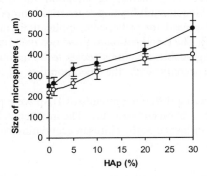

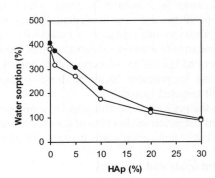

Fig. 1. Size of chitosan and chitosan-hydroxyapatite microspheres, as determined by optical microscopy (n=20). (●) F$_A$= 0.30; (○) F$_A$=0.14.

Fig. 2. Water sorption of chitosan-hydroxyapatite microspheres. (●) F$_A$= 0.30; (○) F$_A$=0.14.

SEM studies revealed that spherically-shaped chitosan-HAp microspheres were obtained. Chitosan microspheres showed a honeycomb-like surface structure (Fig. 3a), while an open pore structure with interconnectivity was observed in cross-section (Fig. 3i).

Increasing HAp content promoted a more homogeneous size distribution as well as smoother and increasingly spherically-shaped microspheres (Figs. 3a-d). Analyses of surfaces as well as cross-sections of the composite microspheres showed a homogeneous distribution of the HAp particles within the polymeric matrix. HAp particles were well embedded in the chitosan matrix. The average pore size decreased with the increase in HAp amount. Pore sizes ranged from ca. 50-100 μm in chitosan microspheres to less than 0.5 μm in chitosan-30%HAp microspheres. Similarly to what was observed concerning chitosan, cross-sections of composite microspheres (Figs. 3j-l) showed considerable interconnectivity, as opposite to the surface (Figs. 3f-h). Apparently, the degrees of acetylation tested did not significantly affect the morphology of the chitosan-HAp microspheres.

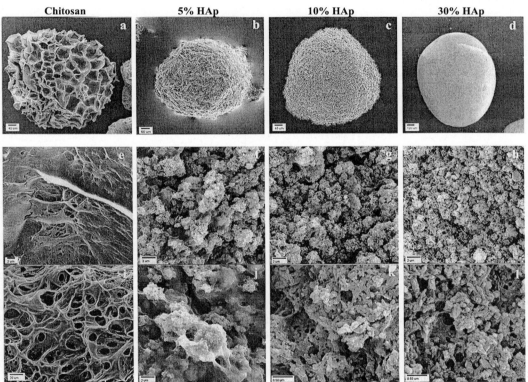

Fig. 3. Chitosan (F_A=0.30)-hydroxyapatite microspheres: a), e) and i) chitosan; b), f) and j) 5% HAp (w/v); c), g) and k) 10% HAp (w/v); d), h) and l) 30% HAp (w/v). Micrographs e) through h) represent higher magnifications at the surface, and i) through l) represent fracture surfaces.

The FTIR spectra of chitosan powders and chitosan microspheres are presented in Fig. 4. The spectra of the initial chitosan showed characteristic peaks of amide I at 1656 cm⁻¹, N-H deformation in NH₂ at 1597 cm⁻¹, and of amide III band at 1321 cm⁻¹. Deacetylation led to a decrease in the absorbance of the amide I peak, and to an increase in the absorbance in the N-H deformation peak from glucosamine units, as expected, resulting in a broad peak centered at 1646 cm⁻¹. In the spectra of chitosan microspheres, for both chitosans, a increase in the absorbance of the 1155 cm⁻¹ peak was observed, together with the emergence of a new peak at 1225 cm⁻¹, indicating the presence of tripolyphosphoric ions, and revealing that some crosslinking also took place during bead formation. In addition, the N-H deformation vibration shifted to lower frequencies, leading to the emergence of a new peak at

1551 cm^{-1}, which was more obvious in the case of microspheres prepared with deacetylated chitosan. Hydrogen bonding is known to lead to shifting of the N-H deformation band to higher frequencies [7]. As this shift was also described for chitosan scaffolds [8], it may be related to the establishment of weaker hydrogen bonding in the liophilizate, compared to the existent in the original chitosan.

The spectra of chitosan/HAp microspheres, is shown in Fig. 5. A standard HAp was used for comparison. The incorporation of HAp lead to the emergence of characteristic peaks assigned to HAp bands, namely the v_1 PO_4 narrow peak at 961 cm^{-1}, the v_3 PO_4 peak at 1034 cm^{-1}, the v_4 PO_4 peaks at 602 and 565 cm^{-1}, and the O-H peak at 632 cm^{-1}. The peaks at 1455 and 1422 cm^{-1} were assigned to the presence of carbonates, already present in HAp. As the HAp amount increased, overlapping of chitosan bands occurred.

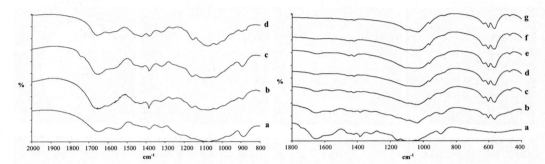

Fig. 4. FTIR spectra of: a) chitosan ([F$_A$]=0.30); b) chitosan microspheres; c) deacetylated chitosan; d) microspheres of deacetylated chitosan.

Fig. 5. FTIR spectra of: a) chitosan ([F$_A$]=0.30); chitosan-HAp composites with varying % w/v HAp: b) 1; c) 5; d) 10; e) 20; f) 30; and g) HAp.

Conclusions

Chitosan-HAp porous spherically-shaped microspheres were successfully prepared using tripolyphosphate as coagulating agent. Chitosan deacetylation promoted a decrease in microsphere diameter. Increasing HAp content promoted an increase in size, a more homogeneous size distribution, and smother and increasingly spherically-shaped surfaces. The water sorption decreased with increasing HAp contents. A homogeneous distribution of the HAp particles within the polymeric matrix was observed, and the ceramic was well embedded in the chitosan matrix. Cross-sections of microspheres revealed considerable interconnectivity, as opposed to their surface.

References

[1] JS Temenoff, AG Mikos: Biomaterials Vol. 21 (2000), p. 2405.
[2] DN-S Hon: *In* S Dumitriu: Polysaccharides in Medical Applications (Marcel Dekker, New York, 1996). p. 631-649.
[3] H Aoki: Medical Applications of Hydroxyapatite (Ishiyaku EuroAmerica, Tokyo, 1994).
[4] A Baxter, M Dillon, KDA Taylor, GAF Roberts: Int J Biol Macromol Vol. 14 (1992), p. 166-169.
[5] J Brugnerotto, J Lizardi, FM Goycoolea, W Arguelles-Monal, J Desbrières, M Rinaudo: Polymer Vol. 42 (2001), p. 3569-3580.
[6] F-L Mi, S-S Shyu, S-T Lee, T-B Wong: J Appl Polym Sci Vol. 37 (1999), p.1551-1564.
[7] G Socrates: Infrared and Raman Characteristics Groups frequencies (John Wiley & Sons, New York, 2001).
[8] Y Zhang, M Zhang: J Biomed Mater Res Vol. 62 (2202), p. 378-386.

Key Engineering Materials Vols. 254-256(2004) pp. 577-580
online at http://www.scientific.net
© *2004 Trans Tech Publications, Switzerland*

In vitro Mineralisation of Chitosan Membranes Carrying Phosphate Functionalities

I.F. Amaral[1,2], P.L. Granja[1] and M.A. Barbosa[1,2]

[1] INEB, Instituto de Engenharia Biomédica, Laboratório de Biomateriais, R. Campo Alegre 823, 4150-180 Porto, Portugal – iamaral@ineb.up.pt

[2] Universidade do Porto, Faculdade de Engenharia, Dept. Eng. Metalúrgica e de Materiais

Keywords: Chitosan, phosphorylated chitosan, calcium phosphates

Abstract. Squid chitosan membranes were phosphorylated through the $H_3PO_4/P_2O_5/Et_3PO_4$/butanol method. P-chitosan membranes were immersed in $Ca(OH)_2$ or $NaOH$ solutions, in order to obtain the Na or the Ca salts, respectively. These materials were investigated regarding their ability to nucleate calcium phosphates, under simulated physiologic conditions. SEM-EDS studies revealed the presence of a calcium phosphate mineral layer all over the surface of P-chitosan membranes, after incubation in $Ca(OH)_2$ solution. The release of ionically bound phosphate functionalities under alkaline conditions, possibly contributed to the formation of calcium phosphate precursor sites, due to the chelation of calcium ions from solution. During the immersion in Simulated Body Fluid (SBF), a multilayer porous mineral structure composed of poorly crystalline carbonated apatite was formed on the surface of these membranes, as shown by EDS and ATR-FTIR analysis. Unmodified membranes and P-chitosan membranes pre-incubated in $NaOH$ solution did not mineralise.

Introduction

The deposition of calcium compounds on polymeric biomaterials may be desirable or not, depending on the application in envisaged. In orthopaedics, the presence of a calcium phosphate interface is often desirable, not only to ensure bone bonding to the implant, but also to be used as a sustained drug delivery system. Therefore, several techniques have been developed in order to turn polymeric biomaterials with limited or no tendency to mineralise into mineralising ones. Blends with bioactive ceramics and biomimetic coatings are among the most frequent ones. The grafting of negatively charged groups, such as phosphates, is another well-known strategy to induce the formation of apatite layers in simulated plasma solutions. This approach has been applied to several natural polymers including chitin and chitosan.

Chitosan, obtained by *N*-deacetylation of chitin, is a biodegradable polysaccharide presently considered for a wide number of biomedical applications, ranging from 3-D matrices for tissue engineering to sutures, wound dressings and gene and drug delivery vehicles. Phosphorylation of chitin and chitosan is described as an effective way of inducing mineralisation, after pre-incubation in calcium-containing solutions [1]. In the present work, an alternative phosphorylation method was used to functionalise chitosan membranes [2]. The ability of the resulting phosphorylated membranes to nucleate calcium phosphates under simulated physiologic conditions, was investigated.

Materials and Methods

Materials. Chitosan powder from squid pens was kindly provided by France-Chitine (Marseille-France), and subsequently purified by the reprecipitation method. The regenerated chitosan presented a molar fraction of *N*-acetylated units of 0.39, as determined by FT-IR following the method of Baxter *et al.* [3], and an average viscosity molecular weight (M_V) of $2.48{\times}10^6$ Da, estimated from the intrinsic viscosity according to Wang *et al.* [4]. All other reagents were of analytical grade.

Preparation of membranes. Clear and transparent membranes of ca. 50 μm thick were prepared by solvent casting, from 1% (w/v) solutions of purified chitosan in 1% (v/v) aqueous CH_3COOH. The resulting membranes were deprotonated in 0.5 M NaOH and thoroughly washed with distilled and deionised water (DDW). Finally, they were cut into 20×60 mm^2 strips and kept in absolute ethanol prior to chemical modification.

Chemical modification. Surface phosphorylation was carried out at 30°C by the $H_3PO_4/P_2O_5/Et_3PO_4$/hexanol method with minor modifications [2], for a period of 8h. After chemical treatment, membranes were rinsed with ethanol and suspended twice in this solvent for 30 min. Finally, they were dialysed against DDW for 24h. The surface P at.% was estimated by XPS spectroscopy.

***In vitro* mineralisation studies.** Unmodified and P-chitosan membranes were sterilized in 70% EtOH before further use. Prior to mineralisation assays, P-chitosan membranes were immersed in saturated $Ca(OH)_2$ or in 23 mM NaOH solutions for 3 days, solutions being renewed every 24h in order to obtain the Na or the Ca salts, respectively. Chitosan membranes were equally immersed in $Ca(OH)_2$, since unmodified chitosan is reported to absorb minute amounts of calcium [5]. Subsequently, each sample was immersed in 40 mL of sterile SBF at 37°C, for periods up to 28 days, SBF being renewed every 2 days. Polyethylene screwtop flasks were used. At the end of each immersion period, the strips were washed twice with DDW and dried under reduced pressure at 60°C. The calcium concentration in the supernatants was analysed by atomic absorption spectroscopy (AAS), using a nitrous oxide-acetylene flame and an ionisation buffer (2000 ppm KCl) in order to remove chemical interferences.

Surface characterisation. Surface mineralisation was assessed by Scanning Electron Microscopy (SEM) equipped with an energy dispersive spectroscope (EDS) microanalysis system, Attenuated Total Reflectance-Fourier Transform Infrared (ATR-FTIR) spectroscopy, using the SplitPea[TM] accessory, and X-ray diffraction (XRD).

Results and Discussion

Phosphorylated membranes presented a 2.87±0.13 P at.%, as shown by XPS studies. AAS analysis of the supernatants showed that P-chitosan membranes significantly adsorbed more calcium than unmodified membranes, during incubation in $Ca(OH)_2$ solution; values of 0.617 and 0.099 mg·cm^{-2} were found, respectively. These results are in accordance with the expected, since P-chitosan is known to be a better calcium ion chelator than chitosan [5]. SEM studies showed the presence of a mineral layer all over the surface of P-chitosan membranes, after incubation in $Ca(OH)_2$ solution (Fig. 1a). EDS analysis revealed a high average Ca/P ratio, namely of 2.42±0.18, probably due to the presence of $Ca(OH)_2$. The present phosphorylated membranes have phosphate groups ionically bound to protonated chitosan amine functionalities, besides having covalently bound phosphates (results to be reported in a future paper). Under alkaline conditions, these ions are released and, together with the presence of covalently bound phosphates, possibly contribute to the anchorage of calcium ions, and consequently to the formation of calcium phosphate precursor sites. After immersion in SBF, several mineral layers were observed in samples immersed for 1 day, presenting an average Ca/P ratio of 1.73±0.01. Samples corresponding to longer immersion periods presented thicker porous layers (Fig. 1b), with average Ca/P ratios in the range typical of apatites, namely 1.65, independently of the immersion time. After 7 days of immersion in SBF, the mineral layer presented a thickness of about 6 μm, as observed in cross-sections (pictures not shown). In addition, Ca was also found in the inner of the membrane, as expected, since the inner of the membrane is also phosphorylated. No mineral formation could be observed on unmodified chitosan membranes, nor on NaOH-treated P-chitosan membranes, suggesting that Na ions were not exchanged by Ca ions from SBF.

Fig. 1. SEM micrograph of P-chitosan membrane: (a) after pre-incubation in Ca(OH)$_2$ and (b) after pre-incubation in Ca(OH)$_2$ and subsequent immersion in SBF for 14 days, showing different mineral layers. The bar corresponds to 6 μm.

AAS analyses of the supernatants agree with the SEM results, revealing practically no calcium uptake for all materials, except for P-chitosan membranes pre-incubated in Ca(OH)$_2$, which presented an uptake of 0.818 mg·cm^{-2} after 28 days of immersion in SBF (Fig. 2). In the case of chitosan pre-incubated in Ca(OH)$_2$, a release of calcium was observed, which suggests that the few

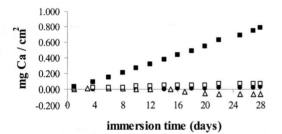

Fig. 2. Cumulative calcium uptake as determined by AAS, during immersion in SBF of the samples: (•) chitosan; (Δ) chitosan pre-incubated in Ca(OH)$_2$; (□) P-chitosan pre-incubated in NaOH; (■) P-chitosan pre-incubated in Ca(OH)$_2$.

anchored calcium ions were probably exchanged with other ions in solution.

The ATR-FTIR spectra of P-chitosan revealed the presence of a new peak at 1220 cm^{-1}, assigned to P=O asymmetric stretching, due to phosphorylation. An increase in the absorbance of the peak at 1064 cm^{-1} was also observed and was assigned to the C–O–P stretching in phosphate esters, overlapping the C–O stretching vibrations in chitosan ether groups. Protonation of chitosan amine functionalities is suggested by the presence of two peaks, both attributed to NH$_3^+$ groups, namely the antisymmetrical deformation at 1630 cm^{-1} and the symmetric deformation at 1533 cm^{-1}. After incubation in Ca(OH)$_2$, the ATR-FTIR spectra revealed the apatitic nature of the mineral layer previously observed by SEM. The spectra of these samples evidenced peaks characteristic of phosphate vibrations in apatitic structures, overlaying those of the phosphorylated substrate. The intense band at 1026 cm^{-1} was assigned to the ν_3 PO$_4$ vibration, the peak at 962 cm^{-1} was assigned to the ν_1 PO$_4$ vibration, and the peaks at 601 and 562 cm^{-1} to the ν_4 PO$_4$ vibration (Fig. 3). In addition, peaks at 1417 and 873 cm^{-1} were also observed, and were assigned to the ν_3 an ν_2 CO$_3$ vibration of carbonate groups, respectively, suggesting the presence of a carbonated apatite. The presence of carbonate groups certainly contributed to the high Ca/P ratio found for these mineralised samples. Fig. 4 shows the spectra obtained for the different materials after 28 days of immersion in SBF, compared to a standard hydroxyapatite sample. The characteristic peaks of phosphate vibrations were only observed in mineralised samples, in accordance to SEM-EDS and AAS studies. The presence of band splitting in the ν_4 PO$_4$ region indicates that the minerals formed were not amorphous. However, the characteristic hydroxyapatite peak assigned to OH groups was not observed. Peaks assigned to carbonate species vibrations were also present, pointing out the growth of a carbonated apatite, *in vitro*.

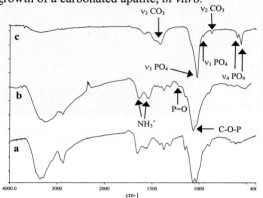

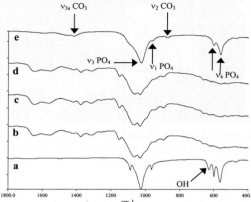

Fig. 3. ATR-FTIR spectra of: (a) chitosan membrane; (b) P-chitosan membrane and (c) P-chitosan membrane after pre-incubation in $Ca(OH)_2$.

Fig. 4. ATR-FTIR spectra of samples, after immersion in SBF for 28 days: (a) control hydroxyapatite; (b) chitosan membrane (c) chitosan membrane pre-incubated in $Ca(OH)_2$; (d) P-chitosan membrane pre-incubated in NaOH and (e) P-chitosan membrane pre-incubated in $Ca(OH)_2$.

The XRD pattern of the mineral formed on P-chitosan pre-incubated in $Ca(OH)_2$ revealed the presence of the main characteristic peaks of crystalline HAp (Fig. 5), although less defined, indicating a poorly crystalline apatite.

Conclusions

A Ca-P mineral was formed on phosphorylated chitosan membranes prepared by the $H_3PO_4/P_2O_5/Et_3PO_4$/hexanol method, after soaking in $Ca(OH)_2$. The subsequent immersion in SBF led to the growth of a partially carbonated and poorly crystalline apatite. The sodium salt of P-chitosan and the unmodified chitosan membranes did not mineralise under the same conditions.

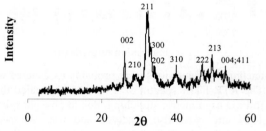

Fig. 5. X-ray diffraction pattern of powder formed on P-chitosan membranes pre-incubated in $Ca(OH)_2$, and immerged in SBF for 28 days.

References

[1] H.K. Varma, Y. Yokogawa, F.F. Espinosa, Y. Kawamoto, K. Nishizawa, F. Nagata and T.Kameyama: Biomaterials Vol. 20 (1999), p. 879.

[2] P.L. Granja, L. Pouysegu, M. Petraud, B. De Jeso, C. Baquey and M. A. Barbosa: J Appl Polym Sci Vol. 82 (2001) p. 3341.

[3] A. Baxter, M. Dillon, K.D.A. Taylor and G.A.F. Roberts: Int J Biol Macromol Vol. 14 (1992), p. 166.

[4] W.Wang, S.Q. Bo, S.Q. Li and W. Qin: Int J Biol Macromo. Vol 13 (1991), p. 281.

[5] N. Nishi, Y. Maekita, S. Nishimura, O. Hasegawa and S. Tokura: Int J Biol Macromol Vol. 9 (1987), p.109.

Key Engineering Materials Vols. 254-256(2004) pp. 581-584
online at http://www.scientific.net
© *2004 Trans Tech Publications, Switzerland*

In vitro Bioactivity in Glass-ceramic / PMMA-co-EHA Composites

B. J. M. Leite Ferreira[1], M. G. G. M. Duarte[2], M. H. Gil[2], R. N. Correia[3]
J. Román [4] and M. Vallet-Regí[4]

[1] Departamento de Engenharia Cerâmica e do Vidro, Universidade de Aveiro, Campus Universitário de Santiago, 3810-193 Aveiro, Portugal, e-mail: bleite@dq.ua.pt

[2] Departamento de Engenharia Química, Faculdade de Ciências e Tecnologia da Universidade de Coimbra, Pólo II - Pinhal de Marrocos, 3030 Coimbra, Portugal, e-mail: garieladuarte@solis.pt

[3] CICECO e Departamento de Engenharia Cerâmica e do Vidro, Universidade de Aveiro, Campus Universitário de Santiago, 3810-193 Aveiro, Portugal, e-mail: rcorreia@cv.ua.pt

[4] Departamento de Química Inorgánica y Bioinorgánica. Facultad de Farmacia. UCM. 28040 Madrid. Spain.

Keywords: bioactive bone substitute, bioactive glass-ceramic, poly(methylmethacrylate) based resin, composites, *in vitro* bioactivity, apatite layer.

Abstract. We developed new composites that consist of poly(methylmethacrylate)-co-(ethyl-hexyl-acrylate) (PMMA-co-EHA) filled with a glass-ceramic (mol%) (70 SiO_2 – 30 CaO). The *in vitro* bioactivity was assessed by determining the changes in surface morphology and composition after soaking in simulated body fluid (SBF) for periods of up to 21 days at 37° C. X-ray diffraction (XRD) and scanning electron microscopy coupled with X-ray energy dispersive spectroscopy (SEM-EDS) after different soaking periods confirmed the growth of apatite-like deposits after 3 days. The deposits consisted of spherical aggregates of acicular crystallites.

Introduction

Polymethylmethacrylate (PMMA) bone cement has been widely used in orthopedic surgery for prosthetic fixation since it was introduced by Charnley. However, an unresolved problem with using PMMA bone cement is a thickening of the intervening fibrous tissue layer that leads to aseptic loosening of the cement in some cases [1]. To improve fixation of PMMA with the host bone, various composites with bioactive materials have been developed and studied [2, 3]. A number of attemps have been made at filling PMMA matrix with calcium phosphates [2] and with bioactive glass [3]. The short-term results obtained are encouraging and suggest that this bioactive materials form an apatite surface layer *in vivo* [1] or in contact with simulated plasma (SBF) *in vitro* [4]. This layer is considered to be responsible for the bonding of bioactive ceramics to the host bone [3]. The aim of this work is to follow the formation of the apatite layer on the surface of bioactive glass-ceramic/PMMA-co-EHA composites.

Materials and Methods

Preparation of composites - Methylmethacrylate (MMA) and ethyl-hexyl-acrylate (EHA) were obtained from Aldrich Chemical Company. Bioactive glass with composition $70SiO_2$ – $30CaO$ (mol%) was obtained by a sol-gel route and calcined at 900° C, for 1 hour, in air, milled and sieved to less than 90 μm particle size. The resultant powder consists of a glass phase and crystalline pseudowollastonite. Benzoyl peroxide (PBO) was obtained from Merck. Only the MMA was purified, extraction of hydroquinone, all the other reagents were used as received. Glass-ceramic/PMMA-co-EHA composites were prepared by addition of the monomers to 50 wt% of the ceramic filler. Benzoyl peroxide was added to the monomer mixture in a ratio of 0,5 mol% as a polymerization initiator. Five types of composites that differed only in the proportions of MMA and EHA were studied (Table 1).

Table 1: Proportions of MMA and EHA in composite matrices.

Composite identification	MMA [wt%]	EHA [wt%]
Composite 1	80	20
Composite 2	70	30
Composite 3	60	40
Composite 4	50	50
Composite 5	100	0

Assessment of *in vitro* bioactivity - Pieces of 5 mm x 5 mm x 5 mm were surface ground, mounted vertically and soaked in 15 mL of *tris*-buffered SBF in sterile polyethylene containers maintained at 37 °C. The SBF solution had a similar composition to that of human plasma, and was previously filtered trough a Millipore 0,22 μm system. Soaking periods were 1, 3, 7, 14 and 21 days. The concentrations of phosphorous (P), silicon (Si), and calcium (Ca), as well as the pH of the solution, were determined for each period by inductively coupled plasma (ICP) spectroscopy. Formation of the apatite layer was identified by X-ray diffraction (XRD) and scanning electron microscopy coupled with X-ray energy dispersive spectroscopy (SEM-EDS).

Results and discussion

Changes in SBF composition - Changes in the concentrations of calcium, silicate, and phosphorous ions in SBF due to immersion of the composites are shown in Fig. 1. Along the first 3 days there is a rapid increase in Si due to filler dissolution. A concomitant deposition of calcium phosphate occurs, as shown by the fall in P concentration (Fig. 1-c). The balance somewhat favours dissolution, since Ca concentration rises moderately in the same period. From 3 to 14 days the concentration profiles are more stable, with comparatively minor changes. This suggests a reformulation of the deposit - perhaps with morphological implications - rather than apposition. A later increase in soluble Ca and P (14 to 21 days) is thought to result from detachment of portions of the deposit.

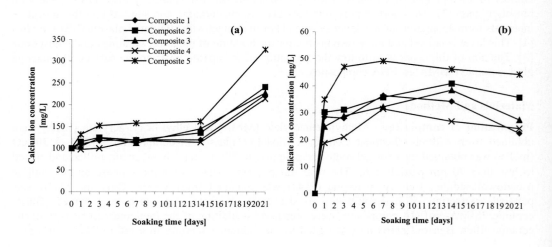

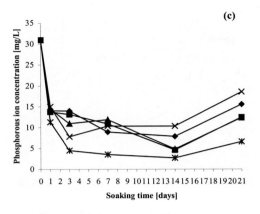

(c)

Fig. 1: Variation of ionic concentrations of (a) calcium, (b) silicate, and (c) phosphorous in simulated body fluid after 21 days of soaking bioactive glass-ceramic/PMMA-co-EHA composites.

Formation of apatite layer - Formation of calcium phosphate deposits on soaking in SBF is evidenced by XRD and SEM-EDS. XRD patterns are shown in Fig. 2(a). After 1 day in SBF the XRD pattern of the surface is characteristic of pseudowollastonite [6] with a certain amount of amorphous phase. At 21 days soaking there is an intensity decrease of the pseudowollastonite peaks and the appearance of a broad band characteristic of apatite [7]. EDS of composite surfaces (Fig 2-b) shows pronounced increases in Ca and P from 1 to 3 days and a corresponding attenuation of the Si signal, indicating calcium phosphate deposition [4]. This deposition slows down between 3 and 21 days.

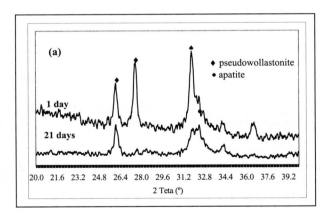

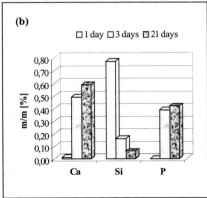

Fig. 2: Cu Kα XRD (a) and (b) EDS patterns of the composites during soaking.

SEM images (Fig. 3) show an almost complete coverage of the substrate after 3 days, by a material whose morphology suggests apatite. After soaking for 21 days, a layer of spherical particles fully covers the composite. The spherical particles consist of numerous acicular crystallites. In the whole, the analyses also demonstrate the formation of an apatite-like layer similar for the five composites studied.

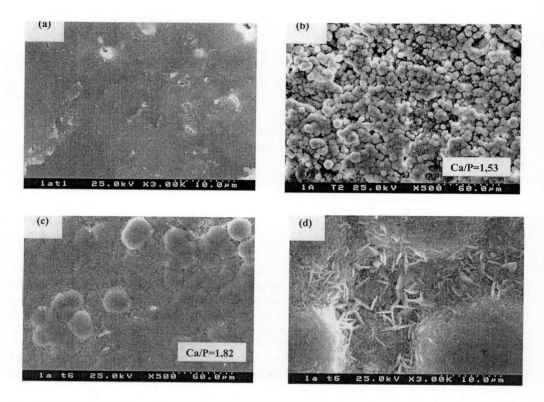

Fig. 3: SEM micrographs of the composites surface after soaking in SBF for (a) 1 day, (b) 3 days, and (c) 21 days. (d) is a magnification of (c). The Ca/P molar ratio was obtained for EDS.

Conclusions

New bioactive composites of PMMA-co-EHA matrix filled with 70 SiO$_2$ – 30 CaO glass-ceramic have been developed. The combined application of XRD and SEM-EDS techniques allowed monitoring the formation of a mineralised deposit following immersion in SBF. This layer is thought to progressively evolve towards a carbonate-substituted, P-deficient apatite.

References
[1] Yoshifumi Okada, Keiichi Kawanabe, Hiroshi Fujita, Ken Nishio, takashi Nakamura: J. Biomed. Mater. Res. Vol. 47 (1999), p. 353-359.
[2] D.T. Beruto, S.A. Mezzasalma, M. Capurro, R. Botter, P. Cirillo: J. Biomed. Mater. Res. Vol. 49 (2000), p. 498-505.
[3] Weam F. Mousa, Masahiko Kobayashi, Shuichi Shinzato, Masaki Kamimura, Masashi Neo, Satoru Yoshihara, Takashi Nakamura: Biomaterials Vol. 21, (2000), p. 2137-2146.
[4] M. Vallet-Regí, A.M. Romero, C.V. Ragel, R.Z. Legeros: J. Biomed. Mater. Res. Vol. 44, (1999), p. 416-421.
[5] X-ray powder data file, ASTM 74-0874.
[6] X-ray powder data file, ASTM 09-0432.

Key Engineering Materials Vols. 254-256(2004) pp. 585-588
online at http://www.scientific.net
© *2004 Trans Tech Publications, Switzerland*

In-vitro Dissolution Characteristics of Calcium Phosphate/Calcium Sulphate Based Hybrid Biomaterials

C. P. Cleere[1], G. M. Insley[2], M. E. Murphy[3], P. N. Maher[4] and A. M. Murphy[5]

[1,2,3,4] Stryker Howmedica Osteonics, Raheen Business Park, Limerick, Ireland.
Cillín.Cleere@emea.strykercorp.com, Gerard.Insley@emea.strykercorp.com,
Matthew.Murphy@emea.strykercorp.com, Peter.Maher@emea.strykercorp.com
[5] Limerick Institute of Technology. Myolish Park, Myolish, Limerick, Ireland.
Ann.Murphy@LIT.ie

Keywords: Calcium sulphate, calcium phosphate, *in-vitro* dissolution, calcium liberation

Abstract. The purpose of this study was to characterize the *in-vitro* dissolution properties of varying percentages of calcium phosphate/calcium sulphate based hybrid biomaterials for possible application in large osseous defect filling. Atomic absorption spectroscopy (AAS), percentage weight loss and macroscopic imaging analysis were carried out on the various formulation tablets at specific time points over a one-month period. The formulations used are outlined in table one below.

Introduction

Customising the properties of synthetic bone substitute materials is a method currently being used by scientists to fill the need for new solutions to surgical problems of bone loss and defect filling. Material properties such as dissolution and mechanical strength can be customised to suit the particular surgical application. Calcium sulphate is a proven biomaterial with a long clinical history [1,2,3] and due to its relatively quick resorption rate [4], is more suited to smaller non-load bearing defects. For calcium sulphate to be considered for larger osseous defect filling, its material properties would need to be tailored to suit the needs of such a defect, such as, stronger mechanical properties and a slower dissolution rate (longer resonance time) while still providing an optimum environment for new bone formation to occur. By adding calcium phosphate, a more stable matrix [5,6], at various percentages to the calcium sulphate material, it is believed more favourable *in-vitro* characteristics can be achieved.

Materials and Methods

A series of different formulation tablets were manufactured for feasibility studies. These tablets consisted of different percentage by weight blends of calcium sulphate hemi-hydrate and calcium phosphate powders, which were then tabletised into three and five mm diameter ($\varnothing$) tablets, see table 1. Four different formulation tablets, (5 mm $\varnothing$), were used in this experiment, with five tablets analysed per three-day time points over a period of one month.

In-vitro dissolution consisted of agitating the samples at 200 cycles per minute in Locke Ringer solution (Oxoid Ringer tablets) while maintaining a constant temperature of 37.4°C $\pm$ 0.2°C

Table 1. Tablet formulations

Formulation	% Calcium sulphate	% Calcium phosphate	% Polyvinyl povidone (PVP)	% Stearic acid
A	98	0	1	1
B	83.3	14.7	1	1
C	63.7	34.3	1	1
D	44.1	53.9	1	1

The following methods were used to characterize the tablets during the dissolution trial,
- Atomic absorption analysis of the dissolution solution for calcium liberated.
- Percentage weight loss of the tablets with time.
- Macroscopic analysis of tablet surface following dissolution.
- Dimetral tensile testing of tablets.

Results and Discussion

Atomic absorption spectroscopy (AAS) analysis was carried out on the dissolution solution of each formulation tablet to determine the quantity of calcium being liberated. Figure 1 shows a background concentration of 90 ppm calcium already existed in Ringer solution, representing the average physiological calcium concentration of blood. All formulations demonstrated similar trends in increased calcium loss with time, both formulation C and D demonstrated the least calcium liberated during dissolution as was expected due to the higher concentration of stable calcium phosphate. [2,3] By day 36, formulation D had liberated 655.4 ppm calcium, compared to 831 ppm liberated by formulation A, a difference of 175.6 ppm. A saturation of the surrounding Locke Ringer solution occurred between days 16-20. At day 20 the Ringer solution was replenished with fresh stock.

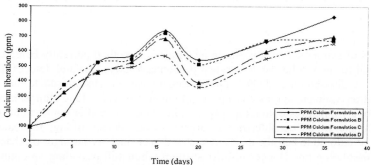

Fig. 1 Calcium liberated with time

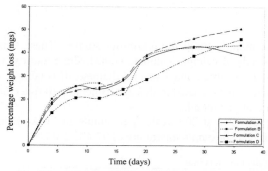

Fig..2 Percentage weight loss with time

Percentage weight loss was recorded for each formulation at the relevant time points to determine the percentage of original weight lost during dissolution. *In-vitro* weight loss simulates the biological resorption of the biomaterial by the relevant bone cells. Biomaterials, which lose high percentages of weight quickly *in-vitro*, tend to be rapidly resorbed by the body *in-vivo*.

From figure 2, it is observed that formulations A, B and C demonstrated similar trends in increased weight loss, with exception to formulation B, which demonstrated similar increase until a sharp drop in weight loss was observed at day six. This is a trend not seen in any of the other formulations, and was most probably due to a statistical deviation in the tablet weight.

Formulation D demonstrated the lowest percentage weight loss on average, throughout the 36-day period. By day 20, it had lost on average 10% less weight than the other formulations. By day 28, both formulations A and B had begun to stagnate, while C and D increased in weight loss steadily. By day 36 formulations A and B has lost approximately 39 and 43% of their total weight, where as C and D lost approximately 50 and 46% respectfully.

This is an unexpected result as both tablets A and B would be expected to demonstrate less stability *in-vitro* due to their low percentage of calcium phosphate. Both tablets C and particularly D demonstrated a more constant weight loss due to their higher percentage of stable calcium phosphate.

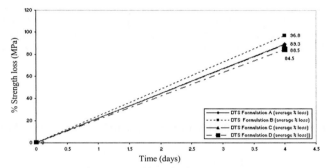

Fig. 3 Mechanical strength loss with time

Dimetral tensile strength (DTS) analysis was carried out on all four formulations at time zero to serve as a benchmark for the strength results. Results at time zero ranged from 2.2 to 2.4 MPa. By day four a dramatic loss in DTS was observed, between 85-97 % strength loss was recorded for the four different formulations, the most plausible explanation for the high strength loss is the dissolution of the glossy steric acid coating (1% w/w) covering all these tablets, see figures 4-7. No formulation demonstrated superior mechanical strengths, and therefore any trends seen thus far in both calcium liberation and percentage weight loss were due to the materials intrinsic properties as opposed to the mechanical properties instilled during the manufacturing process.

Macroscopic analysis

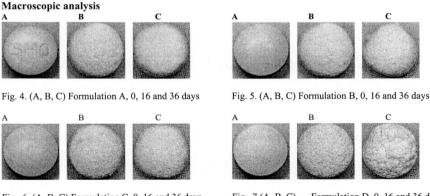

Fig. 4. (A, B, C) Formulation A, 0, 16 and 36 days   Fig. 5. (A, B, C) Formulation B, 0, 16 and 36 days

Fig.. 6. (A, B, C) Formulation C, 0, 16 and 36 days   Fig.. 7 (A, B, C)   Formulation D, 0, 16 and 36 days

Macroscopic image analysis was carried out on all four formulations to demonstrate visually the *in-vitro* dissolution process. Figures 4-7 show the macroscopic images of the four formulation tablets at time 0, 16 and 36 days. The macroscopic images demonstrate the loss of the steric acid coating, which led to the dramatic loss in mechanical strength.

As the *in-vitro* dissolution continues the sharp edges of the tablets noted at time zero were found to smoothen off to a more rounded finish with a distinct powder like surface finish remaining.

As demonstrated in figure 7, formulation D at time 36 days had a very "pitted" surface, which was noted regularly for that formulation. This may be due again to the higher percentage of more stable calcium phosphate, which was left behind after dissolution of the calcium sulphate percentage. Formulations A-C showed the same degree of surface erosion and smoothening of tablet edge surface. This is an expected result as these three formulations demonstrated similar calcium liberation and percentage weight loss properties.

Conclusion

The aim of this study was to determine the *in-vitro* dissolution characteristics of a number of different formulation biomaterial tablets for possible application in large osseous defect filling. Of the four formulation tablets used in this study, the tablets possessing the highest percentage of calcium phosphate, namely C and D, did demonstrate increased *in-vitro* stability compared to the less stable formulations.

References

[1] Dressmann H, Beiter Klin Chir, 9 (1892) 804-810[2] Koffman S, Zentralbl Chir, 45 (1925) 817-818

[3] Coetzee AS, Arch Otolaryngol 106(7) (1980) 405-405

[4] Khalid A and Ruhaimi A, Int.J. Of Oral and Maxillofacial implants, 15 (2000), 859-864

[5] Quintiles dissolution study, Report Reference number: ACR/00/23

[6] Quintiles dissolution study. Report reference number: YOC00101/01

Key Engineering Materials Vols. 254-256(2004) pp. 589-592
online at http://www.scientific.net
© 2004 Trans Tech Publications, Switzerland

Hybridized Hydroxyapatite Bioactive Bone Substitutes

Kleber de Arruda Almeida, Alvaro Antonio Alencar de Queiroz

Departamento de Física e Química, Instituto de Ciências, Universidade Federal de Itajubá (UNIFEI). Av. BPS, 1303, 37.500-903, Itajubá-MG, Brasil. E-mail: kleberalmeida@unifei.edu.br, alencar@unifei.edu.br.

Keywords: Poly(ε-caprolactone), hybridized hydroxyapatite, bioactive composites, osteoinduction, biodegradable composite

Abstract. The aim of this study was to evaluate the osteoinduction activity of the composite poly(ε-caprolactone)/Hydroxyapatite ($PCLI_2/HA$). The bioactive $PCLI_2/HA$ composite was made after bulk polymerization of ε-caprolactone in the presence of HA using I_2 as a catalyst to form a macroporous polyester ($PCLI_2$) having a weight-average molecular weight of 45,500. Bactericidal activities of the $PCLI_2/HA$ were assessed against *Escherichia coli* and *Staphylococcus aureus*. The tested $PCLI_2/HA$ showed more antimicrobial activity against *E. coli* and was less active against *S. aureus*. After sterilization by gamma rays from a ^{60}Co source at 25 kGy the composite $PCLI_2/HA$ was implanted in rabbit tibia cavities ($\phi = 3mm$) and quantitative data for histomorphometric evaluation was collected after microscopic observation. The newly developed HA ceramic with macroporous $PCLI_2$ improves bone tissue osteogenisis after 3 weeks of implantation. The histological analysis after $PCLI_2/HA$ implantation revealed an osteoinductive material that supported bone cell growth suggesting a potentially bioactive composite

Introduction

A number of ceramic analogs of bone, such as hydroxyapatite and tricalcium phosphate, have shown to be promising as alternative graft materials due to their exceptionally good tissue compatibility and adhesion to native bone.

In experimental studies hydroxyapatite (HA) has shown excellent biocompatibility and bone repair properties, but some difficulties for application into periodontal tissue defects have been reported [1].

A variety of synthetic bone grafts from composites based on biodegradable polymers and hydroxyapatite (HA) has been the subject of considerable scientific and clinical interest, improving the clinical applicability of bioceramics in dentistry and medicine [2].

The biodegradable polymer/ceramics composites have gained acceptance in dentistry and medicine for repairing bone and periodontal defects with advantages relatively to autografts and allografts such as unlimited supply, low cost and absence of immunogenicity [3].

The use of biodegradable polymers, such as aliphatic polyesters, to obtain homogeneous injectable composite materials has become a common practice in dental surgery. In this sense, poly (e-caprolactone) (PCL) have been used in several medical applications, but as such they do not enhance tissue regeneration.

The main objective of this work was to develop and test novel biologically active composites based on hydroxyapatite and macroporous PCL with potential for use in periodontal and ortoephedic implants. The experiments are focused on the synthesis and biologic response of bone to the $PCLI_2$ /HA composite.

Experimental

The monomer ε-caprolactone (ε-CL, Sigma) was polymerized in bulk in the presence of HA (Intralock) using I_2 as a catalyst to form a macroporous polyester (PCL) having a weight-average molecular weight of 45,500 [4]. The composite $PCLI_2$/HA was sterilized by gamma rays from a ^{60}Co source at 25 kGy.

The osteoinductive property of the $PCLI_2$/HA composites were evaluated after implantation in rabbit tibia (cavities of φ = 3mm). Seven male adult rabbits (1.5-2.3 kg) were anesthesized by administration of ketamine cloridate, Ketalar® and xilazine, Rompun® (3.0 mL, 2g/100 mL) for insertion of the breathing tube. Isoflurane was given for anesthesia management, using a low flow closed system technique. The post anesthetic recovery was uneventful, the animals were conscious in about 40 minutes after disconnecting the system. At the end of both implantation periods (20 days), all animals were sacrificed by injecting an overdose of pentobarbitalsodium (Nembutal®). After sacrificing the animals, the tibias were removed and fixed in 10% w/w buffered formalin solution (pH 7.4). After dehydration with ethyl alcohol, the samples were decalcificated in formic acid (5 %m/m) and sectioned by microtome technique (5 μm). The plane of tibia sectioning was along the axis of the implanted material. The sections were stained with haematoxylin/eosine and histologically evaluated. The osteoinductives properties of the $PCLI_2$/HA composite was compared with grafts of blood and pure hydroxyapatite as reference materials. Quantitative data for histomorphometric evaluation were collected after microscopic observation of the stained bone sections.

The *in vitro* antimicrobial activities against *Escherichia coli* and *Staphylococcus aureus* were determined on powdery samples of $PCLI_2$/HA composite by the cut plug method on agar [5].

Results and Discussion

The $PCLI_2$/HA composite must have interconnection and pore sizes that are closely controlled during the synthesis of macroporous PCL by bulk polymerization of ε-CL by iodine charge transfer complex.

Figure 1 shows SEM micrographs of $PCLI_2$ (Fig. 1A) and $PCLI_2$/HA composite (Fig. 1B) obtained by iodine charge transfer bulk polymerization of the monomer (ε-CL). The synthesized $PCLI_2$ displayed particles with pore diameter larger than 50 nm, forming an interconnected sponge-like structure. The macropores observed in the particles were closely circular and had rather uniform sizes.

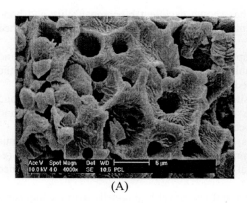 

(A)   (B)

Fig. 1- SEM micrographs of the composite $PCLI_2$/HA obtained by ε-CL iodine bulk Polymerization. Macroporous structure of $PCLI_2$ (A) and the composite $PCLI_2$/HA indicating the bioceramic presence (B).

The use of the I_2 in the formation of the PCLI$_2$/HA composite is very important due to its known anti-microbial activity. The I_2 acting quickly entering the plasmatic membrane, acts by breaking the chemical links of proteins and molecules of DNA. However, we believe that the toxicity of I2 will be controlled, because it is forming a complex with (ε-CL), thus we will have a controlled release of the I_2.

The antibacterial capacity of PCLI$_2$/HA against *E. coli* and *S. aureus* was explored by cut plug method and viable cell counting methods. The capability of the prepared polymer to inhibit the growth of the tested micro-organisms on solid media is shown in Figure 2(A). The macroporous composite PCLI$_2$/HA inhibited the growth of the tested bacteria with an increasing in inhibitory effects in the order *S. aureus* < *E. coli*.

Fig.2- (II): Shows the histological analysis after 3 weeks of the PCLI$_2$/HA composite implantation. The newly developed macroporous composite PCLI$_2$/HA appears to be a promising material for bone tissue repair.

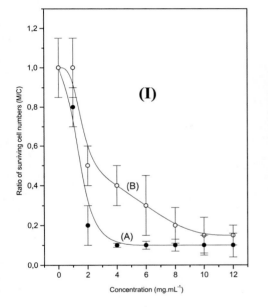
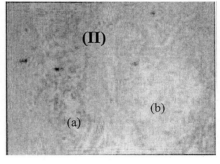

Fig. 2- (I): Growth inhibition of PCLI$_2$/HA for *E.coli (A) and S. aureus* (B), (II): Histological analysis after 3 weeks implantation. New bone formation (a), growth marrow bone (b).

References

[1] V.C. Barney, M.P. Levin, D.F. Adams: J. Periodontol Vol. 57 (1986), p. 764.

[2] C.M. Agrawal: *J.Biomed.Mater.Res.* Vol. 55 (2001), p. 141.

[3] C.N. Cornell: Orthop.Clin.North Am. Vol. 30 (1999), p.591.

[4] A.A.A. de Queiroz, E.J. França, G.A Abraham, J.San Roman: J.Polym.Sci.Polym.Phys.Edn. Vol. 40 (2002), p. 714.

[5] A.A.A. de Queiroz, O.Z. Higa, G.A Abraham, J.San Roman: J.Biomed.Mater.Res. (2003). Submitted for publication.

Key Engineering Materials Vols. 254-256(2004) pp. 593-596
online at http://www.scientific.net
© *2004 Trans Tech Publications, Switzerland*

Phase Mapping: A Novel Design Approach for the Production of Calcium Phosphate-Collagen Biocomposites

A. K. Lynn[1], R. E. Cameron[1], S. M. Best[1], R. A. Brooks[2],
N. Rushton[2], W. Bonfield[1]

[1] Cambridge Centre for Medical Materials, Department of Materials Science and Metallurgy, University of Cambridge, Pembroke Street, Cambridge CB2 3QZ, United Kingdom
akl28@cam.ac.uk

[2] Orthopaedic Research Unit, University of Cambridge, Box 180, Addenbrooke's Hospital, Hills Road, Cambridge CB2 2QQ, United Kingdom
pa.orthopaedics@cam.ac.uk

Keywords: collagen, calcium phosphate, biocomposite, phase mapping, synthesis methodology

Abstract. This paper describes a novel method for the design of calcium phosphate-collagen biocomposites. Employing a rigorous approach based on the phase mapping of calcium phosphate stability regions, this methodology has been developed as a tool through which the compositions of calcium phosphate-collagen biocomposites can be systematically predicted and controlled. The present study details the development of the calcium phosphate phase map, and demonstrates its applicability to the production of biocomposites of i) brushite and collagen, ii) brushite and gelatin, iii) hydroxyapatite and collagen, and iv) hydroxyapatite and gelatin.

Introduction

Biocomposites of calcium phosphate (CaP) and collagen have long been of interest for their potential for use as bone substitutes. While a wide range of methods including mechanical mixing [1], precipitation of CaP on insoluble collagen [2], and co-precipitation [3] have been developed for their production, a comprehensive methodology for the design and control of the stoichiometry of the CaP phase of such composites has yet to be reported.

Variation of the CaP phases present in bone substitutes has been shown to produce significant differences in both implant dissolution and bone apposition rates, and thus biocomposites of collagen and various calcium phosphates ranging from brushite [4] to octacalcium phosphate [5] to hydroxyapatite [6] have been produced to tailor *in vivo* behaviour. Similarly, variations in bone mineral formation and degradation rates between collagen and its denatured form, gelatin, have also been reported [7]. These have led to the production of biocomposites of CaP and both collagen [4-6] and gelatin [8].

The present investigation details the development of a method through which the design of CaP – collagen biocomposites can be systematised. The methodology is based on the production of phase maps which can allow the production of tailored composites using appropriate pH-temperature combinations

Materials and Methods

Phase Mapping. Phase mapping of CaP stability regions was carried out by the x-ray diffraction (XRD) analysis of the products of a dropwise reaction of a soluble calcium source (calcium nitrate; $Ca(NO_3)_2.4H_2O$; Sigma; 1.0M) and a soluble phosphate source (diammonium hydrogen orthophosphate; $(NH_4)_2HPO_4$; BDH; 0.375M) at a Ca:P molar ratio of 1.33.

Temperature and pH were kept constant throughout both the reaction and ageing period using an oil bath and ammonia solution (NH_3; BDH), respectively. After ageing, the reaction products were filtered and the filtrate re-suspended in deionised water to remove any soluble by-products such as ammonium nitrate. The reaction products were then aged for 24 hours.

This reaction was carried out at 36 combinations of temperature and pH, with each experimental trial conducted independently. XRD analysis was carried out on bulk powder specimens of each reaction product using a Phillips PW3020 x-ray diffractometer with scans performed from 2-80° 2θ using Cu-Kα radiation. Resultant peaks were then matched to PDF files using Phillips XPert Plus software to identify the phases present. The results were mapped on a temperature-pH chart to illustrate the stable calcium phosphate phases under each set of conditions.

Composite Production. To demonstrate its utility for the design of CaP-collagen biocomposites, the CaP phase map was used in combination with knowledge of the denaturation temperature of collagen to select appropriate combinations of pH and temperature for the synthesis of brushite/collagen (3.2, 37°C), gelatin/brushite (3.2, 50°C), collagen/ apatite (Ap) (8.0, 37°C) and gelatin/ apatite (Ap) (8.0, 50°C) co-precipitated biocomposites.

Composites were produced by the simultaneous dropwise addition of:
i)   1.0L of a 0.081M calcium nitrate solution and
ii)   1.0L of a solution consisting of 2.0g atelocollagen (Japan Meat Packers Inc.) dissolved in 0.046M H_3PO_4
to 1.0L of continuously stirred deionised water at a selected temperature, and adjusted to the appropriate pH. Following complete addition of these two solutions, the resulting co-precipitates of CaP and collagen were aged for 24 hours under continuous stirring, while the pH was maintained at the appropriate level by titration with ammonia.

XRD analysis to identify the CaP phases present was performed by placing a small amount of each co-precipitate on a silicon wafer and performing measurements under the same parameters described above. Differential scanning calorimetry (DSC) was carried out using a Perkin-Elmer DSC 7 differential scanning calorimeter to identify the presence of either collagen or gelatin in each composite on the basis of the presence or absence of the collagen/gelatin phase transition peak. DSC was performed on the composites prepared at each of the four pH-temperature combinations, as well as on the as-received atelocollagen.

Results

Fig. 1 shows the CaP phase map schematically representing the phases found at each of the 36 combinations of pH and temperature examined.

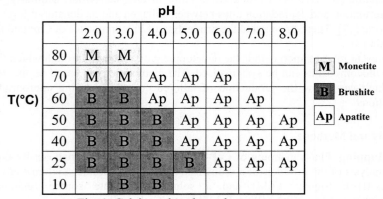

Fig. 1: Calcium phosphate phase map.

DSC of the as-received atelocollagen revealed a collagen/gelatin transition temperature of 41°C. Superimposition of this transition temperature onto the phase map in Fig. 1 produced a composite phase map shown in Fig. 2. for use in the design of CaP/collagen biocomposites.

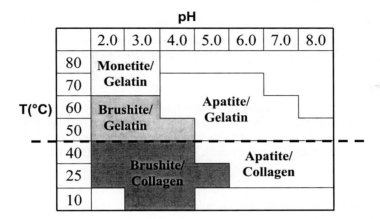

Fig. 2: Composite phase map for the calcium phosphate/collagen system.

XRD analysis of the composites produced at the four combinations of temperature and pH confirmed that for both composites formed at pH 3.2, the sole calcium phosphate phase present was brushite. For the two formed at pH 8.0, the lone phase present was apatite. Furthermore, results from DSC analysis revealed the presence of the collagen/gelatin transition peak in the two composites formed at 37°C, but not in those formed at 50°C. These results confirmed that both the organic and inorganic constituents of the four biocomposites produced corresponded to the composite phase map (Fig. 2) and thus to their intended design.

Discussion

The correlation between the observed compositions of the composites produced at the four respective temperature-pH combinations and the phase region predicted by the composite phase map (Fig. 2) illustrates the power of phase mapping as a tool for the design of biocomposites. By coupling the CaP phase map (Fig. 1) with a basic knowledge of the denaturation behaviour of collagen, the compositions of both the organic and inorganic constituents of biocomposites can be predicted and controlled. Further experimental trials at temperature-pH combinations near the phase boundaries in Fig. 1 will allow further refinement of the phase map. This will increase its predictive capability to allow more accurate control of composition at conditions around these boundaries.

The formation of metastable intermediate phases such as octacalcium phosphate, which do not generally form via direct precipitation, has been observed to occur on collagen fibrils [9]. While the phase mapping method does not predict such phenomena, it should be noted that the formation of such metastable phases is thought to occur as a result of their low surface energies relative to more thermodynamically stable phases [10]. It has been shown that these metastable phases convert readily to their more stable forms [11] and it is these stable states that the phase mapping approach is able to predict. Thus, its application is most appropriate for composites whose production methods include extended ageing periods or other means through which conversion to thermodynamically stable CaP species is promoted.

Although the stable phases in the CaP system depend strongly on pH and temperature, the phase mapping approach is equally suitable for the control of other parameters affecting phase stability including ionic concentration and stirring speed. Therefore, the phase mapping approach can be applied to virtually any composite system to provide simple, visual tools through which the design of biocomposites can be systematised.

Conclusions

A novel method for the design of calcium phosphate/collagen biocomposites has been developed, and its utility for the selection of synthesis parameters during the production of brushite/collagen, apatite/collagen, brushite/gelatin, and apatite/gelatin composites has been demonstrated.

Acknowledgements

The authors would like to acknowledge Dr. Judith Juhasz (University of Cambridge, U.K.) and Dr. Mamoru Aizawa (Meiji University, Japan) for their contributions to this paper. Funding for this work has been provided by the Cambridge-MIT Institute (CMI), the Universities UK Overseas Research Scheme (ORS), the Cambridge Commonwealth Trust and St. John's College, Cambridge.

References

[1] M.R. Bet, G. Goissis, A.M.D. Phelps: Quimica Nova Vol. 20 (1997) p. 475.

[2] A. John, L. Hong, Y. Ikada, Y. Tabata: J. Biomaterial Sci. – Polymer Ed. Vol. 12 (2001) p. 689.

[3] M. Kikuchi, S. Itoh, S. Ichinose, K. Shinomiya, J. Tanaka: Biomaterials Vol. 22 (2001) p. 1705.

[4] K.I. Clarke, S.E. Graves, A.T.C. Wong, J.T. Triffitt, M.J.O. Francis, J.T. Czernuszka: J. Mater. Sci.- Mater. Med. Vol. 4 (1993), P. 107.

[5] Y. Sasano, S. Kamakura, M. Nakamura, O. Suzuki, I. Mizoguchi, H. Akita, M. Kagayama: Anat. Rec. Vol. 242 (1995), p. 40.

[6] S. Itoh, M. Kikuchi, K. Takakuda, Y. Koyama, H.N. Matsumoto, S. Ichinose, J. Tanaka, T. Kawauchi, K. Shinomiya: J. Biomed. Mater. Res. Vol. 54 (2001) p. 445.

[7] N.C. Blumenthal, V. Cosma, E. Gomes: Calcified Tissue Int. Vol. 48 (1991) p. 440.

[8] M.C. Chang, C.C. Ko, W.H. Douglas: Biomaterials Vol. 24 (2003) p. 2853.

[9] M. Iijima, Y. Moriwaki, Y. Kuboki: J. Crystal Growth Vol. 137 (1994) p. 553.

[10] W.J. Wu, G.H. Nancollas: Pure Appl. Chem. Vol. 70 (1998) p. 1867.

[11] S. Graham, P.W. Brown: J. Crystal Growth Vol. 165 (1996) p. 106.

V.   DENTAL AND ORTHOPAEDIC APPLICATIONS

Key Engineering Materials Vols. 254-256(2004) pp. 599-602
online at http://www.scientific.net
© *2004 Trans Tech Publications, Switzerland*

pH Changes Induced by Bioactive Glass Ionomer Cements

H. Yli-Urpo, E. Söderling, P.K. Vallittu and T. Närhi

Department of Prosthetic Dentistry and Biomaterials Research, Institute of Dentistry, University of
Turku, Lemminkäisenkatu 2, 20520 Turku, Finland,
helena.yli-urpo@utu.fi

Keywords: S53P4 bioactive glass; dental materials, glass-ionomer cements, pH

Abstract. The aim of this study was to examine pH-changes in the immersion fluid of a glass ionomer cement (GIC) – bioactive glass (BAG) composite. Glass ionomer cement (GI) (GC Fuji II, GC Corporation, Tokyo, Japan) and resin modified glass ionomer cement (LCGI) (GC Fuji II, GC Corporation, Tokyo, Japan), were mixed with different quantities (10wt-% and 30wt-% of the total powder weight) of BAG (granule size<45µm) (S53P4, Vivoxid Ltd, Turku, Finland). GICs without BAG were used as a control. After mixing the powders for the test specimens, the appropriate liquid was added to the powder mixture, and discoid specimens (n=3) were prepared. The specimens were immersed individually in distilled water at a constant temperature of 37°C. The pH of the immersion liquid was measured electrometrically after 1 h, 6 h, 24 h, 2 d, 7 d and 14 d from the beginning of the immersion. The immediate pH effects of the material suspensions were also determined. The study showed that the more BAG added to GICs, the higher the increase in pH. The pH effect of LC30BAG was the highest among all study materials. Within the limitations of this study, it can be said that the GIC-BAG composites can increase the pH in an aqueous environment around the materials. The increase in pH may affect mineralizing properties of the GIC-BAG composites and possess antimicrobial effects.

Introduction

Glass ionomer cements (GICs), referred also as glass polyalkenoate cements, have been used for nearly 30 years as filling materials in various tooth restorations. Conventional glass ionomer cements contain fluoroaluminosilicate glass, and they set with polymer acid, e.g. polycarboxylic acid due to an acid-base reaction [1]. In an acidic environment glass granules release Ca^{2+} ions, which bond with OH^- groups. After the setting reaction has completed the material is hard and insoluble in the human body. Resin reinforced glass-ionomer cements have been developed in order to improve materials' mechanical and handling properties [2]. Several studies demonstrate that GICs are able to inhibit bacterial growth *in vitro*, and that they have antibacterial activity also *in vivo*. Their antibacterial properties have been suggested to exist due to their ability to release F^- ions resulting in high fluoride concentration and / or on the low initial pH of the cement [3, 4]. The ability to release F^- ions improves the remineralization and reduces the solubility of dentin and enamel to some extent, as well as prevents bacterial attachment to filling surfaces.

Bioactive glass (BAG), first introduced by Hench et al in 1972 [5], are surface-active glasses to which bone minerals are able to bond chemically. Because of their strong bond with living bone, BAGs have been used as bone substitute materials in many different clinical conditions e.g. in orthopedics and in dentistry [6]. They dissolve in aqueous environment, which causes an increase in pH in their surrounding. BAGs have been shown to improve human dentin mineralization both *in vivo* and *in vitro* [7]. They have also been shown to have a broad antimicrobial effect against oral pathogens [8].

In our studies, we have developed bioactive GIC-BAG composites and studied their antimicrobial activity and mineralization properties. The aim of this study was to determine the effects of GIC-BAG composites to the pH of distilled water during two weeks immersion period. The immediate pH effects of the material suspensions were also determined *in vitro*.

Materials and Methods

In this experiment two different kinds of commercially available GIC were used: conventional cure GIC (code: GI) (Batch number: 206251, GC Fuji II, GC Corporation, Tokyo, Japan) and resin-modified light-curing GIC (code: LCGI) (Batch number: 205011, Fuji II LC, GC Corporation, Tokyo, Japan). The material consisted of powder and liquid. A commercially available bioactive glass (code: BAG) (S53P4, Batch number: ABMS53-7-00, Vivoxid Ltd, Turku, Finland) was used in this experiment. The composition of the BAG by weight is SiO_2 53%, Na_2O 23%, CaO 20% and P_2O_5 4%.

The BAG powder with a particle size of <45μm (average particle size, 20μm), was added to the GIC powder. Two different ratios of BAG and GIC powder (10-wt % and 30-wt %) were used. GIC powders without BAG particles were used in fabrication of control specimens. Description of the GIC/BAG powder ratio and powder to liquid ratio is given in Table 1.

Table 1: Weight ratios (wt-%) of GI, LCGI and BAG powders used in the test specimens

Group	GI	LCGI	BAG
GI	100		
GI10BAG	90		10
GI30BAG	70		30
LC		100	
LC10BAG		90	10
LC30BAG		70	30

BAG and GIC powders were measured in 20 ml Falcon plastic test tubes and mixed in a Coulter mixer (Luton, England) for 10 minutes to even the filler particle distribution. The powder was then mixed with poly-acrylic-acid of GI and diacrylate resin-poly-acrylic-acid mixture of LCGI according to the manufacturer's instructions. The mixed materials were packed into metallic molds and gently compressed between glass plates to form discoid specimens (thickness 1.0 mm, diameter 5.5 mm). The specimens made of LCGI cement were cured with a visible light-curing device (ESPE Elipar Highlight, Seefeld, Germany) (470 nm wavelength, light intensity 690 mW/cm^2) for 40 s. The specimens were stored dry for 1-5 days before the immersion. All the specimens were prepared at room temperature (21±1°C) and relative humidity of 55%.

The test specimens were immersed individually in test tubes in 20ml of distilled water in a shaking water bath with a shaking speed of 70 revolutions per minute, at a constant temperature of 37°C (Grant OLS 200, Cambridge, England). All the test materials were present in triplicate. At time points of 1 h, 6 h, 24h, 2 d, 7 d and 14 d from the beginning of immersion the pH was measured electrometrically (PHM80 Portable pH-Meter, Radiometer A/S, Copenhagen, Denmark). Statistical analysis was performed with StatView for Windows program (SAS Institute, Cary, NC, USA). Repeated measures analysis of variance was used to evaluate the differences in pH among the study materials in terms of dissolution time. Pairwise comparisons among the materials were made with Fisher's PLSD test. The immediate pH effects of the materials were also tested using material suspensions. The test specimens were homogenized to powder form with a MM-type Retsch-laboratory mixer (Retsch GmbH & Co, Haan, Germany). The ground-material powders (50 mg) were suspended in 1ml of distilled water and the pH was measured electrometrically (PHM80 Portable pH-Meter, Radiometer A/S, Copenhagen, Denmark) after 15 min.

Results

In the immersion test a significant increase in pH over time was found for all GICs containing BAG ($p < 0.0001$). Change in pH was associated with the type of GIC and the amount of BAG in the samples (Fig. 1). The more BAG added to GICs, the higher the increase in pH ($p < 0.0001$). Furthermore, LCGI based materials with BAG achieved higher increase in pH than GI based materials with BAG. In the immersion test the biggest increase was found for LC30BAG. No significant pH changes were found for GI or LC cements as such.

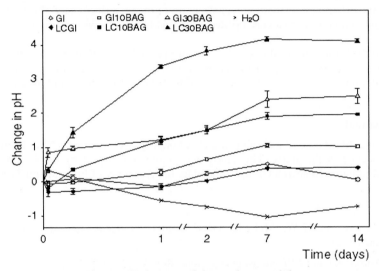

Fig. 1: pH changes of the study materials

Experiment with the material suspensions also confirmed that the more BAG was added to GIC, the higher pH value was determined (Table 2). When comparing the conventional and resin-reinforced GIC together, the pH effect of LC30BAG was the highest among the material suspensions.

Table 2: The immediate pH of the study material slurries

Material	GI	GI10BAG	GI30BAG
pH	6,2	7,4	8,8

Material	LC	LC10BAG	LC30BAG
pH	4,6	5,8	8

Discussion and Conclusions

The presence of BAG increases the pH of the surroundings due to the ion dissolution [8]. In the literature is suggested that the apatite does not form in simulated body fluid at pH 4 or pH 6 [9]. This suggests that also the mineralizing effects of BAG might decrease with the decreasing pH [10]. The immediate pH effects of the material suspensions in this study showed that the more BAG added to GIC, the higher pH effect was determined. This was found on both conventional curing GICs and resin-modified light-curing GICs with BAG. Similar results were also found when the GIC-BAG composites were studied in a set form: They increased the pH of distilled water during the first 2 d of immersion and from approximately 2 d up to 14 d the pH stayed on a high level. Furthermore, the biggest increase in pH surrounding the set material samples and the highest pH effect of the material suspensions was found for the LC30BAG. The most probable reason for the pH effects is the different composition and structure of the materials tested. In our previous studies GI30BAG was found out to be a very brittle material [11], which can explain the pH effects seen in this study.

Within the limitations of this study, it can be said that the GIC-BAG composites increase the pH values in an aqueous environment. These results suggest that the increase in pH may affect positively the mineralizing properties of the GIC-BAG composites and possess antimicrobial effects.

Acknowledgements

The authors would like to thank biomedical laboratory technologist Oona Kalo for excellent technical assistance in the study. This study was conducted in part by the Centre of Excellence "Nano and Biopolymer Research Group" funded by the Academy of Finland and TEKES.

References

[1]   A.D. Wilson and J.W. McLean: *Glass Ionomer Cement* (Quintessence Publishing Co, Chicago 1988).

[2]   J.M. Antonucci and J.W. Stansbury: J. Dent. Res. Vol. 68 (1989), p.251

[3]   C.J. Palenik, M.J. Behnen, J.C. Setcos and C.H. Miller: Dent. Mat. Vol. 8 (1992), p.16.

[4]   R.A. Barkhordar, D. Kempler, R.R.B. Pelzer and M.M. Stark: Dent. Mat. Vol. 5 (1989), p. 281

[5]   L.L. Hench, R.J. Splinter, W.C. Allen and T.K. Greenlee: J. Biomed. Mat. Res. Symp Vol. 2 (1971), p. 117

[6]   J. Wilson, A. Yli-Urpo and R.P. Happonen: *An introduction to bioceramics* (World Scientific Publishing Co, Singapore 1993).

[7]   J. Salonen, U. Tuominen and Ö.H. Andersson: Proc. Finn. Dent. Society (1996), p. 25

[8]   P. Stoor, E. Söderling and J. Salonen: Acta Odontol. Scand. Vol. 56 (1998), p. 161

[9]   T. Kokubo, H. Kushitani, S. Sakka, T. Kitsugi and T. Yamamuro: J. Biomed. Mat. Res. Vol. 24 (1990), p. 721

[10]   P. Stoor, E. Söderling, O.H. Andersson and A. Yli-Urpo: Bioceramics Vol. 8 (1995), p. 253

[11]   H. Yli-Urpo, A.P. Forsback, M. Väkiparta, P.K. Vallittu, T. Närhi: Submitted for publication.

Key Engineering Materials Vols. 254-256(2004) pp. 603-606
online at http://www.scientific.net
© 2004 Trans Tech Publications, Switzerland

Chiral Biomineralization: Epitaxial and Helical Growth of Calcium Carbonate Through Selective Binding of Phosphoserine Containing Polypeptides, and the Dental Application onto Apatite

Hiroyuki Yamamoto[1], Mina Yamaguchi[2], Tetsunori Sugawara[1], Yukie Suwa[1], Kousaku Ohkawa[1], Hisaaki Shinji[2] and Shigeaki Kurata[3]

[1][1]Institute of High Polymer Research, Faculty of Textile Science and Technology, Shinshu University, Ueda 386-8567, Japan, E-mail hyihpr2@giptc.shinshu-u.ac.jp

[2] Department of Pediatric Dentistry and [3]Department of Dental Materials, Kanagawa Dental College, Yokosuka 238-8580, Japan, E-mail yamagumi@kdcnet.ac.jp

Keywords: chiral biomineralization, phosphoserine polypeptides, epitaxial growth, morphology of $CaCO_3$ crystals, adsorption onto apatite

Introduction

Biomineralization is the hybrid forming reaction of inorganic salts (calcium, silicon, barium, strontium and iron) and organic materials by organisms in nature [1]. As examples, teeth, bone and shells are biominerals. Most biomineralization processes are initiated by the secretion of specific matrix molecules at an interface [2]. The precise structure and polymorphs of the biominerals are controlled by a small amount of bio-macromolecules, such as proteins and polysaccharides, contained in the minerals [2,3]. The marine shell matrix proteins were shown to be highly acidic by amino acid analysis which found that 90% of the amino acids were comprised of approximately equimolar amounts of Asp, Gly, and Ser. The majority of the Ser residues in these proteins were present as *O*-phospho-L-serine [Ser(P)] [4]. In vertebrates, a rat dentin protein, phosphophoryn, has been recently revealed, and it has repetitive blocks of -Asp-Ser-Ser- or -Ser-Asp-, in which the Ser residues are mostly phosphorylated [5]. Thus the Ser(P) and Asp are essentially required for the matrix proteins in biominerals. We present here the main effects of Ser(P) containing polypeptides on the crystallization of calcium carbonate ($CaCO_3$), and in part a fundamental study for new pediatric dentinal applications using Ser(P) and Ser(P)-containing polypeptides.

Materials and Methods

The Ser(P) containing polypeptides were synthesized by the method reported in our previous articles [6,7], and used in the biomineralization experiments.

Solutions supersaturated with respect to $CaCO_3$ were prepared by mixing aqueous solutions of $CaCl_2$ dihydrate and $NaHCO_3$. These gave the final $CaCO_3$ concentrations of 5.0, 10, 15, and 20 mM. The aqueous poly[Ser(P)] or copoly[Ser(P)^x Asp^y] solution (2.5, 5.0, 10 mM residue) was added to the supersaturated $CaCO_3$ solutions. The mixed solutions were allowed to stand under static conditions for 3 weeks. The resulting $CaCO_3$ precipitates were characterized by optical microscopy and SEM. The $CaCO_3$ polymorphs were determined using an FTIR with KBr pellets.

Hydroxyapatite was used for the dental adsorption experiments. The amounts of adsorbed

---

[1] Correspondence and requests for materials should be addressed to H. Y. (e-mail: hyihpr2@giptc.shinshu-u.ac.jp).

Ser(P)-related biomaterials were determined by GPC.

Results and Discussion

CaCO₃ crystallization: epitaxial and helical growth.

In the absence of additives, a rhombohedral calcite was obtained at the air/water interface (Fig. 1a). These observations were not significantly changed by varying the total $CaCO_3$ concentrations ranging from 5.0 to 20 mM. When the poly[Ser(P)] was used as an additive, two types of crystals were obtained depending on the concentrations. One is a calcite crystal with rough surfaces, and the other is an aggregate of spherical vaterites (Fig. 1b). The Ca^{2+}/polymer residue molar ratios in the mixed solutions between 10 to 60 produced the calcite with rough surfaces. The quantitative analysis of the phosphate group in this calcite using ammonium molybdate showed that the calcite particles contained about 1.0 wt% of Ser(P) residues in the crystal. On the other hand, by increasing the concentration of Ca^{2+}, the Ca^{2+}/polymer residue molar ratios between 60 to 80 produced the spherical vaterite. Quantitatively, the spherical vaterite contained about 0.6 wt% of the Ser(P) residues.

Peptides and model proteins containing Ser(P) played an important role in controlling the morphology of the biominerals. In copoly[Ser(P)^x Asp^y] (x : y = 75 : 25, 50 : 50, 25 : 75), the spherical vaterite (Fig. 1b), the epitaxial calcite (Fig. 1c) and the helical structure with a clockwise P twined spiral morphology (Fig. 1d) were obtained. For example, the epitaxial calcite is formed in the presence of copoly[Ser(P)^x Asp^y] (x : y = 75:25, 50:50, 25:75) when the Ca^{2+}/polymer residue molar ratios are 20, 40, and 60. The epitaxial calcite consisted of nanofilaments with about a 70 nm width, parallel to one another, and contained about 0.9 wt% of the polymer residues in the crystal.

The effects of the ratios between Ca^{2+} and the copolypeptide residues on the biomineralized $CaCO_3$ morphology and polymorph were investigated. The experimental results suggested the following findings: a) the increased Ca^{2+}/polymer residue molar ratios in the mixed solutions produce the spherical vaterite, and b) the decreased ratio of Ser(P) in copoly[Ser(P)^x Asp^y] tended to form the epitaxial calcite. Most interestingly, when the Ca^{2+}/copoly[Ser(P)^{75} Asp^{25}] residue molar ratio in the mixed solution was 10, a $CaCO_3$ helical structure with a clockwise P twined spiral morphology was obtained (Fig. 1d). These molar ratios of 10 for the organic compound and calcium carbonate are almost the same in which a shellfish forms a shell. The helical crystals were calcite and had a regular repetition structure.

When the organic macromolecules modified the crystal shape during the growing process, the inorganic crystals are laid down in orderly arrays in association with a matrix of macromolecules playing the role of a crystal template [8]. Our results will serve as a clue on a molecular level to clarify why almost all gastropod shell's twist sense is in the clockwise P-direction.

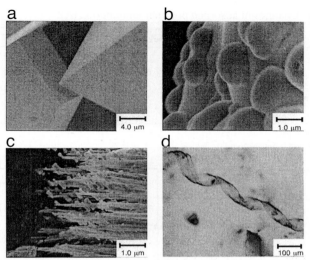

Fig. 1 Optical and SEM micrographs of CaCO$_3$ crystals; a) in the absence of additives,
b)-d) the presence of poly[Ser(P)] and copoly[Ser(P)^x Asp^y]. a-c, SEM; d, optical.

Study on the new method of caries prevention: Adsorption onto apatite.
For pediatric dental treatment, the pit and fissure sealant is one of the important caries preventive treatments. Dental caries are easily formed as pits and fissures in immature teeth enamel. We decided to develop new pit and fissure sealant materials without drying and demineralization using a calcification mechanism of Ser(P). This article describes the fundamental study of the reaction of Ser(P) and apatite.

The amount of Ser(P) or polymers adsorbed onto HAP was measured as follows. One ml of 500. g/ml Ser(P) (Sigma) was added to 50 mg hydroxyapatite (HAP:Wako Pure Chem. Ind.) in a 2 ml plastic tube. Five samples were similarly prepared. The samples were stirred for 1, 2, 6, 12 and 24 hours, and then centrifuged at 1500 rpm for 10 minutes. The clear supernatants were filtered by Millex®-GV (Millipore). The amount of Ser(P) remaining without adsorbing onto the HAP was analyzed by the gel permeation chromatography (GPC) (C-R4A:Shimadzu) using an Asahi Pak GS-220. The amount of the polymers -phosvitin, poly[Ser(P)] and copoly[Ser(P)^{50} Asp^{50}], and poly(Asp)- was determined in the same way using GPC with an Asahi Pak GS-510 [9].

Changes in the adsorbed amounts of the polymers and the adsorption rate onto HAP are shown in Table 1 and Fig. 2, respectively. The relationship between the Ser(P) contents and the amount of adsorption of copoly[Ser(P) Asp] onto HAP is shown in Fig. 3. The amount of the polymers adsorbed on HAP increased with the increasing Ser(P) residues in the polymers. The highest adsorption amount of phosvitin, which contains lesser about 30% Ser(P), was higher than that of the copoly[Ser(P) Asp] systems (Table 1). These adsorption characteristics of phosvitin on HAP may be caused by its higher molecular weight and/or other amino acids in the protein. From the present adsorption results, new pit and fissure sealant materials without drying may be promising using the adsorption mechanism of Ser(P) in synthetic model proteins.

An additional study on the calcification between the Ser(P) containing model proteins and HAP to focus on the caries preventive treatments will be reported in the near future.

Table 1   Changes in adsorbed polymer amount on HAP

Sample	Reaction time (hour)					
	0	1	2	6	12	24
Phosphoserine	0 (0)	114 (23)	121 (24)	60 (12)	94 (19)	40 (8)
Phosvitin	0 (0)	453 (91)	450 (90)	500 (100)	500 (100)	500 (100)
Poly[Ser(P)]	0 (0)	500 100)	500 (100)	500 (100)	500 (100)	500 (100)
Copoly[Ser(P)^{50} Asp^{50}]	0 (0)	390 (78)	371 (74)	364 (73)	376 (75)	368 (74)
Poly(Asp)	0 (0)	254 (51)	254 (51)	265 (53)	270 (54)	275 (55)

1) Starting quantity : 500 μg

2) Adsorption amount : μg (adsorption ratio : %)

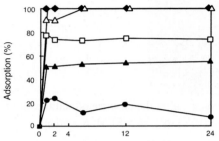

Fig. 2   Adsorption rate of the polymers on HAP: ◆ poly[Ser(P)]; △ phosvitin; □ copoly[Ser(P)^{50} Asp^{50}]; ▲ poly(Asp); ● phosphoserine.

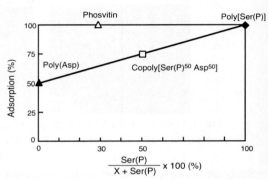

Fig. 3   Effect of contents of Ser(P) residues in the polymers on the adsorption of the polymers on HAP: X, amino acid(s).

Acknowledgement

This work was supported by Grant-in-Aid for Scientific Research (No. 12450330 and No. 13555178) by the Ministry of Education, Culture, Sports, Science and Technology of Japan.

References

[1] S. Mann: Biomineralization: Principles and Concepts in Bioinorganic Materials Chemistry (Oxford Univ Press, UK 2001).

[2] K. Wada and I. Kobatashi, Eds: Biomineralization and Hard Tissue of Marine Organisms, (Toukaidaigaku shuppankai, Japan 1996).

[3] G. Falini, S. Albeck, S. Weiner and L. Addadi: Science Vol. 271 (1996), p. 67-69.

[4] K.W. Rusenko, J.E. Donachy and A.P. Wheeler: Surface Reactive Peptides and Polymers: Discovery and Commercialization (Amer Chem Soc Press, US 1991).

[5] A. George et al.: J. Biol. Chem. Vol. 271 (1996), p. 32869-32873.

[6] K. Ohkawa, A. Saitoh and H. Yamamoto: Macromol. Rapid Commun. Vol. 20 (1999), p. 619-621.

[7] H. Yamamoto, A. Saitoh and K. Ohkawa: Macromol. Biosc. Vol. 3 (2003), p. 354-363.

[8] S. Mann: Nature Vol. 332 (1988), p. 119-124.

[9] N. Kurosaka: Jpn. J. Conservative Dentistry Vol. 44 (2001), p. 390-400.

Key Engineering Materials Vols. 254-256(2004) pp. 607-610
online at http://www.scientific.net
© 2004 Trans Tech Publications, Switzerland

A New Knee Prosthesis with Bisurface Femoral Component Made of

Zirconia-Ceramic (Report 2)

T. Nakamura, E. Oonishi, T. Yasuda, Y. Nakagawa

Department of Orthopaedic Surgery, Kyoto University School of Medicine, Japan,
ntaka@kuhp.kyoto-u.ac.jp

Keywords: TKA, zirconia, alumina, bisurface

Abstract. To avoid fractures of alumina ceramic, we developed a total knee prosthesis with zirconia ceramic femoral componen,t designated bisurface knee prosthesis, by the cooperation of Kyoto University and Kyocera Co. The prosthesis has a unique ball-and-socket joint in the midposterior portion of the femoral and tibial components, which functions as a posterior stabilizing cam mechanism and causes femoral rollback. We reported at Bioceramic 14 that the new zirconia prosthesis worked as well as alumina one in a short term. In this paper we reported the results of 4-year follow-up. We had good results for clinical and X-ray examinations at 4 year-follow-up without any revision.

Introduction

Two years ago, we reported new knee prosthesis made of zirconia ceramics for femoral component [1]. Now we have implanted 128 joints in 115 patients. We will describe 33 joints, which were followed for more than 2 years after operation. For total joint replacement the reduction of polyethylene wear is most important for its long-term durability as well as for total hip replacement. The wear rates of polyethylene against alumina ceramic and zirconia ceramic are known to be lower than that against metal. [2,3] Recent advances in fine-ceramic technology produce ceramics of high purity and good mechanical properties. For total hip replacement alumina or zirconia ceramic heads are used to reduce the wear debris. The femoral head of 22 mm diameter has been developed and used in clinic. For total knee replacement ceramic will be a useful material for the reduction of polyethylene wear. However, total knee prosthesis has complicated shape compared with femoral head of THA and it needs development of technology to make zirconia femoral component for TKA to withstand mechanical loads. We developed total knee prosthesis with alumina ceramic femoral component, designated bisurface knee prosthesis by the cooperation of Kyoto University and Kyocera Co. The bisurface knee prosthesis was designed in 1989 to improve knee flexion without affecting the durability of the prosthesis. The prosthesis has a unique ball-and-socket joint in the midposterior portion of its tibio-femoral joint, which functions as a posterior stabilizing cam mechanism and causes femoral rollback. We reported that from our 4- to 8-year follow-up studies, the bisurface knee with alumina ceramic femoral component provided an excellent clinical result[4]. However, five fractures of alumina ceramics have been reported to occur by hitting with a hammer during operation by the company. To avoid fractures, we developed new zirconia bisurface knee and began to use it instead of alumina bisurface knee because zirconia has better mechanical properties than alumina. In this paper we will report results of 2 to 4 year's follow-up.

Knee Prosthesis and Patients

The bisurface knee has been reported. [1]. Its specialty is that the femoral component of bisurface knee was made of pure zirconia ceramics. Kyocera has developed tetragonal stabilized zirconia ceramic with high strength and toughness. The zirconia ceramic achieves sufficient stability of the tetragonal crystal phase by optimizing grain size and adding specific additives like yttrium in appropriate concentration. The grain size of zirconia ceramic is 0.3 μm, one third of the size of high purity alumina. Zirconia ceramic has a flexural strength almost double that of alumina. Thus, the zirconia ceramic gives higher fracture safety. We tested their characteristics of lubrication against polyethylene using pin-on disc and knee joint simulator. The polyethylene wear against zirconia were as small as that against alumina ceramics *in vitro* [3]. The tibial component is made of UHMWPE insert and titanium metal tray. The UHMWPE insert of thickness from 9 mm to 18 mm is prepared. Five sizes of prosthesis are made as extra small, small, standard, large and extra large. The prosthesis are fixed with bone cement and cruciate ligaments are sacrificed.

Patients were operated in June 1999 to Jun 2001. Clinical results were evaluated using Japanese Orthopedic Association Knee Score (JOA score). JOA score of a normal knee joint is 100 points (Pain:30, MMT 20, ROM 12, ADL 8,Walk 20). Loosening was evaluated as movement of prosthesis or clear zone of more than 2 mm thick around implant.

Results

One hundred sixteen zirconia bisurface knees have been implanted in our department. In these, 33 knees of 29 patients have been followed-up beyond two years and are objects for this study. The average of follow-up was 3.1 years (2 years - 4.1 years). Diagnosis of thirteen patients was osteoarthritis and that of 16 patients was rheumatoid arthritis. One knee of RA patient was a revision from total condylar type. Other patients have primary TKA. Preoperatively the averages of JOA score of OA and RA knees were 58 and 49, respectively. At final follow-up, their averages of JOA score of OA and RA patients were improved to 80 and 79, respectively. The radiological examination revealed no loosening and no implant failure. Averages of preoperative and postoperative knee flexion were 114 (65-144) and 117(80-140), respectively. Difference of knee flexion between OA and RA patients was not detected. All the patients satisfied with TKA.

Discussion

Brittleness is a weak point of ceramic materials. According to the manufacturer's information of over 2000 Bisurface knee operations, five fractures of alumina ceramics have occurred by hitting with a hammer during operation. Thus, we have employed zirconia ceramic as the replacement of alumina ceramic because zirconia ceramic is mechanically twice as strong as alumina ceramic. We began to use the bisurface knees with zirconia ceramic since 1999. We did not have any implant fracture. Previously we reported their two-year follow-up and those knees showed good results as well as bisurface knees with alumina ceramic and no fracture of zirconia occurred at operation. In this paper the results between two and four years after operation were examined. The results were good and no fracture of ceramics was observed. These results suggested zirconia femoral component has worked well.

For ceramic material it is very important to make a design to avoid the concentration of stress. The bisurface knee was designed to have a ball and socket surface at the posterior one-third in a tibio-femoral joint. This joint acts as a posterior stabilizer as well as for knee bending [5]. This conformation of a ball and socket joint is thought to be proper for ceramic surface to avoid stress compared with a conventional posterior stabilized design, because we have already shown good results using alumina bisurface knee during more than 13 years.

In vitro zirconia ceramic reduced polyethylene wear compared with metal. *In vivo* effect of zirconia to reduce polyethylene wear can be expected. However, a long-term study should be performed to confirm reduction of polyethylene wear by using zirconia ceramic in clinic.

References

[1] P.Kumar, M.Oka, K.Ikeuchi, K.Shimizu, T.Yamamuro, H.Okumura, Y.Kotoura1: J Biomed Mater Res Vol. 25(1991), p.813-828

[2] T.Nakamura, M.Akagi, T.Yasuda, Y.Nakagawa, M.Shimizu: *Bioceramics 14* (Trans Tech Publication, 2001)

[3] M.Ueno, KIkeuchi, T Nakamura, M Akagi: *Bioceramics,* Vol. 15 (Trans Tech Publication, 2002)

[4] M.Akagi,T.Nakamura, Y.Matsusue, T.Ueo, K.Nishijyo, E.Ohnishi: J Bone Joint Surg. Vol.82-A, (2000), p.1626-33

[5] M.Akagi, T.Ueo, H.Takagi, C.Hamanishi, T.Nakamura: J Arthroplasty. Vol 17 (2002), p.627-34

Key Engineering Materials Vols. 254-256(2004) pp.611-614
online at http://www.scientific.net
© *2004 Trans Tech Publications, Switzerland*

Effects of Polymer Molecular Weight and Ceramic Particle Size on Flexural Properties of Hydroxyapatite Reinforced Polyethylene

M.Wang[1,2], L.Y.Leung[2], P.K.Lai[2] and W.Bonfield[3]

[1] Medical Engineering and Mechanical Engineering, Faculty of Engineering

University of Hong Kong, Hong Kong

[2] Rehabilitation Engineering Centre, Hong Kong Polytechnic University, Hong Kong

[3] Department of Materials Science and Metallurgy, University of Cambridge, Cambridge, U.K.

Email: memwang@hku.hk

Keywords: hydroxyapatite, polyethylene, composite, molecular weight, ceramic particle size, flexural properties

Abstract: Hydroxyapatite (HA) reinforced polyethylene (PE) was the first bioactive ceramic-polymer composite developed for bone substitution on the basis of mimicking the structure and matching properties of bone. Both biological and mechanical properties of this material have been extensive studied. To understand the behaviour and determine mechanical properties of HA/PE composite under the bending condition, flexural tests of the composite were conducted in this investigation. It was found that an increase in HA volume percentage led to a corresponding increase in flexural modulus of the composite. Flexural modulus values were at the same levels as those of tensile modulus at different HA volume percentages. Generally, a larger HA particle size in the composite resulted in a lower flexural modulus of the composite. The fracture mechanism of HA/PE composite in the bending mode was the same as that for other hard particle filled polymers.

Introduction

Particulate bioactive ceramic reinforced polymers have been investigated as analogue bone replacement materials since bone itself is a natural composite material [1]. Bonfield *et al* pioneered the research in hydroxyapatite reinforced polyethylene (HA/PE) in the 1980s [2] and this composite has already been used as bone substitutes in human bodies [3, 4]. Investigations into basic mechanical properties of HA/PE had mainly concentrated on its tensile, torsional and viscoelastic properties [5-8]. Even though dynamic mechanical analysis (DMA) was used in the bending mode of testing [7, 8], conventional bending tests had not been conducted on HA/PE composite. Considering loading conditions of bone-substituting materials in the human body, it is thus necessary to assess flexural properties of HA/PE since load-bearing bones in the body commonly encounter bending stress in addition to other types of stresses. It has also been found that tensile and torsional properties of HA/PE composite were affected by a number of materials factors [6, 7]. In this study, effects of polymer molecular weight and ceramic particle size on flexural behaviour and properties of the composite were systematically investigated.

Materials and Methods

Synthetic HA (Two grades: P88 and P81B, both from Plasma Biotal, UK) was incorporated into a high density polyethylene (HDPE: M_w = 270,000, from BP Chemicals, UK) and a cross-linkable polyethylene (XPE: M_w = 200,000, from DePuy, UK), respectively, to produce HA/PE composite. The average particle size of P88 HA and P81B HA was 4.14 and 7.32μm, respectively. Other properties of raw materials can be found elsewhere [6, 7]. The production of HA/PE composite followed a standardised procedure [5]. Composite containing 0, 10, 20, 30, 40 and 45vol% of HA was tested. Test specimens conforming to ASTM D790 were made from composite plates (around 4mm in thickness) through normal cutting and machining processes.

Flexural testing was conducted on a universal testing machine (Instron 5569, Instron, USA). A three-point bending fixture was used and the testing speed was 0.5mm/min. Flexural modulus and flexural strength were determined for the composite. For composite containing high amounts of HA (normally greater than 40vol%), tests proceeded until specimens had fractured. For specimens that did not fracture, tests were stopped after peak stresses had been passed. During bending tests, deformation of specimens was monitored and photos were taken at regular intervals. After flexural testing, fracture surfaces of tested specimens were examined under a Leica Stereoscan 440 scanning electron microscope (SEM). Densities of the four series of the composite were also measured using a Shimadzu analytical balance equipped with a density kit. (Density measurements were conducted to check uniformity of test specimens cut from different parts of composite plates.)

Results and Discussion

Fig.1 displays measured densities of the four series of HA/PE composite. Standard deviations of density measurements for composite of each composition were small, indicating the uniformity of composite plates produced. Furthermore, measured densities were close to theoretical values obtained using the "Rule of Mixtures", suggesting non-presence of voids in these plates.

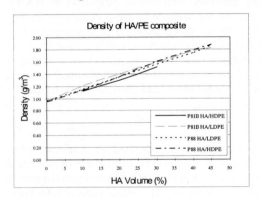

Fig. 1: Measured density of the four series of HA/PE composite

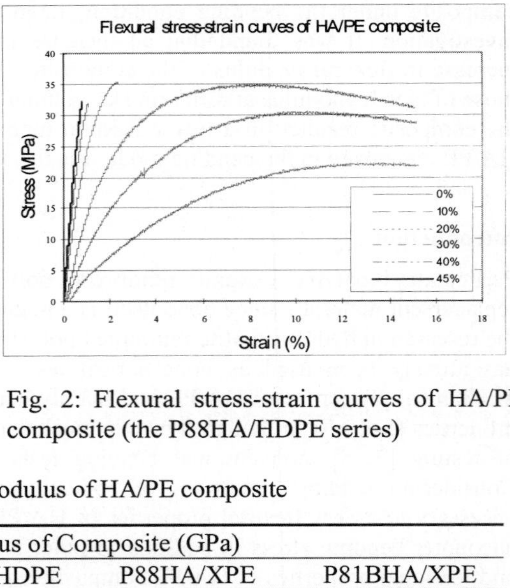

Fig. 2: Flexural stress-strain curves of HA/PE composite (the P88HA/HDPE series)

Table 1   Flexural modulus of HA/PE composite

HA Volume (vol.%)	Modulus of Composite (GPa)			
	P88HA/HDPE	P81BHA/HDPE	P88HA/XPE	P81BHA/XPE
0	0.44	0.44	0.93	0.93
10	1.03	0.71	1.45	2.11
20	1.55	1.16	2.25	2.80
30	3.17	2.43	3.27	3.52
40	4.30	----	5.17	4.90
45	4.92	----	6.74	6.32

For all four series of the composite, it was found that flexural modulus of the composite increased with an increase in ceramic volume percentage (Fig.2 and Table 1). This trend was also observed for tensile, torsional and storage moduli of the composite [5-8]. Furthermore, flexural modulus values were at the same levels as those of tensile modulus at different compositions (Fig.3). It was also evident that composite flexural modulus was affected by the matrix and reinforcement (Table 1 and Fig.3). The flexural strength of HA/PE composite was normally within the range of 20-40MPa, which are generally greater than tensile strength values at corresponding HA volume percentages of the same composite series.

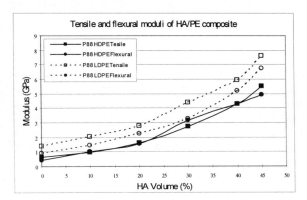

Fig. 3: Tensile and flexural moduli of HA/PE composite

Fig. 4: HA/PE specimen during flexural testing

(a)

(b)

Fig. 5: HA/PE specimens after flexural testing
(a) 30vol% P81BHA/HDPE
(b) 45vol% P88HA/HDPE

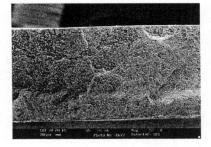

Fig. 6: Flexural fracture surface of HA/PE (30vol% P81BHA/HDPE)

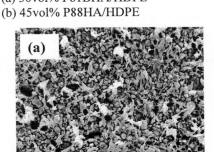

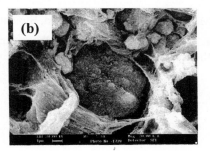

Fig. 7: Flexural fracture surface of HA/PE (45vol% P88HA/HDPE) (a) general view; (b) interface

Even with high HA percentages, the composite still exhibited considerable ductility in the bending mode (Fig.4). Post-flexural testing examination revealed straight crack propagation paths in composite specimens (Fig.5), although at the microscopic level, composite having lower HA percentages exhibited long polymer fibrils on fracture surfaces (Fig.6). Composite having HDPE as the matrix was more ductile than that with the XPE matrix. SEM examination revealed even distributions of HA particles in the composite (Fig.7a). Debonding took place between HA particles and the polymer matrix and hence gaps were observed at the HA-polymer interface (Fig.7b). It appears that the bending fracture mechanism was the same as that for tensile fracture for the HA/PE composite [7].

Conclusions

For HA/PE composite developed for hard tissue substitution, polymer molecular weight and ceramic particle size have significant influences on flexural properties of the composite. An increase in HA volume percentage leads to a corresponding increase in flexural modulus of the composite. Flexural modulus values of the composite are at the same levels as those of tensile modulus at different HA volume percentages. Generally, a larger HA particle size results in a lower flexural modulus of the composite. The fracture mechanism of HA/PE composite in the bending mode is the same as that for other hard particle filled polymers.

Acknowledgements

The authors would like to thank the Jockey Club Rehabilitation Engineering Centre, Hong Kong Polytechnic University, for supporting the research work reported in this paper. Funding from the UK EPSRC (and also SRC and SERC) for developing HA/PE composite is gratefully acknowledged.

References

[1] M.Wang, *MRS Symp. Proc. 724: Biological and Biomimetic Materials – Properties to Function*, San Francisco, USA, 2002, 83-93

[2] W.Bonfield, M.Wang, K.E.Tanner, Acta Met., 46 (1998), 2509-2518

[3] J.L.Dornhoffer, The Laryngoscope, 108 (1998), 531-536

[4] R.E.Swain, M.Wang, B.Beale, W.Bonfield, Biomed. Eng. Appl. Basis & Commu., 11 (1999), 315-320

[5] M.Wang, D.Porter, W.Bonfield, Brit. Ceram. Trans., 93 (1994), 91-95

[6] M.Wang, C.Berry, M.Braden, W.Bonfield, J. Mater. Sci. Mater. Med., 9 (1998), 621-624

[7] M.Wang, R.Joseph, W.Bonfield, Biomaterials, 19 (1998), 2357-2366

[8] F.Tang, C.H.Ng, M.Wang, Proc. Intel Conf. Thermophysical Properties of Materials, Singapore, 1999, 502-508

Key Engineering Materials Vols. 254-256(2004) pp. 615-618
online at http://www.scientific.net
© *2004 Trans Tech Publications, Switzerland*

Diffusion of Ions From a Calcium Phosphate Cement for Dental Root Canal Treatment and Filling

S. Munier, H. El Briak, D. Durand and P. Boudeville

Laboratoire de Chimie Générale et Minérale, Faculté de Pharmacie, 15 Avenue Charles Flahault, BP 14491, 34093 Montpellier cedex 5, France, e-mail : boudevil@univ-montp1.fr

Keywords : Ion diffusion, calcium phosphate cement, calcium oxide, dicalcium phosphate dihydrate, root canal filler.

Abstract. Calcium phosphate cements were developed from dicalcium phosphate dihydrate and calcium oxide giving, after setting, mixtures of hydroxyapatite and calcium hydroxide. The objective of these cements is to replace calcium hydroxide pastes that are currently used in dentistry but only as temporary material. They must keep the advantages of calcium hydroxide (antimicrobial activity due to hydroxyl ion release, induction of hard tissues due to calcium ion release) without its drawbacks (solubility, low and slow hardening, retraction on drying). The release of the different ions OH^-, PO_4^{3-}, Na^+ and Ca^{2+} from cements with Ca/P ratios of 1.67, 2 and 2.5 was studied and results confirm the antimicrobial potency of these cements.

Introduction

Calcium hydroxide is currently used in dentistry for endodontic treatments such as pulp capping, root canal treatment, periapical lesions or apexification [1]. The main advantages of calcium hydroxide are: (i) an antimicrobial activity due to the alkaline environment following its dissolution because most of bacteria do not resist a pH over 9.5; (ii) an anti-inflammatory effect by resolution of exudates; (iii) the promotion of the reparative dentinal bridge or the apical plug and, (iv) it does not inhibit the polymerization of restorative materials unlike zinc oxide-eugenol based cements. It presents also some drawbacks. As pulp capping agent, it provokes a deep pulp necrosis and the quality of the dentinal bridge is poor. Moreover, its slight solubility, its low and slow hardening and its retraction by drying make that it is only used as temporary material for root canal filling given the lack of tightness. Our objective is to prepare calcium phosphate cements (CPCs) that keep the advantages of calcium hydroxide but suppress or minimize its drawbacks.

Mixing dicalcium phosphate dihydrate (DCPD) and calcium oxide with a sodium phosphate (NaP) buffer, pH 7, 0.75 M as liquid phase, we obtained a CPC that expands, has better mechanical properties than calcium hydroxide pastes and with rheological properties that comply with it being used for root canal treatment and filling [2]. Given the Ca/P ratio in the starting powder, the setting reaction leads to hydroxyapatite (HA) (Eq. 1) or a mixture of HA and calcium hydroxide (Eq. 2).

$$6\ CaHPO_4 \cdot 2\ H_2O\ +\ 4\ CaO\ \rightarrow\ Ca_{10}(PO_4)_6(OH)_2\ +\ 14\ H_2O \qquad (Ca/P = 1.67) \quad (1)$$
$$6\ CaHPO_4 \cdot 2\ H_2O\ +\ 9\ CaO\ \rightarrow\ Ca_{10}(PO_4)_6(OH)_2\ +\ 5\ Ca(OH)_2 + 9\ H_2O \qquad (Ca/P = 2.5) \quad (2)$$

We studied the release over time from these cements of different ions such as hydroxyl that ensure the antimicrobial potency or calcium, sodium and phosphate that can induce the mineralization of an apical barrier.

Material and methods

All chemicals were reagent grade. CaO (from Aldrich) was heated at 900°C for 2 h to remove H_2O and CO_2 and stored in a vacuum desiccator. After heating, average particle size was around 7 µm.

DCPD (from Fluka), average particle size 8μm, was used as received. Sodium phosphate buffer pH 7, 0.75 M was prepared from NaH_2PO_4 and $Na_2HPO_4 \cdot 12H_2O$ (from Fluka) by dissolving 0.75 and 4.48 g respectively into a 25-ml gauged flask.

DCPD and CaO in variable proportions depending on the desired Ca/P ratio (1.67, 2.0 and 2.5) were mixed with the NaP buffer (L/P = 0.6 ml g^{-1}) on a glass plate at 22 ± 1°C by successive fractions (1/6) as for zinc phosphate dental cements. Cylindrical samples (8.5 mm diameter x 5 mm height) were molded and stored for 1 h at 37°C, 100% RH for hardening and obtaining samples easy to handle. Samples were removed from the molds and then immersed into 10 ml of distilled water. The elution solutions were changed every day during 3 days and their ionic content analyzed: OH^- by pH-metric titration (Tacussel PHM 210), Na^+ (from the NaP buffer) and Ca^{2+} by atomic absorption spectrophotometry (PerkinElmer AA300) and PO_4^{3-} by UV-Visible spectrophotometry (Beckman Model 24) at 630 nm (phosphomolybdate complex reduced by ascorbic acid). For each cement, two samples were prepared and tested. Only mean values will be given in the results and the mean deviation was less than 5% of the mean value.

Results and Discussion

The concentration, in millimoles per liter (mM), of the different ions in the elution solutions at day 1, 2 and 3 for the three cements (Ca/P = 1.67, 2 and 2.5) are gathered in Table 1. In Table 2 we reported the cumulated masses (in mg) of the ions released into the solutions after 3 days from the three cements and, for comparison, the total masses of ions in each cement sample of 450 ± 5 mg. After complete reaction, the interstitial water was around 45% w/w (Eqs. 1, 2 and L/P = 0.6 ml g^{-1}).

Table 1. pH and ionic composition (in mM) of the elution solutions at days 1, 2 and 3 in contact with the three cements

	pH	Ca^{2+}	Na^+	OH^-	PO_4^{3-}
Ca/P = 1.67					
day 1	7.35	0.39	10.98		6.72
day 2	7.17	0.22	3.72		3.07
day 3	7.0	0.31	1.62		0.74
Ca/P = 2					
day 1	12.13	0.81	14.17	12.8	0.055
day 2	12.40	5.39	2.17	12.4	0.01
day 3	12.22	5.09	2.45	7	0.006
Ca/P = 2.5					
day 1	12.45	5.3	13.63	20.4	0.014
day 2	12.58	6.34	1.29	18.6	0.015
day 3	12.57	6.54	0.7	14.4	0.013

Table 2. Total amount of ions released after 3 days and in each cement samples

Ca/P ratio	Amount of ions released after 3 days (mg)				Total amount of ions in each sample (mg)			
	Ca^{2+}	Na^+	PO_4^{3-}	OH^-	Ca^{2+}	Na^+	PO_4^{3-}	OH^-
1.67	0.37	3.75	10.0	≈ 0	94.36	4.85	134.13	≈ 0
2.0	4.52	4.32	0.07	5.47	103.76	4.85	123.33	13.28
2.5	7.27	3.59	0.04	9.08	115.52	4.85	109.82	29.16

The pH of the solutions increased logically with the Ca/P ratio in the cement (Table 1, second column). These values did not change significantly at day 1, 2 and 3. From the concentrations of the different ions, an iterative calculation of the ionic strength and of the activity coefficients confirmed the electroneutrality of the solutions.

The Na^+ release was over 50% at day 1 (Table 1) and represented, after 3 days, 75-90% of the total amount releasable (Table 2). The release of Ca^{2+} and OH^- increased logically with the Ca/P ratio in the cement but decreased slightly over time unlike phosphate ions. The phosphate ion release was important only for the cement with Ca/P = 1.67. For this cement, given the pH value of the elution solution, the Na^+ and Ca^{2+} concentrations found, the phosphate concentration in the solution complies with an equimolar mixture of $NaH_2PO_4 + Na_2HPO_4$. These results lead to the following remarks.

1) Na^+ and PO_4^{3-} release

* Since the release of sodium ions from the hardening cements is almost total after 3 days, whatever the Ca/P ratio in the cement, it can be stated that ions diffuse toward the solution, not only from the surface, but also from the bulk of the cement samples through the interstitial water. Environmental scanning electron microscopy examination of the surface and of a cross section of cement samples shows microporosities that contain water and facilitate the ionic diffusion from the cement bulk (Fig. 1).

* Since the NaP buffer used as liquid phase is almost totally released after 3 days for the cement with Ca/P = 1.67, it does not seem to participate to the setting reaction although it shortens the setting time and increases the compressive strength of these cements [2]. For cements with Ca/P = 2 and 2.5 the phosphate release was very low because the phosphate ions, coming from the NaP buffer, react with CaO in excess to give HA.

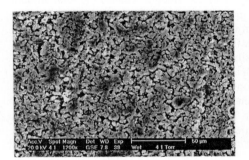 

Fig. 1. ESEM pictures of the surface (left) and of the cross section (right) of a cement sample after 18-h hardening. (white arrows indicate porosities; bar = 50 μm).

2) OH^- and Ca^{2+} release

* For Ca/P = 2 and 2.5, the hydroxyl ion release is associated with the sodium and calcium ion release the first day and then principally with calcium ions the other days. Because calcium hydroxide is more soluble than hydroxyapatite even in basic medium, the elution solutions were principally composed of NaOH and $Ca(OH)_2$ that ensure the alkalinity of the environment and thus the antimicrobial property of the cement [2,3].

* The OH^- release decreases over time (Table 1) and after 3 days the cumulated amount is around a third of the total amount releasable from the cement (Table 2). From these results we can deduce a release half-life time ($t_{1/2}$) of 6 days. Thus the time required for a complete release of OH^- would be around 1 month (6 x $t_{1/2}$) ensuring, *a priori*, a long term

antimicrobial activity. This point will have to be confirmed by new experiments for longer periods.

Are these *in vitro* results transposable to *in vivo* conditions? These tests were performed with a solution-to-cement ratio of 22 (10 ml / 0,45 g). When the cement will be inserted into a root canal, this ratio will be lower, probably around 1 or 2 and the time for complete dissolution of $Ca(OH)_2$ will be longer. Moreover, solutions were changed every day, which would not happen in a root canal after pulpectomy. Because the solubility of calcium hydroxide decreases when pH increases (10^{-1} M at pH 7 to 10^{-5} M at pH 11 [4]), the dissolution will be lower when a pH of 12 will be reached. Inversely, body fluids are buffered and will lower pH increase. But, after root canal filling and alkalinisation of the physiological liquids contained in the dentinal tubules [3], the single contact between cement and living tissues will be through the apex before the formation of the apical plug. Thus we can consider that the release of hydroxyl ions from the cement will be rapidly low and slow, reaching equilibrium, without supply of buffered liquids from outside. This situation arises when there are leakages around the restorative material in the coronal part of the tooth that allow bacterial percolation. The reserve of hydroxyl ions would be sufficient to resist microbial invasion, ensuring the antimicrobial protection on the long term. *Ex vivo* experiments are in progress, using the two-chamber method, to confirm this hypothesis and the *in vivo* antimicrobial potency of this kind of cement.

Conclusion

This work demonstrates that the diffusion of the different ions arises both from the surface and the bulk of the cement through the interstitial water. This point is of paramount importance to ensure a long antimicrobial activity of these cements because the total amount of calcium hydroxide present in the cement is mobilizable. In cements with Ca/P $\geq$ 2, the precipitated hydroxyapatite network ensures their solidity and acts as a drug delivery system releasing slowly the calcium and phosphate ions that induce the mineralization of the apical barrier after pulp removal and more especially the hydroxyl ions responsible of the antimicrobial activity as it was demonstrated with an agar diffusion test [3]. This DCPD-CaO-based cement with a Ca/P ratio of 2.5 that contains, after setting, 27% w/w of calcium hydroxide is a good candidate to replace calcium hydroxide pastes for root canal treatment and could be used as definitive root canal sealer, keeping its antimicrobial potency during a long period without loss in its mechanical properties.

References

[1] C. Ricci and V. Travert. Rev. Fr. Endodon. 6 (1987) 45-73.
[2] H. El Briak, D. Durand, J. Nurit, S. Munier, B. Pauvert and P. Boudeville. J. Biomed. ater. Res, Appl Biomater. 63 (2002) 447-53.
[3] M. Kouassi. Gutta-percha et nouveaux ciments phosphocalciques à usage endodontique : comportement en présence de micro-organismes bucco-dentaires. Thesis, Paris V, 2002.
[4] K.S. TenHuisen and P.W. Brown. J. Biomed. Mater. Res. 36 (1997) 233-41.

Key Engineering Materials Vols. 254-256(2004) pp. 619-622
online at http://www.scientific.net
© *2004 Trans Tech Publications, Switzerland*

Biomimetic Coatings vs. Collagen Sponges as a Carrier for BMP-2: A Comparison of the Osteogenic Responses Triggered *in vivo* Using an Ectopic Rat Model

Y. Liu,[1,2] E. B. Hunziker[2], C. Van de Vaal,[1] K. de Groot[1]

[1] Biomaterials Research Group, Leiden University, IsoTis, S. A. The Netherlands

[2] ITI Research Institute, University of Bern, Switzerland, Email: Maria.liu@isotis.com

Keywords: Biomimetic; Calcium phosphate; Coating; BMP-2; Collagen, Titanium; Implant; Osteoinductive.

Abstract. We have developed a new carrier system for bone morphogenetic protein 2 (BMP-2). This agent has been successfully incorporated into biomimetic calcium phosphate implant coatings without losing its osteoinductive potency. Numerous materials have been tested for their efficiency in carrying BMP-2 and in promoting its osteoinductive effects at both ectopic and orthotopic sites. In the present study, we compare the osteoinductive effects of BMP-2 delivered by biomimetic coatings and by classical collagen sponges *in vivo* using an ectopic rat model. Titanium-alloy discs, either bearing a co-precipitated layer of calcium phosphate and BMP-2 (1.7 µg/disc) or inserted into collagen sponges impregnated with the same drug (10 µg/sponge), were implanted subcutaneously in the dorsal region of rats and the bone-formation process monitored after 2 and 5 weeks. After 2 weeks, the net volume of bone formed in the biomimetic coating group (5.8 mm^3/disc) was greater (p < 0.01) than in the collagen sponge one (2.3 mm^3/sponge). By the end of the fifth week, the net volume of bone deposited had increased significantly (p < 0.001) in the biomimetic coating group (11.6 mm^3/disc), whereas in the collagen sponge one it remained unchanged (3.0 mm^3/sponge; p = 0.82). As a carrier for BMP-2 and in facilitating its osteogenic effects, biomimetic calcium phosphate coatings offer a distinct advantage over the classical collagen sponge.

Introduction

To be efficacious at a biological target site, BMP-2 must be delivered locally by a suitable carrier system. Numerous materials have been tested for their efficiency in carrying BMP-2 and in conducing to its osteoinductive effects at both ectopic and orthotopic sites [1-4]. Amongst these materials, collagen, synthetic or natural ceramics, demineralized bone matrix and polyglycolic acid have been most frequently investigated. Using each of these carrier materials, the BMP-2-release profile is similar and characteristically biphasic, most of the drug being liberated during an initial rapid burst of a few hours duration, the remainder being delivered more gradually over a period of several days. None of these carrier materials meets all of the requirements of an ideal osteoinductive system. We have recently investigated the potential of biomimetic calcium phosphate implant coatings to act as a carrier for BMP-2 and to facilitate its osteoinductive effects in an ectopic rat model, with very encouraging results [5]. Biomimetic coatings afford a vehicle for delivering BMP-2 slowly but at a steady and sufficiently high level to support and sustain local osteogenic activities *in vivo*.

In the present study, we compared the osteoinductive effects of BMP-2 delivered either by biomimetic coatings or by classical collagen sponges in an ectopic rat model, the bone-formation process being monitored histologically and histomorphometrically 2 and 5 weeks after implantation.

Materials and Methods

Titanium-alloy (Ti6Al4V) discs (1cm in diameter) were coated biomimetically with a co-precipitated layer of calcium phosphate and BMP-2, as previously described [6]. Briefly, the discs were immersed first in concentrated simulated body fluid under high-nucleation conditions for 24 hours at 37°C, and then in a supersaturated solution of calcium phosphate (pH 7.4) containing BMP-2 (10 µg/ml) for 48 hours at 37°C. The amount of BMP-2 incorporated into each coating (determined by ELISA) was 1.7 µg per disc.

Absorbable collagen sponges [Helistat® (2 x 2 cm)] were immersed in a buffer solution of BMP-2 (20 µg/ml), and soaked up a drug concentration of 10 µg per sponge. A sterile titanium-alloy disc (of the same kind used in the biomimetic coating experiments) was inserted into each sponge to facilitate its localization *in vivo*. All implants were prepared under sterile conditions and inserted subcutaneously in the dorsal region of rats [5]. Implants, together with the surrounding tissue, were retrieved after 2 and 5 weeks, fixed in 10% formaldehyde for several days at ambient temperature, dehydrated in ethanol, and embedded in methylmethacrylate. Using a systematic random sampling protocol, 300-µm-thick sections were cut from each embedded sample, polished, and surface-stained with basic Fuchsine, McNeil's Tetrachrome and Toluidine blue O for evaluation in the light microscope. Using a well-defined systematic random sampling protocol, eight digital light microscopic images taken at two different magnifications were obtained per section and printed in colour. The histomorphometric analysis of bone formation was performed on these coloured prints using the point and intersection counting methodologies elaborated by Cruz-Orive *et al.*[7] and Gunderson *et al.*[8]. Differences between the two groups at a particular sampling time and within the same group at each sampling time were analyzed statistically using the *t*-test, the level of significance being set at p = 0.05.

Results and Discussion

Ectopic bone formation was observed in each experimental group of animals (Fig. 1). Bone tissue was deposited directly upon the coated implant or collagen sponge surface as well as within vicinal connective tissue and occurred by an intramembranous (direct) growth mechanism. No cartilage tissue formation was observed in either group. Foreign-body giant cells were actively involved in the degradation of both biomimetic coatings and collagen sponges. However, the coverage of implant surfaces with foreign-body giant cells was lower in the biomimetic coating group than in the collagen sponge one. Furthermore, this lower coverage decreased between the 2- and 5-week junctures, whereas the population associated with collagen sponges remained fairly constant. In the biomimetic coating group, bone-marrow tissue was sometimes observed to contact the implant surface directly - a situation that was never encountered in the collagen sponge group. This finding indicates that the biomimetic coatings were highly biocompatible, since bone marrow contains an abundance of immunocompetent cells which are very sensitive to foreign material.

The net volume of bone tissue laid down during the 5-week follow-up period was substantially greater in the biomimetic coating group than in the collagen sponge one (Fig. 2), even though the amount of BMP-2 incorporated (1.7 µg/sample) was much lower than that with which collagen sponges were impregnated (10 µg/sample). In the biomimetic coating group, the net volume of bone formed increased from 5.8 mm^3/disc at 2 weeks to 11.6mm^3/disc at 5 weeks (p < 0.001). In the collagen sponge group, the net volumes of bone formed at the 2-week (2.3 mm^3/sponge) and 5-week (3.0 mm^3/sponge) junctures did not differ significantly from each other (p = 0.82).

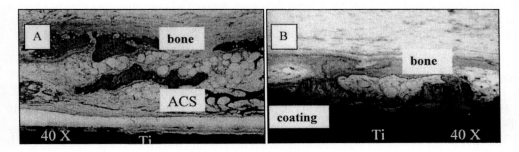

Fig. 1. Ectopic bone formation induced by BMP-2 released from collagen sponges (A) or biomimetic calcium phosphate coatings (B) 5 weeks after implantation. Ti: titanium-alloy implant; ACS: collagen sponge.

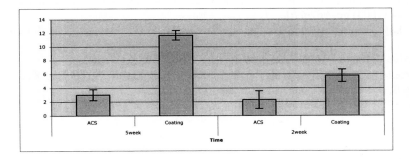

Fig. 2. Comparison of the net volume of bone formed per collagen sponge (impregnated with 10 µg of BMP-2/sample) or biomimetic coating (1.7µg of BMP-2 incorporated/sample) at the 2- and 5-week sampling times. Mean values (± SEM) are represented; n = 6 for each group and time point.

This finding is deemed to be attributable to differences in the BMP-2-release profiles: most of the drug associated with collagen sponges is released rapidly during a short time span (of a few hours duration), whereas BMP-2 incorporated into biomimetic calcium phosphate coatings is liberated more slowly, at a lower but sustained level, over several weeks. BMP-2 is liberated gradually because it forms an integral part of the inorganic latticework. BMP-2 soaked up by collagen sponges is released in a single rapid burst, and the drug, being highly water-soluble, diffuses away from the implantation site too rapidly to be capable of exerting a sustained osteogenic effect. In addition to these physical considerations, the release of BMP-2 from biomimetic coatings will of course be influenced by cell-mediated coating degradation.

Conclusions

Titanium-alloy implants bearing a biomimetically co-precipitated layer of calcium phosphate and BMP-2 are mechanically stable, highly biocompatible, osteoconductive and osteoinductive. Our results demonstrate that BMP-2 was released at a level that not only induced but sustained osteogenic activity until the end of the 5-week follow-up period. Collagen sponges carrying a higher load of BMP-2 were less efficient. As a carrier for BMP-2 and in facilitating its osteogenic effects, biomimetic calcium phosphate coatings offer a distinct advantage over the classic collagen sponge.

References

[1] H. Uludag, D.D. Augusta, R. Palmer, G. Timony and J.Wozney: J Biomed. Mater Res, vol 46 (1999), pp193-202

[2] Y. Sasano, I. Mizoguchi, I. Takahashi, M. Kagayama, T. Saito and Y. Kuboki. Anat Rec, 247: (2001) 472.

[3] I. Alam, I. Asahina, K. Ohmamiuda, K. Takahashi, S. Yokota and S. Enomoto S. Biomaterials. 22:1643-51. 2001.

[4] H. Uludag, D. D'Augusta, J. Golden, L. Li, G. Timony, R. Riedel and J.M. Wozney. J Biomed. Mater Res, 2000, 50:227-238.

[5] Y. Liu, E. B. Hunziker, P.Layrolle, K. de Groot, Bioceramics 15, Key Engineering Materials Vols. 240-244 (2003) pp. 667-670.

[6] Liu Y, Layrolle P, de Bruijn J, van Blitterswijk C, de Groot K: *J Biomed Mater Res* 2001, 57:327-335.

[7] Cruz-Orive L.M and Weibel E.R: *Am J Physiol* 1990, 258:L148-156.

[8] Gundersen H.J, Bendtsen TF, Korbo L, Marcussen N, Moller A, Nielsen K, Nyengaard JR, Pakkenberg B, Sorensen FB, Vesterby A, et al.: *Apmis* 1988, 96:379-394.

Key Engineering Materials Vols. 254-256(2004) pp. 623-626
online at http://www.scientific.net
© *2004 Trans Tech Publications, Switzerland*

A Randomised RSA Study of Peri-Apatite™ HA coating of a Knee Prosthesis

Ulrik Hansson [1] and Sören Toksvig-Larsen[2]

[1,2]Dept of Orthopedics, Lund University hospital S221 85 Lund, Sweden. ulrik.hansson@ort.lu.se and soren.toksvig-larsen@ort.lu.se

Keywords: Knee prosthesis, hydroxyapatite, cementless, radiostereometry

Abstract: Fifty knees scheduled for a total knee prosthesis were randomized in two groups. Both groups had a noncemented fixation of the porous-coated implants but one group received a prosthesis with ceramic coating on the porous surface of the implant. The knees were evaluated with RSA technique regarding micromotion the first two years. There were less micromotion in the coated group. The results indicate that the technique for ceramic coating here studied provides a safe and stable fixation of the implants.

Introduction

Hydroxyapatite (HA) coating of prosthetic joint components is believed to enhance the stability and the longevity of the prosthesis -bone fixation [1]. To measure the implants stability, radiostereophotogrammetry (RSA) can be used. With this technique movements between the implant and host bone can be measured with an accuracy of 0,2 mm [2, 3]. RSA is an invaluable diagnostic tool allowing clinicians the ability to predict the fixation longevity of different prosthetic surfaces *in- vivo* after a relatively short term of implantation[4].

Hydroxyapatite ceramic coatings have mostly been plasma sprayed onto the fixation surface of the implant. Plasma spraying is a high temperature process in which powder raw materials are introduced into a high temperature high velocity stream of combusting gases. The powders are melted by the flame and propelled towards the substrate where they deposit and instantly solidify to form a HA coating. Plasma spraying is largely a line of sight technique and as such has difficulty in covering three-dimensional coatings such as porous beaded fixation surfaces typically used for fixation surfaces on many knee prosthesis. A new solution-deposition technique developed to coat hydroxyapatite (Peri-Apatite™-PA) onto porous coating fixation surfaces has been developed. The coating is formed through a chemical based precipitation process, which occurs at low temperature and physiological pH and is very similar to the apatite or mineral phase found in natural bone. The objective of this study was to access the clinical performance of this coating, regarding its ability to stimulate an endurable and stable implant fixation.

Materials and methods

Fifty knees, scheduled for TKR were randomised in two groups. The Duracon modular total knee, non-cemented, porous coated with a cruciform stem was used. One group received the prosthesis with the PA coating and the other with porous coating only. The femoral components had the same fixation surface as the tibial components in each group. The coating was done using a solution deposition technique, described elsewhere [5], which has the ability to coat all exposed surfaces of the beaded ingrowth substrate.

The patients were allowed full weight bearing starting from the 1st postoperative day. All patients were followed clinically and with RSA analysis. The prostheses were analysed with RSA examinations at day 2-3, at 6 weeks, 3 and 6 months, 1 and 2 years. The RSA analysis was done with segment motion along and around the three coordinal axes, and point motion. In point motion the migration of the marker that moves the most was recognised as the maximal total point motion (MTPM). The prostheses were classified as stable or continuously migrating, depending on the amount of migration between the first and second year.

Results

Largest movements of the rigid body of the tibial component were found along the longitudinal axis (y-axis). (Table 1.) There was a lower migration pattern for the PA group. However, the MTPM at 1 and 2 years showed no clear statistically significant difference between the two groups, even though the migration was lower for the PA-group.

Table 1. Mean Y-translation (subsidence) in mm.

	Periapatite	Porous
6 weeks	0.48	0.77*
3 months	0.54	0.92*
6 months	0.57	1.02*
1 year	0.57	1.02*
2 years	0.62	0.92

Table 2. MTPM in mm

	Periapatite	Porous
6 weeks	1.12	1.52*
3 months	1.27	1.76
6 months	1.36	1.93*
1 year	1.46	1.84
2 years	1.58	1.92

* $p < 0.05$

Discussion

This study has demonstrated a benefit in using the hydroxyapatite coatings for cementless fixation of total knees. In all cases measured the hydroxyapatite coated tibial components migrated less in both the y translation (subsidence) values and the MTPM values. Even though this difference is not statistically significant the short-term benefits to patients are clear. Increased stability means better implant / bone fixation which will make these implants more resistant to loosening in the long term. The study is currently being completed out to 5 years to verify this.

The solution deposited hydroxyapatite used in this study is unique in that all the surface area of the beaded fixation substrate was fully coated. The coating did in fact perform similar to plasma sprayed hydroxapatite. Earlier studies with the same prosthetic design have shown a migration pattern of the same magnitude. In a previous study using a plasma prayed hydroxyapatite coating, slightly lower migration values were measured [6], however this study was done using a smaller patient population and it is not possible to compare the results. Theoretically a coating that provides a full hydroxapatite coating on all layers, not just the top layer, of an ingrowth surface has to have an advantage with regard to bone ingrowth and fixation.

Conclusion

Hydroxapatite coatings are recommended for cementless fixation of total knee arthroplasty. The solution deposited hydroxyapatite coating technique studied in this RSA trial provides a stable and safe fixation of the implant.

References

[1] I.Onsten, et al., Hydroxyapatite augmentation of the porous coating improves fixation of tibial components. A randomised RSA study in 116 patients. J Bone Joint Surg Br, 1998. 80(3): p. 417-25.

[2] G. Selvik, Roentgen stereophotogrammetric analysis. Acta Radiol, 1990. 31(2): p. 113-26.

[3] L. Ryd, et al., Tibial component fixation in knee arthroplasty. Clin Orthop, 1986(213): p. 141-9.

[4] L. Ryd, et al., Roentgen stereophotogrammetric analysis as a predictor of mechanical loosening of knee prostheses. J Bone Joint Surg Br, 1995. 77(3): p. 377-83.

[5] B. R. Constanz, G. C. Osaka, Hydroxyapatite prosthesis coatings. United States Patent 5,164,187,1992

[6] S. Toksvig-Larsen, et al., Hydroxyapatite-enhanced tibial prosthetic fixation. Clin Orthop, 2000(370): p. 192-200.

Key Engineering Materials Vols. 254-256(2004) pp. 627-630
online at http://www.scientific.net
© *2004 Trans Tech Publications, Switzerland*

Manufacture of Paste Opaque Porcelains Using Glycols as a Solvent and Evaluation of their Physical Properties

Kwang-Mahn Kim[1], Sa-Hak Kim[2], Dae-Jin Ko[3], Kyoung-Nam[1]

[1] Dept. of Dental Biomaterials & Bioengineering, Yonsei University, Seoul 120-752, Korea,
kmkim@yumc.yonsei.ac.kr

[2] Dept. of Dental Lab. Technology, Dong-u College, Sokcho 217-070, Korea, kdtpksh@unitel.co.kr

[3] Institute of Alphadent Technology, Paju 413-850 Korea, djkoh@alphadent.co.kr

Keywords: Paste opaque porcelain, metal-ceramic system, properties, glycol, dental materials

Abstract. Paste opaque porcelains are used in metal-ceramic systems in order to laminate opaque porcelains in thin layers, interrupt the metal color effectively, and enhance workability. We manufactured the paste opaque porcelains using Propylene Glycol (PG) and Butylene Glycol (BG) as a solvent, and compared viscosity, the coefficient of thermal expansion, debonding/crack-initiation strength to metal, SEM observations of the interface between porcelain and metal, and color with those of the commercial products (Duceram Plus, Duceram GmbH; VMK 95, Vita Co.; Noritake EX-3, Noritake Co.) used in the clinical field. The viscosity of test materials was measured using viscometer at 25°C. The coefficient of thermal expansion and debonding/crack-initiation strength to metal were measured by ISO 9693:1999, and the color was evaluated by a CIELab system. The rheological property of tested materials showed pseudo-plastic behavior. The coefficient of thermal expansion ($\times 10^{-6} K^{-1}$) of the tested material ranged from 13.3 to 14.3 while debonding/crack-initiation strength (MPa) ranged from 35 to 41. ISO requires debonding/crack-initiation to be over 25 MPa. All test groups adhere effectively onto the metal surface in the SEM observations. And the colors of tested materials were distinguishable from each other despite of having similar shades (A3).

Introduction

Ceramics used for porcelain fused to metal crowns are basically composed of opaque porcelains, dentin porcelains and enamel porcelains. The opaque porcelains play an important role in the bonding between metal and porcelain. Meanwhile, the opaque layer intercepts light transmission, reflects and partially scatters light in porcelain fused to metal systems. Even though the thickness of the layer is only 0.2~0.3 mm, it has a great effect on naturalness and elegance of the final porcelain restorations. The main components of opaque porcelains are feldspar, quartz and clay, which are basically the same compositions as body and enamel porcelains. However, a large amount of the components such as tin oxide or titanium oxide are included so that the metal color is interrupted and opaqueness is obtained after firing. It is also processed into small particle powder so that it can be well fused [1]. There were some reports on the relationship between bonding strength and component element [2, 3] and on the influence on the color of the final restoration material by the opaque porcelains [4, 5]. Test methods and requirements on metal-ceramic systems were specified by ISO Specification 9693:1999 [6].

Materials and Methods

Powder opaque porcelain used for paste opaque porcelain was CeraMax A30 (Alphadent Co., Ltd., Korea), which was ground by alumina ball mill so that its maximum size was below 50 .m. An organic solvent, comprising of 49% propylene glycol, 49% butylenes glycol and 2% glycerine by weight, was used to turn the porcelain powder into a paste. While adding organic solvent to the fine-ground opaque porcelain powder, paste opaque porcelain was made by mixing evenly for 4 hours with Automatic Mortar (Samhwa Ceramics, Korea). The ratio between powder and solvent was

7:3 by weight. In order to compare the physical properties of self-made paste opaque porcelain, Duceram Plus (Ducera Dental GmbH, Germany) and VMK 95 (Vita Co., Germany) and Noritake EX-3 (Noritake Co., Limited, Japan) were used for control group paste opaque porcelain, and were marked as DU, VM, NO, respectively. Self-made opaque porcelain was marked as EX.

Each pastes' viscosity was evaluated at 25 °C with a viscometer (DV-III, Brookfield, U.S.A.). In order to measure the coefficient of thermal expansion, each specimen was made with a 4 mm diameter and 5 mm height by molding and firing. The specimens were measured using a thermo-mechanical analyzer (TMA 2940, TA Instrument, U.S.A.) at temperatures ranging from 30 °C to 500 °C. In order to make the specimen for debonding/crack-initiation strength tests, metal substructures (25 mm × 3 mm × 0.5 mm) were cast with Ni-Cr base metal (VeraBond, AlbaDent, U.S.A.) and degassed at 980 °C for 5 min. Each specimen was built up to 8 mm length and 0.2 mm thickness and fired according to schedule so that the paste opaque porcelain was made symmetrical in the center with a 3 mm width. Body porcelain was built up and fired so that the thickness of the total porcelain including opaque porcelain was 1.1 mm. It was trimmed by disc and glazing treatment was conducted as instructed by the manufacturer (Fig. 1). Debonding/crack-initiation strength was measured with Universal Testing Machine (Instron 6022, Instron Co., England) using 3-Point Bending Test mode. At this time, span length between holders were 20 mm, and the contact sites of both the holder and pressing piston to the specimen were made to a round shape with 1.0 mm radius. The amount of load applied was measured when the sample was broken by applying the load at the cross-head speed of (1.5 ± 0.5) mm/min. Breaking load F_{fail} (N) for each 6 specimens were measured. Debonding/crack-initiation strength was calculated by multiplying constant k to the breakage load F_{fail} (Eq. 1). At this time, the constant k can be obtained from Fig. 2, and is determined by metal plate thickness d_M and the metal's Young's modulus E_M. The Young's modulus of the alloy used in this study was 209.6 GPa.

Debonding/crack-initiation Strength, $T_b = k \times F_{fail}$.   (1)

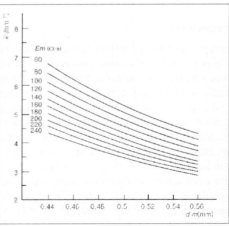

Fig. 1. Schematic showings the sample and jig for debonding/crack-initiation strength test.

Fig 2. Diagram to determine the coefficient k as a function of metal substrate thickness d_M and Young's modulus E_M of the metallic material (ISO 9693:1999).

Fractured sites were observed using SEM. Metal specimens for color evaluation were shaped by diamond disc and alumina so that each specimen measured (2 cm × 2 cm) with 0.5 mm thickness. And then, oxidation film treatment was carried out for each specimen at 980 °C for 5 minutes. A designed plastic sheet was used so that the opaque porcelain is laminated in 0.2 mm thickness and 11 mm diameter. After placing the prepared plastic sheet on the metal plate, the opaque porcelain was applied and excess opaque porcelain was removed so that it can be laminated with constant thickness and shape. Then, the plastic sheet was carefully removed, and porcelain-metal was fired according to the firing schedule. The specimen was placed on an 8 mm round color sensing plate of Spectrophotometer (CM-2500d, Minolta, Japan.) and its color was measured in the SCI (Specular Component Included). The values of L^*, a^*, b^* were measured. The data of debonding/crack-initiation strength and color were compared statistically by Kruskal-Wallis test and Tukey's grouping test (p=0.05).

Results

All pastes opaque porcelains are pseudo-plastic materials and DU has the highest viscosity in all the rate of shear (Fig. 3). The results of thermal expansion coefficient, debonding/crack-initiation strength and color evaluation are in Table 1. There is no significant difference between the test groups in the test of debonding/crack-initiation strength. The requirement value of ISO specification is over 25 MPa, so all test groups were satisfactory. The value of L^* of EX was 86.89 ± 0.63, which was brighter than the control group, and the brightness decreased in the order of VM, NO, and DU. The value of a^* showed positive values except for EX, and the value of b^* was that DU showed 22.45 ± 0.52, which had the highest value. So, the color of the fired opaque porcelain differed from one another despite having similar shades.

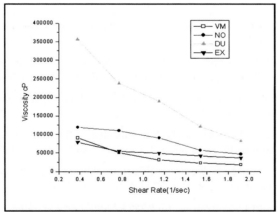

Fig. 3. The viscosity of test materials shows that all materials have pseudo-plastic properties.

Table 1. The results of thermal expansion coefficient, debonding/crack-initiation strength and color evaluation

	a [X10^{-6}/°C]	debonding/crack initiation strength [MPa]	color L^*	a^*	b^*
EX	14.0	35.53±4.37[a]	86.9±0.6[a]	-2.6±0.1[d]	6.7±0.8[d]
DU	13.9	40.88±5.15[a]	76.7±0.4[c]	2.8±0.4[c]	22.5±0.5[a]
VM	14.3	39.43±4.23[a]	82.7±0.2[b]	4.5±0.1[b]	15.3±0.2[c]
NO	13.3	35.39±3.61[a]	76.8±0.1[c]	5.2±0.2[a]	16.1±0.4[b]

Identical letters indicate no significant difference (p>0.05).

SEM photography of fractured sites was as Fig. 4. All fired opaque porcelains intimately adhere to metal surface. EX and VM, it was observed that a lot of bubbles were in the opaque layer.

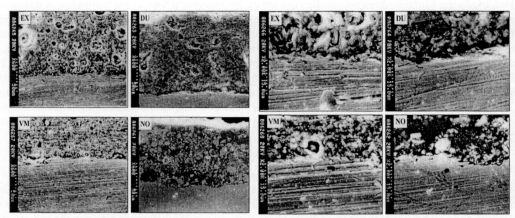

Fig. 4. SEM photography of fractured site shows intimate adhesion between porcelains and metal.

Conclusions

Glycols can be used as a convenient solvent in making paste opaque porcelains with good bonding to metal and reproducible properties.

References

[1] Kim, D.J. Ko, S.H. Kim, et al.: Analysis of Opaque Porcelain Paste, Kor. Res. Soc. Dent. Mater. Vol. 27 (2000), p. 237-243.

[2] K.J. Anusavice, J.A. Horner and C.W. Fairhurst: Adherence controlling elements in ceramic-metal system, I. Precious alloy, J. Dent. Res. Vol. 56 (1977), p 1045-1051.

[3] J.M. Meyer: Porcelain-metal bonding in dentistry, Encyclopedia of materials science and engineering Vol 5 (1986), p. 3825-3830.

[4] A. Obregon, R.J. Goodkind and W.B. Schwabacher: Effect of opaque and porcelain surface texture on the color ceramometal restoration, J. Prosthet. Dent. Vol. 46 (1981), p. 330-340.

[5] P. Yaman, S.R. Qazi and M.E. Razzoog: Effect of adding opaque porcelain on the final color of porcelain laminates, J. Prosthet. Dent. Vol. 77 (1997), p. 136-140.

[6] International Standard Organization: *ISO 9693 Metal-ceramic dental restorative systems*, ISO (1999).

Key Engineering Materials Vols. 254-256(2004) pp. 631-634
online at http://www.scientific.net
© *2004 Trans Tech Publications, Switzerland*

Successful Osteoinduction by Macroporous Calcium Metaphosphate Ceramic-Osteoblastic Cell Complex Implantation

S.H. Oh[1], S.Y. Kim[1], E.K. Park[2], S.Y. Kim[2], J.H. Chung[3], H.M. Ryoo[4], K.H. Kim[4], and H.I. Shin[4]

[1]School of Materials Engineering, Yeungnam University, Daegu, Korea
[2]School of Medicine, Kyungpook National University, Daegu, Korea
[3]School of Medicine, Yeungnam University, Daegu, Korea
[4]School of Dentistry, Kyungpook National University, Daegu, Korea, hishin@knu.ac.kr

Keywords: Macroporous calcium metaphosphate, scaffold, osteoinduction, ceramic-cell complex

Abstract: To evaluate as a scaffold for guided bone regeneration, the macroporous calcium metaphosphate ceramics, having 250μm or 450μm average pore size in an interconnected framework of structured blocks, were implanted into skid mice subcutaneous pouches for 3 weeks in a form of macroporous calcium metaphosphate ceramic-osteogenic cell complex. The macroporous calcium metaphosphate ceramics allowed appropriate cellular attachment and proliferation with osteogenic differentiation. In addition, the macroporous calcium metaphosphate ceramic-cell complex induced ectopic bone formation effectively along the inner surface of the interconnecting frame forming a macroporous structure. These findings suggest that macroporous calcium metaphosphate ceramic can be an ideal scaffolding material for guided bone regeneration in terms of an excellent delivery vehicle for osteogenic cells. Macroporous calcium metaphospate ceramic also possesses excellent biocompatibility, osteoconductivity, and controlled biodegradibility.

Introduction

Bone defects occur in a wide variety of clinical situations, and their reconstruction to provide mechanical integrity to the skeleton is a necessary step in the patient's rehabilitation. Although the autograft is the current gold standard for bone reconstruction, it has several potential drawbacks such as limited availability, significant donor site morbidity, wound complications, and discomfort. The alternative allografts also have a potential risk of disease transmission with significantly less osteogenic activity [1,2]. To minimize these limitations of natural bone graft materials for hard tissue repair and replacement, various synthetic biomaterials were developed. With the growing demands of bioactive materials for orthopaedic as well as maxillofacial surgery, the ultilization of calcium hydroxyapatite and tricalcium phosphate as fillers, spacer, and bone graft substitutes has received great attention during past two decades, primarily because of their biocompatibility, bioactivity, and osteoconduction characteristics with respect to host tissue[3]. The primary goal of these graft materials is to facilitate bone healing in numerous situations whereby healing normally would not occur [4]. To address this problem, novel materials, cellular transplantation and bioactive molecule delivery are being explored alone and in various combinations. Scaffold-guided tissue regeneration is one enabling technology which involves seeding highly porous biodegradable scaffolds with donor cells and /or growth factors, then culturing and implanting the scaffolds to induce and direct the growth of new tissue.

In this study, we evaluated biodegradable macroporous calcium metaphosphate ceramic blocks as a scaffold for guided bone regeneration which allows for appropriate cellular attachment and proliferation with osteogenic differentiation by evaluating ectopic bone formation ability after implantation of cell-matrix construct into skid mice subcutaneous pouches for 3 weeks.

Materials and Methods

The macroporous calcium metaphosphate ceramic blocks were prepared by a sponge method with a 250 um or 450 um average pore size in an interconnected framework of structured blocks, respectively. The biocompatibility of macroporous calcium metaphosphate ceramic blocks was evaluated by the standard agar overlay cytotoxic assay. In brief, MC3T3-E1 cells were plated at 1.25×10^6 viable cells/6-cm tissue culture dish with 5ml α-MEM media and incubated until confluent. Culture medium is then replaced with 5ml mixed solution of 0.6% agar and 2X α-MEM supplemented with 20% FBS. After solidification, 2ml of 0.01% neutral red was added and stained for 15 mins. The test samples were place on top of the agar layer and incubated for 24 hr at 37°C in 5% CO_2. 24 hr at 37°C in 5% CO_2. The cells were then scored for cytotoxic effect. To evaluate the efficiency of macroporous calcium metaphosphate ceramic blocks as a scaffolding material for the delivery of osteogenic cells, the macroporous calcium metaphosphate ceramic blocks were placed in a 24-well culture plate and cultured with seeded primary mouse primary calvarial osteoblastic cells (1×10^4/well) in osteogenic media (α-MEM supplemented with 10 % FBS, 50 μg/ml ascorbic acid, 10^{-7} M dexamethasone and 10^{-2} M β-glycerophosphate) for 2 weeks. These cell-calcium metaphosphate ceramic block complexes were implanted into skid mouse subcutaneous pouches. At one, two, and three weeks after implantation, the samples were harvested and prepared for routine LM observation. The degree of ectopic bone formation was evaluated.

Results

The agar-overlay assay revealed that macroporous calcium metaphosphate ceramic blocks caused neither distaining nor lysis of cultured neutral red stained MC3T3-E1 cells, indicating noncytotoxic as negative control, silicon. Whereas the postive control copper revealed severe cytotoxicity. Unlike the silicon and macroporous calcium metaphosphate ceramic block, copper caused lysis of MC3T3-E1 cells with distaining of neural red (Fig. 1).

Fig. 1. Photomicrograph of agar-overlay assay. (a) Positive control; copper, (b) Negative control; silicon, (c) Macroporous calcium metaphosphate ceramic. (Neutral red staining, x100).

Both types of macroporous calcium metaphosphate ceramic blocks with 250μm or 450μm average pore size allowed appropriate cellular attachment and proliferation with osteogenic differentiation enhancing the potential for manipulating local repair in a beneficial manner. In addition, the average pore sizes of 250μm and 450 μm were enough for the ingrowth of vascularized connective tissue, which is necessary for the stromal support of carried osteoblasts.

All implanted macroporous calcium metaphosphate ceramic-osteoblastic cell complexes in the skid mouse subcutaneous pouches showed excellent biocompatibility. From two weeks after implantation, they induced ectopic bone formation effectively along the inner surface of the interconnecting frame forming a macroporous structure with well aligned functional osteoblasts (Fig. 2). The newly formed bone around the macroporous calcium metaphosphate ceramic framework was directly fused without insertion of fibrous connective tissue. On the other hand, the implanted macroporous calcium metaphosphate ceramic blocks without cellular preparation did not induce ectopic bone formation.

The macroporous calcium metaphosphate ceramic blocks implanted intosubcutaneous pouches without cells were surrounded by loose delicate connective tissue and the inner side of porous structure was filled with edematous loose connective tissue with mild inflammatory cell infiltration at early stage. With time, however, the fibrotic tissue surrounding the implants became more dense with an increase of collagenous bundles and an ingrowth of vascular fibrous connective tissue through the porous spaces was more prominent. Further more, there was no inflammatory or foreign body reaction through out the observation periods. These findings indicate the excellent biocompatibility of macroporous calcium metaphosphate ceramic blocks.

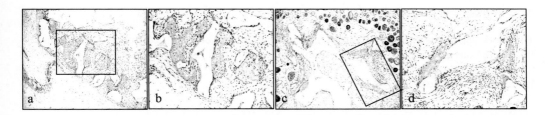

Fig.2 Photomicrograph of ectopic osteoinduction by macroporous calcium metaphosphate ceramic block-osteogenic cell complexes implanted into the subcutaneous pouches of skid mice at 3 weeks. (a) Macroporous CMP with 250um average pore size, x100, (b) Inset of (a), x200, (c) Macroporous CMP with 450um average pore size, x100, (d) Inset of (a), x200, H&E staining.

Discussion

The ideal synthetic bone graft substitute should mimic the native bone in both mechanical and osteogenic properties. In addition, the ideal bone substitute should have structural integrity, provide a framework for host bone formation and act as a delivery system for factors that are important in regulating local bone response [1,5]. Promising structural integrity, a stable framework for host bone formation and an appropriate delivery system are important factors which will maximize repair of large bone defects. Scaffold-guided tissue regeneration is one of the therapeutic approaches to facilitate bone healing in numerous situations in which healing normally would not occur. Bone can be repaired or regenerated by the three processes; osteoinduction, osteoconduction, and osteogenesis. Ostoeinduction is defined as the ability to stimulate the proliferation and differentiationof pluripotential mesenchymal cells[4]. To be an osteoinductive material, it must induce bone formation in a non-bony site. In establishing a direct biological bond between bioactive substrata and adjacent tissue, there must be cellular, molecular, and ionic activity between the tissue and implant surface [6]. Biocompatible matrices are currently being developed to stimaulate osteogenesis via osteoconduction and to promote osteoinduction by using osteognic growth factors, mesenchymal cell implants, and genetically engineered cells [4,7,8]. The macroporous calcium metaphosphate has been known to satisfy a range of goals related to strength, toughness, osteoconductivity, controlled degradation, and complete biocompatibility. In our study, the macroporous calcium metaphosphate ceramics with an average pore size of 250 μm or 450 μm in an interconnected framework of structured blocks worked well as a scaffold for attaching primary calvarial osteoblastic cells and allowing their active proliferation with osteoblastic differentiation. The implanted macroporous calcium metaphosphate ceramic-osteoblastic complex blocks in the skid mouse subcutaneous pouches induced ectopic bone formation effectively along the inner surface of the interconnecting frame forming a macroporous structure. These porous structures were large enough to allow for vascular ingrowth and bone deposition.

Summary

In conclusion, with the growing demands of bioactive materials for orthopaedic as well as maxillofacial surgery, the utilization of macroporous calcium metaphosphate ceramic blocks with an average pore size of 250 to 450 µm may be an appropriate scaffolding material which acts as not only delivery vehicles for cells and gowrh factors important in regulating local bone response, but also as osteoconductive matrices in stimulating direct bone deposition which maximizes the repair of bone defects.

Acknowledgement

This work was supported by a Grant-in-Aid for Next-Generation New Technology Development Programs from the Korea Ministry of Commerce, Industry and Energy(N11-A08-1402-04-1-3).

Literature References

[1] W. R. Moore, S. E. Graves, and G. I. Bain: ANZ. J. Surg. Vol. 71(2001), p. 354

[2] Y.M. Lee, Y.J. Seol, Y.T. Lim, et al.: J. Biomed. Mater. Res. Vol. 54(2000), p. 216

[3] C.J. Damien and J.R. Parsons: J. Appl. Biomaterials, Vol. 2(1990), p. 187

[4] G.A. Jelm, H. Dayoub, and J.A. Jane, Jr.: Neurosurg. Focus, Vol. 10(2001), p. 1

[5] J.F. Keating and M. M. McQuen: J. Bone Joint Surg. Br. Vol. 82-B(2001), p. 3

[6] J.E. Block and M.R. Thorn: Calcif. Tissue Int. Vol. 66(2000), p. 234

[7] P.J. Boyan, C.H. Lohmann, J.Romero, et al.: Clin. Plast. Surg. Vol. 26(1999), p.629

[8] J.B. Brunski, D.A. Puleo, A. Nanci: Int. J. Oral Maxillofac.Implants, Vol. 15(2000), p. 15

Key Engineering Materials Vols. 254-256(2004) pp. 635-638
online at http://www.scientific.net

Do Calcium Zirconium Phosphate Ceramics Inhibit Mineralization?

U. Gross,[1] Ch. Müller-Mai,[2] G. Berger[3], and U. Ploska[3]

[1]Institute of Pathology and [2]Department of Traumatology and Reconstructive Surgery, Klinikum B. Franklin, Free University of Berlin, Hindenburgdamm 30, D 12200 Berlin, Germany.
e-mail: ugross@zedat.fu-berlin.de
[3]Federal Institute for Materials Research and Testing, Unter den Eichen 87, D 12200 Berlin, Germany. e-mail: georg.berger@bam.de

Keywords: Calcium zirconium phosphate ceramics, rabbit model, tissue response, impaired mineralisation, inhibition of matrix vesicles.

Abstract. The solubility of a new calcium zirconium phosphate ceramics was found reduced compared to hydroxyapatite in *in-vitro* studies. Therefore, and for technical reasons cylinders of this new material were implanted into the distal femur epiphysis of female Chinchilla rabbits and investigated after 7, 28 and 84 days postoperatively in sawn sections and TEM. The cellular and tissue responses were characterized by the development of some macrophages in the soft tissue at the interface and by primary trabecular bone approaching the interface and being mostly made up of osteoid, i.e. non mineralised bone matrix, in a seam up to 150 μm thick. The reason for this impaired mineralization is not yet clear. From further in vitro testing and additional transmission electron microscopy it is speculated that the inhibited development of matrix vesicles could be responsible for this process.

Introduction

Recently the system $CaTi_xZr_{4-x}(PO_4)_6$ with x = 0-4 was described regarding solubility in a circulating fluid chamber and compared to a $CaTi_4(PO_4)_6$ system [1, 2]. The solubility tests of these materials were favourable, i.e. the solubility less than the solubility of hydroxyapatite. Animal tests carried out with $CaZr_4(PO_4)_6$ showed direct bonding of bone and implant [3]. The tissue response of material C6Z3P8 (mol%: 46 CaO, 23 ZrO_2, and 31 P_2O_5), however, demonstrated zones of non mineralised bone precursor, i.e. osteoid instead of mineralised bone [4]. Therefore, the material studied in a rabbit model was additionally investigated using transmission electron microscopy. Furthermore, the in vitro behaviour of the new material was studied in the well known simulated body fluid system of Kokubo regarding deposition of apatite and other calcium phosphate moieties under the influence of Zr ions.

Methods

The preparation of the C6Z3P8 sample is described in [2]. The crystallized material was crushed and milled (grain size <45 μm). The powder was uniaxially pressed to tablets of 15 mm in diameter and 3 mm in thickness. Before soaking in simulated body fluid the specimens were washed with ethanol, dried for 1 hr at 60°C, cooled down to room temperature and weighed thereafter. The specimens were kept for 7 and 28 days at 37°C in Kokubo's simulated body fluid pH 7. Thereafter, the specimens were rinsed with ethanol, dried and weighed again after cooling down. The TF-XRD method was used to investigate the surface of the specimens.
Cylinders of the new C6Z3P8 ceramics were made with a diameter of 4 mm and a length of 8 mm, sterilized by dry heat, and implanted under sterile conditions bilaterally into a hole, diameter 4 mm,

depth 8 mm, drilled into the distal femur of adult female Chinchilla rabbits, body weight 3350-4100 g. Six animals per time period of 7, 28, and 84 days were used giving a total of 18 individuals. The implants with surrounding bone were removed after sacrificing the animals in general anesthesia, fixed in formaldehyde solution, dehydrated in a series of ethanol solutions, and embedded in PMMA blocks, sectioned using a diamond saw, polished, and stained using Giemsa and von Kossa Paragon staining solutions. The evaluation was performed using light microscopy in transmitted and reflected light. Furthermore, one specimen in the neighbourhood of implants from each time period was divided into small pieces of tissue for TEM [5]. The material was fixed in 4 % glutaraldehyde solution, post fixed in OsO_4, dehydrated in ethanol solutions (70, 80, 90, 96, and three times 100 %), soaked in propylene oxide and embedded in Epon resin, sectioned with a diamond knife using a Reichert ultra microtome. The ultra thin sections were stained using lead citrate and uranyl acetate and investigated in a transmission electron microscope (Leitz, Oberkochen, Germany) at a magnification of 20000 and 30000, respectively, for detection of matrix vesicles. At least 10 frames from mineralising pre-bone adjacent to the interface with the implant material were taken and compared to corresponding sites and time intervals of control material $Ca_2KNa(PO_4)_2$ that was investigated in the same way in a previous experiment. Matrix vesicles were spherical, at the periphery tri-lamellar structures of the following types: amorphous, crystalline, ruptured, calco-spheritic in the neighbourhood of mineralising fronts. The number of matrix vesicles per field width were estimated. A thorough statistical analysis was not possible due to the insufficient number of specimens derived from only one animal per time interval.

Results

The light microscopic findings were demonstrated, discussed and published [1]. After 7 and 28 days postoperatively the development of non mineralised bone at the interface had increased and was covering large areas. This very impressive seam of non mineralised bone, i.e. osteoid, was built up at the interface with a thickness of up to 150 μm. It contained some osteoblasts exhibiting a normal aspect (Figure 1). Otherwise, there was loose connective tissue with some macrophages and debris of bone, i.e. remains of the bone fragments produced during the drilling of the hole.

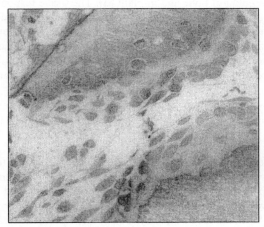

Fig. 1: Semi thin section of proliferating osteoblasts near the implant displaying broad area of non mineralised intercellular matrix (osteoid). Richardson stain. Field width 180 μm.

In high magnification (30000 x) there were less matrix vesicles of early stage, i.e. empty and amorphous vesicles, than usually seen in mineralization of primary bone (Figure 2). The number of empty, amorphous, crystal and ruptured vesicles seemed to be less frequent than usual. The TEM pictures after 28 days were similar to those after 7 days.

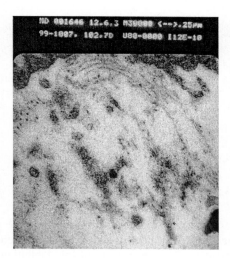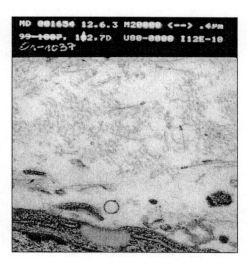

Fig. 2: TEM of matrix vesicles with empty, amorphous, crystal, and ruptured types in control specimen (left) and zirconium oxide specimen (right) 7 days after implantation. Magnification 30000 and 20000 times, respectively.

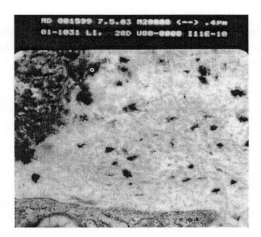

Fig. 3: Calcifying front (left), mineralised fibers and mainly crystal, and ruptured matrix vesicles near the surface of an osteoblast (bottom) 28 days postoperatively. Magnification 20000 times.

As seen in Fig. 4 no hydroxyapatite was formed on the surface of the ceramic specimens stored in SBF solution for 7 and 28 days, respectively. All specimens showed the same XRD spectra generated by TF-XRD.

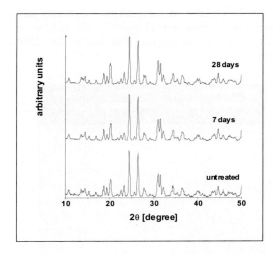

Fig. 4: TF-XRD spectra of the C6Z3P8
material soaked in SBF solution
compared with an untreated specimen

Discussion

According to available data, there is inhibition of mineralization of bone at the interface of calcium zirconium phosphate ceramics of the described composition. The mechanism of this inhibition is not yet clearly elucidated. There must be an interference between chemical species leaching from the calcium zirconium phosphate ceramics and the development or function of matrix vesicle mineralization or autocatalytic apatite deposition at bone surfaces. Similar features were published for leaching products of glass-ceramics of the Ceravital type3 [6]. Recent data speak in favour of effects on matrix vesicles [7].

Conclusions

There are additional arguments from in vitro experiments speaking in favour of the inhibition of mineralization by Zr containing material. The mechanism can be attributed to inhibition of matrix vesicle development and function.

Acknowledgements

Financial support by the German Research Foundation (DFG), Bonn, and technical support by M. Dilger-Rein and H. Renz are gratefully acknowledged.

References

[1] U. Ploska and G. Berger: Biomaterials Vol. 18 (1997), pp. 1671-1675.
[2] U. Ploska: Diss. (PhD) (2000), RWTH Aachen, Germany.
[3] S. Smukler-Moncler, G. Dalcusi, N. Passuti and C. Deudon: Biomaterial-Tissue Interfaces, J. Doherty et al. (Eds), Advances in Biomaterials Vol. 10 (Elsevier Science Publishers B.V., 1992)
[4] U. Gross, Ch. Mueller-Mai, G. Berger, U. Ploska: Key Engineering Materials Vol. 240-242 (2003), pp. 629-632.
[5] U. Gross: *Processing* in: Handbook of Biomaterials Evaluation, Ed.A.von Recum, Taylor & Francis, Philadelphia, USA, 1999, pp.716-717.
[6] U. Gross and V. Strunz: J. Biomed. Mater. Res., Vol.19 (1985), pp. 251-271.
[7] C.H. Lohmann, D.D. Dean et al.: Biomaterials, Vol. 23 (2002), pp.1855-1863.

Key Engineering Materials Vols. 254-256(2004) pp. 639-642
online at http://www.scientific.net
© *2004 Trans Tech Publications, Switzerland*

Effect of Acetabular Cup Position on the Contact Mechanics of Ceramic-on-Ceramic Hip Joint Replacements

M M Mak and Z M Jin

School of Engineering, Design and Technology, University of Bradford, BD7 1DP, England,
z.m.jin@bradford.ac.uk

Keywords: Ceramic-on-ceramic, alumina, total hip arthroplasty, finite element analysis, contact mechanics

Abstract. The importance of the position of an alumina acetabular cup in close contact with an alumina femoral head was examined in this study using the finite element method. The effects of a wide range of inclination angles from 45° to 84° and anteversion angles from 0° to 25° were investigated on the predicted contact mechanics at the articulating surfaces. For the range of implantation angles considered in this study, no edge contact was observed at the rim of the acetabular cup, and the effect of the acetabular cup position on the predicted maximum contact pressure was found to be very small (less than 10%). These findings suggest that the edge contact and associated stripe wear observed clinically in ceramic-on-ceramic hip implants should be related to other mechanisms such as micro-separation, rather than to the position of the acetabular cup directly.

Introduction

The major long-term concern in modern total hip arthroplasty is the adverse biological reaction and loosening due to wear particles produced at the articulating surfaces over the working life of the implant. Ceramic-on-ceramic bearing couples using alumina have been shown to produce extremely low wear in artificial hip joints and rare incidence of osteolysis. It is generally perceived that the position of the acetabular cup can have a dramatic influence on the clinical outcome. For example, steep cup inclination angles of more than 55° have been shown to be associated with high wear rates from clinical retrieved Mittelmeier prostheses [1]. However, the simulator study by Nevelos et al [2] has shown that the inclination angle up to 60° has no appreciable effect on the wear rates. It can be expected that steep inclination angles can potentially lead to the edge contact at the rim of the cup, not only elevating the contact stresses and blocking the lubricant into the contact, but also increasing the rotational torque. The purpose of this study was to investigate the effect of inclination and anteversion angles of an alumina ceramic cup in contact with an alumina ceramic femoral head.

Material and Methods

An alumina femoral head, in articulation with an alumina acetabular insert attached to a polyethylene backing, as shown in Fig. 1a, was considered in this study. The polyethylene backing was secured in a metallic shell for the purpose of fixation. The radius of the femoral head was fixed at 14 mm. The radial clearance between the femoral head and the acetabular cup was assumed to be 0.04 mm. The thickness of the ceramic insert and the polyethylene backing was assumed to be 5 mm each. The implantation position of the acetabular cup was characterised by both the inclination angle (lateral tilt) and the anteversion angle. Inclination angles from 45° to 84° and anteversion angles from 0° to 25° were investigated.

Two 3D finite elements models were developed to simulate the contact between the two articulating surfaces for the configuration shown in Fig. 1a. The full model shown in Fig. 1b consisted of an anatomic pelvic bone with both cortical and cancellous regions created from CT scans (Yew et al 2002) [3], the metallic fixation shell, the polyethylene backing and both the alumina acetabular and femoral components, meshed with approximately 64,000 8-noded brick, 6-noded wedge and shell (for cortical bone) elements. The interface between the pelvic bone and the metallic fixation shell was assumed to be full boned and also at the interface between the ceramic insert and the polyethylene backing. Fig. 1c shows the simplified model, where both the pelvic bone and the metallic fixation shell were neglected, since the elastic modulus for the metallic backing is at least two orders of magnitude of that of polyethylene, and consequently the outside of the polyethylene backing was fully constrained. The simplified model consisted of approximately 40,000 8-noded brick and 6-noded wedge elements. Bonded condition was again assumed at the interface between the ceramic insert and the polyethylene backing. The elastic modulus and Poisson's ratio for alumina were chosen to be 380 GPa and 0.26 respectively; 1 GPa and 0.4 for the polyethylene backing; 110 GPa and 0.3 for the titanium fixation shell; 17 GPa and 0.3 for cortical bone and 800MPa and 0.3 for cancellous bone. A displacement through the centres of both the cup and the head, corresponding to an equivalent vertical load of 2500N (±1%), was applied to both models.

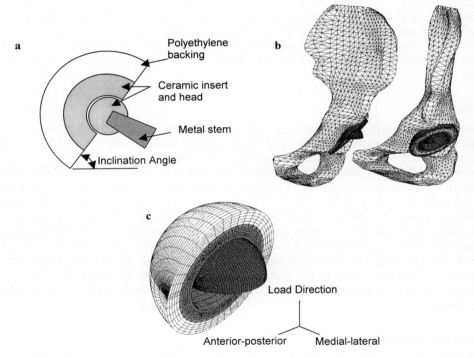

Fig. 1 Models for contact mechanics analysis of ceramic-on-ceramic hip implants a) Schematic model; b) Full finite element model with pelvic bone; c) Simplified finite element model.

Results

Good agreement of the predicted contact pressure was obtained between the full and the simple models shown in Fig. 1b and 1c. The maximum contact pressure obtained from the full model was 84.2 MPa, as compared with 82.8 MPa from the simple model. Therefore the simplified finite element model was used for subsequent parametric analyses on the cup implantation angle.

Fig. 2a to 2d show the contact pressure distribution, viewed from the vertical loading axis, for different inclination and anteversion angles. Table 1 summarises the maximum contact pressure. It is clear that both the inclination and the anteversion angles considered in the present study have little effect on the predicted maximum contact pressure. No edge contact was observed for all the cases considered, even for an extreme case with an inclination angle of 84° and an anteversion angle of 25°.

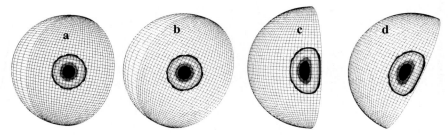

Fig. 2 Contact pressure distributions for different inclination and anteversion angles a) 45° and 0°; b) 45° and 25°; c) 84° and 0° and d) 84° and 25°.

		Inclination		Angles	
		45°	60°	75°	84°
Ante-	0°	81.6	80.5	78.5	77.6
Version	10°	81.8	80.6	78.7	77.6
Angles	15°	82.8	80.6	78.7	77.6
	25°	82.8	80.9	78.9	77.9

Table 1 Maximum contact pressure (in MPa) for different inclination and anteversion angles

It is also interesting to note that with increases in inclination angles, the maximum contact pressure decreases slightly due to slight increases in contact areas, and the area of contact is shifted towards the cup edge. Fig. 3a and 3b shows the contact pressure distributions for different inclination angles in the anterior-posterior and medial-lateral directions respectively. It is clear that the dimension of the contact area in the anterior-posterior direction increases, while the dimension in the medial-lateral direction slightly decreases, resulting in an elliptical shape as shown in Fig. 2d.

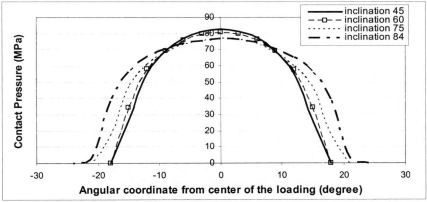

Fig. 3a Contact pressure distribution in the anterial-posterior direction for varies inclination angle (anterversion 0^0)

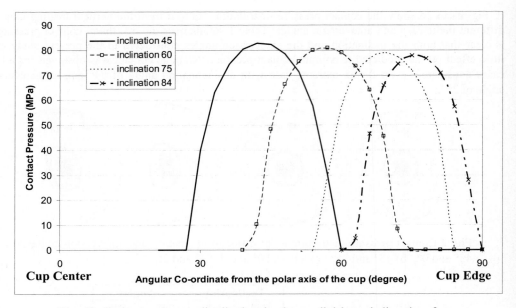

Fig. 3b Contact pressure distribution in the medial-lateral direction for varies inclination angles (anterversion 0^0)

Discussion and conclusion

It appears that the implantation position of alumina cups alone is not directly responsible for causing edge contact and elevating contact stresses at the rim, even for an inclination angle as high as 84°. However, it should be pointed out that a steep inclination angle is more likely to cause edge contact due to micro-separation [4]. For example, for a cup positioned with an inclination angle of 45°, edge contact occurs if micro-separation exceeds two times radial clearance or 80 μm for the case consider in the present study. However, for an inclination angle of 84°, micro-separation of 44 μm may be sufficient to cause edge contact. Furthermore, a smaller clearance between the cup and the head may also be more prone to edge contact for a given implantation position.

Acknowledgement

This work was supported by the Engineering and Physical Sciences Research Council (EPSRC, UK).

References

[1] J. E. Nevelos, E. Ingham, C. Doyle, J. Fisher and A. B. Nevelos: Biomaterials, Vol. 20, Issue 19 (1999), p. 1833-1840.
[2] J. E. Nevelos, E. Ingham, C. Doyle and J. Fisher: 45th ORS Transactions Vol.24 (1999), p.857.
[3] Yew, A., Ensaff, H. and Jin, Z. M. (2002). What caused equatorial contact in McKee-Farrar metal-on-metal hip implants? Proceedings IMechE, International Conference, Engineers & Surgeons - Joined at the Hip, London, UK.
[4] M. M. Mak, A.A. Besong, Z. M. Jin and J. Fisher: Proceedings of IMECH E Part H Journal of Engineering in Medicine, Vol. 216, No. 6 (2002), p. 403-408(6).

Key Engineering Materials Vols. 254-256(2004) pp. 643-646
online at http://www.scientific.net
© *2004 Trans Tech Publications, Switzerland*

Excellent Bone Ingrowth into HA Granules Filled in Acetabular Massive Bone Defect under Weight bearing Condition

H. Oonishi[1], S. C. Kim[1], H. Dohkawa[1], Y.Doiguchi[1], Y. Takao[1] and

K. Oomamiuda[2]

[1]H. Oonishi Memorial Joint Replacement Institute, Tominaga Hospital, 1-4-48, Minato-machi, Naniwa-ku, Osaka, 556-0017, Japan
oons-h@ga2.so-net.ne.jp
[2]Sumitomo Osaka Cement Co., LTD., 585, toyotomi-cho, Funabashi-shi, Chiba, 274-8601, Japan
koomamiuda@sits.soc.co.jp

Keywords: Revision total hip arthroplasty, bone defect, hydroxyapatite, histology

Abstract. We have used HA granules in acetabular massive bone defect in revision THA since 1985. Because as HA is osteoconductive and crystal HA is not resorbable, new bone formation will continue forever. In clinical cases, porous HA granules were filled densely into massive acetabular bone defects under weight bearing condition. Whole HA granule masses of approximately 2.0 and 2.5 cm in thickness were retrieved at revision THA. Non-decalcified specimens were observed. As a control, in non-weight bearing condition, HA granules were filled densely into the cavity made at the proximal end of the tibia of mature goats. At one year and six months the goat was sacrificed. It was very difficult to drill into the HA granule mass and to cut the mass with a chisel. A large amount of bone ingrowth from the acetabulum were found in the whole spaces. In a control a large amount of bone ingrowth were obtained into the spaces only at the peripheral area of the cavity to approximately 5 mm in depth. A large amount of bone growth to the deep area can be expected in weight bearing areas. However, in non-weight bearing areas, it can't be expected.

Introduction

We have used HA granules in acetabular massive bone defect in revision total hip arthroplasty since 1985 (Fig.1). Because it was difficult to obtain allografts in good quality in Japan. Moreover, as HA is osteoconductive and crystal HA is not resorbable, new bone formation will continue forever after osteoporosis due to aging (Fig.2) [1,2,3,4].

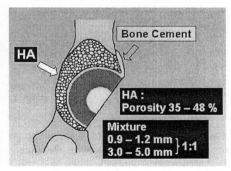

Fig.1: Scheme of revision THA

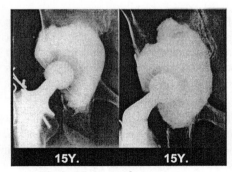

Fig.2: 15 years after surgery

Materials and Methods

In clinical cases, porous HA granules (35-48% porosity) of sizes 0.9 to 1.2 mm (G-4) and 3.0 to 5.0 mm (G-6) (Boneceram-P ™ Sumitomo Osaka Cement Co. Ltd.) mixed in the same ratio were filled densely into massive acetabular bone defects at revision total hip arthroplasty. HA granules were filled under weight bearing condition.

As Kerboull cross plates as cup supporter broke in two cases, HA granule masses of approximately 2.0 and 2.5 cm in thickness were retrieved at two and two and half years after revision total hip arthroplasty, respectively. They contained whole thickness of HA granules.

Non-decalcified ground thin specimens stained by Toluidine blue were observed by optical microscopy and non-decalcified ground blocks were observed by back scattered electron image.

As a control, in non-weight bearing condition, HA granules, which were used in clinical cases, were filled densely into the cavity at the proximal end of the tibia of mature goats. The cavities were made by 2.0 cm in diameter and 10 cm in length. The proximal part of the tibia was cancellous bone and distal part was cortical bone. At one year and six months the goat was sacrificed.

Results

In clinical cases, in weight bearing condition, at the 2nd revision total hip arthroplasty, it was very difficult to drill into the HA granule mass and to cut the mass with a chisel. In histological studies, a large amount of dense bone ingrowth from the base of dense bone and a large amount of cancellous bone ingrowth from the base of cancellous bone were found in the whole spaces between HA granules, respectively. New bone adhered to HA directly (Fig.3, 4).

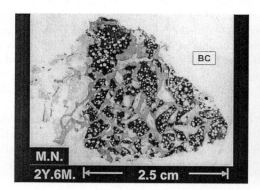
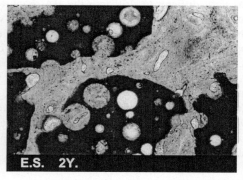

Fig.3 HA granule mass retrieved two years and six months after revision THA. Right area contacted bone cement. Left area contacted acetabular bone. The thickness of HA granules mass was 2.5 cm.

Fig.4. Dense bone ingrowth

In animal experiments, as a control in non-weight bearing condition (Fig.5), a large amount of dense bone ingrowth from the base of dense bone (Fig.6) and cancellous bone ingrowth from the base of cancellous bone (Fig.7) were obtained into the spaces of HA granules at the peripheral area of the cavity to approximately 5 mm in depth.

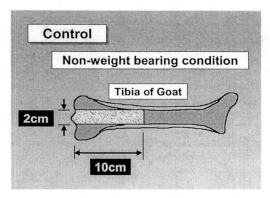

Fig.5 Scheme of animal experiment as a control in non-weight bearing condition

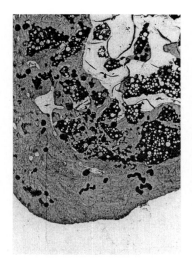

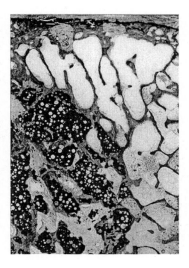

Fig.6. Animal experiment 1 and 1/2 years after implantation. HA granules were impacted into cortical bone area.

Fig.7. Animal experiment 1 and 1/2 years after implantation. HA granules were impacted into cancellous bone area.

However, only a very small amount of bone ingrowth was observed at the central area of 1 cm in depth and some of the HA granules dropped while making histological specimens.

As a result, a large amount of new bone growth was found in the spaces of HA granules over 2.5 cm in depth in weight bearing areas in clinical cases. However, in non-weight bearing areas in animal experiments, only a small amount of new bone ingrowth was found at the center of 1 cm in depth.

Discussion

A large amount of new bone growth was found in the spaces of HA granules over 2.5 cm in depth in weight bearing areas in clinical cases. The HA granule masses were very strong and stable. However, in non-weight bearing areas in animal experiments, a large amount of bone ingrowth could be expected only at the peripheral area of the cavity to approximately 5 mm in depth.

Conclusion

Filling of HA granules in acetabular massive bone defect in revision THA was very effective.

References

[1] H. Oonishi: Biomaterials, Vol.12 (1991), p.171
[2] H. Oonishi, L.L. Hench, J. Wilson: J. Biomed. Mater. Res., Vol. 44(1999), p.31
[3] H. Oonishi, L.L. Hench, J. Wilson: J. Biomed. Mater. Res., Vol. 51(2000), p.37
[4] H. Oonishi, H. Iwaki, N. Kin: J. Bone Joint Surg. [Br], Vol.79-B, (1997), p.87

Key Engineering Materials Vols. 254-256(2004) pp. 647-650
online at http://www.scientific.net
© 2004 Trans Tech Publications, Switzerland

Optimum Fixation at Bone / Bone Cement Interface by Interposing HA Granules (IBBC)

H. Oonishi[1], S. C. Kim[1], H. Dohkawa[1], Y.Doiguchi[1], Y. Takao[1] and K. Oomamiuda[2]

[1]H. Oonishi Memorial Joint Replacement Institute, Tominaga Hospital, 1-4-48, Minato-machi, Naniwa-ku, Osaka, 556-0017, Japan
oons-h@ga2.so-net.ne.jp

[2]Sumitomo Osaka Cement Co., LTD., 585, toyotomi-cho, Funabashi-shi, Chiba, 274-8601, Japan
koomamiuda@sits.soc.co.jp

Keywords: total joint arthroplasty, bone cement, hydroxyapatite, histology

Abstract. It would be a revolutionary idea to interpose unresorbable osteoconductive HA at bone and bone cement interface by expecting chemical bonding of HA with bone and osteoconduction forever to prevent radiolucent line and loosening. As a surgical procedure, less than two layers of HA granules of 300 to 500 micron in diameter were smeared on the bone surface just before the cement insertion (Interface Bioactive Bone Cement : IBBC). In animal experiments, at one week, bone ingrowth began into one to two layers of HA granules. At two to three weeks, bone ingrowth completed. The bonding strength in the case of IBBC without anchor holes at six weeks attained to 50% of non-IBBC and IBBC with anchor holes and showed the same tendency as HA coating on the smooth surface. In clinical cases, the majority of HA granules were incorporated into dense cortical bone and cancellous bone connected to adjacent dense cortical bone and cancellous bone, respectively. The shape and sizes of HA granules were not changed at 17 years. In conventional bone cement (Non-IBBC) and cementless fixation, the spaces will appear at the bone interface due to aging of bone. As unresorbable crystalline HA is used in IBBC and HA is osteoconductive, at present enduring osteoconduction could be expected in only IBBC.

Introduction

Long term after total joint arthroplasty spaces will appear at the interface of bone cement and bone or cementless fixation after occurrence of osteoporosis due to aging of the bone. It would be a revolutionary idea to interpose unresorbable osteoconductive HA at bone and bone cement interface by expecting chemical bonding of HA with bone and osteoconduction forever [1, 2, 3].

Materials and Methods

As a surgical procedure, less than two layers of HA granules of 300 to 500 micron in diameter were smeared on the bone surface just before the cement insertion (Interface Bioactive Bone Cement : IBBC) (Fig.1). Since 1984, IBBC has been used in total hip arthroplasty (THA) and total knee arthroplasty (TKA). In animal experiments, for histological studies, IBBC was performed into the drill hole made in the femoral condyles of mature rabbits (Fig.2).

IBBC   Surgical Procedure

Less than two layers of HA (300~500 μ m)
are interposed onto the bone surface
just before the cement fixation.

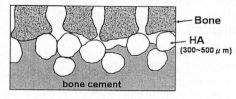

— Bone

— HA
(300~500 μ m)

bone cement

Fig.1

IBBC   Animal experiments

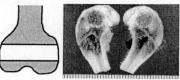

Femoral condyle of mature rabbit
{ Bonding strength : push out test
  Histology

Fig.2

For bonding strength test, holes for cementing of 10 mm in diameter were made on the tibia of goats (Fig.3).   Two models were made; a hole for cementing with several small anchor holes on the wall of the hole, and a hole without anchor holes.   In two models IBBC and Non-IBBC were performed.   These bonding strengths were compared with the materials with HA coating on the Ti column.

Bonding Strength with Bone

Push-out Test

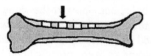

Tibia of Mature Goat

Fig.3

The goats were sacrificed from 0 to six months after surgery, and the push-out test was performed.

In clinical cases, five specimens containing well-fixed bone and bone cement interface were retrieved at 1, 2, 6, 10, 14 and 17 years postoperatively due to late infection and loosening between the socket metal back and the cement layer.   Non-decalcified hard tissue specimens were made and observed by an optical microscopy and a back-scattered electron image.

Results

In animal experiments, at one week bone ingrowth began into one to two layers of HA granules (Fig.4).   At two to three weeks, bone ingrowth completed (Fig.5). At six weeks and three years, bone growth behaviors were not different. The bonding strength in the case of IBBC without anchor holes at six weeks attained to 50% of non-IBBC and IBBC with anchor holes. The bonding strengths in IBBC without anchor holes and HA coating on the smooth surface showed the same tendency (Fig.6).

In clinical cases, the majority of HA granules were incorporated into dense cortical bone (Fig.7) and cancellous bone (Fig.8) connected to adjacent dense cortical bone and cancellous bone, respectively. Both histological and back-scattered electron microscopic examinations demonstrated that a convoluted bone and bone cement interface was well maintained and the shape and sizes of HA granules were not changed at 17 years.

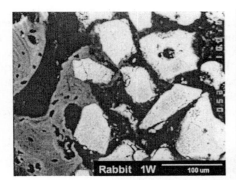

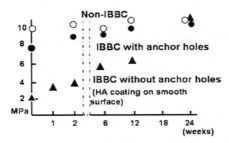

Fig.4: Animal experiment
IBBC was performed into the drill hole
made on the femoral condyle of mature rabbit.
One week after surgery.

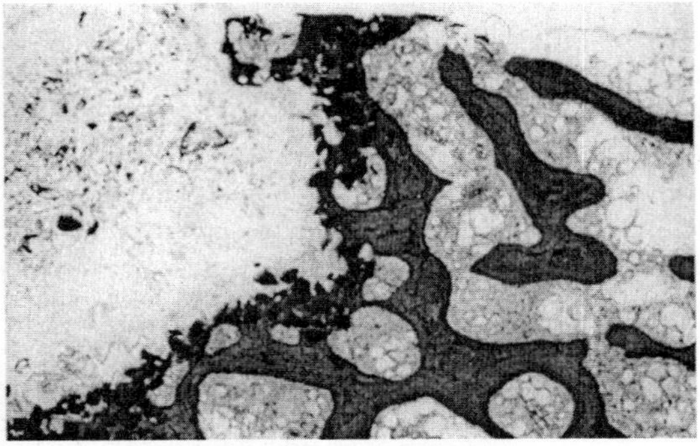

Fig.5: Animal experiment
Tree years after surgery

Fig.6 Bonding strength with bone (Push-out Test)

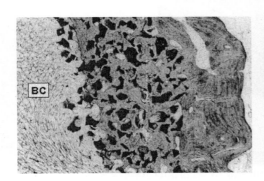

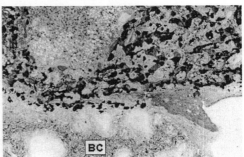

Fig.7 Clinical case, four years after THA. This material was retrieved from the acetabular dense bone of OA due to loosening between metal block socket and bone cement.

Fig.8 Clinical case, four years after surgery. This material was retrieved from the atrophic cancellous bone of the acetabulum of OA due to loosening between metal back socket and bone cement.

Discussion

In conventional bone cement (Non-IBBC) and cementless fixation of porous coating, the spaces will appear at the bone interface after osteoporosis due to aging of bone.   Even in HA coating, as coated HA consists of crystal and amorphous HA, HA will be absorbed after more than ten years.

 As unresorbable crystal HA is used in IBBC and HA is osteoconductive, at present enduring osetoconduction at the interface between bone and bone cement could be expected in only IBBC even after osteoporosis due to aging of the bone (Fig.9).

 Radiographically, the appearance rates of radiolucent line, stress shielding and osteolysis were lowest in IBBC.

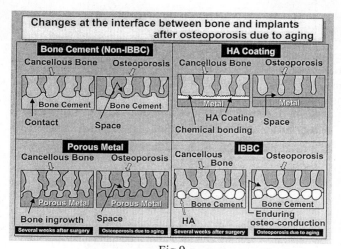

Fig.9

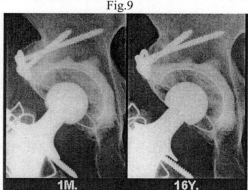

Fig.10 Radiographs of one month and 16 years after surgery
Radiolucent line, osteolysis and stress shielding are not seen.

Conclusion

Enduring osteoconduction at the bone interface could be expected in only IBBC.

Reference

[1] H. Oonishi, S. Kushitani, M. Aono: Bioceramics-1, Vol.1 (1989), p.102
[2] H. Oonishi, Y. Kadoya, H. Iwaki: J. Appl. Biomat. , Vol.53 (2000), p.174
[3] H. Oonishi, Y. Kadoya, H. Iwaki: J. of Arthroplasty, Vol.16,6 (2001), p.784

Key Engineering Materials Vols. 254-256(2004) pp. 651-654
online at http://www.scientific.net
© *2004 Trans Tech Publications, Switzerland*

Biomechanical Analysis of Hydroxyapatite-Coated External Fixation Pins Removed from Osteoporotic Trochanteric Fracture Patients

F. Pegreffi[1], M. Romagnoli[1], A. Moroni[1], S. Giannini[1]

[1] Department of Orthopaedic Surgery, Rizzoli Orthopaedic Institute, Bologna University,

Via C. Pupilli 1 – 40136, Bologna, Italy, a.moroni@ior.it

Keywords: Biomaterials, hydroxyapatite, external fixation pin, osteoporotic trochanteric fracture

Abstract. Elderly osteoporotic patients with pertrochanteric fractures are usually managed with DHS or Gamma Nails. Investigators have observed in these patients an higher intraoperative number of complications caused by a longer surgical time, and a consistent blood loss during surgery. External fixation represents a minimally invasive treatment option for elderly osteoporotic trochanteric fracture patients. This appealing method in osteoporotic bone is a challenge because of the potential for pin loosening and infection. Because of this, pins coated with hydroxyapatite are recommended in these cases.
We wanted to determine whether or not similar improvement in fixation strength could be achieved in a highly loaded weight-bearing situation such as osteoporotic trochanteric fracture.

Introduction

External fixation is a viable treatment option for elderly osteoporotic trochanteric fracture patients [1] largely because it is minimally invasive, operative time is short, blood loss minimal and hospital stay brief. Another advantage is that the loading mode can be adjusted during treatment.
This minimally invasive treatment has appeal, but of surgical concern is the potential for pin loosening and infection [3].
For this reason, pins coated with hydroxyapatite are recommended for fixation in osteoporotic bone. Previous weight-bearing and non-weight-bearing studies, both in normal and osteoporotic bone, have shown faster fixation, higher bonding between host bone and the implant, as well as the so called "sealing effect".
In a recent clinical study of osteoporotic wrist fracture patients, hydroxyapatite-coated pins were better fixed than similar standard pins [2, 4].

Materials and Methods

Ten consecutive osteoporotic patients with trochanteric fracture were selected. Inclusion criteria were: female, age ≥ 65 years; AO type A1 or A2 femoral fracture; bone mineral density (BMD) at the contralateral hip lower than -2.5 T score. Patients were treated with a pertrochanteric fixator and four 5-6mmØ OsteoTite hydroxyapatite-coated tapered pins (Orthofix, Bussolengo, Italy). Pins were numbered 1 through 4, proximal to distal. Two pins were inserted into the femoral head (positions #1 and #2) and two into the proximal femoral shaft (positions #3 and #4). Insertion torque was measured with a torque measurement system. Weight-bearing was as tolerated. After the fractures healed (three months after surgery) the pins were removed with Orthofix instrumentation and the extraction torque measured. After removal, pins were immediately fixed in 2.5 glutaraldehyde 0.1M sodium cocodylate buffer, ph 7.2 for 3h at 4°C, postfixed by 1% osmium tetroxide for 1h at 4°C. They were dehydrated in a graded ethanol series, air dried and gold sputter-coated for ultrastructural examination by Cambridge Stereoscan SEMs.

Results

Average patient age was 82±7 years. Average BMD (bone mineral density) was 538±77. There were five A1 fractures and five A2 fractures. All fractures healed and no fixation failed. Mean pin insertion torque was 1967±1254 Nmm, mean pin extraction torque was 2770±1710 Nmm (p=0.001). Pin insertion and extraction torque were lower for pin positions #1 and #2 compared with pin positions #3 and #4 (p<0.0005) (Fig. 1).

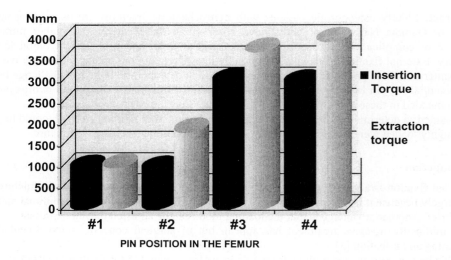

Fig. 1. Pin insertion and extraction torque 3 months after implantation

After removal, the surface of the pins looked different than identical hydroxyapatite-coated pins that had not been implanted. Microscopic analysis of the extracted pins showed bone fragments attached to the coating and no exposure of the metallic substrate. In several areas among the threads, bone particles still attached to the hydroxyapatite-coating were observed (Fig. 2). These particles were more frequently seen in pins taken from positions #1 and #2 than in pins taken from positions #3 and #4.

Femoral neck shaft angle (FNSA) was 130°±3 postoperatively and 129°±3 at 6 months. No significant differences in FNSA at post-op and 6 months were seen.

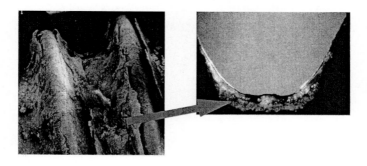

Fig. 2. Microscopic analysis of the extracted pins showed bone fragments attached to the coating and no exposure of the metallic substrate

Discussion and Conclusions

This is the first quantitative weight-bearing clinical study showing that hydroxyapatite-coated pins are well-fixed in osteoporotic bone. Pin insertion torque was shown to relate to pin position. The lowest pin insertion torque was found in pin positions #1 and #2. These pins were implanted into the cancellous bone of the femoral head, where bone density is lower than that of the femoral shaft. This finding confirms that pin insertion torque is influenced by bone density. Our study shows that hydroxyapatite-coated pins improve bone/pin fixation over time, as shown by an average pin extraction torque greater than the corresponding insertion torque (p=0.001). Reduction was maintained, as there were no significant differences between FNSA at post-op and at 6 months. The hydroxyapatite-coated pins were well-fixed and osteointegrated, despite the challenging biomechanical conditions of this study. Fracture fixation in mechanically weak bone is inherently problematic, thus, we consider these results a major achievement in osteoporotic bone fixation.

References

[1] Vossinakis IC and Badras LS: J. Bone Joint Surg. (Br) Vol. 84 (2002), p. 23-29.

[2] Moroni A, Toksvig-Larsen S, Maltarello MC and Giannini S: J. Bone Joint Surg. (Am) Vol. 4 (1998), p. 547-554.

[3] Moroni A, Heikkila J, Magyar G, Toksvig-Larsen S and Giannini S: Clin. Orthop. Vol. 388 (2001), p. 209-17.

[4] Moroni A, Faldini C, Marchetti S, Manca M, Consoli V and Giannini S: J. Bone Joint Surg. (Am) Vol. 83 (2001), p. 717-21.

Key Engineering Materials Vols. 254-256(2004) pp. 655-658
online at http://www.scientific.net
© 2004 Trans Tech Publications, Switzerland

A Radiological Follow-up Study of Plasma Sprayed Bovine Hydroxyapatite (BHA) Coatings

S. Ozsoy[1], K. Altunatmaz [1], .S. Ozyegin[2], F.N. Oktar[3], T. Yazici[4], O. Bayrak[2],

E. Demirkesen[4], D. Toykan[4]

[1]Department of Surgery, Istanbul University, Istanbul, Turkey,

serhatozsoy@yahoo.com, altunatmaz@hotmail.com

[2]Department of Dental Technology, School of Health Related Professions, Marmara University, Istanbul, Turkey, ozyeginsevgi@yahoo.com

[3]Department of Industrial Engineering, Marmara University, Istanbul, Turkey, foktar@gmx.net

[4]Metallurgical & Materials Science Department, Istanbul Technical University, tokayyazici@hotmail.com, demirkesen@itu.edu.tr, deryaderya@hotmail.com

Keywords: Bovine hydroxyapatite, plasma spraying, femoral implant, animal studies

Abstract. In this study a short follow-up study of plasma sprayed stainless steel femoral implants in dogs was performed. Because of its excellent biologic properties, hydroxyapatite (HA) has been required widely in orthopaedic surgery. In this study HA has been used on metal in a form of a thin coating. The HA material has been derived from calcined (850°C) bovine bone. The bovine HA (BHA) which was produced had nearly the same mechanical properties when compared with the a commercial type HA (CHA) that was investigated in a recent research, that is available in literature There were not any vital complications observed postoperatively after 1.5 month of implantation. The long-term follow-up investigations of this study should be carried out in the future.

Introduction

The biologic characteristics of hydroxyapatite (HA)-coated implants are well known. However, over the years both experimental and clinical studies have demonstrated less than ideal and/or variable results achieved using HA-coated devices. Despite excellent early clinical results, concern has developed regarding the long-term results, especially the consequences of potential HA loss and/or separation from the metal substrate. Questions arise as to the nature of the variable performance of the HA coatings since commercially available coatings are required to satisfy guidelines established by the Food and Drug Administration (FDA). The clinical experience of HA-coated orthopaedic prostheses has been recent and limited [1]. Manufacturing of HA powders were carried out with complicated methods like sol & gel, gas atomisation, the rapid spinning cup method, the rotating-electrode process, rapid solidification and the plasma atomisation methods [2]. However, these methods are costly. But the production of HA from bovine bones seems to be economical compared to the methods as mentioned above. Since domestic animals such as dogs and cats have limited life expectation compared to humans, there have been very limited HA coated femoral implants on the commercial market for veterinary. There have been many methods described in literature for the implantation of femoral implants. Femoral stem adaptations have usually been stabilized by cementation. However the usage of cement leads to aseptic loosening, implant corruption, migration and infection. Moreover, complications arise when using PMMA. Thus, in order to minimize these complications and to increase the implant-bone interface strength, animal implants have been sprayed with HA in order to achieve a more porous surface. Nowadays such porous coated implants are being used in the veterinary area to decrease implant complications [3].

The purpose of this study is to prepare a reliable stable commercial good for the veterinary market and cutting the prices down in order to supply a better life with superior quality for domestic animals. Many methods have been used to fix femoral implant into the femur.

Methods

BHA powder was used as the main powder in this study. The BHA material was derived from bovine femurs with calcination method at 850°C [4]. BHA was carefully ground with a blade grinder. Dog femoral implants were machined in different sizes, from 316 grade stainless steel. Femoral heads were die casted. Acetabulum was manufactured with high-density propylene. One ends of the implants were mounted on a round steel plate which was held by a lathe machine for continuous turning movement. This movement enables controllable continuous plasma spraying process to tailor the desired coating thicknesses. BHA powder (particle size was around ~200-150μm) was sprayed on femoral implants. Type 3MB Metco Plasma Flame Spray Gun was used. The surgical operation of two dogs, having degenerative osteoarthritis on hip joints, which could not be cured with conventional methods, has been carried out. The manufacturer company had also manufactured a special surgery kit, which could be used for dogs. Dogs were operated under general anaesthesia. After the operation control x-rays were taken. Consequently, x-rays were taken in one and tree month periods (case 1 was shown at Fig. 1 and 2).

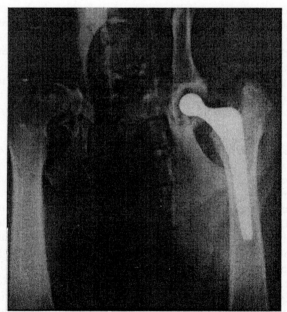

Fig. 1: (Case 1) Postoperative radiography, no screwing procedure was applied with the interlocking system. A 13 year aged male German Shepherd, bilaterally advanced hip dysplasia present (during the manufacturing process no marker was placed around the acetabular implant. This is the reason why this part cannot be seen from the radiography image).

In this study 13-14 aged two German Shepherds that had advanced osteoarthritis, resulting form hip dysplasia, were examined. They were anaesthetized by usuing Rompun (Xylazine Hydrochlorur, 2 mg/kg Rompun i.m.) [Bayer Inc. Istanbul, Turkey] and Ketalar (Ketamin hydrochlorur, 5mg/kg i.v.) [Eczacibasi Inc. Istanbul, Turkey]. After applying Ketalar, the dogs were entubated and Forane (Isoflorane 2.5 %) [Abbott Inc. Queenborough, England] were used for permitting inhalation during the anaesthesia. For hip joint and proximal femoral approach, craniolateral incision was applied. Following the exposure of the hip joint area, femoral head was excised. With the drilling of the femora a femoral canal was formed in order to insert the test implants. It was seen that, there was enough place for femoral prosthesis. Afterwards the acetabulum was reamed and the polyethylene cup was fixed with cement. After the test implants were removed, the femoral implants were inserted carefully into the medullar canal without cementation. Suitable femoral heads for the modular system were fixed to the femoral system by press-fit application and inserted in the acetabular cup. Soft tissues were sutured routinely. For postoperative protection, the dogs were treated with cephazoline sodium for ten days. Patients were kept in their cages for the first two weeks at the postoperative period. Afterwards the dogs were let to have limited exercise by being kept under control with their collars for 1.5 month.

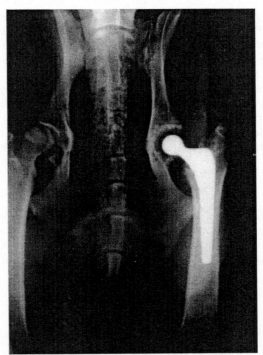

Fig. 2 (Case 1): The radiography of the case 1 postoperatively after 1.5 month. Placement is at normversion. A very slight loosening was seen at the acetubular cement at the dorsal. No abnormalities detected at the femoral stem.

Results and Discussion

Post surgery AP and ML control x-rays were taken at intervals of one and three months. The x-ray analysis did not show any loosening of the implants. After surgery the skin healed naturally. No postoperative infection was observed. After the 1-2 days of the surgery, patients were able to use their legs partially. After 5-7 days they were able to use their limbs very freely. Patient owners pointed out that they were very satisfied from the clinical application. There was not any postoperative infection, aseptic loosening, implant corruption, luxation and other negative signs at the late post-operative period. No osteolysis was detected around the femoral stem.

Before this clinical study the mechanical properties of BHA coatings was tested. This test was done following ASTM tensile testing standards (F-633) for plasma spraying. The tensile strength for BHA coatings that Oktar and Goller had measured was found to be 9.33 MPa (with bond-coatings 10.55 MPa) [5]. Same authors had also used the commercially accepted (for biomedical purposes) HA powder (CHA) at a former study. This officially accepted powder is produced by Sulzer Inc. (Switzerland) in a trademark as AMDRY 6021 and is and widely used as a bioceramic powder for biomedical coating purposes. Oktar and Goller had found the tensile strength to be 1.04 MPa for this CHA and 11.6 MPa for bond-coated CHA [6]. Moreover prior to this study the biocompatibility of this BHA was tested using cell culture methods. It was seen that BHA had the same biological properties of CHA [7].

Summary

In this study bond-coatings were used because the biocompatibility with such coatings was not tested. Since it is known that this BHA is biocompatible [7] there was not any hesitation to use it instead of commercially available AMDRY 6021 HA (CHA). The tensile strength value for CHA was measured as 11.6 MPa and 9.33 MPa for BHA. This difference in strengths could be decreased with adding other reinforcement materials such as titanium and bioglass to BHA powder.

It is also an advantage to use calcined BHA prepared from bovine bone because that it is very economical and easy to produce. Future studies of biocompatibility for bond-coatings should be investigated. There was not any infection evidence observed from the x-rays taken after 3 months of implantation. Thus, BHA plasma sprayed femoral implants are suitable to be used in animal subjects prior to human studies.

Short time results indicate that radiological and clinical evaluations of BHA coated implants present hopeful results. However, long termed clinical and radiological investigations should also be carried out.

Acknowledgement

Authors would like to thank Yahya Yucel from Serbay Orthopedics Inc. who has manufactured the dog implants and the surgical instruments without gaining any profit. The authors would also like to thank Yasar Celik and Turgut Halamoglu from Senkron Ceramic Coating Inc. (www.senkronmetal.com) for the plasma spraying process and Asst. Prof. Gultekin Goller for his kindest supports.

References

[1] J.E. Dalton, S.D. Cook: J. Biomed. Mater. Res. Vol. 29 (1995), p. 239.

[2] M. Entezarian, F. Allaire, et al.: JOM , Vol 48 (1996), p. 53.

[3] K.S. Schulz: Vet. Surg. Vol. 29 (2000), p. 578.

[4] F.N. Oktar, K. Kesenci and E. Piskin: Artif Cell, Blood Sub Immob Biotech, Vol. 27 (1999), p. 367.

[5] F.N. Oktar, G. Goller, et al.: (submitted for XII International Materials Research Congress, August 17-21, 2003, Cancun, Mexico accepted as an oral presentation) Plasma sprayed bovine hydroxyapatite (BHA) coatings.

[6] F.N. Oktar, G. Goller, et al.: Key Eng Mater, Vol. 240-242 (2003), p. 315.

[7] P. Valeiro, F.N. Oktar, et al.: (submitted for 16 Bioceramics, November 6-9, 2003, Porto, Portugal accepted as an oral presentation) Biocompatibility evaluation of dentine, enamel and bovine derived hydroxyapatite.

Key Engineering Materials Vols. 254-256(2004) pp. 659-662
online at http://www.scientific.net

Study of a Silicate Cement for Dental Applications

Pereira S[1], Cavalheiro J[2,3], Branco R[1], Afonso A[1], Vasconcelos M[1]

[1] Faculdade de Medicina Dentária da Universidade do Porto R. Dr. Manuel Pereira da Silva
4200-393 Porto Portugal, fmdup@mail.pt
[2] INEB- Instituto de Engenharia Biomédica, R. do Campo Alegre 823, 4150-180 Porto, Portugal
[3] Universidade do Porto, Faculdade de Engenharia, Departamento de Engenharia Metalúrgica e
Materiais, R Roberto Frias Porto, Portugal, jcavalheiro@fe.up.pt

Keywords: Dental cements, ceramics

Abstract. Dental cements have a wide field of applications in clinical practice. The use as a base material is a very important one in dentistry. The aim of this study is to assess some relevant properties of a silicate cement (SC) developed for dental applications, comparing it with two other cements already commercialised, a zinc phosphate cement De Trey Zinc Cement Improved® (DT) and ProRootMTA®(MTA). Their compressive strength, dimensional changes and weight variation in water after setting, according to ISO standards were compared. The mean compressive strength of DT, MTA and SC were, respectively, 134MPa, 31MPa and 46MPa. There were no statistically significant differences for the three tested cements at the various time intervals, in respect to their weight variation or dimensional changes. However, there was a little loss of weight for DT (0.15%) and for MTA (3%) and a gain of 3% for the SC. Regarding the dimensional changes, all materials undergo little expansion, more significant in the MTA and in the SC samples of 2,07% and 0,4%, respectively. The new Silicate Cement, presented enough compressive strength, low weight variation, very small dimensional changes and good adaptation to the surface of dental cavity. For these reasons this new cement could be a valuable proposal as a base material for dental applications.

Introduction

Dental cements are fast setting pastes obtained by mixing solid-liquid components. One of the most important applications is as base material [1-5] to support other restorative dental materials, such as dental amalgams or composites. The continuous search for new cements for these clinical applications, with improved mechanical properties, namely higher compressive strength, small dimensional changes and weight variation in water after setting and with a very important feature such as the absence of microleakage, is nowadays a very important field of dental research [6-9].

The aim of this study is to assess the mechanical properties and dimensional variation of a silicate cement (SC) developed for dental applications, comparing it with two others cement already commercialised. The study of the ability of the cement to prevent marginal infiltration was achieved on a previous work, where the results were very promising [10].

Materials and Methods

A zinc phosphate cement De Trey Zinc Cement Improved® (DT), a commercial cement ProRootMTA® (MTA), and a new Silicate Cement (SC) were studied.
A sintered ceramic with the following chemical composition (%w/w): SiO_2 19.9, Al_2O_3 5.8, Fe_2O_3 3.7, CaO 65.1, MgO_2 2.0, $CaSO_4$ 1.1, $K_2O1.4$, Na_2O 0.1, was reduced to powder, using an agate mill. The mixture of commercial cements was done according to manufacture's instructions, and the SC was mixed with water (5,9/1 w/w), which provides an initial setting time of 10-11 min.

Their compressive strength, dimensional changes and weight variation in water after setting, according ISO standards, were compared [11].

The structure of the SC adapted to a dental cavity was observed using scanning electronic microscopy (SEM) after 7 days of immersion in a Simulated Body Fluid, where the Ca normally used to prepare the solution was changed by Sr.

For the compressive strength tests 10 cylindrical specimens of each cement were prepared in a split stainless-steel mould and then they were maintained 24h in distilled water at 37°C before the compressive tests. These tests were made in an universal testing machine using a crosshead speed of 0.750m/min.

For the dimensional changes and weight variation in water tests, four cylindrical specimens of each cement were prepared and maintained in an oven during 5h at 37°C. Each sample was weighed on an analytical balance and measured with an electronic micrometer and then stored individually in glass bottles containing 50ml of distilled water at 37°C, during 19h. The superficial water was removed and the samples were weighed again and measured. The same procedure was performed at 24h, 1, 2, 3, 4, 5, 6, 7 and 8 weeks. Results were analysed statistically by t-test and the Kruskal-Wallis test.

Typical class I cavities were prepared in freshly extracted human teeth. After previous cleaning and disinfection procedures with 0.5% chlorhexidine, the SC cement was used to fill the cavity. The teeth after immersion during 7 days in a simulated body fluid prepared with Sr salts, were sliced using a diamond disc and the interface cement/tooth was observed on SEM.

Results and Discussion

The mean compressive strength of DT, MTA and SC were, respectively, 134MPa, 31MPa and 46MPa. The DT, as expected, presented higher compressive strength than the other two cements. However the compressive strength of the new cement was significantly greater than that of the MTA (p<0.001) at 24h, but less than DT cement, as it can be observed in Table 1.

Table 1-Compressive strength of the three cements tested

	SC	MTA	DT
Mean (MPa)	46.40	31.43	134.33
Std. Dev.	6.15	5.75	16.89

There were no statistically significant differences for the three tested cements at the various time intervals, in respect to their weight variation or dimensional changes. However, there was a little loss of weight for the zinc phosphate cement (0.15%) and for MTA (3%) and again of 3% for the SC, as it can be observed in Table 2.

Regarding dimensional changes, all materials undergo little expansion, more significant in the MTA and in the SC samples of 2,07% and 0,4%, respectively, presented in Table 3.

Table 2-Weight variation of the SC cement over 8 weeks

Week	0	1	2	3	4	5	6	7	8	Fin.Var.(%)
Mean (g)	0.171	0.173	0.173	0.175	0.176	0.176	0.176	0.176	0.176	+2.92
Median	0.168	0.171	0.171	0.173	0.173	0.174	0.174	0.174	0.174	+3.57
Maximum	0.184	0.186	0.186	0.191	0.191	0.191	0.191	0.191	0.191	+3.80
Minimum	0.163	0.165	0.165	0.166	0.166	0.166	0.166	0.167	0.166	+1.84

Table 3- Dimensional changes of the SC cement over 8 weeks

Week	0	1	2	3	4	5	6	7	8	Fin.Var.(%)
Mean (mm)	5.991	5.997	6.00	6.015	6.013	6.015	6,013	6.011	6.013	+3.67
Median	5.949	5.952	5.948	5.970	5.966	5.967	5.966	5.964	5.966	+2.85
Maximum	6.349	6.349	6.380	6.383	6.381	6.386	6.385	6.379	6.380	+4.88
Minimum	5.719	5.737	5.738	5.739	5.739	5.742	5.738	5.739	5.741	+3.84

High strength, low weight variation and no significant dimensional changes, are desirable for any base material. The Silicate Cement compressive strength is superior to that of the MTA. The dimensional changes and weight variation in water tests did not show statistically significant differences. However, the slight expansion of SC is important and can explain the excellent sealing ability of this material performed in another study of dye leakage, where the Silicate Cement presented better results than the other two cements.

The SEM observations (Fig. 1) of the new cement applied on a dental cavity after immersion in a simulated body fluid prepared with Sr salts, show a progressive precipitation of Sr phosphates in the surface in contact with the solution, proving the bioactivity of the material and the possibility of a positive evolution of its sealing ability. The material is in close contact with dental tissue, without gaps (arrows), which explains its good sealing behaviour achieved in a previous work [10].

Fig.1 After immersion in a simulated physiological serum, the cement (SC) was able to improve its sealing ability (z1- Sr precipitates in the surface) and a good adaptation to the tooth (T)   cavity ( )

Conclusions

The new Silicate Cement presented enough compressive strength, low weight variation, very small dimensional changes and good sealing ability. For these reasons this new cement seems to be a valuable proposal as a base material for dental applications.

References

[1] Craig RG, Rovers JM. *Restorative Dental Materials*. 9th Ed. St Louis: Mosby; 2002, p 593-627.
[2] Gladwin M, Bagby M. *Clinical Aspects of Dental Materials*. First Edition. Philadelphia: Lippincott Williams & Wilkins; 2000, p 81-92.
[3] Leinfelder KF. *Changing Restorative Traditions: The Use of Bases and Liners*. JADA 1994; 125:p 65-67.
[4] Macchi RL. *Mat. Dentales*. 3ª Ed., Buenos Aires: Edit. Médica P.americana S.A.; 2000, p129-135
[5] Wise DL et al. *Encyclopedic Handbook of Biomaterials and Bioengineering*. New York: Marcel Dekver; 1995, p 1398-1411.
[6] Gordan VV, Mjör IA, Hucke RD, Smith GE. *Effect of Different Liner Treatments on Postoperative Sensivity of Amalgam Restorations*. Quintessence Int 1999; 30:55-59.
[7] Irie M, Nakai H. *Effect of Immersion in Water on Linear Expansion and Strength of Three Base /Liner Materials*. Dent Mater 1995; 14:70-77.
[8] Leevailoj C, Cochran MA, Matis BA, Moore BK, Platt JA. *Microleakage of Posterior Packable Resin Composites With and Without Flowable Liners*. Oper Dent 2001; 26:302-307.
[9] Weiner RS, Weiner LK, Kugel G. *Teaching the Use of Bases and Liners : A Survey of the North American Dental Schools*. JADA 1996; 127:1640-1645.
[10] Pereira S, Branco R, Cavalheiro J, Afonso A, Vasconcelos M. *Sealing ability of a new dental base material*; Bioeng 2003.
[11] International Organisation for Standardization. *Specification for Dental Water-Based Cements*. ISO 9917; 1991(F).

Key Engineering Materials Vols. 254-256(2004) pp. 663-666
online at http://www.scientific.net

Surface Analysis of Explanted Alumina-Alumina Bearings

G. M. Insley[1] and R. M. Streicher[2]

[1] Stryker Howmedica Osteonics, Raheen Business Park, Limerick, Ireland.
gerard.insley@emea.strykercorp.com
[2] Stryker SA, Florastrasse 13, CH-8800 Thalwil, Switzerland.
robert.streicher@strykereurope.com

Keywords: alumina, wear, total joint arthroplasty, microseparation.

Abstract. The purpose of this study was to characterize wear scars on explanted third generation ceramic-on-ceramic components. Macroscopy, light microscopy and scanning electron microscopy (SEM) was carried on the wear scars to determine their extent and the material failure mechanism that led to their formation. The analysis shows that the wear scars were formed through high stress, point contact of the head on the rim of the cup. This type of damage occurs due to microseparation of the head and cup during specific patient activities.

Introduction

Alumina ceramic-on-ceramic articulation is one of the best choices for hip joint bearings for young and active patients. Their low wear, less than 0.5 mm³ per 10^6 cycles *in vitro* and the benign periprosthetic tissue reaction to the small amount of debris produced make them an ideal biomaterial for this application [1-3]. Even though the normal wear mechanism of these bearings is generally understood to be slight relief polishing; recently a number of wear scars were reported on a series of explanted alumina-on-alumina components [4, 5]. The ceramics components concerned were part of a larger series of more than 1,500 successful implants done by mainly one surgeon at the Orthopaedic Hospital Sydney, Australia. The majority of the bearings were removed during a routine re-operation for a variety of clinical conditions such as psoas tendonitis, periprosthetic fracture and infection. The surgeon noticed that the surface of the heads had a slight loss of polish in some areas and this coincided with a similar mark on the cup. This is the first time that wear scars of this type have been reported on modern third generation alumina components. The objective of this study was to characterize the wear scars in terms of their extent and the material failure mechanism that led to their formation. This data will further add to the baseline of knowledge about ceramic articulation for artificial implants and will be essential in validating simulator testing for ceramic-on-ceramic components.

Materials and Methods

Sixteen explanted alumina-alumina couples were analysed in this study. Eleven components had evidence of wear scars (11 heads and 8 liners). The remaining components, which showed no wear scars, were used as controls.

Macro examination. The following dimensional and angular measurements of the scars were carried out to correlate the data with the available clinical information:

On the heads:

- the length of the scars and their maximum width,
- the latitude angle of the centre of the scar relative to the head equator ('inclination' angle, see figure 1 (a)).
- the angle of the long axis of the scar to a line of latitude running through the centre of the scar ('tilt' angle, see figure 1 (b)).

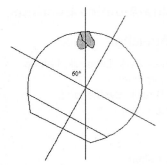

Fig. 1 (a) Latitude angle of scar

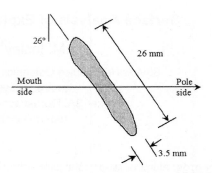

Fig. 1 (b) Tilt angle of scar

On the cups:
- The circumferential length of the scars and their maximum width.

For examination at higher magnification, a conventional optical microscope was used. Various lenses were used to image the scars.

For scanning electron microscopy, a number of representative samples were chosen. The heads were mounted on the SEM stub using graphite adhesive, and the area of interest was lightly sputter-coated with gold/palladium to provide conductivity.

Results and Discussion

The following features appeared to be common to all explanted heads and cups showing localised wear scars:

The worn areas on the heads are usually well defined with sharp boundaries, especially at the ends and on the trunnion side of the scar, see figure 2 (a) and (b). There tended to be more damage outside the main scar on the pole side than on the trunnion side, see figure 2 (b).

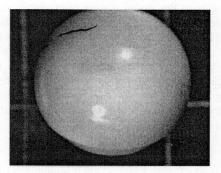

Fig. 2 (a) Macrograph of scar on alumina head

Fig. 2 (b) Micrograph of well-defined boundary

The worn areas on the cups were lens-shaped, generally narrower than on the matching head, and were usually located along the 'blend' line between the highly polished hemispherical bearing surface and the less well-polished convexly curved chamfer near the cup edge, see figure 2 (c). In a number of cases, it was apparent that a very narrow scar existed over an extended length of the blend line outside the main lens-shaped scar, in one or both directions. In extreme cases, the scar was found to extend beyond this convexly curved region onto the conical chamfer region, along the outer blend line, see figure 2 (d).

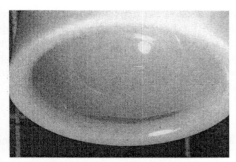
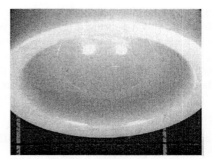

Fig. 2 (c) Macrograph of narrow scar   Fig. 2 (d) Macrograph of narrow scar

The lengths of the worn areas on the head and cup are similar, and increase in size with implantation time, suggesting that the initial non-conforming contact of the head on the cup rim becomes conforming with time and/or number of cycles of movement.

Both heads and cups exhibited scratches in the form of lines of pits both within and outside the main scar areas. Scratches were most prevalent on components with more pronounced scars. These could be parallel to the length of the scars, and/or at a steep angle across them. They were often in parallel sets.

There was a wide variation in the tilt angle of the scar to the lines of latitude on the head. Generally, the direction correlated with whether the implant was on the patient's left or right side. All scars showed a tilt direction that was retroverted, i.e. tilting backwards in respect to the body. There was also a wide variation in inclination angle of the head scars to the head equator, from 0° to 60°.

The SEM analysis of the wear scars showed that they all comprised of grain pull-out with clear facetted sides to the remaining pits. In some cases, there was evidence of subsequent smoothing over of the surface, almost a partial re-polishing of the surface, with some fine-scale debris being trapped in the pits, see figure 3.

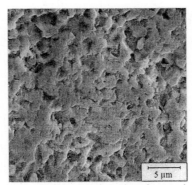

Fig. 3 Head scar centre region showing grain pull-out

Conclusions

Out of the sixteen explants examined, eleven showed evidence of wear scars. The components examined with a stripe scar on the head had a corresponding wear scar on the rim of the cup. The SEM analysis confirms that the initial intense wear process consists of individual grain 'pluck-out' followed later by groups of unsupported grains breaking away. Evidence of some repolishing of the pitted surface was also noted at high magnifications. This pitting type of wear, followed by a later smoothing process, is typical of highly localized contacts occurring initially between the head and cup rim over a defined area. As these contacts conform geometrically with increments in the size of the scar, the wear process becomes less severe and can even lead to partial repolishing of the damaged surface. In conclusion, the wear scars examined on these explanted alumina-alumina couples were formed through high stress, point contact of the head on the rim of the cup. This, coupled with a lack of lubrication at this point initiated the reported wear scars without leading to either catastrophic failure or super-critical wear values. This type of contact occurs, *in vivo*, through microseparation of the head and cup and this mechanism needs to be incorporated into modern simulator testing of alumina-on-alumina components and also other hard-on-hard bearings.

Acknowledgements

The authors wish to acknowledge the staff and technicians at the University of Bath for their contribution and assistance in this work.

References

[1] Amstutz, H. C., Campbell, P., Kossovsky, N. and Clarke, I. C.: Clin. Orthop. Vol. 276 (1992), p.7.

[2] Maloney, W. J., Jasty, M. and Harris, W. H.: J. Bone and Joint Surg. Vol. 72B (1990), p.966.

[3] Schmalzreid, T. P., Jasty, M. and Harris W. H.: J. Bone and Joint Surg. Vol.74 (1992), p.849.

[4] Nevelos, J. E., Ingham, E., Doyle, C., Streicher, R., Nevelos, A. B., Walter, W. and Fisher, J.: J. Arthro. Vol.15 (2000), p.793.

[5] Stewart, T. Tipper, J., Streicher, R., Ingham, E. and Fisher, J.: J. Mater. Sci. (In Press).

Key Engineering Materials Vols. 254-256(2004) pp. 667-670
online at http://www.scientific.net
© 2004 Trans Tech Publications, Switzerland

Tricalcium-Phosphate/Hydroxyapatite Bone Graft Extender for Use in Impaction Grafting Revision Surgery – An in Vitro Study in Human Femora

E.H. van Haaren[1, 2], T.H. Smit[1, 3], K. Phipps[4], P.I.J.M. Wuisman[1, 2], G. Blunn[4], G.M. Insley[6,] I.C. Heyligers[, 2, 5]

[1] Skeletal Tissue Engineering Group Amsterdam, VU University Medical Center, P.O. Box 7057, 1007 MB Amsterdam, The Netherlands, e-mail: EH.vanHaaren@vumc.nl
[2] Department of Orthopaedic Surgery, VU University Medical Center, P.O. Box 7057, 1007 MB Amsterdam, The Netherlands
[3] Department of Physics and Medical Technology, VU University Medical Center, P.O. Box 7057, 1007 MB Amsterdam, The Netherlands, e-mail: th.smit@vumc.nl
[4] The Center for Biomedical Engineering, Institute of Orthopaedics and Musculo-Skeletal Science, University College London, Stanmore, United Kingdom
[5] Department of Orthopaedic Surgery, Atrium Medical Center, P.O. Box 4446, 6401 CX Heerlen, The Netherlands
[6] Stryker Howmedica Osteonics, Raheen Business Park, Limerick, Ireland
gerard.insley@emea.strykercorp.com

Keywords: Hip, surgical revision, bone grafting, calcium phosphates, materials testing

Abstract. Impacted morsellised allografts have successfully been used to address the problem of poor bone stock in revision surgery. However, concern exists about pathogen transmission, high cost and shortage of supply of donor bone. Bone graft extenders, such as tricalciumphosphate (TCP) and hydroxyapatite (HA), have been developed to minimize the use of donor bone. In a human cadaver model we evaluated the surgical and mechanical feasibility of a TCP/HA bone graft extender during impaction grafting revision surgery.

TCP/HA allograft mix increased the risk of producing a fissure in the femur during the impaction procedure, but provided a higher initial mechanical stability as compared to bone graft alone (subsidence ratio graft : TCP/HA = 2.34).

If surgeons are properly trained, this type of graft extender can be viable for impaction grafting revision surgery.

Introduction

Revision of total hip arthoplasty (THA) has become a mainstay of orthopaedic practice. This growth can be attributed to an increase of primary cases as well as a higher life expectancy of mankind [1]. A major challenge in revision surgery is the loss of bone stock in either the acetabulum or the femoral shaft. To date, this problem of poor bone stock has been successfully addressed by the use of morsellised bone allografts [2-7].

However, there are concerns associated with the use of donor bone: risk of pathogen transmission [8], high costs, and shortage of supply [9,10]. Galea *et al.* estimated that the demand of bone graft already exceeds the supply in the UK [11]. Thus, new products, known as bone graft extenders have been developed to decrease dependency on donor bone. Examples are tricalciumphosphate (TCP) and hydroxyapatite (HA), both with proven biocompatibility and the ability to act as an osteoconductive material [4, 12-14]. These materials can best be combined with donor bone in a 50:50 weight relation thus effectively reducing the required amount of donor bone by half. To date, this mixture has not been tested in the femur. Therefore, the purpose of this study was to compare a 50:50 mixture of allograft and TCP/HA to 100% allograft in a simulated revision of a THA.

Materials and Methods

Study design: Fourteen femora from seven human cadavers were used in this study. All femora underwent the same preparation to simulate the state of segmental bone loss encountered during a revision total hip arthroplasty. This involved performing an osteotomy through the collum as in a primary case, followed by removal of all spongiosa. The right femora were designated to the experimental group (equal amounts of TCP/HA bone graft extender and donor bone) and the left femora served as controls (100% human allograft bone). The characteristics of the TCP/HA bone graft extender (BoneSave™, Stryker® Howmedica Osteonics) were: particle size 2.0 - 4.0 mm; pore size 400 – 600 μm; interconnectivity pore size < 100 μm and porosity of 50%. Standard materials were used to perform the operations *in vitro*, consisting of a bone plug, a central guide wire, Simplex bone cement (Stryker® Howmedica Osteonics) and Exeter stems (Stryker® Howmedica Osteonics).

Mechanical testing: During the impaction procedure, applied impaction force and the number of strikes were measured with a specially modified sliding hammer [15,16]. Measurement frequency was 200 kHz.

After the surgical procedure, the distal ends of all the femora were embedded in bismuth with the artificial head positioned precisely above the center of the knee joint. To test mechanical stability the femora were placed in a hydraulic material testing device (Instron 8872, Instron® Corporation) and loaded under a compressive sinusoidal force of 400 to 2,000 N. This load was applied *via* a flat plate, allowing a horizontal shift of the femoral head under bending of the specimen. The loading frequency was 6 Hz and all specimens underwent a total of 50,000 cycles. The test was paused after 1,000 and 10,000 cycles for a rest period of 5 minutes. The rest periods were used to quantify total deformation and its components elastic and plastic deformation. The elastic deformation was defined as the amount of recovery during the rest period. By subtracting the elastic deformation from the total deformation, the plastic deformation (also referred as subsidence) was calculated.

Statistics: The Wilcoxon paired test and Mann-Whitney tests were used to analyse the impaction forces. All group data from the mechanical tests were analysed using a paired, two tailed Student t-test. Significance was set at $P \leq 0.05$.

Results

Surgical procedure: In specimens 1, 2 and 4 of the experimental group, a fissure developed in the proximal femur. No fissures were observed in the control group. The fissures were treated with three cerclage wires around the proximal cortex of the femur, after which the femoral stems could be inserted as planned.

The mean amount of allograft used in the control group was 75.1 gram (range 62.9-90.6 gram, SD 11.0). In the experimental group a mean amount of 70.7 gram (range 52.1-84.5 gram, SD 11.3) of graft was used. The amount of graft in the 100% graft group was significantly higher than in the TCP/HA mix group ($P= 0.05$). The amount of donor bone was decreased on average by 52.9 % (range 48.7 – 58.6 %).

Impaction force: The average impaction force in control group was 13.0 kN (95%-CI 10.8-15.2 kN) and 12.0 kN (95%-CI 8.2-15.8 kN) in the experimental group. There was no significant difference between the two groups (Wilcoxon test $P> 0.05$). The average number of impacts in the 100% graft group was 230 and in the TCP/HA mix group 263. The impaction forces of specimen one of the experimental group were significant higher as compared to the rest of the group (Mann Whitney tests $P <0.0001$).

Mechanical testing: The femurs reconstructed with 100% allograft produced higher plastic deformation after 10,000 cycles. Average subsidence in the control group was 2.31 mm (SD 1.89

mm) and 0.99 mm (SD 0.62 mm) in the experimental group. This difference was statistically significant (P= 0.048). Also, the femurs of the control group demonstrated greater variation in subsidence as shown by a higher standard deviation (table 1). The ratio of the average subsidence of the control over the experimental group is 2.34. This indicates that subsidence was more than two times greater in the experimental group.

Table 1: Plastic deformation (in mm) and ratio of graft over TCP/HA mix after 10,000 cycles.

Cadaver number	Graft	TCP/HA mix	Ratio Gr/TCP
1	0.13	0.21	0.63
2	2.34	1.59	1.47
3	3.74	1.72	2.18
4	0.89	0.49	1.82
5	5.53	1.54	3.59
6	0.84	0.49	1.72
7	2.69	0.86	3.12
Average	2.31 (SD 1.89)	0.99 (SD 0.62)	2.34

After the 1,000 as well as after the 9,000 cycle run (cumulative 10,000), the elastic deformation in the control group was higher than in the experimental group.
All postoperative radiographs showed well-positioned prostheses.

Discussion

Impacted bone grafting has shown good results in addressing bone defects in revision hip arthroplasties [2, 4-7]. However, the limited availability of donor bone and the risk of disease transmission necessitate the development of substitutes. Synthetic ceramics, for example TCP, HA and biphasic TCP/HA particles, have been suggested for this purpose. In experimental studies these materials have shown complete healing of bone defects [17,18,19].

TCP/HA mixed with bone (50:50 weight mixture) cannot be compressed as much as allograft bone, hardly deforms under loading, is much stiffer than bone graft, and shows almost no visco-elastic behaviour (elastic modus of 135 N/mm^2 for human bone grafts and 522 N/mm^2 for 80:20 TCP/HA) [20]. This is expressed in the lower amount of graft used in the experimental group as compared to the control group. Also, these characteristics of the ceramics may result in a greater force transmission to the cortex during the impaction procedure. Indeed, three out of the seven femurs operated with the TCP/HA / allograft mix developed a fissure during the proximal impaction procedure, whereas no fissure occurred in any of the specimens in the 100% allograft group. Consequently, we propose that smaller steps must be taken during the impaction procedure, that is, less material must be impacted with less force to avoid a high stress to the cortex. Another solution for addressing the increased risk of fissure development could lie in adapting the currently available impaction tools. For example, the sliding hammer can be adjusted with a lighter hammerhead or a shorter sliding rod. Also, surgeons should be informed and trained before they use TCP/HA mixtures with impaction grafting. Furthermore, we identified that there was an important learning curve associated with the use of the TCP/HA allograft mixture. This is shown by the fact that the impaction forces in specimen one of the experimental group were significantly higher compared to the rest, and data analyses showed that the operating surgeon adapted the amount of force used over time.

When revision hip surgery is performed, the initial mechanical stability is essential for bone healing and biological fixation. The plastic deformation (i.e. subsidence) in the allograft group after 10,000 cycles was statistically significantly higher than in the TCP/HA mix group. An average subsidence ratio of the control over the experimental group of 2.34 was calculated, indicating that the subsidence in the allograft group was more than twice as high than in the TCP/HA mix group. The standard deviation, an indicator of variation, of the graft group was three times higher. The subsidence in the TCP/HA mix group was thus also more reproducible. An explanation for the higher stability may lie in the earlier mentioned higher elastic modus of the TCP/HA particles as compared to allograft. Another explanation for this finding may lie in a greater cement penetration through the inter-particle space of TCP/HA particles as suggested by Bolder et al. [21], resulting in more solid construction.

In addition, the elastic deformation in the control group demonstrated to be greater than the experimental group after 1,000 and 10,000 cycles. These observations suggest less reversible deformation in the TCP/HA mix group as compared to the graft group.

Conclusions

In our study it has been shown that the use of TCP/HA as a graft extender gives a high initial mechanical stability compared to bone graft alone. Therefore, from a biomechanical point of view, we conclude that this type of TCP/HA is a viable graft extender for the use in impaction grafting of the femur. However, it is very important that surgeons are aware that TCP/HA mixtures need to be handled differently than 100% allograft. We recommend that surgeons receive thorough training, and that the impaction tools are adapted for the use with TCP/HA mixtures.

Acknowledgements
The authors like to thank Klaas Boshuizen for his technical assistance and Kimi Uegaki for critically reading the manuscript.

References
[1] H.Malchau, P.Herberts, T.Eisler and others: J.Bone Joint Surg.Am. Vol. 84-A Suppl 2:2-20. (2002), p. 2
[2] G.A.Gie, L.Linder, R.S.Ling and others: J.Bone Joint Surg.Br. Vol. 75 (1993), p. 14
[3] F.Piccaluga, D.Gonzalez, V, J.C.Encinas Fernandez and others: J.Bone Joint Surg.Br. Vol. 84 (2002), p. 544
[4] B.W.Schreurs, T.J.Slooff, P.Buma and others: J.Bone Joint Surg.Br. Vol. 80 (1998), p. 391
[5] M.Lind, N.Krarup, S.Mikkelsen and others: J.Arthroplasty Vol. 17 (2002), p. 158
[6] B.W.Schreurs, T.J.Slooff, J.W.Gardeniers and others: Clin.Orthop. (2001), p. 202
[7] G.B.Fetzer, J.J.Callaghan, J.E.Templeton and others: J.Arthroplasty Vol. 16 (2001), p. 195
[8] S.Sugihara, A.D.van Ginkel, T.U.Jiya and others: J.Bone Joint Surg.Br. Vol. 81 (1999), p. 336
[9] S.D.Cook, S.L.Salkeld, and A.B.Prewett: Spine Vol. 20 (1995), p. 1338
[10] E.U.Conrad, D.R.Gretch, K.R.Obermeyer and others: J.Bone Joint Surg.Am. Vol. 77 (1995), p. 214
[11] G.Galea, D.Kopman, and B.J.Graham: J.Bone Joint Surg.Br. Vol. 80 (1998), p. 595
[12] A.M.Gatti, D.Zaffe, and G.P.Poli: Biomaterials Vol. 11 (1990), p. 513
[13] T.Kitsugi, T.Yamamuro, T.Nakamura and others: Biomaterials Vol. 14 (1993), p. 216
[14] G.Zambonin and M.Grano: Biomaterials Vol. 16 (1995), p. 397
[15] K.Phipps, J.Saksena, S.Muirhead- Allwood and others: EORS 12 th Annual Meeting, Lausanne, Switzerland (2002), p.
[16] H.Oonishi: Biomaterials Vol. 12 (1991), p. 171
[17] I.Manjubala, M.Sivakumar, R.V.Sureshkumar and others: J.Biomed.Mater.Res. Vol. 63 (2002), p. 200
[18] K.D.Johnson, K.E.Frierson, T.S.Keller and others: J.Orthop.Res. Vol. 14 (1996), p. 351
[19] K.Nagahara, M.Isogai, K.Shibata and others: Int.J.Oral Maxillofac.Implants. Vol. 7 (1992), p. 72
[20] N.Verdonschot, C.T.van Hal, B.W.Schreurs and others: J.Biomed.Mater.Res. Vol. 58 (2001), p. 599
[21] S.B.Bolder, N.Verdonschot, B.W.Schreurs and others: Biomaterials Vol. 23 (2002), p. 659

Key Engineering Materials Vols. 254-256(2004) pp. 671-674
online at http://www.scientific.net
© 2004 Trans Tech Publications, Switzerland

Bone-bonding Ability of Zirconia Coated with Titanium and Hydroxyapatite under Load-bearing Conditions

T. Suzuki[1], S. Fujibayashi[1], Y. Nakagawa[1], I. Noda[2], T. Nakamura[1]

[1] Department of Orthopaedic Surgery, Kyoto University Graduate School of Medicine,
Shogoin-Kawaharacho 54, Sakyo-ku, Kyoto 606-8507, Japan
Email: tsts@kuhp.kyoto-u.ac.jp
[2] Bioceram Division, Kyocera Corp., Gamocho-Kawai 10-1, Gamogun, Shiga 529-1595, Japan

Keywords: Zirconia, cementless, load-bearing, total knee arthroplasty

Abstract. We evaluated the bone-bonding ability of HTOZ (hydroxyapatite and titanium on zirconia) under load-bearing conditions in animal experiments, as a pre-clinical study. Three kinds of specimens, HTOZ, HTOC (HA and Ti on Co-Cr alloy) and TOC (Ti on Co-Cr alloy), were implanted into the weight-bearing portion of the femoral condyles of six beagle dogs. Femurs were extracted 4 and 12 weeks after the implantation and examined mechanically, by pullout testing, and histologically by toluidine blue staining, SEM, and calculation of affinity index. The interfacial shear strengths (Mean ± SD) of HTOZ, HTOC, and TOC groups were 4.42 ± 0.453, 3.90 ± 0.903, and 4.08 ± 0.790 MPa at 4 weeks, and 6.82 ± 2.64, 6.00 ± 1.88, and 6.63 ± 1.63 MPa at 12 weeks. There were no significant differences in the interfacial shear strengths between the three groups at both times. Affinity indices (Mean ± SD) obtained from SEM images of the HTOZ, HTOC, and TOC groups were 49.6 ± 6.52, 43.3 ± 10.43, and 23.7 ± 3.95% at 4 weeks, and 55.0 ± 6.72, 51.5 ± 3.07, and 28.6 ± 4.09% at 12 weeks. HA-coated implants (HTOZ, HTOC) had significantly higher affinity indices than non-HA-coated implants (TOC) at both times. HTOZ has bone-bonding ability equivalent to HTOC and TOC. HTOZ is an excellent material for components of cementless joint prostheses.

Introduction

Compared with metallic materials, using ceramic materials such as zirconia as the bearing parts of joint prostheses can reduce the wear volume of polyethylene. However, ceramic femoral components have not been applied as widely in total knee arthroplasty as in femoral heads for total hip arthroplasty. One of the reasons for this is that there are no ceramic femoral components of the cementless type because it is technically difficult to effect the surface treatment as a porous coating by which ceramics can directly bond to bone. To solve this problem we have developed a new composite material for cementless ceramic components: hydroxyapatite (HA) and titanium on zirconia (HTOZ). The basic mechanical properties and biological safety of this material have been already reported [1, 2]. In the present study, we evaluated the bone-bonding ability of HTOZ under load-bearing conditions in animal experiments, as a pre-clinical study.

Materials and Methods

Three kinds of trapezoid-shaped implants (8.5 × 8.5 × 4.3 mm) including HTOZ, HTOC and TOC were prepared. HTOZ consists of zirconia ceramics that satisfy the ISO standard criteria (ISO 13356) as substrate, titanium (Ti) as a deep coating layer and HA as a superficial coating layer. HTOC (HA and Ti on Co-Cr alloy) consists of Co-Cr alloy as substrate, Ti as a deep coating layer and HA as a superficial coating layer. TOC (Ti on Co-Cr alloy) consists of Co-Cr alloy as substrate, and Ti as the coating layer. (Fig. 1) Ti coating was performed using the inert gas-shielded arc spray method [2]. The peak thickness of the Ti coating layer was determined to be 500 μm, and the surface roughness (Rmax) was approximately 360 μm. HA coating of the Ti coating layer (HTOZ, HTOC) was performed using the flame spray method. The thickness of the HA coating layer was less than 50μm (mean 20μm) (Fig. 2).

Fig. 1 Three kinds of trapezoid-shaped implants (8.5 × 8.5 × 4.3 mm) (HTOZ, HTOC, TOC)

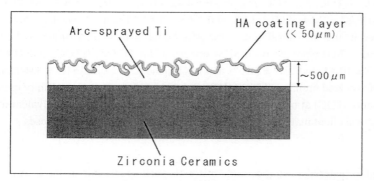

Fig. 2 Cross-section of HTOZ (hydroxyapatite and titanium on zirconia), consisting of zirconia ceramic as substrate, titanium (Ti) as the deep coating layer and hydroxyapatite (HA) as the superficial coating layer.

Implants were conventionally sterilized with ethylene oxide gas and implanted into the condyles of the femurs of six adult beagle dogs weighing 9.0–11.0 kg. The surgical methods have been described previously [3, 4]. Briefly, the anesthetized dogs were placed in the supine position and the knee was exposed via a medial parapatellar approach. The implantation sites in the weight-bearing portion of the medial and lateral femoral condyles were prepared using a specially designed broach with cutting surfaces, followed by a surgical electronic drill. An implant was then inserted into the hole by tapping in for press fitting. Both femurs were extracted following euthanasia at either 4 or 12 weeks after implantation and examined mechanically, by pullout testing, and histologically, by toluidine blue staining, SEM, and calculation of affinity index. At each time point, nine implants (3 HTOZ, 3 HTOC, 3 TOC) were selected at random from 12 implants, for pullout tests, and the other three (1 HTOZ, 1

HTOC, 1 TOC) were used for histological examinations only.

The pull-out tests were performed at a cross-head speed of 2.0 mm/min using an Instron-type autograph, taking care to ensure that the line of action of the pull-out force was parallel to the long axis of the implant. The detaching failure load was measured when the implant was detached from the bone. The failure load values were divided by the bone-implant interface area to obtain the interfacial shear strength. For the histological examination, new bone formation on the coated surface was evaluated by toluidine blue staining and SEM images. To calculate the affinity index from SEM photographs, the length of bone in direct contact with the coated surface with no intervening soft tissue was divided by the total length of the coated surface, and this value was multiplied by 100. Data were recorded as mean ± standard deviation and analyzed using a one-way factorial ANOVA with Fisher's PLSD testing as the *post hoc* test. Differences at $p < 0.05$ were considered to be statistically significant.

Results

The interfacial shear strengths (Mean ± SD) of the HTOZ, HTOC, and TOC implants, obtained by pull-out testing, were 4.42 ± 0.453, 3.90 ± 0.903, and 4.08 ± 0.790 MPa, respectively, at 4 weeks, and 6.82 ± 2.64, 6.00 ± 1.88, and 6.63 ± 1.63 MPa, respectively, at 12 weeks. There was no significant difference in the interfacial shear strengths between the three groups at both times. Histologically, new bone formation was observed on the surface of the coating layer of HTOZ, HTOC, and TOC implants in SEM and toluidine blue staining images at both 4 and 12 weeks. Affinity indices (Mean ± SD) obtained from the SEM images of the HTOZ, HTOC, and TOC implants were 49.6 ± 6.52, 43.3 ± 10.43, 23.7 ± 3.95% respectively at 4 weeks, and 55.0 ± 6.72, 51.5 ± 3.07, 28.6 ± 4.09% respectively at 12 weeks. At both times, the HA-coated implants, HTOZ and HTOC, had significantly higher affinity indices than the non-HA-coated implant (TOC).

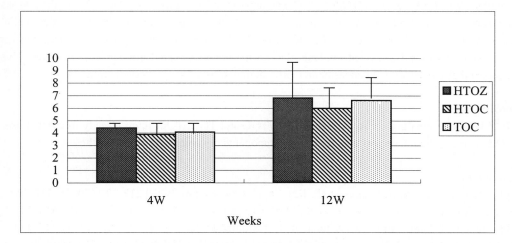

Fig. 3 Interfacial shear strengths (MPa) between bone and implants in pull-out tests at 4 and 12 weeks. There was no significant difference in the interfacial shear strength between the three groups at both times.

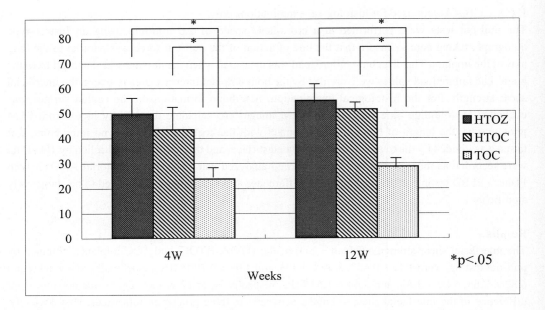

Fig. 4 Affinity indices obtained from SEM images of each implant at 4 and 12 weeks. HA-coated implants (HTOZ, HTOC) have significantly higher affinity indices than the non-HA-coated implant (TOC) at both times.

Discussion

The interfacial shear strength of all three types of implant were sufficiently high at both time points to consider that the implants bonded to bone. However, there was no significant difference in the shear strength between HA-coated (HTOZ, HTOC) and non-HA-coated implants (TOC). One of the reasons for this observation could be that the shear strength at pull-out test partially depend on the quality (maturity) of bone. Affinity indices obtained from SEM images showed significantly greater direct bone ingrowth to the HA-coated implants than to the non-HA-coated implant at both times. This suggested that HA coating enhances bone ingrowth due to its good osteoconduction [5].

Conclusions

HTOZ has bone-bonding ability equivalent to HTOC and TOC, TOC being already in clinical use. HTOZ is an excellent composite material for components of cementless joint prostheses.

References

[1] I. Noda, S. Masuda, H. Kitano, et al.: Key Engineering Materials Vol.218-220 (2002), 577-580
[2] A. Fujisawa, I. Noda, Y. Nishino, et al.: Material Science and Engineering C2 (1995), 151-157
[3] H. Takagi, T. Yamamuro, K. Hyakuna, et al.: J Biomed Mater Res. Vol.23, A2 (1989), 161-181
[4] Z.L. Li, T. Kitsugi, T. Yamamuro, et al.: J Biomed Mater Res. Vol.29 (1995), 1081-1088
[5] Y. Nakashima, K. Hayashi, T. Inadome, et al.: J Biomed Mater Res. Vol.35 (1997), 287-298

Key Engineering Materials Vols. 254-256(2004) pp. 675-678
online at http://www.scientific.net
© 2004 Trans Tech Publications, Switzerland

Next Generation Ceramics Based on Zirconia Toughened Alumina for Hip Joint Prostheses

G. M. Insley[1] and R. M. Streicher[2]

[1] Stryker Howmedica Osteonics, Raheen Business Park, Limerick, Ireland.
gerard.insley@emea.strykercorp.com
[2] Stryker SA, Florastrasse 13, CH-8800 Thalwil, Switzerland.
robert.streicher@strykereurope.com

Keywords: alumina, wear, total joint arthroplasty, microseparation

Abstract. Zirconia toughened alumina (ZTA) was tested as a candidate fourth generation alumina based ceramics for the bearing surfaces of orthopaedic hip joint replacements. The results show the ceramic has superior strength and fracture toughness to third generation alumina currently used for hip replacements. The material is chemically stable with no ageing noted in any of the testing. The wear testing showed the material to be as wear resistant as alumina in standard simulation mode. The ZTA ceramic wore significantly less than the alumina in the severe microseparation wear testing. Zirconia toughened alumina is a promising ceramic for the next generation of fracture and wear resistant ceramic articulation, for hip and eventually knee joint prostheses.

Introduction

Alumina exhibits excellent hardness and wear properties, however it is a brittle material with an inherent risk of fracture that even though is low, 1 in 2500 [1], this can be improved. The objective of this study was to investigate the feasibility of a new alumina based ceramic with improved toughness for artificial hip joint articulation against PE, itself or alumina whilst maintaining all other properties such as hardness, stability and chemical inertness. This objective can be achieved through the addition of zirconia to alumina. Addition of approximately 25% Zirconia to Alumina (Zirconia Toughened Alumina, ZTA) during the manufacturing process is promising to achieve the objectives [2].

Materials and Methods

Two candidate Zirconia Toughened Alumina (ZTA) ceramics were obtained from two manufacturers. The samples were supplied in the form of flat polished coupons, 45 x 4 x 3 mm bars and ball heads as well as inserts for modular cups with 28 mm internal diameter. For the mechanical evaluation 10 samples per test and for the wear test three samples each were used.

One ZTA (NZTA) had a composition of 75% Alumina and 25% Zirconia, the other one, (CZTA) had a composition of 74% Alumina, 24% Zirconia and 1% mixed oxides. To characterize the two new ZTAs several methods were used. Alumina served for all tests as a reference. Mechanical testing involved: hardness (HV), flexural strength (ASTM C1161) and indentation fracture toughness determination. X-ray diffraction (XRD) was used to measure the crystalline phase composition of the ZTAs and was used to monitor any transformation during ageing. Determination of stability was conducted by accelerated ageing (5 hrs at 134°C in a steam autoclave, equivalent to 20 years in vivo [3]) and real time ageing for one year in Ringer's solution at 37°C. Wear simulator testing and friction testing was carried out applying a six-station physiological hip simulator, described elsewhere. The simulator testing was done using standard conditions and in micro-separation mode that has been shown to achieve similar wear pattern as retrievals [4]. All test data was analyzed by descriptive statistics where applicable.

Results and Discussion

The mechanical test results and friction results obtained for the unaged ZTA specimens are summarized in table 1.

Table 1. Summary of mechanical test results for the candidate ceramics.

Test	Units	Alumina	CZTA	NZTA
Flexural Strength	MPa	466 ± 106	1203 ± 101	800 ± 131
Fracture toughness	MPa*m^{½}	2.78	4.1	4.1
Hardness	HV(30)	2041 ± 70	2014 ± 84	2043 ± 60
Friction factor	(f)	0.005	0.004	-

The two ZTA ceramics differ in the crystalline form of Zirconia seen at the surface when measured by XRD. NZTA contains Zirconia in a purely tetragonal crystalline form with no measurable monoclinic phase present, figure 1(a). CZTA contains Zirconia with up to 35% of the monoclinic phase present at the surface. This is shown as a peak at both 28.2° and 31.5° 2 theta, figure1(b). No further transformation in the Zirconia phase was observed after accelerated aging and up to 12 months real time aging for both ZTAs. This indicates the ZTA is a chemically stable ceramic.

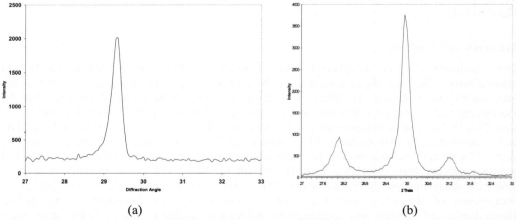

(a)   (b)

Fig.1(a) XRD spectrum for NZTA, 1(b) XRD spectrum for CZTA showing monoclinic peaks at 28.2° and 31.5° 2 theta.

The wear testing, in standard hip joint simulator mode, showed that both ZTA - ZTA couples articulating against themselves have lower wear than even alumina – alumina couples which has been reported before. Using a micro separation simulator set-up, similar results for both ZTA - ZTA combinations were obtained. Although the overall wear rate was increased in this mode, it also reached a steady state after 0.5 million cycles. The ZTAs wore in this test set-up again less than the alumina – alumina combination.

Conclusion

Zirconia Toughened Alumina is a promising ceramic for the next generation of fracture and wear resistant ceramic articulation for hip and eventually knee joint prostheses.

Acknowledgements

The Department of Mechanical Engineering, University of Leeds.

References

[1] Willmann, G, Orthopaedics International, vol. 2, (1994), pg. 66
[2] Wang, J, Stevens, R, Journal of Materials Science, vol. 24, (1989), pg.3421.
[3] Cales, B, Stefani, Y, Lilley, E, Journal of Biomedical Materials Research, vol. 28, (1994) p 619.
[4] Nevelos, J. E., Ingham, E., Doyle, C., Streicher, R., Nevelos, A. B., Walter, W. and Fisher, J.: J. Arthro. Vol.15 (2000), p.793.

Key Engineering Materials Vols. 254-256(2004) pp. 679-682
online at http://www.scientific.net
© *2004 Trans Tech Publications, Switzerland*

Silica/Calcium Phosphate Sol-Gel Derived Bone Grafting Material – From Animal Tests to First Clinical Experience

T. Traykova[1], R. Bötcher[2], H.-G. Neumann[3], K.-O. Henkel[4] ,V. Bienengraeber[4], Th.Gerber[1],

[1]Fachbereich Physik der Universitaet Rostock, University Platz 3; 18051 Rostock, Germany
trayko@physik1.uni-rostock.de
[2]Ohrdruf 99885, Germany
[3]DOT GmbH, Charles-Darwin Ring 1a, 18059 Rostock, Germany
[4]Klinik und Poliklinik fuer Mund-, Kiefer- und Gesichtschirurgie, Universitaet Rostock,
Strempelstr.13, 18057 Rostock, Germany

Keywords: Calcium phosphates, sol-gel method, bone formation, clinical study

Abstract. Calcium phosphate ceramics have attracted the attention of researchers, working in the field of synthetic bone grafts, because of their high biocompatibility, structural similarity with natural bone and fast biodegradation when implanted *in vivo*. The applications of calcium phosphate ceramics are constantly increasing and the areas of their usage are expanding. Together with the traditional applications in orthopaedics and dentistry, calcium phosphates are now successfully employed in new areas such as ophthalmology for making an artificial eye.

This paper describes the comparative *in vivo* study between three commercially available bioceramics (Endobone®, Cerasorb®, Targobone®), and a laboratory developed monophasic and biphasic calcium phosphate silica-based ceramics. The fast degradation and contribution in new bone formation of our ceramics, have given us confidence to proceed with clinical study, which is still continuing. From the primary results of the clinical survey we once again can prove the good material/bone integration and fast tissue growth around the medical intervention.

Introduction

The bioceramics market is expanding rapidly in parallel with the higher demands for bone graft substitute materials. Since the golden standard, the autograft, is still preferred by many surgeons, the developed bioceramics have to be highly bioactive, to posses osteogenic potential, to degrade easily in biological fluids and not to cause immunological inflammations. Therefore, the variety of the specially synthesized calcium phosphate ceramics to be used as bone graft substitute materials is very big. Although, all these ceramics are of the same family, there are differences between the product properties, such as crystallinity, porosity, phase purity, mechanical and thermal stability, etc., which are dependent on the required application of the material. We have designed a novel kind of calcium phosphate ceramic with high level of silica, which can be used as bone filler. Some previous investigations utilising these ceramics have reported good biodegradation, cell-mediated resorption and ability to support bone ingrowth [1]. A parallel *in vivo* study was performed, in which Endobone®, Cerasorb®, Targobone® were used. These commercial materials are well known on the market, however the results of their *in vivo* performances did not report biodegradation, substitution and bone formation in overcritically created defects. Thus, the aim of this study was to compare the osteoconductive potential between our designed ceramics and the commercial ceramics while they were implanted in lower jaws of mini pigs for eight months. The primary results from an ongoing clinical study with our bioceramics are as well included.

Materials and methods

The biomaterials, designed in our laboratory were prepared according to procedure described in [2]. Two kinds of bioceramics were prepared: monophasic bioceramic consisting of hydroxyapatite, and biphasic calcium phosphate ceramic, which consists of hydroxyapatite/tricalcium phosphate with 60/40 weight ratio. Both ceramics contain high level of silica: the monophasic 24 wt. %, and the biphasic 16 wt. %. Due to the presence of silica gel a small shrinkage during heating of the ceramics occurred, which led to the formation of macropores. In general, both ceramics have well expressed porosity (between 60 and 80%), with pore sizes varying from nanometer to milimeter range.

The commercial ceramics used for the study were Endobone®, Cerasorb® and Targobone®. Endobone® is a sintered hydroxyapatite, Cerasorb® is a beta-tricalcium phosphate, and Targobone® is an undenaturated bovine protein. All materials were sterilised prior the surgical procedure at 180°C for 1 hour.

For the *in vivo* study; adult, Goettinger mini pigs were used. Under general anaesthesia overcritical defects (5 cm^3) were created in the lower left jaw of the pigs. These defects were subsequently filled with the bioceramics. The comparative study continued eight months, afterward the animals were sacrificed and thin sections of the implanted sites were prepared for histological, X-ray and microscopy observations.

Results and Discussion

In a previous paper [3], we have mentioned about the structure of the newly designed ceramic material, which has an open crystal structure with three kinds of pores. It has been shown that fast resorption of our ceramics occurred within five weeks of implantation. The materials resorption has been proven to be a cell-mediated process. It has been reported as well, that the materials resorption done by osteoclasts is intimately related with new tissue growth (Fig. 1). Eight months after the surgical intervention, the defects in lower jaw look clinically healed when our designed ceramics were used (Fig. 2). Not the same is the picture when the commercial ceramics were involved.

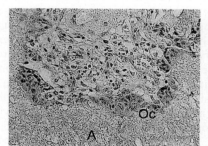

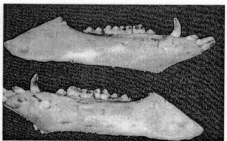

Fig. 1: Osteoclasts (Oc) digesting the ceramic surface (A), HE staining

Fig. 2: Clinically healed defect in 8 months after implantation of designed ceramic

In Figure 3 one can clearly see that the jaw defect, which was filled with Endobone® is clinically closed. Although, the material was substituted by new grown bone, it did not degrade (as indicated by the tweezers in Fig.3). Even worse situation appeared when Targobone® was implanted (Fig. 4). The defect is clinically not closed and appeared similar to the empty defect, used as a control group. Nevertheless biodegradation of Targobone® occurred, it did not guide new bone formation. Likewise, Cerasorb® have shown non-closed defect in eight months, although it has undergone degradation (Fig. 5). The Cerasorb® degradation has not been followed by bone formation.

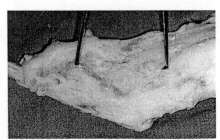 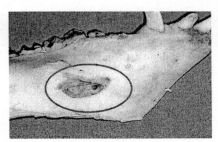

Fig. 3: Endobone® remained nondegraded in the defect (as indicated by tweezers)

Fig. 4: Fast degradation of Targobone® without subsequent bone formation, open defect in 8 months

In contrast, an image of the defect filled with our designed ceramics, appeared clinically healed (Fig. 2). There were as well negligible differences in the material remains between our designed ceramics. For example, monophasic ceramic has shown the fastest resorption and best results, therefore only 0.84 % of it were found as leftovers. In addition the results with biphasic ceramic have shown also good bone formation, but nearly 3,5 % of the materials remained in the defect. We believe that the variations are due to the higher content of silica in the monophasic ceramic composition.

Due to the materials characteristics there are clear differences in the *in vivo* behaviour between used implants. All three commercial ceramics did not support the new bone growth. Some of them like Targobone® degraded very fast, but no subsequent bone formation followed and the wound remained unclosed. We therefore, consider this case as the worst. Similar *in vivo* behaviour has been performed by Cerasorb®. Endobone® remained nondegraded in the defect after 8 months.

Biphasic calcium phosphate silica-based ceramic as a monolithic porous body was used in the clinical study. It was implanted as space filler in lower jaws, where direct contact between the ceramic and titanium dental rods occurred (Fig.6). Due to the relatively low mechanical strength of the ceramic material, it was possible to shape it additionally so that to fill all the empty spaces in the defect. The first X-ray examinations demonstrate good integration between the bioceramic and bone, as well fast tissue growth similar to the observed bone growth in the animal experiments. During the clinical study, which is still ongoing, no complications have been observed.

The silica presence in our bioceramics can be the factor, which led to their involvement in the bone formation process in lower jaws of mini pigs. On the other side, our ceramics as non-sintered possess very high level of porosity, which has been indicated as one of the main requirements for osseointegration. All these have proven once again the beneficial role of the designed mono- and biphasic calcium phosphate ceramics as bone graft substitutes as well for human beings. The *in vivo* performance of Endobone®, Cerasorb® and Targobone® has shown poor material degradation and inability of these materials to enhance bone formation in overcritical jaw defects.

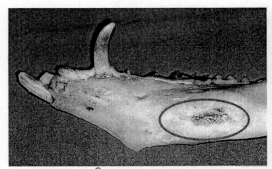

Fig. 5: Cerasorb® has left unclosed defect due to inability to support bone growth

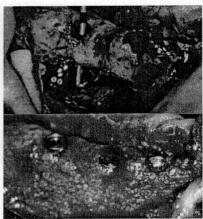

Fig. 6: Clinical study: an empty defect in the jaw (upper part), and bioceramics in contact with dental implants (lower part)

Summary

The comparative study have shown that only monophasic and biphasic calcium phosphate silica-based ceramics have supported the new tissue growth in the surgically created jaw defects. In the ongoing clinical study there is a clear evidence for the involvement of biphasic calcium phosphate silica-based ceramic in the bone formation.

References

[1] Th. Gerber, T. Traykova, K.-O. Henkel, V. Bienegraeber, M. Witt, J. Koewitz: Key Engineering Mat. Vols. 240-242 (2003), p.411

[2] Th. Gerber, T. Traykova, K.-O. Henkel, V. Bienengräber, J Sol-Gel Sci Techn, 26 (2003), p.1173

[3] Th. Gerber, T. Traykova, K.-O. Henkel, V. Bienegraeber: Key Engineering Mat. Vols. 218-220 (2002), p.399

Key Engineering Materials Vols. 254-256(2004) pp. 683-686
online at http://www.scientific.net
© *2004 Trans Tech Publications, Switzerland*

The Effect of Dental Grinding and Sandblasting on the Biaxial Flexural Strength and Weibull modulus of Tetragonal Zirconia

Tomaž Kosmač

Jožef Stefan Institute, Jamova 39, 1001 Ljubljana, Slovenia, tomaz.kosmac@ijs.si

Keywords: tetragonal zirconia; dental grinding; sandblasting; aging; strength and reliability

Abstract. The effects of dental grinding, sandblasting and aging on the biaxial flexural strength of Y-TZP ceramics containing 3 mol % yttria were evaluated. Dental grinding at a high rotation speed lowers the mean strength and reliability, whereas sandblasting may provide a powerful tool for surface strengthening. The coarse-grained materials were less susceptible to grinding induced defects, presumably due to higher fracture toughness. During aging at 140 ^{0}C for 24 hours, an about 100 μm thick layer has transformed to monoclinic on the surface of the standard-grade Y-TZP ceramics resulting in considerable strength degradation, whereas the "corrosion resistant" materials exhibited a much higher resistance toward the low-temperature degradation. Furthermore, pre-existing monoclinic zirconia in the surface of the sandblasted Y-TZP hinders the propagation of the diffusion-controlled transformation during subsequent exposure to aqueous environments.

Introduction

Growing demand for esthetic restorations in dentistry has led to the development of tooth-colored, metal-free systems. In most cases, yttria partially stabilized tetragonal zirconia (Y-TZP) is used as a core material, owing to its biocompatibility, acceptable esthetic appearance and superior mechanical properties. Compared to other dental ceramics, the superior strength, fracture toughness and damage tolerance of Y-TZP are due to a stress-induced transformation toughening mechanism operating in this particular class of ceramics [1]. Y-TZP is currently used as a core material in full-ceramic post-and-core systems, bridges, implant superstructures and orthodontic brackets [2-5]. These dental restorations are produced by dry- or wet-shaping of ceramic green bodies which are than sintered to high density. In order to achieve the perfect fit between the prosthetic work and the prepared tooth structure a final adjustment by dental grinding is usually required. Because Y-TZP ceramics exhibit a stress-induced transformation, the surface of the ground tetragonal zirconia ceramic will be transformed, i.e. constrained, and also damaged, which will influence the mechanical properties and reliability of the material. It has been shown that grinding induced surface flaws largely depend on the grinding parameters, such as diamond grit size, cutting depth, feed velocity and use of water coolant, and may become strength-determining if their length exceeds the depth of grinding-induced surface compressive layer [6-8]. Since sandblasting is commonly used to improve the bond between the luting agent and the prosthetic material, the effect of impact flaws on the reliability and properties of Y-TZP ceramics has been investigated too. Substantial strength improvement has been reported with this mechanical surface treatment indicating that impact-induced flaws are less detrimental to the strength of the tetragonal zirconia ceramics [7,8]. While most current dental ceramics exhibit an excellent chemical durability in situ, the transformability of Y-TZP ceramics under isothermal conditions requires attention. When exposed to an aqueous environment above 100 °C over long periods of time, Y-TZP ceramics start transforming spontaneously into the monoclinic structure [9]. This t-m transformation is diffusion-controlled and is accompanied by extensive micro cracking which ultimately leads to strength degradation [10,11].

The purpose of this study was to evaluate the effect of grinding and sandblasting on biaxial flexural strength and Weibull modulus of several Y-TZP ceramics, differing in the chemical composition and mean grain size. In addition, the susceptibility of pristine and mechanically treated Y-TZP ceramics to degradation in aqueous environment was investigated.

Experimental work

Materials were fabricated from commercially available ready-to-press Y-TZP powders containing 3 mol% yttria in the solid solution, using uniaxial dry pressing and pressureless sintering in air. Two standard grades (TZ-3YB and TZ-3YSB) were of the same chemical composition but they differed in their mean grain size, whereas the third material (TZ-3YSB-E) also contained a 0.25 % alumina addition to suppress the t-m transformation during aging. Starting powders and the main characteristics of the sintered materials are listed in Table 1. Also listed in Table 1 are some characteristics of an yttria-coated Y-TZP material (ZN-101) which has been included into the ageing experiments. The relative density of sintered specimens exceeded 99 % of the theoretical value and they were 100 % tetragonal. The difference in the mean strength between materials was statistically insignificant whereas the differences in the mean indentation toughness were found to be significant.

Table 1:   Starting Y-TZP powders, sintering conditions and main characteristics of sintered ceramics.

Starting powder grade	Sintering conditions	Mean grain size [μm]	Flexural strength [MPa (SD)]	K_{Ic} [MPa m$^{1/2}$ (SD)]
TZ – 3YB[*]	1500°C/2h	0.31	1021 (89.5)	4.45 (0.05)
TZ – 3YSB[*]	1550°C/4h	0.44	914 (58.3)	4.76 (0.11)
TZ – 3YSB-E[*]	1450°C/4h	0.51	1080 (75)	5.08 (0.10)
TZ – 3YSB-E[*]	1550°C/4h	0.59	990 (111)	5.18 (0.12)
ZN-101**	<1400°C	~0.25	n.m.	n.m.

* Tosoh, Tokyo, Japan
** CeramTech, Plochingen, Germany

After firing, the top surface of the disc-shaped specimens (15.5 ± 0.03 mm in diameter and 1.5 ± 0.03 mm thick) was subjected to a different surface treatment. A coarse grit (150 μm) and a fine grit (50 μm) diamond burr were chosen for dry and wet surface grinding, in order to simulate clinical conditions. The grinding load of about 100 g was exerted by finger pressure, the grinding speed was 150,000 rpm. For sandblasting, discs were mounted in a sample holder at a distance of 30 mm from the tip of the sandblaster unit, equipped with a nozzle of 5 mm in diameter. Samples were sandblasted for 15 seconds with 110 μm fused alumina particles at 4 bars. The stability of pristine and mechanically treated materials in an aqueous environment was tested under the conditions specified in ISO 6872:1996(E) for testing the chemical solubility of dental ceramics. After extraction by refluxing for 16 hours, specimens were analysed by XRD for phase composition and tested for strength. Aging of pristine and mechanically treated materials in an aqueous environment was performed under isothermal conditions at 140 °C for 24 hours. After aging, specimens were analysed by XRD for phase composition, whereas optical microscopy of polished cross-sections was used to determine the thickness of the transformed monoclinic zirconia.

Biaxial flexural strength measurements were performed according to ISO 6872 at a loading rate of 1 mm/min. Surface-treated specimens were fractured with the surface treated side under tension. The load to failure was recorded for each disc and the flexural strength was calculated using the equations of Wachtman et al. [12]. The variability of the flexural strength values was analyzed using the two-parameter Weibull distribution function.

Results and Discussion

The mean values of biaxial flexural strength, the respective standard deviation and Weibull modulus, m (which is the slope of the $\ln(\ln 1/1-P$ vs. $\ln \sigma$ plots) are graphically represented in Fig. 1.

Dental grinding evidently lowered the mean strength and reliability, whereas sandblasting provided a powerful tool for strengthening, though at the expense of somewhat lower reliability. Also, there is a strong grain-size dependence of the damage tolerance upon grinding. A fractographic examination indicated that failure of the ground samples originated from radial cracks extending several tens of μm (up to 50 μm) from the grinding grooves into the bulk of the material, whereas the failure of the sandblasted samples was initiated from a lateral crack linked to subsurface cracks [13]. The amount of transformed monoclinic zirconia on the surface of ground specimens was negligible, and so was the contribution of dental grinding to the surface strengthening. In contrast to grinding, sandblasting is capable of transforming a larger amount of zirconia at the surface of Y-TZP ceramics indicating lower temperatures during this operation. Surface flaws, which are introduced by sandblasting, do not seem to be strength determining, otherwise the strength of the material would have been reduced instead of being increased. The counteracting effect of dental grinding and sandblasting on flexural strength can thus be explained by considering two competing factors influencing the strength of surface treated Y-TZP ceramics: residual surface compressive stresses, which contribute to strengthening, and mechanically induced surface flaws, which cause strength degradation.

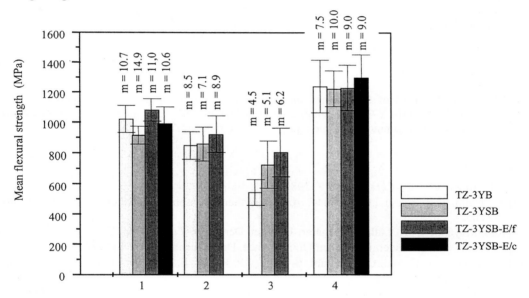

Fig. 1:   Mean biaxial flexural strength values for as-sintered and surface treated Y-TZP ceramics. 1- As sintered, 2- Dry ground (50 μm diamond burr), 3- Dry ground (150 μm diamond burr), 4- Sandblasted. Error bars represent one SD from the mean, m = Weibull modulus.

Although there was no measurable mass loss upon extraction by refluxing, about 8-12 % of monoclinic zirconia was detected on the extracted surfaces. However, the extraction time was not long enough to reveal statistically significant differences between materials and the extracting media and strength degradation has not yet occurred. These results are in line with other recently published studies on the chemical stability of tetragonal zirconia in an aqueous environment [14]. After autoclaving at 140 ^{0}C for 24 hours, however, an about 100 μm thick layer transformed on the surface of the standard-grade Y-TZP ceramics, resulting in almost 50 % strength reduction. The two "corrosion resistant" ceramics (E-grade from Tosoh and the ZN-101 material from CeramTech) exhibited a considerably higher corrosion resistance (Fig. 3a) and the no significant strength reduction was observed. Furthermore, as shown in Fig. 3b, pre-existing monoclinic zirconia in the surface of the sandblasted Y-TZP hinders the propagation of the diffusion-controlled transformation during subsequent exposure to aqueous environments.

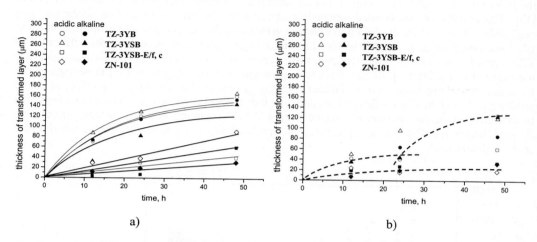

a)   b)

Fig. 2   The thickness of transformed monoclinic layer on the surface of as-sintered (a) and sandblasted (b) Y-TZP ceramics after autoclaving at 140°C in an acidic (4% acetic acid) and alkaline (ammonia solution, pH = 9.5) environment.

References

[1] E.C. Subbarao: Adv. Ceram., Vol. 3 (1981) pp.1.

[2] O. Keith, R.P. Kusy, J.Q. Whitley: Am. J. Orthod. Dentofacial. Orthop., Vol. 106 (1994) pp.605.

[3] K.H. Meyenberg , H. Lüthy, P. Schärer: J. Esthet. Dent. Vol. 7 (1995), pp.73.

[4] A. Wohlwend, S. Studer, P. Schärer: Quintessence. Dent. Technol., Vol. 1 (1997) pp.63.

[5] R. Luthardt, V. Herold, O. Sandkuhl, B. Reitz, J.P. Knaak, E. Lenz: Dtsch. Zahnarztl. Z., Vol. 53 (1998) pp.280.

[6] R.G. Luthardt, M. Holzhuter, O. Sandkuhl, V. Herold, J.D. Schnapp, E. Kuhlisch, M. Walter: J. Dent. Res., Vol. 81 (2002), pp. 487.

[7] T. Kosmač, Č. Oblak, P. Jevnikar, N. Funduk, L. Marion: Dent. Mater., Vol. 15 (1999), p. 426

[8] T. Kosmač, Č. Oblak, P. Jevnikar, N. Funduk, L. Marion: J. Biomed. Materi. Res., Vol. 53 (2000), pp. 304.

[9] T. Sato.,M. Shimada: J. Am. Ceram. Soc. Vol. 68 (1985), pp. 356.

[10] D.J. Kim: J. Euro. Ceram. Soc., Vol.17 (1997), pp. 897.

[11] J. L. Drummond, J. Am. Ceram. Soc., Vol. 72 (1989), pp. 675.

[12] J.B. Wachtman, W. Capps, J. Mandel: J. Mater. Sci., Vol. 7 (1972), pp.188.

[13] T. Kosmač, Key Eng. Mater., Vol. 223 (2002), pp. 181.

[14] B.I. Ardlin, Dent. Mater., Vol. 18 (2002), pp. 590.

Key Engineering Materials Vols. 254-256(2004) pp. 687-690
online at http://www.scientific.net
© *2004 Trans Tech Publications, Switzerland*

Rapid Prototyping Applications in the Treatment of Craniomaxillofacial Deformities - Utilization of Bioceramics

J. V. L. Silva[1], M. F. Gouvêia[1], A. Santa Barbara[1], E. Meurer[2], C.A.C. Zavaglia[3]

[1] Division for Product Development, Renato Archer Research Center P.O. Box 6162, 13083-970, Campinas/SP, Brazil, jorge.silva@cenpra.gov.br

[2] Post-graduating Program in Oral and Maxillofacial Surgery, PUCRS, Porto Alegre /RS, Brazil, emeurer@matrix.com.br

[3] Department of Materials Engineering, Mechanical Engineering Faculty, State University of Campinas, P.O. Box 6122, 13083-970, Campinas/SP, Brazil, czavaglia@uol.com.br

Keywords: Rapid prototyping, customized implants, biomaterials

Abstract. The aim of this article is to develop a methodology for the selection of suitable biomaterials for the treatment of craniofacial deformities using rapid prototyping as a support tool. This research is carried out by a multidisciplinary team of electronic, chemical and computer engineers, and surgeons. The main motivation for the team is to put available, disseminate and integrate computer systems, methodologies, and rapid prototyping usage in order to reduce postoperative costs and risks from the insufficiency of information. It is also useful for the effective surgical planning and for the selection of the suitable biomaterial to create a customized implant for the patient. The present work dealt with bioceramics for the production of customized implants considering small injuries.

Introduction

The indication for reconstruction of the cranial defects after trauma, tumor or infection are aesthetic reasons and protection of intracranial structures. Several techniques and materials have been described for fit these defects. The material must be biocompatible, strong and stable in the body and permit Computerized Tomography (CT) and Magnetic Resonance Imaging (MRI) diagnostics. Metal, acrylic resin, autogenous or lyophilized bone, have been used for this deal. Many problems are associated with these implants. Some of these problems are antigenic reactions of foreign body, irritation, infection and bone resorption. Researchers are still looking for the ideal material for cranioplasty. A new paradigm appeared with the development of bioceramics. Nowadays some bioceramics with acceptable mechanical properties are specifically designed for fit bone defects. Some of their advantages are easy manipulation, injectability, adaptability and no exothermic reaction.

A big problem in the surgical reconstruction either the material choice is the evaluation of the defect to be reconstructed. Not only the size of the defect is important to reestablish form and function, but also its three-dimensional shape.

In the last decades the computer hardware and software development associated to a significant growing in the mechanic precision and materials availability make possible the realization of virtual models in precise touch-and-feel models quickly creating the concept of rapid prototyping (RP) or solid freeform fabrication – SFF [1]. This technology became disseminated in many domains of knowledge as architecture, paleontology, biochemical and medical applications.

In the medical application field the data of a specific part of human body can be manipulated as tridimensional data and used to construct a biomodel in one of the many, rapid prototyping system, e.g. stereolithography, selective laser sintering and fused deposition modeling. This data set can be obtained from any medical imaging modality. In the cases discussed in this article the datasets utilized are obtained from a CT scanner that is more suitable for bone reconstruction for the surgical planning instead of soft tissue visualization.

The Methodology is as follows: 1) Dataset acquisition of the patient through CT; 2) Image processing and visualization using the *InVesalius* software and reconstruction of the anatomical structure of interest of the patient in a virtual 3D model and design of the mold for the implant; 3) Build a physical model in a rapid prototyping system from the virtual 3D model reconstructed; 4) Surgical planning, rehearsal on the model and selection of the more suitable biomaterial for the case; 5) Surgical intervention; 6) Follow up of the results in the postoperative. [1, 6]

Methodology

To carry out this work some diverse technologies were used and developed. The Methodology followed the steps described bellow:

Acquisition of the skull image using a spiral CT scanner. The dataset acquisition has to follow a basic protocol in order to have high quality images that will influence enormously in the final quality of the overall process. The Slice thickness shall be as thinner as possible – 1-2 mm is a good trade-off in a skull study. The slice depth shall be less than the thickness and a FOV – Field of View - with resolution of 512X512 and 25 Cm yields a good dataset. High contrast and none filters shall be used [1, 6].

The second step is the medical image processing to retrieve the structure of interest. This step is accomplished by the use of *InVesalius* Software developed at CenPRA to import data from medical scanners in DICOM [2] - Digital Imaging Communication – format that is an International standard for the interoperability among medical equipment and export 2D and 3D data in many formats. One of these formats exported is the STL a "de facto" standard to export data to rapid prototyping systems. The quality of the 3D model from the region of interest is of great importance for the surgical planning. To do so the InVesalius import DICOM dataset of a patient organizing it in series. The user can choose the slices that will be used and automatic correction of gantry/tilt and elimination of duplicated slices is done. The user can also visualize the 2D images in reconstructed sagittal, coronal, axial and oblique planes. The *InVesalius* is based on VTK [3] and Phyton [4] to render a friendless interface running in a regular PC. It implements the most efficient algorithms for tissue segmentation, slice interpolation, rendering and extraction of structures [Fig. 1a].

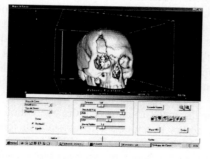 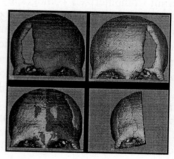

(a)   (b)

Fig.1. (a) View of the 3D model of the patient skull rendered with the *InVesalius* software; (b) Mirroring of the contra lateral morphology to generate a mold.

To produce a mold for implant customization it is necessary to use the STL file of the structure of interest, in this case a skull, generated using *InVesalius* [Fig. 1a]. This file is imported into a

CAD – Computer-Aided Design – System, mirrored using the contralateral and by means of booleans operations the skull hole is filled generating a model of the implant [Fig. 1b]. With the model of the implant that fills the hole it is used another CAD system to create a tool with cavity to conform this part in a bioceramic material to be implanted. Until this step all of operations are done virtually in a computer [Fig. 2a].

In the third step the skull models, molds and the parts that cover the skull holes were reproduced in a physical model using the rapid prototyping SLS- Selective Laser Sintering – process [Fig. 2b]. All of the rapid prototyping processes are based on paradigm of model construction growing layer-by-layer. The SLS process is based on a CO_2 Laser that imprints each layer continuously in a surface covered by a thin layer of preheated polymer until sintering the whole object. The figure 2 shows the skull, mold – to reproduce the implant, and model of the implant in its place. The result is a mold for the implant customization.

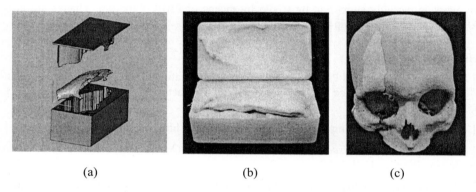

(a)   (b)   (c)

Fig. 2. (a) The mold is created in a CAD software; (b) Rapid prototyping technology is used to build this mold and the implant replica; (c) Using the mold or the biomodel the implant can be build in several materials, including bioceramics. The anatomy is reconstructed perfectly.

The fourth step is the surgical planning and surgical intervention. In the past, the exact fitting of the cranioplasty could not be achieved due to a lack of accuracy of the manufacturing process, resulting in prolonged operation time and in some cases poor results. In order to enhance the surgical planning the virtual and physical biomodel are very useful. They can improve the communication between surgical team members, radiologist, doctors and the patient. The mirror imaging techniques can also be used to design a mould from the contra-lateral side of the skull and a precise fitting implant can be fabricated to re-establish skull contours. [5]

This biomodels can be used as an accurate anatomic guide, increasing safety, indicating the osteotomy sites for the surgery, minimizing blood loss, help to fitting miniplates, increase the understanding of the procedure and promote better results. All these advantages permit reduce patient's discomfort, the number of reconstructive surgeries, improving the speed of patient recovery and by the end save costs for the national health system.

Final considerations

Bioceramics can be used in these reconstruction surgeries. Calcium phosphate ceramics like hydroxyapatite and β- tricalcium phosphate are indicated for small defects. For larger defects it's possible to use zirconia, zirconia-hydroxyapatite and zirconia-alumina composites. Another possibility is to use of carbon fibers reinforced calcium phosphate cement [7]. Some of these bioceramics are been tested now, for these application.

With the possibility to inject bioceramics in the molds building by rapid prototyping technologies we can obtain a perfect shaped biocompatible implant. The bioceramic also can be shaped directly in a biomodel with the same result, and in the future maybe we can use bioceramics as a material to build implants directly in a layer by layer manufacturing machine.

Association of CAD-CAM technologies and bioceramics has a great potential. The use of these two technologies can provide a great advance in the management of craniomaxillofacial deformities and has a radical impact on the reduction of the time spent in surgeries, in the cost of a perioperative procedure and in the final function and aesthetic result. Finally, and the most important, it provides the best physical, psychological and social recovery, to the patient.

Acknowledgements

This work is partially supported by CNPq by means of PIBIC Program. Thanks are also due to FAPERGS for its scholarship program.

References

[1] J.V.L Silva, *et al.* Rapid Prototyping: Concepts, Applications and Potential Utilization in Brazil, 15th International Conference in Cad/Cam Robotics and Factories of the Future, Águas de Lindóia, Sp, Brazil, 1999.

[2] DICOM Standards Committee. DICOM Home Page. <http://medical.nema.org/>

[3] Schroeder *et al. The Visualization Toolkit: An Object-Oriented Approach to 3D Graphics*. 2ed, New Jersey: Prentice Hall, 1998.

[4] M. Lutz Programming Python: *Solutions for Python Programmers*. 2ed. Sebastopol, CA: O'Reilly & Associates, 2001.

[5] M. Perry, *et al:* The use of computer-generated three-dimensional models in orbital reconstruction. British Journal of Oral and Maxillofacial Surgery, Edinburgh, Vol. 36, n. 4, p. 275-284, Aug. 1998

[6] E. Meurer, *et al* : Rapid Prototyping Biomodels in Oral and Maxillofacial Surgery, PUCRS, Porto Alegre, RS, Brazil, 2002. 237p.

[7] L.A. Santos: Desenvolvimento de Cimento de Fosfato de Cálcio Reforçado por Fibras par Uso na Área Médico-Odontológica, tese (doutorado), DEMA, FEM, UNICAMP, Campinas, SP, Brazil, 2001, 243p.

Key Engineering Materials Vols. 254-256(2004) pp. 691-694
online at http://www.scientific.net
© *2004 Trans Tech Publications, Switzerland*

Clinical Investigation on Bioactive Glass Particles
for Dental Bone Defects

A.M. Gatti, E. Monari, L. Simonetti

INFM- Laboratory of Biomaterials, Dip. di Neuroscienze, Testa Collo, Riabilitazione, University of Modena and Reggio Emilia, Via del Pozzo, 71-4100 Modena Italy, gatti @unimo.it

Keywords: Dentistry, bioactive glasses, bone defect.

Abstract. The research investigates the bioactive behaviour of a glass used as granules in dental bone defects. The materials was inserted in post-extractive defects to induce a bone growth for following dental implants. During the surgical manouvre of implantation a bone biopsy was extracted and analyzed hystologically and under Electron Scanning Microscope. The observations showed the integration of the granules in a new bone grown environment. Some of them appeared already degraded after six months from implantation.

Introduction

Bioactive materials are the most interesting materials of 80's. The first of them was the so-called Bioglass that was developed at the University of Florida by Prof. L.L. Hench, considered the father of active glasses [1, 2]. He, first, developed glasses with specific compositions that, when put in contact with biological fluids, cells or tissues, can degrade and the released products can interact positively with the surrounding environment creating the boundary conditions for the new bone deposition. There is the creation of a CaP rich layer that induces the deposition of osteoblasts and a bone formation in a shorter time than the natural one. The original silicate disappears and becomes a calcium-phosphate matrix that will be remodeled by bone. Others researchers developed other similar glasses to emphasize the bioactivity and propose new applications of this class of materials. [3, 4, 5, 6]

Materials and Methods

Bioglass sterilized granules were implanted in 4 human post-extractive defects in jaw for bone augmentation for following dental implants. The bioglass (Perioglass©, Novabone Ltd. Aluchua, USA) has the following composition in weight %: 45 SiO_2, 24.5 Na_2O, 24.5 CaO, 6 P_2O_5 and the granule size ranges from 150 to 600 µm. After 6 months the zone was drilled for the implantation of a titanium device and a bone biopsy was taken away.

Every patient was informed about all the procedures and signed the Informed Consensus.

X-Ray radiographs were carried out before and after the filling of the defect and after 6 months at the time of the implant insertion, in order to verify the behavior of the material and its degradation. The biopsies in the implantation zones were fixed in 4% buffered para-formaldheyde and dehydrated in ethanol. After embedding in poly-methyl-methacrylate, the samples were sectioned by a diamond saw (Accutom, by Struers, Denmark) in 200- µm thick sections for the Scanning Electron Microscopy (ESEM-QUANTA by FEI Company, The Netherlands). The SEM observations were performed in secondary emitted and backscattered mode at high and low vacuum. Elemental analyses and x-ray dot maps were carried out with an Energy Dispersive System (EDS, by EDAX, USA) in order to detect the elements of glass granules after implantation. For the histological observations 20-µm thin sections were obtained with an automatic microtome (Leitz, Germany). These non-decalcified sections were stained with hematoxylin-eosin, suitable to show the calcium deposition and observed under optical microscope (Nikon, Japan) to observe the interface bone-glass granule created.

Results

The four patients, treated with glass granules in post-extractive bone defects (Fig. 1a), after 6 months, did not show any clinical evidence of inflammation. The X-ray images showed the complete integration of the material in all the cases. From the X-ray density a bone remodelling is suspected. In fact the X-ray images of the filled area, carried out immediately after the implantation (Fig. 1 b), showed a greater X-ray opacity than that after the 6 month implantation (Fig. 1c).

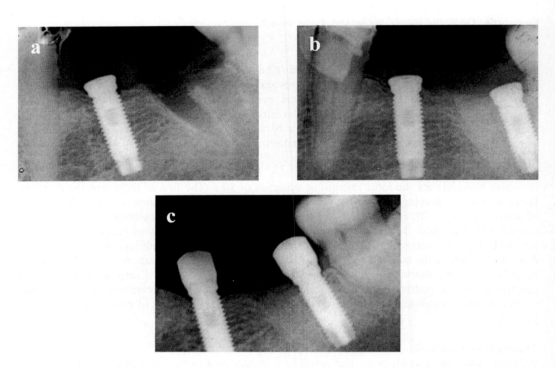

Fig. 1. X-ray images of the different steps of implantation of Perioglass©. a=post-extractive site, b= after the bone filling with glass granules, c=the same site view after 6-month implantation.

The histological observations of the bone biopsy after 6 months showed the glass granules surrounded by reactive connective tissue. Histological images show there is no inflammation in the site of implantation. No macrophages or giant cells are present. (Fig. 2).

Some islands of trabecular bone grew directly on the surface of the glass particles. The low magnification SEM observations showed bone on the apical part with zones of mineralization. The granules appear whiter since they are electronically denser than the bone. On the ridge, islands of trabecular bone grew directly on the granules. In one degraded granule the formation of a pouch was noted and inside that a tissue with the same electronic density of newly formed bone was observed.

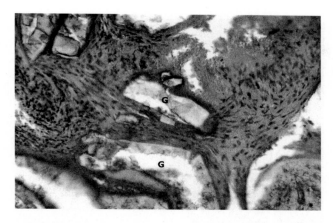

Fig. 2 – Histological view of some glass granules in the middle of the biopsy
(Original magnification 40x, hematoxylin-eosin)

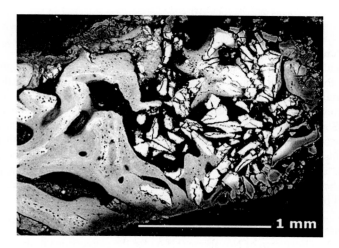

Fig. 3 - SEM image of the bone biopsy with the granules at low magnification. Some of them are
completely surrounded by new formed bone and show fractures running along their core.

Fig. 3 shows the SEM image of the bone biopsy with the granules at low magnification. Some
granules appear to be surrounded by the new formed bone creating an intimate interface with that.
Some granules appear homogenously coloured (white); in others two different zones are visible (an
outer dark layer and a whiter core. (Fig. 4).The EDS analyses showed that some granules degraded
and are already transformed in Calcium-phosphate compounds; others, still surrounded by soft
tissue, show a core with the original composition and an outer layer rich in Calcium and Phosphorus
that interact with the cells creating the natural matrix for the osteocytes deposition.
In another case the new bone formation was less than the presented case.

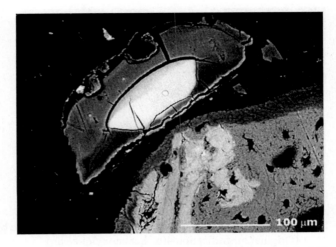

Fig. 4 – SEM image of a glass granule and bone. The granule shows two different zones: a not-degraded core with the original composition and an outer layer composed of Ca and P.

Discussion and Conclusions

The present clinical application of bioactive glass granules for the bone augmentation in post extractive defects for a following implant insertion resulted very successful. After 6 months from implantation a certain quantity of newly formed bone was detected in the site of the bone defect. The histological view showed the absence of inflammatory reactions around the degraded granules. The SEM and EDS analyses explained how the bioactivity of this material works. The granules help to maintain the space immediately after their implantation, but also they start to degrade releasing ions. A contra diffusion from the biological environment starts to compensate the glass diffusion. A calcium-phosphate rich layer is created in the surface of the granules that helps the osteocytes colonization and the new bone formation, chemically connected to the material. A remodelling of this phosphate will occur giving the disappearance of the original material.

Acknowledgements

The authors thank Dr. Greenspan of Novabone (USA) for the suggestions in the developing of this research.

References

[1] L.L. Hench, R.J:Splinter, W.C.Allen, T.K.Greenlee, J.Biomed Mater.Res. Vol. 5, (1972), p117.
[2] Hench LL. In: Mitchell J, ed. *Bone Grafts and Bone Substitutes*, (W.B. Saunders Company, 1992)
[3] A.M. Gatti, L.L. Hench, T. Yamamuro, O. Andersson , Cells and Materials, Vol. 3(3), (1993), p 283.
[4] A.M. Gatti, G. Valdrè, O. Andersson, Biomaterials, Vol. 15(3), (1994) p.208.
[5] A.M. Gatti, G. Valdrè, A. Tombesi, J. of Biomedical Mater. Res., Vol. 31(4),(1996) p.475.
[6] A.M.Gatti, E. Monari, *New Biomedical Materials: Basic and Applied Studies*, Ed. Haris, P.I., Chapman D., IOS Press Amsterdam, (1998), p73.

Key Engineering Materials Vols. 254-256(2004) pp. 695-698
online at http://www.scientific.net

Apatite Remineralization: *In Vivo* Long Term Study In Dental Tissue

R. L. Mourão[1]; H. S Mansur[2]; F.R Tay[3] and L. DLanza[1]

[1]Department of Restorative Dentistry, Federal University of Minas Gerais, Brazil,

[2]Department of Metallurgy and Material Engineering, Federal University of Minas Gerais, Brazil,

[3]Department of Children's Dentistry and Orthodontics and Oral Biology, Faculty Dentistry, The University of Hong Kong,

hmansur@demet.ufmg.br

Keywords: Dental tissue, TEM, dental materials, electron microscopy

Abstract. Sites of incomplete resin infiltration in acid-etched dentin expedite water sorption, leaching of hydrophilic resin components, or hydrolytic degradation of the denuded collagen matrices that subsequently result in deterioration of these bonds after aging. Akimoto [1], recently reported that these sites can be remineralized *in vivo* in primate teeth after a 6-month period. The objective of this study was to validate these findings using *in vivo* human specimens. Unexposed, buccal class V preparations with enamel cavosurface margins were performed in 16 human premolars scheduled for orthodontic extractions, under a protocol approved by the ethical committee, of Federal University of Minas Gerais, Brazil. Half of the samples were acid-etched with phosphoric acid for 15s, intentionally air-dried for 10 s and dry-bonded (over-dried group OD), while the other half was acid-etched for 60 s and then moist-bonded (over-etched group OE), to create zones of incomplete resin infiltration within the acid-etched dentin. The cavities were bonded using Single Bond® (3M) and restored with Protect Linner F® (Kuraray). Half of the restored specimens were retrieved after 10 min, and the other half at 6 months, was used the ammoniacal silver nitrate pH 9.5 as a tracer and than the samples were processed to be examined with transmission electron microscope (TEM). Stained, demineralized sections (SS) revealed 2 mm thick hybrid layers in OD and 13 mm thick hybrid layers in OE. Incompletely infiltrated regions within the acid-etched dentin were identified by zones of the extensive silver deposition from unstained, undemineralized sections (UUS), OD and OE. Similar zones were seen in 10 min and 6-month specimens, with no ultrastructural evidence of apatite deposition within the hybrid layers. The results showed no evidence of remineralization, besides that, we could speculate that collagen fibrils without the protection of adhesive resin, storage in oral environment, could be digested by enzymes as metalloproteinases (MMPs)[2]. The long time evaluation added to *in vivo* research can give us these answers or may be a better approach. It is concluded that incompletely infiltrated, acid-etched human dentin did not remineralize *in vivo* after a 6-month period.

Introduction

Biomaterials have experienced a very impressive growth in the last 2 to 3 decades. The dental and bone tissue replacements have been the driving force to novel materials, mainly based on bioceramics, due to hidroxyapatite compatibility to these hard tissues. Despite many achievements in this field, a major problem is yet to be fully understood: the hard tissue-biomaterial interface. In the present work, we have moved a step further on the apatite remineralization process at the dentin-restoration interface. Most adhesive interface studies have involved transmission electron microscope (TEM) demonstration of the penetration of adhesive resins into demineralized dentin surfaces with subsequent creations of hybrid layer [3]. It had been observed the presence of nanoleakage in bond interfaces, first notes by Sano, [4,5]. Nanoleakage is a term that describes the diffusion of small ions or molecules within the hybrid layer in the absence of gap formation using a tracer as ammoniacal silver nitrate. We still do not know exactly, the behavior and its consequence of these voids inside the hybrid layer with a long-term evaluation. In this study we intend to evaluate the *in vivo* dentin-resin interface, during a 6 months period of time.

Materials and Methods

Samples of 16 erupted pre-molars, non-carious, not fractured or restored were selected in 7 patients that needed the extraction for orthodontics procedures, besides that, they must be scheduled for the extraction, 6 month after finishing the restorative procedures. Another 2 pre-molars were storage in T-chloramines at $4^{o}C$ for a period of 30 days to do the *in vitro* control group. This *in vivo* research was approved by ethical committee of Faculty of Dentistry of Federal University of Minas Gerais. Before start the restorative procedures the patient were anesthetized and were used rubber dam for all specimens. Class V was performed on the lingual surface, with enamel margins, was used a diamond point n° 2294 with a stop to standardize the depth of the cavity, 2 mm, under air/water refrigeration. The specimens were divided into 3 groups. The 2 *in vitro* specimens were prepared as manufacturer instructions. The other 16 *in vivo* specimens were divided in 2 groups of 8 teeth each: Group 2 – over-etched technique, the cavities were acid-etched for 60 sec, rinsed for 30 sec with water and the excesses carefully removed by reabsorbing paper. Group 3- over-dry technique the cavities were acid-etched for 60 sec, rinsed for 30 sec and over dry with an air syringe for 10sec. For all groups were used Uni-Etch Bisco, phosphoric acid 32%, Protec Liner Kuraray, incrementally as a restorative material. The adhesive and restorative resin were light-cured for 20 sec (each increment), using Optilux Demetron® with source intensity of 500 ± 50 mW/cm^2. In groups 2 and 3, half of the specimens (4) were retrieved 10 min after finishing the restorative procedures and the other half (4) 6 months later. The specimens were stored in glutaraldehyde in phosphate buffer solution (2.5% PBS/0.1M, pH=7.3). The samples were prepared for TEM analysis. The specimens were sectioned in strips approximately 2 mm wide, perpendicular to bond interface using a diamond-impregnate copper disc under copious water supply. The surface was coated with nail varnish, except for the 1mm around the interface and immersed in 50% (w/v) of ammoniacal silver nitrate solution for 24h, then were soaked in photo developing solution under fluorescent light for 8h. The strip was completely demineralized in an aqueous solution of 0.1 M ethylene diamine tetra-acetic acid (EDTA) that was buffered with sodium hydroxide to a pH of 7. After demineralization, the specimens were fixed in karnovisky`s fixative(2.5% glutaraldehyde and 2% paraformaldehyde in 0.1 M cacodylate buffer, pH 7.3) for a minimum of 4h and rinsed thoroughly with 0.1M sodium cacodylate buffer. Post-fixation was performed with 1% osmium tetroxide in 0.1 M cacodylate buffer, (pH 7) for 1h at room temperature. The bonded dentin strips were then rinsed three times in cacodylate buffer. They were dehydrated in an ascending ethanol series (30% to 100%), immersed in propylene oxide as a transition fluid, embedded in epoxy resin (TAAB 812 resin, TAAB laboratories, Aldermaston,UK). Two 2x2 mm blocks were trimmed and re-embedded in epoxy resin to ensure proper orientation of the resin dentin interface. Following initial screening of all the semi-thin sections from each group, representative 1x1 mm ultra-thin sections about 90 nm thick were prepared with an ultramicrotome (Reichert Ultracut S, Leica, Vienna, Austria) using a diamond Knife and collected on 100-mesh copper grids. Samples were double stained with 2% uranyl acetate for 10 min followed by Reynold`s lead citrate for an additional 5 min. After drying, they were examined with a TEM (JEM-100, JEOL, Akishima, Japan) operating at 80KV [6]. Both unstained and stained, demineralized ultra-thin sections were examined with TEM. Photomicrographs were collected and evaluated regarding to the dental tissue-composite resin interface and morphology. Resin-infiltrate hybrid layer was investigated depending on acid etching time and the dentin substrate under wet or dry conditions.

Results and Discussion

Representative TEM micrographs of demineralized specimens from Single Bond are illustrated, where sections were unstained (Fig.1, 2, 3, 4, and Fig.6) and stained (Fig.5 and Fig.7). Th e evaluation of formation of hybrid layer was better exposed with the use of (SS) [7]. On the stained photomicrographs from group OE, it was noted a thicker hybrid layer, 10-12μm, (Fig.3 and Fig.6) compared to control group (CG), (Fig.1) and to OD group 5-6μm (Fig.2 and Fig.4). The reason for that was the use of a phosphoric acid (32%) for a longer period of time, 60 sec, than the recommended by the manufacturer, causing a deeper demineralization [8,9]. The results of over-etched, *in vitro* specimens in literature, done in the same conditions as our research, shows an even thicker hybrid layer, 15-16μm, compared to our *in vivo* results. We can speculate that *in vivo* experiments could have been neutralized by the acid, much faster than *in vitro* conditions, caused by the permeability released from pulp-dentin complex. In contrary in OD group the period used for demineralization was 15s, the one recommended by the manufacturer, but the use of the air syringe really hard collapsed the dentin matrix collagen and the adhesive resin could not infiltrate completely, forming a thin hybrid layer. In both groups we wanted to leave an incompletely demineralized dentin to compare the behavior of this region 6 months later in mouth environment, under oclusal strength, expose to enzymes and mastigatory stress [4,5,10,11].

In unstained undemineralized sections (UUS) was seen the presence of reticular nanoleakage pattern in all situations, OD, OE and control group (Figs.1, 2, 3, 4 and Fig.6). The adhesive system was not able to infiltrate completely in any situation. The higher density of nanoleakage was observed in OE group 6 months evaluation (Fig.6). We may assume that the phosphoric acid reduces the pH in the interface causing the activation mechanism of metalloproteinases (MMP) enzymes capable to digest collagen (pH-dependent). These MMPs are a family of enzymes, which, in concept, are capable of degrading collagen [12]. These authors have noted that such enzymes were activated at low pH 4.5, where the acid etched solution used in bonding procedure is responsible for such activation. We also saw the spotted nanoleakage pattern, only observed with the use of ammoniacal silver nitrate tracer [13] (Fig.1, Fig.6 and Fig.2). This kind of nanoleakage is different from the one defined by Sano in 1994, this one shows in hydrophilic domains the incomplete light cure of the resin.

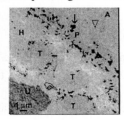
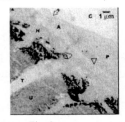
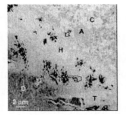

Fig. 1   Fig. 2   Fig. 3

Fig. 1 Control group (UUS), the lowest density of voids (pointer). Fig. 2 (OD) group, 10 min retrieved time. It can be observed a thin hybrid layer (H), between the arrows, and a high density of nanoleakage reticular pattern. Fig. 3 (UUS). OE group, 10 min retrieved time. A thick hybrid layer, between the arrows (H) a high density of reticular pattern of nanoleakage, voids, (pointer). (black arrow) spotted nanoleakage Pattern (Hydrophilic domains), adhesive layer (A), resin composite(C), dentin (U) dentinal tubule (T).

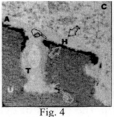 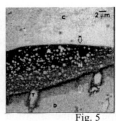 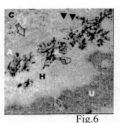 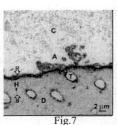

| Fig. 4 | Fig. 5 | Fig.6 | Fig.7 |

Fig.4 (UUS). OD group, 6 month retrieved time. It can be observed a thin hybrid layer, between the arrows, reticular pattern of nanoleakage, voids, (pointer). Fig.5. (SS) shows a thin Hybrid layer (H) and (P) polyalcenoic copolymer acid. Fig. 6 (UUS). OE group, 6 month retrieved time. A thick hybrid layer, between the arrows (H) a high density of reticular pattern of nanoleakage, voids, (pointer), (black arrow) spotted nanoleakage Pattern (Hydrophilic domains). Fig. 7 (SS) shows a thick hybrid layer. Adhesive layer (A), resin composite(C), dentin (U).

Conclusions
The adhesive system used was not able to infiltrate completely into demineralized dentin in any situation investigated. By changing the treatment of the dentin substrate we increased the density of nanoleakage in the hybrid layer. No evidence of apatite deposition was found.

References

[1] N.Akimoto, G.Yokoyama, K. Ohmori, S. Suzuki, A. Kohno, C. F. Cox: Quintessence Int., Vol.32 (2001), p. 561.

[2]M. Hashimoto, H.Ohno, M. Kaga, K. Endo, H. Sano, H, Oguchi: Dent. Mater., Vol.79 (2000), p.1385.

[3] F. R. Tay; A. J. Gwinnet, K. M.Pang, S. H. Y. Wei: J. Dent. Res., Vol. 75 (1996), p. 1034.

[4]H. Sano, M.Yoshiyama, S. Ebisu, M.F. Burrow, T.Takatsu, B.Ciucchi, R.Carvalho, D. H. Pashley: Oper. Dent., Vol. 20 (1995), p. 160.

[5]H.Sano: Oper. Dent. Vol.20 (1994), p.18.

[6]F. R.Tay, M. K. Moulding, D. H. Pashley : J. Adhes. Dent. ,Vol. 1 (1999), p. 103.

[7] R. L.Mourão, H. S. Mansur, F. R. Tay, M. H. Santos, L. D.Lanza: Bioceramics 15/ 15th International Synphosium on Ceramics in Medicine; Key Engineering Materials, Vol. 240 (2002), p. 357.

[8] R. L. Mourão, F. R.Tay, D. H. Pashley: Annual Meeting & Exhibition of the AADR, 32, San Antonio. (CD-ROM of abstracts n.1201); J. Dent. Res. Vol. 82, (2003), Special Issue A.

[9] M. R. O.Carrilho. *Efeito da armazenagem sobre as propriedades mecânicas da união resina composta/dentina e seus constituintes.* (2002). 130 p. Thesis (PhD in Dental Material) – Faculty of Dentistry, University of São Paulo, São Paulo.

[10]H. Li, M. F. Burrown, M. J. Tyas: Oper. Dent., Vol.26 (2001), p.609.

[11]P. Spencer, J. R. Swafford: Quintessence Int., Vol.32 (1999), p. 561.

[12]L. Tjäderhane, H. Larvaja, T. Sorsa, V. J. Uitto, T. Larmas, Salo: J. Dent. Res., Vol. 77 (1998), p.1622.

[13]F. R. Tay, D. H. Pashley, M.Yoshiyama: J. Dent. Res., Vol. 81(2002), p. 472.

Key Engineering Materials Vols. 254-256(2004) pp. 699-702
online at http://www.scientific.net
© *2004 Trans Tech Publications, Switzerland*

Zirconia/Alumina Composite Dental Implant Abutments

Dae-Joon Kim[1], Jung-Suk Han[2*], Sun-Hyung Lee[2], Jae-Ho Yang[2], Deuk Yong Lee[3]

[1] Department of Advanced Materials Engineering, Sejong University, Seoul 143-747, Korea

djkim@sejong.ac.kr

[2] Department of Prosthodontics & Dental Research Center, College of Dentistry, Seoul National University, Seoul 110-749, Korea

*proshan@unitel.co.kr

[3] Daelim College of Technology, Anyang 431-715, Korea

dylee@daelim.ac.kr

Keywords: dental implants, abutments, ceramic composites, clinical applications

Abstract. (Y,Nb)-TZP/20 vol% Al_2O_3 composite was utilized as dental implant abutments. Biaxial strength and fracture toughness of the composite were 820 MPa and 8.5 $MPam^{1/2}$, respectively. A total of 55 ceramic abutments were connected to implant fixtures for single and short span fixed prostheses for a period of 2 years. Neither fracture nor screw loosening of abutments was observed so far.

Introduction

Dental implants have been successfully used as one of the viable treatment modalities in dentistry [1]. The dental implant system consists of a fixture and an abutment. The fixture part is placed into the bone bed surgically and left for several months of undisturbed healing period. After second surgery, the abutments are connected to fixtures to replace the crown portion of the teeth. Traditionally, titanium and titanium alloy have been used for the abutments. Recently, however, ceramic abutments are introduced to solve esthetic problems of metal abutments [2]. Alumina has been used for this purpose, but it has limited applications due to the fracture during preparation for desired shapes and connection procedures as a result of relatively inferior mechanical properties. Although zirconia possesses higher mechanical properties than alumina, it experiences the low temperature degradation during steam sterilization and its color is too white to be esthetic in the dental application. On the other hand, zirconia/alumina composites have suitable mechanical properties, biocompatibility, and esthetics for the implant abutments [3]. The purpose of this study was to evaluate the clinical performance of the loaded single and short span fixed prostheses supported by the zirconia/alumina composite abutments.

Materials and Methods

Powder preparation procedure of the zirconia/alumina composite abutments (ZirAce, Acucera, Korea) was delineated elsewhere [3]. Zirconia/ 20 vol% alumina composite powders were die-pressed into disks, and then isostatically pressed at 140 MPa. The green compacts were sintered for 2 h at 1600°C in air. Both flexural strength and fracture toughness of disk specimens were measured using a flat-on-three-ball biaxial-fixture [4] and toughness was determined by the indentation-strength method [5]. Low temperature degradation was determined by detecting m-ZrO_2 phase on aged specimens using x-ray diffractometer.

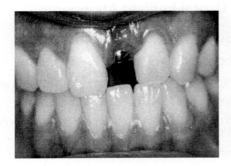

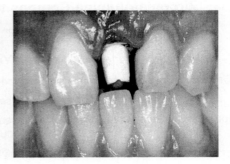

Fig. 1. Titanium healing abutment was connected to implant fixture after second stage surgery.

Fig. 2. Patient restored with a ZirAce abutment (a maxillary left central incisor).

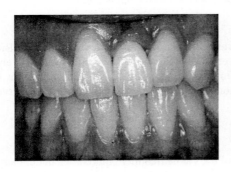

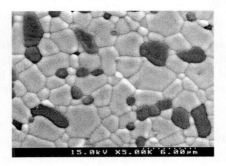

Fig. 3 Clinical view of cemented final all ceramic restoration on the abutment.

Fig. 4. Scanning electron micrograph of zirconia/alumina composite abutment surface.

A total of 40 patients received the ceramic abutments to restore single and short span fixed implant prostheses. Among them, 29 patients (17 female and 12 male) were treated with single tooth implant restorations. The implant replaced 5 missing incisors, 8 premolars, and 4 molars in the maxilla, and one canine, 4 premolars, 7 molars in the mandible. The rest of patients, 8 female and 3 male, were treated with short span fixed implant prostheses.

Only regular size implant fixtures were selected for ZirAce abutment connection. Healing abutments were connected at second stage surgery (Fig 1). After soft tissue healing, fixture level impression was taken and working cast was fabricated. ZirAce abutments were connected onto the fixture replicas and prepared with a high speed diamond burr under copious water spray in the laboratory and finished intraorally (Fig 2). Single abutments were connected to fixtures using titanium fixation screws with a 30 Ncm torque and multiple abutments were connected with a 20 Ncm torque force. The first abutment was placed in September 2001 and the last one in December 2002.

Provisional restoration was fabricated and functioned for a couple of months. Final prostheses were permanently cemented on the prepared 55 ceramic abutments with resin modified glass ionomer cement (Fig 3). Periodical recall check was performed at 3, 6 months and one year after delivery of final prostheses. All complications as noted by the patient or diagnosed by the clinician were recorded and peri-implant soft tissue evaluation, standard periapical radiograph, mobility test, and evaluation of esthetics were performed.

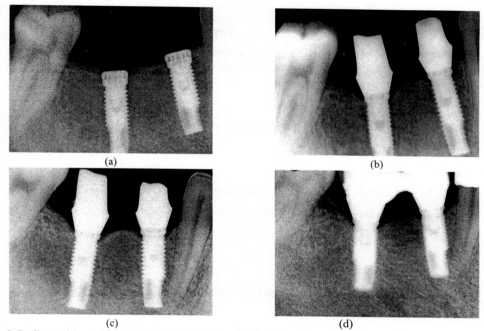

(a)

(b)

(c)

(d)

Fig. 5. Radiographic verification of stability of marginal bone level: (a) after first surgery; (b) 6 months later after second surgery; (c) 11 months later after loading; (d) 20 months later after loading.

Results

Microstructure of the composite abutment containing 20 vol% alumina is shown in Fig. 4, where dark grains are alumina and the light ones are zirconia. Sintered density, biaxial strength, and fracture toughness of the zirconia/alumina composite were 5.53 g/cm^3, 820 MPa, and 8.5 MPam$^{1/2}$, respectively. No degradation was observed after autoclave treatment for 10 h at 200°C and 1.6 MPa water vapor pressure.

A total of 55 ceramic abutments were connected to implants for single and short span fixed prostheses. One titanium implant fixture was failed after several months of loading. However, there was no screw loosening between abutment and fixture. Remaining 54 ceramic abutments showed neither fracture nor screw loosening so far. No adverse soft tissue reaction was observed and reported by patients. Marginal bone level was stable during the observation period (Fig. 5). No mobility was recorded for 54 implants. All patients were satisfied with function and esthetics.

Discussion and Conclusion

Tooth-colored zirconia/alumina composite abutments were introduced to replace titanium abutments for esthetic reasons in the present study. They have excellent mechanical and biological properties [3] so that the abutments can be used for both single tooth restorations and multiple units. Although alumina ceramic abutments also demonstrated promising clinical results for single and short span fixed restorations, several abutment fractures were reported [6,7]. After 1 year loading, about 7% percent (2 out of 34) of alumina abutments had fractured on single tooth replacement cases and about 2% (1 out of 53) had fractured on short span fixed partial dentures after 2 years loading. On the other hand, there was neither fracture nor chipping of the zirconia/alumina composite abutments during preparation and prosthetic procedures because of their high flexural strength and fracture toughness. Furthermore, neither screw loosening nor fracture of abutments was detected during observation period. Thus, clinicians and technicians feel more comfortable to handle ZirAce abutment than other ceramic abutment

The clinical outcome of the ZirAce abutments was favorable within the relatively short observation period. It showed a possibility as a routine abutment for single and short span fixed implant prostheses although a long-term clinical result is needed to make definite prognosis for the zirconia/alumina composite abutment.

Acknowledgement.
This work was supported by the Korea Ministry of Health and Welfare under contract No. 01-PJ4-PG4-01VN02-0036.

References

[1] R. Adell, U. Lekholm, B. Rockler and P.L. Branemark: Int. J. Oral Surg. Vol. 10 (1981), p. 387

[2] V. Prestipino and A. Ingber: J. Esthet. Dent. Vol. 8 (1996), p. 255

[3] D-J. Kim, M-H. Lee, D. Lee and J-S. Han: J. Biomed. Mater. Res. (Appl. Biomater.) Vol. 53 (2000), p. 438

[4] ASTM Standard F39-78 (1996), p. 446

[5] G.R. Anstis, P. Chantikul, B.R. Lawn and D.B. Marshall: J. Am. Ceram. Soc. Vol. 64 (1981), p. 533

[6] B. Andersson, A. Taylor, B. Lang, H. Scheller, P. Sharer and J. Sorensen: Int. J. Prosthodont. Vol. 12 (2001), p.432

[7] B. Andersson, P. Sharer, M. Simion and C. Bergstrom: Int. J. Prosthodont. Vol. 12 (1999), p. 318

Key Engineering Materials Vols. 254-256(2004) pp. 703-706
online at http://www.scientific.net

The Wear Resistance Testing of Biomaterials Used for Implants

Radek Sedlacek[1], Jana Rosenkrancova[2]

[1] Czech Technical University, Faculty of Mechanical Engineering, Department of Mechanics, Technicka 4, 166 07 Prague, Czech Republic, sedlacek@biomed.fsid.cvut.cz

[2] Czech Technical University, Faculty of Mechanical Engineering, Department of Mechanics, Technicka 4, 166 07 Prague, Czech Republic, rosenkra@biomed.fsid.cvut.cz

Keywords: biomechanics, tribology, ceramics, implant

Abstract. This article deals with very specific wear resistance testing of biomaterials used for surgical implants. This type of testing is very important for evaluation of the pertinence of different type of biomaterials. The special wear resistance test is called "Ring On Disc". The experiment was carried out in "Laboratory of Biomechanics of Man" with two types of materials - zirconia ceramics and PEEK (PolyEtherEtherKetone). The main parameter of the test – the wear volume – was determined.

Introduction

No known surgical implant material has ever been shown to be completely free of adverse reactions in the human body. However, long-time clinical experience of use of biomaterials has shown that an acceptable level of biological response can be expected, when the material is used in appropriate applications.

This article deals with very specific wear resistance testing of the bio-compatible and bio-stable materials used for surgical implants. Abrasion is an indispensable parameter for evaluation of the mechanical properties. This type of testing is very important for appreciation of new directions at the joint replacement design (for example in total knee replacement). The investigation in this area is based on a new combination of biomaterials. The experiments were carried out in collaboration with the company Walter Corporation. They are developing and producing bone-substitute biomaterials and implants in Czech Republic.

Materials and Methods

The PEEK (PolyEtherEtherKetone) and ceramics (based on yttria-stabilized tetragonal zirconia Y-TZP) are a new combination of materials used for surgical implants. The special wear resistance tests, called "Ring On Disc", were carried out with three pairs of these new materials (300 hours of testing). The experiment was executed according to ISO 6474:1994(E) [1]. This International Standard deals with evaluation of bio-materials based on highly pure and stabilized bio-ceramics. These materials are used for production of bone spacers, bone replacement and components of orthopaedic joint prostheses. The standard requires a long-time mechanical testing at which a complete volume of worn material is evaluated. The test conditions, requirements on the testing system and specimens' preparation are closely determined. The testing objectivity is ensured by the procedure for the specimens' treatment and their evaluation.

The method is based on loading and rotating two pieces from biomaterials (Figs. 1, 2). A ring from zirconia ceramics is loaded onto a flat plate from PEEK. The axial load that is applied on the ring is constant all the time. The ring is rotated through an arc of $\pm 25°$ at a frequency of (1 ± 0.1) Hz for a given period of time (100 ± 1) hours. Distilled water is used as the surrounding medium.

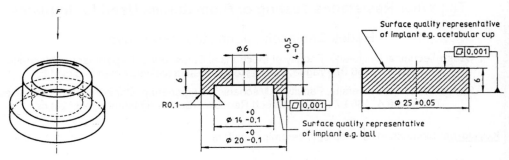

Fig 1: Schematic diagram

Fig 2: Geometry of ring and disc test pieces with dimensions

Special jigs for fixing both specimens during testing were developed and produced. These jigs have to be able to undergo oscillatory rotation of the ring specimen about fixed axis using a sinusoidal or near-sinusoidal rate of change of angle. The disc-holding device is equipped with the special joint to ensure that the plane of the disc surface coincides with the plane of the ring surface at all times during the test.

The experiment was carried out on the top quality testing system MTS 858 MINI BIONIX placed in "Laboratory of Biomechanics of Man" at the Czech Technical University in Prague.

As a measure of wear resistance the volume of the wear track on the disc is used. The wear track cross-sectional area is analyzed from the measured profile. The volume is calculated from this area.

The profile measurement of the tested specimen was carried out using a specially adapted assembly. To determine the vertical position of points on the disc the digital drift sight MAHR EXTRAMESS 2001 was used, with the accuracy of 0.2 µm, placed in a sufficiently stiff stand. A positioning cross-table (ZEISS), containing a make-up piece (in which the disc was inserted), served for the disc shifting. The cross-table is movable in two axes by means of two micrometric screws. The shifting sensitivity is 0.01 mm. Measured data were registered in a table prepared in advance.

Results and Discussion

The axial load, which was applied on the specimens during the experiment, was analyzed after finishing testing part. The values are shown in Table 1.

Table 1: Evaluation of axial load

The demand of the international standard [N]	Real value of axial load [N]
1500±10	1500±0.09

Parameters, determined after measuring profile of the tested disc No. 1 from recorded data, are shown in Table 2. The final parameters, evaluated from the profile of each disc, were calculated (see Table 3).

Table 2: Parameters evaluated the profile of the disc No. 1

No of measured track	1	2	3	4	5	6
Inner radius of the track r_i [mm]	7.1	7.1	7.5	7.5	7.2	7.1
Outer radius of the track r_o [mm]	9.3	9.5	9.9	9.9	9.7	9.5
Vertical position of points out of track h_o [mm]	-0.005	-0.009	-0.027	-0.041	-0.031	-0.009
Vertical position of points in track h_i [mm]	-0.081	-0.074	-0.086	-0.091	-0.083	-0.069
Height of track h [mm]	0.076	0.065	0.059	0.050	0.052	0.060
Track cross-sectional area A [mm^2]	0.167	0.156	0.141	0.120	0.131	0.144

Table 3: Averaged parameters of each disc

Disc No.	r_i [mm]	r_o [mm]	A [mm^2]	V [mm^3]
1	7.3	9.6	0.143	7.59
2	7.4	9.6	0.139	7.42
3	7.3	9.5	0.146	7.71

The test was executed with only three pairs of specimens, but that means 300 hours of testing. We obtained the information about wear resistance for this combination of materials. The resulting wear volume indicates the amount of material that is being lost during loading of the bone substitute implant in human body. This parameter describes one of the mechanical properties and shows that this new combination of biomaterials is suitable as surgical implants. The finished parameters obtained in this test - the wear volume and standard deviation - were calculated (see Table 4).

Table 4: Resulting volume

Wear volume [mm^3]	Standard deviation [mm^3]
7.57	±0.21

Conclusion

These experiments describe the behavior of the tested biomaterials during loading. In the near future the full test will be completed, as required in the international standard – that means with five pairs of specimens from zirconia ceramics and PEEK. A database of these parameters for other combinations of bone-substitute materials will be created.

Acknowledgements

This research has been supported by the Ministry of Education project No. MSM 210000012.

References

[1] Sedlacek, R – Rosenkrancova, J.: Methodics for testing of Mechanical Properties of Ceramic Materials. Workshop 2001, CTU Prague, p. 788-789.

[2] Sedlacek, R – Rosenkrancova, J.: Mechanical Testing of Wear Resistance According to ISO 6474:1994(E) with Ceramics and UHMWPE. EMBEC 2002, Vienna, p. 254-255.

[3] ISO 6474:1994(E) Implants for surgery – Ceramic materials based on high purity alumina. 1994

Key Engineering Materials Vols. 254-256(2004) pp. 707-710
online at http://www.scientific.net
© *2004 Trans Tech Publications, Switzerland*

Finite Element Analysis of Ceramic Dental Implants Incorporated into the Human Mandible

Andy H. Choi[1], Richard C. Conway[2] and Besim Ben-Nissan[1]

[1] Department of Chemistry, Materials and Forensic Science, University of Technology, Sydney, P.O. Box 123, Broadway, N.S.W., 2007, Australia, Andy.H.Choi@uts.edu.au, Besim.Ben-Nissan@uts.edu.au

[2] Department of Oral and Maxillo-facial Surgery, Westmead Hospital, Westmead, N.S.W., 2145, Australia, Richardcconway@optusnet.com.au

Keywords: dental implants, finite element analysis, clenching

Abstract. Three-dimensional finite element analysis (FEA) was used to evaluate the influence of the design of dental implants on distortion and stresses (the maximum and minimum principal, and Von Mises) acting in the dental implant and the surrounding osseous structures during clenching. The model geometry was derived from physical measurements taken of an average size human mandible. Implants of various lengths and root configurations were embedded into the first molar region of the right mandible. Four dental implant materials have been selected for this analysis: Titanium-aluminum-vanadium (Ti-6Al-4V), cobalt-chromium-molybdenum (Co-Cr-Mo), alumina, and zirconia. All of the materials used in this analysis were idealized as homogeneous, isotropic and linear elastic. The complete model 3D-model was completed using STRAND7 and consists of 700 solid elements and 3738 nodes. The results indicated that the stresses acting in the implants are determined by several factors such as implant length, root configurations, and the implant material.

Introduction

Implants have been used to support dental prostheses for many decades, but they have not always enjoyed a favorable reputation. This situation has changed dramatically with the development of endosseous osseointegrated dental implants. They are the nearest equivalent replacement to the natural tooth, and are therefore a useful addition in the management of patients who have missing teeth because of disease, trauma or developmental anomalies.

The clinical success of an implant is largely determined by the manner in which the mechanical stresses are transferred from implant to the surrounding bone without generating forces of a magnitude that would jeopardize the longevity of implant and prosthesis [1]. Carefully planned functional occlusal loading will result in maintenance of osseointegration and possibly increased bone to implant contact. In contrast, excessive loading may lead to bone loss and/or component failure.

Methods for the evaluation of stress around dental implant systems include photoelasticity, finite element analysis and strain measurement on bone surfaces. The finite element method offers several advantages, including accurate representation of complex geometries, easy model modification, and representation of the internal state of stress and other mechanical quantities [2 – 5].

The purpose of this study is to determine the influence of implant length and root configuration on the distribution of stress and distortion in and around the dental implants during functional movements by the use of three-dimensional finite element analysis.

Materials and Method

A dry human mandible has been used to define the geometry of the model. The mandible was cast in a polyester resin and cross-sectioned into 28 sections. The Palette CAD database software was used to reconstruct the digitized co-ordinate data "wire frame" structure. The final construction of the three-dimensional geometry was then completed by creating the symmetry. Finite element analyses were carried out under non-linear static conditions. Both the modeling and analysis were performed with a commercially available finite element analysis package, STRAND7 (G+D Computing, Australia). For this analysis three nodes on the symmetry plane were fixed in space by the use of spring elements. The model is completely free to deform removing the approximation caused by over-restraining the model.

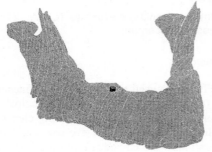

Fig. 1 Dental implant incorporated into the three-dimensional finite element model of a human mandible

The implants were embedded in the first molar region of the right mandible. Eight implants with a diameter of 3.26mm and lengths of 8 and 12mm were incorporated in the model. The implant angle was also modified to study the effects of varying taper angles of 3, 6, and 9°. Implants were modeled as a solid structure with abutments that were 2mm high.

Titanium-aluminum-vanadium (Ti-6Al-4V), cobalt-chromium-molybdenum (Co-Cr-Mo), alumina, and zirconia were used as implant materials. The properties of cortical and cancellous bone [6 - 8] and all the implant materials used in this analysis are given in Table 1. All the materials in this model were assumed to be isotropic, homogeneous, and linearly elastic. The finite element model of the mandible with an implant embedded into the first molar region consisted of 700 solid elements and 3738 nodes (Fig. 1).

Table 1   Physical properties of materials incorporated in the model [6 – 8] (The 1 direction is radial, the 2 direction is circumferential and the 3 direction is longitudinal directions).

Material	Young's Modulus (GPa)			Poisson's Ratio		
	E_1	E_2	E_3	ν_{12}	ν_{23}	ν_{13}
Cortical Bone	6.9	8.2	17.3	0.315	0.325	0.310
Cancellous Bone	0.32	0.39	0.96	0.3	0.3	0.3

	Young's Modulus (GPa)	Poisson's Ratio
Titanium alloy	114	0.34
Cobalt-chromium alloy	210	0.29
Alumina (Al$_2$O$_3$)	420	0.27
Zirconia (PSZ)	210	0.31

The muscle forces considered in this analysis were selected from various investigators [9, 10]. All the forces were assumed to be symmetrical with respect to the mid-line and to have an equal magnitude on the right and left side of the mandible (Table 2).

Table 2 Averaged calculated force (N) acting on the mandible during clenching (for one side only).

Muscles	Force (N)
Masseter	340.0
Temporalis	528.6
Medial Pterygoid	191.4
Lateral Pterygoid	378.0
Joint Reaction Force	471.9
Bite Force	403.7
Openers	155.0

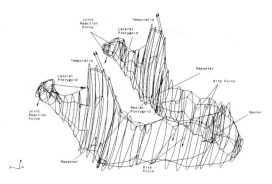

Fig. 2 Calculated forces and co-ordinate system acting on the mandible during clenching.

Results

The distribution of displacements around the dental implants was slightly higher with the 12mm implants. For all four implant materials, displacement at the implant-bone interface increased gradually from the root of the implant in the cancellous bone to a maximum at the abutment of the implant above the cortical bone. The displacement measured in the osseous structure was found to be lower with the inclusion of dental implants.

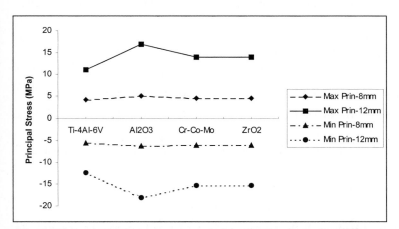

Fig. 3 The principal stresses recorded in the implants for different lengths and material properties.

In this study, the maximum principal stresses were tensile and minimum principal stresses were compressive. The tensile stress was mainly at the back of the implants, whereas the compressive stress acted towards the front. Tensile stress was observed mainly on the buccal side while compressive stress on the lingual side of the bone. The distribution of principal stresses in the dental implants and around the osseous structure (both cortical and cancellous bone) was slightly higher with alumina implants. Increase in the stress values was observed for the 12mm implants. Figure 3 shows the maximum and minimum principal stresses acting in the dental implants with various lengths and material properties.

The results on the distribution of stresses in the implants with different root taper angle for all the materials investigated shows that the highest tensile stress for 8mm implants occur at 3° taper and the lowest tensile stress was recorded at 6° taper. For 12mm implants, the tensile stress increases as the root taper angle increases. This study also demonstrates that with increasing implant root taper angle, the minimum principal (compressive) and Von Mises stresses acting around the implant increases. The results of the studies on the influence of root taper angle on the stress distribution around the alumina implant are summarized in Table 3.

Table 3   Influence of root taper angle of the implant with various lengths on the stresses acting in the alumina implant (σ_1, tensile stress; σ_3, compressive stress).

Taper Angle	8mm Implants		12mm Implants	
(Deg)	σ_1 (MPa)	σ_3 (MPa)	σ_1 (MPa)	σ_3 (MPa)
3	4.72	-6.29	17.31	-18.55
6	4.48	-6.29	18.09	-19.34
9	4.54	-6.34	18.78	-20.02

Conclusion

Finite element modeling of the human mandible incorporating a dental implant and the analysis of stress and distortion distribution in and around the implants has been studied. Some factors such as (1) the length of the implant embedded into the cancellous bone region, (2) the implant root taper angle, (3) the magnitude and direction of the forces applied on the implants, and (4) the mechanical properties of the dental implant used appeared to influence the stresses acting in and around the dental implants and its surrounding bone structure. This study concludes that the best implant combination to use during clenching is an 8mm-titanium alloy implant with a 6-degree taper angle.

References

[1] R. Skalak: J. Prosthet. Dent. Vol. 49 (1983), p.843

[2] S. Lin, S. Shi, R.Z. LeGeros and J.P. LeGeros: Implant Dentistry Vol. 9 (2000), p.53

[3] K. Akca and H. Iplikcioglu: Int. J. Oral Maxillofac. Implants Vol. 16 (2001), p.722

[4] G. Menicucci, A. Mossolov, M. Mozzati and G. Preti: Clin. Oral Impl. Res. Vol. 13 (2002), p.334

[5] T. Nagasao, M. Kobayashi, Y. Tsuchiya, T. Kaneko and T. Nakajima: J. Cranio-Maxillofac. Surg. Vol. 30 (2002), p.170

[6] F.J. Arendts and C. Sigolotto: Biomed. Technik. Vol. 34 (1989), p.248

[7] F.J. Arendts and C. Sigolotto: Biomed. Technik. Vol. 35 (1990), p.123

[8] C.H. Turner, S.C. Cowin, J.Y. Rho, R.B. Ashman and J.C. Rice: J. Biomechanics Vol. 23 (1990), p.549

[9] J.C. Barbenel: J. Biomechanics Vol. 1 (1974), p.19

[10]G.J. Pruim, H.J. De Jongh and J.J. Ten Bosch: J. Biomechanics Vol. 13 (1980), p.755

VI. INTERACTIONS WITH CELLS AND TISSUES

Key Engineering Materials Vols. 254-256(2004) pp. 713-716
online at http://www.scientific.net
© *2004 Trans Tech Publications, Switzerland*

Quantitative Analysis of Osteoprotegerin and RANKL Expression in Osteoblast Grown on Different Calcium Phosphate Ceramics

Chaoyuan Wang[1,2], Yourong Duan[1], Boban Markovic[3], James Barbara[4], C. Rolfe Howlett[4], Xingdong Zhang[1] and Hala Zreiqat[4]

[1] Engineering Research Center in Biomaterials, Sichuan University, Chengdu, Sichuan Province 610064 P. R. China, wangcyscu2001@yahoo.com
[2] Yangtze Fisheries Institute, Chinese Academy of Fishery Science, Jingzhou, Hubei Province, 434000, P. R. China
[3] Chemical Safety and Toxicology Laboratories, School of Safety Science, the University of New South Wales, Sydney 2052, Australia
[4] Bone Biomaterial Unit, Department of Pathology, School of Medical Science, University of New South Wales, Sydney 2052, Australia

Keywords: Calcium phosphate ceramics, phase composition, OPG, RANKL, osteoblast

Abstract: Calcium phosphate ceramics are well-known biomaterials used in orthopaedic and dental applications as bone substitute. In this study, SaOS-2 cells were grown on calcium phosphate ceramics with different phase composition for 3 and 6 days, RANKL and OPG mRNA expression were detected with quantitative in situ hybridization technique. Result showed that Saos-2 grown on tricalcium phosphate and a biphasic calcium phosphate ceramics with high tricalcium phosphate ratio expressed higher RANKL mRNA after 6 days culture. There is no significant difference in OPG expression on different surfaces after 3 days and 6 days culture. It can be concluded that tricalcium phosphate ceramics may induce more bone remodeling than hydroxyapatite when implanted *in vivo*.

Introduction

Due to their differences in chemical composition, commonly used calcium phosphate ceramics can be sorted as pure hydroxyapatite (HA), pure tricalcium phosphate (TCP) and biphasic calcium phosphate (BCP) with different HA/TCP ratio. Experimental evidence has suggested that some calcium phosphate ceramics possess osteoinductivity [1].

Bone remodeling is a natural process that occurs in bone metabolism throughout life. It is clear that osteoblasts are the major cells responsible for new bone formation while osteoclasts are responsible for bone resorption. Bone metabolism is the balance of new bone formation and bone resorption. In 1981, Rodan et al proposed a hypothesis that osteoblasts might play a key role in osteoclast genesis and bone resorption [2]. After that, experimental evidences have accumulated to demonstrate that osteoblastic cells were essential for osteoclast formation by secreting soluble factors and also by signaling to osteoclast progenitors through cell to cell contact [3]. The discovery of the function of osteoprotegerin (OPG) and RANKL has paved the way to understand the mechanism of osteoblast-stimulated osteoclast differentiation and bone remodeling [4,5].

Until now, little information is available concerning how materials property of bioceramics may regulate OPG and RANKL expression. This paper aims at investigating the effect of different phase composition of calcium phosphate ceramics on the expression of OPG and RANKL using Quantitative *in situ* hybridization (QISH).

Materials and Method

Calcium phosphate ceramic disks preparation: Calcium phosphate powder with different phase composition were pressed under 15 MPa for 3 minutes and sintered at 1200°C. The calcium phosphate powders were: hydroxyapatite (HA), biphasic calcium phosphate 1 (BCP1, HA/TCP: 70/30), biphasic calcium phosphate 2 (BCP2, HA/TCP: 35/65) and tricalcium phosphate (TCP). Disks were grinded with sandpaper and polished Roughness of the disks was measured with profilometer. The average roughness (Ra) was 35±10nm (n=6) for HA, BCP1 and BCP2 ceramics and 80±15nm for TCP (n=6). Disks were washed, dried and autoclaved.

RANKL and OPG mRNA detection: mRNA hybridization was done with the method described by H. Zreiqat [6]. Briefly, Saos-2 cells (ATCC HTB85) were seeded onto the disks at a concentration of $3x10^4$ cells /cm^2 and were harvested in 3 and 6 days respectively. $1x10^4$ cells were then aliquoted to each well in a 96-well plate. Plates were centrifuged at 3000 rpm for 10 minutes to precipitate cells. Medium inside the wells was vacuum aspirated using a fine needle. Plates were then placed in a fan forced incubator at 50°C for 30 minutes to fix the cells to the wells.

Photobiotin labeled cDNA probes for RANKL and OPG (kindly provided by Dr. Haynes DR, Department of Pathology, University of Adelaide, SA, Australia) were used for the detection of mRNA expression in Saos-2. cDNA labeling was done with the method on the kit's instruction (Bresatec, Australia). 50µL of labeled probes were added to each well and hybridized at 42°C overnight. After posthybridization washing, streptavidin conjugated alkaline phosphate (Dakopatts) and *para*-nitrophenyl phosphate (*p*-NPP) substrate was added. The optical density of the yellow color (*p*-NP) was read by an ELISA plate reader with the software of GENESIS \ protocol \ mrna. prt & mrna. plt at 405nm. GAPDH gene (kindly provided by Dr. Hoggs, New South Wales, Australia) was used as internal control for mRNA expression detection.

Statistical analysis: Student's paired t-test was applied to determine statistical significance, and a *p* value of <0.05 represented a significant difference.

Result

As shown in Fig.1, surface morphology is different on different surfaces. HA had the biggest crystal size (larger than 2µm), while the crystal size of BCPs was smaller. Interestingly, TCP surface morphology showed that crystal particles were fused under 1200°C sintering temperature and pores with a diameter of approximately 2µm were evenly distributed on the whole surface.

As shown in Fig. 2, after 3 days culture, no significant differences in the expression of OPG and RANKL mRNA was detected among the different surfaces. At 6 days culture, OPG mRNA expression was comparable for all different substrates tested. On the other hand, SaOS-2 grown on TCP and BCP2 expressed more RANKL mRNA than that on HA and BCP1 (p<0.05).

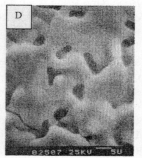

Fig. 1. Surface morphology and microstructure of calcium phosphate ceramics with different phase composition. A: HA; B: BCP1; C: BCP2; D: TCP

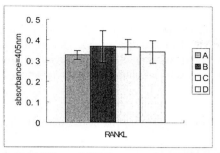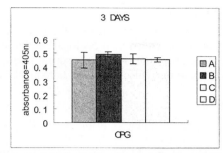

Fig. 2. RANKL and OPG expression of Saos-2 grown on calcium phosphate ceramics with different phase composition for 3 days. A = HA, B = BCP1, C = BCP2. D = TCP. Means of three wells ± SD.

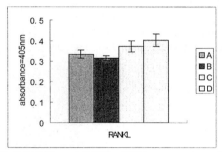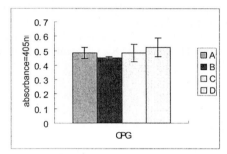

Fig. 3. RANKL and OPG mRNA expression of Saos-2 grown on calcium phosphate ceramics with different phase composition for 6 days. A = HA, B = BCP1, C = BCP2, D = TCP. Means of three wells ± SD.

Discussion

It has been reported that chemistry, energy and topography of an implant surface were important factors in inducing biological responses in vivo [7,8]. The differences in crystal particle size, chemistry and surface microstructure might have played an important role in inducing different cell behavior.

In this study, OPG mRNA level seemed not to be regulated by different surfaces. But after 6 day in vitro culture, RANKL mRNA expression was higher on TCP and BCP2. This might imply that TCP and BCP2 could stimulate more bone turnover when they were implanted *in vivo*.

RANKL/RANK/OPG pathway is very important in bone remodeling. The mechanism of osteoblast-stimulated osteoclast differentiation is that, by binding to RANK, a specific receptor of RANKL on the preosteoclastic cell membrane, RANKL can stimulate preosteoclastic cell differentiation. OPG is a soluble cytokine secreted by osteoblasts and served as a decoy receptor of RANKL. By competitively binding to RANKL, OPG inhibits RANKL to bind to RANK on preosteoclastic cell and therefore to inhibit osteoclast differentiation.

Conclusion

Tricalcium phosphate ceramics and biphasic calcium phosphate with high tricalcium phosphate ratio induce osteoblasts to express more RANKL mRNA expression than pure hydroxyapatite and biphasic calcium phosphate ceramics with low tricalcium phosphate ratio. No significant differences can be detected among different surfaces in terms of OPG mRNA expression. Tricalcium phosphate ceramics may induce more bone remodeling than hrdroxyapatite when implanted in vivo.

Acknowledgements

This work is financially supported by the State Key Science Research Project of China (Contract No. G1999064760) and Australian National Health and Medical Research Council. The author thanks China Scholarship Council for sponsoring Chaoyuan Wang to complete this cooperative work in Australia. The authors also thank Dr. Jianxin Wang for providing calcium phosphate powders.

References

[1] Yuan H, Li Y, Yang Z, *et al*: *Biomedical Materials Research in the Far East* (III). Kyoto: KoBunshi Kankokai; 1997. p 228-229
[2] Rodan GA, Martin TJ. Calcified Tissue Int. 1981; 33:349-351
[3] Takahashi N, Akatsu T, Uadagawa N, *et al*. Endocrinology 1988; 123:2600–2602
[4] Yasuda H, Shima N, Nakagawa N, *et al*. Proc Natl Acad Sci USA 1998; 95:3597–3602
[5] Kong Y-Y, Yoshida H, Sarosi I, *et al*. Nature 1999; 397:315–323
[6] Zreiqat, H. Markovic B. Walsh W.R. *et al*. J Biomed Mater Res 1996, 33(4): 217-23
[7] B.D. Ratner, A.B. Johnston, and T.S. Lenk. J. Biomed. Mater. Res., 1987 21(A1), 59-90
[8] Zreiqat H, Standard OC, Gengenbach T, et al. Cells and Materials, 1996; 6:45-56

Key Engineering Materials Vols. 254-256(2004) pp. 717-720
online at http://www.scientific.net
© *2004 Trans Tech Publications, Switzerland*

Properties of Two Biological Glasses Used as Metallic Prosthesis Coatings and After an Implantation in Body

Y. Barbotteau, J.L. Irigaray, E. Chassot, G. Guibert and E. Jallot

Université Blaise Pascal, Clermont-Ferrand II
Laboratoire de Physique Corpusculaire, IN2P3-CNRS
63177 Aubière cedex, France
email : yves.barbotteau@qse.tohoku.ac.jp

Keywords: biological glass, coating, Ti6Al4V, corrosion, nuclear microscopy.

Abstract. We carried out electron and nuclear microscopy in order to study the *in vivo* behaviour of two biological glasses used as Ti6Al4V prosthesis coating. First glass is bioactive and improves the osseointegration of the prosthesis. Second is bioinert and it is employed like a cement. Nuclear microscopy enables us to prove that two glasses are an effective barrier against corrosion when they are placed between living medium and the metallic prosthesis.

Introduction

Biomaterials and their associated medical devices are an important future technology, both in terms of their health care capabilities and the opportunities they present for wealth creation. Biomaterials development is essentially a multidisciplinary activity.

In orthopaedic or dental surgery, most prostheses are made with metal or metallic alloys. A successful long-term implant requires biocompatibility, toughness, strength, corrosion resistance, wear resistance and fracture resistance [1]. Three groups of metal prevail for clinical application [2]: stainless steels, cobalt-based alloys and titanium or titanium alloys.

Nevertheless, metal, in contact with body fluids, forms corrosion products which cause different tissue reactions according to the metal. A metallic contamination surrounding tissues may play a major role in aseptic loosening of the implant. A way to prevent this problem of corrosion is to use a biomaterial coating. A layer thickness of few tens of micrometers allows to isolate the implant from the corrosive environment during some months, to confer the layer biocompatibility on the prosthesis and to keep the mechanical properties of the supporting metal. Such as hydroxyapatites ceramics, biological glasses are good candidates for coating metallic prostheses. In contact with cell tissue, they show *in vivo* and *in vitro* biocompatibility, no inflammatory and no toxic processes. They can be bioactive or bioinert according to their composition in oxides. The bioactive fixation is defined as an interfacial bond between implant and tissues by means of formation of a biologically active hydroxyapatite layer on the implant surface [3].

We have studied two different glasses marked off BVA and BVH. Both glasses are used as coating of titanium alloy prostheses (Ti6Al4V). Several methods have been performed in these studies such as nuclear and physico-chemical techniques. The implant with its coating is in contact either with trabecular bone or with lacuna site.

Materials

Glasses are manufactured [4] by melting the components and heating them between 1300°C and 1600°C. They are cast, crushed and transformed into powder of grain suitable to the spraying process. They are deposited on sand blasted Ti6Al4V cylinders, 18 mm in length and 4 mm in diameter. The compositions are shown on table 1.

Table 1: Composition (% wt.) of biological glasses: BVA & BVH.

	SiO$_2$	Na$_2$O	CaO	P$_2$O$_5$	K$_2$O	Al$_2$O$_3$	MgO
BVA	47%	19%	20%	7%	5%	1%	1%
BVH	71%	14%	9%	--	1%	2%	3%

The coating consists of small granules linked together in an heterogeneous and non continuous form which provides a favorable surface for the attachment of bone tissue. The thickness ranges from 30 to 100 μm.

The rods are implanted into lateral femoral epiphisis of sheep by manual pressure. The animals are sacrified at 3, 6 and 12 months after implantation. The samples are embedded in PMMA resin and cut in transversal sections of 400 μm thickness.

Methods

Scanning Electron Microscopy (SEM), Scanning Transmission Electron Microscopy (STEM) and Energy Dispersive X-ray Spectroscopy (EDXS) are used firstly for a qualitative and quantitative analysis of the coatings according to the implantation duration. For STEM and EDXS, some samples are cut into ultra-thin sections of about 100 nm with a diamond ultra-microtome. Bony tissues and coating are also studied simoultaneously by Particle Induced X-ray Emission (PIXE) and Rutherford Backscattering Spectrometry (RBS) with nuclear micro-probe (protons of energy 3 MeV, intensity 600 pA and 5 μm beam diameter). PIXE is performed to identify trace elements and to cartography their repartition and RBS to determine organic matrix composition and to measure the electric charge received by the sample during irradiation. This experimental setup has been published elsewhere [5][6].

Results and discussion

With BVA and when the coating is in contact with trabecular bone, there is gel formation relatively dense and homogeneous. When the coating is in contact with lacuna site, this gel has different densities and is heterogeneous. The BVA glass proved to be bioactive: it is transformed into a silicon gel with incorporation of protein and trace elements: Zn and Sr. This gel disappears gradually and is replaced by neoformed bone at 6 months after implantation. These transformations lead to a better osseointegration of the coated implant than uncoated prostheses [7]. Bony contact perimeter is increased, then BVA glass permits to limit micro-motion of the implant.

The BVH glass is bioinert: there is no gel formation. Its composition is constant versus time. However, the formation of a 2 μm thickness interface, induced by plasma spray coating process, weakens the inter-granular connections. This fact results in the fragmentation of the coating and the migration of glass particles through the lacunar network of surrounding bone. Its expected function is to protect metallic prosthesis against corrosion: BVH coating is used like a cement.

In table 2, we present mean titanium concentration values measured in surrounding bone of the implant for each kind of coating and for the three implantation durations. In the case of uncoated implants, we found that after a period of establishment, the titanium contamination in bony tissue is constant (about 1200 ppm). This result is consistent with the observation of Agins [8]. We consider that the two glasses protect efficiently the implanted metallic prostheses against the corrosion since we do not detect titanium in bone directly in contact with glass (table 2). In the same time, coating glasses can be contaminated themselves by metallic elements coming from the metallic prosthesis.

Table 2: Titanium concentration in ppm (LOD = Limit Of Detection).

	3 months	6 months	12 months
bone directly in contact with prosthesis initially coated with BVA	--	285 ± 14	888 ± 86
bone directly in contact with prosthesis initially coated with BVH	--	387 ± 246	1307 ± 167
bone directly in contact with the glass (BVA or BVH)	< LOD	< LOD	< LOD
bone directly in contact with prosthesis initially uncoated	651 ± 26	1221 ± 99	1134 ± 67

In both glass coating, we have detected titanium contamination. But, as mentioned above, BVA glass completely disappears between 3 and 6 months after implantation and BVH glass is gradually detached from the prosthesis. Then a large part of prosthesis surface becomes uncoated and the corrosion begins. And after 12 months of implantation, the titanium amounts in bony tissue directly in contact with metal of uncoated and initially coated prosthesis are similar. The protective effect of glasses is thus effective as a long time as glasses remain in place. If we look to the contamination away from the edge of the implant, we observe that it is concentrated in the first 100 μm and that it decreases fastly after some hundred micrometers (fig. 1).

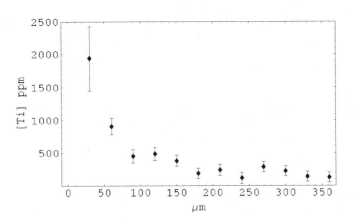

Fig. 1: Profile of titanium contamination of bony tissue surrounding an uncoated implant after 12 months of implantation.

Conclusion

The two glasses have different behaviour during the first months after their implantation in the body. However, they prevent metallic contamination of surrounding tissues. One year after implantation, BVA is resorbed and BVH is destroyed into debris and the contamination from metallic prosthesis began.

It is shown that the bioactive glass BVA coating enhances the osseointegration of a metallic implant: this is an advantage in the first months after surgery. On the other hand, the plasma spraying at high temperature can induce surface modification and we look after an alternative technique.

References

[1] Parr GR, Gardner LK, Toth RW. J. Prosthet. Dent. 1985; 54: 410-414.

[2] Breme J. *In Criteria for the bioinertness of metals for osseointegrated implants.* G Heimke ed. Osseo-integrated implants, CRC Press, Boca Raton 1990; 31-80.

[3] Hench LL. Biomaterials 1998;19: 1419-1423.

[4] *Construzione e Gestione Dati Bioingegneria* -- Viale dell'industria 5 -- 20070 Livraga (MI) -- Italy.

[5] Llabador Y, Moretto Ph, Guegan H. Nucl. Inst. and Meth. 1993;B77: 123-127.

[6] Irigaray JL, Oudadesse H, Brun V. Biomaterials 2001;22: 629-640.

[7] Ducheyne P, Qiu Q. Biomaterials 1999;20: 2287-2303.

[8] Agins HJ, Alock NW, Bansal M, Salvati EA, Wilson PD, Bullonugh PG. J. Bone Jt Surg. 1988;70A: 347-356.

Key Engineering Materials Vols. 254-256(2004) pp. 721-724
online at http://www.scientific.net
© 2004 Trans Tech Publications, Switzerland

Comparative Study of the Biomineral Behavior of Plasma-Sprayed Coatings in Dynamic Porous Environment

Q.Y. Zhang[1,2], J.Y. Chen[1], C.L. Deng[1], Y. Cao[1], J.M. Feng[1] and X.D. Zhang[1]

[1] Engineering research center in biomaterials, Sichuan University, China, jychen@scu.edu.cn

[2] College of chemical engineering, Sichuan University, China, qyzhang71@hotmail.com

Keywords: Plasma-sprayed, coatings, biomineral behavior, porous environment, dynamic

Abstract. The formation of bone-like apatite is always seen as one of characteristics of bioactivity. In this study, a novel model was established to simulate the porous structure environment to investigate the biomineralization behavior of the plasma-sprayed coatings. The coating with two different thermal treatments (120 °C for 6h in water vapor [0.15MPa] and 650 °C for 1h in vacuum) and as-received coating were immersed in three different immersion systems. The results showed that mainly needle-like biomineral layer was formed on samples treated at 120°C in water vapor, and round-shaped crystallites were observed on samples treated at 650°C in vacuum. These differences are caused by the difference in the composition, dissolution ability of the coatings. The difference in morphology of the biomineral layer in different model was governed by the ion concentration in local microenvironment.

Introduction

It has been reported that porous calcium phosphate (Ca-P) ceramics possessed good osteoinductivity, while dense Ca-P ceramics did not show any osteoinductivity *in vivo* [1,2]. Therefore, the geometry of materials seems to play an important role in promoting their bioactivity [3,4]. The plasma-sprayed coatings are widely applied in clinic because they can enhance the early bone bonding *in vivo*. However, limited by line-of-sight, the plasma-sprayed coating technique is not applicable on porous structures. In this study, a novel model was established to simulate the porous structure environment in order to investigate the biomineralization behavior of plasma-sprayed coatings. The ability to form biomineralized apatite is always considered an evaluation of the bioactivity of the materials *in vitro*.

Materials and methods

A METCO MN plasma spraying system with AR2000 Robot was used. The substrates for coatings were 8×6×2mm titanium plates with a coating about 70 μm thick. The as-received coatings were post-treated as follows: 120 °C for 6h in water vapor (0.15MPa) and 650 °C for 1h in vacuum (denoted as 120 treatment and 650 treatment respectively). The composition and crystallinity of the coating were assessed by X-ray diffraction (XRD). After post-treatment, samples were placed in a porous model and immersed in dynamic SBF solution. The porous models were made by titanium

Fig.1 The photo of the porous model.

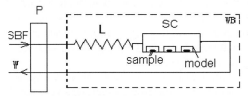

Fig.2 Diagram of the dynamic immersion system.
P: peristaltic pump, SC: soaking cell, L: equilibration tube, W: waste, WB: water bath (37 °C).

with two grooves occluding as shown in Fig 1. The four sides of the model were notched to facilitate liquid exchange in the chamber. Samples were placed in the calcium phosphate-coated porous model or uncoated model for 7days. Samples immersed in the same dynamic SBF solution without the porous model were used as control.

The SBF solution was prepared after Kokubo's recipe [5]. The dynamic immersion system was shown in Fig.2, the rate of flowing was 2ml/min/100ml.

Results and Discussion

After thermal post-treatment, the plasma-sprayed coatings have similar crystallinity of approximately 88%, but the 650 treatment coatings have some additional phases such as TCP, TTCP, and the crystal size is smaller than 120 treatment. The profile of the XRD is shown in Fig.3. The morphology of the biomineral layer on the samples is shown in Fig.4.

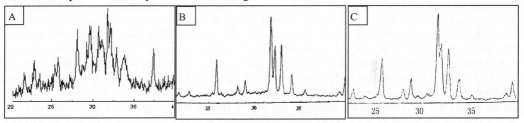

Fig.3 The XRD profile of samples.
A: as-received coating; B: 120 treatment; C: 650 treatment

Under coated or uncoated model or without model, all the coatings with 120 treatment could form needle-like crystal on their surface. But under the coated model there are more crystallites on the surface, and additional round-shaped crystallites were also found to be evenly distributed on the whole surface. Some of the needle-like crystals grew from the interstice of the coatings, some of them grew directly on the surface, and the size of them is about 5-10 μm. whereas on the 120 treatment coating under uncoated model, only needle-like crystal growth could be seen, and the amount of crystal formed was not as many as in coated model, but the crystals size (20 μm) was bigger than that in coated model. Needle-like crystals could be found on the surface of control group, but they grew only in the interstice of the coating surface, and the size of them is also about 10 μm. The same sample in different immersion environments showed different biomineralization behavior. In the coated model, which represented porous ceramics, samples were in an environment rich in Ca and P ions, furthermore, in porous condition, the exchange of ions was comparatively slow and thus, sufficient Ca and P resources were available for the growth of crystals. So the biomineral layer formed on the surface was needle-like, and round-shaped crystallites. As competition in the crystal growth took place, the crystals were not fully grown and thus resulted in short needle-like crystal formation. In the uncoated model, the local Ca and P ions were not as rich as in coated model, therefore there were only needle-liked crystals formed and obviously the biomineral layer was less than that in coated model, but the needle-like crystal grew more sufficiently because of less nucleation sites. In the control group, the local concentration of Ca and P ions at the interstice of coatings maintained at relatively high level and thus, crystal only grew in the interstice, and grew vertically to the surface.

As for 650 treatment and as-received samples, the biomineralization behavior in different immersion model was different, which was very similar to the changes occurred in sample with 120 treatment. The results were compared in table 1.

The behaviors of the three kinds of thermal treated samples in three different immersion conditions were obviously different. In the porous condition, with or without coating, the microenvironment is

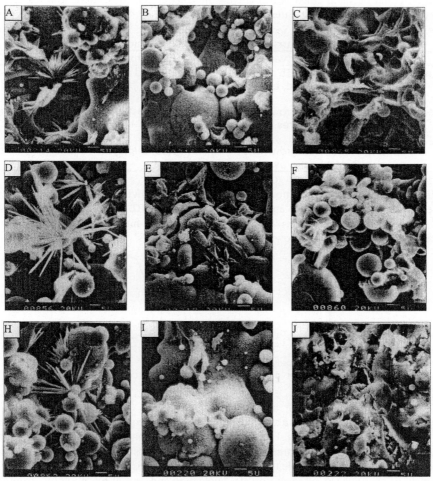

Fig.4 Morphology of the biomineralized layer.
Samples of 120 treatment, 650 treatment and as-received were immersed in flowing system in the coated model (A, B, C), uncoated model (D, E, F) and without model (H, I, J)

beneficial for biomineralization. The crystal size and morphology of the newly formed crystallites on different samples were governed by composition, solubility of the coating and the ion concentration of the microenvironment. The solubility of the 120 treatment was the least, so the local ions concentration was comparatively at a low level. In the three immersion conditions, biomineralization of 120 treatment samples are characterized by needle-like crystallites. In coated porous model, round-shaped crystallites formed due to the high local concentration of calcium and phosphate ions and mineral layer appeared on both the interstice and all surfaces. The solubility of the 650 treatment samples was higher than that of 120 treatment samples because of the additional phases and thus, the crystallites grew in round-shaped, a way that consumed more calcium and phosphate ions. Although the much higher solubility of the untreated samples, no crystal growth could be observed in non-porous environment, the reason is that under flowing condition, the coating dissolved so fast that there were no stable sites for nucleation. While in porous environment, new crystal phase formation could be found because the ion exchange would be slow in the porous model, so the local ion concentration could be maintained at a relatively high level and thus, there was enough time for nucleation and the growth of crystallites.

Table 1. The comparison of the morphology of biomineral coating

	120 °C, water vapor	650 °C, vacuum	untreated
coated model	Needle-like (~5 μ) + round-shaped, evenly distributed	Round-shaped	Sheet-shaped and interconnected
uncoated model	Needle-like grown out rfrom micropores (~20 μ)	Closely connected round-shaped + grain-like	A few, Round-shaped
control	Needle-like grown out from micropores	A few, Round-shaped + grain-like	No significant crystal growth

The difference of biomineralization behaviors is shown obviously in SEM photos, but this difference cannot be seen in XRD and FT-IR profile. From FT-IR spectrum, the peaks around 1460cm^{-1}, 1420cm^{-1} were characterized as carbonate, which further proved the newly formed biomineralized layer, as the Fig.5 shown.

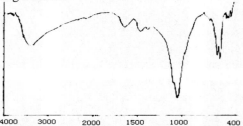

Fig. 5. The representative of FT-IR spectrum of the coating after immersion

Conclusion

The formation of bone-like apatite is always believed to be a prerequsite of osteoinduction and one of characteristics of bioactivity *in vitro*. From this work, we can conclude that the porous structure is benefit for the formation of the biomineral layer because the most crystal formation is observed in coated model than in the others, while the least crystal formation is found in control group. Different post-treatment of coatings can affect the crystal nucleation and crystal morphology. From the result above, we may conclude that calcium phosphate coating in porous structure environment has better bioactivity than that in non-porous environment.

Acknowledgements

This work was financially supported by the Key Basic Research Project of China, (Contract No.G1999064760)

References

[1] Zhang XD, Yuan HP, de Groot K., Workshop, 6th World Biomaterials Congress, Hawwai, USA, May14-20,2000
[2] Yuan HP, Yang ZJ, Li YB et al., J Mater.Sci.Mater.Med., 1998;9:721-726
[3] Maniatopoulos C, Pilliar RM, Smith DC, J.Biomed. Mater. Res 20:1309-1333,1986
[4] Pilliar RM, Implant Dent 1998,7(4):305-314
[5] Kokubo T, Ito S, Yamamuro T et al. J.Biomed.Mater.Res., 1990;24:331-343

Key Engineering Materials Vols. 254-256(2004) pp. 725-728
online at http://www.scientific.net
© 2004 Trans Tech Publications, Switzerland

Tissue Responses to Titanium with Different Surface Characteristics

after Subcutaneously Implanted in Rabbits

Y. Wu, B.C. Yang, C.Y. Bao, J.Y. Chen and X.D. Zhang

Engineering Research Center in Biomaterials, Sichuan University, Chengdu, 610064 China
e-mail: yaowu_amanda1970@sina.com

Keywords: subcutaneously implanted titanium, soft tissue responses, attachment strength, fibrous capsule, calcium phosphate

Abstract. The objective of this study was to investigate the soft tissue responses to titanium with different surface characteristics. Titanium bars with three kinds of surface were available to be anchored in the femur bone of rabbits and then penetrated into the soft tissue just under the skin: hydroxyapatite (HA) coated Ti, anode-oxidized Ti and sandblasted Ti as control. After 4 weeks and 8 weeks, the implants with surrounding soft tissue were retrieved and processed mechanically and histologically. All samples were encapsulated with quite thin fibrous capsules and little inflammatory reaction was observed except the sandblasted bars. At 4 weeks postoperation, the histological reaction and the attachment strength of the anode oxidized group was similar to the HA coated group. But at 8 weeks postoperation, the histological response around the HA coated group was better than at the anode oxidized group and the attachment strength was almost three times more. The sandblasted bars attached the soft tissue so weakly that it was too difficult to test the push-out strength. The SEM-EDX micrographs showed that a layer of calcium phosphate emerged at the interfaces between soft tissue and implants except sandblasted Ti, so we can infer that the layer of calcium phosphate plays a very important role in improving the soft tissue integration with the titanium.

Introduction

For percutaneous devices (PD) anchored in bone, the major failure modes are marsupialization and infection, which are resulted from skin downgrowth and mechanical damage [1]. In order to avoid these phenomena occurring, the implant's surface could be bioactivated to facilitate its firm attachment and bonding with the soft tissue under the skin, which is one of the main methods to inhibit the epithelial downgrowth. [2,3] Titanium and titanium alloys are increasingly being used as PD implants as a result of their excellent bulk properties and biocompatibility. [3,4] But up to now, there are few reports available concerning soft tissue responses to the bioactivated surface of Ti. In this paper, three kinds of titanium bars with different surfaces characteristics were subcutaneously implanted in rabbits to study their soft tissue responses, including the attachment strength of implant-tissue interface.

Materials and Methods

Commercially pure titanium was cut into cylinder bars $\Phi3$ x12mm. Half part of the bar along the long axis was screwed and the other part was modified with two different bioactivation methods or sandblasted. The first group of bars, defined as HAC group, was plasma sprayed with hydroxyapatite (HA) and then treated under 125°C in water vapor for 6 hours. The crystallinity of HA coating was approximately 80%. The second group of bars, defined as AO group, was anode oxidized in the electrolyte solution of 1M H_2SO_4 under 155V for 1 minute. The third group of bars, defined as SB group, was just sandblasted by 400μm Al_2O_3 to obtain a rough surface as control. The surface morphology of the bars was analyzed by scanning electron microscopy (SEM) and their surface was analyzed with X-ray diffraction (XRD).

Twelve mature healthy male rabbits weighing from 2.5 to 3.5 kg were chosen. Before surgical operation, Titanium bars were sterilized with high-pressure water for 30 minutes. The rabbits were anesthetized by an intravenous injection of pentobarbital sodium (2%ml/Kg). The bars were subcutaneously implanted into the rabbits. The part with screws was anchored in the femur bone completely and the other part was placed in muscle and other connective tissue just under the skin. 4 and 8 weeks after implantation, the rabbits were sacrificed and only the implants with the surrounding soft tissue were harvested. Some of the bars were used for the push-out mechanical test (SHIMADZN AG-10TA, JAPAN) and the others were fixed in 10% formalin, HE stained for the histological observation under light microscopy. The interface between tissue and implant was analyzed by scanning electrical microscopy and energy dispersion x-ray spectroscopy (SEM-EDX)

Results and Discussion

After plasma sprayed and subsequent water vapor treatment, the HA coating was about 50-60μm thick, with crystallinity above 80%. (Fig. 1A, B) After anode oxidized, the surface of Ti became uniform microporous and the XRD showed that lot of the TiO_2 layer appeared rutile and the other was anatase. The Ti peak was reflected from the underlying substrate. (Fig.2A, B)

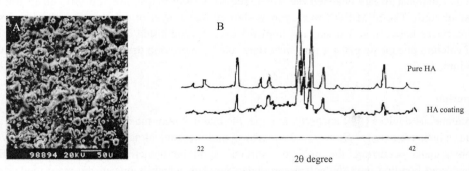

Fig.1   SEM photograph of HAC Ti (A) and the XRD patterns of the coating (B)

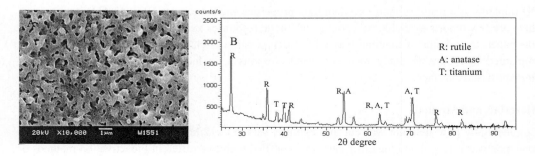

Fig.2 SEM photograph of AO Ti (A) and the XRD patterns of the surface (B)

The attachment strengths of Ti with different surface characteristics are shown in table 1. From the table, we can see that 4 weeks postoperation, the attachment strength of the AO group was similar to the HAC group all about 0.15Mpa. At the end of 8 weeks, the attachment strength of both of the AO and HAC groups were higher than those of 4 weeks, but the strength of HAC group was almost three times more than the AO group. The sandblasted bars attached the soft tissue so weakly that no push-out strength could be determined.

Table 1   Attachment strength of Ti with different surface characteristics   [x 10^6Pa]

	Time[week]	Sample 1	Sample 2	Sample 3	Sample 4	Sample 5	Average
HAC	4	0.17	0.32	0.14	0.14	0.09	0.17
	8	2.64	1.98	2.16	2.47	1.52	2.15
AO	4	0.14	0.09	0.17	0.14	0.11	0.13
	8	0.61	0.69	1.36	0.49	0.21	0.67
SB	8	-	-	-	-	-	-

These results can be conformed by the histological sections. At 4 weeks postoperation, the thickness of the fibrous capsule around the HAC group and AO group were equally thin and a few inflammatory cells were present (Fig 3A, B), but the thickness of the fibrous capsule around the sand blasted group was so thick that the soft tissue could not adhere the implant surface firmly (Fig 3C), so the attachment strength of SB was defined as 0 Pa. After 8 weeks, although the capsule's thickness of the HAC group and the AO group became thicker than that of 4 weeks to some different extent (Fig.4A, B), the attachment strength of both of the groups became higher. Meanwhile, Fig.4 demonstrated that the fibrous capsule adjacent to the implant surface became rougher than at 4 weeks and inflammatory cells were hardly found.

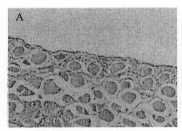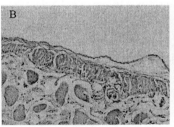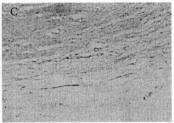

Fig.3   Histological sections of the interface between the soft tissue and the different Ti after 4 weeks.
A: HA coated Ti   B: anode oxidized Ti   C: sand-blasted Ti

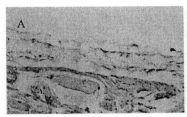

Fig.4   Histological sections of the fibrous capsule interfaced on different Ti after 8 weeks.
A: HA coated Ti   B: anode oxidized Ti

Through the observation of SEM-EDX (Fig.5A, B), Ca and P peak clearly appeared at the interface of tissue-AO implant, while none appeared at the interface of tissue-SB implant. According to the previous works, HA coating can degrade gradually and induce carbonate apatite to form at the interface of tissue-implant [5], and the AO bioactive Ti can also induce carbonate apatite on the surface [6] *in vitro*. So, it was assumed that the layer of calcium phosphate was taking part in the adherence between the soft tissue and the implant, which may induce the fibrous tissue or collagen fiber ingrowth into the structure of CaP layer. The HA coating's ability to induce the formation of carbonated apatite maybe stronger than that of the bioactive anode oxidized surface *in vivo*. So, at the early beginning of the implantation of 4 weeks, the difference between the two groups was not significant, but it became clear after 8 weeks. Thus, we can deduce that the formation of the layer of CaP plays a very important role in the tissue response at the tissue-implant interface.

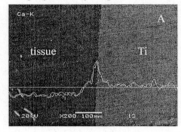 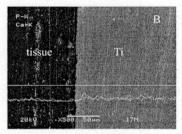

Fig.5   SEM-EDX photograph about Ca and P on the interface on the different Ti at 8 weeks
A: anode oxidized Ti   B: sandblasted Ti

Conclusions

Accordingly, the CaP layer plays a very important role in the attachment between the soft tissue and the implant, which may induce the fibrous tissue or collagen fiber ingrowth into its structure.

Acknowledgement

This work is supported by the National High Technology Research and Development Program of China (863 program), No.:2001AA326010.

References

[1] Andreas F.Von Recum: Journal of Biomedical Materials Research, Vol.18 (1984), p. 323-336

[2] B.Chehroudi, T.R.L.Gould and D.M.Brunette: Journal of Biomedical Materials Research, Vol. 26 (1992), p.493-515

[3] J.A.Jansen, Y.G.C.J.Paquay, J.P.C.M.van der Waerden: Biomaterials, Vol. 28(1994), p.1047-1054

[4] M.Gerritsen, Y.G.C.J.Paquay, J.A.Jansen: Journal of Materials Science: Materials in Medicine, Vol. 9 (1008), p. 523-528

[5] J.A.Jansen, J.P.C.M.Van der Waerden, J.G.C.Wolke: Journal of Materials Science: Materials in Medicine, Vol. 4 (1993), p. p.466-470

[6] B.C.Yang, Y.Wu, M.Tang, J.Y.Cheng, X.D.Zhang: Chinese Materials Conference 2002,10, D69

Key Engineering Materials Vols. 254-256(2004) pp. 729-732
online at http://www.scientific.net
© 2004 Trans Tech Publications, Switzerland

Effect of Electrophoretic Apatite Coating on Osseointegration of Titanium Dental Implants

Cristina C. Almeida[1], Lídia A. Sena[1], Alexandre M. Rossi[2], Marcelo Pinto[3], Carlos A. Muller[3] and Gloria A. Soares[1]

[1] Metallurgy and Materials Dep., UFRJ, P.O.Box 68505, Rio de Janeiro, 21941-972, RJ, Brazil, gloria@ufrj.br

[2] CBPF, Rua Dr. Xavier Sigaud, 150, Rio de Janeiro, 22290-180, RJ, Brazil

[3] IOC/FIOCRUZ, Av. Brasil, 4365, Rio de Janeiro, 21045-900, RJ, Brazil

Keywords: hydroxyapatite, coating, osseointegragion, dental implants.

Abstract. This paper discusses the *in vivo* behavior of HA-coated titanium implants compared with non-coated implants. HA coating was produced by using electrophoresis. Coatings were characterized using scanning electron microscopy (SEM) and the bone-implant contact length was used as the measure of osseointegration. Results show significantly higher percentage of bone-implant contact for hydroxyapatite coated than for non-coated condition. The obtained results allow us to conclude that coating by electrophoresis appears to be an attractive process to coat metallic implants with an osteoconductive material like hydroxyapatite, for small or medium scale production.

Introduction

Titanium is considered a suitable metal for permanent implants, in dental and orthopedic areas. In order to accelerate bone formation around the implant, several surface treatments [1, 2] including bioactive coatings [3-9] were developed. As synthetic hydroxyapatite is very similar to the mineral part of bone, coating metallic implants with hydroxyapatite (HA) or other calcium phosphates ceramics (CPC) may result in optimized properties by coupling mechanical strength with osteoconductive behavior. These coatings, usually produced by plasma spray techniques, can be obtained by using alternative processes, like biomimetic process [5], electrolytic deposition [6] or electrophoretic process [7-9].

The electrophoretic process seems to be a good alternative choice. The time required for coating is very short and the process presents high reproducibility [7]. Although bone formation around an implant is a complex process, several researchers measure osseointegration as the percentage of direct bone-contact implant in *in vivo* tests [2, 8, 10]. This methodology allows the comparison between different surface finishing. In this paper the osseointegration of HA electrophoretically coated Ti implants was compared to non-coated titanium ones, using rabbit as the animal model.

Methods

Screw dental implants made from commercially pure titanium (c.p. Ti) were coated with synthetic hydroxyapatite (HA) with Ca/P ratio equal to 1.66 ± 0.04 by using electrophoresis. A precipitation wet method was used to synthesise the hydroxyapatite powder. The coating methodology was developed by Sena *et al.* [7] following Zhitomirsky and Gal-Or [9] procedure. Titanium implants were used as cathode, and platinum as anode in an electrophoretic cell, with electrodes 40 mm apart as schematised on Figure 1. HA suspension in ethanol was prepared using hydrochloric acid as dispersing agent. The process was carried out at 24 V for 3 minutes. The HA coating was characterised by scanning electron microscopy (SEM) and energy dispersive spectroscopy (EDS).

In vivo tests were carried out through the insertion of both non-coated and HA-coated implants into rabbits' tibia, following approved protocols. A total of six New Zealand rabbits, adults of both sexes, weighing between 3.2 and 4kg were used in the study. Each animal received two implants, one of commercially pure titanium and the other one coated with hydroxyapatite, both inserted in the proximal tibial methaphysis. Three animals received c.p. Ti implants in the proximal tibial methaphysis and three received HA-coated implants in the distal tibial methaphysis. In the remaining animals the opposite procedure was used. Three animals were sacrificed after two months and the other three after three months.

Histomorphometric analysis was performed by processing back-scattered electron (BSE) images with 100x magnification for both side of each section. This methodology was successfully applied by Gotfredsen *et al.* [2] and Vidigal Jr. *et al.* [11]. Digital images were processed with the aid of an image analysis software and the percentage of direct bone contact was calculated as the length of implant in contact with bone divided by the total length of implant.

Results and Discussion

SEM micrograph of HA-coated implants shows an homogenous layers of hydroxyapatite with "crackled" appearance due to the volumetric contraction due to the calcination process, Fig. 2. Figure 3 shows the Ti screw implanted on rabbit's tibia with different magnifications. In these images obtained with backscattered electrons (BSE), the gray phase was identified as calcium-phosphorus rich, while the titanium appears white, due to the fact that titanium has a greater atomic weight than phosphate phase. SEM image of bone section of HA-coated implant (Fig. 3b) shows an almost continuous Ca-P-rich layer with thickness varying from 4 to 8 μm well adhered to titanium implant [7].

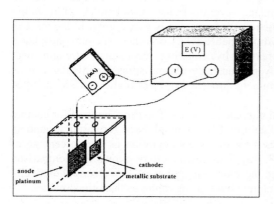

Fig.1 Schematic drawing of electrophoresis

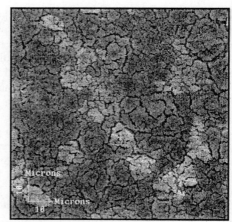

Fig.2 SEM image of HA-coated implant surface (original magnification 1,000 x)

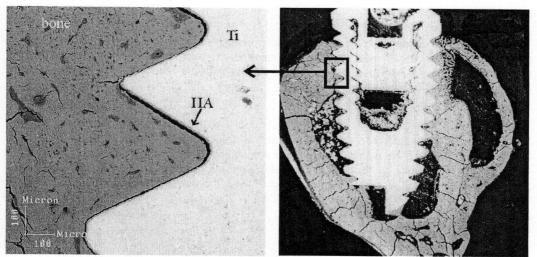

Fig. 3   BSE images of HA-coated Ti implant placed on rabbit's tibia

Figure 4 showed the means and standard deviations of the percentage of bone-implant contact for each animal, considering   the three best consecutive threads. Non-significant differences ($\alpha=0,05\%$) were observed when comparing the results obtained after 2 or 3 months, in accordance with the findings of some authors [2, 8]. Consequently, the results were grouped - independent of osseointegration time - and the percentage of bone-implant contact of c. p. Ti and HA-coated implant was compared, Figure 5. Student t-test showed significant differences ($\alpha=0,05\%$) between non-coated and HA-coated implants for the three best consecutive threads. The superior performance of HA-coated implants regarding the bone-implant contact percentage (Figure 5) is also in good agreement with other researchers results [2, 10], although the number of implants and animals used in this study cannot unequivocally support this conclusion.

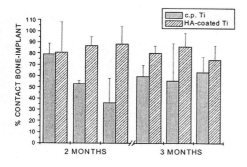

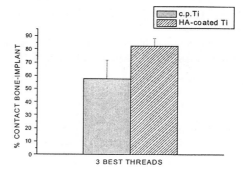

Fig. 4 Bone-implant contact for all animals, considering the three best consecutive threads

Fig. 5 The percentage of bone-implant contact for the three best consecutive threads of both conditions

Conclusions

The obtained results allow us to conclude that electrophoresis appears to be an attractive process to coat metallic implants with an osteoconductive material like hydroxyapatite, for small or medium scale production. However, additional tests must still be carried out before process commercialisation.

Acknowledgements

Financial support was received from several Brazilian agencies like CAPES, CNPq, FAPERJ and RENAMI/CNPq. The authors would also like to acknowledge FIOCRUZ where the surgeries were carried out. In addition, the authors are grateful to thank Neodent (Curitiba, PR, Brazil) for providing titanium implants. This research is part of the Molecular and Interfaces Nanotechnology Research Network (RENAMI), supported by CNPq.

References

[1] Z. Strnad, J. Strnad, C. Povysil and K. Urban: The Intern. JOMI 15 (2000), p. 483-490.

[2] K. Gotfredsen, A. Wennerberg, C. Johansson, et al.: Journ.Biom.Mater.Research 29 (1995), p.1223-1231.

[3] L.A. Sena, N.C.C. Rocha, M.C. Andrade and G.A. Soares: Surface and Coatings Technology 60 (2002), p.1-7.

[4] W.R. Lacefield: Implant Dentistry 7 (1998), p. 315-320.

[5] T. Kokubo, F. Miyaji, H. Min-Kim and T. Nakamura: Journal of the American Ceramic Society 79 (1996), p. 1127-29.

[6] M.H. Prado da Silva, J.H.C. Lima, G.A. Soares, et al.: Surface and Coatings Technology 137 (2001), p.270-276.

[7] L.A. Sena, M.C. Andrade, A.M. Rossi and G.A. Soares: Journ of Biomedical Mater Research 60 (2002), p. 1-7.

[8] M. Gottlander, C.B. Johansson, A. Wennerberg, T. Albrektsson, S. Randin and P. Ducheyne: Biomaterials 18 (1997), p. 551-57.

[9] L. Zhitomirsky and L. Gal-Or: Journal of Materials Science: Materials in Medicine, 8 (1997), p.213-19.

[10] G.M. Vidigal Jr., L.C.A. Aragones, A.Campos Jr. and M. Groisman: Imp. Dent. 8 (1999), p.295-302.

[11] G.M. Vidigal Jr., M.S. Sader and G.A. Soares: Proceedings of the XVIII Brazilian Congress for Microscopy and Microanalysis (2001), p. 9-10.

Key Engineering Materials Vols. 254-256(2004) pp. 733-736
online at http://www.scientific.net
© 2004 Trans Tech Publications, Switzerland

The Dualism of Nacre

Evelyne Lopez, Christian Milet, Meriem Lamghari, Lucilia Pereira Mouries, Sandrine Borzeix and Sophie Berland

Muséum National d'Histoire Naturelle

Département des Milieux et Peuplements Aquatiques USM 0401

UMR CNRS 5178 "Biologie des Organismes Marins et Ecosystèmes"

7, rue Cuvier 75231 Paris cedex 05

email : lopez@mnhn.fr, milet@mnhn.fr, berland@mnhn.fr

Keywords: Biomaterial, nacre, bone cells, osteogenesis, organic matrix, signal molecules

Abstract. The nacreous part of the shell of *Pinctada maxima*, bivalve mollusk, was used in these studies. *In vivo* studies, carried out on adult sheep, showed that implanted pieces of nacre pass bone acceptance. Nacre implants were not subjected to intolerance reaction and the recipient bone provided with nacre underwent a sequence of bone regeneration within an osteoprogenitor rich cell layer. Newly formed bone and nacre welded into a dual biomineralized unit. For *in vitro* studies, the water soluble organic matrix was extracted from powdered nacre by a gentle non-decalcifying process. Three mammalian cell types, fibroblasts (human), bone marrow stromal cells (rat) and pre-osteoblasts (mouse) were used to characterize the effect of nacre water soluble matrix (WSM) on mammal cell recruitment and differentiation in the osteogenic pathway. Alkaline phosphatase activity and osteocalcine expression, markers of cell stimulation and osteoblastic differentiation, were analysed in the culture models. *In vitro* studies provided evidence for the presence, in nacre organic matrix, of signal molecules responsible for the recruitment of mammal cells in the osteogenic pathway and bone cell activation undergoing a complete sequence of mineralization.

Introduction

Nacre, mother of pearl, offers a model for studying biomineralization control processes and can provide support for bone regeneration [1].

A material of fragile aspect, nacre however results from a multiscale organization yielding mechanical strengh properties. Nacre microstructure consists of alternative sheets of mineral, CaCO3 crystallized under aragonite form, and organic matrix.

Low in rate, 1 to 5%, the organic matrix is essential to nacre properties. It provides strong resistance to deformation [2] and matrix-mineral relationships manage crystal growth [3].

Bone and nacre share a matrix supported and controlled biomineralization process. They are the result of a mineral crystallization deposited on an organic matrix scaffold: calcium phosphate in hydroxyapatite form in bone and calcium carbonate in aragonite form in nacre. This matrix is laid by special cells: osteoblasts in vertebrates and epithelial mantle cells in mollusks. Bone and shell undergo self reparation when damaged, evidence for the presence in the two mineralized structures of signal molecules, targeting the cells involved in bone or shell renewal.

Materials and methods

Biomaterial. Mother of pearl was obtained from the inner shell layer of Pinctada. When used under powder form, nacre was ground to particles 50-150μm size range at the Centre de Transfert de Technologie Ceramique, Limoges, France.

In vivo studies. Studies were done on sheep. Nacre was implanted under sterile surgical conditions within experimental bone defects. Solid nacre implants were placed in the femur epiphysis [4]. Nacre powder was used to fill cavities prepared in lumbar vertebrae [5].

Extraction of water soluble matrix (WSM). Nacre powder was suspended according to the procedure described in patent n°FR951225, and the resulting supernatent, WSM, was added to the culture media.

In vitro studies. MRC5, human fetal lung fibroblasts, obtained from the Institute for medical research (London); primary cultures of bone marrow cells from femurs of young male wistar rats, MC3T3-E1, preosteoblasts cell line from mouse calvaria (Dr Kumeggawa kind supply, Josai Dental University, Sakado, Japan) and MG63 cell line, (human osteosarcoma, European Collecion of Cell Cultures, Salisbury, UK) model of immature osteoblasts. Cultures were processed as described in [6]. Effects of WSM on the cells were evaluated by mesurement of Alkaline phosphatase activity, early marker of cell differentiation in the osteogenic lineage, osteocalcine levels and detection of mineralized nodule using Von Kossa staining. WSM activity was compared to that of standards osteogenesis inducers, Dexamethasone (DEX), BMP-2, ß-glycerol-phosphate (BGP), and ascorbic acid (AA).

Results

Nacre exhibits dual biodegradability properties. When implanted as replacement bone device, nacre pieces have a persistance in bone tissue without general shape alteration. On another hand, nacre undergoes biodissolution when implanted under powder form (fig. 1a). The biodissolution process is however initiated in implanted pieces of nacre and lead to surface erosion at the edge of the implant (fig. 1b). In any case, new bone formation occured after nacre implantation. Histological analysis of undecalcified samples showed that osteogenesis began within an intervening activated cell layer. Osteogenesis resulted in a direct contact between newly formed bone an nacre, providing the anchoring of the implant (fig. 1b).

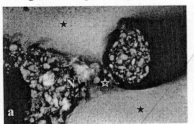 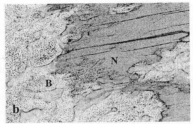

Fig. 1. Light microscopy. Undecalcified sections of trabecular bone provided by a nacre implant, under powder form (a) or as a raw piece of nacre (b).

a - Microradography showing new bone trabeculae formation (I) within the host bone (H). Nacre (N) is being gradually dissolved.

b - Direct bone (B) nacre (N) interface providing the bone implant anchoring. Nacre piece underwent surface erosion. Toluidine blue staining.

X-ray diffraction electron microspy analysis established that the first step of nacre osteogenic effect occured via the establishment of a calcium and phosphorus front (fig. 2). Ca and P, constitutive elements of bone mineral phasis, show a dense distribution pattern at the edge of nacre within the host tissues. The Ca-P rich cell layer indicates that Nacre cellular activation targets osteogenesis.

Fig. 2. Electron microscopy and energy dispersive X-ray analysis at the interface of bone replacement nacre implant (X 1200).
a- Scanning image showing nacre implant (N) and mineralization clusters in the host tissue (™) and at the edge of nacre (∏).
b- Phosphorus rich front at the edge of the nacre implant (⌌) within the host tissues (™).

The organic matrix of nacre was supposed to be the source of signal molecules diffusible from the implanted nacre. The water soluble components of the organic matrix were extracted from nacre following a gentle patented process. No demineralization step was performed. The organic matrix released by this process exhibits paradoxical features. It is highly hydrophobic and rich in glycine and alanine, more than 65% of the total amino acid composition. The large proportion of hydrophobic amino acids results in a very low C/HP value (0.29).

To test the hypothesis that WSM was responible for nacre bioactivity on bone, in vitro experiments were done using the mammalian cell types involved in bone regeneration. Cell types were chosen to provide a set of osteogenic lineage, from precursors to predifferenciated bone cells. Bone marrow cells, MRC5 fibroblasts, MC3T3-E1 and MG63, respectively preosteoblasts and immature osteoblasts were used.

Bone marrow cells supplemented with WSM in combination with standards inducers BGP, AA and DEX, started to mineralize after 14 days of culture (fig. 3b). In the meanwhile, osteocalcine level increased consistently, corroborating the maturation in bone forming cells of WSM treated bone marrow cells (fig. 3).

MRC5 fibroblasts response to WSM treatment was an early increase in alkaline phosphatase activity when compared with differentiating factors (DEX and BMP-2) effects. Alkaline phosphatase activity and osteocalcine measurements showed that osteoblastic cells were induced to maturation (MG63), up to mineralization in MC3T3-E1 cultures.

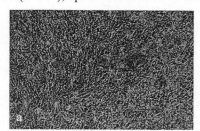
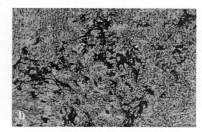

Calcium staining at day 14 using Von Kossa technique.
a – Control. b – mineral clusters (dark) in WSM treated cells.

Day 14 assay	Control	WSM treated
Osteocalcine (ng/mg prot)	2646,5 ± 495,5	4651,8 ± 310,2 *

Fig. 3: WSM effect on mineralisation in bone marrow stromal cells cultures

The water-soluble matrix isolated from *Pinctada maxima* nacre contains the signals responsible for the biological activity of the whole nacre. Indeed, this matrix acts in particular on bone cell differentiation, up until the final step of mineralization.

Discussion

Nacre can provide biomaterials for bone regeneration. The objective to reach in bone reconstruction is the biomechanical quality of the newly formed bone or the bone provided by a replacement implant. Implant fixation implies to obtain strong binding while preserving the physiological conditions and normal stress pattern. Our studies gave evidence that bone and nacre can form a hybride interactive system, transient or sustained since durability of implanted nacre is shape-related [4,5]. Nacre dual composition is responsible for mechanical strength and bioactivity. Organic matrix of nacre also supports duality. The organic framework is involved in stress shielding [2] and in the activation of the cells responsible for biomineralization [6]. Matrix proteins characterization attempts also gave rise to paradoxical findings. The pattern of a water solubilized fraction shows a profil conventionnally classified hydrophobic [7]. Nacre water soluble matrix encloses diffusible signal molecules that can target bone forming cells. The guidelines for the reasons why nacre can pass bone acceptance come from nacre matrix bioactivity on bone cell lineage and on cell recruitment in the osteogenesis pathway .

Because the interactions resulting in controlled crystallization are tailored to biological function, the understanding of bone nacre interactions would therefore be of interest in structural biology, medicine and engineering technology. Nacre, highly organized organic and mineral material hides clues in mechanical design and bioactivity supply and thus provides a challenge to biomimicry.

References

[1] P. Westbroek and F. Marin. Nature vol 392 (1998), pp 861-862

[2] A.G. Evans., Z. Suo, R.Z. Wang, I.A. Aksay, M.Y. He and J.W. Hutchinson.. J. Mater. Res. Vol 16 (2001), pp 2475-2484

[3] G. Falini, S. Albeck, S. Weiner and L. Addadi. Science vol 271 (1996), pp 67-69

[4] G. Atlan, O. Delattre, S. Berland, A.Le Faou, G. Nabias G, D. Cot and E. Lopez. Biomaterials vol 20 (1999),pp 1017-1022

[5] M. Lamghari, S. Berland, A. Laurent, H. Huet and E. Lopez. Biomaterials vol 22 (2001), pp 555-562

[6] L. Pereira-Mouries, M.J. Almeida, C. Milet, S. Berland and E. Lopez. Comparative Biochemistry and Physiology vol 132 (2002), pp 217-229

[7] L. Pereira-Mouries, M.J. Almeida, C. Ribeiro, J. Peduzzi, M. Barthelemy C. Milet and E. Lopez. European Journal of Biochemistry, vol 269 (2002), pp 4994-5003

Key Engineering Materials Vols. 254-256(2004) pp. 737-740
online at http://www.scientific.net
© 2004 Trans Tech Publications, Switzerland

Osseointegration of Grit-Blasted and Bioactive Titanium Implants: Histomorphometry in Minipigs

C. Aparicio[1], F.J. Gil[1], U. Thams[2], F. Muñoz[3], A. Padrós[4] and J.A. Planell[1]

[1] Research Center of Biomedical Engineering (CREB), Technical University of Catalonia, Avda. Diagonal 647, 08028 Barcelona, Spain. conrado.aparicio@upc.es

[2] Department of Animal Pathology II, University Complutense of Madrid, Ciudad Universitaria, 28040 Madrid, Spain.

[3] Veterinary Hospital Rof Codina, University of Santiago de Compostela, Estrada da Granxa s/n, 27002 Lugo, Spain.

[4] Instituto Padrós. Martí i Julià, 6-8. 08034 Barcelona, Spain.

Keywords: bioactive titanium, roughness, grit blasting, histomorphometry, osseointegration, dental implant.

Abstract. In order to improve osseointegration of commercially pure titanium implants one of the main influencing factors is surface quality, which refers to their topographical and physicochemical features. A new two-step (2S) surface treatment has been previously developed. The method consists of 1) grit blasting the surface in order to roughen it and 2) thermochemical treatment in order to obtain a bioactive metal surface. This new 2S-treatment has been *in vivo* evaluated and compared with as-machined (Ctr), acid etched (E) and grit-blasted (R) implants by means of an histomorphometric quantification (percentage of the total implant surface in direct contact with new bone) after 2, 4, 6 and 10 weeks of implantation into the mandibles of minipigs. The results showed that the rough and bioactive-c.p. Ti implants obtained by a new two-step surface treatment accelerate their osseointegration compared to grit-blasted, acid etched and as-machined implants, which make them preferential candidates as immediate-loaded dental implants. These implants developed *in vivo* onto their surfaces a layer of a calcium phosphate, which is probably apatite.

Introduction

Dental and orthopaedic implants are widely made of commercially pure titanium (c.p. Ti) and its alloys. The appropriate interactions between titanium implants and bony tissue results in mechanical anchorage of the implant but without chemical bond because titanium is a bioinert material that leads a very thin soft-tissue layer formation around its surface in clinical use [1]. In order to improve implant-bone fixation, i.e. osseointegration, physicochemical and topographical surface quality of c.p. Ti dental implants is one of the most influencing factors [2].

It is known that a firmer fixation can be achieved with a rough titanium surface than with a smooth one [3]. It is also known that this is more relevant at long implantation times when the bone will have completely penetrated in surface microroughness and, consequently, mechanical interlocking improved.

On the other hand, several surface chemical treatments have been developed in order to obtain a bioactive titanium surface that will be able to exhibit bone-bonding ability [4-6]. In Kokubo *et al.* method [4], an alkali and heat treatment is carried out on titanium in order to obtain a dense and tightly-bonded amorphous sodium-titanate gel on its surface. This gel can react with psysiological fluids *in vivo* [7] or with a Ca-P supersaturated acellular simulated-body fluid (SBF) *in vitro* [4]. As a result, an homogeneous apatite layer is formed on the metal substrate. This apatite layer is an essential requirement for artificial materials to bond to living bone in the body [8].

In previous studies, a new two-step method (2S) for obtaining bioactive rough titanium surfaces was developed for improving short-term (due to its bioactivity) and long-term (due to its roughness) osseointegration [9]. This 2S-method consists on: 1) Grit blasting (GB) on titanium surface in order to roughen it; and 2) Thermo-chemical treatment (TCh) in order to obtain a titanate-surface with bone bonding ability. It was also demonstrated that a rough and bioactive surface enhances *in vitro* adhesion and differentiation activity of human osteoblasts cells [10,11].

The aim of this work is to evaluate the *in vivo* short and mid-term response of these rough and bioactive c.p. Ti dental implants compared to as-machined and rough (grit-blasted or acid-etched) implants. It is hypothesised that 2S-treated c.p. Ti implants will develop *in vivo* a layer of apatite onto their surfaces, showing their bioactivity and accelerating their osseointegration. A histomorphometric quantification of the direct contact between new bone and implants inserted into mandibles of minipigs and *in vivo* bioactivity evaluation have been carried out.

Materials and methods

Implants. Four different surface-qualities of screw-shaped c.p.-Ti dental implants were prepared in the laboratory:

1.- as-machined (Ctr),

2.- acid-etched (E): with 0,15-M fluorhidric acid during 15 s at room temperature.

3.- rough-GB (R): with Al_2O_3-particles of 600 μm in mean size and 0,25 MPa blasting-pressure.

4.- rough-GB and bioactive-TCh (2S): blasting in the same way as R-implants + TCh-treatment developed by Kokubo *et al.* [4].

Animal study. The implants were implanted adequately distributed along both mandibles of 4 animals, with a total of 20 implants per minipig: 4 Ctr-, 4 E-, 8 R- and 4 2S-implants. The minipigs were females and 5-7 year-old. The animals were powder-fed during all the experiment.

The implants were inserted into the mandibles after a 2-month period without teeth. The surgical procedure was carried out with the animals anesthetised and following the same steps as for implantation into human mandibles.

One minipig was sacrificed at each time of evaluation, i.e. at weeks 2, 4, 6 and 10 after implantation.

Histomorphometry. Block sections containing the implants were dehydrated in alcohol and embedded in a photopolymerizable resin. Longitudinal 100 μm-thick sections were sawed, polished, and stained with toluidine blue.

As a measure of the osseointegration of the implants, the percentage of the total implant surface in direct contact with new bone (% direct contact) was measured in 160 images (2 per implant). The images (x100) were obtained with an optical microscope (Olympus CH30) and an acquisition software (PCTV Vision 1.96), and finally analysed with an image-analyser software (MicroImage 4.0). The evaluation was carried out along all the threads of the implant.

Evaluation of bioactivity. Blocks with Ctr-, E-, R- or 2S-implants were observed with an Scanning Electron Microscope (SEM) (Jeol JSM 6400) and chemically analysed by means of X-Ray dispersive energy (EDS) (Lynk Analytical LZ-5).

Results and Discussion

There were 3, 5, 2 and none failed implants for Ctr-, E-, R- and 2S-surfaces, respectively.

Fig. 1 summarises the histomorphometric quantification of the osseointegration of the implants.

The Ctr-implants had significant-statistically (Fisher's test; 0,05) lower percentage of direct contact than the R- and 2S-implants from week 4, which demonstrates that highly rough ($R_a \approx 4$ µm) surfaces (bioactive or not) improve bone integration, as compared to those as-machined ($R_a \approx 0.7$ µm). However, E-implants ($R_a \approx 1.5$ µm) had not higher values than as-machined implants until 10 weeks of implantation.

2S-treated implants had significant statistically differences in percentage of direct contact (Fisher's test; 0,05) between weeks 2 and 4, maintaining the % direct contact in week 4 until

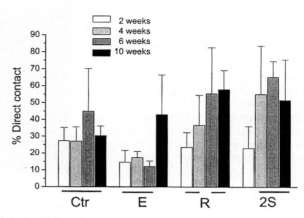

Fig. 1: % of the total implant surface in direct contact with new bone for the four different types of surfaces and all the times of implantation studied. Straight lines bellow the columns join groups with non-significant statistically differences.

the end of the experiment without significant changes (% direct contact ≈ 55 %). However, R-treated implants only showed significant statistically differences in percentage of direct contact between weeks 4 and 6, reaching at the end of the study the same value as 2S-implants. These results indicate that an acceleration of osseointegration at short-times of implantation as brief as 4 weeks is achieved.

At 2 weeks of implantation the amount of bone in contact with the surface of all kind of implants is mainly because of the contact between the threads and the original machined bone during surgical procedure (Fig. 2, a). At 4 weeks of implantation newly immature bone was formed in closed contact with 2S-implants (Fig. 2, b), whereas at 6 weeks of implantation all the bone in contact with 2S- implant is new organised bone (Fig. 2, c). However, immature bone can be seen in R-implants after 6 weeks of implantation. This shows qualitatively the different rates of formation of bone around the different types of implants.

Fig. 2: Histologies (x100) of 2S-implants after a) 2 weeks (showing position of the implant after surgical procedure), b) 4 weeks (showing immature bone in direct contact with implant) and c) 6 weeks (showing well-organised new bone in direct contact with implant) of implantation.

SEM analysis of the interfaces between bone and implant of the rough and bioactive implants confirmed the development of a calcium-phosphate layer (Fig. 3). The Ca/P-ratio obtained with EDS for the layer was 2,20 and for bone was 2,16. Despite the high Ca/P-ratio measured (probably obtained because EDS peaks for P and Au are very close) this result could confirm that the calcium-phosphate layer is apatite. This apatite-layer may be the responsible for accelerating osseointegration of the rough and bioactive implants compared to those that only were rough.

Conclusions

Rough and bioactive c.p. Ti implants obtained by a new two-step surface treatment accelerate their osseointegration compared to grit-blasted, acid etched and as-machined implants. This acceleration of the osseointegration makes 2S-implants preferential candidates as immediately-loaded dental implants. These implants developed *in vivo* onto their surfaces a layer of a calcium phosphate, which is probably apatite.

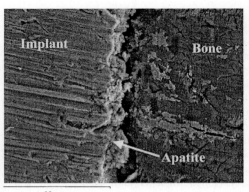

60μm

Fig. 3: SEM image of a 2S-implant after 6 weeks of implantation showing a cristalline calcium-phosphate layer at the implant-bone interface

Acknowledgements

The authors would like to acknowledge Klockner Implants, S.L. for providing the implants used in this work. The authors would also acknowledge Drs. J.M. Arano, E. Pedemonte, A. Sanromán, J.M. Manero, M. López and S. Stigson for their help in different steps during the development of this study.

References

[1] P. Thomsen, C. Larsson, L.E. Ericsson, L. Sennerby, J. Lausmaa and B. Kasemo: J. Mater. Sci.: Mater. Med. 8 (1997), p. 653-665.
[2] T. Albrektsson, P.I. Brånemark, B.O. Hansson and J. Lindström: Acta Orthop. Scand. 52 (1981), p. 155-170.
[3] A. Wennerberg: On surface roughness and implant incorporation (PhD Thesis, University of Göteborg, Sweden 1996).
[4] T. Kokubo, F Miyaji and H.M. Kim: J. Amer. Ceram. Soc. 79 (1996), p. 1127-1129.
[5] C. Ohtsuki, H Iida, S. Hayakawa and A. Osaka: J. Biomed. Mater. Res. 39 (1998), p. 141-152.
[6] P. Li, I. Kangasniemim, K. de Groot and T. Kokubo: J. Amer. Ceram. Soc. 77 (1994), p. 1307-1312.
[7] W. Q. Yan, T. Nakamura, M. Kobayashi, H.M. Kim, F. Miyaji, T. Kokubo: J. Biomed. Mater. Res. 37 (1997), p. 265-275.
[8] T. Kokubo: Anales de Química Int. Ed. 93 (1997), p. S49-S55.
[9] C. Aparicio, F.J. Gil, M. Nilsson, J.M. Manero and J.A. Planell: Transactions of the Sixth World Biomaterials Congress. Vol. III (2000), p.1050.
[10] C. Aparicio, F.J. Gil, J.A. Planell and E. Engel: J. Mater. Sci.: Mater. Med. 13 (2002), p. 1050-1111.
[11] F.J. Gil, A. Padrós, J.M. Manero, C. Aparicio, M. Nilsson and J.A. Planell: Mater. Sci. Eng. C 22 (2002), p. 53-60.

Key Engineering Materials Vols. 254-256(2004) pp. 741-744
online at http://www.scientific.net

Mechanism of Apatite Formation on Anodically Oxidized Titanium Metal in Simulated Body Fluid

Hyun-Min Kim,[1] Hideki Kaneko,[2] Masakazu Kawashita[2] Tadashi Kokubo,[3] and Takashi Nakamura[4]

[1]Department of Ceramic Engineering, School of Advanced Materials Engineering, Yonsei University
134 Shinchon-dong, Seodaemun-gu, Seoul 120-749, Korea (hmkim@yonsei.ac.kr)
[2]Department of Material Chemistry, Graduate School of Engineering, Kyoto University
Yoshida, Sakyo-ku, Kyoto 606-8501, Japan
[3]Research Institute of Science and Technology, Chubu University
1200 Matsumoto-cho, Kasugai-shi, Aichi 487-8501, Japan
[4]Department of Orthopaedic Surgery, Graduate School of Medicine, Kyoto University
Shogoin, Sakyo-ku, Kyoto 606-8507, Japan

Keywords: Titanium, anodic oxidation, titania, bioactivity, apatite, SBF, TEM-EDX

Abstract. Mechanism of apatite formation on anodically oxidized titanium metal in a simulated body fluid was investigated by XPS and TEM observation. The anodically oxidized metal was found to have rutile and anatase titania with a large number of Ti-OH groups on its surface. On immersion in SBF, the metal formed a bonelike apatite on its surface through formations of an amorphous calcium titanate and an amorphous calcium phosphate. The formation of the calcium titanate was induced by the Ti-OH groups, which reveals negative charge to interact selectively with positively charged calcium ions in the fluid. The calcium titanate is postulated to reveal positive charge, thereby interacting with the negatively charged phosphate ions in the fluid to form the calcium phosphate, which eventually crystallized into bonelike apatite.

Introduction

An essential requirement for an artificial material to bond to living bone is the formation of a biologically active bonelike apatite layer on its surface in a body environment [1]. Titania hydrogel has been documented to induce formation of bonelike apatite formation on its surface in body environment [2]. Based upon these fundamentals, titanium metal has been subjected to anodic oxidation (AO) to form a bioactive titania on its surface. When the AO was applied to the metal at spark discharge field in some specific aqueous electrolytes such as H_2SO_4, the treated metal formed a bonelike apatite on its surface in SBF and in vivo, through which integrates with living bone [1, 3]. While this indicates that the AO is an effective method to prepare bioactive surface in situ on titanium metal, the characteristics of the AO surface governing the formation of apatite is not yet clear.

In this study, we investigated the mechanism of apatite formation on an AO titanium metal in SBF. The surface composition and structure of the AO metal were surveyed as a function of soaking time in the simulated body fluid (SBF) using X-ray photoelectron spectroscopy (XPS) and transmission electron microscopy (TEM) attached with energy dispersive X-ray spectrometry (EDX). The process of apatite formation on the AO metal was discussed in terms of electrostatic interaction between the surface of the metal and ions in the SBF.

Materials and Methods

Plates ($10 \times 10 \times 1$ mm^3) or mesh-grids (3 mmø) of pure titanium metal (Ti > 99.8%, Nilaco Co., Tokyo, Japan) were subjected to AO by being soaked in 1M-H$_2$SO$_4$ solution and applied with electrical field of 150 V for 1 min at room temperature. The AO specimens of plates and grids were then gently washed with distilled water, and dried at 40°C for 24 h. They were soaked at 36.5°C in 30 ml of SBF [4] with a pH (7.40) and ionic concentrations (Na$^+$ 142, K$^+$ 5.0, Mg^{2+} 1.5, Ca^{2+} 2.5, Cl$^-$ 148.0, HCO$_3^-$ 4.2, HPO$_4^{2-}$ 1.0, SO$_4^{2-}$ 0.5 mM) nearly equal to those in human blood plasma.

After soaking for various periods, the plate and grid specimens were subjected to XPS (Model 5500, ULVAC-PHI Co., Ltd., Chigasaki, Japan) and TEM (JEM-2000FXIII, JEOL, Co., Tokyo, Japan) attached with EDX (Voyager III, NORAN Instruments, Inc., Middleton, WI), respectively. The analyses were performed adopting reference materials of untreated titanium metal, titania gel prepared by sol-gel process, tricalcium phosphate (Ca$_3$(PO$_4$)$_2$; Taihei Chem., Osaka, Japan) and hydroxyapatite (Ca$_{10}$(PO$_4$)$_6$(OH)$_2$; Taihei Chem.).

Results and Discussion

Figure 1 shows the O$_{1s}$ Ti$_{2p}$, Ca$_{2p}$ and P$_{2p}$ XPS spectra of the surfaces of AO titanium metals soaked in SBF for various periods. The AO surface before soaking in SBF (0 h) revealed O$_{1s}$ and Ti$_{2p}$ peaks ascribed to Ti-O and Ti-OH bonds. After soaking in SBF for 0.5 h, the surface revealed Ca$_{2p}$ peak in addition to the O$_{1s}$ and Ti$_{2p}$ peaks. While the Ti-OH peak appeared to slightly increase in O$_{1s}$ spectrum, binding energy positions were apparently unchanged. After soaking for 24 h, the surface revealed P$_{2p}$ peak in addition to the O$_{1s}$ Ti$_{2p}$ and Ca$_{2p}$ peaks. The binding energy position of O$_{1s}$ spectrum shifted to those ascribed to P-O and P-OH bonds. Thereafter by increasing soaking time to 72 h, both the Ca$_{2p}$ and P$_{2p}$ peaks increased appreciably their intensities.

Figure 2 shows the TEM-EDX profiles of the surfaces of AO titanium metals soaked in SBF for various periods. The profile together with electron diffraction pattern indicates that the AO surface of the metal (0 h) is composed of porous titania containing spherical nano-crystals of anatase and rutile (TiO$_2$). After soaking in SBF for 0.5 h, the surface titania was apparently unchanged in morphology and diffraction pattern, but the EDX detected Ca in addition to O and Ti. After 24 h, new particle-like nano-deposits were observed to precipitate on the surface titania, where EDX detected Ca and P with Ca/P atomic ratio of 1.39, but electron diffraction found no resolvable pattern. After 72 h, new aggregates of needle-like deposits were observed to precipitate on the surface, where EDX detected Na and Mg in addition to Ti, O, Ca and P, and the Ca/P ratio of 1.58. This aggregates revealed electron diffraction pattern ascribed to apatite.

The surface of AO titanium metal was shown above being composed of nano-crystalline anatase and rutile, which on immersion in SBF forms a thin amorphous calcium titanate within 0.5 h, thereon an amorphous calcium phosphate with a low Ca/P ratio within 24 h, and eventually the crystalline apatite with bonelike composition and structure within 72 h. A distinctive feature of the titania on the AO titanium metal is abundance in Ti-OH groups able to induce the formation of apatite [1, 2]. Therefore, the AO titanium metal appears to induce the apatite on its surface in the SBF via the following process.

Immediately after soaking in SBF, the Ti-OH groups of the surface anatase and rutile on the metal are dissociated into negatively charged units of Ti-O$^-$ owing to its isoelectric point considerably lower than the pH of SBF, i.e., 7.40. These Ti-O$^-$ units combine with the positively charged calcium ions in the fluid to form an amorphous calcium titanate, which is postulated to reveal positive charge. This calcium titanate thereby combines with the negatively charged phosphate ions in the fluid to form an amorphous calcium phosphate. The amorphous calcium

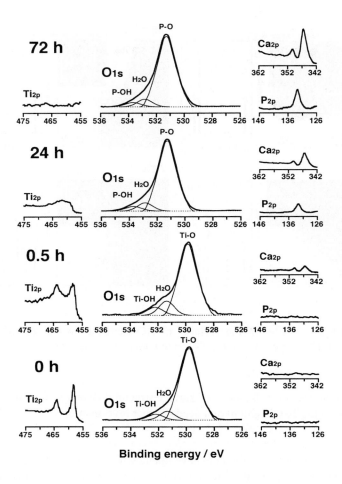

Fig.1. O_{1s} Ti_{2p}, Ca_{2p} and P_{2p} XPS spectra of the surfaces of AO titanium metals soaked in SBF for various periods.

phosphate crystallizes ions into bonelike apatite incorporating Na, Mg and carbonate, owing to a high solubility in SBF.

In fact, this mechanism of apatite formation was early documented with empirical proof concerning different type of bioactive titanium metal, which has been prepared by forming a sodium titanate on the metal through NaOH and heat treatment [5]. In this case, the sodium titanate forms the Ti-OH groups by ion exchange of Na^+ ions with H_3O^+ ions in the SBF to progress the apatite formation through the same surface structural change as discussed above, i.e., through formations of an amorphous calcium titanate and an amorphous calcium titanate. In this context, the Ti-OH groups induce the apatite formation through the same electrostatic process, in which they interact early with the calcium ions and later with the phosphate ions in the fluid. The difference

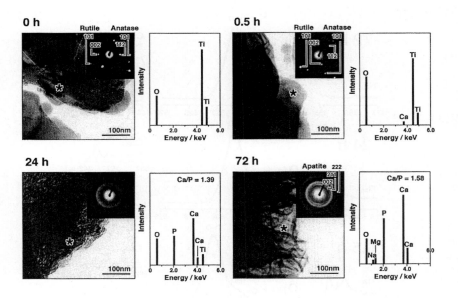

Fig.2. TEM-EDX profiles of the surfaces of AO titanium metals soaked in SBF for various periods (*: Center of electron diffraction and EDX analysis).

appears to be the kinetics along the same the surface structural change, which is speculated to relate with arrangements of the Ti-OH groups and needs further investigation.

Conclusions

The anodically oxidized metal was found to have titania with a large number of Ti-OH groups on its surface. On immersion in SBF, the metal forms a bonelike apatite through formations of an amorphous calcium titanate and an amorphous calcium phosphate. The formation of the calcium titanate is induced by the Ti-OH groups, which reveals negative charge to interact with calcium ions in the fluid. The amorphous calcium titanate is postulated to reveal positive charge, thereby interacting with the phosphate ions in the fluid to form the amorphous calcium phosphate, which eventually crystallized into bonelike apatite.

References

[1] T. Kokubo, H.-M. Kim and M. Kawashita: Biomaterials Vol.24 (2003), p.2161-2175.
[2] P. Li, C. Ohtsuki, T. Kokubo, K. Nakanishi, N. Soga, T. Nakamura, T. Yamamuro and K. de Groot: J. Biomed. Mater. Res. Vol. 28 (1994), p.7.
[3] B. Liang, S. Fujibayashi, J. Tamura, M. Neo, H.-M. Kim, M. Uchida, T. Kokubo and T. Nakamura: Bioceramics Vol.15 (2002), p.923.
[4] T. Kokubo, H. Kushitan, S. Sakka, T. Kitsugi and T. Yamamuro: J. Biomed. Mater. Res.Vol.24 (1990), p. 721.
[5] H. Takadama, H.-M. Kim, T. Kokubo and T. Nakamura: J. Biomed. Mater. Res. Vol.57 (2001), p.441-448.

Key Engineering Materials Vols. 254-256(2004) pp. 745-748
online at http://www.scientific.net
© 2004 Trans Tech Publications, Switzerland

Osteogenetic Effect on Cortical Bone of Cultured Bone/Ceramics Implants

N. Satoh[1], T. Yoshikawa[1,2], A. Muneyasu[1], J. Iida[1], A, Nonomura[2] and Y. Takakura[1]

Departments of [1]Orthopaedic Surgery and [2]Diagnostic pathology, Nara Medical University, Kashihara Nara 634-8522, Japan, tyoshi@naramed-u.ac.jp

Keywords: marrow cell, dexamethasone, hydroxyapatite, osteogenesis.

Abstract. Hydroxyapatite(HA) has been reported to have a good affinity for cancellous bone with high osteoblastic activity, binding directly to such bone. However, little work has been done concerning the strength of binding to cortical bone with a low cellular activity. Bio-artificial bone with a high osteogenetic capacity can be produced by combining cultured bone tissue with an artificial bone material. We examined this bio-artificial bone for its osteogenetic capacity around the bone tissue to evaluate whether the bone is applicable to osteogenetic treatment. Bone marrow cells were collected from the femurs of 7-week-old male Fischer rats, placed into two T75 flasks, and incubated in standard medium. After 2 weeks in primary culture, the cells were seeded on a porous hydroxyapatite block, and incubated in an osteogenic medium prepared by adding 10 nM dexamethasone, ascorbic acid and b-glycerophosphate. to the standard medium. Two weeks later, cultured HA constructs were implanted onto the cortical bone of the femur. At four or eight weeks after implantation, the femur was removed and radiographically and histologically examined for osteogenesis. The bio-artificial bone of HA impregnated with cultured marrow cells bound firmly to the femoral cortex both radiographically and histologically, and calcification indicating new bone formation was observed around the HA. The use of this bio-artificial bone makes it possible to perform osteogenetic treatment without damage to autogenous bone, avoiding pain at the site of bone collection and deformity of the pelvis, and reducing the burden on the patient.

Introduction

Many studies reported that when marrow cells were cultured in a medium containing compounds such as dexamethasone bone-like tissue formed, and that this cultured bone tissue possessed a calcified collagen matrix, contained osteocalcin (a bone specific protein) and exhibited bone morphogenetic protein activity [1-4]. The results of biochemical and gene expression analyses showed that the osteoblast activity of cultured cells was high. We have reported that it is possible to prepare artificial bone with a potent osteogenic potential by culturing artificial bone materials with functionally active bone tissue [5-10]. We have also documented that the osteoblast activity of such cultured bone constructs is similar to that of cancellous bone [11,12].

Autologous cancellous bone is used for spinal fixation and treatment of pseudoarthrosis due to its high osteogenic potential. Cancellous bone is often harvested from the ilium, but this process can result in pelvic deformation, and there are limitations on the amount of bone that can be harvested. Furthermore, patients often experience uncomfortable levels of pain at the site of bone harvest. On the other hand, cultured bone constructs, with properties comparable to cancellous bone, can be produced by harvesting marrow cells using bone marrow aspiration, which is a minimally invasive procedure. When a bone fracture damages the periosteum, pseudoarthrosis is likely to occur with cortical bone due to low cellular activity. Therefore, we investigated the osteogenic potential of cultured bone constructs grafted onto cortical bone and investigated whether cultured bone constructs could be used for bone regeneration in the treatment of pseudoarthrosis.

Materials and Methods

Marrow cells were collected from the femurs of 7-week-old male Fisher rats, placed into two T75 flasks, and cultured in a standard medium (minimal essential medium (MEM) containing 15% calf serum and antibiotics). After two weeks of incubation, cultured cells were detached using trypsin, and a concentrated cell suspension was incubated porous hydroxyapatite with (3x3x3mm, Inter por 500, made by Interpore Co,Ltd, USA) blocks in an osteogenic medium (above-mentioned standard medium containing 10 nM dexamethasone, 82 ug/ml vitamin C phosphate, and beta-glycerophosphate). After two weeks of incubation, the resulting bone constructs were grafted onto cortical bone by placing an incision on the periosteum of the shaft of the right femur in eight syngeneic 7-week-old rats. For comparison, plain HA blocks were grafted onto cortical bone of the left femur in the same manner. Four rats were sacrificed at 4 weeks after implantation, and the other four rats were sacrificed at 8 weeks after implantation. Femurs were removed and fixed in a 10% formalin solution. Each fixed femur was analyzed radiologically to investigate osteogenesis. Osteogenesis was also histologically analyzed by decalcifying samples and staining sections with hematoxylin and eosin.

Results

Radiological analyses revealed that at 4 weeks after implantation, calcification indicative of new bone formation was not present on the surface of cortical bone in any of the four HA grafts. With two grafts, a radiolucent area was seen between the cortical bone and HA. On the other hand, with all four cultured bone constructs, there was favorable continuity with the cortical bone, and calcification, indicative of new bone formation, was continuous with the cortical bone around the construct [Fig.1]. At 8 weeks after implantation, one HA graft came loose when the femur was excised. Findings indicative of osteogenesis were not seen around any of the four HA grafts. With three of the four cultured bone constructs, favorable continuity with the cortical bone was seen, and calcification, indicative of new bone formation, was continuous with the cortical bone. With the remaining construct, while clear signs of osteogenesis were not seen, hardening of the surrounding cortical bone and favorable continuity between the cortical bone and HA were observed.

The results of histological analyses showed that, at 4 weeks after implantation, fibrous tissue was present between the cortical bone and HA in all four HA grafts, and there was no osteogenesis in or around the HA pores. On the other hand, with cultured HA constructs, osteogenesis was seen in HA pores and was continuous with the cortical bone tissue [Fig.2].

Therefore, the above radiological and histological findings show that cultured bone constructs can effectively facilitate bone regeneration when grafted onto cortical bone.

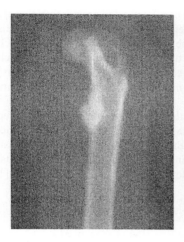
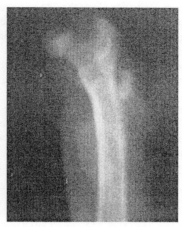

Fig.1. X rays of cultured cultured HA construct (left) and HA alone (right) at 4 weeks after implantation.

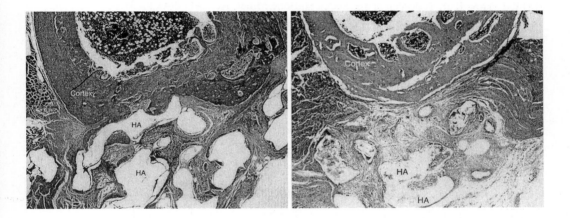

Fig.2. Histological findings of cultured HA construct (left) and HA alone (right) at 4 weeks after implantation.

Discussion

HA possesses a high degree of affinity towards cancellous bone, which has a high level of osteoblast activity. However, there have not been many studies on the bond between HA and cortical bone, which has a low level of osteoblast activity. In the present study, when a plain HA block was grafted onto cortical bone, it did not bind strongly to the cortical bone and the level of osteogenesis was low. On the other hand, when an HA block covered by cultured bone tissue was grafted, it bound strongly to the cortical bone, and findings indicative of osteogenesis were observed. Because bone regeneration around the cortical bone is required for spinal fixation and the treatment of pseudoarthrosis, cultured bone constructs may prove to be useful. Based on these findings, we are in the process of conducting a study on bone regeneration using cultured bone constructs for spinal fixation and the treatment of pseudoarthrosis.

Pseudoarthrosis is generally treated by implanting a cancellous bone graft harvested from the ilium in order to facilitate bone regeneration at the affected area. However, bone harvesting from the ilium is very painful, and the severity of pain at the site of bone harvest is often greater than that at the grafted site, thus lowering patient QOL. Furthermore, it has been reported that pain at the harvest site persists for more than one year in 25% of cases [13]. Bone harvesting can also lead to pelvic deformation, and in women with little subcutaneous fat tissue, surgical scarring and pelvic deformation are notable and esthetically undesirable. In men, cancellous bone is harvested from the ilium where the belt is worn, and as a result, bone harvesting makes it difficult for some patients to wear pants.

Marrow cells contain some osteogenic cells. Because these cells can be harvested by bone marrow aspiration, which is a minimally invasive procedure, harvesting is possible under local anesthesia during an outpatient visit. By using artificial bone cultured with marrow cells, it is possible to perform bone regeneration therapy without scarifying autologous bone and causing harvest-induced pain or pelvic deformation, thus reducing the level of physical stress to patients [14].

References

[1] C. Maniatopoulos, J. Sodek, and A.H. Melcher: Cell Tissue Res. 254 (1988), p. 317.

[2] J.E. Davies, R. Chernecky, B. Lowenberg, et al.: Cells Mater. 1 (1991), p.3.

[3] T. Yoshikawa, S.A. Peel, J.R. Gladstone, et al.: Biomed. Mater. Eng. 7 (1997) , p369.

[4] K-L Yao , R.Jr. Todescan and J. Sodek: J. Bone Miner. Res. 9 (1994), p.231.

[5] T. Yoshikawa, H. Ohgushi, Y. Dohi, et al.: Biomed. Mater. Eng. 7 (1997), p.49.

[6] T. Yoshikawa, H. Ohgushi and S. Tamai: J. Biomed. Mater. Res. 32 (1996), p.481.

[7] T. Yoshikawa, H. Ohgushi, Akahane M, et al.: J. Biomed. Mater. Res. 41(1998), p.568.

[8] T. Yoshikawa, H. Nakajima, E. Yamada, et al.: J. Bone Miner. Res. 15 (2000), p. 1147.

[9] T. Yoshikawa, H. Ohgushi, H. Nakajima, et al.: Transplantation. 69(2000), p.128.

[10] T. Yoshikawa, H. Ohgushi, T. Uemura, et al.: Biomed. Mater. Eng. 8 (1998), p.311.

[11] T. Yoshikawa, T. Noshi, H. Mitsuno, et al.: Mat. Sci. Eng. C-Biomim., 17 (2001), p.19.

[12] T. Yoshikawa : Mat. Sci. Eng. C-Biomim. 13(2000), p.29-37.

[13] S.S. Reuben, P. Vieira, S. Faruqi, et al.: Anesthesiology 95 (2001), p390.

[14] T. Yoshikawa , T. Ohmura, Y. Sen et al.:Bioceramics Vol.15. (2003), p383.

Key Engineering Materials Vols. 254-256(2004) pp. 749-752
online at http://www.scientific.net
© *2004 Trans Tech Publications, Switzerland*

Effect of Enamel Matrix Protein-Coated Bioactive Glasses on the Proliferation and Differentiation of the Osteogenic MC3T3.E1 Cell Line

S. Hattar[1], A. Asselin[1], D. Greenspan[2], M. Oboeuf[1], A. Berdal[1] and J-M Sautier[1]

[1]Laboratoire de Biologie Orofaciale et Pathologie, INSERM U 0110, Université Paris 7, UFR d'Odontologie, Institut Biomédical des Cordeliers, Esc. E - 2è étage, 15-21 rue de l'Ecole de Médecine, F-75270 Paris Cedex 06, France. E-mail : sautier@ccr.jussieu.fr
[2]US Biomaterials Corporation, Novamin Technology, Inc, 13709 Progress Blvd. # 23, Alachua, Florida, USA

Keywords: Bioactive glass, enamel matrix protein, osteoblast, *in vitro*.

Abstract. Bioactive glasses, osteoproductive materials, have received considerable attention as a bone graft substitute in the treatment of periodontal loss. More recent strategies for achieving a predictable periodontal regeneration include the use of enamel matrix proteins (EMP). However, the osteoblastic response to these materials has not yet been clearly understood. The aim of our study is to examine the effects of these two materials on the behaviour of osteoblastic cells *in vitro*. For this purpose, cells of the murine osteoblastic cell line MC3T3.E1, were seeded in contact with 45S5® bioactive glass granules coated or not with Emdogain®, and a less reactive glass with 60wt% silica content (60S). Phase contrast microscopy was used to evaluate the evolution of the cultures. The bone/biomaterial interface as well as tissue ultra-structural features were visualized using transmission electron microscopy. Our results have shown that the granules supported the growth and maturation of osteoblastic cells. Zones of early differentiation and mineralization were observed in all three cultures, while occurring at an earlier stage in cultures with the bioactive glasses, compared to 60S cultures. Transmission electron microscopic (TEM) observations have shown metabolically active cells in contact with granules. Furthermore, The tissue/biomaterial interface was established *via* an electron dense layer located at the periphery of bioactive granules. In conclusion, these findings indicate that bioactive glasses alone or combined to enamel matrix proteins have the ability to support the growth and enhance the differentiation of osteoblast-like cells *in vitro*. Our findings would have potential clinical applications in cases of severe bone loss, when both osteoconductive and osteoproductive properties are required to assure an optimal periodontal regeneration.

Introduction

In dentistry, the supporting tissues including cementum, periodontal ligament and alveolar bone can be lost following periodontal disease. Among different regenerative techniques, the use of autogeneous bone, considered by many to be the gold standard, has its limitations. Consequently, this led the clinicians to look for other synthetic grafting materials or biologic mediators, among others, bioactive ceramics have a common property: their capacity to form a hydroxycarbonate apatite (HCA) layer on their surfaces once exposed to simulated body fluids or implanted *in vivo*. One promising material are bioactive glasses (e.g.Perioglas®) widely used in the treatment of periodontal bone defects.

The biologic mediators involved in the process of regeneration are gaining importance as viable alternatives to existing regenerative techniques. For instance, it is well established that enamel matrix proteins (EMP) are involved in the formation of acellular cementum, periodontal ligament and alveolar bone. The only commercially available biological mediator used in treatment of intrabony defects is an EMP derived from the developing tooth germ of fetal pigs (e.g.Emdogain®) [1]. It has been shown that these proteins induce proliferation, protein production and nodule formation of periodontal ligament cells *in vitro*.

While the recruitment of cells such as osteoblasts is essential for the formation of new attachment, little information has been reported with regard to the specific effects of Bioglasses or EMD on cell behaviour and osteoblastic differentiation, or their mechanism of action. The aim of our study was to investigate the possible effect of these biomaterials at a morphological and cellular level, by cultivating a mouse osteoblastic cell line, MC3T3.E1, in contact with bioactive granules. In addition, we evaluated the effect of combining the bioactive glasses with EMP, to see whether this combination may further enhance the cellular response.

Materials and Methods

Material Preparation

Bioactive glass melt type class A (45S5® bioglass), were kindly provided by USBiomaterials Corporation (Alachua, FL, USA). The composition is (in wt %): 45% SiO_2, 24.5% Na_2O, 24.5% CaO and 6% of P_2O_5. In each experiment, two types of bioactive granules were used: 45S5 bioactive granules (*BG*), 45S5® granules coated with enamel matrix proteins Emdogain® (kindly provided by Biora AB, Malmö, Sweden). This was performed by incubating the granules in a carbonate buffer at a concentration of 100µg/ml Emdogain, for 24 hours at 33°C (*BG/EMD*). The third substrate was 60S glass (*60S*), known to be less reactive served as the control for our experiments.

Cell Culture Conditions

MC3T3.E1 osteoblast-like cells, kindly provided by Dr RT Franceschi (Ann Arbor USA), were used in our experiments. The cells were grown in α-MEM, supplemented with ascorbic acid, ß-glycerophosphate, and 10% FCS (Hyclone®). The cells were maintained at 37°C in a humidified atmosphere at 5% CO_2 in air. At confluence, the cells were passaged with trypsin, counted on Malassez and plated at a density of 2.10^4 cell/cm^2 in 60 mm Ø petri dishes. The granules were added 24 hours later (20mg/ culture dish).

Transmission Electron Microscopy

The cell cultures were washed in α-MEM 0% FCS, fixed in Karnovsky solution for 1 hour. After several rinses in sodium cacodylate buffer (pH 7.4), the cultures were post fixed for 1hr in osmium tetroxide diluted in 0.2M sodium cacodylate buffer. The cells were then dehydrated in graded series of ethanol, and left overnight in a mixture of absolute ethanol and epon 1:1. The next day the cells were embedded in Epon-Araldite, and incubated at 60°C for 1 day. Semi-thin sections were cut with a diamond knife, mounted on glass slides, stained with methylene blue (Azur II), then examined under light microscopy for orientation purposes. Ultrathin sections were performed, collected on copper grids, and stained with 5% uranyl acetate in water for 4 minutes and lead citrate for 2 minutes. The sections were then examined under a TEM (Philips CM-12)

Results

Morphological Analysis: Cellular Proliferation and Mineralization

Phase contrast microscopy showed that the cells proliferated and reached confluence by day 3. By the end of the first week, first evidence of osteoblast differentiation was seen, in the form of cellular condensations especially around the particles. At first, the condensations represented discrete patches, then increased in number to become continuous afterwards. This was noticed earlier and to a higher degree in the cultures of *BG* and *BG/EMD* when compared to *60S* (Fig.1). The next weeks, mineralization was seen first around the granules, then extended all through the culture dishes. This was more advanced in the cultures of *BG* and *BG/EMD*.

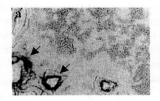

Fig. 1: Observations by contrast phase microscopy at day 10, cultures of *BG* (Fig. 1a), *BG/EMD* (Fig. 1b), *60S* (Fig. 1c). Arrows: Bioglass granules.

Our observations by TEM showed metabolically active cells, with well developed RER, large nuclei and occasional chromatin masses. The undecalcified sections produced a break down of the material, but we could observe an electron dense layer that corresponds to the periphery of the granules (Fig. 2a). This layer represents the famous bioactive layer rich in Ca-P to which multiple cellular attachments could be observed (Fig. 2b). The osteoid matrix was composed of densely-packed collagen fibres and the mineralized matrix was in contact with the electron dense layer located at the periphery of the glass (Fig. 2c).

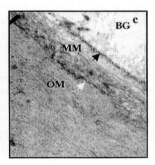

Fig. 2: Observations by TEM at day 20, showing cultures in contact with bioactive glasses coated with Emdogain (*BG/EMD*). BG: bioglass granules. Black arrow= bioactive layer. MM= mineralized matrix. OM= osteoid matrix. White arrow= mineralized front.

Discussion

A number of earlier studies have shown that bioactive glasses can promote proliferation, differentiation and mineralization of bone cells [2,3]. However, little is known on the cellular and molecular events responsible for their activity. On the other hand, data considering the effect of EMD on osteoblasts is scarce, because their major use has been advocated in the treatment of periodontal disease, most of *in vitro* studies were conducted on periodontal ligament cells. For these reason, the purpose of our study was to investigate the ability of Bioglasses and enamel matrix proteins to influence the proliferation and differentiation of a mouse pre-osteoblastic cell line.

Our results demonstrated that the bioactive glasses provide a favourable environment for the growth and proliferation of MC3T3.E1 cells. The observations obtained by light microscopy have shown that the cells proliferated, and formed nodules around and at distance from the glass particles. Since these glasses can and do support osteoblast growth and phenotype, it would be an appealing idea to develop surfaces on which cells can proliferate and differentiate to the extent of forming mature bone tissue *in vitro* [4]. Recent studies have demonstrated that bioactive glasses can modulate the genetic expression of osteoblasts, and support the osteoblastic phenotype *in vitro* [5,6]. In our study, the mineralization and nodule formation was seen earlier and to a larger extent in the cultures of *BG* when compared to cultures of less reactive glasses 60S. Furthermore, and in support of our morphological observations, the expression of major osteoblastic markers such as bonesialoprotein and osteocalcin seemed enhanced in cultures with bioactive granules (data not shown). These late osteoblast markers characterize mature cells of the osteoblastic lineage, actively producing mineralized tissue.

Biomimetism enhances cellular behaviour on bioactive glasses by means of modifying the surface properties of the material. In an attempt to mimic *in vivo* situations, and to optimize cell function, the BG granules in our study were first preincubated in culture medium [7]. On the other hand, EMD under *in vitro* conditions form protein aggregates. Since they adsorb both to hydroxyapatite and collagen [8], and considering the fact that BG forms a superficial HCA layer, we hypothesized that a combination of EMD on BG provides a three dimensional substratum that is favourable for cell function. In fact, cells cultures of *BG/EMD* showed also earlier and more extensive bone nodule formation when compared to other cultures. Indeed, clinical studies have demonstrated that combination of EMD with alloplastic graft materials results in more favourable enhancement of periodontal regeneration [9]. Our *in vitro* data, in accordance with these clinical studies, demonstrated that a combination of *BG/EMD* promoted osteoblast differentiation and mineralization not only at a morphological level but at a molecular one as well. In fact, higher levels of osteoblastic markers especially OC and BSP were detected when a combination of both BG and EMD was used (data not shown). Although EMD is not osteoinductive [10], it could enhance the attachment of cells on biomaterials, probably through cell surface integrins [11].

In conclusion, our preliminary results, confirmed earlier studies demonstrating osteoproductive capacities of bioactive glasses. In addition, the combination of an osteoproductive material and enamel matrix proteins, each one acting at a different level and in conjunction, could further enhance the cellular response. Enamel matrix proteins in the bioactive layer could favour the cellular attachment and regulate crystal formation, hence result in overgrowth of mineralized tissues.

References

[1] L. Hammarström, L. Heijl , S. Gestrelius: J Clin Periodontol Vol. 24 (1997), p.669

[2] W.C.A. Vrouwenvelder, C.G. Groot, K. Groot: Biomaterials Vol. 13 (1992), p.382

[3] C. Loty, J.M. Sautier, M.T. Tan, M. Oboeuf, E. Jallot, H. Boulekbache, D. Greenspan, N. Forest: J Bone Miner Res Vol. 16 (2001), p. 231

[4] A. El-Ghannam, P. Ducheyne, I.M. Shapiro: J Biomed Mater Res Vol. 29 (1995), p. 359

[5] E.A.B. Effah Kaufmann, P. Ducheyne, I.M. Shapiro: Tissue Eng Vol. 6 (2000), p. 19

[6] S. Hattar, A. Berdal, A. Asselin, S. Loty, D.C. Greenspan, J-M. Sautier: European Cells and Materials Vol. 4 (2002), p. 61

[7] A. El-Ghannam, P. Ducheyne, I.M. Shapiro: Biomaterials Vol. 18 (1997), p. 295

[8] S. Gestrelius, C. Andersson, D. Lindström, L. Hammarström, M. Sommerman: Clin Periodontol Vol. 24 (1997), p. 685

[9] P.M. Camargo, V. Lekovic, M. Weinlaender, N. Vasilic, E.B. Kenny, M. Madzarevic: J Clin Periodontol Vol. 28 (2001), p. 1016

[10] B.D. Boyan, T.C. Weesner, C.H. Lohmann, D. Andreacchio, D.L. Carnes, D.D.Dean, D.L. Cochran, Z. Schwartz: J Periodontol Vol. 71 (2000), p. 1278

[11] M.T.M. Van der Pauw, V. Everts, W. Beertsen J Periodont Res Vol. 37 (2002), p. 317

Key Engineering Materials Vols. 254-256(2004) pp. 753-756
online at http://www.scientific.net

Apatite-Forming Ability of Calcium Phosphate Glass-Ceramics Improved by Autoclaving

T. Kasuga, T. Fujimoto, C. Wang and M. Nogami

Graduate School of Engineering, Nagoya Institute of Technology

Gokiso-cho, Showa-ku, Nagoya 466-8555, Japan

kasugato@mse.nitech.ac.jp

Keywords: Calcium phosphate, glass-ceramic, simulated body fluid, apatite, autoclave

Abstract. Calcium phosphate invert glass-ceramic with a composition of $60CaO$-$30P_2O_5$-$3TiO_2$-$7Na_2O$ in mol% has an apatite-forming ability in simulated body fluid (SBF) at 37 °C; on its surface, the bonelike apatite forms after 20 days of soaking in SBF. In the present work the glass-ceramic was autoclaved in SBF or water at 121 °C prior to the soaking in SBF at 37 °C. Bonelike apatite started to form on the autoclaved glass-ceramic after 3 days of soaking in SBF at 37 °C; the apatite-forming ability could be drastically improved even after autoclaving in water at 121 °C. Although the residual glassy phase around the surface dissolved during the autoclaving in water, titanium ion in the phase did stay around the surface. X-ray photoelectron spectroscopic analysis suggested that the hydrated titania groups form newly around the surface during the autoclaving.

Introduction

We developed silica-free calcium pyrophosphate glass-ceramics with a small amount of glassy phase containing TiO_2 and Na_2O [1]. The typical composition was determined to be $60CaO$-$30P_2O_5$-$3TiO_2$-$7Na_2O$ in mol%. When the glass was heated at 800~850 °C, they crystallized, resulting in formation of glass-ceramic containing crystalline phases such as β-$Ca_3(PO_4)_2$ (β-TCP) and β-$Ca_2P_2O_7$ (β-CPP) [2]. Bonelike apatite formation was observed on the glass-ceramic after 14~20 days of soaking in simulated body fluid (SBF); the glass-ceramic was suggested to show bioactivity. It seems difficult, however, to form the apatite by soaking in SBF within 10 days [3].

Surface modification for enhancement of the apatite-forming ability may induce high bioactivity of the glass-ceramic. Body fluid or SBF is already highly supersaturated concerning the apatite. The apatite formation on the glass-ceramic would be associated with the presence of plentiful inducers for the apatite nucleation around the surface [4]. Some groups such as Si-OH, Ti-OH, or COOH are known as the promising candidates for supplying inducers for the apatite formation [5]. A hydrothermal process may be one of the promising methods for introducing the sufficient hydrated functional groups as the inducers for apatite formation around the surface of the glass-ceramic. In the present work, for the first time, the apatite-forming ability of the glass-ceramic was found to be drastically enhanced after autoclaving in water or SBF.

Materials and Methods

A phosphate glass was prepared using a molar composition of $60CaO$-$30P_2O_5$-$3TiO_2$-$7Na_2O$. The batch mixture was melted in a platinum crucible at 1350 °C for 0.5 h. The melt was poured onto a stainless-steel plate and quickly pressed using an iron plate, resulting in formation of glasses with thickness of 0.5~1 mm. The obtained glass plates were reheated at 800 °C for 1 h to be crystallized. The resulting glass-ceramic plates with a surface area of 50~200 mm^2 were autoclaved in 50 mL of SBF or water (pH~5.1; DW) at 121 °C for 20 min. The treated glass-ceramic plates were then soaked in 100 mL of SBF at 37 °C. For these experiments, c-SBF [6] (consisting of 2.5 mM of Ca^{2+}, 142.0

mM of Na$^+$, 1.5 mM of Mg^{2+}, 5.0 mM of K$^+$, 148.8 mM of Cl$^-$, 4.2 mM of HCO$_3^-$, 1.0 mM of HPO$_4^{2-}$ and 0.5 mM of SO$_4^{2-}$) that included 50 mM of (CH$_2$OH)$_3$CNH$_2$ and 45.0 mM of HCl was used.

After the soaking, the surface was examined by thin-film X-ray diffractometry (TF-XRD) at a glazing angle of 1 ° and observed by scanning electron microscopy (SEM) incorporating x-ray microanalysis using energy dispersive spectrometry (EDS). The surfaces before and after the autoclaving in DW were discussed using X-ray photoelectron spectra (XPS). XPS spectra of the glass-ceramics were measured under an ultra high vacuum (~10^{-7} Pa) using monochromatized Al-Kα X-ray irradiation (15 kV, 10 mA). The binding energy was normalized to the C$_{1s}$ energy.

Results and Discussion

Figure 1 shows SEM photos of the surface of the glass-ceramics before and after the autoclaving in SBF or DW. As reported earlier [2], the glass-ceramic consists of calcium phosphate grains of 0.5~2 μm and the residual glassy matrix phase that appears dark in Fig. 1(a). When the glass-ceramic was autoclaved in SBF (Fig. 1(b)), its surface was completely covered by numerous crystallites of diverse morphologies, which were regarded as hydroxyapatite, judged from the TF-XRD pattern (not shown here). As shown in Fig. 1(c), after the autoclaving in water, no new products appear on the glass-ceramic and the glassy phase seems to slightly decrease in amount.

Figure 2 shows SEM photos of the surface of the glass-ceramics after 10 days of soaking in SBF at 37 °C. Obviously, on the glass-ceramic without the autoclaving, no products formed (Fig. 2(a)), and leaf-like bonelike apatite formed on the autoclaved glass-ceramics and covered their surfaces completely (Fig. 2(b, c)). As shown in Figs. 1(b) and 2(b), although hydroxyapatite had formed on the glass-ceramic surface after the SBF-autoclaving, bonelike apatite seen after soaking in SBF at 37 °C is different from the previous one, which could be distinguished from their morphology. The bonelike apatite formation in SBF at 37 °C is suggested to be enhanced by the apatite which forms

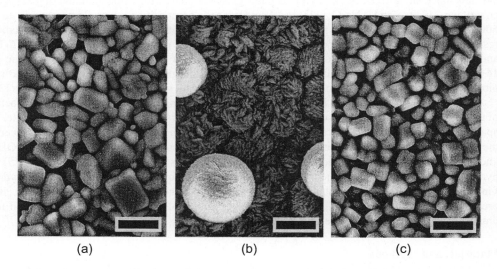

(a)   (b)   (c)

Fig. 1. SEM photos of the glass-ceramic surface (a) before and (b, c) after the autoclaving at 121 °C for 20 min. (b) and (c); the samples autoclaved in DW and in SBF, respectively. Bar scale is 2 μm.

during the SBF-autoclaving. Meanwhile, even when the DW-autoclaved sample was soaked in SBF at 37 °C, bonelike apatite formed within 10 days (Fig. 2(c)); further experiments showed that the apatite formation of the DW-autoclaved glass-ceramic started to occur after 3 days of soaking in SBF at 37 °C. Thus, the autoclaving using SBF or DW was found to be an effective method for enhancement of the bonelike apatite formation on the glass-ceramic. Especially, the DW-autoclaving is a very simple and satisfactory method for preparation of the glass-ceramic on which bonelike apatite forms within 10 days during the soaking in SBF at 37 °C.

   Figure 3 shows XPS Ti_{2p} spectra of the glass-ceramics before and after the DW-autoclaving. Before the autoclaving, the peaks are seen at 459.6 eV and 465.5 eV (Fig. 3(a)), which correspond to those observed in an XPS spectrum of $60CaO-30P_2O_5-10TiO_2$ glass, as shown in Fig. 3(c); titanium ion is included in the residual glassy phase in the glass-ceramic. On the other hand, after the autoclaving, the Ti_{2p} peaks at 459.6 eV and 465.5 eV disappear and the peaks newly appear at 458.5 eV and 464.2 eV (Fig. 3(d)), which are close to those from a Ti–O bonding in TiO_2 crystal [7], as shown in Fig. 3(d). The changes in the peak positions are suggested to originate from bond breaking around titanium ions in the glassy phase with accompanying dissolution of the phase to form a new Ti–O bonding at the surface during the autoclaving. Although Na_{1s} spectra before and after the autoclaving are not shown here, the peak intensity decreases drastically after the autoclaving; this change shows that a large amount of Na^+ ion in the glassy phase around the surface dissolves into water during the autoclaving. No significant changes in the spectra of Ca_{2p} and P_{2p} were seen. Some oxygen states bonded to phosphorus or titanium ions were suggested to be superimposed in the O_{1s} peaks. It was difficult, however, to deconvolute the spectra.

   Almost no titanium ions would dissolve into the water at pH~5 and the ions may form the hydrated group around the surface. When the autoclaved sample is soaked in SBF, the negative-charged, hydrated titania groups around the surface would induce nucleation of the apatite. As a result, the apatite formation in SBF would be enhanced.

(a)   (b)   (c)

Fig. 2. SEM photos of the glass-ceramics after 10 days of soaking in SBF. (a); the sample without autoclaving. (b) and (c); the samples autoclaved in DW and SBF, respectively. Bar scale is 2 μm.

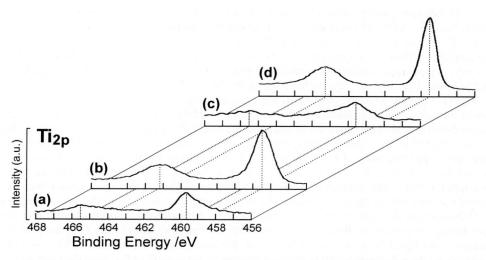

Fig. 3. Ti$_{2p}$ XPS spectra of the glass-ceramics before and after the autoclaving. (a);
the sample before the autoclaving, and (b); the sample autoclaved in DW. (c) and
(d); 60CaO-30P$_2$O$_5$-10TiO$_2$ glass and commercial powders TiO$_2$ (including 80%
anatase and 20% rutile), respectively, which are shown as examples of typical XPS
patterns for Ti–O bonds.

Conclusion

The apatite-forming ability in SBF at 37 °C of the calcium phosphate glass-ceramic was drastically
enhanced by the autoclaving in SBF or DW at 121 °C prior to the soaking in SBF at 37°C. Especially,
the DW-autoclaving is a very simple and satisfactory method for preparation of the glass-ceramic on
which bonelike apatite forms within 10 days. The negative-charged, hydrated titania groups, which
form easily around the glass-ceramic surface during the autoclaving, would induce nucleation of
bonelike apatite. As a result, the apatite is suggested to form on the glass-ceramic in SBF at 37 °C in
a short period.

Acknowledgement

This work was supported in part by a Grant-in-Aid for Scientific Research from Japan Society for the
Promotion of Science and grant from the NITECH 21st Century COE program "World Ceramics
Center for Environmental Harmony."

References

[1] T. Kasuga, Y. Abe: J. Non-Cryst. Solids, Vol. 243 (1999), p. 70.
[2] T. Kasuga, M. Sawada, , M. Nogami, Y. Abe,: Biomater., Vol.20 (1999), p. 1415.
[3] T. Kasuga, T. Mizuno, M. Watanabe, M. Nogami, M. Niinomi: Biomater., Vol. 22, (2001), p.
577.
[4] P. Li, C. Ohtsuki, T. Kokubo, K. Nakanishi, N. Soga, K. de Groot: J. Biomed. Mater. Res., Vol.
28 (1994) p. 7.
[5] T. Kokubo: Acta Mater., Vol. 46 (1998), p. 2519.
[6] H.-M. Kim, T. Miyazaki, T. Kokubo, T. Nakamura: Key Eng. Mater. (Bioceramics 13), Vol.
192-195 (2001), p.47.
[7] R.N.S. Sodhi, A. Weninger, J.E .Davis: J. Vac. Sci. Technol., Vol. A9 (1991), p. 1329.

Key Engineering Materials Vols. 254-256(2004) pp. 757-760
online at http://www.scientific.net
© *2004 Trans Tech Publications, Switzerland*

Radiological and Histological Examination of Gap Healing on Plasma Sprayed HA Coating Surface

Y. Cao[1,2], W. Lu[2], B. Zhang[1], L.P.Wang[1] ,C.Y. Bao[1*] , B. C.Yang[1], C.Y. Wang[1], J.Y.Chen[1], J.M.Feng[1], S.P.Chow[2], X.D.Zhang[1]

1. Engineering Research Center for Biomaterials, Sichuan University, P.R. China, 610064, E-Mail: caowanghui@yahoo.com.cn
2. Department of Orthopedic Surgery, the University of Hong Kong, Hong Kong

Keywords: hydroxyapatite, coating by plasma spray, implant, gap healing

Abstract. Bone formation on hydroxyapatite coatings in the presence of gaps is important for clinical applications. Samples of pure titanium and of this material coated with hydroxyapatite by plasma spraying were implanted in dogs. The implants were surrounded by gaps of 2mm, and the follow-up period was 12 weeks. Histological examination revealed that gaps could be bridged by bone if the hydroxyapatite coating was applied, and pure titanium implants were surrounded by fibrous tissue with no bone contact at all. In clinical trials, hip implants were HA-coated by plasma spraying and implanted in human femurs. There was a 2mm gap between host bone and implanted device on the top. After 1, 6, 12, 24 months, the hip implant was observed by radiological examination, and showed that one month after operation, part of the gap had been replaced by new bone. After 6 months, about 80% of gap was filled by new bone, and after 12 and 24 months, the percentage increased to over 90%.

Introduction

Hydroxyapatite (HA), the main inorganic component of human bone, has been applied in surgery as a bone substitute and showed excellent biocompatibility as well as bone-bonding capacity [1]. However, bulk HA ceramics cannot be used in load-bearing situations because of their poor mechanical properties. Hence, HA coating on metal or alloy implants have been developed to overcome HA drawbacks. One method of applying ceramic coating to metallic substrates is the plasma-spraying process. For plasma-sprayed HA-coated implants, good results have been reported with respect to bone adaptation and binding strength of the HA-coated implant-bone interface. Many clinical reports on HA-coated implants have indicated high success rates [2-5].

The healing of gaps around implants, called 'bridge-like binding', is one of the major advantages of HA coatings over uncoated implants [6]. Gaps between an implant and bone arise in clinical implant surgery because the procedure is far more complicated than in laboratory circumstances. Extreme gaps can be encountered during revision surgery. However, the maximum gap size that allows bone apposition on HA coating [3,6] is still controversial. In this study, histological and radiological examination were used to investigate whether HA coating can "bridge" 2mm gaps in dog model and human hip implants.

Materials and Methods
Implant materials

The HA coatings were prepared by plasma-spraying technique [2] on commercial pure titanium cylinders at Sichuan University. The HA powders for coatings were wet-synthesized and sintered at 1250^{0}C for 2 hours, then ground and sieved. The mean size of particle was ≤ 70μm. X-ray examination showed that the powder was pure HA [2]. Cylindrical implants were designed (Fig. 1-A). The diameter of one side (A) is 6mm, while the other side (B) is 2mm, and the length is 15mm.The samples were polished and sandblasted with Al_2O_3 and cleaned. Plasma spraying of HA coating was done with METCO equipment. Only side B was coated, while side A was uncoated. The as-received coating was kept in water vapor at 125^0C, with a pressure of 0.15Mpa, for 6 hours. Then, the sample was dried at 80^0C for 2 hours [2]. The HA coating composition is shown in Table 1.

Table 1. HA coating composition

	HA (%)	Other phases (%)	Crystallinity (%)
HA powder	100	0	100
As-received coating	82	18	26
Post treated coating	96	4	88

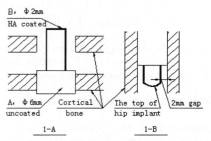

Fig.1. The model of implant for animal trial (1-A) and human hip implant (1-B)

Surgical procedure

Six male adult and healthy dogs were used as the animal model in this study. Animals were anaesthetized with an intra-abdominal injection of 2.5% sodium pentobarbital. The limbs were shaved and disinfected, and lateral skin incision and blunt dissection next to the quadriceps femoral muscle exposed the femora. The periosteum was incised and reflected and four holes were drilled in the lateral cortex under cooling with saline solution. Each hole was drilled in two steps by drills with increasing diameters up to 6mm, with a distance between holes of 2cm. The implantation was performed by press fitting, and one side (A) contacted tightly with cortical bone, while the other side (B) did not contact host bone; there were a 2mm gap between cortical bone and implants (Fig.1-A). As control specimens, pure Ti cylinders were implanted.

After implantation periods of 2, 4, 12 weeks, the animals were sacrificed and the implants were excised by sawing the femora in transversal segments. The samples were fixed in a series ethanol (70, 80, 90, and 100% ethanol) and embedded in methyl methacrylate (MMA). Thin sections ($50\mu m$) were made with a diamond saw. The section was stained with HE.

Using a light microscope coupled to a Vidas Image Analysis System, the percentage of the implant surface covered by the bony tissue was measured.

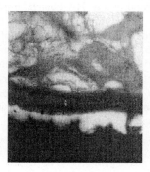

Fig.2. Pure Ti implants after 2 weeks; a thick layer of fibrous tissue between implant and host bone.

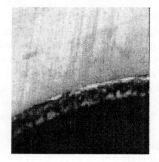

Fig.3. HA coating implant after 2 weeks; a thin layer of fibrous tissue between HA coating and host bone.

Hip implants was designed and HA-coated by plasma spraying. Then, the device was implanted in human femur. There was a 2mm gap between host bone and implanted device on the top (shown in Fig.1-B). After 1, 6, 12, 24 months, the hip implant was observed by radiology.

X-ray diffraction (XRD) also was employed to analyse the phase composition and structure.

Fig.4. HA-coated implanted after 12 weeks; direct bone contact established on HA coating.

Fig.5. Pure Ti implanted after 12 weeks; a thick layer of fibrous tissue between implant and host bone.

Results

(1) Histological observation in animal model: Histological results showed that there was a layer of fibrous tissue between bone and implants at end B after 2 weeks for all specimens, but the fibrous tissue around pure Ti specimens (Fig. 2) was thicker than that around HA coated specimens (Fig. 3). After 4 weeks, new bone was observed around HA coating and approximately 63% of the original gap was replaced by new bone, while in pure Ti no bone formation in the gap was observed. At 12 weeks, 88% of the gap was filled with new bone in HA-coated samples (Fig. 4), while only fibrous tissue was observed in the gap in control group (Fig. 5).

(2) Clinical radiological results: 1 month after operation, the X-ray photo showed that part of the gap has been replaced by new bone (Fig. 6, Fig.7-A.). After 6 months, about 80% of the gap was filled by new bone, and after 12 and 24 months, the percentage increased to over 90% (Fig. 7-B, 7-C).

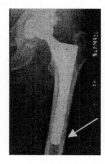 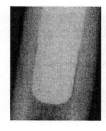  

7-A   7-B   7-C

Fig. 6. The radiological picture of human hip implant after one month.

Fig.7.The magnified radiological picture of the arrow part after one month (7-A), 6 months (7-B) and 24 months (7-C) implantation.

Discussion

Fixation of arthroplasty implants to bone is one of the most significant factors in order to obtain a satisfactory clinical outcome. However, most of the artificial joints being used now, are made of bioinert materials such as titanium, CoCr alloy, etc. They do not form a chemical bond with bony tissue. Therefore, they must be fixed to bone by mechanical interlocking in clinical practice, and the possibility of loosening over a long period may become a critical problem.

It has been demonstrated that there was a layer of fibrous tissue between bone and the uncoated titanium surface if the gap is more than 0.25mm [7]. J. A. M. Clement et al [6] have reported that direct bone apposition between bone and HA coating in goats could be observed if the gaps were up to 1mm. But fibrous tissue would interpose between bone and the HA coated implant surface if the gaps was 2mm. Others have observed gap healing up to 3mm in rabbit model [3]. However, this result might have been explained by osteoconduction [1]. In our experiment, to prevent osteoconduction, samples were made press-fit at end A (uncoated) only, and with non-press-fit (end B) coated with HA. There was directly new bone apposition between host bone and HA coating at 2mm gaps in dog model. The difference in results may be due to the different animal model used [8].

Generally, in surgery, it is impossible to meet the demands by optimal implant insertion. Even with the best tool and skills, gaps will still arise between the implant and host bone. Calcium phosphate coatings significantly increase the tolerance for inaccuracy during the insertion of an non-cemented prosthesis, thus promising great success rate of the HA coating of implants in clinical applications [7,9].

Conclusions

The plasma spraying HA coating can "bridge" 2mm gaps between host bone and implant in dog model and clinical application.

Acknowledgments

This work was supported by the National High Technology Research and Development Program of China ("863" Program, Number: 2001AA326010)

References

[1] B. S. Chang, C. K. Lee, K. S. Hong, H. J. Youn, H. S .Ryu, S. S. Chung, K. W.Park. Biomaterials, 21 (2000) 1291-1298.
[2] Yang Cao, Jie Weng, Jiyong Chen, Jiaming Feng, Zongjian Yang and Xingdong Zhang Biomaterials 17 (1996) 419-424.
[3] S.H.Maxian, J.P.Zawadsky, and M.G. Dunn. J. Biomed. Mater. Res., 28 (1994) 1311-1319.
[4] K.de Groot, R.Geesink, and C.P.A.T. Klein. J.Biomed. Mater.Res.,21 (1987)1375-1381.
[5] S.D.Cook,J.F.Kay,K.A.Thoma, and M.Jarcho. Int.   J.Oral Maxilloface. Impl., 2 (1987)15-22.
[6] J.A.M.Clemens, C.P.A.T.Klein, R.C.Vriesde, P.M.Rozing, K.de Groot. J. Biomed. Mater. Res, 40 (1998) 341-349.
[7] L.Carlsson, T.Röstland, and B.Albrektsson. Acta Orthop. Scand., 59(1988) 272-275.
[8] Zongjian Yang, Huipin Yuan, Weidong Tong, Ping Zou, Weiqun Chen and Xingdong Zhang. Biomaterials, 17 (1996) 2131-2137.
[9] K. Søballe, E. Stender-Hansen, H. Brockstedt-Rasmussen, V. Hjortdal, G. I. Juhle, C. M. Pedersen, I. Hvid, and C. Bünger. Clin. Orthop. Rel. Res.,272 (1991) 300-307.

*Corresponding Author: Dr.C.Y.Bao

Key Engineering Materials Vols. 254-256(2004) pp. 761-764
online at http://www.scientific.net
© 2004 Trans Tech Publications, Switzerland

α-Tricalcium Phosphate Cements and the Granules to Dental Pulp and Periapical Tissue

Masataka Yoshikawa[1] and Tadao Toda[2]

[1] 5-17, Otemae 1-chome, Chuo-ku, Osaka 540-0008, JAPAN, yosikawa@cc.osaka-dent.ac.jp

[2] 5-17, Otemae 1-chome, Chuo-ku, Osaka 540-0008, JAPAN, Toda@cc.osaka-dent.ac.jp

Keywords: α-TCP cement, granules, dental pulp, periapical tissue, hard tissue formation, tissue irritation, latex beads

Abstract. α-tricalcium phosphate (α-TCP) granules and α-TCP cements kneaded with high or low concentration of organic acid solution were respectively placed on the dental pulp surface after pulp extirpation to estimate pulp tissue responses and hard tissue formation in rat mandibular first molar canals. The tissue responses in the residual pulp in the middle and apical portion of the root canals and periapical area were histopathologically estimated 1, 3 and 5 weeks postoperatively. α-TCP granules and α-TCP cement kneaded with a low concentration of citric acid solution induced acute inflammation in the pulp tissue. α-TCP cements kneaded with high concentration of citric acid solution induce hard tissue deposition at the apical portion of root canal. It is concluded that fine granules dispersed from the material in the tissue, even though it is a biocompatible material, caused inflammatory responses.

Introduction

In this study, tissue responses and hard tissue formation in tooth root canal were estimated using α-tricalcium phosphate (α-TCP) cements kneaded with high/low concentration of citric acid solution in comparison with α-TCP granules. The use of α-TCP cement as a root canal filling material or a bone filling material in periapical lesion is recommended. This is mainly because of its biocompatibility and hard tissue inductivity [1,2]. However, the material probably stimulates the residual pulp in the apical portion of the root canal or periapical tissue considering the existence of fine particles [3] and acidic components [4] in the cement. In this *in vivo* study, we focused on pulp and/or periapical tissue responses to fine granules on the surface layer of α-TCP cement compared with responses to chemically stable spherical beads.

Materials and Methods

Experimental materials. α-TCP granules were prepared from $CaHPO_4 \cdot 2H_2O$ and $CaCO_3$ by dry synthetic methods. The α-TCP granules ranging from 0.6 to 46.0 μm were obtained by calcining an equimolar mixture of $Ca_2P_2O_7$ and CaO at 1,400°C. Citric acid solution was used as a liquid phase of α-TCP cement. α-TCP granules were kneaded with 36.0 % or 2.5 % citric acid (powder and liquid ratio = 1.4 g/ml), and used for histological examination. As control materials, spherical polycarbonate beads measuring 1, 6 or 25 μm in diameter (latex beads: Latex beads, POLYSCIENCES) were used.

***In vivo* examinations.** This study was performed under the Guidelines for Animal Experimentation at Osaka Dental University. Access cavities were prepared in mandibular first molars of 7-week-old male SD rats. The tooth pulp was extirpated at the root canal orifice or close to the apical foramen in Groups 1, 2 and 3, and removed at the root canal orifice in Groups 4, 5 and 6. All preparations were performed under general anesthesia. α-TCP granules (Group 1), cement composed of α-TCP granules and 36.0 % citric acid solution (Cement-H: Group 2), and another one kneaded with 2.5 % citric acid solution (Cement-L: Group 3) were respectively placed in the root canal of the teeth. As control materials, latex beads measuring 1 μm (Group 4), 6 μm (Group 5) and 25 μm (Group 6) in diameter were respectively placed in the root canal in the same manner as α-TCP materials. The access cavities were hermetically sealed with a light curing resin. The mandibles were dissected after euthanasia with an anesthetic overdose to the rats 1, 3 and 5 weeks postoperatively. The specimens were fixed, decalcified and embedded in paraffin. Serially sectioned 6μm-thick specimens were made and stained with hematoxylin and eosin, and examined histopathologically under an optical microscope.

Results

One week postoperatively, in all experimental groups, resorption of periapical alveolar bone was observed when the pulp was extirpated at the apical portion of root canal close to the apical foramen.

Group 1. Severe inflammatory responses were seen 1 and 3 weeks postoperatively. Accumulation of polymorpho-nuclear leucocytes (polymorphs) was seen in the residual pulp adjacent to the cement (Fig. 1). Accumulation of polymorphs was also seen in the residual pulp tissue at the apical portion when the radicular pulp was extirpated near the apical foramen. Periapical alveolar bone resorption and infiltration of polymorphs around the apex were also found. Inflammatory responses in the pulp and periapical tissue were still observed 5 weeks postoperatively.

Group 2. Cement-H caused severe inflammatory reaction in both pulp tissue and periapical tissue 1 week after pulp extirpation close to the apical foramen. As shown in Fig. 2, cementum-like hard tissue was newly formed at the root apex 3 weeks postoperatively. There were no inflammatory reactions such as accumulation of polymorphs in the residual pulp tissue adjacent to Cement-H and in the periapical region at this period. Alveolar bone was reconstructed. Five weeks postoperatively, the apical foramen was closed with a newly formed cementum-like hard tissue and reconstruction of periapical alveolar bone was recognized. The periapical region looked normal with alveolar bone reconstruction. When the cement was placed on the amputated pulp surface at the orifice, hard tissue was deposited on the root canal wall 3 weeks postoperatively, and the root canal was completely closed by hard tissue 5 weeks postoperatively.

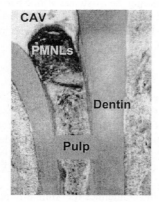

Fig. 1   α-TCP granules applied pulp
3 weeks postoperatively
CAV: Access cavity
PMNLs: Accumulated polymorphs
H.E. stain (Orig. Mag. ×18)

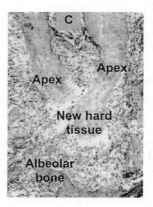

Fig. 2   Apical canal and periapical region
Cement-H (C) is in the root canal.
3 week postoperatively
Apex: Root apex
H.E. stain (Orig. Mag. ×25)

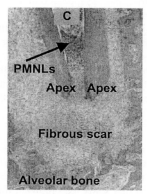

Fig. 3 Apical canal and periapical region
Cement-L (C) is in the root canal.
3 weeks postoperatively
PMNLs: Accumulated polymorphs
Apex: Root apex
H.E. stain (Orig. Mag. ×18)

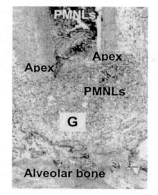

Fig. 4 Apical canal and periapical region
Latex beads (1 μm) were placed in the
root canal after pulp extirpation.
3 weeks postoperatively
PMNLs: Accumulated polymorphs
G: Granulaton tissue   Apex: Root apex
H.E. stain (Orig. Mag. ×20)

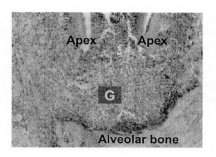

Fig. 5 Apical canal and periapical region
Latex beads (6 μm) were placed in the
root canal after pulp extirpation.
3 weeks postoperatively
G: Granulaton tissue   Apex: Root apex
H.E. stain (Orig. Mag. ×20)

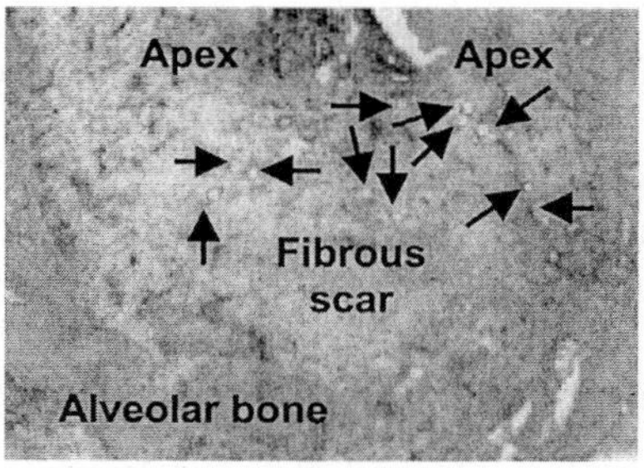

Fig. 6 Root apex and periapical region
Latex beads (25 μm) were placed in the
root canal after pulp extirpation.
5 weeks postoperatively
Arrows: 25 μm latex beads
H.E. stain (Orig. Mag. ×18)

Group 3. One week postoperatively, Cement-L caused inflammatory responses in the surface layer of radicular pulp after extirpation of coronal pulp. There were no inflammatory responses observed in the residual pulp except for the presence of polymorphs in the pulp adjacent to Cement-L 5 weeks postoperatively. When pulp was extirpated closely at the apical foramen, polymorphs accumulated in the residual pulp 3 weeks postoperatively (Fig. 3). Polymorphs in the amputated residual pulp disappeared in the periapical region 5 weeks postoperatively. Periapical alveolar bone resorption was still observed.

Group 4. Latex beads (1 μm in diameter) in the root canal induced accumulation of polymorphs not only in the residual pulp but also in the periapical tissue. These findings were observed 1 and 3 weeks postoperatively (Fig. 4). Five weeks postoperatively, polymorphs had disappeared in these areas and fibrous scar was seen in the periapical lesion with alveolar bone resorption.

Group 5. Mild inflammatory responses were seen 1 and 3 weeks postoperatively. Three weeks postoperatively, many macrophages were conspicuously seen in the periapical lesion (Fig. 5). There was no reconstruction of alveolar bone 3 and 5 weeks postoperatively.

Group 6. A small number of latex beads (25 μm in diameter) were present with no inflammatory responses around the beads and in the periapical region. Neither polymorphs nor macrophages were seen 1, 3 and 5 weeks postoperatively. Resorption of periapical alveolar bone was still observed 5 weeks postoperatively (Fig. 6).

Discussion

In this study, it was confirmed that cement composed of α-TCP granules and 36.0 % citric acid solution induced hard tissue in the pulp and around the apical portion of the tooth root. These findings support the results of our previous study [5]. The findings in this study also indicate that calcium phosphates do not always show biocompatibility. As chemically stable latex beads of 1 μm in diameter induced polymorphs, granules dispersed from the surface of calcium phosphate cement may also induce polymorphs. Calcium phosphate cement is expected to demonstrate excellent biocompatibility and osseous healing ability. A low concentration of citric acid solution hardens α-TCP granules by the process of coagulation. Numerous micro α-TCP granules might scatter in the tissue and induce polymorphs. The α-TCP granules may biologically resolve because of their low crystallinity. However, α-TCP granules caused the accumulation of polymorphs. Though there is a possibility that citric acid solution injures tissue because of high acidity, it may be desirable for α-TCP granules to be kneaded with a high concentration of citric acid solution in order to achieve a hard set. Then, the cement may induce hard tissue deposition or reconstruction of alveolar bone.

Conclusion

Micro granules of calcium phosphate cement may induce inflammatory responses in the tissue. It is concluded that α-TCP cement must set immediately not to scatter granules in the tissue.

Acknowledgements

This study was performed in the Laboratory Animal Facilities and Photograph-Processing Facilities, Institute of Dental Research, Osaka Dental University. This study was supported in part by a 2001 Grant-in-Aid for Scientific Research (C) (No. C:13672016) and a 2002 Grant-in-Aid for Scientific Research (C) (No. C:14571833) from the Japan Society for the Promotion of Science.

References

[1] K. Ikami, M. Iwaku and H. Ozawa: Arch. Histol. Cytol. Vol. 53 (1990), p.227

[2] M. Yoshikawa, S. Hayami, T. Toda and Y. Mandai: Bioceramics Vol. 14 (2001), p. 353

[3] Y. Miyamoto, K. Ishikawa, M. Takechi, T. Toh, T. Yuasa, M. Nagayama and K. Suzuki: J Biomed. Mater. Res. Vol. 48 (1999), p.36

[4] M. Yoshikawa, H. Oonishi, Y. Mandai, K. Minamigawa and T. Toda: Bioceramics Vol. 10 (1997), p. 361

[5] M. Yoshikawa, T. Toda, Y. Mandai and H. Oonishi: Bioceramics Vol. 13 (2000), p. 841

Key Engineering Materials Vols. 254-256(2004) pp. 765-768
online at http://www.scientific.net
© *2004 Trans Tech Publications, Switzerland*

Surface Modification of 58S Bioactive Gel-Glass with an Aminosilane

R.S. Pryce and L.L. Hench

Tissue Engineering Centre, Department of Materials, Imperial College, London, UK. SW7 2AZ
russell.pryce@imperial.ac.uk

Keywords: Bioactive glasses, surface modification, aminosilane, APTS, sol-gel, dissolution kinetics, in-vitro bioactivity.

Abstract. 58S bioactive gel-glasses containing amino-functional ($-NH_2$) moieties have been prepared by means of a surface modification process using an organosilane (3-aminopropyltriethoxysilane, APTS) dissolved in an organic solvent (toluene). Textural characterisation using nitrogen sorption revealed that the surface modified gel-glasses retain the mesoporous structure, but with a reduced surface area, pore diameter and pore volume as APTS concentration increases. However, the initial loading concentration of the aminosilane onto the surface of the 58S gel-glass does not seem to strongly influence the final coverage; rather it is the availability of surface silanols that provide active sites for chemical bonding with the aminosilane. Dissolution of the modified bioactive gel-glasses was assessed using inductively coupled plasma-optical emission spectroscopy, whilst bioactivity was examined using Fourier transform infrared spectroscopy. The in-vitro studies indicate that the surface modified NH58S gel-glasses retain the bioactivity of the unmodified 58S powder, exhibiting similar dissolution profiles and HCA formation. It has been shown that surface modification of bioactive glasses with an aminosilane can produce an amino-functionalised biomaterial, which could be used as reaction handles for further modification.

Introduction

The production of bioactive glasses using the sol-gel process has expanded the potential applications of glass-ceramics in the field of biomaterials. The mesoporous nature of the sol-gel structure enables the incorporation of specific stimuli into the glass for delivery at the local site of implantation [1,2]. Incorporating specific functional group onto the surface of the gel-glass can also increase the site reactivity. One such group is the amino-functional group ($-NH_2$), which has an important chemical role in the adsorption of proteins onto biomaterials. The substitution of surface silanols ($\equiv Si-OH$) with amino groups on the surface of a mesoporous bioactive gel-glass enables the combination of chemical patterning and bioactive properties to result in a potentially superior material with which to examine the controlled sorption of specific proteins at the surface.

The use of aminofunctional silane compounds to chemically modify silica has been commonly employed for a wide variety of applications in science and industry, with the method of modification being well studied [3,4]. Aminosilanes self-catalyse their reaction with silica gel in dry conditions resulting in a siloxane bridge between the aminosilane and the silica substrate. They also simplify the chemical modification process since only three components (silica, aminosilane and solvent) are required.

Previous work has provided a glimpse into the surface modification of bioactive gel-glasses with APTS, using an initial loading of 5% APTS/toluene with 58S bioactive foams [5]. This study aims to further this investigation into the characterisation of bioactive gel-glass materials prior to protein adsorption studies. Therefore the objective of this research was to modify the ternary 58S bioactive gel-glass with an organosilane, 3-aminopropyltriethoxysilane (APTS) and subsequently characterise the textural and bioactive properties of the modified gel-glass as a function of loading of APTS on the surface.

Methods

Preparation of NH58S powders. The 58S gel glass (60% SiO_2, 36% CaO and 4% P_2O_5, in mole percent) was prepared via the sol-gel technique starting from tetraethyl orthosilicate, triethyl phosphate and calcium nitrate tetrahydrate (with nitric acid as catalyst) in the standard way [6]. The stabilised glass was ground in a jar mill and sieved to obtain a range of 90-710 μm prior to modification. Surface modification of the powder was carried out by immersing 1g of the 58S powder in APTS/toluene solution (either 0%, 0.5%, 1%, 2.5%, 5% or 10% APTS). The samples were agitated constantly at 37°C for 3 hours. The powder was then rinsed with toluene to remove any unreacted organosilanes before being dried at 100°C for 12 hours.

Textural characterisation. The textural characterisation of the modified gel-glass powders was performed on a six-port Quantachrome AS6 Autosorb gas sorption system using N_2 gas (assumed cross-sectional area of $0.162nm^2$). Prior to N_2 sorption, accurately measured powder samples were degassed at 100°C for 15 hours to remove any physically adsorbed molecules from their surfaces. The specific surface areas were estimated in relation to the masses of the degassed samples using the BET theory. Pore size analysis were computed from the desorption data using the BJH method.

Bioactivity. 0.075g samples of the modified gel-glass powders were immersed in 50ml of simulated body fluid (SBF) in large conical flasks and placed in an orbital shaker at 175rpm and 37°C, for periods ranging from 30 minutes up to 1 week. The surfaces of the specimens were characterised using Fourier transform infrared spectroscopy (FTIR) whilst the gel-glass dissolution products were analysed using inductively coupled plasma-optical emission spectroscopy (ICP).

Results and Discussion

Textural characterisation. Figure 1 shows the textural profile as a function of concentration, with the 58S gel-glass represented by the dotted line. Surface modification of the bioactive gel-glass with APTS resulted in complete coverage even at low concentrations of APTS (0.5%). Increasing the APTS concentration during loading resulted in a very gradual increase in final loading of the gel-glass surface (not shown).

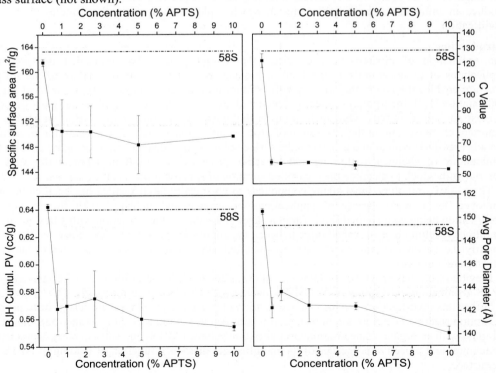

Fig. 1: Textural characterisation as a function of APTS concentration

The decrease in surface area, pore size diameter and the increase in surface loading of APTS can be attributed to the presence of trace quantities of water, since only monolayer coverage would otherwise be expected in fully dry conditions [4].

Dissolution profiles. Fig. 2 shows the dissolution profiles of 58S, 0.5NH58S and 10%NH58S. A notable decrease in the concentration of phosphorus is seen within the first 2 hours for all samples, indicating that the phosphorus in solution is being depleted as calcium is released from the gel-glasses. After 8 hours, the concentration of phosphorus in the SBF is negligible, preventing further deposition of a calcium phosphate (CaO-P_2O_5) layer.

Calcium release from the gel-glasses shows a rapid burst within the first 30 minutes (as high as 257ppm for 0%NH58S) which then decreases with time within the first 15 hours. This is due to reaction with phosphorus in the solution. Between 15 and 24 hours, there is a slight surge in calcium concentration that gradually decreases as time increases up to 1 week. There seems to be little correlation between calcium concentration and APTS concentration within the first 24 hours, but by 1 week, increased loading seems to diminish the observed final calcium concentration with 10%NH58S showing the lowest calcium concentration. All glasses exhibit a similar trend of rapid release of silica within the first 4 hours (between 150 to 160ppm Si) after which a plateau is reached. There does not seem to be noticeable relationship between APTS concentration and silica release with all modified glasses showing similar trends; however increased APTS loading resulted in an overall increase in silicon release from the glass.

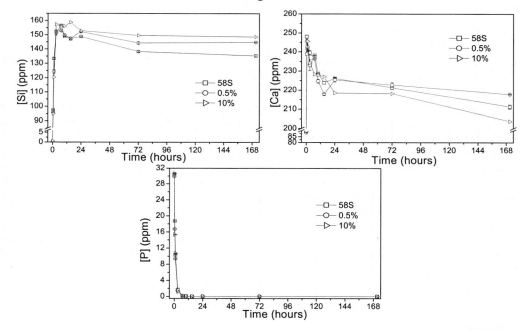

Fig. 2. Dissolution profiles (Si, Ca and P) for 58S, 0.5%NH58 and 10%NH58 gel-glass powders.

Bioactivity. Analysis of the hydroxyapatite layer formed on the surface of the powders indicated a well developed layer at a rate comparable to the original 58S gel-glass. Fig. 3 shows FTIR traces for 5%NH58S as a function of reaction time in SBF, as well as traces for all powders after 7 hours immersion in SBF. As can be seen, the development of a well-defined P-O doublet at 571 and 604cm^{-1} has occurred by 30 minutes with the peaks sharpening and increasing in intensity. The band at 1060cm^{-1} is due to the P-O stretch vibration in the PO_4 tetrahedra of crystalline HCA. Carbonate bands at about 1420-1500 cm^{-1} and 874cm^{-1} also indicate the formation of HCA on the surface [7]. The presence of -NH vibrational bands were occluded by the presence of the stronger carbonate and phosphate bands at 1600 and 1060cm^{-1} respectively; other means were used to confirm their presence (Raman spectroscopy and Ninhydrin assay, not shown).

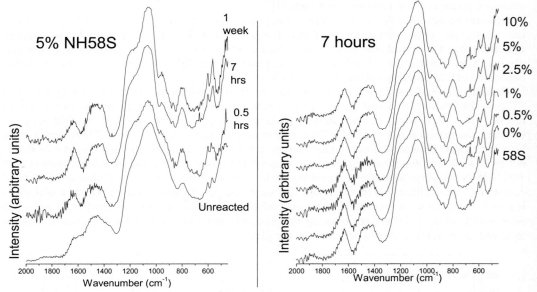

Fig. 3. FTIR traces of 5% NH58S (unreacted, 0.5 hours, 7 hours and 1 week immersed in SBF) and all modified gel-glasses immersed for 7 hours in SBF.

Conclusions
Surface modification of the 58S bioactive gel-glass with APTS results in the incorporation of aminopropyl functional groups onto the gel-glass via the formation of siloxane bonds at surface silanols. The initial aminosilane concentration does not seem to be a significant variable, as fast adsorption occurs for all concentrations, with only a gradual increase in the polysiloxane layer once monolayer coverage has occurred due to the presence of trace water. The in-vitro studies in SBF indicate that the surface modified NH58S gel-glasses retain the bioactivity of the unmodified 58S powder, exhibiting similar dissolution profiles and HCA formation. Thus, surface modification of bioactive glasses with aminosilanes provides a means to chemical pattern the gel-glass surface. The attached amino-functional groups can subsequently be used as reaction handles for further modification, to tailor for specific applications. Further studies are underway to determine the interactions of the modified bioactive glass in physiological fluids (and in particular protein absorption) using Raman spectroscopy.

Acknowledgements
The authors acknowledge the support of the March of Dimes Birth Defects Foundation.

References
[1] L.L. Hench and J.K. West: *Life Chemistry Reports Vol.* 13 (1996) p.187
[2] M.M. Pereira and L.L. Hench: Journal of Sol-gel Science and Technology Vol. 7 (1996), p.59.
[3] E.P. Plueddemann: *Silane coupling agents* (Plenum Press, New York 1991).
[4] E.F. Vansant, P. Van Der Voort and K.C. Vrancken: *Characterization and Chemical Modification of the Silica Surface* (Elsevier, Amsterdam 1995).
[5] R.F.S. Lenza, W.L. Vasconcelos, J.R. Jones et al. Journal of Materials Science–Materials in Medicine. Vol. 13 (2002) p.837.
[6] R. Li, A.E. Clark and L.L. Hench: Journal of Applied Biomaterials Vol. 2 (1991) p.231.
[7] A. Stoch, W. Jastrzebski, A. Brozek et al: Journal of Molecular Structure. Vol. 511-512 (1999) p. 735.

Key Engineering Materials Vols. 254-256(2004) pp. 769-772
online at http://www.scientific.net
© 2004 Trans Tech Publications, Switzerland

Study of Osteoblasts Mineralisation *In-Vitro* by Raman Micro-Spectroscopy

Ioan Notingher, Julie E. Gough and Larry L. Hench

Department of Materials, Imperial College London, Exhibition Road, London SW7 2AZ, United Kingdom, email: i.notingher@imperial.ac.uk

Keywords: Raman spectroscopy, osteoblasts, mineralisation

Abstract. The effect of bioinert and bioactive materials on the formation and mineralisation of bone nodules *in-vitro* was studied by Raman spectroscopy. Human primary osteoblasts were cultured on bioinert fused silica and bioactive 45S5 Bioglass® for 10 days. No mineralisation was detected in both cases when the culture media was not supplemented with beta-glycerophosphate. When beta-glycerophosphate was added to the culture media, hydroxyapatite was detected in the Raman spectra for both substrates. However, when osteoblasts were cultured on fused silica, the intensity of HA peak was considerably lower compared to the values measured when the cells were cultured on 45S5 Bioglass®. Bone nodules were observed only when 45S5 Bioglass® was used as substrate and the mineralisation was indicated by very intense HA peaks.

Introduction

The failure of prostheses made of bioinert materials following implantation in patients with bone failure led to the development of a new generation of materials. These materials interact and bond with living tissue or stimulate self-regeneration of the diseased bone [1]. Another alternative for bone repair is the implantation of engineered bone built *in-vitro* by culturing bone cells harvested from the patient on three dimensional scaffolds made of bioactive materials [1]. In both cases, the understanding of the interaction between human bone cells and the bioactive materials is an important factor in the development of the materials. Previous studies indicated that bioactive materials have a significant effect on the cell behavior, such as cell attachment, proliferation, differentiation, mineralisation or cell death [2,3].

Bioactive glasses based on silica, phosphate and calcium and sodium oxides, were reported by Hench et al [4], as the first man made materials to bind to living tissue. The bioactive glasses develop a hydroxyapatite layer at the surface when immersed in aqueous solutions on which bone cells attach, proliferate, differentiate and form new bone. These effects are due to the active role of the released ions on these cells, such as activation of genes related to cell attachment, proliferation and differentiation [2,3].

Experiments designed to follow in time the behavior of cells cultured on biomaterials *in-vitro* are the first stage for testing and designing of bioactive materials. However, time-course experiments are difficult to be carried out because most biological assays are invasive, need labels, cell fixation or cell lysis.

Raman spectroscopy is a well established analytical technique and it is based on the inelastic scattering of electromagnetic radiation by molecules. The spectral shifts induced to the excitation radiation depend on the vibrational frequencies of the molecules, therefore a Raman spectrum is a fingerprint of the chemical composition of the sample. Cultures of living cells can be investigated *in-situ* and in real-time by Raman spectroscopy using near-infrared lasers due to the low Raman signal and low absorption of water in this spectral region [5,6]. Raman spectroscopy proved sensitive to detect changes at a molecular level due to cell differentiation (not published), cell cycle and cell death [6]. Studies of *in-vitro* mineralisation of mouse bone cells cultured on bioinert materials were also reported [7]. In this study, we used Raman spectroscopy to compare *in-situ* the effect of bioinert materials (fused silica) and bioactive glass (45S5 Bioglass®) on the formation and mineralisation of primary human osteoblasts *in-vitro*.

Materials and Methods

Cell Culture. Human primary osteoblasts (HOBs) were isolated as described previously from femoral heads after hip replacement surgery [8]. The isolated cells were cultured on fused silica squares 5x5mm (Helma Ltd., UK) and on 15 mm diameter 45S5 Bioglass® discs (US Biomaterials, US) with and without beta-glycerophosphate (BGP) (Sigma, UK) supplemented culture medium (Dulbecco's Modified Eagles Medium (DMEM) containing 10% foetal bovine serum (FBS) with 2% penicyllin/streptomycin and 0.85mM ascorbic acid) at 37°C, 5% CO_2 for 10 days.

Raman Spectroscopy. Raman spectra were measured with a RM 2000 Renishaw micro-spectrometer equipped with a high power 785nm laser (250mW). A long working distance 63x magnification water immersion objective was used. The cells were maintained in phosphate-buffered saline (PBS) during the experiments. Spectral collection time was 40 sec.

Results and Discussion

Raman spectra of primary human osteoblasts cells cultured for 10 days on fused silica substrates in culture media with and without beta-glycerophosphate (BGP) are shown in Fig. 1.

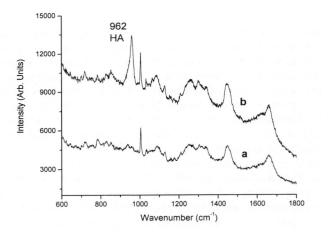

Fig. 1. Raman spectra of primary human osteoblasts cultured on fused silica substrates:
(a) without BGP, (b) with BGP

The measured Raman spectra of human osteoblasts cultured on bioinert fused silica substrates are similar to the spectra reported for other cell types [5,6]. The spectra show peaks corresponding to vibrational modes of all cell components, nucleic acids, proteins, carbohydrates and lipids. A tentative peak assignment based on published literature is given in Table 1 [5-7,9]. From all cell components, proteins (1660 cm^{-1} Amide I, 1200-1300 cm^{-1} Amide III, 1005 phenylalanine, 938 cm^{-1} backbone C-C stretching) and DNA (1095 cm^{-1} phosphodioxy group PO_2^-, 788 cm^{-1} phosphodiester bond O-P-O) have the strongest contribution to the Raman spectra of osteoblasts cultured on fused silica in absence of BGP (Fig.1(a)).

Table 1. Peak assignment for Raman spectra of human osteoblasts
(A, G, C, T-adenine, guanine, cytosine, thymine; str-stretching) [5-7,9]

Peak (cm^{-1})	Nucleic acids	Proteins	Carbohydrates	Lipids	Hydroxy Apatite
1660		Amide I		C=C str	
1449		CH def	CH def	CH def	
1300⌐	A, G	Amide III		=CH def	
1230⌐					
1095⌐	PO$_2^-$ str		C-C, C-O str	Chain C-C str	
1005		Phe			
962					PO$_4^{3-}$ asym
938		C-C str α-helix			str
854		Tyr			
813	O-P-O RNA				
788	O-P-O DNA			O-P-O	
782	U, C, T				

When the culture media was supplemented with BGP, at 10 days the osteoblasts formed a confluent layer but no bone nodules were observed. However, in the Raman spectra a new peak was present at 962 cm^{-1}, which is specific to the PO$_4^{3-}$ group in hydroxyapatite (HA) [9] (Fig.1(b)). The amount of HA can be estimated relative to the amount of organic components in the Raman spectra by computing the ratio between the intensities of the HA 962 cm^{-1} and CH deformation 1449 cm^{-1} peaks. For the osteoblasts cultured on fused silica and supplemented with BGP, the maximum value of this ratio was $I_{962}/I_{1449} \cong 1.75$.

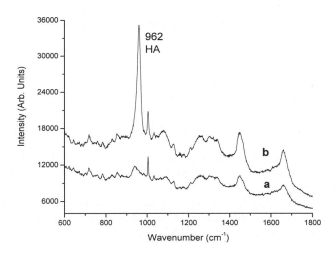

Fig. 2. Raman spectra of primary human osteoblasts cultured on 45S5 Bioglass®:
(a) without BGP, (b) with BGP

The presence of the HA peak in the Raman spectra of human osteoblasts fed with BGP and cultured on fused silica suggests that even though bone nodules did not formed, a few HA crystals formed from the precipitation components in the culture medium.

Fig. 2 shows the Raman spectra of human osteoblasts cultured on 45S5 Bioglass® discs. If the culture media was not supplemented with BGP, no bone nodules and no mineralisation was observed. The absence of mineralisation was confirmed by the absence of the 962 cm^{-1} HA peak in the Raman spectra (Fig. 2(a)). If the culture media was supplemented with BGP and the osteoblasts were grown on 45S5 Bioglass®, several bone nodules started to form after 10 days and large HA peaks was observed at 962 cm^{-1} in the Raman spectra, as shown in Fig.2(b). The magnitude of the HA peak was much higher than for fused silica substrates, the ratio between the HA peak and the 1449 cm^{-1} CH deformation peak of proteins was found to be higher than 3.5. The high intensity of the HA peak in the Raman spectra confirms that 45S5 Bioglass® has a different effect on the human osteoblasts compared to fused silica. The dissolution ions released by the 45S5 Bioglass® increase the proliferation and differentiation rate of the cells and led to formation and mineralisation of bone nodules, which is consistent with results of Refs 2 and 3.

Conclusions

Raman spectroscopy can be used to monitor and quantify non-invasively in real-time the mineralisation of primary human osteoblasts *in-vitro* cultured on various substrates. The formation of hydroxyapatite is indicated by the strong phosphate group at peak at 962 cm^{-1}. No mineralisation was observed when osteoblasts were cultured on fused silica and 45S5 Bioglass® without supplementation with BGP after 10 days. When the culture media was supplemented with BGP, hydroxyapatite was detected both on fused silica and 45S5 Bioglass®. However, no formation of bone nodules was observed on fused silica. The amount of HA formed was estimated relative to the magnitude of the protein peaks. The intensity of the HA was more than 2-fold higher when 45S5 Bioglass® was used as substrate. These results show that 45S5 Bioglass® leads to formation and mineralisation of bone nodules as early as 10 days.

Acknowledgements

The authors thank US Defence Advanced Research Projects (Contract No. N66001-C-8041) and the UK Medical Research Council for financial support.

References

[1] L.L. Hench, J.M. Polak: Science 295 (2002), p. 1014
[2] I.D. Xynos, M.V. Hukkanen, J.J. Batten, L.D. Buttery, L.L. Hench: Calcif. Tissue Int. 67 (2000), p. 321
[3] I.D. Xynos, A.J. Edgar, L.D. Buttery, L.L. Hench, J.M. Polak: Biochem. Byophys. Res. Commun. 276 (2000), p. 461
[4] L.L. Hench, R.J. Splinter, W.C. Alle, T.K. Greenlee: J. Biomed. Mater. Res. 74 (1971), p.1478
[5] G. J. Puppels, F. F. de Mul, C. Otto, J. Greve, M. Robert-Nicoud, D. J. Arndt-Jovin, T. M. Jovin: Nature 347 (1990), p. 301
[6] I. Notingher, S. Verrier, S. Haque, J.M. Polak, L.L. Hench: Biospectrosc0py 72 (2003), p. 230
[7] C.P. Tarnowski, M.A. Ignelzi Jr., M.D. Morris: J. Bone Miner. Res. 17 (2002), p. 1118
[8] J.E. Wergedal, D.J. Baylink: Proceedings of the Society for Experimental Biology and Medicine 176 (1984), p. 60
[9] J.C. Elliot ed.: Structure and Chemistry of the Apatites and other Calcium Orthophosphates, Studies in Inorganic Chemistry 18, (1994), Elsevier Science

Key Engineering Materials Vols. 254-256(2004) pp. 773-776
online at http://www.scientific.net
© 2004 Trans Tech Publications, Switzerland

Characteristic of Osteoblast Vacuole Formation in the Presence of Ionic Products from BG60S Dissolution

P. Valério[1], A. M. Goes[2], M. M. Pereira[3], M. F. Leite[1]

[1] Department of Phyisiology and Biophysics,

[2] Department of Biochemistry and Immunology,

[3] Department of Metallurgical Engineering,

Federal University of Minas Gerais, Belo Horizonte, Brazil

patricia.valério@terra.com.br

Keywords: Vacuole, bioactive glass, silicon, collagen, alkaline phosphatase

Abstract. Bioactive ceramics developed during the past few decades have interesting properties from the biological standpoint, but their effects on cellular events remain partially unknown. In the current work, we investigated morphology changes of rat primary culture osteoblasts in contact with ionic products from the dissolution of a bioactive glass with 60% of silica (BG60S). We observed that osteoblasts cultured with BG60S showed vacuole formation. We also found that high silicon concentration could induce cellular vacuole formation. Additionally, energy dispersive spectroscopy analysis indicated that vacuole contains 75% more silicon than other regions in the cell, outside the vacuole. We further found that vacuole formation was not related to cell degeneration nor to lysosomatic activity. Together, our results indicate that osteoblast vacuole formation was due to high silicon contents in the dissolution of BG60S.

Introduction

Considering that osteoblasts are the cells that support the synthesis, secretion and mineralization of extracellular bone matrix, the investigation of the their behavior in the presence of bioceramics is important to evaluate biocompatibility. The bioceramics composition, cristallinity, and porosity are characteristics directly related to cell physiology (1) and the ionic products from bioceramics dissolution are responsible for alterations in osteoblast proliferation and secretion capability (2). In our previous studies we demonstrated that osteoblasts in the presence of ionic products from a bioactive glass with 60% of silicon (BG60S) dissolution have a higher proliferation and collagen secretion, when compared to control and biphasic calcium phosphate (BCP). We also demonstrated that osteoblasts in the presence of the ionic products from BG60S dissolution show a large number of cytoplasmatic vacuoles and this vacuole formation is related to silicon content of the ionic products (3). In this work, our aim is to investigate the characteristics of these vacuoles.

Material and Methods

Material

In the present study was used a ionic product obtained from the dissolution of a bioactive glass with 60% of silica BG60S. The composition of BG60S (in weight %): silica 60%, calcium 35% phophate 5%.. The bioceramic was used in an approximately particle size of 38μm and suspended in a 1/100 ratio w/v in culture medium, shaked and filtered. Cell culture medium RPMI (Sigma), Dulbecco´s phosphate buffered saline, trypsin-EDTA, fetal bovine serum (Gibco), crude bacterial collagenase (Boehringer), Mytotracker, Lysotracker, Fluo-4 and Phallotoxin (Molecular Probes), T 25 culture flasks and multdish 24 well (Nunc Products).

Methods

To investigate the vacuole characteristics, Mytotracker (500 nM), Lysotracker (100 nM), Fluo-4/AM (100μM) and Phalloidin (6,6 μM) were added to osteoblats that had been incubated with the ionic products from the dissolution of BG60S. To investigate the presence of apoptotic cells we used propidium iodide probe which is based on the property of this fluorescent dye to stain apoptotic cells nucleous. For this assay, cells were rinsed twice with PBS and incubated for 5 min with 50 μl/ml propidium iodide. Osteoblasts were observed by fluorescence microscope or confocal microscope using the proper filter for each assay. For (EDS) energy dispersive spectroscopy analysis, osteoblasts were platted in stainless plates and put in contact with medium containing ionic products from BG60S dissolution for 72 hours. The same condition was done using micro cover glasses considering that it was not possible to evaluate vacuole formation by optical microscopy in the metal plates. The cells were fixed by 3,7% phormaldeid for 5 min followed by 70% ethanol for 5 min and slowly dried (4). The specimens were then analysed by EDS using punctual analysis at an operating voltage of 15 Kv and the silicon ratio was determined inside and outside the vacuole.

Results

To investigate the vacuole characteristics we used fluorescent selective probes to label and track organelles, search for calcium containing compartments, and check for apoptotic cells.

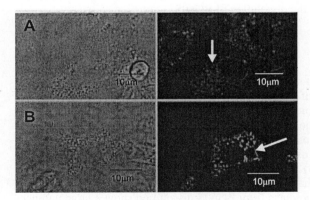

Fig. 1.Organelles staining.
  A. Mytotracker probe indicated that vacuoles were not hypertrophic mitochondria.
  B. Lysotracker probe showed that they were not lysosome.s
Left panels show transmission images and right panels the fluorescent organelles.Arrows show vacuoles

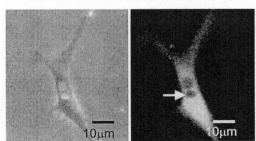

Fig. 2. Free calcium staining.
 Fluo- 4/AM probe was used to evidentiate intracellular free calcium. No fluorescent calcium was present inside vacuole.
Left panel show transmission image of a vacuolated osteoblast and right the fluorescent calcium. Arrow shows a big vacuole without calcium.

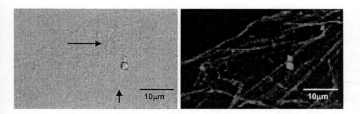

Fig. 3. Actin staining. When actin cytoskeleton was evidentiated, no actin was present n the vacuole walls.Left panel shows transmission image and rigth the fluorescent actin. Arrows show vacuoles.

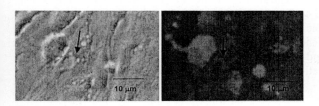

Fig. 4. Apoptotic cells staining. Vacuolated cells showed no staining nucleous indicating apoptosis.Left panel shows a transmission image and right the fluorescent apoptotic nucleous. Arrows show vacuoles.

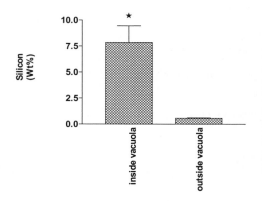

Fig. 5. Energy Dispersive Spectroscopy analysis. Osteoblasts plated in stainless metal plates were put in contact with ionic products from BG60S dissolution and after 72 hours analysed by EDS analysis. Graphic shows high percentage of silicon inside vacuoles when compared to other regions in the cell. Result represents: Mean ± SD of 4 different measurements ($P< 0.05$).

The Mitotracker and Lysotracker probes showed that vacuole were neither hypertrophic mitochondria nor lysosomes (Fig. 1A, Fig. 1B). The Fluo-4/AM probe indicated that the vacuole were not intracellular vesicles containing high calcium concentration (Fig. 2). We also found that there was no actin staining on the vacuole walls (Fig. 3), suggesting that the vacuoles are not endocytic vesicles. Additionally, we found that osteoblasts containing vacuoles are not apoptotic cells (Fig. 4). We then investigated the vacuole contents, using an EDS analysis. Our results indicated that the percentage of silicon inside the vacuole was 75% higher than the percentage in other regions of the cell, outside the vacuole (Fig. 5).

Discussion

Vacuole is an intracellular structure delimited by a membrane (5) and in some cases vacuole formation occurs by a pinocytotic process with internalization of plasma membrane (6). Despite vacuolization are commonly associated to cell death (7), we did not find correlation between vacuole formation and apoptosis or cell degeneration. Instead of this we have found previously, higher cell proliferation when compared to control (3). Vacuole formation has been related to changes in actin cytoskeleton (8, 9, 10), but when we evidentiated actin filaments we did not find any correlation between actin cytoskeleton and osteoblast vacuole formation. It can be explained by the evidentiation of other mechanisms involved in vesicle motor complex such as kinesin or microtubules (11,12). It leads us to the speculation that other cytoskeleton constituent may be involved in osteoblast vacuolization. Vacuole formation was also described as a mechanism of molecule transportation inside many kind of cells and frequently associated with increasing in collagen production (13, 14, 15, 16,), and these vacuoles in some cases were associated to a lysosomal origin and in other cases to storage vesicles (17). In our studies, the lysosomatic vesicle characteristic was discarded, but our results strongly support the idea that associates high collagen secretion and vacuolization. When we performed the EDS analysis, we found a high percentage of silicon inside vacuole, when compared to other cell regions. Considering that it is already demonstrated that fibroblast, a cell with the same embryologic origin of osteoblasts, internalize silicon without detectable cell damage (18) and also considering that there is a positive correlation between silicon concentration and collagen concentration in tissues (19), we can speculate that the internalization of silicon may stimulate the collagen production and cross linking and the formed collagen could be in storage inside vacuole. On the other hand we can also speculate that the internalized silicon forms a silica gel core inside the cell, since it was already described this gel core formation when silicon reacts with body fluid in vivo (20). Both speculations are in accord with our previous finding of osteoblast vacuole formation in the presence of pure silica dissolution (3).

References

[1] R Langer. Molecular Therapy. Vol 1 (2000) p 12-15

[2] I D Xynos et al. Byochem Biophis Res. Vol 276 (2000) p 461-467

[3] P Valerio et al. Key Eng Mat. Vol 250 (2003) p 699-702

[4] S N Silva et al. J Biomed Mater Res. Vol 65 (2003) p 475-481

[5] E Holtzman. Cells and Organelles. CBS college publishing.(1983)

[6] G E Davis. Exp Cell Res. Vol 224 (1996) p 39-51

[7] C M Serre et al. J Biomed Mat Res. Vol 42 (1998) p 623-626

[8] S Shibata et al. Bone. Vol 14 (1993) p 35-40

[9] T L Herring et al. J Membr Biol. Vol 171 (1999) p 151-159

[10] A Aderem et al. Annu Rev Immunol. Vol 17 (1999) p 593-613

[11] G J Hyde et al. Cell Motil Cytoskeleton. Vol 42 (1999) p 114-124

[12] A S De Pina et al. Micros Res Tech. Vol 47 (1999) p 93-106

[13] S Takano et al. The Anat Rec. Vol 263 (2001) p 127-138

[14] S G Katz. Tissue Cell. Vol 27 (1995) p 713-721

[15] Y S Lee et al. Ann Acad Med Singapure. Vol 24 (1995) p 902-905

[16] A Gritli-Lindi et al. Calcif Tissue Int. Vol 57 (1995) p 178-184

[17] L B Creemers et al. Matrix Biol. Vol 16 (1998) p 575-584

[18] T Baroni et al. J Investg Med. Vol 49 (2001) p 146-156

[19] M R Calommed et al. Biol Trace Elem Res. Vol 56 (1997) p 153-165

[20] W Lai et al. Bioceramics 11. (1998) p 383-386

Key Engineering Materials Vols. 254-256(2004) pp. 777-780
online at http://www.scientific.net

Evaluation of Osteoblasts Viability, Alkaline Phosphatase Production and Collagen Secretion in the Presence of TiHA

P. Valério[1], F. N. Oktar[2], G. Goller[3], A. M. Goes[4] and M. F. Leite[1]

[1] Federal University of Minas Gerais, Department of Phyisiology and Biophysics, Brazil

[2] Marmara University, Campus of Goztepe, Industrial Engineering Department Istambul, Turkey

[3] Istambul Technical University, Metallurgical & Material Engineering Department, Turkey

[4] Federal University of Minas Gerais, Department of Biochemistry and Immunology, Brazil

patricia.valerio@terra.com.br

Keywords: Titanium, collagen, alkaline phosphatase, osteoblast, hydroxyapatite

Abstract. Titanium reinforced hydroxyapatite (TiHA), prepared in a 5% and 10% ratio and synterized at 1200^0C and 1300^0C, were analysed. 5% TiHA synterized at 1300^0C showed significant increase in collagen production, when compared to the others TiHA and to control.

Introduction

Biocomposite materials have been developed in order to combine bioactivity of ceramics and mechanical properties of metals (1). Hydroxyapatite (HA) is known for its weakness and brittles. When titanium is added to HA, an improvement of the biomaterial mechanical properties occurs (2). However, the proportion of Ti and HA in each biocomposite can alter its characteristics. The sintered temperatures can also alter the biological properties and therefore, they need to be characterized. Considering that calvarie osteoblast primary culture is a well stabilished model to investigate biocompatibility, in this study it is evaluated the osteoblast viability, alkaline phosphatase production and collagen secretion in the presence of four different TiHA biocomposites.

Material

Four types of titanium reinforced hydroxyapatite were used in the present study. Cell culture medium RPMI (Sigma), Dulbecco´s phosphate buffered saline, trypsin-EDTA, fetal bovine serum (Gibco), crude bacterial collagenase (Boehringer), T 25 culture flasks and multdish 24 well (Nunc Products), MTT, NBT-BCIP assay (Sigma), Sircol kit.

Methods

TiHA was prepared mixing 5 and 10 wt % of Ti powder with HA. The 2 specimens were subjected to sintering at temperatures up to 1200^0C and 1300^0C. This process resulted in 4 different biocomposites (2) that were processed in large particles ($\pm$ 50μm). Primary cultures of osteoblasts were prepared from sequential enzymatic digestion of calvarie obtained from 1-5 days old Wistar rats. Briefly, after cut into small pieces, the calvarie bone was digested with trypsin 1% and four times with collagenase 2%. The supernatant of the three last washes were centrifuged at 1400 g for 5 min, the pellet ressuspended in culture medium and plated. After confluence the cells were replicated and used in passage 2.

Culture medium containing each TiHA powder was put in contact with osteoblasts that have been plated in 5 x 10⁴ cell density at passage 2. After 72 h of incubation the viability and proliferation was measured by MTT method that is based on the capacity of viable cell to metabolize tetrazolium

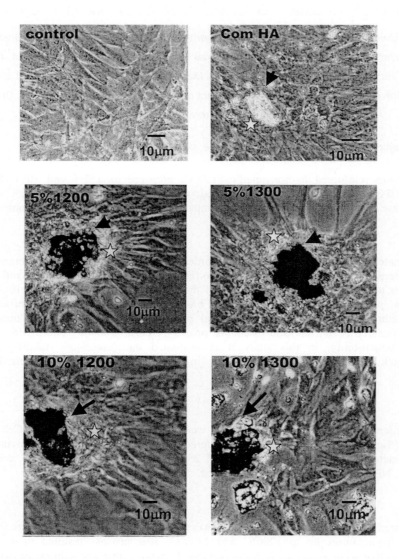

Fig. 1. Osteoblast morphology Photomicrograph 400 X. Under optical microscopy, osteoblasts showed no morphological changes in the presence of all biocomposites and commercial hydroxyapatite (arrows show biomaterials crystals), when compared to control. Osteoblasts seem to adhere to the biomaterial (pointed by stars).

to formazan crystals, a purple dye that can be solubilized and measured by optical density. Alkaline phosphatase production was analysed by NBT-BCIP assay. Nitrobluetetrazolium is activated by the alkaline phosphatase secreted by osteoblasts and after some reactions, spread a blue dye that may be solubilized and measured by optical density. Collagen secretion was measured using SIRCOL method. This assay is based on the capacity of Syrius red dye to bind to the end of collagen molecule and precipitate it. After solubilization, collagen can be quantificated using linear regression from samples of known concentrations of collagen. Data were analysed statistically by variance test ANOVA, using Bonferroni's post-test.

Results

Osteoblasts morphology showed no change in the presence of any biocomposite, neither in the presence of commercial HA, when compared to control. Osteoblasts seem to adhere to biomaterials crystals as showed in Fig.1. The viability and proliferation of osteoblasts were not altered in the presence of any biocomposite, when compared to control and commercial HA (Fig. 1). Alkaline phosphatase production decreased plus than 20% in the presence of all TiHA as well as commercial HA, when compared to control (Fig. 2) Collagen secretion had a significat increasing in the presence of 5% TiHA syntered at $1300^{0}C$ (Fig. 3), when compared to the other biocomposites (75%), commercial HA (35%), and control (100%).

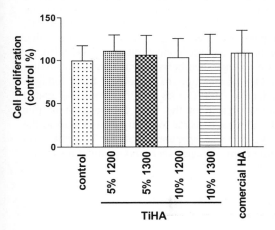

Fig. 2. Cell proliferation. 1 X 10^5 osteoblasts were plated in the presence of granules from the four TiHA samples and a commercial HA. After 72 h of incubation, proliferation was evaluated by MTT assay. Osteoblasts showed no proliferation difference in the presence of all bioceramics, when compared to control. Results represent Mean ± SD of triplicates from 3 separate experiments (P<0.05).

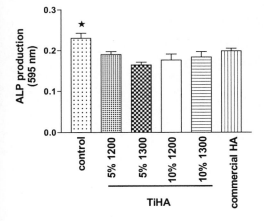

Fig. 3. Alkaline Phosphatase production. 1 X 10^5 osteoblasts were plated in the presence of granules from the four TiHA samples and a commercial HA. After 72 h of incubation, alkaline phosphatase production was measured by NBT-BCIP assay.Osteoblasts in the presence of all bioceramics showed a decreasing in alkaline phosphatase production, when compared to control. Results represent Mean ± SD of triplicates from 3 separate experiments (P<0.05).

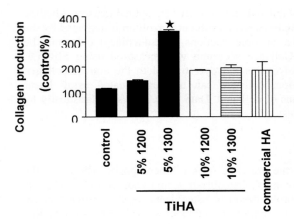

Fig. 4. Collagen production. 1×10^5 osteoblasts were plated in the presence of granules from the four TiHA samples and a commercial HA. After 72 h of incubation, the supernatant of each culture were analised to measure collagen secretion. Osteoblasts in presence of 5% TiHA sinterized at 1300^0 C have higher collagen secretion when compared to the other biocomposites and comercial HA. Results represent Mean ± SD of triplicates from 3 separate experiments ($P<0.05$)

Discussion

The biomaterial substrates markedly influence the temporal sequence of matrix protein secretion (3). Ceramics structure and topography can alter the osteoblast function (4). It is also known that the synthesizing method can alter the biomaterial crystallites morphology (5). When sintered at 1300^0C, 5%TiHA shows a high proportion of TCP when compared to 1200^0C (2). It has been demonstrated a correlation between collagen production by osteoblasts and the concentration of TCP (6). In addition, it is known that the presence of TCP increase collagen synthesis and that particles smaller than 10μm do not effect collagen production (7). Our observation of osteoblast high collagen production in the presence of the biocomposite sinterized at 1300^0C with 5% of Ti could be related to a possible adequate particle size and TCP amount present in the final composition of this biocomposite. As it was also demonstrated that TiO_2 / HA ratio seems to be directed related to the stimulation of osteoblasts markers expression (8), we can also speculate that the Ti / HA ratio associated to crystallite morphology could be an important collagen secretion stimulating factor, requiring therefore, further investigations.

References

[1] L Hench. J Am Ceram Soc. Vol 81(1998) p 1705-1728

[2] G Goller *et al.* Key Eng Mat. Vol 240 (2003) p 619-622

[3] R Z LeGeros, R G Craig. J Biomed Mater Res. Vol 15 (1999) p 585-594

[4] K S TenHuisen, P W Brown. Biomaterial. Vol 19 (1998) p 2209-2217

[5] T J Webster *et al.* J Biomed Mater Res. Vol 51 (2000) p 475-483

[6] A Ehara. Biomaterial. Vol 24 (2003) p 831-836

[7] D Pioleti. Biomaterial. Vol 21 (2000) p 1103-1114

[8] P A Ramires *et al.* Biomaterial. Vol 22 (2001) p 1467-1474

Key Engineering Materials Vols. 254-256(2004) pp. 781-784
online at http://www.scientific.net

A549 Lung Carcinoma Cells: Binary vs. Ternary Bioactive Gel-Glasses

P.Saravanapavan [1], S. Verrier [1,2] and L.L. Hench [1]

[1] Tissue Engineering Centre, Department of Materials, Imperial College London, UK
p.pavan@imperial.ac.uk
[2] Tissue Engineering & Regenerative Medicine Centre, Faculty of Medicine, Imperial College London, UK

Keywords: calcium silicate, attachment, proliferation, lung cells, bioactive glass, sol-gel

Abstract. The binary gel-glasses in the $CaO-SiO_2$ system are being studied for use as a bioceramic material and hence have to be compared to other materials that are similar in properties and are in use either for clinical or research purposes. In this paper, the attachment, proliferation and the morphology of human lung epithelial cells are investigated using substrates of two compositions (one with and another without phosphate). The results of this study indicate that the ternary 58S bioactive gel-glass substrates offer favourable conditions for A549 cell adhesion and proliferation.

Introduction

Ternary bioactive gel-glasses in the $CaO-SiO_2-P_2O_5$ system have been widely studied since 1991 [1]. Both S70C30 and 58S gel-glass compositions are Class A bioactive [2,3] and have been compared to melt-derived 45S5 Bioglass®, which has been in clinical use for twenty years.

The aim of the investigation reported herein is to compare cell adhesion and proliferation properties of binary and ternary substrates using human epithelial-like lung carcinoma cells (A549). Porous scaffolds of both compositions are being developed for a lung cell based bio-photonics toxicity detector system.

Method

Binary and ternary gel-glass monoliths (composition as given in Table 1) were prepared using the sol-gel technique with tetraethyl orthosilicate, triethyl phosphate and calcium nitrate tetrahydrate as starting materials [4]. Physical characterization of porous substrates consisted of pore size and textural analysis using N_2 sorption.

Table 1. Nominal composition of the monoliths (mol %)

	SiO_2	CaO	P_2O_5
S70C30	70	30	-
58S	60	36	4

Human lung epithelial (A549) cells (purchased from ECACC, UK) were cultured in F12K medium supplemented with 10% foetal calf serum (FCS; Invitrogen) at 37°C in 5% CO_2 and in humidified atmosphere. Cell seeding on S70C30 foams and monoliths was performed as previously described [5]. The cells were seeded at the same density (10 000 cells / sample) on each sample, and incubated at 37°C in 5% CO_2 for various periods of time. The medium was changed twice a week.

A549 adhesion was assessed using WST-1 test (Roche, Mannheim, Germany). Two hours after cell seeding on materials or control plastic, non-adherent cells were removed by gently washing twice in PBS. WST-1 substrate (tetrazolium salt) diluted to 10% in F12K medium was added to A549 culture and incubated at 37°C in 5% CO_2 atmosphere. After 3 hours, the absorbance (A) was measured at 450 nm using a MRX ELISA reader (Dynatech Laboratories). Results are expressed in percentage of cell adhesion relative to cell adhesion onto cell culture plastic surface (positive control). Background absorbance was obtained by incubation of the WST-1 on substrate in cell culture medium only or on material only. Cell proliferation was investigated after incubation for 2 hours, 1, 3, 7, 15 and 28 days before analysis. The same method was used to follow cell proliferation and the results are expressed as optical density units. All biological samples were repeated 4 times, and experiments were done 3 times. The U Mann-Whitney test was applied to results. Values with $p<0.05$ were considered as significant (*) and $p<0.01$ is shown as (**).

Cell morphology was studied using a JSM LV 5610 scanning electron microscopy. The substrates were gently washed in phosphate buffer after different incubation periods. They were fixed using 2.5% glutaraldehyde in 0.1M phosphate buffer for 40 minutes at 4°C and dehydrated through increasing concentrations of ethanol and hexamthyldisilasane (HMDS). The specimens were vacuum dried and sputter coated with gold. Image analysis was performed using 10kV accelerating voltage.

Results

N_2 sorption analysis on the monoliths indicates that both substrates are mesoporous in nature (see Table 2) with interconnecting, non-perfect cylindrical pores of a narrow pore size distribution.

Table 2. Textural parameters

	S70C30	58S
Surface Area	136 m^{2}g^{-1}	151 m^{2}g^{-1}
Pore Volume	0.71 cm^{3}g^{-1}	0.45 cm^{3}g^{-1}
Average Pore	210 Å	132 Å

After 2 hours of cell incubation, about 70% of cell adhesion was obtained on both substrates. No significant differences were observed between cell adhesion to S70C30 and to 58S (Fig. 1).

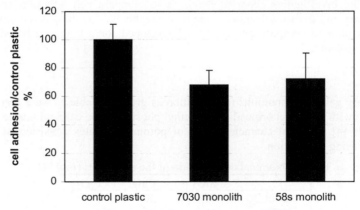

Fig. 1: A549 cell adhesion on gel-glass substrates.

Cell proliferation studies showed (Fig. 2) a higher proliferation rate on the 58S monoliths compared to the S70C30. A typical cell proliferation curve was obtained on plastic surfaces with the cells reaching a plateau of cell proliferation after 2 weeks. On both gel-glass substrates, the cells remain viable and proliferate even after 4 weeks as needed for the toxicity detector system.

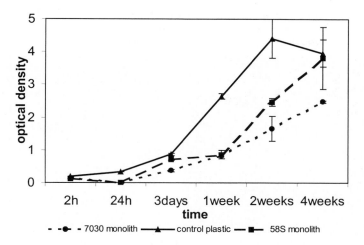

Fig. 2: A549 cell proliferation on gel-glass substrates.

The morphology of the A549 cells was observed after 2, 24 and 72 hour incubation periods (micrographs in Fig. 3 and Fig. 4). It is clear from the 2-hr micrograph of the 58S substrate that cells have attached to the surface and started to proliferate (the cell seen in the micrograph is at the end of mitosis). Only attachment of the cells was observed on the S70C30 surfaces.

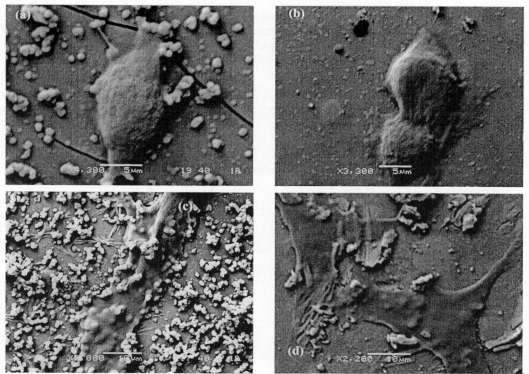

Fig. 3: A549 cell morphology on gel-glass substrates (a and c on S70C30 and b and d on 58S) after 2 hours (a and b) and 24 hours (c and d) of incubation.

By 24 hours the proliferation and spreading of cells were observed on both surfaces. It is clear from these SEM micrographs, cell attachment and proliferation on the S70C30 substrates was slower compared to 58S (Fig. 4).

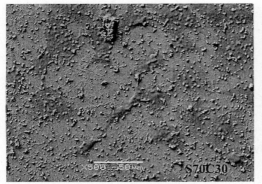 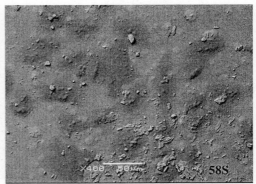

Fig. 4: A549 cell morphology on gel-glass substrates after 72 hours of incubation on S70C30 and 58S substrates.

Discussion and conclusions

The results of this study indicate that the ternary 58S bioactive gel-glass substrates offer favourable conditions for A549 cell proliferation and viability. This may be due to the rapid dissolution as well as the nucleation of amorphous calcium phosphate that take place at the surface of the ternary gel-glass. It is also possible that the mesoporous texture (smaller pore size and volume for 58S) may have had an influence on cell attachment and proliferation. However, this is not fully understood. Further analyses are underway to optimise lung cell lifetimes on gel-glass substrates.

Acknowledgements

The authors wish to thank EPSRC (UK), MRC and DARPA for their support.

Reference

[1]. R. Li, A.E. Clark, L.L. Hench: J Appl Biomater. 2 (1991), p. 231-232.

[2]. P. Saravanapavan, L.L. Hench: J Biomed. Mater. Res. 54: (2001), p. 608-618.

[3]. P. Saravanapavan, J.R. Jones, R.S. Pryce, L.L. Hench: J Biomed Mater Res. 66(2002), p. 110-119.

[4]. P. Saravanapavan, L.L. Hench: J Non-Crys Solids 318 (2003), p. 1-26.

[5]. J. Amedee, R. Bareille, R. Jeandot et al.: Biomaterials 15 (1994), p. 1029-1031

Key Engineering Materials Vols. 254-256(2004) pp. 785-788
online at http://www.scientific.net
© 2004 Trans Tech Publications, Switzerland

Indirect Cytotoxicity Evaluation of Soluble Silica, Calcium, Phosphate and Silver Ions

P.Saravanapavan, J. Selvakumaran, and L.L. Hench

Tissue Engineering Centre, Department of Materials, Imperial College London, London SW7 2AZ, UK

p.pavan@imperial.ac.uk

Keywords: cytotoxicity, soluble ions, bioactive glass, osteoblasts, fibroblasts

Abstract. New classes of materials are being designed to interact specifically with mammalian cells to control their behavior and subsequently direct the repair or regeneration of organ specific tissues. It has been reported that ionic products released by melt-derived 45S5 Bioglass® activated several families of genes that induced new bone formation. In this study, the cytotoxicity of soluble silica, calcium, silver and phosphate ions on osteoblasts is investigated. It is concluded that the amount of soluble ions, such as silica, calcium, phosphate and silver, in the culture medium can influence the cellular viability in both a positive and negative manner and a combination of Si and Ca ions induce cell proliferation.

Introduction

Bioactive gel-glasses and most other glass-ceramic materials considered for biological applications possess one common characteristic, which is the generation of a carbonated hydroxyapatite layer that is equivalent chemically and structurally to the biological mineral phase of bone [1]. However, it is not known exactly how the formation of this layer affects the surrounding cells, particularly since this process is known to involve the release of significant amount of calcium and silicate ions.

Recent investigations by Xynos *et al.* [2] concluded that ionic products released by melt-derived 45S5 Bioglass® activated several families of genes that induced new bone formation. During this investigation Bioglass® particles were immersed in Dulbecco's modified eagle medium (DMEM) for 24 hours. The culture media was supplemented with the filtrate 48 hours after cell seeding. Nonetheless, this study did not take into consideration the biphasic profile of glass dissolution (that is the rate of dissolution is rapid at short times and slow thereafter) or the cytotoxicity of the said ions when saturation is achieved.

This study is aimed to evaluate the cytotoxicity of substances leached by sol-gel derived third generation calcium silicate ($CaO-SiO_2$), calcium silicate phosphate ($CaO-SiO_2-P_2O_5$) and silver-doped calcium silicate ($CaO-SiO_2-Ag_2O$) bioactive glasses. Both Ca_2^+ and SiO_3^- ions are known to influence the metabolism of osteoblastic cells; the role of PO_4^- ions is thought to aid the mineralisation process and tissue bonding mechanism; Ag^+ ions offer antimicrobial properties. The indirect toxicity evaluation was performed by an extraction method, according to International standard organization (ISO). The cell viability was assessed by mitochondrial activity (MTT assay) and membrane integrity (neutral red (NR) uptake by viable cells).

Method

Materials. Unary, binary and ternary gel-glass monoliths (compositions listed in Table 1) were prepared using the sol-gel technique as previously published [3].

Extract preparation. The indirect cytotoxicity evaluation was performed by an extraction method. The extracts were prepared according to International Standard Organisation (ISO 10993-5). Under sterile conditions, samples of S100, S70C30, 58S and Ag-S70C30 monoliths were immersed in the culture medium (DMEM) for different time periods (from 10 minutes to 24 hours) at 37 °C in a humidified atmosphere of 5% CO_2 and 95% air, without agitation. The amount of Si, Ca and P ions in solution was quantified using ICP-OES.

Table 1. Nominal composition of the monoliths

	SiO$_2$	CaO	P$_2$O$_5$	Ag$_2$O
S100	100	-	-	-
S70C30	70	30	-	-
58S	60	36	4	-
Ag-S70C30	70	28	-	2

Extracts containing known amounts of single ions (either Si, Ca or P) were also prepared using standard solutions that are available for elemental analysis (Sigma Aldrich chemicals).

Cytotoxicity assays. Human osteosarcoma (MG-63) cells were cultured in DMEM supplemented with 10% foetal bovine serum (FBS), 2mM L-Glutamine, 1% non-essential amino acids (NEAA), and 1% antibiotics. The cell viability was assessed by the conversion of tetrazolium salt (MTT) by mitochondrial dehydrogenase enzymes to form formazan product and the uptake and retention of neutral red (NR) by lysosomes in cells with intact membrane. Cells were grown in 96-well plates to form sub-confluent layers before the treatment with extracts. Different groups were tested: positive control (0.1% Triton X-100), negative control (complete culture medium) and culture media supplemented with extracts and standards prepared at different time periods. Four samples per group were tested in each experiment with two independent experiments (n=8).

Results

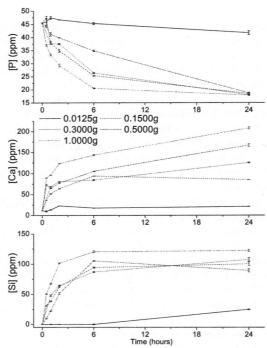

Fig. 1: Si, Ca and P ion release in 100 ml DMEM over 24 hour immersion period

Figure 1 shows the ionic release from 58S bioactive gel-glasses when immersed in 100 ml of DMEM for 24 hrs. The dissolution profile of the gel-glass is similar to that obtained for the other compositions. As the dose concentration increases (from 0.0125 g/100ml to 1.00 g/100ml) the amount of calcium, silica and silver ions in solution increases with increasing time, whereas the amount of phosphate ions decreases. This is due to the different stages of glass dissolution: exchange of alkali ions, network breakdown and calcium phosphate precipitation; all of which lead to the formation of HCA layer formation and tissue bonding.

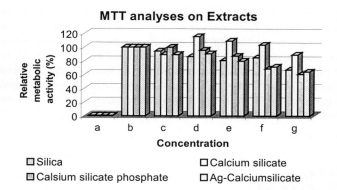

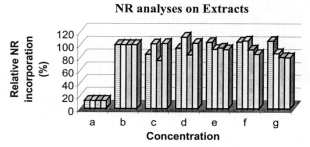

Fig. 2: Results from the MTT assay and NR assay on MG63 cells using extracts; (a – positive control (0.1% TritonX-100); b – negative control (complete culture media) c-g corresponds to exposure time given in tables)

The viability of osteoblast cells treated with extracts measured by MTT and NR assays are shown in Fig.2. Cells were exposed to complete DMEM supplemented with extracts for 24 hours. The levels of mitochondrial activity are generally lower for cells treated with extracts of S100, Ag-S70C30 and 58S compared to cells treated with extracts of S70C30. The metabolic activity of cells treated with extracts of S70C30 from 30 minutes to 6 hours of exposure is higher than the control cells. The metabolic activity of cells treated with extracts of S70C30 show an initial increase up to 30 minutes exposure time and a steady decline after 30 minutes. Whereas cells treated with S100, Ag-S70C30 and 58S show decline in metabolic activity with increasing exposure time after 10 minutes. Overall the extracts of bioactive gel-glasses had no significant effect on the membrane integrity as seen from NR incorporation. The viability of cells treated with S70C30 is the highest and higher than the control cells after 30 minutes exposure.

The viability of osteoblast cells treated with standards measured by MTT and NR assays are shown in Fig.3. Complete DMEM was supplemented with varying amounts of 1000 ppm standards of soluble Si, Ca, P and Ag. Levels of mitochondrial activity decrease with increasing concentration of soluble ions.

Discussion and conclusions

Various *in vitro* and *in vivo* studies have demonstrated the capacity of bioactive ceramics to induce an hydroxyapatite like layer indispensable for bone-biomaterial interaction. However, *in vitro* cytotoxicity studies have been rarely performed. The bioactive gel-glasses investigated in this paper, are able to induce several changes. As the pH increases, Ca_{2+}, SiO_3^-, PO_4^- and Ag_+ ions are released into the environment leading to apatite like layer formation [1]. These changes can potentially lead to cellular toxicity. This work was performed to study the impact of different soluble ions on the cellular metabolism. It is clear that ionic release during the first 30 minutes from the S70C30 induces cell proliferation. This is not seen in 58S extracts, leading to the conclusion that there is a critical calcium ion concentration that is responsible for the behaviour observed. In 58S the Ca_{2+} precipitates as a Ca-P layer so the final ionic concentration is lower in the prepared extracts. However, the results from the MTT and NR assays for the standards do not show increased cell proliferation with Ca ions. This is probably due to the fact that a combination of Si and Ca ions are responsible for the results observed with S70C30.

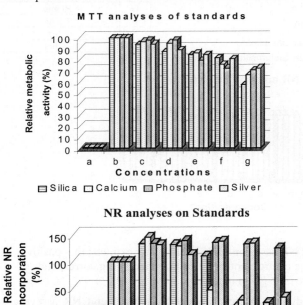

Key	Si	Ca	P	Ag
A	Positive control			
B	Negative control			
C	10ppm	15ppm	10ppm	1ppm
D	25ppm	50ppm	15ppm	5ppm
E	50ppm	100ppm	25ppm	10ppm
F	100ppm	150ppm	50ppm	15ppm
G	200ppm	200ppm	100ppm	20ppm

Fig. 3: Results from the MTT assay and NR assay on MG63 cells using; (a – positive control (0.1% TritonX-100); b – negative control (complete culture media) c-g corresponds to concentrations given in tables)

Acknowledgements

The authors wish to thank EPSRC (UK) for their support of this research.

References

[1]. L. L. Hench J. Amer. Ceram. Soc. (1991) 74:1487-1510
[2]. I.D. Xynos, A. J. Edgar, L. D. Buttery et al. (2000) Biochem. and Biophys. Res. Comm., 276: 461-465.
[3]. P. Saravanapavan, L.L. Hench (2003) J Non-Crys Solids 318: 1-26.

Key Engineering Materials Vols. 254-256(2004) pp. 789-792
online at http://www.scientific.net
© *2004 Trans Tech Publications, Switzerland*

Bioactive Glass Coatings on Ti6Al4V Promote the Tight Apposition of Newly-Formed Bone *In-Vivo*

A. Merolli[1], C. Gabbi[2], M. Santin[3], B. Locardi[4], P. Tranquilli Leali[1]

[1]Universita' Cattolica, largo Gemelli 8, 00168 Roma, ITALY, antoniomerolli@tiscali.it
[2]Universita' di Parma, Dip. Salute Animale, via del Taglio 8, 43100 Parma, ITALY
[3]University of Brighton, School of Pharmacy, Brighton, UK
[4]Stazione sperimentale del Vetro, Murano VE, ITALY

Keywords: *in-vivo* study, bioactive glass, coatings, bone tissue

Abstract. The rationale for a degradable bioactive glass coating is to lead the bone to appose gradually to the metal. Two formulations of bioactive glasses, already described in the literature, have been studied: bg A and bg F. A sodium-calcium-silicate non-bioactive glass was sprayed as a control. Young adult New Zealand White rabbits were selected as animal model. A hole was drilled from the femoral intercondylar groove and a Ti6Al4V coated cylinder was implanted. Retrievals took place at 1, 2, 4, 6, 8, 10 months. For all the samples and for both preparations of bioactive glass, it was noticed that bone was in tight apposition with the coating. As time progressed, pictures were found where bone showed characters of physiological remodelling (newly formed bone substituting areas of bone resorption) close to the coating. At the interface between bone and bioactive glass coating the apposition was so tight that it was not possible to discern a clear demarcation, even at higher BSEM magnification (more than 2500X). A second key feature in the behaviour of the bioactive glass coatings was their gradual degradation and the eventual apposition of bone directly to Ti6Al4V.

Introduction

The rationale for a degradable bioactive glass coating is to lead the bone to appose gradually to the metal [1]. This process should happen without the production, at its end, of bulky non-degradable particles as those observed with the fragmentation of the crystalline phase of hydroxyapatite coatings [2]. In this paper Authors report on the *in-vivo* comparison, in the rabbit femur, between the response to a couple of formulations of bioactive glass coatings and the response to a non-bioactive glass coating.

Materials and Methods

Two formulations of bioactive glasses, already described in the literature, have been studied: namely bg A and bg F. Bg A has 6% P_2O_5 and a ratio 1:1 CaO/Na_2O. Bg F has 7% P_2O_5 and a ratio 9:1 fra CaO/Na_2O. A sodium-calcium-silicate, non-bioactive glass, already described in literature as glass H, lacking P_2O_5 and resulting to be a biocompatible material, was sprayed as a control [3,4,5]. Glass H releases 10 times less SiO_2 and Na_2O in solution after 96 hours, in comparison with bg A. Fusion of the glasses has been obtained in a furnace at 1350 °C, then solidified glass has been pulverized and plasma-sprayed on Ti6Al4V cylinders of 10mm in length and 3mm in diameter. Young adult New Zealand White rabbits, weighing about 2700 g, were selected as animal model. Site selected was the distal femoral canal (meta-epiphyseal region). A hole was drilled from the intercondylar groove after access has been gained to articular cavity of the knee by a lateral parapatellar approach. The cylinder was inserted along the main axis of the femur. Thirty-six animals were implanted. Retrievals took place at 1, 2, 4, 6, 8, 10 months. The retrieved femur was embedded in PMMA.

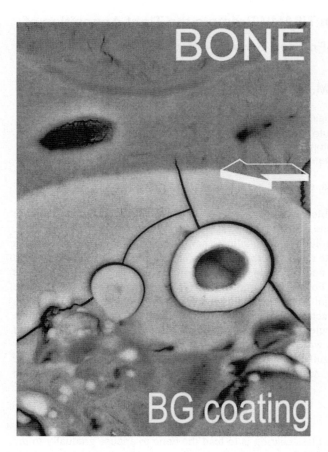

Fig. 1. Bone (note the black area of an osteocytic lacuna on the left) was in tight apposition with the bioactive glass coating (note a small solid droplet and a bigger empty one, embedded in the coating); at the interface between bone and bioactive glass coating (arrow) the apposition was so tight that it was not possible to discern a clear demarcation, even at higher magnification (Back Scattered Electron Microscopy (BSEM), 2000X, after 4 months).

Sections of 100 micron of thickness were taken by a rotating diamond-saw microtome (Leitz Wetzlar). Then followed the analysis by polarized light microscopy. After sectioning, two blocks were obtained; a face of one block was ground, sputter-coated with gold and analyzed by Back Scattered Electron Microscopy (Jeol JSM 6310) and Electron Dispersive Analysis (Isis, Oxford Instruments Ltd).

Results

For all the samples of both preparations of bioactive glass, it was noticed that bone was in tight apposition with the coating (Fig.1). As time progressed, pictures were found where bone showed characters of physiological remodelling (newly formed bone substituting areas of bone resorption) close to the coating. At the interface between bone and bioactive glass coating the apposition was so tight that it was not possible to discern a clear demarcation, even at higher magnification (more than 2500X). A second key feature in the behaviour of the bioactive glass coatings, was their gradual

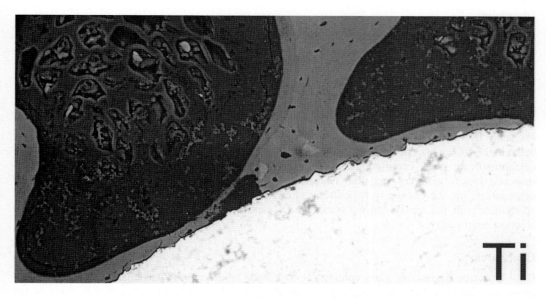

Fig. 2. After 10 months bone was found directly apposed to the metal (Ti) in more than half of the samples (BSEM, 150X, after 10 months).

degradation with time. At 10 months, bone was found directly apposed to the metal in more than half of the samples (Fig. 2). On the opposite, the non-bioactive glass coating used as a control showed pictures of complete integrity at any time examined (Fig. 3). It did not elicit an adverse reaction and, actually, bone was able to growth in its proximity but a clear demarcation remained.

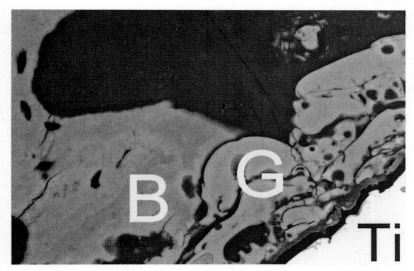

Fig. 3. The non-bioactive glass coating used as a control (G) showed pictures of complete integrity at any time examined; it did not elicit an adverse reaction and, actually, bone (B) was able to growth in its proximity but a clear demarcation remained (BSEM, 500X, after 10 months).

Discussion

Since the early work of Larry Hench it has been appreciated that a peculiar composition of glass may be "bioactive" and, eventually, promote the growth of newly formed bone in tight apposition with the glass itself [6]. Authors show that this characteristic is retained when a bioactive glass is used as a plasma-sprayed coating on a titanium alloy metallic substrate.

The exchange of ions between the glass and the biological environment is involved in the mechanism of the bioactive behaviour and has been described in several steps by Hench [7]. Anyway, other glass compositions (or virtually any glass composition) may promote an ion-exchange in a provided suitable reactive environment but this does not means that a "bioactive" behaviour will ensue. In this paper Authors demonstrate that *even if* a non-bioactive sodium-calcium-silicate glass coating may elicit a favourable response when implanted in bone, this response lacks two very peculiar features of the behaviour of bioactive glass coatings: *a)* degradation, i.e. the change in structure at microscopic level during time; *b)* promotion of a tight apposition with the newly formed bone, which cannot be resolved even by high magnification electron microscopy.

Conclusions

Only bioactive degradable glass coatings elicit a bone response which is characterized by a tight apposition between bone and coating and by the gradual degradation process that leads to the disruption and partial resorption of the material and the eventual apposition of bone directly to Ti6Al4V substrate.

References

[1] A. Merolli, P.L. Guidi, L. Gianotta, P. Tranquilli Leali, in Ohgushi et Al. eds.:"Bioceramics 12" (World Scientific, Singapore 1999), p. 597

[2] P. Tranquilli Leali, A. Merolli, O. Palmacci C. Gabbi, A. Cacchioli, G. Gonizzi: J Mater Sci Mater Me*d* , Vol 5, (1994), p. 345

[3] C. Gabbi, A. Cacchioli, B. Locardi, E. Guadagnino: Biomaterials, Vol 16, (1995), p. 515

[4] A. Merolli, P. Tranquilli Leali, P.L. Guidi, C. Gabbi: J Mater Sci Mater Med, Vol 11, (2000), p. 219

[5] A. Merolli, A. Cacchioli, L. Giannotta, P. Tranquilli Leali: J Mater Sci Mater Med, Vol 12, (2001), p. 727

[6] L.L. Hench, R.J. Splinter, W.C. Allen, T.K. Greenlee: J Biomed Mat Res Symp (1971), p. 117

[7] L.L. Hench, O.H. Andersson, in Hench et Al. Eds.:"An Introduction to Bioceramics" (Reed Healthcare Comm, Singapore 1993) 41

Key Engineering Materials Vols. 254-256(2004) pp. 793-796
online at http://www.scientific.net
© 2004 Trans Tech Publications, Switzerland

Orientation of Human Osteoblast Cells on Biphasic Calcium Phosphates Tablets with Undulated Topography

Euler A. dos Santos[1], Adriana B. R. Linhares[2], Alexandre M. Rossi[3], Marcos Farina[2], Gloria A. Soares[1]

[1] Dep. de Eng. Metal. e de Materiais, UFRJ, P.O.Box 68505, Rio de Janeiro, 21941-972, RJ, Brasil, gloria@ufrj.br

[2] Laboratório de Biomineralização, ICB, UFRJ, Rio de Janeiro, 21941-590, RJ, Brasil, farina@anato.ufrj.br

[3] CBPF, Rua Dr. Xavier Sigaud, 150, Rio de Janeiro, 22290-180, RJ, Brasil, rossi@cbpf.br

Keywords: hydroxyapatite, α-tricalcium phosphate, surface topography, composites, human osteoblast culture.

Abstract. In this work the *in vitro* behavior of human osteoblast cells on the undulated surfaces of biphasic calcium phosphate tablets was investigated. The tablets were produced by uni-axial press and undulations occupied just only half of the surface area. Chemical and physical characterization was performed by scanning electron microscopy (SEM), X-ray diffraction (XRD) and Fourier-transform infrared spectroscopy (FT-IR). XRD and FT-IR analysis revealed presence of hydroxyapatite (HA) and α-tricalcium phosphate (α-TCP) in significant proportion. Undulations heights and widths were variable. These superficial topographical differences induced different cellular arrangements confirming their influence on the cells orientation.

Introduction

The process of cell adhesion depends on the initial contact of cells with the biomaterial surface. All the subsequent processes of cell growth, migration and differentiation are significantly influenced by this first stage. It is known that the chemical surface composition [1] and physical factors as crystallinity, particle size, porosity and topography [2], also induce distinct cellular response. The surface of the biomaterial can influence cell reaction through changes in the cytoskeleton. Actin microfilaments are involved in the formation of cell processes, cell shape, and cell attachment [3]. Cell orientations to long grooves produced in materials surfaces confirm the influence of the topography on cell behavior [4]. This orientation phenomenon can be a very important factor in scaffold design inducing cells migration and the production of an organized tissue similar to that observed in the natural tissues.

Hydroxyapatite (HA) and tricalcium phosphate (TCP) have been widely used as bone-substitute ceramics. HA is more stable than α-tricalcium phosphate (α-TCP) and β-tricalcium phosphate (β-TCP) under physiological conditions, as it has a lower solubility and slower resorption kinetics [5]. Thus, when fast bone remodeling is desired, a HA/TCP composite can be selected.

Methods

The calcium-deficient HA powder (Ca/P = 1.58 ± 0.02) was compacted by uni-axial press under a load of 500Kg. Parallel undulations were machined on half of the matrix, with the rest of the surface remaining flat. Tablets with 7.6 mm of diameter and 1.0 mm height were produced using

100 mg of powder per tablet. The tablets were calcinated at 1150°C for 2h in order to increase mechanical strength and also produce a biphasic (HA/TCP) structure. After calcination, the tablets were characterized by scanning electron microscopy (SEM), X-ray diffraction (XRD) and Fourier-transform infrared spectroscopy (FT-IR). The tablets topography was estimated by performing image analysis on SEM images.

Human osteoblast cells were cultured in Dulbecco's modified Eagle's medium (DMEM) containing 10% fetal bovine serum (FBS) in an incubator at 37 °C under a humidified atmosphere of 95% air and 5% CO_2. Approximately 10^4 cells of fourth passage were seeded on tablets previously immersed for 1h in the supplemented medium. Cell morphology, cell migration and cell orientation were monitored at 4, 7 and 11th days of culture by SEM.

Results and Discussion

After calcination, a shrinkage of approximately 21% on the tablets diameter was observed. Figure 1 shows a SEM image of the undulated region of a tablet. Undulations contours were obtained from lateral views of polished tablets by image processing and revealed heights varying from 10 to 85 μm and widths varying from 87 to 235 μm (contour line depicted in Fig. 1). In addition, the undulations are not evenly distributed on the surface. Despite of these dimensional variations the radius of curvature of each undulation was similar. By XRD analysis the presence of both HA (2θ = 31.8°) and α-TCP (2θ = 30.7°) was confirmed (Fig. 2). The quantities of each phase on calcinated tablets were determined with the aid of a calibration curve [6] and were equal to 78.2% of HA and 21.8% of α-TCP. The absorption characteristic bands of the groups PO_4^{3-} (571, 602, 632 and 962 cm^{-1}) and characteristic additional bands of TCP (944 and 971 cm^{-1}) were identified on the FT-IR spectrum (not shown).

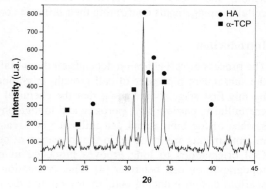

Fig. 1 Details of the superficial undulations and their contour.

Fig. 2 XRD pattern of the biphasic tablets after calcination at 1150°C for 2h.

After 4 days, the cells were completely attached and spread on the material surface without being confluent. In the fourth day, cells on the flat portion of the surface did not show any particular orientation (Fig. 3). Despite of the undulations size (significantly higher than the cell dimensions), oriented cells were easily observed on the wavy portion (Fig. 4-7).

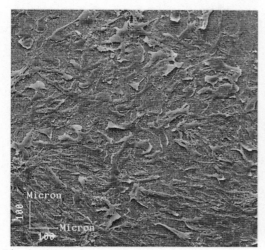

Fig. 3 SEM micrograph showing cells on the tablet flat portion, without any preferential orientation.

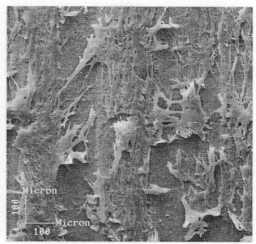

Fig. 4 SEM micrograph showing oriented cells along the undulation axis.

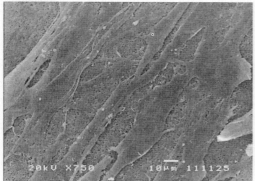

Fig. 5 SEM micrograph showing oriented cells on undulation.

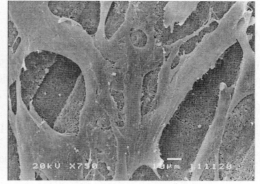

Fig. 6 SEM micrograph showing cellular bridges between two undulations.

The cells migrated from the depressions to undulation crests. This preference for the undulation crests was evident once few cells were observed on the regions between two undulations. Bridges of up to 60 μm between two neighbor undulations could be seen (Fig. 6-7). After 7 and 11 days, several cellular layers were observed. Cells on the upper layers did not present a particular orientation relatively to the composite geometry, probably because these cells did not make direct contact with the biomaterial.

 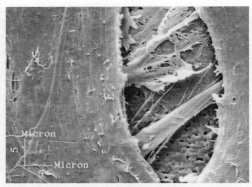

Fig. 7 SEM image showing the limit between
both flat and wavy part of a tablet.

Fig. 8 SEM image showing cells in the upper
layer oriented differently from those in direct
contact with the composite surface.

Conclusions

This work shows that osteoblasts may be influenced by much larger topographical differences in
undulated surfaces than those described in the literature for grooved surfaces [7-8]. Short-term
cultures are being carried out and are expected to contribute to the understanding of such behaviour.

Acknowledgements

The authors would like to thank, CAPES, FAPERJ and FUJB Brazilian Agencies. This research is
partly supported by the Instituto do Milênio de Bioengenharia Tecidual/CNPq.

References

[1] P. Kasten, R. Luginbhl, M. van Griensven, T. Barkhausen, C. Krettek, M. Bohner, U. Bosch:
    Biomaterials, v. 24 (2003), p. 2593-2693.
[2] D. D. Deligianni, N. D. Katsala, P. G. Koutsoukos, Y. F. Missirlis: Biomaterials. v. 22 (2001),
    p. 87-96.
[3] L.A. Amos, W.B. Amos: *Molecular Biology - molecules of the cytoskeleton*, Hampshire;
    MacMillan, 1991, p. 30.
[4] M. J. Dalby, M. O. Riehle, S. J. Yarwood, C. D. W. Wilkinson, A. S. G. Curtis: Experimental
    Cell Research, v. 284 (2003), p. 274-282.
[5] S.V. Dorozhkin, M. Epple: Angew. Chem. Int., v. 41 (2002), p. 3130-3146.
[6] J. F. Oliveira, P. F. Aguiar, A. M. Rossi, G. A Soares: Artificial Organs, v. 27 (2003), p. 386-
    391.
[7] K. Matsuzaka, X. F. Walboomers, M. Yoshinari, T. Inoue, J. A. Jansen: Biomaterials, v. 24
    (2003), p. 2711-2719.
[8] K. Anselme: Biomaterials, v. 21 (2000), p. 667-681.

Key Engineering Materials Vols. 254-256(2004) pp. 797-800
online at http://www.scientific.net

Changes in Distant Organs in Response to Local Osteogenic Growth Factors Delivered by Intraosseous Implants: a Histological Evaluation

Elsie Damien[1] and Peter A Revell

[1] Osteoarticular Research, Dept. of Histopathology, Royal Free Campus,

University of London, London, NW3 2PF, UK. email: edamien@rfc.ucl.ac.uk

Keywords: bioactive; biomaterial; bone ingrowth; drug delivery; hydroxyapatite implant; insulin like growth factor; inflammation; osteogenesis

Abstract. Osteogenic growth factors are added to enhance osteointegration and osteogenesis of synthetic bone substitutes to improve clinical outcome. Reactions to particles of wear debris from implanted material could lead to bone resorption similar to resorption around a total joint prosthesis and also to inflammatory responses in distant organs. Porous hydroxyapatite (HA) scaffolds pretreated with insulin like growth factor (IGF) -I (0.5 (LD) or 3.0 (HD) µg/implant) and IGF-II (0.5 µg/implant) were implanted *in vivo* in rabbit femur to enhance their bioactivity and bone bonding properties. Heart, kidney, liver, lung, lymph nodes and spleen were collected during systematic postmortem examination at 1, 3 and 5 weeks postimplantation for quantitative and qualitative analysis. Local wound healing of the periprosthetic bone and the responses in the distant internal organs were characterised using light microscopy and electron microscopy.

All tissues from implanted groups except heart and kidneys exhibited an increase in the cellularity at week 1 and 3. In the lung, there was also evidence of lympho-proliferation and aggregation in the IGF groups and presence of exudates in the IGF-1 HD group. The hypertrophy and hyperplasia appeared to be growth factor, dose and time dependent with IGF-1 LD <IGF-II <IGF-1 HD. At 5 weeks the tissues appeared as regenerated to the normal level with regression of the lymphocytic infiltration. We report for the first time, that local delivery of IGF-I and -II by intraosseous HA implant to enhance osteointegration produced no adverse persistent effects in important internal organs.

Introduction

Synthetic hydroxyapatite is used as a bone replacement material for osseous defects or for coating of metal implants due to its bioactivity and similarity in composition to bone mineral [1, 2]. The clinical success of bone graft material depends upon the integration of the implant with surrounding tissue. Addition of growth factors may enhance osteointegration of HA scaffold by supporting and stimulating bone formation within the implant and at the interface. Wear debris from implanted material could cause bone resorption similar to resorption around a total joint prosthesis and to inflammatory responses in distant organs [3]. It is a disadvantage that bioceramics may produce local and systemic inflammatory responses. To ensure that bone graft substitute can perform the expected physiological functions, one must understand the systemic effects in response to the implant material and the drugs it delivers locally. IGF-I and -II, osteogenic growth factors produced endogenously in bone and also produced and expressed in tissues other than bone, were used successfully to enhance osteointegration [4-6]. Both IGFs could act as paracrine or autocrine factors in the regulation of bone-forming process and targeted disruption of the genes impair the growth of many organs [7-10].

The study reported here was performed to find systemic effects in response to intraosseous implantation of biodegradable porous HA ceramic with IGF-I and -II. It was reported that HA served as an effective scaffold for bone ingrowth when implanted in rabbit femur and also served as excellent IGF-I and -II delivery system. The addition of IGFs enhanced the periprosthetic tissue response and so improved bone healing resulting in better integration and bonding of the HA implant with the existing bone [11, 12]. We reported that pretreatment of HA implant with IGF-I and -II significantly increased bone healing, new bone formation and bone mineralisation rate in the cancellous bone in the rabbit femur compared with HA alone or sham operation at 1, 3 and 5 weeks post-implantation [12, 13]. It is hypothesised that locally delivered IGF-I and -II to enhance the bioactivity and osseointegration of HA will not produce any persistent adverse visceral or systemic effects. The purpose of this study was to investigate the prevalence of systemic effects and histopathological responses to intraosseously implanted HA and the IGFs it delivered locally.

Materials and Methods
Implants, growth factors, animal models and implantation procedure: Cylindrical porous HA ($Ca_{10}(PO_4)_6(OH)_2$) implants (4.5±0.4mm x 6.6±1.2mm) alone or with IGF-I or IGF-II, 0.5 or IGF-I, 3.0 µg/implant (R&D Systems Europe Ltd), were implanted bilaterally in defects made in the lower femur of NZW rabbits (2.7±0.2Kg) under approved general anaesthesia protocol and Government licences [4, 8]. Sham operated and normal rabbits served as controls. Body weight and general health of all experimental animals were checked during the study.

Retrieval and processing of tissues: Heart, kidney, liver, lung, lymph nodes and spleen were collected during systematic postmortem examination at 1, 3 and 5 weeks postimplantation for quantitative and qualitative histological analysis and comparison from basal, sham, HA, HA+IGF-I LD, HA+IGF-I HD, and HA+IGF-II. Tissue samples were fixed in formal alcohol (70% ethanol and formaldehyde) for 4 days. After fixation, samples were dehydrated using ascending grades of alcohol and embedded in paraffin wax.

Haematoxylin and eosin staining of tissues: Local wound healing of the periprosthetic bone and the responses in the distant internal organs were characterised using light microscopy (Axiocam Digital Image system, Zeiss) of 'haematoxylin and eosin' stained sections and also by electron microscopy (n=6/treatment/time point). Haematoxylin was prepared by dissolving 1 gm of Mayer's haematoxylin in 1000ml distilled water with added potassium (50gm), sodium iodate (0.2gm) followed by 1gm citric acid and 50gm chloral hydrate. Sections were stained with haematoxylin for 6 minutes, excess stain rinsed off with water, and counterstained with 1% aqueous eosin. Excess eosin was rinsed off with water and sections were allowed to air dry before mounting.

Statistical analysis: The means and standard deviations were calculated for each group and the statistical significance of difference was determined using the 'student t test'.

Results
Basal and sham controls: There was no pathology found in any of the tissues examined from normal or sham animals.

HA: All tissue specimens exhibited an increase in the cellularity at 1 and 3 week postimplantation.

HA+ IGF-I and -II: In addition to the generalised increase in the cellularity in all tissues examined, in the lung there was also evidence of a lympho-proliferation in all groups and the aggregation of lymphocytes into nodules not associated with vessels, bronchi or bronchioles at 1 and 3 week postimplantation. 3 weeks specimens from all groups showed a marked increase in cellularity with high numbers of lymphoid infiltration in the lung. These appeared dose dependent with IGF-1 LD <IGF-II <IGF-1 HD. It also appeared that the proliferated lymphocytes were infiltrating the parenchyma of the lung. Marked thickening of the alveolar walls was also evident (Fig. 1). Spleens demonstrated an increase in cellular activity with the formation of new germinal centres and hypertrophied cells (Fig. 2). The hypertrophy and hyperplasia appeared to be growth factor, dose and time dependent. In the livers of these animals, it was noted that there was centri lobular degenerative change of the hepatocytes of lobules and venous congestion within the sinusoids. There also appeared to be an increase in the number of apoptotic cells with condensed nuclei. The livers of the IGF-1 HD treated groups appeared more vascular and having hypertrophied hepatocytes with varying nuclear sizes. There were also a larger number of apoptotic cells visible. A cell count was done to compare cellularity within the specimen groups. Mean cell count revealed a significant dose dependent hyperplasia of hepatocytes, normal 320±12, IGF-1 LD 410±16, IGF-1 HD 340±10 cells per high power field. Tissue exudate was also noted in the IGF-1 HD group, as was the presence of many polymorphs. Heart and kidneys showed normal histology. At 5 weeks post implantation all tissues appeared normal with a regression of the lymphocytic

infiltration and reduction in inflammatory responses. Appearance of lung was comparable with that of normal (Fig. 3). The overall cellularity of the parenchyma of the liver and lungs was however increased above that of the control animals. The spleens were no longer overactive and the livers appeared to have stopped degenerating (Fig. 4).

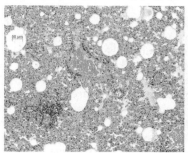 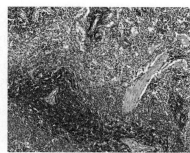

Fig. 1: Lung from 3 week IGF-I HD.   Fig. 2: Spleen from 3 week IGF-I HD.

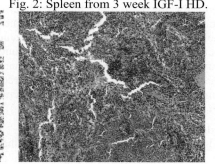

Fig. 3: Lung from 5 week IGF-I HD.   Fig. 4: Spleen from 5 week IGF-I HD.

Discussion

The observations in soft tissue sections from experimental rabbits implanted with hydroxyapatite impregnated with IGF-1 and -II compared with those of age and sex matched normal, sham operated and HA alone implanted rabbits are presented here. These studies were carried out to assess whether HA could cause any adverse systemic reactions when used in conjunction with IGFs, a known mitogen for osteoblasts, in guided enhancement of bone tissue regeneration [14]. The tissue responses to the HA implants and IGFs were assessed with respect to the number of foreign body giant cells, cellular infiltration, the mitotic figures, cellularity, apoptotic cells, and general inflammatory responses. The rationale for including sham and normal rabbits were to see if any of the observed histological effects could be attributed to the procedure itself. In the lung it was seen that normal rabbits have small aggregations of lymphoid tissue associated with the bronchi and bronchioles. It also gave an idea of the range of cellularity for normal rabbit tissues.

Upon introduction into the human body, biomaterials initially trigger a foreign-body inflammatory response. Furthermore, the wear debris associated with materials used for orthopaedic implants and dental implants can cause the body to mount an inflammatory response. This involves the production of phagocytic macrophages that ingest the foreign material while simultaneously producing cytokines that serve as chemotactic agents for an amplified immune response [15]. The objective of this study was to investigate the effect of HA alone, HA+IGF-I LD, IGF-I HD and IGF-II on the inflammatory reactions expressed in internal distant organs. The major histopathological findings were an initial, up to 3 weeks postimplantation, distant visceral hyperplastic reaction in spleen, and lymph nodes and increased inflammatory responses and cellularity in liver and lung. There were small aggregates of lymphocytes and macrophages occurring as infiltrates with some polymorphs in lung, liver, spleen and lymph nodes in tissues from both HA and HA+IGFs implanted groups. This hypertrophy and hyperplasia is in correlation with the recruitment, migration, attachment, proliferation, and differentiation of mesenchymal cells into new bone in the periprosthetic bone and within the implant at week 1 and 3. Macrophages, capable of ingesting HA particles, treated with HA showed irregular cytoplasmic borders with an increase in vacuolisation leading to cell lysis [15].

Visceral hyperplastic reaction but no hepato- or splenomegaly was observed with the doses used in this study. Lung parenchyma showed polymorph and lymphocyte infiltration and presence of exudates in the IGF-1 HD. Liver specimens demonstrated degenerative changes and venous congestion within the sinusoids. IGF-1 HD showed increased vascularity, hypertrophied hepatocytes with varying nuclear sizes and an increased number of apoptotic cells with condensed nucleus. These histological observations correspond to the increase in active proliferation, differentiation and matrix secretion of osteoblasts at these time points within the implants [11]. After 5 weeks the tissues appeared to have returned to normal level with a regression of the lymphocytic infiltration and degeneration. These findings are in keeping with the histological evidence of osteoblast differentiation and mineralisation of the newly formed bone matrix in and around the implant.

We report for the first time, that local delivery of IGF-I and IGF-II by HA to enhance osteointegration produced no adverse persistent inflammatory responses in important internal organs. These results provide a basis for further studies on the future clinical use of IGFs to enhance bone formation in response to prosthetics without any adverse systemic effects. The inference is that the supplementation with IGF-I will enhance the bioactivity and osteointegration of the biodegradable hydroxyapatite implant thereby reducing the healing time, which will have major economic importance in healthcare if the postoperative patient recovery period can be accelerated in clinical situations.

Acknowledgements

Acknowledgements are due to EPSRC, UK for the financial support, Dr Hing for the generous gift of implants and Mr MacInnes for expert technical assistance.

References

[1] T. Yamamoto, T. Onga, T. Marui and K. Mizuno: J Bone Joint Surg Br. 82 (2000), 1117-20.
[2] R.E. Holmes, R.W. Bucholz and V. Mooney: J Orthop Res. 5 (1987), 114-21.
[3] S. Saeed, E. Damien and P. Revell: Proceedings of the European Society for Biomaterials. (2002), P109.
[4] E. Damien, K. Hing, T. MacInnes and P. A. Revell: Proceedings of the European Society for Biomaterials. (2001), T140.
[5] W.H. Daughaday and Rotwein: Endocrine reviews. 10 (1989), 68-91.
[6] D. Shinar, N. Endo, D. Haperin, G. Rodan and M. Weinreb: Endocrinology. 132 (1993), 1158-67.
[7] D. Bikle, J. Harris, B. Halloran, C. Roberts, D. LeRoith and E. Morey: Am J Physiol. 276 (1994), E278-86.
[8] C.J. Rosen and L.R. Donahue: Proc Soc Exp Biol Med. 219 (1998), 1-7.
[9] T.M. DeChiara, A. Efstratiadis and E. J. Robertson: Nature. 345 (1990), 78-80.
[10] A. Ward, P. Bates, R. Fisher, L. Richardson and C. Graham: Proc Natl Acad Sci. USA. 91 (1994), 10365-69.
[11] E. Damien, K. Hing, S. Saeed and P.A. Revell: J Biomed Mater Res. (2003), (in press).
[12] E. Damien, T. MacInnes and P.A. Revell: Bone. (2001), 28: S140.
[13] E. Damien, K. Hing, T. MacInnes and P.A. Revell: The J Pathology. 193 (2001), 6A.
[14] E. Damien, J.S. Price and L.E. Lanyon: J Bone Miner Res. 15 (2000), 2169-77.
[15] L. Bell, H. Benghuzzi, M. Tucci and Z. Cason: Biomed Sci Instrum. 37 (2001), 161-66.

Key Engineering Materials Vols. 254-256(2004) pp. 801-804
online at http://www.scientific.net
© 2004 Trans Tech Publications, Switzerland

Experimental Study on Construction of Vascularized Bone Graft with Osteoinductive Calcium Phosphate Ceramics *In Vivo*

C Bao[1,2], H Fan[1], C Deng[1], Y Cao[1]*, Y Tan[1], Q Zhang[1], X Zhang[1]

[1] Engineering Research Center in Biomaterials, Sichuan University, Chengdu, 610064, China

[2] West China School of stomatology, Sichuan University, Chengdu, 610064, China

*Corresponding author, cybao9933@sina.com

Keywords: bone graft, osteoinduction, calcium phosphate ceramics, tissue engineering, *in vivo*

Abstract. This study was to aimed to explore a new method for fabricating a bone substitute, which is constructed of vascularized bone graft with osteoinductive calcium phosphate ceramics *in vivo*. A total of 32 small and 16 big ceramic cylinders were prepared, and bone-like apatite layer was formed on the surface and the pore walls of ceramics in revised simulated body fluid (RSBF). The smaller ceramics cylinders were implanted in dorsal muscles of dogs to evaluate the osteoinductive capacity. The bigger cylinders were mixed with fresh bone tissue and were implanted near a branch of femoral blood vessel in the dog's leg muscles to construct vascularized bone grafts. After implantation for 6 and 12 weeks, the specimens were harvested. The specimens were evaluated by blood vessel staining, histological observation, tetracycline fluorescence labeling, ^{99m}Tc–MDP SPECT and mechanical tests. The results showed that bone formation was found in all the samples and bone grafts had good blood supply. This result indicated that this kind of calcium phosphate ceramics had a good osteoinductivity and that vascularized bone grafts could be constructed with osteoinductive calcium phosphate ceramics *in vivo*.

Introduction

Bone defect is a common disease in clinic and to date, there is no satisfactory method to treat the segmental bone defect. At present, autograft, allograft, xenograft and alloplasm materials could be used for repairing bone defects. Autograft is the best bone substitute, but it causes a secondary wound. Allograft and xenograft may bring immune response and infectious disease. Alloplasm material have been developed as a major bone substitute [1]. An ideal bone substitute should possess good biological and mechanical properties. The biological properties of an ideal bone substitute include biocompatibility, osteoconductivity, osteoinductivity, osteogenesis capability. In 1980s, the appearance of tissue engineering brought about opportunities for repairing bone defects. Great achievements have been obtained in this field in the last 20 years [2]. But there are still some challenges ahead for bone tissue engineering [3]. Until now, no tissue engineering bone composed of scaffold and cultured cells *in vitro* has been permitted to be applied in clinic [4].

Calcium phosphates ceramics are frequently used as bone substitute materials because of their similarity to the mineral phase of bone, absence of antigenicity, and excellent osteoconductivity. Some studies have proved that calcium phosphate ceramics with special structure could induce bone formation in soft tissue [5]. Based on the exploration of the osteoinductive mechanism, Zhang [6] proposed a new concept of bone tissue engineering *in vivo*, i.e., bone grafts could be constructed directly with osteoinductive biomaterials singly or osteoinductive biomaterials mixed with fresh bone tissue in some areas of the body. This study was conducted to explore a new method for fabricating bone substitute, which constructed vascularized bone graft with osteoinductive calcium phosphate *in vivo*.

Materials and Methods

Preparation of the ceramics Calcium phosphate ceramics was prepared by H_2O_2 foaming method and sintered at 1250°C for 3 hours with wet-synthetic calcium phosphate powder. The chemistry of the ceramics was analyzed by X-ray diffraction (XRD). The average pore size of the ceramics was tested by SEM. The porosity of the ceramics was determined by the Archimedes water displacement methods. A total of 16 ceramic cylinders were prepared with 10 mm in diameter and 20 mm and 30 mm in length respectively, Φ=2 mm canal in the center. Other 32 ceramic cylinders with 5 mm in diameter and 8 mm in length were also prepared. After being cleaned with de-ionized water by ultrasonication, all ceramics were soaked in dynamic revised simulated body fluid (RSBF) for 72 hours [7]. The microstructure of ceramics was analyzed by SEM, and the formation of hydroxycarbonate apatite layer on the surface was observed by Infrared spectroscopy (IR).

Animal experiments. Eight dogs were used in this study. After anaesthesia with 3% sodium pentobarbital, the four smaller ceramics cylinders were implanted in dorsal muscles of each dog to evaluate the osteoinductive capacity, and the bigger cylinders mixed with marrow and cancellous bone were implanted near a branch of femoral blood vessel in the leg muscles to construct vascularized bone grafts (fig 1).

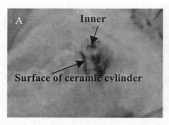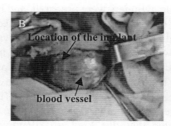

Fig.1 Construction of vascularized bone graft. (A) ceramic cylinder mixed with fresh marrow and cancellous bone; (B) the picture of the implanted location; (C) formation of muscle pouch and implantation of ceramics composite

Samples analysis. Tetracycline (50mg/kg) and ^{99m}Tc–MDP (1mCi/kg) were injected to all dogs in the vein a week and four hours, respectively, before sacrifice. Four dogs were sacrificed and the specimens were harvested at 6 and 12 weeks after implantation. The specimens were detected by blood vessel staining, histological observation, tetracycline fluorescence labeling, ^{99m}Tc–MDP Single Photon Electronic Computer Tomography (SPECT) and mechanical property testing. Blood vessels of bone grafts were stained with 3% methylthioninium chloride before sacrifice. The samples for histological observation were fixed, and decalcified thin sections were made and stained with HE. The samples for tetracycline fluorescence labelling were fixed, dehydrated and embedded in methyl methacrylate. 20μm undecalcified sections were made. A total of 12 ceramic cylinders with 10 mm in diameter and 20 mm in length were used for compression stress tests, and the same numbers of ceramic cylinders with 30 mm in length for bend stress test before and after implantation.

Results and discussion

Osteoinductive ceramics. The XRD analysis showed that the ceramics consists of HA/α-TCP (60:40). SEM test showed that the average pore size was about 200μm. The porosity of the ceramics was 60%. After immersion for 72h in RSBF, SEM observation showed that bone-like apatite layer was formed on the surface of the ceramics (Fig 2A). FT-IR spectrum showed the characteristic peaks of calcium phosphate ceramics, PO_4^{3-}(~1000 cm^{-1}) and OH$^-$ (~3568 cm^{-1},630 cm^{-1}). The characteristic peaks of CO_3^{2-} (~1400 cm^{-1}) further confirmed that bone-like apatite was formed on the surface of the ceramics (Fig 2B). Histological observation showed that bone formation was found in all the ceramics implanted in dorsal muscles (Fig. 3).

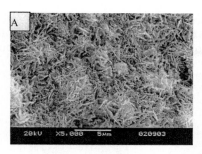
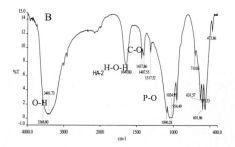

Fig 2. Character of osteoinductive ceramics (A) bone-like apatite formed on the surface of the ceramics (SEM) (B) the spectrum of FT-IR

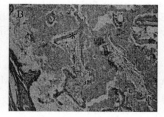
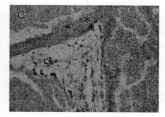

Fig 3. Decalcified light micrograph after the ceramics were implanted in dorsal muscle for 12 weeks, *New forming bone. Original magnification (A) 40×; (B) 100×; (C) 400×

The chemical composition and physical structure of ceramics are the major factors to initiate the osteoinduction of biomaterials and determine their osteoinductive capacity. It is believed that formation of bone-like apatite is an important factor for osteoinduction. Porous biphasic calcium phosphate ceramics were used in this study, and a bone-like apatite layer was formed on the surface of the ceramics and their pore walls. The histological observation revealed that bone formation was found in the ceramics implanted in dorsal muscles, i.e. this kind of calcium phosphate ceramics has a good osteoinductivity.

Vascularized bone grafts. After implantation for 6 and 12 weeks, the result of vascular staining with 3% methylthioninium chloride showed that these bone grafts turned blue. This suggested that these bone grafts had good blood supply. Histological observation showed new bone formation in the surface pores of ceramics which were mixed with marrow and cancellous bone. Fibrous tissue and bone-like tissue formed in the central canal of ceramic cylinders, and few inflammatory cells were observed (Fig. 4).

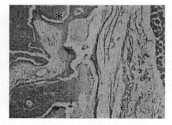

Fig. 4. Light micrographs of decalcified sections after 6 weeks intramuscular implantation of ceramics composite. *New forming bone. (A) fibrous and bone-like tissue ingrows the central canal. Original magnification 40×; (B) new bone formation in the surface pore of ceramics, 40×; (C) 100×

Tetracycline labelling revealed that fluorescence in the central canal of ceramic cylinders was weak while strong fluorescence in the surface pore of ceramics (Fig. 5 A, B). Tetracycline as a vital stain, fluoresced when exposed to ultraviolet light, thus revealing new bone deposition [8].

Fig.5C is the photo of ^{99m}Tc–MDP SPECT. The color distribution in implanted ceramics and in femoral diaphysis of nature bone was coincident, which revealed that the density at implanted ceramic is equal to that of the nature bone. SPECT test showed that the newly formed bone could metabolize in the body [9].

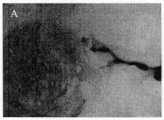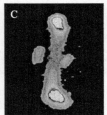

Fig. 5. Light micrographs of tetracycline labelling and SPECT after 6 weeks intramuscular implantation of ceramics composite. (A) weak fluorescence in the central canal, 40×;(B) strong fluorescence in the surface pore of ceramics, 40×;(C) ^{99m}Tc–MDP SPECT of the implants and natural bone.

Compression stresses of ceramic cylinders varied from 4.25 MPa to 6.32 MPa at 6 weeks, and 8.46 MPa at 12 weeks. Bend stresses changed from the original 4.13 MPa to 6.25 MPa at 6 weeks, and 6.90MPa at 12 weeks.

The results of histological staining, tetracycline labelling and ^{99m}Tc–MDP SPECT demonstrated that vascularized bone grafts with considerable size could be constructed with osteoinductive calcium phosphate ceramics *in vivo*.

Conclusion

These results indicated that the calcium phosphate ceramics used in the present work have good osteoinductivity and thus vascularized bone grafts with considerable size could be constructed with osteoinductive calcium phosphate ceramics *in vivo*.

Acknowledgments

This work was supported by the Key Basic Research Project of China (Contract NO G1999064760)

References

[1] Schnurer SM, Gopp U, Kuhn KD: Orthopade.Vol. 32 (2003), p. 2

[2] Robert F, Service: Science. Vol. 289 (2000), p. 1498

[3] Linda G, Griffith, Gail Naughton: Science. Vol. 276 (1997), p .181

[4] Dan Ferber: Science. Vol. 284 (1999), p. 423

[5] Yuan HP, de Bruijin JD, Li YB: J MATER. SCI,MATER .MED. Vol. 12 (2001), p. 7

[6] X.D Zhang, H.P Yuan, K.de Groot: Biomaterials With Inteinsic Osteoinductivity (The 6th World Biomaterials Congress, Hawaii, USA, 2000)

[7] Chunlin Deng, Yourong Duan, Jiyong Chen. Key Engineering Materials. Vol. 240-242 (2003), p. 81

[8] Treharne RW, Brighton CT: Clin Orthop Relat Res. Vol. 140 (1979), p. 240

[9] Kobayashi W, Kobayashi M, Nakayama K: Science. Vol. 289 (2000), p. 1467

Key Engineering Materials Vols. 254-256(2004) pp. 805-808
online at http://www.scientific.net
© 2004 Trans Tech Publications, Switzerland

Neural Cells on Iridium Oxide

I-S. Lee[1]*, J-C. Park[2], G. H. Lee[3], W. S. Seo[4], Y-H. Lee[5],

K-Y. Lee[6], J. K. Kim[7], F-Z. Cui[8]

[1]Atomic-scale Surface Science Research Center, Yonsei University, Seoul 120-749, Korea.
inseop@yonsei.ac.kr
[2]Department of Medical Engineering, Yonsei University College of Medicine, Seoul 120-752, Korea
[3]Korea Institute of Machinery & Materials, Chang-Won 641-010, Korea
[4]Korea Institute of Ceramic Engineering & Technology, Seoul 153-023, Korea
[5]Department of Electronic Engineering, Kangwon University, Kyungnam 200-701, Korea
[6]Bioengineering Research Center, Sejong University, Seoul 143-747, Korea
[7]Department of Biomedical Engineering, Inje University, Kyungnam 621-749, Korea
[8]Department of Materials Science & Engineering, Tsinghua University, Beijing 100084, China

Keywords: Neuron, Iridium oxide, charge injection, e-beam evaporation, ar ion beam, stimulation

Abstract. Iridium oxide was investigated as a material for the stimulating neural electrode. Iridium oxide was formed by potential sweep of iridium film that was deposited on either Si wafer or silicone rubber with electron-beam evaporation. The rate of iridium oxide formation was dependent on the upper and lower limits of potential sweep. The higher thickness of iridium oxide produced the higher charge injection due to the reversible valence transition of iridium within oxide. Embryonic cortical neural cells formed neurofilament after 4-day culture on iridium oxide, which indicated neural cells could adhere and survive on iridium oxide.

Introduction

Interaction between neurons and electrodes is very important for stimulation or signal collection. The restoration of a variety of physiological functions by electrically activating nerves serving paralyzed muscles has been of particular interest. The first application of functional electrical stimulation (FES) was by Liberson [1] who stimulated the peroneal nerve in adult hemiplegia to correct drop foot during the swing phase of locomotion. For the rehabilitation of locomotion in paraplegics and hemiplegics, thirty-two electrodes are required and even more stimulations are demanded for the refined control [2]. A major problem limiting the widespread application FES has been the lack of stimulating electrodes that can be used for long-term precise multipoint stimulation of nerves. Practical application requires miniaturized electrode geometry, which still demands high charge injection capability over the suitable range in potential. Recently, iridium oxide was investigated as a candidate electrode material [3-6]. Iridium oxide shows electrochromic behavior, where a change of color takes place between oxidative and reductive state of iridum. The color of iridium oxide starts changing from metallic to blue around 0.4 V (SHE) in 0.1 M H_2SO_4, become darker and remains dark blue above near 0.9 V (SHE). The bleaching process occurrs during the cathodic sweep. The dark blue became lightened around 1.0 V (SHE) and no blue color could be observed below 0.4 V (SHE). The double proton-electron injection mechanism has been proposed to explain the electrochromism of the anodic iridium oxide film by Gottesfeld et al. [7]. They postulated the following reaction to describe the film conversion processes:

$$Ir(OH)_n \leftrightarrow Ir_x(OH)_{n-x} + xH^+ + xe^-$$

In this mechanism, during the coloration process, electrons are removed from the oxide across the metal-oxide interface by application of a suitable anodic potential to the metal substrate. Charge repulsion causes an equivalent amount of mobile positive charge carriers (protons) to be ejected across the oxide-electrolyte interface, thus preserving electroneutrality inside the oxide. Similary, during bleaching, electrons are injected into the oxide film from the metal while the compensating positive charge is injected at the oxide-electrolyte interface. The injection mechanism of iridium oxide is even more beneficial for stimulating neural electrode since neuron communication is inherently electrical and chemical in nature. Since the charge injection is related to the valence transition of iridium oxide, the thicker oxide produces the higher charge injection as shown in Figure 1 [5].

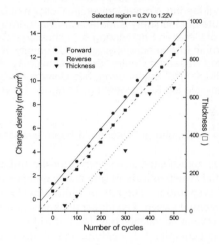

Fig. 1 Charge densities of iridium oxide

Under the identical stimulating action of potential the higher charge produced by the thicker oxide is very important property since the charge injection can be modulated and optimized for the stimulation of different nerve fiber simultaneously. In this study, iridium oxide was grown from iridium film formed by electron-beam evaporation with Ar ion beam assist and was characterized in term of charge injection capability, morphology, and neural cell interaction.

Materials and Methods

Iridium was deposited on either Si wafer or silicone rubber plate. An electron beam gun (Telemark, Fremont, CA, USA) for evaporation and a Mark □ TM end-hall type ion gun (Commonwealth Scientific, Alexandria, VA, USA) for ion assist was employed. After chamber was vacuumed down to the usual base pressure of 2×10^{-7} Torr, a vapor flux of Ir atoms is generated with an electron beam evaporator and deposited on a rotating substrate with the deposition rate of 0.3 Å /s with an Ar ion beam assist. Iridium oxide was grown on iridium film by cyclic voltammetry with an EG&G Princeton Applied Research Model 273 potentiostat. In 0.1M H_2SO_4, the voltage was scanned linearly with a triangular waveform between 0.0 V and 1.45 V versus the standard hydrogen electrode (SHE) at a rate of 100 mV/s, and the resultant current density was plotted on a linear scale. A sample was mounted in a specially constructed Teflon^{TM} sample holder and the details can be found elsewhere [6]. The morphology was observed with an SEM (JSM-5310, JEOL, Tokyo, Japan) and a high resolution TEM (JEM-4010, JEOL, Tokyo, Japan).

Results and Discussion

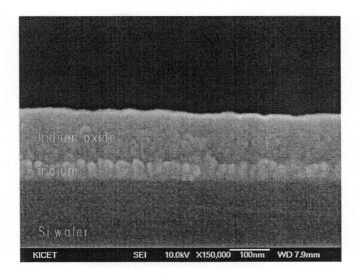

Fig. 2 SEM micrograph showing cross section of iridium oxide grown on iridium film

Figure 2 shows the cross sectional view of scanning electron micrograph of iridium oxide grown on iridium film. The thickness of the as-deposited iridium film was 580 Å measured by surface profiler, and the thickness of the iridium oxide formed was 1300 Å and the thickness of the remained iridium film was 420 Å, which were approximately estimated from Figure 2. The thickness of 160 Å iridium film was oxidized to the thickness of 1300 Å iridium oxide by cycling of iridium in 0.1M H_2SO_4 with a triangular waveform between 0.0 V and 1.45 V versus the standard hydrogen electrode (SHE) at a rate of 100 mV/s. The rate of iridium oxide formation from iridium, called activation rate, is very dependent on the upper and lower limits of the activating potential sweep. The charge density of 12 mC was obtained by cycling to 400 N with the upper and lower limits of -0.0 V (SHE) and 1.45 V (SHE) respectively, but the number of cycle was increased to 1000 N with increasing both limits by 30 mV and it was decreased to 340 N with decreasing both limits by 30 mV as shown in Table 1. It could be said the activation rate is higher if the number of cycle is lower to reach the intended charge density. The effect of potential limits on the activation rate was assumed to relate to the irreversible formation of iridium oxide.

Table 1. Effect of potential sweep limits on the activation rate

| Potential, V (SHE) | | Number of cycle to inject |
Upper limit	Lower limit	the charge density of 12 mC, N
0.0	1.45	400
0.03	1.48	1000
-0.03	1.42	340

Figure 3 shows SEM photomicrograph of embryonic cortical neural cells cultured for 4-day on iridium oxide without applying any electrical signal. Cells were prepared from the rat (Sprague-Dawley) cortex by mechanical dissociation of the 18 day old embryo. The formation of neurofilament indicates neural cells could adhere and survive on iridium oxide. The network of neurons and extending axons are expected to form more actively with inducing charge injection, which are currently investigating.

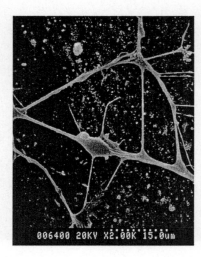

Fig. 3 SEM micrograph of embryonic cortical neural cells
cultured for 4-day on iridium oxide

Conclusions

The 1300Å thick iridium oxide was formed from the thickness of 160 Å iridium by potential sweep of iridium in 0.1M H_2SO_4 with a triangular waveform between 0.0 V (SHE) and 1.45 V (SHE) at a rate of 100 mV/s. Embryonic neural cells prepared from the rat cortex formed neurofilament on iridium oxide after 4-day culture.

Acknowledgements

This study was partially supported by a Grant (M1-0214-00-0064) of the Ministry of Science and Technology and partially by Korea Research Foundation through the Research Institute for Medical Instruments and Rehabilitation Engineering at Yonsei University, Wonju.

References

[1] W.T. Liberson, H.J. Holmquest, D. Scott and A. Dow: Arch Phys. Med. Rehabil. Vol. 42 (1961), p. 101.
[2] E.B. Marsolais and R. Kobetic: J. Bone Joint Surg. Vol. 69-A (1987) p. 728.
[3] L.S. Lobblee, J.L.Lefko and S.B. Brummer: J. Electerochem Soc. Vol. 130 (1983), p. 731.
[4] I-S. Lee, R.A. Buchanan and J.M. Willaims: J. Bio Meter Res. Vol. 25 (1991), p. 1039.
[5] I-S. Lee, C-N. Whang, K. Choi, M-S. Choo and Y-H. Lee: Biomaterials Vol. 23 (2002), p.2375.
[6] I-S. Lee, C-N. Whang, J-C. Park, D-H. Lee and W-S. Seo: Biomaterials Vol. 24 (2003), p.2225.
[7] S. Gottesfeld, J.D.E. McIntyre, G. Beni and J.L. Shay: Appl Phys Lett. Vol. 33 (1978), p. 208

Key Engineering Materials Vols. 254-256(2004) pp. 809-812
online at http://www.scientific.net

The Effect of Surface Treatment and Corrosive Etching on Flexural Strength of a Dental Porcelain

E.M. Reis[1], W.C. Jansen[2], M.M. Pereira[3], R. Giovani[2], P.R. Cetlin[3]

[1]Federal University of Minas Gerais, Dental School, Department of Dentistry, Brazil
(emoisesreis@aol.com)
[2]Federal University of Minas Gerais, Dental School, Department of Dentistry, Brazil
[3]Federal University of Minas Gerais, Eng. School, Metallurgical Eng. and Materials Dept., Brazil

Keywords: Strength, dental porcelain, polishing, glazing, corrosion

Abstract. The present study was aimed to assess the flexural strength of the porcelain Duceram Plus ® (Degussa GmbH & Co), through variation of surface treatment and exposure or not to corrosive environment. Fifty six samples with standardized dimensions were fabricated on refractory boxes and divided into four groups according to the surface treatment. Group 1 remained only ground. Group 2 was submitted to natural glazing by heating the specimen to the glazing temperature. The samples from group 3 were submitted to overglazing using the porcelain Duceram Plus Glaze Material®. Samples from group 4 were polished using silicon rubber and diamond paste of 3 and 6μm grit. Half of the specimens in each group were stored for 16h at 80°C in a 4% acetic acid solution (ISO 6872). Tukey's test (p<0,05) showed that the samples not submitted to corrosion, overglazing and polishing provided an increased strength in relation to the other groups. In the samples submitted to corrosion, overglazing showed a higher strength than polishing, although both of them did not show significant differences when coMPared to the other groups. Student's t test (p<0,05) showed that glazed and overglazed porcelains did not have their strengths altered due to corrosion effects. Corrosion showed to diminish strength of samples submitted to polishing and enhanced strength of the samples submitted to grinding.

Introduction

Porcelain in dentistry is one of the most widely used restoration materials for anterior and posterior teeth. Despite its outstanding aesthetic quality and hardness, the most serious problem of this material is the susceptibility to fractures. Limiting factors involved in its strength include inner porosity, superficial roughness and microcracks. Such "structural defects" constitute areas of stress, which may be generated by condensation, sinterization, wear and abrasion during the chewing process, and also by the treatment given to the surface [1]. Irregular surfaces may enable the beginning and spread of cracks in the ceramic material, even under low magnitude stress, as far as complete fracture happens. Beside its mechanic strength, an adequate chemical strength is desirable, once the oral cavity is a conductive medium to form corrosion products [2]. The exact mechanism of corrosion in porcelain is complex and not fully understood. Literature data has reported that the corrosive phenomenon may lead to an increased roughness on the surface and weakening of the ceramic [3], whereas other investigators, surprisingly, have remarked enhanced values of strength, suggesting a possible ion exchange mechanism between the ceramic material and the corrosive environment [4].

Materials and Methods

The commercial porcelain Duceram Plus® (Degussa, GMbH &Co., Germany) was used. The production of samples was carried out on refractory boxes fit for *Ducera-Lay Superfit®* (Degussa, GMbH &Co) porcelains. Before porcelain application, the refractory boxes were submitted to sinterization at a temperature of 1100°C for five minutes, according to the manufacturers instructions. Application and condensation of porcelain were performed by using constant vibrations and drying with absorbent paper. The heating stage was carried out under vacuum conditions at 910°C for 90 seconds. Fast cooling was performed in all porcelains as well as 4 heating stages for each sample. We performed a complete flattening of the porcelain, present outside the refractory box, with silicate carbonate sand paper numbered 120 and 220 with the help of a emery machine at a speed of 300 to 400 rpm.

The 56 samples under study were produced and randomly divided into 4 groups, with 14 samples each, according to the surface treatment given. Group 1 was submitted to abrasive wear, as previously described, without any other surface treatment. Group 2 was submitted to a glazing, following the manufacturer's instructions. The samples were heated with no vacuum until 940°C for 90 seconds. Overglazing was undertaken in Group 3, using glaze porcelain, according to the manufacture's directions. A fine layer of the material was applied to the outer surface in order to get a uniform shape. These samples were heated with no vacuum until 890°C for 90 seconds. To polish the samples from Group 4, we used abrasive silicone rubbers, specific for finishing and polishing, together with diamond pastes with granulation of 3 and 6μm. Samples were obtained as bars and their dimensions were standardized in accordance with ISO 6872 (1995) for three-point flexural test: 1.2 (± 0.2) mm x 4 (± 0.25) mm x 20 mm (minimum). Dimensions were gagged by the average of 3 measures taken from the central region of the sample with a caliper ruler, accuracy of 0.01 mm.

Afterwards, half of the samples from each group, totaling 28 specimens, were stored in 4% acetic acid solution at a temperature of 80 °C for 16 hours, following ISO 6872 recommendations for porcelain solubility tests. The flexural three-points tests was undertaken in Instron 5582 machine, with a 100 N-loaded cell, at a distance of 15 mm between supports of 1.6 mm diameter each (ISO 6872), and the treated surface was down positioned during the assay. At a load speed of 1 mm/minute, all samples were fractured and maximum stress values were obtained.

Results and Discussion

Maximum stress fracture measured for the samples with different surface treatments are presented in Table 1 and Figure 1. Through analysis of variance ($p<0,05$), we concluded that there was a significant difference regarding stress values to fracture among the samples under study, both the group that was assigned corrosive etching ($p= 0.035$) and the group that did not suffer corrosion ($p= 0.002$).

Table 1: Maximum fracture stress of samples with different surface treatments.

Surface treatments	Descriptive measures (MPa)				
	N	Mínimum	Máximum	Mean	S.d
Abrasive wear	7	48,30	63,32	57,96	4,78
Glazing	7	46,81	67,80	56,65	7,39
Overglazing	7	59,32	83,95	69,04	8,65
Polishing	7	63,22	72,80	67,47	3,22

Tukey test ($p<0,05$) applied to analyze the group that was not submitted to corrosive etching, showed that the polished and overglazed porcelains had maximum stress averages significantly higher than the averages obtained from the glazed and worn porcelains, whose values did not differ among themselves. Higher strength values from the overglazed porcelains might be explained by the capacity of the ceramic of low temperature fusion, applied as a more fluid paste, to flow into the surface cracks, working as a sealing substance and avoiding their spread [5]. Furthermore, a difference in coefficient of thermal expansion (C.E.T.L), between the overglazing layer and subjacent ceramic, could promote residual compressive stress on the sample surfaces, which would inhibit crack spread and would enhance the specimens strength [6]. Polishing of ceramic materials also led to an increased strength and a possible explanation for such result is that the polishing process promotes both flattening of irregular surfaces and a residual compressive stress on the ceramic surface, filling superficial cracks and then avoiding their spread. Although glazing also may generate a residual compressive stress under the surface ceramic [7], glazed specimens showed a significantly lower strength than polished and overglazed samples. It is likely that glazing is an effective process, but it depends on a previous finish given to the surface. Glazing itself may not be enough to diminish superficial roughness and enhance strength [8]. Low strength values of worn specimens might be explained by the roughness produced by sand paper used for finish and flattening of the specimen surface. Indeed, literature data has reported that porcelains after corrosive etching showed to have less strength and more roughness than those after some sort of superficial treatment [9].

Results from the group submitted to corrosion showed that overglazing (Table 2, Figure 1) demonstrated higher average values to fracture, though not statistically different from average values from those specimens after wearing and glazing. However, in the group including porcelains after corrosion, polished specimens showed to have significantly lower values than those overglazed.

Table 2: Maximum fracture stress for samples with different surface treatments and submitted to corrosive etching.

Surface treatments	N	Descriptive measures (MPa)			
		Mínimum	Máximum	Mean	S.d
Abrasive Wear	7	64,22	71,52	67,90	2,61
Glazing	7	49,74	67,84	62,55	6,66
Overglazing	7	58,67	84,60	70,63	8,98
Polishing	7	51,08	67,38	61,26	5,59

In order to evaluate the influence of exposure to a corrosive medium on strength, after each kind of surface treatment, Student t test was carried out at a significance level of 5%. Concerning abrasive wear, average before corrosive etching was 57.96 MPa with a standard deviation of 4.78 MPa. After corrosion, standard deviation decreased to 2.61 MPa and the average enhanced to 67.90 MPa; such difference was highly significant ($p < 0.001$). Inversely, polished specimens had their standard deviation enhanced and the average dropped from 67.47 MPa to 61.26 MPa, which showed to be significantly lower ($p=0.031$). Regarding glazing and overglazing, there was no significant difference in maximum stress measurements between the group under corrosive etching and the one not submitted to corrosion ($p>0.05$). It was expected that specimens from the worn group, before corrosion, showed a higher level of roughness, in as much as this group had not been submitted to any surface treatment. Corrosive solution could preferably etch the top of superficial irregularities, which would present a higher surface energy than the valleys, and then a sort of "chemical polishing" would be enabled. As a consequence, a higher flattening could be expected for this group after corrosion together with flexural strength enhancement. As for the polished samples, flatter surfaces would be expected before corrosion. In this case, there would be a preferable etching to the vitreous phase of the porcelain, which could expose the crystalline stage and then generate more roughness and consequent diminished flexural strength. Concerning glazed and overglazed porcelains it is likely that corrosion may have happened homogeneously on those surfaces, which could be explained by the greater amount of vitreous phase in such regions. More uniform corrosion would lead to less roughness and would not have a markedly effect on strength.

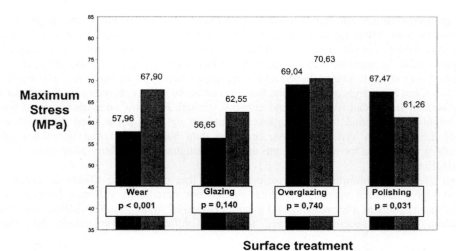

Surface treatment

Fig. 1: Influence of corrosive etching on the flexural strength of dental porcelain with different surface treatments.

Conclusions

Under the experimental conditions described here, we may conclude:

a) Strength of polished and overglazed samples, not submitted to corrosion, showed to be significantly higher than those values obtained from worn and glazed specimens;

b) Strength values among samples after corrosion showed no difference, except for those submitted to overglazing, which showed higher values than those from polished specimens;

c) Glazed and overglazed specimens did not have their flexural strength values changed after corrosion. Corrosive medium was able to decrease flexural strength of polished specimens, but not of the worn ones, which showed to have higher strength after corrosion;

d) Regarding flexural strength, overglazing showed better performance when compared to all the other surface treatments under study, in as much as these samples showed higher strength values before corrosion, which did not change after corrosive exposure.

References

[1] H. Baharav *et al.*: J. Prosthet. Dent., v.81, n.5 (1999), p.515.

[2] K.J. Anusavice and N.Z. Zhang: Dent. Mater., v.13, n.1 (1997), p.13.

[3] K.J. Anusavice and N.Z. Zhang: J. Dent. Res., v.77, n.7 (1998), p.1553.

[4] N.L Jestel; M.D. Morris and W.J. O'Brien: Dent. Mater., v.14, n.5 (1998), p.375.

[5] S.E. Brackett *et al.*: J. Prosthet. Dent., v.61, n.4 (1989), p.446.

[6] R. Giordano, S. Campbell and R. Pober: J. Prosthet. Dent., v.71, n.5 (1994), p.468.

[7] R. Samuel *et al:* J. Am. Ceram. Soc., v. 712, n.10 (1989), p.1960.

[8] M. Zalkind, S. Lauer and N. Stern: J. Prosthet. Dent., v.55, n.1 (1986), p.30.

[9] R. Giordano, M. Cima. and R. Pober: Int. J. Prosthodont., v.8, n.4 (1995), p.311.

Key Engineering Materials Vols. 254-256(2004) pp. 813-816
online at http://www.scientific.net
© 2004 Trans Tech Publications, Switzerland

Osteoblast Responses to Sintered and Tapecast Bioactive Glass

J E Gough[1], DC Clupper[2] and LL Hench[2]

[1] Manchester Materials Science Centre, UMIST and The University of Manchester, Grosvenor Street, Manchester M1 7HS, UK. E-mail: j.gough@umist.ac.uk.

[2] Department of Materials, Imperial College London, South Kensington Campus, London SW7 2BP, UK. E-mail: l.hench@imperial.ac.uk.

Keywords: Tapecast, bioactive glass, osteoblast, nodule, mineralisation, cell death, hydroxyapatite.

Abstract. The manufacture of tape cast bioactive glasses allows the build up of layers and therefore the production of complex shapes. This therefore has applications to tissue repair where specific shapes are required such as repair of craniofacial defects. Tape cast discs sintered at temperatures from 800°C to 1000°C for 3 or 6 hours were used to culture primary human osteoblasts. Hydroxyapatite formation, cell attachment, cell death, and nodule formation were studied. It was found that cell death and nodule formation were dependent on Si release concentrations and the development of the HA layer.

Introduction

Bioactive glasses have applications to tissue engineering and bone repair due to their ability to bond to bone *in vivo* via a hydroxyapatite (HA) surface layer [1, 2]. The bioactivity is due to formation of a crystalline hydroxyapatite surface layer, similar in structure and composition to that of the inorganic portion of bone after incubation in body fluid [3]. The formation of this surface layer has been reported on tape cast and sintered bioactive glass ceramics by incubation in simulated body fluid (SBF) [4, 5]. Using tape cast bioactive glasses allows the build up of layers and therefore the production of complex shapes, and therefore has applications to tissue engineering. The release profiles of tapecast bioceramics has been shown to vary according to sintering temperature and time [5]. Therefore cell attachment, cell death, and nodule formation were studied directly on tapecast surfaces and also in response to dissolution products from the tapecast bioceramics to determine effects of released products and interaction with the HA layer on the tapecast surface. The formation of calcium phosphate and HA on the disc surfaces was also analysed using FTIR.

Methods
In vitro bioactivity testing
Discs were incubated in DMEM culture medium for 2, 24 hours and 14 days. Samples were soaked in acetone for 15 seconds and allowed to dry. Samples were then analysed by FTIR and Raman spectroscopy.
FTIR
Fourier transform infrared spectroscopic (Genesis II FTIR, Spectronic Unicam) analysis was conducted using KBr pellets (1:100 weight ratio). Scans were taken between 1600-400 cm^{-1}.
Inductively coupled plasma optical emission spectroscopy (ICP-OES)
Ionic concentrations of released sodium (Na) and silicon (Si) were determined using ICP-OES (3580B ICP Analyser, Applied Research Laboratories, IM35xx ICP Manager software, Micro-Active Australia Pty Ltd). After immersion of tapecast discs in culture medium (DMEM) for various time periods at 37°C in static conditions (to mimic cell culture conditions) the samples were diluted 1:10 in a 2N HNO$_3$ matrix. Three integrations were performed per sample and the mean and the standard deviation was calculated.

Cell Culture

Human primary osteoblasts (HOBs) were isolated as described previously from the femoral head and cultured in Dulbecco's modified eagles medium (DMEM) containing 10% fetal bovine serum with 1% glutamine, 2% penicillin/streptomycin, and 50μg/ml ascorbic acid. Cells were seeded onto material surfaces at a density of 40,000 cells/cm2. At certain time points cells were fixed in 4% paraformaldehyde and stained with propidium iodide to identify apoptotic, necrotic and viable cells. Cells were also fixed to observe nodule formation.

Results

After 2 hr in culture media, FTIR spectra of the 800C, 900C and 1000(3)C samples each appeared nearly identical to the as-processed spectra, with no detectable HA formation (image not shown). In contrast, FTIR spectral analysis indicated an HA layer began forming on the 1000(6) sample surface after only 2 hr in DMEM (sharp peaks near 598 and 573 cm^{-1}). Hydroxyapatite formation was clearly evident in the 900, 1000(3) and 1000(6) samples after 24 hours (598, 573, and 565 cm^{-1}). A semi-crystalline calcium phosphate phase formed on the 800 surface. The formed surface layers continued to increase in thickness after 14 d, as indicated by the decrease in the Si-O bending peaks near 450 cm^{-1}.

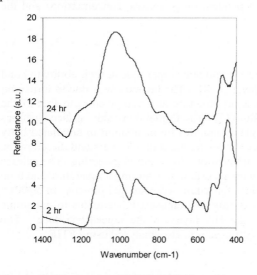

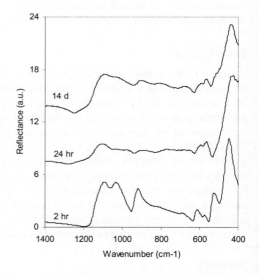

Fig. 1. FTIR spectra of 800C TCS bioactive glass following 2 hr and 24 immersion.

Fig. 2. FTIR spectra of 900C TSC bioactive glass following 2 hr, 24 hr and 14 d immersion.

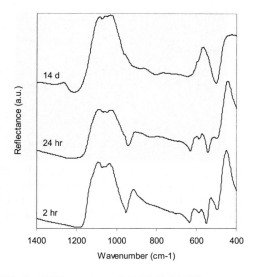

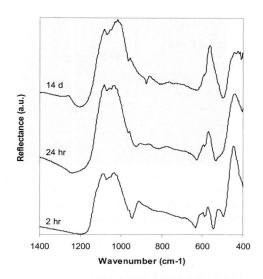

Fig. 3. FTIR spectra of 1000(3)C TCS
bioactive glass following 2 hr, 24 hr and 14 d
immersion.

Fig. 4. FTIR spectra of 1000(6)C TCS
bioactive glass following 2 hr, 24 hr and 14 d
immersion.

Release of Si and Na ions (shown in Table 1) followed the pattern 800(3)> 900(3)> 1000(3)>
1000(6). Release of Ca and P ions were not significantly different. Culture in dissolution products
from the tapecast discs resulted in apoptosis, necrosis or viable cells, as shown in Table 2,
depending on concentration of Si released. Nodule formation was only observed on the 1000(3)
sample.

Table 1. Si release / Na release (ug/ml) over time.

Time/Sample	24	48	72	120	168	192
800(3)	413 / 3012	699 / 3841	1011 / 4896	1410 / 6778	1665 / 7777	1792 / 8166
900(3)	136 / 423	143 / 539	179 / 539	269 / 1043	444 / 1491	531 / 1727
1000(3)	102 / 227	102 / 318	102 / 318	163 / 454	199 / 511	199 / 566
1000(6)	89 / 242	89 / 323	89 / 323	144 / 561	176 / 713	176 / 746

Table 2. Percentage of apoptotic, necrotic and viable cells after culture in dissolution products from
tapecast bioglass for 24 hours.

Sample	% Apoptotic	% Necrotic	% Viable
DMEM	4	0	96
800(3)	22	78	0
900(3)	79	8	13
1000(3)	13	0	87
1000(6)	91	6	3

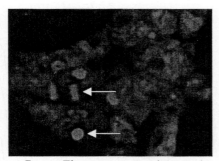

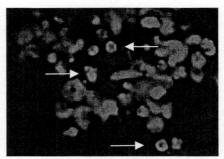

Fig. 7. Fluorescence micrograph of propidium iodide stained cells cultured on tapecast 1000(3). Mitotic cells are denoted by arrows. Top arrow shows cell in anaphase stage of cell cycle.

Fig. 8. Fluorescence micrograph of propidium iodide stained cells cultured on tapecast 1000(6). Arrows show examples of apoptotic cells where nuclear material has condensed.

Discussion and Conclusions

Previously studies of tape cast bioactive glass-ceramic revealed that the ionic release in Tris buffer followed the relation: 800C > 900C > 1000(3)C > 1000(6)C [4]. This study showed a similar trend under cell culture conditions. The pattern of cell death appears to be related to levels of ion release. The 800 sample released the highest rates of Na and Si and it was determined that this caused the high levels of cell death. The author has previously demonstrated Si release from 58S bioactive glass foams and apoptosis [6]. The greatest number of nodules formed were on the 1000(3) sample. Hydroxyapatite formation in DMEM was similar to that formed in Tris buffer and SBF, although the process was slightly slower in culture medium. The most rapid HA formation was observed on the 1000(6) sample, which also resulted in high levels of apoptosis. The results suggest that hydroxyapatite formation should not be too rapid, nor should excessive ion leaching occur; i.e. a medium-level bioactivity rate may be ideal for osteoblast survival, proliferation and nodule formation.

References

[1] L.L. Hench and O. Anderson: *An introduction to bioceramics*. (Singapore World Scientific, 1993).

[2] T. Kitsugi, T. Yamamuro, T. Nakamura, S. Higashi, Y. Kakutani, K. Hyakuna, S. Ito, T. Kokubo, M. Takagi and T. Shibuya: J Biomed Mater Res Vol 20 (1986), p. 1295.

[3] L.L. Hench: J Am. Ceram. Soc. Vol 74 (1991), p. 1487.

[4] D.C. Clupper, J.J.Mecholsky Jr, G.P. LaTorre and D.C. Greenspan: Biomaterials Vol 23 (2001), p. 2599.

[5] D.C. Clupper DC, J.J Mecholsky Jr,G.P. LaTorre and D.C. Greenspan: J Biomed Mater Res Vol 57 (2001), p. 532.

[6] J.E. Gough, J.R. Jones and L.L Hench. *Submitted*.

Key Engineering Materials Vols. 254-256(2004) pp. 817-820
online at http://www.scientific.net
© 2004 Trans Tech Publications, Switzerland

Chondrocyte Maturation on Biphasic Calcium Phosphate Scaffold: A Preliminary Study

Cristina C. Teixeira[1], Racquel Z. LeGeros[2], Claudia Karkia[2], and Yelena Nemelivsky[1]

[1]Department of Basic Science & Craniofacial Biology, New York University College of Dentistry, NY, USA, ct40@nyu.edu

[2]Department of Biomaterials, New York University College of Dentistry, NY, USA, rzl1@nyu.edu

Keywords: chondrocytes, biphasic calcium phosphate, maturation, collagen, alkaline phosphatase

Abstract. In recent years artificial bone replacement materials have been developed that can be coated with growth factors and loaded with osteoblasts to induce osteogenesis. All these attempts from the bone tissue engineering field disregard the fact that skeletal development can proceed by two pathways: intramembranous or endochondral bone formation. Here we propose to mimic nature by creating an *in vitro* cartilage template that could after implantation *in vivo*, remodel into bone, as is normally carried out during the process of endochondral bone formation. The specific aims of this study are: (1) to establish a method of growing chondrocytes in a well characterized biphasic calcium phosphate (BCP) scaffold, and (2) to induce chondrocyte hypertrophy and cartilage matrix deposition in cells growing in the BCP scaffold. <u>Materials and Methods:</u> Chondrocytes isolated from 18-day chick embryo tibial growth plates were grown on particles of macroporous biphasic calcium phosphate (MBCP®, *Biomatlante, France*). Chondrocytes were grown in culture continuously for 3 weeks, and fed every day with Dulbecco's modified Eagle medium (GIBCO) containing 10% Nu serum (Fisher), 2mM L-glutamine, 100 U/ml penicillin/streptomycin, and 50 µg/ml ascorbic acid. After one week, cultures were treated daily with 100 nM all trans-retinoic acid (RA), to induce chondrocyte maturation, and extracellular matrix synthesis. Chondrocyte proliferation was observed using scanning electron microscopy (SEM) on appropriately prepared specimens. Alkaline phosphatase (AP) activity was measured spectrophotometrically. Levels of proteoglycans and levels of type X collagen were determined using Alcian blue staining and Western blot analysis, respectively. <u>Results:</u> Chondrocytes attached, and proliferated on the BCP scaffold. Levels of AP, proteoglycans, and type X collagen increased in the presence of RA. An unexpected observation was made regarding changes in chondrocyte morphology in the presence of RA. Chondrocytes grown on BCP scaffold assumed an elongated morphology in an intricate net of collagenous proteins, while cells grown on tissue culture plate surfaces maintained a polygonal and flat morphology. <u>Conclusion:</u> Results from this study demonstrated for the first time the proliferation, maturation of chondrocytes, and cartilage matrix deposition on a macroporous calcium phosphate scaffold, and the potential of such a scaffold in tissue engineering through the endochondral bone formation mechanism.

Introduction

Bone grafting has been used for many years to restore areas deficient in bone due to trauma, growth defects or pathology. Human derived bone grafts were historically the only osteoinductive replacement materials, providing cellular elements and growth factors necessary for osteogenesis. In recent years, a number of artificial bone replacement materials have been developed that can be coated with growth factors and loaded with osteoblasts to induce bone formation after implantation *in vivo*. Until now, the focus has been to engineer bone grafting materials from direct differentiation of osteoprogenitor cells, all attempts geared toward the creation of intramembranous bone, disregarding the fact that there are two distinct mechanisms of skeletal development: intramembranous and endochondral bone formation [1]. Intramembranous bone forms by differentiation of mesenchymal cells directly into osteoblasts, while during endochondral ossification there is the gradual and partial replacement of a cartilage model by bone. The residual cartilage acts as a growth plate or an articular surface. The long bones, pelvis, vertebral column, base of the skull and mandible are formed through the endochondral mechanism [1].

Here, we propose to mimic nature, by creating a cartilage template of the correct size and shape that, after implantation *in vivo*, will remodel into the required bone, as is normally carried out in the body during the process of endochondral bone formation. Our approach to the field of bone tissue engineering, presents significant advantages including the resistance of chondrocytes to low oxygen [2, 3] and the fact that these cells provide the signaling system for vascular invasion and osteogenesis [4, 5]. These characteristics will allow the generation of an improved osteoconductive and osteoinductive scaffold. By successful inducing endochondral bone formation from an artificially created cartilage model, we will be able to address situations where current bone grafting techniques have failed to correct severe bony defects. In the future, a large cleft palate, a severely resorbed dental alveolar ridge, or a large bone defect caused by cancer, may be repaired by placing a cartilaginous scaffold (artificially created) in the area, to provide the template and induce new bone formation. This new bone will be indistinguishable from natural bone, will present the same properties, respond to loads and have the potential to grow, just as the patient own bone.

There are two specific aims in the current investigation: 1) to establish a method of growing chondrocytes in a well characterized BCP scaffold, 2) to induce chondrocyte maturation and cartilage matrix deposition in cells growing in the BCP scaffold.

Materials and Methods

The material we chose to use in this project is macroporous biphasic calcium phosphate (Biomatlante France) consisting of a mixture of 60% Hydroxyapatite and 40% ß-tricalcium phosphate. This material has been extensively characterized and shown to be biocompatible and an adequate bone replacement scaffold [6-8]. Chondrocytes were isolated from 18 day chick embryo tibia, using a method described by Rajpurohit et al [2] and plated in flat bottom polystyrene 100mm tissue culture dishes. After one week, chondrocytes were sub cultured (0.125 million cells/well) into 24-well tissue culture plates containing the BCP particles, no attempt being made to seed the cells into the scaffold. Chondrocytes were grown continuously without further subculturing for 3 weeks at a temperature of 37° C and at a 5% carbon dioxide atmosphere. Cultures were fed every day with Dulbecco's modified Eagle medium (GIBCO) containing 10% Nu serum (Fisher), 2mM L-glutamine, 50 µg/ml ascorbic acid, and 100 U/ml penicillin/ streptomycin. After one week, cultures were treated daily with 100 nM all-trans retinoic acid to induce chondrocyte maturation, and matrix synthesis. To access chondrocyte attachment and proliferation in the BCP scaffold, the samples were examined by scanning electron microscopy at the end of each week, and photographed. To evaluate chondrocyte maturation, we investigated the expression of two important markers of chondrocyte hypertrophy: alkaline phosphatase activity and type X collagen protein levels. Alkaline phosphatase activity was measured using a method described by Leboy et al [9], and western blot analysis was performed using an antibody against chick type X collagen. Protein content in each sample was analyzed using the DC Protein Assay (BioRad, CA) according to manufacturers instructions. In addition, AP staining and Alcian blue staining were performed to visualize levels of this enzyme activity and proteoglycans deposited on the BCP scaffold.

Results

Results show that chondrocytes attach and divide on the BCP scaffold. Scanning electron microscopy demonstrates that at the end of the first week in culture chondrocytes became confluent or almost confluent on the scaffold. After 3 weeks in culture, a continuous layer of polygonal chondrocytes covered the BCP particle surface and pores (data not shown). Once it was established that chondrocytes could attach and proliferate in the BCP surface, cell maturation was induced by adding retinoic acid (RA) to the cultures. Chondrocytes were allowed to proliferate for one week

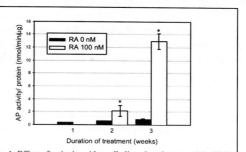

Fig. 1: Effect of retinoic acid on alkaline phosphatase activity. Tibial chondrocytes growing on BCP were treated with RA (100 nM) for 2 weeks, and chondrocytes collected for alkaline phosphatase analysis. Note that RA causes a time dependent increase in the enzymatic activity. * Significantly different from control. (RA 0nM)

before treatment with 100 nM RA was initiated. After two weeks of daily RA treatment, BCP particles were collected and used for different analysis. Macroscopic analysis was performed, by staining the beads to visualize both proteoglycan production and AP activity. RA treatment caused a marked increase in AP activity evidenced by deep red coloration. An increase in

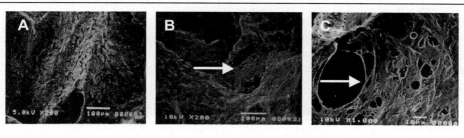

Fig. 2: Scanning electron microscopy of chondrocytes after 3 weeks in culture. Chondrocytes were grown to confluence for 1 week, and then treated with RA for 2 additional weeks. The samples were fixed and dehydrated, sputter coated with gold, viewed using SEM (Joel JSM 5400, Tokyo, Japan) and photographed. Note that the cells grown in the presence of 100nM of RA (B,C) secreted a matrix rich in collagen fibers (arrow in C) that even covered the pores (arrow in B). In the absence of RA (A) the chondrocytes presented the characteristic polygonal shape and formed a thin cellular layer.

proteoglycans deposition on the extracellular matrix was characterized by the presence of an intense blue staining (data not shown). The level of AP activity was also measured by spectophotometry. A significant increase in the alkaline phosphatase activity was observed in the presence of RA in the

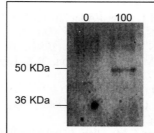

Fig. 3: Western Blot analysis of type X collagen in tibial chondrocytes grown on BCP. Tibial chondrocytes were treated with RA (0 and 100 nM) for 2 weeks. After extraction, proteins were transferred onto a nitrocellulose membrane and immunostained with antibody against type X chick collagen. Note that RA caused an increase in the protein level of type X collagen in chondrocytes.

culture media, and this increase in enzymatic activity was dependent on the duration of treatment (Fig. 1).

Scanning electron microscopy performed at the end of the 3rd week in culture showed that in the absence of RA, chondrocytes form a confluent layer over the BCP surface (Fig. 2A) with a polygonal and flat cell morphology.   Treatment with the retinoid induced formation of a thick extracellular matrix over the particle surface (Fig. 2B) and over the pores (arrow in Fig. 2B). Chondrocytes are embedded in this matrix composed of numerous fibers creating an intricate net of collagenous proteins (white arrow on Fig. 2C). Western blot analysis confirmed an increase in the protein level of type X collagen when chondrocytes were treated with the retinoid (Fig. 3).

Discussion

While attempts to create an in vitro cartilage scaffold have been geared towards articular cartilage replacement therapies, here we propose the creation of an *in vitro* calcified cartilage scaffold as a bone replacement material.   To the best of our knowledge, this is the first report of chondrocytes being grown on a mineralized scaffold, specifically BCP. This material has been used as a scaffold for bone grafting but always using osteoblasts as the cell precursor.   Over the last 25 years calcium phosphate materials have been well tested clinically in craniofacial bone replacement therapies [10-12]. We reasoned that due to its characteristics, the material would be a suitable template for endochondral bone formation, by first allowing attachment and maturation of chondrocytes, and later supporting and inducing osteoblast ingrowth. Experiments described here reveal that indeed chondrocytes attach and proliferate on the BCP scaffold.   After three weeks of continuous cell culture, the beads became completely covered by a continuous layer of cells and extracellular matrix.   To test the hypothesis that chondrocytes grown in an already mineralized surface can undergo hypertrophy and maturation, we treated the cells with retinoic acid. Our results show an increase in 2 important markers of maturation, AP activity and type X collagen expression, in the presence of this agent.

In conclusion, we have successfully established a method of growing and inducing chondrocyte maturation on a mineralized template, BCP. Further studies will be conducted to test the suitability of this template as an intermediate in endochondral bone formation.

References

[1]   S.D. Howell, and D.D. Dean: Disorders of Bone and Mineral Metabolism (1992), p. 313-353, Raven Press.
[2]   R. Rajpurohit, C.J.   Koch, Z. Tao, C.C. Teixeira, and I.M. Shapiro: Journal of Cellular Physiology 168 (1996), 424-432.
[3]   E. Schipani, H.E. Ryan, S. Didrickson, T. Kobayashi, M. Knight, and R.S. Johnson: Genes Dev 15 (2001), 2865-76.
[4]   C. Maes, P. Carmeliet, K. Moermans, I. Stockmans, N. Smets, D. Collen, R. Bouillon, and G. Carmeliet: Mech Dev 111(2002), 61-73.
[5]   W. Petersen, M. Tsokos, and T. Pufe: J Anat 201 (2002), 153-7.
[6]   R.Z. LeGeros, J.R. Parsons, G. Daculsi, F. Driessens, D.   Lee, S.T. Liu, S. Metsger, D. Peterson, and M. Walker: Ann N Y Acad Sci 523 (1988), 268-71.
[7]   G. Daculsi, N. Passuti, S. Martin, C. Deudon, R.Z. Legeros, and S. Raher: J Biomed Mater Res 24 (1990), 379-96.
[8]   R.Z. LeGeros: Clin Mater 14 (1993), 65-88.
[9]   P.S. Leboy, L. Vaias, B. Uschmann, E. Golub, S.L. Adams, and M. Pacifici: Journal of Biological Chemistry 264 (1989), 17281-17286.
[10] T.E. Butts, L.J. Peterson, and C.M. Allen: J Oral Maxillofac Surg 47 (1989), 475-9.
[11] S.F. Hulbert, S.J. Morrison, and J.J. Klawitter: J Biomed Mater Res 6 (1972), 347-74.
[12] L.M. Wolford, R.W. Wardrop, and J.M. Hartog: J Oral Maxillofac Surg 45 (1987), 1034-42.

Key Engineering Materials Vols. 254-256(2004) pp. 821-824
online at http://www.scientific.net
© *2004 Trans Tech Publications, Switzerland*

In Vitro Mineralisation of Human Bone Marrow Cells Cultured on Bonelike®

M. A. Costa[1], M. Gutierres[2], L. Almeida[2], M. A. Lopes[3,4], J. D. Santos[3,4], M. H. Fernandes[5]

[1]ICBAS - Instituto de Ciências Biomédicas de Abel Salazar da Universidade do Porto, Largo Abel Salazar, 4000 Porto, Portugal
[2]Serviço de Ortopedia do Hospital de S. João, Largo Hernáni Monteiro, 4200 Porto, Portugal
[3]FEUP - Faculdade de Engenharia da Universidade do Porto, DEMM, Rua Dr Roberto Frias, 4200 Porto, Portugal
[4]INEB - Instituto de Engenharia Biomédica, Laboratório de Biomateriais, Rua do Campo Alegre 823, 4150 Porto, Portugal
[5]FMDUP - Faculdade de Medicina Dentária da Universidade do Porto, Rua Dr Manuel Pereira da Silva, 4200 Porto, Portugal, mhrf@portugalmail.pt

Keywords: Bonelike®, human bone marrow cells, *in vitro* mineralisation

Abstract. Bonelike® is a $CaO-P_2O_5$ based glass-reinforced hydroxyapatite (HA) designed to mimic the inorganic composition of the bone tissue. This work evaluates the response of human bone marrow cells to Bonelike® concerning cell proliferation and osteoblast differentiation. HA was used as control material. Results showed that Bonelike® allowed the proliferation of bone marrow cells and their complete differentiation, as evidenced by the formation of cell-mediated mineralisation. In comparison with HA, Bonelike® had a positive effect on the expression of alkaline phosphatase and also on the formation of a mineralised matrix, two osteoblast markers.

Introduction

Bonelike® is a synthetic hydroxyapatite (HA) that is sintered in the presence of $CaO-P_2O_5$-based glasses using a patented process [1]. This synthetic bone graft was designed to improve the mechanical properties of calcium phosphate ceramics and mimic the inorganic composition of bone tissue. The physicochemical and mechanical behaviour of Bonelike® have been extensively reported in literature [1-3].

Previous *in vitro* biological studies showed that glass-reinforced HA composites allow the proliferation of MG63 osteoblast-like cells and human bone marrow cells and the expression of osteoblast markers [4-6]. Also, *in vivo* studies performed in a rabbit model demonstrated that Bonelike® composites induced earlier new bone formation around implants than HA [7].

Recently, a composite prepared by the addition (4 wt %) of a glass with the composition of $65P_2O_5-15CaO-10CaF_2-10Na_2O$ (in % mol) to HA was subject of clinical trials in implantology and maxillofacial surgery [8]. This study demonstrated extensive new bone formation around implanted granules and continuous replacement by new bone. Osteoblasts are the cells responsible for the formation of the bone tissue at the bone/material interface and the present work evaluates the response of human bone marrow cells to Bonelike® composite, with the same chemical composition, concerning cell proliferation and osteoblast differentiation. HA was used as control material.

Materials and methods

For the preparation of Bonelike®, a P_2O_5-based glass with the chemical composition of $65P_2O_5$-$15CaO$-$10CaF_2$-$10Na_2O$, in mol %, was prepared from reagent grade chemicals by conventional techniques. The glass was then milled and added to HA powder in a proportion of 4.0 wt % using methanol as solvent medium. Disc samples were then prepared by uniaxial pressing at 288 MPa and sintered at 1300 °C for 1h followed by natural cooling inside the furnace. Detailed description of Bonelike® preparation was previously reported [1]. For cell culture studies, discs were polished down to 1 μm, ultrasonically cleaned and sterilised before testing. HA samples were also prepared and used as control material.

Human bone marrow cells (first subculture) were cultured on the surface of Bonelike® and HA material samples for 28 days in α-MEM, 10 % foetal bovine serum, 50 μg/ml gentamicin and 2.5 μg/ml fungizone, and supplemented with 50 μg/ml ascorbic acid, 10 mM β-glycerophosphate and 10 nM dexamethasone, experimental conditions that favour the complete expression of the osteoblast phenotype [9]. As reference control, cells were cultured in parallel on standard tissue culture plastic plates. Material samples were also incubated in the absence of cells in the same experimental conditions as those of the seeded materials. Control cultures and seeded materials were characterised at days 1, 3, 7, 14, 21 and 28 for cell viability/proliferation (MTT assay), total protein content and alkaline phosphatase (ALP) activity, using procedures previously described in detail [6,7,10,11]. Control and seeded materials were observed by scanning electron microscopy (SEM). The *in vitro* biological studies were performed at the *Faculdade de Medicina Dentária da Universidade do Porto*.

Results and discussion

Biochemical evaluation of the seeded Bonelike® showed that human bone marrow cells proliferated during approximately three weeks and attained a stationary phase afterwards, as shown by the results concerning the MTT reduction and total protein content (Figs. 1 A and B). Cells expressed ALP, especially during the second and third weeks of culture (Fig. 1C). Seeded HA presented a similar behaviour concerning cell growth but lower (20 to 30 %) ALP activity than that measured on seeded Bonelike®.

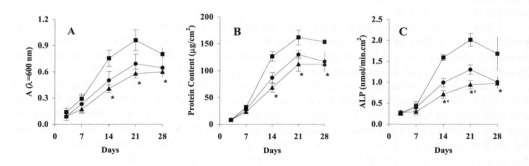

Fig.1. Cell viability/proliferation (A), total protein content (B) and alkaline phosphatase (ALP) activity (C) of human bone marrow cells grown on Bonelike® (●) and HA (▲). *Significantly different from cultures performed on standard tissue culture plates (■); 'Significantly different from cultures performed on seeded Bonelike®.

SEM observation of Bonelike® samples incubated in the absence of cells showed significant surface modifications resulting from the dissolution of bioresorbable phases. This process led to the formation of different size microcavities that became progressively larger with the incubation time (Fig. 2A). Seeded Bonelike® was partially covered with bone marrow cells that successfully adapted to the surface irregularities of the material surface and were able to grow towards inside the cavities, as shown in Figs. 2B for a 21-day culture. At day 28, abundant globular mineralised structures incorporated into a network of fibres were present, Fig. 2C. It is worthwhile to note that the formation of the mineralised matrix was especially related with the surface cavities resulting from the dissolution of bioresorbable phases. A detailed SEM micrograph showed exuberant cell-mediated mineralisation inside these cavities.

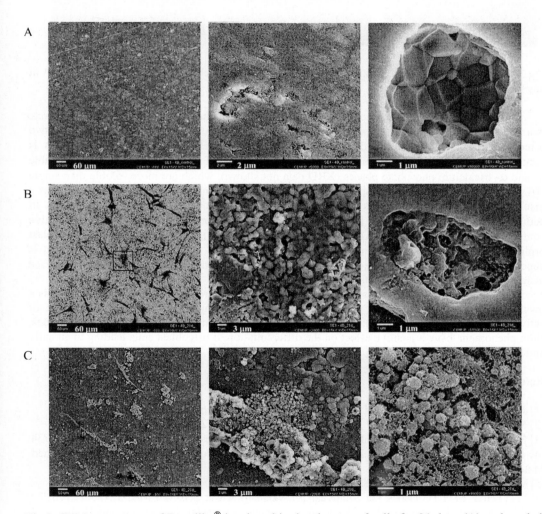

Fig.2. SEM appearance of Bonelike® incubated in the absence of cells for 21 days (A) and seeded Bonelike® at days 21 (B) and 28 (C).

HA material samples incubated in the absence of bone marrow cells showed a relatively stable surface during the 28 days incubation, Fig.3A. Seeded HA presented cell proliferation in some areas of the material surface but under the present experimental conditions, formation of mineralised structures associated with the cell growth was not clearly observed, Fig. 3B.

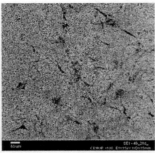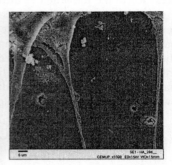

Fig.3. SEM appearance of HA incubated in the absence of cells for 21 days (A) and seeded HA at days 21 (B) and 28 (C).

This work showed that, in comparison with HA, Bonelike® had a positive effect on bone marrow cells concerning the expression of ALP and formation of a mineralised matrix, two osteoblast markers. The improved biological performance of Bonelike® compared with HA is related with its chemical composition. Bonelike® is composed of an HA matrix with bioresorbable β- and α- tricalcium phosphate phases (XRD data not shown), which are more soluble than single HA and liberate Ca and P ionic species to the local environment. Surface changes occurred as a result of these ionic interactions appear to significantly induce osteoblast differentiation. In addition, the presence of fluoride ions in the composition of Bonelike® may also have a positive contribution on the performance of this material, as fluoride is known to stimulate osteoblast proliferation [10]. However, the ionic release from Bonelike® was carefully controlled and is dependent on the content of bioresorbable phases since a strong dissolution can cause high surface instability, which negatively affects cell adhesion.

The results reported in this work suggested that the improved bioactivity of Bonelike is a result of very important surface changes that occur after incubation, namely ionic release of activities species (Ca, P and F to the local environment) and surface topography modifications. Both phenomena have a positive effect on the differentiation of osteoblast cells leading to an early and strong bone /material bonding.

References

[1] J.D. Santos, J.K. Campbell and G.H. Winston: European Patent N° WO0068164, 2000
[2] M.A. Lopes, J.C. Knowles and J.D. Santos: Biomaterials 21 (2000), p. 1905
[3] M.A. Lopes, F.J. Monteiro and J.D. Santos: J. Biomed. Mater. Res.:App. Biomater. 48 (1999), p. 734
[4] M.A. Lopes, J. C. Knowles, L. Kuru, J.D. Santos, F.J. Monteiro and I. Olsen: J. Biomed. Mater. Res. 41 (1998), p. 649
[5] M.A. Lopes, J. C. Knowles, J.D. Santos, F.J. Monteiro and I. Olsen: Biomaterials 21 (2000), p. 1165
[6] M.P. Ferraz, M.H. Fernandes, J.D. Santos and F.J. Monteiro: J. Mat. Sci.: Mat. Med. 12 (2001), p. 629
[7] M.A. Lopes, J.D. Santos, F.J. Monteiro, C. Ohtsuki, A.Osaka, S. Kaneko and H. Inoue: J. Biomed. Mater. Res. 54 (2001), p. 463
[8] F. Duarte, J. D. Santos and A. Afonso: Advanced Materials Forum, in press
[9] M.J. Coelho and M.H. Fernandes: Biomaterials 21 (2000), p. 1095
[10] M. Kleerekoper: Principles of Bone Biology, Academic Press (1996), p. 1053

Key Engineering Materials Vols. 254-256(2004) pp. 825-828
online at http://www.scientific.net
© *2004 Trans Tech Publications, Switzerland*

Biological Activity of Two Glass Ceramics in the Meta- and Pyrophosphate Region: a Comparative Study

A. G. Dias[1,2], M. A. Costa[3], M. A. Lopes[1,2], J. D. Santos[1,2], M. H. Fernandes[4]

[1]FEUP - Faculdade de Engenharia da Universidade do Porto, DEMM, Rua Dr. Roberto Frias, 4200 Porto, Portugal
[2]INEB - Instituto de Engenharia Biomédica, Laboratório de Biomateriais, Rua do Campo Alegre 823, 4150 Porto, Portugal
[3]ICBAS - Instituto de Ciências Biomédicas de Abel Salazar da Universidade do Porto, Largo Abel Salazar, 4000 Porto, Portugal
[4]FMDUP - Faculdade de Medicina Dentária da Universidade do Porto, Rua Dr. Manuel Pereira da Silva, 4200 Porto, Portugal, mhrf@portugalmail.pt

Keywords: Biodegradable glass ceramics, MG63 osteoblast-like cells

Abstract. Based upon the $CaO-P_2O_5$ glass system, two glass ceramics were prepared in the meta- and pyrophosphate regions. The present work describes preliminary results concerning the biological activity of MK5B ($45CaO-45P_2O_5-5MgO-5K_2O$, in mol %) and MT13B ($45CaO-37P_2O_5-5MgO-13TiO_2$, in mol %) using MG63 osteoblast-like cells. Both materials supported cell growth and proliferation which increased with the incubation time. However, cell growth was significantly lower on seeded MK5B samples. SEM observation showed that cell adhesion and spreading were hampered on this material. Evident degradation of the material surface was observed with simultaneous cell proliferation and material precipitation. By contrast, MT13B presented a relatively stable surface throughout the culture time. Results suggest that the *in vitro* biological behaviour of MK5B and MT13B is related with differences in the degradation rates of the two materials during the culture time.

Introduction

Several efforts have been made to obtain allografts for bone regeneration processes with controlled biodegradability. For several medical applications e.g. implantology these allografts should be capable of being degraded with simultaneously new bone regeneration without gaps formation at bone/implant interface. Based upon the $CaO-P_2O_5$ glass system two glass ceramics were prepared in the meta- and pyrophosphate regions [1-3]. K_2O and TiO_2 oxides were also added to promote the precipitation of biodegradable and bioactive phases. By working in the meta- and pyrophosphate regions it is possible to obtain calcium phosphate biomaterials with tailored degradability and enough mechanical strength and therefore to modulate their behaviour to specific clinical applications. The present work aims at analysing the biological activity of the two glass ceramics using MG63 osteoblast-like cells.

Materials and methods

Glass ceramics MK5B ($45CaO-45P_2O_5-5MgO-5K_2O$, in mol %) and MT13B ($45CaO-37P_2O_5-5MgO-13TiO_2$, in mol %) were prepared by controlled crystallisation and powder sintering technique, respectively. Sintering and crystallisation heat-treatments were conducted according to differential thermal analysis results. Crystallisation of MK5B was performed using two-step heat-treatment of nucleation followed by crystal growth to obtain volume crystallisation [1] and MT13B was sintered at 703 °C [1,2]. For cell culture studies, samples were ground down to 1000 mesh of SiC paper and sterilised by autoclaving.

Ceramic material samples were seeded with MG63 osteoblast-like cells (10^4 cell/cm^2) and cultured for 9 days in α-MEM supplemented with 10% foetal bovine serum, 50 µg/ml ascorbic acid, 50 µg/ml gentamicin and 2.5 µg/ml fungizone. Control cultures were performed on standard plastic tissue culture plates. Control and seeded materials were characterised at days 1, 3, and 6 for cell viability/proliferation and observed by scanning electron microscopy (SEM).

Cell proliferation was evaluated by the MTT assay – reduction of 3-[4,5-dimethylthiazol-2-yl]-2,5-diphenyltetrasodium bromide (MTT) to a purple formazan product by living cells. Cultures were incubated with 0.5 mg/ml of MTT during the last four hours of the culture period tested; the material samples were transferred to a new plate, the formed formazan salt was dissolved with dimethylsulphoxide and the absorbance was measured at 600 nm. Results were normalised in terms of macroscopic surface area and expressed as A.cm^{-2}.

The *in vitro* biological studies were performed at the *Faculdade de Medicina Dentária da Universidade do Porto*.

Results and discussion

Both materials supported the growth of the osteoblast-like MG63 cells and cell proliferation increased with the incubation time. The values for the MTT reduction measured in MK5B samples were significantly lower than those observed in MT13B samples, throughout the incubation time (Fig. 1A). The macroscopic appearance of the seeded samples after the incubation with the MTT showed an evident difference on the cell colonization of the two materials (Fig. 1B). SEM observation showed that, at day 1, seeded MK5B (Fig. 2) presented few cells with a round morphology, reflecting a difficulty on cell adhesion and spreading; at day 6, the material surface was only partially colonised and extensive material degradation and precipitation were evident. By contrast, on seeded MT13B (Fig. 3), cells were well spread at 1 day incubation time and covered the material surface on day 6, adapting to the material topography.

Previous published results using X-ray diffraction analysis have shown the presence of α- and β- $Ca_2P_2O_7$, $CaTi_4(PO_4)_6$ and TiP_2O_7 phases in the MT13B glass ceramic and $KCa(PO_3)_3$, β-$Ca(PO_3)_2$, β-$Ca_2P_2O_7$, $Ca_4P_6O_{19}$ phases in the MK5B glass ceramic [1,2]. Both $KCa(PO_3)_3$ and β-$Ca(PO_3)_2$ are quite soluble in normal physiological conditions, as previously demonstrated by *in vitro* testing[1,3], while $CaTi_4(PO_4)_6$ is almost insoluble. The interference of MK5B surface on cell behaviour in the experiments performed without pre-immersion can be explained by the several phenomena. Firstly, the instability in the topography of this glass ceramic as a consequence of its surface degradation may have negatively interfered with the initial cell adhesion process. Also, the pyrophosphate groups firstly released to the medium may have then occupied potential sites for the protein adhesion [4] which precedes cell adhesion, causing a detrimental effect on cells anchorage. Finally, previous *in vitro* degradation studies have shown that this glass ceramic causes pH changes [1], which is also deleterious for cell adhesion and proliferation, especially in closed *in vitro* systems. These three phenomena decisively contributed to the less favourable biological behaviour of MK5B glass ceramic compared to MT13B.

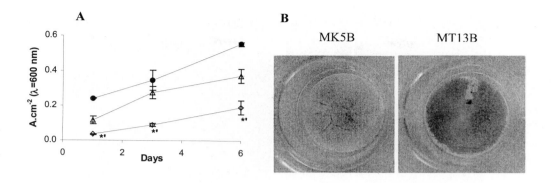

Fig. 1. MTT reduction by MG63 cells cultured on MK5B (◊) and MT13B (Δ) material samples.
A: Cell viability/proliferation; *Significantly different from cultures performed on standard tissue
culture plates (●); 'Significantly different from cultures performed on seeded MT13B.
B: Macroscopic view of seeded MK5B and MT13B after 4 h incubation with MTT, at day 6.

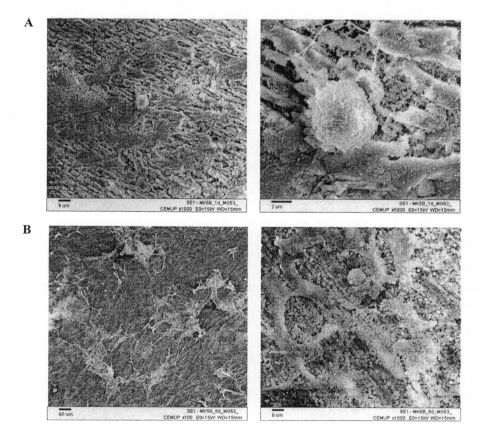

Fig. 2. SEM appearance of seeded MK5B - A, 1 day; B - 6 days.

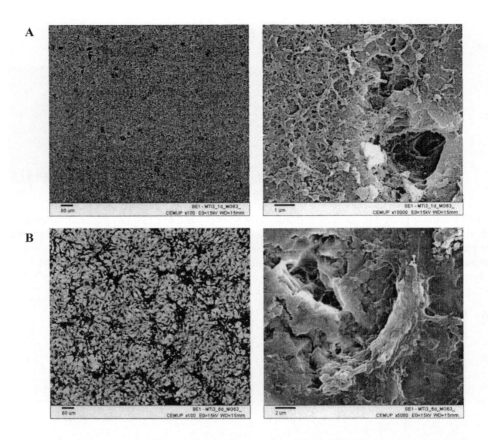

Fig. 3. SEM appearance of seeded MT13B - A, 1 day; B, 6 days.

In conclusion, both MK5B and MT13B glass ceramics allow the growth of MG63 osteoblast-like cells. The high dissolution rate of MK5B glass ceramic appears to impair *in vitro* cell adhesion.

Acknowledgements

The authors wish to acknowledge the grant Ref: PRAXIS XXI/BD/21458/99 financed by FCT (Fundação para a Ciência e Tecnologia).

References

[1] A. G. Dias, M. A. Lopes, I. Gibson, J. D. Santos, J. Non-Crystal. Solids, *in press*.

[2] A. G. Dias, M. A. Lopes, K. Tsuru, S. Hayakawa, J. D. Santos, A. Osaka, Physiscs and Chemistry of Glasses, *in press*.

[3] A. G. Dias, M. A. Lopes, J. D. Santos, Advance Materials Science, *in press*.

[4] M. Rykke, G. Rolla, T. Sonju: Scand. J. Dent. Res. Vol 96 (1988), p. 517-22.

Key Engineering Materials Vols. 254-256(2004) pp. 829-832
online at http://www.scientific.net
© 2004 Trans Tech Publications, Switzerland

In Vivo Evaluation of Hydroxyapatite and Carbonated Hydroxyapatite Fillers

C. Mangano[1], A. Scarano[1], R. Martinetti[2], L.Dolcini[2]

[1] Università degli Studi "G. d'Annunzio" , Via dei Vestini, 11 Chieti, Italy,

[2] FIN-CERAMICA, Via Ravagnana 186, Faenza I – 48018, Italy

roberta.martinetti@fin-ceramicafaenza.com

Keywords: Carbonated-hydroxyapatite, hydroxyapatite, bioactive ceramics, clinical behaviour

Abstract. Carbonated hydroxyapatite (CHA) obtained by synthetic route is more comparable to biological apatite in respect to hydroxyapatite (HA) because hard tissues minerals contains carbonate ions [1]. The aim of this study was to compare two calcium phosphate fillers (CHA and. HA granules) with similar crystallinity regarding their performance *in vivo*: newly bone formation and bioresorption behaviours were investigated. Both granules, after 4 months promote newly bone formation even if CHA granules are in contact and surrounded with mature bone; besides CHA granules surface show more evident bioresorption in respect to HA.

Introduction

Calcium-phosphate materials have been modified in their physical, chemical, and structural features, to improve the interaction of these materials with biological regenerative processes. The range of available calcium-phosphate-based materials is great, as are the variations in their composition, Ca/P ratio, structure, surface, etc. They may be presented in the form of powder, granules, or dense or porous blocks and their solubility degree vary from very low to very high, depending on the impurity, Ca/P ratio, porosity, granulometry, etc. Autologous bone is undoubtedly one of the best choices for bone reconstruction, however, limitations of volume, morbidity of the donor area, and other factors often inhibit or impede its use, particularly when there are limitations related to revascularization and infection in the implanted region, or difficulty stabilizing the material, which can lead to the decomposition of the graft. In large bone defects, the risks of using auto-homo, or xenografts increase considerably. These risks include the possibility of disease transmission (e.g., HIV, hepatitis, prions), and the possibility of immunological responses that will affect the regenerative process.

Hydroxyapatite (HA) exhibits high potential for application in the fields of medicine and dentistry because of its similarity with bone mineral phase. In recent years by introducing osteogenic cells and bone morphogenetic proteins (BMPs) into the biomaterials, many efforts were made to make biomaterials osteoinductive, while certain calcium phosphate biomaterials were reported to be osteoinductive, because they can induce bone formation in extraskeletal sites without additional osteogenic cells or bone morphogenetic proteins. Being the main inorganic constitution of hard tissues calcium phosphates have been attractive in hard tissue repair for a long time. Fundamental studies and clinical applications in the past three decades have demonstrated that calcium phosphate biomaterials (hydroxyapatite tricalcium phosphate) and calcium phosphate-based biomaterials (carbonated hydroxyapatite) are biocompatible and osteoconductive. HA is not toxic or adverse to tissues, it is able to promote chemical bonds with live bone tissue promoting bone regeneration. HA has been widely used as a bone substitute because of its excellent biocompatibility and osteoconductivity. Recent reports on ectopic bone formation of calcium phosphate biomaterials showed that osteoconduction might be an intrinsic property of calcium phosphate biomaterials [2, 3]. Ripamonti [2] reported bone formation in coral-derived HA implanted in muscles of baboons, and others animal models. Carbonated hydroxyapatite (CHA) is more similar to biological apatite than stoichiometric HA since human bone contains up to 8 wt% carbonate ions. References studies have shown that both precipitated and sintered CHA have higher solubility that HA [4]. The carbonate ion is one of the major ingredients in bone apatite and chemically modifies the properties of apatite also in term of solubility. This study reports *in vivo* evaluation data comparing this two kinds of filling materials based respectively on HA and CHA.

Materials and Methods

HA and CHA powders were synthetised by wet methods using Ca, P and, in particular for CHA, CO_3^{2-} sources [5, 6, 7], after synthesis HA and CHA have been washed, filtered and dried and used for granulation.

HA and CHA granules were investigated: X-Ray diffraction analysis (CuKα radiation, Rigaku Miniflex), IR spectroscopy (Perkin Elmer FT-IR mod.1600 spectrometer, spectral range 4000-400cm^{-1}, KBr pellets as supports), elemental analysis of Ca, P, Na, Mg, Pb, Hg, As, Cd and C (ICP) and morphology of the surface (SEM, Leica, Cambridge).

In vivo trials were performed. The specimens for both fillers and surrounding tissues were achieved from patients after 4 months, washed in saline solution and immediately fixed in 4% parapholmaldehyde and 0.1% glutaraldehyde in 0.15 M cacodylate buffer at 4 °C and pH 7.4, to be processed for histology. The specimens were processed to obtain thin ground sections according to the cutting-grinding system. Briefly the specimens were dehydrated in an ascending series of alcohols and embedded in a glycolmethacrylate resin (Technovit 7200 VLC, Kulzer, Germany). After polymerization the specimens were sectioned with a diamond saw at a thickness of about 200 μm and ground down to about 30 μm. About 3 sections were cut for each bone sample in a way parallel to the major axis. After polishing, the slides were stained with basic fuchsin-methylene blue and with basic fuchsin- toluidine blue and were observed with a Leitz Laborlux microscope (Leitz, Wetzlar, Germany) in normal light.

Results

X-ray diffraction analysis of powders revealed no foreign phases besides CHA and HA for respectively sample; traces showed also that powders are characterized by low crystallinity degree (30 – 40 vol %). The carbonate content of CHA granules was determined from the c:a ratio (lattice parameters): the c:a tends to 0,75. IR spectra show that CHA were B-type and that the carbonate bands were produced between 1410 and 1460 cm^{-1} and at 870 cm^{-1} in agreement with literature data [1, 2]

Table 1. Comparative data in term of elemental analysis of the two powders

Samples	Ca / P	CO$_3$ [wt %]	Na [%]	Mg [%]	Pb [ppm]	Hg [ppm]	As [ppm]	Cd [ppm]
HA	1,66	-	0,000233	0,00185	< 1	< 1	< 1	< 1
CHA	1,90	14	2,5	-	< 1	< 1	< 1	< 1

The specimens were achieved and investigated after 4 months: new bone was detected close to HA granules (Fig.1 and 3); in the newly bone wide osteocitarie cavities have been observed and the bioresorption behaviour is not evident at this stage. On the other hand, the CHA granules after 4 months (Fig. 2 and 4) have been surrounded with mature bone that result in contact with the granules surface that show an evident bioresorption.

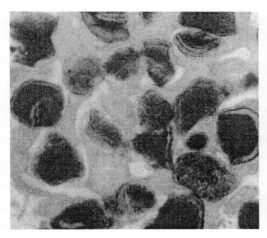

Fig. 1. HA granules

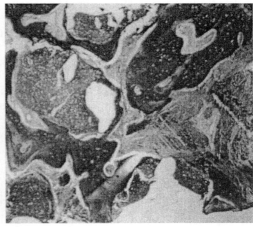

Fig. 2. CHA granules

Fig. 3. HA granules

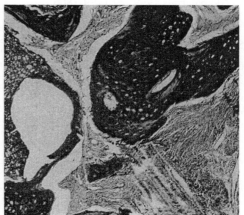

Fig. 4. CHA granules

Quantitative analysis between histological sections showed for HA that bone was present for 18%, connective tissue 36%, HA particles 46%; while for CHA bone was 30%, connective tissue 40% and CHA particles 30%

Discussion and Conclusions

The advantages of using synthetic materials include: their unlimited supply, the purity of the material, the absence of biological contamination and most importantly, our knowledge of their potential and limitations *in vivo,* which makes their behaviour more predictable and the process safer. In addition to these considerations, the possibility of using such a bioceramics as a vehicle for drug delivery may contribute to the biological regenerative process. The purity and quality control under which the material is synthesized eliminates the risk of contamination from infectious-contagious diseases (e.g, HIV, hepatitis, viruses), immunological reactions, and unknown or little known diseases such as those caused by prions.

Two important feature of calcium phosphate biomaterials are their biocompatibility and the solubility degree at a rate similar to bone growth. Bioresorption appear to be mainly regulated by solution and cell-mediated degradation.

Both fillers, HA and CHA, have shown good biocompatibility with bone and a direct contact with newly formed bone was observed.

However bone defects, filled with CHA have shown a faster healing process than with HA granules. The CHA after 4 months are surrounded by more mature formed bone in respect to HA granules and showed to be also more soluble in respect to HA samples.

This findings suggest that CHA would be more suitable as a bioresorbable bone substitute in respect to HA.

References

[1] R.Z. LeGeros, M. Tung, Carves Res., 17. 19-29 (1983)

[2] Ripamonti U. Osteoinduction in porous hydroxyapatite implanted in deterotopic sites of different animal models: Biomaterials 1996;17:31-35

[3] Yuan H.: A preliminary study on osteoinduction of two kinds of calcium phosphate ceramics. Biomaterials 1999; 20:1799-1806.

[4] Jake E. Barralet : 2000 Society for Biomat. Sixth World Biomat., p. 1385

[5] F.C.M. Driessen: *Bioceramics of Calcium Phosphates*, K. de Groot (ed.), 1983, pp 1- 32

[6] Dj.M. Maric, P.F. Meier and S.K. Estreicher: Mater. Sci. Forum, Vol. 83-87 (1992), p. 119

[7] M.A. Green: *Advanced Bioceramic Processing* ,Trans Tech Publications, Switzerland 1987.

Key Engineering Materials Vols. 254-256(2004) pp. 833-836
online at http://www.scientific.net
© 2004 Trans Tech Publications, Switzerland

Processing of Ca-P Ceramics, Surface Characteristics and Biological Performance

S. Cazalbou[1], C. Bastié[1], G. Chatainier[1] N. Theilgaard[2], N. Svendsen[2], R. Martinetti[3], L. Dolcini[3], J. Hamblin[4], G. Stewart[4], L. Di Silvio[5], N. Gurav[5], R. Quarto[6], S. Overgaard[7], B. Zippor[7], A. Lamure[1], C. Combes[1], C. Rey[1].

[1]CIRIMAT, ENSIACET, 118 route de Narbonne, 31077 Toulouse Cedex, France. crey@cict.fr
[2] Danish Technological Institute, Gregersensvej, 2630 Taastrup, Denmark.
[3] FIN-CERAMICA FAENZA srl, Via Ravegnana 186, Faenza I – 48018, Italy.
[4] Hi-Por Ceramics ltd, Stopes Road Stannington Sheffield S6 6BW, United Kingdom.
[5] IIRC in Biomedical Materials,Institute of Orthopaedics (UCL), Brockley Hill, Stanmore, Middlesex HA7 4LP, United Kingdom.
[6] Dip. Oncologia, Biologia, Genetica - Universita' di Genova, Italy.
[7] Department of Orthopaedics, Aarhus University Hospital, 8000 Aarhus, Denmark.

Keywords: XPS, surface energy, apatite, tri-calcium phosphate.

Abstract. Surface pollution by Mg and carbon has been identified by X-Ray Photoelectron Spectroscopy (XPS) analysis on monophasic (100% stoichiometric hydroxyapatite) or biphasic (70% hydroxyapatite, 30% β-tricalcium phosphate) ceramics obtained by sintering of Ca-P powder of standard purity. Magnesium was the main impurity, although it is present only as traces in synthesised Ca-P powders, it concentrates on the surface of the ceramic during the sintering process due to its poor solubility in HA. Carbon pollution is common on all materials and generally assigned to the adsorption of atmospheric volatile organic compounds, although other types of pollution may occur. These surface species alter the surface energy of the ceramics which has been shown to be related to cell adhesion. The change in surface composition and physical properties can interfere with the sintering process as well as with the biological behaviour of the ceramics and a special effort should be made to control these events at all levels from processing to handling, storage and packaging.

Introduction

The purity of HA (hydroxyapatite) and HA-TCP (hydroxyapatite-β-tricalcium phosphate) ceramics are crucial for their biological performances. It is well known that apatites can take up many trace elements exhibiting a biological effect as substitutes of the main lattice ions, but also as surface contaminants due to the high specific surface area of the powders generally used for sintering. Although the purity of Ca-P ceramics has to comply with existing standards, only the average level of a few toxic trace elements (namely, As, Cd, Hg, and Pb) must be under specific limits; other trace elements which may have a biological effect such as Mg, one of the major impurities of calcium salts, have rarely been considered. The occurrence of a heterogeneous distribution of toxic as well as beneficial elements between the surface and the bulk of the ceramics is generally not investigated. However, ceramic processing often alters the location of trace elements inside crystals, at grain boundaries or on the surface of ceramics, thus affecting their biological behaviour. The aim of this preliminary work was to analyse the modifications of surface characteristics of Ca-P ceramics related solely to processing. From the same batch of powder several sintered samples (dense HA or porous HA and HA-TCP) were produced by three different organisations and were characterised by XPS and contact angle measurements. The biological performance of porous samples was assessed by human osteoblast cell culture and implantation in mice.

Materials and Methods

Powder characteristics. The powder samples were prepared by standard methods involving precipitation in aqueous media from commercial calcium and phosphate salts. The main impurity was Mg (0.24%) followed by Sr(0.016%). These contaminants essentially originated from the Ca salts. The powders complied with existing ASTM and CEN standards relative to the amount of potential toxic elements.

Sintered samples. All three producers (referenced as A, B and C) used the same sintering program involving preheating for binder removal at 600 °C, sintering at 1200 °C and free cooling. Porosity was close to 5% for dense samples and 80% for porous samples (open, interconnected pores). One dense sample and three porous samples from each producer were analysed.

XPS determinations were performed on an ESCALAB 2000 (VG Scientific, East Grinstead, UK), and using an Al K_α source. The atomic ratios of the elements analysed were determined using the Scofield sensitivity factors. The XPS and surface energy determination were done on the same dense pellets (with a 10% relative accuracy). The XPS data reported for porous ceramics correspond to the average measurements in three different samples. The porous samples were split with a scalpel blade and the analyses performed on the inside part to avoid unnecessarily taking into account pollution by contact.

Contact Angle measurements were performed on dense pellets on a Digidrop (GBX, Lyon, France) with water, ethylene glycol and diiodomethane. The data were interpreted according to the theory of Owens-Wendt.

Biological evaluation.

-   **Cell cultures.** Porous samples were seeded in triplicate with human osteoblast cells. Proliferation was estimated from alamar blue assay DNA at 1,2,4 and 7 days.
-   **Implantations in mice.** The porous samples (3x3x3mm cubes) were implanted subcutaneously with or without expanded human bone marrow stromal cells in immunodeficient mice. The formation of bone was assessed by histology on decalcified sections.

Results

Dense ceramics (HA). The results of XPS determination on dense pellets are reported in table 1:

Table 1: XPS analysis of dense ceramics (atomic ratios)

Sample	P/Ca	O/Ca	Mg/Ca	C/Ca
Powder	0.64	2.44	0	2.14
Ceramic A	0.67	2.51	0.20	0.82
Ceramic B	0.58	2.48	0.10	1.46
Ceramic C	0.62	2.32	0.09	0.10
Pure HA (theoretical values)	0.60	2.60	0	0

The data reveal a high proportion of Mg ions on the surface of the sintered ceramic not detected on the powder. The carbon peak corresponds to aliphatic carbon contaminants always present on any material. The amount of C varies widely depending on the samples. The oxygen values always appeared slightly lower than the theoretical ones.

These differences in surface composition lead to different surface energies (Table 2). The dispersive component was almost constant for all samples but the polar component was found to vary strongly. It was minimum for the sample with the highest carbon contamination.

Grazing angle X-ray diffraction did not reveal any foreign phase on the surface of the ceramics.

Table 2: Surface energy determined by contact angle measurements [mJ.m^{-2}]

Sample	Total Energy	Polar component	Dispersive component
Ceramic A	30.3	3.3	26.9
Ceramic B	27.2	0.3	27.1
Ceramic C	40.4	9.6	30.8

Porous ceramics. The results of XPS determinations are reported in table 3.

Table 3: XPS analysis of porous samples (atomic ratios)

Sample	P/Ca	O/Ca	Mg/Ca	C/Ca
HA ceramic A	0.73 ± 0.07	2.60 ± 0.09	0.40 ± 0.05	0.84 ± 0.04
HA ceramic B	0.645 ± 0.04	2.48 ± 0.13	0.37 ± 0.03	0,83 ± 0.03
HA ceramic C	0.645 ± 0.007	2.369 ± 0.008	0.25 ± 0.02	0.67 ± 0.04
HA-TCP ceramic A	0.66 ± 0.07	2.49 ± 0.10	0.46 ± 0.11	0,74 ± 0.14
HA-TCP ceramic B	0.647 ± 0.03	2.611 ± 0.06	0.373 ± 0.04	0.547 ± 0.02
HA-TCP ceramic C	0.653 ± 0.003	2.58 ± 0.10	0.161 ± 0.03	0.481 ± 0.14

The proportion of Mg appears even more important than in the dense pellets and depends markedly on the makers. The surface concentration seems similar for HA and HA-TCP samples except for the producer C where it is lower in HA-TCP mixtures. The carbon pollution seems generally higher for HA than HA-TCP ceramics.

Osteoblast cell cultures. Variable but consistent data were obtained depending on the sample batches. Some diferences were observed in cell response, but only at the early stages of cell culture (4 days). Porous samples analogous to those analysed by XPS did not exhibit any significant differences in cell attachment or behaviour.

Implantations in mice. The porous implants did not induce any inflammatory reaction. Bone formation was noticed in the pores of all samples implanted with marrow cells after only 2 weeks implantation. No difference appeared depending on the samples origin.

Discussion

The surface pollution observed in these ceramics is rather common, its origin however seems different, as does its potential biological effect.

Effect of surface pollutants. Surface pollutants are responsible for the variations in surface energy and also for surface heterogeneities. It has been shown that surface energy, especially the polar component, is crucial for the adhesion of osteoblast and osteoclast cells [1-2]. The presence of aliphatic organic substances especially may delay the initial adhesion of cells. Magnesium on apatite surfaces, on the contrary, seems to favour osteoblast primary adhesion [3] although this effect is controversial [4]. In our case it shall be observed that porous ceramics with more magnesium are also those which have more aliphatic carbon, however, the differences observed between porous samples were small enough and no significant differences were detected in cell culture experiments. Similarly implantations in mice did not reveal significant differences.

Origin of surface pollution. Mg is the main ionic surface pollutant of the ceramics and this ion already exists in the powder as the major impurity. Its concentration on the surface of the ceramic could be related to its very low solubility in the calcium phosphate hydroxyapatite [5]. As sintering increases the size of apatite crystals and reduces their specific surface area, the concentration at the surface of ceramics and probably at grain boundaries can be very high. Mg ions, however, are much more soluble in β-TCP, therefore, they could diffuse into this phase, in biphasic ceramics, and their surface concentration should be lower. This is not generally observed except for biphasic ceramics C, it may be concluded that in most cases the diffusion of Mg into β-TCP does not occur during the sintering time. Mg is known to inhibit apatite crystal growth. It may affect sintering and generally the grain size of sintered Mg-substituted HA is smaller than those in pure HA [6]. The role of these impurities in the mechanical properties of the ceramics and their degradation *in vivo*, have yet to be studied. The large and consistent differences observed between manufacturers is rather puzzling as the same powders and the same sintering programs were used. An external source of pollution could be suspected but the analysis of binders did not show any significant Mg content. Other sources of Mg pollution such as talc from gloves or moulding substances have also been excluded. As the global content of Mg in all ceramics seems constant to originate in the powder, its variable concentration on the surface is probably related to faint alterations in the sintering process depending on the makers even thought the samples appear homogeneous and constant. Another observation is the higher concentration of surface Mg in porous than in dense samples. This phenomenon could be related to differences in the sintering process, but it can also be explained by the architecture of the samples.

Carbon is also a frequent pollutant of surfaces easily detected by XPS. Obviously just after heating in air the ceramic should not have any carbon on its surface and the main source of pollution comes from the atmosphere and organic materials in contact with the ceramic after cooling. Volatile organic components may be found in any place and have various sources, the gathering of these substances on the surface of ceramics is related to their surface properties and seems to be favoured on HA compared to biphasic surfaces. In addition direct contact pollution may arise from handling with gloves or bare hands (forceps should be recommended); and also from abrasion of plastic packages by the ceramic (most ceramics are packaged in plastic bags or containers). The surface contact does not however alter the interior of the porous ceramics where carbon pollution is essentially due to volatile organic components. This characteristic appears to be variable but also rather reproducible for each producer and the distribution of pollutants then seems homogeneous. The surface pollution by contact, inversely, is heterogeneous. All these surface impurities should be controlled even if they have little impact on the biological performance.

Acknowledgments

This work was carried out within the frame of a European Community Project, PORELEASE, (GROWTH Contract No G5RD-CT-1999-00044, Project No GRD1-1999-10590).

References

[1] S.A. Redey, M. Nardin,D. Bernache-Assolant, C.Rey, P. Delannoy, L. Sedel, P.J. Marie: J. Biomed. Mat. Res. Vol 50 (2000), p. 353

[2] S.A. Redey, S. Razzouk, S; C. Rey, D. Bernache-Assollant, G. Leroy, M. Nardin, G. Cournot: J. Biomed. Mat. Res. Vol 45 (1999), p. 140.

[3] Y. Yamasaki, Y. Yoshida, M. Okazaki, A. Shimazu, A. Uchida, T. Kubo, H. Akagawa, Y. Hamada, J. Takahashi, N. Matsuura: J. Biomed. Mat. Res.Vol 62 (2002), p. 99

[4] C.M. Serre, M. Papillard, P. Chavassieux, J.C. Vogel, G. Boivin: J. Biomed. Mat. Res. Vol 50 (1998), p. 626

[5] S. Ben Abdelkader, I. Khattech, C. Rey and M. Jemal: Thermochim. Acta. Vol. 376 (2001), p. 25.

[6] C. Ergun, Th. Webster, R. Bizios, R.H. Doremus: J. Biomed. Mat. Res. Vol 59 (2002), p. 305

Key Engineering Materials Vols. 254-256(2004) pp. 837-840
online at http://www.scientific.net
© 2004 Trans Tech Publications, Switzerland

Biocompatibility Evaluation of Dentine, Enamel and Bone Derived Hydroxyapatite

P.Valerio[1], F. N. Oktar[2], L.S. Ozyegin[3],G. Goller[4], A. M.Goes[5] , M. F. Leite[1]

[1] Federal University of Minas Gerais, Department of Phyisiology and Biophysics, Brazil,
patricia.valerio@terra.com.br

[2] Marmara University, Department of Industrial Engineering, Istanbul, Turkey

[3] Marmara University, Department of Dental Technology, Istanbul, Turkey

[4] Istanbul Technical University, Department of Metallurgical & Material Engineering, Turkey

[5] Federal University of Minas Gerais, Department of Biochemistry and Immunology, Brazil

Keywords: Natural hydroxyapatite, osteoblast, collagen, alkaline phosphatase, biocompatibility

Abstract. The biocompatibily of dentine, enamel and bone derived hydroxyapatite was studied using primary culture of osteoblasts. The cell viability, collagen and alkaline phosphatase production and cell morphology were analyzed in the presence of biomaterials powder and compared to control and to osteoblasts in the presence of synthetic hydroxyapatite. Osteoblasts in the presence of bone derived hydroxyapatite showed increased proliferation and collagen production. Dentine and enamel derived hydroxyapatite did not alter cell behavior when compared to control or to synthetic hydroxyapatite.

Introduction

Hydroxyapatite (HA) is the most investigated calcium phosphate among biomaterials certainly because this mineral is the main inorganic phase of bone. Porous hydroxyapatite has already been used as a bone substitute due to its similarity with this mineral part of the bone [1]. However, natural HA crystals are relatively small compared to synthetic one, and it is already known that the crystal structure can alter the dissolution rate and inffluence the cell behavior in the presence of the biomaterial [2]. Natural HA can be produced from natural sources like corals, sea-algae and human / animal mineralized tissues [3], and, because of their different stoichiometry, the solubility properties of natural HA are different from those of commercial HA [4]. The aim of this study was to investigate if these different characteristics could alter some biocompatibility parameters. Thus, cell viability, alkaline phosphatase production and collagen secretion were evaluated. Natural HA obtained from human dentine, human enamel and bovine bone were compared to a commercial one.

Material

Natural HA derived from freshly obtained human teeth (dentine and enamel) and bovine bone, were deproteinized, calcinated, grinded and sieved to 40-50 μm particle size [3]. Cell culture medium composed of RPMI (Sigma) Dulbecco´s phosphate buffered saline, trypsin-EDTA, fetal bovine serum (Gibco), crude bacterial collagenase (Boehringer), MTT, NBT-BCIP (Sigma), Sircol Kit.

Methods

Osteoblasts were prepared from calvarie of 1-5 days old Wistar rats by using a sucessive enzymatic digestion method. Briefly, after cut into small pieces, the calvarie bone was digested with trypsin 1% and four times with collagenase 2%. The supernatant of the three last washes were centrifuged at 1400 g for 5 min, the pellet ressuspended in culture medium and plated. After confluence the cells were replicated and used in passage 2.

Medium containing each HA powder, was put in contact with osteoblasts that have been plated at 1 X 10^5 cell density. The cell / particle ratio was 10/1 and was controled visualy under optical microscopy. The experiments were performed 72 hours after incubation. Morphological changes were investigated under light microdcopy. Viability was assayed by MTT method that is based on the capacity of viable cell to metabolize tetrazolium to formazan crystals, a purple dye that can be solubilized and measured by optical density. Alkaline phosphatase production was analysed by NBT-BCIP assay. Nitrobluetetrazolium is activated by the alkaline phosphatase secreted by osteoblasts and after some reactions, spread a blue dye that may be solubilized and measured by optical density. Collagen secretion was measured using SIRCOL method. This assay is based on the capacity of Syrius red dye to bind to the end of collagen molecule and precipitate it. After solubilization, collagen can be quantificated using linear regression from samples of known concentrations of collagen. Data were analysed statistically by variance test ANOVA, using Bonferroni's post-test.

Results
Under light microscopy, there were no osteoblasts morphological changes observed each 24 hours, during 3 days (Fig.1). Of all the bioceramics tested, crystal rejection by the cells was not observed. The viability and proliferation of osteoblasts had a significant increasing in the presence of bone HA, when compared to control and commercial HA (Fig. 2). Alkaline phosphase production had no statistical significant alteration in the presence of all natural HA as well as commercial HA (Fig. 3) Collagen secretion by osteoblasts was higher in the presence of bone derived HA, when compared to the other natural HA, commercial HA and control (Fig. 4).

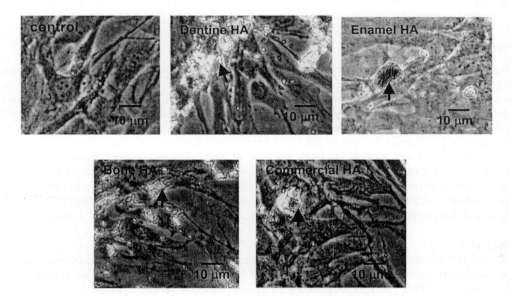

Fig. 1. Osteoblast morphology. Under ligth microscopy, osteoblasts did not show any morphological changes in the presence of all tested bioceramics, when compared to control. Panels show images of 72 h after incubation. Granules of all bioceramics were not rejected by the cells (arrows). Photomicrograph (X 400).

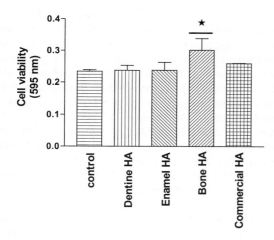

Fig. 2. Osteoblast viability. 1×10^5 osteoblasts were plated in the presence of granules from dentine, enamel and bone derived HA and commercial HA. After 72 h of incubation, cellular viability was evaluated by MTT assay. Osteoblasts showed increased viability in the presence of bone derived HA, when compared to HA derived from the others sources and to control. Results represent Mean ± SD of triplicates from 3 separate experiments ($P<0.05$)

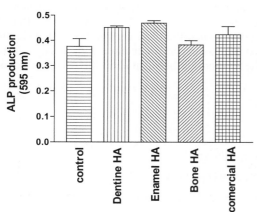

Fig. 3. Alkaline phosphatase production. 1×10^5 osteoblasts were plated in the presence of granules from dentine, enamel and bone derived HA and commercial HA. After 72 h of incubation, alkaline phosphatase production was evaluated by NBT-BCIP assay. No statistical significant difference was found when comparing all the HA to control. Results represent Mean ± SD of triplicates from 3 separate experiments ($P<0.05$)

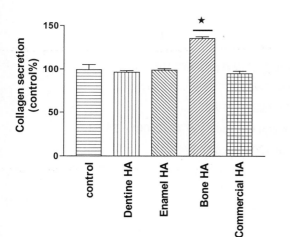

Fig. 4. Collagen production. 1×10^5 osteoblasts were plated in the presence of granules from dentine, enamel and bone derived HA and commercial HA. After 72 h. of incubation collagen secretion was measured by SIRCOL assay. Osteoblasts in the presence of bone derived HA showed higher collagen secretion, when compared to the HA from the others sources and to control (*). Results represent Mean ± SD of triplicates from 3 separate experiments ($P<0.05$)

Discussion

An increased collagen production in the presence of bone HA was observed. It was already demonstrated that secreted collagen could be drawn into the nanoscopic crystallites of HA [5]. So, this finding could be justified by the fibrillar organization of crystallites, with a periodic spacing of them, showed by the apatite crystallites in mature mammalian bone, completely different from synthetic HA [6]. The level of crystallinity in HA is thought to be responsible for its degradation in the physiologic milieu [7] and lower crystallinity might be favorable to cell proliferation [8]. This is in accord to the observation of increased proliferation in the presence of bone derived HA. However, dentine and enamel HA did not show the same characteristic. It can be explained in comparison with previous studies of crystallinity differences between tooth derived HA and bone derived HA, where the authors partially attributed this difference to the incorporation of Mg and minor elements, during processing [9]. So, it can be speculated that these chemical and structural differences could also be related to the absence of increased collagen secretion in the presence of dentine and enamel HA.

It is known that processing parameters such as Ca/P ratio of start material, calcination temperature and other elements incorporation influence the characteristic of calcium phosphates [1,10]. It is also known that calcium phophates ceramics can be transformed trough a sequence of reactions, which include disolution, precipitation and ion exchange [2,11]. Considering this, the results obtained from this study supports the hypothesis that the bone derived HA tested can be a suitable material to biological application, when high collagen production is necessary. Further investigation is important to better understand the relationship between natural HA crystallites and osteoblast behavior

References.

[1] J F Oliveira et al. Artif Organs Vol 27 (2003) p 406-411

[2] P Ducheyne et al. J Biomed Mater Res. Vol 27 (1993) p 25-34

[3] F Oktar et al. Art Cells Blood Immob Biothec. Vol 49 (1991) p 367-379

[4] E Moreno. Calcif Tissue Int. Vol 49 (1991) p 6-13

[5] M Olszta et al. Calcif Tissue Int. (2003) in press

[6] V Rosen et al. Biomaterials. Vol 23 (2002) p 921-928

[7] Y L Chang et al. J Oral Maxillofac Surg. Vol 57 (1999) p 1096-1108

[8] L Chou et al. Biomaterials. Vol 10 (1999) p 977-985

[9] W Zhao et al. Zhonghua Kou Qiang. Vol 37 (2002) p 219-221

[10] G Penel et al. Bull Group Int Rech Sci Stomatol Odontol. Vol 42 (2000) p 55-63

[11] R Tang et al. J Colloid Interface Sci. Vol 260 (2003) p 379-384

Key Engineering Materials Vols. 254-256(2004) pp. 841-844
online at http://www.scientific.net
© 2004 Trans Tech Publications, Switzerland

Effects of Bioactive Glass 60S and Biphasic Calcium Phosphate on Human Peripheral Blood Mononuclear Cells

C.C.P. Silva[1], A. Bozzi[2], M.M. Pereira[3], A.M. Goes[2], M.F. Leite[1]

[1] Department of Physiology and Biophysics, UFMG, Belo Horizonte, Brazil, pcarolc@hotmail.com

[2] Department of Immunology, UFMG, Belo Horizonte, Brazil

[3] Department of Metallurgical Engineering, UFMG, Belo Horizonte, Brazil

Keywords: biomaterials, flow cytometry, peripheral blood mononuclear cells, cytokines

Abstract. During the last decade, bioactive materials, including synthetic hydroxiapatite and bioactive glass ceramics, have been applied to repair bone defects showing promising results [1,2]. The purpose of this study is to characterize the population of peripheral blood mononuclear cells (PBMC), to determine its proliferation, cellular viability, and the profile of cytokines secreted by these cells, when in the presence of bioactive glass with 60% of silica (BG60S) or biphasic calcium phosphate (BCP). Our data indicate no induction of an inflammatory process when PBMC were cultured with these bioceramics, since the cell population and the profile of cytokines secreted remained similar or smaller than control.

Introduction

Implant materials are extensively used for replacing diseased or injured hard tissues. A key issue in biomaterial research is the biocompatibility of the materials used for implantation. The implant material surface is in intimate contact with the living tissue and its biocompatibility is determined in large part by surface properties of the materials, with have a direct effect on cellular responses [3]. Materials like bioceramics can induce hypersensitivity, chronic inflammation and immunostimulation. The cell-mediate immune response is critical for the induction of these effects and evaluation of cytokines is considered one of the most important indicators to evaluate it. Moreover, the soluble mediators released from immune cells regulate bone metabolism and for this reason can bring about abnormal resorption around the implant [4]. Monocytes and macrophages are largely implicated in phagocytosis and degradation of bioceramics [5]. Moreover, many studies evaluating tissue at the interface between implant and bone have demonstrated an association between macrophages and areas of bone resorption [6]. Therefore, the purpose of this study is to characterize the population of peripheral blood mononuclear cells (PBMC), to determine its proliferation, cellular viability, and the profile of cytokines secreted by these cells, when in the presence of bioactive glass with 60% of silica (BG60S) or biphasic calcium phosphate (BCP).

Methodology

Cell preparation – The PBMC from healthy volunteers were isolated from heparinized blood by Ficoll-diatrizoate density gradient centrifugation [7]. The cells were cultured in RPMI 1640, supplemented with 1.6 % 1-glutamine, 100 units/ml of penicilin G sodium, 100 µg/ml of streptomycin sulfate, 0.25 µg/ml of amphotericin B, 0.06 mg/ml of gentamicin and 10% of fetal bovine serum, and used throughout the experiments.

Cell proliferation and viability assay – PBMC were cultured in 96-well flat-bottomed plates at the density of 3×10^5 cells/well in a final volume of 200 µl. The cells were cultured with either, 50µg/ml particles or ionic products from the dissolution of bioceramics BCP or BG60S [8]. The cultures were also performed with medium alone, as negative control, or PHA (1µg/25µl), as positive control. The cells were maintained in culture at 37°C in an atmosphere of 5% CO2. For the last 18h of incubation, cells were pulsed with 0.5µCi/well of [methyl-3H] thymidine. On day 5 after culture initiation, cells were harvested onto glass fiber paper and the incorporated radioativity was measured by a liquid scintillation β-counter. The results are expressed as counts per minute (cpm). The MTT assay was used as a measure of cell viability. MTT is a pale yellow substrate (3-[4,5-dimethylthiazol-2-yl]-2,5-diphenyltetrasodium bromide), which is reduced, by living cells, to a dark blue formazan reaction product [9]. After 5 days of culture, 20μl of MTT (5mg/ml) was added to each well and incubated at 37°C for 2h. At the end of the assay, the blue formazan reaction product was dissolved by addition 70μl of SDS 10%/HCL and incubated at 37°C for 18h. The absorbance was measured at 595.

Cytokine assay – For in vitro cytokine measurements, PBMC were plated as described above. Supernadant fluids were harvested at 120h and cytokine levels for, IL-2, IL-4, IL-5 and IFN-γ were assayed by ELISA using commercially detection kits following the instructions of the manufacturer. Flow cytometry – 3×10^5 cells were cultured with medium alone, PHA or ionics products from dissolution of the bioceramics. One fraction of the cells were marked after isolation with PE-conjugated antibodies (mouse anti-human CD4, mouse anti-human CD8, mouse anti-human CD28, mouse anti-human CD86, mouse anti-human CTLA-4, mouse anti-human HLA-DR and mouse anti-human IgG1) and FITC-conjugated antibodies (mouse anti-human CD3, mouse anti-human CD4, mouse anti-human CD8, mouse anti-human IgG1, mouse anti-human CD33 and mouse anti-human CD19).The other PBMC portion were marked after 5 days of culture with the same antibodies. The light scattering properties and the fluorescence of the cells stained with FITC and PE were measured on a FACScan flow cytometer. The excitation source was an argon-ion laser emitting a 488nm beam at 15 mW. Analyses were performed on 20000 cells. The FSC and SSC were measured on linear scales of 1024 channels, while green fluorescence (FITC) and red fluorescence (PE) emission was detected on a logarithmic scale of four dacades of log. Data were colleted with CELLQuest Software and analysed with WinMDI. Data were analyzed statistically by variance test ANOVA, using Bonferroni`s post-test.

Results

Our results showed no significative difference on cell sub-populations (Fig. 1) and activation (Fig. 2) between PBMC cultured with either medium, or ionic products from the dissolution of BG60S or BCP.

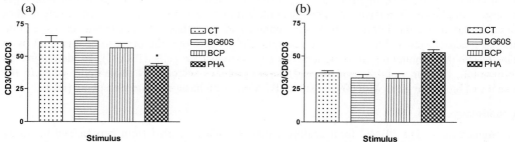

Fig. 1 (a) Proportion of lymphocyte T populations (CD3/CD4); (b) proportion of lymphocyte T population (CD3/CD8). The figure represent mean ± SD of 4 separate experiments; * p < 0.05.

The cells were responsive, since PBMC cultured with PHA showed change on proportion of lymphocyte T populations (CD3/CD4 and CD3/CD8) (Fig. 1), and increased activation of lymphocyte T helper (CD4/HLADR) (Fig. 2), compared with control and the bioceramics.

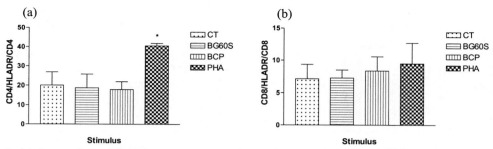

Fig. 2 (a) Proportion of lymphocyte T populations activated (CD4/HLADR); (b) proportion of lymphocyte T populations activated (CD8/HLADR). The figure represent mean ± SD of 4 separate experiments; * p < 0.05.

We also found that PBMC cultured with BG60S or BCP proliferated 55% and 20% less, compared to control (Fig. 3). However, the ionics products from the dissolution of BG60S caused an increase of 24% on cellular viability compared to cells cultured with control and BCP (Fig. 4).

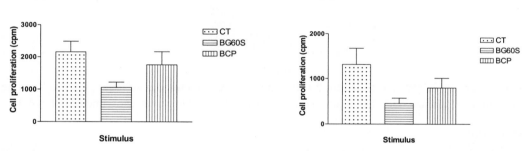

Fig. 3 (a) Cell proliferation when PBMC were cultured with particles of BG60S or BCP; (b) cell proliferation when PBMC were cultured with ionics products from the dissolution of BG60S or BCP. The figure represent mean ± SD of triplicates from 5 separate experiments; p < 0.05.

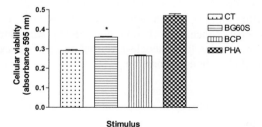

Fig. 4 Cellular viability when PBMC were cultured with ionics products from the dissolution of BG60S or BCP. The figure represent mean ± SD of triplicates from 5 separate experiments; * p < 0.05.

We then investigated the profile of cytokines secreted by PBMC. We found that IL-4 and IL-5 release did not change significantly after the exposure of PBMCs to the bioceramics (Fig. 5). However, IFN-γ was reduced significantly (80%) compared to control when mononuclear cells were cultured with BG60S ionic products (Fig. 6).

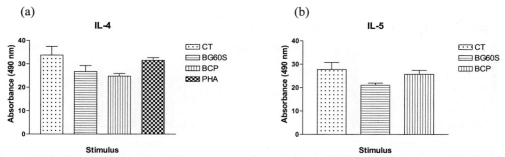

Fig. 5 (a) Profile of IL-4 secreted by PBMC when this cells cultured with particles of BG60S or BCP; (b) profile of IL-5 secreted by PBMC when this cells cultured with particles of BG60S or BCP. The figure represent mean ± SD of 4 separate experiments; $p < 0.05$.

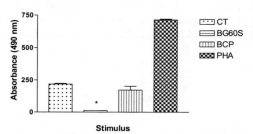

Fig. 6 – Profile of IFN-γ secreted by PBMC when this cells cultured with ionic products from the dissolution of biomaterials. The figure represent mean ± SD of 4 separate experiments; * $p < 0.05$.

Discussion

Blood cells are among the first cell population that interacts with an implant. Depending on the cell type recruited around the implant area and the cytokines released, the inflammatory response can occur in different intensity. Our data suggested no induction of an inflammatory process when PBMC were cultured with BG60S and BCP powders, or with the ionic products from their dissolution, since the cell population and the profile of cytokines secreted remained similar or smaller than control. Contrary to our findings, other authors had shown an increase on TNF-α in macrophages when exposed to 45S5 Bioglass [10]. Since it is known that IFN-γ stimulates superoxide generation by white cells, that are necessary for osteoclast formation [5], we can speculate that an implant with BG60S would promote less bone absorption.

References

[1] L. Hench, J. K. West: Life Chem. Rep. Vol.13 (1996), p. 187-241
[2] T. Kokubo: Biomaterials Vol. 12 (1991), p.155-163
[3] H. Zeng et al.: Biomaterials Vol. 20 (1999) p. 377-384
[4] D. Granchi et al.: Biomaterials Vol. 21 (2000) p. 1789-1795
[5] M. Benahmed et al.: Biomaterials vol 17 (1996) p 2173-2178
[6] I. Catelas et al.: Biomaterials Vol. 20 (1999) p. 625-630
[7] C. A. Almeida, A. M. Góes: Parasitology International vol 48 (2000) p 255-264
[8] L. Hench. J.: Am. Ceram. Soc. Vol 81 (1998) p. 1704-1728
[9] M. P. Ferraz at al.: Biomaterials Vol. 21 (2000) p. 813-820
[10] M. Bosetti et al.: J. Biomed. Mater. Res. Vol.60 (2002) p. 79-85

Key Engineering Materials Vols. 254-256(2004) pp. 845-848
online at http://www.scientific.net
© *2004 Trans Tech Publications, Switzerland*

Biological and Physical-Chemical Characterization of Phase Pure HA and SI-Substituted Hydroxyapatite by Different Microscopy Techniques

C.M. Botelho[1,2], R. A. Brooks[3], S.M. Best[4], M. A. Lopes[1,2], J.D. Santos[1,2], N. Rushton[3] and W. Bonfield[4]

[1]INEB- Instituto de Engenharia Biomédica, Laboratório de Biomateriais, Rua do Campo Alegre, 823, 4150-180 Porto, Portugal
[2]FEUP- Faculdade de Engenharia da Universidade do Porto, DEMM, Rua Dr. Roberto Frias, 4200-465 Porto, Portugal, jdsantos@fe.up.pt
[3]Orthopaedic Research Unit, Box 180, Addenbrooks Hospital, Hills Road Cambridge, CB2 2QQ.
[4]Department of Materials Science and Metallurgy, University of Cambridge, CB2 3 QZ, UK.

Keywords: Silicon-substituted hydroxyapatite, human osteoblast cells, confocal microscopy, fluorescence microscopy, atomic force microscopy and environmental electron scanning microscopy.

Abstract. Two different microscopy techniques were used to investigate the response of human osteoblasts to hydroxyapatite (HA) and silicon substituted hydroxyapatite (Si-HA), namely, fluorescence and confocal microscopy The changes in the surface of HA and Si-HA, after incubation for different periods of time in simulated body fluid, were assed using atomic force microscopy and environmental electron scanning microscopy. Cell proliferation was higher on Si-HA compared to HA. In addition more focal points of adhesion were seen in Si-HA than on HA. Using atomic force microscopy and environmental scanning electron microscopy it was possible to observe changes in the surface of both materials, namely dissolution features and the formation of an apatite layer. These findings support the results of a previous study, which showed that Si-HA had a higher dissolution with the preferential release of silicon into the medium and this fact may account for the changes observed in the cell behaviour.

Introduction

In the 1970´s several groups demonstrated that bone mineralisation requires a minimum concentration of soluble silicon [1-4]. Hench reported that the deterioration in the proliferation and function of osteoblasts due to osteopenia and osteoporosis is related to the loss of biologically available silicon [5] and Keeting reported that, bone cells in culture proliferate more rapidly in the presence of soluble silicon [6]. A recent study by Reffit demonstrated that physiological concentrations of soluble silicon stimulate collagen type I synthesis in human osteoblast-like cells and promote osteoblastic differentiation [7]. These studies clearly demonstrate the possible advantage of incorporation of silicon into the lattice of biomaterials intended for applications in bone tissue regeneration. Therefore to take advantage of the positive biological effect of silicon, Si-substituted hydroxyapatite (Si-HA) has been developed using a chemical precipitation route [8,9]. Patel et al demonstrated that the *in vivo* bioactivity of hydroxyapatite is significantly enhanced with the addition of silicate ions into the lattice of hydroxyapatite [10]. Gibson et al [11] also demonstrated that Si-HA increases the metabolic activity of human osteosarcoma cells. The aim of this work was to assess the effect of Si-HA on human osteoblast (HOB) cells and to correlate these findings with the rate of dissolution of the Si-HA. The techniques used to analyse the effect of silicon on dissolution of the material and on the cells were: confocal microscopy (CM), fluorescence microscopy (FM), environmental scanning electron microscopy (ESEM) and tapping mode atomic force microscopy (AFM).

Materials and Methods

The preparation of phase pure HA and Si-HA is fully described elsewhere [8,9]. The powders were uniaxially pressed to discs and sintered at 1300 °C for 2 hours. The HOB cells were grown in McCoy's 5A modified medium, supplemented with 10% foetal calf serum, 30 µgml^{-1} Vitamin C, and 1 % L-Glutamine at 37 °C. Cells of passage number 5 were seeded onto two different materials, HA and 0.8 wt% Si-HA at a density of 10 000 cells/cm^2. After 3 and 7 days of culture, the cells were fixed in 4 % formaldehyde/phosphate buffer saline solution, pH 7.4 at room temperature for 15 min. The staining procedure is described elsewhere [12]. After staining the samples were mounted with Vectashield fluorescent mountant and viewed using both FM and CM. At each time point the lactate dehydrogenase (LDH) release was determined, using an enzyme assay kit (Promega, UK). LDH is a cytosolic enzyme present within all mammalian cells [13]. If the cell membrane is damaged there is a leakage of LDH into the medium, through this method is possible to have an accurate measure of cell membrane integrity and cell viability [14]. To determine the changes on the surface of HA and Si-HA, samples of both materials were incubated in simulated body fluid (SBF), at 37 °C, for different periods of time. Changes were assessed by AFM and ESEM. AFM imaging was performed using a Digital Instruments NanoScope III and ESEM signals were collected in low vacuum mode, using an off axis gaseous secondary electron detector, the beam intensity used was 10 kV.

Results

The number of cells on the 0.8 wt% Si-HA was significantly higher than on phase pure HA, which could indicate that the silicon enhances the proliferation of HOB cells (Figure 1a), and there was no significant variation in the LDH release with time. When the LDH was expressed per cell, this parameter was significantly lower on Si-HA than on HA, and this could indicate that there are more viable cells on the surface of Si-HA than on the HA surface (Figure 1b).

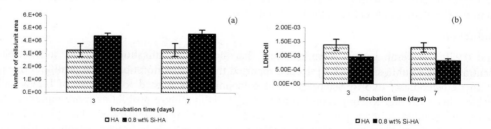

Fig. 1 – Variation of the number of cells (a) and LDH/Cell release (b) on phase pure HA and Si-HA with time, (mean ± standard deviation, n=3).

Using FM it was possible to view the distribution and number of cells on the HA and Si-HA. The number of HOB cell on the Si-HA surface appeared to be higher and more closely packed than on HA, (Figure 2 a,b and Figure 3 a,b)

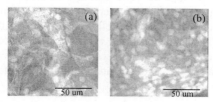
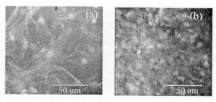

Fig. 2 – FM images of HOB onto HA (a) and Si-HA (b) after 3(a) days of incubation.

Fig. 3 – FM images of HOB onto HA (a) and Si-HA (b) after 7(a) days of incubation.

The cell cytoskeleton and vinculin in focal adhesion complexes were visualized using confocal microscopy to image the cells on the different materials. At day 3, cells were spread on the substrates as shown in Figure 4 a,b. Although at day 7 the cells seemed to be less spread on both materials (Figure 5 a,b). There appeared to be more focal points of adhesion on the Si-HA than on the HA.

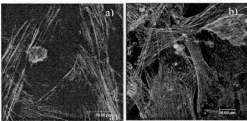

Fig. 4– CM image of HOB on HA (a) and Si-HA (b) after a period of 3 days of incubation.

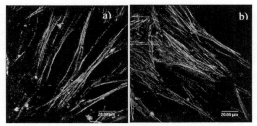

Fig. 5 – CM image of HOB on HA (a) and Si-HA (b) after a period of 7 days of incubation.

Both AFM and ESEM revealed that the dissolution of the Si-HA started earlier than the dissolution of HA. After 7 days of incubation the Si-HA surface was partially covered by an apatite layer (Figure 6b), indicating, that there had been extensive dissolution of Si-HA at earlier time points [15], while for HA significant changes were only detected after 14 days of incubation (results not shown). AFM was used to observe small features in the surface of both materials. At day 1, dissolution features were visible at the surface of Si-HA, but not on the HA surface (Figure 7 a,b) and it was not possible to observe significant dissolution features on the HA surface at 7 days.

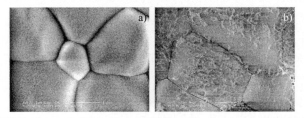

Fig. 6 – ESEM image of HA (a) and Si-HA (b) after incubation in SBF for 7 days

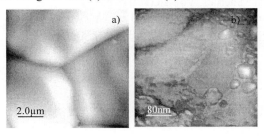

Fig. 7 – AFM image of HA (a) and Si-HA (b) after incubation in SBF for 1 day

Discussion

These results show an increase in proliferation and an increase in the number of viable HOB grown on Si-HA compared to HA. These results complement those of Gibson et al [11], who showed that Si-HA increase the activity of osteoblast-like cells. The adhesion of cells to a substrate involves extracellular matrix proteins, cell membrane proteins and cytoskeleton proteins which interact together to induce signal transduction, promoting the action of transcription factors and consequently regulating gene expression [16]. The external face of focal contacts present specific proteins such as integrins, these molecules bind to specific ligands. Bone cells can adhere to the substrate through $\alpha_2\beta_1$ integrin that binds to collagen [16]. According to the literature, there is a preferential dissolution of silicon in the Si-HA and we have shown early dissolution of this material, so there will be soluble silicon in the medium available to cells [17]. Reffit [7] showed that soluble silicon increases the production of collagen, and this could be an explanation for the increased number of focal points of adhesion, observed on the HOB cell cultured on the Si-HA allowing the cells to attach and spread more extensively.

Conclusions

Si-HA increases the proliferation of human osteoblast cells, resulting in more viable cells on the surface compared to HA. The increase in the number of focal points of adhesion may be due to the release of silicon to the medium stimulating the production of collagen by HOB cells. However the mechanism by which the silicon increases the collagen production and increases the proliferation is not still fully understood and more studies are required.

Acknowledgements

The authors wish to acknowledge the financial support of ref. SFRH/BD/6173 grant, the project entitled "Revestimento de Hidroxiapatite modificada com silício para aplicações biomédicas" ref. POCTI/CTM/49238/2002 financed by FCT (Fundação para a Ciência e Tecnologia).

References

[1] E. M. Carlisle, Fed. Proc., Vol. 43, (1984), p. 680.
[2] E. M. Carlisle, in: Silicon Biochemistry, (D. Evered & M. O'Connor, eds.), Wiley, N. York 1986
[3] K. Schwarz, in: Biochemistry of Silicon and Related Problems, N. York, (1978), p .207
[4] K. Schwarz, and D.B. Milne, Nature, Vol. 239, (1972), p. 333
[5] L.L. Hench, in: Sol-Gel Silica Properties, Processing and technology Transfer, (1999), chapter 10 Biological Implications, p .116-163
[6] P.E. Keeting, et al, J. Bone Mineral Res. Vol. 7, (1992), p. 1281
[7] D. M. Reffit et al, Bone, Vol. 32, (2003) p.127
[8] L.J. Jha, et al, Worldwide patent, PCT/GB97/02325 and US Patent Serial N° 09/147773, (1999).
[9] I.R. Gibson et al, J. Biomed. Mat. Res., Vol. 44, (1999), p.422
[10] Patel et al, J. of Mat. Sci.: Mat. in Med., Vol. 13 (2002) p. 1199
[11] I.R. Gibson et al, in Bioceramics, Vol. 12, (1999), p. 191
[12] M. J. Dalby et al, Biomaterials, Vol. 24 (2003), p. 927
[13] M. Allen et al, Clinical Materials, Vol. 16 (1994), p. 189
[14] T. Rae et al, J. Biomed. Mater. Res, Vol. 11, (1977), p. 839
[15] C.M. Botelho et al, Advanced Materials Fórum, in Press
[16] K. Anselme, Biomaterials, Vol. 21 (2000), p. 667
[17] C. M. Botelho et al, J. Mater. Sci:Mater. Med, Vol. 13 (2002), p. 1123

Key Engineering Materials Vols. 254-256(2004) pp. 849-852
online at http://www.scientific.net
© *2004 Trans Tech Publications, Switzerland*

Osseous Cell Response to Electrostatic Stimulations of Poled Hydroxyapatite Ceramics in Canine Diaphyses

Satoshi Nakamura, Miho Nakamura, Takayuki Kobayashi, Yasutaka Sekijima, Shohei Kasugai* and Kimihiro Yamashita

Institute of Biomaterials & Bioengineering, Tokyo Medical & Dental University,
2-3-10 Kanda-Surugadai, Chiyoda, Tokyo 101-0062 Japan, nakamura.bcr@tmd.ac.jp,
miho.bcr@tmd.ac.jp, kobayashi@vet.ne.jp, sekijima.bcr@tmd.ac.jp, yama-k.bcr@tmd.ac.jp

*Masticatory Function Control, Tokyo Meical & Dental University,
1-5-45 Yushima, Bunkyo, Tokyo 113-8549 Japan, kas.mfc@tmd.ac.jp.

Keywords: poling, electrostatic stimulation, surface charge, hydroxyapatite, osteoconductivity

Abstract. Cell responses of electrically charged surfaces of poled hydroxyapatite (HA) ceramics were investigated by implantation in wide defects of canine femora, compared with the uncharged surfaces. The HA ceramic specimens were poled in a dc field in air at 300-400°C. Base on the thermo stimulated current measurements, the stored charge and the half–value period of HA ceramics polarized at 300°C and 1.0 kVcm^{-1} were 4.2 μCcm^{-2} and 7.3×10^9 s (at 37.0°C), respectively. Although the non-polarized HA ceramic surfaces were still isolated from osteoid tissues by dominant fibrin multi-layers 7 days after the implantation, newly formed bone layers of 0.01-0.02 mm in thickness contacted the negatively charged surfaces without any inclusion. On the contrary, The osteoid tissues surrounded by osteoblastic cells occupied the gap between the positively charged surface and the cortical bone. We have demonstrated that the polarized HA ceramic had a large and long durable charge suitable for biomedical applications and that the HA surface charges induced by electrical polarization stimulated osseous cells and promoted bone reconstruction.

Introduction

Surface charges of biomaterials are recognized as one of the important factors to determine cell responses [1, 2]. The cells receiving the stimulation exhibited various modulated reactions, such as migration, alteration of differentiation, cell phase, and extracellular matrix secretion. We have more recently disclosed that exclusively large surface charges were inducible on hydroxyapatite (HA) ceramics by proton transport polarization [3, 4] and demonstrated that the negatively charged surface of the HA ceramics enhanced their osteobonding ability in canine femora [5, 6]. In the present study, the cell responses of the negatively and the positively charged surfaces of the electrically poled HA ceramics were investigated by implantation in wide defects of canine femora, compared with the uncharged surfaces.

Materials and Methods

HA powder was precipitated from calcium hydroxide aqueous suspension and phosphoric acid solution. The HA powder, calcined at 850°C, was pressed in a mold at 200 MPa. Dense HAp ceramics prepared by sintering at 1250°C for 2 h in saturated water vapor were employed as the specimens. The HA ceramic specimens with a size of 5.0×8.0×1.0 mm^3 were electrically poled in a dc field of 1.0 kVcm^{-1} with a pair of Pt electrodes in air at 300-400°C for 1 h (Fig. 1a). The confirmation of the poling charges of the samples chosen at random was examined by the thermally stimulated current (TSC) method using a handmade measurement cell [4].

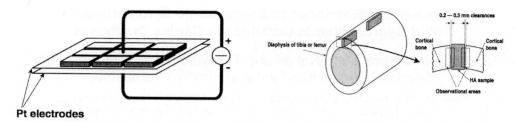

Pt electrodes

Fig. 1 Schematic illustrations of setups of electrical poling procedures (a: left) and implantation geometries (b: right).

The samples thus polarized at 300°C were implanted in the femoral and tibial diaphyses after being sterilized by the ethylene-oxide gas method. The experiments were carefully completed by veterinarians in accordance with the Guideline for Animal Experimentation (Tokyo Medical & Dental University) as well as the Guide for the Care and Use of Laboratory Animals (National Institutes of Health Pub. No. 85-23, Rev. 1985). Six male beagle dogs weighing 12-13 kg were inhalationally anesthetized with isoflurane using tracheal tubes. We exposed the femola by lateral luxation and bored rectangular through-holes on the lateral faces of the bones with a 0.7-mm dental fisher bur (Fig. 1b). The gaps between the observational faces of the samples and the cut cortical bone faces were fixed at 0.2 – 0.5 mm. The samples were rigidly held by the friction between the lateral faces of the samples and the bone faces [5]. The bones containing the samples harvested at 7, 14 and 28 days after the implantation were histologically observed.

Results

The magnitude and time durability of the polarization charge were estimated from the TSC spectra measured at 1 day after the polarization [4]. The peak temperature (T_{peak}) and maximum current density (J_{max}) increased with increasing polarization temperature, as shown in Table 1. The increase in the polarization temperature raised the calculated stored charge (Q). The peak temperature (T_{peak}) giving the maximum current widely varied from 295 to 420°C, depending on the polarization temperature. This result suggests that the polarization of HA was not due to a phase transition but by mass transport. The relaxation time (τ) is described by the Arrhenius law:

$$\tau(T) = \tau_0 \exp(H / kT), \tag{1}$$

where H is the activation energy, τ_0 is a pre-exponential factor and k is Boltzmann's constant. The activation energies of 0.84-0.89 eV for the HA polarized in a field of 1.0 kVcm^{-1} were almost independent of the polarization temperature. The obtained τ_0 increased with increasing polarization temperature. Polarization charge (Q) is given by a function of time:

$$Q(t) = Q_0 \exp(- t / \tau), \tag{2}$$

Where Q_0 is the initial polarization charge. The calculated half–value period (t_{50}) of HA polarized at 300°C and 1.0 kVcm^{-1} was 7.3×10^9 s (ca. 230 years) at the biological temperature of 37.0°C. Therefore, the polarized HA ceramic had a long durable charge for biomedical applications [4].

Table 1   Effects of polarization temperature on depolarization parameters, activation energy and relaxation time obtained from TSC spectra.

Polarization	Depolarization					
T_p [°C]	T_{peak} [°C]	J_{max} [nAcm^{-2}]	Q [μCcm^{-2}]	H [eV]	τ_0 [ms]	$\square_{50}$ [s]
300	375	3.06	4.2	0.87	0.077	7.3×10^9
350	386	5.33	6.8	0.89	0.068	1.4×10^{10}
400	420	7.87	14.9	0.84	0.43	1.4×10^{10}

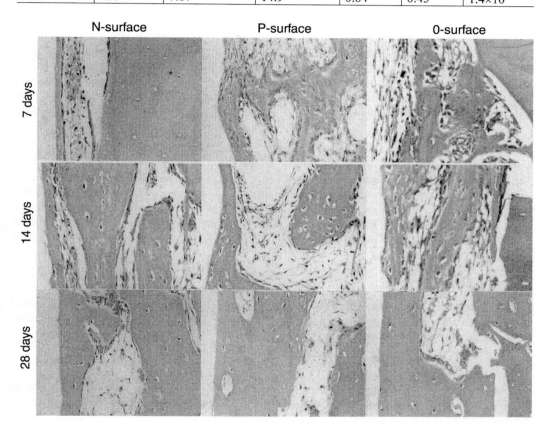

Fig.2   H.-E.-stained histological sections of Bone formation and cell responses in wide gap between HA ceramics and cortical bones .

Newly formed bone layers of 0.01-0.02 mm in thickness contacted the negatively charged (N) surfaces without any inclusion 7 days after the implantation. The bone layers were accompanied by mono-layered osteoblastic cells on the bone surfaces and maturing osseous cells in the bones. The osteoid tissues surrounded by osteoblastic cells occupied the gap between the positively charged (P) surface and the cortical bone 7 days postoperatively. Although a small part of the osteoids directly contacted the P-surface, almost all of the osteoids were isolated from the ceramic surface by fibrin granulation tissues. The layer-structured osteoblasts were ubiquitous at the margins of the osteoids and significantly larger than those in the N-surfaces and had a square shape associated with the bone forming stage. The non-polarized HA ceramic surfaces (0-surfaces) were still isolated from osteoid tissues by dominant fibrin multi-layers.   In all groups, no phyagocytic reaction with multinucleated giant cells was found in the gap areas 7 days after the implantation.

At 14 days, the newly formed bone layers directly bonding to the N-surfaces thickened 2-3 times compared with that on day 7. All of the newly formed bones included many osseous cells. A few multinucleated giant cells associated with osteoclasts were ubiquitous in the vicinity of the cut cortical bone surfaces, while no Howship's lacuna was found. The proximity of the P-surface was occupied by the osteoids surrounded by osteoblast layers and fibrous granulation tissues. A few multinucleated giant cells were found to be identifiable as an osteoclast at the margins of the osteoids.

At 28 days, the newly formed bones surrounded by the osteoblast layers were matured in the N- and the P-surfaces. The multinucleated giant cells with Howship's lacunae were found at the margins of the newly formed bones over the entire region of the gaps. In the vicinity of the 0-surface, the newly formed bones were found without any significant ubiquitousness in the gaps. Parts of the bones were in direct contact with the 0-suface. Although the osteoblasts at the fringes of the newly formed bones, the sporadic osteoclasts, and the osseous cells in the osteoids exhibited similar morphologies to those in the charged surfaces, the amount of the newly formed bones in the 0-surfaces was significantly smaller than in the N- and the P-surfaces, based on of low magnified observations.

Discussion

The bone reconstruction on the ceramic and the cut cortical bone surfaces was promoted in the vicinity of the N-surfaces. For the P-surface, the formation of the bones derived from osteoids was more predominant than the bone formation in direct contact with the ceramic surfaces. Although it was not clear that the osteobonding ability on the P-surface was suppressed, in the proximity of the P-surfaces, the activities of the osteoblastic cells surrounding the osteoids were stimulated by the positive charges.

Conclusion

The HA surface charges induced by electrical polarization stimulated osseous cells and promoted bone reconstruction in the entire gap region, regardless of the polarity of the charge, while the reconstruction process depended on the polarity.

Acknowledgements

This work was partly supported by a Grants-in-Aid from the Japan Society for the Promotion of Science (#15360338), the Mitsubishi Foundation, the Murata Scientific Foundation, a grant for Development of Advanced Medical Technology from the Ministry of Education, Science, Sports and Culture of Japan.

References

[1] J.E. Davies, B. Causton, Y. Bovell, K. Davy and C.S. Sturt: Biomaterials Vol. 7 (1986), p.231-233.

[2] M. Krukowski, D.J. Simmons, A. Summerfield, P. Osdoby: J Bone Min Res Vol. 3 (1988), p.165-171.

[3] K. Yamashita, N. Oikawa and T. Umegaki: Chem Mater Vol. 8 (1996), p.2697-2700.

[4] S. Nakamura, H. Takeda and K. Yamashita: J Appl Phys Vol. 89 (2001), p. 5386-5392.

[5] T. Kobayashi, S. Nakamura and K. Yamashita: J Biomed Mater Res Vol. 57 (2001), p. 477-484.

[6] S. Nakamura, T. Kobayashi and K. Yamashita: J Biomed Mater Res Vol. 61 (2002), p. 593-599.

Key Engineering Materials Vols. 254-256(2004) pp. 853-856
online at http://www.scientific.net

Platelet Adhesion on Metal Oxide Layers

S. Takemoto, T. Yamamoto, K. Tsuru, S. Hayakawa, A. Osaka

Biomaterials Laboratory, Faculty of Engineering, Okayama University

3-1-1, Tsushima-naka, Okayama, 700-8530, Japan, kanji@biotech.okayama-u.ac.jp

Keywords: Titanium oxide layer, platelet adhesion, sol-gel coating

Abstract. This study was concerned with blood compatibility of titanium oxide layers on stainless steel. The titanium oxide layers were prepared through sol-gel process by dip-coating of tetraethyltitanate solution and heated at 500°C. The crystal phase, thickness and wettability of the oxide were characterized. The blood compatibility was evaluated in term of platelet adhesion using human platelet rich plasma. Consequently, with increase in the thickness of the titanium oxide layers, the number of platelet adhered on the stainless steel coated with titanium oxide layer decreased rapidly. When the thickness of titanium oxide layers on stainless steel grew more than 150 nm, the number of adherent platelets decreased less than 10% in comparison with that on non-coated stainless steel. The titanium oxide layers indicated to be more hydrophilic than non-coated stainless steel. In conclusion, the thicker and more hydrophilic titanium oxide layer on stainless steel appears to inhibit platelet adhesion.

Introduction

Blood-contacting biomedical materials for artificial heart casing, artificial heart valves, and cardiovascular stents are required to have good blood compatibility. Among those materials, cardiovascular stents are mostly made of metal such as stainless steel and titanium-nickel. However, these metallic materials activate the intrinsic coagulation system and platelet-adhesion to induce thrombosis. It is common to coat an anticoagulant like heparin on those stents to prevent thrombosis. However, a few investigations found [1,2] the heparin coatings are insufficient for sustained blood compatibility. Since titanium oxide is known to be blood compatible [3-14], it is reasonable to expect that the titanium oxide coatings should provide anti-thrombogenic surfaces and prevent thrombosis. In the present experiment an attempt was made to prepare a crack-free TiO_2 layer on stainless steel and to prevent adhesion of platelets onto the coated stainless steel substrates. The platelet adhesion was discussed and correlated to the physico-chemical properties of the titanium oxide layer, such as crystalline phases, thickness and wettability.

Materials and Methods

Substrates 10x10x0.1 mm in size were cut from a sheet of commercial SUS316L and rinsed three times with acetone for 5 min in an ultrasonic cleaner. The titania sols for the coatings were prepared through the sol-gel procedure starting from reagent grade tetraethyltitanate ($Ti(OEt)_4$), ethanol (EtOH), distilled water (H_2O) and hydrochloric acid (HCl). Table 1 shows the mixing ratios of the chemicals for coating. The resultant sols were denoted as E10, E20 or E50 after the amount of EtOH. The sols were dip-coated on the substrate where the pull-up speed was 2 mm/s. The substrates were then heated at 500°C for 10 min in an electric furnace. The TiO_2 layers prepared by iterating the coating and heating processes for 1, 3 or 5 times were denoted as D1, D3 or D5, respectively. Surface

Table 1 Mixing ratios of starting chemicals for coating (molar ratio)

	$Ti(OEt)_4$	EtOH	HCl	H_2O
E10	1	10	0.2	2
E20	1	20	0.2	2
E50	1	50	0.2	2

texture was observed under a scanning electron microscope (SEM). Crystalline phases of the TiO_2-coated substrates were characterized by thin-film X-ray diffraction (TF-XRD) equipped with a thin-film attachment (the glancing angle was 1°). The thickness of the TiO_2 layers was estimated using the ultra violet and visible (UV-Vis) reflection spectra (reflection angle was 5°) and the refractive index for anatase (2.561) was used for calculation of the oxide thickness [15]. Static contact angle toward distilled water was measured with an automatic contact angle meter. Platelet rich plasma (PRP) was taken from human whole blood containing 3.8 % citrate acid solutions (blood : citrate acid = 9 : 1) after centrifuged at 400g for 5 min. The TiO_2-coated substrates were placed in 24 well microplates, and 1.0 ml of PRP poured to each well. After incubated for 30 min at 36.5°C, PRP was taken away form the wells. The substrates in the wells were gently rinsed with phosphate buffer solution (PBS), the platelets that adhered specifically on the surface were fixed, dehydrated and dried under a freeze dryer in overnight. The sample surface was observed under a SEM. Five areas were randomly chosen on the surface and the number of the adhered platelets was counted to give the average values per unit area and standard deviation. The significance of the difference among those average values was statistically evaluated by the one-way analysis of variation (ANOVA).

Results

The SEM observation indicated that the TiO_2 layer derived from sol E10 had many cracks after being heated at 500°C for 10 min. However, the TiO_2-layer from sol E20 or E50 was crack-free when heated at 500°C for 10 min. The crack-free titania layers derived from sols E20 and E50 and heated at 500°C was used for further characterization. TF-XRD patterns of the TiO_2-coated substrates derived from sol E20 or E50, and subsequently heated at 500°C for 10 min indicated that the peak intensity of anatase (JCPDS card: 21-1272) was increased with increasing number of dip-coating (not shown here). TiO_2 layer prepared from sol E20 had stronger X-ray intensity for anatase than that from E50. The thickness of titania layers and static contact angle toward distilled water on SUS316L substrate are listed in Table 2. In the case of coating with E20, the E20_D1 had about 130 nm in thickness of TiO_2 layer, and E20_D5 had about 520 nm. That is, a TiO_2 layer could increase about 100 nm each coating immersion. In the case of coating with E50, a TiO_2 layer could increase about 50 nm each coating immersion. The value of static contact angle for the TiO_2-coated substrate was 20-30° except for E50_D1 (50°). Coated TiO_2 layer on the substrates, was found to reduce the contact angle when compared with the SUS316L substrate without the TiO_2 layer. That is, the surface on SUS316L substrate became hydrophilic after being coated with TiO_2 layer. SEM photographs of SUS316L substrate with or without titanium oxide layer after contact with PRP for 30 min indicated that many adherent platelets could be observed on SUS316L substrate without coating layer. Some platelets aggregated and spread their pseudopodia (not shown here). On the other hand, the TiO_2 coated SUS316L substrates were observed to reduce the number of adherent platelets. Table 3 shows the

Table 2   Thickness of titanium oxide layers and static contact angle toward distilled water on the SUS316L substrate with titanium oxide layer. See text for the sample.

Sample	Thickness /nm	Contact angle /°
SUS316L	—	83 (±7)
E20_D1	133 (±12)	27 (±4)
E20_D3	289 (± 7)	25 (±4)
E20_D5	517 (±67)	22 (±4)
E50_D1	67 (±23)	50 (±5)
E50_D3	161 (±16)	32 (±5)
E50_D5	235 (±27)	27 (±6)

— : no titanium oxide layer

Table 3   The number of platelets adhered on each sample after contacted with PRP for 30 mim. The number of platelets adhered on the SUS316L substrates without TiO_2 coating was shown as 100%. See text for the sample.

Sample	Normalized number of adherent platelets (SD)
SUS316L	100
(±12)	
E20_D1	21 (±2)
E20_D3	4 (±3)
E20_D5	5 (±2)
E50_D1	62 (±2)
E50_D3	3 (±1)

number of platelets adhered on each sample after contact with PRP for 30 min. Here, the number of platelets adhered on the SUS316L substrates without TiO_2 coating was shown as 100%. The number of platelets for E50_D1 was about 60% as much as that for SUS316L substrates without TiO_2 layer. In addition, the number of platelets adhered on E50_D3 and E50_D5 was 3 ± 1 and 6 ± 1 %, respectively. The number of immersions influenced the number of adherent platelets. The difference in the number of adherent platelets with the number of immersion times was statistically significant. Similar trend could be observed for TiO_2 layer prepared in composition of E20. Consequently, TiO_2 layer on SUS316L substrates was effective for protection of platelet adhesion.

Discussion

The platelet adhesion has often been correlated to the thickness of TiO_2 layer in Fig. 1 where the number of adherent platelets is plotted as a function of the thickness of TiO_2 layer. Here, the number of platelets adhered on the SUS316L substrate without TiO_2 layer was shown as 100%. With the increase in thickness of the TiO_2 layer, the number of adherent platelets reduced rapidly for thickness up to 150 nm. The number of adherent platelets kept a constant value for films with thickness of more than 180 nm. The difference in the number of platelet adhesion with thickness of TiO_2 layer was statistically of significance. The result of our experiment clearly suggested that the TiO_2 layer produced by sol-gel method inhibited platelet adhesion. Nygren et al. reported the effect of titanium oxide layer on titanium metal. Thick oxide layer (39 nm) had less amount of platelets adhesion than thin one (4 nm) [8, 9]. Sunny and Sharma mentioned the relation between the thickness of titanium oxide layer and the amount of fibrinogen adsorbed on titanium substrate oxidized by anodic oxidation [5]. They reported that titanium oxide layer more than 40 nm in thickness suppressed the adsorption of fibrinogen. These results indicated that titanium oxide layer inhibited platelet adhesion and adsorption of fibrinogen. In previous study, we investigated the platelet adhesion on titanium substrate oxidized by hydrogen peroxide solution or heating, and concluded that the thickness of titanium oxide layer was related to the decrease in platelet adhesion [14]. The results obtained in this and previous studies agree with other research on the inhibition of platelet adhesion on thick titanium oxide layer. However, the thickness of titanium oxide layer, which inhibited platelet adhesion and adsorption of fibrinogen, was different in our result compared to previous studies. This will have to be discussed in detail in the future. Wettability of TiO_2 layer on SUS316L substrate could be considered as factor related to platelet adhesion. Tables 2 and 3 indicate that less platelet adhered on 20-30° in contact angle of TiO_2-coated 316L substrate. Ikada et al. reported that adhesion of cells and platelets was well related to wettability of polymer substrates [16]. The surfaces with the contact

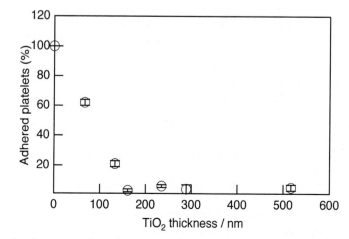

Fig. 1 Normalized amount of platelets adhered on the SUS316L substrate coated with titanium oxide layer as a function of the thickness of TiO_2 layer. The number of platelets adhered on the SUS316L substrate without TiO_2 layer was shown as 100%.

angles less than 30° and more than 90° were not favorable for cell and platelet adhesion. However, they indicated that surface composition and functional group on the surface governed the wettability. In this study, there were no remains of the solvent (EtOH) and catalyst (HCl) in TiO_2 layer, according to infrared analysis and X-ray photoelectron spectroscopy (not shown here). Surface compositions and functional groups of TiO_2-coated SUS316L substrates may not be affected by contact angle. Hydrophilic surface of TiO_2-coated SUS316L substrate may also affect the inhibition of platelet adhesion.

Conclusion

Titanium oxide layers on SUS316L substrate were prepared through sol-gel procedure by dip-coating and heating up to 500°C. The wettability of titanium oxide layers indicated them to be more hydrophilic than SUS316L substrate, and the crystalline phase of titanium oxide was anatase. The thickness of titanium oxide layers could be controlled by changing the number of coating immersions and the composition of solution. The platelet adhesion on titanium oxide layers was discussed in terms of thickness and wettability of titanium oxide layers. The number of adherent platelet decreased rapidly until the thickness of titanium oxide layer reached to 150 nm, and that value was less than 10% in comparison with that on stainless steel without titanium oxide coating. Therefore, the technique of titanium oxide coating is useful to decrease the number of platelet adhesion onto titanium oxide layers coated stainless steel and can be widely applied to the medical materials to contact with blood.

Acknowledgements

This study was supported by the Industrial Technology Research Grant Program (02A47022a) in '02 from the New Energy and Industrial Technology Development Organization (NEDO) of Japan.

References

[1]   K. Christensen, R. Karsson, H. Emanuelsson, G. Elgue and A. Larsson, Biomaterials, Vol. 22 (2001), p. 349-355

[2]   P. Klement, YJ. Du, L. Berry, M. Andrew and AKC. Chan, Biomaterials, Vol. 23 (2002), p. 527-535

[3]   B. Wälivarra, I. Lundström and P. Tengvall, Clin. Mater., Vol. 12 (1993), p. 141-148

[4]   B. Wälivarra, BO. Aronsson, M. Rodahl, J. Lausmaa and P. Tengvall, Biomaterials, Vol. 15 (1994), p. 827-834

[5]   MC. Sunny and CP. Sharma, J. Biomater. Appl.,Vol 6 (1991), p. 89-98

[6]   JY. Park and JE. Davies, Clin. Oral Impl. Res., Vol. 11 (2000), p. 530-539

[7]   JY. Park, CH. Gemmell and JE. Davies, Biomaterials, Vol. 22 (2001), p. 2671-2682

[8]   H. Nygren, C. Eriksson and J. Lausmaa, J. Lab. Clin. Med., Vol. 129 (1997), p. 35-46

[9]   C. Eriksson, J. Lausmaa and H. Nygren, Biomaterials, Vol. 22 (2001), p. 1987-1996

[10] N. Huang, P. Yang, X. Cheng, Y. Leng, X. Zheng, G. Cai, Z. Zhen, F. Zhang, Y. Chen, X. Liu and T. Xi, Biomaterials, Vol. 19 (1998), p. 771-776

[11] N. Huang, YU. Chen, JM. Luo, J. Yi, R. Lu, J. Xiao, ZN. Xue and XH. Liu, J. Biomater. Appl., Vol. 8 (1994), p. 404-412

[12] S. Takemoto, K. Tsuru, S. Hayakawa, A. Osaka and S. Takashima, Bioceramics, Vol. 13 (2000), p. 35-38

[13] S. Takemoto, K. Tsuru, S. Hayakawa, A. Osaka and S. Takashima, J. Sol-Gel Sci. Technol., Vol. 21 (2001), p. 97-104

[14] S. Takemoto, T. Yamamoto, K. Tsuru, S. Hayakawa, A. Osaka and S. Takashima, Biomaterials, submitted.

[15] S. Sakka, K. Kamiya and Y. Yoko, "Sol-Gel Prepareation and Properties of Fibers and Coating Films" in ACS symposium series, editors M. Zeldin, KJ. Wynne, HR. Allcock, Vol. 360 (1988), p. 345-352

[16] Y. Tamada and Y. Ikada, Polym, Vol. 34 (1993), p. 2208-2212

Key Engineering Materials Vols. 254-256(2004) pp. 857-860
online at http://www.scientific.net
© 2004 Trans Tech Publications, Switzerland

In Vitro Cytocompatibility of Osteoblastic Cells
Cultured on Chitosan-Organosiloxane Hybrid Membrane

Y. Shirosaki[1], K. Tsuru[1], S. Hayakawa[1], A. Osaka[1], M. A. Lopes[2,3],
J. D. Santos[2,3] and M. H. Fernandes[4]

[1] Biomaterial Laboratory, Faculty of Engineering, Okayama University, Tsushima, Okayama-shi, 700-8530, Japan, e-mail:osaka@cc.okayama-u.ac.jp

[2] FEUP-Faculdade de Engenharia da Universidade do Porto, DEMM, Rua Dr Roberto Frias, 4200-465, Porto, Portugal

[3] INEB-Instituto de Engenharia Biomedica, Lab. de Biomateriais, Rua do Compo Alegre, 823, 4150-180, Porto, Portugal

[4] FMDUP-Faculdade de Medicina Dentaria, Universidade do Porto, Rua Dr Manuel Pereira da Silva, 4200-393, Porto, Portugal

Keywords: chitosan, chitosan-inorganic hybrid, γ-glycidoxypropyltrimethoxysilane (GPSM), osteoblastic cell (MG63), cytocompatibility

Abstract. Chitosan-silicate hybrids were synthesized using γ-glycidoxypropyltrimethoxysilane (GPSM) as the agent to cross-link the chitosan chains. Fourier-transform infrared (FT-IR) spectroscopy was used to analyze the structures of the hybrids. The swelling ability and the cytocompatibility of the hybrids were investigated as a function of the GPSM concentration. The swelling of the hybrids was suppressed by addition of GPSM. MG63 human osteoblastic cells were cultured on the hybrids. The adhesion and proliferation of the MG63 cells cultured on the chitosan-GPSM hybrid surface were improved comparing to that on chitosan membrane, regardless the GPSM concentration.

Introduction

Tissue engineering approach to repairing, complementing and regenerating damaged tissues depends on polymer scaffolds that serve to support and reinforce the regenerating tissue. A number of natural and synthetic polymers are currently employed as tissue scaffolds. Chitosan, a mucopolysaccharide having structural characteristics similar to glycosamines, is obtained from alkali-deacetylation of chitin derived from exoskeleton of crustaceans [1]. Chitosan is biodegradable, biocompatible, non-antigenic and non-toxic [1]. For most of medical applications, the polysaccharide network of chitosan should be crosslinked in order to improve its mechanical properties and control its biodegradability. Various reagents have been used such as epoxy compounds, formaldehyde, and glutaraldehyde [2]. Note that these cross-linking agents are all highly cytotoxic and may impair the biocompatibility of the crosslinked biomaterials [3]. Ren et al. already reported that the hybrids of gelatin and γ-glycidoxypropyltrimethoxysilane (GPSM, $CH_2OCHCH_2OCH_2CH_2Si(OCH_3)_3$) were bioactive and cytocompatibile in vitro [4]. In this study, chitosan-organosiloxane hybrids were prepared using GPSM. Fourier-transform infrared (FT-IR) and ^{29}Si NMR spectroscopies, contact angle measurements and swelling ability were used to assess the structure of the hybrids. Cytotoxicity of the hybrids was investigated using MG63 human osteoblastic cells.

Materials and Methods

Materials preparation
An appropriate amount of chitosan (Aldrich®, high molecular weight) was dissolved in 0.25 M acetic acid aqueous solution to attain a concentration of 2 (w/v) %. Then, appropriate amounts of GPSM (Aldrich®) were added to the chitosan solution. Table 1 indicates the composition of the hybrids and sample reference. After stirring, the resultant solutions were spread on the bottom of polyethylene cases, and stand overnight at room temperature. Afterwards, they were kept at 60 °C for 2 days to yield hybrid membranes. The chitosan membrane was soaked in 0.25N sodium hydroxide and washed with distilled water to remove excess acid. For sterilization, all membrane obtained were autoclaved in phosphate-buffered saline (PBS) solution, pH7.4, at 121 °C for 20 min.

Table 1 Starting composition of chitosan-silicate hybrids (molar ratio).

	Ch	ChG1	ChG5	ChG10	ChG15	ChG20
chitosan	1.0	1.0	1.0	1.0	1.0	1.0
GPSM	0	0.1	0.5	1.0	1.5	2.0

Structural analysis
The FT-IR spectra of the hybrids were measured by the KBr method on a JASCO FT-IR300 spectrometer. The data were recorded by the accumulation of 200 scans with a resolution of 2 cm^{-1}. ^{29}Si cross-polarization (CP) magic-angle spinning (MAS)-NMR spectra were recorded at 7.05 T (tesla) on a Varian ^{UNITY}INOVA300 FT-NMR spectrometer, equipped with a CP/MAS probe. The samples were placed in a zirconia sample tube. The sample spinning speed at the magic angle to external field was 5.0 kHz. The ^{29}Si CP MAS-NMR spectra were measured at 59.6 MHz with 10.0 μs ($\pi/2$) pulses, 10 s recycle delays and 1500 μs contact time. The signals for about 3400 pulses were accumulated. The chemical shift is represented in δ (ppm) by convention. Polydimethylsilane (PDMS: δ=-34 ppm with respect to tetramethylsilane: δ=0 ppm) was used as the secondary external reference.

The wettability was assessed by the sessile drop method, using distilled water as the testing liquid and an automatic contact angle meter (Kyowa Interface Science, Model CA-V). To determine the swelling ability of the hybrids were soaked in PBS for different periods of time. Wet samples were wiped with a piece of filter paper to remove excess liquid and reweighed. The amount of adsorbed water was calculated according to Eq. 1, where Wa and Wb stand for weight after and before being soaked in PBS solution, respectively.

$$\text{Water uptake (\%)} = 100 \, (Wa-Wb) / Wb \qquad (1)$$

Biological assesment
MG63 cells were cultured at 37 °C in an atmosphere of humidified 5 % CO_2 in air. α-modified Eagle's medium (α-MEM) that contained 10% fetal bovine serum (FBS), 1 % penicillin and streptomycin solution (GIBCO BRL 15140-122), 2.5 μgml^{-1} fungizone and 50 μgml^{-1} ascorbic acid was used as the culturing medium. MG63 cells were cultured (10^4 cellcm^{-2}) up to 6 days in control conditions (absence of materials, standard plastic culture plates) and on the surface of the hybrids. Control cultures and the seeded specimens were evaluated throughout the incubation time at days 1, 3, 6 for cell viability / proliferation and observed by scanning electron microscopy (SEM). MTT assay was used to estimate cell viability / proliferation. Control cultures and the seeded specimens were incubated with 0.5 mgml^{-1} of MTT during the last 4 hours of the culture period tested; the medium

was then decanted, formazan salts were dissolved with dimethylsulphoxide and the absorbance was measured at 600nm. Results were compared in terms of macroscopic surface area. For SEM observation, the specimens were fixed with 1.5 % glutaraldehyde in 0.14 M sodium cacodylate buffer (pH 7.3), then dehydrated in graded alcohols, critical-point dried, sputter-coated with gold and analysed in a JEOL JSM 6301F SEM equipped with an energy dispersive X-ray microanalyser (Voyager XRMA System, Noran Instruments).

Results

Fig. 1 shows the FT-IR spectra of Ch and ChG10 as well as the spectra of GPSM monomer and GPSM homopolymer solutions used as references. GPSM monomer showed three characteristic peaks at 2840, 1192 and 914 cm^{-1}; these peaks were assigned to represent the methoxy group (2840 cm^{-1}) and the epoxide group (1192 and 914 cm^{-1}). The spectrum of GPSM homopolymer showed v(-OH) and v(Si-O-Si) bands at 3400 and 1000~1100 cm^{-1}, respectively. For Ch and ChG10, specimens bands at 1650 and 1565 cm^{-1} were detected and correspond to amide bands I and II, respectively, and are characteristic of the chitosan structure. The Si-O bands for ChG10 were found at the same region as those for GPSM homopolymer. The spectra of other hybrids were similar to ChG10 (data not shown).

The contact angle obtained using distilled water decreased slightly due to the modification of chitosan with GPSM (data not shown). Fig. 2 shows the swelling of the hybrids after being soaked in PBS for various periods. The hybrid membranes were unable to imbibe as much PBS solution during 27 h of hydration as the chitosan membrane. Ch became saturated after being soaked in PBS for 7 h. After 27 h of hydration, ChG1 achieved a water uptake of 50 % compared to 65 % for the Ch. As the content of GPSM increased, swelling decreased in the following order: ChG1 (50 %) > ChG5 (37 %) >ChG10 (30%) >ChG15 (23 %) >ChG20 (18 %).

MG 63 cells grew better on the hybrids than on the chitosan membrane. The values for MTT reduction were higher on the hybrids than those on the chitosan membrane (data not shown). Fig. 3 shows light microscopy photographs of the MG63 cells cultured on the hybrids for 1, 3 and 6 days after the reaction with the MTT reagent. Seeded specimens showed cell attachment at 1 day and cell growth afterwards. At day 6, cell growth on the hybrids was significantly higher than that observed on the chitosan membrane. The content of GPSM appeared not to influence cell attachment and proliferation. SEM observation of the specimens provided the same information (data not shown).

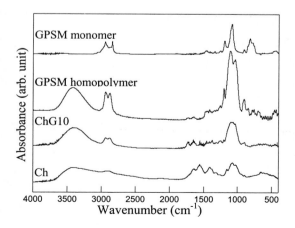

Fig. 1 FT-IR spectra of the specimens.

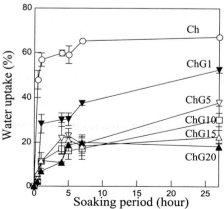

Fig. 2 Swelling of the hybrids after being soaked in PBS for various periods.

Discussion and Conclusion

During the preparation of chitosan-GPSM hybrids, the epoxy group of GPSM reacted with amine groups of chitosan chains and the methoxy groups of GPSM were hydrolyzed to form Si-O-Si bonds. The wettability of the hybrid surfaces and the percentage of water uptake were found to increase with the content of GPSM as a consequence of the polycondensation of GPSM in the hybrids. Most of crosslinking agents are highly cytotoxic and may impair the biocompatibility of the crosslinked biomaterials. However, the present hybrids were all cytocompatible with MG63 osteoblastic cells, regardless to the GPSM content. Therefore, GPSM is suitable as a crosslinking agent of chitosan and it is expected that these chitosan-GPSM hybrids find application on bone regeneration.

| Ch | ChG1 | ChG5 | ChG10 | ChG15 | ChG20 |

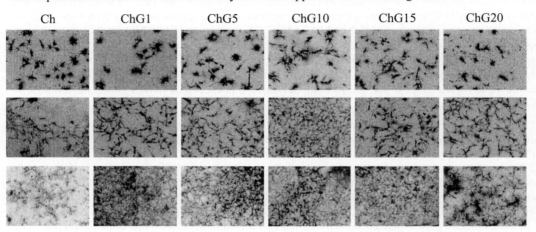

Fig. 3 Light microscopy photographs of MG63 cells cultured on the hybrids for 1, 3 and 6 days (x40). MTT reduction by living cells; top row : 1day, middle row : 3 days, bottom row : 6 days

Acknowledgements

The Okayama Original Entrepreneur Seeds Excavation Enterprise Grant.

References

[1] T., Chandy, P. C. Sharma: Biomat Art Cells Art Org, Vol. 18 (1990), p. 1-24.

[2] S. R. Jameela, A. Jayakrishnan, Biomaterials, Vol. 16 (1995), p. 769-775.

[3] C. Nishi, N. Nakajima, Y. Ikada, J Biomed Mater Res. Vol. 29 (1995), p. 829-834.

[4] L. Ren, K. Tsuru, S. Hayajkawa, A. Osaka, Biomaterials, Vol. 23(2002), p. 4765-4773.

Key Engineering Materials Vols. 254-256(2004) pp. 861-864
online at http://www.scientific.net
© 2004 Trans Tech Publications, Switzerland

Hydroxyapatite Ceramic Particles As Material For Transfection

Patrick Frayssinet, Eliane Jean, Nicole Rouquet

Urodelia, Le Gaillard, Rte de St Thomas, 31470 St Lys, France
(patrick.frayssinet@wanadoo.fr).

Keywords: transfection, calcium phosphate ceramics, tissue culture

Abstract. Hydroxylapatite ceramics were developed for cell transfection *in vitro* and *in vivo*. They were tested at the contact of isolated bone cells and bone tissue culture. This experiment showed that calcium phosphate matrices are able to transfect isolated cells and cells in a tissue culture with a good efficiency.

Introduction

The introduction of a gene (transfection) into mammalian cells is a key-step in gene therapy. Different strategies can be applied to put a gene in an organism. The cells can be transfected either *in vitro* or *in vivo*. When isolated cells are transfected *in vitro*, they must then be reimplanted in the organism, however, in this case, a wide range of transfection tools can be used. *In vivo* transfection is limited to the use of viral vectors such as adeno or retrovirusesbut this may present several disadvantages or side effects which can be disastrous. Adenoviruses produce proteins which can trigger immune reactions. They have already been responsible for the death of patients. Furthermore, the expression of a gene transfected using a viral vector is transient and can be brought to an end when an immune reaction against the viral proteins occurs. It must also be pointed out that selection of the cells transfected by the virus is very poor and transfection efficiency can be dependent on the cell stage.

There is thus a need for a biocompatible carrier, which might transduce cells both *in vitro* and *in vivo* with a good efficiency and with a good or partial cell selectivity. Polycationic matrices have been developed for this aim. These polymeric matrices allow the transfection of cells involved in healing reactions in order to transiently produce factors which improve healing[1].

We have developed calcium phosphate ceramics which can play the same role and which can be degraded within a few weeks after implantation. Calcium phosphate ceramics are widely used in human surgery: they are integrated when implanted in bone tissue, they only induce a very mild foreign body reaction when implanted in soft tissue and they are totally degradable. The dissolution of grain boundaries releases ceramic grains which are then dissolved in the low pH compartment of the cells. These ceramics were used to transduce isolated cells *in vitro* and cells grown in tissue culture with a plasmid bearing a galactosidase gene, which was evidenced histochemically.

Materials and methods

Ceramic characteristics: The hydroxylapatite powder (> 98% hydroxylapatite)was obtained by precipitation. The particle size was 80-125 µm, the surface area was 0.62 m^2/gr. The surface potential was − 35 mV. The powders were hydrophobic.

Plasmid adsorption on the ceramic surface: The ceramic particles (2 mg) were placed under a phosphate buffer (0.12M, pH 6.8) for two hours at 60°C. The powder was then washed in demineralised water and incubated in phosphate buffer (0.12M, pH 6.8) containing 10 µg of the plasmid bearing a galactosidase gene (pCMVβ - Clontech, Palo Alto)

Cell culture: Newborn rat calvariae were dissected out and cut into 2-3 mm pieces which were placed at the bottom of Petri dishes. They were incubated in culture medium made of DMEM supplemented with 10% foetal calf serum and glutamine (20 mM) until a confluent cell layer was obtained. Then, they were suspended in the culture medium using a trypsin solution before being plated in Petri dishes at a 10^5 cells/ml (3 ml). Once the cells had adhered to the dishes, 1mg of powder was introduced in the culture for 48 h.

Tissue culture: Newborn rat tibias were aseptically dissected out and cut under the proximal metaphysis. The tibia extremity was deposited at the surface of an agar gel (1% in DMEM) at the interface between the air and the culture medium (DMEM supplemented with 10% fetal calf serum). A few µg of powder were deposited on the explant using a Pasteur pipette for the whole culture period. The galactosidase activity was checked at 48 h and three weeks of culture.

Histological analysis: The cells and the tissue samples were fixed in absolute ethanol. The tissue samples were then embedded in ethyl methacrylate and 5 µm sections were performed. The cells and tissue sections were immersed in X-gal solution at 37°C for two hours (100 mM sodium phosphate pH 7.3, 1.3 mM $MgCl_2$, 3 mM $K_3Fe(CN)_6$, 3mM $K_4Fe(CN)_6$ and 1mg/ml X-Gal). The cells and sections were observed with a light microscope.

Control: Cells and sections of tissues which were not in contact with the carrier bearing the plasmid were also immersed in the X-gal solution before being examined under a light microscope.

Results

Cell culture. Numerous cells showed a galactosidase activity at 3 days (fig. 1). The cells in the immediate vicinity of the particles were almost all labelled while the density of labelling decreased with distance from the particles. Between 30% and 64% of the cells were coloured blue. At three weeks, the number of cells showing a galactosidase activity was comparable to the previous period.

Tissue culture. At three days, only a few cells were labelled around the calcium phosphate particles. By three weeks, all the samples appeared uniformly blue indicating that a great number of cells expressed the galactosidase gene. Histological sections showed that the majority of the bone marrow cells were dead. The osteoblasts of the cancellous bone constituting the metaphysis at this age lost their cuboidal appearance and became stellar. All the cells present in the cancellous bone were labelled blue while in the bone marrow none were (fig. 2). The chondroblasts and the perichondral cells of the epiphysis were also labelled.

Discussion

This experiment showed that calcium phosphate matrices are able to transfect isolated cells and cells in a tissue culture with a good efficiency. The negative surface charge of the particles did not impair immobilisation of plasmids at the particle surface nor plasmid transduction into the cells. These ceramic matrices differ from the polycationic matrices made of polymers whose activity is based on their positive surface charge[2] responsible for the adsorption of the negatively charged DNA molecules. However, plasmid immobilisation at the ceramic material surface could result from an interaction of the positive groups of the DNA which are not available when DNA is in solution.

The surface modification of calcium phosphates in saline solution is complex. Epitaxial growth of carbonated apatite has been noted [3] [4] when calcium phosphate is implanted in the organism or incubated in medium supersaturated in calcium and phosphorus. It has been demonstrated that in the first hour, dissolution predominates while it is precipitation in the following hours. When organic molecules are present in the medium, the dissolution/precipitation phenomenon is more complex [5].

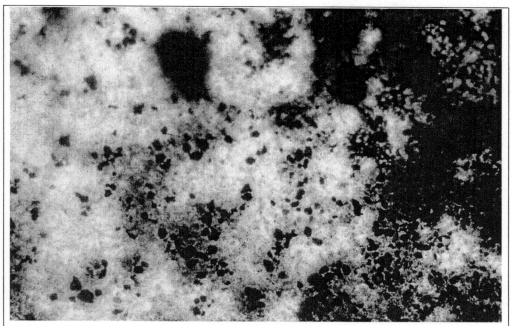

Fig 1. Photomicrograph of a cell culture stained for its galactosidase activity. The density of positive cells is greater in the proximity of the particles.

It is possible that although the phosphate buffer used in this experiment to incubate the plasmid with the ceramics is not supersaturated in calcium and phosphorus, these late elements released during the dissolution process would coprecipitate with DNA at the ceramic surface and in the ceramic micropores. The coprecipitation of DNA with calcium phosphates has been widely used for isolated cell transfection. The coprecipitates are obtained by mixing $CaCl_2$ with phosphate buffers containing the DNA[6].

The cells are transfected at some distance from the particles as shown on isolated cells and in tissue culture. In the tissue culture, cells such as the osteoblasts of the cancellous bone and chondroblasts which have not been in contact with the particles are coloured blue.

Examination of tissue culture sections showed that all the cells were labelled regardless of their cell cycle stage.

Conclusions

Calcium phosphate ceramics under a powder form are able to transfect cells whether they are isolated or not. Although the mechanism is not fully understood, the applications of these properties could be considerable when transient gene expression is needed as for the case healing of various tissues or vaccination. The calcium phosphate ceramic particles are inexpensive, degradable and biocompatible. They can induce gene expression in the cells close to the particle and thus show a certain level of selectivity.

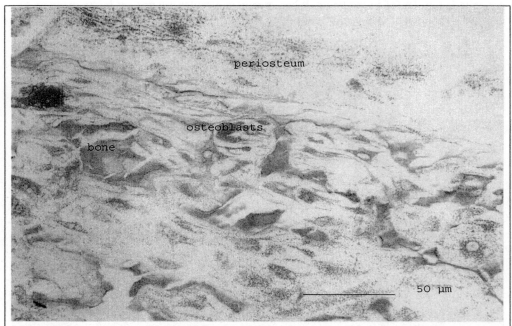

Fig. 2. Cancellous bone of the tibia epiphysis grown for three weeks in the presence of transfecting calcium phosphate particles. All the osteoblasts are stained.

References

[1] J. Bonadio, S. A. Goldstein, and R. J. Levy. Gene therapy for tissue repair and regeneration. Advanced Drug Delivery Reviews 33 (1998), p. 53-69

[2] J. Bonadio, E. Smiley, P. Patil, and S. Goldstein. Localized, direct plasmid gene delivery *in vivo*: prolonged therapy results in reproducible tissue regeneration. Nature Medicine 5 (7) (1999), p. 753-759,

[3] M. M. Monteiro, N. Carbonel Campos da Rocha, A. Malta Rossi, G. de Almeida Soares . Dissolution properties of calcium phosphate granules with different compositions in simulated body fluids. J Biomed Mater Res, 65 (2003), p. 299, 305

[4] M. Heughebaert, R. Z. LeGeros, M. Gineste, and A. Guilhem. Hydroxyapatite (HA) ceramics implanted in non-bone-forming sites. Physico-chemical characterization. J Biomed Mat Res 22 (1988), p. 257-268

[5] P. Frayssinet, , N. Rouquet, , J. Fages, , M. Durand, , P.O. Vidalain, , G. Bonel, , Influence of HA-ceramic sintering temperature on the proliferation of cells grown in their contact. Journal of Biomedical Material Research 35 (1997), p. 337-347

[6] E. T. Schenborn and V. Goiffon. *Calcium phosphate transfection of mammalian cultured cells.* edited by M. J. Tymms, Totowa, NJ: Humana Press Inc, 2000, p. 135-144.

Key Engineering Materials Vols. 254-256(2004) pp. 865-868
online at http://www.scientific.net
© 2004 Trans Tech Publications, Switzerland

Platelet Adhesion on Titania Film Prepared from Interaction of Ni-Ti Alloy with Hydrogen Peroxide Solution

Satoshi Hayakawa, Kanji Tsuru and Akiyoshi Osaka

Biomaterials Laboratory, Faculty of Engineering, Okayama University

3-1-1, Tsushima-naka, Okayama, 700-8530, Japan, satoshi@cc.okayama-u.ac.jp

Keywords: Ni-Ti alloy, titania, hydrogen peroxide, platelet adhesion.

Abstract. Titania films were prepared from interaction of Ni-Ti shape memory alloy with hydrogen peroxide solution followed by heating at 400°C. The titania films consisted of crystalline phase of TiO_2 (anatase). The *in vitro* platelet adhesion property of the titania films was examined. The titania films on Ni-Ti alloy substrates adhered less platelets than Ni-Ti alloy substrate without treatment. This indicates that the titania film can suppress the adhesion of platelet. The present chemical modification process with hydrogen peroxide solution is applicable to prevent the Ni-Ti shape memory alloy materials from platelet adhesion when in contact with blood.

Introduction

Ni-Ti alloy has mechanical properties such as shape memory effect and superelasticity. A number of biomedical applications have been proposed such as orthodontic wire, orthopedic implants for osteosynthesis and stents. However, Ni ions, in general, have a cytotoxic effect on endothelium cells. Thus, considerable anxiety is casted about the medium- and long-term behavior of Ni-Ti alloy materials when they are implanted for long periods as coronary stents or bone grafts. One can expect that the surface chemical modifications improve not only the corrosion resistance but also biocompatibility of Ni-Ti alloy. We already reported that the titania film was derived from combined chemical and thermal treatments of titanium substrates [1, 2].

In this study we proposed novel preparation technique of titania film from the interaction of Ni-Ti alloy with hydrogen peroxide solution, and examined the platelet adhesion behavior on the titania films.

Materials and Methods

Commercial Ni-Ti shape memory alloy substrates (5 x 5x 0.5 mm) were washed with acetone for 5 minutes three times in an ultrasonic cleaner. The Ni-Ti alloy substrates were immersed in 20 ml of 6 mass% H_2O_2 solutions with pH 2.5. After being kept at 80°C for 30 min., 1, 2 and 3 hours, they were washed with distilled water for 5 minutes in an ultrasonic cleaner. Then, these samples were heat-treated at 400°C for 1 hour. Their surface structures were examined by thin film X-ray diffraction (TF-XRD: RINT2500, Rigaku, Tokyo, Japan) and scanning electron microscopy (SEM: JSM6300CX, JEOL, Japan). Thier surface compositions were examined by energy dispersive X-ray spectrometer (EDX: DX-4, EDAX, Japan). The concentrations of Ti and Ni ions dissolved from the Ni-Ti alloy substrates into the H_2O_2 solutions were measured by inductively coupled plasma emission spectroscopy (ICP: SPS-7700, Seiko, Japan). Fresh blood was obtained after informed consent from healthy human donors and mixed with 3.8 % sodium citrate. The citrated whole blood was centrifuged at room temperature to obtain platelet rich plasma (PRP). The Ni-Ti alloy substrates were placed in a polystyrene 24-well plate dish. Then, 1 mL of PRP was added into each Ni-Ti alloy substrate well and incubated for 30 min at 36.5°C. The Ni-Ti alloy substrates were rinsed with phosphate buffer solution (PBS). The platelets adhered on the Ni-Ti alloy substrates were fixed in PBS containing 2 % glutaraldehyde and then in a 1 % osmic acid aqueous solution for 1 h at 4°C. After fixation, the Ni-Ti alloy substrates were dehydrated with *t*-butanol and freeze-dried overnight.

The surface density of adhered platelets were observed with SEM photographs after sputter-coated with gold.

Results

Fig. 1 shows the TF-XRD patterns of the Ni-Ti substrate with no treatment (denoted hereafter NT) and the Ni-Ti substrates treated with 6 mass% H_2O_2 with pH 2.5 at 80°C for 30 min. to 3 hours and subsequently heat-treated at 400°C for 1 hour (denoted hereafter as CT). The XRD patterns of NT and CT indicated the presence of crystalline phases of Ni-Ti(JCPDS #18-0899) and TiO(JCPDS #02-1196), while the XRD patterns of CT indicated the presence of crystalline phase of titania (anatase, JCPDS #21-1272). The EDX analysis indicated that the atomic ratios of Ti/Ni of NT and CT(3h) were 0.98 and 1.65, respectively.

Fig. 2 shows the change in the concentrations of Ti and Ni dissolved in the H_2O_2 solutions as a function of the chemical treatment time. The Ni concentration increased with the treatment time, while the Ti concentration was almost constant after 30 min.

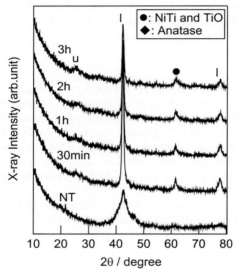

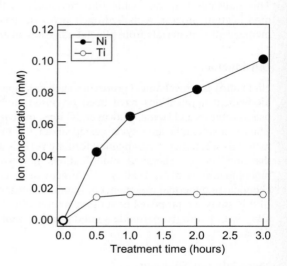

Fig. 1 TF-XRD patterns of the Ni-Ti substrate before and after treated with 6 mass% H_2O_2 with pH 2.5 at 80°C for 30 min. to 3 hours and subsequently heat-treated at 400°C for 1 hour.

Fig. 2 Changes in the concentrations of Ti and Ni ions dissolved in the H_2O_2 solutions as a function of the chemical treatment time.

Figs. 3(a) and (b) show SEM images of NT and CT substrates, respectively. Figs. 4(a) and (b) show SEM images of NT and CT substrates after *in vitro* platelet adhesion experiment, respectively. Note that the CT substrate adhered less platelets than NT substrate.

(a)   (b)

Fig. 3 SEM images of NT (a) and CT substrates (b), respectively.

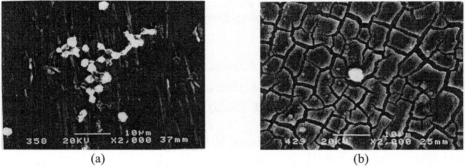

(a)   (b)

Fig. 4 SEM images of NT (a) and CT substrates (b) after *in vitro* platelet adhesion experiment, respectively.

Discussion.

We have already reported the characterization of titania films derived from combined chemical and thermal treatments of titanium substrates.[1, 2] Our recent study [3, 4] showed that titania films are useful in biomedical applications such as orthopedic implants. A series of reactions leading to direct formation of titanium peroxides in the titanium and hydrogen peroxide system (Ti-H_2O_2) are considered on the basis of Tengvall *et al.*[5, 6] : They involve some redox reactions that the derivative species are subsequently subject to follow.

$$Ti(s) + 2H_2O_2 \rightarrow TiOOH(aq) + 1/2O_2(g) + H_3O^+ \qquad (1)$$
$$TiOOH + 1/2O_2 (g) + H_3O^+ \rightarrow Ti(OH)_4 \qquad (2)$$
$$Ti(OH)_4 \rightarrow TiO_2(aq) + 2H_2O \qquad (3)$$
$$Ti(OH)_4 \rightarrow Ti(OH)_{4-n}^{n+} + nOH^- \qquad (4)$$
$$TiO_2(aq) + 4H^+ \rightarrow Ti(IV)(aq) + 2H_2O \qquad (5)$$

Here, eq. (1) shows that TiOOH is derived in H_2O_2 solution accompanying the formation of protons or oxonium ions. TiOOH(aq) may stay either on the substrate surface or in the solution. In eq. (2) hydrated neutral complex $Ti(OH)_4$(aq) is derived from TiOOH(aq), consuming the oxonium ions, while eq. (3) describes the formation of hydrated titania gel, TiO_2(aq), from $Ti(OH)_4$(aq). Eq. (4) shows that $Ti(OH)_4$ changes to positively charged complex $Ti(OH)_{4-n}^{n+}$. In eq. (5) dissolution of the hydrated titania gel proceeds at the expense of protons or oxonium ions. Thus, pH decreases in eq. (1) while it increases in eqs. (2), (4), and (5). However, the system may involve some important chemical species, including $TiO_3(aq)$, superoxide radical O_2^-, stable complex $TiO_2^- - TiO_2^{2-}$, OH^-, hydroperoxyl (HO_2), hydroperoxide ion (HO_2^-) or TiO_2^{2-} that have not been considered in eqs. (1) through (5). Nevertheless, eqs. (1)-(5) will be used in the following discussion because they well describe the correlation between the possible reactions and the formation of titania films after heat-treatment.

We confirmed that the selective dissolution of Ni ions was accelerated especially in the H_2O_2 solution with pH 2.5. Since Fig. 2 indicated that the Ni ions selectively dissolved into the H_2O_2 solution, the titania-rich phase (Ti/Ni ratio: 1.65) must be left on the surface of the Ni-Ti alloy substrates, resulting in the formation of a titania film due to the reactions eq.(1) through eq. (3). The result of *in vitro* evaluation of platelet adhesion indicated that the crystalline titania film suppressed the adhesion of platelet.

It is concluded that the present modification process with hydrogen peroxide solution is applicable to prevent the Ni-Ti shape memory alloy materials from platelet adhesion when in contact with blood.

Acknowledgements.

This study was supported by the Industrial Technology Research Grant Program (02A47022a) in '02 from the New Energy and Industrial Technology Development Organization (NEDO) of Japan.

References.

[1] S. Hayakawa, K. Shibata, K. Tsuru and A. Osaka, (2001), Bioceramics, Vol. 14, (2001) p. 71.

[2] K. Shibata, K. Tsuru, S. Hayakawa and A. Osaka, (2002), Bioceramics, Vol. 15, (2002), p. 55.

[3] S. Kaneko, S. Takemoto, C. Ohtsuki, T. Ozaki, H. Inoue and A. Osaka , Biomaterials, 22[9], (2001), p. 875.

[4] T. Kim, M. Suzuki, C. Ohtsuki, K. Masuda, H. Tamai, E. Watanabe, A. Osaka, H. Moriya, J. Biomed. Mater. Res., Vol. 64B[1], (2003), p. 19.

[5] P. Tengvall, H. Elwing, and I. Lundström, J. Colloid. Interf. Sci., Vol. 130, (1989), p. 405.

[6] P. Tengvall, E.G.Hörnsten, H. Elwing, and I. Lundström, J. Biomed. Mater. Res., Vol. 24, (1990), p. 319.

Key Engineering Materials Vols. 254-256(2004) pp. 869-872
online at http://www.scientific.net
© 2004 Trans Tech Publications, Switzerland

Orthosilicic Acid Increases Collagen Type I mRNA Expression in Human Bone-Derived Osteoblasts *In Vitro*

M. Q. Arumugam[1,3], D. C. Ireland[2], R. A. Brooks[3], N. Rushton[3], W. Bonfield[1]

[1] Cambridge Centre for Medical Materials, Dept of Materials Science and Metallurgy, Cambridge University, Pembroke St, Cambridge, CB2 3QZ, United Kingdom; mqa20@cam.ac.uk

[2] Bone Research Group, School of Clinical Medicine, Cambridge University, Box 157, Addenbrooke's Hospital, Cambridge CB2 2QQ, United Kingdom; dci20@mole.bio.cam.ac.uk

[3] Orthopaedic Research Unit, Dept of Surgery, Cambridge University, Box 180, Addenbrooke's Hospital, Cambridge CB2 2QQ, United Kingdom

Keywords: Silicon, orthosilicic acid, human osteoblasts, reverse transcriptase-polymerase chain reaction, collagen type I

Abstract. The effect of 0, 5μM, 10μM, 20μM and 50μM orthosilicic acid on collagen type I gene expression in cultured human trabecular bone-derived osteoblasts was investigated. Collagen type I messenger RNA (mRNA) was quantified using real–time reverse transcriptase-polymerase chain reaction (RT-PCR). After 20 hours incubation a significant increase in collagen type I mRNA was seen in cells cultured with 5μM, 10μM and 50μM orthosilicic acid ($p<0.05$) compared to untreated controls. This study suggests that silicon has a stimulatory effect on collagen type I mRNA in human osteoblasts.

Introduction

There is considerable evidence that bone growth is significantly enhanced by the addition of carbonate or silicon into a hydroxyapatite (HA) matrix [1, 2]. The mechanism by which this occurs is therefore increasingly being explored. A previous study by Refitt et al. [3] using MG63 osteosarcoma cells and primary human bone marrow stromal cells, indicated that silicon addition increased collagen type I synthesis in cells treated with 10 and 20μM orthosilicic acid. Collagen type I mRNA levels, however, remained unchanged. We investigated the effect of soluble silicon on human osteoblast cells by studying the effect of increasing concentrations of orthosilicic acid (0-50μM) on collagen type I gene expression in human osteoblast cells isolated from trabecular bone. The collagen type I mRNA were measured using reverse transcriptase-polymerase chain reaction (RT-PCR). Changes in collagen type I mRNA due to either altered transcription or mRNA stabilisation would indicate a direct effect of silicon on cells leading to altered collagen type I production. The following work is based on the assumption that silicon is released from the silicon-substituted hydroxyapatite substrate into the surrounding tissue over time after implantation *in vivo*. As an initial investigation, it is therefore important to consider the effect of silicon on the cells *in vitro*.

Materials and Methods

Two separate experiments were carried out.

Experiment 1

Human osteoblast cells (HOBs), isolated from trabecular bone, were grown in McCoy's medium supplemented with 10% Foetal Calf Serum, 1% glutamine and Vitamin C (30μg/ml). Cells at passage 5 were grown in four 75cm^2 flasks to 50-60% confluence. A 1.0μM stock solution of orthosilicic acid was prepared from sodium silicate and MilliQ ultra high purity water and the pH was adjusted to approx. pH 7.0. This solution was added to each of 3 flasks to give final concentrations of 10μM, 20μM and 50μM in 12ml of medium. As a control, 240μL of ultra high purity water was added to a fourth flask. The cells were incubated for 20 hours, following which medium was removed and the cells solubilised using 7.5ml of Trizol (Invitrogen, Paisley, U.K.) per flask and the lysates were stored at -80°C. Total RNA was prepared from the aqueous phase after addition of 0.2ml chloroform per ml of Trizol used for the original homogenisation. After centrifugation at 3000 RCF for 45 minutes at 2-8°C, the aqueous phase was transferred to fresh tubes and 0.5ml of isopropyl alcohol was added per ml of Trizol. After incubation for 10 minutes at room temperature and centrifugation at 3000 RCF for 30 minutes at 2-8°C, the supernatant was removed from the gel-like pellet and the RNA was washed with 75% ethanol. The tubes were then vortexed and centrifuged at 3000 RCF for a further 20 minutes at 2-8°C. The supernatant was removed and the pellet was air dried for 10 minutes before being re-suspended in 100μL of RNase free water.

Collagen type I mRNA levels were quantified by real-time RT-PCR. RT-PCR is a very sensitive, quantitative method for investigating differences in mRNA levels between samples. It relies on the ability of the PCR reaction to amplify a few molecules of RNA, once DNA copies have been formed by the action of reverse transcriptase. Real time RT-PCR is based on the release of a fluorescent marker when each molecule of DNA is amplified. The number of amplification cycles required to reach a threshold fluorescence value is a measure of the amount of mRNA in the sample. The comparative threshold-cycle method was used to determine mRNA levels [4]. Collagen type I mRNA levels were expressed relative to mRNA for the housekeeping gene, glyceraldehyde-3-phosphate dehydrogenase (GAPDH), which encodes a protein essential for cell metabolism and its message levels remain stable under altered external conditions.

Experiment 2

HOBs isolated from trabecular bone were grown in McCoy's medium supplemented as above. Cells at passage 5 were grown in five 25cm^2 flasks to 50-60% confluence. The 1.0μM stock solution of orthosilicic acid was then added to each flask to give final concentrations of 5μM, 10μM, 20μM and 50μM in 4ml of medium. 80μl of ultra high purity water was added to a fifth flask as a control. As above, the cells were incubated for 20 hours before being solubilised using Trizol, the RNA was then harvested and collagen type I relative to GAPDH mRNA levels were quantified by RT-PCR.

Results

Figures 1a) and 1b) show the morphology of the HOBs prior to and immediately after the addition of orthosilicic acid. Figures 1c) and 1d) show the cells with and without orthosilicic acid after 20 hours of incubation.

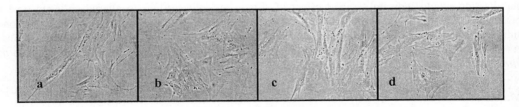

Fig. 1 Optical Microscope images of HOBs before incubation, from left to right; a) without addition of orthosilicic acid, b) with 50μM orthosilicic acid, c) without orthosilicic acid (control) after 20hrs incubation and d) 50μM orthosilicic acid after 20 hours incubation.

The addition of orthosilicic acid at concentrations up to 50μM produced no visible effect on cell morphology after 20 hours as seen in Figure 1. It was therefore concluded that the addition of orthosilicic acid caused no short-term cytoxicity to the cells.

Experiment 1

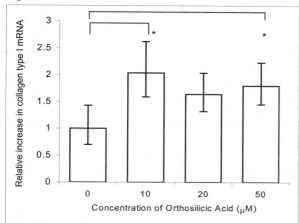

Fig. 2: Collagen type I mRNA levels, following the addition of orthosilicic acid (0-50μM), are shown relative to GAPDH mRNA levels. Results are normalised to levels without silicon. * $p<0.05$ using Satterthwaite's approximate test with the Bonferroni correction [5]. Mean values of $\pm$ S.D. for four replicates are shown.

Experiment 2

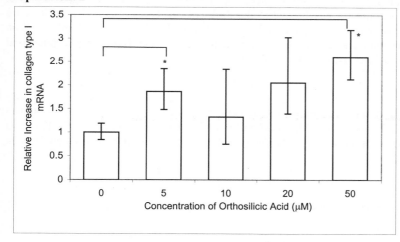

Fig. 3: Collagen type I mRNA levels, following the addition of orthosilicic acid (0-50μM), are shown relative to GAPDH mRNA levels. Results are normalised to levels without silicon. * $p<0.05$ using Satterthwaite's approximate test with the Bonferroni correction [5] Mean values of $\pm$ S.D. for four replicates are shown.

A significant increase in collagen type I mRNA was seen with orthosilicic acid concentrations of 10μM and 50μM (Fig. 2) and 5μM and 50μM (Fig. 3) after 20 hours incubation *in vitro*. The results from experiment 1 suggested that the increase in collagen type I mRNA with the addition of orthosilicic acid was not dose dependent. However, the results from experiment 2 suggested there may be an increase in collagen type I mRNA levels with increasing concentrations of orthosilicic acid. Further experiments are required to investigate this effect in more detail.

Discussion

There has been extensive study on the effects of silicon on bone formation [6-9] with increasing evidence suggesting that silicon is essential in skeletal development. The main source of silicon for humans is the diet, but the bioavailability of silicon from solid foods is not well understood. These studies [6-9] suggest that silicon affects collagen formation and bone mineralisation and it is also known to be involved in the formation of cross-links between collagen and proteoglycans during bone growth.

Work done by Carlisle et al. [10], found that Si was involved in the early stage of bone calcification in physiological conditions. Gibson et al. [11] found that the *in vitro* bioactivity of HA, was substantially enhanced with the incorporation of silicon and that the acceleration in bone apposition for SiHA may partly result from an up-regulation in osteoblast cell metabolism, but the exact mechanism of this is yet to be resolved. Work done by Patel et al. [1], indicated that early *in vivo* bioactivity of HA was significantly improved with the incorporation of silicate ions into the HA structure. The absolute percentage of bone ingrowth, coverage and bone mineral apposition rate for SiHA implants were found to be significantly greater than that of HA at 23 days *in vivo*. This implies that SiHA is an attractive alternative to conventional HA for use as bone substitute.

In contrast to the study by Refitt et al. [3], we found that collagen type I mRNA levels increased significantly with the addition of orthosilicic acid at concentrations of 5, 10 and 50µM.

Refitt et al. found that collagen type I protein synthesis increased in all cells treated with orthosilicic acid at concentrations of 10 and 20 µM although there was no change in the collagen type I mRNA levels in the treated MG63 cells. The authors therefore suggested a post-translational mechanism for the increased production of collagen. Our results however, suggest a mechanism involving increased transcription or mRNA stability. The different results obtained in this study compared to that of Reffit et al. may be due to the methods used to quantify collagen type I mRNA. Fluorescence based detection in real-time RT-PCR is more sensitive that end-product detection on agarose gels. It is possible also that the differences may be due to the different cell types and media. MG63s are highly proliferative osteosarcoma cells, producing lower levels of collagen type I mRNA than HOBs. In addition, Reffit et al. used a culture medium supplemented with β-glycerophosphate and dexamethasone, which stimulates osteoblastic differentiation.

Conclusion

The results of this study showed a significant increase in collagen type I mRNA in human osteoblasts cultured for 20 hours with 5µM, 10µM and 50µM orthosilicic acid ($p<0.05$). This data suggest a stimulatory effect of silicon on collagen type I mRNA levels in human osteoblasts *in vitro*.

Acknowledgements

The author gratefully acknowledges the Cambridge-MIT Institute for funding.

References

[1] N Patel et al., J. Mat. Sci.: Mat. in Med. 13, (2002)
[2] J Merry, PhD Thesis, June 2000, Queen Mary and Westfield College, University of London
[3] D M Reffitt et al., Bone 32 (2003), p127-135
[4] L Fink et al., Nat. Med. 4 (1998), p1329-1333
[5] P Armitage, G Berry, *Statistical Methods in Medical Research*, (Blackwell Science, 1994)
[6] C D Seaborn, F H Nielsen, Biological Trace Element Research 89(3): (2002), p239-250,
[7] K L Tucker et al., Journal of Bone and Mineral Research 16: (2001), S510-S510
[8] R Jugdaohsingh et al., American Journal of Clinical Nutrition 75(5): (2002), p887-893
[9] M C Hott et al., Calcified Tissue International 53(3): p174-179.
[10] E M Carlisle et al., Science 167 & Calc. Tiss. Int. 33 (1981), p27 [11] I Gibson et al., J. Biomed. Res. 44 (1999), p422

Key Engineering Materials Vols. 254-256(2004) pp. 873-876
online at http://www.scientific.net

Mechanical Properties and Biocompatibility
of Surface-Nitrided Titanium for Abrasion Resistant Implant

Fumio Watari, Yutaka Tamura, Atsuro Yokoyama,
Motohiro Uo and Takao Kawasaki [1]
[1] Hokkaido University Graduate School of Dental Medicine, Sapporo 060-8586, JAPAN
watari@den.hokudai.ac.jp

Keywords: Biocompatibility, bone formation, surface modification, implant, titanium, titanium nitride, particles, hardness, abrasion, scratch, ultrasonic scaler

Abstract. Physical properties and biocompatibility of surface-nitrided titanium Ti(N) were investigated to examine its possible use as an abrasion resistant implant material. Mechanical properties of Ti(N) were evaluated by three different tests, Vickers hardness test, Martens scratch test and ultrasonic scaler abrasion test. Vickers hardness of a nitrided layer of 2 μm in thickness was 1300, about ten times higher than that of pure Ti. Martens scratch test showed the high bonding strength of the nitrided layer with matrix Ti. The abrasion test using an ultrasonic scaler showed very small scratch depth and width, demonstrating extremely high abrasion resistance. The dissolved amount of Ti ion in SBF was as low as the detection limit of ICP, and that in the 1% lactic acid showed no significant difference from Ti. The tissue reaction of cylindrical implant in soft tissue of rats showed very little inflammation, and fine particles of 1 μm induced phagocytosis, which was similar to Ti. The implantation in the femora of rat showed the new bone formed in direct contact with implants. Ti(N) with sufficient abrasion resistance and biocompatibility comparable to Ti is promising as an abrasion-resistant implant materials such as abutment in dental implants and artificial joints.

Introduction

Ti is highly corrosion-resistant due to the thin and stable protective oxide layer on surface and one of the most biocompatible metals [1]. In this sense Ti is the nearly ideal material for implant and used most commonly in orthopedics and dentistry. There are, however, not only good properties but also disadvantages. Low abrasion resistance is one of the few shortcomes of Ti. The abraded fine particles produced in the sliding parts of artificial joints often cause inflammation in the surrounding tissue.

As for dental implants it is protruded from the inside of jaw bone to oral cavity [2]. Since the abutment part (mucosa penetration part) is exposed in the oral cavity, plaque and dental calculus tend easily to adhere on it. Removal of them is necessary to obtain a good prognosis throughout the long term maintenance of the implant. Plastic tip is currently used for this purpose to avoid the mechanical scratch on the implant surface, which may cause further plaque adhesion. Abrasion resistance against the scaling treatment is desirable.

Although bulk Ti is stable and biocompatible *in vivo*, minute Ti abrasion powder may cause an inflammatory reaction [3,4]. The development of implant materials with both biocompatibility and abrasion resistance is, therefore, important. Ti nitride is known for its high surface hardness and mechanical strength. The dissolution of Ti ions is also expected to be very low.

In the present study the analysis of surface properties and mechanical properties, including abrasion test and biocompatibility was done for surface-nitrided Ti, abbreviated as Ti(N) in the following text, and compared with Ti for the possible use as an abrasion resistant implant material.

Materials and Methods

Surface nitriding was done on pure Ti plates (10x10x0.5mm, JIS type 1, KOBE STEEL) and rod(1φx7mm, 99.9%, NILACO) under a N_2 atmosphere of 1atm at 850 °C for 7 hours. The products of nitriding were identified by X-ray diffraction(JDX-3500, JEOL). The cross-section of the Ti(N)

was observed by SEM (S-4000N, HITACHI) after the specimen was cut vertically and polished. Surface durability was evaluated by three static and dynamic mechanical tests: Vickers hardness test, Martens scratch test and for the more practical viewpoint a newly developed abrasion test with an ultrasonic dental scaler which is used to remove calculus on teeth in dental clinics. Surface roughness measurement machine (Surf Com 200C, TOKYO SEIMITSU) was applied to measure the depth profiles of indentations made by hardness test and scratch grooves by scratch and abrasion tests. Corrosion resistance was evaluated by dissolution test in SBF and acid solution using ICP(P4010, HITACHI) elemental analysis. Animal implantation test was done in soft and hard tissue of rat with rod and fine particles (3 μm Ti, SOEKAWA CHEMICALS; 1.5 μm TiN, KISHIDA CHEMICAL) to verify biocompatibility.

Results
Mechanical and Surface Properties.   Fig.1 shows the cross-section of nitrided layer with about 2μm thickness observed by SEM.   X-ray diffraction indicated that the nitrided layer is composed of TiN and Ti N. Vickers hardness of Ti(N) was 1308, nearly ten times larger than 146 of Ti. Corrosion test showed that the dissolved amount of Ti ion was as low as the detection limit of ICP in SBF and had no significant difference from Ti in 1% lactic acid.

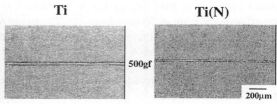

Fig.2. Scratched trace after Martens scratch test (load 500gf) on Ti and Ti(N)

Fig.1. Cross-section of nitrided layer observed by SEM

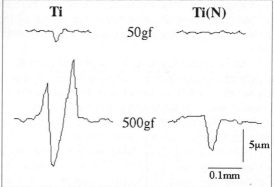

Fig.3. Cross-section profiles of traces after Martens scratch test(load 50, 500gf)

Fig.2 and 3 show the scratched traces and their cross-section profiles after Martens scratch test on Ti and Ti(N), respectively. The trace depth and width increased with an increase of load. Protrusions were recognized on both sides of the trace at loads higher than 250gf for Ti. For Ti(N), however, traces were not observed at loads of 50 and 100gf. Martens scratch test showed that the bonding of nitrided layer to Ti is strong and coherent to matrix Ti, enough to be resistant against being peeled off by scratching.

Fig.4 shows the abrasion test equipment using a dental ultrasonic scaler. Fig.5 shows the cross-sectional profiles after the abrasion test. For the load of 50gf, protrusions by scaling were confirmed on Ti. The trace depth was about 9μm . For Ti(N), a trace was hardly observed, and no change of the surface roughness could be recognized. For 500gf, the protrusions were observed on both sides of the trace on Ti. The trace depth was about 25μm. For Ti(N), the trace was barely observed. The trace depth was about 2.5μm, about one-tenth that of Ti and there were no protrusions. Abraded volume was increased with the load in Ti, while no trace was formed in Ti(N) instead the stainless tip of scaler was abraded. The test showed that abrasion would be negligibly small under the practical conditions of the load 50gf in dental clinics.

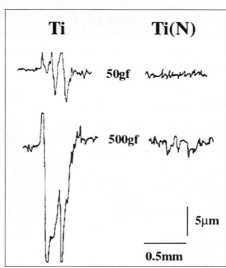

Fig.4. The abrasion test equipment using a dental ultrasonic scaler

Fig.5. Cross-sectional profiles after ultrasonic scaler abrasion test (load 50, 500gf)

Biocompatibility. The tissue reaction of the cylindrical implant of Ti(N) in soft tissue of rats showed little inflammation, and fine particles of 1 μm induced phagocytosis, which was similar to Ti. Fig.6 shows the new bone formation around Ti and Ti(N) in femoral bone marrow after 4 weeks. A 30μm thick layer of newly formed bone was mostly in direct contact with implant surface. Quantitative histopathological analysis for 4 to 8 week implantation showed that the volume ratio of new bone was increased from 22.4 to 42.6% in Ti, 22.0 to 40.8% in Ti(N) and the direct contact ratio of new bone to the implant surface was increased from 51.5 to 67.8% in Ti, 46.6 to 65.8% in Ti(N). There was a significant increase between 4 and 8 weeks. There were no significant differences between Ti and Ti(N) for each period. The fine particles of 3 μm Ti and 1.5 μm TiN inserted in soft tissue induced phagocytosis and inflammation. There was little difference in tissue reaction between Ti and Ti(N). All these results of Ti(N) in corrosion resistance and biocompatibility were nearly equivalent to those of Ti.

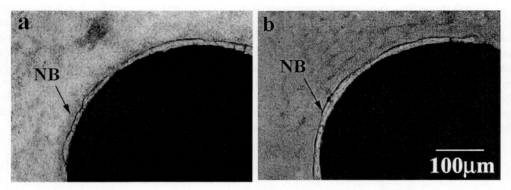

Fig.6. New bone formation around Ti(a) and Ti(N) (b) after 4 week implantation in femoral bone marrow of rat. NB: new bone.

Discussion and Conclusion

Surface-nitrided titanium, Ti(N), showed the extremely high abrasion resistance [5], corrosion resistance and nearly equivalent biocompatibility to Ti in soft and hard tissue in animal implantation test [6]. These facts suggest that Ti(N) would be suitable as abrasion-resistant implant materials and is promising for dental implants especially for abutments. Dentists could use more easily the dental ultrasonic scaler with a stainless tip to remove the plaque and dental calculus instead of the current, careful use with a plastic tip. As a possible target of Ti(N), application in orthopaedics is also suggested for the sliding parts of artificial joints in which abrasion powders cause cytotoxicity.

References

[1] H. Matsuno, A. Yokoyama, F. Watari, M. Uo and T. Kawasaki : Biomaterials. Vol.,22(2001) p. 1253-1262.

[2] F. Watari, A. Yokoyama, F. Saso, M. Uo, and T. Kawasaki: Composites Part B. 28B (1997) 5-11

[3] R. Kumazawa, F. Watari, N. Takashi, Y. Tanimura, M. Uo and Y. Totsuka: Biomaterials, Vol. 23 (2002), p. 3757-3764.

[4] K.Tamura, N.Takashi, R.Kumazawa, F.Watari, Y.Totsuka: Materials Transactions, Vol. 43(12), (2002), p. 3052-3057.

[5] Y.Tamura, A.Yokoyama, F.Watari, M.Uo, T. Kawasaki: Materials Transactions, Vol. 43(12), (2002), p. 3043-3051.

[6] Y.Tamura, A.Yokoyama, F.Watari, T.Kawasaki: Dental Materials Journal: Vol. 21(4), (2002), p. 355-372.

Key Engineering Materials Vols. 254-256(2004) pp. 877-880
online at http://www.scientific.net

Adhesion and Proliferation of Human Osteoblastic Cells
Seeded on Injectable Hydroxyapatite Microspheres

CC Barrias,[1,2] CC Ribeiro,[1,3] MA Barbosa [1,2]

[1] INEB – Instituto de Engenharia Biomédica, Laboratório de Biomateriais, R. Campo Alegre 823, 4150-180 Porto, Portugal, ccbarrias@ineb.up.pt, cribeiro@ineb.up.pt, mbarbosa@ineb.up.pt

[2] Universidade do Porto, Faculdade de Engenharia, Dep. de Eng. Metalúrgica e de Materiais

[3] ISEP – Instituto Superior de Engenharia do Porto, Dep. de Física

Keywords: Injectable microspheres, hydroxyapatite, bone regeneration, osteoblast-like MG63 cells

Abstract. In the present work, the interaction of human osteoblast-like MG63 cells with novel hydroxyapatite (HAp) microspheres was investigated. Cells were seeded on the microspheres and incubated for up to 7 days. The cell-material constructs were visualised by scanning electron microscopy and confocal laser scanning microscopy, and the rate of cell proliferation was estimated using the Neutral Red assay. The results showed that osteoblastic cells were able to adhere and proliferate on the HAp microspheres, which are intended to be used as injectable support-materials for bone regeneration.

Introduction

The use of injectable biomaterials for bone regeneration purposes has received much attention recently [1]. The main advantage of such strategy relies on the possibility of filling defects of different shapes and sizes, while requiring only minimally invasive techniques for implantation, which provide less patient discomfort. Microparticles, suspended on appropriate vehicles, are among the most common forms of injectable materials. Once implanted, they are expected to conform to the irregular implant site, with more or less close packing, and to encourage host cell migration, attachment, proliferation and differentiation. The interstices between the particles, if presenting an appropriate size, may also provide a space for both tissue and vascular ingrowth.

In a previous work, a novel route for the preparation of calcium-phosphate microspheres was described [2,3]. The ceramic granules are first mixed with alginate, which enables the preparation of homogeneous spherical particles through ionotropic gelation in the presence of Ca^{2+}. The spherical particles are subsequently sintered to burn-off the polymer and fuse the ceramic granules, producing pure hydroxyapatite (HAp) microspheres [2,3]. In the present study, the ability of HAp microspheres to support the adhesion and growth of human osteoblastic cells was investigated using the MG63 cell line.

Materials and Methods

HAp microspheres were prepared as previously described [2,3]. Briefly, HAp powder (CAM Implants) pre-heated at 1000°C was mixed with sodium alginate solution (3% w/v) at a ratio of 0.2 w/w and well homogenised. The paste was extruded drop-wise into a 0.1M $CaCl_2$ crosslinking solution, where spherical-shaped particles instantaneously formed and were allowed to harden for 30 min. The size of the microspheres was controlled by regulating the extrusion flow rate using a syringe pump and by applying a coaxial air stream (Encapsulation Unit Var J1, Nisco). At completion of the gelling period, microspheres were recovered and rinsed in water in order to remove the excess $CaCl_2$. Finally, they were dried overnight in a vacuum-oven at 30°C, and then sintered at 1100°C for 1h, with a uniform heating rate of 5°C/min from room temperature. The size of the microspheres was determined using a laser scanner particle size analyzer (Coulter Electronics), and their morphology was analyzed by optical microscopy.

Cell culture studies were performed using MG63 osteoblast-like cells, which express a number of features characteristic of relatively immature osteoblasts. Cells were routinely maintained in α-MEM supplemented with 10% v/v foetal calf serum, 2.5 μg/ml fungizone and 50 μg/ml gentamicine. Microspheres were steam-sterilised (120°C, 20 min) and pre-incubated in culture medium over-night in non-tissue culture plates, to avoid cell adhesion on the bottom of the wells. MG63 cells were seeded on the microspheres at 500 cells/mg and incubated at 37°C in a humidified atmosphere of 5% v/v CO_2 in air for 4h, or 1, 3, 5 and 7 days with the medium being replaced every 2 days. Cell proliferation was estimated using the Neutral Red (NR) assay. Cells seeded on tissue culture grade polystyrene (TCPS) plates were used as a control.

Cell morphology was visualized at different time points by scanning electron microscopy (SEM) and confocal laser scanning microscopy (CLSM).

Results and Discussion

The granulometric analysis of the microspheres (Fig. 1a) revealed that the size distribution of the main population of particles is narrow and follows a normal distribution, with 90% (in volume) of the particles being smaller than 600 μm and having an average diameter around 550 μm.

A second population of microspheres with an average diameter around 1000 μm and corresponding to approximately 4% of the particles was also identified. Sieving could easily eliminate these larger microspheres, which result from occasional particle-particle aggregation during the extrusion process. An optical micrograph of the microspheres is presented in Fig. 1b and illustrates their uniformity.

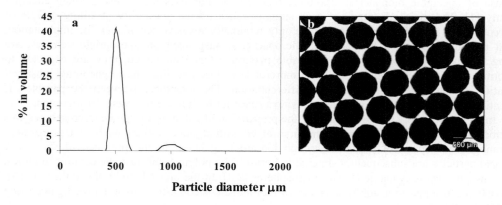

Fig. 1. (a) Particle size distribution and (b) optical microscopy image of the HAp microspheres.

SEM analysis of the cell-microspheres constructs revealed that after an initial period of 4h (Fig 2a), some adherent cells could already be observed on the surface of the microspheres. Because the ability of cells to attach, adhere and spread will influence their capacity to proliferate and differentiate in contact with the material, this initial phase is of critical importance [4]. After 1 day (Fig. 2b, 2c) several cells exhibiting a spindle-like morphology were dispersed on the surface of the microspheres. Some round cells (possibly mitotic cells) were also present. At day 5 (Fig. 2 d-f), cells were well flattened exhibiting numerous filopodial-like extensions and cell-cell contact points.

Finally, at day 7 (Fig 2 g-i) the microspheres were almost completely covered by the cells that formed continuous cell layers in some regions.

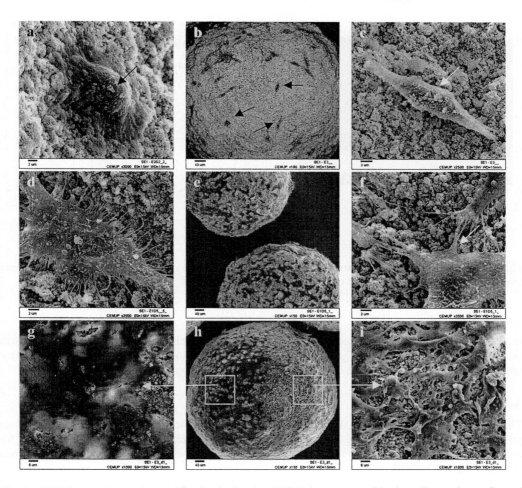

Fig. 2. SEM images (several magnifications) showing MG63 human osteoblastic cells on the surface of hydroxyapatite microspheres after an incubation period of 4h (a), 1 day (b-image obtained with the backscattered electrons mode and c), 5 days (d-f) and 7 days (g-i).

The distribution of cells on the HAp microspheres (day 5) was also visualised using CLSM. (Fig. 3a). Pictures constructed by superimposing images obtained using both the fluorescence and reflection channels showed numerous cells on the surface, a result similar to the one obtained by SEM.

The number of viable cells on the surface of the microspheres was evaluated by performing the NR assay at different time points. This assay is based on the incorporation of a vital colorant by viable cells, and its subsequent fixation at anionic sites on the surface of lysosomal membranes. As any alteration of the membranes will result in a diminished fixation of the colorant, only viable cells will be stained, and so the intensity of the developed coloration allows the indirect quantification of the number of viable cells in the sample. Fig. 3b shows that the number of viable cells, both on the surface of the microspheres and on control TCPS plates, gradually increased along the 7 days in culture, indicating that cells were actively proliferating.

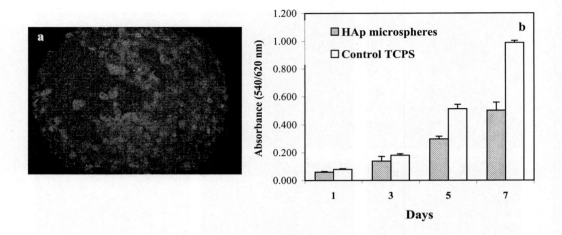

Fig. 3. (a) CLSM image showing osteoblastic cells (light grey areas) on the surface of HAp microspheres (day 5), and (b) proliferation of cells on HAp microspheres estimated by the Neutral red assay.

Conclusions

In this study it was demonstrated that hydroxyapatite microspheres are able to support human osteoblastic cells adhesion and proliferation. Further studies to evaluate the influence of these materials on the expression of the osteoblastic phenotype are in progress.

Acknowledgments

The authors are grateful to Dr. Paula Sampaio (IBMC) for her assistance with CLSM and to programme Praxis XXI from the Portuguese Foundation of Science and Technology (FCT) for awarding Cristina Barrias a scholarship. This work was carried out under contract POCTI/FCB/41523/2001.

References

[1] Temenoff JS and Mikos AG. Injectable biodegradable materials for orthopaedic tissue engineering. Biomaterials 21(2000):2405-2412.

[2] Ribeiro CC, Barrias CC and Barbosa MA. Calcium phosphate-alginate microspheres as protein delivery matrices for bone tissue regeneration. Submitted 2003.

[3] Ribeiro CC, Barrias CC and Barbosa MA A novel route for the preparation of injectable ceramic porous microspheres for bone tissue engineering. Submitted 2003.

[4] Anselme K. Osteoblast adhesion on biomaterials. Biomaterials 21(2000):667-681.

Key Engineering Materials Vols. 254-256(2004) pp. 881-884
online at http://www.scientific.net
© 2004 Trans Tech Publications, Switzerland

Preliminary Radiological *in vivo* Study of Calcium Metaphosphate Coated Ti-Alloy Implants

S.W.Kim[2], S.Oh[1], C.K.You[1], M.W.Ahn[2], K.H. Kim[3], I.K.Kang[4], J.H.Lee[5], S.Kim[1]

[1]School of Metallurgical & Materials Eng., Yeungnam University, Kyongbuk, Korea,
sykim@yumail.ac.kr
[2]Dept of Biomedical Eng., Yeungnam University, Daegu, Korea
[3]College of Dentistry, Kyungpook National University, Daegu, Korea
[4]Dept of Polymer Engineering, Kyungpook National University, Daegu, Korea
[5]Dept of Metallurgical Engineering, Dong-A University, Busan, Korea

Keywords: CMP, biodegradable, Ti-alloy, implant, integration

Abstract. In order to evaluate the bone-implant integration behavior of biodegradable calcium metaphosphate (CMP, $[Ca(PO_3)_2]_n$) coated metallic implant, as-machined, blasted, and blasted /CMP-coated Ti6Al4V screw-type implants were prepared and aseptically implanted into male New Zealand white rabbits. CMP sol was prepared by sol-gel process and coated on each substrate by dip and spin coating. The CMP coated layer was smooth and uniform with fine grains, compared to that of as-machined and as-blasted specimens. Each specimen was inserted into the defects of bilateral intratibial metaphysis bone and then followed up for 1 and 6 weeks. From the radiographs at 1 and 6 weeks after implantation, all the implants were shown to be apparently well integrated with surrounding bone tissue without interfacial fracture, bony resorption, or radiolucent lines. With the combination of histological results, CMP-coated group was noticed that bony bridges were extending from the endosteum onto the implants at 6 weeks after implantation, with the showing good osseo-intergration compared to other two groups.

Introduction

The development of a stable direct bonding between bone and implant surface (osteointegration) is the critical issue for the long-term success of orthopedic and dental implants. The establishment and maintenance of osteointegration depends on wound healing, repairing and remodeling of hard tissues. The tissue response to an implant involves physical factors such as implant design and surface topography, and chemical factors such as composition and structure of the material surface [1]. To improve the implant fixation to a host bone, several strategies have been developed focusing on the surface modification of materials. For example, the physical surface modification of implants in roughness by various techniques has been attracted, because it has been demonstrated that the osteoblastic cells tend to attach more easily to rough surface [4], consequently increasing the bone apposition [5]. Chemical surface modifications have been also realized by covalent attachment of an organic monolayer anchored by a siloxane network [2], and immobilization of specific adhesive peptides like arginine-glycine-aspartic acid-serine (RGDS) [3].

In addition, the implants coated with different bioactive materials such as calcium phosphates, bioactive glasses [6], diamond-like carbon, and amorphous C-N film [7] enhanced the bonding to bone. However, it has been currently reported that the coating layer was sometimes delaminated from the substrates. Therefore, as one of the alternatives to solve this problem, the biodegradable material coating on implants, which may allow the organism to replace the foreign material by new bone tissue in a balanced time schedule, was conductd [8]. Calcium metaphosphate (CMP, $[Ca(PO_3)_2]_n$) is a promising biodegradable material due to its hydrolytic degradation behavior of PO_4^{3-} polymeric chains.

In this study, in order to preliminarily evaluate *in-vivo* bone-implant osseo-integration behavior and stability of biodegradable CMP coated implants radiologically and histologically, Three groups of metallic screws were aseptically implanted into male New Zealand white rabbits and then followed up for 1 and 6 weeks: as-machined, as-blasted, and blasted/CMP-coated Ti6Al4V.

Materials and Methods

CMP sol was prepared by the reaction of $Ca(NO_3)_2 \cdot 4H_2O$ (Aldrich 99%, USA) with $P(OC_2H_5)_3$ (Fluka 97%, Japan) in methyl alcohol using correct amounts to obtain the stoichiometric Ca/P ratio of 0.5. First, Ca-precursor dissolved in methyl alcohol was dehydrated at 180 °C and then anhydrous calcium precursor was dissolved in methyl alcohol in Ar-atmosphere. P-precursor was pre-hydrolyzed for 1 hour with the addition of HCl and H_2O as catalysts, before the reaction with the Ca-precursor. The prepared CMP sol was aged at 40 °C for 48 hours. As-machined (group 1), grit-blasted with CMP powder (75~150μm) (group 2), and blasted and then CMP-coated (group 3) screw-shaped Ti-alloy implants were prepared. CMP sol was coated by dip and spin coating at 5000 rpm for 50 seconds. CMP coated specimens were immediately dried at 70 °C for 12 hours and then heat-treated at 600 °C for 3 hours. All the specimens were sterilized in an autoclave at 120 °C for 10 min before implantation. The surface morphology was observed with SEM and the material phases were identified with XRD. Six groups (as-machined, as-blasted and blasted/ CMP-coated for 1 and 6 weeks, respectively) of skeletally mature male New Zealand white rabbits (2.5 ~ 2.8 Kg) were used for aseptic implantation under general anesthesia induced by Xylaxine (Rumpun, Bayer Korea) 11mg/kg and Ketamine (Ketamine, Yuhan, Korea) 10mg/kg intramuscular injection. Bilateral intratibial metaphysis bone defects were formed by a round burr, each specimen was inserted into the defects and then followed up radiologically and histologically for 1 and 6 weeks.

Results and Discussion

Homogeneous and transparent CMP sol was prepared by sol-gel method. After 48 hour-aging at 40 °C, the appropriate viscosity of CMP sol was obtained for dip and spin coating on Ti-alloy dental implants. The viscosity was 12.7 cP at 27 °C. The surface morphology of the coated and heat-treated CMP layer is smooth and uniform with fine grains (about 100nm), compared to that of as-machined and as-blasted specimens (Fig. 2). The phases of the fine grains were identified with δ-CMP (JCPDS #9-363) from XRD study. The radiography of three different groups at 1 and 6 weeks after implantation showed all the implants were apparently well integrated with surrounding bone tissue, without fractures at the CMP-implant interface, bony resorption, or radiolucent lines.

    (a) as-machined   (b) blasted with CMP   (c) blasted/CMP-coated

Fig. 1 Ti6Al4V screw-type implants with different surface morphology (×12).

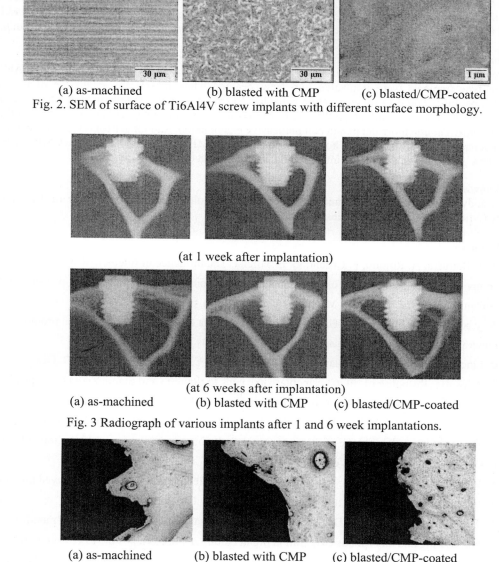

(a) as-machined   (b) blasted with CMP   (c) blasted/CMP-coated

Fig. 2. SEM of surface of Ti6Al4V screw implants with different surface morphology.

(at 1 week after implantation)

(at 6 weeks after implantation)

(a) as-machined   (b) blasted with CMP   (c) blasted/CMP-coated

Fig. 3 Radiograph of various implants after 1 and 6 week implantations.

(a) as-machined   (b) blasted with CMP   (c) blasted/CMP-coated

Fig. 4 Histologies of implants at 6 weeks after implantation.

After 1 week, new bone formation did not appear, however, new bone formation was observed after 6 weeks. In CMP-coated group, it was noticed that bony bridges were extending from the endosteum onto the implants at 6 weeks after implantation, showing good osseo-intergration, compared to the other two groups (Fig. 3).

With a combination of histological results (Fig. 4), a relatively good biocompatibility and osseo-integration of CMP-coated group could be observed without a foreign body reaction or an inflammatory response. However, it was not easy to determine here the exact evidence of resorption and replacement into bone of CMP coated layer by a process of remodeling after integration, because the observation period of 6 weeks might have been short to figure out *in-vivo* behavior. CMP-blasted and CMP coated groups showed better bone formation and integration than the as-machined group.

Conclusion

Three groups of as-machined, grit-blasted with CMP powder, and blasted/CMP coated Ti6Al4V screw-type implants were implanted into New Zealand white rabbits for evaluation of *in-vivo* biocompatibility and osseo-intergadation behavior. Based on apparent radiological and histological examinations, bone-implant integradation of CMP coated group was much better, compared to other two groups. Though more detail and comprehensive examinations and longer follow-up are required to understand the phenotype bone bonding mechanism of CMP layer, it is believed that CMP coating will be a promising candidate for improved osseo-integration of metallic implant.

Acknowledgements

This study was supported by a Grant-in-Aid for Core Technology Development Programs from the Korea Ministry of Commerce, Industry and Energy.

References

[1] P.A. Ramires, A. Romito, F. Cosentino, E. Milella: The influence of titania/hydroxyapatite composite coatings on in vitro osteoblasts behaviour. Biomaterials Vol. 22 (2001), p. 1467-1474.

[2] Sukenik CN, *et. al,*. J Biomed Mater Res Vol. 24 (1990), p.1307-23.

[3] Kay c. Dee, David c. Rueger, Thomas T. Andersen, Rena Bizios: Conditions which promote mineralization at the bone-implant interface. Biomaterials Vol. 17 (1996), p. 209-15.

[4] J. Lincks, B.D. Boyan, C.R. Blanchard, C.H. Lohmann, Y. Liu, D.L. Cochran, D.D. Dean, Z. Schwartz: Response of MG63 osteoblast-like cells to titanium and titanium alloy is dependent on surface roughness and composition. Biomaterials Vol. 19 (1998); p. 2219-32.

[5] Tetsuya Jinno, Victor M. Goldberg, Dwight Davy, Sharon Stevenson: Osseointegration of surface-blasted implants made of titanium alloy cobalt-chromium alloy in a rabbit intramedullary model. J Biomed Mater Res Vol. 42 (1998), p. 20-29.

[6] S. Vercaigne, J.G.C. Wolke, I. Naert, J.A. Jansen: The effect of titanium plasma-sprayed implants on trabecular bone healing in the goat. Biomaterials Vol. 19 (1998), p. 1093-9.

[7] C. Du, X.W. Su, F.Z. Cui, X.D. Zhu: Morphological behaviour of osteoblasts on diamond-like carbon coating and amorphous C-N film in organ culture. Biomaterials Vol. 19 (1998), p. 651-8.

[8] Antia A. Ignatius, Carla Schmidit, Daniela Kaspar, Lutz E. Claes: In vitro biocompatibility of resorbable experimental glass ceramics for bone substitutes. J Biomed Mater Res Vol. 55 (2001), p. 285-294.

VII. NANOTECHNOLOGIES APPLIED TO BIOCERAMICS

Key Engineering Materials Vols. 254-256(2004) pp. 887-890
online at http://www.scientific.net
© 2004 Trans Tech Publications, Switzerland

Preparation of Nano-HAP as Vectors for Targeting Delivery System

Y.R. Duan[1] , Z.R. Zhang*[1] , Y.Huang[1] and C.Y. Wang[2]

1. West China School of Pharmacy, Sichuan University, Chengdu 610041, China ,
    jychen@scu.edu.cn

2. Engineering Research Center in Biomaterials, Sichuan University, Chengdu 610064, China

Keywords: Orthogonal design, nano-Hap vectors, PELGE, nanoparticles.

Abstract. Nanoparticles can be easily transported in vivo and can reach and attack cancer cells. The biodegradable hydroxyapatite(HAp) nanoparticles which strongly inhibited the proliferation of the cancer cell wrapped with degradable copolymer monomethoxy(polyethyleneglycol)-poly (lactide-co- glycolide)-monomethoxy (poly-ethyleneglycol) *(PELGE)*. The nanoparticle preparation method is a critical problem for small-sized particles. HAP-loaded nanoparticles(NP) were fabricated by using a double-emulsion system. Orthogonal design was applied to optimize the preparation technology on the basis of the single factor evaluation. The optimal conditions for preparation HAP-loaded nanoparticle was as follows: 20mg/ml was the concentration of PELGE, volume of inner-phase of HAp was 0.5ml(C=12.5mg/ml), the ratio of DCM/acetone was 3/2 and the concentration of Pluronic® F68 (Poloxamer 188 NF) was 3%. The entrapment efficiency was >88% and particle size less than 500nm. The nanoparticles, as detected by transmission electron microscopy (TEM), have a smooth and spherical surface. The HAP could be loaded into PELGE copolymers. In this study , the HAP nanoparticle-polymer delivery system was established using PELGE polymers as wraping material.

Introduction

It has been long recognized that the efficient use of drugs requires that they should be delivered selectively at the site of cancer cells, preferably at a controlled rate. This is especially true for potent drugs with strong side effects, such as the anticancer drugs. In addition to selectivity, such as beta-ray radioactive elements, requires that they should be protected from in vivo before reaching their cancer cells.Because nanoparticles can be easily transported in vivo, beta-ray source materials are fabricated by adding beta-ray radioactive elements in the nanoparticles and then, wrapped by biocompatible polymeric biomaterials. In the meantime, specific antibodies to cancer cells are added into the particles. When this nanoparticles reach the tumor, they can bind specifically to the target cancer cells and attack them with beta-ray so that their destruction to the surrounding normal tissues is reduced to the least. HAP nanoparticles which inhibited the proliferation of the cancer cell [1] is one of the suitable materials for this purpose. Because of nano-grade hydroxyapatite can be used as the vectors for diagnosis and therapeutic medicine and the radioactive elements and amino- and carboxy-containing medical molecules can be easily adsorbed to the particles, the nanoparticle preparation method is a critical step for this targeting delivery system. PELGEs are excellent in biocompatibility, biodgrability and therefore, PELGE are applied as the wrapping materials to the nano-particles. Two biodegradable PELGE polymers of different molecular weight were selected to prepare nano-HAP targeting delivery system. In this study, orthogonal experiment design was adopted to select the optimal conditions for preparation of nanoparticles by the experiment of the singular factor. This method, namely w/o/w solvent evaporation,was used for the preparation of nano-HAP as vectors to investigate whether biodegradable nanocapsules could be obtained. The results reported could be useful to the rational design of nano-HAP as vectors for targeting delivery system.

Materials and Methods

Preparation of Materials: The HAP were synthesized by slowly dropping the $(NH_4)_2HPO_4$ aqueous solution into the stirred $Ca(NO_3)_2$ aqueous solution. The nanograde needle-like crystals (Fig.2) was observed by TEM. PELGE copolymers were prepared by a melting polymerization process under vacuum using stannous octoate as catalyst. Their composition was characterized by ^{1}H-NMR and their molecular weight and molecular distribution were analysised by gel permeation chromatography (GPC). Two kind copolymers of different molecular weight were used. PELGE1(LA:GA:70:30, PEG 10%) with molecular weight of 12kDa, and PELGE2(LA:GA:70:30, PEG 10%) with molecular weight of 20kDa.

Fabricate Method of Nanoparticles: The nanoparticles were fabricated by entraping HAP nanoparticles with PELGE1 or PELGE2. An aqueous solution of HAP was emulsified in dichloromethane/acetone, in which the copolymer had been dissolved, using probe sonication at 180w for 10 seconds. This w/o (water/oil) emulsion was transferred to an aqueous solution of F68 and the mixture was probe sonicated at 360 W for 30 seconds. The w/o/w emulsion formed was gently stirred at room temperature until the evaporation of the organic phase was completed.

Orthogonal Experiment Design: Orthogonal experiment design was adopted to select the optimal scheme by the experiment of the singular factor. The following six factors: concentration of PELGE, ultrasonic time, concentration of F68, molecular weight of PELGE, stirring time and volume of inner-phase, were selected to investigate their influence on the HAP encapsulation efficiency and particle size.

Characteristic of Nanoparticles: Nanoparticles were analyzed on negative stain electron microscopy using a JEM 1200 EXII electron microscope (Jeol Ltd, Tokyo, Japan). The particle morphology was observed by TEM. The size distribution of nanoparticles was determined by laser diffractometry (Mastersize/2000, Malvern). The samples were diluted with distilled water and measured at room temperature with a scattering angle of 90°.

The amount of HAP in the aqueous phase was quantitatively determined by inductively coupled plasma (ICP). The encapsulation efficiency was calculated by the ratio of the practical HAP entrapment to the theoretical HAP entrapment: practical HAP entrapment (% w/w)/theoretical HAP entrapment (% w/w) x100.

Results and Discussion
Fig.1 shows the ^{1}H-NMR spectra of PELGE triblock copolymers in CDCL$_3$.TEM photographs of the HAP nanograde needle-like crystals is shown in Fig.2.

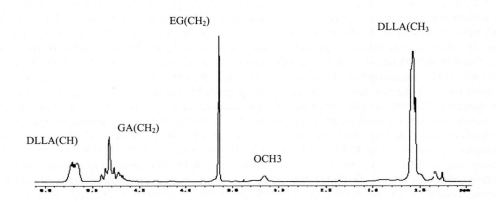

Fig.1.^{1}H-NMR spectra of PELGE1(LA:GA ratio in PLGA was 70:30, PEG was 10%) in CDCL$_3$

The polymer used to form the nanoparticles will also strongly affect the structure, properties and applications of the nanoparticles. As previously stated, PLGA has been the most common polymer used to make biodegradable nanoparticles, however, these are not the optimal carriers for all drug

Fig. 2. TEM photographs of the HAP

delivery applications. PELGEs are excellent in biocompatibility, biodgrability and therefore, In our study, PELGEs are applied as the wrapping materials to fabricate the nano-particles. Pluronic F68 was used as an the stabilizer [2]. The possibility of using Pluronic F68 to produce HAP-PELGE nanospheres is an interesting option because of its well-known acceptability for parenteral administration. The amount of stabilizer used will also have an effect on the properties of the nanoparticles. Most importantly, if the concentration of the stabilizer is too low, aggregation of the polymer droplets will occur. Alternatively, if too much of the stabilizer is used, the drug incorporation could be reduced due to the interaction between the drug and stabilizer.

However, when the stabilizer concentration is between the 'limits', adjusting the concentration can be a means of controlling nanoparticle size [3]. When the concentration of F68 is less than 3%, experimental results show that there is an evident effect of concentration on nanoparticle size. The size decreases with the increase of the concentration of F68. When the concentration of F68 exceeded 3%, the size decreased with increased concentration of F68, however, the entrapment efficiency could be reduced. So, we chose the concentration of F68 at 3% (w%).

A number of parameters, e.g, conditions of the aqueous phase, composition of the organic phase and emulsification conditions, concentration of F68, may influence the nanoparticles size, encapsulation efficiency or both. Optimizing the parameters that greatly affect the particle size is done by keeping the concentration of F68 (3%, w%) while change other parameters, such as, concentration of PELGE, ultrasonication time, molecular weight of PELGE, stirring time etc. These findings showed that ultrasonic time did not directly influence the nanoparticle diameter or encapsulation efficiency and neither the temperature of the external phase and stirring time did.

When nanoparticles were prepared, the mean diameter can be changed. Especially, the nanoparticle size tended to changed when changing the concentration of PELGE, the ratio of solvent (DCM/acetone, v/v) used to dissolve the copolymer and the volume of inner-phase(HAP, C=12.5mg/ml).

The main factors that could affect the particle size and the entrapment efficiency are: concentration of PELGE, composition of the organic phase and the volume of inner-phase, which was denoted as A, B and C. Four points of experimental for each factor were selected(Table 1): the concentration of PELGE (A) is 5~20(mg/ml); the ratio of acetone/DCM is 0/5~5/5, the volume of inner-phase is 0.2ml~1.0ml. In the orthogonal table(table 1), the experiments for the preparation of nanoparticles of optimization is to investigate the effect of these three factors on the particle size and the entrapment efficiency.

Table 1.The point of experimental selection for orthogonal design($L_{16}(4^3)$)

Factor	A	B	C
Level	Concentration of PELGE (mg/ml)	Ratio of Acetone/DCM (V/V)	The volume of inner phase (ml)
1	5	0/5	0.3
2	10	1/4	0.5
3	15	2/3	0.7
4	20	3/2	0.9

After the $L_{16}(4^3)$ table was selected, experiment plans were carried out .These plans were $A_1B_1C_1$, $A_1B_2C_2$, $A_1B_3C_3$, $A_1B_4C_4$, $A_2B_1C_2$, $A_2B_2C_3$, $A_2B_3C_4$, $A_2B_4C_1$, $A_3B_1C_3$, $A_3B_2C_4$, $A_3B_3C_1$, $A_3B_4C_2$, $A_4B_1C_4$, $A_4B_2C_1$, $A_4B_3C_2$, $A_4B_4C_3$.

The particle size of the nanoparticles, entrapment ratio and drug loading were investigated.and discovered that $A_4B_3C_2$ are close to the predicted level. The optimal conditions were found that the concentration of PELGE was at 20mg/ml, volume of inner-phase of HAp was V=0.5ml, the ratio of DCM/acetone was 3/2.

The nanoparticles, as shown in TEM photos (Fig.3), have a smooth and spherical surface. Fig.3a is for HAP-free nanoparticles, 3b is for HAP-loaded nanoparticles wrapped by PELGE1 and 3c is for HAP-loaded nanoparticles wrapped by PELGE2. The size distribution of nanoparticles was determined by laser diffractometry, and particle size ranged from 38 to128 nm

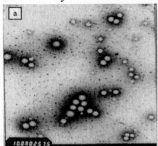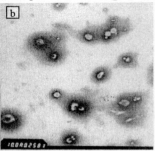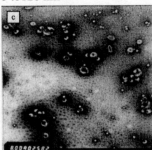

Fig.3. TEM photographs of the HAP-free nanoparticles(a) and HAP-load nanoparticles(3b and 3c)(10,000×)

High concentration of PELGE solution, by including acetone in the organic solution, formed a fine PELGE membrane at the interface of the HAP droplet and a PELGE organic phase during the first W/O emulsion stage. The HAP would be enclosed in a PELGE membrane before the second emulsification. Thus, the loss of HAP in the continuous aqueous phase during the second sonication would be greatly reduced. An increase in HAP entrapment was observed after increasing the concentrations of both HAP and PELGE solutions. However, lower MW PLGA under the same conditions produced much lower entrapment of HAP. This might be because the low molecular weight PLGA formed a weaker film when deposited on the droplet interface which could be easily disrupted later by the second sonication. On the contrary, high MW PLGA formed a stronger film which would not easily be disrupted and therefore would provide a better barrier against HAP leakage. These results suggest that the deposition of higher density PELGE immediately around the HAP droplets is a key factor in preparing nanoparticles with high HAP loads.

The formation mechanism of nanoparticles can be explained by the large interfacial area resulting from emulsification and the gradual reduction of the globule size due to solvent transferation and probably, the interfacial turbulence generated during diffusion. A higher concentration of PELGE was found to rapidly increase the size and polydispersity of nanoparticles. In contrast, an increase in concentration of stabilizer agent was found to moderately reduce the size of the nanoparticles[4].

Conclusion

Nanoparticles were prepared by a double-emulsion adjustment method to reduce the particle size to nano grade. The best condition was obtained by orthogonal design based on the single factor's experiments. The optimal conditions for preparation HAP-loaded nanoparticle was as follows: 20mg/ml was the concentration of PELGE, volume of inner-phase of HAp was 0.5ml (C=12.5mg/ml), the ratio of DCM/acetone was 3/2 and the concentration of F68 was 3%. The entrapment efficiency was >88% and particle size is <500nm. Beta-ray source materials were fabricated by adding beta-ray radioactive elements in the biodegradable hydroxyapatite phosphate (HAP) nanoparticles, which will be reported in another paper.

References

[1] S.P.Li. The Chinese Journal of Nonferrousals. 9(1999), p. 650-654.

[2] Fontana G . Biomaterials 22 (2001), p.2857–2865.

[3] M.L. Hans, A.M. Lowman. Current Opinion in Solid State and Materials Science 6 (2002), p. 319–327.

[4] Kumaresh S. Soppimath , Tejraj M. Aminabhavi , Walter E. Rudzinski, et al. Journal of Controlled Release 43 (1997), p. 197–212.

Key Engineering Materials Vols. 254-256(2004) pp. 891-894
online at http://www.scientific.net
© *2004 Trans Tech Publications, Switzerland*

Nanostructure Of Hydroxyapatite Coatings Sprayed In Argon Plasma

Elena I. Suvorova[1], Yury D. Khamchukov[2] and Philippe-A. Buffat[3]

[1] Institute of Crystallography RAS, Leninsky pr., 59, Moscow 119333, Russia,
Suvorova@ns.crys.ras.ru or Elena.Suvorova@epfl.ch

[2] Institute of Technical Acoustics, Ludnikov pr., 13, Vitebsk 210023, Belarus, ita@vitebsk.by

[3] Centre Interdisciplinaire de Microscopie Electronique, MXC, EPFL Lausanne, Switzerland,
philippe.buffat@epfl.ch

Keywords: High resolution transmission electron microscopy, nanostructure, hydroxyapatite, plasma sprayed coatings.

Abstract. High resolution transmission electron microscopy (HRTEM) and X-ray energy dispersive spectroscopy (EDS) were applied to study the structure of hydroxyapatite (HAP) coatings deposited by plasma spraying on different substrates (Cu, Cr, Ni, NaCl, BaF_2). Coatings were obtained with a 13.56 MHz RF argon plasma at 1 Pa by placing a HAP target as the RF electrode. The typical rate of coating formation was about 1.0 - 1.5 nm/s and the total thickness was 350 – 400 nm. For all coating conditions and substrates investigated here, coatings are constituted of single-phase HAP polyhedral nanocrystals of 3-20 nm size randomly oriented to each other and without texture relatively to the substrate.

Introduction

Plasma spraying now is widely used for manufacturing hydroxyapatite coatings on different substrates for implants and successful experience in clinical application was reported more than ten years ago already [1]. However, it is believed that such coatings can be accompanied with several calcium phosphates in uncertain proportions and particle sizes. This may lead to rejection and inflammation effects. Unsuccessful implantations are observed from time to time, mainly having to do with the degradation and fatigue behavior of the coatings. Microstructure (nanostructure) of deposited layers plays a significant role in the process of biointegration.

Thus the availability, efficiency and reliability of materials in orthopedics, prosthesis and pharmacopoeia require an accurate control of the microstructure with relevant methods of characterization. In particular only analytical transmission electron microscopy can provide the necessary morphological, crystallographic and chemical information to understanding the mechanical, chemical and bio-integration properties of thin coatings observed during wear, corrosion in bio-environment and after implantation. Model coatings were deposited on very different substrates (Cu, Cr, Ni, NaCl, BaF_2) to investigate their influence on microstructure.

Materials and Methods

The initial HAP powder used for target preparation was obtained by precipitation in aqueous solutions following the chemical reaction:

$$10Ca(NO_3)_2 + 6(NH_4)_2HPO_4 + 8NH_3 + 2H_2O = Ca_{10}(OH)_2(PO_4)_6 + 20NH_4NO_3$$

Hydroxyapatite precipitation was washed, dried and annealed at 950°C. 58mm HAP targets were manufactured by mixing HAP with polyvinyl alcohol, then pressing and annealing at 500°C.

The argon plasma at 1 Pa was created in quartz reactor with a RF generator (13.56MHz) between a grounded titanium electrode and a hydroxyapatite target as the RF electrode. The target and the substrate surfaces were parallel and the distance between them was about 10 mm. The approximate temperature read on a thermocouple lying close to the substrate surface was 100°C. The typical rate of the coating formation was about 1.5 nm/s.

HAP suffers from irradiation damage under ionizing beams. Thus EDS chemical analysis was performed with electron nano-probes down to a few nanometers in diameter for short counting times, leading to high spatial resolution, with low composition accuracy or with longer counting times on larger areas for better accuracy, when the coating looked uniform.

Observation to study the layer morphology and the structure down to the atomic level was carried out in a Philips CM300UT FEG transmission electron microscope (300 kV field emission gun, 0.65 mm spherical aberration, 1.2 mm chromatic aberration, 0.17 nm resolution at Scherzer defocus and 0.11 nm limit of resolution). The images were recorded on a Gatan 797 slow scan CCD camera (1024 pixels x 1024 pixels x 14 bits) and processed with the Gatan Digital Micrograph 3.6.1 software (Gatan, Inc., Pleasanton, CA), including Fourier filtering.

Phase identification of the samples was carried out by analyzing electron diffraction patterns taken on micron-scale areas and Fast Fourier Transforms (FFT) of HRTEM images in sub-micron and nanometer-scale areas.

The interpretation of the HRTEM images and electron diffraction patterns were performed with the help of the JEMS software package for images simulation and diffraction calculations [2]. Diffraction patterns and images were calculated for the electron-optical parameters given above and the structural data [3] of the hexagonal hydroxyapatite phase (unit cell $P6_3/m$ with parameters a=0.942 and c=0.688 nm). The best match of these images on experimental micrographs provided unambiguous phase assessments. In addition to HAP, 7 possible calcium phosphate modifications were considered.

Results

TEM observation of cross-sectioned samples showed that a 350-400-nm thick calcium phosphate layer was formed. Coatings on all substrates looked dense, uniform with a low surface roughness (<10 nm). The smoothest coatings were obtained on the atomically flat faces of NaCl and BaF_2 single crystals. Figures 1a and 2a demonstrate cross section and plan-view of samples.

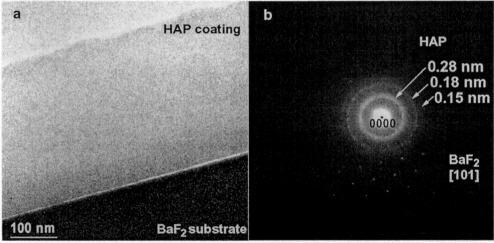

Fig. 1. Cross section of the coating on the BaF_2 substrate (a) and the corresponding selected area diffraction pattern (b) where broaden rings belong to the HAP coating and spots are coming from the BaF_2 substrate.

The presence of Ca and P in the coating was determined from EDS patterns (Fig.2b). Only qualitative information was considered because the pretty fast irradiation damage leads to phosphorous loss and thus to unreliable composition concentrations when the necessary electron dose for statistical relevancy was attained (Fig. 3).

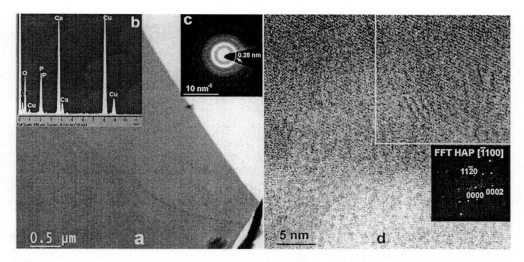

Fig. 2. Plan-view of HAP coating (a) with EDS pattern (b) and SAED pattern (c). HRTEM image (d) and FFT were taken close to the $[\bar{1}100]$ zone axis of a HAP nanocrystal from the coating.

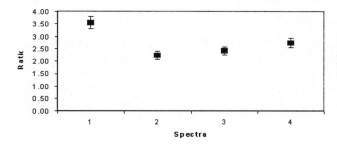

Fig. 3. The Ca/P ratio in different points of the coating determined from EDS patterns.

Selected areas electron diffraction (SAED) patterns of the coating and underlying substrate taken from areas of about one micron in diameter behave diffuse reflections (Figs.1b and 2c) due to size effect broadening [4, 5]. Also only 3-4 visible rings appeared due to large structure factors of reflections given by {0002}, {$21\bar{3}1$} - {$03\bar{3}0$}, {$22\bar{4}2$} - {$3\bar{2}\bar{1}3$}, and {$24\bar{6}0$} – {$05\bar{5}2$} atomic plane groups. Taking into account the interplanar spacings and distribution of reflection intensity, the rings were indexed within the HAP lattice. Nevertheless, it cannot be excluded that they may also be formed by other phases. Thus these SAED patterns are unsuitable for accurate phase determination.

Microdiffraction from nanoareas only helped to assess the local crystalline character of the coatings, but also could not provide better phase identification because the beam convergence broadens the reflection, in addition to the already reported size effect in SAED and the high density of the electron beam induces fast irradiation damage.

The highest spatially resolved information was obtained from HRTEM (Figs. 2d and 4) images showing that the layer contained nanocrystals of 3-20 nm size randomly oriented relatively to each other. Indexing the diffractograms obtained by Fourier Transform on a few nanometers areas (FFT in Fig.2d and a, b, c and d boxes in Fig.4) lead only to hydroxyapatite, and no other crystalline phase is present in the coating.

The size of HAP crystals increases with distance from the interface. Close to it the coating structure is very disordered nearby the interface (within some10 nm). It looks most likely to be formed by HAP nanocrystals with sizes comparable to their unit cells, rather than by some undefined amorphous material.

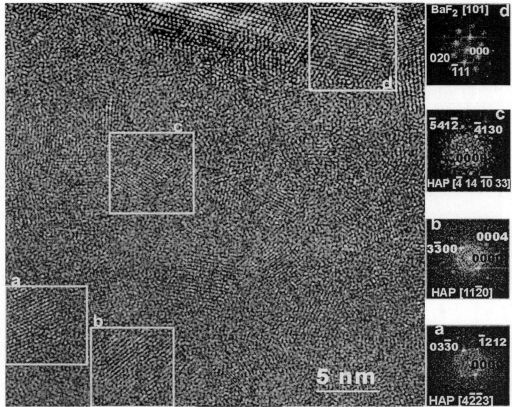

Fig. 4. HRTEM image of cross section from the HAP coating on the BaF2 substrate. Individual HAP nanocrystals are selected by white squares a, b, c. The corresponding diffractograms a, b, and c confirm the phase composition of the nanocrystals and provide their orientations. Diffractogram d was taken from the substrate (white square d) along the zone axis close to [101].

Conclusions

This work proves the ability of HRTEM to study calcium phosphate coatings. Under the investigated conditions of plasma spraying only crystalline hydroxyapatite was found. The coatings contain nanocrystals (up to a few nm) in random orientation. No substrate effect on phase composition or orientation was found.

This work was supported by Russia-Belarus Foundation for Basic Research, grant N 02-02-81009.

References

[1] G. Willmann: Advanced Engineering Materials Vol.1 (1999), p.95.

[2] P.Stadelmann: http://cimewww.epfl.ch/people/Stadelmann/jemsWebSite/jems.html

[3] M.I. Kay, R.A. Young, A.S. Posner: Nature Vol. 204 (1964), p. 1050.

[4] E.I. Suvorova, P.A. Buffat: J.Microscopy Vol. 196, Part.1 (1999), p. 46.

[5] E.I. Suvorova, P.A. Buffat: Crystallography Reports Vol.46, No5, (2001), p.722.

Key Engineering Materials Vols. 254-256(2004) pp. 895-898
online at http://www.scientific.net
© *2004 Trans Tech Publications, Switzerland*

Modulating the Nanotopography of Apatites

M. Bohner

Dr Robert Mathys Foundation, Bischmattstrasse 12, CH-2544 Bettlach, Switzerland;
marc.bohner@rms-foundation.ch.

Keywords: cement, calcium phosphate, specific surface area, osteoinductivity, calcium-deficient hydroxyapatite.

Abstract. Various additives were added to α-tricalcium phosphate - water mixtures to identify their effect on the specific surface area (SSA) of the resulting hardened cement blocks. Results indicated that the specific surface area could be modified in the range of 15 to 40 m^2/g. The most potent additives were calcium sulfate dihydrate (CSD) and calcium carbonate (CC) powders.

Introduction

Calcium phosphate materials are known to be osteoconductive bone substitutes, i.e. bone forms in the bone substitute when it is in close apposition to bone. A decade ago, a study of Ripamonti [1] suggested that coral-derived apatites could also be osteoinductive, i.e. bone can form within the bone substitute even though the bone substitute is in a bone ectopic site. Since then, there has been numerous studies showing that apatites and calcium phosphate materials can be osteoinductive [2-17]. Nevertheless, there is so far no clear understanding for this phenomenon. Factors such as calcium phosphate chemistry [2,5,13,15-16], porosity [3,5,10,12], pore size [6,12], pore shape [12], implant location (intramuscular or subcutaneous, back or thigh) [5,9,11], implant type (granule or block) [6], pre-hardened or injected cement [11], block shape [8], implantation time [1-6,9-11,15], and animal type [5,7,9] have been tested. Generally, more bone has been found (i) after longer implantation times [1-6,9-10,12,15], (ii) in less resorbable calcium phosphates [2,13,15-16], (iii) in baboons, dogs and pigs (rather than rabbits, mice and rats) [7,9], (iv) in more microporous materials [3,5,10,12], (v) in macropores [11], in particular macropore concavities [12], (vi) in blocks (rather than granules) [6], and (vii) intramuscularly (rather than subcutaneously) [9].

Until now, most efforts made on the material side have been focused on the effect of composition, micro- and macrotopography. Little has been done to assess the effect of nanotopography despite the fact that bone consists of calcium phosphate nanocrystals (rather than microcrystals). Moreover, it has been speculated that an increase of SSA of the bone substitute can promote osteoinductivity [1]. Therefore, the goal of the present study is to investigate methods for modifying the SSA of apatite blocks.

Materials and Methods

As a sintering process would erase nanotopographies, calcium phosphate cements are used to produce apatite blocks. Apatite blocks were obtained by hydrolyzing α-TCP powder in an aqueous solution. A wide range of synthesis conditions were tested: (i) the SSA of the starting α-TCP powder was adjusted from to 0.8, 0.9, 1.1 and 1.8 m^2/g by grinding the initial α-TCP powder for variable times; (ii) the liquid to powder ratio of the mixtures was varied between 0.43 and 0.50 mL/g; (iii) precipitated hydroxyapatite (PHA; SSA = 58 m^2/g), calcium carbonate (CC; SSA = 7.1 m^2/g) and calcium sulfate dihydrate (CSD; SSA = 4.5 m^2/g) powders were added to replace part of the α-TCP powder (up to 28 w%); (iv) the phosphate concentration ($Na_2HPO_4 \cdot 2H_2O$; DSHPD) of the mixing solution was changed from 0 to 0.4 M; and (v) the samples were incubated for one, two and/or 16 days. The choice of the latter synthesis conditions is supported by the fact that the particle size of α-TCP powder as well as the addition of CSD, DSHPD and CC modify the setting time [18-19]. Short setting times are associated with high supersaturations which should lead to small crystal sizes and hence large SSA. The experiments were performed according to several series of measurements. All experiments were randomized and repeated at least twice. More details can be found elsewhere [18]. Some of the experiments were made according to a factorial design of experiments where the factors were: (A) Specific surface area of α-TCP (related to three different grinding times: 7, 15 and 60 min); (B) Ratio between α-TCP and CSD (0.00, 0.37, 0.74 and 1.11 g CSD for 4.00, 3.63, 3.36 and 2.89 g α-TCP); (C) $Na_2HPO_4 \cdot 2H_2O$ concentration in the mixing liquid (0.1/0.2 M); (D) Addition of PHA (0.00/0.10 g); (E) Ageing time (1 or 2 days).

Chemicals. α-TCP powder was obtained from Mathys Medical (Bettlach, Switzerland). Its specific surface area was 0.78 ± 0.03 m^2/g (95% confidence level). This powder was ground for 7, 15 and 60 min (Planetary mill, Pulverisette 5, Fritsch, Germany) to obtain finer powders with a SSA of 0.93 ± 0.04, 1.07 ± 0.03, and 1.84 ± 0.07 m^2/g (95% confidence level). The powder-to-liquid ratio of the cements was 2.32 g/mL except in a few cases where the P/L ratio was varied (from 2.32 to 2.00 g/mL). According to the x-ray diffraction (XRD) pattern, the purity of the α-TCP powder was superior to 98%. Small amounts of HA were detected. CSD was purchased from Fluka (Switzerland; Art No 21244). CC and PHA were purchased from Merck (Germany; Art No 2196 and 102076, respectively). These chemicals had a (crystalline) purity close to 100% as assessed by XRD. The phosphate buffer solution (PBS) was made using demineralized water and the following salts: 9.0 g/L DSHPD (Fluka, Switzerland; Art No 71645), 1.79 g/L potassium di-hydrogen phosphate (Fluka, Art No 4873), 4.5 g/L sodium chloride (Merck, Art No 1.06400), and 0.02 g/L sodium azide (Fluka, Art No 71290).

Mixing and preparation. Each sample containing 4.0 g powder (α-TCP + additives) and the solution was mixed for 45 s with a spatula in a small beaker. The paste was then placed into one or two syringes (only in the factorial design) whose tip had been previously cut off (diameter: 12.5 mm). All experiments were performed at $23 \pm 2°C$. Fifteen minutes after setting, the samples were placed into 10 mL of PBS solution at 37°C. In the factorial design of experiments, half of the samples were incubated for 1 day, whereas the other half was incubated for two days. In the subsequent series, all samples were incubated for two days. Each sample containing CC was cut into two parts. The first part was characterized, whereas the second part was incubated for two additonal weeks. After incubation, the samples were taken out and dried in air at 37°C. One day later, the samples were dried at 110°C until a constant weight was reached.

Characterizations. The specific surface area of the dried samples was measured using nitrogen adsorption and applying the BET theory (Gemini 2360, Micromeritics, USA). The cement samples were initially ground, but as no difference was measured between sample chunks and ground specimens, all further measurements were carried out using intact specimens. The weight of each specimen varied in the range of 0.2 to 0.7 g. For XRD measurements, the samples were ground by hand with a pestle and a mortar. The powder was packed in a cavity in an aluminium sample holder. The measurements were done on a Philips PW1800 X-ray powder diffractometer (XRD) using Ni-filtered Cu-Kα radiation (40kV, 30mA) and an automatic divergence slit. The investigation range was from 4 to 40 degrees (2θ) with 1 second per step (0.02 degrees 2θ). Some of the samples were investigated by scanning electron microscopy (Cambridge S360). However, the size of the crystals was too small to note

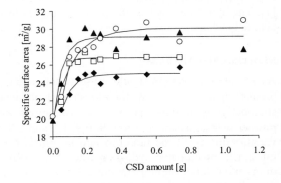

Fig. 1: Effect of CSD and phosphate concentration on SSA. (o) 0.05 M; (σ) 0.1 M; (o) 0.2 M; ($\blacklozenge$) 0.3 M. Each point corresponds to an average of two measurements. Incubation time: 2 days.

morphological changes. Therefore, further evaluation of the samples via SEM was stopped.

Results

The results of the factorial design of experiment suggested that apart from a large increase of SSA due to the addition of CSD, the investigated factors did not have any significant effect on the SSA, or if they did have a significant effect (at $p < 0.01$), it was very limited ($< 10\%$). This was in particular seen with an increase of the phosphate concentration, the SSA of α-TCP, the incubation time and the addition of PHA. All those factors had a large effect on the cement setting time [18].

The SSA of the cement increased with an increase of the CSD amount (Fig 1). The effect of CSD was mostly limited to small amounts (< 0.2 g) (Fig 1). The position of the plateau of SSA varied according to the phosphate concentration. An intermediate phosphate concentration (0.1 - 0.2 M) led to the highest plateau value.

The addition of CC provoked a linear increase of the SSA (Fig 2) and the α-TCP content of the samples after two days of incubation (α-TCP/HA 100% peak intensity ratio close to 1 with 0.2g CC). The presence of large α-TCP remnants motivated the use of a subsequent 2-week-long incubation. Despite the disappearance of α-TCP in all but one sample (0.2 g CC), no additional SSA modification was measured (Fig 2).

A small but significant increase of the SSA was observed with an increase of the amount of mixing liquid (Fig 3). Slightly lower SSA values were measured with 0.74g CSD compared to 0.37g CSD.

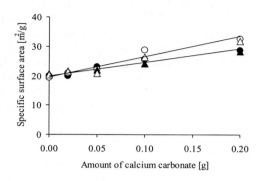

Fig. 2: Effect of CC amount on SSA: (•) 0.37g CSD, 2 days of incubation; (o) 0.74g CSD, 2d; (σ) 0.37g CSD, 16d; (Δ) 0.74g CSD, 16d. Each point corresponds to an average of two measurements.

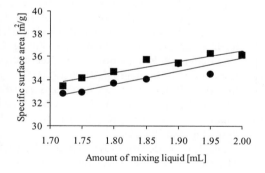

Fig. 3: Effect of the amount of mixing liquid on SSA. CSD amount: (v) 0.37g CSD; (•) 0.74g CSD. Each point corresponds to an average of two measurements. Two days of incubation.

Discussion

The SSA of all compositions were in the range of 15 to 40 m^2/g. No comparison could be made with other apatite cements because no information could be found in the literature. For comparison, our values are much larger than the SSA of most non-cementious bone substitute materials [20], but still much lower than the SSA of the apatite crystals present in bone (80-100 m^2/g). Despite this large difference, the present study not only shows that large variations of the SSA can be attained but also how these variations can be obtained.

Surprisingly, there was very little correlation between setting time and SSA values [18], even though large SSA values are expected with fast setting reactions. This discrepancy might be explained by the fact that the setting time measurements are only related to the initial part of the setting reaction, whereas the precipitation of apatite crystals takes place during the whole setting reaction.

Interestingly, the SSA of samples containing CC did not vary between 2 and 16 days of incubation, (Fig 2) even though large amounts of α-TCP disappeared within this time period. This suggests that long incubation times promote the growth and not the nucleation of apatite crystals.

This study represents a first step in the direction of bone substitutes with high surface areas. The SSA plays an important role in the interactions between body fluids and material. However, the crystal size and shape are also expected to play an inportant role, because cell adhesion and proliferation is strongly influenced by nanotopographies [21], and by nanoparticle shape [22]. Therefore, further experiments are required to fully understand the present results and to evaluate ways to modify the crystal size and shape.

Conclusion

The specific surface of apatite blocks can be easily modified by changing the composition of the initial composition. The most efficient way to increase the SSA is to add a small amount of CSD (3-5%) or increasing amounts of CC. The use of an adequate cement composition and an adequate block synthesis technique [23-25] should enable the obtention of apatite blocks with controlled nano-, micro- and macrotopography. It would then be possible to test the effect of apatite nanotopography on osteoinductivity.

References

[1]   Ripamonti U. J Bone Joint Surg Am 1991;73[5]:692-703.
[2]   Ripamonti U, Ma SS, Reddi AH. Plast Reconstr Surg 1992;89[4]:731-9; discussion 740.
[3]   Yamasaki H, Sakai H. Biomaterials 1992;13[5]:308-312.
[4]   Ripamonti U, Van den Heever B, van Wyk J. Matrix 1993;13[6]:491-502.
[5]   Klein C, de Groot K, Chen W, Li Y, Zhang X. Biomaterials 1994;15[1]:31-4.
[6]   Van Eeden SP, Ripamonti U. Plast Reconstr Surg 1994;93[5]:959-66.
[7]   Ripamonti U. Biomaterials 1996;17:31-5.
[8]   Magan A, Ripamonti U. J Craniofac Surg 1996;7[1]:71-8.
[9]   Yang Z, Yuan H, Tong W, Zou P, Chen W, Zhang X. Biomaterials 1996;17[22]:2131-7.
[10]   Yuan H, Kurashina K, de Bruijn JD, Li Y, de Groot K, Zhang X. Biomaterials 1999;20[19]:1799-806.
[11]   Yuan H, Li Y, de Bruijn JD, de Groot K, Zhang X. Biomaterials 2000;21[12]:1283-90.
[12]   Thomas ME, Richter PW, Crooks J, Ripamonti U. Key Eng Mater 2001;192-95:441-4.
[13]   Yuan H, Yang Z, De Bruijn JD, de Groot K, Zhang X. Biomaterials 2001;22[19]:2617-23.
[14]   Yang ZJ, Yuan H, Ping Z, Tong W, Qu S, Zhang XD. J Mater Sci Mater Med 1997;8:697-701.
[15]   Yuan H, De Bruijn JD, Li Y, Feng J, Yang Z, De Groot K, Zhang X. J Mater Sci Mater Med 2001;12:7-13.
[16]   Kurashina K, Kurita H, Wu Q, Ohtsuka A, Kobayashi H. Biomaterials 2002;23:407-12.
[17]   Ooms EM, Egglezos EA, Wolke JGC, Jansen JA. Biomaterials 2003;24:749-57.
[18]   Bohner M. Biomaterials, Accepted.
[19]   Bohner M. Dr Robert Mathys Foundation. Internal report.
[20]   Weibrich G, Trettin R, Gnoth SH, Goetz H, Duschner H, Wagner W. Mund Kiefer GesichtsChir 2000;4:148-152.
[21]   Curtis ASG and Wilkinson C. Biomaterials, 1997;18[24]:1573-83.
[22]   Laquerriere P, Grandjean-Laquerriere A, Jallot E, Balossier G, Frayssinet P, Guenounou M. Biomaterials 2003;24[16]:2739-47.
[23]   Bohner M. Key Eng Mater 2001;192-195:765-768.
[24]   Barralet JE, Grover L, Gaunt T, Wright AJ, Gibson IR. Biomaterials 2002;23[15]:3063-72.
[25]   Takagi S, Chow LC. J Mater Sci Mater Med 2001;12:135-9.

Key Engineering Materials Vols. 254-256(2004) pp. 899-902
online at http://www.scientific.net

In Vitro Evaluation Of Zirconia Nanopowders

S.Braccini[1], C.Leonelli[1], G.Lusvardi[2], G.Malavasi[2], L.Menabue[2]

[1]Department of Materials and Environmental Engineering,University of Modena and Reggio Emilia,
Via Vignolese 905, 41100 Modena, Italy

[2]Department of Chemistry, University of Modena and Reggio Emilia,
Via G.Campi 183, 41100 Modena, Italy, lusvardi.gigliola@unimore.it

Keywords: zirconia, nanopowders, in vitro behaviour, microwave, bioactivity

Abstract. ZrO_2 is used for a long time as biomaterial. Nanopowders of ZrO_2 are prepared *via* a microwave assisted hydrothermal synthesis and morphological studies carried out by TEM. The powders are tested *in vitro* and this test does not alter the dimension of the particles.

Introduction

Recent advances in microwave technology have met the challenge of providing faster, cleaner, safer, more reproducible, and more accurate nanopowder preparation alternatives [1-3]. The microwave cavity is nowadays functionalized with temperature and pressure sensors, corrosion-resistant coating of all metal surfaces, and advanced feedback power control have transformed the instrumentation into an easy-to-use automated system with a high sample throughput. Vessel design allows control of temperatures up to 300°C and pressure up to 800 psi [4], and some continuous flow reactor are also available on the market. The aim of this paper is to present the use of a microwave digestor to prepare nanopowders via hydrothermal route. In the field of biomaterials the researchers (due to some problems with the traditional materials) are looking for the design of biomaterials with surface properties similar to physiological bone (grain sizes in the nanometric range [5]); this would aid in the formation of new bone at the tissue/biomaterial interface and therefore improve implant efficacy. With the advent of nanostructured materials (materials with grain sizes less than 100 nm in at least one direction [6]) it may now be possible to prepare materials which simulate the surface properties of physiological bone. The use of zirconia ceramics, with several advantages over other ceramic materials started about twenty years ago [7]. Zirconia exhibits phase transformation at high temperatures with a volume expansion; stresses generated by the expansion originate cracks in the zirconia ceramics and the addition of stabilizing oxides (CaO, MgO, Y_2O_3) inhibits this effect.

In this work we report the characterisation and the *in vitro* study of nanometric particles of ZrO_2 in terms of chemical and phase stability.

Experimental

Nanocrystalline ZrO_2 powders were prepared starting from a $ZrOCl_2 \cdot 8H_2O$ (RPE, Carlo Erba, Milan, Italy) aqueous solution 0.5 M [8,9]. The solutions were neutralized with NaOH 1 M to pH 10. The hydrothermal syntheses were conducted using a MW reactor (mod. MDS 200, CEM Corp.), operating at 2.45 GHz and at an optimized power level automatically adjusted to the maximum pressure of 180 psi (corresponding to 185°C) for 2 hrs. The solid phase was separated from the solution by filtering and washed free of salts with distilled water. Transformation from $Zr(OH)_4$ to ZrO_2 was tested by thermogravimetric analyses (TGA, Netzch 402). Crystalline phases were analyzed before and after *in vitro* tests with a powder diffractometer (PW3710 Philips) operating with the CuKa radiation. The data were collected in the 2θ angular range 20-60° (2θ) (qualitative analysis) or 3-100° (2θ) (quantitative analysis). The quantitative X-ray diffraction analysis was performed using the combined Rietveld and reference intensity ratio (RIR) methods [10-14] (by GSAS [15]). Moreover, the nanopowders were investigated by means of transmission electron microscopy (TEM, JEM 2010, JEOL). The *in vitro* tests were performed leaving for 30 days the nanopowders in three different solutions: distilled water, SBF (simulated body fluid [16]) and SBF added of citric acid (0.08M), at 60°C. This temperature was chosen to better evaluate powder behaviour under the most extreme conditions. Leaching solutions were subjected to elemental analysis by inductively coupled plasma spectrometer (ICP, Spectroflame Modula, Spectro).

Results and Discussion

The thermogravimetric analysis (Figure 1) shows two weight losses: the first correlated to water and the second correlated to the loss of OH groups. The hydrothermal synthesis of zirconia could be explained by the following mechanism:

$$[Zr_4O_{(8-x)}(OH)_{2x}] \longrightarrow ZrO_{(2-y/2)}(OH)_y \longrightarrow t\text{-}ZrO_2$$

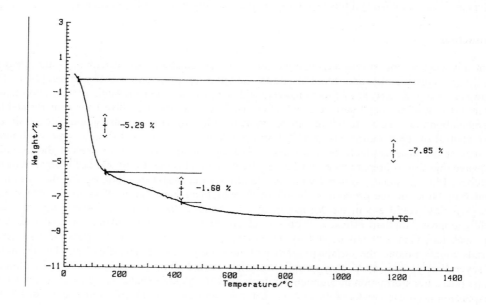

Fig.1. TGA curve of investigated powders

X-ray diffraction analysis reveals that the nanopowders are constituted of ZrO_2 in the tetragonal and monoclinic forms (ICCD file nos.17-923, 37-1484). The quantitative analysis by Rietveld-RIR method indicates also the presence of amorphous phase and the results are (wt%): tetragonal phase 36(1), monoclinic phase 22 (1) and amorphous 42 (1)

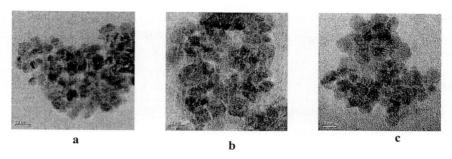

a   b   c

Fig. 2. TEM micrograph before (a) and after *in vitro* test: (b) SBF and SBF with citric acid (c)

TEM investigation reveals that the nanopowders (Figure 2a) are monodispersed, spherical-shaped with dimension range of about 10 nm; some particles exhibit sharp interference fringe patterns that monitor a relatively high crystalline order. Preliminary characterization performed by TEM electron diffraction patterns reveals an intermediate crystallisation degree which can be related to the TGA weight losses. The *in vitro* tests do not alter the dimension of the particles: their reactivity (evaluated by the release of zirconium ion to the solution) is varying and in the subsequent order: SBF (with acid citric)>SBF>H_2O. These results are also confirmed by the XRD spectra corresponding to those solutions (Figure 3): the peak intensity decreases in the same order, but their position is unchanged. There is no evidence of transformation of the tetragonal phase into the monoclinic that could inhibit mechanical properties of the powders [7].

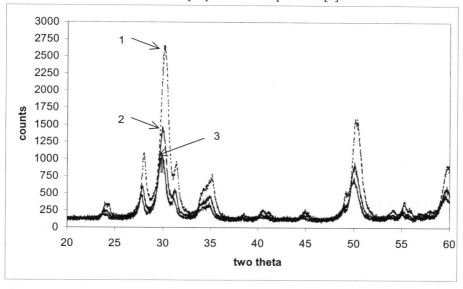

Fig. 3. XRD powders pattern of ZrO_2 before (1) and after *in vitro* test SBF (2) and SBF with citric acid (3)

Acknowledgements

This work is fulfilled under support of MIUR, PRIN 2001-2003 "Synthesis of nanopowders under microwave irradiation".

References

[1]   L.B. Gilman, W.G. Engelhart, *Spectroscopy*, Vol 4, (1989), 14.

[2]   H. Matusiewicz, *Anal. Chem.*, Vol 66, (1994), 751.

[3]   H. Matusiewicz, *High-Pressure Closed-Vessel Systems* in Microwave-Enhanced Chemistry, Eds. H.M. Kingston, S.J. Haswell, Am.Chem. Soc. Washington, DC, USA, (1997).

[4]   W. Lautenschläger, T. Schweizer, *Labor Praxis*, Vol 14, (1990), 376.

[5]   Kaplan FS, Hayes WC, Keaven TM, Einhorn TA, Iannotti JP in: Simon SR, editor. Orthopaedic basic science. Columbus OH:American Academy of Orthopaedic Surgeons, (1994),Vol 127

[6]   Siegel RW, Scientific American, Vol 275, (1996), 42-7.

[7]   C.Piconi, G.Maccauro Vol 20, (1999), 1-25

[8]   F. Bondioli, C. Leonelli, C. Siligardi, G.C. Pellacani, S. Komarneni, Report from the 8th International Symposium on "Microwave and High Frequency Processing", Springer Verlag, (2002).

[9]   F. Bondioli, A.M.Ferrari, C. Leonelli, C. Siligardi, G.C. Pellacani , *J.Am.Ceram.Soc.,* Vol 84[11], (2001), 2728.

[10]   H. M. Rietveld, Acta Cryst., Vol 22, (1967), 151.

[11]   H.M. Rietveld, J. Appl. Cryst., Vol 2, (1969) 65.

[12]   R.L.Snyder, Pow. Diff., Vol 7, (1992), 186.

[13]   M.Bellotto, C.Cristiani, Mater. Sci. Forum, Vol 79, (1996), 745.

[14]   A.Gualtieri, Pow. Diff., Vol 11, (1996), 1.

[15]   A.C.Larson, R.B.Von Dreele, Report LAUR LANL 86-748, Los Alamos National Laboratory, New Mexico,1997.

[16]   T. Kokubo, H. Kushitani, S. Sakka, T. Kitsugi, T. Yamamuro, J. Biomed. Mater. Res. Vol 24 (1990) 721.

Key Engineering Materials Vols. 254-256(2004) pp. 903-906
online at http://www.scientific.net

Preparation and Characterization of Calcium Phosphate Nanoparticles

C.M.Manuel[1],M.Foster[2],F.J.Monteiro[1,3],M.P.Ferraz[1,4],R.H.Doremus[2],R.Bizios[5]

[1]INEB – Instituto de Engenharia Biomédica, Lab. de Biomateriais, Universidade do Porto,
Rua do Campo Alegre, 823, 4150 -180 Porto, Portugal, cmmanuel@fe.up.pt
[2]Rensselaer Poly. Institute, Materials Science and Engineering Dep., Troy, NY, 12180-3590
USA, fosterm5@rpi.edu, doremr@rpi.edu
[3]Universidade do Porto, Faculdade de Engenharia, Dep. de Engenharia Metalúrgica e Materiais,
Rua Dr. Roberto Frias, 4200-465 Porto, Portugal E-mail: fjmont@fe.up.pt
[4]Universidade Fernando Pessoa, Faculdade de Ciências da Saúde, Rua Carlos da Maia, 296,
4200-150 Porto, Portugal, mpferraz@ufp.up.pt
[5]Rensselaer Poly. Institute, Department of Biomedical Engineering, Troy, NY, USA, bizios@rpi.edu

Keywords: nanoparticles, hydroxyapatite, tricalcium phosphate, wet chemical precipitation

Abstract. In this work, CaP nanoparticles were prepared by wet chemical precipitation using three different approaches and were characterised and quantified in terms of their composition. Different precipitating, ageing and drying temperatures were applied and affected HA to TCP ratio. Sintering induced formation of large agglomerates and affected the relative composition, specifically, increased concentrations of TCP as proved by SEM, FTIR and XRD. Different reactants also affected chemical final product composition. Contacts with fibroblasts and osteoblasts revealed better adhesion to sintered substrates.

Introduction

Calcium phosphate-based ceramics have proved to be attractive materials for orthopaedic applications. Among these bioceramics, particular attention has been given to hydroxyapatite and tricalcium phosphate due to their bioactivity and biocompatibility. The search for materials similar to the main mineral component of bone led to the development of nanoparticles. New trends in this field focus on simulating the mineral structure of bone by calcium phosphate materials resulting from agglomerated nanoparticles. The present study used a wet chemical precipitation method to prepare hydroxyapatite and tricalcium phosphate [1-3] and characterized their physico-chemical properties and cytocompatibility.

Materials and Methods

Preparation and characterization of nanoparticles. HA nanoparticles were prepared by wet precipitation under stirring and nitrogen bubbling, using three different methods. In the first case (A), orthophosphoric acid (H_3PO_4) was added to a solution of calcium hydroxide ($Ca(OH)_2$) and lactic acid ($CH_3CH(OH)COOH$) till the Ca/P ratio was 1.67. The pH of the homogeneous solution was adjusted to 10 by adding ammonium hydroxide (NH_4OH). Crystal growth took place at room temperature (Sample A1) and at 4°C (Sample A2) for 24 hours. After drying (T=60°C), part of each material batch was sintered at 1100°C for 1 h. In the second case (B), H_3PO_4 was added to a solution of calcium carbonate ($CaCO_3$) and acetic acid (CH_3COOH) till the Ca/P ratio was 1.67. Crystallisation started after ammonium hydroxide was added at room temperature and pH = 10. Crystal growth took place at room temperature over 24 hours. After drying (T=60°C), part of each material batch was sintered at 1100°C for 1 hour (sample B1). The same procedure was applied in the preparation of sample B2 but crystal formation took place at 90°C for 3 hours. Sample B3 was prepared using the methods described for sample B1, but the resultant material in this case was dried at 110°C. In the third approach (C), various amounts of $Ca(NO_3)_2 \cdot 4H_2O$ but the same amount of $(NH_4)_2HPO_4$ were used to obtain several (in the range 1.45 to 1.75) Ca/P ratios (samples C1 to C6). These precipitates were dried and sintered at 1100°C. The dried samples were analysed before and after sintering. Elemental composition and the presence of impurities were determined by x-ray dispersive spectroscopy (EDS) for the non-sintered (ns) and sintered (s) samples. The surfaces of these materials were examined by scanning electron microscopy (SEM). The crystalline phases before and after heat treatment were characterised by x-ray diffraction (XRD). For infrared spectroscopy measurements (FTIR), the specimens were diluted 100 fold with KBr powder. The HA and TCP compositions were determined with the help of a calibration curve prepared using mixtures of known HA/TCP ratios.

Cell Attachment. Cultured cells, both rat calvaria osteoblasts (isolated and characterized according to established methods [4] at population number 3) and rat-skin fibroblasts (purchased from, and characterized by, the American Type Culture Collection; population number 14-16), in Dulbecco's modified Eagle's medium (containing 10% FBS) were seeded (3500 $cell/cm^2$) per substrate and allowed to adhere in a 37°C humidified, 5% CO_2/95% air environment for 4 hours. Adherent cells were stained with Hoechst 33342 and counted using fluorescence microscopy. Numerical data were statistically analyzed using ANOVA and the Tukey test; significance was considered at $p < 0.05$.

Results and Discussion

Elemental analysis. The Ca/P ratios of samples A1 and A2 (Table 1) were slightly different from the intended 1.67 and 1.5, respectively, probably due to heterogeneities in the material and presence of contaminant (such as magnesium) ions.

Table 1. Ca/P ratios of the non-sintered (ns) and the sintered (s) material

Experiment	ns	s
A1	1.47	1.76
A2	1.60	1.64

Particle morphology of non-sintered (ns) and sintered (s) material. The representative scanning electron micrographs of Fig.1 illustrate the materials revealing two distinct morphologies: spherical particles (with diameters in the range or 60–150 nm) on the non-sintered (Fig.1a), and agglomerates that formed both porous (Fig.1b) and non-porous arrangements on the sintered substrates with grain size with diameter in the range 400–600 nm.

Fig.1 Surface morphology (SEM) of (a) non-sintered and (b) sintered materials.

Crystallinity of the non-sintered (ns) and sintered (s) materials. FTIR and XRD analyses (Figs 2 and 3) revealed that non-sintered materials contained mainly poorly crystallized HA with carbonate ions (peaks at 874, 1420 e 1485 cm^{-1} of the FTIR spectra) [5].

Small amounts of dicalcium phosphate dihydrate (peak at 1646 cm^{-1}) and tricalcium phosphate (peaks at 947 and 1122 cm^{-1}) were also present. In contrast, the sintered materials contained non-carbonated HA and TCP. Strong peaks at 1212 and 935 cm^{-1} and peak at 725 cm^{-1} were attributed to β-calcium pyrophosphate in sintered sample s - B3.

The ageing and drying temperatures affected the HA/TCP ratios; ageing at lower temperature resulted in increase TCP content while drying temperature of 110°C promoted the formation of calcium pyrophosphate (s-B3).

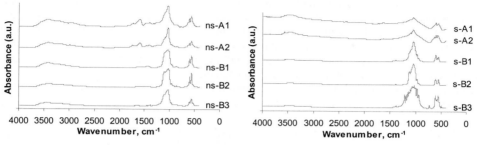

Fig. 2 FTIR spectra of non-sintered (left) and sintered (right) materials.

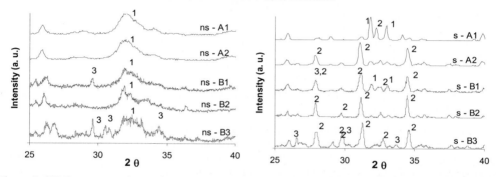

Figure 3. XRD patterns of non-sintered (left) and sintered (right) materials. The indicated structures are: (1)-HA;(2)-TCP;(3)-CaHPO$_4$.2H$_2$O.

The Ca/P ratios and percentage HA composition were determined (Table 2) using a calibration curve prepared with the XRD analysis data of various HA/TCP mixtures of known composition. The ratios of the highest peak intensities of the two components (HA and TCP) were used to calculate the HA percentage; the correlation between the experimental and theoretical results (Figure 4) was good (R = 0,9957) and proved to be effective in quantifying relative compositions.

The reagents composition and the different temperatures used for precipitation, ageing and drying affected the HA to TCP ratios. As observed from the analysis of relative HA weight percentages (Table 2), ageing at lower temperature (4°C), precipitating at higher temperature (90°C) or drying at higher temperature (110°C) resulted in a relative decrease of HA content.

Table 2. Ca/P ratios and HA percentage

Sample	Ca/P Ratio	Relat. % HA
A1	1.67	78
A2	1.67	9
B1	1.67	19
B2	1.67	10
B3	1.67	5
C1	1.45	21
C2	1.50	38
C3	1.60	57
C4	1.67	100
C5	1.70	100
C6	1.75	100

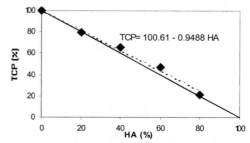

Fig. 4 Correlation between TCP and HA amounts in samples of known composition. Solid line = theoretical prediction. Dotted line = trend line. Black squares = experimental data.

For C samples, calcium phosphates with Ca/P ratios less than that of hydroxyapatite (1.67) contained TCP. As the ratio increased from 1.45 to 1.67 the percentage of hydroxyapatite increased too. Traces of CaO were found for the 1.75 ratio. Comparing the three different precipitation approaches using Ca/P=1.67, a higher percentage of HA was obtained following the C method.

Cell Attachment. Osteoblasts and fibroblasts attached similarly to the three non-sintered B substrates. Also these cells attached similarly to the three sintered B substrates (Fig. 5). Attachment on the sintered substrates, however, was significantly better than on non-sintered ones. It is possible that pH changes in the vicinity of these less crystalline, and subsequently more soluble, substrates create a micro-environment not favouring cell attachment.

There were no significant differences observed in terms of the different materials (B1, B2 and B3), both for non-sintered and sintered substrates.

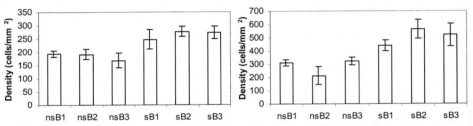

Fig. 5 Adhesion of osteoblasts (left) and fibroblast (right) to non-sintered (ns) and sintered (s) B substrates. Values are mean ± STD; n=3

Same conclusions may be drawn for fibroblasts adhesion on the surface of tested materials.

These results apparently indicated that the chemical structure did not affect cell adhesion.

Comparing osteoblasts and fibroblasts behaviour in contact with these materials, lower adhesion of the former cells is statistically significant except for non-sintered B2.

Conclusions

The wet chemical precipitation method used to prepare nanoparticles proved to be successful. Optimization of the experimental conditions is, however, required in order to obtain pure and homogeneous HA and to tailor the HA/TCP ratios.

Apart from inducing formation of larger agglomerates and crystals, sintering of the nanoparticles also affected their chemical composition by increasing TCP formation.

The aging and drying temperatures also affected both the HA/TCP ratio and particles size.

Sintered materials prepared according to method B, promoted cell attachment, a pre-requisite of subsequent functions of anchorage dependent cells.

Acknowledgements

The authors acknowledge the financial support in the USA of the Nanoscale Science and Engineering Initiative of the National Science Foundation under NSF Award Number DMR-0117792, and in Portugal, FCT's project POCTI/FCB/41523/2002.

References

[1] CM Manuel, MP Ferraz, FJ Monteiro: Key Engineering Materials (2003), 240-242.
[2] M Jarcho, CH Bolen, MB Thomas, J Bobick and RH Doremus: J Mat Sci (1976), 2027-2035.
[3] M Jarcho, RL Salsbury, MB Thomas and RH Doremus: J Mat Sci (1979), 142-150.
[4] DA Puleo, LA Holleran, RH Doremus, R Bizios: J Biomed Mat Res (1991), 25:711-723.
[5] B.O.Fowler, E.C. Moreno, W.E. Brown (1966) Arch. Oral Biol. 11:477-92.

Key Engineering Materials Vols. 254-256(2004) pp. 907-910
online at http://www.scientific.net
© 2004 Trans Tech Publications, Switzerland

Modulation of Defense Cell Functions by Nano-particles *in vitro*

Marilena Lucarelli[1], Emanuela Monari[2], Antonietta M. Gatti[2] and Diana Boraschi[1]

[1] Istituto di Tecnologie Biomediche, CNR, Pisa, Italy, email borasc@tin.it

[2] Laboratorio di Biomateriali, Università di Modena, Italy, email gatti@unimo.it

Keywords: Nano-particles, inflammation, defense, epithelial cells, monocytes, TLR receptors, interleukin 1.

Abstract. The interaction of nano-particles derived from ceramic objects and handicrafts with the human body has been investigated with a particular emphasis to defense mechanisms. Macrophage (professional) and epithelial (non-professional) cells were exposed *in vitro* to TiO_2, SiO_2, or ZrO_2 nano-particles, and their vitality and inflammatory/immune activation checked. All nano-particles showed effects on cell activation, although different in nature and extent. In particular, SiO_2 nano-particles directly induced significant inflammatory activation of macrophages, as judged by production of the inflammatory cytokines IL-1β and TNF-α, and TLR modulation. Thus, it is suggested that exposure to ceramic nano-particles can contribute to the induction of inflammation-related pathologies.

Introduction

The body is protected from foreign attacks both by professional and non-professional defense cells, which interact with invading agents and contribute to their disposal. Mechanical damage due to particulate agents of different nature usually leads to necrotic cell damage and a consequent potent inflammatory reaction, eventually leading to the elimination of the particle. In other cases, foreign particles cause severe inflammation, fibrosis and macrophage apoptosis, leading to chronic pathology [1]. Nano-particles do not induce cell damage and necrosis, due to their reduced size, but they may influence in other ways the biological functions of defense cells. In this study, nano-particles of titanium oxide (TiO_2), silicon oxide (SiO_2), and zirconium oxide (ZrO_2) have been examined for their ability to modulate the innate defense functions of human macrophages (the U937 myelomonocytic cell line) and epithelial cells (the liver-derived cell line HepG2), two cell types involved in the response to mineral dust [2]. The parameters examined were the expression of the defense receptors of the TLR/IL-1R family [3], and the production of inflammatory cytokines.

Materials and Methods

Cell Lines. The human cell lines of epithelial origin HepG2 (hepatocellular carcinoma) and the human myelomonocytic cell line U937 (histiocytic lymphoma) were used. Cells were grown to sub-confluency and passaged every two-three days. All media and supplements were tested as endotoxin-free. Cultures were routinely tested for mycoplasma contamination.

Nano-particles. Nano-particles of SiO_2, TiO_2, and ZrO_2 were prepared by flame spray pyrolysis, i.e. by combusting metallo-organic alcohol solutions with oxygen at 1200°-2000°C, followed by rapid quenching, to produce unagglomerated single particle nano-powders. Average nano-particle size is 70 nm for SiO_2 (range 20-160 nm), and 15 nm for TiO_2 (range 4-40 nm). The size of ZrO_2 nano-particles is in the range of 5-30 nm.

Cell Survival. Cells were plated at $2x10^5$ cells/well of Cluster[24] plates and allowed to adhere for 24 h (epithelial cells) or exposed to 10 nM PMA for 24 h (U937, to induce macrophage differentiation and adherence). Cells were then washed and exposed to increasing doses of nano-particles for different times. Proliferation was then evaluated with the XTT assay, whereas the number of residual cells was evaluated by fixing and staining cell monolayers with crystal violet, and the number of necrotic cells was assessed by trypan blue dye exclusion.

TLR/IL-1R Expression. Semi-quantitative RT-PCR was set up with specific primer pair for TLR receptors (TLR1, TLR2, TLR3, TLR4, TLR5, TLR6, TLR7, TLR8, TLR9, TLR10), TLR4 accessory chains (CD14, MD-2), IL-1 receptors (IL-1RI, IL-1RII, IL-1RAcP), IL-18 receptors (IL-18Rα, IL-18Rβ, IL-18BPa). mRNA weas retro-transcribed and amplified for 20, 23, 26, 29, 32, 35, 38, 40, 43, 46 cycles. Densitometric values were compared with those obtained for the reference gene HPRT. mRNA expression was calculated in arbitrary units, taking the HPRT value as 100.

Assay of cytokine production. Cells were plated at 2×10^5 cells/well of Cluster[24] plates and allowed to adhere for 24 h (HepG2) or exposed to 10 nM PMA for 24 h (U937). Cells were then washed and exposed to nano-particles for 24 h. Presence of the soluble mediators IL-1β and TNF-α was evaluated in culture supernatants by specific ELISA.

Results

Toxicity studies. Exposure of cells to increasing doses of nano-particles for variable length of time did not cause cell death nor inhibition of cell proliferation. The lack of toxic/anti-proliferative/pro-apoptotic effects was evident at doses as high as 400 μg of nano-particles/10^6 cells for up to 96 hours (data not shown). This dose was subsequently used for functional experiments.

Expression of TLR/IL-1R. Expression of TLR/IL-1R receptors has been evaluated with a semi-quantitative RT-PCR assay in resting cells and upon exposure to nano-particles. Data in Table 1 show that in resting conditions the professional defense cells (macrophages, i.e. PMA-differentiated U937 cells) express, though at different levels, all the TLR chains except TLR3, all the IL-1/IL-18 receptor chains and the TLR4 accessory chains CD14 and MD-2. On the other hand, liver epithelial

Table 1. Expression of TLR/IL-1R

Expressed gene	mRNA expression (units) in					
	U937			HepG2		
	m	SiO₂	TiO₂	m	SiO₂	TiO₂
TLR1	22.7	28.2	31.8	20.6	16.6	19.0
TLR2	83.8	59.6*	45.7*	7.5	n.t.	n.t.
TLR3	0.0	0.0	0.0	4.3	n.t.	n.t.
TLR4	47.2	n.t.	n.t.	28.0	n.t.	n.t.
TLR5	74.6	17.5*	37.7*	57.5	89.4*	25.1*
TLR6	88.5	35.9*	60.7*	67.6	50.9	n.t.
TLR7	28.8	n.t.	n.t.	+	n.t.	n.t.
TLR8	31.0	n.t.	n.t.	+	n.t.	n.t.
TLR9	30.9	72.4*	79.0*	32.7	66.2*	11.7*
TLR10	+	n.t.	n.t.	15.9	n.t.	n.t.
IL-1RI	31.9	n.t.	n.t.	67.6	n.t.	n.t.
IL-1RII	10.9	n.t.	n.t.	80.6	n.t.	n.t.
IL-1RAcP	114.8	n.t.	n.t.	102.6	n.t.	n.t.
IL-18Rα	+	n.t.	n.t.	43.7	n.t.	n.t.
IL-18Rβ	+	n.t.	n.t.	0.0	n.t.	n.t.
IL-18BPa	+	n.t.	n.t.	+	n.t.	n.t.
CD14	+	n.t.	n.t.	0.0	n.t.	n.t.
MD-2	+	n.t.	n.t.	0.0	n.t.	n.t.

+ indicates qualitatively positive expression. * significant difference from medium control. n.t. = not tested

cells express all TLR (including TLR3), IL-1R and IL-18R chains (except IL-18Rβ), but do not express the TLR4 accessory chains. The modulation of some genes was examined following exposure of cells for 18 h to SiO_2 or TiO_2 nano-particles (400 $\mu g/10^6$ cells). It is interesting to see that nano-particles could have different effects on expression of different TLR genes in the same cells, and different effects on the same TLR gene in different cells. In fact, SiO_2 nano-particles had opposite effects on TLR5 expression in macrophages (inhibition) as compared to liver cells (enhancement), whereas on TLR9 expression the effect was of enhancement in both cell types. TiO_2 nano-particles may have effects different from those of SiO_2 nano-particles: inhibitory effects on TLR5 expression in either cell type, and opposite effect on TLR9 expression (enhancement in macrophages, inhibition in liver cells).

Production of IL-1β and TNF-α. The production of the inflammatory and defense mediators IL-1β and TNF-α was evaluated in PMA-differentiated U937 cells cultured for 24 h in the absence or in the presence of nano-particles. As shown in Table 2, the low level of basal IL-1β and TNF-α production of macrophages was not significantly affected by exposure to TiO_2 nano-particles, whereas it was significantly enhanced by SiO_2 nano-particles and, to a lower extent, by ZrO_2 nano-particles. The ability of nano-particles to modulate macrophage activation induced by the bacterial component LPS was then assessed. When exposed to an optimal concentration of LPS, macrophages are activated to produce large amounts of IL-1β and TNF-α (Table 2). Co-exposure to TiO_2 nano-particles does not significantly affect the LPS-mediated macrophage activation. On the other hand, macrophages exposed to LPS in the presence of SiO_2 nano-particles produced amounts of IL-1β and TNF-α significantly higher than the sum of those induced by exposure to either stimuli alone, indicating synergistic induction. Nano-particles of ZrO_2 had a selective enhancing effect on LPS-induced IL-1β production, whereas no significant effect was detected on TNF-α production.

Table 2. Production of IL-1β and TNF-α

Treatment of macrophages	Cytokine production (pg/10^6 cells/24 h) by macrophages exposed to nano-particles			
	none	TiO$_2$	SiO$_2$	ZrO$_2$
IL-1β: medium	7.0 ± 2.8	9.1 ± 0.5	896.1 ± 38.0*	17.1 ± 3.7*
LPS	60.8 ± 11.6*	45.8 ± 4.6*	2,253.7 ± 79.1**	214.7 ± 21.3**
TNF-α: medium	0.9 ± 0.1	1.5 ± 1.1	20.5 ± 0.9*	1.9 ± 0.8
LPS	27.1 ± 2.4*	19.8 ± 0.4*	190.3 ± 0.1**	26.0 ± 0.1*

* significant difference from respective medium control.
** significant difference from LPS alone or nano-particles alone.

Discussion

Ceramical micro- and nano-particles can be found as environmental contaminants of industrial locations and factories, and can be released into the body by worn prosthetic implants (e.g. dental bridges). Whereas the inflammatory effect of micro-particles is well-documented, and depends on their size (that limits their diffusion and hampers disposal by macrophages, causing tissue necrosis, localised inflammation, granuloma formation and fibrosis), the possible activity of nano-particles on cell function is largely unknown. At variance with particles of larger size, nano-particles can enter the blood stream and diffuse throughout the body, and localise in different organs. Their effect on the function of both professional and non-professional innate defense cells has been examined, in order to evaluate the possible consequences of exposure to nano-particles on the host ability to cope with environmentally adverse conditions (e.g. exposure to infectious agents). Innate defense mechanisms represent the first line of protection of the body from external invasion. Macrophages (professional scavengers and inflammatory cells) and epithelial cells of the respiratory and gastrointestinal tract (non-professional barrier/defense cells) are the first cells entering in contact with foreign particles and are constitutively able to initiate the innate defense response. This response includes, upon contact with and physical uptake of the foreign particles, production of inflammatory mediators which initiate and amplify the inflammatory and immune reaction eventually leading to elimination of the foreign body. The reduced size of nano-particles is expected to fail triggering a normal inflammatory response, leading to hypothesize that these can be "inactive" as far as physiological function of cells are concerned. In this study, the possible effects of ceramic nano-particles (TiO_2 SiO_2, and ZrO_2) on professional and non-professional innate defense cells has been examined in vitro, using as cellular models two continuous human cell lines, one of myelomonocytic origin (U937) and one of epithelial origin (HepG2). As expected, none of the nano-particles could induce necrosis of cells in culture. In fact, neither their proliferation rate nor their survival in culture was affected by exposure to up to 400 μg nano-particles/10^6 cells for up to 96 hours. Despite the lack of effect on cell survival/proliferation, the possibility was examined that ceramic nano-particles could modulate the defensive function of innate defense cells. This was evaluated by two parameters: the expression of receptors of the TLR/IL-1R family (a class of receptors of pivotal importance in the mechanisms of activation of the innate inflammatory/immune response), and the production of the inflammatory/immune cytokines IL-1β and TNF-α. Preliminary results show that ceramic nano-particles have significant effects on TLR/IL-1R expression, suggesting that they are likely to affect cell reactivity to infectious triggers by altering innate receptor expression. For instance, SiO_2 nano-particles enhance expression of both TLR5 and TLR9 in liver cells, thus likely amplifying the inflammatory response mediated by these receptors. In macrophages, SiO_2 nano-particles have enhancing effects on the expression of TLR9, but can decrease expression of TLR2, TLR5, and TLR6. Regarding cytokine production, whereas TiO_2 nano-particles do not show any significant activity, ZrO_2 nano-particles could selectively increase the basal and the LPS-induced production of IL-1β, being without effect on TNF-α production, whereas SiO_2 nano-particles dramatically induce the production and release of IL-1β and TNF-α, either alone or in synergy with LPS. This is in agreement with previous data showing strong pulmonary inflammation and apoptosis caused by large silica particles in vivo, as opposed to weak and transient effect caused by TiO_2 particles [4, 5]. Thus, despite the lack of macroscopic, mechanical inflammatory effects, ceramic nano-particles (in particular SiO_2) can induce a significant inflammatory response. Therefore, exposure to SiO_2 nano-particles may result in the pathological consequence of chronic inflammation, with unregulated amplification of immune responses and risk of autoimmune derangements [6].

Acknowledgements

This work was supported by EU contract QLK4-CT-2001-00147 (NANO-PATHOLOGY). Diana Boraschi was also supported by a grant from AIRC (Associazione Italiana Ricerca sul Cancro, Milano, Italy) and by the FIRB project "NIRAM" of the Italian MIUR.

References

[1] K.D. Srivastava, W.N. Rom, T.A. Gordon and K.M. Tchou-Wong: Am. J. Respir. Crit. Care Med. Vol. 165-4 (2002), p. 527
[2] F. Tao and L. Kobzik: Am. J. Respir. Cell Mol. Biol. Vol. 26-4 (2002), p. 499
[3] G.M. Barton and R. Medzhitov: Curr. Top. Microbiol. Immunol. Vol. 270 (2002), p. 81
[4] D.D. Zhang, M.A. Hartsky and D.B. Warheit: Exp. Lung Res. Vol. 28-8 (2002), p. 641
[5] K.E. Driscoll, R.C. Lindenschmidt, J.K. Maurer, L. Perkins, M. Perkins and J. Higgins: Toxicol. Appl. Pharmacol. Vol. 111-2 (1991), p. 201.
[6] C.G. Parks, K. Conrad and G.S. Cooper: Environ. Health Perspect. Vol. 107-5 (1999), p. 793.

Key Engineering Materials Vols. 254-256(2004) pp. 911-914
online at http://www.scientific.net

Micropatterning of Apatite by Using CaO-SiO$_2$ Based Glass Powder Dispersed Solution

T.Matsumoto[1], N.Ozawa[1]and T.Yao[1]

[1] Graduated school of Energy Science, Kyoto University

Yoshida, Sakyo-ku, Kyoto, 606-8501 Japan

yao@energy.kyoto-uc.ac.jp

Keywords: Apatite, micropattern, biomimetic method, resist pattern, transcription, simulated body fluid

Abstract. Combination of biomimetic method and transcription of resist pattern was made to form apatite micropattern. A resist pattern printed substrate was suspended in CaO-SiO$_2$ based glass (CaO 44.7, SiO$_2$ 34.0, P$_2$O$_5$ 16.2, MgO 4.6, CaF$_2$ 0.5 mass%) powder dispersed simulated body fluid (SBF) with ion concentration nearly equal to human blood plasma, and then soaked in SBF. Apatite nuclei were seeded on whole the surface of the substrate. Next the resist material was dissolved off with the apatite nuclei just formed on it. Then, the substrate was soaked in SBF to grow the remaining nuclei. As the result, an apatite micropattern transcribing the resist pattern was obtained. This method is promising for producing multifunctional materials having bioaffinity

Introduction

It is thought that various smart biomaterials utilizing the bioaffinity of apatite can be developed by forming apatite micro pattern. For example, by using the affinity of apatite to living cells, pattern of living cells can be obtained by culturing them on the apatite pattern. This will be applicable for developing a cellular biosensing system. Because apatite can adsorb biomolecules such as proteins, it will also be used for capturing of biomolecules in biosensing devices. Moreover, advanced multifunctional biomaterials can be produced by combination of the bioaffinity of apatite with mechanical, electric, magnetic or optical properties of other materials in micron scale. Previously, we have presented that combination of biomimetic method and transcription of resist pattern was made to form apatite micropattern. [1-4] In this method, apatite nuclei were seeded on a resist pattern printed substrate by setting in contact with CaO-SiO$_2$ based glass in a simulated body fluid (SBF) (Na$^+$ 142.0, K$^+$ 7.5, Ca^{2+} 2.5, Mg^{2+} 1.5, HCO$_3^-$ 4.2, Cl$^-$148.0, HPO$_4^{2-}$ 1.0, SO$_4^{2-}$ 0.5 mmol dm^{-3} pH=7.25) with inorganic ion concentrations nearly equal to human blood plasma. In this study, in order to simplify the seeding process, a resist pattern printed substrate was suspended in CaO-SiO$_2$ based glass (CaO 44.7, SiO$_2$ 34.0, P$_2$O$_5$ 16.2, MgO 4.6, CaF$_2$ 0.5 mass%) powder dispersed solution. Next the resist material was dissolved off with the apatite nuclei just formed on it. Then, the substrate was soaked in SBF to grow the remaining nuclei.

Methods

Polyethersulfone (PESF) plates were used as substrates. Resist patterns were formed on the substrate as follows. Positive type resist material(AZ1500, Clariant Japan Corp.) was coated on the whole surface of the substrate by using spin coater, and the coating was pre-baked at 100°C for 1min on a hotplate. A photomask was put on the surface and ultraviolet rays were irradiated. By soaking in developing solution, the region of the resist material exposed to ultraviolet rays was dissolved off and the resist pattern was obtained on the surface of the substrate. After the development, the substrate was washed with distilled water and postbaked at 100°C for 1min on a hotplate to fix the resist material. The resist pattern printed PESF plate was cut into 10 x 10 x 0.2 mm^3 in size and used as substarate.

Chemical regants of $CaCO_3$, SiO_2, $CaPO_4$, $CaHPO_4 \cdot 2H_2O$, MgO, and CaF_2 were mixed to obtain nominal composition of MgO:4.6, CaO:44.7, SiO_2:34.0, P_2O_5:16.2, and CaF_2:0.5 wt%. The mixture was melted in a platinum crucible at 1450°C for 2 h in an electric furnace. The melt was poured onto a stainless steel plate to obtain a glass. This glass is referred to as glass G, hereafter. Glass G was crushed with a laboratory planetaly-type ball mill (Model P7, Feritsch Co., Idar-Oberstein, Germany) and sieved to obtain particles under 150μm in diameter. A simulated body fluid (SBF). An SBF, a so-called Kokubo solution [5], with ion concentration nearly equal to those of human plasma was prepared by dissolving reagant-grade NaCl, $NaHCO_3$, KCl, $K_2HPO_4 \cdot 3H_2O$, $MgCl_2 \cdot 6H_2O$, $CaCl_2$, and Na_2SO_4 into ion exchanged and distilled water in a polystyrene bottle and buffering at pH=7.25 with tris(hydroxymethyl)aminomethane [$(CH_2OH)_3CNH_2$] and hydrochloric acid (HCl) at 36.5°C.

Glass G powder under 150μm in diameter was dispersed in 30 ml of SBF, in which the substrate was suspended at 36.5°C for 1 d. Apatite nuclei were seeded on whole the surface of the substrate by this process. Then, the substrate was soaked in acetone shaking gently for 1min. The resist material was dissolved off with the apatite nuclei just formed on it. Next, the substrate was soaked in 30 ml of SBF for 4 d to grow the apatite nuclei. SBF was renewed every 2 d. After soaking, the surface of the substrate was analyzed by thin film X-ray diffraction (TF-XRD, RINT 2500, Rigaku, Japan) and scanning electron microscopy (SEM: ESEM-2700, Nicon, Japan).

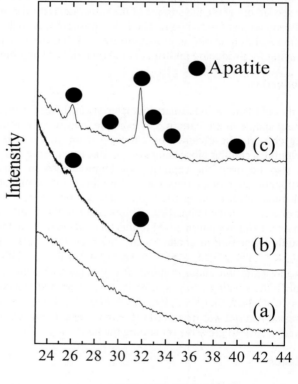

Results and Discussion

Fig. 1 shows TF-XRD patterns of the substrate suspended in 30 ml of the glass G powder dispersed SBF at 36.5°C for 1 d, soaked in acetone for 1min, and then soaked in 30 ml of SBF for 0 d, 2 d and 4 d, respectively. After the soak in SBF for 2d, characteristic XRD peaks attributed to hydroxyl apatite were observed. The peak intensities increased for 4 d soaking. This indicates that the apatite nuclei seeded on the substrate by the soak in glass G powder dispersed SBF grew by the subsequent soaking in SBF. Fig. 2 shows SEM micrograph of line patterns of the substrate suspended

Fig. 1 TF-XRD patterns of the substrate suspended in 30 ml of the glass G powder dispersed SBF at 36.5°C for 1 d, soaked in acetone for 1min, and then soaked in 30 ml of SBF for (a) 0 d, (b) 2 d and (c) 4 d.

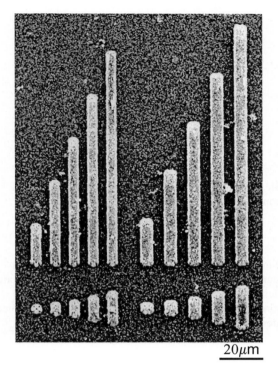

$20\mu m$

Fig. 2 SEM micrograph of line pattern of the substrate suspended in 30 ml of the glass G powder dispersed SBF at for 1 d, soaked in acetone for 1min, and then soaked in 30 ml of SBF for 4 d.

in 30 ml of the glass G powder dispersed SBF for 1 d, soaked in acetone for 1min, and then soaked in SBF 2 d. A dense apatite thin film grew at region where the nuclei remained, and a micropattern clearly transcribing the resist pattern was formed. Fig. 3(a) and (b) are the pictures of the positive type and negative type apatite pattern. As shown in Fig. 3, apatite-existing part and apatite-not-existing part are clearly separated, and the boundary is straight line. The corner of the bending line was clearly outlined, a $5\mu m$ square pattern was also formed by apatite. Fig. 4 shows the magnification picture of submicron scale apatite pattern. As is shown in Fig.4 (a), rings $2\mu m$ in diameter formed with lines of about 500nm width were clearly formed by apatite. In Fig.4 (b), holes about 400nm in diameter were clearly shown.

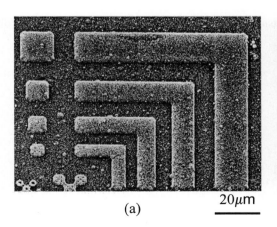
(a) $20\mu m$

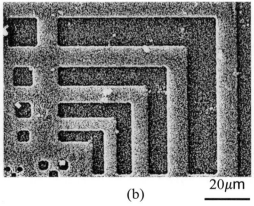
(b) $20\mu m$

Fig. 3 SEM micrograph of (a) the positive type and (b) the negative type apatite pattern.

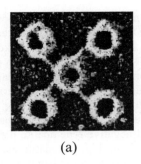
(a)

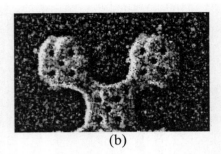
(b)

Fig. 4 Magnification picture of submicron scale apatite pattern, (a) rings 2μm in diameter formed with lines of about 500nm width, (b) holes about 400nm in diameter.

Conclusion

Resist pattern printed substrate was suspended in glass G powder dispersed SBF. This method is effective for seeding apatite nuclei on the substarte. After removing the resist material, the substrate was soaked in SBF to grow the remaining nuclei. Minute patterns such as lines, blocks, rings ans holes were well formed. It was possible to obtain apatite patterns with submicron scale. This method is promising for producing multifunctional materials having bioaffinity.

References

[1] T. Yao: Japan Patent, Publication No. 2001-294411, (2001).

[2] N.Ozawa and T.Yao: Bioceramics, Vol. 14, (2002), p.123.

[3] N.Ozawa and T.Yao: J. Biomed. Mater. Res., Vol. 62, (2002), p.579.

[4] N.Ozawa and T.Yao: Solid State Ionics, Vol. 151, (2002), p.79.

[5] T. Kokubo, H. Kushitani, S. Sakka, T. Kitsugi and T. Yamamuro: J. Biomed. Mater. Res., Vol. 24, (1990), p.721.

Key Engineering Materials Vols. 254-256(2004) pp. 915-918
online at http://www.scientific.net
© 2004 Trans Tech Publications, Switzerland

Microstructural Changes of Single-crystal Apatite Fibres during Heat Treatment

M. Aizawa[1], A. E. Porter[2], S. M. Best[2] and W. Bonfield[2]

[1]Department of Industrial Chemistry, School of Science and Technology, Meiji University, 1-1-1 Higashimita, Tama-ku, Kawasaki, Japan 214-8571; mamorua@isc.meiji.ac.jp
[2]Department of Materials Science and Metallurgy, University of Cambridge, Pembroke street, Cambridge, CB2 3QZ, UK

Keywords: Hydroxyapatite, apatite fiber, strain, defects, transmission electron microscope

Abstract. The apatite fibres were prepared using a homogeneous precipitation route. The as-prepared fibres were composed of carbonate-containing apatite with preferred orientation along the c-axis direction. The fibres were highly strained and were composed of domains that preferentially oriented themselves with the c-axis parallel to the surface of the substrate. When the as-synthesized fibres were heated at 800°C to 1200°C for 1 h, the domain structure of the fibres disappeared but they remained highly strained. In the cases of the apatite fibres heated at 1000°C and 1200°C, their microstructures changed during sintering to form some grain boundaries and dislocations. The apatite fibres heated at 1200°C for 1 h contained many angular voids formed by releasing the carbonate groups during sintering.

Introduction

Hydroxyapatite (HAp) has widely been applied as a biomaterial for substituting human hard tissues. We have successfully synthesized single-crystal apatite fibres by a homogeneous precipitation method [1] and then characterized specific properties of the apatite fibres by high-resolution transmission electron microscopy (HR-TEM) [2]. The as-synthesized apatite fibres were composed of carbonate-containing apatite with a long-axis dimension of 60-100 μm. The fibres have a preferred orientation in the c-axis direction and therefore developed the a-plane of the HAp crystal. In addition, the fibres were highly strained and were composed of domains that preferentially oriented themselves with the c-axis parallel to the surface of the substrate. A sintering process is needed to fabricate high-performance biomaterials, such as porous ceramics [1], bioactive HAp/polymer hybrids [3, 4] and scaffolds for tissue engineering of bone [5, 6], from our current apatite fibres. Therefore, it is important to elucidate how the microstructure of the apatite fibres changes during heat treatment. The objectives of this investigation were to examine the microstructure of the apatite fibres heated at 800°C, 1000°C and 1200°C for 1 h using several characterization techniques and to compare these microstructures to those of apatite fibres that had not been heat-treated.

Materials and Methods

Apatite fibres were prepared by the same process as given in Ref [1]. The resulting apatite-fibre sheets were heated at 800°C, 1000°C and 1200°C for 1 h at a ramp-rate of 10°C•min⁻¹. The heated apatite fibres were characterized by X-ray diffractometry (XRD), Fourier transform infrared spectrometry (FT-IR), scanning electron microscopy (SEM) and TEM. For TEM observation, bright and dark field imaging, combined with selected area electron diffraction (SAED), was performed on a Jeol CX200 TEM operated at 200 kV. TEM samples were prepared by dispersing the fibers in ethanol and collecting them onto lacey carbon, Cu mesh TEM grids.

Results and Discussion

Figure 1 shows the XRD patterns of the apatite fibres heated at 1200°C for 1h.   In this XRD pattern, the (100), (200) and (300) reflections of the apatite phase were more intense than those of a typical HAp listed in JCPDS card #9-432.   These reflections were also more intense than those arising from the apatite fibres before heat treatment.   The XRD patterns of the apatite fibres heated at 800°C and 1000°C for 1 h were almost the same as that of fibres heated at 1200°C. These results indicate that the present apatite fibres retained their preferred (h00) orientation

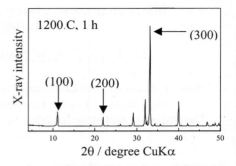

Fig. 1 XRD pattern of the heated apatite fibres.

after heat treatment.   The crystallinity of the apatite fibres was enhanced after heat treatment.

The FT-IR spectra of the apatite fibres heated at 800°C (a) and 1200°C (b) for 1 h are shown in Fig. 2.   The FT-IR spectra of the apatite fibres heated at 800°C and 1000°C indicated that those absorptions were assigned to typical carbonate-containing HAp.   The absorptions after heat treatment were sharper than those of the apatite fibres before heat treatment.   However, the absorptions assigned to carbonate group decreased in the case of the spectra of apatite fibres heated at 1200°C for 1 h.   This is may be due to the release of carbonate ions from the apatite crystals during sintering.

Figure 3 shows an SEM micrograph of the apatite fibres heated at 1200°C as typical particle morphology.   The apatite fibres retained their original morphology after heating.   Sintering of the fibres to each other was not revealed by SEM of the fibres heated at 800°C and 1000°C; however, the edge of fibres had become rounded after heating at 1200°C.   In addition, the relatively smaller fibres had become fused with the large fibres by sintering process.

The results illustrated so far describe characterization of the apatite fibres heated at 800°C, 1000°C and 1200°C for 1 h by XRD, FT-IR and SEM.   These techniques are effective to make the average properties of the heated apatite fibres clear; however, those techniques are not suitable to examine the specific property of apatite fibres.   Thus we characterized the specific microstructural property of the heated apatite fibres using TEM.

Typical results from TEM observations are shown in Fig. 4 and these indicate the morphology of apatite fibres heated at 1000°C for 1 h.   The overviews of the heated fibres are shown in Fig. 4(a) and the microstructures of A, B and C areas in low magnification image are examined using bright-field image and SAED techniques of TEM; the result for C area is shown in Fig. 4(b, c).   In the bright-field images (Fig. 4(b)), black bent-contours were

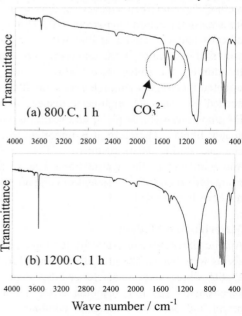

Fig. 2 FT-IR spectra of the heated apatite fibres.

presented across a short-axis of the heated fibres. The contrast from these contours was also observed in dark-field imaging mode. Tilting experiments confirmed that this contrast arose due to strain in the fibres, i.e. that they are bend contours and not defects. It was found from the microstructure observation that the apatite fibres were still highly stained after heat treatment. In the SAED modes (Fig. 4(c)), the diffraction pattern from point B clearly showed a diffraction pattern corresponding to an apatite structure with high crystallinity. The bundle-like domain structure observed in the apatite fibres before heating disappeared after heat treatment at 1000°C for 1h, as well as the case of the apatite fibres heated at 800°C for 1 h.

Fig. 3 SEM micrographs of the apatite fibres heated at 1200°C for 1 h.

Although the typical microstructure of the fibres heated at 1000°C for 1 h are shown in Fig. 4, the present fibres also contained some dislocations and grain boundaries. Figure 5 shows the microstructure including grain boundary and dislocation in the apatite fibres heated at 1000°C for 1h. In a bright-field image mode, we could observe both the grain boundary and loop-shaped dislocations. The contrast from these contours was also observed in dark-field imaging mode. We confirm that these contours will arise due to defects in the fibres by performing tilting experiments. The SAED pattern showed the high-order Laue zone (HOLZ) based on the bending of apatite crystals.

In the case of the apatite fibres heated at 1200°C for 1 h, a large number of angular voids were present in the fibres, together with some strains and grain boundaries. The SAED pattern showed clear spots correspond to an apatite structure with high crystallinity. In addition, the bundle-like domain in the fibre before heating disappeared after heat treatment at 1200°C for 1h, as well as the case of the apatite fibres heated at 800°C and 1000°C for 1 h.

Figure 6 shows the ultrastructure including grain boundary and some voids in the apatite fibres heated at 1200°C for 1h. In the bright-field image mode of a grain boundary (Fig.6 (a)), we could observe the crevasse across a short-axis of fibre. We confirm that this contour arise due to grain boundary in the fibres by tilting experiments. The contrast from bent contours was also observed in dark-field imaging mode (Fig. 6 (b)). Thus, the apatite fibres have highly strained structure. The SAED pattern showed the HOLZ of apatite crystal (Fig. 6(c)). On the other hand, angular voids were also present in the heated fibre (Fig. 6 (d)). This SAED pattern also showed the

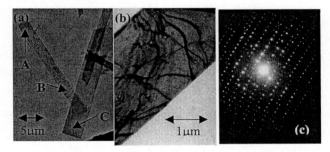

Fig. 4 TEM results of the apatite fibres heated at 1000°C for 1 h. (a) Low magnification, (b) bright-field image and (c) SAED pattern of B area.

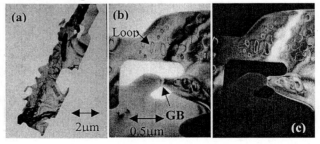

Fig. 5 TEM results of the apatite fibres heated at 1000°C for 1 h. (a) Low magnification, (b) bright-, (c) dark-field images. GB: grain boundary

HOLZ of apatite crystals (Fig. 6 (e)). The formation of lots of voids may be due to the release of carbonate groups during sintering from the apatite structure, as shown in the FT-IR spectra (Fig. 2).

Conclusion

The apatite fibres before heat treatment were composed of carbonate-containing apatite with preferred orientation in the c-axis direction. In addition, the fibres were highly strained and were composed of domains that preferentially oriented themselves with the c-axis parallel to the surface of the substrate. When the fibres were heated at 800°C to 1200°C for 1 h, the domain structure disappeared and but they remained highly strained. In the cases of the apatite fibres heated at 1000°C and 1200°C, significant microstructural changes were observed during sintering and the presence of some grain boundaries and dislocations were noted. The apatite fibres heated at 1200°C for 1 h contained many voids. We suggest that these were formed by the release of the carbonate groups during sintering.

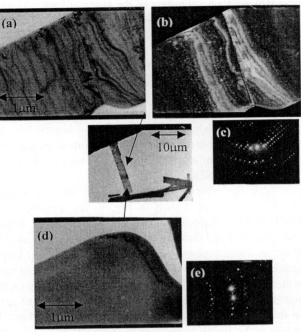

Fig. 6 TEM results of the apatite fibres heated at 1200°C for 1 h. (a, b, c) grain boundary, (d, e) voids; (a, d) bright-, (b) dark-field images, (c, e) SAED pattern.

References

[1] M. Aizawa, F. S. Howell, K. Itatani, Y. Yokogawa, K. Nishizawa, M. Toriyama and T. Kameyama, (2000) J. Ceram. Soc. Jpn. 108: 249-253.

[2] M. Aizawa, A. E. Porter, S. M. Best, W. Bonfield, Key Engineering Materials, 240-242, 509-512(2003).

[3] M. Aizawa, Y. Tsuchiya, K. Itatani, H. Suemasu, A. Nozue and I. Okada, Bioceramics 12 (1999) 453-456.

[4] M. Aizawa, M. Ito, K. Itatani, H. Suemasu, A. Nozue, I. Okada, M. Matsumoto, M. Ishikawa, H. Matsumoto and Y. Toyama, (2002) Key Engineer. Mater. 218-220: 465-468.

[5] M. Aizawa, H. Ueno, K. Itatani, (1999) Material Integration 12: 75-77.

[6] M. Aizawa, H. Shinoda, H. Uchida, K. Itatani, I. Okada, M. Matsumoto, H. Morisue, H. Matsumoto and Y. Toyama, (2003) Key Engineer. Mater. 240-242: 647-650.

Key Engineering Materials Vols. 254-256(2004) pp. 919-922
online at http://www.scientific.net

Effects of Micro/Nano Particle Size on Cell Function and Morphology

K. Tamura[1], N. Takashi[1], T. Akasaka[2], I. D. Roska[2], M. Uo[2], Y. Totsuka[1], F. Watari[2]

[1] Department of Oral and Maxillofacial Surgery, Graduate School of Dental Medicine,
tam@den.hokudai.ac.jp
[2] Dental Materials and Engineering, Graduate School of Dental Medicine,
Hokkaido University, kita 13 nishi 8, kita-ku, Sapporo, 060-8586, Japan

Keywords: particle, biomaterial, titanium, titanium dioxide, carbon nanotube, biocompatibility

Abstract. The cytotoxicity of micro/nano particles in Ti, TiO and carbon nanotube was investigated by in vitro biochemical analyses using human neutrophils. The particles smaller and larger than the neutrophils were used to determine the relationship between cell and particle size with respect to cytotoxicity. As the particle size decreased, the cell survival rate was decreased and, with the good corresponding relation to this, the value of lactate dehydrogenase (LDH), which is the indication of cell disruption, was increased. The release of superoxide anion showed the increasing tendency. Proinflammatory cytokines were detected distinctly for $3\mu m$ or smaller particles and very little in more than $10\mu m$, which is closely related to the phagocytosis by neutrophils. ICP elemental analysis showed that the dissolution from Ti particles was below detection limit. Micro and nano particles stimulated the cell reactions according to the results of the human neutrophil functional tests. As the particle size was smaller, the inflammation was pronounced. The fine particles less than $3\mu m$ caused distinctly the inflammation in the surrounding tissue. All these results indicated that the cytotoxicity was induced due to the physical size effect of particles, which is different from the ionic dissolution effect. The clinical phenomenon confirmed the result obtained in vitro cell tests. The neutrophils stimulated by fine particles may cause the inflammatory cascade and harm the surrounding tissue.

Introduction

Ti and its alloys are the commonly used material in plastic surgery because it is one of the most biocompatible metals [1,2]. Ti is highly corrosion-resistant at ambient temperature due to its thin and stable protective oxide layer formed on its surface. In this sense Ti is the ideal metallic material for implant [3,4]. However, it is clinically reported that the abraded fine titanium particles produced in sliding parts of artificial joints often caused inflammation in the surrounding tissue [1,2]. However, little is known about the effect of micro/nano particles on cellular function and the relevance between *in vivo* and *in vitro* findings. The purpose of this study is to analyze the vital reactions of human neutrophils to the Ti, TiO_2 particles and carbon nanotubes and their size effect.

Materials and probe cells

The dependence of cytotoxicity on particles size in titanium, titanium oxide (TiO_2) and carbon nanotubes was investigated by biochemical functional analysis and by microscopic observation of cellular morphology.

Particles. The Ti, TiO_2 and carbon nanotubes particles colloid solutions were prepared. ICP elemental analysis showed that the dissolution from Ti particles were below detection limit. The carbon nano-tubes is 99% purity. The solubilization distributed processing was carried. The average diameter and size distribution of the TiO_2 particles and carbon nanotubes were determined by electron microscopy (SEM) and by laser scattering particle distribution analyzer (SALD-7000, Shimadzu).

The various sizes of Ti, TiO_2 and carbon nanotubes particles were mixed with HBSS (Hanks' balanced salt solution). The colloid solutions were adjusted to PH 6.8 by 1N NaOH solution, sterilized by autoclave and dispersed by sonicator [5,7].

Cells. Human peripheral blood was obtained from healthy volunteers in our group. Neutrophils were separated from the blood using the 6% isotonic sodium chloride containing the hydroxyethyl starch and lymphocyte isolation solution (Ficoll-Hypaque)[6,7]. The cells were maintained in HBSS. After particles were kept dispersed, neutrophils were added, and incubated in a humidified atmosphere of 5% carbon dioxide at 37 °C for 60 minutes. The experiments were performed using cells within 3 hours after collection of blood and the cell density was adjusted to 10^6 cells / ml [7].

Methods

Cell survival rate, leakage of lactate dehydrogenase (LDH), product of superoxide anion, and release of cytokines of tumor necrosis factor-alpha (TNF-alpha), interleukine-1 beta (IL-1beta) were measured to analyze biochemical reaction.

(1) Cell survival rate: The cell stained with trypan blue population was counted under an optical microscope using Thomas' hemacytometer. The number of vital cell in the control specimen was considered as 100%.

(2) LDH activity: The LDH values of samples were measured using the lactate dehydrogenase C test kit (Wako Pure Chemical Industries) and by spectrophotometry.

(3) Superoxide anion production: Superoxide anion (O_2^-) was assayed by measuring the superoxide dismutase-inhibiting reduction of equine ferricytochrome C (550 nm). The reaction was promoted by adding 1.39mM PMA (phorbol 12-myristate 13-acetate) [7].

(4) Cytokine release: TNF-alpha and IL-1 beta in the supernatant was measured using ELISA kits (Endogen) [7].

The values are expressed as means +/- standard deviation (n=6). Data were analyzed by Student's-t test with the level of significance set at 5%.

The pathological and morphological changes were observed by optical microscopy and SEM.

Results

Ti particles diameter were 3, 10, 50, and 150 μm and TiO_2 particles were 0.05, 0.5, and 3 μm in average size, as confirmed by SEM and the particle distribution analyzer. The SEM image shows carbon nanotubes with diameter of 20 nm and length of 100 nm.

The significant difference of the survival rate from control was observed in all nano particles (Fig.1). The Ti micro particles showed clearly the size dependency. The cell survival rate decreased, when the particle size became smaller. The nanoparticle of TiO_2 also showed the similar tendency. The lowest mean value of the survival rate was 84.6% in the size 50 nm for TiO_2 particles, which were the smallest particles in this study.

LDH showed the tendency to increase as the particle size became smaller (Fig.2). The LDH level of 147.2 Wroblewski unit was significantly higher in the 50 nm than the other larger sizes.

The neutrophils stimulated by the 3 μm or less particles showed the large productions of superoxide anion. The other larger size particles were slightly higher than control solutions (Fig.3).

There is a clear difference in the emission of inflammatory cytokines for 3 μm and 10 μm (Fig.4). The distinct release of TNF- alpha was observed in 3 μm or less particles. There was no statistical difference of cytokines under 3 μm. The IL-1beta showed the similar

SEM observation revealed the degenerative changes in the morphology of neutrophils (Fig.5). The activated neutrophils extended some pseudopods and phagocytized particles into the cytoplasm. 50nm TiO_2 particles and carbon nanotube induced the morphological change of neutrophils. After 6 hours, the atrophied and destroyed neutrophils were observed.

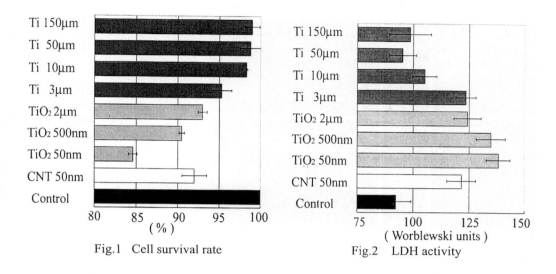

Fig.1　Cell survival rate

Fig.2　LDH activity

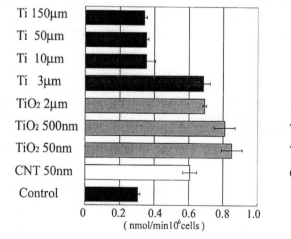

Fig.3 Superoxide anion

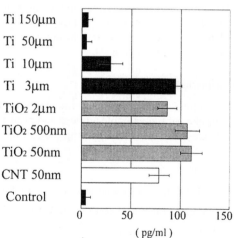

Fig.4　TNF-alpha

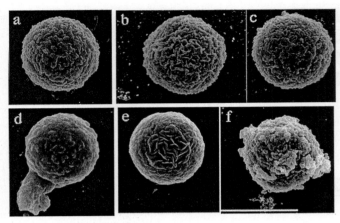

Fig.5
SEM images of morphological change of neutrophils
(Bar = 10 μm)
a: Neutrophil with HBSS
b: Activated by TiO$_2$ particles
c: Activated by Carbon nanotubes
d: Extension of pseudopod and phagocytized particles
e: Atrophied neutrophil
f: Destroyed membrane

Discussions

The study clearly showed the cytotoxicity due to the particle size effect in Ti, TiO$_2$ and CNT. Micro-nano particles may cause cytotoxicity, although the macroscopic size is quite biocompatible. The increased superoxide content in vivo may affect the cell circumference. The chemical mediators, TNF- alpha and IL-1beta, may induce the inflammatory cascade to affect tissue and organ. The effect is further pronounced by phagocytosis when particles are smaller than cells [7]. The clinical inflammation reaction around the abrasion powders can be well understand by the results obtained in cell functional *in vitro* tests.

Acknowledgements

The authors thank Dr. Takafumi Domon, Dr. Ami Fukui, Dr. Kiyomi Tsuji and Dr. Yoshinobu Nodasaka for the valuable advices and preparation of SEM observation, and Dr. Wenzhi Hugetu and Pr. Kazuyuki Touji for supplying the carbon nanotubes.

References

[1] K. Takamura, T. Yamada and Y. Sugioka, et al. : J Biomed Mater Res. Vol 28 (1994), p. 583-9.

[2] Young-kyun Yeo, Seoug-cheul Lim, et al. : J Oral Maxillofac. Surg. Vol 55 (1997), p. 322-326.

[3] A. Rosenberg, K.W. Grätz and H.F. Sailer: J. oral Maxillofac. Surg. Vol 22 (1993), p. 185-188.

[4] Y. Tanimura, F. Watari, et al. : J of Oral and Maxillofacial Surgery. Vol 46(11) (2000), p. 750.

[5] M. Uo, F. Watari, A. Yokoyama, T. Kawasaki, et al. : Biomaterials. Vol 20 (1999), p. 47-55.

[6] F. Takesita, H. Tkata, Y. Ayukawa and T. Suetsugu. : Biomaterials. Vol 18 (1997), p. 21-5.

[7] Kazuchika Tamura, Fumio Watari, et al. : Materials Transactions Vol. 43(12)(2002), p.3052-3057

Key Engineering Materials Vols. 254-256(2004) pp. 923-926
online at http://www.scientific.net

Comparative Study of Sonochemical Synthesized ß-TCP- and BCP-Nanoparticles

M. de Campos[1,2,3], F. A. Müller[1], A. H. A. Bressiani[2], J. C. Bressiani[2] and P. Greil[1]

[1] University of Erlangen-Nuremberg, Department of Materials Science-Glass and Ceramics, Martensstr.5-D91058-Erlangen, Germany, fmueller@ww.uni-erlangen.de

[2] IPEN–DMC, Department of Materials Engineering, Travessa R 400, Cidade Universitária Armando Salles de Oliveira, 05508-900, São Paulo, SP, Brazil, abressia@net.ipen.br

[3] University of Mogi das Cruzes, Technological Research Center, Av. Dr. Cândido Xavier de Almeida Souza 200, 08780-911, Mogi das Cruzes, SP, Brazil, magali@umc.br

Keywords: Tricalcium phosphate, biphasic calcium phosphate, nanopowders, ultrasound, sonochemical synthesis.

Abstract. Hydroxyapatite (HA), beta tricalcium phosphate (ß-TCP) and biphasic calcium phosphate (BCP) are successfully used materials for biomedical applications due to their ability to promote bone integration. In this work calcium phosphate nanopowders were produced. Their morphology and crystalline phases were investigated in dependence of ultrasonic irradiation and addition of D-glucose and glycerol in the neutralization method. The results show that the neutralization method was effective to obtain ß-TCP and BCP with adjustable Ca:P ratios. The morphology and crystallinity of synthesized nanopowders are significantly affected by ultrasonic irradiation and additives. The grain size of the obtained nanopowders was decreased from 300nm to 50 nm by assistance of ultrasound.

Introduction

The interest in calcium phosphates for biomedical applications is increasing because theses ceramics exhibit high potential in the fields of orthopedic and dentistry, due to their chemical similarity to the mineral phase of bone tissue and their ability to promote bone integration. The most widely used calcium phosphates are hydroxyapatite, ($Ca_{10}(PO_4)_6(OH)_2$, HA) and beta tricalcium phosphate, ($Ca_3(PO_4)_2$, ß-TCP). They have been widely applied as biomaterial in the form of powder, granules, dense or porous bodies, and coatings for metallic and polymeric implants [1]. In the field of biomedical applications the rate of solubility is a very important factor. ß-TCP shows *in vivo* solubility 3-12 times higher than HA. Biphasic calcium phosphate (BCP), consisting of a mixture of HA and ß–TCP, has been considered as an ideal bone substitute due to its controllable degradability. It has been demonstrated that the bioactivity of these ceramics may be controlled by manipulating the HA/ß-TCP ratio [2]. The properties of the HA, ß-TCP and BCP powders such as crystal morphology, crystallinity, thermal stability, and solubility have been shown to be dependent on the route of fabrication. In a previous work synthesis by a neutralization method was reported to be suitable to obtain HA [3]. Ultrasonic irradiation was convincing as a promising tool to assist the synthetic reactions to prepare fine ceramic powders [4]. The aim of this work was to study the influence of ultrasonic irradiation and addition of D-glucose and glycerol on the synthesis, crystallinity and morphology of ß-TCP and BCP nanopowders.

Materials and Methods

Synthesis of calcium phosphate powders by neutralization was achieved by the addition of a 0.3 M aqueous solution of orthophosphoric acid (H_3PO_4) to 0.1 M aqueous suspension of calcium hydroxide (Ca(OH)$_2$) at room temperature. The reagents were mixed with a Ca/P molar ratio of 1.50 during magnetic stirring (N), or ultrasonic irradiation (-US). The aqueous suspension of Ca(OH)$_2$ was irradiated by ultrasound at two frequencies, 40 or 50 KHz. Additionally additives were used to prepare calcium phosphate, i.e. 40 wt% D-glucose (GLU) and 40 v% glycerol (GLY), respectively. The slurry was aged for 24h, filtrated, dried at 85 °C for 24h, and finally calcinated at 1000 °C for 3h. A white nanopowder was obtained. The main techniques used for powder characterization were Fourier transform infrared spectroscopy (FTIR), to determine the bands corresponding to functional groups, X-ray diffractometry (XRD) to obtain the crystalline phases and scanning electron microscopy (SEM) to evaluate the powder morphology.

Results and Dicussion

The synthesis of calcium phosphates by the neutralization methods with magnetic stirring (N) leads to the formation of an amorphous calcium phosphate (ACP) with a Ca/P molar ratio of 1.5. The results of the HT-XRD analysis reveal the presence of ACP in a temperature range between 100 and 600 °C, fig. 1. At room temperature ACP is only obtained under conditions of high supersaturation from solutions containing only calcium and phosphorous ions. ACP obtained in this manner was determined to have approximate formula Ca$_3$(PO$_4$)$_{1.87}$(HPO$_4$)$_{0.2}$nH$_2$O [5]. At temperatures above 600 °C ACP transforms into ß-TCP, fig. 1, which is in good agreement with studies showing that pure ß-TCP cannot be directly obtained from aqueous systems but by calcination of ACP powders above 700 °C [5].

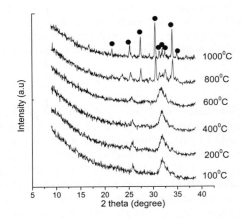

Fig. 1: HT-XRD patterns for the transformation of ACP produced by neutralization method into ß-TCP powders. (•) ß-TCP phase

The powders produced by neutralization method consisted of homogeneous spherical particles with an average grain size of 300 nm. Fig. 2 (A) and (C) show the morphologies of the powders dried at 85 °C for 24h and calcinated at 1000 °C for 3h, respectively. The specific surface area of the calcinated powders was detected to be 1.9 m^2/g. Ultrasound irradiation in neutralization method change the morphology, fig. 2(B). Calcination at 1000 °C for 3h leads to homogeneous spherical nanopowders with an average crystallite size of 50 nm and a specific surface area of 4.2 m^2/g, fig. 2 (D). The significant decrease in grain size can be assumed due to higher available energies during ultrasound irradiation, resulting in a disturbed growth of nuclei to their critical size.

Fig. 3 compares the calcinated powder XRD patterns obtained from the samples synthesized with and without assistance of ultrasound with a frequency of 50 KHz. From the N-patterns, no other crystalline phases than HA and ß-TCP were detected, while the powders produced by N-US presented mainly HA phase and secondary ß-TCP phase. Sonochemistry derives principally from acoustic cavitation, growth, and implosive collapse of bubbles in liquids. Cavitation serves as a means of concentrating the diffuse energy of sound [6]. The shock wave produces energy which could be able to generate ions and create free radicals. This can stimulate the reactivity of chemical species, resulting in the effective acceleration of heterogeneous reactions between liquid and solid reactants. Therefore the collapsing microbubbles may generate an energy which can rupture the molecular ionic bond and create reactive species as hydroxyl radicals [7]. Fig.4 shows FTIR spectra of CaP powders after calcination at 1000 °C for 3h. The OH absorption band at 633 cm^{-1} which is characteristic for HA phase is absent in N but present in the N-US.

Fig.2: SEM micrographs of powders after drying at 85 °C for 24h (A) N and (B) N-US; and after calcination at 1000 °C for 3h (C) N and (D) N-US.

The influence of certain parameters such as frequency and viscosity by change of the concentration of additives was evaluated with a frequency of 40 KHz. In this experiment, the viscosity increased with the content of glucose (10-40 wt%) and glycerol (10-40 v%). It was observed that the viscosity impedes chemical activity of OH° by decreasing the formation of HA. The propagation of ultrasonic waves in water leads to H_2O sonolysis. One of the main consequences is the production of radical species, especially the very strong oxidizing hydroxyl radical OH° [8]. However, it seems that the viscosity impedes chemical reactivity by decreasing the fluctuations in bubble diameter. The limit for formation of HA at 40 KHz were 40 wt% of GLU and 40 v% of GLY. New series of sonochemical experiments were repeated with this concentration of additives at a frequency of 50 KHz. Fig. 3 (GLU and GLU-US) shows powders with 40 wt% D-glucose additive which consist of ß-TCP phase, independent of magnetic stirring or ultrasound irradiation. Powders produced with 40 v% glycerol with magnetic stirring presented mainly ß-TCP phase with traces of HA, fig. 3 (GLY). HA phase was found as secondary phase in the ß-TCP produced by 40 v% glycerol, fig. 3 (GLY-US). GLY-US shows the OH absorption band at 633 cm^{-1}, characteristic of HA phase, fig. 4. The ultrasound frequency as well as the viscosity of the medium determine the type and intensity of cavitation.

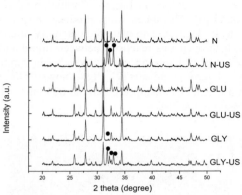

Fig. 3: XRD patterns of the powders, obtained and calcinated at 1000 °C for 3h. (●) HA phase.

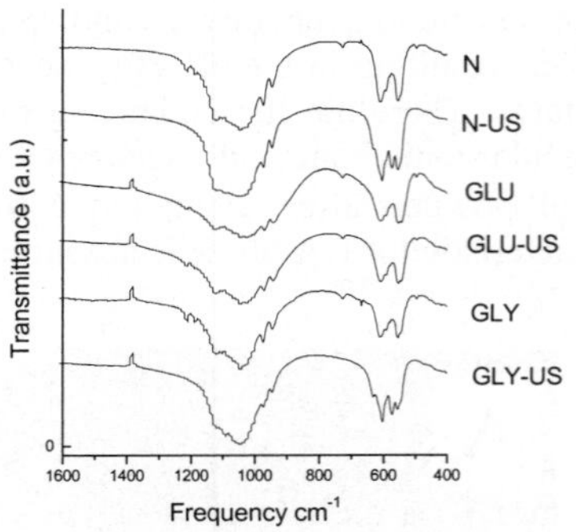

Fig.4: FTIR absorption spectra of powders obtained and calcinated at 1000 °C for 3h.

Conclusions

Precipitation of calcium phosphates by neutralization method was effective to obtain ß-TCP powder after calcination above 600 °C. The major contribution of this work is to emphasize that there is considerable potential in increasing the efficiency of sonochemical reactions in relation:

(i)   with the production of nanopowders of calcium phosphate with a significant decreased grain size by neutralization method and ultrasound irradiation (50 KHz);

(ii)   with the control of the formation of ß-TCP and HA by addition of D-glucose or glycerol to the solution, which leads to an increased viscosity, a minimized occurrence of microbubbles and thus prevents the formation of HA due to a reduced amount of OH° radicals;

(iii) with the production of BCP with adjustable ß-TCP/HA ratios which are important to control the bioactivity and the solubility of the composite.

Acknowledgements

The authors are grateful to CAPES/DAAD-PROBRAL Brazil-Germany for the financial support.

References

[1] L. L. Hench, J. Wilson (editors). *An Introduction to Bioceramics - Advanced Series in Ceramics* 1, World Scientific, Singapore (1993).

[2] G. Daculsi, R. Z. LeGeros, E. Nery, K. Lynch, J. Kalbfleisch,: J. Biomed. Mater. Res. Vol.23 (1989), p. 883.

[3] M. de Campos de; A. H. A Bressiani,, Key Eng. Mater. 218-220 (2002) 171.

[4] W. Kim; F. Saito: Ultrason. Sonochem. 8 (2001) 85.

[5] R. Z. LeGeros: Calcium Phosphates in Oral Biology and Medicine – Monographs in Oral Science 15; edited by H. M. Myers, Karger, Basel (1991).

[6] B. Li; Y. Yie: Ultrason. Sonochem. 6 (1999) 217.

[7] D. Dunn, W. D. O'Brien (editors), Ultrasonic Biophysics - Benchmark Papers in Acoustic, 7, Dowden, Hutchingon & Ross, Stroudsburg (1976).

[8] M. H. Entezari, P. Kruus, Ultrason. Sonochem. 3 (1996) 19.

Key Engineering Materials Vols. 254-256(2004) pp. 927-930
online at http://www.scientific.net
© *2004 Trans Tech Publications, Switzerland*

Specific Characteristics of Wet Nanocrystalline Apatites. Consequences on Biomaterials and Bone Tissue

Diane Eichert[1], Hocine Sfihi[2], Christèle Combes[1], Christian Rey[1]

[1] CIRIMAT, UMR CNRS 5085, ENSIACET, 118 route de Narbonne, 31077 Toulouse Cedex, France. crey@ensiacet.fr

[2] Laboratoire de Physique Quantique, UMR CNRS 7143, ESPCI, 10 rue Vauquelin, 75231 Paris Cedex, France. hocine.sfihi@espci.fr

Keywords: apatite, nanocrystals, bone mineral

Abstract. The analysis of wet nanocrystalline apatites by spectroscopic techniques (FTIR, solid state NMR) reveals fine, transient, structural details testifying for the existence of a structured hydrated layer on the surface of nanocrystals probably related although not identical, to the structure of octacalcium phosphate. On drying this layer loses its very fragile structure and gives a disordered assembly of ions corresponding to non-apatitic environments. The existence of a hydrated layer with mobile ionic entities due to an enhanced surface reactivity, gives nanocrystalline apatites very specific and interesting properties, which are used by living organisms in mineralised tissue and which can be utilised in material science. They allow strong intercrystalline interactions, specific adsorption properties and adhesion to other surfaces.

Introduction

Despite a wide occurrence in mineralised tissues and involvement in the biological activity of several bioceramics and bioglasses, the composition, structure an properties of poorly crystalline apatites (PCA) are not very well known. In addition to their nanocrystalline nature, the main difficulty is their high reactivity inducing composition and structural evolution. As usual in materials science, most investigations concerning PCA have been done with physical techniques involving "dry" stabilised samples often obtained by lyophilisation. When nanocrystals are involved, however, drying may considerably alter the surface energy and can even totally destabilise the structural organisation. As most PCA are prepared and used in wet media, we investigated the effect of an aqueous environment on the structure of nanocrystalline apatites and studied the possible alterations on drying and their consequences on our understanding of biomaterials and bone.

Materials and methods

Sample preparation. The nanocrystalline apatites were prepared by double decomposition between a calcium nitrate solution ($Ca(NO_3)_2$, $4H_2O$ 17,7 g in 250 ml deionised water) and an ammonium phosphate solution (($NH_4)_2HPO_4$, 40g in 500 ml deionized water). The calcium solution was rapidly poured into the phosphate solution. The precipitate was filtered and washed immediately. Part was analysed wet and another part after freeze drying.

Fourier Transform Infrared Spectroscopy (FTIR). The FTIR spectra of the wet gel-like samples were obtained by transmission through a polyethylene membrane coated with the gel on a Perkin-Elmer 1760 FTIR spectrometer. The lyophilised samples were analysed by transmission using a KBr pellet. The spectral treatments were performed with Grams 32 (Galactic, Salem, NH, USA).

Solid-state Nuclear Magnetic Resonance Spectroscopy (NMR). The ^{31}P MAS and CP-MAS NMR measurements were made at 202.47 MHz on Bruker ASX 500 spectrometer. The gel-like wet sample or the lyophilised samples were put in a 4mm zirconia rotor and spun at 10 kHz.

Results

The Chemical analysis of the samples just after precipitation, indicated a very low Ca/P ratio (Ca/P = 1.37) and a high amount of HPO_4^{2-} ions (27 % of total P), very close to the values of Octacalcium phosphate (OCP): Ca/P = 1,33 and HPO_4^{2-} = 33% of total P.

FTIR spectra of the wet and lyophilised PCA samples are shown in figure 1. The spectrum of the wet sample shows distinct thin peaks enhanced by self-deconvolution. Several bands or shoulders

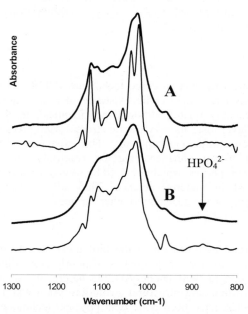

Fig. 1: FTIR spectra of wet and freeze-dried Poorly crystalline apatite.

The wet sample (A, bold)) shows thin specific bands. The HPO_4^{2-} bands appears very broad and barely visible. Self-deconvolution (thin line) reveals distinc bands in close positions to those of Octacalcium phosphate except for bands at 915 and 1195 cm^{-1}.

The freeze-dried sample (B) shows broader bands but the HPO_4^{2-} band is clearly seen (arrow). Self-deconvolution indicates the presence of weak bands of the original structure, with a few distorsions.

may thus be detected and compared to those of well crystallised Ca-P salts. A strong analogy is observed with OCP [1] although some very specific bands are missing. In contrast, the spectrum of the same freeze dried sample shows broad bands lacking fine structure. Resolution enhancement partly restores the wet sample spectrum with significant band shifts. One of the most intriguing observations is the weakness of the HPO_4^{2-} band assigned to P-OH stretching at 870 cm^{-1} in the wet precipitate, despite a high concentration of these species attested by chemical analysis. Paradoxically, this band is more clearly visible in the spectrum of the freeze-dried sample.

Solid-state ^{31}P MAS NMR spectra of wet and lyophilized samples are shown in figure 2-a. For comparison the OCP spectrum is also given. Like for FTIR, the spectrum of the wet sample is more resolved than that of the lyophilized one, but less resolved that of OCP. The line at ~ 1.4 ppm, assigned to HPO_4^{2-}, clearly appearing in wet sample spectrum, is at the same position as that of HPO_4^{2-} ions in DCPD. However, the apatitic phosphate groups chemical shift is the same as that observed for these groups in OCP, even though there is a larger dispersion of the chemical shift in the wet sample. This distribution can be explained by the fact that the apatitic PO_4^{3-} are locally better organised in OCP than in the wet sample. Note that some of these PO_4^{3-} groups disappear on drying (line at ~ 3.7). In addition the $^1H \rightarrow ^{31}P$ CP-MAS NMR results (Figure 2-b) reveal that the apatitic PO_4^{3-} are in closer vicinity to the protons (water molecules) in the wet sample than in OCP.

X-ray diffraction patterns of wet and freeze-dried samples seem essentially similar (data not shown) and are characteristic of nanocrystalline apatites, the specific very intense 100 peak of OCP was not detected.

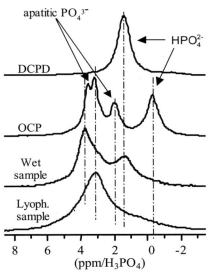

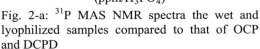

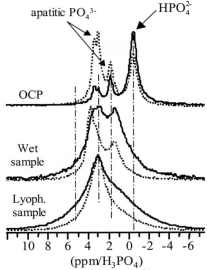

Fig. 2-a: ^{31}P MAS NMR spectra the wet and lyophilized samples compared to that of OCP and DCPD

Fig. 2-b: Comparison of ^{1}H$\rightarrow$^{31}P CP-MAS (contact time 1 ms) and ^{31}P MAS (dotted lines) NMR spectra of

Discussion

Spectroscopic data reveal a significant alteration of nanocrystalline apatites on drying. This phenomenon is partly reversible and rehydrated lyophilised samples exhibited the characteristics features of wet samples although somewhat attenuated. The spectral features observed appear very transient and fade after a few hours of ageing in the wet media. After 6 hours, the spectra appear very close to those of lyophilised samples. These observations do not appear to be due to technical artefacts, the use of different sampling techniques on FTIR (transmission through the wet sample or attenuated total reflectance), for example, gave the same results and the spectra of well known, well crystallised calcium phosphates, as well as those of matured nanocrystalline apatites were not modified in the wet state. Thus, the wet nanocrystalline apatite seems to exhibit transient but specific features existing only in the presence of water. These data indicate that nanocrystals are strongly sensitive to the media in which they are immersed, due to their very high specific surface area and the role of surface events on their global structural stability. It therefore seems appropriate to characterise these structure in the media in which they are used.

The chemical composition of the precipitate corresponds to OCP and the analogy of the FTIR spectrum of the wet precipitate with OCP is striking, however NMR revealed only a weak analogy and the presence of OCP was not confirmed by XRD. This failure could be due to the existence of very thin crystals inducing a considerable broadening of the X-ray peaks and XRD on such nanocrystals cannot be considered as an efficient tool for precise structural determination. However, a close examination of the FTIR and NMR spectra also shows several anomalies which prevent unambiguous identification of OCP. It can be noticed that several alterations of well established 3-dimensional structures have already been observed in nanocrystals [3] and it might be suggested that the wet precipitate could exhibit such an altered OCP structure. OCP can be described as a layered structure composed of alternating apatite sheets and hydrated sheets. The spectral data suggest that the main differences between the spectra of well crystallised OCP and the nanocrystalline phase reside in HPO_4^{2-} species belonging to the hydrated layer. The NMR peak due to HPO_4^{2-} in the nanocrystals is in-between two peaks of OCP, suggesting some mobility of the ions averaged to a medium position or, as previously suggested by Yesiniovski [2], a hopping of the proton from one phosphate group to another, also giving an averaged position, like for phosphate ions in solution. This undetermined environment of the HPO_4^{2-} ion would also lead to broad FTIR bands and could explain the absence of strong HPO_4^{2-} bands in the fresh precipitate and its stronger intensity in the lyophilised samples where the removal of water forbids ion mobility and/or proton hopping.

The existence of a transient hydrated layer has several consequences on the reactivity of nanocrystalline apatites. The most important is probably the ability of the loosely bound ions of the hydrated layer to participate in fast exchange reactions with ions from the solution. This exchange behaviour involves ions normally participating in the apatitic structure such as phosphate, carbonate, calcium, strontium, but also ions which do not penetrate into the apatitic structure, such as magnesium for example, and are confined to the hydrated layer. This property considerably increases the ion reservoir role of bone mineral. Another interesting property of the hydrated layer is to allow ionic segments of proteins to exchange and interact with mineral ions. The adsorption properties of nanocrystalline apatites have seldom been investigated, it is clear however that the number of adsorption sites and the adsorption affinities are related to the extension of the hydrated layer [4]. Such interactions with collagen could be responsible for the mechanical properties of bone and many reports have clearly established the alteration of the mechanical properties of bone on drying. The hydrated layer could also permit direct crystal-crystal interactions and could explain the "crystal fusion" phenomena described in bone and enamel.

Materials scientists can also take advantage of the surface properties of apatite nanocrystals. Thus, low-temperature ceramics can be formed from the slow drying of aqueous suspensions of nanocrystalline apatites. Once the crystals have become close enough packed, water evaporation from their hydrated layer brings the mineral ions in direct contact and their junction is then irreversible. The main problem is the strong shrinkage that occurs during drying and the residual tensions in the ceramic which weaken the mechanical properties [5]. On a similar basis however, composite materials can be built involving charged polymeric substances with improved mechanical properties. The direct interaction between nanocrystals is also probably involved in "biomimetic cements" where the hardened body is composed of nanocrystal platelets. The ion mobility in the hydrated layer could also explain the extraordinary ability of the nanocrystals to bind to almost any substrate. This primary bonding can then be tightened on drying as water release allows direct interaction between ions of the hydrated layer and ions or dipolar groups on the substrate. Low-temperature coatings of nanocrystalline apatites on titanium have thus been shown to involve a magnesium-rich layer, probably corresponding to a hydrated layer at the surface of nanocrystals [6].

Conclusion

The transient, structured surface layer on apatite nanocrystals offers an opportunity to build a wide variety of materials (dense bodies, coatings, composites) which has already been exploited without full understanding of the mechanisms involved. The control of the surface layer and of its evolution on maturation is probably a key factor for the improvement of low-temperature biomaterials based on calcium phosphates.

References:

[1] B.O. Fowler, M. Markovic, W.E. Brown. Chem. Mater. Vol. 5 (1993), p. 1417.
[2] J.P. Yesinovski. *"Calcium phosphates in biological and industrial systems"*. Eds Amjad Z., Kluwer Academic Publishers, Boston. (1997), p.103.
[3] S. Ram, S. Rana. Mat. Letters. Vol. 42 (2000), p.52
[4] S. Ouizat, A. Barroug, A. Legrouri A, C. Rey. Mater. Res. Bul. Vol. 34 (1999), p. 2279.
[5] S. Sarda, A. Tofighi, M.C. Hobatho, D. Lee, C. Rey. Phos. Res. Bull. Vol. 10 (1999), p. 208.
[6] F. Barrere, C.A. van Blitterswijk, K. de Groot, P. Layrolle. Biomaterials Vol.23 (2002), p.2211

Key Engineering Materials Vols. 254-256(2004) pp. 931-934
online at http://www.scientific.net

Influence of Water Addition on the Kinetics of Mechanochemical Synthesis of Hydroxyapatites from DCPD+CaO

H. El Briak-BenAbdeslam[1], C. Mochales[2], J.A Planell[2], M.P. Ginebra[2] and P. Boudeville[1]

[1] Laboratoire Chimie Générale et Minérale, Faculté de Pharmacie, Université Montpellier I, 15, Avenue Chales Flahault, BP 14491, 34093 Montpellier cedex 5, France.
e-mail : boudevil@univ-montp1.fr

[2] Research Center in Biomedical Engineering, Biomaterials Division, Department of Materials Science and Metallurgy, Technological University of Catalonia (UPC), Avda. Diagonal 647, E-08028 Barcelona, Spain. e-mail : maria.pau.ginebra@upc.es

Keywords: mechanosynthesis, biphasic calcium phosphate ceramics, hydroxyapatite, calcium oxide, dicalcium phosphate dihydrate, kinetics, water addition

Abstract. Mechanosynthesis is a new possible route to obtain calcium phosphates of biological interest. In most of papers concerning the mechanosynthesis of apatites, grinding was performed under wet conditions. The influence of the water content of the slurry on the kinetic parameters of the dicalcium phosphate dihydrate – calcium oxide reaction mechanically activated was investigated varying the powder-to-water ratio and the calcium-to-phosphate ratio in the powder. Ground powders before and after heating at 950°C for 2 h were analyzed by XRD. The addition of water decreases, not linearly, the reaction rate and, after heating, increases the amount of α and β-tricalcium phosphate to the detriment of hydroxyapatite for a given calcium-to-phosphate ratio. Dry grinding seems preferable to wet grinding to obtain bioceramics with the expected Ca/P ratio.

Introduction

Calcium phosphate or biphasic calcium phosphate (BCP) ceramics can be prepared by sintering a calcium deficient or stoichiometric hydroxyapatite (CDHA or HA respectively) obtained by precipitation or hydrothermal reaction. Mechanosynthesis is a new possible route to obtain HA or CDHA [1-3]. Mechanochemical synthesis can be performed under wet or dry conditions. Under wet conditions, a slurry of the starting materials, with a powder-to-liquid ratio ranging generally from 10 to 40% w/w, was subjected to mechanical attrition. Under dry conditions, starting powders are directly ground without addition of liquid. In most of papers or patents concerning the mechanochemical preparation of calcium phosphates, mechanosynthesis was performed under wet condition [1,2,4 and references therein]. Thus we investigated the influence of the water addition on the mechanochemical reaction kinetics of dicalcium phosphate dihydrate (DCPD) with calcium oxide (CaO) (Eq. 1 with $3 \leq x \leq 4$).

$$6 \text{ CaHPO}_4 \cdot 2\text{H}_2\text{O} + x \text{ CaO} \rightarrow \text{Ca}_{(6+x)}(\text{HPO}_4)_{(4-x)}(\text{PO}_4)_{6-(4-x)} + (14-(4-x)) \text{ H}_2\text{O} \qquad (1)$$

Materials and Methods

CaO (from Aldrich) was heated at 900°C for 2 h to remove H_2O and CO_2 and stored in a vacuum desiccator. After heating, median particle size d_{50} was around 7 μm (d_{10}–d_{90} = 2–40 μm; calculated specific surface area = 4.3 m^2 g^{-1}, Mastersizer, Malvern Instruments). DCPD (from Fluka), median particle size 8 μm (1.6–27 μm; calculated specific surface area, 3.5 m^2 g^{-1}) was used as received. Slurries were prepared by adding various volumes of distilled water (from 5 to 100 ml) to 15 g of a mixture of DCPD and CaO each weighted to obtain the desired molar calcium-to-phosphate ratio (Ca/P = 1.5, 1.6 or 1.67). They were ground in a planetary ball mill (Retsch Instruments: porcelain vial eccentricity on the rotating sun disc 3.65 cm) at a rotation velocity of 350 rpm with 5 porcelain balls, 2.4 cm in diameter (total mass 119 g and surface area 98.5 cm²).

To study the reaction kinetics, after different intervals, 200-500 μl of the suspension was taken, filtered onto a fritted glass filtering flask (porosity 4) under vacuum, washed with acetone for water removing and dried in a stove at 37°C for 1 h. The DCPD content in the dried powder at the different intervals was determined either by differential scanning calorimetry (DSC6 PerkinElmer), value of the enthalpy change of the endotherm between 170-210°C corresponding to the DCPD dehydration into DCPA, or by X-ray diffraction (XRD), surface area of the DCPD peak at $5.80°\theta$ (plane 0.2.0) after baseline subtraction [5,6]. The rate constant (k) of DCPD disappearance was given by the slope of the $\ln(\text{DCPD}) = f(t)$ plot (Figs. 1 and 2) and the final reaction time (t_f) was determined with a phenolphthalein test ($\varphi\varphi$ test) [6]. In this test, some grains of powder are poured in water containing $\varphi\varphi$ and a pink color of the solution or of the grains indicates the presence of unreacted calcium hydroxide. After complete reaction (negative $\varphi\varphi$ test) the as-prepared powders and powders after heating at 950°C for 2 h were analyzed by XRD (Philips PW3830X, horizontal goniometer CGR, Cu $K_{\alpha 1}$ = 1.5405 Å, Ni filter) from 2 to $30°\theta$, by $0.02°\theta$, 20 acquisitions, acquisition delay 500 ms.

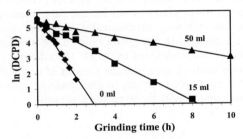

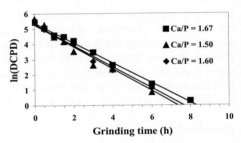

Fig. 1. Variations in ln(DCPD) with the grinding time and with the amount of water added to 15 g of powder with Ca/P = 1.67

Fig. 2. Variations in ln(DCPD) with the grinding time and with the Ca/P ratio. Water added 15 ml to 15 g of powder.

Results and Discussion

The rate constant k of DCPD disappearance and the final reaction time (t_f) for the different volumes of water added and Ca/P ratios are given in Table 1. The mechanochemical reaction of DCPD with CaO was always slower (plots 15 and 50 ml in Fig. 1) when water was added than under dry conditions (Plot 0 ml in Fig. 1).

Table 1: k and t_f obtained for different amounts of water added to 15 g of powder with different Ca/P ratios

Water added (ml)	0		5	10		15		30	50	100
Powder (% w/w)		100	75	67		50		33	23	13
Ca/P	1.50	1.60	1.67	1.67	1.67	1.50	1.60	1.67	1.67	1.67
k (h⁻¹)	1.65	1.68	1.51	0.28	0.38	0.72	0.70	0.66	0.41	0.24	0.16
t_f (h)	2.6	3.8	10	18	12	5	7	14	32	nd	nd

nd : not determined

The variations of k and t_f with the amount of water added were not linear. With water, the best compromise was obtained for a powder percentage of 50% (w/w) (Table 1). For 50 and 100 ml of water added to 15 g of powder with Ca/P =1.67, because the reaction was not achieved after 2 days of grinding, the final reaction time was not determined.

In presence of water, the mechanochemical reaction followed the same mechanism as under dry conditions [6]. It proceeds in two well differentiated stages: reaction of DCPD with CaO or $Ca(OH)_2$ to give CDHA with Ca/P ≈ 1.5, and then $Ca(OH)_2$ in excess reacts with CDHA 1.5 leading to a CDHA with a greater Ca/P ratio. We can also observe that during wet grinding, the DCPD disappearance rate was independent on the Ca/P ratio (Fig. 2) as we noted for dry grinding [6], which was not for the final reaction time that increased not linearly with the Ca/P ratio.

After complete reaction (negative φφ test), whatever the amount of water added and the Ca/P ratio, XRD patterns of the as-prepared powders showed the profile of an apatite with a poor crystallinity (Fig. 3). After heating at 950°C for 2 h, the XRD pattern profiles varied with the Ca/P ratio but the final phase(s) obtained did not comply with the expected composition for Ca/P 1.60 and 1.67. Indeed, for Ca/P = 1.67, we would expect to obtain after heating only pure hydroxyapatite, which was nearly so under dry grinding but not under wet milling (Fig. 4). Note that for Ca/P = 1.50, under dry or wet conditions, we obtained after heating only β-TCP, without α-calcium pyrophosphate.

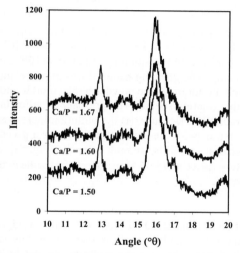

Fig. 3. XRD patterns of the ground powders with different Ca/P ratios (powder 15 g + water 15 ml).

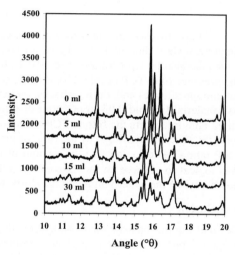

Fig. 4. : XRD patterns after heating at 950°C for 2 h of powders (Ca/P = 1.67) ground with different volumes of water added to 15 g of powder.

Wet mechanosynthesized powders were composed after heating of α and β-TCP and HA in proportions depending on the water content. Because the main diffraction peaks of α -TCP (plane (0 3 4) at 15.36°θ) and β-TCP (plane (0 2 10) at 15.50°θ) overlap, we calculated the surface area of these two peaks together after baseline subtraction. In the same way, we calculated the whole surface area of the HA peaks at 15.82 and 16.04°θ (planes (2 1 1) and (1 1 2)). The variations in the surface area of the α+β-TCP peaks and of the HA peaks are represented in Fig 5. The proportion of α and β-TCP increased with the amount of water because, certainly, more calcium than phosphate remained dissolved in the liquid phase leading to CDHA. The same observation was reported by Avvakumov et al. [1] and Shuk et al. [2].

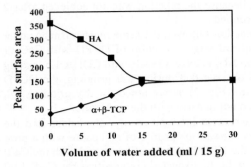

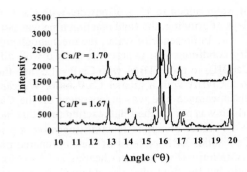

Fig. 5. Variations in the surface area of the α+β-TCP and HA peaks with the amount of water added to 15 g of powder (Ca/P = 1.67).

Fig. 6. XRD patterns of powders with theoretical Ca/P of 1.67 and 1.7 ground under dry conditions after heating at 950°C for 2h. (β =β-TCP).

Pure hydroxyapatite was only obtained by dry milling a DCPD-CaO mixture with a theoretical Ca/P ratio of 1.70 (Fig. 6) for reasons that have been discussed in a previous paper [6]. Given the heating temperature used, 950°C, below the β to α-TCP transition temperature (1160°C), the presence of α-TCP beside β-TCP could seem surprising. The grinding material was in porcelain (aluminum silicate) and, during shocks, there was a slight pollution of the powder by the mortar and ball material (around 0.03 % per hour of grinding at this rotation speed). The presence of silica in slight amount lowers the β to α-TCP transition temperature [7]. Because, (i) the pollution increases with the time of grinding, (ii) the time of grinding increases with the amount of water for a given Ca/P ratio (Table 1) and (iii) the amount of β-TCP increases with the amount of water (Fig. 5), it was logical to observe an increase of the α-TCP content (peak at 15.36°θ in Fig. 3) in the heated powder when the water volume in the slurry was increased.

Conclusion

Dry and wet mechanosynthesis both allow the preparation of nano-sized CDHAs that can be used to prepare biphasic calcium phosphate ceramics. But from these results, we think that dry milling is preferable to wet grinding because (i) dry milling is faster, lowering the pollution by the mill material, (ii) the final powder has the expected Ca/P [3] and (iii) the powder can be directly used without filtering and drying steps to prepare bioceramics.

References

[1] E. Avvakumov, M. Senna and N. Kosova. Soft mechanosynthesis: a basis for new chemical technologies. Kluwer Academic Publishers, Boston, 2001.
[2] P. Shuk, L. Suchanek, T. Hao, E. Gulliver, R.E. Riman, M. Senna, K.S. TenHuisen and V.F. Janas. J. Mater. Res. Vol 16 (2001) 1231-4.
[3] P. Boudeville, B. Pauvert, M.P. Ginebra, E. Fernandez and J.A. Planell. *Bioceramics 13* (Trans Tech Publications Switzerland 2000) 115-8.
[4] Y. Hakamatsuka, H. Irie, S. Kawamura and M. Toriyama. US Patent 5322675, 1994
[5] S Serraj, P. Boudeville, B. Pauvert and A. Terol. J. Biomed. Mater. Res. Vol 55 (2001) 566-75.
[6] H. El Briak-BenAbdeslam, C. Mochales, M.P. Ginebra, J. Nurit, J.A. Planell and P. Boudeville. J. Biomed Mater Res 2003, to appear.
[7] R.W. Nurse, J.H. Welch and W. Gutt. J. Chem. Soc. Vol 220 (1959) 1077-83.

VIII.   POROUS BIOCERAMICS

Key Engineering Materials Vols. 254-256(2004) pp. 937-940
online at http://www.scientific.net

Manufacturing of Biocompatible TiO₂-Surface-Structures with a Water Based Tape Casting

Julia Will[1], Stephan Zuegner, Håvard Haugen[2], Ursula Hopfner, Joachim Aigner and Erich Wintermantel

Central Institute for Medical Engineering ZIMT, Technische Universität München, Germany,
[1]will@zimt.tum.de, [2]haugen@zimt.tum.de

Keywords: TiO₂, scaffold, porous ceramics

Abstract. TiO₂ a bioactive material, as it has been demonstrated in several studies. Tape casting is a well established method of high-quality ceramic sheet production. Little work has been done in the field of combining this method and TiO₂ for biomedical applications. This paper shows a water-based manufacturing method for thin ceramic scaffolds with an adjustable micro- and macro surface structure (pore diameters $0.37 - 0.96$ µm and $50 - 150$ µm), which is needed for cell attachments.

Introduction

One of the most promising biocompatible materials has been proven in previous studies to be TiO₂ [1,2]. This material has shown particular biocompatible properties, where scaffolds were implanted in rats for 55 weeks without any signs of inflammatory responses or encapsuling [3]. Tape casting is an ample proven procedure for the manufacturing of thin high-quality ceramic membranes, panels and tapes with high density and very smooth surfaces [4-6]. The objective of this work was the production of TiO₂-tapes with adjustable thickness, surface-structure, mechanical strength and reproducibility for the application as scaffold.

Method

Manufacturing of a TiO₂ tape ceramic. Basic elements of the fluid manufacturing process are the ceramic powder and the solvent. Additives are mostly dispersants (=surface active agents to avoid an excessive increasing of agglomerates), binders (= cohesion of the green body and placeholder for a micro structure in the sintered ceramic), plasticizers (=flexibiliser of the polymeric binder) and organic particles (=placeholder for a adjustable macro structure). To achieve a feasible slurry, screening-tests with nine selected organic and inorganic solvents and seven dispersions in combination with different binders, plasticizers and placeholders were performed (see Table 1). A quantitative examining of useful combinations was applied to find practical formulas for the manufacturing process. Main selection criteria were strong pseudo-plastic behaviour, homogeneous mixing of the slurry, good adhesion on the tape-carrier and drying/sintering without crack initiation [5,6]. The slurry analysis were performed by Electronic Sonic Amplitude (ESA), rheology, grindometry and SEM / optical microscope. Regarding the surface of the sintered ceramic, two branches were chosen: One is a smooth tape with only micro structured porosity. The other one has a rough surface, produced by burn-out of polypropylene particles (placeholders) to create an additional macro structure. Prototypes should be produced in various thickness to achieve an idea of crack-free producibility. Shaping of the green body was realized by a non-continuous single blade tape casting system. To ascertain the influence of drying conditions, tests with different atmospheres were accomplished - the required sinter parameters were measured with a dilatometer.

Characterization for scaffold purposes. The characterisation of the structure was performed using mercury porosimetry and SEM-analyses. Mechanical stability was tested with a modified 3-Point-Flexural-Test by measuring the maximal force before breaking and using the idealized formula for the calculation of flexural stress of beams with a constant profile as illustrated in Fig. 1. A micrometer in combination with SEM-analyses to avoid erroneous measurements by mechanically destroying the surface structures was chosen to measure the different thickness of the tape prototypes. Finally, a preliminary cell culture test was performed with a 3T3 fibroblastic cell line.

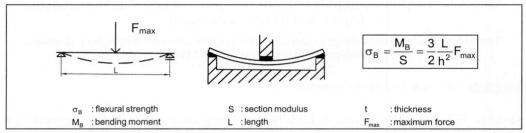

$$\sigma_B = \frac{M_B}{S} = \frac{3}{2}\frac{L}{h^2}F_{max}$$

| σ_B | : flexural strength | S | : section modulus | t | : thickness |
| M_B | : bending moment | L | : length | F_{max} | : maximum force |

Fig. 1: 3-Point Flexural Test modified by polymer supports

Results

Manufacturing and additive selection. Utilising the results of extensive screening tests, water was selected as the best economical and technical solvent. For additives, glycerine, polyphosphatester, acrylat and – for the rough surface version – polyurethane particles were chosen (illustrated in Table 1). A decomposition of the dispersion in the time frame of processing was not detectable. The pseudo-plastic behaviour of the slurry, needed for a high producing precision, was sufficiently given by the viscosity drop from 3.75 / 2.15 Pa*s (slurry of rough/smooth tape) for nearly unstressed slurry, down to 0.16 / 0.09 Pa*s at a stress rate of 400 s^{-1} (Fig. 2). Using a polyurethane foil as tape-carrier the adhesion was sufficient to cast tapes with almost constant thickness over the full width. Tapes with maximal 0.80 mm and minimal 0.40 / 0.08 mm thickness (rough / smooth tape) were produced.

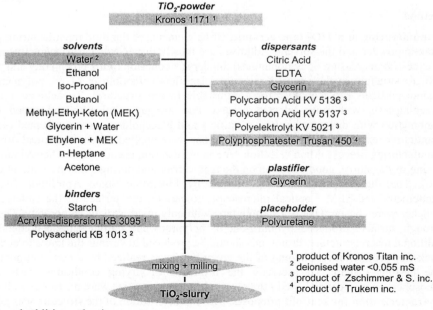

Table 1: Solvent and additive selection

After drying in a conditioned humid atmosphere (20 hours, 80% humidity and 22°C) and sintering, crack free, flat and plane tapes were obtained. For TiO_2 sintering conditions a rising temperature of 0.25°K/min up to 1150°C and cautious cooling down was applied. The dilatometer curve is displayed in Fig. 3.

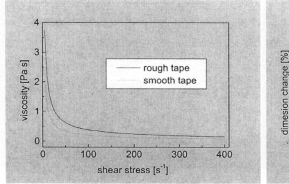

Fig. 2: Rheological data of different slurries   Fig. 3: Dilatometer analyses of TiO_2

Characterization for scaffold purposes. Porosity measurements showed a pore diameter of 0.37 – 0.96 µm and a macro structure with pore sizes of 50 – 150 µm as showed in Fig. 4 to 7. The latter region is adjusted through the polyurethane particles (placeholders).

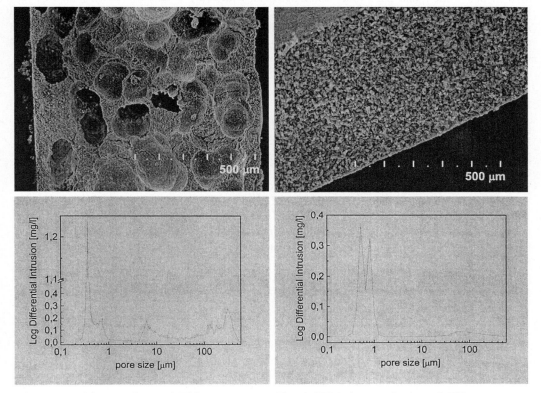

Fig. 4: SEM picture of a rough TiO_2-tape   Fig. 6: SEM picture of a smooth TiO_2-tape

Fig. 5: Pore size data of a rough TiO_2-tape   Fig. 7: Pore size data of a smooth TiO_2-tape

The flexural mechanical stability of the sintered ceramic is on average 73 N/mm^2 with a standard deviation of 2.3 (samples with 0.72 – 0.77 mm thickness and a section modulus between 1.2 and 1.9 mm^3). Initial cell culture tests showed that fibroblasts adhere on both surfaces (Fig. 8 and 9).

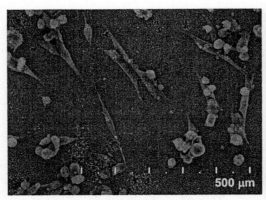
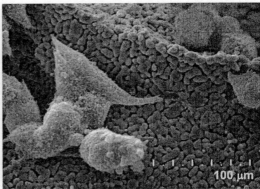

Fig. 8: SEM picture of fibroblasts on a smooth surface

Fig. 9: SEM picture of fibroblasts on a rough surface

Discussion and Conclusion

While most industrial Tape Casting processes of oxide-ceramics use organic solvents, fundamental tests in this work found pure water combined with commonly used dispersions, plastifiers and binders to be the optimal base to provide – mixed with TiO$_2$ ceramic powder – an exact adapted slurry for the tape-casting process. It was shown that flat, crack free and plane TiO$_2$-Tape-Scaffolds can be produced with adjustable micro and macro surface structure, thickness and strength. Positive cell culture tests show a potential application for scaffolds.

References

[1] Nygren, H., Tengvall, P. and Lundstrom, I., The initial reactions of TiO$_2$ with blood. Journal of Biomedical Material Research, 34, 4 (1997), p. 487-492.

[2] Wintermantel, E., Biomaterials, human tolerance and integration. Der Chirurg, 70, 8 (1999), 847-857.

[3] Wintermantel, E., et al., Angiopolarity of cell carriers, In: Angiogenesis: Principles-Science-Technology-Medicine, Steiner, R., Weisz, B., and Langer, R., Editors. Birkhaeuser Verlag, Basel 1992.

[4] Shanefield, D.J., Organic additives ceramic processing. 1995: Kluwer Academic Publishers.

[5] Mistler, R.E. and Twiname, E.R., Tape Casting: Theory and Practice. The American Ceramic Society, 2000.

[6] Hellebrand, H., Tape Casting, In: Cahn, R.W., Haasen, P., Kramer, E.J. and Brook, R.J., Materials Science and Technology: Processing of Ceramics Part 1 (Volume 17A), VCH Publishers Inc., New York, 1996.

Key Engineering Materials Vols. 254-256(2004) pp. 941-944
online at http://www.scientific.net

Processing and Characterisation of a Potential TiO$_2$ Scaffold

Håvard Haugen[1], Julia Will[2], Anne Köhler, Joachim Aigner and Erich Wintermantel

Central Institute for Medical Engineering ZIMT, Technische Universität München, Germany,
[1]haugen@zimt.tum.de, [2] will@zimt.tum.de

Keywords: TiO$_2$, scaffold, porous ceramics

Abstract. The Schwartzwalder process was chosen for the production of ceramic TiO$_2$ scaffolds and showed a fully open structure with a permeability for water of 39 %. The window sizes were 445 μm (45 ppi scaffolds) and 380 μm for the 60 ppi scaffolds. The porosity of all scaffolds was above 78 % (n=8). It was shown that scaffolds can be produced with defined pore sizes, shape and architecture, which is a requirement for scaffold production. The macro- and microarchitecture was reproducible. Hence a reproducible ceramic scaffold processing method has been established.

The interconnectivity of the pores in the scaffold was tested with a novel method. For the tests a new device was constructed where the permeability was linked to the degree of interconnectivity. Results from the permeability measurements in the mercury intrusion meter and permeability tester show that increasing pore size increases the rate of permeability. The tortuosity, which was measured in the mercury intrusion meter, was several factors higher for 60 ppi scaffolds compared to 45 ppi.

Introduction

One principle of tissue engineering is to harvest cells, expand this cell population *in vitro,* if necessary, and seed them onto a supporting three-dimensional scaffold, where the cells grow into a complete tissue or organ. For most clinical applications, the choice of scaffold material and structure is crucial [1].

In order to achieve a high cell density within the scaffold, the material has to have a high surface area to volume ratio. The pores must be open and large enough such that the cells can migrate into the scaffolds. When cells have attached to the material surface there must be enough space and channels to allow for nutrient delivery, waste removal, exclusion of material or cells and protein transport, which is only obtainable with an interconnected network of pores [2]. Biological responses to implanted scaffolds are also influenced by scaffold design factor such as three-dimensional microarchitecture [3]. In addition to the structural properties of the material, physical properties of the material surface for cell attachment are essential.

TiO$_2$ has been proven in several studies to be bioactive [4,5]. When this material was implanted in rats for 55 weeks no inflammatory responses or encapsulation were found [6].

The objective of this work was to produce ceramic scaffolds with defined macro-, micro- and nano-structures and to show the possible application of ceramic scaffolds for cell cultures.

Materials and Methods

Polymer Scaffolds

Fully reticulated polyester based polyurethane foams of 45 and 60 ppi (Bulbren S, Eurofoam GmbH, Wiesbaden, Germany) were used in this study. The foams were supplied in large plates of 5 mm thickness, and were cut by stamping with a cylindrical metal blade to diameters of 10 mm.

Scaffold Fabrication

The polymers foams were impregnated with the following suspension: 88,4 g of TiO_2 powder (Kronos TiO_2 No 1171, Kronos Titan GmbH, Leverkusen, Germany) with a median grain size of 0.3 μm mixed with 25,8 g deionised water. Additional powder was added to the suspension in several steps up to a volume content of 57 vol.% (133,4 g) after milling for an hour in a ball mill. 0,13 g of a polysaccharide binder (Product KB 1013, Zschimmer & Schwarz GmbH, Lahnstein, Germany) was further added and finally homogenised in the ball mill for 10 min.

The cylindrical shaped polymer foams were then dipped into the ceramic slurry, and later squeezed with the assistant of a hand roll, leaving only a thin layer of slurry. The samples were then placed onto a porous ceramic plate and dried at room temperature for at least 18 hours. The heating schedule for the burnout of the polymer and the sintering of the ceramic part was chosen as follows: slowly heating to 450°C with 0,5 K/min, one hour holding time at 450°C, heating to 1150°C with 3 K/min, cooling to room temperature with 5 K/min.

Characterisation

Pore Size Distribution and Porosity

The pore size distributions were measured by mercury intrusion porosimetry (Autopore IV, Micromeretics, Norcoss, GA, USA). A solid penetrometer with 6 mL bulb volume (model 07-044506-01, Micromeritics, Norcoss, GA, USA) was used. The intrusion chamber was then filled with mercury at a pressure of 3,45 kPa. The samples were penetrated with mercury until the total intrusion volume reached a plateau at a pressure of 420 MPa. The mean pore diameter of the sample was calculated from the measured pore diameter distribution, and the porosity was obtained from the total intruded volume.

Micro- and Macrostructure Analysis

The scaffolds were sputter coated with gold (Cool Sputter, Bal-Tec AG, Balzers, Liechtenstein) for 40 seconds at current of 40 mA before examination in a Scanning Electron Microscope (SEM) (Hitachi S-3500-N, Hitachi High Technologies, Japan) under high vacuum with current range between 5-15 kV.

Permeability

The permeability was measured with a self-design permeability-meter. Water was used as flow media. The sample was placed in fitting consisting of two half shells. The percolation rate was measured with a zero flow rate with no sample set as 100%. 15 samples were tested 10 times each.

In order to validate the experimental results described above, four sample were tested for permeability using the mercury intrusion porosimetry. Katz and Thompson have applied the percolation theory to laminar flow for porous media in order to develop a mercury porosimetry-based prediction for the ratio of the fluid permeability k to the electrical conductivity σ of a porous material [7]. Their main result is the prediction stated in equation 1.

$$\frac{k}{\sigma} = \frac{cd_c^2}{\sigma_o}$$

(Eq. 1)

Cell Adhesion Test

After the sterilisation in an autoclave the scaffolds (n=2) were seeded with the 3T3 fibroblastic cell line (American Type Culture Collection, Manassas VA, USA) and cultured in Dulbecco´s Modified Eagle Medium DMEM (Biochrom AG, Berlin, Germany). The cells were then seeded at concentrations of 5×10^4 and 5×10^5 in a volume of 100 μL culture medium directly on the scaffolds. The scaffolds were later sputtered and examined in a scanning electron microscope according to procedure previously described.

Results
Foam structure

The macrostructure of the sintered scaffolds is characterised by a fully open structure for both ppi-numbers, although a very low number of closed cells is always present. Three types of porous structure was be found: a fine-scale porosity within the ceramic part, a triangular pore in the centre of the struts and large pores which are also called "windows".

Pore Size Distribution and Porosity

The pore size distribution of 45 and 60 ppi scaffolds (Fig. 1) can be described by three regions: (1) a fine-scale portion within the solid portion. 0,5-1,7 µm for the 45 ppi scaffolds and 0,5-1,2 µm for the 60 ppi scaffolds; (2) pores resulting of the pyrolysis of the polymer struts with pore sizes from 12-45 µm with a maximum pore size at 0.15 µm (45 ppi) and 7-25 µm with a maximum pore size at 19 µm (60 ppi). These pores are in the size of the original struts; (3) pores with a diameter of 385-700 µm and 310-385 µm, respectively. The last region correspond to the macroscopic, which should serve as channels for the metabolism products.

The three different pore sizes could also be observed in the SEM. However, the smallest sized pores had a large diameter then measured with the porosimetry, which show an typical error for the mercury intrusion porosimetry. The overall porosity was 74 %.

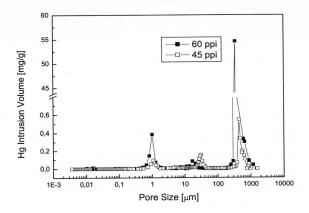

Fig. 1: Pore size distribution from mercury intrusion porosimetry of sintered TiO$_2$ scaffold from 60 ppi foam.

Permeability

The permeability rate depends of the foam's density. 45 ppi scaffolds had a high value of permeability and was measured to 39%. Its permeability decrease slightly with increasing density. The 60 ppi scaffolds show, however, a strong correlation between decreasing permeability and increasing density. Hence, the higher the density the lower the permeability rate.

The 60 ppi scaffolds also showed lower permeability values, obviously as the smaller pore size increases the resistance to the flow media. The tortuosity, which was measured in the mercury intrusion meter, was several factors higher for 60 ppi scaffolds compared to 45 ppi and therefore do also explain the lower permeability for higher ppi-numbers.

Permeability measurements (n=4) of the mercury intrusion meter revealed values of 50 mDarcy for 60 ppi scaffolds and 153 mDarcy for 45 ppi scaffolds calculated by equation 1, and verify the experimental results from the self-design permeability meter.

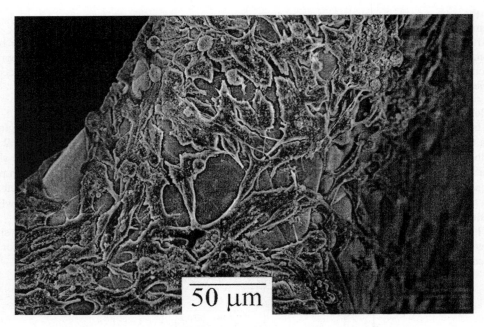

Fig. 2: Fibroblasts adhere and circumstantially cover a scaffold's strut of the TiO₂ surface.

Discussion and Conclusion

It was shown that scaffolds can be produced with defined pore sizes, shape and architecture, which is a requirement for scaffold production. The pore size is easily adjusted by using polymer foams with different ppi-numbers. The scaffolds were open porous, clearly permeable as the water permeability was 39 % and had above porosity 78 %. The macro- and microarchitecture was reproducible. Hence a reproducible ceramic scaffold processing method has been established.

The cell seeding showed that fibroblasts adhered to the TiO₂ surface and were alive. The adhesion of fibroblasts can be viewed in figure 2.

References

[1] R.C. Thomson et al., Scaffold processing. In: R. Lanza, R. Langer and J. Vacanti (Editors), Principles of Tissue Engineering. (Academic Press, San Diego, 2000).
[2] M.C. Peters and D.J. Mooney: Mater. Sci. Forum, Vol. 250, (1997), p. 43-52
[3] K. Whang et al.: Tissue. Eng., Vol. 5, 1, (1999), p. 35-51
[4] H. Nygren, P. Tengvall and I. Lundstrom: J. Biomed. Mat. Res., Vol. 34, 4, (1997), p. 487-492
[5] E. Wintermantel et al.: Der Chirurg, Vol. 70, 8, (1999), p. 847-857.
[6] E. Wintermantel et al., Angiopolarity of cell carriers, Directional angiogenesis in resorbable liver cell transplantation devices. In: R. Steiner, B. Weisz and R. Langer (Editors), Angiogenesis: Principles-Science-Technology-Medicine. (Birkhaeuser Verlag, Basel, 1992).
[7] A.J. Katz and A.H. Thompson: Phys. Rev., Series B, Vol. 34, 11, (1986), p. 8179–8181

Key Engineering Materials Vols. 254-256(2004) pp. 945-948
online at http://www.scientific.net
© 2004 Trans Tech Publications, Switzerland

Development of a New Calcium Phosphate Glass Ceramic Porous Scaffold for Guided Bone Regeneration

M. Navarro[1], S. Del Valle[1], M.P. Ginebra[1] ,S. Martínez[2] and J.A. Planell[1]

[1] Research Center of Biomedical Engineering (CREB), Technical University of Catalonia, Avda. Diagonal 647, 08028 Barcelona, Spain, maria.pau.ginebra@upc.es; melba.navarro@upc.es

[2] Dept. Crystallography, University of Barcelona, Marti i Franqués s/n, 08028 Barcelona, Spain

Keywords: Calcium phosphate glass, porous glass ceramic, scaffolds.

Abstract. In this work, different porous scaffolds were developed using calcium phosphate glass particles in the system (P_2O_5-CaO-Na_2O-TiO_2) and H_2O_2 as foaming agent. The different 3D structures were obtained by varying the foaming agent content and the thermal treatment. The microstructure and morphology of the specimens were observed by SEM (Scanning Electron Microscopy) and the percentage of porosity was quantified by image analysis. XRD (X Ray Diffraction) was used to study the formation of new crystalline phases. The results showed that a increase in the amount of H_2O_2 solution mixed with the glass particles increased both, the number of pores and their size. Indeed, the percentage of porosity ranged between 35% and 55% and the pore size varied from 100 to 500μm for the samples with 40 and 60% w/w of H_2O_2 respectively. The thermal treatments induced the formation of crystalline phases in the material. After sintering at 540°C a vitreous phase was the predominant, while at 570°C, the percentage of crystalline phases was higher The crystalline phases detected were meta and pyrophosphates. The pores sizes and interconnectivity obtained were in good agreement with those demanded for bone engineering scaffolds.

Introduction

Nowadays, the study of new resorbable porous scaffolds for guided bone regeneration is acquiring great importance, given the limitations that bone grafts present such as the material availability, and the risk of inducing transmissible diseases among others.

It is well known that one of the main features this kind of materials must accomplish is the reabsorbability at controlled rates, to match that of tissue repair. Calcium phosphate glasses are particularly suitable for this application, since they have a composition similar to the mineral phase of bone, and their dissolution rate can be precisely modulated by modifying their chemical composition as it has been demostrated in previous studies [1].

Another characteristic of paramount importance in the development of Tissue Engineering scaffolds is a high interconnected macroporosity, to ensure cell colonization and flow transport of nutrients and metabolic waste.

Several methods for obtaining these porous scaffolds have been proposed. Some of them are based in the introduction of porogenic agents, foaming agents or emulsifiers. Other approaches use the casting or moulding technique, through the impregnation of a porous polymer foam with ceramic emulsions [2-8].

The scope of this work is to develop glass ceramic porous scaffolds from a calcium phosphate glass in the system P_2O_5-CaO-Na_2O-TiO_2, whose in vitro biocompatibility was assessed in a former study [9]. In these scaffolds, the kinetics of resorption is expected to be controlled by the chemical composition, percentage of crystalline phases and porosity.

Materials and Methods

Preparation of the scaffolds For the elaboration of the porous scaffolds, calcium phosphate glass particles with the following molar composition, $44.5\%P_2O_5$-$44.5\%CaO$-$6\%Na_2O$-$5\%TiO_2$ were used. For the preparation of the glasses $CaCO_3$, $NH_4H_2PO_4$, Na_2CO_3 and TiO_2 were used as raw materials. The glasses were melted in a Platinum crucible at 1350°C for 3 hours, and then rapidly quenched on a metallic plate preheated at 350°C. Finally, the glasses were annealed at their Tg (533°C) and slowly cooled. The glass was pulverized (< 30µm) in an agate ball mill.

DTA analysis of the glass particles was performed to obtain the values of the glass transition temperature(T_g) as well as the crystallization temperature (T_c) in order to know the temperature range for the different thermal treatments. Both, Tg and Tc were measured in a differential thermal analyzer (DTA; Netzch) at a heating rate of 10°C/min and are reproducible to ± 5°C. The analysis was conducted up to 1200°C.

Glass powder was mixed with different percentages of a H_2O_2 solution (10% vol/vol). Once mixed, the slurry was cast in teflon molds and stored into a furnace at 60°C for 2.5 h. Afterwards, different thermal treatments were applied, at temperatures slighly above the glass transition temperature of the glass to avoid full crystalisation of the samples. The variables studied were the following : (i) the amount of H_2O_2 solution in the slurry: 40, 50 and 60% by wheight; (ii) the sintering temperature: 540 and 570°C; (iii) the sintering time: 2, 3 and 4 hours.

Characterization of the scaffolds The porosity, morphology and microstructure of the samples were analyzed by scanning electron microscopy (SEM)(Jeol JSM-6400) and an image analyzer system (Omnimet V5.1). The devitrification or formation of new crystalline phases was evaluated by X Ray powder Diffraction (XRD). XRD measurements were performed in a Bragg-Brentano θ/2θ Siemens D-500 diffractometer with CuKα radiation. The starting and the final 2θ angles were 4 and 70° respectively. The step size was 0.05° 2θ and the measuring time 3s per step.

Results and Discussion

SEM analysis showed the microstructure and morphology of the different samples (see Fig. 1a and 1b). The influence of the H_2O_2 content in the porosity, interconnectivity and pore size could be observed. The samples with 40% w/w of H_2O_2 showed a structure formed mainly by laminar pores which could be formed by the union of several smaller pores. The specimens with 50% w/w of H_2O_2 displayed a porous structure very similar to that of the 40% samples. In the case of the 60% H_2O_2 constructs, pores with rounded shape and larger size were the most common. In addition, a higher level of interconnectivity than in the materials with 40% and 50% w/w of H_2O_2 was observed.

The image analysis showed an increment in the macroporosity percentage as the H_2O_2 content added into the samples increased. The specimens sintered at both temperatures (540° and 570°C) showed similar morphologies. Neither the time of sintering had any significant influence on the amount of macroporosity. The different percentages are shown in Table 1. The macroporosity varied from 40% to 55% for the samples with 40% and 60% w/w of H_2O_2 respectively.

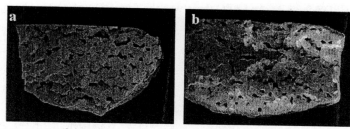

Fig. 1. Macrostructure of the 40% sample (a) and the 50% sample (b) after sintering at 570°C for 2h

Table 1. Macroporosity variation with H_2O_2 content. The value was calculated as an average for all the series having the same H_2O_2 content. (value ± SD)

H_2O_2 %wt	Macroporosity %
40	40 ± 5
50	45 ± 5
60	55 ± 5

SEM images showed that there was a relevant difference in the degree of sintering between the samples heated during 2 and 3 hours at 540°C. In the samples sintered for 2 hours, the glass particles were slighly sintered, and the formation of necks between the different particles was scarce. However, the specimens sintered for 3 hours displayed the formation of multiple bonds and necks between the particles. The union between the glass particles due to the sintering process, introduced a change in the microporosity of the specimens structure.

The samples heated during 4 hours at 540°C, showed both, microstructure and macrostructure very similar to that obtained after the 3 hours treatment.

The specimens sintered at 570°C revealed, in general, a sintering level comparable to that of the samples treated at 540°C for 3 and 4 h, independently of the sintering time (2, 3 and 4 hours).

The pore size distribution in number for the samples with different thermal treatments and different amounts of the foaming agent is shown in Figure 2a. The distribution of the number of pores was similar between the specimens with the same H_2O_2 content. There was an increment in the pore size according to the percentage of H_2O_2 incorporated. In this way, the 40% samples presented a higher number of pores between 100-200μm while the 50% and 60% samples presented pores over 500μm. However, the pore size distribution for the 50% samples was more homogeneous and the number of small pores was higher than in the 60% specimens.

The thermal treatment induced the formation of crystalline phases in the material, as revealed by X ray diffraction (see Fig. 2b). The nature and quantity of these phases depended on the temperatures and times of the thermal treatments performed. Thus, after sintering at 540°C for 3 hours, the material was mainly formed by a vitreous phase, with a small percentage of crystalline phases. However, at 570°C, the percentage of crystalline phases increased, independently of the amount of foaming agent incorporated in the slurry. The crystalline phases detected were meta and pyrophosphates. After the treatment at 540°C the resultant phases were δ- $Ca(PO_3)_2$ (JCPDS-9-363) and α-Ca$_2$P$_2$O$_7$ (JCPDS-9-345). At 570°C there was the formation of a new phase, β-Ca$_2$P$_2$O$_7$ can be observed.

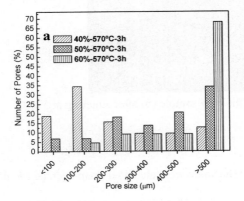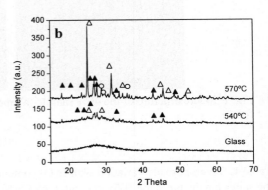

Fig. 2. a) Pore size distribution in number for the specimens containing 40, 50 and 60% (w/w) of H$_2$O$_2$ sintered at 570°C for 3 h. b) X ray diffractograms of the 60 % (w/w) samples processed at 540 and 570°C after 3 h compared with the glass diffractogram. The main peaks are highlighted, $\triangle$ δ-Ca(PO$_3$)$_2$, $\blacktriangle$ α-Ca$_2$P$_2$O$_7$ and o β-Ca$_2$P$_2$O$_7$

Conclusions

Macroporous calcium phosphate glass ceramics can be obtained by foaming a suspension of glass particles with H$_2$O$_2$. The percentage of porosity, pore sizes and percentages of crystallinity can be controlled by varying the amount of H$_2$O$_2$ incorporated and the thermal treatments applied. The pores sizes and interconnectivity obtained in this work are in good agreement with those demanded for bone engineering scaffolds. Thus, these constructs seem to be a good option as scaffolds for biomedical applications.

References

[1] M.Navarro, M.P.Ginebra, J.Clément, S. Martínez, G. Avila and J.A. Planell: J. Ame.Ceram.Soc. Vol.87-6(2003)

[2] A. El-Ghannam, P. Ducheyne and I.M. Shapiro: J.Biomed.Mater.Res. Vol. 29 (1995), p.359

[3] L. Dean-Mo: J.Mater.Sci:Mater in Med. Vol. 8 (1997), p.227

[4] P. Sepulveda, J.R. Jones and L. Hench: J.Biomed.Mater.Res. Vol.59 (2002), p.340

[5] C. Tuck and R.G. Evans: J.Mater.Sci.Letters. Vol.18 (1999), p.1003

[6] J. Tian: J.Mater.Sci. Vol.36 (2001), p.3061

[7] G. Gong, A. Abdelouas and W. Lutze: J.Biomed.Mater.Res. Vol.54 (2001), p.320

[8]H.Yuan, J. De Bruijn, X. Zhang, C. Van Blitterswijk and K. De Groot: J.Biomed.Mater.Res(Appl.Biomat.) Vol. 58 (2001), p.270

[9] M. Navarro, M.P. Ginebra and J.A.Planell: J.Biomed.Mater.Res. In Press.

Key Engineering Materials Vols. 254-256(2004) pp. 949-952
online at http://www.scientific.net

Fabrication and Characterization of Porous Hydroxyapatite and Biphasic Calcium Phosphate Ceramic as Bone Substitutes

Nurşen Koç[1], Muharrem Timuçin[2], Feza Korkusuz[3]

[1]METU, Metallurgical and Materials Eng. Dept., Ankara-06531 TURKEY, nkoc@metu.edu.tr
[2]METU, Metallurgical and Materials Eng. Dept., Ankara-06531 TURKEY, timucin@metu.edu.tr
[3]METU, Health Center., Ankara-06531 TURKEY, feza@metu.edu.tr

Keywords: Hydroxyapatite, biphasic calcium phosphate, porous ceramic, slip casting.

Abstract. In the present study, the calcium phosphate powders (HAP, HAP-TCP) were produced by a precipitation technique involving the use of aqueous solutions of calcium nitrate and diammonium hydrogen phosphate. Preforms of calcium phosphate ceramics were produced by using a modified slip casting technique. Porosity in the cast and in the sintered ceramic was obtained through custom made PMMA beads ~200-300 μm size. The green ceramic samples were prepared by suspending the apatite powder and PMMA beads in an aqueous medium stabilized with an acrylic deflocculant. Porous ceramics were obtained by sintering the polymer-free preforms for 2 hours at 1000°C to 1200°C. Pore size and pore size distribution as revealed by SEM analyses were appropriate for repair of cortical defects.

Introduction

Calcium phosphate is the major constituent of calcified tissue of living bodies. Hence, synthetic bone substitutes based on calcium phosphates are frequently used for the repairing of damaged bone due to disease and for fabricating matrix as drug delivery devices. They are biocompatible, bioactive and osteoconductive [1]. Rate of new bone formation and rate of dissolution depends on the Ca/P ratio, implantation site and the microstructure of ceramics. Porous calcium phosphate ceramics have been manufactured by using several techniques. Uniaxial or isostatic compaction of powders that contain granules of organics, impregnation of polymeric sponge, and foaming process are among common methods of green ceramic production. Present work was conducted for studying the production of porous calcium phosphate ceramics by the slip casting process.

Materials and Methods

The calcium phosphate precursors were produced by a precipitation technique involving the use of aqueous solutions of calcium nitrate and diammonium hydrogen phosphate at 60 °C under argon atmosphere. At the start of precursor production, first 1lt 0.2 M of diammonium hydrogen phosphate solution was introduced into the reaction vessel. The reactor was fitted with a tube condenser and the system was heated to 60 °C. Temperature of the solution was measured and kept constant during the process. Next, 1 lt. 0.32 mole calcium nitrate solution was added dropwise into the hot diammonium hydrogen phosphate solution. The reaction vessel was placed on a magnetic stirrer and the contents were stirred vigorously to eliminate any possible chemical inhomogenity by refluxing for 15 hours. Throughout the process, with the aid of a 12.5% solution of ammonia the pH of the system was set at 11 for HAP precursor, and at 8 for biphasic calcium phosphate precursor. The precipitates were separated from the mother liquor by centrifugal filtration at 8000 rpm and washed in the centrifuge with deionized water twice, and then dried at 100°C for 24 hours. The conversion of the precursor into HAP-TCP was monitored and verified by qualitative and quantitative powder XRD analyses using a Rigaku D-MAX/B powder diffractometer with mono-chromatic CuKα radiation at 40 kV and 20 mA.

Preforms of calcium phosphate ceramics were produced by using a modified slip casting technique. Slips were prepared at solid loadings of 55 and 60 weight percent. DI water was used as the suspending medium for the powders, weak powder agglomerates were eliminated by ultrasonic agitation. In order to enhance slip stability a commercial organic deflocculant, Dolapix PC33 (Zchimmer&Schwarz GmbH and Co., Germany) was introduced in amounts of 0.5-3 wt%, based on the weight of the powder. Rheological characteristics of slips were determined by measuring their viscosities and flow behavior in terms of shear stress versus shear rate. These parameters were measured by using a rotating-spindle Brookfield RV Digital Viscometer (Brookfield Engineering, Stoughton MA, USA). Controlled porosity in the cast and in the sintered ceramic was obtained through costum made PMMA beads ~200-300 μm size. The green ceramic samples were prepared by mixing the calcium phosphate slip and PMMA beads. The slip was poured into teflon rings placed on absorbent plaster blocks. The resultant cakes were dried at room temperature for 24 hours. Porous ceramics were obtained by sintering the polymer-free preforms for 2 hours at temperatures1000°C to 1200°C. Density and total porosity of porous ceramics were determined by liquid displacement method using xylene. Pore size and pore size distributions in sintered ceramics were examined under a JEOL 6400 Scanning Electron Microscope.

Results and Discussion

The XRD patterns of calcium phosphate precursors after calcinations at different temperatures are given in Figure 1. The diffraction pattern of dried precursors contained apatite peaks only. The XRD patterns of the precursors fired at progressively higher calcination temperatures revealed that TCP and HAP were developed fully at 800 °C. The ratio of β-TCP to HAP in biphasic calcium phosphate powders was determined as 40 to 60.

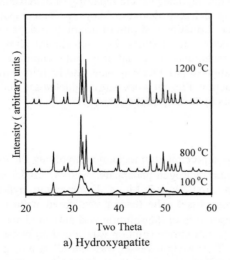 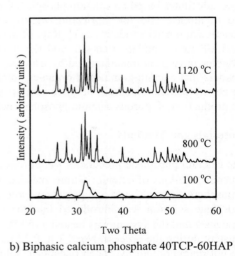

a) Hydroxyapatite

b) Biphasic calcium phosphate 40TCP-60HAP

Fig. 1. X-ray diffraction patterns of hydroxyapatite and biphasic calcium phosphate precursor heated at temperature indicated on the patterns.

The results of apparent viscosities of the ceramic slips measured at a constant shear rate of 18,6 s^{-1} as a function of dispersant concentration are presented in Figure 2. It can be observed that 1 wt % dispersant was sufficient for both biphasic calcium phosphate and hydroxyapatite powders. The flow behavior of slips prepared from these powders calcined 800 °C, as expressed in terms of shear rate and shear stress are shown in Figure 3. The slip had relatively low viscosity and its flow could be characterized as ideal-like. The results of slip optimization experiments revealed that Dolapix PC 33 was a suitable dispersant for calcium phosphate powders.

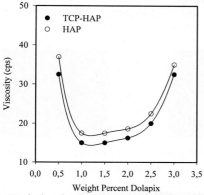

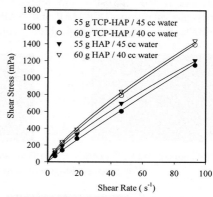

Fig. 2. Variation in the viscosity of 55 wt % slip with dispersant concentrations.

Fig. 3. Flow behaviour of 55 and 60 wt % slip at 1 wt % dolapix addition.

The sintering behaviour of porous ceramics are given in Figure 4 as a function of sintering temperature. The sintering temperature of biphasic ceramic was limited by 1120 °C in order to avoid the transformation of β-TCP into α form. Maximum sintering temperature for HAP ceramics was 1200 °C in order to maintain the hydroxyl group in the structure.

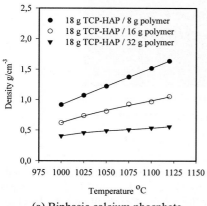

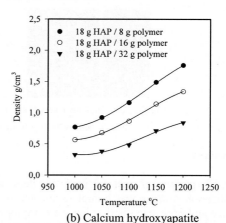

(a) Biphasic calcium phosphate

(b) Calcium hydroxyapatite

Fig. 4. Density of porous calcium phosphate ceramics as a function of sintering temperatures.

The morphologies of porous biphasic ceramics sintered at 1120°C prepared according to optimal slip condition and sintered at 1120 °C are shown in Figure 5 as an example. The examination of the fracture surface revealed that the pores were uniformly distributed throughout the entire ceramic. Micrographs (a), (b) and (c) in Figure 5 belong to the slips prepared from 18 gram calcium phosphate powder containing 8, 16 and 32 gram polymer, respectively. The average size of large pores in the sintered ceramic was around 190 microns, as determined by areal analysis of the cross section in accordance with the suggestion of Fullman [2]. Although the optimal pore diameter may depend on the location of the defect to be filled, a minimum pore size of 100 μm is necessary for porous implant materials[3], so porous calcium phosphate ceramics produced in this study are suitable for new bone formation. An interconnected pore network was generated with the aid of smaller pores linking the larger ones and providing the continuity of the pore channels. The diameters of the interconnections varied between 20 to 80 microns. As the amount of polymer beads was raised, the size and amount of interconnections increased proportionately. SEM micrographs of porous TCP-HAP and HAP ceramics shown in Figure 6, indicated that selected sintering temperatures produced a highly coherent matrices.

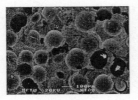 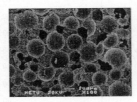 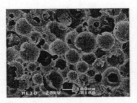

a)Total porosity: 48±1 vol. percent   b) Total porosity: 66±1 vol. percent   c) Total porosity: 82±2 vol. percent

Fig. 5. Morphologies of porous biphasic calcium phosphate ceramics.

a) TCP-HAP; Inside the pore X 4000   c) HAP; Inside the pore X 4000

b) TCP-HAP; Ceramic matrix X 4000   d) HAP; Ceramic matrix X 4000

Fig. 6. Matrices of porous calcium phosphate ceramics

Conclusions

The XRD pattern of the calcined powders and that of the sintered ceramic indicated that crystallization process was complete and the products could be used as artificial bone materials. Pore size and pore size distribution were appropriate for the repair of cortical defects. In-vivo tests performed with ceramic samples containing 66 volume percent porosity revealed that the porous ceramics allowed natural tissue integration within a reasonably short period following implantation [4].

References
[1] Larry L. Hench: J. Am. Ceram. Soc.,81 (1998), p.705
[2] R.L. Fullman: Trans. AIME, 197 (1953), p.447
[3] G. Daculsi, N. Passuti: Biomaterials,11 (1990), p. 86
[4] C. Balçık, A. Şenköylü, N. Koç, M. Timuçin, P. Korkusuz, F Korkusuz: J. Arthroplasty & Arthroscopic Surgery 14 (2003), p. 39

Key Engineering Materials Vols. 254-256(2004) pp. 953-956
online at http://www.scientific.net
© *2004 Trans Tech Publications, Switzerland*

Osteoinduction of Bioactive Titanium Metal

Shunsuke Fujibayashi[1], Masashi Neo[1], Hyun-Min Kim[2], Tadashi Kokubo[3], and Takashi Nakamura[1]

[1]Department of Orthopaedic Surgery, Graduate School of Medicine, Kyoto University, 54 Shogoin Kawahara-cho, Sakyo-ku, Kyoto 606-8507, Japan
email: shfuji@kuhp.kyoto-u.ac.jp
[2]Department of Ceramic Engineering School of Advanced Materials Engineering, Yonsei University, 134, Shinchon-dong, Seodaemun-gu, Seoul 120-749, Korea
[3]Research Institute for Science and Technology, Chubu University, 1200 Matsumoto-cho, Kasugai 487-8501, Japan

Keywords: Osteoinduction, titanium, porous, metal surface treatment

Abstract. Four types of titanium implants were implanted in the dorsal muscles of mature beagle dogs, and were examined histologically after periods of 3 and 12 months. Chemically and thermally treated titanium (bioactive titanium), as well as pure titanium, was implanted either as porous blocks or as fibre mesh cylinders. Bone formation was found only in the chemically and thermally treated porous block implants removed after 12 months. The present study shows that even a non-soluble metal that contains no calcium or phosphorus can be an osteoinductive material when treated to form an appropriate macrostructure and microstructure.

Introduction

Certain calcium phosphate biomaterials have recently been reported to be osteoinductive if they possess a specific porous structure [1,2]. These calcium phosphate biomaterials induce bone formation at extra-skeletal sites without the need for additional osteogenic cells or bone morphogenetic protein. However, the mechanism of osteoinduction by calcium phosphate ceramics is not clear. Titanium metal is considered a bioinert material, and is used for scaffolds when loaded with BMP to induce ectopic bone formation. In a previous study, we showed that titanium metal could be converted into an osteoconductive material through specific chemical and thermal treatments [3]. Although the bioactive titanium has an osteoconductive ability, it does not possess any osteoinductive ability.

Recently, we have developed an interconnective porous titanium block using a plasma-spray technique, and this porous titanium has been successfully subjected to chemical and thermal treatments [4]. In this paper, we describe our development from titanium metal of an osteoinductive material that contains no calcium and phosphorus, and discuss the mechanism of material-induced osteogenesis, along with the relationship between osteoconduction and osteoinduction.

Material and Methods

Four types of titanium implants were prepared. Bioactive titanium was prepared by a chemical treatment (5 M aqueous NaOH solution at 60 °C for 24 h), followed by a hot water treatment (distilled water at 40 °C for 48 h), followed by a thermal treatment (at 600 °C for 1 h). Pure titanium was used as a control. These materials were implanted as porous blocks (5 x 5 x 7 mm^3, porosity = 40–60%, pore size = 300–500 μm) and titanium fibre mesh cylinders (diameter = 4 mm, length = 11 mm, porosity = 40–60%, pore size = 50–450 μm). The porous titanium blocks were manufactured by plasma spraying method. The titanium fibre mesh implants were manufactured by compacting a single 250-μm fibre into a die to a porosity of 50%, followed by vacuum sintering. Before implantation, the bioactive ability of the samples was examined by soaking them in a simulated body fluid.

Four types of samples were implanted in the dorsal muscles of mature beagle dogs for periods of 3 and 12 months. Six beagle dogs were used in this study. Twenty-four samples were implanted. Undecalcified sections were examined by SEM and light microscopy.

Results
SEM examination of the samples before implantation revealed that two types of porous macrostructure were present in the block and cylinder implants (see Figures 1-A and 1-B). The porous macrostructure of the block-type implant was more complex than that of the cylinder-type implant. High magnification revealed that micro-porous structures were recognizable in the macro-pores of both the chemically and thermally treated titanium implants (Figures 1-C and 1-D). On the other hand, the micro-surface of both of the non-treated titanium implants was smooth (Figures 1-E and 1-F). The average porosity of the porous titanium samples was calculated to be 38.8%.

After the titanium blocks had been soaked in the SBF, apatite deposits could be recognized on all the chemically and thermally treated titanium blocks and cylinders within a seven-day period. This result indicates that the chemically and thermally treated titanium implants possessed an in vitro apatite forming ability and an in vivo osteoconductive ability.

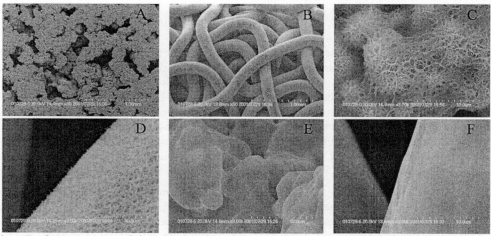

Fig.1 Surface structure of four kinds of implant. A: Macro-porous structure of porous block, B: Macro-porous structure of fibre mesh cylinder, C: Micro-porous structure of porous block after chemical and thermal treatments, D: Micro-porous structure of the fibre mesh cylinder after chemical and thermal treatments, E: Surface of the porous block without any surface treatment, F: Surface of the fibre mesh cylinder without any surface treatment.

Both SEM and optical microscopy showed that no bone formation had occurred in all the samples removed after three months. After 12 months, however, bone formation was found in the chemically and thermally treated porous block implants. Bone formation was observed in all three porous bioactive titanium samples after 12 months (Table 1). Optical microscopy showed mineralized, newly formed bone that stained well, and had a lamellar structure at the surface of the inner-pores

(Figure 2). SEM and EDX analysis showed that new bone had bonded to the titanium surface directly, and that it contained calcium and phosphorus.

Although bone formation was found in all the bioactive porous titanium samples harvested at 12 months, only a small mass of bone had formed at the centre of the implants. In all, 16.2±7.5% new bone had formed in the pores that covered 23.2±5.5% of implant area.

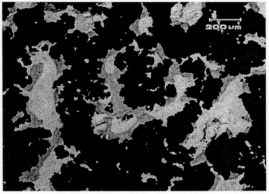

Fig. 2 New bone formation into the pore.

Table 1. Bone formation by four kind of titanium implants in muscle of dog.

	3 months	12 months
TP	0/3	0/3
TF	0/3	0/3
BTP	0/3	3/3 (100%)
BTF	0/3	0/3

TP: non-treated porous titanium, TF: non-treated titanium fiber mesh, BTP: treated porous titanium, BTF: treated titanium fiber mesh

Discussion

The porous bioactive titanium prepared using specific chemical and thermal treatments induced bone formation at non-osseous sites without the need for additional osteogenic cells or osteoinductive agents.

From the results of our study, we can first hypothesize that the complex macro-porous structure plays an important role in material-induced osteogenesis. Although the chemically and thermally treated cylinders showed apatite formation in vitro, they did not show osteoinduction in vivo. A second point to note is that, although the non-treated blocks also possessed a similar macro-porous interconnective structure, they did not exhibit an osteoinductive nature. The surface 3D micro-porous structure, which was formed by the chemical and thermal treatments, plays an important role in osteogenesis. The interconnective porous structure may play an important role in osteogenesis. In this work, the interconnection of the porous structure of the porous titanium block fabricated using a plasma spray technique was established from the histological examinations.

Although many hypotheses could be postulated from the results of this study and the osteoinduction of other calcium phosphate ceramics, the true mechanism of osteoinduction by porous bioactive titanium is not understood. This paper also represents the first report of interconnective structured titanium metal. In general, the fabrication of porous metals is difficult, and this is especially so for titanium. The porous structure is useful for tissue regeneration in relation to its use as a scaffold for growth factors or osteogenic cells. However, the mechanical properties of porous ceramics are too poor for clinical use under load-bearing conditions. Porous titanium has a high enough mechanical strength for use under load-bearing conditions.

References

 [1] Ripamonti U: The morphogenesis of bone in replicas of porous hydroxyapatite obtained from conversion of calcium carbonate exoskeletons of coral, J. Bone Joint Surg. 73(A), (1991), p. 692-703

[2] Yuan H et al. A preliminary study on osteoinduction of two kinds of calcium phosphate ceramics. Biomaterials 20, (1999), p.1799-1806

[3] Kokubo T, Miyaji F, Kim HM, Nakamura T. Spontaneous apatite formation on chemically surface treated Ti. J Amer Ceram Soc. 79, (1996), p. 1127–1129

[4] Kim HM, Kokubo T, Fujibayashi S, Nishiguchi S, Nakamura T. Bioactive macroporous titanium surface layer on titanium substrate. J Biomed Mater Res. 52, (2000), p. 553–557

Key Engineering Materials Vols. 254-256(2004) pp. 957-960
online at http://www.scientific.net
© *2004 Trans Tech Publications, Switzerland*

Calcium Phosphate Porous Scaffolds From Natural Materials

Luis M. Rodríguez-Lorenzo[1], Kārlis A. Gross[2]

School of Physics and Materials Engineering, Building 69, Monash Uni., VIC 3800, Australia

[1] luis.rodriguez-lorenzo@spme.monash.edu.au; [2] karlis.gross@spme.monash.edu.au

Keywords: Porous scaffolds, tissue engineering, calcium phosphates

Abstract. American oak, Quercus Alba, was selected as a template for the development of a porous scaffold suitable for tissue engineering applications. The selection was based on the pore size of 150 µm and cell wall thickness. Pyrolysis of the wood at 800°C produces an amorphous carbon template, which maintains the original porous structure. Successive soaking of the carbon template in calcium and phosphorous containing solutions followed by calcination and sintering processes yields a calcium phosphate scaffold with analogous pore structure to the original template and chemical composition analogous to natural hard tissues that make them promising materials for cell colonization.

Introduction

The design of scaffolds for tissue engineering must allow ease of handling in the preparation and transportation of porous bodies, high porosity to allow rapid vascular tissue ingrowth and space for growth of differentiated specialized extracellular matrices. Scaffolds must have the capacity to be trimmed or sculpted to exactly fit every surgical site and exhibit a rapid and reproducible degradation or dissolution process in every insertion site [1].

Wood is the most ancient and still most widely used structural material in the world [2]. Like bone, wood present contains elongated pores, which facilitates fluid transportation in a given direction. The central cavity known as the lumen can occupy up to 80% in volume in early wood. In comparison with bone, the pore volume relative to the structural unit volume in the tracheid is higher with lumen orientated in one specific direction. Small openings in the wall of the tracheid allow nutrient transport through a membrane to neighboring tracheids or rays. Rays are similar to tracheids but are orientated horizontally and directed towards the center of the wood. Pits are the smallest pores and offer a means of water transport. The surface area of the pores in wood is lower than in bone. Pore connectivity is also different since the transport mechanism is different in both tissues. Chemically, wood is a composite of cellulose, hemicellulose and lignin as major constituents with minor additions of fat, oil, wax, resin, sugar, minerals and alkaloids [3]. Previous work has shown that the lumen in some woods offers a suitable site for hard tissue abutment [4]. Heating wood in a non oxidizing atmosphere at temperatures above 600°C results in decomposition of the polyaromatic constituents to form a carbon residue which can reproduce the original structure [5, 6]

The purpose of this work is to use wood as a template to obtain a carbon scaffold which subsequently will be transformed into a calcium phosphate skeleton which resembles the original structure of the wood and therefore posses the appropriate structure needed for tissue engineering.

Methods

Quercus Alba (American Oak, QA) has been selected for this work. The characterization was carried out by thermogravimetrical analysis (TGA) in nitrogen atmosphere with a Setaram TG92, true density was determined with a Micromeritics AccuPyc 1330 (Norcross GA, USA) helium gas pycnometer from an average of 15 measurements, bulk density was determined according to the

Archimedes principle by immersion in water from an average of four measurements per material, specific surface area by the BET method. The 5 point method with nitrogen was performed using a Micromeritics Gemini 2360 (Norcross GA, USA), mercury intrusion porosimetry (MIP) in a Micrometrics Autopore III from 0.003 up to 413.685 MPa and scanning electron microscopy (SEM) in a Hitachi S570. Wood was pyrolysed at 800°C in a tubular furnace under N_2 flow after extracting the air up to 4.5 x 10^{-2} ats. Pyrolysed wood was characterized by X-ray diffraction (XRD), patterns were obtained using a Rigaku Geigerflex diffractometer, using copper K_α radiation at 22.5 mA and 40 kV passing through a 0.15° receiver and 1° divergence slit. A scan rate of 1 second per step and a step size of 0.05 degrees were selected over a 10-60 two-theta range. Archimedes's density, MIP and SEM were performed using the same conditions described for wood.

Carbon residue was then successively vacuum impregnated in $Ca(NO_3)_2.4H_2O$, and K_2HPO_4 or in $Ca(NO_3)_2.4H_2O$ and KF/K_2HPO_4. The specimens were then calcined at 500°C and sintered at 1100°C. Calcium phosphate scaffolds have been characterized by XRD and optical microscopy in a Olympus SZ40 and the carbon content of the scaffolds was analyzed in the Campbell microanalytical laboratory of the Department of Chemistry of the University of Otago, Dunedin, N.Z.

Results and discussion

Quercus alba was selected based upon its cell diameter and cell wall thickness [7]. It belongs to the group of Angiosperm dicotiledoneous or hardwood and its composition includes cellulose (38-43 wt%), hemicellulose (19-26 wt%) and lignin (25-34 wt%) and its structure contains a great variety of cell types, including thick-walled fibres [8]. Fig. 1 left displays a transversal cut showing the vessels or pores with a diameter of about 150 μm, which provide the pathway for the conduction of water and solutes along the trunk of the living tree. Fig. 1 middle-left shows a detail of the tracheids and the thickness of the walls. Other analysis have shown a density of 1.487 (2) g/cm^3 and a bulk density of 0.620(10) g/cm^3 that according with the formula taken from reference [9] yields a 58% total porosity.

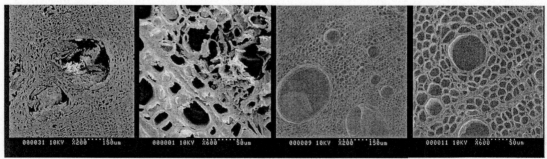

Fig. 1 Left and middle left, transversal cut of Quercus Alba showing the pores and traqueids that form its structure. Middle right and right, tranversal cut of a pyrolysed specimen showing the same structural components

Figure 2 displays the TGA analysis showing a 85% total weight loss as a result of the biopolymer degradation. The three steps can be associated with the decomposition of hemicellulose, cellulose and lignin respectively. The thermal degradation of the key biopolymer of wood, cellulose, can be divided in two steps, low temperature, which consist in the releasing of water to yield dehydrocellullose followed by the elimination of CO and CO_2 above 260°C. The second step represents carbonization at temperatures above 600°C that breaks down the carbon chains of the biopolymer structures and forms graphitic structures [10]. These studies determined the temperature for pyrolysis to be 800°C.

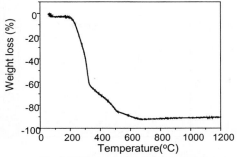

Fig.2 TGA analysis of Quercus Alba

After the heat treatment at that temperature, the total transformation into an amorphous graphitic carbon template has been confirmed by the XRD pattern displayed in fig 3 left. The loss in appearance of wood together with the change in colour occurs upon heating. Morphological characteristic of the pyrolysed specimens can be examined in figs 1 middle right and right. The substitution of the polymer components by the more rigid graphitic yields a clean surface for the SEM examination of the structural components. This analysis enables to appreciate similar elements than in the original wood i.e. vessels of more than a hundred microns, tracheids and pits connecting vessels. The pore size distribution curve shown in Fig 3 right reveals four maxima centered at 150, 30, 1 and 0.1 μm with a total porosity of 60%. This is a promising pore size distribution for cell infiltration, since pores greater than 120 μm are necessary for cellular settlement and the ingrowth of blood vessels while pores in the range 20-50 μm are necessary to allow the physiological liquid to reach the cell and supply nutritional factors and to drain the metabolic substances [11]. This is akin to the lacunae in bone where osteocytes are located.

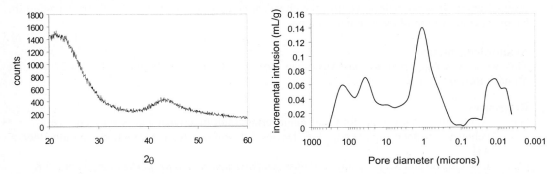

Fig. 3 Left, XRD pattern of a carbon template. Right, MIP pore size distribution of a carbon template

The appearance changes after soaking the specimens in the successive solutions and the calcination and sintering processes producing a white body that still resembles the original wood structure. XRD analysis, displayed in fig 4, show that the scaffold is made now of an apatite phase when the fluoride solution has been used but a tricalcium phosphate when the carbon scaffold has been soaked in a solution without fluoride. The difference can be related with the mechanism of reaction. Calcium ions react with acid phosphate groups and then release the proton to yield the phosphate group characteristic of calcium phosphate apatites and absorb, either hydroxyl groups or fluoride when available. It seems that the smaller size of the fluoride ion in comparison with hydroxyl groups promotes a faster accommodation [12] into the apatite structure than the larger hydroxyl groups. When the reaction is incompleted, a calcium deficient apatite is formed that on heating above 800°C yields tricalcium phosphate, as displayed in fig 4 left.

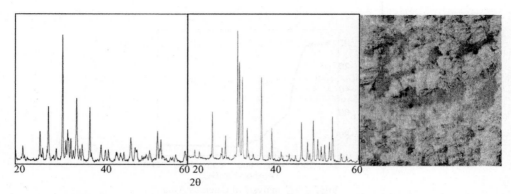

Fig. 4 Left, XRD pattern of a specimen soaked in $Ca(NO_3)_2.4H_2O$, + K_2HPO_4. Middle, XRD pattern of a specimen soaked in $Ca(NO_3)_2.4H_2O$ + KF/K_2HPO_4. Right, optical micrograph of the latter specimen

The carbon analysis of the scaffolds yielded a 2.00 (0.10) Wt% carbonate content for the fluoride containing scaffold, shown in fig 4 centre, illustrating the formation of a carbonate fluoride containing apatite that resembles in composition the mineral component of natural hard tissues while the optical microscopy picture (fig 4 right) taken on a longitudinal cut of the scaffold permits to appreciate the longitudinal pores resembling the original pore structure of the wood used as a template.

Conclusions

Quercus Alba wood is a promising template for the fabrication of scaffolds for tissue engineering. The pyrolysis of the wood at $800^{\circ}C$ yields an amorphous carbon structure that can be used to develop a carbonate containing calcium phosphate scaffold that maintains the pore structure of the original wood. It is possible to control the chemical composition of the scaffolds by changing the chemical composition of the soaking solution.

Acknowledgements

This work was supported by the Australian Research Council with grant F10017027 and the Monash Research Fund.

References

[1]   A. I. Caplan, "New logic for tissue engineering: multifunctional and biosmart delivery vehicles," in Bone engineering, J. E. Davies, Ed.: em squared incorporated, Toronto 1999, p. 441.

[2]   L. J. Gibson and M. F. Ashby, *Cellular solids: structure and properties* Pergamon, U.K. 1988.

[3]   P. Greil, T. Lifka, A. Kaindl, J Eur ceram Soc, vol. 18, (1998) p. 1961

[4]   K.A. Gross, E. Ezerietis, J Biomed Mater Res, vol 64 (2003), p. 672.

[5]   F. Shafizadeh, Y. Sekiguchi, Carbon, vol. 21, (1983) p. 511

[6]   P. Ehrburger, J. Lahaye, Carbon, vol. 20, (1982) p. 433,

[7]   P. Greil, J Eur ceram Soc, vol. 21, (2001), p. 105.

[8]   B. G. Butterfield, B. A. Meylan, *Three-dimensional structure of wood. An ultrastructural approach*, 2nd ed. Chapman and Hall, U.K.1980.

[9]   M. Aizawa, F. S. Howel, K. Itatani, Y.et al. J ceram soc Japan, vol. 108, (2000). p. 249,

[10]   H. Sieber, C. Hoffman, A. Kaindl, P. Greil, Advanced eng mater, vol. 2, (2000).p. 105,

[11]   A. Ravaglioli, A. Krajewski, Mater Scie Forum, vol. 250, (1997).p. 221,

[12]   L. M. Rodriguez-Lorenzo, J. N. Hart, and K. A. Gross, J Phys Chem B,. In press,

Key Engineering Materials Vols. 254-256(2004) pp. 961-964
online at http://www.scientific.net
© *2004 Trans Tech Publications, Switzerland*

Thermally Sprayed Scaffolds for Tissue Engineering Applications

Kārlis A. Gross[1] and Luis M. Rodríguez-Lorenzo[2]

School of Physics and Materials Engineering, Building 69, Monash Uni., VIC 3800, Australia.

[1] karlis.gross@spme.monash.edu.au; [2] luis.rodriguez-lorenzo@spme.monash.edu.au

Keywords: scaffolds, scaffold template, tissue engineering, thermal spraying, pore connectivity

Abstract. Scaffolds for tissue engineering need to exhibit specific pore characteristics for optimized fluid flow and tissue growth. Scaffolds were successfully manufactured by using a template made from sintering flame sprayed salt particles. Good interconnectivity of large pores was achieved between spherical pores, with an additional of submicron pores.

Introduction

The range of approaches used to produce porous inorganic scaffolds include gel-casting, layering of phosphate solutions on polymer beads and the utilization of biological porous structures . Pore size is typically established for biological structures, such as for corals, and typically exhibits a monomodal pore size distribution. The species of coral will dictate the pore volume, pore size and connecting passage geometry. Manufactured porous scaffolds typically exhibit a range in interconnectivity, pore volume and pore size. Limited attention has been devoted to the control of pore size and pore morphology for tissue engineering situations, with more emphasis on pore volume. The most attention is being directed to gel-cast structure and structures made by coating calcium phosphates onto polymer beads, where the polymer is typically removed by heat treatment. Limited knowledge is available on the requirement of pore size, pore volume, pore geometry for tissue engineering applications. It is known that sufficient free volume needs to be provided for entry of nutrients to stem cells seeded in the scaffold, removal of waste products, and growth of tissue. For stress loading applications, added demands for strength are placed on structural member geometry and strut thickness.

A new approach for designing porosity in tissue engineering scaffolds includes the use of thermal spraying. This approach will be shown to provide an interconnected pore structure with rounded interconnected pores.

Methods

Sodium chloride salt was ball milled for 2 minutes at 300rpm and then sieved with a vibratory stack of sieves for 20 minutes to obtain powder within particle size ranges of 80-100, 100-150, 150-300 and >300μm.

Salt was injected into a Metco 5P flame spray torch at a rate of 1 g/min. Powder injected into the flame was accelerated with the flame, transferred into a long metallic tube and collected in a stainless steel bucket. Salt of each particle size was assessed for the degree of melting in a gas flame. Salt, with a particle size that produced spherical particles was used in further studies. Source powder was used as a comparison for each further step of scaffold production.

Salt was placed in an alumina crucible and sintered at 780 °C for 2 hours A fracture surface of the connected salt template was examined by scanning electron microscopy to reveal the neck formation between the salt particles.

A suspension of 17 grams of apatite in 25 ml of pure ethanol was prepared with mechanical and ultrasonic agitation. Sodium chloride scaffolds were soaked for 5 minutes in the apatite suspension in an ultrasonic bath. The sodium chloride with the apatite slurry was dried in at 100 °C for 3 hours. Poly(L-lactic acid), a weight of 0.2g, was dissolved in 6mL of CH_3Cl and introduced into the apatite filled salt template. The solvent was allowed to evaporate at room temperature for 24 hour and then in vacuum at -100kPa for an additional 24 hours. The embedded scaffold was soaked in deionized water for 30 minutes and the obtained scaffold fractured for examination in the SEM.

Results and discussion

The flame within the conventional thermal spray torch produced different degrees of melting for the particle size range investigated. It was found that particles less than 80 μm were spherical, as were the particles within the 100-150 μm particle size range, Fig. 1. Larger particles, within the size range of 150-300 μm experienced some melting within the flame that produced rounded edges, but particles greater than 300 μm did not exhibit any melting at all. The map indicated the maximum particle size that can be made spherical in an acetylene/oxygen flame of a flame spray torch.

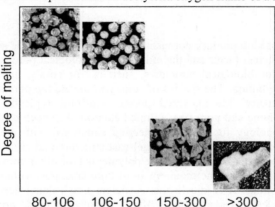

Fig 1. Degree of melting for particles of different size

The spherical shape can form from different degrees of melting, Fig. 2. Case 1 represents a scenario where particles are completely molten and cool to form a sphere. Slightly larger particles, Case 2, only have a molten shell, however, this is sufficient to impart the change in morphology.

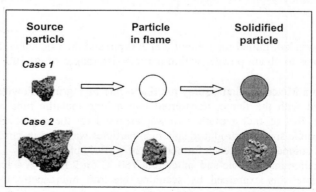

Fig 2. Melting of particles showing the original microstructure in the core of larger particles

Sintering of the salt scaffolds produced material transfer from one particle to the next. The evaporation-condensation sintering mechanism that occurs with halides [2] has formed a mesa-like or butte-like topography often seen in geological structures [3]. The contours formed in these particles appears to be directed to the closest neighbouring particle. Closely arranged particles can lead to bridging of salt particles, however, the separation between the salt particles in the initial packing generally determines the locations for neck formation. Formation of necks can occur between asperities, a line contact and contact of particle faces This produces a wide range in neck size between angular particles. A neck size of 50 microns is observed in Figure 3, that possibly occured from contact of two edges. The neck size between spherical particles is larger, Figure 3 right, and will produce a smaller neck size distribution within the salt template.

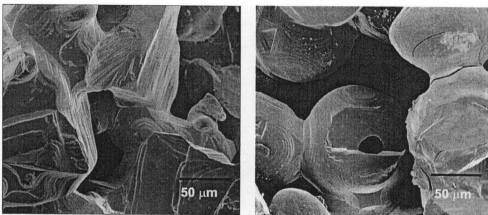

Fig. 3. Template of sintered (left) angular and (right) spherical particles

The scaffold produced by entrainment of apatite and polylactic acid into the salt template resulted in effective filling of the large connective pore channels within the scaffold, Figure 4. Angular salt particles produced angular pores within the scaffold. The use of spherical pores provided rounded pores interconnected by the neck regions created during the sintering treatment of the salt particles.

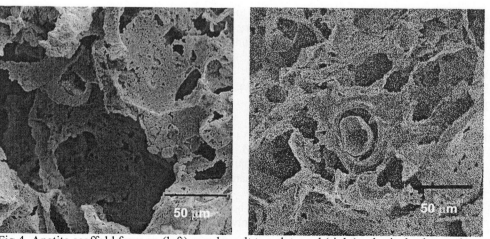

Fig 4. Apatite scaffold from an (left) angular salt template and (right) spherical salt template

The large pores in the scaffold reproduced a negative mould of the interconnected salt template, but smaller pores resulted from the characteristics of the materials introduced into the scaffold. Submicron apatite crystallite packed together to produce submicron pores of size ranging from a fraction of the crystallite size to several times the size of a crystallite. This submicron size shows promise to provide nutrients and a channel for waste removal that typically occurs through canaliculi, from the lacunae where osteocytes are typically resident in bone [4].

Nonconnecting salt particles in the salt template will lower the degree of interconnectivity, and the lower packing efficiency of angular particles suggests that pore interconnectivity in the resulting scaffold will be lower. The fluid transfer within a scaffold produced from angular particles will thus be irregular. The submicron pores within the scaffold offer promise to house a cell, that will provide a "cave-like" site for slow exchange of fluids. The polylactic acid component added to the salt template in a more fluid form provided reinforcement, Fig. 5. The excellent ability to follow the details of the contours are clearly seen in the negative of a contoured structure, as previously described as a "mesa-like" or "butte-like" structure.

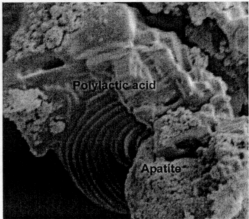

Fig 5. Scaffold showing precise reproduction of salt template morphology

Conclusions

Thermally sprayed salt particles, when sintered to form an interconnected template, provided the ability to finely replicate the rounded like pore morphology and good interconnectivity for a carbonated hydroxyfluorapatite/ polylactic acid composite..

Acknowledgements
This work was supported by the Australian Research Council with grant F10017027 and the Monash research fund

References
[1] K.A. Gross and L.M. Rodríguez-Lorenzo LM. Interconnected pore networks from sintered salt particles for scaffolds in tissue engineering applications. J Austr Ceram Soc, 2003, Accepted.
[2] W.D. Kingery, H.K. Bowen and D.R. Uhlmann, *Introduction to Ceramics* (John Wiley, New York 1976), p. 473.
[3] A. Holmes, *Principles of Physical Geology* (Thomas Nelson, Melbourne 1972), p. 569.
[4] K.S. Saladin, *Anatomy and Physiology* (McGraw Hill, New York 2001), p. 235.

Key Engineering Materials Vols. 254-256(2004) pp. 965-968
online at http://www.scientific.net
© *2004 Trans Tech Publications, Switzerland*

Fabrication of α-Tricalcium Phosphate Porous Body Having a Uniform Pore Size Distribution

M. Kitamura, C. Ohtsuki, S. Ogata, M. Kamitakahara and M. Tanihara

Graduate School of Materials Science, Nara Institute of Science and Technology,
8916-5, Takayama-cho, Ikoma-shi, Nara 630-0192, Japan
kitamako@ms.aist-nara.ac.jp

Keywords: α-tricalcium phosphate, porous ceramics, bioresorbable ceramics, pore size distribution, compressive strength, dissolution property

Abstract. An α-tricalcium phosphate (TCP) porous body with a continuous 10-50 μm pore structure was fabricated from slurry composed of β-TCP, potato starch and ultra-pure water through a conventional sintering process. The effect of molding methods on porous structures and mechanical properties of porous body were investigated. Two types of molding methods were applied to the fabrication; i.e. direct casting in an alumina crucible (CP method) and holding with polyurethane sponge (SP method). The CP method resulted in a porous body with high uniformity in pore size distribution of continuous pores around 10-50 μm, while the SP method produced bimodal-type distribution of pores around 10-50 μm and 150 μm. The high uniformity brought about a higher compressive strength of the porous body synthesized by the CP method than that synthesized by the SP method. Porosity of the body could be controlled from 55% to 75% by the content of starch in the slurry. The compressive strength of the porous body synthesized by the CP method increased inversely to its porosity. On the other hand, no significant increase was observed on the compressive strength for the porous body synthesized by the SP method, even when the content of starch in the slurry decreased. This is attributed to macro-cracks in the porous body synthesized by the SP method. Increase in porosity accelerated dissolution of α-TCP body in buffered solution (pH=4). These results indicate that the CP method gives a porous α-TCP body with superior mechanical property as well as controlled porosity.

Introduction

Tricalcium phosphate (TCP) is a popular bioresorbable ceramics [1,2,3]. α-TCP phase is thermodynamically stable at temperatures above 1100 °C, and shows higher solubility than β-TCP phase. Porous α-TCP with continuous pores is therefore expected to be a useful material for bone regeneration by combination with drugs and osteoinductive factors [4,5]. High porosity may result in poor mechanical properties, which relates to less workability on implantation. It is important to control pore size distribution to achieve functions as a drug delivery carrier as well as high workability. In this study, α-TCP porous ceramics were synthesized through a conventional sintering process by different molding methods; direct casting in alumina crucible (CP method) and holding with polyurethane sponge (SP method). Pore structure and mechanical properties were investigated in the bodies prepared by these two molding methods.

Materials and Methods

Commercial β-TCP powder was mixed with potato starch at 10, 25 or 50 mass%. Ultra-pure water was added to the powder mixture, at water/powder mass ratio of 0.875. The slurry was molded either by the CP method or the SP method. In the CP method, the slurry was directly poured into an alumina crucible of 66 mm in diameter and 54 mm in height. In the SP method, the slurry was held in a 15 x 15 x 15 mm polyurethane sponge, with a continuous 1000 μm-diameter pore, by dipping the sponge in the slurry. The molded samples were then dried at 60 °C for 1 hour, followed by heating to 1000 °C at a rate of 5 °C/min, and kept at 1000 °C for 3 hours in air to burn-off the organic component. After cooling to room temperature, the sample was heated again to 1400 °C at a rate of 5 °C/min, and kept at 1400 °C for 12 hours. The samples were then allowed to cool to room temperature under natural cooling in the furnace. The porous bodies synthesized by the CP method and the SP method were noted as CPm and SPm respectively, where m means starch content (mass%) in the starting powder mixture. Porosity and pore size distribution of CPm and SPm were measured by mercury intrusion porosimetry. The compressive strength of the samples was measured with an Instron testing machine with a crosshead speed of 20 mm/min. Dissolution of the porous body was estimated by immersion of 1 g of a sample in 30 mL of a buffered solution (pH=4), which contained potassium hydrogen phthalate of 0.05 mol/L and NaCl of 0.142 mol/L. The concentration of calcium and phosphate ions was measured for the solution after soaking for 24 hours at 36.5 °C using inductively coupled plasma emission spectroscopy.

Results
Fig. 1 shows SEM images of porous ceramic bodies, CP50 and SP50. Both samples have a porous body with a continuous pore of around 10-50 μm, whereas macro-cracks more than 100 μm appeared in SP50 under observation at lower magnification. A remarkable difference was observed in pore size distribution between CP50 and SP50, as shown on Fig. 2. Two peaks were detected on SP50 at about 10-50 μm and 150 μm, while only one peak in CP50 at around 10-50 μm. The porosity of the porous body in CP50 is 73% which is similar to that in SP50, 74%. CP50 showed about 5 times as large compressive strength as that of SP50, as given on Table 1.

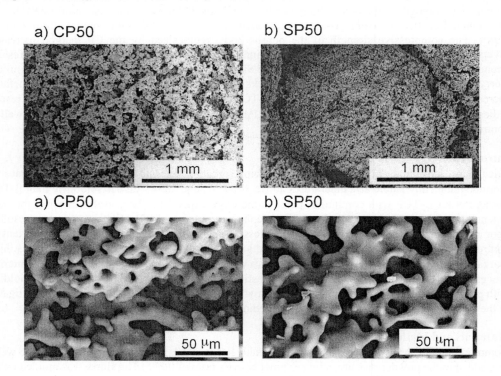

Fig. 1. SEM images of the porous ceramics bodies of a) CP50 and b) SP50

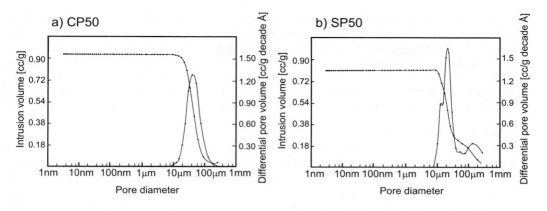

Fig. 2. Pore size distribution of porous ceramic bodies, a) CP50 and b) SP50.

Table 1. Properties of porous ceramic bodies of CP50 and SP50.

	Porosity [%]	Ave. pore diameter [μm]	Compressive strength [MPa]
CP50	73.8	39.4	0.90
SP50	72.9	24.5	0.14

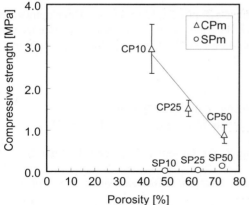

Fig. 3. Relationship between compressive strength and porosity of CPm and SPm.

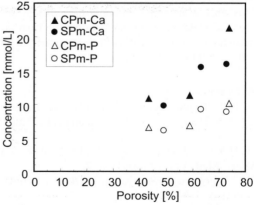

Fig. 4. Change in elemental concentrations of buffered solution at pH4.0 due to immersion of CPm and SPm.

Fig. 3 shows the relationship between the compressive strength and porosity of CPm and SPm. Porosity was controlled, ranging approximately from 40% to 75%, depending on the content of starch in the powder mixture. The compressive strength of CPm increased to about 3 MPa being inversely proportional to porosity, while that of SPm showed low values below 0.2 MPa independent of their porosities. The low values of compressive strength of the SPm specimens are attributed to macro-cracks that could be observed under SEM. Changes in elemental concentrations of the buffered solution at pH4.0, due to immersion of the CPm and SPm specimens, are shown in Fig. 4. Increases in calcium (Ca) and phosphorus (P) concentrations are attributed to dissolution of Ca and phosphate ions from the specimens. Dissolution of the specimens increased with increasing porosity of the bodies irrespective of molding methods. Namely, the dissolution of Ca and phosphate was governed by the total porosity of the specimens. Ca/P molar ratio of the dissolved ions was about 1.5.

Discussion and Conclusions
Pore size distribution in CP50 shows a narrow peak in the range from 10-50 μm, whereas that in SP50 is bimodal with peaks at about 10-50 μm and 150-500 μm. A peak in pore size around 150 μm in SP50 is attributed to thermal decomposition of the polyurethane sponge because of the similarity in framework size. The continuous pores of 10-50 μm are caused by starch in the powder mixture that may produce a pore during burning out and also act as binder between sponge and slurry during fabrication before sintering. The CP method gives a uniform pore size distribution free from macro-cracks. Such a uniform structure without macro-cracks produces higher compressive strength of the body synthesized by the CP method than that synthesized by the SP method. The higher compressive strength and uniform pore size distribution of CP50 leads to high workability. The compressive strength for the body synthesized by the SP method shows a lower value irrespective of its porosity, since it has macro-cracks caused by burning-out of the polyurethane sponge. A remarkable increase in the concentration of Ca and phosphate ions in buffered solution was produced by increasing porosity from 60% to 75%. The rate of dissolution is mainly governed by the porosity of the body irrespective of the method on fabrication. Consequently, the casting process with potato starch (CP method) is a superior process for fabrication of α-TCP body having high uniformity of pore size distribution and higher compressive strength. The porosity of the body can be easily controlled by the amount of starch in the starting powder mixture. The porosity of the porous body mainly governs the rate of dissolution of the porous body.

Acknowledgements
This work was conducted under the auspices of the research project, "Technology Development for Medical Materials Merging Genome Information and Materials Science", in the Kansai Science City Innovative Cluster Creation Project, supported by the Ministry of Education, Culture, Sports, Science and Technology of Japan.

References
[1] B. V. Rejda, J. G. Peelen and K. De. Groot: J. Bioeng. Vol. 1 (1977), p. 93.
[2] D. S. Metsger, T. D. Driskell and J. R. Paulsrud: J. Am. Dent. Assoc. Vol. 105 (1982), p. 1035.
[3] H.Yuan , J. D. De. Bruijn, Y. Li, J. Feng, Z. Yang, K. De. Groot and X. Zhang: J. Mater. Sci. Mater. Med. Vol. 12 (2001), p. 7.
[4] M. Itokazu, T. Sugiyama, T. Ohno, E. Wada and Y. Katagiri: J. Biomed. Mater. Res. Vol. 39 (1998), p. 536.
[5] T. Gao, T. S. Lindholm, A. Marttinen and M. R. Urist: Int. Orthop. Vol. 20 (1996), p. 321.

Key Engineering Materials Vols. 254-256(2004) pp. 969-972
online at http://www.scientific.net
© 2004 Trans Tech Publications, Switzerland

Bioactive Behaviors of Porous Si-Substituted Hydroxyapatite Derived from Coral

S.R. Kim[a], J.H. Lee[b], Y.T. Kim[b], S. J. Jung[a], Y. J. Lee[a], H. Song[a] and

Y.H.Kim[a*]

[a]Korea Institute of Ceramic Engineering and Technology, Seoul, Korea

[b]Department of Orthopedic Surgery, University of Ulsan College of Medicine, ASAN Medical Center, Korea

*yhkokim@kicet.re.kr

Keywords: Hydroxyapatite, silicon substitution, coral, bioactive behavior, porous

Abstract. A porous silicon-substituted hydroxyapatite has been prepared using natural coral as a calcium source to obtain a biomaterial having an improved biocompatibility. From the XRD analysis, it was confirmed that the single-phase hydroxyapatite containing silicon has formed without revealing the presence of extra phases related to silicon dioxide or other calcium phosphate species. Silicon content was 0.369% by weight. EDS investigation confirmed the presence of silicon in the framework of hydroxyapatite structure. Based on in-vivo test, Si-substituted porous hydroxyapatite can be considered a useful material for bone implants.

Introduction

The most important properties with respect to the use of hydroxyapatite as a biomaterial for filling bone defects are depend on the high porosity and the interconnectivity pore system of the materials. A natural coral has a porous structure with all pores interconnected throughout the skeleton and structure resembles that of trabecular bone. Specially, Porites genus of natural coral is expected to be an excellent starting material to synthesize porous hydroxyapatite since its microporous structure resembles that of bone. In recent, hydroxyapatite derived from coral has been used extensively in clinical applications.

Ion-substituted hydroxyapatite can be used for a potential biological material in the form of porous body, granule, and coating material on metal alloy substrates to improve biocompatibility. Especially, incorporation of silicon into hydroxyapatite structure is of great interest since they play an important role in developing artificial bone [1-7].

In this study, preparation of Si-substituted hydroxyapatite derived from coral was investigated using a repeated treatment by hydrothermal and solvothermal methods.

Materials and Methods

Block of coral was immersed in 4% sodium hypochlorite to remove organic part. After thoroughly washing with distilled water, block of coral was put in 2M $(NH_4)_2HPO_4$ in a Teflon® lined hydrothermal bomb and heated for 16h at 180°C. Then block of coral was transferred into silicon acetate saturated acetone solution and heated for 24h at 180°C in Teflon® lined hydrothermal bomb. Finally, to obtain single phase hydroxyapatite, the sample was treated hydrothermally again at 180°C for 24h. After thoroughly washing with distilled water, block of coral was dried at 90°C.

The phase transformations of natural coral at different reaction process were observed using X-ray powder diffractometer (MAC Science Co., Ltd.). Cu Kα radiation was used at the operating condition of 40kv and 20mA. Microstructure of natural coral was examined using a Scanning Electron Microscopy (JOEL, JSM-6700F).

The bony ingrowth into the sample was macroscopically observed and the histopathological analysis at the interface between the sample and the cancellous bones was performed. For in-vivo test of the Si-substituted hydroxyapatite derived from coral, a hole (5mm in diameter and 7mm in length) was made through the lateral femoral condyle of New Zealand white rabbit and hemostasis was conducted using gauze. The sample was pushed into the hole and the rabbits were euthanized at 3 and 6 weeks. The bony ingrowth on the area surrounding the sample was then macroscopically observed.

Results

Fig. 1 shows scanning electron microscopy data of the porites coral genus obtained from the coast of Indonesia. Pore size distributions are ranged from 200µm to 300µm. This coral exhibits a porous structure with all pores interconnected throughout the skeleton and structure resembles that of trabecular bone. The observed X-ray powder diffraction pattern of the coral sample is indexed for aragonite phase on the basis of JCPDS Card No. 24- 2005 (Fig. 2). Conversion of coral into Si-substituted hydroxyapatite was carried out by repeated treatments of hydrothermal and solvothermal methods. For the first step, natural 2M $(NH_4)_2HPO_4$ in the Teflon® lined hydrothermal bomb and heated for 16h at 180°C. After the treatment of first step, X-ray powder diffraction of the sample shows the mixture phases of aragonite, tricalcium phosphate and hydroxyapatite.

As a second step, then block of mixed phases was transferred into silicon acetate saturated acetone solution and heated for 24h at 180°C in Teflon lined hydrothermal bomb. After the treatment of second step, X-ray powder diffraction data of the sample represents that the mixture phases of tricalcium phosphate and hydroxyapatite. Finally, to obtain single phase hydroxyapatite, the sample was treated hydrothermally again at 180°C for 24h.

From the XRD analysis, it was confirmed that the single-phase hydroxyapatite containing silicon has formed without revealing the presence of extra phases related to silicon dioxide or other calcium phosphate species. Average of silicon content determined by ICP was about 0.369% by weight. (Table 1) EDS investigation confirmed the presence of silicon in the framework of hydroxyapatite structure. The silicon-substituted hydroxyapatite that prepared from natural coral possesses uniformly permeable micropore and about 70% uniform pore volume.

Based on in-vivo test, Si-substituted hydroxyapatite can be considered a useful material for bone implants. From the macroscopic observation of bony ingrowth into the sample, symptoms of necrosis and discolorations were not observed on the sites surrounding the implanted Si-HA samples as shown in Figure X. Moreover, the boundary between the sample and the cancellous bone clearly appears at 3 weeks (Fig. 3(a)) but it completely disappears at 6 weeks(Fig. 3(b)). This result indicates that the sample was resolved and new bone completely grew into the sample after 6 weeks.

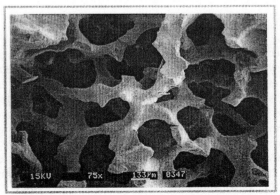

Fig. 1 SEM photograph of Si- HA derived from coral

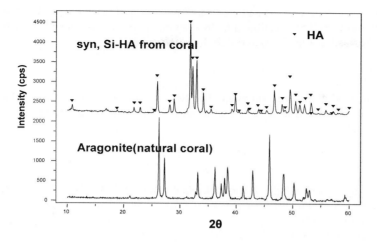

Fig. 2 X-ray powder diffraction data of Coral and Si- HA derived from coral

Table 1. Chemical analysis data of the samples

Minerals	Ca	Si	Mg	Fe
wt %	38.380	0.369	0.093	0.001

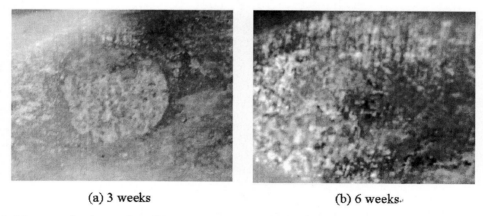

(a) 3 weeks   (b) 6 weeks

Fig. 3. Macroscopic observation of bony ingrowth into the Si-substituted hydroxyapatite sample.

Discussion and Conclusions

To obtain porous hydroxyapatite block having an improved biocompatibility, Si-substituted hydroxyapatite has been prepared using natural coral as a starting material. From the XRD, it was confirmed that the major-phase hydroxyapatite with silicon level 0.37wt% was synthesized. The Si-substituted hydroxyapatite derived from natural coral exhibits a porous structure with all pores interconnected throughout the skeleton and structure resembles that of trabecular bone. Pore size distributions are ranged from 200μm to 300μm.

From in-vivo test, symptoms of necrosis and discolorations were not macroscopically observed and new bone completely grew into Si-substituted hydroxyapatite at 6 weeks. Therefore, Si-substituted hydroxyapatite may be used as a useful material for bone implants.

Acknowledgements

Financial support of this work was provided by Next Generation Project (Development of Biological Hybrid Materials) of Ministry of Commerce, Industry and Energy, Korea under contract No. N11-A08-1402-07-1-3.

References

[1] I.R. Gibson, S.M. Best, W. Bonfield: J. Biomed. Mater. Res,. Vol. 44 (4) (1999), p. 422
[2] S.R. Kim, J.H. Lee, Y.T. Kim, D.H. Riu, S.J. Jung,Y.J.Lee, S.C. Chung andY.H.Kim: Biomaterial Vol. 24 (2003) p. 1389
[3] M.Okazaki: Biomaterials Vol. 16 (1995) p. 703
[4] E.Bertoni, A. Bigi, G. Cojazzi, M. Gandolfi, S. Panzavolta. N. Rover: J. Inorg. Biochem. Vol.72 (1998) p.29
[5] D. Skrtic, J.M. Antonucci, E.D. Eanes, R.T. Brunworth: J. Biomed. Mater. Res. Vol 59 (2002) p. 597.
[6] I. Mayer, J.D.B. Featherstone: J. Crystal. Growth. Vol. 219 (2000) p98.
[7] J.G. Lopez, R. Pomes, C.O. Vedova, R. Vina, G. Punte: J. Raman. Spectrosc. Vol. 32 (2001) p.255

Key Engineering Materials Vols. 254-256(2004) pp. 973-976
online at http://www.scientific.net

Porous Bioactive Glasses with Controlled Mechanical Strength

Linda Fröberg[1], Leena Hupa[1] and Mikko Hupa[1]

[1]Åbo Akademi University, Process Chemistry Center, Biskopsgatan 8, 20500 Turku, Finland,
lfroberg@abo.fi.

Keywords: bioactive glass, sintering, porous implant, mechanical strength

Abstract. Porous implants were sintered of three different grain size fractions of two established bioactive glasses. The sintering parameters for the implants were carefully recorded. The porosity of the implants was measured from SEM-images. The compression strength of the implants was measured with a Crush Tester. The data was used to find the sintering parameters desired in the manufacture of bioactive glasses needed to obtain a specific porosity and mechanical strength.

Introduction

The goal of the work was to study the densification and concurrent evolution of mechanical strength of porous implants manufactured by sintering crushed bioactive glasses. The formation of a firm bonding of bone is enhanced when using a porous implant structure [1]. Melt-derived porous bioactive glass implants have successfully been manufactured by sintering spherical particles of a narrow size fraction [2]. The porosity of the implant is regulated by the sintering parameters time and temperature, while the maximum pore size depends on the particle fraction used to manufacture the spheres. However, the method used to manufacture the spheres is restricted only to relatively small particles with maximum diameter less than roughly 300 μm. In some applications the desired pore texture is larger. In this work the porosity and mechanical strength of implants made by sintering two crushed bioactive glasses of three different grain size fractions larger than 300 μm were measured.

Materials

Sintering of glasses takes place through viscous flow above their transformation temperature. Any devitrification process slows down the sintering, and inhibits the bioactivity of the glass. The experimental glasses were chosen according to devitrification measurements performed for four established bioactive glasses [3]. The glasses could be divided into two groups, i.e. glasses that devitrify easily just above their transformation temperature, and glasses that devitrify at relatively high temperatures. The bioactive glasses Glass 13-93 and Glass 1-98 are supposed to be interesting materials for manufacture of porous implants [3].

The implants were sintered using three different fractions of Glass 1-98 and Glass 13-93. The glasses were melted from analytical grade raw materials and Belgian quartz sand according to a standard procedure for achieving homogeneous glasses. The chemical composition of the glasses is given in Table 1. The final glasses were crushed and screened into the desired fractions. The fractions 300-500 μm, 500-800 μm, and 800-1000 μm were used for the sintering experiments.

Table 1. The chemical composition of the experimental glasses in wt-% (mol-%).

Glass Code	SiO_2	Na_2O	K_2O	MgO	CaO	B_2O_3	P_2O_5
1-98	53 (53,8)	6 (5,9)	11 (7,1)	5 (7,5)	22 (23,9)	1 (0,9)	2 (0,9)
13-93	53 (54,6)	6 (6,0)	12 (7,9)	5 (7,7)	20 (22,1)	0 (0)	4 (1,7)

Methods

The test implants were made by sintering the different fractions into cylindrical bodies (length and diameter 5 mm) in a graphite mold containing 16 sites. The sintering was performed in an electrical furnace in a nitrogen atmosphere. The sintering temperatures 680, 690, 710, and 740 °C were selected between the transformation and devitrification temperatures of the glasses. The transformation temperature of the glasses is roughly 580 °C. The devitrification temperature is around 850 ° C for Glass 13-93 and 880° C for Glass 1-98 [3]. Time recording for the sintering process started when the temperature of the mold reached 650 °C. The temperature was measured with a separate thermocouple attached to the mold. The sintering times at the selected temperatures varied between 5 and 1200 minutes. Most of the samples were sintered for 5 to 13 different times, and at least three times were used for each sample.

Altogether more than 1300 cylinders were studied. Porosity as well as pore size was calculated from cross-sectional surfaces of the cylinders using SEM images (LEO 1530). To assure that no devitrification had taken place during the sintering process, several cylinders were studied by XRD (X'pert by Philips). The compression strength was measured with a Crush Tester (AB Lorentzen & Wettre).

Results

For both glasses the densification was faster when using smaller grain size fractions and higher temperatures. When the glasses were sintered at the same conditions, Glass 1-98 had clearly lower porosity. The densification degree as a function of sintering time for the fraction 500-800 μm is given in Figure 1. The curves in the figure are trend lines based on six samples measured for each selected combination of sintering time and temperature. The figure clearly shows that the higher the sintering temperature the faster the densification. The results for the highest experimental sintering temperature 740 °C are not shown in the figure. At this temperature it was difficult to accurately adjust the porosity.

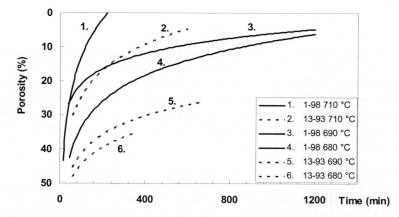

Fig. 1. Porosity of the implants as a function of sintering time for Glass 1-98 and Glass 13-93 at 680, 690 and 710 °C. Implants were sintered of fraction 500-800 μm.

At all temperatures Glass 1-98 sintered faster than Glass 13-93. The sintering times needed at different temperatures to achieve a porosity of 30 % were for Glass 1-98 and Glass 13-93 following: at 680 °C 140 and 730 min, at 690 °C 70 and 360 min, and at 710 °C 20 and 80 min. At porosities > 20 % the standard deviation is less than 5%. The standard deviation is somewhat higher at the lower porosities due to the low number of pores in each sample. XRD-analyses showed that none of the glasses devitrified when sintered with the chosen parameters.

A smaller grain fraction leads to a denser structure than a larger fraction if sintered under the same conditions. Also the pore size is less for implants sintered of smaller fractions. Typical SEM-images showing the cross section of cylinders sintered from different grain fractions are given in Figure 2. The cylinders were sintered of Glass 1-98 for 1 hour at 710 °C. The porosity of Glass 1-98 increases from 19 % (S.D. 0,9) for the smallest fraction to 22 % (S.D. 2,3) for the middle fraction, and further to 36 % (S.D. 11,1) for the largest fraction (Fig. 2).

Fig. 2. SEM-images showing the cross section of cylinders sintered of Glass 1-98 for one hour at 710 °C. 1) 300-500 µm, 2) 500-800 µm, and 3) 800-1000 µm. The white line equals to 2 mm.

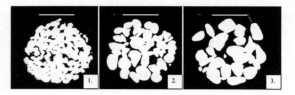

The sintering parameters for the desired porosity were determined from the SEM-images of the cylinders sintered at different time-temperature combinations. Figure 3 gives the sintering parameters as a Time-Temperature-Porosity-diagram (TTP) of the grain fraction 500-800 µm for both experimental glasses.

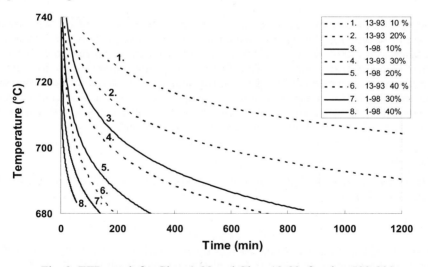

Fig. 3. TTP-graph for Glass 1-98 and Glass 13-93, fraction 500-800 µm.

The glasses have the same average compression strength for each densification level independent of the grain size or the glass composition used for sintering the cones. The sintering time at a specific temperature and porosity should, however always be adjusted according to the grain size and glass composition, c.f. Figure 3. The average strength at different porosity levels were: 10 % - 36,2 N/mm^2 (S.D. 11,4), 20 % - 21,2 N/mm^2 (S.D. 5,4), 30 % - 10,2 N/mm^2 (S.D. 3,6).

Discussion

Implants with controlled porosity can be sintered from crushed glass. The irregular particles of crushed glass start to round at the sharp edges above transformation temperature. After neck-formation the sintering proceeds similarly to that of spherical particles. The final porosity and maximum pore size depend on the grain size of the initial raw material, and on the chosen sintering parameters. Longer sintering times lead to a denser implant structure for a constant sintering temperature. Smaller particles sinter faster than larger ones due to the differences in surface energy for different sized particles [4]. As the sintering of glass takes place through viscous flow, the sintering parameters giving a desired porosity depend on the viscosity of the glass, and thus on its chemical composition. The observed faster sintering of Glass 1-98 compared to Glass 13-93 can be explained by the slightly lower content of network forming oxides and thus a lower viscosity at given temperature for Glass 1-98. The compression strength depends primarily on the total porosity of the glass. The strength is more than tripled when the porosity is decreased from 30 to 10 %. The strength is further increased at lower porosity values. However, a closed pore structure starts to form at a porosity level lower than roughly 10 %. By choosing the sintering parameters suitably the porosity and the compression strength can be controlled within certain limits. However, its should be noted that the actual strength is always controlled by the more or less random orientation of the glass particles in the sintering mold, and thus the distribution and size of the pores in the final implant.

Conclusions

Crushed glass of a known grain size distribution can be sintered into implants with controlled porosity by careful selection of the sintering parameters. At the same porosity level crushed glass with a small grain size (300-500 µm) leads to a structure with smaller sized pores but a larger number of them than a coarser grain size. The sintering range and optimum sintering parameters depend on the chemical composition of the glasses, and should be measured for each composition. In general, the sintering is slower and the control of the porosity is easier at low temperatures. The compression strength of the sintered implants depends on their total porosity rather than on the chemical composition of the glass or on the average pore size of the implant.

Acknowledgements

This work was funded by grants from the National Technology Agency (TEKES) and the Graduate School of Materials Research.

References

[1] K.H Karlsson, H. Ylänen and H.T. Aro: Ceramics International. Vol. 26 (2000), p.897-900.
[2] H. Ylänen, K.H. Karlsson, A. Itälä and H.T. Aro: J. Non-Cryst Solids. Vol. 275 (2000), p. 107-115.
[3] H. Arstila, E. Vedel, L. Hupa, H. Ylänen and M. Hupa: Key Engineering Materials, *accepted*.
[4] M.O Prado, E.D Zanotto and R. Müller: Non-Cryst Solids. Vol. 279 (2001), p. 169-178.

Key Engineering Materials Vols. 254-256(2004) pp. 977-980
online at http://www.scientific.net
© *2004 Trans Tech Publications, Switzerland*

Development of Hydroxyapatite Ceramics with Tailored Pore Structure

Ulrike Deisinger[1], Frauke Stenzel[2] and Günter Ziegler[1,2]

[1] Friedrich-Baur-Research Institute for Biomaterials, University of Bayreuth, 95440 Bayreuth, Germany, e-mail: ulrike.deisinger@fbi-biomaterialien.de

[2] BioCer EntwicklungsGmbH, Bayreuth, Germany

Keywords: hydroxyapatite ceramics, interconnecting porosity, slip casting, bone substitute materials, rapid prototyping

Abstract. The porosity of a hydroxyapatite ceramic was tailored by transferring polymeric pore models into ceramic forms via the slip casting technique. The rheological properties of the hydroxyapatite slurries play an important role in facilitating the casting operation. The optimised slurry with a solids content of 60 wt% hydroxyapatite had a slight shear thickening flow behaviour and a low viscosity. Via impregnating of polymeric pore models and dip coating of polymeric foams ceramic green bodies were fabricated. After sintering at 1250 °C the polymer was burnt out and a porous ceramic was achieved. This porous ceramic could be reasonably handled and had a very high interconnecting porosity of 91 – 96 vol%. The pore size varied between 300 and 800 µm. The density value of the bulk hydroxyapatite ceramic covered the range between 94 and 96 % th.d.. The characteristics of the porous ceramics could be varied between high and undirected porosity on the one hand, and lower porosity with defined pore channels in the three dimensional directions on the other hand.

Introduction

There is an increasing clinical requirement for bone graft material, because there are many possible applications such as revision hip surgery, defect filling after e.g. a tumor surgery, or reconstructive orthopaedic surgery. As hydroxyapatite closely resembles the mineral phase of natural bone and is highly biocompatible, it is among other calcium phosphates widely used as bone substitute material [1]. Synthetic hydroxyapatite can be reproducibly processed, and additionally, provides no danger of infections. A large number of research activities has been dealing with the fabrication of synthetic hydroxyapatite ceramics [2, 3, 4, 5].

In this work the porosity of a hydroxyapatite ceramic was tailored, so that the resulting bone substitute material is adapted to the requirements of the implantation site. The tailoring of the pore structure was carried out by using commercially available polymeric foams and polymeric pore models, which were produced via rapid prototyping. These polymeric models were coated or impregnated with hydroxyapatite using the slip casting technique.

This work is part of the research network ForTePro (Bayerische Forschungsstiftung, Germany), in which scientists of different research areas, particularly medical doctors, develop individually adapted implants for bone and cartilage defects, which will be cultivated with the patients own cells.

Materials and Methods

For producing the slurry the commercially available hydroxyapatite powder from Merck, Germany, was used. The specific surface area was characterised by the BET method (model Gemini 2370, micromeritics, Germany). The mean particle size d_{50} was determined with a laser particle size analyser (model Granulomètre 850, Cilas Alcatel, France). The powder was slowly added to distilled water under constant stirring. The dispersant agent was a solution of a natrium salt of an acrylic acid copolymer. Additionally, a surfactant for improved wettability and a binder were used.

The viscosity of the slurries was measured in a rotational rheometer (model Rheolab MC 100, Physica, Germany) with shear rates from 0 to 800 1/s.

Porous hydroxyapatite ceramic was fabricated by impregnating or coating polymeric pore models, e.g. rapid prototyping models or polyurethane (PU) foams (KURETA, Germany) of different pore sizes 30, 35 and 45 ppi (pores per inch), with the hydroxyapatite slurry. The resulting structure was dried and then heat-treated at 1250 °C. The pyrolyses behaviour of the rapid prototyping models and PU foams was characterised by thermo-gravimetric measurements (model STA 409, Netzsch, Germany).

The porosity was determined with a helium pyknometer (model Accu-Pyk 1330, micromeritics, Germany) and a scanning electron microscope (model Jeol 6400, Jeol, Japan). To analyse the phase composition of the ceramic a X-ray diffractometer (model XRD 3000P, Seifert, Germany) was used.

For comparison dense hydroxyapatite cylinders were prepared by casting the slurry into plaster moulds. After drying the resulting green bodies were sintered at 1250 °C. Green density was determined by geometrical weight-volume evaluation. Sintering density was measured by the Archimedes' method. The shrinkage and sintering start temperature were determined with a dilatometer (model 402 E/7, Netzsch, Germany).

Results and Discussion

The commercially available hydroxyapatite powder has a specific surface area of 64 m²/g. Using this powder only slurries with a relatively low solids content of 35 wt% and a high viscosity could be prepared. After calcination of the powder at 900 °C the specific surface area was reduced to 12.5 m²/g. Using the calcined powder slurries with a high solids content from 55 to 65 wt% were produced. The flow behaviour of these slurries is affected by the character and the amount of the dispersant (Fig. 1). Other additives like the binder can change the viscosity as well, but to a minor degree. The binder used in this work reduced the viscosity. The slurries showed a shear thickening flow behaviour, whereas the viscosity increased with rising solids content. As an optimum between the two desired properties low viscosity and high solids content, slurries with a solids content of 60 wt% were chosen for further experiments. The mean particle diameter in the stabilized slurries is approximately 2.8 – 3.2 µm. No sedimentation occurred for several hours.

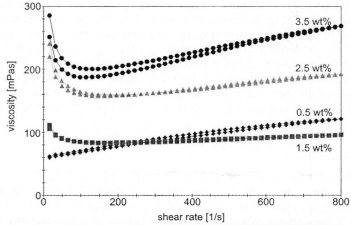

Fig. 1: Flow behaviour and viscosity of the hydroxyapatite slurry as a function of the amount of dispersant agent.

Green and sintering density values were determined with cast dense cylinders. As can be seen from Table 1, these values decrease slightly with increasing dispersant amount. Dilatometer measurements exhibit a mean shrinkage of the cylinders of 20.6 %. The sintering started at approx. 950 °C. From X-ray diffraction patterns of the dense samples sintered at 1250 °C it is evident that the final product consists of pure hydroxyapatite. No other phase was detected.

Table 1: Green and sintering density depending on the amount of the dispersant agent.

dispersant content [wt%]	green density [% th.d.]	sintering density [% th.d.]
0.5	47.0	96.5
1.5	47.5	94.6
2.5	46.6	94.3
3.5	45.9	94.3

Error: for green density: +/- 1.5 %, for sintering density: +/- 0.9 %

With the optimised slurry rapid prototyping polymeric models were impregnated and PU foams were coated. Subsequently, the polymer was pyrolysed and the ceramic sintered with a defined temperature profile. The thermo-gravimetric measurement revealed that the PU foam decomposes in two steps. The weight loss started at 210 °C, and at 650 °C the foam was almost completely removed. In the case of the rapid prototyping models the polymer burn out was finished at 700 °C. The porosity of the resulting ceramic is very high (91 to 96 vol% - Table 2). The ceramic has density values between 94 and 96 % th.d. which are comparable with the density of the bulk ceramic cylinders (see Table 1).

Table 2: Porosity of the PU foams and the hydroxyapatite (HAp) ceramic.

PU foam	Total porosity of PU foam [vol%]	Total porosity of HAp ceramic [vol%]	Density of bulk HAp [% th.d.]
30 ppi	98.4	91.3	94.8
35 ppi	97.9	96.4	95.5
45 ppi	97.9	96.2	96.2

Scanning electron microscopy images (Fig. 2) show the pore structure of the ceramic prepared from the PU foam. In all cases the pores were interconnected, only the pore size varied from 300 to 800 μm in diameter. The struts had very few cracks. Thus, the fabricated hydroxyapatite ceramic foam could be reasonably handled, although the struts of the ceramic were hollow. While the ceramics based on the PU foam had a high and undirected porosity, the rapid prototyping polymeric models resulted in bodies with a lower porosity, but defined pore channels in the three dimensional directions. In this way the porosity of the hydroxyapatite ceramic could be varied over a broad range.

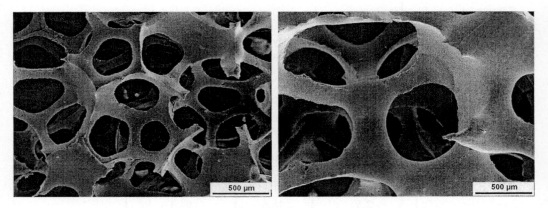

Fig. 2: SEM images of the sintered hydroxyapatite ceramics with different pore sizes (from left to right formed from PU foam with 45 and 30 ppi).

Conclusions

Hydroxyapatite ceramics could be fabricated by transferring polymeric pore models into ceramic forms using hydroxyapatite slurries with optimised characteristics. The porous ceramics revealed a uniformly distributed open and interconnected macroporosity. By using different types of polymeric pore models with diverse grades of porosity (50 to 96 vol%) and pore sizes (400 to 900 µm) as basic materials the pore structure of the bioceramic (undirected pores as well as defined pore channels) could be tailored according to the requirements of the specific implantation site. As a next step the porous ceramic scaffolds will be used for cell cultivation.

Acknowledgements

This work has been supported by the Forschungsverbund ForTePro initiated by the Bayerische Forschungsstiftung (Bavaria, Germany).

References

[1] R. Schnettler and E. Markgraf: *Knochenersatzmaterialien und Wachstumsfaktoren* (Georg Thieme Verlag, Stuttgart, 1997)

[2] D.-M. Liu: Ceramics International Vol. 24 (1998), p. 441-446

[3] M. Milosevski, J. Bossert, D. Milosevski and N. Gruevska: Ceramics International Vol. 25 (1999), p. 693-696

[4] J. Tian and J. Tian: Journal of Materials Science Vol. 36 (2001), p. 3061-3066

[5] M. Fabbri, G.C. Celotti and A. Ravaglioli: Biomaterials Vol. 16 (1995), p. 225-228

Key Engineering Materials Vols. 254-256(2004) pp. 981-984
online at http://www.scientific.net

Optimising the Strength of Macroporous Bioactive Glass Scaffolds

Julian R. Jones[1], Lisa M. Ehrenfried[2], Larry L. Hench[1]

[1] Department of Materials, Imperial College London, South Kensington Campus, London, SW7 2BP UK. julian.r.jones@imperial.ac.uk, l.hench@imperial.ac.uk.

[2] Zentralinstitut, fur Medizinitechnik der Technischen Universitat Munchen, Germany. Lisa_Ehrenfried@gmx.de

Keywords: Macroporous, scaffold, bioactive glass, strength.

Abstract. Resorbable 3D macroporous bioactive scaffolds have been produced for tissue engineering applications by foaming sol-gel-derived bioactive glasses of the 70S30C (70mol% SiO_2, 30mol% CaO) composition. The foams exhibit a hierarchical structure with interconnected macropores (10-500µm in diameter), which provide the potential for tissue ingrowth, and mesopores (2-50nm in diameter) that enhance bioactivity and resorbability. The effect of sintering temperature on the compressive strength of bioactive glass scaffolds was investigated. Compressive strength generally increased as sintering temperature increased to 800°C, due to a decrease in the textural mesopore diameter and a thickening of the macropore walls. The compressive strength of the foams did not increase as sintering temperature increased from 800°C to 1000°C, even though the mesoporosity was eliminated. The modal interconnected macropore diameter decreased as sintering temperature increased, but remained above 100µm until the sintering temperature exceeded 800°C. Sintering the bioactive glass foam at 800°C was the optimum treatment for the production of a macroporous scaffold for tissue engineering.

Introduction

A possible solution to the shortage of bone-graft donors is an artificial scaffold that can stimulate restoration of damaged or diseased bone to natural form and function [1]. Resorbable 3D macroporous bioactive scaffolds have been produced by foaming sol-gel-derived bioactive glasses of the 70S30C (70mol% SiO_2, 30mol% CaO) composition [2]. The foams exhibit hierarchical pore structures similar to that of trabecular bone with modal interconnected macropore diameters in excess of 100µm, which provide the potential for tissue ingrowth and vascularisation. The scaffolds have the potential to bond to bone [2, 3] and to release ions that will stimulate the genes in osteoblasts to promote bone regeneration [4]. The final step in the standard sol-gel processing of binary bioactive glasses was sintering at 600°C [2, 5] and previous work reported that the scaffolds exhibited low compressive strength (0.2-1.4MPa) [6]. The aim of this work was to investigate how changing the sintering temperature (final stage of the foam synthesis) affected the structure and compressive strength of the foams.

Materials and Methods

Foams were prepared using sol-gel-derived bioactive glass of the 70S30C composition (70mol% SiO_2, 30mol% CaO). Sol preparation involved mixing of the reagents in order; deionised water, 2N nitric acid, tetraethyl orthosilicate (TEOS), and calcium nitrate [2]. 50ml aliquots of sol were foamed by vigorous agitation with the addition of 0.5ml surfactant (Teepol) and 1.5ml HF (gelation catalyst). As the gelling point was approached the solution was cast into moulds. The samples were aged, dried and thermally stabilised at 600°C [2].

Foams were then sintered at 700°C, 800°C and 1000°C for 2 hours. The foams were characterized using SEM (JEOL, JSM 5610LV), X-ray diffraction (XRD) spectroscopy, mercury porosimetry (Quantachrome Poremaster 33) and nitrogen sorption analysis (Quantachrome AS6) [2, 5]. XRD spectra were collected using a Philips PW1700 series automated spectrometer, using a step scanning method (step size of 0.04°) at 40kV, 40mA. Differential thermal analysis (DTA, Stanton Redcroft STA-780) of a foam sample was carried out in air using a heating rate of 10°C/ minute to monitor phase transformations as a function of temperature. The foam cylinders (height = 9mm, diameter = 27mm) were compression tested (parallel plate test), using an Instron with a crosshead velocity of 0.5mm/minute. Five samples were used for each sintering temperature and mean values taken.

Results

For tissue engineering applications, the modal interconnected pore diameter (D_{mode}) must be greater than 100µm, to allow tissue ingrowth and eventually vascularisation. Fig. 1a shows interconnected macropore distributions, as a function of sintering temperature (T_s), obtained from mercury intrusion porosimetry. The vertical axis is a derivative of the volume of mercury intruded into the foam relative to the interconnect pore diameter. Fig. 1a shows that D_{mode} decreased as T_s increased. D_{mode} decreased from 122µm at T_s = 600°C to 113µm at T_s=700°C, to 98µm at T_s= 800°C and 87µm at T_s = 1000°C. Therefore heating the scaffold above 800°C reduced D_{mode} below the 100µm required for tissue engineering applications.

Fig. 1b shows textural pore size distributions as a function of (T_s), obtained from desorption branch of the nitrogen sorption isotherms using BJH analysis. The vertical axis is a derivative of the volume of nitrogen adsorbed into the surface of the foam at each pore diameter. The modal textural pore diameter (d_{mode}) was 17.1 nm for the foams sintered at 600°C. The textural pore size was therefore in the mesoporous range (2-50nm). As T_s increased from 600° to 700°C, the d_{mode} value was 17.5nm, which was not a significant change as a change of 0.4 nm was within the accuracy limits of the nitrogen sorption analysis technique (± 0.5nm). As T_s increased to 800°C, d_{mode} decreased from approximately 17nm to 12.3 nm. As T_s increased to 1000°C, d_{mode} decreased values below the detection limits of BJH analysis (<1nm), i.e. although interconnected macropores were present after sintering at 1000°C, the mesoporosity was negligible and the struts between the macropores were fully dense.

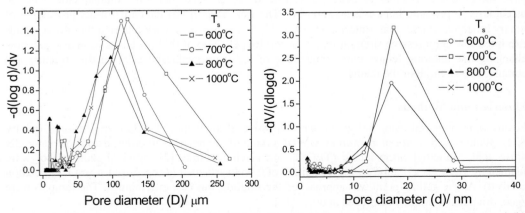

Fig. 1. (a) Interconnected macropore size distribution from mercury intrusion porosimetry and (b) textural mesopore size distributions from nitrogen BET sorption analysis.

Fig. 2 shows SEM micrographs of sol-gel derived bioactive foam scaffolds sintered at (a) 600°C
and (b) 800°C. Fig. 2a shows that foams sintered under standard procedures at 600°C exhibited
macropores in excess of 500μm in diameter connected by pore windows of up to 200μm diameter.
Fig. 2b shows that after sintering at 800°C the macropores were reduced in diameter, but the
interconnecting pore windows were high in number and confirms that the interconnected pore
diameter was in the region of 100μm.

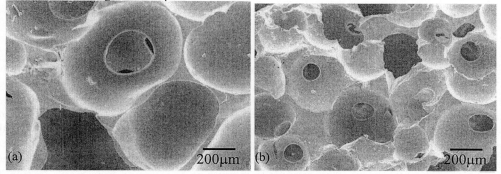

Fig.2. SEM micrographs of sol-gel derived foam scaffolds sintered at (a) 600°C and (b) 800°C.

The load/ displacement curves (not shown) were typical for compression of porous ceramics. Fig. 3
shows a graph of maximum compressive strength. Table 1 shows a summary of data. Figure 3 and
Table 1 show that as sintering temperature increased from 600°C to 800°C, the mean σ_{max} increased
from 0.34MPa to 2.26MPa and E increased from 0.93MPa to 22.6MPa. The increase in stiffness
and strength is due to a thickening and densification (reduction in mesoporosity) of the macropore
walls. As T_s increased to 1000°C, σ_{max} did not increase and E decreased to 18.2MPa.

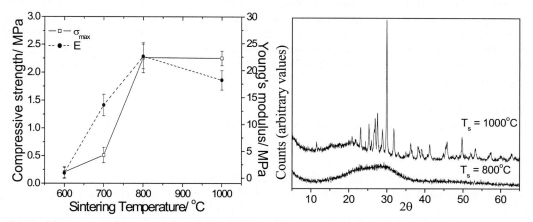

Fig. 3. (a) Compressive strength and Young's modulus as a function of sintering temperature. (b)
XRD traces of bioactive glass foams sintered at 800°C and 1000°C.

No further increase in strength was achieved as T_s increased from 800°C to 1000°C due to changes
in the glass structure. Differential thermal analysis (DTA) traces (not shown) of the foam revealed
that the glass transition temperature of the foams was 844°C. This means that above 844°C the
glass structure started to become more ordered and crystallization may have started. This was
confirmed by the XRD spectra (Fig. 3b).

Table 1: Data summary; T_s = sintering temperature, D = interconnected pore window diameter from d = textural pore diameter, mercury intrusion porosimetry, σ_{max} = maximum compressive strength, E = Young's modulus.

T_s [°C]	ρ_b [gcm^{-3}]	Modal D [μm]	Modal d [nm]	σ_{max} [MPa]	E [MPa]
600	0.25	122	17.1	0.34	9.7
700	0.41	113	17.5	0.9	24.3
800	0.51	98	12.3	2.45	25.9
1000	0.58	87	-	2.31	41.9

Fig. 3b shows XRD spectra collected from foams sintered at 800°C and 1000°C. Spectra collected at T_s = 800°C contained no crystalline peaks and exhibited an amorphous halo commonly observed for glasses, however spectra collected at T_s = 1000°C contained peaks that were attributed to wollastonite ($CaSiO_3$) [7] but also still exhibited an amorphous halo. DTA traces of the foams showed that total crystallization of the glass should occur at 1050°C. Therefore, at 1000°C, the foam was a mixture of glass and wollastonite phases.

Fig. 1b showed that the textural porosity of the struts of the foam was eliminated at 1000°C, which should have caused the compressive strength to increase as T_s increased from 800°C to 1000°C, however the phase transformation of the glass to wollastonite created a mixture of phases effectively creating a crystalline material with a high concentration of microstructural defects and causing a mismatch in mechanical properties, preventing an increase in σ_{max}.

Conclusion

The compressive strength of bioactive foam scaffolds increased with sintering temperature as temperature increased to 800°C due to a thickening of the struts between macropores. However at 1000°C, which was above the glass transition temperature of the gel-derived bioactive glass, the glass underwent a phase transition to wollastonite and the modal interconnected macropore diameter decreased to 87μm. The compressive strength did not increase from the value at 800°C. To obtain a bioactive glass scaffold of optimum strength and with a pore structure suitable for tissue engineering, the foams should be sintered at 800°C.

Acknowledgements
Lloyds Tercentenary Foundation, EPSRC and MRC (UK).

References

[1] J.R. Jones, L.L. Hench J. Mat. Sci. Technol. Vol. 17 (2001), p. 891.
[2] P. Sepulveda, J.R. Jones, L.L. Hench, J. Biomed. Mater. Res. Vol. 59 (2) (2002), p. 340.
[3] L. L. Hench J. Am. Ceram. Soc, Vol. 74 (7) (1991), p. 1487.
[4] I. Xynos, A.J. Edgar, L.D.K. Buttery, J. Polak, Biochem. Biophys. Res. Com. Vol. 277 (2000), p. 604.
[5] P. Saravanapavan, L.L. Hench, J. Biomed. Mater. Res. Vol. 54 2001p. 608.
[6] J.R. Jones, L.L. Hench, Key Eng Mat. Vol.s 220-242 (2003), p. 209.
[7] Powder Diffraction File – Inorganic phases. International Centre for Diffraction Data, USA. 2000

Key Engineering Materials Vols. 254-256(2004) pp. 985-988
online at http://www.scientific.net
© *2004 Trans Tech Publications, Switzerland*

Osteoblast Nodule Formation and Mineralisation on Foamed 58S Bioactive Glass

JE Gough[1], JR Jones[2] and LL Hench[2]

[1] Manchester Materials Science Centre, UMIST and The University of Manchester, Grosvenor Street, Manchester M1 7HS, UK. E-mail: j.gough@umist.ac.uk.

[2] Department of Materials, Imperial College London, South Kensington Campus, London SW7 2BP, UK. E-mail: julian.r.jones@imperial.ac.uk, l.hench@imperial.ac.uk.

Keywords: Osteoblast, bioactive glass, ion release, mineralisation, apoptosis.

Abstract. In this study we analysed human osteoblast responses to a porous bioactive glass scaffold. Attachment, spreading and formation of mineralised nodules in response to culture on the bioactive glass were analysed. Dissolution products are a key feature of bioactive glasses and these were measured by inductively coupled plasma optical emission spectroscopy to determine effects of both the glass surface and ion release. Osteoblasts attached and proliferated on the foams as demonstrated by scanning electron microscopy. Nodule formation was also observed in the pores of the glass and also in conditioned medium containing dissolution products at certain concentrations and these nodules were shown to be mineralised by alizarin red staining.

Introduction

Bioactive glasses have gained much interest for bone tissue engineering due to their ability to upregulate certain osteoblast genes [1]. Tissue engineering has excellent potential to repair diseased or damaged tissue through regeneration of tissue rather than replacement. Third generation materials that are bioactive, resorbable and stimulate specific desired cell responses are now being investigated [2]. Tissue engineering aims to regenerate damaged tissue using cell seeded resorbable materials. This study describes responses of human primary osteoblasts to foamed 58S bioactive glass. This bioactive glass is being investigated for bone tissue engineering. Nodule formation and mineralisation are characteristics of osteoblasts during bone formation on biomaterials for tissue repair. The objectives of this study were to demonstrate nodule formation and mineralisation by scanning electron microscopy and alizarin red staining, a histochemical stain commonly used to detect calcium deposits in osteoblast cultures.

Methods
Glass Synthesis
Colloidal solutions (sols) of 58S composition (60mol% SiO_2, 36mol% CaO, 4mol% P_2O_5) were prepared by mixing (in order); distilled water, 2N nitric acid, tetraethyl orthosilicate (TEOS), triethyl phosphate (TEP) and calcium nitrate [3]. Aliquots of 50ml of sol were foamed by vigorous agitation to produce porous scaffolds: 1.5ml of an acidic catalyst (0.5wt% HF) was added as a gelling agent, to reduce the gelling time to 7 minutes and 1.5ml of Teepol®, a detergent containing a low concentration mixture of anionic and nonionic surfactants was added as a foaming agent. As the gelling point was approached the solution was cast into molds. Samples were aged (60°C), dried (130°C) and thermally stabilized (600°C) according to published procedures [4].
Foamed samples measured 25mm diameter x 10mm deep and had a geometrical bulk density of 0.28-0.32g/cm³. Porosity and surface area values were previously determined [4].
For cell culture studies, samples were used as above or cut to approximately 10mm x 10mm x 5mm using a scalpel and placed into the wells of a 6 well plate containing 8 mls medium.

Inductively coupled plasma optical emission spectroscopy (ICP-OES)

Ionic concentrations of released calcium (Ca), phosphorous (P) and silicon (Si) were determined using ICP-OES (3580B ICP Analyser, Applied Research Laboratories, IM35xx ICP Manager software, Micro-Active Australia Pty Ltd). After immersion of foams in culture medium (DMEM) for 24 hours at 37°C in static conditions (to mimic cell culture conditions) the samples were diluted 1:10 in a 2N HNO_3 matrix. Three integrations were performed per sample and the mean and the standard deviation was calculated. The instrument detection limits for the elements of interest were 0.10 (Ca), 0.20 (P) and 0.05 (Si) ppm.

Osteoblast Culture

Human osteoblast cells (HOBs) were isolated as described previously from femoral heads after total hip replacement surgery [5, 6]. Bone fragments of approximately 3mm x 3mm were removed and washed several times in phosphate buffered saline (PBS) to remove blood cells and debris with a final wash in culture medium. Fragments were then placed in tissue culture flasks in complete Dulbecco's Modified Eagles Medium (DMEM) containing 10% Foetal Bovine Serum (FBS) with 1% glutamine, 2% penicillin/streptomycin and 0.85mM ascorbic acid. The fragments were incubated at 37°C in a humidified incubator with 5% CO_2. Medium was changed every 2 days and after approximately 4 weeks in culture, bone fragments were discarded and cells harvested using trypsin EDTA. Cells were seeded onto materials or Thermanox discs as positive controls, at a density of 80,000 cells/cm^2.

Culture in Conditioned Medium

Conditioned medium was made by incubating foam samples in DMEM for 24 hours. Cells were then cultured in this conditioned medium neat, or diluted 1:1 or 1:4 with DMEM.

Scanning Electron Microscopy

Cells were cultured on foams or Thermanox discs and at required time-points fixed in 1.5% glutaraldehyde for 30 minutes at 4°C. Cells were then post-fixed in 1% osmium tetroxide for 1 hour at 4°C. Cells were dehydrated through a series of increasing concentrations of ethanol and dried using hexamethyldisilazane (HMDS). Samples were sputter coated with gold and viewed using a Cambridge Stereoscan S360 scanning electron microscope operated at 10kV.

Mineralisation

Whether mineralisation of nodules occurred was determined using alizarin red staining. As the dye binds to calcium, and calcium is present in the material making it difficult to visualise true positive staining, nodules were trypsinised from the foams, and then fixed in 70% ethanol and stained. Cells were washed several times in distilled water and viewed under the light microscope.

Alizarin red staining was also performed after culture in conditioned medium.

Results

The scaffolds exhibited a hierarchical interconnected macropore network, with a texture consisting of mesopores with a mean diameter of ~20 nm (nitrogen sorption) and modal interconnected pore diameter of ~100μm (mercury porosimetry) which is ideal for bone ingrowth [4].

Cells attached to the foam surface and within pores as shown in Fig. 1 A, B and C. Dorsal ruffles and cell projections or filopodia were observed. By 10 days of culture, nodules had begun to form within pores of the foam as shown in Fig 1. D.

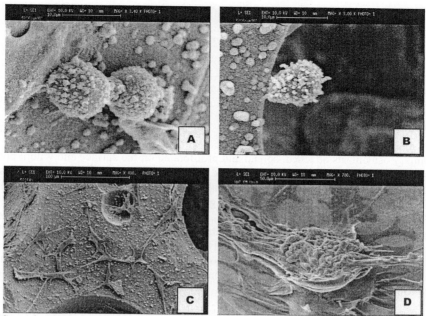

Fig. 1. Scanning electron micrographs of human osteoblasts cultured on 58S porous foamed scaffold. A = 90 minute culture, cells attaching within pore. B = 90 minute culture, cell attaching to edge of pore. C = 4 hour culture, cells attaching to material surface and bridging smaller pores. D = nodule formed within pore after 10 days culture.

These nodules positively stained with alizarin red (Fig 2.), as did nodules that formed after culture in 1:4 conditioned medium:DMEM. Nodule formation was not observed after culture in neat dissolution product or 1:1 conditioned medium:DMEM, or control DMEM.

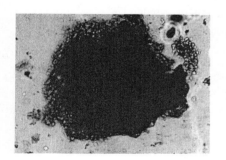

Fig. 2. Nodule stained with alizarin red after 10 days culture on 58S bioactive glass foam. Positive dark red staining demonstrates calcium deposits within nodule i.e. mineralisation.

Table 1 shows ICP-OES results demonstrating that the 1:4 ratio contained approximately 48µg/ml Si, whereas the neat dissolution products contained approximately 230µg/ml and the 1:1 ratio contained approximately 120µg/ml.

Table 1. Concentration of released Si (µg/ml) in conditioned medium (DMEM) from 58S foam after 24 hours.

Sample	Si Concentration	+/- SD
58S	230	45
1:1 58S:DMEM	120	18
1:4 58S:DMEM	48	8
DMEM	0	N/a

Discussion and Conclusions

Human primary osteoblasts attached, spread and proliferated on the surface and within pores of the 58S bioactive glass foamed scaffold. Mineralized nodule formation, a characteristic of osteoblasts in culture, occurred within pores of the foam and also in a 1:4 ratio of conditioned medium:DMEM, which contained approximately 48µg/ml Si. Higher concentrations of Si in the culture medium appeared to prevent nodule formation occurring, as did control medium, which contained no Si. These are exciting results as traditionally osteoblast cultures require supplementation with "mineralizing agents" dexamethasone and beta-glycerophosphate to cause mineralisation [7-9]. In this study we have demonstrated mineralized nodule formation without these agents but in the presence of certain concentrations of Si. Further investigations are underway determining effects of Si concentration on other osteoblast parameters.

References

[1] I.D. Xynos, A.J. Edgar L.D.K. Buttery and J.M. Polak: Biochem. Biophys. Res. Com. Vol 277 (2000), p.604.

[2] L.L. Hench and J.M. Polak: Science Vol 295 (2002), p. 1014.

[3] R. Li, A.E. Clark and L.L. Hench: J Appl Biomater Vol 2 (1991), p. 231.

[4] P. Sepulveda, J.R. Jones and L.L. Hench: J Biomed Mater Res Vol 59 (2002), p. 340.

[5] J.E. Wergedal and D.J. Baylink: Proc Soc Expt Bio Med Vol 176 (1984), p. 60.

[6] L. Di-Silvio: PhD Thesis 1995. Inst Orth. UCL Medical School, London UK.

[7] R. Bizios: Biotechnology and Bioengineering Vol 43 (1994), p. 582.

[8] J.E. Aubin: Biochem Cell Biol Vol 76 (1998), p. 899.

[9] H. Ohgushi, Y. Dohi, T. Katuda, S. Tamai, S. Tabata and Y. Suwa: J Biomed Mater Res Vol 32 (1996), p. 333.

Key Engineering Materials Vols. 254-256(2004) pp. 989-992
online at http://www.scientific.net
© 2004 Trans Tech Publications, Switzerland

Macroporous Bioactive Glasses

Victoria J. Shirtliff and Larry L. Hench

Tissue Engineering Centre, Department of Materials, Imperial College London, Exhibition Road, London SW7 2AZ, UK

victoria.shirtliff@imperial.ac.uk, l.hench@imperial.ac.uk

Keywords: bioactive glass, scaffold, macroporous, foam, sol-gel.

Abstract. Glass foams with composition 67% SiO_2, 33% CaO (S67C33) were produced by foaming the sol during the sol-gel process. Nitrogen gas sorption and mercury intrusion porosimetry revealed the foams' hierarchical structure of interconnected macropores as large as 200 μm and a mesoporous texture, suggesting their suitability as scaffolds for tissue engineering. *In vitro* dissolution in simulated body fluid showed that although an amorphous phosphate material was initially deposited on the surface of the glass foams, it did not develop into crystalline hydroxycarbonate apatite. The solution was depleted of phosphorous ions within 12 hours and calcite was found to be the major component on the surface of the foams. The experimental conditions for *in vitro* dissolution of these macroporous glass foams should be optimised for further cellular or *in vivo* studies to be carried out.

Introduction

The discovery, in 1969, of a four-component glass which could bond to living tissue [1] has enabled the development of many novel prostheses but also allowed the treatment of damaged or diseased tissue without simple replacement. Bioactive glasses make possible the regeneration of tissue as their ionic dissolution products have been found to stimulate genes which control cell growth and proliferation [2]. Bioactive glasses produced via the sol-gel process exhibit an inherent porosity (pores of 2-50 nm diameter) which enhances their bioactivity and resorbability. A larger surface area permits greater release of ionic dissolution products. Sol-gel glasses in the simple binary CaO-SiO_2 system have been found to be bioactive [3, 4] and have also been produced with macropores (>100 μm) [5] large enough to allow for cell penetration, tissue ingrowth, vascularisation and nutrient delivery to the growing tissue, making them suitable for tissue engineering applications.

This study assesses the potential of macroporous sol-gel glasses of the composition 67% SiO_2, 33% CaO (S67C33), previously unstudied, for use as tissue engineering scaffolds. Mercury intrusion porosimetry is used to measure the diameter of pore throats, i.e. interconnections within a porous network, up to 200 μm. *In vitro* formation of a hydroxycarbonate apatite (HCA) layer on the surface of a material in aqueous solution is generally accepted to be indicative of *in vivo* bioactivity.

Experimental Procedure

Foam preparation. The sol for composition S67C33 was prepared as follows. The following compounds were added to 159.19 ml deionised water in sequential order: 26.53 ml 2N nitric acid (HNO_3), 164.13 ml tetraethyl orthosilicate (TEOS) and 85.67g of calcium nitrate salt [$Ca(NO_3)_2 \cdot 4H_2O$]. The solution was stirred for 1 hour after the additions of TEOS and calcium nitrate. Quantities of 50ml of the prepared sol were foamed by adding approximately 0.3 ml of Teepol (a diluted mixture of 10-15 vol% anionic and <5 vol% non-ionic surfactants), 1.5ml HF (0.1 wt%) to catalyse the polycondensation reaction and by employing vigorous mechanical agitation. As the viscosity of the sol began to increase more rapidly, the foam was cast into polymethyl pentene (PMP) moulds where gelation subsequently took place. The containers were hermetically sealed, aged at 60 °C for 72 hours, dried at 130 °C for 48 hours and thermally stabilised at 600 °C for 22 hours, according to established procedures [5].

Textural analysis. Prior to analysis specimens were degassed under vacuum at 150 °C for 15 hours to remove physically adsorbed moisture and impurities from their surfaces. Helium pycnometry was used to ascertain the skeletal (true) density (ρ_s) of the foams, to enable the calculation of their porosity using the following equation:

$$V_V = 1 - (\rho_b / \rho_s) \qquad (1)$$

where V_V is the volume fraction porosity and ρ_b is bulk density, measured using digital Vernier calipers and a digital four-decimal place balance. Nitrogen sorption was employed to determine the specific surface area of the foams, using the BET method, the average mesopore diameter (mean as well as modal, calculated by applying the BJH model to the desorption data) and total pore volume due to mesopores. Mercury intrusion porosimetry was used to obtain macropore size distributions for the foams.

Dissolution analysis. Foam cubes of 7 x 7 x 7 mm^3 were prepared and each soaked in 50 ml of simulated body fluid (SBF) [6], in 100 ml conical flasks, at 37 °C and 175 rpm for different time periods to assess their ability to form a surface layer of HCA *in vitro*. This was carried out in triplicate for each time period. The ionic concentrations of Si, C and P in the reacted SBF solutions were measured using inductively coupled plasma (ICP) analysis and the pH of these solutions was measured at 37 °C. The surfaces of the foams were examined using Fourier transformed infra-red (FTIR) spectroscopy, normalising to the spectrum of a blank KBr pellet, scanning electron microscopy with energy-dispersive X-ray (SEM-EDX) analysis and X-ray diffraction (XRD) analysis. XRD analysis was performed with a Philips PW1710 series diffractometer, using the step scanning method with Cu$_{K\alpha}$ radiation, operated at 40 kV and 40 mA with a 0.040° 2θ step in the range of 5-60° 2θ and a count rate of 5 seconds per step.

Results and Discussion

Textural features. Foams with a bulk density of 0.25 g cm^{-3} were chosen for analysis in this study. Table 1 summarises the textural characteristics of these foams. The volume fraction porosity of the foams was found to be 89 %. Nitrogen sorption and mercury porosimetry analyses were performed in triplicate and the results are displayed as the mean and standard deviation of the three values obtained for each textural feature. The values obtained in this study are consistent with those presented in the literature for sol-gel derived glasses and foams in the CaO-SiO$_2$ system [5, 7]. The mean mesopore size is larger than the modal value due to the presence of large mesopores and macropores. Figure 1 shows a typical macropore size distribution for the foams, obtained by mercury intrusion porosimetry. It illustrates the wide range of pore sizes within the foams as well as the presence of pores as large as 200 μm (limit of detection for the instrument). SEM images not shown) revealed the presence of pores greater than 200 μm and the pore interconnections were evident.

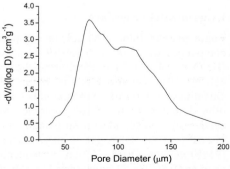

Table 1. Textural characteristics of S67C33 foams

Textural Feature	Mean	S. D.
Bulk density [g cm^{-3}]	0.25	-
Skeletal density [g cm^{-3}]	2.2713	-
Volume fraction porosity [%]	89.0	-
BET surface area [m^2 g^{-1}]	128.9	5.4
Average pore diameter [Å]		
Mean	233.0	34.2
Modal (BJH)	173.1	26.4
Pore volume [cm^3 g^{-1}]	0.7761	0.0914
Modal macropore diameter [μm]	77.1	28.2

Fig. 1. Macropore size distribution for S67C33 foam

In vitro bioactivity assessment. During the first 3 hours of immersion in SBF the foams release Si and Ca ions, accompanied by an increase in pH, as shown in Figure 2. After 3 hours Si ions continue to be released but a reduction in the concentration of Ca and P ions in solution is seen, suggesting that a Ca-P material is being deposited on the surface of the foam. After 12 hours the solution is depleted of P ions and Si and Ca release continues.

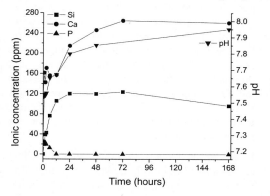

Fig. 2.   Variation in Si, Ca and P ion concentrations and pH with immersion time in SBF for S67C33 foams

The SEM micrographs (not shown) revealed that a number of small round particles have developed on the surface of the foam after 2 hours immersion in SBF. The EDX spectrum of the surface at 2 hours (not shown) indicates the presence of phosphorous and the EDX spectrum of one of the small particles reveals a Ca:P ratio of 1.7, suggesting that the particles may have a similar composition to that of the mineral apatite in human bone, the Ca:P ratio of which ranges between 1.6 and 1.7. After 12 hours of immersion the particles have increased in number but after 48 hours a material with unusual morphologies was deposited on the surface. The corresponding EDX spectra show reduced concentrations of Ca and P on the foam surface, with respect to Si. This suggests that the material deposited after 48 hours may be rich in Si.

The FTIR spectra of the foam surface, shown in Figure 3, suggest that after 1 hour an amorphous phosphate material was deposited. After 2 hours a peak indicating the presence of a carbonate group can be seen. However the amorphous phosphate is not seen to develop into crystalline HCA.

Instead, after 12 hours, the peaks representing silica continue to resolve and grow in intensity while the phosphate and carbonate group peaks grow smaller.

XRD analysis was performed on an unreacted foam and on foams after 2 hours and 72 hours of immersion in SBF. The spectrum of the unreacted specimen, shown in Figure 4 confirms the absence of any crystalline phase as no diffraction maxima are observed and only a broad band between 25 and 40 $°2\theta$ was detected. The spectrum after 2 hours of immersion does not reveal the presence of any crystalline HCA. The major crystalline component on the surface of the foam was found to be calcium carbonate (calcite, $CaCO_3$). The spectra after 2 and 72 hours of immersion are very similar. It was expected for the diffraction maxima in the spectrum after 72 hours of immersion to

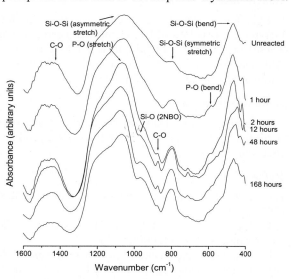

Fig. 3. FTIR spectra for S67C33 foam surface before and after *in vitro* dissolution

have higher intensities however the more spherical nature of this specimen resulted in a broadening of the peaks, hence comparisons between peak intensities in each spectrum cannot be made.

Both the EDX and FTIR spectra confirm that phosphorous is being taken up onto the surface of the material from solution, however XRD analysis suggests that this is not in a crystalline phase. It is likely that, at some point, deposition of calcite became preferential to the deposition of calcium phosphate, possibly due to the experimental conditions. The specimens in dissolution for time periods longer than 12 hours were prone to disintegration at the vortices. It is therefore possible that the crystalline layer is thinner on the specimens in dissolution for longer time periods, hence the growing Si peaks and diminishing carbonate and phosphate peaks in the FTIR spectra for post-12 hour specimens. Unfortunately it is not possible to verify this using relative XRD peak intensities for the reason stated above.

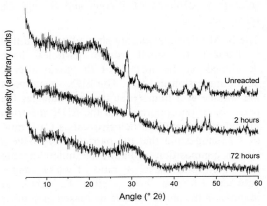

Fig. 4. XRD spectra for S67C33 foam surface before and after *in vitro* dissolution

In this *in vitro* study the foams were not Class A bioactive, although powder and monolith forms of sol-gel glasses of similar compositions have been, previously [4, 7]. These results indicate that optimisation of test conditions may be required, to prevent damage to the foam. Variables include: ratio of surface area of foam to volume SBF employed, size of conical flask used and degree of agitation. Under the current experimental conditions the depletion of P ions from solution within 12 hours advocates the transfer of foam specimens to fresh SBF at regular time intervals for studies longer than 12 hours, to allow a thick crystalline layer of HCA to develop.

Conclusions

Sol-gel foams have been successfully produced for the composition S67C33, which exhibit a hierarchical structure of interconnected macropores as large as 200 μm and a mesoporous texture. These macroporous sol-gel glasses hold promise for use as scaffolds for tissue engineering. Although they were not found to be Class A bioactive, *in vitro* experimental conditions could be optimised to enhance HCA formation.

Acknowledgements

The authors wish to thank Dr Priya Pavan for performing ICP analysis, Mr Richard Sweeney for performing XRD analysis and the Engineering and Physical Sciences Research Council (UK) for its financial support.

References

[1]   L. L. Hench, R. J. Splinter, W. C. Allen and T. K. Greenlee, Jr.: J. Biomed. Mater. Res. Vol. 2 (1971), p. 117
[2]   I. D. Xynos, A. J. Edgar, L. D. Buttery, L. L. Hench and J. M. Polak: Biochem. and Biophys. Res. Comm. Vol. 276 (2000), p. 461
[3]   P. Saravanapavan and L. L. Hench: J. Biomed. Mater. Res. Vol. 54 (2001), p. 608
[4]   P. Saravanapavan, J. R. Jones, R. S. Pryce and L. L. Hench: J. Biomed. Mater. Res. Vol. 66A (2003), p. 110
[5]   P. Sepulveda, J. R. Jones and L. L. Hench: J. Biomed. Mater. Res. Vol. 59 (2002), p. 340
[6]   T. Kokubo, H. Kushitani and S. Sakka: J. Biomed. Mater. Res. Vol. 24 (1990), p. 721
[7]   P. Saravanapavan and L. L. Hench: J. Non-Cryst. Solids Vol. 318 (2003), p. 14

Key Engineering Materials Vols. 254-256(2004) pp. 993-996
online at http://www.scientific.net
© 2004 Trans Tech Publications, Switzerland

In-vitro and *in-vivo* Validation of a New Bonesubstitute for Loaded Orthopaedic Applications

G. M. Insley[1] and R. M. Streicher[2]

[1] Stryker Howmedica Osteonics, Raheen Business Park, Limerick, Ireland.
gerard.insley@emea.strykercorp.com
[2] Stryker SA, Florastrasse 13, CH-8800 Thalwil, Switzerland.
robert.streicher@strykereurope.com

Keywords: calcium phosphate, hydroxyapatite, tricalcium phosphate, impaction grafting.

Abstract. A highly sintered HA/TCP granule, with 50% surface porosity in two size ranges, has been proven suitable for impaction grafting in femoral and acetabular revision surgery through an extensive series of novel test methods specific to the intended clinical application.

Introduction

Impaction grafting as a surgical technique for revision hip arthroplasty has been reported extensively in the literature [1, 2]. The procedure involves the impaction of fresh frozen allograft chips into the femoral or acetabular cavity after the implant has been removed cleaned and prepared. The resulting impacted bone bed allows normal cementation and replacement with a standard stem and cup designed for this application. One of the main issues with this technique involves the allograft used. The technique typically requires two to three allograft femoral heads that are subsequently passed through a bone mill to produce bone chips. Allograft supply is due to outstrip demand [3] in Europe. Also the allograft quality is inconsistent which can affect the clinical outcomes and there exist also immunological concerns. The objective of this study was to validate a synthetic allograft extender for impaction grafting application. A new, fully sintered calcium phosphate based granule was evaluated for clinical applications in areas where loading is important i.e. impaction grafting for femoral and acetabular revision surgery.

Materials and Methods

HA/TCP (25%: 75%) granules in two size ranges, 2-4 mm and 4-6 mm were characterized in both novel laboratory mechanical test models and relevant animal models in order to asses their suitability as bone extenders for impaction grafting clinical applications.

Mechanical testing for femoral impaction grafting involved fatigue testing 50:50 mixtures of allograft bone and synthetic granules impacted under standardized conditions in a tube /cone set up using a relevant hip stem to apply the loading. The stems were loaded for up to 5,000 cycles, peak loads up to 1.2 kN and subsidence of greater than 5 mm was considered a failure. The effect of granule size, granule sintering temperature and porosity level on the amount of hip stem subsidence was measured [4].

Mechanical testing for acetabular impaction grafting involved testing the initial stability of mixes of allograft and synthetic granules using an acetabular model in sawbones with both cavitary and combined defects [5]. Primary cemented cups and reconstructions using 100% human cancellous chips were used as controls. The cemented acetabular cup model was loaded (3 kN) on a servo-hydraulic machine at 45° inclination; RSA (radiostereophotogrammetry) marker beads were used to measure the subsequent subsidence level of the cup.

In-vivo testing involved using a fully loaded hip hemi-arthroplasty sheep model. Four randomised groups of eight skeletally mature sheep in each case were implanted with a sheep hip stem (Exeter™) with impaction grafted mixtures as follows: 100% pure allograft, 50% allograft/50% synthetic bone substitute (HA/TCP = 25%:75%), 50% allograft/50% synthetic bone substitute (HA/TCP = 75%:25%) and 10% allograft/90% synthetic bone substitute. The prosthesis and graft were fully loaded in the animal model for 24 months with analysis every six months including force plate measurement, XRD and DEXA analysis. Limited histology was performed on representative sheep from each group [6].

Results and Discussion

Mechanical testing results from the *in-vitro* femoral stem model showed that the impaction graft is more stable, i.e. less stem subsidence occurs when mixtures of allograft and synthetic granules are used, when compared to using allograft alone, see figure 1. The results demonstrate that the best combination of granule properties for impaction grafting in the femur are highly sintered, 2-4 mm in size with 50% surface porosity and a composition of 25%:75% HA: TCP. The consistency of the subsidence results when using the synthetic granules is remarkable when compared to the higher variability seen in the allograft groups, see figure 1.

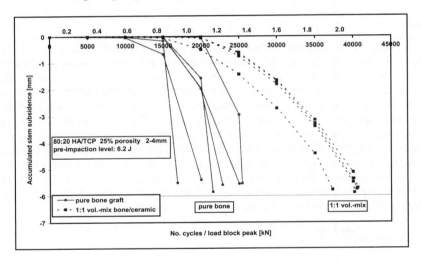

Fig. 1 Mechanical test results, femoral model

Mechanical testing using the acetabular *in-vitro* model demonstrated similar results with the allograft/granule mixtures being more stable than using allograft alone. Excessive cement penetration into the graft was observed when 100% synthetic granules were used. This may lead to problems with bone remodelling and subsequent stability of the cup. The cups with 100% allograft produced the most subsidence.

The loaded sheep hip model showed at 24 months post operatively that there were no significant differences in radiographic, DEXA or force plate analysis (see figure 2) between the four treatment groups. This would imply, using functional outcomes, that mixtures of synthetic granules and allografts are as effective as allograft alone in impaction grafting of the femur. The limited histological analysis shows that the synthetic granules are remaining with beneficial bone remodelling around the graft materials. New bone was seen growing up to all the synthetic granules with no gaps or fibrous tissue evident, see figure 3.

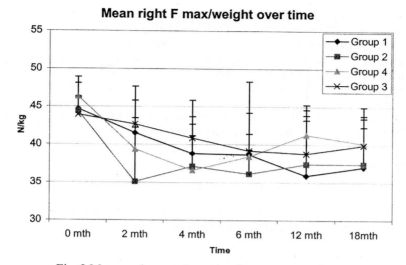

Fig. 2 Mean peak ground reaction forces F max over time

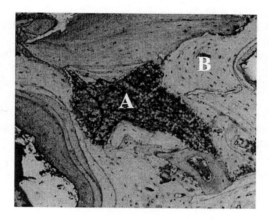

Fig. 3 Photograph of histology section of synthetic TCP:HA granule (A) surrounded by new bone
(B) with no evidence of gaps or fibrous tissue

Conclusions

The studies reported here confirm the safety and effectiveness of using a synthetic TCP:HA granule
(BoneSave™) for the specific clinical application of impaction grafting.

The *in-vitro* laboratory testing using validated test models demonstrated both increased
stability of the graft and less user variance on the subsequent subsidence measured. The highly
sintered, highly crystalline HA:TCP - 80%:20% granules with 50% surface porosity were found to
be the most suitable for this application. The 2-4 mm granules gave the highest stability for the
femoral impaction grafting and the 4-6 mm granules are more suitable for filling the larger defects
encountered in acetabular impaction grafting.

The sheep study confirmed the *in-vitro* laboratory model results with immediate stability of
the graft post operatively assured in all animals. The force plate analysis showed that all sheep in
the synthetic mix groups were loading the operated limb as much as the unoperated limb. Long-

term stability and biological stability was assured with the grafts at 24 months implantation showing no significant differences in terms of radiographic and DEXA outcomes. The histological analysis demonstrated excellent bone ingrowth into the synthetic grafts.

Extensive *in-vitro* and *in-vivo* testing is required to validate synthetic bone substitutes for orthopaedic applications. Furthermore these tests need to be specific to the intended application before all the requirements for both short term and longer term performance in the clinical application can be assured.

A highly sintered HA/TCP granule with 50% surface porosity in two size ranges has been proven suitable for impaction grafting femoral and acetabular revision surgery.

Acknowledgements

The authors wish to acknowledge the following technical centres for their contribution to this work:
- University of Bath, UK.
- Orthopaedic Research Laboratory, University Medical Centre Nijmegen, Netherlands.
- The Royal Veterinary College and Orthopaedic Institute, UCL, UK.

References

[1]   G. A. Gie, L. Linder, R. S. M. Ling, J. P. Simon, T. J. Slooff and A. J. Timperley: J Bone and Joint Surg Vol. 73 (1993), p.14-21.

[2]   B. W. Schreurs, T. J. Slooff, P. Buma, J. W. Gardeniers and R. Huiskes: J Bone and Joint Surg Vol. 80 (1998), p.391-395.

[3]   G. Galea, D. Kopman and B. J. M. Graham: Supply and demand of bone allograft for revision hip surgery in Scotland: J Bone and Joint Surg Vol. 80 (1998), p.595-599.

[4]   B. Grimm, A. W. Miles and I. G. Turner: J. Mater Sci.: Mater. Med. Vol. 12 (2001), p.929-934.

[5]   S. B. T. Bolder, N. Verdonschot, B. W. Shruers and P. Buma: Biomat. Vol. 23 (2002), p.659-666.

[6]   Blom, A. W., Cunningham, J. L., Lawes, T. L., Hughes, G., Learmonth, I. D., Goodship, A. E.: 4th combined EORS, 1-2 June 2001, Rhodes, Greece.

Key Engineering Materials Vols. 254-256(2004) pp. 997-1000
online at http://www.scientific.net

Porous Hydroxyapatite and Glass Reinforced Hydroxyapatite for Controlled Release of Sodium Ampicillin

A.C. Queiroz[1,2], S. Teixeira[1,3], J.D. Santos[1,3], F.J. Monteiro[1,3]

[1] INEB - Instituto de Engenharia Biomédica, Laboratório de Biomateriais, Rua do Campo Alegre, 823, 4150-180 Porto, Portugal; email: aqueiroz@ineb.up.pt
[2] Escola Superior de Tecnologia e Gestão, 4901 Viana do Castelo Codex, Portugal
[3] Faculdade de Engenharia da Universidade do Porto, Departamento de Engenharia Metalúrgica e Materiais, Rua Dr. Roberto Frias, 4200-465 Porto, Portugal

Keywords: Hydroxyapatite, glass reinforced hydroxyapatite, porous, drug delivery

Abstract. Porous hydroxyapatite (HA) and glass reinforced hydroxyapatite (GRHA) with adequate macro and microporous structure were developed, aiming at being used as drug delivery carrier of antibiotics, for the *in situ* treatment of periodontitis. Materials were characterised by XRD and FTIR, presenting no changes from similar dense materials. Mercury intrusion porosimetry revealed micropores of less than 1 μm, accounting for 15% of the total porosity. Compression tests have shown close values for HA and GRHA, with the former showing slightly higher values of strength. Ampicillin adsorption was more effective on porous than on dense HA, and was similar for HA and GRHA.

Introduction

Periodontitis is an oral disease that promotes, in its most severe form, maxillar alveolar bone loss [1, 2]. Current systemic approach leads to the use of long-lasting treatments with very high dosages of antibiotics and anti-inflammatory drugs. This may induce drug resistance of oral and medial pathogens to common antibiotics [3] and particularly when they are present locally in less than the minimum inhibitory concentration (MIC). A local delivery system should provide the locally required amount of antibiotic, and this should naturally correspond to a significant decrease, when compared to a systemic treatment. Also the period of treatment should decrease [3]. To maximise local antibiotic delivery, for a system where the drug is adsorbed, the surface area of the drug releasing agent must be maximised. For example, polymer porous scaffolds have been widely used for soft tissue replacement and drug delivery.

By proposing the use of a porous bioactive ceramic scaffold, the aim is not only to locally deliver a specific drug, in this case sodium ampicillin, a wide spectrum antibiotic, but also, after delivery, to continue acting by inducing cell attachment, osseointegration and vascularisation, promoting rapid healing of the bone tissue after an efficient treatment of the periodontitis infection.

In this work both dense and highly porous structures of hydroxyapatite (HA) and glass-reinforced hydroxyapatite (GRHA) were obtained. Hydroxyapatite (HA) is a well-known bioceramic extensively used in medical applications. Glass-reinforced hydroxyapatite (GRHA) [4] has been found to present higher mechanical strength than HA, both as coatings [5] and dense materials, and also to be able to degrade faster in contact with living tissues, due to the presence of controlled amounts of β-TCP, eventually leading to a faster response from the host bone.

Materials and Methods

Dense materials were obtained in disc shapes by uniaxially pressing at 180 MPa, followed by sintering at 1200 °C, and then milled and sieved until granules of 250-850 μm were obtained. Porous HA and GRHA samples with approximately 5x5x5 mm were obtained by immersing polyurethane (PU) foams in the slurries respectively containing HA and GRHA, at 50 °C. The slurries were obtained by mixing, commercially pure HA (P-120 Plasma Biotal) or GRHA powders, water and a tensoactive agent (LM3 commercial detergent) at pre-established proportions (HA (g): water (ml):tensoactive agent (ml)). After immersion, the sponges were squeezed, for excess removal, and dried overnight. PU was removed by burn-out method as previously described [6]. The best results were obtained with the 6:6:0.2 ratio for both materials.

Materials were characterised by SEM, FTIR with split pea accessory and XRD. Microporosity was characterised by mercury intrusion porosimetry, and compressive strength tests were performed in a Lloyd LR 30 K tensile testing equipment, with a load cell of 5 KN.

Samples were then tested for drug adsorption and releasing capability. Both dense granules (D) and porous (P66) samples were put in contact with a 10 mg/mL sodium ampicillin solution for 24 hours, at 37°C, with continuous agitation of 250 rpm. The amount of sodium ampicillin adsorbed was measured from the supernatant, using UV spectroscopy at 230 nm. Adsorption in similar sets of porous samples (VP66, VP64) were carried out in vacuum for 15 min. at 37 °C, and following a sequence similar to the previously described one, but being agitated at 110 rpm.

Results and Discussion

SEM observations of the ceramic sponges revealed an interconnective macroporous structure where very well organised macro and micro porosity was observed (Fig.1). These features were found for both HA and GRHA, with this last presenting some less organised aspects, namely more irregular distribution of micropores, as result of the glass liquid sintering and reaction with HA, occuring during sintering cycle. This kind of structure is considered to be adequate, as the macroporosity induces osteoconductivity and allows for cells to enter the structure, and nutrients to reach all the cells by vascularisation, while microporosity allows improved cell adhesion [7].

The results of mercury intrusion porosimetry have confirmed that besides macroporosity, there is microporosity, as detected by SEM, with the majority of micropores occuring at diameters less than 1 μm. Micropores are responsible for approximately 15 % of the total porosity. Previous work had established macroporosity as being around 73% and corresponding to pore size between 150-400 μm [8]. Pores distribution for HA and GRHA were similar.

Both FTIR and XRD analyses for porous HA and GRHA presented similar features to those previously found [8] with the same dense materials, indicating that the presence of the tensoactive agent and PU sponge did not affect the phase composition.

Compression tests were carried out on porous HA (with two different (HA:W:TA) ratios) and GRHA (Fig.2). For similar contents of ceramic powders, HA was more resistant than GRHA, in particular withstanding higher stress for the highest values of strain achieved in both cases.

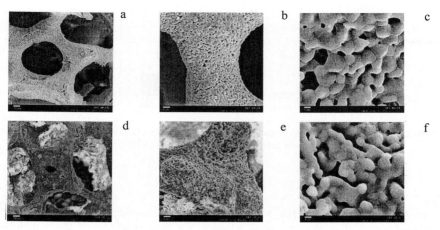

Fig 1. Porous 6:4:0.2 (HA:W:TA) structures of HA a) b) and c) and GRHA d) e) and f). (SEM images).

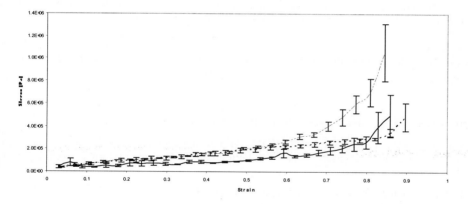

Fig 2- Compression test for HA 6:6:0.2(- — -), HA 6:4:0.2(- - - - -), and GRHA (——)

The adsorption studies, including for the porous samples an initial stage of adsorption in vacuum, have shown that both granular and porous material could adsorb sodium ampicillin, but adsorption was much more effective, in the case of HA sponge, due to the significantly increased of surface area put in contact with the ampicillin containing solution (Fig. 3). Adsorption onto GRHA followed the same trend as in the case of HA samples.

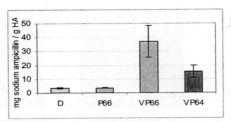

Fig. 3. Sodium ampicillin adsorption onto dense HA granules (D); porous 6:6:0.2 (P66) and vacuum adsorption onto porous 6:6:0.2 (VP66) and 6:4:0.2 (VP64).

Conclusions

Porous structures of both HA and GRHA, being composed of interconnected macropores, associated to a well distributed range of micropores seem adequate for drug releasing of this antibiotic. Their morphologies seem to adapt well to be used as scaffolds for bone tissue to invade and proliferate, if adequate conditions are created.

Porous HA and GRHA showed to be able to adsorb larger amounts of antibiotic than granules, becoming a better candidate material to be used as a scaffold for drug delivery in the case of periodontitis. HA and GRHA did not present very relevant differences.

Acknowledgements

The authors would like to acknowledge FCT Project "TEXMED", ref. POCTI/FCB/41402/2001 for financial support.

References

[1] Krejci CB, Bissada NF, Farah C, Greenwell H.:J Periodontol 1987; 58: 521-528.

[2] RA Seymour, Heasman PA, MacGregor IDM. The pathogenesis of periodontal disease. In: MacGregor IDM, editor. Drugs, Diseases, and the Periodontium. Oxford: Oxford University Press, 1992. p. 1-10.

[3] Steinberg D, Friedman M, Soskolne A, Sela MN: J Periodontol 1990; 61: 393-398.

[4] Santos JD, Silva PL, Knowles JC, Talal S, Monteiro FJ: J Mater Sci: Mater Med 1996; 7: 187-189.

[5] Ferraz MP, Monteiro FJ, Santos JD: J Biomed Mat Res 1999; 45: 376-383.

[6] Queiroz A.C.,Teixeira S.,Santos J.D.,Monteiro F.J.: Key Engineering Materials, submitted

[7] Nishihara K: Clin Materials 1993; 12: 159-167.

[8] Queiroz AC, Santos JD, Monteiro FJ, Gibson IR, Knowles JC: Biomaterials 2001; 22: 1393-1400.

Key Engineering Materials Vols. 254-256(2004) pp. 1001-1004
online at http://www.scientific.net

Fabrication of Low Temperature Hydroxyapatite Foams

A. Almirall[1], G. Larrecq[1], J.A. Delgado[1], S. Martínez[2], M.P Ginebra[1], J.A. Planell[1]

[1] Research Centre in Biomedical Engineering (CREB). Biomaterials Division. Department of Materials Science and Metallurgy. Technical University of Catalonia (UPC), Av. Diagonal 647, E08028 Barcelona, Spain. e-mail: amisel.almirall@upc.es; maria.pau.ginebra@upc.es

[2] Dept. Crystallography, University of Barcelona, Barcelona, Martí i Franquès s/n, 08028-Barcelona, Spain.

Keywords: calcium phosphate cement, porosity, scaffolds, bone tissue engineering

Abstract. Bone tissue engineering requires appropriate scaffold materials with a high-interconnected macroporosity. In this work a novel two-step method, based in the foaming of a calcium phosphate cement paste by the addition of hydrogen peroxide, and its subsequent hydrolysis to a calcium deficient hydroxyapatite (CDHA) is presented. The size of the interconnected macropores ranged between 50 μm and 2 mm, reaching the total porosity a maximum value of 66 %. The foaming capacity of the H_2O_2 solution was strongly influenced by the particle size of the α-tricalcium phosphate (α-TCP) powder. The size of the macropores increased with increasing L/P ratio. As expected, the compressive strength of the apatitic foams decreased with increasing porosity, ranging between 2 and 9 MPa.

Introduction

One of the most promising approaches to the problem of bone regeneration and repair is bone tissue engineering (BTE). To guide in vitro or in vivo tissue regeneration, it is necessary to obtain appropriate scaffold materials with a high-interconnected macroporosity. For this application, several porous ceramic manufacturing techniques have been proposed [1-3]. However, the majority of these methods are based on the production of high temperature apatites, which are known to be hardly resorbable in normal physiological conditions. In this work an alternative route is proposed to develop macroporous calcium phosphate scaffolds at low temperature, from calcium phosphate cements. Consolidation is obtained not by sintering, but through a low temperature setting reaction. This approach has several advantages: the cementing reaction takes place at low temperature and therefore the final product is a low temperature precipitated hydroxyapatite, chemically more similar to the biological apatites, and with a much higher specific surface than that of a sintered hydroxyapatite. All these factors contribute to bring this material a higher reactivity, as compared to a ceramic hydroxyapatite. In addition, the low temperature processing allows the introduction of drugs, proteins or signaling molecules into the material.

Materials and Methods

To prepare the foams, an α-TCP powder obtained by solid state reaction at 1400°C, which contained a 2 wt% PHA as a seed material [4], was mixed with a 10 vol% aqueous solutions of hydrogen peroxide (H_2O_2) as foaming agent. The parameters studied were: (i) The liquid to powder ratio of the mixture (L/P = 0.32 and 0.38 ml/g) and (ii) the particle size of the α-TCP powder: two powders were studied, a coarse powder (average size 5.6 μm) and a fine powder (average size 2.2 μm). In order to evaluate the efficiency of foaming process, materials with no hydrogen peroxide on the liquid phase were also prepared. The experimental design is shown in Table 1.

Table 1. Experimental design.

SERIES	Vol % H$_2$O$_2$ in the liquid phase	Liquid/Powder (ml/g)	Average particle size of the α-TCP powder (μm)
Cement 1	0	0.32	5.56
Foam 1	10	0.32	5.56
Cement 2	0	0.38	5.56
Foam 2	10	0.38	5.56
Cement 3	0	0.32	2.21
Foam 3	10	0.32	2.21

After mixing, the paste was poured in Teflon moulds, and kept at 60°C for 2h. The decomposition of the hydrogen peroxide at this temperature produced the foaming of the paste. Afterwards, the specimens were immersed in Ringer's solution (0.9 wt% NaCl) at 37°C for 15 days.

The micro, macro and total porosity of the samples were evaluated by an indirect method, proposed by Takagi and Chow [5], and based in the measurement of the apparent density of samples. This parameter was measured by immersion in Hg, applying the Archimedes principle.

The morphology of the specimens was observed by SEM, and the pore size distribution was evaluated by image analysis using the Image Analysis software Omnimet 1.5, on micrographs obtained by Light microscopy at a magnification of x40.

The foamed and unfoamed specimens were analyzed in terms of the phases present by X-ray diffraction (XRD) and infrared spectroscopy (IR). The compressive strength (CS) was evaluated in cylindrical specimens (φ = 6 mm, height = 12 mm), in a Universal Testing Machine MTS 858 Bionix, at a crosshead speed of 1 mm.min^{-1}

Results and Discussion

In all the series studied, the addition of H$_2$O$_2$ in the liquid phase resulted in the foaming of the paste when it was stored at 60°C. Porous dry bodies were obtained after 2 hours. The foamed structure was maintained after immersing the specimens in Ringer's solution. The structural integrity of the samples was not destroyed by the diffusion of the liquid through the foamed α-TCP powder. SEM observations prior to immersion in Ringer's solution indicated that no CDHA precipitation was produced during the foaming step. It was only after immersion in Ringer's solution that the hydrolysis of α-TCP started. Porous specimens obtained after 15 days of immersion on Ringer's solution are shown in Fig. 1.

The formation of a homogeneous foamed structure, with interconnected macroporosity was observed in Foams 1 and 2. In contrast, Foam 3 showed a much lower porosity. The use of a fine powder reduced considerably both the porosity and the pore size, indicating that, within the ranges studied, the particle size of the starting powder was the factor that had a most significant effect on the foaming ability of the H$_2$O$_2$ solution. The characteristic needle-like entangled CDHA microcrystals were observed by SEM, Fig. 1 d), with a microporous structure superimposed to the macroporous foam. Indeed, the apatitic phase obtained by this cementitious reaction was more similar to the biologic one, in terms of chemical composition, crystallinity and specific surface. All these features can contribute to greatly increase the reactivity of the material.

According to Image Analysis measurements, in the most porous material (Foam 2), the main part of the pore volume was occupied by pores with a diameter larger than 600 μm. In contrast, in Foam 1 the pore size was smaller, being the greater part of the pore volume occupied by pores in the range of 100-450 μm.

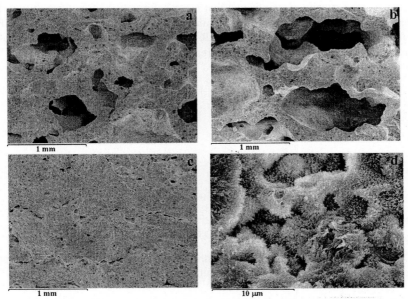

Fig. 1. Scanning Electron Microscopy of a) Foam 1, b) Foam 2, c) Foam 3 and d) Foam 1 at higher magnifications.

The micro, macro and total porosity, as well as the compressive strength of the samples are reported in Table 2. The macroporosity reached a maximum value of 35.3% for Foam 2, prepared with the coarse powder and L/P = 0.38 ml/g. Again, the particle size of the starting powder had the most significant effect on the created macroporosity. By comparing Foams 1 and 2, it was observed that an increase of L/P ratio increased the quantity and size of the pores. A high intrinsic microporosity was observed in the unfoamed specimens.

Table 2. Micro, macro and total porosity, and compressive strength of the different series studied. (Standard deviation between brackets)

SERIES	Microporosity (%)	Macroporosity (%)	Total porosity (%)	Compressive Strength (MPa)
Cement 1	46.3 (0.6)	-	46.3 (0.6)	29.54 (5.16)[a]
Foam 1	33.9 (1.3)	26.8 (2.8)	60.7 (1.5)	2.69 (0.91)[b]
Cement 2	48.1 (0.3)	-	48.1 (0.3)	28.33 (3.24)[a]
Foam 2	31.1 (1.6)	35.3 (3.3)	66.4 (1.7)	2.18 (0.54)[b]
Cement 3	45.1 (0.5)	-	45.1 (0.5)	33.36 (1.01)[a]
Foam 3	40.1 (0.6)	11.0 (0.8)	51.1 (0.7)	8.81 (1.16)

[a,b]The differences are not statistically significant (p>0.05)

Therefore, it is possible to say that, in the CDHA low temperature foams, the total porosity is generated by two different and well defined mechanisms: on one side, the macroporosity is introduced by the foaming agent, and on the other side the setting of the CPC, creates some intrinsic microporosity, like the one present in the unfoamed materials.

As expected, the introduction of macroporosity in the foams resulted in a strong decrease of the mechanical properties (Table 2), the highest values corresponding to the cement series, followed by the less macroporous construct (Foam 3).

XRD of the foamed specimens confirmed that no transformation took place during the foaming step, and that the foamed structure transformed into a CDHA by hydrolysis of α-TCP (Fig. 2) after immersion in Ringer's solution. Both XRD and IR spectra indicated that the foaming with H_2O_2 did not have any significant effect on the hydrolysis reaction of the α-TCP.

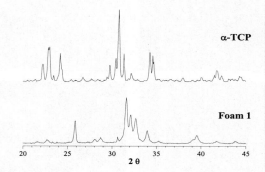

Fig. 2. XRD patterns of the α-TCP before hydrolysis and Foam 1 after 15 days of immersion in Ringer's solution at 37°C.

Conclusions

Macroporous calcium deficient hydroxyapatite scaffolds can be obtained at low temperature by foaming and hydrolysis of an α-TCP cement, using H_2O_2 solution as the liquid phase. Macro and microporosity can be controlled by different parameters such as the particle size distribution of the powder, and the liquid to powder ratio.

Acknowledgements

The authors thank the Science and Technology Spanish Ministry for funding this work through project CICYT MAT2002-04297.

References

[1] R. White, J. Weber, E. White: Science Vol. 176 (1972), p.922

[2] J. Tian, J. Tian: J Mater Sci Vol. 36 (2001), p.3061.

[3] S.H. Li, J.R. De Wijn, P. Layrolle, K. De Groot: J. Biomed. Mater. Res. Vol. 61 (2002), p.109

[4] M.P. Ginebra, E. Fernandez, E.A.P. De Maeyer, R.M.H. Verbeeck, M.G. Boltong, J. Ginebra, F.C.M. Driessens, J.A. Planell: J. Dent. Res. Vol. 76 (1997), p.905.

[5] S. Takagi, L.C. Chow: J Mater Sci: Mater Med Vol. 12 (2001), p.135

Key Engineering Materials Vols. 254-256(2004) pp. 1005-1008
online at http://www.scientific.net

Osteoinductive Properties of
Micro Macroporous Biphasic Calcium Phosphate Bioceramics

Guy Daculsi, Pierre Layrolle

EM INSERM 998 03 Research Center on Materials of Biological Interest

Dental Faculty, Nantes University, place Alexis Ricordeau, 44042 Nantes France,
gdaculsi@sante.univ-nantes.fr

Keywords: Bioceramics, Biphasic calcium phosphate, micropores, macropores, bone ingrowth, osteoinduction, osteoconduction

Abstract. We have developed 17 years ago, with the collaboration of Lynch, Nery, and Legeros in USA, a bioactive concept based on biphasic calcium phosphate ceramics (BCP). The concept is determined by an optimum balance of the more stable phase of HA and more soluble TCP. The material is soluble and gradually dissolves in the body, seeding new bone formation as it releases calcium and phosphate ions into the biological medium.

The bioactive concept based on the dissolution/transformation processes of HA and TCP with a specific microstructure (micropore) and macrostructure (mesopores and macropores) represents a dynamic process, including physico-chemical processes, crystal/proteins interactions, cells and tissue colonization, bone remodelling, finally contributing to ingrowth at the expense of the MBCP. The microstructure of such material is achieved with low temperature sintering conditions preserving the microstructure. The material has shown osteoconductive properties largely reported in the literature, but its osteionductive properties have not been explored and documented for such bioceramics (Triosite® Zimmer). This paper presents retrospectives series of animal data (rats, rabbits, dogs, cats) demonstrating that promotion of the mineralization into micropores occurred simultaneously with osteoid and bone ingrowth into mesopores and macropores in non bony sites.

Introduction

Synthetic calcium phosphate bioceramics are attracting more interest in bone reconstruction due to their unlimited availability, excellent biocompatibilty, osteo-conductivity and even more recently reported osteo-inductivity. Macroporous biphasic calcium phosphate MBCP, a mixture of hydroxyapatite (HA) and beta tricalcium phosphate (β-TCP) was developed 20 years ago by Lynch, Nery, Legeros and Daculsi [1]. MBCP bioceramics have been largely used as bone substitutes and represent the largest bibliography on this concept of bioceramics. However, data on its potential osteoinductive properties have not been reported to date. The mechanism of osteoinduction by biomaterials is not understood and the amount of induced bone might be too limited for the reconstruction of large skeletal defects. It has been shown recently that some calcium phosphate bioceramics and coatings on metal implants can induce ectopic bone formation after implantation in muscles of different animals.[2-10] These biomaterials have demonstrated the ability to induce bone intramuscularly within their porosity in 6-12 weeks without the addition of osteogenic cells or growth factors prior to implantation. Implanted in critical size osseous defects, these osteoinductive biomaterials have demonstrated superior bone healing capacities than conventional bioceramics and coatings. Previous investigations from our group and others laboratories have shown that the implants should exhibit two features to induce ectopic bone: (i) a calcium phosphate surface with micropores and (ii) a macroporous structure. This paper reports retrospectives data on biological crystals and bone formation in non bony sites in different animal models from small to large one using various methods of investigation and analysis.

Materials and Methods

The material used was a mixture of hydroxyapatite (HA) and beta tricalcium phosphate (β-TCP) with a weight ratio of 60/40 % HA/β-TCP, 30 % of microporosity of less than 1 micron and a specific surface area of the crystal of 4 ± 0.02 m^2/g. The macroporosity was 50 % with a range of macropores of 150 to 700 with a mean of 400 microns (Triosite™ Zimmer, MBCP™ Biomatlante France). Light microscopy, X rays microradiographies, scanning electron microscopy SEM, electron microprobe EDS, transmission electron microscopy TEM and high resolution Hr TEM, Fourier transformed Infrared spectroscopy FTIR, Selected area electron diffraction SAED have been applied to samples collected after implantation in non bony sites of small animals (36 Rats and 72 rabbits) and large ones (24 dogs, 2 cats). The materials were calibrated cylinders or granules implanted in subcutaneous areas or in paravertebral muscles. Positive controls for osteoinduction have been performed in rats and dogs using the same samples associated with autologous bone marrow. This type of artificial bone has been used from a long time in preclinical and in clinical studies.

Results

The biodegradation of MBCP included the dissolution of the individual HA or ß-TCP crystals. The proportion of HA to ß-TCP crystals in MBCP appeared greater after implantation than the initial content of 60/40 % and resulted from the higher reactivity or solubility of ß-TCP compared to HA. Formation of microcrystals with Ca/P ratios similar to those of bone apatite crystals was observed after implantation.

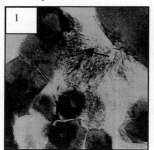

The abundance of these biological microcrystals was directly related to the initial ß-TCP/HA ratio in the BCP: higher was the ß-TCP content, greater was the abundance of biological apatite precipitation (fig.1). This dissolution precipitation process is observed both *in vitro* an *in vivo*, after subcutaneous fibrous tissue or intra muscular implantations and obviously in bony areas. The MBCP materials induced similar responses from cells coming at the implantation site in soft tissue as well as in bony sites.

Fig.1 : Biological apatite precipitation (arrow) into micropores and at the calcium phosphate crystal surface

These materials allow cell attachment, proliferation and expression. The first biological events after BCP ceramics implantation are biological fluid diffusion, followed by cells colonization. In early steps, these cells are macrophages, followed by mesenchymal stem cells, fibroblasts, and angiogenic cells.

In animal experiment with rabbits, we have not observed osteoid or bone ingrowth after either subcutaneous or intramuscular implantations (fig.2). We have not found evidence of ectopic bone formation induced by MBCP bioceramics in rats and rabbits. In these small animals, *de novo* bone formation was only observed when bone marrow was added to the MBCP.

In contrary, ectopic bone formations have been observed in large animals like cats or dogs. Ectopic bone with osteoblasts and osteoids has been found in some macropores always in the deeper macropores and not preferentially in the outer surface after 3-7 months of implantation in muscles (fig.3). The ectopic bone formation is still limited to a small amount of macropores (range 1 to 6%, $4.3\% \pm 2$). EDS analysis of the ectopic bone formed in the inner macropores indicated Ca and P contents and SAED patterns corroborated the formation of biological apatite crystals similar to those found in osseous sites. With time of implantation, the thin bone trabecula covering the macropores in dogs (fig.4) appeared more thicker, occupying largely the core of the macropores (fig.5).

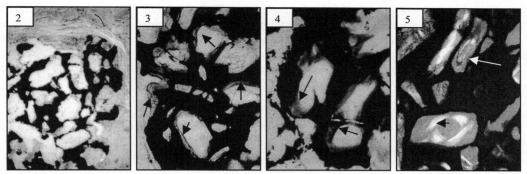

Fig.2: Soft tissue colonization without osteoid formation into macropores after rabbit sub-cutaneous implantation during 1 month; light microscopy using Masson staining.

Fig.3: Bone ingrowth (arrow) into macropores after 3 months of intramuscular implantation in cat. Light microscopy using Movat's staining

Fig.4: Bone ingrowth (arrow) into macropores after 2 months of intramuscular implantation in dog. Light microscopy using Movat's staining

Fig.5: Bone ingrowth (arrow) into macropores after 7 months of intramuscular implantation in dog. Light polarized microscopy.

Discussion

The mechanism of bone induction by biomaterials is not elucidated to date. Several hypotheses have been proposed to explain this intriguing biological property. (i) Klaas de Groot [11] has proposed that some calcium phosphate ceramics can concentrate bone growth factors from body fluids which will trigger stem cells to form bone tissue. (ii) Others have thought that a low oxygen tension in the central region of implants might provoke a dedifferentiation of pericytes from blood microvessels into osteoblasts, (iii) it has been also postulated that the nanostructured rough surface or the surface charge of implants might cause the asymmetrical division of stem cells into osteoblasts ; (iv) finally, the mesenchymal cells might be triggered by the local high level of calcium ions and recognize the bone-like apatite layer formed in vivo by dissolution-reprecipitation to differentiate into osteoblasts producing bone. However, all of these hypotheses are based on the triggering of stem cells and osteogenic differenciation due to local CaP environments resulting from micro and macropore structures of materials. Thus, the term 'Osteoinduction' is questionable to define these materials but undoubtedly the induction of mineralization has been perfectly demonstrated.

By Using HR TEM [12] we have shown for the first time that the formation of these microcrystals after implantation was non-specific, i.e., not related to implantation sites, types of implantation, and nature of CaP ceramics. We think that the formation of biological apatite microcrystals *in vivo* can be the main factor for explaining the "osteoinductive properties" of these materials. Certain bioceramics have the ability to promote mineralization on their surface or inside inner spaces and that should be the first event in the formation of bone tissue. The coalescing interfacial zone of biological apatite micro crystals provides a scaffold for bone-cell adhesion and further bone formation. The resorbing process involves dissolution of calcium phosphate crystals and then a precipitation of carbonated apatite needle-like crystallites into the micropores near by the dissolving crystals. The coalescing zone constitutes the new biomaterial/bone interface, suitable for undifferentiated osteogenic cells spreading. In addition, the macropores were able to develop localised three dimensionnal cells organisation in cooperation with angiogenesis [1].

If micro / macroporous bioceramics can induce ectopic bone formation within their macropores in few weeks after implantation in non bony sites, this intriguing biological property is not the main clinical outcome for bone substitutes usually used in bony environments. On the opposite, bone tissue engineering might require highly bioactive materials to be used as scaffolds for seeding, culturing, differentiating and transplanting bone marrow cells and thus providing an autologous bone graft for reconstructive surgery.

Conclusion

The control of HA/ ß -TCP ratio as well as micropores and high crystalline surface area with specific low sintering temperature process, promotes a high release of Ca and P ions into the body fluid and thus, induces the precipitation of biological apatite microcrystals within macropores of MBCP bioceramics. This early mineralization and biological apatite microcrystals precipitation seems to be a key factor for the induction of bone in ectopic sites. However, this process remains limited to a low amount of macropores within the implants. Ectopic bone formation induced by biomaterials seems to be related to the recruitment and differentiation of STEM cells and not to be the result of a process of transduction which include de-differenciation and newly phenotype for the cells.

Acknowledgments

The individual and collaborative studies were supported by research grants from the INSERM U225, CJF 93-05 and E 99-03 and CNRS EP 59 [Dr. G. Daculsi, Director] and from NIH-NIDR of Health Nos. DE04123 and DE07223 and special Calcium Phosphate Research Funds [Dr. R.Z. LeGeros, Principal Investigator]. We thank Zimmer France and Biomatlante France to provide us in Calcium phosphate bioceramics and to support some of the animal experiments.

References

[1] Daculsi G., Laboux O., Malard O., Weiss P. J. Mater. Sci. Mater Med (2003), 14: 195-200

[2] Yamasaki H, Sakai H. Biomaterials (1992) 13(5) 308-12

[3] Ripamonti U. Biomaterials (1996) 17(1), 31-5

[4] Yuan H, Kurashina K, de Bruijn JD, Li Y, de Groot K, Zhang X. Biomaterials (1999) 20(19) 1799-806

[5] Yuan H, Li, Y, de Bruijn JD, de Groot K, Zhang X. Biomaterials (2000) 21(12) 1283-90

[6] Yuan H, Yang Z, de Bruijn JD, de Groot K, Zhang X. (2001) 22(19) 2617-23

[7] Yuan H, de Bruijn JD, Zhang X, van Blitterswijk CA, de Groot K. J Biomed Mater Res (2001) 58(3) 270-6

[8] de Bruijn JD, Yuan H, Dekker R, Layrolle P, de Groot K, van Blitterswijk CA.. In: "Bone Engineering" edited by J.E. Davies, Part V, Chapter 38, p. 421-431 (2000).

[9] Barrère F, van der Valk CM, Dalmeijer RAJ, Meijer G, van Blitterswijk CA, de Groot K, Layrolle P. J Biomed Mater Res (2003) 66A(4) 779-88

[10] Gosain AK, Song L, Riordan P, Amarante MT, Nagy PG, Wilson CR, Toth JM, Ricci JL. Plast Reconstr Surg (2002) 109(2) 619-30

[11] de Groot K. Carriers that concentrate native bone morphogenetic proteins in vivo. *Tiss Eng* (1998) 4(4) 337-41

[12] Daculsi G., LeGeros R.Z., Heugheubaert M., Barbieux. Calcif Tissue Int (1990) 46: 20-27.

Key Engineering Materials Vols. 254-256(2004) pp. 1009-1012
online at http://www.scientific.net
© 2004 Trans Tech Publications, Switzerland

Multi-scale Structure and Growth of Nacre: a New Model for Bioceramics

Marthe Rousseau[1], Evelyne Lopez[1] , Alain Couté[2] , Gérard Mascarel[2],
David C. Smith[3], Roger Naslain[4] and Xavier Bourrat[4]

[1] Muséum National d'Histoire Naturelle, Département des Milieux et Peuplements aquatiques, UMR 5178: CNRS-MNHN: Biologie des Organismes Marins et Ecosystèmes, 7, rue Cuvier 75231 Paris Cedex 05 France, rousseam@gmx.net, lopez@mnhn.fr

[2] Muséum National d'Histoire Naturelle, Département Régulations Développement et Diversité Moléculaire USM 505 Ecosystèmes et Interactions toxiques, 12, rue Buffon 75231 Paris Cedex 05 France, acoute@mnhn.fr, mascarel@mnhn.fr

[3] Muséum National d'Histoire Naturelle, Département de l'Histoire de la Terre, Bâtiment de Minéralogie, 61, rue Buffon 75231 Paris Cedex 05 France, smith@mnhn.fr

[4] Université de Bordeaux, Laboratoire des composites thermo-structuraux, UMR 5801 :CNRS SNECMA-CEA-UB1, 3, allée de la Boétie 33600 Pessac France, naslain@lctcs.u-bordeaux.fr, bourrat@lcts.u-bordeaux.fr

Keywords: biomaterial, nacre, structure, crystal growth, organic matrix, biofilm.

Abstract. Nacre is the internal lustrous 'mother of pearl' layer of many molluscan shells. The structure is a brick and mortar arrangement: the bricks are flat polygonal crystals of aragonite and the mortar consists of organic compounds.The biological mineralization of composites such as the molluscan shell has generally been thought to be directed by preformed organic arrays of proteins or others biopolymers.The possibility that the organic matrix behaves as a template for crystal formation by heteroepitaxial growth, is examined. The structure and growth of nacre of *Pinctada margaritifera* and *Pinctada maxima* are studied at different scales.We propose a new model in nacre growth and maturation. Understanding the molecular mechanisms that regulate biomineralization may thus provide practical routes to the synthesis of new high-performance composite materials.

Nacreous shell growth was studied by electron microscopy after fixation of immature nacre. It shows that the first step in crystal formation takes place in a biofilm, as evidenced by Laser Raman Spectroscopy.This biofilm is composed of organic matrix and calcium carbonate in the aragonite crystal form. Crystalin nuclei grow in the biofilm as an aligned sequence, starting with a row of intitial nucleation centres, gradually increasing concentrically in size and fusing together to form polygons in the next highest nacreous level.

In *Pinctada* biomineralization takes place in a biopolymer that has the ability to initiate nucleation and to control orientation in the aragonite crystal form and growth. Also is stressed the existence of several levels or steps of mineralization, simultaneously, as a key mechanism for the thickening of the shell. This natural bioceramic is studied with the aim of using it as a biomaterial in bone repair.

Introduction

The active role of proteins in biomineralization is fundamental [1-3]. It represents a source of inspiration for future nanotechnology with a bottom-up approach. The brick and mortar ordering of nacre has already inspired the toughening of ceramic materials by co-processing rigid ceramic as silicon carbide and supple interlayers as boron nitride [4].

The interdigitating brickwork array of tablets, specific, in bivalves ("sheet nacre") is not the only interesting aspect of nacre structure. The bio-crystal itsef is a composite. It has not only the mineral structure of aragonite but possesses intracrystalline organic material [5]. Among others things we aim to understand the role of the organic template [6], already validated by in-vitro experiments.

Incipient shell nacre (*Pinctada margaritifera*) was analyzed by electron microscopy at the interface on the mantle-side. Experimental observations indicate the key role of the step-like growing front in sheet-like nacre.

Material and methods

Preparation of growing shell surface. A *Pinctada margaritifera* oyster was killed. It was fixed in 70% ethanol, then dehydrated progressively, and finally the fat was removed with xylene. At this stage the oyster was mounted in methyl methacrylat. Some sections were cut in the shell (100μm thick) with a diamond draw to conserve the intimate mineral and organic structure.

SEM examination. After coating the samples in gold observations with SEM were systematically carried out using a JEOL JSM-840A. An Hitachi 4500 FEG was also used at lower voltage for some finer details.

Raman spectroscopy. Specimens for laser Raman spectroscopy were analysed as follows: objective x50, green argon laser excitation at 514.53 nm, laser power: 100 to 200 mW, multichannel detection in the range of 131-1201cm^{-1} with two batteries of 1024 diodes, slit width 100-150μm, 5 counts of 20 seconds and peak position calibration corrected to standard diamond at 1332 cm^{-1}.

Results

Contrary to usual SEM observations, Fig.1 gives a view of the growing nacre layers on the side of the epithelial tissue. After fixation, dehydratation and mounting of the animal with the shell, the latter was removed from the animal. What is shown in Fig.1 is therefore a fracture surface in-between the mantel and mature nacre, as seen by SEM on the mantel side.

It is important to analyze this preparation as a fracture surface, which results from the crack propagation produced by the removal of the mineral shell. It can be seen that the fracture surface follows the interlaminar matrix inbetween two layers. The result is the "garden terrace" appearance revealed by the fracture in this interface, in relation to the morphogenesis of the bio-mineralization. The fracture propagation (stable fracture) and the different marking on the preparation indicate that the film that forms close to the epithelium is in a viscous state and becomes more and more rigid or mineralized moving towards the mature nacre. This coincides with an increasing percentage volume of aragonite tablets that progressively grow until biocrystallization is completed and only residual film can be found inbetween some tablets.

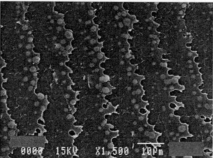

Fig. 1. SEM growing surface image of the mantle: fracture surface after fixing, mounting and separating the mantel from the shell of the oyster showing the incipient nacre layers on the mantel side.

The nucleus is located, in most cases, at the level of the intercrystalline matrix of the underlying layer. Also, nucleation is seen to initiate when tablets of the underlying layer complete biocrystallization and come into contact with each other. The state of maturation of the underlying layer is thus critical in controlling the brick wall-like nacre struture. Nucleation of new tablets is well controlled to keep statistically constant in density: 11 nuclei for 100 μm^2. This makes 1 nucleus for approximately 10 μm^2, which is consistent with the size of the tablets in *Pinctada margaritifera* nacreous layer.

The opposite side of that presented in Fig.1 can be observed in Fig.2. It is easy to distinguish aragonite tablets growing inside the film. The location of two comparative Raman analysis are obtained on growing white tablets (star) and on the film in the surrounding area (circle). The two spectra exhibit the peak of aragonite. The film also shows the presence of organic mater with a lot of fluoresence: background increasing with wavenumber.

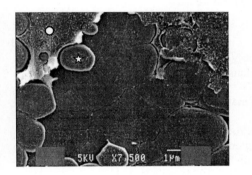 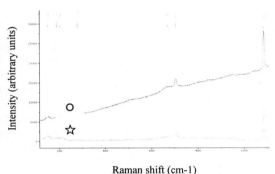

Raman shift (cm-1)

Fig. 2. Growing surface on nacre-side: Raman characterization of nuclei (star) and film (circle). The film exhibits the same characteristic peaks of aragonite but with the presence of organic matter (increasing background of fluorescence).

Discussion

Nacre layered structure is developed in three steps.

(i)   The discharge of a mixture of organic and ionic species by some specialized cells of the mantel.
(ii) Each layer of this fluid forms a gel coating at the interface between the mantel and the previous layer already formed.
(iii) This film further undergoes a self-ordering until its right dehydratation. Self-ordering means numerous irreversible interactions determined by the amount of the different species at the starting point and the precise succession of molecular interactions (polymerization, self-assembling of proteins or crystallization and mineralization). Nevertheless some external signal are seen to be necessary for the film to complete its structure perfectly.

This work proposes in Fig.3 a global understanding taking into account the well known step-like growth of shell nacre by introducing the model of self-ordering of nacre layers within the film of extrapallial fluid secreted by the mantel.
Further experiments are now in progress to explain the way nucleation occurs.

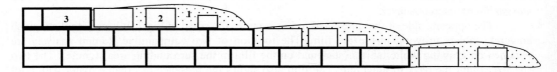

Fig. 3. Step-like growing front of sheet-nacre. It shows the simultaneous multilayered growth of shell:1) extrapallial fluid in films or compartments, 2) incipient nacre developing by self-ordering of the different constituents within each compartment and 3) mature nacre

References
[1]   G. Falini, S. Albeck, S.Weiner, L. Addadi. Science Vol 271 5 (1996), pp 67-69
[2]   A.M. Belcher, X.H. Wu, R.J. Christensen, P.K. Hansma, G.D. Stucky, D.E. Morse. Nature Vol 381 (1996), pp 56-58
[3]   Q.L. Feng, H. B. Li, F.Z. Cui, H.D. Li. Journal of materials science letters 18 (1999), pp1547-1549
[4]   R. Naslain, R. Pailler, X. Bourrat, F. Heurtevent. Proceed. ECCM-8 (J.Girelli-Visconti, ed.) vol.4 (1998), pp 191-199, Woodhead Publ., Abington, Cambridge, UK
[5]   N. Watabe. J. Ultrastructure Research 12 (1965), pp 351-370
[6]   S. Weiner, W. Traub . Phil. Trans. R. Soc. Lond. B 304 (1984), pp 425-434

Key Engineering Materials Vols. 254-256(2004) pp. 1013-1016
online at http://www.scientific.net
© 2004 Trans Tech Publications, Switzerland

In-vitro Calcium Phosphate Formation on Cellulose – Based Materials

Lenka Jonášová, Frank A. Müller, Heino Sieber and Peter Greil

Department of Materials Science – Glass and Ceramics, University of Erlangen-Nuremberg, Martensstr. 5, 910 58 Erlangen, Germany, jonasova@ww.uni-erlangen.de

Keywords: cellulose, sol-gel coating, calcium phosphate, simulated body fluid

Abstract. CaO-SiO$_2$ sol-gel coatings were deposited on natural cellulose-based polymers with a 3D porous network structure. Changes in the sample surface after coating and subsequent soaking in simulated body fluid (SBF) were determined by SEM – EDX, FT-IR and X-ray diffraction analysis. Sample - solution interactions were quantified on the basis of gravimetric and solution analysis. Within 3 days a homogeneous calcium phosphate layer was deposited on the sample surface. Furthermore, the porous structure of the samples was maintained. The Ca/P ratio in the deposited layer was 1.63, which is close to that in hydroxyl carbonated apatite (HCA). At the beginning of Ca-P layer precipitation calcium leaches out from the gel resulting in the formation of Si-OH groups in the surface. The Si-OH groups serve as favorable sites for Ca^{2+} and (PO$_4$)$^{3-}$ absorption from SBF resulting in the formation of a Ca-P enriched layer, which grows by consumption of calcium and phosphorus from the surrounding fluid. Sol-gel coatings of porous structures can be used to enhance osteointegration of porous bone replacement materials and scaffolds for bone tissue engineering.

Introduction

Biomaterials are manufactured from all kinds of man-made materials, including polymers, ceramics and metals as well as their composites. However, none of them can serve as perfectly as the living tissue for a replacement. If used as bone repairing or replacing material, metal implants will cause stress shielding. The low modulus of synthetic polymers limits their clinical application in bone reconstruction. For ceramic materials, the low fracture toughness is not favorable for bone-repairing material, thus the desire to search for new biomaterials is beneficial. Natural cellulose based materials possess mechanical properties close to those of human bone [1]. The biocompatibility of cellulose and its derivatives is well established [2]. Nevertheless, a complete mineralization could not be observed after implantation [3, 4]. The aim of the present work was to suggest a simple treatment that would induce calcium phosphate formation and mineralization in the surface of biological cellulose-based structures via a biomimetic process.

Materials and Methods

Luffa aegyptiaca sponge was used as a source of cellulose. This natural material with a fibrous network, obtained from the matured dried fruit of *Luffa aegyptiaca* (syn. *Luffa cylindrica*), was reported to consist of cellulose, hemicellulose and small amount of mannan and galactan [5, 6, 7]. Pieces of Luffa approximately 15 mm in length were first extracted in a 2:1 mixture of toluene and ethanol for 24 hours by Soxleth method and than dried for 24 hours at 110°C. The dried samples were dip-coated with a sol containing 20 mol% CaO and 80 mol % SiO$_2$ [8]. The coated specimens were kept 3 days at room temperature, than 3 days at 70°C and subsequently 3 days at 120°C for hydrolysis, aging and drying of the coating, respectively.

The coated *Luffa aegyptiaca* substrates were soaked in simulated body fluid (SBF) with ion concentration nearly equal to those of human blood plasma [9]. The samples were exposed to the solutions under static conditions in a biological thermostat at 37 °C for 3, 5, 7 and 10 days. For each time period three samples were soaked. The ratio of sample surface area to soaking solution volume S/V was 0.05 cm^{-1}. After soaking in SBF the samples were washed gently with distilled water and dried at room temperature.

Changes in the sample surface after treatment and soaking in SBF were determined by SEM – EDX (Cambridge Instruments, USA) on carbon sputtered samples, FT-IR (Nicolet Impact 420T, USA) using KBr tablets and X-ray diffraction analysis (XRD, Siemens DIFFRAC 500, Germany). Sample - solution interactions were quantified on the basis of gravimetric and solution analysis. The changes in the concentration of phosphorus and calcium in the solution were determined using inductively coupled plasma optical emission spectrometry (ICP-OES: Modula, Spectro, Germany). The measurement error for phosphorus and calcium concentration was determined to ± 1.9 mg/l and ± 5.1 mg/l, respectively.

Results and Discussion

Fig. 1a shows SEM micrographs of the *Luffa aegyptiaca* coated with CaO-SiO$_2$ gel. The pore size and distribution of the Luffa sponge are maintained. A new surface layer with spherical particles was observed after 3 days in SBF (Fig.1b,c). The macrostructure of the Luffa sponge remained unchanged (Fig.1b). The precipitated layer was homogenous (Fig.1b) and covered all around the sample (Fig.1c). Its thickness was approximately 10 μm. Using EDX, calcium and silicon were detected in the sample surface resulting from the gel coating (Fig 2a). EDX analysis confirmed phosphorus and calcium to be present in the surface after 3 days in SBF (Fig.2b). The calcium to phosphorus molar ratio was about 1.63, which is close to that of hydoxyapatite (Ca$_{10}$(PO$_4$)$_6$(OH)$_2$: Ca/P=1.67).

a)　　　　　　　　　b)　　　　　　　　　c)

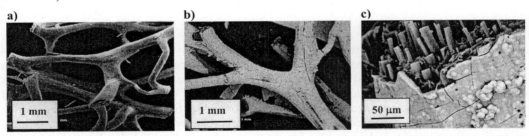

Fig. 1 SEM micrographs of *Luffa aegyptiaca* coated with CaO-SiO$_2$ gel a) before soaking in SBF and b, c) after 3 days in SBF.

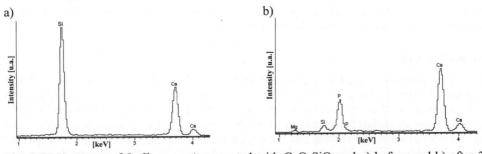

Fig. 2 EDX pattern of *Luffa aegyptiaca* coated with CaO-SiO$_2$ gel a) before and b) after 3 days in SBF.

Changes in calcium and phosphorus concentration after exposure of coated Luffa samples in SBF allow to suppose the formation of a Ca-P enriched surface layer. Fig. 3a shows the time dependence of calcium and phosphorus concentration after soaking of *Luffa aegyptiaca* coated with CaO-SiO$_2$ gel in SBF compared to SBF without samples, assigned as P-blank, Ca-blank. The results are average values from three samples in every time period. The calcium concentration in SBF with samples was for every soaking time higher than that measured in SBF without samples. It is necessary to take into account that the Ca-concentration is affected by transfer of this component from the CaO-SiO$_2$ coating into solution. However, from the decreasing calcium concentration in the solution with increasing time can be assumed a calcium absorption in the sample surface. On the other hand, the phosphorus concentration in SBF is not influenced by transfer of this component from the gel into SBF. The concentration of phosphorus was lower than in the blank solution for every soaking time. These results indicate calcium phosphates precipitation in the surface of gel coated Luffa.

The results of gravimetric analysis (Fig.3b) are in agreement with those obtained by solution analysis. The results are average values from three samples in every time period. From the decrease in the sample weight it can be assumed the dissolution of the coating, which corresponds with the increase of calcium concentration in SBF after soaking of Luffa substrate due to the dissolving gel layer. However, after 7 days in SBF the sample weight starts to increase again. It can be supposed that a new Ca-P layer precipitates on the sample surface.

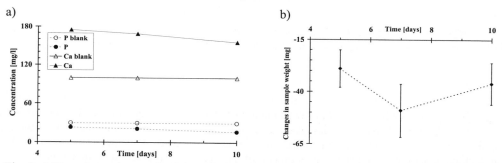

Fig. 3 a) Time dependence of calcium (Ca) and phosphorus (P) concentrations in SBF after soaking of CaO-SiO$_2$ gel coated *Luffa aegyptiaca*; b) Gravimetric analysis of *Luffa aegyptiaca* coated with CaO-SiO$_2$ gel after soaking in SBF.

Fig. 4a shows the X-ray diffraction analysis of the Luffa sample coated with CaO-SiO$_2$ gel after 10 days in SBF. The X-ray pattern corresponds to an amorphous material with very low crystallinity. In Fig. 4b the FT-IR spectra of CaO-SiO$_2$ coated Luffa before soaking and after 5 days in 1.5 SBF are compared. The 460 cm^{-1} vibration can be assigned to the Si-O-Si and O-Si-O bending modes. The 830 cm^{-1} and 780 cm^{-1} absorbtion bands correspond to the stretching mode of O-Si-O bonds. The 960 cm^{-1} vibration is associated with Si-O-Ca bonds containing non-bridging oxygen. The doublet formed at 1080 cm^{-1} and 1240 cm^{-1} can be assigned to Si-O-Si symmetric stretching. The band at 1380 cm^{-1} corresponds to the stretching vibration of (NO$_3^-$), which results from the preparation of the sol using Ca(NO$_3$)$_2$.4H$_2$O and HNO$_3$. At 1630 cm^{-1} the vibration of H$_2$O was detected [8, 10]. After 5 days in SBF (Fig.4b)-B) a new band at 568 cm^{-1} appeared, which corresponds to the P-O antisymmetric vibration mode. In addition, a variation in the shape of the band at 1080 cm^{-1} was observed due to the presence of the P-O antisymmetric stress band at 1040 cm^{-1}. Composed and overlapped bands in the 960-1200 cm^{-1} range resulted from the P-O antisymmetric stretching vibration, which indicates a deviation of phosphate ions from the ideal tetrahedral structure [11]. Furthermore, the recessive vibration of Si-O-Ca at 960 cm^{-1} shows destruction of O-Ca bonds and leaching of calcium, which is in agreement with the increase of calcium concentration in the soaking solution (Fig.3a). As a consequence of calcium leaching, silanol groups (Si-OH) from the gel coating form in the sample surface. The Si-OH groups were described as favorable sites for Ca^{2+} and $(PO_4)^{3-}$ absorption resulting in the formation of a Ca-P enriched layer [12]. This deposited layer can grow up by consumption of calcium and phosphorus from the surrounding fluid.

a)

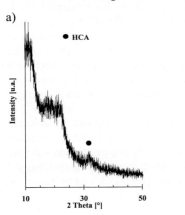

b)

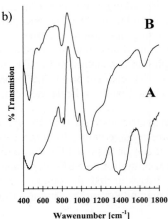

Fig. 4 a) XRD analysis of the *Luffa aegyptiaca* coated with CaO-SiO$_2$ gel after 10 days in SBF; b) FTIR spectra of the CaO-SiO$_2$ gel coating: A- before and B- after soaking in SBF for 5 days.

Conclusion

The coating of the natural cellulose-based cellular tissue of *Luffa aegyptiaca* with CaO-SiO$_2$ gel and subsequently soaking in SBF resulted in the formation of calcium phosphate in the material surface within 3 days. The Ca-P layer was homogeneous and the 3D porous network of the natural tissue was maintained. The coating of biocompatible porous structures with bioactive gels using sol-gel technique can be used for development of porous bone replacement materials or for mineralization of tissue engineering scaffolds.

References

[1] M. Martson, J. Viljanto, T. Hurme and P. Saukko: Eur Surg Res 30 (1998), p. 426
[2] D. Chauveaux, C. Barbie, X. Barthe, C. Baquey and J. Poustis: Clin Mater 5 (1990), p. 251
[3] T. Miyamoto, S. Takahashi, H. Ito, H. Inagaki and Y. Noishiki: J Biomed Mater Res 23 (1989), p. 125
[4] G. Franz: Adv Polym Sci 76 (1986), p. 1
[5] Y. Liu, M. Seki, H. Tanaka and S. Furusaki: J Fermen and Bioengin 85 (1998), p. 416
[6] S. Masuda: Cellulose Industry 3 (1927), p. 321
[7] M.R. Vignon, C. Gey: Carbohydrate Research 307 (1998), p.107
[8] I. Izquiero-Barba, A.J. Salinas and M. Vallet-Regí: J Biomed Mater Res 47 (1999), p. 243
[9] L. Jonášová, A. Helebrant and L. Šanda: Ceramics – Silikáty 46 (2002), p. 9
[10] P. Saravanapavan, L.L. Hench: J Non-Cryst Sol 318 (2003), p. 1
[11] A. Stoch, W. Jastrzebski, A. Brozek, J. Stoch, J. Szatraniec, B. Trybalska, G knita: J Molec Str 555 (2000), p. 375
[12] L.L. Hench: J Am Ceram Soc 74 (1991), p. 1487

Key Engineering Materials Vols. 254-256(2004) pp. 1017-1020
online at http://www.scientific.net
© *2004 Trans Tech Publications, Switzerland*

Performances of Hydroxyapatite Porosity in Contact with Cells and Tissues

A. Ravaglioli[1], A. Krajewski[1], M. Mazzocchi[1], R. Martinetti[2], L. Dolcini[2]

[1] ISTEC – CNR Via Granarolo 64, Faenza I-48018, ITALY, ravaglioli@irtec1.irtec.bo.cnr.it
[2] FIN-CERAMICA FAENZA s.r.l. Via Ravegnana 186, Faenza I-48018, ITALY

Keywords: porous bioceramics, calcium phosphate, release of biologically active molecules

Abstract. HA is the major component of the bone tissue (about 70%). Synthetic HA is totally bio-compatible, non-toxic and osteoconductive. The particulate form of HA promotes ingrowth and attachment of bone. The HA scaffold can be produced by several methods obtaining blocks, spheres and/or microspheres. The scaffolds were prepared and impregnated with different drug studying the kinetics of the release in time. The results are presented on scaffolds with controlled pore size used as containers for targeted drug delivery of prolonged pharmacokinetics. It is reported that to control the drug release kinetic it is necessary the addition of biocompatible surfactants and/or organic matrices to exploit the role of bi-modal porosity. From the information acquired, scaffold for bone regeneration have been also produced; in further applications cells can be inserted in the HA architecture to regenerate new bone in a shorter time.

Introduction

The scientific research is going on towards new strategies to deliver drugs from devices implanted into the body to assure a suitable constant flux of pharmacological substances for a prefixed interval of time. Ceramic hydroxyapatite is a very suitable material to produce capsule for drug delivery. However, the releasing surfaces (not always of the same nature) must have tough walls with suitable microstructure and interconnecting pores of different dimensions and morphology. The best characteristics can be achieved through specific technological processes (slip casting, extrusion or slurry expansion), the drying and the sintering.

The authors carried out experiments focusing on the porous structure of ceramics based on hydroxyapatite at a controlled porosity for drug release, as cells carriers for tissue engineering and as scaffold architectures to repair large bone defects. A fallout from these knowledge involves other particular applications such as the production of preformed suitable scaffolds able to recover bone tissue when injured for trauma or degenerative diseases.

Materials and Methods

Commercial HA powder of medical grade (Riedel de Haën) was calcined at 900°C for 1 h to reduce its specific surface from 61 to 5 m^2g^{-1} (the mean particles size from 0.35 top 0.89 μm), as well as its reactivity in sintering. Two different types of capsule were prepared with different methods to be adaptable to different kinds of release.

A first type of capsule, used for sodium hydrocortisone acetate (Na-HCA) (1) or with biphosphonates (2), was realized by imbibition of cellulose sponges in an HA aqueous slurry and then sintered (samples A, Table 1).

A second type, to be utilized for the release of Paxitaxel drug (3), was realized by extrusion of cylindrical porous walls to give rise to reservoir-shaped capsules. Granular polyvinyl-butyral (PVB, B-76, Monsanto Co.) was used to produce the porous walls. To cap the cylindrical porous walls to constitute the reservoir device, two thin HA circular disks were glued (with a suitable process) at each base of the porous cylinder. The porosity of the sintered samples was measured by Hg-porosimeter (Mod. 200, Carlo Erba, Milan), and/or examined by scanning electron microscopy (SEM – Cambridge, Stereoscan 360).

Results

All porous devices have bimodal porosity. Those of Figures 1, coming from the samples utilised for release tests of hydrocortisone acetate (Na-HCA) and bisphosphonates, have the main physical parameters reported in Table 1 and the porosity distribution of Figure 3.

Table 1 – Parameters of characterizing HA devices for drug delivery.

Samples	Total porosity	Density	Compressive Strength	Firing Temperature
	(%)	(g·cm^{-3})	(MPa)	(°C)
(a)	62.06	1.2	9.9($\pm$ 1.3)	1170
(b)	40.74	2.0	36.5($\pm$ 11.4)	1280

The amount of hydrocortisone acetate (Na-HCA) released from these HA samples resulted a bit high during the first hours (Figures 4, a and b); the release amount of bisphosphonates loaded on the same bimodal porous capsules was relatively high during the first 20 days, giving also a long term release of the agent.

The reservoir capsules loaded with Paclitaxel with GrEL medium (a biocompatible molecular surfactants in the case of lipophilic drug) for a prolonged drug delivery showed a more calibrated release, at least up to 10 hours (2). Microstruture of these capsules is reported in microphotos of Figures 2.

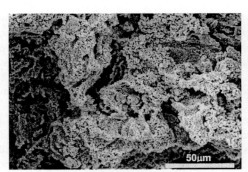

Fig. 1 a - Porosity Image (SEM macro sample a)

Fig. 1 b - Porosity Image (SEM micro sample a)

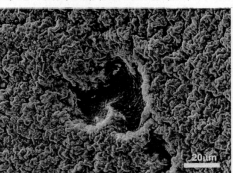

Fig. 2 a - Porosity Image (SEM macro sample b)

Fig. 2 b - Porosity Image (SEM micro sample b)

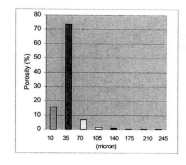

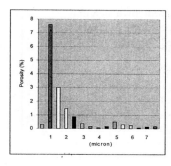

Fig. 3 a (left) - Sample (a) Total Porosity Distribution

Fig. 3 b - (right) Sample (a) 0 – 10 μm porosity distribution

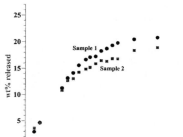

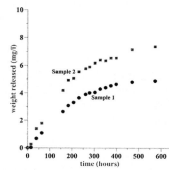

Fig. 4 a (left) - Sample (a) Trend of the weight percent release of Na-HCA in the time in 100 ml of salt solution.

Fig. 4 b (right) - Sample (a) Concentration vs. time during the release.

Sample 1 and sample 2 refer to two different types of cellulose sponges used to produce the capsule.

Through the porosity of these last capsule (Figure 5) has a different morphology, its size distribution is similar to that of the first type of device (Figure 3, a and b).

Discussion and Conclusions

Making reference to the HA releasing devices it was ascertained the role, in particular, of the small pores (<5μm, able to control the diffusion process) and the outstanding one of the bimodal porosity morphology as well.

While for the biphosphonates the release occurs steadily, controlled and lasting over a long period of time, in the case of the other releasing drugs, as Paclitaxel, the addition of biocompatible surfactants and/or organic matrices is needed to trim their outlet from the bimodal porosity of the walls.

Fig. 5 – Photo of a capsule used for Paclitaxel release. Photo width corresponds to 18 mm.

Fallout from these results

From the results and experiences obtained from the release of all devices a sufficient amount of information on the mechanisms of molecular and microparticles intrusion on pores was achieved to promote a different experience. Bone graft devices, characterised by bimodal porosity, have been produced and tested to be loaded with cells and pharmacological principles to rebuild parts of bone. As concerns the porous bimodal hydroxyapatite ceramic used as scaffold loaded with cells (in comparison to porous hydroxyapatite implant without cells), a significant improvement in bone repair was achieved. A decisive role is anyway played by the total porosity (still arranged in a bimodal distribution) in the osteointegration mechanism that occurs during the cellular spreading of the seeded cells in the porosity of the hydroxyapatite device after implantation as bone substitute. The porous structure allows new bone tissue ingrowth and vascularisation, with incoming of fluids with a high rate of new bone formation for a faster repair of large bone loss (4).

Relative to the development of a calcium phosphate device characterised by a bimodal porosity, obtained through imbibition of cellulose sponges and used as scaffold for bone re-colonisation, it was ascertained that seeded osteoprogenitor cells are able to differentiate and regenerate new bone in a shorter time inside such material. For instance, these bone graft devices, characterised by bimodal porosity, have been also widely used to solve large bone defects and also in neurosurgery for reconstruction of cranial lacunae.

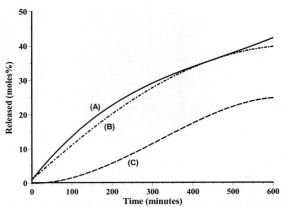

Fig. 6 - Kinetic release curves of Paxitaxel from the indicated capsule in saline solution: (A) without any mixing mean; (B) mixed with 50% CrEL and 50% Ethanol; (C) mixed with 90% CrEL and 10% Ethanol. The reported Release% refers to the amount initially present inside the capsule.

Acknowledgements

The authors are indebted with Prof.s C.M. Camaggi and E. Strocchi (Pharmacokinetics ANT Laboratory, Dept. of Organic Chemistry, Industrial Chemistry, University of Bologna) for the tests carried out with Paxitaxel.

References

[1] R. Martinetti, L. Dolcini, A. Krajewski, A. Ravaglioli, L. Montanari; *"An antiphlogistic release study from hydroxyapatite porous ceramic devices"*, Proc. 6th Meeting & Seminar on Ceramics, Cells and Tissues, Editors A. Ravaglioli and A. Krajewski, Published by ISTEC-CNR (Faenza, 1999), pp. 75-79.

[2] H. Denissen, E. Van Beek, R. Martinetti, C. Klein, E. Vanderzee, A. Ravaglioli; "Net shaped hydroxyapatite implants for release of agents modulating periodontal-like tissues", J. Period. Res. 32 (1977) 40-46.

[3] A. Krajewski, A. Ravaglioli, M. Mazzocchi, E. Strocchi, C.M. Camaggi; "Study of the release kinetics on some anticancer substances from porous ceramic capsule", Proc. 17th European Soc. for Biomaterials, Conference on Biomaterials September 11-14, 2002 (Barcelona).

[4] A. Facchini, G. Lisignoli, B. Gigolo, A. Krajewski, M. Mazzocchi, C. Capitani, A. Ravaglioli; "Porous ceramics structured for bone-cartilage implants", Proc. ECERS Meeting-Seminar, June 30-July 3, 2003 (Istanbul).

Key Engineering Materials Vols. 254-256(2004) pp. 1021-1024
online at http://www.scientific.net
© *2004 Trans Tech Publications, Switzerland*

Porous Triphasic Calcium Phosphate Bioceramics

J.S.V. Albuquerque[1], R.E.F.Q. Nogueira[1], T.D. Pinheiro da Silva[2], D.O. Lima[2] M.H. Prado da Silva[1]

[1] Universidade Federal do Ceará, Centro de Tecnologia, Dept° Eng. Mecânica e de Produção, Bl.714, Campus PICI, 60455-760, Fortaleza, CE, svalbuquerque@yahoo.com.br

[2] Universidade Federal do Ceará, Centro de Tecnologia, Dept° Eng. Química, Bl.709, Campus PICI, 60455-760, Fortaleza, CE, thaudeze@yahoo.com.br

Keywords: Porous bioceramics, alpha-TCP, beta-TCP, hydroxyapatite, nanoparticles.

Abstract. In the present study, calcium phosphate powders were produced by precipitation in aqueous solution. Porous discs were produced by organic additives incorporation. The final microstructure consisted of a triphasic bioceramic after sintering. The produced material is a good candidate to be used as bone-filler. In the present study a detailed study of the sintering temperature to produce triphasic calcium phosphate ceramics is presented.

Introduction

Calcium phosphate bioceramics can be used in applications where bone ingrowth is intended. In this application, the ideal material is the one with a degradation ratio similar to that of new bone formation. It is well established that pure and crystalline hydroxyapatite, $Ca_{10}(PO_4)_6(OH)_2$ has a low degradation ratio *in vivo*. For this reason, recent studies have pointed to the design of multiphasic calcium phosphate bioceramics as bone fillers [1-3]. These materials contain other calcium phosphates with higher solubility when compared to pure hydroxyapatite. Tricalcium phosphates are bioactive ceramics that can be associated to hydroxyapatite. These bioceramics exhibit polimorphism: α-$Ca_3(PO_4)_2$ (alpha-TCP) and β- $Ca_3(PO_4)_2$ (beta-TCP). In temperatures above 1300°C, beta-TCP is likely to decompose in alpha-TCP [4,5].

Porous bioceramics are suitable for being used as bone-fillers. In this application, pores are requested to have diameters above 100μm to allow proper vascularisation of the newly formed bone [6-7]. In recent studies, porous triphasic bioceramics were produced by dry pressing and sintering a patented glass-reinforced hydroxyapatite [8]. In the present study, similar results were obtained by a simple method.

Materials and Methods

In this study, apatite nanoparticles were produced by aqueous precipitation. The starting solution was 0.3M H_3PO_4, 0.5M $Ca(OH)_2$, 1M CH_3CHCO_2HOH. The pH value of the solution was adjusted to pH=10 by NH_4OH addition. The suspensions were left overnight for ageing. The suspension was then vacuum filtered and washed in deionised water to remove NH_4OH. The powders were dried in an oven at 100°C overnight. Porous specimens were produced using a dry method developed in previous studies [1,2]. In this study, wax spheres (Licowax®, Clariant) with two different particle sizes were used to produce porous structures: 0.8mm (90%>800μm) and 0.3mm (90%>300μm). Porous discs with 65 vol% of Licowax® were produced by dry mixing spheres to the powders and green bodies were produced by uniaxially pressing the mixture at 40 MPa. The discs were heated at 550°C in a muffle furnace at a heating rate of 0.5°C/min to burn the organic additives. Final sintering was performed at a heating rate of 4°C/min to consolidate the porous structures. Five different temperatures were used to assess the microstructural evolution: 900°C, 1100°C, 1250°C, 1300°C and 1350°C. Phases were identified with a Philips X'Pert X-ray

Diffraction (XRD) System operating with CuKa (l=0,1542nm) with 40kV and 40mA. Sintered powders and porous specimens were analysed by XRD. Scanning electron microscopy (SEM) analysis was an important tool to assess the morphology and interconnectivity of the pores. SEM analysis was performed with a scanning electron microscope (SEM) Philips model XL-30.

Results

XRD analysis showed no difference between the patterns for sintered powders and porous specimens. Sintering at 1350°C showed the presence of alpha-TCP. Table 1 shows phase composition for each sintering temperature.

Table 1 – XRD analysis of porous specimens sintered at different temperatures.

Sintering temperature	Phases present
900°C	HA + β-TCP
1100°C	HA + β-TCP
1250°C	HA + β-TCP
1300°C	HA + β-TCP
1350°C	HA + α-TCP + β-TCP

Figure 1 shows XRD pattern corresponding to a porous specimen sintered at 1350°C showing the presence of HA + α-TCP + β-TCP.

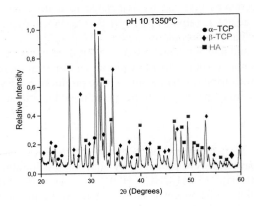

Fig. 1 – XRD pattern after sintering at 1350°C showing the presence of HA, α-TCP and β-TCP.

Porous samples were produced by sintering at 1350°C. Scanning electron microscopy showed that 65 vol% of Licowax® provided interconnected pores with pore diameters above 100μm, irrespective of the particle diameter used. Figure 1 corresponds to SEM observation of a porous disc produced by a mixture of spheres with the two different particle sizes used (50-50 wt%).

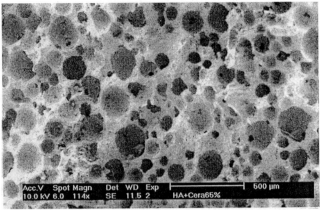

Fig. 2 – SEM analysis showing interconnected pores with pore sizes above 100µm.

Discussion

In previous studies [1, 8], a good interconnectivity was obtained with 65 vol% Licowax® to glass-reinforce hydroxyapatite sintered at 1300°C. The final microstructure consisted of HA + α+ β. In this study, the sintering temperature to produce a triphasic bioceramic was found to be 1350°C where the powder used was produced by precipitation from aqueous solution. Alpha-TCP presence occurred after sintering powders obtained at pH=10 at 1350°C during 1 hour. A possible explanation for the lower sintering temperature required for having triphasic bioceramics when using GR-HA powders rely on the sintering mechanism. The glass used in the studies with GR-HA has a low melting point and sintering occurs with the presence of liquid phase, which causes a better interaction between particles and thus favors phase transformation. A different trial was performed, regarding the particle size of the organic additives. A mixture of Licowax with different particle diameters produced the interconnected structure shown in Figure 2.

Conclusion

Porous triphasic bioceramics were produced. The specimens showed interconnected pores with pore sizes above 100µm. The results of this study are promising and present a simple alternative to the use of composite materials for the production of triphasic bioceramics.

Acknowledgements

The authors would like to thank CNPq through PROFIX programme 540191/-1-9, Andre Galdino (DEMP-UFC) and Dr. Lisiane Navarro (UFCG) for their help in the XRD sintering.

References

[1] M. H. Prado da Silva, A. F. Lemos, J. M. F. Ferreira, J. D. Santos, J. Non-Cryst. Sol., Vol. 304 (2002), p. 286.

[2] J.D. Santos, R.L. Reis, F.J. Monteiro, J.C. Knowles, G.W. Hastings, J. Mater. Sci: Mat. Med. 6 (1995) p. 348.

[3] J. S. V. Albuquerque, J. V. Ferreira Neto, J. I. L. Almeida Jr., D. O. Lima, R. E. F.Q. Nogueira, M. H. Prado da Silva, Key Eng. Materials, Vol. 240 (2003) p.23.

[4] R. LeGeros, 1991, Calcium Phosphates in Oral Biology and Medicine (Karger, N. Y., USA), Vol. 5 (1991) p. 15.

[5] L.L Hench, J. Wilson,. "An Introduction to Bioceramics" (1 ed., Gainesville, USA, World Scientific), Vol.1 (1993), p. 154.

[6] A.F. Lemos, J.M.F. Ferreira, Mater. Sci. and Eng. Vol. 11 (2000) p. 35.

[7] Lyckfeldt, J.M.F. Ferreira, J. Eur. Ceram. Soc. Vol. 18 (1998) p. 131.

[8] M. H. Prado da Silva, A. F. Lemos, J. M. F. Ferreira and J. D. Santos, Key Eng. Materials, Vol. 230 (2002), p. 483.

Key Engineering Materials Vols. 254-256(2004) pp. 1025-1028
online at http://www.scientific.net

Macro Porous Hydroxyapatite with Designed Pores

Erik Adolfsson

Swedish Ceramic Institute, Box 5402, 402 29 Göteborg, Sweden, ea@sci.se

Keywords: Hydroxyapatite, macro porous, free form fabrication, slip casting

Abstract. Hydroxyapatite suspensions were prepared by ball milling and consolidated by slip casting. The suspensions were used to produce both solid and macro porous materials with designed pore channels by infiltrating free form fabricated moulds. Materials with three different pore shapes were produced, cylindrical shaped pore channels, square shaped pore channels and one scaffold consisting of cylindrical rods (negative of cylindrical pores). The size of the continuous and interconnected pore channels was around 500μm and the total macro porosity was between 48-65 % when sintered at 1300°C. The solid materials reached sintered densities around 99% when sintered at 1250°C or above.

Introduction

Since it was shown that a certain pore size was required for bone ingrowth to take place in the pores [1], a lot of work has been performed with the aim to prepare ceramics with a suitable pore sizes for bone regeneration. The interconnections between the pores are another important characteristic of the porous materials to consider and these may also require a certain size not to inhibit the bone tissues to continue to grow into the adjacent pores. The interconnections are usually more difficult to control during the material fabrication than the pore size and pore volumes. This has inspired different attempts to use moulds that correspond to the desired macro porosity. By infiltrating the mould with a ceramic suspension, the shape of the mould was converted to macro porosity in the green body when sacrificed. The technique has been used to produce replicates of cancellous bone [2]. The produced porosity will then depend on the bone sample used and to obtain a more isotropic macro porous material, spherical polymer beads were heated to partially fuse together, which allowed the size of both the pores and interconnectivity to be controlled [3]. This is also possible with free form fabrication where the moulds are built layer by layer based on a CAD-model, which also has been used to produce designed macroporous scaffolds [4,5].

The aim of the work was to evaluate the possibilities not only to produce a macroporous scaffold of hydroxyapatite with continuous and interconnected pore channels, but also to design the shape of the pore channels.

Materials and Methods

Hydroxyapatite powder (Plasma Biotal) was heat-treated at 900°C and used to prepare ceramic suspensions with a solids loading of 48vol% by ball milling. A poly acrylic acid (Dispex A40) was used as dispersant and the viscosity was measured with a rheometer (Stress tech). To reduce the brittleness of the unfired ceramics, an addition of 3% binder (Monowlith LDM7651S) was made to the suspensions, which were consolidated by slip casting (colloidal filtration) in plastic moulds on a block of plaster where the excess of water was drained from the suspension until green bodies were formed. The sintering performance of solid slip cast materials was evaluated at temperatures between 900°C and 1300°C

The macro porosity was designed by a CAD tool and the moulds corresponding to the porosity were built with free form fabrication (Model Maker) equipment using an inkjet printing principle. The moulds were built with a layer thickness of approximately 50 μm of a thermoplastic material (Protobuild) surrounded by a supporting wax based material (proto support), allowing overhangs to be built. The moulds were infiltrated by the ceramic suspension and thereafter consolidated by slip casting. The mould material had a melting temperature of around 100°C and were sacrificed during sintering by using a heating rate of 1°C/minute up to 600°C and 5°C/minute up to the sintering temperature, that was kept for 2 hours before temperature was decreased by 5°C/minute down to room temperature. The bulk density of the sintered material was measured by Archimedes principle and the macroporosity was calculated from the weight and geometrical dimensions of the macro porous structures. The microstructure of the sintered material was studied by scanning electron microscopy and x-ray diffraction was used to confirm phase composition, where no other phases were detected.

Results and discussion

When ceramics are shaped from suspensions, high solids loading and a low viscosity are desired [6]. Suspensions with a solids loading of 48 vol% with a viscosity that showed a shear thinning behaviour (fig. 1) were prepared to cast both solid and macroporous materials. To produce a ceramic green body from the suspension, a suitable consolidation process that transforms the suspension from a liquid to a solid state before the mould can be removed is required. One such consolidation process that has been used for dense cylindrical shaped samples of hydroxyapatite is slip casting [7] where a porous mould removes the excess of water in the suspension. This process allows the dispersed particles in the suspension to reach a high packing density, which facilitates the sintering performance of the slip cast materials. The densification was found to be initiated at temperatures above 1050°C, where the density increased from slightly above 60 vol% until almost fully dense materials were obtained at temperatures above 1250°C (fig. 2).

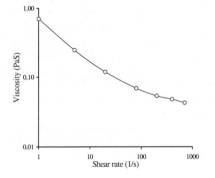
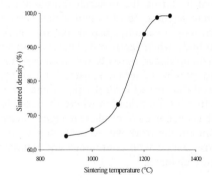

Fig. 1 Viscosity of the prepared suspensions with a solids loading of 48vol%.

Fig. 2 Sintered density of slip cast hydroxyapatite.

Models of the desired macroporosities, with three different shaped pore channels where drawn with a CAD tool and used to produce moulds using a free form fabrication technique. Both the pore size and the distance between the pores were in all cases 600µm in the CAD model. The shape of the mould was transferred to the green body by infiltrating the mould with the hydroxyapatite suspension, and finally converted to macroporosity by a slow temperature increase to allow the mould material to melt. The different hydroxyapatite scaffolds produced had either a square or cylindrical shaped pore channels that were continuous and interconnected and one scaffold was formed of cylindrical rods, corresponding to the negative of the scaffold with cylindrical pore channels (Fig 3). The pore size, after sintering, was around 500μm and the total macro porosities were between 48 and 65vol% (table 1). The reason for the discrepancies between the macro porosity in the sintered scaffolds and calculated porosities corresponding to the differently shaped pore shapes that where around 7-8% was not know.

The viscosity of the suspension was sufficiently low to allow the pore system in the moulds to be infiltrated, however, even the small reduction of the cross section area from the square to cylindrical shaped pores gave a noticeably increased resistance of the suspensions to enter the pore channels. This indicates that an increased viscosity or reduced pore size in the moulds would cause difficulties related to the infiltration with the procedure used. The scaffold fabrication technique has then a lower limit of the dimensions of the structure in the scaffold that can be produced since these are determined by the pore size in the mould that has to be filled. A slight reduction of the minimum size of the ceramic structure would however be possible to obtain by a vacuum assisted infiltration.

Fig. 3 Macro porous scaffold consisting of rods, sintered at 1300°C (the side of the cube is around 6.8mm).

Table 1 Macro porosity of the prepared scaffolds sintered at 1300°C and for the differently shaped pore shapes in the CAD-models

Pore shape	Macro porosity	
	CAD-model (Vol%)	Sintered scaffold (Vol%)
Cylindrical shaped pore channels (Fig. 5a)	41	48
Square shaped pore channels (Fig 5b)	50	58
Scaffold formed of cylindrical rods (Fig. 5c)	58	65

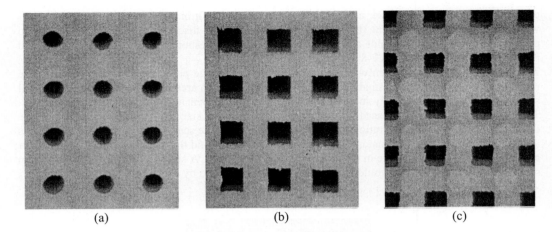

(a)   (b)   (c)

Fig. 4 Macro porous hydroxyapatite sintered at 1300°C, (a) with cylindrical pore channels, (b) square shaped pore channels and (c) a scaffold consisting of cylindrical rods (the negative of (a)).

Conclusion

The hydroxyapatite obtained from the slip cast suspensions gave a favourable sintering performance and almost fully dense materials were reached when sintered at 1250°C. By infiltrating free form fabricated moulds with the suspensions, macroporosities consisting of continuous and interconnected pore channels with defined shapes were produced.

The combination of slip casting and free form fabricated moulds gives a shaping process where hydroxyapatite components of almost any desirable shape can be produced and sintered to high densities. The fabrication technique would then be suitable for preparation of designed materials that can be used as research tools.

Acknowledgement

Matts Andersson at Nobel Biocare AB is thanked for financial support.

References

[1] J.J. Klawitter and S.F. Hulbert: J. Biomed. Mater. Res. Vol. 5 (1971) p. 161

[2] D.C. Tancred, B.A.O. McCormac and A.J. Carr: Biomaterials Vol. 19 (1998) p. 2303

[3] O. Richart, M. Descamps and A. Liebetrau: Key Eng. Mater. Vol. 218-220 (2002) p. 9

[4] T.-M.G. Chu, J.W. Halloran, S.J. Hollister and S.E. Feinberg: J. Mater. Sci. Med. Vol. 12 (2001) p. 471

[5] S. Bose, J. Darsell, H.L. Hosick, L. Yang, D.K. Sarkar and A. Bandyopandhyay: J. Mater. Sci. Med. Vol. 13 (2002) p. 23

[6] F. Lelièvre, D. Bernach-Assollant and T. Chatier: J. Mater. Sci. Med. Vol. 7 (1996) p. 489

[7] L.M. Rodriguez-Lorenzo, M. Vallet-Regi and J.M.F. Ferreira: Biomaterials Vol. 22 p. 1847

Key Engineering Materials Vols. 254-256(2004) pp. 1029-1032
online at http://www.scientific.net
© 2004 Trans Tech Publications, Switzerland

Extensive Studies on Biomorphic SiC Ceramics Properties for Medical Applications

P. González[1], J.P. Borrajo[1], J. Serra[1], S. Liste[1], S. Chiussi[1], B. León[1], K. Semmelmann[2], A. de Carlos[2], F.M. Varela-Feria[3], J. Martínez-Fernández[3], A.R. de Arellano-López[3]

[1]Universidad de Vigo, Dpto. Física Aplicada, Lagoas-Marcosende, 36200 Vigo, Spain, pglez@uvigo.es

[2]Universidad de Vigo, Dpto. Bioquímica, Genética e Inmunología, Lagoas-Marcosende, 36200 Vigo, Spain

[3]Universidad de Sevilla, Dpto. Física de la Materia Condensada, Apdo. 1065, 41080 Sevilla, Spain

Keywords: silicon carbide, bioceramics, biomaterials, EDS, XPS, XRF

Abstract. Biomorphic silicon carbide ceramics are light, tough and high-strength materials with interesting biomedical applications. The fabrication method of the biomorphic SiC is based in the infiltration of molten-Si in carbon preforms with open porosity. The final product is a biostructure formed by a tangle of SiC fibers. This innovative process allows the fabrication of complex shapes and the tailoring of SiC ceramics with optimised properties and controllable microstructures that will match the biomechanical requirements of the natural host tissue. An interdisciplinary approach of the biomorphic SiC fabricated from beech, sapelly and eucalyptus is presented. Their mechanical properties, microstructure and chemical composition were evaluated. The biocompatible behaviour of these materials has been tested *in vitro*.

Introduction

In the last decades many materials have been developed and improved for medical applications, such as metals, ceramics, glasses, polymers and composites. In particular, there has been a recent interest in the research and development of biomaterials from natural biological structures with an open porosity. The challenge of the implant technology is to get lighter materials, with enhanced mechanical properties, wear resistance and with better biological response.

Biomorphic silicon carbide ceramics are very promising as a natural base material for dental and orthopaedic implants due to their unique mechanical and microstructural properties. This innovative material is produced by molten-Si infiltration of carbon templates obtained by controlled pyrolysis of wood [1]. This innovation introduces two beneficial factors: first, the versatility for the fabrication of complex shapes, because it will only require the shaping of the wood template, and secondly, the high strength and toughness associated with the fibrous nature of the wood. The biodiversity of the natural grown wood structures offers a large variety of templates with different porosity and morphologies [2]. The SiC ceramics retain the microstructural details of the biostructure derived carbon preforms and it allows the tailoring of a wide range of SiC ceramics with optimised microstructure and properties close to those of the tissue to be repaired. The final product is a light, tough and high-strength material, with controllable microstructure, that mimics the starting wood fibrous structure that has been perfected by natural evolution.

Presented in this work is an extensive study on the physico-chemical and mechanical properties, as well as the biocompatible behaviour of biomorphic SiC ceramics obtained from three selected woods.

Materials and methods

Biomorphic SiC has been obtained from beech, sapelly and eucalyptus woods. The samples are pyrolyzed in an argon atmosphere at 1000°C with well-controlled heating and cooling ramps. The porous carbon preforms are then infiltrated with liquid silicon in vacuum. A subsequent spontaneous reaction forms SiC [1].

Room temperature crushing strength experiments were performed using a universal mechanical testing machine. Compressive rates were $2x10^{-5}$ s^{-1} on 5x2x2 mm^3 samples tested along their longer axis. Specimens were cut for testing both parallel (axial) and perpendicular (radial) to the direction of growth of the wood.

In order to assess possible adverse physiological effects in the body, *in vitro* tests were carried out by soaking the SiC ceramics in 1,5 ml of Simulated Body Fluid (SBF) [3] at 37°C for 1 week. Chemical analyses of the SiC ceramics and the fluids for the identification of the composition and the trace element content by Energy Dispersive X-Ray Spectroscopy (EDS), X-Ray Fluorescence Spectrometry (XRF), X-Ray Photoelectron Spectroscopy (XPS) and Inductively Coupled Plasma Mass Spectrometry (ICP-MS) were performed. Cytotoxicity was assessed performing a solvent extraction test according to ISO 10993 standards, using PVC and Thermanox as positive and negative controls respectively. MG-63 osteoblast-like cells were grown to confluent layers in 96-well tissue culture plates and tested against different concentrations of the extracts for 24 hours. The cell viability was assessed by determining the reduction of MTT measuring the absorbance in each of the wells.

Results and discussion

Figure 1 shows the microstructure of the three Bio-SiC materials used in this study. Different porous size and porosity grade can be observed. The biomorphic microstructure resembles the cellular microstructure of bone, and this type of large porosity may contribute to the vasculature of the repaired tissue around the implant, providing a route for migration of osteogenic cells and, consequently, new bone formation within the implant.

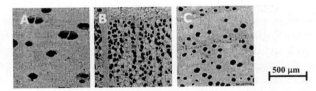

Fig. 1. SEM cross-sections of Bio-SiC materials from (A) sapelly, (B) beech and (C) eucalyptus wood.

Table 1 presents the densities of the woods, the carbon preforms and the density of the SiC skeleton, which controls the mechanical properties of the material. Error bars are assessed to be less than 5%.

Depending on the amount of remaining Si filling the porosity of Bio-SiC, the bulk density of the material can vary. When all porosity is filled with Si these three Bio-SiC materials reach a density of approximately 2.0 $g.cm^{-3}$. Recent developments have shown that unreacted Si can be removed by appropriate chemical treatment. If all porosity is empty, then the density is that shown in Table 1 as SiC.

Table 1. Densities of Bio-SiC

	Density ($g.cm^{-3}$)		
	Wood	Carbon	SiC
Sapelly	0.61	0.36	1.15
Beech	0.70	0.48	1.42
Eucalyptus	0.79	0.58	1.57

As demonstrated, an important advantage of this process is the possibility to tune the density of this material in a wide range (1,15 − 2 g.cm^{-3}). Biomorphic SiC is defined as a light material, because its density is very low in comparison with the literature values of the theoretical density of SiC (3.21 g.cm^{-3}) and both medical-grade Ti (4,51 g.cm^{-3}) and Ti-alloy (4,42 g.cm^{-3}).

The room-temperature crushing strengths of the three types of Bio-SiC, in axial and perpendicular direction are included in Table 2. For comparison purposes, literature crushing strengths of cortical bone, and both medical-grade Ti and Ti-alloy are shown.

As observed, the microstructure and mechanical properties of the material can be tailored by an appropriate wood precursor selection. From the material science point of view, the bulk properties of this material will match the biomechanical requirements of a particular type of natural host tissue that should be repaired.

Concerning the chemical composition of these ceramics, EDS analyses show that this biomorphic material is a composite of a SiC skeleton with small amounts of unreacted Si and C. Recent studies conclude that the residual silicon does not present adverse physiological effects in the body; Si is safely excreted through the urine and no accumulation was found distributed in the major organs [4]. Table 3 shows the XRF and XPS analyses of the three SiC types. The elements identified by both techniques are marked in grey. The measurements reveal the majority presence of Si and C. Small amounts of Ca were found in all samples and, in some cases, Al, Cu and Na. Other trace elements were also identified in very low concentrations.

The biocompatible behaviour of this material has been tested by soaking the SiC ceramics in Simulated Body Fluid. Table 4 shows the ICP-MS analyses of the SBF fluids after one-week immersion of the Bio-SiC materials. The SBF element concentrations used as reference values are also shown. Only weak signals related to the release of Si and

Table 2. Crushing Strength results

	Strength (MPa)	
	Axial	Radial
Bone [6]	195	135
Ti [7]	310	--
Ti-6Al-4V [7]	880	--
Sapelly	1070 ± 200	105 ± 20
Beech	1160 ± 300	530 ± 100
Eucalyptus	1175 ± 400	165 ± 60

Table 3. Chemical composition of Bio-SiC

Element	Bio-SiC type					
	Beech		Sapelly		Eucalyptus	
	XRF	XPS	XRF	XPS	XRF	XPS
Al	X		X	X	X	X
B				X		X
C	X	X	X	X	X	X
Ca	X	X	X	X	X	X
Cr					X	
Cu	X		X		X	X
Fe	X		X		X	
K	X		X		X	
La					X	
Mn					X	
N		X		X		X
Na	X		X		X	X
O		X		X		X
P	X		X		X	
S	X		X		X	
Si	X	X	X	X	X	X
Ti	X		X		X	
Zn					X	

Table 4. Release test of Bio-SiC (mg/l)

Element	SBF	Beech	Sapelly	Eucalyptus
Al	0,014	0,007	0,005	0,004
Ca	93	104	97,8	86,4
Cu	0,01	0,375	0,097	0,166
Fe	<0,005	<0,005	<0,005	<0,005
P	37,1	32,4	33,8	32,3
S	4,62	6,33	5,3	4,5
Si	0,369	5,76	4,69	0,829
Ti	<0,005	<0,005	<0,005	<0,005

Cu until reaching concentration values in the order of ppm were observed. Most of the elements remain unchanged within experimental error. Therefore, it can be concluded that no important dissolution rate of these materials is observed and no adverse physiological effects in the body are expected.

Figure 2 summarises the result of a representative solvent extraction test. Cells incubated with the biomorphic SiC extracts at different concentrations showed viability rates similar to those obtained for the negative control. Only a slight decrease in cell activity could be observed when the eucalyptus and sapelly extract concentrations were 100%.

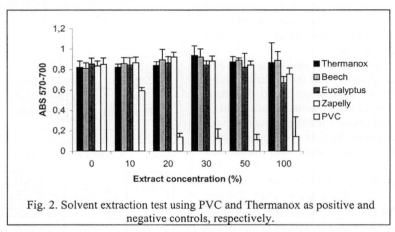

Fig. 2. Solvent extraction test using PVC and Thermanox as positive and negative controls, respectively.

This interdisciplinary study demonstrates that biomorphic SiC ceramics are promising materials for medical applications. Moreover, these ceramics can be successfully coated with a bioactive glass layer [5], which will improve the biological response of this new product.

Conclusions

Biomorphic SiC ceramics appear as an interesting alternative to Ti implants, by showing higher biomechanical requirements and lower density. The *in vitro* biocompatible behaviour of the biomorphic SiC obtained from three selected woods is demonstrated.

Acknowledgements

The authors wish to thank Ángeles Fernández for her collaboration during the experimental work. This work was supported by Xunta de Galicia (PGIDT02PXIC30302PN), Universidad de Vigo (645021908 and 6452I106) and Ministerio de Ciencia y Tecnología (MAT2001-3434).

References

[1] M. Singh: Ceram. Sci. Eng. Proc. Vol. 21 (2000), p. 39.
[2] F.M. Varela-Feria, J. Martínez-Fernández, A.R. de Arellano López and M. Singh: Ceramics Engineering Science Proceedings Vol. 23 (2002), p. 681.
[3] T. Kokubo, M. Tanahashi, T. Yao, M. Minoda, T. Miyamoto, T. Nakamura and T. Yamamuro: Bioceramics Vol. 6 (1993), p. 327.
[4] W. Lai, J. Garino and P. Ducheyne: Biomaterials Vol. 23 (2002), p. 213.
[5] P. González, J. Serra, S. Liste, S. Chiussi, B. Léon, M. Pérez-Amor, J. Martínez-Fernández, A. R. de Arellano-López and F.M. Varela-Feria: Biomaterials (2003) in press.
[6] L.J. Gibson, M.J. Ashby: *Cellular Solids* (Pergamon Press, Oxford 1988)
[7] P.L. Mangonon: *The Principles of Materials Selection for Engineering Design* (Prentice-Hall, First Edition, Upper Saddle River, New Jersey 1999)

Key Engineering Materials Vols. 254-256(2004) pp. 1033-1036
online at http://www.scientific.net
© *2004 Trans Tech Publications, Switzerland*

New Method for the Incorporation of Soluble Bioactive Glasses to Reinforce Porous HA Structures

A. F. Lemos[1,2,3], J. D. Santos[2,3] and J. M. F. Ferreira[1]

[1]. Dept. of Ceramic and Glass Engineering, University of Aveiro, Campus de Santiago, 3810-193 Aveiro, Portugal, ilemos@cv.ua.pt, jmf@cv.ua.pt
[2]. Dept. of Metalurgic and Materials Engineering, Engineering Faculty of the University of Porto, Rua Dr. Roberto Frias, 4200-465 Porto, Portugal, jdsantos@fe.up.pt
[3]. Institute of Biomedical Engineering, University of Porto, Rua do Campo Alegre 823, 4150-180 Porto, Portugal

Keywords: Macroporous structures, bioactive glass, liquid phase sintering, reinforced hydroxyapatite

Abstract. The most widely used calcium-phosphate ceramic in clinical applications is hydroxyapatite, $Ca_{10}(PO_4)_6(OH)_2$, however its dissolution rate in the human body after implantation is too low to achieve good results and it has poor mechanical properties. Recently, glass reinforced hydroxyapatite composites (GR-HA) have been developed, in order to improve the mechanical properties and to increase the similarity between implants and bone. The addition of a P_2O_5 based glass to a hydroxyapatite matrix, promotes the liquid phase sintering, improving the mechanical properties of the final bodies, and enabling the formation of two other calcium-phosphate phases, which have a higher dissolution rate, more compatible with bone regeneration. However, some of these glass compositions are soluble in water, making their incorporation in aqueous slurry impossible. Therefore, the aim of the present work is to study a new alternative method for the incorporation of soluble P_2O_5 based glass into macroporous structures obtained by different methods.

Introduction

The most widely used calcium-phosphate ceramic in clinical applications is hydroxyapatite, $Ca_{10}(PO_4)_6(OH)_2$, due to it excellent biocompatibility, bioactivity and osteoconduction characteristics [1-3]. However, many studies have indicated that the dissolution rate of HA in the human body after implantation is too low to achieve good results. On the other hand, the dissolution rate of β-tricalcium phosphate, β-TCP, is too fast for bone bonding. The development of biphasic calcium-phosphate ceramics composed of HA and β-TCP, enables to achieve a suitable dissolution rate of the implant materials, through the HA/ β-TCP ratio control [4, 5]. Another limitation of HA ceramics is their insufficient mechanical reliability for biomedical applications, where high-load strength is required.

It is well established that improved mechanical properties of HA and better chemical similarity between implanted HA materials and bone can be obtained through liquid phase sintering route using CaO-P_2O_5 glasses as a sintering aid [6, 7]. It was demonstrated that during sintering CaO-P_2O_5 glass reacts with HA forming β-TCP, which then might transform into α-TCP at higher temperatures. The relative proportions between β and α-TCP phases in the final microstructure depend upon several experimental factors, including the glass content and composition. The CaO-P_2O_5 glass presents high hygroscopic properties and is water soluble, hindering its incorporation into the aqueous powders suspensions due to its strong coagulating effect and the consequent accentuated increase in viscosity.

Therefore, this study aims at developing a new alternative method for the incorporation of soluble CaO-P_2O_5 glass into macroporous structures obtained by different methods, which involves the impregnation of pre-sintered ceramic bodies with a glass solution of a suitable concentration tailored to the pore volume fraction of the pre-sintered bodies.

Experimental Procedure

Materials. The materials used in this work were: Hydroxyapatite P215S (PlasmaBiotal, UK); ammonium polycarbonate (Targon 1128, BK Ladenburg, Germany), as a deflocculant to disperse HA suspensions; albumin, chicken egg (A) (dried egg white, grade II, Sigma Aldrich Chemie, Germany); starch, used both as consolidator agent and pore former - TRECOMEX AET1 (Lyckeby Starkelsen AB, Sweden), which is an etherficated potato starch modified by hydroxyl-propylation and cross-linking; foam-bath concentrate - FBC (Dibel, S. A., Porto, Portugal) as foaming agent; supplementary foaming agent - Sodium Lauril Sulphate – SLS (Texapon k 12, V. P., Lisbon, Portugal).

Glass Preparation. To reinforce the HA macroporous structure, a glass was prepared by melting suitable proportions of P_2O_5, $CaCO_3$, and CaF_2 in order to have the following final molar composition: $75P_2O_5$-$15CaO$-$10CaF_2$. The glass was poured in cold water to get a frit, which was then milled in ethanol until a mean particle diameter lower than 14 μm has been obtained.

Slip Preparation. A stock HA suspension was prepared with 60-vol.% solids loading and stabilised using 0.4-wt.% Targon 1128, according to the previously described [8].

Samples Preparation. The porous microstructures were prepared by using two different methods: (1) a new direct consolidation method that takes advantage of the foaming capability of egg white and of its gelatinization properties on heating, thus acting simultaneously both as pore former and as consolidator agent; (2) combination of foaming and starch consolidation methods.

Several samples were prepared according to the first method using different mass proportions between HA and A. The albumin solutions were firstly beaten and then added to aliquots of the stock HA suspension. Ceramic parts were consolidated by pouring the beaten mixtures into closed moulds followed by consolidation at 80°C for one hour. The samples were demoulded, dried and pre-sintered according to the following heating schedule: 1°C/min from 20°C up to 550°C, a holding time of 2 hours at this temperature, followed by a heating rate of 4°C/min up to 1100°C, 1 hour holding at the maximum temperature and free cooling down inside the furnace.

The samples prepared following the second method included different mass proportions of HA, FBC and SLS. 10-vol.% starch relative to HA, was added to the stock HA suspension prior the incorporation of the foaming agents. The final mixtures were poured into a closed mould and consolidated at 80°C for one hour. The pre-sintering was made following the same heating schedule as described above. Table 1 lists the sample codes and the amounts and types of foaming agents used to prepare the samples. The amounts of foaming agents are expressed as weight fractions relative to the mass of water present in the suspension. The porosity of the pre-sintered samples was evaluated in order to estimate the amounts of glass solution that might be absorbed and to tailor the glass solution concentrations to the respective porous structures. Based on these measurements, glass solutions with the required concentrations were prepared in order to incorporate about 4-wt.% of glass. These solutions were prepared under heating and magnetic stirring. The samples were immersed in the boiling glass solution for different time periods (2, 5 and 10 minutes) and then removed. The mass of glass incorporated was measured by the differences in the weight, before and after immersion/drying. Finally the samples were sintered at 1300°C for one hour, with a heating rate of 4°C/min.

Samples Characterization. The apparent density was evaluated from the weight and dimensions of the samples. XRD analyses (Rigaku, Tokyo, Japan) were carried out in the sintered samples to evaluate the final phases present in the composites. Scanning electron microscopy analysis (S-4100,

Table 1. Sample codes and amounts and types of foaming agents

Sample Code	Egg White [wt.%]	FBC [wt.%]	SLS [wt.%]
I	-	1.25	1.25
II	-	8	2
III	-	3.3	-
IV	5	-	-

HITACHI, Tokyo, Japan) performed on fracture surfaces of the samples were used to characterize the microstructure and qualitatively assess pore size and pore morphology.

Results and Discussion

Fig. 1 presents the microstructures of pure HA porous samples after sintering at 1300°C for one hour. This is to give an idea about the type of microstructures that would be obtained in absence of glass. It can be seen that the size of pores and their interconnectivity strongly depend on samples formulation. The size of pores is higher for lower total added amounts of foaming agents. On the other hand, the presence of SLS plays an important role in the establishment of pore interconnectivity (samples I and II). When egg white is used as foaming agent the porous

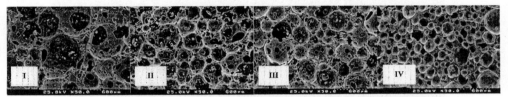

Fig. 1 Microstructures of the samples I, II, III and IV sintered at 1300 °C.

microstructure is mainly formed by closed pores of smaller size in comparison with the other samples.

Comparison of Fig. 1 and Table 2, shows that despite samples I and II have similar porosity values, the degree of pores interconnectivity is very superior in sample II. This fact can explain the superior amount of glass incorporated by sample II, as presented in Table 2, since the glass can enter through the entire pores network. In fact, pore interconnectivity seems to play the most important role in the glass impregnation in comparison with total porosity. The sample IV consisting mainly of closed pores is the one that presents the slowest impregnation kinetics and requires the longest time to impregnate the desired amount.

Table 2. Estimated values of porosity and amount of glass incorporated

Sample Code	Porosity [%]	Amount of Glass Incorporated [wt.%]		
		2 minutes	5 minutes	10 minutes
I	89.36	4	7	9
II	89.36	6	8	10
III	84.70	3	5	7
IV	75.58	2	3	5

The experimental results presented in Table 2 demonstrated that the amount of glass phase incorporated by this process can be suitably controlled by playing with the features of porous microstructure, the glass concentration and the immersion time.

Fig. 2 shows different details of the surface of the pores of impregnated samples, which are featured by the concentration of the reinforcing glass at the grain boundaries and at the surface of pores (samples I and III), as well as in the thinner parts of the struts (sample II), promoting high local densification levels. During drying of the pre-sintered and glass solution soaked samples, the liquid phase will tend to migrate to the evaporating surfaces where the soluble glassy phase after reaching a supersaturation concentration will start to precipitate preferentially. During sintering, the glassy phase will enhance the densification in the above referred parts, closing eventual microcracks existing in the struts, and therefore exerting an overall reinforcing effect of the porous structure. In the case of sample IV, mainly featured by close porosity, the impregnated glass tends to mostly concentrate onto the accessible pore surfaces given rise to higher local concentrations that change locally the pore morphology. Furthermore, the overall enrichment of the surface of pores in glassy phase gave rise to a partial transformation of HA into α and β-TCP phases as revealed by XRD analysis (not shown). This means that besides reinforcing the porous microstructure, this method of incorporating the glassy phase might also enhance the dissolution rate of the implant after the first contact with the host.

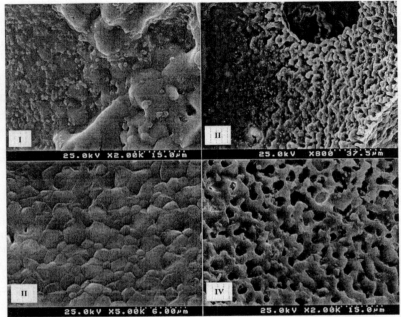

Fig. 2 Microstructures of the samples I, II, III and IV with glass incorporated (soaked on the glass solution for 5 minutes), sintered at 1300 °C.

Conclusions

The results obtained and discussed along this work show the feasibility of incorporating water soluble P_2O_5 based glass into GR-HA composite porous microstructures consolidated from aqueous slurries. Moreover, the new method disclosed transforms what could be considered a serious unsolvable drawback into other practical and functional advantages concerning the enhancement of the low dissolution rate of HA based materials to a most suitable level compatible with the bone regeneration rate. Therefore, this innovative approach has a two fold advantages: (i) the reinforcement of the porous ceramics for biomedical applications specially in the target sites; (ii) offers the possibility for tailoring the dissolution rate of the implant by partial transformation of HA into more soluble α and β-TCP phases.

References

[1] K. de Groot: *Bioceramics of Calcium Phosphates* (CRC Press, Boca Raton, FL 1983);

[2] L. L. Hench and J. Wilson: *An Introduction to Bioceramics* (World Scientific, UK 1993);

[3] L. L. Hench: J. Am. Ceram. Soc. Vol. 81 (1998), p. 1075;

[4] M. Kohri, K. Miki, D. E. Waite, H. Nakajima and T. Okabe: Biomaterials Vol. 14 (1993), p. 299;

[5] N. Krivak and A. C. Tas: J. Am. Ceram. Soc. Vol. 81 (1998), p. 2245;

[6] J. D. Santos, J. C. Knowles, R. L. Reis, F. J. Monteiro and G. W. Hastings: Biomaterials Vol. 15 (1994), p. 5;

[7] M. A. Lopes, J. C. Knowles, J. D. Santos, F. J. Monteiro and I. Olsen: J. Biomed. Mater. Res. Vol. 41 (1998), p. 649;

[8] A. F. Lemos, J. D. Santos and J. M. F. Ferreira: to be published in a special volume of Advanced Materials Forum (2003).

Key Engineering Materials Vols. 254-256(2004) pp. 1037-1040
online at http://www.scientific.net
© 2004 Trans Tech Publications, Switzerland

Designing of Bioceramics with Bonelike Structures Tailored for Different Orthopaedic Applications

A. F. Lemos[1,2,3] and J. M. F. Ferreira[1]

[1] Departamento de Engenharia Cerâmica e do Vidro, CICECO, Universidade de Aveiro, Campus de Santiago, 3810-193 Aveiro, Portugal, ilemos@cv.ua.pt, jmf@cv.ua.pt;
[2] Faculdade de Engenharia da Universidade do Porto, Rua Dr. Roberto Frias, 4200-465 Porto, Portugal;
[3] Instituto de Engenharia Biomédica, Laboratório de Biomateriais, Rua do Campo Alegre 823, 4150-180 Porto, Portugal.

Keywords: Hydroxyapatite, consolidation techniques, porous ceramics, bioceramics.

Abstract. Innovative methods for the production of porous bioceramics for implants with tailored pore volume fractions, pore sizes, pore size distributions and pore interconnectivity have been developed. The average pore sizes can vary from a few micrometers to more than one millimeter, therefore fulfilling all the needs required for components used in implantology. The methods combine in a suitable way the knowledge coming from different areas in the field of colloid and interface science, namely, on the processing of ceramic materials by traditional colloidal shaping techniques such as slip casting with a number of new direct consolidation techniques, which transform the fluid hydroxyapatite (HA) based suspensions into rigid bodies, and particularly the expertise in the production of porous ceramics by different techniques. Slip casting was preferably used for designing the cortical part of the bone, while direct consolidation techniques were used to "freeze" tailored foamy structures or suspensions containing different kinds of organic inclusions of suitable sizes and amounts that generate pores after burning out.

Introduction

There has been much interest in the development of porous synthetic bone replacement materials for the filling of both load bearing and non-load bearing osseous defects since the demonstration of improved biocompatibility of macroporous materials as compared with dense bodies [1]. Biocompatibility of HA-based materials, namely glass reinforced hydroxyapatite (GR-HA) materials have been already well established [2,3]. Consequently there is an increasing interest in the development of porous HA-based ceramic bodies with different but complementary purposes: restoration of vascularity and complete penetration of osseous tissue throughout the repaired site, applications such as scaffolds for tissue engineering, and systems for controlled delivery of drugs. The main morphological requisites for allowing bone ingrowth are the existence of open and interconnected porous, with pore diameters larger than 100 μm for proper vascularisation [4], while the main shortcoming for spreading the use of these kinds of materials is the difficulty of tailoring the porous microstructures for the desired applications. One approach involves the incorporation in the green bodies of some organic inclusions that will burn out during sintering leaving the corresponding pores in the sintered material [5]. In this case, one has to play with the size and amount of the organic inclusions to tailor the porous microstructure. The use of polymeric sponges as templates is another interesting method that was first patented by Schwartzwalder and Somers [6]. This method involves coating flexible open-cell polymer foams with slurries of ceramic particles. The polymer is burned out and the ceramic is sintered to yield a replica of the original foam. The structural properties of the ceramic foams are controlled by the characteristics of the polymer foam, such as porosity and pore size distribution, as well as the amount of slurry impregnated on the original foam and shrinkage taking place during densification. Such parameters will influence their relative density and the morphology, size and distribution of cell walls. Building a foamy structure into a suspension is another way to obtain porous ceramics and glass ceramics [7]. However, the consolidation of the as formed foam, i.e., its transformation into a rigid network, is a critical step, which can not be accomplished by using the traditional shaping methods involving liquid removal, demanding other techniques based on different setting mechanisms, which have been reviewed before [8]. This new set of "direct consolidation techniques" can make use of non-porous moulds and offer incomparable advantages in terms of shaping capability, i.e., near-net shaping and shape complexity. In fact, another important concern is the fabrication of ceramic parts with precise sizes and shapes that match in the implant sites. Recently, a big breakthrough has been made with the discovery of "starch consolidation" [9]. This technique is based on the formulation of a high concentrated suspension containing the ceramic particles and starch granules, which is then poured into a non-porous mould. Upon heating up to temperatures in the range of 55-80°C, the starch acts swell by water uptake, therefore acting as a consolidator agent, exhibiting also good binding properties that enable green machining of the ceramic parts. The pore volume fraction could be controlled within a wide range by the proposition of the starch in the mixture. However, the size of the pores is limited to the size of the starting starch granules and does not satisfy the requirements of a few hundred of microns for bone ingrowth.

This paper presents a series of innovative approaches to overcome most of the existing difficulties in terms of microstructural design and shaping capability. By suitably combining the knowledge gained in all the previous techniques, which enable to engineering implants with complex shapes and tailored porous microstructures. Namely, trabecular bone microstructures have been obtained by setting foamy structures by starch consolidation, protein consolidation or by Direct Coagulation Casting (DCC), or by using the polymeric sponge method. Slip casting from suspensions containing the required amounts of starch granules was preferably used for designing the cortical part of the bone. The combination of all these possibilities would enable to design HA-based ceramic bodies with structures mimicking those of cortical and/or trabecular bones.

Materials and Methods

A synthetic medical grade hydroxyapatite powder HA (P215S, Plasma Biotal Limited, UK) was used. Well dispersed aqueous suspensions were prepared with solids loadings as high as 60-vol.% following a suitable deagglomeration procedure, in the presence of a suitable amount of an ammonium polycarbonate, (Targon 1128, BK Ladenburg, Germany), which revealed to be a very efficient dispersant for HA powders [10]. The as prepared suspensions could then be used to consolidate simple or complex shaped ceramic products by slip casting and/or starch consolidation [9] for obtaining microstructures resembling that of cortical bone. Designing of macroporous microstructures resembling that of trabecular bone was accomplished by suitably combining the microstructural capabilities of the foaming method or the polymeric sponge method [7], with the shaping capabilities of direct consolidation methods (starch consolidation [9], protein consolidation method [11] and gel casting with polysaccharides [8]). The consolidator agents included an etherficated potato starch modified by hydroxy-propylation and cross linking, Trecomex AET1 (Lyckeby Stärkelsen AB, Sweden), albumin, chicken egg (A) (dried egg white, grade II, Sigma Aldrich Chemie, Germany), and a 80/20 mixture of two polysaccharide powders: agar (Fluka, Mr 3000-9000) and locust bean gum, i.e. the galactomannan extracted from seeds of *Ceratonia siliqua* (Sigma, USA), or a home made partially water soluble bioglass.

Results

The pictures presented in Fig. 1 reflect the major advantage of the polymeric sponge method, namely the easy control of pores size and amount of porosity of the final structure. The effectiveness of this method is strongly determined by the rheological characteristics of the suspension, particularly by the solids content and rheological behaviour, since the sponge must be impregnated with the suspension. Simultaneously, a setting agent (bioglass partially soluble in water or the 80/20 mixture of polysaccharides) must be added to the suspension in order to avoid its drainage and the consequent gradient accumulation of suspension. The result obtained after drying and burn-out of the sponge is a macroporous structure with a near total interconnection between pores. As the samples shown were cut from bigger blocks, complex shapes are difficult to obtain through this method.

Fig. 1. Macroporous HA structures obtained by the polymeric sponge method accompanied by the respective template sponge. The pictures show structures with extremely different pores sizes.

In order to obtain more complex shapes, combination of well-known methods like slip casting, foaming, starch consolidation and others, enables to tailor the porous structures according to the requirements of the site of implantation. Figs. 2 and 3 give examples of some interesting combinations.

Fig. 2A shows porous microstructures obtained by adding starch granules and wax spheres as pore formers. Such additives insure a high control of pore volume fraction and pores' characteristics, in spite of a limited capability of establishing pore interconnectivity. Combining the foaming and starch consolidation methods (Fig. 2B) enables to prepare macroporous hydroxyapatite structures with pores larger than 100 μm, which are believed to be suitable for allowing bone ingrowth. The porous structure can be tailored according to the final application by varying the fraction of the foaming agents. Contrarily to the established knowledge, the structures obtained by the foaming method are open and the pores left by the burning out of starch granules enhance pore interconnectivity.

Fig. 2. Macroporous HA structures of different shapes, obtained by different methods: (A) Starch consolidation with wax spheres; (B) Starch consolidation and foaming.

Fig. 3A demonstrates how the different consolidation techniques can be suitably combined to produce structures resembling the different parts of the bone, showing their potential for tissue engineering.

Fig. 3. Macroporous HA structures of different shapes, obtained by combining different methods: (A) 1 - Slip casting + polymeric sponge; 2,4 - Polymeric sponge, and 3 – Slip casting + starch consolidation + foaming; (B) Polymeric sponge + DCC using a partially water soluble glass (cylinder), or polysaccharides (block) as setting agents. Samples A1 and A3 present outer high density and inner macroporosity.

Conclusions

The results presented in this work show that ceramic parts mimicking the cortical or trabecular bone structures could be prepared by suitably combining slip casting from suspensions containing a suitable volume fraction of starch granules, or by using the polymeric sponge or the foaming methods combined with suitable setting agents. Furthermore, the complete bone structure (cortical + trabecular) can also be designed by combining all the above techniques.

Acknowledgments

The first author is grateful to Fundação para a Ciência e a Tecnologia of Portugal for the grant SFRH/BD/8755/2002

References

[1] K. A. Hing, S. M. Best, K. E. Tanner, W. Bondfield, P. A. Revell: J. Mater. Sci: Mater Med. Vol. 8 (1997), P. 731.
[2] J. D. Santos, J. C. Knowles, R. L. Reis, F. J. Monteiro and G. W. Hastings: Biomaterials Vol. 15 (1994), p. 5.
[3] M. A. Lopes, J. C. Knowles, J. D. Santos, F. J. Monteiro and I. Olsen: J. Biomed. Mater. Res. Vol. 41 (1998), p. 649.
[4] E. Kon, A. Muraglia, A. Corsi, P. Bianco, M. Marcacci, I. Martin, A. Boyde, I. Ruspantini, P. Chistolini, M. Rocca, R. Giardino, R. Cancedda, R. Quarto: J. Biomed. Mater. Res. Vol. 49 [3] (2000), p. 328.
[5] M. H. Prado da Silva, A. F. Lemos, M. A. Lopes, J. M. F. Ferreira and J. D. Santos: Journal of Non-Crystalline Solids Vol. 304 (2002), p. 286.
[6] K. Schwartzwalder, A.V. Somers: U.S. Patent nº 3 090 094, (1963).
[7] J. Saggio-Woyanski, C. E. Scott, W. P. Minnear: Amer. Ceram. Soc. Bull. Vol. 71[11] (1992), p. 1674.
[8] S. M. Olhero, G. Tarì, M. A. Coimbra and J. M. F. Ferreira: J. Eur. Ceram. Soc. Vol. 20 [4] (2000), p. 423.
[9] O. Lyckfeldt, J.M.F. Ferreira: J. Eur. Ceram. Soc. Vol. 18 (1998), p. 131.
[10] A. F. Lemos, J. D. Santos and J. M. F. Ferreira: in press Advanced Materials Forum (2003).
[11] O. Lyckfeldt, J. Brandt and S. Lesca: J. Eur. Ceram. Soc. Vol. 20 [14-15] (2000), p. 2551.

Key Engineering Materials Vols. 254-256(2004) pp. 1041-1044
online at http://www.scientific.net

Combining Foaming and Starch Consolidation Methods to Develop Macroporous Hydroxyapatite Implants

A. F. Lemos[1,2,3] and J. M. F. Ferreira[1]

[1] Departamento de Engenharia Cerâmica e do Vidro, CICECO, Universidade de Aveiro, Campus de Santiago, 3810-193 Aveiro, Portugal, ilemos@cv.ua.pt, jmf@cv.ua.pt;

[2] Faculdade de Engenharia da Universidade do Porto, Rua Dr. Roberto Frias, 4200-465 Porto, Portugal;

[3] Instituto de Engenharia Biomédica, Laboratório de Biomateriais, Rua do Campo Alegre 823, 4150-180 Porto, Portugal.

Keywords: starch consolidation, foaming method, hydroxyapatite, porous ceramics, bioceramics.

Abstract. Hydroxyapatite (HA) is the calcium-phosphate material with composition closest to that of bone, what makes it suitable for osseous implant purposes. This material was used to produce macroporous structures with pores larger than 100 μm, which are believed to be suitable for allowing bone ingrowth. The macroporous structures were generated and consolidated by combining foaming and starch consolidation methods. The porous structures could be tailored according to the final application by varying the proportion of different foaming agents, foam-bath concentrate (FBC) and sodium lauril sulphate (SLS). Playing with these proportions it was also possible to improve foam stability and model the size of pores and pore interconnections in order to reproduce the pore structure of natural bone.

Introduction

Hydroxyapatite (HA) is a material widely used in orthopaedic implants because of its biocompatibility and osteoconductivity [1,2]. HA has been extensively used in applications such as non-load-bearing implant material and coatings on metal implants for load-bearing applications. In porous form, the surface area of HA bodies is greatly increased, which allows more cells or tissue to be carried into the implant, in comparison with dense HA. For bone ingrowth, pores larger than 100 μm in diameter are required, which must be interconnected to guarantee the supply and the circulation of the necessary nutrients, through the ingrowth of fibrous tissue, vascular tissue and bone tissue [3]. It is generally agreed that it is very difficult to produce porous HA with a structure similar to human bone, because the control of the porous structure is a critical point. One way to overcome these difficulties is to combine the capability of microstructural design of the foaming method and the shaping capability of starch consolidation. The foaming method is based on the addition of a foaming agent to the ceramic suspension and foam is formed, through agitation. After liquid phase removal, the foam gives rise to a porous structure, which is said to be mainly composed by closed pores that remain after sintering [4]. In the manufacture of porous bioceramics by using foaming methods, the consolidation of the foamy structure built in a slurry state into a rigid network is probably the most critical step, which cannot be achieved by using the traditional shaping techniques involving liquid removal. This transformation could be carried out by a new set of emerging "direct consolidation techniques", which are based on different setting mechanisms [5-11]. In these techniques, the particulate structure of the ceramic slips is consolidated without powder compaction or removal of liquid, thus preserving the homogeneity achieved in the slurry state. The use of starch granules as consolidator agent enables the control of the overall porosity and enlarges the pore size distribution, and also allows for a closer control of the final dimensions of the ceramic parts [12].

The aim of the present work is to design macroporous HA bioceramics with a structure similar to human bone by combining foaming and starch consolidation methods.

Experimental Procedure

Materials. The materials used in this work were: Hydroxyapatite P215S (PlasmaBiotal, UK); ammonium polycarbonate (Targon 1128, BK Ladenburg, Germany) as a deflocculant to disperse HA suspensions; starch, used both as consolidator agent and pore former - TRECOMEX AET1 (Lyckeby Starkelsen AB, Sweden), which is an etherified potato starch modified by hydroxyl-propylation and cross-linking; foam-bath concentrate - FBC (Dibel, S. A., Porto, Portugal) as foaming agent; supplementary foaming agent - Sodium Lauril Sulphate – SLS (Texapon k 12, V. P., Lisbon, Portugal).

Slip Preparation. A stock HA suspension was prepared with 60-vol.% solids loading and stabilised using 0.4-wt.% Targon 1128, according to what was previously described [13]. The starch was added to the stock suspension prior to the incorporation of the foaming agents, in a proportion of 10-vol.% relative to HA.

Samples Preparation. To obtain bodies with different porous microstructures, several samples were prepared with different mass proportions between HA, FBC and SLS. The amounts of foaming agents are expressed as weight fractions relative to the mass of water present in the suspension and are presented in Table 1.

Table 1. Sample codes, total added foaming agent (TFA) and relative fractions of FBC and SLS

Sample code	I	II	III	IV	A82	A64	A55	B82	B64	B55	C82	C64	C55
TFA [wt.%]	1.7	3.3	6.7	10	2.5	2.5	2.5	5	5	5	10	10	10
FBC [%]	100	100	100	100	80	60	50	80	60	50	80	60	50
SLS [%]	-	-	-	-	20	40	50	20	40	50	20	40	50

The final mixtures were poured into a closed mould, consolidated at 80°C for 1 hour, de-moulded and dried at 60°C for 24 hours, and then at 110°C for further 24 hours. The dried samples were sintered according to the following heating schedule: 1°C/min from 100°C up to 550°C, a holding time of 2 hours at this temperature, followed by a heating rate of 4°C/min up to 1300°C, 1 hour holding at the maximum temperature and free cooling down inside the furnace. The densities of sintered bodies were determined by measuring the weight and apparent volume of the samples. Microstructural observations of the overall porous structures were performed on fracture surfaces by using a scanning electron microscope (SEM, S-4100, HITACHI, Tokyo, Japan).

Results and Discussion

The overall porosity of the samples, estimated from the apparent density data, is presented in Fig. 1.

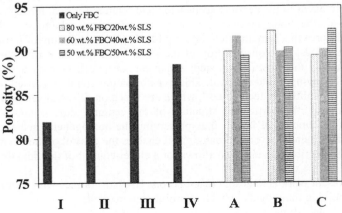

Fig.1. Estimated values of total porosity based on the determined apparent density

It is possible to observe that although FBC alone enables to obtain high porosity values, which increase with increasing added amounts of this foaming agent, its combination with SLS produces a synergetic effect, enhancing the total amount of porosity, even for lower total added amounts of foaming agents. Apparently, only small changes in the overall porosity seem to occur for TFA > 2.5 wt.% in the range of the FBC/SLS proportions tested.

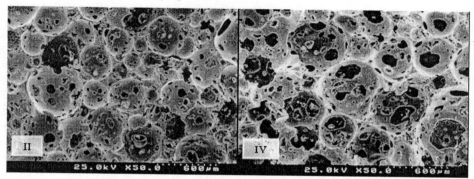

Fig. 2. Microstructure of the samples II and IV, sintered at 1300°C.

Fig. 2 shows the microstructures of the sintered porous samples obtained by adding just FBC. It can be observed that over 90 % of the total porosity consists of open and interconnected pores. This means that the fraction of closed pores is relatively modest if one considers that foaming methods tend to form essentially closed porosity [4]. These results suggest that the pores left by the burning out of starch granules enhance pore interconnectivity. This also proves the suitability of combining the foaming method with the starch consolidation method to achieve the desired goals in terms of porous microstructure. However, pore size and pore interconnectivity also depend on the added amount of foaming agent. In fact, increasing the amount of FBC from 3.3 wt.% (sample II) to 10 wt.% (sample IV) increases the size of pores and enhances pores interconnectivity.

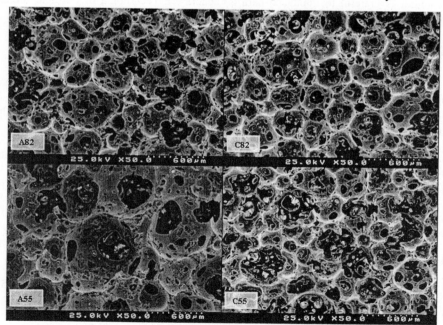

Fig. 3. Microstructure of the samples A82, A55, C82 and C55, sintered at 1300°C.

Although the SLS tends to originate abundant foam, its use alone revealed to be unsuitable because of the very low stability of the formed foam, which quickly fades.

Adding small amounts of SLS to the suspension with FBC greatly improved the foam stability and also contributed to significantly increase the size of air bubbles formed as well as the degree of their interconnections, as can be observed in Fig. 3. However, excess amounts of SLS imparted accentuated viscoelastic-type properties to the suspensions making handling and pouring very difficult.

The comparison between the samples A (TFA = 2.5 wt.%) and C (TFA = 10 wt.%), shows that increasing the total amount of foaming agents generates more pores (i.e. the average size of pores decreases), although this increase seems not to affect the size of the pore interconnections. The latter are mainly affected by the proportion of SLS, as can be clearly seen by comparison of the two pairs of microstructures (top and bottom) in Figure 3. These results allow to conclude that higher amounts of foaming agent produce more pores, and the addition of SLS promotes the increase of the number and the size of pore interconnections.

Conclusions

- Combining foaming and starch consolidation methods enables the preparation of macroporous hydroxyapatite structures with pores larger than 100 μm, which are believed to be suitable for allowing bone ingrowth;
- Using SLS as supplementary foaming agent enables to improve foam stability and to appropriately increase the size of pores and pore interconnections;
- The porous structure can be tailored according to the final application by varying the fraction of SLS and FBC.

Acknowledgments

The first author is grateful to Fundação para a Ciência e a Tecnologia of Portugal for the grant SFRH/BD/8755/2002

References

[1]   K. de Groot: *Bioceramics of Calcium Phosphates* (CRC Press, Boca Raton, FL 1983).
[2]   L. L. Hench and J. Wilson: *An Introduction to Bioceramics* (World Scientific, London, UK 1993).
[3]   H. Ohgushi, M. Okumura, T. Yoshikawa, K. Inoue, N. Senpuku, S. Tamai and E. C. Shors: J. Biomed. Mater. Res. Vol. 26 (1992), p. 885.
[4]   W. P. Minnear: *Processing of Foamed Ceramics, Ceramic Transactions, Forming Science and Technology for Ceramics Vol. 26* (ed. M.J. Cima, The American Ceramic Society, USA 1992).
[5]   O. Omatete, M. A. Janney, and R. A. Strehlow: Am. Ceram Soc.Bull. Vol. 70 (1991), p. 1641.
[6]   T. Kosmac, S. Novak, and M. Sajko: *Fourth Euro Ceramics, Vol. 1, Basic Science: Developments* (ed. C. Galassi. Gruppo Editoriale Faenza Editrice S.p.a., Italy 1995).
[7]   T. J. Graule, F. H. Baader, and L. J. Gauckler: cfi/Ber. DKH. Vol. 71 (1994), p. 317.
[8]   B. E. Novich, C. A. Sundback and R. W. Adams: *Ceramic Transactions, Vol. 26* (ed. M. J. Cima. American Ceramic Society, USA 1992).
[9]   R. D. Rives : *U. S. Patent* 4113480 (1976).
[10] O. Lycfeldt, J. Brandt, and S. Lesca : J. Eur. Ceram. Soc. Vol. 20 [14-15] (2000), p. 2551.
[11] S. M. Olhero, G. Tari, M. A. Coimbra and J. M. F. Ferreira: J. Eur. Ceram. Soc. Vol. 20 [4] (2000), p. 423.
[12] O. Lycfeldt, and J. M. F. Ferreira: J. Eur. Ceram. Soc. Vol. 18 (1998), p. 131.
[13] A. F. Lemos, J. D. Santos and J. M. F. Ferreira: in press Advanced Materials Forum (2003).

Key Engineering Materials Vols. 254-256(2004) pp. 1045-1048
online at http://www.scientific.net

The Valences of Egg White for Designing Smart Porous Bioceramics: as Foaming and Consolidation Agent

A. F. Lemos[1,2,3] and J. M. F. Ferreira[1]

1. Ceramic and Glass Engineering Department, CICECO, University of Aveiro, Campus de Santiago, 3810-193 Aveiro, Portugal, ilemos@cv.ua.pt, jmf@cv.ua.pt;
2. Engineering Faculty of University of Porto, Rua Dr. Roberto Frias, 4200-465 Porto, Portugal;
3. Biomedical Engineering Institute, Rua do Campo Alegre 823, 4150-180 Porto, Portugal.

Keywords: Egg-white, hydroxyapatite, consolidation technique, porous ceramics, bioceramics

Abstract. Porous hydroxyapatite bioceramics were developed by a new direct consolidation method based on the foaming capability of egg white and on its gelling properties on heating, thus acting both as pore former and consolidator agent. The bubble forming capability of fresh egg white, due to the presence of a globular protein – albumen, allows the formation of a stable foam when added to a ceramic suspension. Adding a sufficient amount of the globular protein to the suspension revealed to be crucial to control either the pore volume fraction as well as the gelling behaviour and the rigidity of the green bodies. This consolidation method can be performed in nonporous moulds, enabling materials with controlled porosity to be obtained by varying the egg white content in the slip composition. After drying, burnout and sintering, materials with ultimate porosities close to 75 % were obtained.

Introduction

Calcium-phosphate bioceramics have received considerable attention as bone-graft substitutes due to their biocompatibility, bioactivity and osteoconduction characteristics, being hydroxyapatite $Ca_{10}(PO_4)_6(OH)_2$ (HA) the most widely used [1].

Recently, most of the efforts have been made towards the development of porous bioceramics for implantation purposes, namely porous hydroxyapatite, since the porous network allows the tissue to infiltrate, which further enhances the tissue-implant attachment [2]. These porous materials should have the proper pore size, morphology and interconnectivity. Although, there is no agreement concerning a desirable specific value for pore size, it is well established that the pores must be interconnected.

One of the methods to produce porous materials is the foaming method. However, this method comprises a critical step, i.e., the consolidation of the foamy structure built in a slurry state into a rigid network, which cannot be achieved by using the traditional shaping techniques involving liquid removal. This transformation could be carried out by a new set of emerging "direct consolidation techniques", which are based on different setting mechanisms such as polymerisation reactions (*gel casting*) [3], polycondensation reactions (*hydrolysis assisted solidification, HAS*) [4], destabilisation of the suspension (*direct coagulation casting, DCC*) [5], freezing (*quick set*) [6], or the gelling properties of starch [7], methylcellulose [8] or proteins [9] that occur on heating, or the gelation of other polyaccharides on cooling [10] to create a three-dimensional network. In these methods, the particulate structure of the ceramic slips is consolidated without powder compaction or removal of liquid, thus preserving the homogeneity achieved in the slurry state.

The use of proteins, including bovine serum albumin (BSA), whey proteins concentrate (WPC), and albumin, as gelling agents to consolidate dense green bodies has been reported recently by Lyckfeldt *et al* [9]. The amounts of albumin added to the ceramic slurries were restricted to around 5-8-wt.% in order to keep suitable gel strength values and avoid accentuated thickening effects. Antifoaming agents had to be used to avoid pore formation in the dense bodies. Dhara *et al* [11] referred that the use of egg white ovalbumin as consolidator agent has many advantages in comparison with systems such as the MAM/MBAM used in aqueous gelcasting [12], namely, it does not need any chemical additives to initiate and complete gelation and is non-toxic, biodegradable, cheap, and widely available. Heating the ovalbumin (protein) to near 80°C causes denaturation of the protein structure involving unfolding of the coiled protein structure or alteration in the nature of the folded structure [13]. As a result of protein denaturation, the fluid slurry turned into a gel with ceramic particles bound in the gel structure [9]. The gelled green ceramic body was rigid and strong and could be easily removed from the mould.

The aim of the present work is make use of this new direct consolidation method for producing porous hydroxyapatite by taking advantage of the foaming capability of egg white and of its gelling properties on heating, thus acting simultaneously both as pore former and as consolidator agent.

Experimental Procedure

Materials. The materials used in this work were: Hydroxyapatite P215S (PlasmaBiotal, UK); ammonium polycarbonate (Targon 1128, BK Ladenburg, Germany), as a deflocculant to disperse HA suspensions; albumin, chicken egg (A) (dried egg white, grade II, Sigma Aldrich Chemie, Germany).

Slip Preparation. A stock HA suspension was prepared with 60-vol.% solids loading and stabilised using 0.4-wt.% Targon 1128, according to the previously described [14].

Foams Characterization. Before the preparation of the porous samples several tests were made to evaluate accurately the gelling temperature of the albumin and its foam capability and stability. These tests included rheological measurements of protein solutions beaten for 5 minutes with various concentrations (5-wt.% to 20-wt.% relative to mass of water). Rheological measurements were carried out in a C-VOR rheometer (Bohlin Instruments, USA).

Samples preparation. To obtain bodies with different porous microstructures, several samples were prepared with different mass proportions between HA and A. First were prepared the albumin solutions, which after being beaten were added to aliquots of the stock suspension. The rheological properties of the final mixtures were evaluated. Ceramic parts were consolidated by pouring the beaten mixtures into closed moulds followed by consolidation at a suitable temperature in the range of 60-80°C (for one hour), according to the added amounts of egg white. The samples were demoulded, dried and sintered according to the following heating schedule: 1°C/min from 20°C up to 550°C, a holding time of 2 hours at this temperature, followed by a heating rate of 4°C/min up to 1300°C, 1 hour holding at the maximum temperature and free cooling down inside the furnace.

The apparent density was evaluated from the weight and dimensions of the samples. Microstructural observations of the overall porous structures were performed on fracture surfaces by using a SEM (S-4100, HITACHI, Japan).

Results and Discussion

To characterize the gelling behaviour, the protein solutions beaten for 5 minutes, were subjected to oscillation at a constant frequency (1Hz) and strain (10^{-3}) during an increasing temperature sweep (5°C/min) from 25 to 90 °C. The results presented in Fig. 1 shown that the gelling temperature does

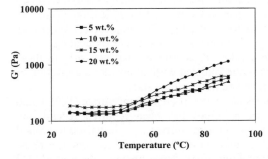

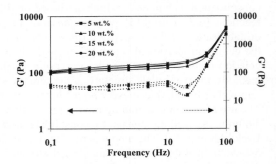

Fig. 1. Elastic modulus (G'), at a constant strain (10^{-3}), versus temperature, of protein beaten solutions.

Fig. 2. Elastic modulus (G') and viscous modulus (G''), at a constant temperature (25 °C), versus frequency, of protein beaten solutions.

not change significantly with increasing amounts of protein from about 5-20-wt.%, with the gelatinization process starting at around 50 °C for all the cases, as revealed by the increase of the elastic modulus (G') at this temperature. G´ tends to increase with increasing amounts of added A, but the differences became more salient in the presence of 20-wt.% protein beaten solution. The stability of the foam during handling and stirring at room temperature is of paramount importance because it determines the mechanical conditions under which the foam can be used before collapsing. Fig. 2 shows the results of frequency sweeps performed at 25°C. It can be seen that G'>G'' along all the frequency range, indicating that the solid-like characteristics of the foams predominate over the viscous ones. The same increasing trend, as referred to for Fig. 1 is also observed, although the changes with A concentration are less significant, since no gelatinization takes place at this temperature.

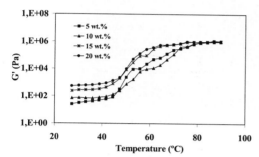

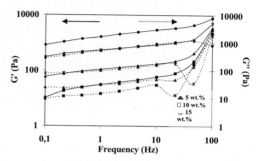

Fig. 3. Elastic modulus (G'), at a constant strain (10^{-3}), versus temperature, of hydroxyapatite suspensions with different amounts of beaten protein solutions.

Fig. 4. Elastic modulus (G') and viscous modulus (G''), at a constant temperature (25 °C), versus frequency, of hydroxyapatite suspensions with different amounts of beaten protein solutions.

The incorporation of the beaten protein solutions into the hydroxyapatite suspensions confers to the mixed systems a dramatic increase of solid-like characteristics, as shown in Fig. 3. The gelatinization process occurs faster and at a lower temperature. However, for all added amounts of beaten protein G' reaches a plateau value of about 1MPa at about 80°C, which is about three orders of magnitude higher compared with the maximum valued (20-wt.% A) presented in Fig. 1. The results presented in Fig. 4 reveal that the viscoelastic properties of the mixed systems are very sensitive to the concentration of beaten protein solutions. In fact the G' and G'' curves become more discrete when compared with Fig. 2. This confirms that the incorporation of the inorganic component enhances the consistency of the systems at room temperature, as expected, as well as the gel strength after heat treatment.

The average values of the results of apparent density, as evaluated from the weight and dimensions of the samples sintered at 1300°C, are reported in the Table 1. The estimated values of porosity are also presented as percentage relative to the true density of hydroxyapatite powder. Fig. 5 shows the porous microstructures of the samples A5, A10, A15 and A20 where a range of pore sizes suitable for bone ingrowth can be observed.

Comparison of the data from Table 1 and Fig. 5, it can be concluded that macroporous structures with pore volume fractions as high as ≈ 75-80% can be prepared by the present method.
The average size of the pores tends to decrease and the degree of pore interconnectivity increases with increasing amounts of added protein.

Table 1. Apparent density and porosity values of the porous samples prepared

Sample Code	Apparent Density [g/cm³]	Porosity [%]
A5	0.80	74.65
A10	0.69	78.14
A15	0.69	78.14
A20	0.68	78.45

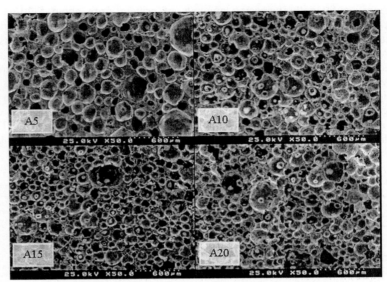

Fig. 5. Microstructures of the samples A5, A10, A15 and A20
sintered at 1300°C.

Conclusions

Adding egg white to a concentrated ceramic suspension enables to build a foamy structure, which can be directly consolidated into rigid and highly porous green body by heat-treating it at moderate temperatures. The size of the pores and the degree of interconnectivity could be modelled by the added amount of egg white. This method shows the potential for tailoring the porous microstructures for the applications. Interconnected pores with sizes in the range of a few hundred microns, suitable for bone ingrowth could be easily prepared.

Acknowledgments

The first author is grateful to Fundação para a Ciência e a Tecnologia of Portugal for the grant SFRH/BD/8755/2002

References

[1] K. De Groot: Biomaterials Vol. 1 (1980), p. 47;
[2] H. Ohgushi, M. Okumura, T. Yoshikawa, K. Inoue, N. Senpuku, S. Tamai, and E. C. Shors: J. Biomed. Mater. Res. Vol. 26 (1992), p. 885;
[3] O. Omatete, M. A. Janney, and R. A. Strehlow: Am. Ceram Soc.Bull. Vol. 70 (1991), p. 1641;
[4] T. Kosmac, S. Novak, and M. Sajko: Fourth Euro Ceramics Vol. 1, Basic Science: Developments (ed. C. Galassi. Gruppo Editoriale Faenza Editrice S.p.a., Italy 1995);
[5] T. J. Graule, F. H. Baader, and L. J. Gauckler: cfi/Ber. DKH. Vol. 71 (1994), p. 317;
[6] B. E. Novich, C. A. Sundback and R. W. Adams: Ceramic Transactions Vol. 26 (ed. M. J. Cima. American Ceramic Society, USA 1992).
[7] O. Lycfeldt, and J. M. F. Ferreira: J. Eur. Ceram. Soc. Vol. 18 (1998), p. 131;
[8] R. D. Rives, U. S. Patent 4113480 (1976).
[9] O. Lycfeldt, J. Brandt, and S. Lesca: J. Eur. Ceram. Soc. Vol. 20 [14-15] (2000), p. 2551;
[10] S. M. Olhero, G. Tari, M. A. Coimbra and J. M. F. Ferreira: J. Europ. Ceram. Soc. Vol. 20 [4] (2000), p. 423;
[11] S. Dhara, and P. Bhargava: J. Am. Ceram. Soc. Vol. 84 [12] (2001), p. 3048;
[12] M. A. Janney, O. Omatete, C. Walls, S. Nunn, R. Ogle, and G. Westmoreland: J. Am. Ceram. Soc. Vol. 81 [3] (1998), p. 581;
[13] R. Lumry, and H. Eyring: J. Phys. Chem. Vol. 58 (1954), p. 110;
[14] A. F. Lemos, J. D. Santos and J. M. F. Ferreira: to be published in a special volume of Advanced Materials Forum (2003).

IX.   TISSUE ENGINEERING SCAFFOLDS

Key Engineering Materials Vols. 254-256(2004) pp. 1051-1054
online at http://www.scientific.net

Osteogenic Potential of Cryopreserved/ thawed Human Bone Marrow-derived Mesenchymal Stem Cells

Motohiro Hirose[1], Noriko Kotobuki[1], Hiroko Machida[1], Shigeyuki Kitamura[1], Yoshinori Takakura[2], and Hajime Ohgushi[1]

[1]Tissue Engineering Research Center (TERC),
National Institute of Advanced Industrial Science and Technology (AIST),
3-11-46 Nakoji, Amagasaki, Hyogo 661-0974, Japan
[2]Department of Orthopedic Surgery, Nara Medical University,
840 Shijo-cho, Kashihara, Nara 634-8522, Japan
e-mail: motohiro-hirose@aist.go.jp

Keywords: Cryopreservation, mesenchymal stem cell, bone formation, tissue engineering

Abstract. Cryopreserved human bone marrow-derived mesenchymal stem cells were thawed and cultured on tissue culture polystylene dishes. Viability of the cells immediately after thawing was 82%. The cells treated with dexamethasone formed abundant mineralized nodules at their extracellular regions after about culture day 14, while the cultures without dexamethasone never exhibited bone formation. The calcein uptake into the bone matrices was detected in the dexamethasone treated cells with the increase in fluorescent intensity with culture periods but not in the non-treated cells. These results suggest that human bone marrow-derived mesenchymal stem cells retain bone-forming capability even after cryopreservation/ thawing.

Introduction

With the rapid advancement in tissue engineering, cultured cells will be used for construction of regenerative tissues in multiple applications. As cryopreservation technology of cultured cells has progressed [1, 2], cryopreserved cells have been expected to become cell sources for fabrication of regenerative tissues and have potentially important implications for the clinical applications toward regenerative medicine due to the limitation of fresh donor cells. We have previously reported that primary cultures of mesenchymal stem cells derived from bone marrow cells could differentiate into osteoblasts by the treatment of dexamethasone (Dex), which formed bone matrices with abundant minerals on various ceramic surfaces [3-6]. Recently, we have developed the monitoring system of mineralization processes by cultured cells using Calcium binding fluorescent dyes [7, 8]. The calcein uptake was also quantified *in situ* by using an image analyzer and the uptake of calcein well correlated with the Calcium content of the mineralized bone matrix formed by cultured cells.

In this report, we examined the viability and calcein uptake into extracellular regions of cryopreserved/ thawed human bone marrow-derived mesenchymal stem cells (hMSCs) in culture to assess whether the hMSCs after freezing and thawing could retain the bone-forming capability.

Methods

Culture of Cryoperserved/thawed hMSCs.

Cryopreserved hMSCs were thawed in MEM-. supplemented with 15% FBS. Subcultures at a cell density of 2×10^4 cells/well in Falcon 24-well tissue culture polystyrene (TCPS) dishes were performed with 1 mL of MEM-α supplemented with 15% FBS, 10 mM ß-glycerophosphate (Calbiochem), 0.07 mM ascorbic acid 2-phosphate (Wako Pure Chemical), and antibiotics with or without 100 nM Dex under a 5% CO_2 atmosphere at 37°C for 25 days.

Quantitative fluorescence analysis of calcein uptake in cryopreserved /thawed hMSCs.
To enable the assay of the calcein uptake, calcein (Dojindo) at a final concentration of 1 μg/mL was added to subcultures of hMSCs. The cells were washed twice with Calcium-free phosphate-buffered saline (PBS, Invitrogen), after which 1 mL of the culture medium was added. The fluorescence of the incorporated calcein was visualized and quantified by using an image analyzer (Amersham Pharmacia Biotech) and was also observed by using a fluorescent microscope (Olympus) at each time point during the subsequent culture periods (day 4 through day 25).

Cell viability assay
Cell viability was assessed with the LIVE/DEAD Viability assay kit (Molecular Probes) based on a simultaneous determination of living and dead cells with two probes, calcein-AM for intracellular esterase activity and ethidium homodiner-1 (EthD-1) for plasma membrane integrity [9]. Briefly, cryopreserved hMSCs were thawed in MEM-. containing 15% FBS and the working solution (2 mM of calcein-AM and 5mM of EthD-1 in PBS) was added directly to the cell suspension. After a 15-min incubation at 37°C, the stained cells were observed under a fluorescent microscope.

Results and Discussion

Viability of cryopreserved hMSCs immediately after thawing.
Cryopreserved hMSCs immediately after thawing were plated on TCPS dishes in the presence of the LIVE/DEAD Viability assay solution. Intense green fluorescence was observed in almost all of the cell cytoplasms (Fig. 1b), while red staining of nuclei was seen in few cells (Fig. 1c). The viable cells were found to be 82% of the total cells by cell counting under fluorescent microscopy. These results show that even after the cryopreservation of hMSCs, cell viability is well retained.

Osteogenic capability of cryopreserved/thawed hMSCs in culture.
Cryopreserved/ thawed hMSCs were then subcultured in the presence or absence of Dex. Under these conditions, the Dex induces primary cultures of hMSCs into osteoblasts that produce a mineralized matrix about 14 days after the start of the subculture as previously reported by our group. In order to check whether the cryopreserved hMSCs could have osteogenic potential, we observed bone-forming ability of the cells after thawing by culturing the cells with or without Dex for long culture periods. Phase-contrast microscopy demonstrated that the cells cultured in the presence of Dex showed abundant mineral deposition around the cells to form mineralized nodular aggregates (Fig. 2b) comparable to primary cultured hMSCs (data not shown). In contrast, non-treated cells did not exhibit bone-forming ability (Fig. 2d).
    Recently, we have reported a novel quantitative method for mineralization by cultured cells utilizing a Calcium-binding fluorescent dye of calcein [7, 8]. Calcein was found to be specifically incorporated and deposited into extracellular bone matrices evidenced by the co-staining with Alizarin Red S. Advanced characteristics of the method included the monitoring of the mineralization of the same specimens of cultured cells in a time dependent manner due to continuous cultivation without fixation processes of cell layers. To evaluate the degree of mineralization by cryopreserved/ thawed hMSCs, we added calcein to the culture medium and observed the cells with a fluorescent microscope and then quantified the fluorescent intensity of deposited calcein. Fluorescent microscopy demonstrated that the cells treated with Dex could have the ability of mineralization and calcein uptake (Fig. 2a). By contrast, the cultures without Dex did not display matrix formation and the calcein uptake was negligible (Fig. 2c). Measurements to quantify the amount of calcein deposited into mineralized matrices of the cells were performed using the image analyzer. Increase in fluorescent intensity as a result of the calcein uptake was clearly determined only in the cells treated with Dex as culture days proceeded (data not shown).

Taken together, the results clearly showed that cryopreserved/ thawed hMSCs maintained the Dex-dependent osteogenic potential. Furthermore, the cells were morphologically, phenotypically and functionally comparable with primary hMSCs. The facts that hMSCs could possess the differentiated activity into osteoblasts even after cryopreservation mean the clinical significance because one of the main issues we should address to realize regenerative medicine using living cells is to secure sufficient cells with high differentiated potential. Since our recent results showed that cryopreserved/ thawed hMSCs formed bone matrices on certain types of bioceramics such as alumina ceramics and hydroxyapatite scaffolds [10, 11], we believe that cryoperserved hMSCs could become a promising candidate of cell sources for fabrication of regenerative bone tissues.

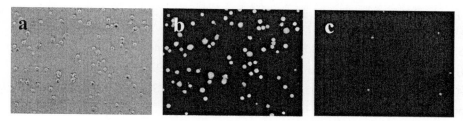

Fig. 1. Cell viability of cryopreserved hMSCs immediately after thawing .

Cryopreserved hMSCs were thawed in the presence of the LIVE/DEAD Viability assay solution. The cells with green fluorescent signals (living cells) were 82% of the total cells (b), while few dead cells were observed (c). Phase-contrast microscopy is shown in (a). The photomicrographs show the same microscopical region.

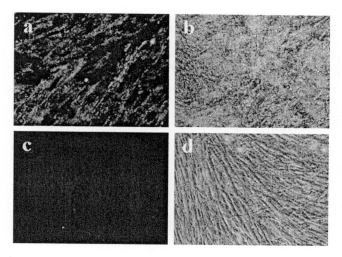

Fig. 2. Incorporation of calcein into extracellular matrices of cryopreserved/thawed hMSCs.

Cryopreserved hMSCs were thawed and cultured in the presence (a, b) or absence (c, d) of Dex with addition of calcein. At culture day 21, the cells were observed with a fluorescent microscope. Calcein-uptake into extracellular regions occurred in the Dex-treated cells (a), while did not occur in the non-treated cells at all (c). Phase-contrast micrographs of (b) and (d) show the same microscopical regions of (a) and (c), respectively.

Acknowledgments

This work was done by Three-Dimensional Tissue Module Project, METI (A Millennium Project) and in part supported by the R&D Projects in "Advanced Support System for Endoscopic and Other Minimally Invasive Surgery" entrusted from the New Energy and Industrial Technology Development Organization (NEDO) to the Japan Fine Ceramics Center. The author (M. Hirose) is grateful for a fellowship from New Energy and Industrial Technology Development Organization (NEDO) of Japan.

References

[1] E.E. Spurr, N.E. Wiggins, K.A. Marsden, R.M. Lowenthal and S.J. Ragg: Cryobiology 44:210-217 (2002)

[2] S.P. Bruder, N. Jaiswal and S.E. Haynesworth: J Cell Biochem 64:278-294 (1997)

[3] H. Ohgushi and A.I. Caplan: J Biomed Mater Res 48:913-927 (1999)

[4] H. Ohgushi, Y. Dohi, T. Katuda, S. Tamai, S. Tabata and Y. Suwa: J Biomed Mater Res 32:333-340 (1996)

[5] H. Ohgushi, Y. Dohi, T. Yoshikawa, S. Tamai, S. Tabata, K. Okumura and T. Shibuya: J Biomed Mater Res 32:341-348 (1996)

[6] H. Ohgushi, T. Yoshikawa, H. Nakajima, S. Tamai, Y. Dohi and K. Okunaga: J Biomed Mater Res 44:381-388 (1999)

[7] E. Uchimura, H. Machida, N. Kotobuki, T. Kihara, S. Kitamura, M. Ikeuchi, M. Hirose, J. Miyake and H. Ohgushi: Calcif Tissue Int, to appear

[8] M. Hirose, N. Kotobuki, H. Machida, E. Uchimura and H. Ohgushi: Key Eng Mater 240:715-718 (2003)

[9] N.G. Papadopoulos, G.V.Z. Dedoussis, G. Spanakos, A.D. Gritzapis, C.N. Baxevanis and M. Papamichail: J Immunol Meth 177:101-111 (1994)

[10] H. Ohgushi, H. Machida, M. Ikeuchi, T. Tateishi, Y. Tohma, Y. Tanaka and Y. Takakura: Key Eng Mater 240:651-654 (2003)

[11] S. Kitamura, M. Hirose, H. Funaoka, Y. Takakura, H. Ito and H. Ohgushi: in preparation

Key Engineering Materials Vols. 254-256(2004) pp. 1055-1058
online at http://www.scientific.net
© *2004 Trans Tech Publications, Switzerland*

In Vitro Osteogenic Activity of Rat Bone Marrow Derived Mesenchymal Stem Cells Cultured on Transparent Hydroxyapatite Ceramics

Noriko Kotobuki[1], Daisuke Kawagoe[2], Hirotaka Fujimori[2], Seishi Goto[2],

Koji Ioku[2], and Hajime Ohgushi[1]

[1]Tissue Engineering Research Center (TERC),

National Institute of Advanced Industrial Science and Technology (AIST),

3-11-46 Nakoji, Amagasaki, Hyogo 661-0974, Japan

n.kotobuki@aist.go.jp

[2] Division of Applied Medical Engineering Science, Graduate School of Medicine,

Yamaguchi University, 2-16-1 Tokiwadai, Ube, Yamaguchi 755-8611, Japan

ioku@po.cc.yamaguchi-u.ac.jp

Keywords: Transparent hydroxyapatite ceramic, mesenchymal stem cell, rat bone marrow

Abstract. Direct observation of cultured cells on various materials has benefit for assessment of fundamental cellular functions including cell attachment, spreading, proliferation and differentiation on the materials. For this purpose, we made transparent hydroxyapatite (tHA) ceramics, where cultured rat bone marrow-derived mesenchymal stem cells (rMSCs) could be visible by phase contrast microscopy immediately after plating. The seeded rMSCs adhered and proliferated on the tHA ceramics. Furtheremore, the cells cultured on the tHA ceramics were able to differentiate into osteoblasts under osteogenic conditions and showed comparable osteogenic activity to the cells cultured on tissue culture polystyrene dishes. The results suggested that cultured cells could be monitored in a time dependent manner using the tHA ceramics.

Introduction

Tissue engineering is the recent attractive technology creating living tissues or organs utilizing cultured cells and their scaffolds *in vitro*. Hydroxyapatite (HA) ceramics can play important roles in providing three-dimensional scaffolds for adhesion, proliferation, and differentiation of the cultured cells especially bone-related cells. We have previously reported that bone marrow derived mesenchymal stem cells (MSCs) could be cultured on HA ceramics [1-4]. Importantly, the cultured MSCs further differentiated into osteoblasts under osteogenic conditions to fabricate bone matrices on the HA ceramic surfaces. The cultured osteoblast/matrix constructs showed *in vivo* osteogenic capability evidenced by new bone formation after *in vivo* implantation. Therefore, we defined the constructs as *regenerative cultured bone tissues* [5]. When the cells were cultured on tissue culture polystyrene (TCPS) dishes known to be suitable culture surfaces for cells, they could be easily observed by light microscopy. However, the cells cultured on various types of ceramics with three-dimensional structures are usually hard to be observed by light microscopy because of their opacity. In order to overcome the shortcoming, we made transparent hydroxyapatite (tHA) ceramics to observe cultured cells directly by light microscopy [6]. In this report, we observed the rat MSCs (rMSCs) cultured on the tHA ceramics by phase contrast microscopy and checked the osteogenic activity of the rMSCs on the tHA ceramics under osteogenic conditions by biochemical analyses.

Materials and Methods

Ceramics preparation:

A fine powder of HA (High-purity grade) was utilized as a starting material. About 1g of this powder was poured into a graphite mold (inner diameter: 15 mm), and then sintered by the spark plasma sintering method (Sumitomo Coal Mining). Temperatures of the samples during sintering were measured by thermocouples of Rh/ Pt-Pt, which was inserted into a wall of the graphite mold to measure the sample temperature. The samples were pressed uniaxialy under 10 MPa, and then they were heated at 800°C, 900°C and 1000°C for 10 min with a heating rate of 25°C/ min. The ceramics were polished finely with using a paste containing Al_2O_3 fine particles smaller than 0.5 mm in size.

Culture methods:

Rat bone marrow cell plugs were obtained from the SD female, 7-week-old rats and were flushed out by culture medium; minimum essential medium (MEM) containing 15% fetal bovine serum (FBS) and antibiotics. The MSCs from rat bone marrow were primarily cultured up to 80% confluent in culture flasks and resuspended to 5×10^5 cells/ mL in culture medium following harvested using 0.05% trypsin/ 0.53 mM EDTA. The cell suspension was applied on sterilized tHA ceramics (5 mm in diameter X 3 mm in thickness). These cells/ tHA composites were cultured with osteogenic medium containing 10 nM dexamethasone (Dex), 10 mM beta-glycerophosphate and 0.28 mM ascorbic acid 2-phosphate. Then culture medium were changed 2-3 times per week. During the culture periods, the cell morphology was detected by phase contrast microscopy.

Alkaline phosphatase (AP) activity staining:

The cell/ tHA composites were washed with phosphate buffer saline (PBS) and preserved in 4% paraformardehyde. Then fixed cells/ tHA composites were soaked in 0.1% naphthol AS-MX phosphate and 0.1% fast red violet LB salt in 56 mM 2-amino-2-methyl-1, 3-propanediol for 10 min. Following wash step with PBS, stained HA were observed by microscopy.

Results and Discussions

Scanning electron microscopy showed that the grain structures of the calcined tHA surface was fine (less than 1 μm) (data not shown).

When rMSCs were seeded on tHA ceramics, we found that the cells actively adhered and proliferated on the ceramics as well as on TCPS dishes by phase contrast microscopy (Fig. 1). Morphology of the cells cultured on the surfaces of tHA ceramics showed round shape in early phase and became spindle like shape in late phase. These morphological changes are similar to that on the surfaces of TCPS dishes. Furthermore, the rMSCs could differentiate into osteoblasts to form extracellular bone matrices around cell surfaces in the presence of Dex. After 2 weeks cultivation, alkaline phosphatase (AP) activity of the rMSCs cultured on tHA ceramics were measured. Both tHA ceramics and TCPS dishes in Dex-containing medium were stained in red color, which give a sign of high AP activity (Fig. 2, high AP activity is shown as black area).

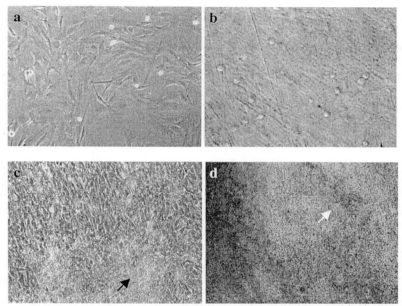

Fig. 1: Phase contrast microscopy of rat bone marrow derived mesenchymal stem cells (rMSCs) cultured on TCPS dishes (a, c) and tHA ceramics (b, d). The cells spread after 1 day culture (a, b), and differentiated into osteoblasts after 14 days culture (c, d). Arrows show the typical regions of extracellular mineralization.

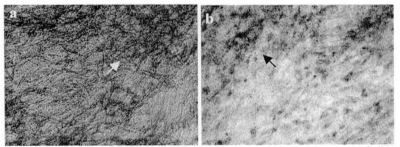

Fig. 2: AP activity staining of rMSCs cultured on TCPS dishes (a) and tHA ceramics (b) after 14 days culture in the presence of 10 nM Dex. Arrows show the typical stained regions.

As described above, the rMSCs could be cultured on tHA ceramic surfaces, moreover the living cells on tHA ceramics were simply observed by phase contrast microscopy without fixation step at any time of the sequencing culture. The present results showed that the cells could attach on tHA ceramic surfaces and the shape of the cells cultured on tHA ceramics was similar to that on TCPS dishes. Furthermore, the tHA ceramic surfaces had the capability to support osteogenic differentiation of the rMSCs into osteoblasts and the differentiation cascade could be easily detected by simple observation using light microscopy, suggesting that tHA ceramic surfaces showed equal characteristics to TCPS dishes.

For observation of living cells, we have taken advantage of transparent ceramics such as alumina ceramics [7]. HA ceramics have high biocompatible characteristics and support cell adhesion. However, there are a lot of questions to be clarified regarding behaviors of cultured cells on the HA ceramics because of inability to observe the cultured cells in a time dependent way. In this paper, we succeeded in creating tHA ceramics to monitor living cultured cells on tHA ceramics. We believe that the tHA ceramics would hold promise to accelerate development of hard tissue engineering.

Acknowledgments

This work was done by Three-Dimensional Tissue Module Project, METI (A Millennium Project) and in part supported by the R&D Projects in "Advanced Support System for Endoscopic and Other Minimally Invasive Surgery" entrusted from the New Energy and Industrial Technology Development Organization (NEDO) to the Japan Fine Ceramics Center.

References

[1] H. Ohgushi, Y. Dohi, T. Katuda, S. Tamai, S. Tabata and Y. Suwa: J Biomed Mater Res, Vol. 32 (1996), p. 333-340

[2] H. Ohgushi, Y. Dohi, T. Yoshikawa, S. Tamai, S. Tabata, K. Okumura and T. Shibuya: J Biomed Mater Res, Vol. 32(1996), p. 341-348

[3] H. Ohgushi, T. Yoshikawa, H. Nakajima, S. Tamai, Y. Dohi and K. Okunaga: J Biomed Mater Res, Vol. 44 (1999), p. 381-388

[4] H. Ohgushi and A. I. Caplan: J Biomed Mater Res, Vol. 48 (1999), p. 913-927

[5] E. Uchimura, H. Machida, N. Kotobuki, T. Kihara, S. Kitamura, M. Ikeuchi, M. Hirose, J. Miyake and H. Ohgushi: Calcif Tissue Int, to appear

[6] K. Ioku, D. Kawagoe, H. Toya, H. Fujimori, S. Goto, K. IShida, A. Mikuni and H. Mae: Trans Mater Res Soc J, Vol. 27 (2002), p. 447-449

[7] H. Ohgushi, H. Machida, M. Ikeuchi, T. Tateishi, Y. Tohma, Y. Tanaka and Y. Takakura: Key Eng Mater., Vol. 240 (2003), p. 651-654

Key Engineering Materials Vols. 254-256(2004) pp. 1059-1062
online at http://www.scientific.net
© 2004 Trans Tech Publications, Switzerland

The Functional Expression of Human Bone-Derived Cells Grown on Rapidly Resorbable Calcium Phosphate Ceramics

H. Zreiqat[1], G. Berger[2], R.Gildenhaar[2], C. Knabe[1,3]

[+1]Department of Pathology, School of Medical Sciences, UNSW, Sydney 2052, Australia,
H.Zreiqat@unsw.edu.au

[2] Federal Research Institute for Material Research and Testing, Laboratory of Biomaterials, Unter den Eichen 87, 12200 Berlin, FRG,
georg.berger@bam.de; renate.gildenhaar@bam.de,

[1,3]Department of Experimental Dentistry, University Hospital Benjamin Franklin, Germany
Christine.knabe@medizin.fu-berlin.de

Keywords: Calcium phosphate ceramics, human bone-derived cells, cell-biomaterial interactions, bone-substitutes, *in situ* hybridization.

Abstract. The use of biodegradable bone substitutes is advantageous for alveolar ridge augmentation, since it avoids second-site surgery for autograft harvesting. This study examines the effect of novel, rapidly resorbable calcium phosphates on the expression of bone-related genes and proteins by human bone derived cells (HBDC) and compares this behavior to that of tricalciumphosphate (TCP). Test materials were α-TCP, and four materials created from ß-Rhenanite and its derivatives: R1-ß-Rhenanite ($CaNaPO_4$); R1/M2 - composed of $CaNaPO_4$ and $MgNaPO_4$; R1+SiO_2 composed of $CaNaPO_4$ and 9% SiO_2 (wt%); and R17 - $Ca_2KNa(PO_4)_2$. HBDC were grown on the substrata for 3, 5, 7, 14 and 21 days and probed for Type I collagen, osteocalcin, osteopontin, osteonectin, alkaline phosphatase and bone sialoprotein mRNA and proteins. All substrata supported continuous cellular growth for 21 days when R1+SiO_2 and R17 had the highest number of HBDC. At 14 and 21d, cells on R1 and on R1+SiO_2 displayed significantly enhanced expression of all osteogenic proteins. Since all novel calcium phosphates supported cellular proliferation together with expression of bone-related proteins at least as much as TCP, these ceramics can be regarded as potential bone substitutes. R1 and R1+SiO_2 had the most effect on osteoblastic differentiation, thus suggesting that these materials may possess a higher potency to enhance osteogenesis than TCP.

Introduction

Bioactive calcium phosphate ceramics are widely used as bone substitute materials in reconstructive surgery because of their biocompatibility [1,2] and osteoconductivity [1,2]. The use of TCP particles as alloplastic bone graft materials for alveolar ridge augmentation and sinus floor elevation procedures has received increasing attention in implant dentistry [3,4]. However, biodegradation of β-TCP has been reported to be incomplete [4]. In non-load-bearing applications such as alveolar ridge augmentation, a biomaterial used as a bone substitute should be a temporary material serving as a scaffold for bone remodeling and must be rapidly resorbable, especially when endosseous dental implants are to be inserted [3,4]. Thus, considerable efforts have been undertaken to produce rapidly resorbable bone substitute materials which stimulate enhanced bone formation at the interface combined with a high degradation rate. This has led to the development of rapidly resorbable calcium alkali orthophosphate ceramics. These materials have a higher solubility and higher degree of biodegradability than tricalciumphosphate (TCP) [5], and therefore could be near optimum alloplastic materials.

This study examines the effect of novel rapidly resorbable calcium alkali orthophosphates as compared to α-TCP on the expression of bone-related genes and proteins by HBDC.

Materials and Methods

Preparation of Disks. The following four calcium phosphate materials which were created from ß-Rhenanite (CaNaPO$_4$) and its derivatives were tested and compared to α-TCP: ß-Rhenanite (CaNaPO$_4$) was denominated R1. Others resulted from modification of CaNaPO$_4$ using magnesium or potassium phosphate or silicate; viz. were made from CaNaPO$_4$ and 20mol% Mg$_2$SiO$_4$ (material denominated R1/M2); CaNaPO$_4$ and 9% SiO$_2$ (wt%) (material denominated R1+ SiO$_2$); CaKPO$_4$ and CaNaPO$_4$ react by forming Ca$_2$KNa(PO$_4$)$_2$ (material denominated R17). Preparation of all these materials has been described in detail elsewhere [5,6]. All specimens were prepared by compressing powder (powder size 40μm) followed by sintering to form 10-mm diameter discs [6]. The ceramic specimens were sterilized at 300°C for 3 h. Phase transformations do not occur below 600°C. Tissue culture polystyrene served as a control (Co).

HBDC were obtained as described previously by enzymatic digestion [7]. HBDC were cultured in alpha-Minimal Essential Medium, 10% (v/v) fetal calf serum, 2mM L-glutamine, 25mM Hepes Buffer, 30μg/ml penicillin, 100μg/ml streptomycin and 0.1M L-ascorbic acid phosphate. All test materials were preincubated in 500μl of culture medium for 24 hours without cells. HBDC were seeded at a density of 2.83x10^4 cells/cm^2 on the test substrata for 3, 5, 7, 14 and 21 days. At the predetermined time point, cells were harvested from the test surfaces by using 0.01% trypsin/0.02% EDTA and plated at a concentration of 1 x 10^4 into wells of 96-well plates, centrifuged, dried, fixed and probed for various intracellular mRNAs and proteins. The amounts of probe and antibody bound, were quantitatively determined as described previously [6] and the results normalized to the internal control β-actin mRNA and protein. The mRNAs of the following proteins were determined: osteocalcin (OC), osteopontin (OP), osteonectin (ON), alkaline phosphatase (ALP), pro-collagen Iα2 (Col Ia2). Intracellular protein production by HBDC was measured by immunohistochemical techniques; the primary antibodies used were against collagen type I (Col I), ALP, OP, OC, ON and bone sialoprotein (BSP).

Results

All substrates supported continuous cellular growth for 21 days (Fig. 1). At day 14, all test surfaces displayed higher cell numbers than the control (Fig. 1), and by day 21 all novel calcium phosphates had as many or more cells than TCP (Fig. 1). At day 3, HBDC on R1 expressed significantly higher mRNA levels for Col Iα2 compared to cells grown on Co and R17 ($p<0.002$). Also for ALP and ON mRNA were significantly higher on R1 compared to Co, TCP, R17 and R1/M2 ($p<0.02$). Protein production for Col I, ALP, OP, OC, ON, and BSP was higher on R1 compared to all other surfaces tested ($p<0.05$). Cells cultured on R1+SiO2 had significantly higher levels of Col Iα2, ALP and ON mRNAs compared to α-TCP, R17, R1/M2 and the control ($p<0.05$).

At day 14, HBDC grown on R1/M2 and R1+SiO$_2$ expressed significantly higher levels of Col Iα2, ALP and OP mRNAs than on TCP ($p<0.04$). Col I, OP, OC, ON and BSP protein levels were significantly higher for R1 compared to all other surfaces ($p<0.008$) (Fig. 2). At day 14, significantly higher protein levels for Col I, ALP, OP, OC, ON and BSP were expressed on R1+SiO$_2$ compared to Co, TCP, R17, and R1/M2 ($p<0.005$) (Fig. 2).

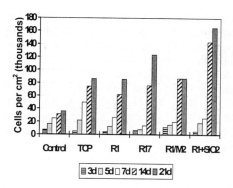

Fig. 1: Number of HBDC cultured over 21 days on different biomaterials

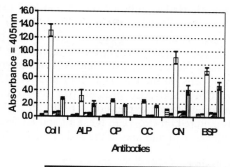

Fig. 2: The expression of osteogenic proteins produced by cells cultured for 14 days on different ceramics.

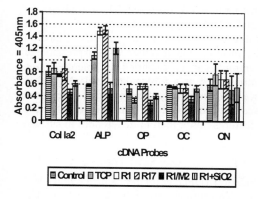

Fig. 3: The expression of osteogenic mRNAs produced by cells cultured for 21 days on different ceramics.

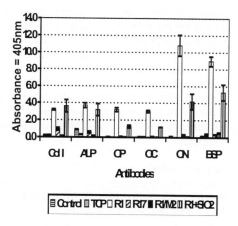

Fig. 4: The expression of osteogenic proteins produced by cells cultured for 21 days on different ceramics.

At day 21, HBDC cultured on R1 and R17 expressed significantly higher levels of ALP, and OP mRNAs (Fig. 3) than cells on TCP ($p<0.003$). At the protein level the pattern observed at day 14 was also maintained at day 21. In detail, protein expression by HBDC cultured on R1 was significantly higher for OP, OC, ON and BSP than in cells on all other substrata ($p<0.01$) (Fig. 4). Protein production for Col I and ALP was significantly higher in cells grown on R1 and R1+SiO2 than in cells on the Co, TCP, R17 and R1/M2 specimens ($p<0.005$), and cells on R17 expressed more Col I and ALP protein than cells on TCP ($p<0.02$) (Fig. 4). The same was true when comparing protein expression for OP, OC, ON and BSP by HBDC grown on R1+SiO2 to that in cells cultured on Co, TCP, R1, R17 and R1/M2 surfaces ($p<0.05$) (Fig. 4). Moreover, R1+SiO2 had the highest cell numbers at the end of the incubation period (Fig. 1).

Discussion
The various calcium phosphates of the study significantly affected cellular growth and the temporal expression of an array of bone-related genes and proteins. R1 had the most effect on the differentiation of HBDC, inducing mRNA and protein expression of osteopontin, osteocalcin, osteonectin, and bone sialoprotein protein suggesting later osteoblast differentiation. These four proteins have been tightly linked to osteoid production and matrix mineralization, suggesting that R1 may posses a higher potency to promote osteogenesis and matrix calcification, than TCP and the other calcium phosphates tested. At 5, 7 and 14 days cell numbers on R1 were lower than on TCP, indicating that stimulation of cell differentiation over proliferation had occurred. These findings are in agreement with those reported previously [6]. R17 and R1/M2 supported the expression of the osteoblastic phenotype mostly to the same or a higher degree than cells grown on TCP. Since all novel calcium alkali orthophosphates supported cellular proliferation together with expression of osteogenic markers at least as much as TCP, these ceramics can be regarded as potential bone substitutes. Hence, their biocompatibilty has been demonstrated at a molecular level. All tested calcium phosphate ceramics displayed a similar surface roughness [6], therefore the differences detected in the amount of osteogenic markers expressed between these surfaces suggest that the differences seen have to be attributed to compositional features rather than to differences in surface roughness

Acknowledgements
This work was funded by the German Research Foundation (DFG grant no. KN 377/2-1) and Australian National Health and Medical Research Council. The authors gratefully acknowledge Dr. Larry Fisher for providing the polyclonal antibodies used in this study and Dr. W. R. Walsh for providing some of the cDNA probes.

References
[1]   M. J. Yaszemski, R. G. Payne, W. C. Hayes, R. Langer, A. C. Mikos: Biomaterials Vol. 17 2 (1996), p. 175.
[2]   J. O. Hollinger, J. Brekke, E. Gruskin, D. Lee: Clin. Orthop. Vol. 324 (1996), p.55
[3]   H. Schliephake, T. Kage: J Biomed Mater Res Vol. 56 (2001), p. 128.
[4]   I. R. Zerbo, A. L. Bronckers, G. L. de Lange, G. J, van Beek, E. H. Burger: Clin Oral Implants Res Vol. 12(2001), p. 384.
[5]   G. Berger, R. Gildenhaar, U. Ploska: In: Wilson J, Hench LL, Greenspan DC: Bioceramics 8. (Butterworth-Heinemann, Oxford 1995)
[6]   C. Knabe, R. Gildenhaar, G. Berger, W. Ostapowicz, R. Fitzner, R. J. Radlanski, U. Gross: Biomaterials Vol 18 (1997), p. 1347.
[7]   H. Zreiqat and C. R. Howlett: J. Biomed. Mater. Res. Vol. 47(1999), p. 360. Web: http://www.ttp.net

Key Engineering Materials Vols. 254-256(2004) pp. 1063-1066
online at http://www.scientific.net
© 2004 Trans Tech Publications, Switzerland

Osteogenic Potential of Multi-layer-Cultured Bone Using Marrow Mesenchymal Cells - For Development of Bio-Artificial Bone

T. Yoshikawa[1], J. Iida[2] and Y. Takakura[2]

Departments of [1]Orthopaedic Surgery and [2]Diagnostic Pathology, Nara Medical University,

Kashihara City, Nara 634-8522, Japan

tyoshi@naramed-u.ac.jp

Keywords: marrow cell, dexamethasone, multiplayer, osteogenesis

Abstract. Bone marrow cells have been reported to contain stem cells and to show bone tissue-regenerating capacity. However, since the osteogenetic capacity of adult human marrow cells is unreliable, the development of a more stable culture technique is needed to apply these cells to clinical osteogenesis. For development of bio-artificial bone, we examined osteogenic potential of multi-layered-cultured bone tissue. Bone marrow cells were collected from the femurs of 7-week-old male Fischer rats, placed into two T75 flasks, and then incubated in standard medium, which was MEM containing 15% bovine fetal serum. Two weeks later, the cultures were trypsinized, and 1/10 of the total number of cells was placed into each of four T75 flasks and incubated in standard medium for multi-layer culture. The remaining cells were seeded into 6-well plates and secondary culture was carried out in an osteogenic medium prepared by adding 10 nM dexamethasone etc. to the standard medium (monolayer group). After one week of secondary culture in the two T75 flasks, cells were trypsinized, seeded again at the same concentration, and then incubated in osteogenic medium for an additional two weeks (two-layer group). After one week of multi-layer culture, the remaining cultured cells were trypsinized, seeded again at the same concentration for further multi-layer culture, and incubated for an additional one week (three-layer group). After three weeks of secondary culture, the ALP activity of the cultured tissue was measured. ALP activity increased with an increase in the number of cell layers and the three-layer cultures showed approximately twice the ALP activity of the monolayer cultures. Thus, formation of a multi-layer culture caused osteoblastic activity to increase. This method could provide more active cells for artificial bone, so that a bone substitute with a greater osteoblastic activity can be obtained more efficiently than by other procedures, even when there is no difference in the number of marrow cells used.

Introduction

Bone marrow cells have been reported to contain stem cells [1] and to show bone tissue-regenerating capacity [2-14]. However, since the osteogenetic capacity of adult human marrow cells is unreliable, the development of a more stable culture technique is needed to apply these cells to clinical osteogenesis. When marrow mesenchymal cells are incubated in medium containing dexamethasone, these cells differentiate into osteogenetic cells that produce bone matrix [4]. In general, proliferation of cultured cells on a culture dish stops when the cells become confluent. In this marrow culture system, however, osteogenetic cells proliferate further while producing bone matrix and become multi-layered, differentiating into osteoblasts and then osteocytes. Thus, there is a transformation from cell culture to tissue culture. This tendency of the cells to form a multi-layer during culture was employed to create artificial bone by guiding the growth of cultured cells to create layers. Here we report on our success in creating artificial bone.

Materials and Methods

Bone marrow cells were collected from the femurs of 7-week-old male Fischer rats, placed into two T75 flasks, and then incubated in standard medium, which was MEM containing 15% bovine fetal serum and an antibiotic. Two weeks later, the cultures were trypsinized, and 1/10 of the total number of cells was placed into each of four T75 flasks and incubated in standard medium for multi-layer culture. The remaining cells were seeded into 6-well plates at a concentration of 10^4 cells/cm^3 and secondary culture was carried out in an osteogenic medium prepared by adding 10 nM dexamethasone, 82 µg/mL vitamin C phosphate, and beta-glycerophosphate to the standard medium (monolayer group). After one week of secondary culture in the two T75 flasks, cells were trypsinized, seeded again at the same concentration, and then incubated in osteogenic medium for an additional two weeks (two-layer group). After one week of multi-layer culture, the remaining cultured cells were trypsinized, seeded again at the same concentration for further multi-layer culture, and incubated for an additional one week (three-layer group) [Fig.1]. After three weeks of secondary culture, the ALP activity of the cultured tissue was measured. As the control, a monolayer of cells was incubated in standard medium [Dex (-) group].

Results and discussion

The ALP activity was 3.0 ± 0.4, 12.4 ± 1.6, 16.4 ± 3.2, and 23.2 ± 2.0 (PN µmol) in the Dex (-), monolayer, two-layer, and three-layer cultures, respectively [Fig.2]. As is apparent from these data, ALP activity increased with an increase in the number of cell layers and the three-layer cultures showed approximately twice the ALP activity of the monolayer cultures. Thus, formation of a multi-layer culture caused osteoblastic activity to increase.

To produce artificial bone, marrow mesenchymal cells can be seeded on an artificial bone substrate and incubated in osteogenic medium, but cellular proliferation is poor on such artificial bone. In contrast, when cells are seeded into T75 flasks, which allow rapid proliferation, a multi-layer structure is formed during incubation. This method could provide more active cells for artificial bone, so that a bone substitute with a greater osteoblastic activity can be obtained more efficiently than by other procedures, even when there is no difference in the number of marrow cells used. In addition, it was found that the osteoblastic activity of artificial bone was at least two-fold greater when estriol was added to the osteogenic medium than when it was not added [15]. (This result will be reported separately). The application of multi-layer culture and addition of estriol allow the activation of artificial bone. This method for creation of activated artificial bone has already been approved by the Ethics Committee of our University and is now under clinical investigation [14].

multiple-layered cultured bone

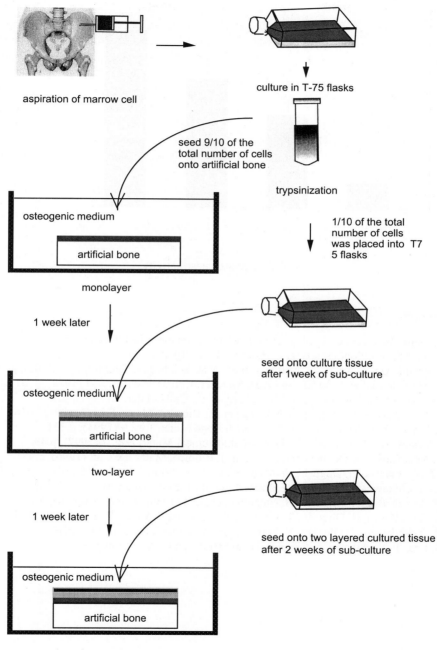

aspiration of marrow cell

culture in T-75 flasks

seed 9/10 of the total number of cells onto artiificial bone

trypsinization

osteogenic medium

artificial bone

monolayer

1/10 of the total number of cells was placed into T7 5 flasks

1 week later

seed onto culture tissue after 1week of sub-culture

osteogenic medium

artificial bone

two-layer

1 week later

seed onto two layered cultured tissue after 2 weeks of sub-culture

osteogenic medium

artificial bone

three-layer cultured bone

Fig.1.   Fabrication of multiple layer cultured bone

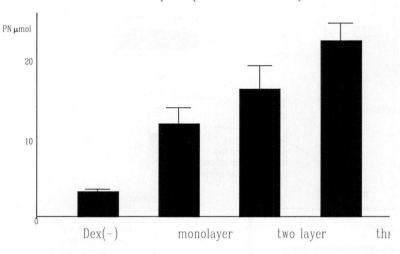

Fig.2. Alkaline phosphatase activity

References

[1] Y. Jiang, B.N. Jahagirdar, R.L. Reinhardt, et al.: Nature 418 (2002), p.41-9.

[2] T. Yoshikawa : Mat. Sci. Eng. C-Biomim. 13(2000), p.29-37.

[3]T. Yoshikawa, T. Noshi, H. Mitsuno, et al.: Mat. Sci. Eng. C-Biomim., 17 (2001), p.19.

[4] C. Maniatopoulos, J. Sodek, and A.H. Melcher: Cell Tissue Res. 254 (1988), p. 317.

[5] J.E. Davies, R. Chernecky, B. Lowenberg, et al.: Cells Mater. 1 (1991), p.3.

[6]T. Yoshikawa,S.A. Peel, J.R. Gladstone, et al.: Biomed. Mater. Eng. 7 (1997) , p369.

[7] K-L Yao , R.Jr. Todescan and J. Sodek: J. Bone Miner. Res. 9 (1994), p.231.

[8]T. Yoshikawa, H. Ohgushi, Y. Dohi, et al.: Biomed. Mater. Eng. 7 (1997), p.49.

[9] T. Yoshikawa, H. Ohgushi and S. Tamai: J. Biomed. Mater. Res. 32 (1996), p.481.

[10] T. Yoshikawa, H. Ohgushi, Akahane M, et al.: J. Biomed. Mater. Res. 41(1998), p.568.

[11]T. Yoshikawa, H. Nakajima, E. Yamada, et al.: J. Bone Miner. Res. 15 (2000), p. 1147.

[12]T. Yoshikawa, H. Ohgushi, H. Nakajima, et al.: Transplantation. 69(2000), p.128.

[13]T. Yoshikawa, H. Ohgushi, T. Uemura, et al.: Biomed. Mater. Eng. 8 (1998), p.311.

[14]T. Yoshikawa , T. Ohmura, Y. Sen et al.:Bioceramics Vol.15. (2003), p383.

[15]J. Iida, T. Yoshikawa, N.Okumura, et al.: Bioceramics Vol.16 in press.

Key Engineering Materials Vols. 254-256(2004) pp. 1067-1070
online at http://www.scientific.net
© 2004 Trans Tech Publications, Switzerland

In vitro Study on the Osteogenetic Capacity of Expanded Human Marrow Mesenchymal Cells - For Development of Advanced Bio-Artificial Bone

K. Miyazaki[1], T. Yoshikawa[1,2], K. Hattori[1], N. Okumura[2], J. Iida[1] and Y. Takakura[1]

Departments of [1]Orthopaedic Surgery and [2]Diagnostic Pathology, Nara Medical University,

Kashihara Nara 634-8522, Japan.

tyoshi@naramed-u.ac.jp

Keywords: human, marrow cell, osteogenesis, dexamethasone, culture

Abstract. Marrow mesenchymal cells contain stem cells and can differentiate into various types of cells with the capacity to regenerate tissue. Cell proliferation can be substantially enhanced by subculture. However, repeated subculture is known to decrease the differentiation potential as proliferation occurs. We examined human marrow mesenchymal cells to determine the number of passages through which these cells maintained their potential to differentiate into osteogenic cells that could be expected to have an efficient intensifying effect on bone regeneration. In a 64-year-old man with lumbar spondylosis, 3 ml of marrow fluid was collected from the ilium at surgery. Primary culture was done in T75 flasks and standard medium (MEM containing 15% bovine fetal serum). Two weeks later, the cultures were trypsinized to prepare a cell suspension. Then 10% of the cell suspension was seeded into a T75 flask and subculture was done in the standard medium. The remaining 90% of the cell suspension was seeded into 6-well plates. Incubation was carried out in an osteogenic medium that was prepared by adding 10 nM dexamethasone, ascorbic acid and beta-glycerophosphate to the standard medium [Dex (+) cultures]. Two weeks later, the calcium level and ALP activity were measured. A similar procedure was repeated up to the fifth subculture and the subcultured marrow mesenchymal cells were examined for their capacity for differentiation into osteogenic cells. There was significant ALP activity in the first to fifth Dex (+) subcultures, cells up to the second subculture were shown to have the potential to differentiate into osteogenic cells. Calcium was detected in the first and second subcultures. These data indicate that a small volume of marrow fluid could be used for extended osteogenetic therapy and that a large amount of artificial bone with stable osteogenetic capacity could be produced.

Introduction

Bone marrow mesenchymal cells include a number of stem cells and these differentiate into various cell types [1]. As a result, bone marrow mesenchymal cells can be used to regenerate tissue and have been applied clinically to regeneration of damaged tissue. We have been investigating the use of bone marrow mesenchymal cells in bone and skin regeneration [2,3]. Marrow cells can be collected and cultured in order to increase the number of bone marrow mesenchymal cells, and as reported by Maniatopoulos et al.[4], osseous tissue can be regenerated by incubating these cultured bone marrow mesenchymal cells in osteogenic media containing compounds such as dexamethasone [4-7]. When this cultured osseous tissue is allowed to form on an artificial bone material, the material can be covered with the cultured osseous tissue and becomes an artificial bone having a comparable osteogenic potential to autologous bone[8-13]. We reported last year that this cultured bone construct could be grafted to treat bone defects or pseudoarthrosis [14], thus proving that it is possible to perform bone regeneration therapy. Furthermore, by attaching cultured osseous tissue to an artificial joint where it meets osseous tissue, an artificial joint having a higher level of bone affinity can be produced.

However, when repairing severely damaged osseous tissue using bone marrow mesenchymal cells, numerous cells are needed. In order to prepare a large cultured bone construct using a small number of marrow cells, it is necessary to multiply the cells. Subculturing facilitates marked increase of cell number. However, some subculturing procedures, such as trypsin treatment, can damage cultured cells. It is also known that repeated subculturing lowers the ability of cells to differentiate, but no detailed studies have yet been conducted. In order to prepare large quantities of cultured bone constructs, it is necessary to determine the efficacy of subculturing in cell multiplication and the effects of subculturing on differentiation into osteogenic cells. Therefore, in the present study, we investigated the ability of subcultured human bone marrow mesenchymal cells to differentiate into osteogenic cells and the osteogenic potential of subcultured osteogenic cells.

Materials and Methods

After obtaining written consent, 3 ml of marrow fluid was collected from the ilium of a 64-year-old man with degenerative lumbar spondylosis who was undergoing surgery under general anesthesia. The marrow fluid was divided into two T75 flasks, and incubated in a minimum essential medium containing fetal calf serum and antibiotics. After two weeks, the cultured cells were treated with trypsin in order to prepare a cell suspension, and one tenth of this suspension was transferred to another T75 flask for subculturing in the standard medium. The remaining nine tenths of the suspension were incubate at a cell density of 10^4 cells/cm^2 on a 6-well plate containing standard medium enriched with glycerophosphoric acid, vitamin C and 10 nM dexamethasone [Dx(+) group]. As a control, cells were incubated using only the standard medium [Dx(-) group]. After two weeks of incubation, subcultured cells were stored frozen at -20°C. This subculturing procedure was repeated five times in order to determine the ability of subcultured bone marrow mesenchymal cells to differentiate into osteogenic cells.

Each frozen cultured tissue sample was thawed, scraped off the plate using a scraper, and placed in a 2 ml tube. After adding 1 ml of 0.2% Nonidet P40, the cells were subjected to ultrasound in order to extract alkaline phosphatase (ALP). Ruptured cells were centrifuged at 13,000 rpm for 10 minutes, and the supernatant was used as an enzymatic solution. To 1 ml of a p-nitrophenyl phosphate solution, 5 ml of the enzymatic solution was added and allowed to react for 30 minutes at 37°C. Next, the reaction was stopped by adding 2 ml of 0.2 N NaOH, and absorbance was measured at 405 nm. A calibration curve was prepared using p-nitrophenol (PN), and enzymatic activity was expressed in PN micro-mol.

To the precipitate, 1 ml of 20% formic acid was added in order to extract calcium. The level of calcium was measured using the OCPC method (Wako Pure Chemical Industries, Ltd.).

Results

The ALP activities for the first through fifth generations in the Dx(+) group were 23.11±1.63, 10.6±0.94, 5.43±0.6, 7.49±0.45, and 4.32±0.09, respectively (PN μmol/well, mean±SD) [Fig.1]. The Ca levels for the first and second generations in the Dx(+) group were 2.71±0.9 and 0.99±0.3, respectively (ng/well, mean±SD), but these remained low after the third generation. On the other hand, for the Dx(-) group, both Ca and ALP levels were low from the first generation. The fact that ALP activity remained relatively high up to the fifth generation for the Dx(+) group suggests the ability of subcultured marrow cells to differentiate into osteogenic cells. As far as calcium production was concerned, the level of Ca remained high up to the second generation, thus suggesting that subcultured cells can produce bone matrix up to the second generation. These findings suggest that subcultured bone marrow mesenchymal cells are capable of regenerating osseous tissue.

Discussion

Cancellous bone, such as that obtained from the ilium, is rich in osteogenic cells, and is often used in autologous bone grafting. However, harvesting bone from the ilium causes severe pain postoperatively, and because the severity of pain at the site of bone harvest is often greater than that at the grafted site, patient QOL is lowered. Management and handling of pain caused by bone harvesting are often difficult for patients and physicians. It has also been reported that pain persists for more than one year after grafting in 25% of cases [15].

Today, bone grafting is commonly performed using allogeneic grafts and artificial bones because large quantities of grafting material can be obtained without damaging the host tissue. However, with allogeneic grafts, there is a risk of unknown infection and immunological rejection, and in some patients, sufficient bone regeneration does not occur. In recent years, numerous biologically compatible artificial bone materials have been developed because they have no risk of unknown infection or immunological rejection. However, artificial bone has no biological function and cannot regenerate bone tissue. Autologous bone grafting is an ideal bone regeneration technique when taking into account factors such as bone regeneration, unknown infection and immunological rejection. However, autologous bone harvesting damages healthy tissue and causes pelvic deformation, and as mentioned above, pain due to bone harvesting and size limitations in bone harvesting must also be considered. On the other hand, a cultured bone construct having comparable function to an autologous bone graft can be prepared by harvesting autologous bone marrow cells by bone marrow aspiration, which is a minimally invasive procedure, and then incubating these marrow cells with an artificial bone material [2]. Hence, with cultured bone constructs, there is no risk of pelvic deformation or harvest site pain, and it is possible to produce a large implant.

At our institution, with approval from the ethics review board, bone regeneration is performed using artificial bones cultured with marrow cells [14]. However, when compared to marrow cells obtained from young rats, the abilities of marrow cells obtained from adult humans to multiply and regenerate osseous tissue are less stable. In actual clinical settings, it is necessary to enhance the tissue regenerating capability of marrow cells, and when treating a broad bone defect, a large quantity of marrow cells is needed. It is possible to markedly increase the number of marrow cells by subculturing, but subculturing has been known to lower the ability of marrow cells to differentiate. The results of the present study showed that the ability of cultured marrow cells to differentiate into osteoblasts and produce bone matrix remained up to the fifth generation and the second generation, respectively. Subculturing to the second generation can increase the number of cells by 10^2 fold, and that to the fifth generation can increase the number of cells by 10^5 fold. Therefore, it is possible to obtain a large number of cells from a small amount of marrow fluid, thus allowing the preparation of a large cultured bone construct having a stable osteogenic potential.

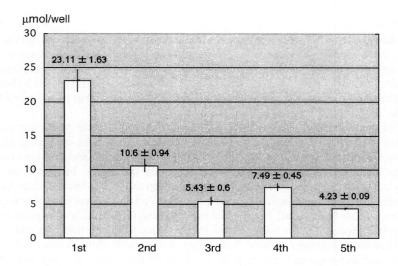

Fig.1. The ALP activities for the first through fifth generation.

References

 [1] Y. Jiang, B.N. Jahagirdar, R.L. Reinhardt, et al.: Nature 418 (2002), p.41-9.

[2] T. Yoshikawa : Mat. Sci. Eng. C-Biomim. 13(2000), p.29-37.

[3]T. Yoshikawa, T. Noshi, H. Mitsuno, et al.: Mat. Sci. Eng. C-Biomim., 17 (2001), p.19.

[4] C. Maniatopoulos, J. Sodek, and A.H. Melcher: Cell Tissue Res. 254 (1988), p. 317.

[5] J.E. Davies, R. Chernecky, B. Lowenberg, et al.: Cells Mater. 1 (1991), p.3.

[6] T. Yoshikawa,S.A. Peel, J.R. Gladstone, et al.: Biomed. Mater. Eng. 7 (1997) , p369.

[7] K-L Yao , R.Jr. Todescan and J. Sodek: J. Bone Miner. Res. 9 (1994), p.231.

[8] T. Yoshikawa, H. Ohgushi, Y. Dohi, et al.: Biomed. Mater. Eng. 7 (1997), p.49.

[9] T. Yoshikawa, H. Ohgushi and S. Tamai: J. Biomed. Mater. Res. 32 (1996), p.481.

[10] T. Yoshikawa, H. Ohgushi, Akahane M, et al.: J. Biomed. Mater. Res. 41(1998), p.568.

[11] T. Yoshikawa, H. Nakajima, E. Yamada, et al.: J. Bone Miner. Res. 15 (2000), p. 1147.

[12] T. Yoshikawa, H. Ohgushi, H. Nakajima, et al.: Transplantation. 69(2000), p.128.

[13] T. Yoshikawa, H. Ohgushi, T. Uemura, et al.: Biomed. Mater. Eng. 8 (1998), p.311.

[14] T. Yoshikawa , T. Ohmura, Y. Sen et al.:Bioceramics Vol.15. (2003), p383.

[15] S.S. Reuben, P. Vieira, S. Faruqi, et al.: Anesthesiology 95 (2001), p390.

Key Engineering Materials Vols. 254-256(2004) pp. 1071-1074
online at http://www.scientific.net
© 2004 Trans Tech Publications, Switzerland

Osteogenic Effect of Genistein on *in vitro* Bone Formation
by Rat Bone Marrow Cell Culture - For Development of Advanced Bio-Artificial Bone

N. Okumura [1,2], T. Yoshikawa [2,3], J. Iida [3], A. Nonomura [2] and Y. Takakura [3]

[1] Center of Dialysis, Tenri Hospital, Tenri City,Nara632-8552,Japan.
Departments of [2]Diagnostic Pathology and [3]Orthopaedic Surgery, Nara Medical University,
Kashihara City, Nara 634-8521, Japan

tyoshi@naramed-u.ac.jp

Keywords: genistein, isoflavone, bone formation, marrow cell, dexamethasone

Abstract. The effect of genistein, a soybean isoflavone, on new bone formation by bone marrow cell culture from mature rats was examined. Bone marrow cells were collected from the femoral diaphysis of 7-week-old Fisher rats, cultured in MEM containing fetal calf serum and then cultured with or without the addition of dexamethasone to the bone-forming medium. Genistein was added at concentrations of 10^{-5}, 10^{-6}, 10^{-7} or 10^{-8} M. *In vitro* bone formation was examined 2 weeks after subculture. When dexamethasone was added to the bone-forming medium, genistein (10^{-7} M and 10^{-8} M) caused a significant increase in the levels of calcium and alkaline phosphatase activity compared with cells not cultured in genistein. In conclusion, genistein was found to promote bone formation at physiological concentrations across species, and thus may be useful as a bone formation-promoting factor.

Introduction

The use of soybean proteins to prevent chronic diseases through various actions has been investigated [1]. Among these soy proteins, isoflavones such as genistein, daidzein, and glycitein have a structure similar to that of estrogens and are known to show a very weak estrogen-like action because of competitive binding to the estrogen receptor [2]. Utilization of these substances has attracted attention for the treatment of arteriosclerosis and osteoporosis in postmenopausal women, as well as for the prevention of prostate cancer. The soybean isoflavones inhibit bone resorption and appear to have bone-specific actions similar to those of estrogens without being carcinogenic for the female genital organs, unlike estrogen, and thus these proteins may possibly be useful for the treatment of osteoporosis. However, the effects of genistein on bone metabolism are unclear. In the present study, we employed an *in vitro* bone formation model using cultured bone marrow cells from rats to investigate the effects of genistein on bone metabolism.

Materials and Methods

1) Bone marrow cells were collected from the femoral diaphysis in seven-week-old male Fisher rats, and were added to a T-75 flask for primary culture. The standard medium was α-minimal essential medium (MEM) containing 15% fetal calf serum and antibiotics (Sigma). After 7 days of culture, a cell suspension was prepared by treatment with 0.1% trypsin.

2) The cell suspension was reseeded into 6-well plates at 10^4 cell/cm^2 and was cultured with or without dexamethasone. Genistein was added to the cultures at concentrations of 10^{-5}, 10^{-6}, 10^{-7}, and 10^{-8} M. After incubation for 2 weeks, the tissue formed in the culture plates was examined macroscopically and microscopically, and the levels of alkaline phosphatase (ALP) activity and Ca were measured. Culture in the presence of 10^{-8} M dexamethasone (Dex, Sigma) was done in standard medium (α-MEM containing 15% fetal calf serum and antibiotics) with 10 mM Na β-glycerophosphate (Merck) and 82 µg/ml ascorbic acid phosphate (Wako Co., Japan). The standard

medium (α-MEM containing 15% fetal calf serum and antibiotics) alone was used for cultures without dexamethasone.

3) Measurement of ALP activity

After culture for 2 weeks, the tissue that formed was collected from the bottom of each well using a scraper and placed in 2-ml Eppendorf tubes. After addition of 1 ml of 0.2% Nonidet P40 and lysis of the cells by ultrasound, centrifugation (4°C, 13,000 rpm, 10 min) was performed and the supernatant was used as the ALP enzyme solution. After addition of 1 ml of assay buffer [0.56 M 2-amino-2-methylpropanol (AMP), 1.0 mM $MgCl_2$, and 10 mM p-nitrophenyl phosphate] to 5 μl of enzyme solution, incubation was done for 30 min at 37°C. Then the reaction was stopped by adding 2 ml of 0.2 N NaOH and the optical density (OD) was measured at 410 nm. ALP activity was expressed as μmol of p-nitrophenol from the calibration curve.

4) Measurement of Ca

1 ml of 20% formic acid was added to the above precipitate and the mixture was decalcified by shaking in a refrigerated chamber for one week (4°C). After decalcification, separation by centrifugation (4°C, 13,000 rpm, 10 min) was performed and the supernatant was used for measurement of Ca by the o-cresolphthalein complexone (OCPC) method (Wako Pure Chemical Industries, Ltd., Japan, code 272-21801).

Results
ALP activity in rat bone marrow cell cultures

With dexamethasone. ALP activity was the highest at genistein concentrations of 10^{-7} or 10^{-8} M, and showed a significant difference from the control culture without genistein in both cases. At 10^{-5} M, genistein caused significant inhibition of ALP activity when compared with the control culture (Fig. 1).

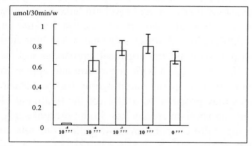

Fig.1   ALP activity of genistein-treated rat bone marrow cell culture with dexamethasone. High activity was observed in cultures with 10^{-7} and 10^{-8} M genistein. (mean±SD)

Without dexamethasone. ALP activity in the control culture without genistein was below the measurable limit. ALP activity was increased significantly at genistein concentrations of 10^{-6} M, 10^{-7} M, and 10^{-8} M when compared with the control culture, but the increase was not significant at 10^{-5} M.

Ca level in rat bone marrow cell cultures

With dexamethasone. At concentrations of 10^{-7} or 10^{-8} M, genistein caused a significant increase of Ca when compared with the control culture. In contrast, 10^{-5} or 10^{-6} M genistein caused a decrease of Ca and there was a significant difference from the control culture (Fig 2).

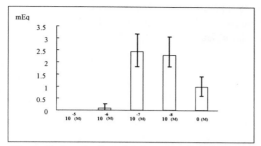

Fig.2 Ca level of genistein-treated rat bone marrow cells cultured with dexamethasone.
High levels were observed in cultures with 10^{-7} and 10^{-8} M genistein. (mean±SD)

Without dexamethasone. Ca was below the measurable limit in all cultures .

Discussion and conclusion
Isoflavones such as genistein, daidzein, and glycitein have a structure similar to that of estrogen and
have been shown to possess very weak estrogen-like activity based on competitive binding to the
estrogen receptor [2]. They are also bone-specific and have no adverse effects on the female genital
organs such as those mentioned above.

The mechanism of action of isoflavones on the bone is similar to that of estrogens and is thought
involve resorption from the gastrointestinal tract and renal tubules via parathyroid hormone, with
estrogens modulating the expression of vitamin D receptors. In recent years, estrogen receptors have
been identified on human and avian osteoclasts [3], and on human, avian, and rabbit osteoblasts, thus
suggesting that estrogen has a direct action on bone [4]. Estrogen may modulate bone resorption via
the regulation of bone-resorbing cytokines such as IL-1, IL-6, and tumor necrosis factor (TNF). In
patients with postmenopausal osteoporosis, production of IL-1 is increased, but this is normalized
when estrogen is administered. When the time course of IL-1, TNF-α, and GM-CSF levels were
examined in ovariectomized women, the levels of these cytokines were found to have increased after
surgery and then returned to normal when estrogen was administered. Because of its increased
production secondary to estrogen deficiency, IL-1 might promote the differentiation of osteoclast
precursors to osteoclasts and thus promote bone resorption. Genistein is known to have an inhibitory
effect on tyrosine kinase and to inhibit bone resorption, in addition to its estrogen-like action, while
these actions are not seen with estrogens.

The increase in the biochemical markers of bone formation at concentrations of 10^{-7} M and 10^{-8} M
genistein, close to the physiological blood concentration in both rats, suggests that genistein can
effectively modulate bone metabolism and may come to be used for the treatment of bone diseases
requiring osteogenesis in the future. We have previously reported that culture of bone marrow cells
with artificial bone can produce bone implants with a high osteogenic potential that can be applied in
surgical bone [5-7]. Because genistein is safe and noncarcinogenic, it should be possible to produce
cultured artificial bone with an even higher osteogenic potential using genistein in such tissue
engineering methods of bone regeneration. Future studies will determine whether genistein is a bone
regeneration-promoting factor.

References

[1] H.Adlercreutz, T .Fotsis and R. Heikkinen: Lancet□, (1982), p.1295.
[2] H.Adlercreutz, B.R.Goldin and S.L. Gorbach: J. Nutr. 125, (1995), p.757.
[3] Y.Ishimi, C. Miyaura and I. Ikegami:Endcrinology 140, (1999), p.1893.
[4] B.S. Komms, C.M.Terpening and D.J.Benz: Science 241, (1998), p.81.

[5] T.Yoshikawa, and H.Ohgushi:Ann..Chir.Gynecol. 88, (1999), p186.
[6] T.Yoshikawa, H.Nakajima, K.Ichijima et al.:J. Bone Miner. Res. 15, (2000),p .1147.
[7] T.Yoshikawa, H.Ohgushi, J.E. Davies et al.:Bone Mater. Eng. 7, (1997), p.49.

Key Engineering Materials Vols. 254-256(2004) pp. 1075-1078
online at http://www.scientific.net
© 2004 Trans Tech Publications, Switzerland

Experience of Osteogenetic Therapy With Advanceded Bio-Artificial Bone - A study in 25 cases

T. Yoshikawa[1,2], Y. Ueda[1], T. Ohmura[1], Y. Sen[1], J. Iida[1], M. Koizumi[1], K. Kawate[1], Y. Takakura[1] and A. Nonomura[2]

Departments of [1]Orthopaedic Surgery and [2]Diagnostic Pathology, Nara Medical University,

Kashihara Nara 634-8522, Japan

tyoshi@naramed-u.ac.jp

Keywords: osteogenesis, marrow cell, dexamehtasone, regeneration

Abstract. Bone tissue was produced when marrow cells were incubated in a medium containing dexamethasone. Bio-artificial bone with a high osteogenetic capacity can be produced by combining such cultured bone tissue with an artificial bone material. With respect to the osteogenetic capacity of adult human marrow cells, however, the potential for proliferation and differentiation varies from individual to individual, involving many factors. We succeeded in producing advanced bio-artificial bone with a greater osteogenetic capacity by this method. In this report, we discuss bone regeneration therapy using activated cultured bone constructs. The activated cultured bone construct was used for 25 bone regeneration treatments in 23 patients. Reasons for bone regeneration were prolonged bone fracture therapy in 11 cases, coxarthrosis in 12 cases, lumbar spondylosis in 1 case and tumor resection in 1 case. In all patients, 10 to 20 ml of bone marrow was collected from the ilium or the tibia and incubated in MEM containing autologous serum or fetal bovine serum and an antibiotic. After two weeks in primary culture, the marrow mesenchymal cells were seeded onto hydroxyapatite, beta-TCP, another ceramic material, or a prosthetic joint and cultured in the osteogenic medium as reported separately. The bio-artificial bone thus obtained was then implanted at the affected site. In patients with pseudoarthrosis, repeat joint replacement, or an osseous defect following tumorectomy, the implanted artificial bone survived and bone regeneration was detected radiographically. Short- and medium-term follow-up has shown that the bone implants were effective in all of the patients. The prosthetic joints which were implanted with cultured bone marrow cells showed good integration at the bone-joint interface.

Introduction

Many studies have reported that marrow cells include cells that are capable of regenerating bone [1-3]. However, in order to apply the osteogenic potential of marrow cells to clinical bone regeneration therapy, it is necessary to understand the mechanisms of tissue regeneration, have in-depth knowledge of medicine, engineering and culturing, and enhance the osteogenic potential of cultured cells. In 1988, Maniatoupoulos et al. reported that osseous tissue could be regenerated by culturing marrow cells in an osteogenic medium, containing compounds such as dexamethasone [4-7]. We have reported that superior bone constructs can be prepared by combining cultured osseous tissue and recently developed artificial bone materials having a high degree of biocompatibility [8-13]. We showed that the osteogenic potential of such cultured bone constructs is comparable to that of cancellous bone grafts, which possess a high degree of osteogenic potential.

However, in humans, even when marrow fluid is collected and cultured, there are individual differences in the number of cells and the level of cellular multiplication, and unlike animal studies, rapid bone regeneration is not assured. Therefore, based on the culture techniques developed by Maniatopoulos et al.[4], we established a new culture technique. Firstly, we found that adding estriol to an osteogenic medium enhanced bone regeneration by more than two-fold *in vitro*, and as a result, we included estriol as an osteogenic factor in the culture medium [14]. Secondly, we found that a large quantity of osteogenic cells can be placed over an artificial bone material by layering, and we have succeeded in preparing cultured bone constructs having more than twice the osteogenic cell activity [15]. With approval from the ethics review board at our university, this improved cultured bone construct has been used clinically [16]. In this report, we discuss bone regeneration therapy using activated cultured bone constructs.

Materials and Methods

An activated cultured bone construct was used for 25 bone regeneration treatments in 23 patients. The average age of the 21 patients was 63 years with a range of 17 to 86 years. There were 10 men and 15 women. Reasons for bone regeneration were prolonged bone fracture therapy in 11 cases, coxarthrosis in 12 cases, lumbar spondylosis in 1 case and tumor resection in 1 case. Fractured bones among the 11 cases of prolonged bone fracture therapy were as follows: femur (n=2), humerus (n=1), spine (n=6), metatarsal (n=1) and clavicle (n=1). Among the 12 cases of coxarthrosis, bone regeneration therapy was performed using a cultured bone construct to treat a bone deficit during the second hip replacement surgery in seven cases, and in the remaining five cases, a cultured bone construct was placed between an artificial joint and osseous tissue.

In all patients, 10-20 ml of marrow fluid was collected from the ilium and then cultured in minimal essential medium [MEM] containing autologous or fetal calf serum, estriol and antibiotics. After two weeks, bone marrow mesenchymal cells were cultured with ceramics (hydroxyapatite (HA) or beta-TCP) or artificial joints in an osteogenic medium containing estriol for about three weeks. At one or two weeks after the start of culturing, another layer of cultured cells was incubated over the existing layer. Cultured bone constructs prepared in this manner were grafted [Fig.1].

Results and Discussion

Postoperative X-rays revealed that the cultured bone constructs, which were grafted to bone defects caused by pseudoarthrosis, total hip replacement or tumor resection, had fused with existing osseous tissue and osteogenesis had occurred [Fig.2]. The results of short-term and intermediate follow-up examinations confirmed efficacy in all cases. When a cultured bone construct was grafted to treat pseudoarthrosis, the bone construct was remodeled three months after grafting, and X-ray findings confirmed favorable bone union. When a cultured bone construct was used with an artificial joint, there was a fine union between the joint and osseous tissue. Graft-related adverse reactions were not seen. Therefore, using the cultured bone constructs we have developed, it is possible to perform bone regeneration therapy without sacrificing autologous bone and inducing harvest-site pain or pelvic deformation, thus reducing physical stress to patients. Furthermore, when used with an artificial joint, the bond and compatibility between the artificial joint and osseous tissue can be increased [Fig.3].

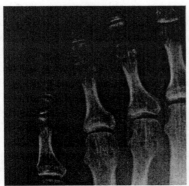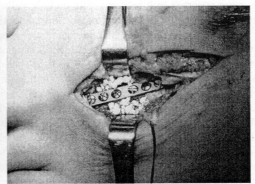

Fig.1. X ray of third metatarsal bone non-union of 26-year-old patient (left). Prefabricated cultured bone constructs were grafted into previous non-union site (right).

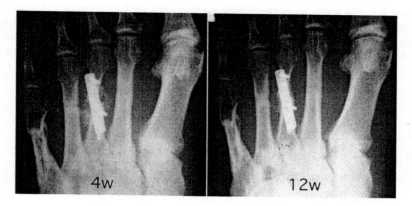

Fig.2. X rays of 4 weeks and 8 weeks after surgery in metatarsal bone non-union of 26- year-old patient.

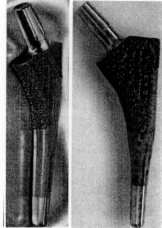

Fig.3. ALP activity stain shows high osteoblastic activity in the surface of stem component (right). The stem shows good integration with bone of medullary cavity.

The cancellous bone collected from bones such as the ilium is rich in osteogenic cells, and as a result, the ilium is most frequently used in autologous bone grafting. However, harvesting bone from the ilium is very painful for patients, and the severity of pain at the site of bone harvest is often greater than that at the grafted site, thus lowering patient QOL [17]. In addition, managing the pain caused by bone harvesting is often difficult for patients and physicians, and it has been reported that pain can persist for more than one year grafting.

Today, bone grafting is done using allogeneic grafts and artificial bones because large quantities of grafting materials can be obtained without damaging the host tissue. However, with allogeneic grafts, there is a risk of unknown infection and immunological rejection, and in some patients, bone regeneration cannot be performed properly. In recent years, numerous biologically compatible artificial bone materials have been developed, and artificial bones have no risk of unknown infection or immunological rejection. However, artificial bones have no biological function and cannot regenerate bone tissue. Autologous bone grafting is ideal with regard to bone regeneration, unknown infections and immunological rejection. However, autologous bone harvesting damages healthy tissue and causes pelvic deformation, and as mentioned above, pain due to bone harvesting and size limitation in bone harvesting need to be taken into account. On the other hand, with cultured bone/HA constructs, autologous bone marrow cells are harvested by a minimally invasive procedure (bone marrow aspiration) and incubated with an artificial bone material to produce a bone construct that is similar to an autologous bone graft [2,3]. Hence, with cultured bone/HA constructs, there is no pelvic deformation or harvest site pain, and it is possible to produce a large implant.

References

[1] Y. Jiang, B.N. Jahagirdar, R.L. Reinhardt, et al.: Nature 418 (2002), p.41-9.

[2] T. Yoshikawa : Mat. Sci. Eng. C-Biomim. 13(2000), p.29-37.

[3]T. Yoshikawa, T. Noshi, H. Mitsuno, et al.: Mat. Sci. Eng. C-Biomim., 17 (2001), p.19.

[4] C. Maniatopoulos, J. Sodek, and A.H. Melcher: Cell Tissue Res. 254 (1988), p. 317.

[5] J.E. Davies, R. Chernecky, B. Lowenberg, et al.: Cells Mater. 1 (1991), p.3.

[6]T. Yoshikawa,S.A. Peel, J.R. Gladstone, et al.: Biomed. Mater. Eng. 7 (1997) , p369.

[7] K-L Yao , R.Jr. Todescan and J. Sodek: J. Bone Miner. Res. 9 (1994), p.231.

[8]T. Yoshikawa, H. Ohgushi, Y. Dohi, et al.: Biomed. Mater. Eng. 7 (1997), p.49.

[9] T. Yoshikawa, H. Ohgushi and S. Tamai: J. Biomed. Mater. Res. 32 (1996), p.481.

[10] T. Yoshikawa, H. Ohgushi, Akahane M, et al.: J. Biomed. Mater. Res. 41(1998), p.568.

[11]T. Yoshikawa, H. Nakajima, E. Yamada, et al.: J. Bone Miner. Res. 15 (2000), p. 1147.

[12]T. Yoshikawa, H. Ohgushi, H. Nakajima, et al.: Transplantation. 69(2000), p.128.

[13]T. Yoshikawa, H. Ohgushi, T. Uemura, et al.: Biomed. Mater. Eng. 8 (1998), p.311.

[14] J. Iida, T. Yoshikawa, N.Okumura, et al.: Bioceramics Vol.16 in press.

[15] T. Yoshikawa, J. Iida, Y. Takakura: Bioceramocs Vol.16 in press.

[16] T. Yoshikawa , T. Ohmura, Y. Sen et al.:Bioceramics Vol.15. (2003), p383.

[17]] S.S. Reuben, P. Vieira, S. Faruqi, et al.: Anesthesiology 95 (2001), p390.

Key Engineering Materials Vols. 254-256(2004) pp. 1079-1082
online at http://www.scientific.net
© *2004 Trans Tech Publications, Switzerland*

Fabrication of Macroporous Scaffold Using Calcium Phosphate Glass for Bone Regeneration

Y.-K. Lee[1,4], Y.S. Park[1], M.C. Kim[1], K.M. Kim[1], K.N. Kim[1,4], S.H. Choi[2,4], C.K. Kim[2,4], H.S. Jung[3,4], C.K. You[5] and R.Z. LeGeros[6]

[1] Dept. & Research Institute of Dental Biomaterials and Bioengineering, Yonsei University College of Dentistry, 134 Shinchon-dong, Seodaemun-ku, Seoul 120-752, Korea, leeyk@yumc.yonsei.ac.kr

[2] Dept. of Periodontics, Yonsei University College of Dentistry, Seoul, Korea

[3] Div. in Histology, Dept. of Oral Biology, Yonsei University College of Dentistry, Seoul, Korea

[4] Brain Korea 21 Project for Medical Science, Yonsei University, Seoul, Korea

[5] School of Materials Engineering, Yeungnam University, Kyongbuk, Korea

[6] New York University College of Dentistry, New York, USA, rzl1@nyu.edu

Keywords: bone, calcium phosphate, scaffold, tissue engineering, MSC, *in vivo* transplantation

Abstract. Numerous techniques have been applied to fabricate three-dimensional scaffolds of high porosity and surface area. In this study, we fabricated three-dimensional macroporous scaffold by polymeric sponge method using calcium phosphate glass in $CaO-CaF_2-P_2O_5-MgO-ZnO$. Calcium phosphate glass slurry was prepared by suspending the glass powder in water, polyvinyl alcohol, polyethylene glycol and dimethyl formamide. Reticulated polyurethane sponges were used as a template and were coated with the prepared slurry by infiltration technique several times. Calcium phosphate glass slurry was homogenously thick coated when the amounts of calcium phosphate glass powder and polyvinyl alcohol were 67 and 8 wt%, respectively. Addition of 10 wt% dimethyl formamide as a drying control chemical additive into a slurry successively prevented microcrack formation. Sintering at 850°C exhibited dense microstructures as well as entire elimination of organic additives. When the macroporous scaffolds were transplanted subcutaneously in the dorsal surface of mice, the vascularized mesenchymal stem cells were formed in the transplanted scaffolds and the regenerated connective tissue was also formed after 4 weeks. These results of the present study suggested that the fabricated macroporous scaffold is expected to be useful scaffold material for osteogenic tissue development.

Introduction

The ideal bone substitute material should be osteoconductive in order to allow as rapid as possible integration with host bone, biodegradable at a preferred rate in order to eventually be replaced by newly formed natural bone, and strong enough to fulfill required load-bearing functions at least during the initial post-implantation period, i.e., before significant bone ingrowth and replacement has occurred. In addition, it should preferably exhibit osteoinductive characteristics in order to encourage rapid new bone formation although this may require the incorporation of biological factors with synthetic materials.

Tissue engineering concepts present an alternative approach to the repair of a damaged hard tissue. Osteoblast stem cells obtained from the patient's hard tissues can be expanded in culture and seeded onto a scaffold that will slowly degrade as the tissue structure grows *in vitro* and *in vivo*. A suitable temporary scaffold material exhibiting adequate mechanical and biological properties is required to enable tissue regeneration by exploiting the body's inherent repair mechanism. The three-dimensional constructed scaffold provides the necessary support for cells to proliferate and maintain their differentiated function, and its architecture defines the ultimate shape of the regenerated bone. The cellular structure must also be designed to satisfy several requirements. High porosity is needed for cell seeding and ingrowth. The several hundred micron-sized pores must be interconnected to allow ingrowth of cells, vascularization and diffusion of nutrients. And the material has to have sufficient mechanical integrity to resist handling during implantation and *in vivo* loading.

Several scaffold materials have been investigated for bone tissue engineering including calcium phosphates and synthetic and natural polymers. Among them, synthetic bioresorbable polymers have attracted increasing attention for their use as tissue engineering scaffolds in the last 10 years, in particular PLA, PGA and PLGA. However, a number of problems have been encountered regarding the use of these polymers in tissue engineering applications due to the release of acidic degradation products leading to inflammatory responses. Another limitation of biodegradable polymers is that they lack of bioactive function, i.e., in particular for bone tissue applications, they do not allow for bone apposition or bonding on the polymer surface. Calcium phosphates are ideal candidates for scaffold materials because the mineral phase of the biological bone has been identified as an apatite structure. However, the clinical application of the traditional calcium phosphates such as HA and TCP has been restricted due to its poor degradability and brittleness. Calcium phosphate glass in the system $CaO-P_2O_5-Al_2O_3$ was firstly developed as a dental crown material and could be expected to extend the application field for hard tissue repair because of its great degradability as well as excellent biocompatibility.

Methods

Calcium phosphate glass (CPG) with Ca/P molar ratio 0.6 was prepared using calcium carbonate, calcium fluoride and phosphoric acid. The molar ratio between calcium oxide and calcium fluoride was fixed at 9. Small amounts of magnesium oxide and zinc oxide were included. The mixture of raw materials was melted in a platinum crucible at $1250^{\circ}C$ and quenched to room temperature.

As-quenched glasses were reduced to powders of around 10 μm and dissolved in distilled water with organic additives such as binder, dispersant and drying chemical control additive (DCCA). Polyvinyl alcohol (PVA), polyethylene glycol (PEG) and dimethyl formamide (DMF) were selected as binder, dispersant and drying chemical control additives, respectively. First, PVA was hydrolyzed and stirred in distilled water at a temperature of $50^{\circ}C$ with various amounts from 2 to 8 wt%. After cooling down to room temperature, PEG was added at 5 wt%, and followed by addition of DMF up to 10 wt%. Preparation of the CPG slurry was completed by dispersing CPG powder into distilled water containing the organic additives at 10 to 67 wt%. Reticulated polyurethane sponges were used as a template in this study (Fig. 1). They were interconnected 3-dimensionally with 50 ppi and were subjected to infiltration process. They were coated with the prepared slurry by infiltration technique several times. Firstly, the temperature raised slowly in order to burn out the sponge entirely, and held at 600. to volatate the organic additives. Then the residual glass was sintered in a rigid form maintaining the original reticulate porous structure at various temperatures from 650 to $900^{\circ}C$.

The maximum compressive strength of each 10 of the porous scaffolds was determined with a universal testing machine (Fig. 2). Rubber plates of 0.2 mm in thickness were placed between each surface of the scaffold and compression punches in order to eliminate the unexpected effect due to heterogeneous horizontal level.

Human mesenchymal stem cells were obtained from surgical specimen and cultured in monolayer. The cultured cells were seeded on the prepared CPG scaffolds and cultured with osteogenic supplement for 1 week. After 4 weeks, they were transplanted subcutaneously into the dorsal surface of immunocompromised beige mice (Fig. 3). The transplants were recovered at 3 weeks post-transplantation, fixed with 10% formalin, decalcified with buffered 10% EDTA, and then embedded in paraffin. Sections were deparaffinized and stained with hematoxylin/eosin and von Kossa. Immunohistochemical staining was performed to detect osteocalcin. The constructs were fixed with glutaldehyde and were examined by electron microscopy.

 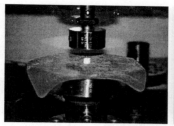 

Fig. 1. Polymeric sponge. Fig. 2. Strength measurement. Fig. 3. Subcutaneous transplantation.

Results

Photographs of the fabricated CPG scaffolds by polymeric sponge method were presented in Fig. 4. When the content of CPG powder was as low as 10 wt%, almost all pores were clogged by a thin film of slurry due to its low surface tension. With increasing the content of CPG powder up to 33 wt%, thin film formation was eliminated, however, the viscosity of the slurry was still low. When the content of CPG powder was fixed, coating efficacy was improved with increasing binder content. The best condition for homogenous-thick coating of the slurry in this study was 67 wt% CPG powder and 8 wt% binder. The next step for porous scaffold fabrication was drying. Without DCCA, lots of cracks were formed and the surface of the coated film was very rough and heterogeneous. Surface was more homogeneous and smoother with 5 wt% DMF as DCCA, however, cracks still remained. When the addition of DMF increased up to 10 wt%, crack formation was absolutely prevented during the drying process. The final step for scaffold fabrication was heat-treatment. The role of heat-treatment is elimination of polymeric sponge and organic additives around 600°C. After that, the residual CPG was sintered at various temperatures from 650 to 900°C. Sintering at 850°C exhibited white color, while sintering at lower temperature presented dark color, which means either the sponge or the additives was remained. The microstructure of CPG powders sintered at 650°C just showed particle contact. With increasing the sintering temperature, the voids between particles decreased. Dense microstructure was formed after sintering at 850°C, without voids and cracks. When the whole process was repeated, the compressive strength increased almost twice.

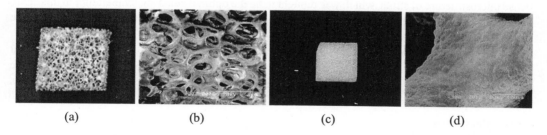

(a)   (b)   (c)   (d)

Fig. 4. Photographs of polymeric sponges (a) infiltrated (×2), (b) dried (×50),
(c) sintered at 850°C (×2) and (d) microstructure after sintering (×1000).

It is well demonstrated that the vascularized mesenchymal stem cells were formed in the transplanted scaffolds and the regenerated connective tissue was also formed after 4 weeks of subcutaneous transplantation in the dorsal surface of mice (Fig. 5).

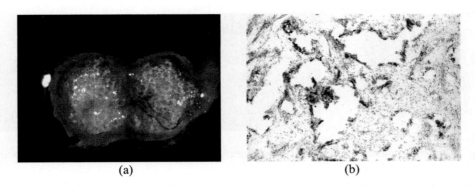

(a)   (b)

Fig. 5. Results of in vivo transplantation showing (a) vascularization and (b) histologic observation.

Discussion and Conclusion

As a preliminary study of the application of the calcium phosphate glass in the system $CaO-CaF_2-P_2O_5-MgO-ZnO$ for bone regeneration by tissue engineering, three-dimensional interconnective porous scaffolds were fabricated using a polymeric sponge as a template. PVA was used as a binder, where much amount of slurry could be coated onto sponge as PVA content increased in the same powder concentration. This might be attributed to increasing thixotropy with increasing PVA content. The mechanical strength of the macroporous scaffold came to considerably decrease due to microcrack formation according to heterogeneous coating thickness, residual internal stress caused from large shrinkage during drying, difference in thermal expansion coefficient between sponge and coated layer, and vapor pressure generated by sponge evaporation during heat-treatment. It was demonstrated that DMF is an excellent candidate to prevent microcrack formation because of its lower surface tension and higher evaporation temperature than that of water. Therefore, DMF still remained between particles even when water volatized entirely and can prevent abrupt shrinkage.

As-dried sponges were heat-treated at temperatures between 650 and 900°C because the glass transition temperature of CPG was 595°C. When sintered at 650°C, they showed gray color due to residual hydrocarbon in sponge. In contrast, sintering at 850°C enabled them to be white colored with dense microstructure of struts, which was thought to be the optimal sintering temperature in this study. Macroporous CPG scaffolds were successively fabricated with several hundred microns of pore size and three-dimensionally interconnected open pore system.

When they were transplanted subcutaneously in the dorsal surface of mice, the vascularized mesenchymal stem cells were formed in the transplanted scaffolds and the regenerated connective tissue was also formed after 4 weeks. These results suggested that the fabricated macroporous scaffold is expectedly useful for osteogenic tissue development.

Acknowledgements

This study was supported by a grant of the Korea Health 21 R&D Project, Ministry of Health & Welfare, Republic of Korea (03-PJ1-PG1-CH8-0001).

References

[1] K. Anselme K et al., Bone, Vol. 25 (1999), p. 51S.
[2] K.D. Johnson et al., J. Orthop. Res. Vol. 14 (1996), p. 351.
[3] D.M. Liu, Ceram. Inter. Vol. 24 (1998), p. 441.
[4] P. Sepulveda P, Am. Ceram. Soc. Bull. Vol. 76 (1997), p. 61.

Key Engineering Materials Vols. 254-256(2004) pp. 1083-1086
online at http://www.scientific.net

Oriented Collagen-Based/Hydroxyapatite Matrices for Articular Cartilage Replacement

Rolf Zehbe[1], Ulrich Gross[2] and Helmut Schubert[3]

[1] TU Berlin, Institute of Materials Science and Technologies, Englische Strasse 20, 10587 Berlin, Germany, rolf.zehbe@tu-berlin.de

[2] FU Berlin, Clinical Centre Benjamin Franklin, Department of Experimental Dentistry and Oral Biology, Aßmannshauser Strasse 4-6, 14197 Berlin, Germany, ugross@zedat.fu-berlin.de

[3] TU Berlin, Institute of Materials Science and Technologies, Englische Strasse 20, 10587 Berlin, Germany, schubert@ms.tu-berlin.de

Keywords: tissue engineering, articular cartilage replacement, mechanical stimulation, hydrogel foam, gelatine, collagen, directional freezing, biomimetic crystallization

Abstract. Mechanical stimulation of articular cartilage is known to be essential for chondrogenesis. In our study we use animal cartilage-bone explants from the femur condylus, which were divided into three experimental groups (loaded, unloaded and controls). Loading was performed in a sterile and physiological environment based on DME-medium. Cell morphology and cell vitality was observed by histological techniques. The knowledge gathered from the mechanical in-vitro study is now used to produce a structured artificial matrix resembling native cartilage in biochemical composition, biomechanical properties and zonal tissue structure. The artificial matrix is then seeded with chondrocytes and loaded with the same parameters we found to be ideal for tissue regeneration. Analysis is performed using histological and immuno-histological methods.

Introduction

Increasing cost factors in social health-care systems are joint diseases [1]. In this medical field the most painful diseases are those, where cartilage and its underlying bone structure are being destroyed. Articular cartilage has considerable capabilities for regeneration, however, the product of regeneration does not fulfil the quality of the original healthy cartilage and subchondral bone. Several attempts were made to transplant or cultivate this tissue, some are more successful than others, but up to now, none have lead to a permanent cure.

To use synthetic materials as cartilage tissue replacement many different properties of the natural system must be considered. Among these parameters are biochemical composition, structural identity (mimicking the zonal morphology) and biomechanical properties. For these materials porous foams based on natural and synthetic polymers were proposed by different authors.

A good overview of technologies for the creation of porous materials for tissue-engineering (fiber-bonding, solvent casting, gas foaming and phase separation) is presented by Mikos et al. [2].

A technology for phase separation of a hydrogel was proposed by Yannas and co-workers [3], who used an axial freezing zone which resulted in an oriented pore-channel structure for axonal regeneration of nerve tissue. A modification of this technology was used in the works of Heschel et al. and Kuberka et al. [4, 5].

Experiments based on collagen/ hyaluronic acid composite structure were carried out by Angele and co-workers, analysing biomechanical stability and cell-differentiation after cell-seeding [6]. Another work on the biomechanical stability of tissue-engineered matrices in vivo was done by Duda and co-workers [7].

Materials and Methods

In this study we used bovine explants from 24 months old female cows. Two loaded, two unloaded and two explants as controls were harvested from the joint.

A fatigue testing machine (Zwick Z005) was used for the cyclic loading of the explants. The load ranged between 0.05 MPa and 2 MPa at a cycle-frequency of 0.125 Hz. Cyclic loading was applied for 72 h, 93 h and 141 h. Two cylindrical explants were placed inside a stainless-steel specimen holder allowing loading against each other. The specimen holder was placed inside a sterile bag, containing DMEM solution and penicillin/ streptomycin. Two heated loading platens were used to ensure constant temperature at 37 °C (Fig. 1).

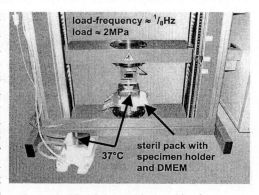

Fig. 1: Modified fatigue testing machine for loading of cartilage-bone-specimens.

After testing, all specimens were cross-linked in formaldehyde solution (5%) with calcium-phosphate-buffer (pH 7.3). The samples were then prepared for histological staining with Hematoxylin-Eosin and Hale-PAS. Evaluation was performed by counting vital cells and observation of cell/matrix morphology.

The acquired data and knowledge of the in vitro explant experiment was used as basis for the synthesis of an oriented and porous matrix structure. Oriented hydrogel foams for cartilage regeneration were produced by a directional freezing process. In this study, collagen-based hydrogels were used to model the biochemical composition of articular cartilage. The apparatus for the directional freezing process was based on a peltier-element as cooling-device, in which the hydrogel was frozen completely in approximately two hours. A novel technique was used to attach a porous hydroxyapatite-layer to the hydrogel resembling the subchondral bone. The frozen collagen-based/hydroxyapatite hydrogel was transferred to a freeze-dryer to eliminate water in a sublimation-process. All equipment was build using materials commonly applied in medical science and was designed to enable sterile handling.

In a first experimental approach, material properties of the synthetic matrices were determined using unconfined and confined compression (sample diameter = 8 mm). After equilibration at a pre-load of 0.003 N, stepwise loads of 0.019 N were applied up to a strain of 15%. The collagen matrices were rinsed in 0.9% NaCl during mechanical tests.

Results

Mechanical Loading of explants. We have found, that a cyclic load resulted in better cell vitality and cell morphology. The cell-quantity in loaded specimens indicated better nutrition of the cells by intermittent compression. The loaded explants were able to retain their zonal structure, whereas unloaded specimens showed more cells in the surface zone and less cells in the middle and deep zone, indicating that there is not sufficient nutrition by diffusion (Fig. 2).

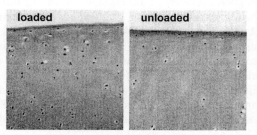

Fig. 2: Difference between loaded and unloaded explants after 141h of continuous testing.

Oriented hydrogel matrices. First experiments to structure hydrogels were carried out using gelatine (Sigma, type A from porcine skin). Experiments using native collagen (Sigma, type I, insoluble, bovine achilles tendon) were done as well, but resulted in matrices with lesser structural stability, because native collagen is highly insoluble and the formed hydrogel consisted mostly of water. In this study we used a novel technique to induce crystallisation of HAp-crystals at one end of the oriented pore-channel structure of the hydrogel. Scanning electron microscopic and light microscopic pictures were made in order to compare the natural cell-alignment with the formed synthetic hydrogel-structure. The left picture in Fig. 3 shows the formed porous HAp-layer.

Fig. 3: synthetic matrix-structure based on gelatine with hydroxyapatite-layer modelling the deep zone (SEM, left) and matrix-structure (SEM) compared to the HE-stain of natural cartilage tissue (right)

To compare the synthetic matrix-structure with the natural collagen fiber network, which is not visible in the histological picture, the reader shall refer to the works of Kääb et al. and Owen et al. [8, 9]. In both works cartilage was studied using the freeze-substitution technique, which does not destroy the fiber arrangement often found in aldehyd-based fixation of tissue. Additional information on the freeze-substitution technique may be found in the work of Gwynn et al. [10]. The natural fiber arrangement is much denser than the synthetic structure shown in Fig. 3. The space between the tubular structure of the matrix should be filled in by the products of cartilage cells that are seeded into the synthetic matrix in later cell-culture experiments. If the porous channels are too small, as in natural tissue, cell-integration and cell-adhesion may be inhibited. Optimisation of the channel diameter is possible by varying the parameters of the directionally freezing process (amount of water of the hydrogel, addition of electrolytes to hydrogel and rate of undercooling). The synthetic matrix with the HAp-layer is then chemically cross-linked by aldehyd-based fixation using glutaraldehyde. Alternatively, physical cross-linking is possible via heat or UV-radiation. In case of aldehyd-based fixation washing in phosphate buffered saline is necessary to ensure safe removal of aldehyde residues, which is measured by HPLC (high pressure liquid chromatography). To enhance later cell-adhesion in the cross-linked synthetic matrix, native collagen II is deposited on the surface of the porous-channels by dipping in a 2% collagen II solution and drying under a laminar flow-workbench.

In mechanical testing, the synthetic matrices showed strain increase caused by fluid efflux (additional 3% of initial strain at each step after 100 s), typical for biphasic materials. Young's moduli were determined to $E_s = 0.67 \pm 0.02$ MPa in unconfined compression, aggregate moduli to $H_A = 0.84 \pm 0.03$ MPa in confined compression and Poisson's ratio $v_s = 0.27$, respectively. Table 1 shows a comparison to values obtained from mechanical testing of native bovine cartilage found by Korhonen et al. [11].

Table 1: Young's Modulus of cell-free synthetic matrices in comparison to native tissue

Reference	Treatment/ remark		Young's Modulus [MPa]	Poisson's Ratio
Korhonen et al. [11]	Confined compression (diameter = 3.7mm) of bovine explants			
		Femur	0.47 ± 0.15	0.26 ± 0.08
		Humerus	1.15 ± 0.43	0.16 ± 0.06
		Patella	0.72 ± 0.19	0.21 ± 0.05
this study	Confined compression (diameter = 8.0mm) of synthetic, cell-free matrix structure		0.67 ± 0.02	0.27

Discussion and Conclusion

In this study we were able to produce an oriented matrix-structure for cartilage replacement based on partially hydrated collagen with a tightly attached hydroxyapatite layer mimicking the subchondral bone. Biochemical properties are optimised for use as cartilage replacement by depositing native collagen II on the pore-channel surface. The mechanical properties of the synthetic matrices are almost identical compared to native bovine cartilage. Nevertheless optimisation is still necessary.

In combination with a perfusion system for the in-vitro cell-seeding experiment and specific growth factors like TGF-β1, we hope to succeed in producing a hybrid of cells and synthetic matrix for the development of regenerative cartilage displaying natural morphology and function.

References

[1] Eckstein F, Reiser M, Englmeier KH, Putz R, Anat Embryol 203 (2001), 147-173

[2] Mikos AG, Temenoff JS, EJB Electronic Journal of Biotechnology 3(2) (2000), 114-119

[3] Yannas IV, Orgill DP, Loree HM, Kirk JF, Chang ASP, Mikic BB, Krarup C, *Patent US 4955893*, 1-10, 1990

[4] Heschel I, Rau G, Patent DE 197 51 031 A1, 1-7, 1999

[5] Kuberka M, Heimburg DV, Schoof H, Heschel I, Rau G, The International Journal of Artificial Organs 25(1) (2002), 67-73

[6] Angele P, Kujat R, Faltermeier H, Schumann D, Müller R, Nerlich M, Biomaterialien 4(1) (2003), 11-18

[7] Duda GN, Haisch A, Endres M, Gebert C, Schroeder D, Hoffmann JE, Sittinger M, J. Biomed Mater Res (Appl Biomater) 53 (2000), 673-677

[8] Kääb MJ, Richards RG, Walther P, Gwynn IA, Nötzli HP, Scanning Microscopy 13(1) (1999), 61-69

[9] Owen GR, Kääb M, Ito K, Scanning Microscopy 13(1) (1999), 83-91

[10] Gwynn IAP, Wade S, Kääb MJ, Owen GRH, Richards RG, Journal of Microscopy 197 (2000), 159-172

[11] Korhonen RK, Laasanen MS, Töyräs J, Rieppo J, Hirvonen J, Helminen HJ, Jurvelin JS, Journal of Biomechanics 35 (2002), 903-909

Key Engineering Materials Vols. 254-256(2004) pp. 1087-1090
online at http://www.scientific.net
© 2004 Trans Tech Publications, Switzerland

Antimicrobial Macroporous Gel-glasses: Dissolution and Cytotoxicity

P.Saravanapavan [1], J. E. Gough [2], J.R. Jones [1] and L.L. Hench [1]

[1] Tissue Engineering Centre, Department of Materials, Imperial College London, UK,
p.pavan@imperial.ac.uk

[2] Manchester Materials Science Centre, University of Manchester Institute of Science and
Technology Manchester, UK, j.gough@umist.ac.uk

Keywords: antimicrobial, macroporous, silver, bioactive glass

Abstract. The incidence of biomaterial-centred infections underlies the need to improve the properties of existing biomaterials. Combining the bioactive properties of calcia-silicate gel-glasses with that of the silver would prevent infections without the use of antibiotic drugs. Inclusion of silver into bioactive gel-glass foam scaffolds is explored using *in vitro* characterization techniques. The amount of silver released from Ag-doped S70C30 foams is well above the minimum bactericidal concentration (0.1 ppm) but below the cytotoxic concentration (1.6 ppm) for human cells. Primary human osteoblasts proliferate on the silver-doped gel-glasses.

Introduction

The incidence of infection following implant surgery has stimulated investigations into the potential antimicrobial properties of bioactive glasses [1]. Research suggests that incorporation of silver in gel-derived bioactive glass (S70C30 and 58S) compositions resulted in antimicrobial properties to the bioactive glass without compromising its bioactivity [2,3].

The aim of this work was to evaluate the *in vitro* bioactivity and cytotoxicity of novel antimicrobial bioactive gel-glass foams. The foams, which were made using previously established procedures by Sepulveda *et al* [4], are designed for orthopedic and cranio-maxillo-facial tissue engineering applications. The bioactivity was investigated by immersing foams in simulated body fluid (SBF) at different time points. Cellular responses were assessed by seeding primary human osteoblasts on the surface of silver-doped S70C30 foam substrates.

Method

Silver doped binary gel-glass foams of composition 70 mol% SiO_2- 28 mol% CaO- 2 mol% Ag_2O were prepared using the sol-gel technique with tetraethyl orthosilicate, silver nitrate and calcium nitrate tetrahydrate as starting materials. Physical characterization of the porous foams consisted of microstructural observation, pore size, and textural analysis.

Segments of the foams were immersed in 50 ml of simulated body fluid (SBF) in large conical flasks, placed in an orbital shaker under 175 rpm at 37 °C, for periods of 10, 20, 30 minutes, 1, 2, 3, 6, 24, 48 and 72 hours. In order to terminate the reactions after the different soaking periods, the foam specimens were removed, rinsed in acetone and dried at 60 °C for 3 hours. The foam surfaces were analysed by Fourier transformed infrared spectroscopy and the filtrate by ICP analysis. Absorbance FTIR spectra were collected using a Mattson Genesis II spectrometer, with a Pike Technologies EsiDiff diffuse reflectance accessory in the range 400 – 4000 cm-1. The crushed foam samples were mixed with KBr (dilution ratio approx. 1:100) to avoid spectral saturation. Filtered extracts were analysed using Inductively Coupled Plasma Optical Emission Spectroscopy (Applied Research Laboratory 3580 B ICP analyser along with ICP Manager software (provided by Micro-Active Australia Pty Ltd)) to obtain the elemental concentrations of Si, Ca, P and Ag. The detection limits were: 0.05 ppm for Si, 0.10 ppm for Ca, 0.20 ppm for P and 0.01 ppm for Ag. The SBF dissolution study was performed in triplicates to ensure reproducibility.

Primary human osteoblasts were cultured under standard conditions in DMEM with 10% FBS, 1% antibiotics, 1% 2mM L-glutamine and 50µg/ml ascorbic acid. The foam samples were fixed to the culture well plates using 2% Agar and pre-incubated in DMEM for 3 days. Cells were seeded onto foams at a density of 80,000/cm2 and left to adhere for 15 minutes before flooding the well plate with complete DMEM. After different incubation periods (90 minutes and 24 hours) the cells were fixed with 1.5% glutaraldehyde and processed for scanning electron microscopy, by dehydrating through increasing concentrations of ethanol with a final drying step in hexamethyldisilazane (HDMS).

Results

The interconnected macro-structure of the foams resembled the hierarchy of trabecular bone, forming a template of intricate shapes that surrounds large spherical pores. The structures contained macropores of 500 μm or larger which were interconnected by smaller (20 – 200 μm) pores (see Fig. 1 and Fig. 2a). The surface of the pore walls and pore cavities contained mesopores (Fig. 2). The textural parameters are listed in Table 1.

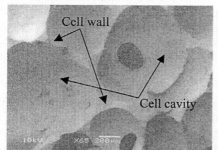 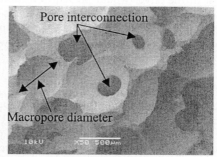

Fig. 1: SEM micrograph of the pores (walls and cavities) in the Ag-S70C30 gel-glass foam.

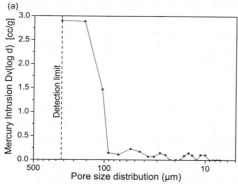 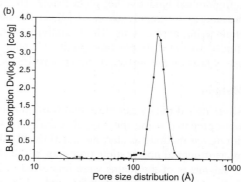

Fig. 2: (a) Macropore distribution using mercury porosimetry (b) BJH pore distribution of mesopores found in the surface of pore walls and cavities.

Table 1: Textural parameters of silver doped S70C30 scaffolds

Surface Area	113.6 m^2/g
Pore Diameter - mesopores	170.3 Å
Pore Diameter - macropores	>200 μm

Fig. 3 shows the dissolution profile for silver-doped S70C30 gel-glass foams. The concentrations of Si, Ca, P and Ag in filtered SBF (in ppm) are plotted against reaction time. The concentration of silicon ions released into solution increases rapidly during the first 24 hours after which equilibrium is reached with the concentrations remaining constant throughout the test period (72 hours). The calcium ions are released more rapidly during the first 2 hours. In contrast, phosphorous ions are depleted from the SBF solution within 24 hours of immersion. These results are similar to bioactive glass dissolution profiles, where the alkali ions are first exchanged before the glass network is broken while a Ca-P rich layer is precipitated onto the surface. The silver ion release is also similar to cation release from bulk surfaces. After an initial surge in Ag+ ion concentration, it is maintained in equilibrium via silver chloride precipitation.

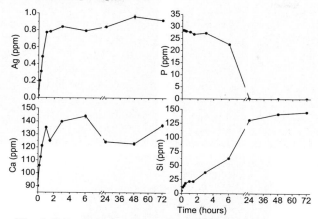

Fig. 3: Dissolution profile of Ag-S70C30 gel-glass foam.

Fig. 4 illustrates the different IR absorption spectra obtained for crushed Ag-S70C30 foams after various reaction time periods in SBF. The un-reacted powder exhibits the bending and stretching vibrations of silica tetrahedra. However, after 1-hour immersion in SBF, the vibrational band for amorphous phosphorous was detected. This protuberance splits into two peaks by 6 hours of reaction indicating the presence of crystalline P-O vibrational modes. The FTIR trace for the 24-hr reaction illustrates well-defined phosphate bending vibrational bands at 610, and 570 cm-1, confirming the formation of a crystalline Ca-P rich layer on the surface of the gel-glass.

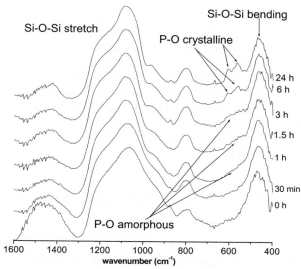

Fig. 4: FTIR absorbance spectra of Ag-S70C30 gel-glass foam for different soaking times.

The SEM micrographs seen in Fig. 5 show primary human osteoblasts cells on the surfaces of these silver-doped calcium silicate gel-glass foams. After 90 minutes of incubation although the majority of the cells are still rounded they have started attaching to the gel-glass surface via filopodia. Few of the cells had the morphology of spreading cells. Cell infiltration into the macropores in the scaffold was also observed. One such cell after 24-hour incubation is shown in Fig. 5.

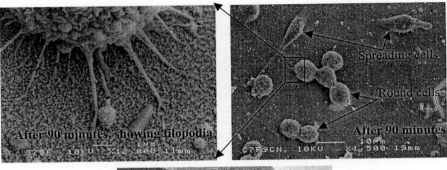

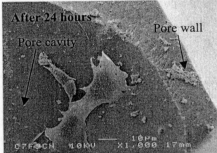

Fig. 5: SEM micrographs of Ag-S70C30 gel-glass foam surfaces showing cell morphology after different incubation periods.

Discussion and conclusions

The dissolution results indicate that the amount of silver release from silver-doped S70C30 foams was well above the minimum bactericidal concentration (0.1 ppm) but below the cytotoxic concentration (1.6 ppm) for cells. The surface characterisation confirms that the presence of silver in the bioactive gel-glass foams does not adversely affect the bioactivity. Cellular studies show that silver-doped S70C30 gel-glass foams provide a favourable environment for osteoblast attachment, spreading and proliferation as shown by qualitative morphological assessment. Further studies are underway to determine bone nodule formation, mineralisation and maintenance of the osteoblast phenotype. This preliminary study shows that antimicrobial calcium silicate gel-glass scaffolds can be used in bone tissue engineering.

Acknowledgements

The authors thank EPSRC (UK), MRC and the Lloyds Tercentenary Foundation for their support of this research.

References

[1]. P. Stoor, E. Soderling, J.I. Salonen: Acta Odontal Scand 56 (1998), p. 161-165.
[2]. M. Bellantone, N.J. Coleman, L.L. Hench: J Biomed Mater Res 51 (2000), p. 484-490.
[3]. P. Saravanapavan, M.H. Patel, L.L. Hench:Key Engineering Materials 240-242 (2003), pp 233
[4]. P. Sepulveda, J.R. Jones, L.L. Hench J Biomed Mater Res 59: (2002), p. 340-348.

Key Engineering Materials Vols. 254-256(2004) pp. 1091-1094
online at http://www.scientific.net
© *2004 Trans Tech Publications, Switzerland*

Mineralised membranes for bone regeneration

Mesquita P[1,2]; Branco R[3], Afonso A[3], Vasconcelos M[3] Cavalheiro J[1,2]

1 INEB- Instituto de Engenharia Biomédica, R. do Campo Alegre 823, 4150-180 Porto, Portugal
2 Universidade do Porto, Faculdade de Engenharia, Departamento de Engenharia Metalúrgica e Materiais, Porto, Portugal, jcavalheiro@fe.up.pt
3 Faculdade de Medicina Dentária da Universidade do Porto, R. Dr. Manuel Pereira da Silva 4200-393 Porto Portugal. fmd_up@mail.pt

Keywords: Biomaterial, membrane, bone regeneration, natural material

Abstract. Guided bone regeneration can be achieved using filling biomaterials and protective membranes. The aim of this work is to study a new membrane prepared from shrimp shells, without modification of the natural concave shape. The "in vitro" bioactivity was evaluated using a simulated body fluid prepared with substitution of calcium by strontium (SBR-Sr). The bioactivity behaviour was increased using a surface treatment with sodium silicate.

The "in vitro" and "in vivo" tests allow to conclude that these membranes are bioactive and can induce bone growth on their surfaces.

Introduction

The bone regeneration of the alveolar ridge can be achieved using ceramic biomaterials to fill bone defects with better results when protective membranes are used [1-4]. Several types of membranes are now available which are not bioactive, except for the slices of human bone collected from cadaver. Besides with the available membranes their fixation is normally difficult.

The aim of this work is to study a new membrane prepared from shrimp shells. The shrimp shells have a multilayer structure, where chitin is associated with calcium carbonate. The shells exhibit an elastic behaviour that can be useful to fix the membrane for dental applications. In a previous work [5] the shear strength with a mean value of $22N/mm^2$ was found, corresponding to the thickness of 0,25mm.

The potential for the utilisation of a mineralised membrane, not only as a protective layer but also as a bone support, will be evaluated.

Materials and methods

The boiled shrimp shells removed and treated with NaOH 3% (w/w) solution at 70°C were sonnicated and used as control. Another group with the same treatment was immersed in a 25 % (w/w) sodium silicate solution during one hour and dried. The samples were immersed during 6h, 2, 4 and 8 days in a simulated body fluid at 37°C, where Ca had been substituted for Sr. The composition of the solution of SBF-Sr was (mmol/L): NaCl 136.8; $NaHCO_3$ 4.2; KCl 3;$KaHPO_4$ 3; H_2O 1; $MgCl_2$ 6; H_2O 1.5; Na_2SO_4 0.5; $SrCl_2$ 2.5; HCl 1M to adjust pH to 7.

A third group of shrimp shells, after the same treatment, was immersed in PBS, a 0.01 M phosphate buffered saline, pH 7, NaCl 0.138M; KCl 0.0027 M (Sigma) and were implanted in the tibial cortical surface of 6 adult Huíla rabbits. The animals were sacrificed 4 weeks after surgery. The bone blocks obtained were immersed in 10% neutral buffered formalin for 24 hours. Afterwards, specimens were dehydrated in a series of alcohols and embedded in a mixture of methylmethacrylate, plastoid N and Perkadox. After polymerization, specimens were sectioned with a diamond saw to a thickness of about $300\mu m$ and ground down to about $40\mu m$ with a polish superfine discs. Slices were then stained with hematoxylin/eosin and solochrome R. Histological characterization was performed under light microscopy.The surface of the samples were analysed using SEM/EDS and FTIR and XPS; P will be expressed as oxide.

Results and discussion

The initial deproteinization treatment produced a reduction of the amount of phosphates on the surface of the external layer, with a decrease of the P percentage from 20.4 to 2.0 %.

The use of SBF Sr was aimed to identify precipitates of Sr under a natural calcified tissue. Before immersion in SBF-Sr solution the control samples exhibited a great difference between the inner and the outer layers. The former were covered with a smooth film organic under a mineralised substract, (Fig. 1a) and the latter show an organised structure of calcium carbonate nodules (Fig 1b) associated with phosphates, which promoted a very fast precipitation during immersion.

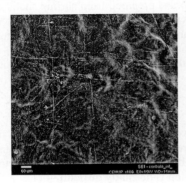 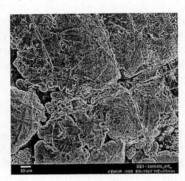

Fig. 1 a- The internal surface of the shell with a poor mineralised tissue with some phosphates
    b- External surface with calcium carbonate nodules, some phosphates and a chitin matrix

The observation of a cut of the control samples showed different layers of tissue (Fig.2). Near the external (convex) surface Z1, we can find a large mineralization with a strong presence of calcium carbonate. The phosphates appear associated to the carbonates on the Z2. Before a zone mineralised with calcium phosphate, the inner layer Z4, we can find the layer Z3, mainly formed by chitin.

The surface analyses of the internal layer confirm the results of observation of Fig. 2 related with the decrease of P from 16.0 on the internal layer to 2.0% on the external (convex) surface. The presence of phosphates and the smooth internal surface will promote a differentiation of the structure and composition of the precipitates formed after immersion in SBF solution.

The precipitation process can be evaluated by the Sr precipitated on the material, depending on the immersion time, because all the Sr will come from the solution.

The precipitation of Sr will occur very fast on the external surface of control samples only a few hours after immersion in SBF (Table 1):

Table 1 Precipitation of Sr on the external layer of control samples after immersion

	control	6h i.	2d i.	4d i.	8d i.
% Sr	0	13.3	7.3	5.2	15.0

Six hours after immersion the Sr precipitates had almost the maximum concentration achieved and then apparently there is a decrease of concentration that could be explained if part of the thick precipitate formed was lost later, during sample preparation.

Fig. 2- Cut of the control shell. Z1- external layer maily calcium carbonates and some phosphates; Z2-calcium phosphates and carbonates; Z3- maily chitin; Z4-internal layer with phosphates

The relationship Ca/P of the outer layer change completely a few hours after immersion, from 65.4 to 35.2 because the carbonate was partially covered by a strontium apatite layer, as it can be concluded observing the superficial variation of P concentration on Table 2:

Table 2. Precipitation of phosphates on the external layer

	control	6h i.	2d i.	4d i.	8d i.
% P	2.0	16.3	14.1	11.8	12.2

Again, as in the Sr results, after 6 h of immersion some decrease of the concentration of the P can be observed. It can now be explained by the occurrence of two phenomena: a simultaneous precipitation of strontium carbonates, with proportional decrease of phosphates, and the possibility of damage and loss of the thick films formed during samples preparation.

The structure of the precipitates was quite different. The precipitation on the internal layer occurred as a smooth continuous film, similar but more mineralised than the initial one (Fig. 1a). On the other hand the external layer precipitates are globular, formed on the surface of the natural inorganic base material. Using a higher magnification than the one used in Fig.1b, in Fig 3a, we could observe the crystals of the precipitate with annular structures, where some rings are formed by carbonates (Z1 and Z3) and other by phosphates (Z2).

The use of silicate allowed a still faster mineralization, even on the less active surface, the inner layer. After 6 h of immersion in SBF-Sr the inner layer treated with silicate was covered with a continuous film of apatite, thicker than the outer layer of the control shells, despite their great potential of nucleation: the concentration of Sr in the treated inner layer was 23.3 % while the more active external surface of the control treatment exhibit only 5.9% after the same time of immersion.

The FTIR confirmed the formation of phosphates as well as Ca carbonate precipitates, detected in FTIR peaks of 1404 and 1035 cm$^{-1.}$

The XPS analysis also confirmed the potential of silicate treatment results, showing an increase of the P atomic concentration from 3.0 to 12.7%, after 6 hours of immersion in SBF solution respectively of the control and the silicate treated samples.

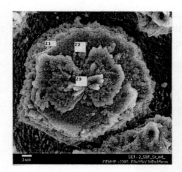 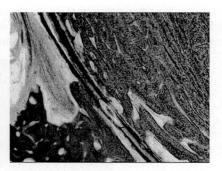

Fig. 3a The natural inorganic structure of the external surface will nucleate the precipitation of carbonates (Z1, Z3) and phosphates (Z2) after immersion in SBF-Sr solution
Fig. 3 b Histology of the rabbit tibia covered with a shell membrane (continuous line), with close contact with new bone (grey) (OM 50x, 4 weeks after implantation).

In vivo results

The *in vivo* tests proved that membrane fixation was easy on convex surfaces, like the rabbit tibia, and that it was possible to observe the formation of osteoid bone growing in close contact with the inner membrane surface, as it can be seen in Fig 3b. The surface of tibia was not scratched before implantation, so the remodelling stimulus comes only from the membrane.

Conclusions

According to the results *in vitro* and *in vivo,* these new membranes are very bioactive providing osteoconduction and they can be useful not only as an elastic cover, but also as a support of the new bone formation.

References

[1]   Baron M, Haas R, Dörtbudak O, Watzek G: Experimentally induced peri-implantitis: a review. Int J Oral Maxillofac Implants 15 (2000), 533-44.

[2]   Hammerle CH, Jung RE, Feloutzis A: A systematic review of the survival of implants in bone sites augmented with barrier membranes (guided bone regeneration) in partially edentulous patients.

J Clin. Periodontol. 29 Suppl 3 (2002), 226-31.

[3]   Brunel G, Brocard D, Duffort JF, Jacquet E, Justumus P, Simonet T, Benque E: Bioabsorbable materials for guided bone regeneration prior to implant placement and 7-year follow-up: report of 14 cases. J Periodontol. 72(2) (2001), 257-64.

[4]   Karapati S, Hugson A, Kugelberg CF: Healing following GTR treatment of bone defects distal to mandibular 2nd molars after removal of impacted 3rd molars. J. Clin. Periodontol. 27 (2000), 325-32

[5]   Simões R, Fonseca L, Vasconcelos M, Afonso A, Cavalheiro J. Modified exoskeleton of shrimp for guided bone regeneration. 17th ESB, Barcelona, Spain, September 2002.

Key Engineering Materials Vols. 254-256(2004) pp. 1095-1098
online at http://www.scientific.net

Inspired Porosity for Cells and Tissues

R. Martinetti[1], L. Dolcini[1], A. Belpassi[1], R. Quarto[2], M. Mastrogiacomo[3], R. Cancedda [2,3], M. Labanti[4]

1 - FIN-CERAMICA FAENZA s.r.l. Via Ravegnana 186, Faenza I-48018, ITALY
roberta.martinetti@fin-ceramicafaenza.com
2 - Dpt. Oncology, Biology and Genetics, Univ. of Genova, ITALY
3 - National Institute for Cancer Research (IST), Genova, ITALY
4 - ENEA Centro Ricerche Nuovi Materiali, Via Ravegnana 186, Faenza I-48018, ITALY

Keywords: Tissue engineering, calcium phosphates, scaffold, bone repair

Abstract. The use of 3D scaffolds provides an adhesive substrate that become the physical support matrix for in vivo tissue regeneration. In the last few years the use of bioengineered 3D scaffolds is becoming the most promising experimental approach for the regeneration of living tissue. Stem cells are used to study the signaling pathways that mediate cell differentiation and to identify optimal microenvironments that support cellular functionality. For tissue engineering applications 3D biomaterials must be able of supporting the functional properties of osteogenic cells. Inorganic devices are particularly relevant for bone regeneration; in particular calcium phosphate ceramics have been shown to interact strongly particularly with bone. The aim of the present work was the design and the production of a 3D Calcium/Phosphate scaffold able to mimics the in vivo environment to induce/promote tissue repair; the scaffold developed shows a defined design able to achieve the required functionality.

Introduction

Porous ceramic are interesting candidate materials for hard tissue engineering; bone is a natural composite in which mineral crystal are enclosed: hydroxyapatite (HA) in bone provides a natural 3D scaffold with organic fibrous material. The current trend of hard tissue engineering is toward the development of bioceramics and in particular porous calcium phosphates are used. In particular porous ceramics can be designed in term of porosity and pore size distribution [1, 2]. The micro and macro architecture of the scaffold is highly dependent on the production process and the aim of the present development is to provide a structure able to host and interact with cells, to induce/promote cells proliferation, differentiation and new bone deposition/formation. With appropriate modification, traditional ceramic manufacturing technique (slip casting) can be used to achieve innovative scaffold structure. In this paper, our achievements in term of characterization parameters of the developed scaffold are reported; the device functionality was also investigated loading the scaffold with stem cells [3, 4], *in vivo* experimental results are described.

Materials and Methods

The porous devices were prepared with an innovative technology using slurry expansion: a slurry with high powder concentration was used and expanded in a known volume to achieve a total porosity close to 80 vol. %; the porosity is characterised by bimodal porous structure and controlled morphology. Physico-chemical and morphological characterisation of the developed 3D scaffold were carried out; closely examination, from the morphological point of view was performed to investigate the design of the micro-macro porosity and the pore size distribution. The porosity distribution was investigate with an Image analyser; the analyses was carried out following ASTM E562 (Leica Imaging System Ltd. Q500MW by Qphase Application); morphology was analysed by SEM,(Leica, Cambridge) while phases investigation and chemical analyses were carried out respectively by X-Ray diffraction analysis (CuKα radiation, Rigaku Miniflex) and   by ICP (Spectroflame Modula, Spectro, equipped with Ultrasonic Nebulizer Cetac U-6000 AT⁺). The scaffold was also investigated from the mechanical point of view: compressive strength and flexural strength tests were carried out. For compressive strength the cross-head rate of the machine was 0,25 mm/ minute, the test was carried out on 15 mm cubes. Flexural strength test was carried out according CEN EN 658-3 standard , the cross-head rate of the machine during the test was 0,5 mm/ minute test.

The scaffold was loaded with osteoprogenitor cells in the presence of a fibrin glue as cell carrier and implanted in immunodeficent mice to allow bone formation; scaffold alone was also implanted as a control. Samples were retrieved 4 and 8 weeks after implantation and bone formation was evaluated using microscopy and quantified by image analysis.

Results

According to X-ray diffraction analysis the starting powder used for the production of hydroxyapatite scaffold resulted to be single phase crystalline HA, corresponding to ICCD card no. 9-432, with purity ≥ 95%. The total porosity of the devices was 80 vol.% (± 3); the porosity distribution results to be of bi-modal type, with pores size mainly in the range 100-200 μm (32 vol.%) and 200-500 μm (40 vol.%) (Fig. 1 and 2).

SEM examination of the scaffold (Fig. 3 and 4) showed a uniform distribution of the macro and micro structure; the interconnection between the pores were also evident (Fig. 3)

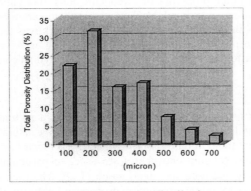

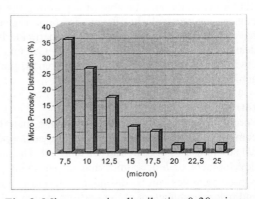

Fig. 1. Total porosity distribution   Fig. 2. Micro porosity distribution 0-30 micron

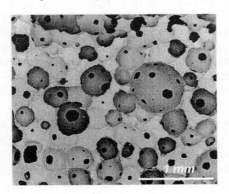

Fig. 3 – HA scaffold macrostructure   Fig. 4 – HA scaffold microstrutture

Fig. 5 report the data collected in term of compressive strength, flexural and Young modulus of the HA device: for compressive strength there is a standard deviation of the value close to ± 0,17 while for flexural strength it is ± 1,08 and ± 0,74 for Young Modulus. Considering the high level of total porosity (80 ± 3 vol%) the mean compressive strength value is higher in respect to the flexural strength: it could be correlated to the isotropic porous structure achieved with the developed production process.

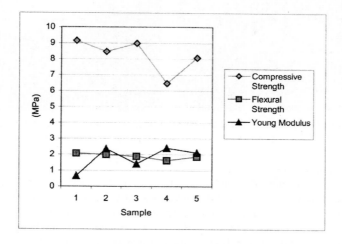

Fig. 5. HA scaffold mechanical data

Both at 4 and 8 weeks, the implanted ceramic samples presented an evident bone formation. In the 8 weeks retrieval (Fig. 6, 7, 8 and 9) newly formed bone had formed an extensive coat over the ceramic internal surface. Bone had covered most of the macro-pore space. Bone tissue was adherent to the vast majority of the free ceramic surfaces suggesting that bone formation occurs as a function of available surface. Therefore the overall bone formed is very prominent in this low density materials.

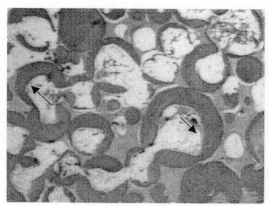

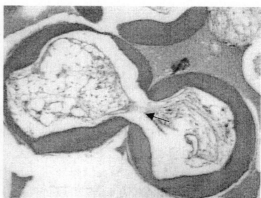

Fig. 6 -Samples after 8 weeks: bone deposition

Fig. 7 – Samples after 8 weeks: pore interconnection

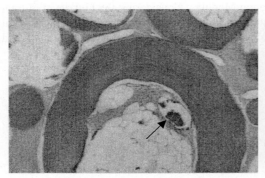 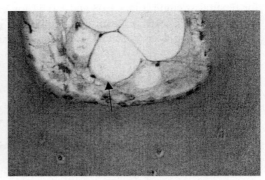

Fig. 8 - Samples after 8 weeks: blood vessel formation is underlined

Fig. 9 - Samples after 8 weeks: starting bone marrow formation

Discussion and Conclusions

Our results suggest that the bi-modal characteristic of the porosity and the low density in hydroxyapatite scaffold result to be useful innovations for tissue engineering; a porosity level of 80 vol. % means also an high surface available ($\cong$ 0,9 m^2/gr) for interaction with cells and bone formation. The osteoprogenitor cells are able, in contact with the high porous hydroxyapatite scaffold to differenciate and generate high amount of new bone, whereas hydroxyapatite alone does not generate any bone.

The development of cell culture technology together with a suitable porous biomaterials may be of future interest for long bone segments substitutions and regeneration; the new device ceramic/cells is able to combine the advantage of the osteogenesis process with the osteoconductivity of the calcium/phosphate material.

Acknowledgments

The authors thank G. Martignani, G.L. Minoccari, L. Pilotti and G. Ronconi (ENEA, Faenza) respectively for porosity determination, mechanical tests, SEM observations and chemical analysis. This work was carried out within the frame of an European Community Project: PORELEASE (Contract N°: G5RD-CT-1999-00044, Project N°: GRD1-1999-10590).

References

[1] J.X.LU, B. Flautre, K. Anselme: Materials in medicine 10(1999) 111-120

[2] I. Martin, A. Muraglia, G. Campanile, R. Cancedda, and R. Quarto, R.: Endocrinology, 138, (1997) 4456-62 .

[3] M. Marcacci, E. Kon, S. Zaffagnini, R. Giardino, M. Rocca, A. Corsi, A. Benvenuti, P. Bianco, R. Quarto, I. Martin, A. Muraglia and R. Cancedda: Calcif Tissue Int, 64, (1999) 83-90 .

[4] P. Chistolini, I. Ruspantini, P. Bianco, A. Corsi, R. Cancedda and R. Quarto: J Mater Sci Mater Med, 10, (1999) 739-742.

[5] R. Martinetti, A. Belpassi, A. Nataloni and C. Piconi (2000), Bioceramics 13, 506-510

Key Engineering Materials Vols. 254-256(2004) pp. 1099-1102
online at http://www.scientific.net
© *2004 Trans Tech Publications, Switzerland*

Repair of Full-thickness Defects in Rabbit Articular Cartilage Using bFGF and Hyaluronan Sponge

T. Yamazaki[1] , J. Tamura[1] , T. Nakamura[1] , Y. Tabata[2] , Y. Matsusue[3]

[1] Department of orthopedic surgery , Kyoto University

Kawahara-cho 54 , Shogoin , Sakyo-ku , Kyoto 606-8507 , Japan

[2] Institute for Frontier Medical Sciences , Kyoto University

53 Kawahara-cho Shogoin , Sakyo-ku Kyoto 606-8507 , Japan

[3] Department of Orthopedic Surgery , Shiga University of Medical Science

Setatukinowa-cho Otsu , Shiga 520-2192 , Japan

Keywords: growth factor, carrier , drug delivery system

Abstract. Basic fibroblast growth factor (bFGF) is reported to be effective for cartilage repair, but its effects are limited without a drug delivery system because bFGF is lost by diffusion. We investigated the effects of exogenous basic fibroblast growth factor (bFGF) with drug delivery system on the repair of full-thickness cartilage defects. bFGF with gelatin microspheres and hyaluronan sponge improved the repair of osteochondral defects in rabbits.

Introduction

Several studies have shown that bFGF is mitogen for chondrocytes, and that it stimulates the synthesis of the cartilagenous matrix [1-4]. But the effects of exogenous bFGF are considered to be limited to only a few days without a drug delivery system. We investigated the effects of exogenous basic fibroblast growth factor (bFGF) with a drug delivery system on the repair of full-thickness cartilage defects, and the profile of bFGF release from bFGF incorporating hyaluronan sponge and gelatin microspheres. In this study hyaluronan sponge is used as carrier, and bFGF is released from gelatin microspheres by degradation. As hyaluronan sponge is well known as biocompatible and absorbable material, it is expected to be a good carrier material.

Materials and Methods

 Gelatin microspheres (average diameter 20 μm) were prepared through glutaraldehyde (2 wt%) crosslinking of a gelatin aqueous solution [5]. The gelatin sample, isolated from bovine bone, had an isoelectric point of 4.9 and a molecular weight of 99 000 (Nitta Gelatin Co., Osaka, Japan).   An aqueous solution of human recombinant bFGF with an isoelectric point of 9.6 (10 mg/ml) was supplied by Kaken Pharmaceutical Co. Ltd., Tokyo, Japan. The hyaluronan sponge was supplied by Denka Co. Ltd., Tokyo, Japan. The hyaluronan sponge was used to keep gelatin microspheres and bFGF in the defects.

72 knees taken from 36 Japanese white rabbits were examined. Four-millimeter diameter drill holes were made in the articular cartilages. The articular defects were divided into 3 groups:
   a) the defect left untreated (group 1)
   b) the defect filled with a hyaluronan sponge (4 mm×4 mm×1mm) (group 2)
   c) the defect filled with a hyaluronan sponge + aqueous solution (20 μl) containing 1μg bFGF and 2 mg geratin microsphere (group 3).
   The animals were killed with an intravenous injection of sodium pentobarbital at 4, 12, 25, and 50 weeks after the operation. The specimens were fixed in 4% paraform aldehyde, decalcified in EDTA, and embedded in paraffin. Sections were stained with toluidine blue and type II collagen antibody.

A semiquantitative scale of histological grading (Table 1) was used to evaluate the sections microscopically. The scale is composed of 6 categories and assigns a score ranging from 0 to 20. The score for normal cartilage is 20 points.

Table1. Histological grading system

A. Cell morphology
4-hyaline cartilage, 3-mostly hyaline cartilage, 2-hyaline and fibrocartilage, 1-mostly fibrocartilage, 0-mostly non-cartilage
B. Matrix staining
4-same as the normal area, 3-slightly reduced, 2-reduced, 1-significantly reduced, 0-none
C. Surface regularity
2-smooth, 1-slightly irregular, 0-irregular
D. Thickness of the cartilage (comparing the cartilagenous thickness of the defect to the surrounding cartilage)
4-100%, 3- 75%, 2- 50%, 1- 25%, 0- 0%
E. Bonding
2-both edges integrated, 1-one edge integrated, 0-both edges not integrated
F. Reconstruction of subchondral bone
4-same as the normal area (100%), 3-slightly lower and/or higher than the normal osteochondral junction (OCJ) (75% and/or 125%), 2-moderately lower and/or higher than the normal OCJ (50% and/or 150%), 1-severely lower and/or higher than the normal OCJ, 0-non-reconstruction

To evaluate the profile of bFGF release from hyaluronan sponge and gelatin microsphere *in vivo*, hyaluronan sponge incorporating gelatin microspheres containing I^{125} labeled-bFGF was implanted into the back subcutis of female DDY mice, 6 weeks of age . These mice were sacrificed at 1,4,7,10, and 14 days after the operation and then radioactivity of the tissue from the back of these mice was counted.

Results

The *in vivo* bFGF release from hyaluronan sponge containing bFGF-impregnated gelatin microspheres is shown in Fig.1. The amount of bFGF implanted in the subcutis of mice decreased about 70% in day 1, and gradually absorbed as the hyaluronan sponge and gelatin microsphere were degraded. Thus, bFGF had been released for 2 weeks and was almost totally absorbed in day 14.

As for rabbit joint cartilage repair, every defect showed, microscopically, a layer of mesenchymal cells at 4 weeks. There were little histological differences between the groups. The deep part of the wound tissue from group 1 had mostly been invaded by fibroblast-like cells, and the matrix was faintly stained with toluidine blue. In the tissues from group 3, the deep part was partially composed of round chondrocytes, and weakly stained with type II collagen antibody. The matrix was stained with toluidine blue, but the superficial part was weakly stained (Fig. 2).

At 12 weeks, the tissues from group 3 shows thick layer of newly formed cartilage (Fig. 3). In the other groups, reconstruction of subchondral bone was poorly observed.

At 25 weeks, the tissues from group 1 could still be distinguished from the surrounding normal cartilage, but the tissues from group 3 resembled normal cartilage (Fig.4). The defects of group 1, 2 had been covered with fibrous tissue. In group 3, the thickness of the metachromatic-stained matrix had increased as compared with 4 and 12 weeks, and resembled the surrounding normal cartilage without any degenerative changes.

At 50 weeks, regenerated cartilage became thinner and matrix staining was lost in groups 1 and 2, and the defects of groups 1 and 2 had been covered with fibrous tissue. But degenerative changes were not so prominent in the defects of group 3 compared with groups 1 and 2 (Fig. 5).

Throughout all periods, there was a tendency for the regenerative cartilage in group 3 to be thicker than that seen in group 1. The histological grading system was used to evaluate the repair of the defects. The score for group 3 increased throughout the experimental period. The score for group 3 was the best in the three groups at 25 weeks, and peaked at 50 weeks (Table 2). The addition of bFGF to the defect induced the formation of a thick cartilage layer composed of chondrocytes and metachromatic-stained matrix after 12 weeks.

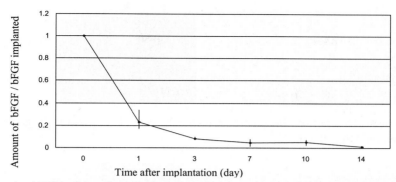

Fig.1. *In vivo* bFGF release from hyaluronic acid sponge containing bFGF-impregnated gelatin microspheres

Table 2. Histological grading score

	12W	25W	50W
Group 1	6.3	9.2	9.6
Group 2	7.4	9.7	10.3
Group 3	10.3	14.6	15.6

Discussion and Conclusions
These findings suggests that exogenous bFGF stimulated the proliferation of chondrocytes, and the accumulation of type II collagen and proteoglycan in the regenerating cartilage *in vivo*. As group 2 did not show worse result compared with group 1 (Table 2), this sponge seems to have no harmful effect in cartilage repair. Compared with a previous study, bFGF with gelatin microspheres and hyaluronan sponge seems to show better results for the repair of full thickness cartilage defect [6].

In conclusion, administration of bFGF with gelatin microspheres and hyaluronan sponge into cartilaginous defects promotes the differentiation of chondrocytes and their matrix synthesis, and improves repair of cartilage *in vivo*. Furthermore, gelatin microspheres and hyaluronan sponge can release bFGF for 2 weeks.

A   B   C
Fig.2: Decalcified sections of specimen implanted after 4 weeks (A- group 1 , B- group 2 , C- group 3 ,×40 , toluidine blue) Arrow head-defect

A   B   C
Fig.3 : Decalcified sections of specimen implanted after 12 weeks (A- group 1 , B- group 2 , C- group 3 , ×40, toluidine blue)

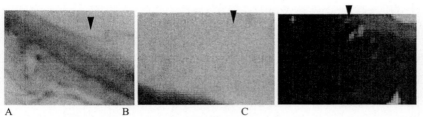

A   B   C

Fig.4 : Decalcified sections of specimen implanted after 25 weeks (A- group 1 , B- group 2 ,
C- group 3 ,×40, toluidine blue)

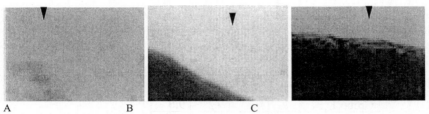

A   B   C

Fig.5 : Decalcified sections of specimen implanted after 50 weeks (A- group 1 , B- group 2 ,
C- group 3 ,×40 , toluidine blue)

References

[1] Cuevas P, Burgos J, Baird A: Basic fibroblast growthfactor (FGF) promotes cartilage repair   in vivo. Biochem Biophys Res Commun 156(1988)   :611-618

[2] Kato Y, Gospodarowicz D: Sulfated proteoglycan synthesis by confluent cultures of rabbit costal chondrocytes grown in the presence of fibroblast growth factor. J Cell Biol 100(1985) :477-485

[3] Kato Y, Nomura Y, Daikuhara Y, Nasu N, Tsuji M, Asada A,Suzuki F   Cartilage-derived factor (CDF) I. Stimulationof proteoglycan synthesis in rat and rabbit costal chondrocytes in culture. Exp Cell Res 130(1980) :73-81

[4] Kato Y, Hiraki Y, Inoue H, Kinoshita M, Yutani Y, Suzuki F: Differential and synergistic actions of somatomedinlike growth factors, fibroblast growth factor and epidermal growth factor in rabbit costal chondrocytes. Eur J Biochem 129(1983) :685-690

[5] Kawai K, Suzuki S, Tabata Y, Ikada Y: Accelerated tissue regeneration through incorporation of basic fibroblast growth factor-impregnated gelatin microspheres into artificial dermis Biomaterials 21 (2000) 489-499

[6] Fujimoto M. Ochi,Y. Kato, Y. Mochizuki,Y. Sumen, Y. Ikuta: Beneficial effect of basic fibroblast growth factor on the repair of full-thickness defects in rabbit articular cartilage Arch Orthop Trauma Surg 119 (1999) :139-145

Key Engineering Materials Vols. 254-256(2004) pp. 1103-1106
online at http://www.scientific.net
© *2004 Trans Tech Publications, Switzerland*

Bioactive Porous Bone Scaffold Coated with Biphasic Calcium Phosphates

Hae-Won Kim[1,2,#], Hyoun-Ee Kim[1] and Jonathan C Knowles[2]

[1] School of Materials Science and Engineering, Seoul National University, 151-742, Seoul, Korea,
hkim@eastman.ucl.ac.uk

[2] Biomaterials and Tissue Engineering, Eastman Dental Institute, University College London,
WC1X 8LD, London, UK

[#] Corresponding Author, email: hkim@eastman.ucl.ac.uk

Keywords: Porous bone scaffold, calcium phosphate coating, zirconia, mechanical property, cell response

Abstract. Calcium phosphates (CaP) were coated on a zirconia (ZrO_2) porous scaffold for hard tissue applications. The ZrO_2 porous body, intended to be a load-bearing part, was fabricated by a polymeric foam reticulate method. On the framework, CaP layers were deposited by a powder slurry method to induce bioactivity and osteoconductivity. The compositions of the coating layers were varied to be biphasic, i. e., hydroxyapatite (HA) with tricalcium phosphate (TCP) and fluorapatite (FA). The coated scaffolds had a porosity of ~ 90 % and pore size of ~500-600 μm. The compressive strength of the coated ZrO_2 scaffolds was ~5 MPa, which was ~7 times higher than those of pure HA. The coating layers had a thickness of ~30 μm and was firmly adhered to the substrate. The adhesive strength of the coating layer was ~20-25 MPa. Human osteoblast-like cells grew and proliferated well on the coated scaffolds. The differentiation of the cells on the coated scaffolds was higher than those on pure ZrO_2 scaffold, confirming improved cell activity via the calcium phosphate coatings.

Introduction

Calcium phosphate (CaP) ceramics, such as hydroxyapatite [HA; $Ca_{10}(PO_4)_6(OH)_2$], fluorapatite [FA; $Ca_{10}(PO_4)_6F_2$], and tri-calcium phosphate [TCP; $Ca_3(PO_4)_2$], have attracted lots of attention as teeth and bone substitutes [1-3]. Their chemical and crystallographic similarities as well as osteoconductivity and bioactivity to hard tissues give them excellent biocompatibility. To be a bone scaffold, the porous structure is favored due to the high specific surface area and blood circulation. However, the poor mechanical properties restricted their applications to the powders, granules, and non-load bearing parts [4]. Therefore, coating systems onto metals and ceramic oxides were employed to improve mechanical properties as well as biocompatibility [5].

On the other hand, zirconia (ZrO_2) is a well-known bioceramic, which possesses excellent mechanical properties. Several researchers have fabricated ZrO_2 bodies for biomedical applications [6]. However, ZrO_2 is not osteoconductive and do not bond directly to bone [3]. Therefore, bioactive materials need to be coated on the surface.

In this study, a porous ZrO_2 body, which was chosen as a framework for load bearing, was coated with CaP layers to enhance biocompatibility and osteoconductivity. In particular, an FA intermediate layer was placed into the CaP layer / ZrO_2 substrate to suppress the reaction between CaP and ZrO_2. The porous ZrO_2 body was fabricated by a polymeric sponge replication technique and the FA and CaP coating layers were obtained by a powder slurry method. The coating system was characterized in terms of structural, mechanical properties, and *in-vitro* cellular responses.

Experimental Procedures

The ZrO_2 slurry was made by mixing a 100 g of ZrO_2 powder (3 mol% Y_2O_3, Cerac) with 6 g of triethyl phosphate and poly vinyl butyl in distilled water. The polyurethane sponge (Customs Foam) was immersed in the slurry then dried at 80 °C. After repeating the dipping/drying step, the foam was heat-treated at 800 °C for 5 h, and subsequently at 1400°C for 3 h. The above replication process was repeated to obtain a porosity of ~90 %. The initial powders for each coating slurry were prepared from HA (Alfa Aesar, USA), TCP (Merck, Germany), and CaF$_2$ (Aldrich, USA). The FA powder was obtained from a reaction between TCP and CaF$_2$, as previously described [7]. The biphasic powders (HA+FA and HA+TCP) were prepared by mixing each powder at an equivalent molar ratio (HA/FA and HA/TCP = 1).

Each coating powder was dissolved in ethanol solution mixed with tri-ethyl phosphate and poly-vinyl butyl. Initially, the ZrO$_2$ scaffold was dip-coated with the FA slurry to form an intermediate layer. After drying and heat-treating, CaP outer layers were coated using the corresponding slurry under the same conditions as the FA pre-coating process. Coating process was repeated to produce a uniform coating layer. The ZrO$_2$ porous scaffolds coated with different CaP compositions are designated as shown in Table 1.

Phase and morphology of the coated scaffolds were analyzed using X-ray diffraction (XRD) patterns and scanning electron microscopy (SEM), respectively. Compressive strength was tested on the porous specimens (5 x 5 x 10 mm) at a crosshead speed of 0.05 mm/min. Adhesive strength of the coating layer was tested with an adhesion testing apparatus (Sebastian V, USA), as described previously [5].

Cellular responses to the scaffolds were assessed in terms of cell proliferation and differentiation. MG63 cells were seeded on the coated scaffolds at a density of 1×10^4 cells/ml. Pure ZrO$_2$ scaffold was also tested for comparison. After culturing for 3 and 7 days, the cells were detached with trypsin-EDTA solution (trypsinization). After centrifuging and washing, the cells were counted using a hemocytometer. Each set of tests was performed in triplicate. Cell growth morphology was observed using SEM after fixing, dehydrating, and critical point drying of the cells. Cell differentiation was assessed in terms of alkaline phosphatase (ALP) activity. MG63 cells were cultured for 10 days, and then trypsinized as described above. After centrifuging, the cells were resuspended with Triton X-100, and disrupted further by freezing / thawing cycles. After centrifugation, the cell lysates were reacted with *p*-nitrophenyl phosphate at pH 10.3 for 60 min. The result-out color product, *p*-nitrophenol, was measured at 410 nm using a spectrophotometer.

Table 1. Designation of scaffolds with different coating composition

Designation	CaP1	CaP2	CaP3
Coating composition	HA+TCP	HA+FA	HA

Results and Discussion

The structures of the ZrO$_2$ porous scaffolds coated with CaP1 layer are shown in Fig. 1. A highly porous structure with interconnected framework was obtained (Fig. 1A). The initial polyurethane foam structure was exactly replicated with perfectly interconnected open pores. As shown in Table 2, the sizes of pores and the stems were about 500-600 µm and 100-200 µm, moreover the porosity was as high as 90 %. This structure is expected to be adequate to allow bone ingrowth as well as to supply blood circulation [8]. At high magnification, CaP1 coating layer is seen to cover the ZrO$_2$ surface uniformly (Fig. 1B). The coating structure was relatively porous with pore sizes of several microns. This micro-porous structure of the coating layer is expected to enhance adhesion of the bone with implants via mechanical interlocking and consequently to promote osseointegration [9]. The cross section view of the scaffold represented a typical double-layered structure (Fig. 1C). The thicknesses of the outer and inner coating layers are ~5 and ~20 µm, respectively. There were no delamination or cracks at the interfaces, indicating tight adhesions among the layers. This ZrO$_2$ scaffold coated with CaP layers had a compressive strength of 5.4 MPa. The value was approximately 7 times higher than that of pure HA scaffold (0.7 MPa). This value confirms the efficacy of ZrO$_2$ as hard tissue applications when considering the compressive strength of human cancellous bone (2-12 MPa) [3]. As shown in the microstructure, the reticulated foam approach is found to be quite effective in obtaining highly porous structures. Moreover, the porosity was easily controllable by repeating the replication process. Especially, during the coating process, a care should be taken to restrain the reaction between CaP layer and ZrO_2 substrate. Previously, the direct contact of HA with ZrO_2 at elevated temperatures caused serious decomposition reactions, producing reaction products of TCP and $CaZrO_3$ [10]. The reactions degrade the mechanical properties of interface, and consequently the biocompatibility of the material [10]. When FA layer was inserted, there was no reaction (data not shown here).

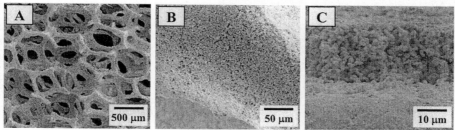

Fig. 1. SEM morphologies of CaP1 coated ZrO_2 scaffold

Table 2. Properties of CaP coated ZrO_2 porous scaffold

Pore size (μm)	Stem size (μm)	Porosity (%)	Compressive Strength (MPa)
500-600	150-200	89.6 (±1.3)	5.42 (±0.45)

The adhesive strength of the coating layer was approximately 20-25 MPa regardless of the coating composition, as shown Table 3. The adhesive strength of the coating layer is the crucial parameter that determines the permanence and durability of the system because a poor bonding results in coating debris and a loss of fixation from the host tissues. The strengths obtained in this study were comparable to those of commercial plasma-spray coatings [3]. This high adhesive strength was deemed to the relaxation of thermal mismatch between the CaP and ZrO_2 (due to porous structure of the coating layer) and also to the chemical inertness of FA with respect to ZrO_2 substrate. However, the intrinsic mechanical properties of the coating layer, such as the toughness and hardness, are expected to be somewhat down regulated due to the micro-porous structure.

Table 3. Adhesive strength of CaP coating layers with respect to ZrO_2 substrate

Scaffolds	CaP1	CaP2	CaP3
Adhesive Strength (MPa)	24.4 (±3)	20.9 (±2.8)	22.1 (±3.2)

Cellular responses of the coated scaffolds were assessed using osteoblast-like MG63 cells. Cell growth morphologies on the CaP1 coated ZrO_2 after culturing for 3 and 7 days are represented in Figs. 2A and B, respectively. When cultured for 3 days, the cell membranes highly spread on the coated surface with an intimate contact, and migrated deep into the large pores, suggesting the osteoconducting characteristics of the porous scaffolds (Fig. 2A). After 7 days, much larger number of cells grew on the surface, covering nearly the whole surface (Fig. 2B). The scaffolds coated with other CaP layers had similar cell growth morphologies to the CaP1 coated sample (data not shown here). The proliferation and differentiation characteristics of the cells were quantified, as shown in Fig. 3. Cell differentiation was evaluated via alkaline phosphatase (ALP) expression level since the ALP activity has been recognized as a marker for the functionality and activity of the osteoblast cells undergoing the differentiation step. Pure ZrO_2 without coating was also tested for comparison. MG63 cells similarly proliferated on both coated scaffolds and pure ZrO_2. In a different manner, the ALP activities of the cells showed higher ALP expression levels on the coated scaffolds compared to pure ZrO_2, confirming that the cells differentiated to a high degree on the coating samples. However, there was little significant difference among the samples depending on coating composition. Even though the cells on ZrO_2 similarly responded in a proliferation level, their activity was highly enhanced with the CaP coatings. In order to understand the biocompatibility of the CaP coated ZrO_2 porous scaffolds completely, *in vivo* experiments are underway.

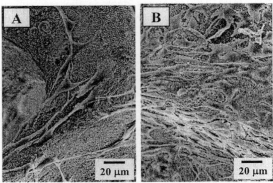

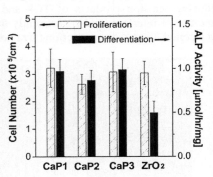

Fig. 2. Cell growth morphology on CaP1 coated ZrO_2 scaffold after culture for (A) 3 and (B) 7 days

Fig. 3. Cell proliferation and ALP activity on CaP1 coated ZrO_2

Summary and Conclusions

Biphasic calcium phosphate layers (HA+TCP and HA+FA) were successfully coated on a strong ZrO_2 porous scaffold for hard tissue applications. The coated scaffolds had highly porous structure with porosity and pore size of about 90 % and 500-600 μm. Compressive strength of the scaffold was superior to that of pure HA. The obtained coating layer was microporous with a thickness of ~30 μm. Adhesive strength of the coating layer was as high as 20-25 MPa. Osteoblast-like MG63 cells on the coating layer grew and spread actively, and showed higher ALP expression levels compared to those on pure ZrO_2, confirming the enhanced cell differentiation on the coating system.

Acknowledgement

This work was supported by the Post-doctoral Fellowship Program of Korea Science & Engineering Foundation (KOSEF)

References

[1] M. Jarcho: Clin. Orthop. Vol. 157 (1981), p. 259-78.
[2] S. F. Hulbert, F. A. Young, R. S. Mathews, J. J. Klawitter, C. D. Talbert, F. H. Stelling: J. Biomed. Mater. Res. Vol. 4 (1970), p. 433-56.
[3] L. L. Hench, E. C. Ethridge: *Biomaterials: An interfacial approach* (Academic Press, New York 1982).
[4] K. de Groot, C. de Putter, P. Smitt, A. Driessen: Sci. Ceram. Vol. 11 (1981), p. 433-7.
[5] H.-W. Kim, Y.-M. Kong, C.-J. Bae, Y.-J. Noh, H.-E. Kim. Biomaterials (in press)
[6] E. W. White, J. N. Weber, D. M. Roy, E. L. Owen, T. T. Chiroff, R. A. White: J. Biomed. Mater. Res. Symp. Vol 6 (1975), p. 23-7.
[7] H.-W. Kim, Y.-J. Noh, Y.-H. Koh, H.-E. Kim, H.-M. Kim J. S. Ko: J. Am. Ceram. Soc. (in press).
[8] E. Tsuruga, H. Takita, H. Itoh, Y. Wakisaka, Y. Kukoki: J. Biochem. Vol. 121 (1977), p. 317-24.
[9] P. Li, K. de Groot, T. Kokubo. J. Sol-Gel Sci. Tech. Vol. 7 (1996), p. 27-34.
[10] H.-W. Kim, Y.-J. Noh, Y.-H. Koh, H.-E. Kim, H.-M. Kim: Biomaterials Vol. 23 (2002), p.4113-21.

Key Engineering Materials Vols. 254-256(2004) pp. 1107-1110
online at http://www.scientific.net

A Self Setting Hydrogel as an Extracellular Synthetic Matrix for Tissue Engineering

P. Weiss, C. Vinatier, J. Guicheux, G. Grimandi, G. Daculsi

EMI 99-03, 1 Place A. Ricordeau, BP 84215, 44042 Nantes Cedex 1, France.
pweiss@sante.univ-nantes.fr

Keywords : biomaterial, hydrogel, rheology, polymer network, tissue engineering.

Abstract. The use of hydrogel for biomaterials or as 3D synthetic matrix for tissue engineering lead to determine the viscoelastic properties of the hydrogels and the moment when the viscous liquid is transformed in a reticulated polymer network. This can be performed by oscillatory rheological study. We developed a new self-reticulating polymer wich is silated hydroxyl propylmethylcellulose (Si-HPMC). This study shows that the parameters (pH, swelling, amount of silane grafting) can be controlled and adjusted using oscillatory rheological measurements. The polymer network can be adapted to provide the best environment for specific cell functions and use it as a scaffold for tissue engineering and as injectable biomaterial for bone repair.

Introduction

The "injectable ceramic suspension" IBS (Injectable Bone Substitute) [1, 2] developed in our laboratory is a composite of calcium phosphate granules with a polymer carrier. The polymeric phase consists of an aqueous solution of hemi-synthetic cellulose derivatives, while the calcium phosphate granules consist of biphasic calcium phosphate (BCP), which is a mixture of hydroxyapatite and β tricalcium phosphate. This IBS product is now ready for industrial development (MBCP Gel™, Biomatlante, Vigneux de Bretagne, France) and human applications. Human pilot clinical trials are now in progress in orthopedics and odontology for bone defect filling in non-loading conditions and bone filling of dental extraction sites.

The current IBS product is a non hardening material limiting its application to well defined cavity to prevent the diffusion of the IBS over the surgical site. To improve the properties of IBS and expand its applications, we developed a new injectable bone substitute, IBS2 [3-6], based on a self-reticulating polymer. This new polymer opens new prospectives in the field of osteoarticular engineering or others. Indeed, it seems commonly accepted today, that polymeric hydrogels are potential scaffolds for three-dimensional culture of cells and/or to make composite formulations for biomaterials. This polymer is silated hydroxyl propylmethylcellulose (Si-HPMC) with an organo-silane. After synthesis the silanised polymer is insoluble in water. The dissolution occurs in a sodium hydroxide solution with a pH of 13. The principle guiding the choice of this polymer for this first study, is its absorbent and viscous liquid character even in the presence of BCP granules and before its injection. In the surgical site, in contact with the biological liquids plugs, and/or with the addition of a specific buffer, a chemical reaction without additive and any system of catalysis allows bridging, by self-setting, between the various macromolecular chains involved (Figure 1). The starting reaction is related to pH modification. This process is pH-dependent, leads to the formation of a more or less dense gel, according to the conditions of polymer synthesis.

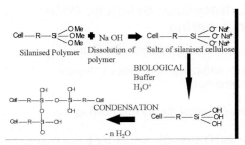

Fig. 1: Representation of dissolution and reticulation process of the self-setting polymer in water.

The synthesis of silanized cellulosic ethers was carried out while grafting in heterogeneous medium of the organosilanes on the polymeric chain of hydroxypropylmethylcellulose (HPMC). In basic medium (fig1), the grafted silanes are ionized and the chains are disjoined, then when the pH decreases the number of protons increases and R-Si-O becomes R-Si-OH very reactive. Covalent bridges with hydroxyls of the cycles of cellulose ethers or the substituents will be created thus forming a three-dimensional network (Figure 2).

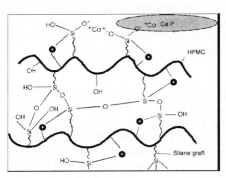

Fig. 2 : Network diagram of the silanised polymer in water with inter-macromolecular and intra-molecular (o) links.

To characterize the network structures, rheological measurements offer a quantitative characterization of the viscoelastic properties of a substance under defined conditions, using the three different types of rheological stresses, i.e. shear flow, elongational flow and oscillatory measurements [8, 9]. The method of choice for characterizing cross-linked structures is dynamic measurements. Here the substance is subjected to a sine oscillation with a defined deformation γ or stress τ, and frequency ω. If the substance shows a viscoelastic behavior, a phase shift δ occurs. If this deformation is sufficiently small, a non-destructive characterization of the substance is possible. By determining the storage and the loss modulus (G', G'') from the oscillatory measurements, the viscous and the elastic properties can be determined separately, and network parameters can be calculated.

Using the phase shift δ and the ratio of the maximum shear stress τ_0 to the maximum amplitude of deformation γ_0, one can calculate a complex oscillatory modulus G^*:

$$G^* = \frac{\tau_0}{\gamma_0} \cdot e^{i\delta} = G' + i \cdot G''$$

The stored elastic deformation energy is expressed by the real component (storage modulus G') and the viscous dissipated energy is expressed by the imaginary component (loss modulus G''). The quotient of the loss modulus G'' and the storage modulus G' gives the proportion of the dissipated energy to the stored energy. It is called the loss factor tanδ.

The aim of this study was to analyse the viscoelastic properties of this hydrogel by dynamic measurements and to determine the moment when the viscous liquid becomes a solid hydrogel and the parameters that change these properties.

Materials and methods

The silated HPMC (si-HPMC) produced was dissolved in NaOH solution at pH equal to 12.8 and with a concentration of 3% (w/w).

The pH of the viscous polymer solution was decreased to around 12.4 before steam sterilization.

HEPES buffer solution, in variable concentrations, was mixed with the polymer solution to obtain a final pH ranging from 7 to 10 and a final polymer concentration between 1 to 2.5%.

Dynamic rheological measurements were then made on a Haake Rheometer (Rheostress 300) using a coni-cylindrical geometry with a diameter of 60 mm and a cone angle of 1°. We used multiwaves procedure with 3 frequencies, 1Hz, 3.2Hz et 10Hz and the imposed stress was 1Pa. The temperature was controlled by an external thermal bath and maintained at 37°. The experiment time was 14 days. To prevent evaporation, oil of paraffin fluid was floated over the exposed sample. The paraffin fluid and sample were immiscible, and there was no anticipated effect on the sample chemistry; no changes in rheological behavior were observed as a result of this treatment.

Oscillation tests measuring parameters G' and G" were performed to study the self setting process, gel point and network parameters.

Results

The gel point (Arrow 1, Figure 3) and the maximum value of conservative modulus G' (Arrow 2, Figure 3) in dynamic rheological measurements show, for this Si-HPMC behaviour, that the viscous liquid becomes a solid hydrogel after 33 minutes and the reticulation was finished after 244 hours with a maximum G' storage modulus at 310 Pa.

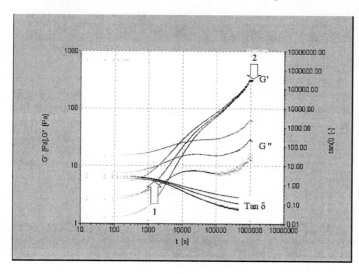

Fig. 3 : Si-HPMC rheometric diagram of E4M (Colorcon-Kent-England) silanized with 14,24 % of 3-glycidoxypropyltrimethoxysilane (GPTMS) graft (0.78 % of Silicium / Dry polymer). Polymer was in 3% NaOH water solution (w/w). Hepes buffer solution volume added was equal of polymer solution volume (1.5 % of dry polymer in the final solution). The pH of this mixture was 7.4

The results also demonstrated that pH catalyses the polycondensation reaction, confirming results reported by Bourges et al. [6] using another method. If pH is near 12, the gel point is about 5 minutes and if pH is near 7, the gel point is about 60minutes.

The maximum value of conservative modulus G' (arrow 2, Figure 3) is linked with the degree of swelling (concentration of dry polymer in the final solution) and/or the quantity of silanol grafted on the polymer. This value is directly linked to the network and the cross-link density.

Discussion

This preliminary oscillatory rheological study confirms that this silanised hydrophilic polymer allows hardening and becomes an hydrogel in up to 60 minutes with a total reaction times of more than 10 days. This polymer provides a network between the macromolecular polysaccharides when pH solution decreases without any adjuvant or catalytic agent. In alkaline medium, the silane functions are ionized and chains are disconnected causing the pH to decrease. As the number of protons increases R-Si-O⁻ Na⁺ becomes R-Si-OH, which is very reactive. Polycondensation occurs and covalent bridges will be created thus forming a three-dimensional network. The parameters (pH, swelling, amount of silane grafting) can be controlled and adjusted using oscillatory rheological measurements. Those properties allow the formation of efficient scaffold for tissue engineering. The polymer network can be adapted to provide the best environment for specific cell functions. For example, osteo-sarcomas (SAOS2 cell line) grow like a nodule in the hydrogel (figure 4) whereas osteoblasts (PAL+ cells, results not shown) from total rat bone marrow grow only on the surface of BCP granules (figure 5). Other preliminary studies with bone cells show the biocompatibility of this hydrogel and the different behavior of the different cell lines used.

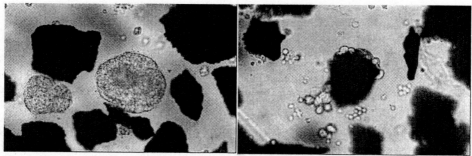

Fig. 4: SaoS2 cells after 24 days of   Fig. 5: Marrow cells after 14 days of culture
Culture in hydrogel and BCP granules.

Conclusion
These results highlight the potential of this hydrogel as a scaffold for tissue engineering and as injectable biomaterials for bone repair.

Acknowledgments
This work was supported by ACI "Technologies pour la Sante 2001-2004" and CPER "Biomatériaux" 2000-2004, Pays de Loire.
Acknowledgments for Dr R.Z. LeGeros for her help and critical reading of this paper, Pr M. Basle and Pr D. Chappard from LHEA Laboratoire Histologie-Embryologie, Faculty of Medicine, 49045 ANGERS cedex, France, for the preliminary bone cell culture studies, and Pr J. -F. Tassin from UMR CNRS 6120 – Le Mans, France for his help in rheological measurements.

References
[1] O. Gauthier, I Khairoun, J Bosco, L Obadia, X Bourges, C Rau, D Magne, JM Bouler, E Aguado, G Daculsi, P Weiss J Biomed Mater Res 66A (2003), 47.
[2] P Weiss, L Obadia, D Magne, X Bourges, C Rau, T Weitkamp, I Khairoun, JM Bouler, D Chappard, O Gauthier, G Daculsi, Biomaterials. (*In press*).
[3] M. Lapkowski, P Weiss, G Daculsi, A Dupraz, CNRS Patent. WO 97/05911, (1995).
[4] X Bourges, P Weiss, A Coudreuse, G Daculsi, G Legeay Biopolymers 63 (2002), 232.
[5] R Turczyn, P Weiss, M Lapkowski, G Daculsi J Biomater Sci Polym Ed 11(2000), 217.
[6] X Bourges, P Weiss, G Daculsi, G Legeay. Adv Colloid Interface 99(2002), 215.
[7] Sau, T. Majewicz, ACS Symp Ser 476, (1992), 265.
[8] D. S. Jones: Inter. J. Pharma. 179 (1999), 167.
[9] K. S. Anseth, C. N. Bowman and L. Brannon-Peppas: Biomaterials 17 (1996), 1649.

Key Engineering Materials Vols. 254-256(2004) pp. 1111-1114
online at http://www.scientific.net

Biomimetic Apatite Formation on Chemically Modified Cellulose Templates

Frank A. Müller, Lenka Jonášová, Peter Cromme,
Cordt Zollfrank and Peter Greil

Department of Materials Science, University of Erlangen Nuernberg,

Martensstr. 5, 910 58 Erlangen, Germany, fmueller@ww.uni-erlangen.de

Keywords: cellulose, Langmuir-Blodgett, calcium phosphate, biomimetic process, SBF

Abstract: Biomimetic precipitation of calcium phosphate phases (Ca-P) from simulated body fluid (1.5 SBF) on the biopolymer cellulose was studied. The basic crystallization behavior was investigated using highly oriented Langmuir-Blodgett films of cellulose activated in $Ca(OH)_2$ solution. Using transmission electron microscopy it was shown that the activated cellulose triggered the formation of nano-cyrstalline Ca-P phases, whereas spherical nano-aggregates were found on pure cellulose films. The formation of two different Ca-P phases, i.e. octa calcium phosphate (OCP) and hydroxy carbonated apatite (HCA) was found after 7 days in 1.5 SBF. $Ca(OH)_2$ pretreated 3D cellulose fabrics (Lyocell®) were homogeneously covered with a 30 μm thick HCA layer after soaking in 1.5 SBF for 21 days. Thus, chemically pretreated cellulose fabrics with adjustable porosity are considered to be capable to build up novel scaffold architectures for tissue engineering.

Introduction

Cellulose is the world's most abundant natural, renewable and biodegradable polymer [1]. Regenerated cellulose products like textile fabrics or felts are of particular interest for tissue engineering applications due to their biocompatibility, mechanical properties similar to those of hard and soft tissue and easy fabrication into a variety of shapes with adjustable interconnecting porosity [2,3]. Nevertheless, their clinical use as bone substitute material is limited because they do not bond directly to bone [4]. The aim of the present study was to develop an innovative process to provide regenerated cellulose with *in vitro* bone bonding ability by initiating biomimetic apatite formation from simulated body fluid (SBF). For optimized deposition of Ca-P phases with respect to osteointegration the effect of chemical activation of the cellulose was evaluated. Defined and oriented thin films of chemically pretreated cellulose on 2D-substrates as well as 3D-structured cellulose templates were used to assess the fundamental chemical processes directing the phase composition and crystal morphology of the deposited Ca-P phases.

Materials and Methods

Highly oriented mono- and multilayers of trimethylsilyether-cellulose (TMS-cellulose) were prepared on carbon coated TEM grids using a commercial Langmuir-Blodgett (LB) film balance (KSV Instruments Ltd, Finland). An average of 10 monolayers was deposited on each TEM grid at a constant surface pressure of 18 mN/m. Regenerated 3D cellulose fabrics (Lyocell®, Lenzing, Austria) with a specific mass of 257 g/m² were used as templates for biomimetic apatite coatings. Pieces of 3D cellulose samples as well as cellulose thin films were treated in supersaturated $Ca(OH)_2$ solutions at 8°C for 72 hours, subsequently washed with distilled water and dried at 50°C for 24 hours.

Biomimetic bone-like apatite (hydroxy carbonated apatite, HCA) formation was evaluated after exposure of chemically treated cellulose and LB-film coated TEM grids to simulated body fluid (1.5 SBF) with composition described elsewhere [5].

The microstructure and composition of biomimetically precipitated layers was determined by scanning electron microscopy / energy dispersive X-ray analysis (SEM/EDX, Cambridge Instruments, USA) on carbon coated samples and by transmission electron microscopy / energy dispersive X-ray spectrometry (TEM, Phillips CM-30, NL / EDS, Oxford ISIS 30, UK). The rate of apatite formation was calculated on the basis of gravimetric analysis and by measuring the residual calcium and phosphorous concentrations after exposure to SBF using inductively coupled plasma optical emission spectrometry (ICP-OES).

Results and Discussion

The LB technique was used to produce well defined, highly ordered cellulose films from TMS-cellulose [6]. The LB films served as a template for the oriented precipitation of Ca-P phases from 1.5 SBF. The morphology of the precipitated Ca-P phases was strongly influenced by the chemical nature of the cellulose template. The hydroxyl groups of the cellulose template triggered the precipitation of amorphous, spherical aggregates of Ca-P-phases, fig. 1a. In contrary, $Ca(OH)_2$ activated cellulose LB-films induced the formation of crystalline needle-like Ca-P phases, Fig. 1b. Electron diffraction proved the nano-crystalline nature of the Ca-P phases due to the observed broad rings. The segmentation of the diffraction rings indicates a preferred alignment of the calcium phosphate nano crystals. However, the evaluation of the d-spacings of the ring diffraction patterns could not clearly clarify whether OCP or HCA were precipitated due to similar d-spacing values. According to EDS analysis a Ca/P molar ratio ranging from 1.33 to 1.63 was found for the crystallized Ca-P phases, which suggests the formation of both, OCP and HCA phases. This observation is consistent with results described in literature where OCP was reported as a precursor in HCA formation [7]. The pretreatment of cellulose in $Ca(OH)_2$ results in a significantly higher amount of precipitated Ca-P phases as well as a high degree of crystallinity together with a preferred orientation of the particles.

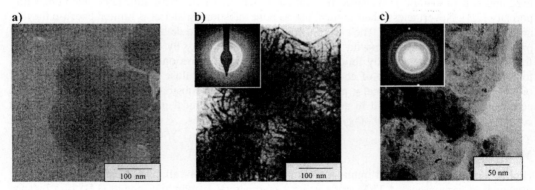

a)   b)   c)

100 nm   100 nm   50 nm

Fig. 1: TEM micrographs of precipitates on a) 2D regenerated cellulose soaked in 1.5 SBF for 3 days b) 2D regenerated cellulose pretreated with $Ca(OH)_2$ and soaked in 1.5 SBF for 3 days, c) 3D cellulose fabrics pretreated with $Ca(OH)_2$ and soaked in 1.5 SBF for 7 days; inset: representative electron diffraction pattern.

Fig. 2a shows a SEM micrograph of the cellulose fabrics treated with Ca(OH)$_2$. Using EDX analysis only calcium was detected in the sample surface resulting from the pretreatment. A surface layer consisting of spherical particles was observed by SEM within 7 days of soaking in 1.5 SBF, fig. 2b. The presence of phosphorus and calcium on the fiber surface was confirmed by EDX analysis.

The morphology of the Ca-P-phases precipitated on the cellulose fabrics after 7 days in 1.5 SBF was analysed by TEM, fig. 1c). The Ca-P-phases found are of a nano-crystalline character forming spherical aggregations, similar to the morphology shown in fig. 2b. Faceted crystals were not observed. Their nano-sized character was proved by electron diffraction. The broad rings in the diffraction pattern indicate a crystallite size smaller than 10 nm. The most prominent rings are assigned to the (002) and (402) planes, corresponding to either OCP, TCP or HCA due to similarity in their d-spacings [8]. However, analyzing the phase composition using EDS exhibited an average Ca/P-ratio of 1.35, which is close to the theoretical one for OCP (1.33). After 21 days in 1.5 SBF an approximately 30 µm thick homogeneous Ca-P-layer was observed using SEM, fig. 2c, d.

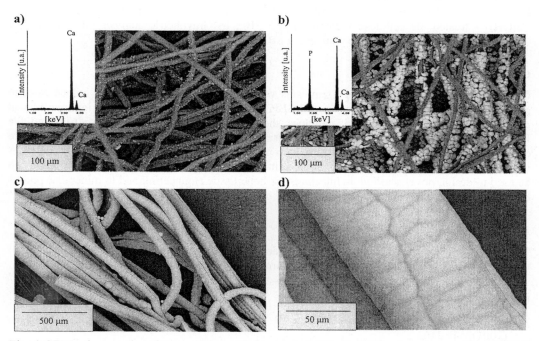

Fig. 2 SEM micrographs of 3D cellulose fabrics treated with Ca(OH)$_2$ a) before soaking in SBF, b) after 7 days in 1.5 SBF and c, d) after 21 days in 1.5 SBF; inset: EDX analysis.

The concentrations of calcium and phosphorus after soaking of Ca(OH)$_2$-treated cellulose substrates in 1.5 SBF were compared with those of SBF without samples, assigned as P-blank and Ca-blank, respectively, fig. 3a. The concentrations of both components were lower than in the blank solution for every soaking time. It may be assumed that a Ca-P enriched surface layer formed in the sample surface. These results are in agreement with those obtained by gravimetric analysis, fig. 3b. The decrease of calcium and phosphorus concentration after 14 and 21 days of soaking in 1.5 SBF corresponds to the increase in sample weight. It can be supposed that a new Ca-P layer precipitate in the surface of chemically treated cellulose fabrics. The Ca-P layer formation can be discussed as follows: The calcium adsorbed after activating in Ca(OH)$_2$ probably leaches out from the surface during soaking in 1.5 SBF and calcium phosphate nucleation is induced due to an increase in the ion activity product of calcium phosphates in the surrounding fluid.

a)

b)

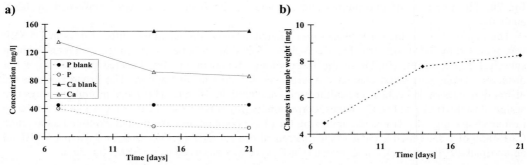

Fig. 3 a) Time dependence of calcium (Ca) and phosphorus (P) concentration after soaking of Ca(OH)$_2$ - treated 3D cellulose substrate in 1.5 SBF; b) Gravimetric analysis after soaking of Ca(OH)$_2$ - treated 3D cellulose substrate in 1.5 SBF.

Conclusion

Chemical pretreatment of cellulosic templates with supersaturated Ca(OH)$_2$ can induce calcium phosphates formation after soaking in 1.5 SBF. The basic crystallization behavior investigated using highly oriented LB-films of cellulose showed that the activated cellulose triggered the formation of nano-cyrstalline Ca-P-phases, whereas spherical nano-aggregates were found on pure cellulose films. Using TEM-EDS the formation of two different Ca-P phases, i.e. OCP and HCA was found after soaking in 1.5 SBF. Porous cellulose structures activated in Ca(OH)$_2$ solution can be used to design novel scaffold architectures for bone tissue engineering and to mineralise porous bone replacement materials.

Acknowledgment

The TEM-work was carried out at the Central Facility for High-Resolution Electron Microscopy of the Friedrich-Alexander University Erlangen-Nürnberg, Germany.

References

[1] P.L.Granja, M.A. Barbosa, L. Pouysége, B. de Jéso, F. Rouais, C. Baquuey: J Mat Sci 36 (2001),   p. 2163
[2] B. Philipp, W. Bock, F. Schierbaum: J Polym Sci Polym Symp 66 (1979), p. 83
[3] T. Miyamoto, S. Takahashi, H. Ito, H. Inagaki, Y. Noishiki: J Biomed Mater Res 23 (1989), p. 125
[4] M. Martson, J. Viljanto, T. Hurme, P. Saukko: Eur Surg Res 30 (1998), p. 426
[5] Jonášová L., Hlaváč J.: *Materials for medical engineering*, EUROMAT Vol.2 (eds. Stallforth H., Revell P.), WILEY-VCH, Weinheim 2000, p. 126
[6] G. Wegner: Ber Bunsenges Phys Chem 95 (1991), p. 1326-1332
[7] N Eidelman, W.E.Brown, J.L. Meyer: J Crys Growth 113 (1991), p. 643
[8] D.G.A. Nelson, G.J. Wood and J.C. Barry: Ultramicroscopy 19 (1986), p. 253

Key Engineering Materials Vols. 254-256(2004) pp. 1115-1118
online at http://www.scientific.net
© 2004 Trans Tech Publications, Switzerland

Preparation of a Composite Membrane Containing Biologically Active Materials

T. Itoh[1], S. Ban[2], T. Watanabe[1], S. Tsuruta[1], T. Kawai[1], H.Nakamura[1]

[1] Aichi-Gakuin University, 2-11, Suemori-Dori, Chikusa-ku, Nagoya, Japan, 464-8651

toshiki@dpc.aichi-gakuin.ac.jp

[2] Kagoshima University, 8-35-1, Skuragaoka, Kagoshima, Japan, 890-8544

sban@denta.hal.kagoshima-u.ac.jp

keywords: Apatite, BMP, PLGA, biodegradable, membrane, composite

Abstract. Oriented needle-like apatite was formed on a pure titanium plate with the hydrothermal electrochemical method. BMP was applied on the titanium plate. Poly-lactic acid/poly-glycolic acid (PLGA) dissolved in dichloromethane was dropped on the titanium substrate. The dried and delaminated PLGA layer containing apatite with BMP was the biologically active membrane. The composite membrane was easy to bend along the shape of the bone defect.

Introduction

Biologically active materials, stem cells and their scaffolds are required to promote new bone formation. Especially in case of dental applications, various diseases such as periodontal disease and severe apical legion cause bone defect. Guided Bone Regeneration and Guided Tissue Regeneration are techniques for the treatment of the bone defects [1, 2]. Biodegradable membranes have been successfully applied to these treatments [3]. Generally, poly-lactic acid, poly-glycolic acid and PLGA are known as materials of biodegradable membranes. The biodegradable membranes are useful materials because of their resorption, usability, and retention property.

On the other hand, it is well known that bone morphogenetic protein (BMP) induces bone formation. PLGA and gelatin are frequently used as the carriers of BMP. We have reported the preparation of biodegradable PLGA membrane containing oriented needle-like apatite [4]. In the present study, BMP was extracted from bovine cortical bone. Oriented needle-like apatite was formed on the titanium plate and biodegradable composite membrane containing oriented needle-like apatite with BMP was prepared.

Materials and Methods
Preparation of BMP

BMP was prepared with the method of Urist [5] and Mizutani [6]. One hundred pieces of the bovine cortical bones removed soft tissue were washed in cold water, frozen in liquid nitrogen and grinded to a particle size of 1 mm^3. The particles of bone were defatted in chloroform and methanol (1:1) for 2 h (chloroform and methanol were evaporated for 12 h after defatted) and decalcified in 0.6 M HCl for 72 h. The decalcified particles of bone were exposed to 2 M $CaCl_2 \cdot 2H_2O$ solution for 1 h, 0.5 M EDTA for 1 h and extracted in 6 M urea containing 0.5 M $CaCl_2 \cdot 2H_2O$ and 1 mM n-ethyl maleimide (n-EM) for 48 h. Extracted solution without particles of bone was dialyzed against distilled water for 72 h. Dialyzed solution was centrifuged for 20 min. The precipitate was dissolved in 1 L of 6 M urea containing 0.5 M $CaCl_2 \cdot 2H_2O$ (without 1 mM n-EM). That solution was dialyzed against 11 L of 0.25 M citrate buffer for 24 h. Dialyzed solution was centrifuged for 20

min.   The final precipitate was defatted in chloroform and methanol (1:1).   Defatted and dried
precipitate was used as BMP.   BMP activity was estimated by the implantation into ddy mice aged 4
weeks.   Capsuled 3mg of BMP was implanted into the thigh muscle pouch of mice.   At 3 weeks
after implantation, all mice were sacrificed and implanted area was observed by roentgen photogram.
The experimental protocol followed the guidelines for animal   experimentation of Aichi-Gakuin
University.

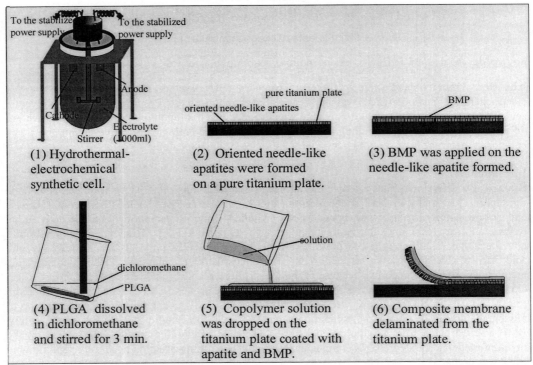

(1) Hydrothermal-electrochemical synthetic cell.

(2) Oriented needle-like apatites were formed on a pure titanium plate.

(3) BMP was applied on the needle-like apatite formed.

(4) PLGA dissolved in dichloromethane and stirred for 3 min.

(5) Copolymer solution was dropped on the titanium plate coated with apatite and BMP.

(6) Composite membrane delaminated from the titanium plate.

Fig. 1. Procedure for preparation of the composite membrane.

Preparation of composite membrane

According to our previous studies [4], needle-like apatites were formed on a pure titanium plate
(15mm x 25mm) with the hydrothermal-electrochemical method in an autoclave with two electrodes.
The anode has a pure platinum plate.   The cathode has a pure titanium plate.   The autoclave was
connected to the stabilized power supply and the heating device.   The electrolyte was prepared by
dissolving given amounts of the reagent-grade chemicals into the distilled water. Table 1 shows the
composition of the electrolyte.   Table 2 shows the electrolytic conditions.

The extracted BMP from bovine cortical bones was applied on the titanium plate covered with
apatites.   One hundred mg of PLGA (GC-membrane large size, GC; Japan) was dissolved in 1.0 ml
of dichloromethane and was stirred for 3 min.   This copolymer solution was dropped on the titanium
plate, which was already covered with apatite and BMP.   After drying at 37°C for 24 h, the composite
membrane was delaminated from the titanium plate.   Figure 1 shows the procedure for preparation of

the composite membrane containing oriented needle-like apatite with BMP. In this study, the composite membrane containing oriented needle-like apatite with BMP was employed as experimental group, and the composite membrane containing BMP alone (without apatite) was employed as control group. Composite membrane containing oriented needle-like apatite with BMP was observed by field emission type scanning electron microscopy (FE-SEM). The tensile strength of both membranes (5 x 5mm) was determined with a measuring instrument (Push-Pull gauge, AIKO Engineering CO. LTD.; Japan).

Table 1 Composition of electolyte

NaCl	137.8mM
K_2HPO_4	1.67mM
$CaCl_2 \cdot 2H_2O$	2.5mM
$(CH_2OH)_3CNH_2$	50mM
HCl	buffer to pH7.2

Table 2 Electrolytic conditions

Current density	12.5mA/cm2
Loading time	1 hour
Electrolyte temp.	100°C
Anode	pure platinum plate
Cathode	pure titanium plate

Results and Discussion

Figure 2 shows the roentgen photogram of the thigh muscle pouch of mice. A radiopaque region was observed around implanted area (arrow). It was found that this BMP had bioactivity.

Figure 3 shows the appearance of the composite membrane of experimental group. The appearance of the experimental membrane was semitransparent. It was found that the semitransparent membrane was created by the apatite needles. BMP was combined with the semitransparent membrane. Figure 4 shows the appearance of the composite membrane of control group. The appearance of the control membrane was transparent. Brown-colored BMP was combined with the transparent membrane.

Figure 5 and Figure 6 show FE-SEM photographs of cross sections and surfaces of the composite membranes containing oriented needle-like apatite. The cross sections showed that the PLGA contained apatite with BMP. The surfaces showed that membrane had pores of varied sizes. It seems that dissolved PLGA evaporated and the dichloromethane disappeared completely. It was suggested that these pores could be utilized for the transportation of various cells and utilized as scaffold for the cell culture.

The tensile strength of the experimental group was 9.9N (SD=2.13). The tensile strength of the control group was 7.1N (SD=0.7).

Composite membrane of both experimental and control groups delaminated from the titanium plate could be bent easily.

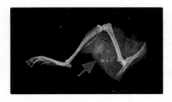

Fig. 2. Roentgen photogram around the implantation area of mice.

Conclusions

A composite membrane was prepared by dropping PLGA on the titanium plate, which was previously covered with apatite and BMP. Eventually this composite membrane contained two types of biologically active materials. The tensile strength of the composite membrane was increased by incorporation of apatite. These results suggest that this membrane could possibly function as a biologically active material.

Fig. 3. Appearance of the composite membrane of experimental group

Fig. 4. Appearance of the composite membrane of control group

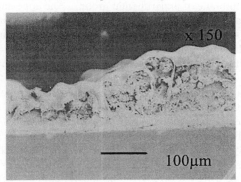

Fig. 5. FE-SEM photograph of cross section of the membrane (x 150).

Fig. 6. FE-SEM photograph of surface of the membrane (x 150).

References

[1] Karring T., Nyman S., Ryndhe J. and Sirirat M.: J. Periodontol. Vol. 11 (1984), p.41-52

[2] Pontoriero R., Nyman S., Lindhe J., Rosenberg E. and Sanavi F.: J. Clin. Periodontol. Vol. 14 (1987), p. 618-620

[3] T. Miki, K. Masaka, Y. Imai, and S. Enomoto: Journal of Cranio-Maxillofacial Surgery Vol. 28 (2000), p. 294-299,

[4] T. Watanabe, S. Ban, S. Tsuruta, T. Ito, T. Kawai and H. Nakamura: Key Engineering Materials Vol. 240-242 (2003), p. 191-194

[5] Urist M.R., Mikulski A. and Lietze A.: Proc Natl. Acad. Sci. USA Vol. 76 (1979), p. 1828-1832

[6] Mizutani H., Urist M. R.: Clin. Orthop. Vol. 171 (1982), p. 213-223

Key Engineering Materials Vols. 254-256(2004) pp. 1119-1122
online at http://www.scientific.net

Effect of Inorganic Polyphosphate on Periodontal Regeneration

T. Shiba[1], Y. Takahashi[2], T. Uematsu[2], Y. Kawazoe[1,3], K. Ooi[3], K. Nasu[2], H. Itoh[1], H. Tanaka[2], M. Yamaoka[2], M. Shindoh[3] and T. Kohgo[3]

[1] Frontier Research Division, Fujirebio Inc., 51, Komiya, Hachioji, Tokyo 192-0031, Japan.
tz-shiba@fujirebio.co.jp

[2] Department of Oral and Maxillofacial Surgery, School of Dentistry, Matsumoto Dental University, Shiojiri, Nagano 399-0781, Japan. uematsu@po.mdu.ac.jp

[3] Graduate School of Dental Medicine, Hokkaido University, Sapporo 060-8586, Japan.
mshindoh@den.hokudai.ac.jp

Keywords: alveolar bone, inorganic polyphosphate, MC-3T3-E1, periodontal regeneration, bone mineralization

Abstract. We designed a new bio-material using sodium polyphosphate [poly(P)] that can be used for acceleration of periodontal regeneration. The effects of poly(P) on cell calcification and periodontal regeneration, including alveolar bone regeneration, were investigated *in vitro* and *in vivo*. Expressions of osteopontin (OPN) and osteocalcin (OC) mRNAs were induced by poly(P) treatment of MC3T3-E1cells. Alkaline phosphatase (ALPase) activity was also enhanced by poly(P), and the cells treated with poly(P) were strongly stained by alizarin red. As an *in vivo* experiment, artificial periodontal pockets were made in the buccal alveolar bone of mandibular first and second molars of each rat. Poly(P) was injected into the artificial pockets over a period of 1 to 3 weeks, and the regeneration of alveolar bone was evaluated histologically. At 1 week after the start of poly(P) treatment, remarkable regeneration of periodontal tissues, including alveolar bone, was observed in the poly(P)-treated group, whereas little regeneration was observed in the control group. The average area of regeneration in the poly(P)-treated group (after 2-weeks treatment) was more than 2-fold larger than that in the control group, suggesting that poly(P) positively stimulates alveolar bone formation. In addition, inflammation of the defected area was diminished and tissue repair was also accelerated in the poly(P)-treated group. Since poly(P) has antibacterial activity against *P. gingivalis* and *S. mutans* and since poly(P) stimulates the growth of fibroblasts by stabilizing fibroblast growth factors (FGFs), poly(P) could be effective for periodontal regeneration.

Introduction

Since more than 80% of people over 40 years of age have periodontal diseases that will cause loss of teeth, this is becoming a very serious problem due to the aging of society. Periodontal diseases are characterized by inflammation of and subsequent damage to or loss of tooth-supporting tissues, including cementum, periodontal ligament and alveolar bone. Increasing evidence that tissues of the periodontium harbor cells with the capacity to regenerate the periodontium has fostered an interest in developing clinical procedures to restore periodontal support by promoting tissue regeneration, i.e., formation of new cementum, periodontal ligament and new bone [1].

Inorganic polyphosphate [poly(P)] is a polymer of orthophosphate (Pi) ranging in size of up to one thousand Pi residues. Poly(P) is spontaneously present in animal cells and tissues and has also been found extracellularly in human blood plasma. The biological functions of poly(P) in mammalian cells have not been elucidated. However, our recent findings revealed that poly(P) could play important roles in tissue regeneration by stabilizing fibroblast growth factors 1 and 2 (FGF-1 and FGF-2) and by increasing their binding affinities to their cell surface receptors [2]. In addition, poly(P) has antibiotic activity that is especially effective against oral bacteria such as *P. gingivalis* and *S. mutans* [3]. Furthermore, since relatively large amounts of poly(P) have been found in normal osteoblasts and normal gingival fibroblasts [4], it may potentially be involved in bone formation and periodontal regeneration. Poly(P) is a completely safe material as evidenced by its use as a food additive all over the world for more than 50 years.

Encouraged by the above-described findings we carried out the present study to determine the applicability of poly(P) as a new material for periodontal regeneration, and we found that poly(P) accelerates bone mineralization and periodontal regeneration.

Materials and Methods

Separation of Long Chain Poly(P). Twenty g of sodium poly(P) (average chain length of about 20 phosphate residues), as specified in the standards for food additives, was dissolved in 200 ml of purified water, and then 32 ml of 96% ethanol was gradually added to the solution. The solution was vigorously agitated and then allowed to stand at room temperature for approximately 30 minutes. Then centrifugation (10,000 x g, 20 minutes, 25°C) was performed to separate the precipitate from the aqueous solution. The aqueous solution fraction was discarded, and 70% ethanol was added to the collected precipitate for washing, and then the precipitate was vacuum-dried. By this process, 9.2 g long chain-poly(P) salt was obtained as a precipitate (average chain length of about 60 phosphate residues) (Fig. 1).

Preparation of Poly(P)-Carboxymethyl Cellulose (CMC) Mixture and Poly(P)-collagen Complex. To use long-chain poly(P) for in *in vivo* experiments, two types of material (poly(P)-CMC mixture and poly(P)-collagen complex) were prepared. The poly(P)-CMC mixture was prepared by mixing long-chain poly(P) with CMC at final concentrations of 1%. For preparation of poly(P)-collagen complex, 5 g of the long-chain poly(P) was dissolved in 50 ml of sterile distilled water. Then, 14.3 g of a chicken-derived atelocollagen was added, thereby generating a gel precipitate. The precipitate was filtered through mesh and then washed with 70% ethanol, yielding complexes with a wet weight of 6.47 g.

Evaluation of Poly(P)-Induced Cell Calcification Using MC3T3-E1 Cells. MC3T3-E1 cells were plated on 60-mm plastic petri dishes at the density of 1×10^5 cells/dish and maintained in α-MEM supplemented with 50 μg/ml of kanamycin and 10% FBS at. After the cells had become confluent, the medium was replaced with α-MEM supplemented with 0.5% FBS and 50 μg/ml kanamycin or the same medium containing 1 mM PO_4 buffer, 1 mM poly(P). The cells were further cultured with or without stimulants, and the medium was replaced every fourth day.

***In vivo* experiments.** Male Wistar rats (8 weeks old; 10 rats in total) were anesthetized. Then, using a 1/2 round bar, approximately 2-mm portions were removed from the buccal alveolar bone crest of the mandibular first and second molars, and artificial periodontal pockets (gingival crevices) were formed. In the treated group (5 rats), approximately 0.1 ml of poly(P)-CMC mixture or poly(P)-collagen complex was injected using a syringe into the pockets. In the control group (5 rats), only CMC or collagen was injected. The injection was given every day from the day following the operation to prepare the periodontal pocket and continued for up to 5 weeks at the longest.

Results and Discussion

Preparation of Long-Chain Poly(P). Since poly(P) with a chain length of more than 20 phosphate residues is likely to be effective in promoting fibroblast growth and FGF stabilization, we prepared long chain-poly(P) with chain length ranging from 20 to 100 phosphate residues and remove the short-chain poly(P) by ethanol precipitation as described in Materials and Methods. As shown in Fig. 1, the average chain length of the resultant long-chain poly(P) was approximately 60 phosphate residues. This long-chain poly(P) was used directly in *in vitro* experiments. For *in vivo* experiments, a mixture of 1% poly(P) with 1% CMC or poly(P)-collagen ion complex (see Materials and Methods) was used.

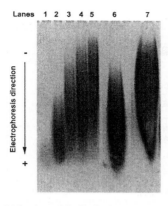

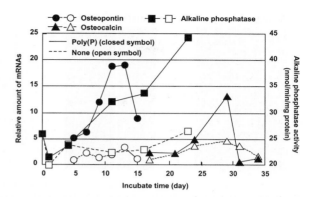

Fig. 1. Molecular weight distribution of poly(P). Molecular weight of the long chaon poly(P) was analyzed by 15% polyacrylamide gel electrophoresis. Lanes 1 to 5 denote chain length markers, which are average chain length of 15 (lane 1), 35 (lane 2), 45 (lane 3) and 65 (lane 4). Lanes 6 and 7 indicate short chain poly(P) before separation and long chain poly(P) after separation, respectively.

Fig. 2. Effect of poly(P) on OPN, OC and ALPase expressions. Levels of OPN and OC expression were measured by determination of their mRNA levels by real-time PCR as described [6]. The mRNA levels were standarlized by the level of glyceraldehyde 3-phosphate dehydrogenase (GAPDH) mRNA as an internal control. ALPase expression was evaluated by its activity [6].

Induction of Calcification of MC3T3-E1 Cells by Poly(P). OPN and OC are matrix gla proteins that are well-known osteoblast differentiation markers [5]. As shown in Fig.2, the level of OPN mRNA expression in poly(P)-treated cells was about 10-fold higher than that in control cells (non-treated cells and phosphate buffer-treated cells) at 10 days after the start of treatment. OC mRNA expression was also induced by poly(P) treatment, and the highest expression level (about 3.5-fold higher than that in control cells) was observed at 27 days. These results indicate that poly(P) positively stimulates calcification of MC3T3-E1 cells.

The level of ALPase activity in poly(P)-treated cells also increased with incubation time (Fig. 2). The fact that no increase in ALPase activity was observed in phosphate buffer-treated cells or non-treated cells indicates that poly(P) has a positive effect on induction of ALPase activity. As increase in the level of ALPase activity in poly(P)-treated cells would result in induction of cell calcification.

Alizarin red staining was performed for visualization of cell calcification,. Cells treated with poly(P) were strongly stained with alizarin red S, and the presence of calcium precipitation was clearly observed. No calcification was observed in the cells treated with phosphate buffer or in the cells that were not treated (data not shown).

Taken together, these *in vitro* results indicate that poly(P) could be a inducer of cell calcification. The effect of poly(P) was also confirmed by normal human osteoblast and odonotoblast cells.

Induction of Alveolar Bone Regeneration by Poly(P) in rats. To confirm the effect of poly(P) on regeneration the periodontal tissue, an experiment on alveolar bone was performed using rats. In this experiment, poly(P)-collagen complex was injected every day into an experimentally made pocket of each rat. Fig. 3 shows samples of hematoxylin/eosin (HE)-stained tissue that had bee injected with the poly(P)-collagen complex or with only collagen. In the group of rats treated withpoly(P) for 5 weeks, significant regeneration of alveolar bones was observed, and the degree of bone reconstruction was also significant. In the collagen-treated group, however, only cured artificial pockets were observed, and no significant increase in reconstruction of alveolar bone was observed. In addition, inflammation of the defected area was drastically diminished in the poly(P)-treated group, and total tissue repair was almost completed within 1 week of poly(P) treatment (data not shown).

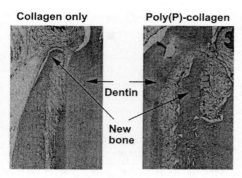

Collagen only   Poly(P)-collagen

Dentin

New bone

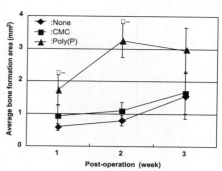

Fig. 3. Effect of poly(P) on alveolar bone regeneration. The rats that had been treated for a period of 5 weeks, and tissues were fixed, decalcified and embedded. Tissue sections were stained by HE and observed under a microscope.

Fig. 4. Changes in the area of the regenerated alveolar bone. Using HE stained tissue samples, the area of the regenerated alveolar bone was measured as the area of bone formation by calculating the sum of the bone areas in the sections using image processing. Asterisks indicate the significant differences between poly(P)-CMC treated group and non-treated group (None) ($P<0.01$ at U-test).

To confirm the poly(P) effect on alveolar bone regeneration, a simple mixture of poly(P) and CMC was used for the same experiment as that described above. Appropriate amounts of the poly(P)-CMC mixture were injected every day into the artificial defects over a period of 1 to 3 weeks. Fig. 4 shows changes in the area of bone formation. The area of regenerated alveolar bone was significantly larger in the poly(P)-treated group than in the control group, which was treated with only CMC, suggesting significantly promoted regeneration of the alveolar bone by poly(P).

According to the results of experiments described above, both poly(P)-collagen complex and poly(P)-CMC mixture were effective for periodontal regeneration, including promotion of new alveolar bone formation and total tissue repair. Although the mechanism is still unclear, poly(P)-stimulated periodontal regeneration might be caused by a combination of the following three factors: (1) acceleration of cell calcification, (2) antibiotic effect against *P. gingivalis* and related oral bacteria and (3) stabilization and modulation of FGFs. We conclude that poly(P) might be a good material for treatment of periodontal diseases.

Acknowledgements:

This work was supported by a Grant-in-Aid for the Creation of Innovations through Business-Academic-public Sector Cooperation (Open Competiton for Development of Innovative Technology), a Grant-in-Aid for Scientific Research on Priority Areas (B) from the Ministry of Education, Culture, Sports, Science and Technology of Japan and a Research on Advanced Medical Technology under Health and Laybour Sciences Research Grants from Ministry of Health, Laybour and Welfare of Japan.

References

[1] S. Gestrelius, C. Andersson, D. Lidström, L. Hammarström and M. Somerman: J. Clin. Periodontol. Vol. 24 (1997), p. 685-692

[2] T. Shiba, D. Nishimura, Y. Kawazoe, Y. Onodera, K. Tsutsumi, R. Nakamura and M. Ohshiro: J. Biol. Chem. Vol. 278 (2003), p. 26788-26792

[3] J. Lee: Newsletter for JADR, http://www.bcasj.or.jp/jadr/Newsletter/jadr

[4] H.C. Schröder: Proc. Mol. Subcell. Biol. Vol. 23 (1999), p. 45-81

[5] G.S. Stein, J.B. Lian, J.L. Stein, A.J. van Wijnen and M. Montecino: Physiol. Rev. Vol. 76 (1996), p. 593-629

[6] Y. Kawazoe, T. Shiba, R. Nakamura, A. Mizuno, K. Tsutsumi, M. Shindoh, T. Kohgo, T. Uematsu, and M. Yamaoka: submitted to J. Bone Miner. Res. (2003)

Key Engineering Materials Vols. 254-256(2004) pp. 1123-1126
online at http://www.scientific.net
© 2004 Trans Tech Publications, Switzerland

PLGA-CMP Composite Scaffold for Articular Joint Resurfacing

J.H. Jeong[1], S.K. Park[2], D.J. Lee[1], Y.M. Moon[1], D.C. Lee[2], H.I Shin[3], and S. Kim[4]

[1] Plastic & Reconstructive Surgery, College of Medicine, Yeungnam Univ., Daegu, Korea

[2] Dept. Orthopedic Surgery, College of Medicine, Yeungnam Univ.

[3] Dept. Oral Pathology, Dentistry, Kyungpook National Univ., Daegu, Korea

[4] School of Material Engineering, Yeungnam Univ., Kyongbuk, Korea

sykim@yumail.ac.kr

Keywords: Biodegradable scaffold, cartilage, chondrocyte, poly DL-lactic-co-glycolic acid, calcium metaphosphate.

Abstract. The treatment of articular cartilage defects originated many approaches for the clinicians to use various biomaterials for its repair and reconstruction. The result of the application of chondrocyte concentrate on the joint defect with periosteal coverage is still controversial. In order to obtain the reliable and durable regeneration of articular joint cartilage, which is firmly fixed to underlying bone, simultaneous regeneration of bone and cartilage was conducted. For autologous implantation, articular cartilage biopsy from non weight-bearing joint surface was dissociated into chondrocytes by enzymatic digestion. The chondrocytes were culture expanded in a routine manner. PLGA disc was seeded with cultured chondrocytes and was attached to a cylindrical CMP block with fibrin glue. A critical size osteochondral defect of 4.5 mm diameter was created on the weight-bearing femoral condylar surface of rabbit knee joint. Then, the PLGA-CMP composite scaffold was implanted into the defects and joint capsule was repaired. At 4 weeks, cartilage regeneration was observed from the surface of the implant. However, below the surface level, active subchondral bone tissue regeneration was incomplete. At 8 weeks, PLGA was replaced by regenerated hyaline cartilage and underlying CMP ceramic was partly degraded and replaced by new bone. These results demonstrated that the PLGA-CMP composite scaffold provides adequate regeneration condition for subchondral bone as well as for the articular cartilage of defected joint.

Introduction

Regeneration of articular cartilage defect is not a simple problem in clinical situation. Although some clinicians apply chondrocytes concentrate on the defective area with periosteal coverage, the result of the procedure is still controversial because the end product of the procedure is fibrocartilage rather than hyaline cartilage [1]. Moreover, the regenerated cartilage layer is easily detached from underlying subchondral bone. For the reliable and durable regeneration of articular joint cartilage, which is firmly fixed to underlying bone, simultaneous regeneration of bone and cartilage might be a possible solution. Recently, the use of various biomaterials such as bioactive glasses, hydroxyapatite (HAp), and their composites, were studied for the regeneration of subchondral cartilage defects in rabbit [2]. In orthopedic surgery, CMP is used as a nontoxic and biocompatible bone substitute which can provide limited preoperative stability together with gradual restoration of the host bone. The porous structure of CMP allows sufficient integration with the host tissue. Unlike HAp and other ceramics, the degradability of CMP can be controlled by changing the composition of potassium and sodium. In previous experiments, HAp-fibrin glue composite scaffolds were investigated to regenerate articular cartilage defects, resulting in fibrocartilage formation due to its instability caused by early absorption of HAp [3].

In this study, we developed a composite scaffold comprising PLGA and CMP for the regeneration of defective joint surface in a New Zealand white rabbit model. The degree of regeneration of hyaline cartilage and subchondral bone was evaluated after 8 weeks of implantation.

Materials and Methods
A total of 16 New Zealand white rabbits were used in this experiment. Anesthesia was done with pentobarbital sodium (Entobar®) and ketamine. Two kinds of materials were used to prepare a composite scaffold. PLGA polymer scaffold with a PLA/PGA ratio of 75:25 was used, having a porosity of 80%, and pore size of 200-250 μm. The salt leaching method was used and the dimension of PLGA scaffold was 5 mm in diameter and 1 mm thick. CMP blocks were ground into cylindrical shape with dimensions of 4.5 mm in diameter and 4.5 mm high.

Experimental procedures were performed in three stages. Stage 1 consisted in harvesting of a small piece of knee joint cartilage from rabbit knee and cell culture. After enzymatic digestion, the chondrocytes were cultured and proliferated for several weeks until an adequate amount of chondrocytes was obtained. At stage 2, PLGA discs were seeded with cultured chondrocytes (20 x 10^7 cells) in a spinner flask filled with 100 mL of chondrogenic medium, at 50 rpm. At stage 3, three days after chondrocyte seeding, the cell seeded PLGA discs were attached to cylindrical CMP blocks with fibrin glue containing chondrocytes. A critical size osteochondral defect of 4.5 mm in diameter was created on the weight bearing femoral condylar surface of knee joint in each rabbit. Then, PLGA-CMP composite scaffold was implanted into the defects and the joint capsule was repaired (Fig. 1).

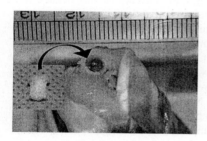

Fig. 1. Composite graft of PLGA-CMP prepared for implantation into the defect of femoral condyle.

Specimens were harvested at 4 and 8 weeks after implantation and observed grossly and microscopically. To evaluate osteochondral regeneration, Villanuevo stain was performed to observe histology sections of bone and cartilage. Resin was embedded into harvested sample in order to maintain good shape of CMP block. Using a high-speed cutter (VC-50, LECO, USA), samples were minimized by trimming the unnecessary parts and attached on the slide glass. Then, those were polished (STRUERS, RotorPol-35) until they have reaches a thickness of 0.3-0.5 μm.

Results and Discussion
The composite scaffold provided adequate regeneration of hyaline cartilage and subchondral bone. At 4 weeks, cartilage regeneration was observed from the surface of the implant (Fig. 2a). However, below the surface level, active bone regeneration was noticed but the regeneration was incomplete and major parts of ceramic scaffold were still observed (Fig. 2b). At 8 weeks, PLGA was replaced by regenerated hyaline cartilage and underlying CMP ceramic was mostly degraded and replaced by new bone. The histological section of the regenerated osteochondral junction looked fairly natural

(Fig. 3). The control sample, which was not filled^a with any implant, showed noticeable crater on the condylar surface grossly and the defect was filled with fibrous scar tissue microscopically (Fig. 4).

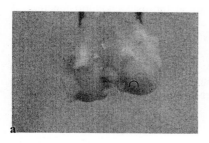
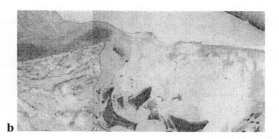

Fig. 2. PLGA-CMP composite scaffold at 4weeks of implantation. The surface of the defect was almostly healed (a), but histological section showed remaining ceramic scaffold and incomplete regeneration of subchondral bone (b).

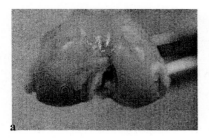
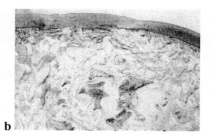

Fig. 3. PLGA-CMP composite scaffold at 8 weeks of implantation. The surface of the defect was completely covered with new cartilage (a) and histologic section showed regeneration of subchondral bone and little trace of ceramic scaffold (b).

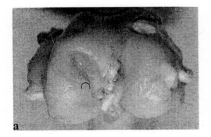
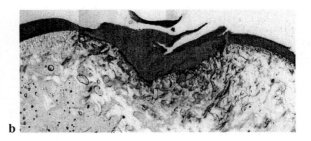

Fig. 4. In the control group, the surface defect was not healed at 4 weeks and chronic cartilage defect was seen at 8 weeks (a). In histological examination, the defect was filled with fibrous scar tissue (b).

Fibrin glue, which combines the two different scaffold together in this study, is a kind of semi-solid natural polymer. Although this is not rigid fixation, this chondrocyte containg semi-solid scaffold provides not only enough stability between the two scaffold materials in this early ambulating animal model, but also a cell containing matrix filling the gaps within the defect.

Physical stress to the knee joint is enormous and we do believe the regenerated cartilage and bone should be tested for mechanical durability and for the stability at the osteochondral junction. A long-term evaluation and mechanical test should be followed.

Conclusions

The PLGA-CMP composite scaffold showed remarkable potential for the regeneration of osteochondral defect of a knee joint. The PLGA enables transplanted chondrocytes to synthesize a new hyaline cartilage tissue within 8 weeks and CMP offered supportive stability and adequate incorporation with host bone.

Acknowledgement

This work was supported by a Grant-in-Aid for Next-generation New Technology Development Programs from the Korea Ministry of Commerce, Industry and Energy.

References

[1] O'Driscoll SW, Fitzsimmons JS: Clin Orthop. Vol. 391,Suppl. (2001), p. 192.

[2] Suominen E, Aho AJ, Vedel E, Kangasniemi I, Uusipaikka E, Yli Urpo A: J Biomed Mater Res Vol. 32 (1996), p. 543-51.

[3] Van Susante JL, Buma P, Homminga GN, van den Berg WB, Veth RP: Biomaterials. 1998 Vol. 19 (1998), p. 2367-74.

Author Index

A

Abad, P. 431
Abrahams, I. 95
Adolfsson, E. 1025
Afonso, A. 565, 659, 1091
Agathopoulos, S. 327
Ahn, M.W. 881
Ahnfelt, N.O. 51, 197
Aigner, J. 937, 941
Aimoli, C.G. 311
Aizawa, M. 915
Akasaka, T. 919
Albuquerque, J.S.V. 1021
Almeida, C.C. 729
Almeida, L. 821
Almirall, A. 253, 1001
Altunatmaz, K. 655
Amaral, I.F. 573, 577
Ambrosio, L. 485
Andersson, J. 557
Annaz, B. 273
Ansell, C. 205, 281
Aparicio, C. 737
Arcos, D. 27
Arias, J. 347, 371, 415
Arikawa, H. 517
Arstila, H. 67
Arts, J.J.C. 221
Arumugam, M.Q. 869
Ashizuka, M. 545
Asselin, A. 749
Axén, N. 265

B

Balas, F. 467
Ban, S. 517, 1115
Bao, C. 7, 725, 757, 801
Barbara, J. 713
Barbosa, M.A. 577, 877
Barbotteau, Y. 717
Barral, R. 323
Barralet, J.E. 79, 205, 249, 277, 281, 297
Barrias, C.C. 573, 877
Bar-Yosef Ofir, P. 11
Bastié, C. 833
Bayrak, O. 655

Belpassi, A. 1095
Ben-Nissan, B. 301, 455, 707
Beppu, M.M. 311
Berdal, A. 749
Berger, G. 71, 407, 411, 635, 1059
Berland, S. 733
Bernstein, A. 407
Bertran, G. 319
Best, S. 371
Best, S.M. 111, 593, 845, 915
Bhadang, K.A. 39
Bibby, J. 335
Bienengraeber, V. 679
Bigi, A. 229
Bizios, R. 903
Björklund, K. 265
Blunn, G. 667
Boch, P. 237
Bohner, M. 895
Boltong, M.G. 161
Bonfield, W. 111, 371, 513, 593
    611, 845, 869, 915
Boraschi, D. 907
Borges, J.P. 573
Born, R. 419
Borojevic, R. 493
Borrajo, J.P. 23, 355, 1029
Borzeix, S. 733
Bötcher, R. 679
Botelho, C.M. 845
Boudeville, P. 103, 107, 615, 931
Bouler, J.-M. 55, 193
Bourrat, X. 1009
Boutinguiza, M. 371
Bozzi, A. 841
Bracci, B. 229
Braccini, S. 899
Braden, M. 513
Branco, R. 659, 1091
Brenner, P. 427
Bressiani, A.H.A. 923
Bressiani, J.C. 923
Brooks, R.A. 593, 845, 869
Bruno, M. 43
Bubb, N. 335
Buckland, T. 273

Buffat, P.-A.. 891
Buma, P. 221
Buth, L.H.O. 427

C

Cabañas, M.V. 363
Cakir, A.F. 463
Calixto, R.. 119
Cameron, R.E. 513, 593
Camiré, C.L. 269
Campos, T.P.R................................. 83, 87
Cancedda, R... 1095
Cao, Y. 721, 757, 801
Cashion, J.D.. 213
Cavalheiro, J. 659, 1091
Cazalbou, S.. 833
Cetlin, P.R... 809
Chang, B.S..................................... 147, 151
Chang, J.-S.................................... 135, 225
Chassot, E. 717
Chatainier, G... 833
Cheang, P.. 315
Chen, J. 7
Chen, J.Y. 131, 339, 351, 721, 725, 757
Chen, Q.Z. 165
Cheng, K. 331, 477
Cheung, K.M.C.. 165
Chin, T.S... 473
Chiussi, S. 23, 355, 1029
Cho, H.M. 47
Cho, K.J. 285
Cho, S.B.. 285
Cho, S.-H... 217
Cho, W.S..................................... 135, 225
Choi, A.H. 707
Choi, J.W. 47, 189
Choi, S.H. 185, 1079
Chow, S.P. 757
Chrzanowski, W. 387
Chung, C.P.. 245
Chung, J.H. 631
Chung, R.J. 473
Chung, Y.-C.. 217
Cleere, C.P. 585
Clupper, D.C... 813
Collía, F. 177
Combes, C. 833, 927
Conway, M. 301
Conway, R.C... 707
Correa, J.C. 431

Correia, R.N. 143, 581
Costa, A.M. 119
Costa, M.A. 821, 825
Couté, A. 1009
Cox, F.P..91
Cromme, P...1111
Cui, F.-Z. 805
Cui, X.-Y. 375, 459

D

Daculsi, G.......................55, 193, 1005, 1107
Damien, E.....................................447, 797
Dard, M. 419
Davy, C.A. 55
De Arellano-López, A.R. 1029
de Arruda Almeida, K. 589
De Aza, P.N. 75
De Aza, S...75
de Campos, M. 923
de Carlos, A..1029
de Groot, K. 619
De Luca, P. 43
de Pedro, J.A. 177
de Queiroz, A.A.A. 589
Deisinger, U. 977
Del Valle, S. 945
Delgado, J.A....................253, 379, 1001
Delubac, C. 427
Demirkesen, E. 655
Deng, C...7, 801
Deng, C.L. 721
Di Silvio, L. 833
Dias, A.G. 825
Diwan, A. 301
Dlanza, L. 695
Dohkawa, H......................................643, 647
Doiguchi, Y. 643, 647
Dolcini, L. 233, 829, 833, 1017, 1095
Doremus, R.H...903
Dorozhkin, S.V..327
Dorozhkina, E.I. 327
dos Santos, E.A. 793
dos Santos, F.F.P..439
Driessens, F.C.M..............................161, 371
Du, P...331, 477
Duan, Y.R.....................131, 351, 713, 887
Duarte, M.G.G.M. 581
Durand, D. 615

E

Eberhardt, A.W. 455
Ehrenfried, L.M. 981
Eichert, D. 927
El Briak, H. 615
El Briak-BenAbdeslam, H. 103, 107, 931
Eloy, R. 427
Engqvist, H. 51, 197, 265
Evans, P. 447

F

Fan, H. 7, 801
Farina, M. 793
Farrar, D.F. 205, 281
Feng, J.M. 721, 757
Fernandes, M.H. 821, 825, 857
Fernandes, M.H.V. 155
Fernández, J. 383
Fernández, M. 177
Ferraz, M.P. 347, 903
Ferreira, J.M.F. 327, 331, 1033
    1037, 1041, 1045
Forsythe, J.S. 529
Foster, M. 903
Frade, J.R. 155
Frayssinet, P. 861
Frèche, M. 201
Fröberg, L. 973
Fujibayashi, S. 671, 953
Fujii, K. 517
Fujimori, H. 19, 1055
Fujimoto, T. 753
Füredi-Milhofer, H. 11
Furutani, Y. 403

G

Gabbi, C. 789
Gaona, M. 383
Garcia, C. 431
Garti, N. 11
Gatti, A.M. 127, 691, 907
Gauthier, O. 55, 193
Gbureck, U. 205, 249, 277, 281
Georgescu, G. 201
Gerber, T. 679
Giannini, S. 651
Gião, D. 347
Gil, F.J. 379, 737
Gil, M.H. 581
Gildenhaar, R. 407, 1059

Ginebra, M.P. 107, 253, 379, 485
    931, 945, 1001
Giovani, R. 809
Girija, E.K. 399
Goes, A.M. 773, 777, 837, 841
Goller, G. 777, 837
Gómez-Ortega, G. 197
Gonzalez, P. 347
González, P. 23, 355, 1029
González-Corchón, A. 177
Goto, K. 173
Goto, S. 19, 1055
Gough, J.E. 769, 813, 985, 1087
Gouvêia, M.F. 687
Granja, P.L. 573, 577
Greenspan, D. 749
Greil, P. 391, 923, 1013, 1111
Grimandi, G. 1107
Gross, K.A. 39, 213, 529, 957, 961
Gross, U. 635, 1083
Grover, L.M. 205, 277
Guibert, G. 717
Guicheux, J. 1107
Guilemany, J.M. 383
Guipont, V. 319
Gurav, N. 833
Gutierres, M. 821

H

Haltia, A.-M. 549
Hamblin, J. 833
Han, G. 331, 477
Han, J.-S. 699
Han, S.B. 245
Hansen, S. 269
Hansson, U. 623
Harmand, M.F. 233
Hashida, T. 395
Hashimoto, M. 115, 569
Hata, K. 497
Hattar, S. 749
Hattori, K. 1067
Hattori, M. 209
Haugen, H. 937, 941
Hawkes, G.E. 95
Hayakawa, S. 443, 853, 857, 865
Hayashi, K. 181
Hein, W. 407
Heino, H. 549

Hench, L.L. 3, 765, 769, 781, 785, 813
   981,985, 989, 1087
Henkel, K.-O. 679
Hermansson, L. 51, 197, 265
Heyligers, I.C. 667
Hijón, N. 363, 481
Hill, R.G. 99
Himeno, T. 139
Hing, K.A. 273
Hirose, M. 1051
Hong, K.S. 147, 151
Hopfner, U. 937
Howlett, C.R. 447
Hsieh, M.F. 473
Hsu, T.J. 473
Huang, J. 111
Huang, Y. 887
Humbert, T. 193
Hunziker, E.B. 619
Hupa, L. 67, 973
Hupa, M. 23, 67, 973
Hutton, A. 205
Hwang, I.-S. 217

I

Iida, J. 745, 1063, 1067, 1071, 1075
Ikoma, T. 561
Inada, H. 59
Inagaki, M. 451
Insley, G.M. 91, 289, 585, 663, 667
   675, 993
Ioku, K. 19, 1055
Ireland, D.C. 869
Irigaray, J.L. 717
Ishikawa, K. 115
Isomoto, S. 181
Ito, A. 541
Itoh, H. 1119
Itoh, S. 561
Itoh, T. 1115
Izquierdo-Barba, I. 359, 363
Izumi, Y. 517

J

Jackson, L.E. 297
Jaecques, S.V.N. 427
Jallot, E. 717
Jansen, W.C. 809
Jean, E. 861
Jeandin, M. 319

Jeong, J.H. 1123
Jeong, Y.K. 47
Jiménez, E. 415
Jin, H.-Z. 375
Jin, Z.M. 639
Jokinen, M. 489, 553, 557
Jonášová, L. 1013, 1111
Jones, J.R. 981, 985, 1087
Jones, T. 261
Jung, H.S. 1079
Jung, S.J. 969

K

Kagami, H. 497
Kameyama, T. 257, 451
Kamitakahara, M. 59, 403, 521, 525, 965
Kaneko, H. 741
Kang, I.K. 881
Kanie, T. 517
Karkia, C. 817
Kasuga, T. 497, 533, 753
Kasugai, S. 849
Kawagoe, D. 1055
Kawai, T. 525, 565, 1115
Kawasaki, S. 443
Kawasaki, T. 873
Kawashita, M. ...139, 375, 459, 467, 521, 741
Kawate, K. 1075
Kawazoe, Y. 1119
Kellomäki, M. 505, 509, 549
Kepenek, B. 463
Khamchukov, Y.D. 891
Khor, K.A. 315, 319
Kiesvaara, J. 489
Kijkowska, R. 127
Kikuchi, M. 15, 561
Kim, C.K. 1079
Kim, C.Y. 305
Kim, D.-J. 699
Kim, H. 147, 151
Kim, H.E. 423, 1103
Kim, H.M. 139, 375, 459, 467, 521
   741, 953
Kim, H.W. 423, 1103
Kim, J.K. 805
Kim, J.L. 189
Kim, K.H. 189, 631, 881
Kim, K.M. 185, 627, 1079
Kim, K.N. 185, 1079
Kim, M.C. 1079

Kim, S. 881, 1123
Kim, S.B. 285
Kim, S.C. 643, 647
Kim, S.H. 627
Kim, S.R. 135, 217, 225, 237, 969
Kim, S.W. 881
Kim, S.Y. 245, 631, 631
Kim, Y.H. 135, 969
Kim, Y.T. 135, 225, 969
Kinoshita, H. 403
Kitamura, M. 965
Kitamura, S. 181, 1051
Klein, C. 391
Klein, S. 221
Knabe, C. 1059
Knowles, J.C. 423, 1103
Ko, D.-J. 627
Kobayashi, N. 537
Kobayashi, T. 849
Koç, N. 949
Kohgo, T. 1119
Köhler, A. 941
Koizumi, M. 1075
Kokubo, T. 173, 467, 521
Kokubo, T. 115, 139, 375, 459
      569, 741, 953
Kolos, E.C. 343
Konagaya, S. 525
Kondo, K. 209
Korkusuz, F. 949
Kortesuo, P. 489
Korventausta, J. 557
Kosmač, T. 683
Kotobuki, N. 1051, 1055
Koyama, Y. 561
Krajewski, A. 1017
Ku, Y. 245
Kurata, S. 603
Kwak, E.K. 47, 189
Kwon, T.G. 47

L

Labanti, M. 1095
Lacout, J.L. 201
Lai, P.K. 611
Lakatos, M. 419
Lambert, C.S. 435
Lamghari, M. 733
Lamure, A. 833
Lamy, B. 55

Larrecq, G. 1001
Latella, B.A. 455
Law, R.V. 99
Layrolle, P. 1005
Lee, C.K. 147, 151
Lee, D.C. 1123
Lee, D.H. 147, 151
Lee, D.J. 1123
Lee, D.Y. 699
Lee, G.H. 805
Lee, I.S. 805
Lee, J.H. 135, 147, 151, 225, 881, 969
Lee, J.K. 217, 225
Lee, K.S. 135, 225
Lee, K.Y. 805
Lee, S.B. 185
Lee, S.H. 699
Lee, S.J. 245
Lee, Y.H. 805
Lee, Y.J. 969
Lee, Y.K. 185, 245, 1079
Lee, Y.M. 245
LeGeros, J.P. 127
LeGeros, R.Z. 127, 185, 817, 1079
Leite, M.F. 773, 777, 837, 841
Leite Ferreira, B.J.M. 581
Lemos, A.F. 1033, 1037, 1041, 1045
Leng, Y. 339
Leon, B. 347
León, B. 23, 355, 371, 415, 1029
Leonelli, C. 899
Leong, J.C.Y. 165
Leung, L.Y. 611
Levenfeld, B. 177
Li, H. 315
Li, Y. 477
Lidgren, L. 269
Lilley, K.J. 281
Lim, H.B. 305
Lima, D.O. 1021
Lind, A. 557
Linden, M. 557
Linhares, A.B.R. 793
Liste, S. 23, 347, 355, 1029
Liu, Y. 79, 619
Llorca, N. 319
Locardi, B. 789
Long, M. 261
Loof, J. 51
Lööf, J. 197

Lopes, M.A. 323, 477, 565, 821
  825, 845, 857
Lopez, E. 733, 1009
Lu, W. 757
Lu, W.W. 165
Lu, X. 339
Lucarelli, M. 907
Lucas, M.-F. 55
Luk, K.D.K. 165
Lusquiños, F. 371
Lusvardi, G. 899
Lynn, A.K. 593

M
Machida, H. 1051
Maeda, H. 497, 533
Magalhães, M.C.F. 143
Maher, P.N. 585
Mak, M.M. 639
Malavasi, G. 899
Manero, J.M. 379
Mangano, C. 829
Mansur, H.S. 695
Manuel, C.M. 903
Marciniak, J. 387
Markovic, B. 713
Marques, P.A.A.P. 143
Martin, A.I. 143
Martín, A.I. 481
Martinetti, R. 233, 829, 833, 1017, 1095
Martínez, S. 379, 945, 1001
Martínez-Fernández, J. 1029
Mascarel, G. 1009
Mastrogiacomo, M. 1095
Matsumoto, H.N. 561
Matsumoto, T. 911
Matsusue, Y. 1099
Matsuya, S. 99
Mavropoulos, E. 123
Mazzocchi, M. 1017
McCarthy, I. 269
Melville, A.J. 529
Menabue, L. 899
Mendes, B.M. 83, 87
Méndez, J.A. 177
Merino, J.M. 481
Merolli, A. 789
Mesquita, P. 1091
Met, C. 367
Meurer, E. 687

Mijares, D.Q. 127
Milet, C. 733
Milev, A. 301
Miyajima, T. 293
Miyata, N. 521
Miyazaki, K. 1067
Miyazaki, T. 59, 285, 403, 525, 545
Mizuno, M. 115, 569
Mochales, C. 107, 931
Monari, E. 691, 907
Monteiro, A.M.R. 439
Monteiro, F.J. 323, 347, 903, 997
Moon, H.J. 185
Moon, Y.M. 1123
Morais, S.M.O. 439
Moreira, J.C. 123
Morejón, L. 379
Moroni, A. 651
Mourão, R.L. 695
Muir, M.M. 343
Mukaida, M. 241
Muller, C.A. 729
Müller, F.A. 391, 923, 1013, 1111
Müller-Mai, C. 635
Mummery, P.M. 335
Muneyasu, A. 745
Munier, S. 615
Muñoz, F. 737
Murphy, A.M. 585
Murphy, M.E. 91, 289, 585

N
Nagata, F. 63, 293, 399
Nakagawa, Y. 607, 671
Nakamura, H. 1115
Nakamura, M. 849
Nakamura, S. 849
Nakamura, T. 139, 173, 241, 375, 459
  467, 521, 607, 671, 741, 953, 1099
Nam, K. 627
Närhi, T. 599
Naslain, R. 1009
Nastro, A. 43
Nasu, K. 1119
Navarro, M. 945
Nawrat, G. 387
Nemelivsky, Y. 817
Neo, M. 241, 953
Neto, J.A.C. 439
Neumann, H.-G. 679

Niemelä, T. 505, 509
Niiranen, H. 505
Noda, I. 671
Nogami, M................................ 497, 533, 753
Nogueira, R.E.F.Q. 1021
Nonomura, A. 745, 1071, 1075
Notingher, I.. 769
Nunes, T.G... 95

O

Oboeuf, M... 749
Ogata, S. 403, 965
Oh, K.O.. 189
Oh, K.S. 47, 189, 217, 237
Oh, S. 881
Oh, S.H. 631
Ohgushi, H........................... 181, 1051, 1055
Ohkawa, K. 603
Ohmura, T.. 1075
Ohsawa, K. 241
Ohta, K... 15
Ohtsuki, C........... 59, 285, 403, 467, 521, 525
545, 565, 965
Oka, M. 169
Oktar, F.N. 655, 777, 837
Okubo, A. 395
Okumura, N. 1067, 1071
Okuyama, M. 209
Oliveira, J.M. 565
Omori, M. 395
Onuma, K... 537
Ooi, K. 1119
Oomamiuda, K................................... 643, 647
Oonishi, E. 607
Oonishi, H... 643, 647
Osaka, A. 443, 853, 857, 865
Otsubo, M. 395
Overgaard, S. 833
Oyane, A..................................... 537, 541
Ozawa, N. 911
Ozsoy, S... 655
Ozyegin, L.S. 837
Ozyegin, S. 655

P

Padilla, S.. 31, 35
Padrós, A. 737
Panzavolta, S. 229
Park, E.K.. 631
Park, J.-C. 805

Park, K.S. 147, 151
Park, S.K. 1123
Park, Y.S. 1079
Patel, M.P... 513
Pauvert, B... 103
Pearson, G. 513
Pegreffi, F... 651
Peláez, A.. 431
Peña, J..27, 359
Pereira, M.M. 773, 809, 841
Pereira, S. 659
Pereira Mouries, L..................................... 733
Pérez Amor, M. 23, 347, 355, 371, 415
Perng, L.H. 473
Persson, T. 265
Phipps, K. 667
Pilet, P. 55
Pinheiro da SIlva, T.D............................. 1021
Pinto, M. 729
Pires, R.A. 95
Planell, J.A.107, 253, 379, 485, 737, 931, 945, 1001
Ploska, U. 71, 411, 635
Pomeroy, M.J. 91
Pons, M... 319
Porter, A.E.. 915
Pou, J. 371
Prado da Silva, M.H. 439, 1021
Pryce, R.S. 765

Q

Quarto, R. 833, 1095
Queiroz, A.C. 997
Queiroz, C.M. 155
Quiniou, L. 367

R

Rahman, F.F. 513
Rajchel, B. 387
Ravaglioli, A. 1017
Reis, E.M.. 809
Revell, P.A. 273, 447, 797
Rey, C....................................... 833, 927
Rhee, S.-H. 501
Rhyu, I.C. 245
Ribeiro, C.C.. 877
Rocha, N.C.C. 123
Rodríguez-Lorenzo, L.M..........529, 957, 961
Roessler, S. 419
Roest, R. 455
Roger, G. 343

Rohanizadeh, R. 343
Rolfe Howlett, C. 713
Romagnoli, M. 651
Román, J. 31, 581
Rosenkrancova, J. 703
Roska, I.D. 919
Rosling, A. 489, 557
Rossi, A.M. 119, 123, 493, 729, 793
Rouquet, N. 861
Rousseau, M. 1009
Rushton, N. 593, 845, 869
Ruys, A.J. 343
Ryoo, H.M. 47, 189, 631
Ryu, H.S. 147, 151

S
Saeed, S. 273
Saffarzadeh, A. 193
Sahre, M. 71
Sainte Catherine, M.C. 367
Saint-Jean, S. 269
Sakaguchi, Y. 525
Salinas, A.J. 143, 481
Salvado, M. 177
San Román, J. 177
Sanginario, V. 485
Santa Barbara, A. 687
Santin, M. 789
Santos, J.D. 323, 477, 565, 821, 825
    845, 857, 997, 1033
Saravanapavan, P. 781, 785, 1087
Satoh, N. 745
Sautier, J.M. 749
Sawamura, T. 209
Scarano, A. 829
Scharnweber, D. 419
Schrooten, J. 427
Schubert, H. 1083
Schultheiss, C. 427
Sedlacek, R. 703
Seker, U.Ö.Ş. 463
Sekijima, Y. 849
Selvakumaran, J. 785
Semmelmann, K. 1029
Sen, Y. 1075
Sena, L.A. 493, 729
Seo, J.H. 147, 151
Seo, W.S. 805
Serra, C. 355
Serra, J. 23, 347, 355, 1029

Serricella, P. 493
Sewing, A. 419
Sfihi, H. 927
Shelton, R.M. 79
Shen, G. 331, 477
Shiba, T. 1119
Shima, M. 169
Shin, H.I. 47, 189, 631, 1123
Shindoh, M. 1119
Shinji, H. 603
Shinomiya, K. 561
Shinzato, S. 173
Shiotsu, Y. 63
Shirosaki, Y. 857
Shirtliff, V.J. 989
Sieber, H. 1013
Sikiric, M. 11
Silva, A.I.N. 573
Silva, C.C.P. 841
Silva, C.H.T. 87
Silva, J.V.L. 687
Silva, R.A. 323
Simonetti, L. 691
Sloten, J.V. 427
Smit, T.H. 667
Smith, D.C. 1009
Soares, G.A. 119, 123, 493, 729, 793
Söderling, E. 599
Song, H. 969
Song, J. 185
Sousa, M. 323
Spiers, K.M. 213
Stamboulis, A. 99
Stenzel, F. 977
Stewart, G. 833
Streicher, R.M. 289, 663, 675, 993
Sugawara, T. 603
Suvorova, E.I. 891
Suwa, Y. 603
Suzuki, T. 671
Svendsen, N. 833
Syoji, D. 561

T
Tabata, Y. 1099
Takadama, H. 115, 569
Takahashi, Y. 1119
Takakuda, K. 561
Takakura, Y. 181, 745, 1051, 1063
    1067, 1071, 1075

Takao, Y. 643, 647
Takashi, N. 919
Takemoto, S. 853
Takeuchi, A. 403
Takigawa, Y. 569
Tamerler, C. 463
Tamura, J. 1099
Tamura, K. 919
Tamura, Y. 873
Tan, Y. 7, 801
Tanaka, H. 1119
Tanaka, J. 15, 561
Tanaka, Y. 181
Taniguchi, A. 181
Tanihara, M. 59, 403, 525, 545, 965
Tavakoli, S.M. 513
Tay, F.R. 695
Teixeira, C.C. 817
Teixeira, S. 997
Teraoka, K. 257
Terol, A. 103
Thams, U. 737
Theilgaard, N. 833
Thian, E.S. 111
Thull, R. 249, 277
Timuçin, M. 949
Toda, M. 19
Toda, T. 761
Tohji, K. 395
Tohma, Y. 181
Toksvig-Larsen, S. 623
Törmälä, P. 505, 509
Totsuka, Y. 919
Toykan, D. 655
Tranquilli Leali, P. 789
Traykova, T. 679
Tsuru, K. 443, 853, 857, 865
Tsuruta, S. 1115

U
Uchida, M. 541
Ueda, M. 497
Ueda, Y. 1075
Uematsu, T. 1119
Uo, M. 873, 919
Ürgen, M. 463
Utech, M. 43
Uzumaki, E.T. 435

V
Vaahtio, M. 489, 553
Vago, R. 301
Väkiparta, M. 553
Valerio, P. 837
Valério, P. 773, 777
Vallet-Regí, M. 27, 31, 35, 143, 359
363, 481, 581
Vallittu, P.K. 553, 599
Van de Vaal, C. 619
van Haaren, E.H. 667
Van Humbeeck, J. 427
Vandenbulcke, L. 367
Varela-Feria, F.M. 1029
Vasconcelos, M. 659, 1091
Vázquez, B. 177
Vedel, E. 67
Verdonschot, N. 221
Verrier, S. 781
Vinatier, C. 1107
Voor, M.J. 221
Vuono, D. 43

W
Walschot, L.H.B. 221
Wang, C. 713, 753
Wang, C.Y. 131, 351, 757, 887
Wang, L.P. 757
Wang, M. 611
Watanabe, M. 63
Watanabe, T. 1115
Watari, F. 873, 919
Weiss, P. 55, 193, 1107
Weng, W. 331, 477
Will, J. 937, 941
Willfahrt, M. 411
Wintermantel, E. 937, 941
Wong, C.T. 165
Wood, D.J. 335
Worch, H. 419
Wright, A.J. 297
Wu, J. 375
Wu, Y. 725
Wuhrer, R. 455
Wuisman, P.I.J.M. 667

X
Xiong, T.-Y. 375

Y

Yamaguchi, M. 603
Yamamoto, H... 603
Yamamoto, T. 853
Yamamuro, T.. 169
Yamaoka, M. 1119
Yamashita, K. 849
Yamazaki, A. 537
Yamazaki, M. 403
Yamazaki, T.. 1099
Yang, B.C. 725, 757
Yang, J.-H.. 699
Yang, S.M... 245
Yao, T. 911
Yasuda, T.. 607
Yazici, T. 655
Ylänen, H... 67
Ylänen, H.O. 23
Yli-Urpo, A............................. 489, 553, 557
Yli-Urpo, H... 599
Yokogawa, Y. 63, 257, 293, 399, 451
Yokoyama, A... 873
Yoshii, S. 169

Yoshikawa, M. 761
Yoshikawa, T. ...745, 1063, 1067, 1071, 1075
Yoshimura, M...19
You, C.K. 881, 1079
Yuda, A. 517

Z

Zak, J. 387
Zavaglia, C.A.C................................435, 687
Zehbe, R. 1083
Zhang, B. 757
Zhang, Q..7, 801
Zhang, Q.Y. 721
Zhang, X..7, 713, 801
Zhang, X.D. 131, 351, 721, 725, 757
Zhang, X.S...447
Zhang, Z.R......................................131, 887
Ziegler, G. 977
Zippor, B. 833
Zollfrank, C. 1111
Zreiqat, H. 713, 1059
Zuegner, S. 937

Keyword Index

A

α-TCP 1021
α-TCP Cement 761
α-Tricalcium Phosphate 225, 229, 297
    793, 965
Abrasion.. 873
Abutments... 699
Accelerated Osteoconduction 273
Acetabular Dysplasia 169
Acetylation.. 311
Adhesion...................................... 155, 455
Adhesive Strength................................... 383
Adsorption onto Apatite 603
AFM.. 537, 845
Aging 683
Aging Test 173
Alkaline Phosphatase 773, 777, 817, 837
Alkoxysilane... 545
Alumina 367, 607, 639, 663, 675
Alumina Ceramics 181
Alveolar Bone....................................... 1119
Aminosilane.. 765
Ammonium Hexafluorophosphate 331
Amorphous Calcium Phosphate 11
Anatase 443, 463
Animal Studies 225, 655
Anodic Oxidation 375, 459, 741
Anodizing 455
Antimicrobial....................................... 1087
Antimicrobial Effect 47
Apatite 99, 115, 131, 375, 459
    501, 521, 525, 533, 541, 545, 569,
    741, 753, 833, 911, 927, 1115
Apatite Coating..................................... 517
Apatite Fiber 915
Apatite Formation.......................... 285, 443
Apatite Layer 155, 581
Apoptosis 985
APTS 765
Ar Ion Beam 805
Articular Cartilage Replacement 1083
Artificial Bone 169
Artificial Bone Grafts 205, 233
Artificial Joint...................................... 181
Atomic Force Microscopy 537, 845
Attachment... 781

Attachment Strength............................... 725
Attritor Milling...................................... 111
Autoclave 753

B

β-TCP 1021
β-Tricalcium Phosphate TCP 237, 281
    297, 509
Bactericidal Effect.................................. 463
Bending Strength.................................... 173
Biaxial Orientation 549
Bioabsorbable....................................... 549
Bioabsorbable Polymer 505
Bioactive...................... 3, 147, 447, 549, 797
Bioactive Behavior 969
Bioactive Bone Cement............. 173, 177, 285
Bioactive Bone Substitute 581
Bioactive Ceramics 27, 233, 829
Bioactive Composites.............................. 589
Bioactive Glass-Ceramic.......................... 581
Bioactive Glasses 23, 67, 177, 355
    427, 505, 691, 749, 765, 773, 781,
    785, 789, 813, 973, 981, 985, 989,
    1033, 1087
Bioactive Materials 379, 467
Bioactive Titanium 737
Bioactivity 23, 59, 71, 75, 115, 155
    335, 439, 451, 501, 545, 553, 569,
    741, 899
Bioceramics 51, 99, 147, 1005, 1029
    1037, 1041, 1045
Biocompatibility................ 135, 241, 265, 407
    837, 873, 919
Biocomposite... 593
Biodegradable....................... 147, 881, 1115
Biodegradable Composite 589
Biodegradable Copolymer........................ 517
Biodegradable Glass Ceramics.................. 825
Biodegradable Scaffold 1123
Biodegradation 245
Biofilm 1009
Biological Glass 717
Biomaterials 43, 55, 387, 407, 435
    651, 687, 733, 797, 841, 919, 1009,
    1029, 1091, 1107
Biomechanics 703

Biomimetic 343, 399, 473, 619
Biomimetic Crystallization.................... 1083
Biomimetic Method.......................... 541, 911
Biomimetic Process 403, 525, 1111
Biomineral Behavior.............................. 721
Biomineralization 311, 327
Bioresorbable Ceramics........................... 965
Biphasic Calcium Phosphate 297, 817, 923
   949, 1005
Biphasic Calcium Phosphate Ceramics 7
   193, 931
Biphasic HA-TCP Ceramic 189
Biphasic Mixtures...................................... 31
Bisurface... 607
Blood Plasma... 115
BMP.. 1115
BMP-2 619
Bone............................. 3, 123, 165, 185, 261
   407, 1079
Bone Bonding.. 447
Bone Cells... 733
Bone Cement 205, 217, 237, 265, 305
   485, 647
Bone Construction 169
Bone Defect 643, 691
Bone Filler 47, 225
Bone Formation 127, 241, 679, 873
   1051, 1071
Bone Graft 221, 801
Bone Graft Substitute 221, 273
Bone Grafting 667
Bone Growth.................................... 319, 513
Bone Ingrowth 797, 1005
Bone Marrow Stromal Cells BMSCs 7
Bone Mineral 927
Bone Mineralization 1119
Bone Regeneration 257, 573, 877, 1091
Bone Repair 1095
Bone Substitute Materials........................ 977
Bone Substitutes 135, 189, 1059
Bone Tissue 789
Bone Tissue Engineering....................... 1001
Bone Tissue Regeneration 561
Bone Void Filler 265, 269
Bone-Like Apatite 7, 139, 327, 351, 497
Bonelike®.. 565, 821
Bovine Hydroxyapatite............................ 655
Bovine Serum Albumin BSA 327, 399
Brain Implants 83
Brushite.. 281, 391

C
Ca-Aluminates...265
Caking...63
Calcification 311
Calcium 261
Calcium Aluminates...........................51, 197
Calcium Carbonate...........................497, 533
Calcium-Deficient Hydroxyapatite 161
   269, 895
Calcium Hydrogen Phosphate
   Dihydrate...11
Calcium Ion...525
Calcium Liberation..................................585
Calcium Metaphosphate..........................1123
Calcium Oxide..........................103, 615, 931
Calcium Phosphate............185, 217, 237, 289
   315, 327, 339, 343, 363, 371, 419,
   537, 577, 585, 593, 619, 667, 679,
   725, 753, 895, 957, 993, 1013,
   1017, 1079, 1095, 1111
Calcium Phosphate Cement 161, 201, 209
   241, 253, 277, 281, 615, 1001
Calcium Phosphate Ceramics..............55, 351
   713, 801, 861, 1059
Calcium Phosphate Coating 1103
Calcium Phosphate Glass 945
Calcium Phytate 11, 11
Calcium Polyphosphate............................245
Calcium Potassium Phosphate 161
Calcium Potassium Sodium Phosphate 161
Calcium Salt 545
Calcium Silicate.......................................781
Calcium Strontium Phosphates 103
Calcium Sulphate 585
Calcium Titanate 411
Calcium Zirconium Phosphate
   Ceramics..635
Cancellous 221
CaO-P$_2$O$_5$ Glass.......................................347
CaO-SiO$_2$...481
CaP Formation..557
CAPS Coatings...319
Carbon Coating 387
Carbon Dioxide 7
Carbon Nanotube..............................395, 919
Carbonate Ion 63
Carbonated Apatite...................................529
Carbonated-Hydroxyapatite 829
Carrier..1099
Cartilage 1123

Cathodic Arc Deposition 463
Cell Attachment 343
Cell Culture 427
Cell Cycle 3
Cell Death 813
Cell Proliferation 135
Cell Response 1103
Cell-Biomaterial Interactions 1059
Cellulose 1013, 1111
Cement 229, 289, 895
Cementless 623, 671
Ceramic Composites 699
Ceramic Particle Size 611
Ceramic-Cell Complex 631
Ceramic-on-Ceramic 639
Ceramics 169, 407, 659, 703
Characterization 55
Charge Injection 805
Chemical Treatment 443
Chemically Bonded Ceramics 51
Chiral Biomineralization 603
Chitin 311, 545
Chitosan 245, 311, 501, 573, 577, 857
Chitosan-Inorganic Hybrid 857
Chondrocyte 817, 1123
Clenching 707
Clinical Applications 699
Clinical Behaviour 829
Clinical Study 679
CMP 881
Coating by Plasma Spray 757
Coating Technologies-Chemical and
   Physical 359
Coatings 355, 371, 407, 419, 435
   451, 467, 619, 717, 721, 729, 789
Coefficients of Friction 367
Collagen 399, 473, 493, 537
   593, 619, 773, 777, 817, 837, 1083
Collagen Type I 869
Colloidal Processing 35
Composite 467, 477, 485, 505, 509, 517
   533, 541, 549, 569, 573, 581, 611,
   793, 1115
Compressive Strength 217, 229, 965
Condensing Techniques 197
Confocal Laser Scanning Microscope 319
Confocal Microscopy 845
Consolidation of Carbon Nanotube 395
Consolidation Techniques 1037, 1045
Contact Mechanics 639

Coral 301, 969
Corrosion 387, 717, 809
Cryopreservation 1051
Crystal 79
Crystal Growth 1009
Crystal Growth Morphology 11
Crystal Orientation 15
Crystallinity 43
Crystallisation 67, 99, 155, 327
Culture 1067
Customized Implants 687
Cytocompatibility 857
Cytokines 841
Cytotoxicity 785

D
DCPD 305
Decay 209
Defects 915
Defense 907
Degradation 553
Dental Cements 659
Dental Grinding 683
Dental Implants 699, 707, 729, 737
Dental Materials 599, 627, 695
Dental Porcelain 809
Dental Pulp 761
Dental Restorative 197
Dental Tissue 695
Dentistry 691
Devitrification 67
Dexamethasone 745, 1063, 1067
   1071, 1075
Diamond Coatings 367
Diamond-Like Carbon 435
Dicalcium Phosphate Dihydrate 103
   615, 931
Diffusion 513
Directional Freezing 1083
Dispersant 35
Dispersion 477
Dissolution 127, 261, 315
Dissolution Kinetics 765
Dissolution Property 965
Dosimetry 83
Drug Adsorbance 529
Drug Delivery 797, 997
Drug Delivery System 1099
Drug Release 489
Drug Release Rate 529

DTA 67
Dynamic 131, 721
Dynamic Revised Simulated Body
    Fluid (Dynamic RSBF) 7

E
E-Beam Evaporation 805
EDS 1029
EDX 139, 355
Egg-White 1045
Elastic Modulus 39
Elastic Property 561
Electrochemistry 419
Electrodeposition 391
Electron Beam Ablation 427
Electron Diffraction 339
Electron Microscopy 695
Electrostatic Stimulation 849
Enamel Matrix Protein 749
Environmental Electron Scanning
    Microscopy 845
Epitaxial Growth 603
Epithelial Cells 907
Ethylene-Vinyl Alcohol Copolymer
    EVOH 541
External Fixation Pin 651

F
Femoral Implant 655
Fibres 343
Fibroblasts 785
Fibrous Capsule 725
Finite Element Analysis 639, 707
Flexural Properties 611
Flexural Strength 431
Flow Cytometry 841
Fluorapatite 335
Fluorescence Microscopy 845
Fluorescence Mode 319
Fluor-Hydroxyapatite 423
Fluoridated Hydroxyapatite 331
Foaming Method 1041
Foams 253, 989
Fosfosal 177
Fracture Toughness 39
Free Form Fabrication 1025
Fretting Wear 383
FTIR 23, 355

G
Gap Healing 757
Gelatin 229
Gelatine 1083
Gelcasting 35
Genes 3
Genistein 1071
Glass 71, 75, 169, 173, 305
Glass Ceramic 335
Glass Reinforced Hydroxyapatite 997
Glass-Ceramics 59, 71, 95, 99
    155, 305, 753
Glass-Ionomer Cements 599
Glazing 809
Glucose 327
Glycol 627
Grain Growth 91
Granules 19, 237, 761
Grit Blasting 737
Growth Factors 3, 1099

H
HA/TCP Ceramics 131
Hard Tissue Formation 761
Hardness 39, 873
High Resolution Transmission
    Electron Microscopy 891
HIP 667
Histological Analysis 565
Histology 643, 647
Histomorphometry 55, 737
Hollow Fibers 391
Hollow Sphere 533
HTM 67
Human 181, 1067
Human Bone Marrow Cells 821
Human Bone-Derived Cells 1059
Human Osteoblast Cells 845
Human Osteoblast Culture 793
Human Osteoblasts 869
Human Parotid Saliva 75
Hybrid 521
Hybridized Hydroxyapatite 589
Hydration Reaction 209
Hydrogel 63, 1107
Hydrogel Foam 1083
Hydrogen Peroxide 443, 865
Hydrothermal 19
Hydrothermal Conversion 301

Hydroxyapatite 15, 19, 31, 35, 43, 63
    83, 87, 91, 103, 111, 119, 123,
    143, 165, 177, 209, 257, 273, 293,
    297, 301, 305, 323, 383, 391, 399,
    403, 419, 423, 439, 447, 473, 477,
    493, 513, 529, 573, 611, 623, 643,
    647, 651, 729, 745, 757, 777, 793,
    813, 829, 849, 877, 891, 903, 915,
    931, 949, 969, 993, 997, 1021,
    1025, 1037, 1041, 1045
Hydroxyapatite Ceramics 977
Hydroxyapatite Coating............................ 395
Hydroxyapatite/Collagen
    Nanocomposite 561
Hydroxyapatite Implant 797
Hydroxyfluorapatite.................................... 39
Hydroxylapatite Cement........................... 221
Hydroxylapatite Coatings Pulsed
    Laser Deposition................................. 415
Hydroxylapatite Thin Films...................... 347
Hyperthermia 213
Hysteresis Heating.................................... 213

I
Ibuprofen 529
Impaction Grafting 993
Impaired Mineralisation 635
Implant................ 87, 619, 703, 757, 873, 881
Implantation Study 135
In Situ Hybridization 241, 1059
In Vitro................ 75, 315, 427, 467, 557, 749
In Vitro Assay.. 31
In Vitro Behaviour 899
In Vitro Bioactivity........... 347, 481, 581, 765
In Vitro Dissolution 585
In Vitro Mineralisation 821
In Vitro Protocols.................................... 143
In Vitro Test.................................... 319, 355
In Vivo.. 801
In Vivo Studies.................................. 565, 789
In Vivo Transplantation 1079
Incorporation 221
Inflammation 797, 907
Inhibition of Matrix Vesicles.................... 635
Injectability... 485
Injectable 573
Injectable Bone Graft................................ 253
Injectable Bone Substitute 55, 193
Injectable Materials for Bone
    Regeneration................................. 55, 205

Injectable Microspheres 877
Inorganic Polyphosphate 1119
Integration 881
Interconnected Defects.............................319
Interconnecting Porosity 977
Interface...165
Interleukin 1 907
Ion Concentration 115
Ion Diffusion 615
Ion Release 557, 985
Iridium Oxide 805
Isoflavone 1071
Insulin Like Growth Factor 797

K
Kinetics.. 931
Knee Prosthesis 623

L
Laminin 541
Langmuir-Blodgett...................................1111
Laser Ablation 347, 355
Laser Cladding 371
Latex Beads 761
Lead..123
Liquid Phase Sintering 1033
Load-Bearing...671
Lung Cells 781

M
Macroaggregate...83
Macropores...1005
Macroporosity 253, 273
Macroporous.................981, 989, 1025, 1087
Macroporous Calcium Metaphosphate......631
Macroporous Structures 189, 1033
Maghemite..213
Magnesium...127, 447
Magnetite...213
Marrow Cell 181, 745, 1063, 1067
    1071, 1075
MAS-NMR...99
Materials Testing......................................667
Maturation 817
MC-3T3-E1 1119
Mechanical Blends 39
Mechanical Properties 51, 147, 197, 205
    277, 281, 289, 301, 415, 521, 1103
Mechanical Stimulation...........................1083
Mechanical Strength................................973

Mechanosynthesis.................... 103, 107, 931
Membrane...................... 1091, 1115
Mesenchymal Stem Cell............... 1051, 1055
Metal Surface Treatment 953
Metal-Ceramic System 627
Methacrylate Polymers 513
MG63 Osteoblast-Like Cells.................... 825
Microhardness 91
Micropattern 911
Micropores.. 1005
Microporosity 273
Microseparation 663, 675
Microsphere...................................... 293, 573
Microstructure 165
Microwave 899
Mineralisation.................. 143, 769, 813, 985
Molecular Weight 611
Monetite.. 439
Monocytes 907
Monte Carlo 83
Morphology 43, 79, 443
Morphology of CaCO$_3$ Crystals 603
Mössbauer Spectroscopy 213
MSC.. 1079
Mullite 335
Multicomposites 557
Multiplayer 1063
Multi-Walled Carbon Nanotube 395

N
Na$_2$O 245
Nacre.. 733, 1009
Nanocoatings 301, 455
Nanocomposite 119, 493
Nanocrystals 927
Nano-HAp Vectors 887
Nanoparticles 887, 903, 907, 1021
Nanopowders 899, 923
Nano-Sized Apatite................................. 439
Nano-Smooth Surfaces 367
Nanostructure.. 891
Natural Hydroxyapatite 837
Natural Material....................................... 1091
Network Modifiers 23
Neuron 805
Niobium 439
Ni-Ti Alloy 865
NMR 95
Nodule 813
Non-Crystalline 185

Novel Biopolymer 553
Nuclear Microscopy 717

O
Octacalcium Phosphate 11, 79
OPG..713
Organic Matrix 733, 1009
Organic-Inorganic Hybrids 473, 481, 545
Orthogonal Design 887
Orthopaedic Implants 435
Orthosilicic Acid 869
Osseointegration.................................729, 737
Osteoblast......................3, 335, 713, 749, 769
     777, 785, 813, 837, 985
Osteoblast Cell 185
Osteoblastic Cell (MG63) 857
Osteoblast-Like MG 63 Cells...................877
Osteoconduction..............................245, 1005
Osteoconductive 509
Osteoconductivity.....................................849
Osteogenesis............................733, 745, 797
     1063, 1067, 1075
Osteoinduction 589, 631, 801, 953, 1005
Osteoinductive..619
Osteoinductivity 895
Osteoporosis 265
Osteoporotic Trochanteric Fracture 651

P
Particles 873, 919
Paste Opaque Porcelain............................627
PDMS 481
PELGE 887
Periapical Tissue 761
Periodontal Regeneration 1119
Peripheral Blood Mononuclear Cells 841
pH 599
Phase Composition...................................713
Phase Mapping 593
Phase Transformation.................11, 249, 277
Phase Transition 151
Phosphate Treatment 455
Phosphates..95
Phosphorylated Chitosan.........................577
Phosphoserine Polypeptides....................603
Phosvitin...537
Photocatalytic Activity.............................463
Pin-on-Disc Friction Tests 367
Plasma Sprayed 721
Plasma Sprayed Coatings.........................891

Plasma Spraying 383, 655
Platelet Adhesion 853, 865
PLGA 473, 565, 1115
PMMA 177, 285
Poling 849
Polishing 809
Poly DL-Lactic-Co-Glycolic Acid 1123
Poly(D,L-lactide) 293, 489
Poly(Lactic Acid) 497, 533
Poly(methylmethacrylate)Based Resin 581
Poly(ε-Caprolactone) 589
Polyacrylic Acid 399
Polyamide Film 525
Polyaspartic Acid 11
Polyethylene 569, 611
Polylactide 477, 509
Poly-l-Glutamic Acid 11
Poly-l-Lysine 11
Polymer Network 1107
Polymers 467
Polymethyl Methacrylate 173
Polysaccharides 209
Polystyrene Sulfonate 11
Pore Connectivity 961
Pore Size Distribution 965
Porogenic Agents 201
Porosity 201, 1001
Porous 953, 969, 997
Porous Bioceramics 1017, 1021
Porous Body 561
Porous Bone Scaffold 1103
Porous Ceramics 19, 257, 937, 941
      949, 965, 1037, 1041, 1045
Porous Composite 497
Porous Environment 721
Porous Glass Ceramic 945
Porous Implant 973
Porous Material 293
Porous Scaffolds 957
Potassium-Containing Apatite 161
pQ-CT 565
Precipitation 11, 79
Preparation Process 115
Proliferation 781
Properties 627
Prosthesis 319
PTMO-Ta$_2$O$_5$ 521
PVA 481

R

Rabbit 135, 225
Rabbit Model 635
Radiological Response 87
Radiostereometry 623
Raman 23
Raman Spectroscopy 769
RANKL 713
Rapid Prototyping 687, 977
Rat 185
Rat Bone Marrow 1055
Reactive Plasma Spraying 451
Regeneration 1075
Reinforced Hydroxyapatite 1033
Release Control 59
Release of Biologically Active
      Molecules 1017
Reliability 683
Resist Pattern 911
Resorbable Bone Substitute 201
Reverse Transcriptase-Polymerase
      Chain Reaction 869
Revised Simulated Body Fluid RSBF 327
Revision Total Hip Arthroplasty 643
RF Sputtering 323
Rheology 1107
Rietveld Analysis 297
Ringer's Solution 367
Root Canal Filler 615
Roughness 737
Rutile 459

S

S53P4 Bioactive Glass 599
Salt Melt 411
Sandblasting 683
Scaffold 253, 631, 937, 941, 945, 961
      981, 989, 1001, 1079, 1095
Scaffold Template 961
Scratch 873
Scratch Test 155
SEM 355
Sericin 403
Serum 139
Signal Molecules 733
Silica Gel Microparticles 489
Silica Xerogel 553
Silicate 107
Silicon 773, 869
Silicon Carbide 1029

Silicon Substitution.................................... 969
Silicon-Substituted Hydroxyapatite.. 135, 845
Silver... 1087
Silver-Doped Hydroxyapatite..................... 47
Simulated Body Fluid SBF........... 71, 75, 115
    131, 139, 147, 285, 335, 339, 343,
    351, 375, 379, 399, 403, 459, 497,
    501, 525, 545, 569, 741, 753, 911,
    1013, 1111
Sintered Body 15
Sintered Hydroxyapatite 139
Sintering................................ 39, 91, 151, 973
Sinus Bone Grafting 193
Skin Terminal 541
Slip Casting................... 111, 949, 977, 1025
Soft Tissue Bonding 427
Soft Tissue Responses 725
Sol-Gel.............. 301, 331, 363, 431, 455, 481
    765, 781, 989
Sol-Gel Coating 423, 853, 1013
Sol-Gel Glasses.................................. 31, 143
Sol-Gel Method 679
Soluble Ions 785
Solvent 477
Sonochemical Synthesis 923
Spark Plasma Sintering SPS 395
Specific Surface Area 895
Splat 315
Starch Consolidation............................... 1041
Stem Cell 181
Stimulation... 805
Strain.. 915
Strength.................................... 683, 809, 981
Strontium Oxide 103
Structure... 1009
Subcutaneously Implanted Titanium 725
Substituted Hydroxyapatite 107
Sulfonic Group 525
Sulfuric Acid Solution 375
Supersaturation 143
Surface Charge 849
Surface Energy 833
Surface Layer.. 411
Surface Modification 387, 467, 501
    765, 873
Surface Reaction....................................... 407
Surface Topography 793
Surgical Revision...................................... 667
Suspension 35
Synthesis Methodology 593

Synthetic Serum 367

T
Tapecast..813
TCP Granule..217
TEM 139, 339, 695
TEM-EDX...741
Test Method..51
Tetragonal Zirconia 683
Thermal Spray...315
Thermal Spraying......................................961
Thermal Treatment 443
Thermodynamic Aspects...........................249
Thin Bioactive Film..................................431
Thin Films 323, 363
Ti-Alloy..881
Ti-6Al-4V Alloy......................367, 411, 717
TiO_2 ....463, 919, 937, 941
TiO_2 Layer...853
Tissue Culture 861
Tissue Engineering..........801, 957, 961, 1051
    1079, 1083, 1095, 1107
Tissue Irritation 761
Tissue Response 635
Titania..............................443, 451, 741, 865
Titania (Anatase) 569
Titanium 443, 451, 455, 619, 741
    777, 873, 919, 953
Titanium Implant......................................423
Titanium Mesh 517
Titanium Metal..................................375, 459
Titanium Nitride................................451, 873
TKA..607
TLR Receptors 907
Topography 443
Toremifene Citrate 489
Total Hip Arthroplasty 639
Total Joint Arthroplasty 647, 663, 675
Total Knee Arthroplasty...........................671
Transcription 911
Transfection..861
Transmission Electron Microscope...165, 915
Transparent Hydroxyapatite Ceramic 1055
Tribology..703
Tricalcium Phosphate........127, 249, 261, 269
    277, 391, 513, 833, 903, 923, 993

U
Ultrasonic Scaler 873
Ultrasound 923

V

Vacuole 773

W

Water Addition 931
Water Uptake 513
Wear 663, 675
Wear Rates 367
Weibull Analysis 431
Wet Chemical Precipitation 903
Wollastonite 151, 379

X

XPS 355, 833, 1029
X-Ray Diffraction 213, 229, 277

X-Ray Powder Diffraction 95
XRD 43
XRF 1029

Y

y-Glycidoxypropyltrimethoxysilane
      GPSM 857

Z

Zeta-Potential 139
Zinc 59, 119
Zirconia 367, 379, 455, 607
      671, 899, 1103
Zirconium 71